INTERNATIONAL ASTRONOMICAL UNION

HIGHLIGHTS OF ASTRONOMY
VOLUME 15

AS PRESENTED AT THE
TWENTY SEVENTH GENERAL ASSEMBLY
RIO DE JANEIRO 3–14 August 2009

COVER ILLUSTRATION

THE FRONTISPIECE OF *Sidereus Nuncius* ("STARRY MESSENGER") BY GALILEO
GALILEI. PUBLISHED IN MARCH 1610, IT WAS THE FIRST SCIENTIFIC REPORT ON OB-
SERVATIONS MADE THROUGH A TELESCOPE, GIVING THE RESULTS OF GALILEO'S EARLY
OBSERVATIONS OF THE MOON, THE STARS, AND THE MOONS OF JUPITER. THIS IS SEEN
AS THE START OF MODERN ASTRONOMY. THE INTERNATIONAL YEAR OF ASTRONOMY
2009 COMMEMORATES GALILEO'S CONTRIBUTIONS TO ASTRONOMY AND SCIENCE.

INTERNATIONAL ASTRONOMICAL UNION
UNION ASTRONOMIQUE INTERNATIONALE

HIGHLIGHTS OF ASTRONOMY VOLUME 15

Editors of Joint Discussions and Special Sessions held at the XXVII General Assembly 2009

JD1: Leon V. E. Koopmans & Tommaso Treu
JD2: Magda Arnaboldi & Ortwin Gerhard
JD3: Tomaso Belloni, Mariano Méndez & Chengmin Zhang
JD4: Margarida S. Cunha, Michael M. Dworetsky & Barry Smalley
JD5: James J. Binney
JD6: Pascale Defraigne & Aleksander Brzezinski
JD7: Elisabete M. de Gouveia Dal Pino & Alejandro Raga
JD8: Dong-Woo Kim & Silvia Pellegrini
JD9: Paolo Molaro & Elisabeth Vangioni-Flam
JD10: Hans-G. Ludwig, Piercarlo Bonifacio & K. N. Nagendra
JD11: Junwei Zhao, Hiromoto Shibahashi & Günther Houdek
JD12: Tommy Wiklund, Volker Bromm, & Bahram Mobasher
JD13: Augusto Daminelli, Theodore R. Gull & Krister E. Nielsen
JD14: Maria R. Cunningham, Carsten Kramer & Vincent Minier
JD15: Elisabete M. de Gouveia Dal Pino & Alex Lazarian
JD16: Sarah E. Gibson & David F. Webb

SpS1: Glenn M. Wahlgren, Hans Ulrich Käufl & Florian Kerber
SpS2: Pedro Russo, Catherine J. Cesarsky & Lars Lindberg Christensen
SpS3: Michael G. Burton
SpS4: Jean-Pierre de Greve, Edward F. Guinan & Magda G. Stavinschi
SpS5: Ray P. Norris & Clive L. N. Ruggles
SpS6: Régis Courtin, Alan P. Boss & Michel Mayor
SpS7: Jane C. Gregório-Hetem & Silvia Alencar
SpS8: Melvin Hoare & Janet E. Drew
SpS9: Terence J. Mahoney
SpS10: Gerard F. Gilmore & Richard T. Schilizzi

INTERNATIONAL ASTRONOMICAL UNION

UNION ASTRONOMIQUE INTERNATIONALE

HIGHLIGHTS OF ASTRONOMY VOLUME 15

AS PRESENTED AT THE
XXVII IAU GENERAL ASSEMBLY
RIO DE JANEIRO, BRAZIL, 2009

Edited by

IAN F. CORBETT
General Secretary

CAMBRIDGE
UNIVERSITY PRESS

Shaftesbury Road, Cambridge CB2 8EA, United Kingdom

One Liberty Plaza, 20th Floor, New York, NY 10006, USA

477 Williamstown Road, Port Melbourne, VIC 3207, Australia

314–321, 3rd Floor, Plot 3, Splendor Forum, Jasola District Centre, New Delhi – 110025, India

103 Penang Road, #05–06/07, Visioncrest Commercial, Singapore 238467

Cambridge University Press is part of Cambridge University Press & Assessment, a department of the University of Cambridge.

We share the University's mission to contribute to society through the pursuit of education, learning and research at the highest international levels of excellence.

www.cambridge.org
Information on this title: www.cambridge.org/9781107005334

First published 2010

A catalogue record for this publication is available from the British Library

ISBN   978-1-107-00533-4   Hardback

Table of Contents

JD8 HOT INTERSTELLAR MATTER IN ELLIPTICAL GALAXIES:

JD11 NEW ADVANCES IN HELIO- AND ASTERO-SEISMOLOGY

SpS 4: ASTRONOMY EDUCATION BETWEEN PAST AND FUTURE

SpS7 YOUNG STARS, BROWN DWARFS, AND PROTOPLANETARY DISKS . . . .

Preface

Highlights of Astronomy 15 records the scientific sessions which took place during the XXVII IAU General Assembly held in the Centro de Convenções SulAmérica during the International Year of Astronomy 2009, in the magnificent city of Rio de Janeiro, Brazil, hosted by the Brazilian Astronomical Society (Sociedad Astromomica Brasiliera, SAB).

The XXVII Assembly offered a rich scientific programme organized by the General Secretary Karel. A. van der Hucht. There were four Invited Discourses, six Symposia (IAUS 262 – 267), 16 Joint Discussions, 10 Special Sessions, and 7 scientific sessions in the course of Divisional Business meetings. A major, and very successful, innovation was the inclusion of Plenary Reviews at the start of the day, the speakers beng drawn from the 6 symposia. This is a feature likely to be repeated at future General Assemblies.

The proceedings of the six IAU GA Symposia IAUS 262 – 267 have been published in the regular *IAU Symposium Proceedings Series* by Cambridge University Press. The Plenary Review corresponding to each symposium will be included in the proceedings of that symposium. I am extremely grateful to the editors and the organizing committees of all the Symposia, Joint Discussions, and Special Sessions for their hard work in producing the manuscripts for the Proceedings and for this volume of Highlights. This would be impossible without their efforts.

The business proceedings of the General Assembly were published in May 2010 by CUP as Transactions of the International Astronomical Union XXVIIB.

Financial support for a limited number of the participants was provided by the IAU, and the invaluable support of all the sponsors is gratefully acknowledged.

It is my pleasure to thank Karel van der Hucht who pulled together the scientific programme, and the staff of the IAU Secretariat, Mme Vivien Reuter and Mme Jana Zilova, for their invaluable assistance in preparing for the General Assembly and then looking after attendees during the meeting, and particularly handling all the grants. I extend particular thanks to James Binney, organiser and editor of Joint Discussion 5, for giving me a powerful LaTeX tool to edit and assemble these Highlights.

Finally, and most importantly, we are all most grateful to Daniela Lazarro, Beatriz Barbuy and every member of the National Organizing Committee, and its sub-committees, for a most memorable XXVII General Assembly.

Ian F. Corbett
IAU General Secretary
Paris, May 2010

Highlights of Astronomy, Volume 15
XXVIIth IAU General Assembly, August 2009
Ian F. Corbett, ed.

MEASURING THE HUBBLE CONSTANT WITH THE HUBBLE SPACE TELESCOPE†

W. L. Freedman, R. C. Kennicutt & J. R. Mould

Abstract. Ten years ago our team completed the Hubble Space Telescope Key Project on the extragalactic distance scale. Cepheids were detected in some 25 galaxies and used to calibrate four secondary distance indicators that reach out into the expansion field beyond the noise of galaxy peculiar velocities. The result was $H_0 = 72 \pm 8$ km s^{-1} Mpc^{-1} and put an end to galaxy distances uncertain by a factor of two. This work has been awarded the Gruber Prize in Cosmology for 2009.

1. Introduction

Our story begins in the mid-1980s with community-wide discussions on Key Projects for the NASA/ESA Hubble Space Telescope. The extragalactic distance scale was the perfect example of a project that would be awarded a generous allocation of time in order that vital scientific goals would be accomplished even if the lifetime of the telescope proved to be short. As Marc Aaronson (Figure 1), the original principal investigator of the project, said in his Pierce Prize Lecture in 1985, "The distance scale path has been a long and torturous one, but with the imminent launch of HST there is good reason to believe that the end is finally in sight."

Figure 1. Jeremy Mould (left) and Marc Aaronson (right) at the van Biesbroeck Prize award ceremony in 1981.

Marc Aaronson died tragically in an accident in 1987, having written a successful proposal for the Key Project, a project designed to shrink the scatter shown in Figure 2 to 10% and put an end to 60 years of debate, commencing with Hubble's estimates in 1929.

† Based on observations made with the NASA/ESA Hubble Space Telescope obtained at the Space Telescope Science Institute which is operated by AURA for NASA.

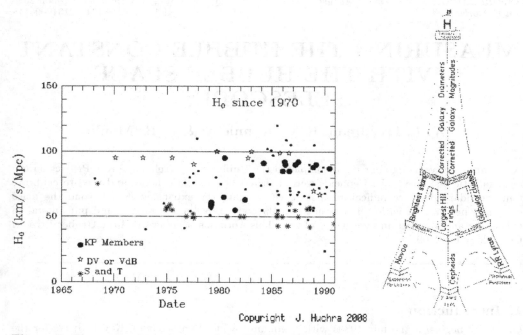

Figure 2. The scatter in estimates of the Hubble Constant showed no sign of convergence in the 1970s and 1980s. On the right is a schematic distance scale by G. de Vaucouleurs, one of the protagonists in the distance scale controversy.

The principal reason for the uncertainty in H_0 is evident in Figure 3. Large scale structure is seen out to distances of 100 Mpc. Ground-based Cepheid distances, however, extended to only 4 Mpc with a "twilight zone" beyond (Figure 4).

Figure 3. Colour coded redshifts show the large scale structure in the nearby Universe (galactic coordinates).

The Key Project's solution to the twilight zone problem was to map Cepheids out to 20 Mpc and calibrate secondary distance indicators within this volume. The secondary indicators would extend the distance scale out to 100 Mpc.

ALLAN SANDAGE AND G. A. TAMMANN

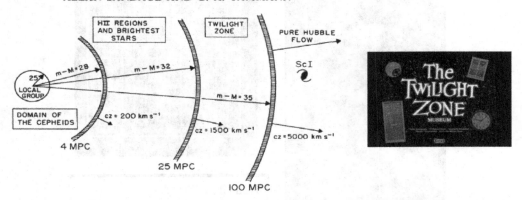

Figure 4. A schematic distance scale by other protagonists of the controversy, A. Sandage & G. Tammann.

An important tenet of the Key Project was to exploit the redundancy of distance indicators, as shown in Figure 5, especially the four secondary distance indicators, the Tully-Fisher relation, surface brightness fluctuations, supernovae, and the fundamental plane.

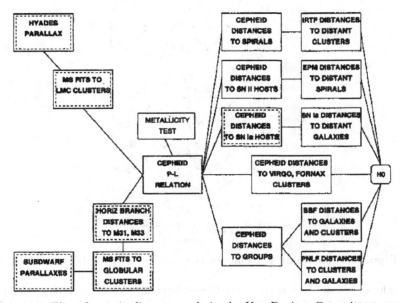

Figure 5. The schematic distance scale in the Key Project Committee report.

2. Cepheids

The Cepheid Period-Luminosity law was discovered by Leavitt early in the twentieth century.

By the end of the century much was understood about the systematics of Cepheids. For example (Madore & Freedman 1991), Cepheid amplitudes are maximum in the blue, and interstellar absorption is minimum at long wavelengths (Figure 7). Therefore, the best strategy for discovering Cepheids is to observe at visible wavelengths; to minimize the effect of dust, luminosities are best measured in the infrared.

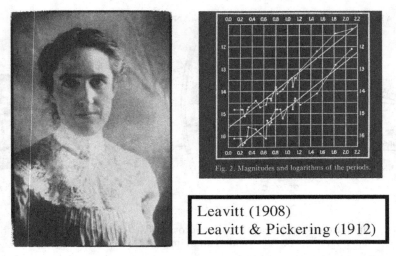

Leavitt (1908)
Leavitt & Pickering (1912)

Figure 6. Henrietta Leavitt and the period-luminosity relation in the Magellanic Clouds.

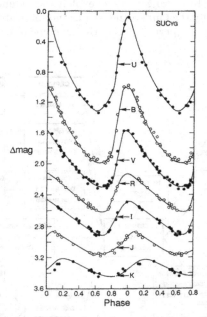

Figure 7. Amplitude as a function of wavelength for Cepheid variables.

HST brought a number of strengths to the Cepheid distance scale: linear detectors, multiwavelength observations, and a planned cadence of observations. Figure 8 shows for M33 (Freedman *et al.* 1991) how an absolute distance modulus is obtained with knowledge of the reddening law.

Figure 9 shows how a power law observing cadence is superior for luminosity measurement to equally spaced observations (Madore & Freedman 2002, 2005).

The results are shown in Figure 10. The light curves for these periods are unmistakably Cepheids. Our observations populated the range 10–100 days in galaxies with distances of order 10 Mpc.

Figure 11 shows a typical field placement for the Key Project in the large face-on spiral galaxy in the Virgo cluster, M100. Twelve V observations were obtained during the sequence and four I observations. Cosmic ray splits were used and a fixed roll angle was adopted. Photometry was

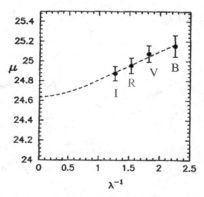

Figure 8. Finding the distance of M33 (Freedman *et al.* 1991).

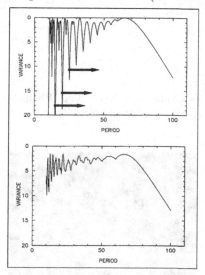

Figure 9. Observing cadence. Upper panel: equally spaced observations. Lower panel: power law cadence.

carried out on the frames using DoPhot and a custom version of DAOPHOT called ALLFRAME (Stetson 1994).

A composite I-band period-luminosity relation for 800 Cepheids in 24 galaxies is shown in Figure 12 (corrected for distance) (Ferrarese *et al.* 2000).

Ferrarese *et al.* (2000) also carried out a comprehensive comparison of Key Project distances with other prominent distance indicators, such as the tip of the red giant branch (e.g. Sakai *et al.* 1997) and surface brightness fluctuations (Tonry *et al.* 2001) (Figure 13) and the planetary nebula luminosity function (Ciardullo & Jacoby 1992) and the globular cluster luminosity function (Secker & Harris 1993) (Figure 14).

3. Metallicity Calibration

The Cepheid period luminosity relation is a straightforward primary distance indicator if the precepts described above are followed. But there is a complication. Cepheids vary in their chemical composition, and the period luminosity relation is affected. We can write

$$(m - M)_{true} = (m - M)_{PL} - \gamma \log(Z/Z_{LMC})$$

where Z is the metallicity of the field and Z_{LMC} is the metallicity of the LMC.

However, theory is not predictive, even about the sign of γ. According to Chiosi, Wood &

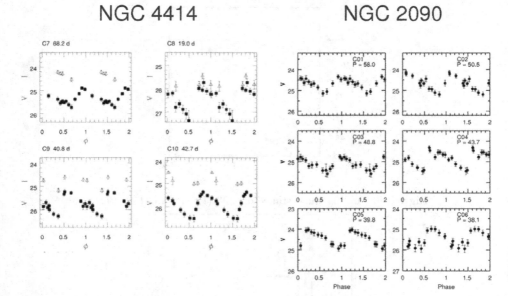

Figure 10. Light curves for Cepheids in NGC 2090 (Phelps *et al.* 1998) and NGC 4414 (Turner *et al.* 1998).

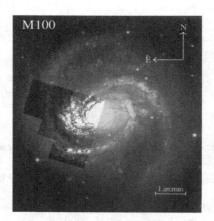

Figure 11. The WFPC2 footprint is shown on an image of M100.

Capitanio (1993) $\gamma_{VI} = -0.11$ mag/dex, but elsewhere we find with different opacities $\gamma_{VI} = +0.06$ mag/dex. The Key Project took an empirical approach, described by Kennicutt (1998). Both an inner and an outer field were observed in M101 (Figure 15), a galaxy with a large abundance gradient (Figure 16). The Cepheid metallicities were assumed to follow the oxygen abundances of nearby HII regions. The inner field modulus was found to be 29.20 ± 0.09 mag, and the outer modulus was 29.36 ± 0.08 mag. There was a factor of five difference in Z. This yielded $\gamma = -0.24 \pm 0.16$. Oxygen abundances were measured for each of the Key Project galaxies by Zaritsky, Kennicutt, and Huchra (1994).

Confirmation of this result comes from a comparison of tip of the red giant branch distances (TRGB) and Cepheid distances by Ferrarese *et al.* (2000) (Figure 17).

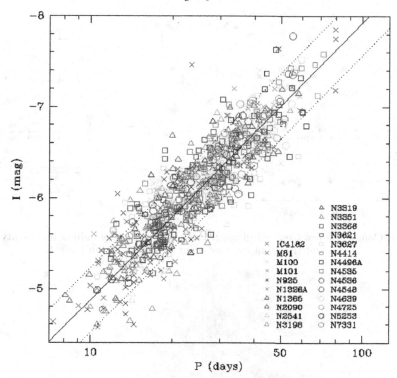

Figure 12. Composite I-band PL relation for the Key Project.

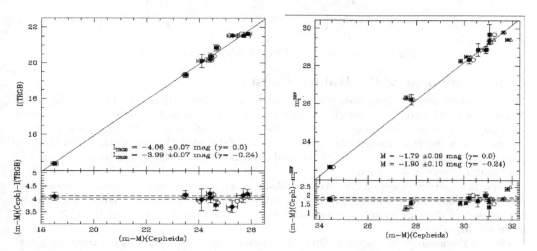

Figure 13. Comparison of Cepheid distances with the tip of the red giant (left) and surface brightness fluctuations (right).

More recent work has strengthened these results. Mould & Sakai (2008, 2009ab) have shown that substitution of TRGB distances for Cepheid distances in secondary distance indicator calibrations returns a Hubble Constant in agreement with the Key Project. And Scowcroft *et al.* (2009) obtained $\gamma_{VI} = -0.26$ in a study of M33.

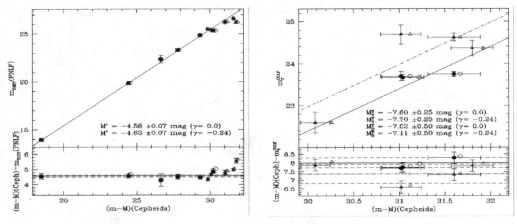

Figure 14. Comparison of Cepheid distances with the planetary nebula luminosity function (left) and the globular cluster luminosity function (right).

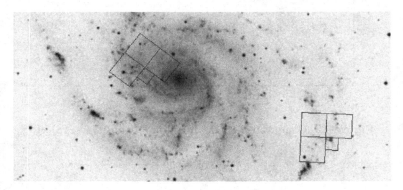

Figure 15. The two WFPC2 fields observed in M101.

4. Measurement of the Hubble Constant

Four secondary distance indicators were calibrated by the Key Project. The first was the Tully-Fisher relation (Figure 18). Sakai *et al.* (2000) obtained $H_0 = 71 \pm 3 \pm 7$ km s^{-1} Mpc^{-1}, where the first uncertainty is the random error and the second uncertainty is the systematic error.

Next comes the fundamental plane. Kelson *et al.* (2000) used Cepheid distances to the Leo, Virgo, and Fornax clusters to calibrate the fundamental plane and the D_n, σ relation (Figure 20), obtaining $H_0 = 78 \pm 5 \pm 9$ km s^{-1} Mpc^{-1}.

Type Ia supernovae were calibrated by Gibson *et al.* (2000), using 6 supernova hosts with measured decline rates, some of them reworked from Saha *et al.* (1999). Figure 21 shows the application of the calibration to supernovae out to 30,000 km s^{-1}, yielding $H_0 = 71 \pm 2 \pm 7$ km s^{-1} Mpc^{-1}.

Finally, Ferrarese *et al.* (2000a) calibrated surface brightness fluctuations in early type galaxies, obtaining $H_0 = 69 \pm 4 \pm 6$ km s^{-1} Mpc^{-1}.

A good summary of the results is provided by Figure 22, which shows the Cepheids and the calibrated distance indicators to a redshift of 0.1, well beyond the effect of local velocity field perturbations. Hubble's (1929) distances are confined to the first tick mark in Figure 23.

The uncertainties in the Hubble Constant remain dominated by systematic errors. The first of these is reddening, and this has been tested in work by Macri *et al.* (2001), who reobserved many of the Key Project Cepheids with the NICMOS infrared camera. Their results for M81 are shown in Figure 24. Overall, the H band distances agreed with the Key Project distances to 1%.

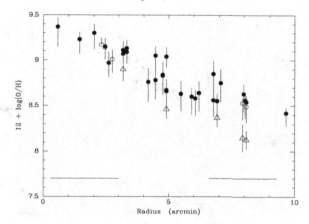

Figure 16. The gradient in oxygen abundance in M101.

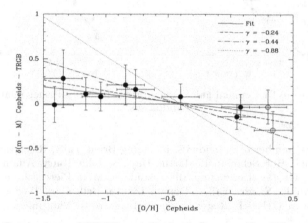

Figure 17. Comparison of TRGB and Cepheid distances for different values of γ.

We have dealt with the second systematic error, metallicity differences with the Large Magellanic Cloud (LMC), but the LMC remains the principal systematic error because of the assumption of its distance modulus, 18.50. A second fiducial distance has subsequently become available in the maser distance of NGC 4258. Herrnstein *et al.* (1999) obtained the distance of this galaxy by fitting a simple kinematic model to the maser radial velocities and VLBI proper motions (7.3 ± 0.4 Mpc). The Cepheid distance is 7.5 ± 0.3 Mpc (Macri *et al.* 2006). In addition, HST trigonometric parallaxes for a sample of Cepheids have become available (Benedict *et al.* 2007), confirming the Key Project period-luminosity calibration, as shown in Figure 25.

The probability distribution for H_0, combining the results of the secondary distance indicators and the full error budget of the Key Project is shown in Figure 26 (Freedman *et al.* 2001).

5. The Team

Most of the team members (many referenced above) are captured in Figure 27. Special roles were played by the individuals depicted in Figure 28 and 29.

Other notable contributions were made by John Graham, Nancy Silbermann, Randy Phelps, Daya Rawson, Fabio Bresolin, Lucas Macri, Bob Hill, Kim Sebo, Paul Harding, Anne Turner, Han Ming Sheng, Shaun Hughes, Charles Prosser, John Huchra, Holland Ford, and Garth Illingworth. Jim Gunn, Sandra Faber and John Hoessel were instrument team liaisons. The team drew

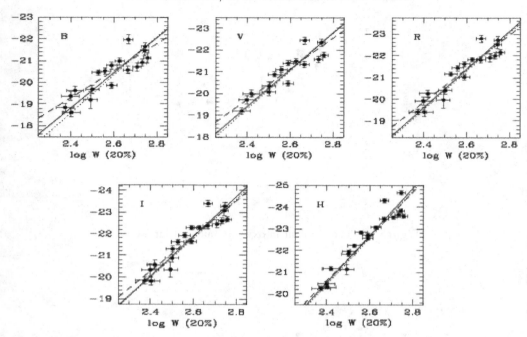

Figure 18. Distances to 21 Cepheid galaxies calibrate the Tully-Fisher relation in the BVRIH bandpasses.

on work from a large number of individuals, including Brent Tully, Riccardo Giovanelli, Mario Hamuy, Mark Phillips, Bob Schommer, Martha Haynes, John Tonry, Adam Riess, Bob Kirshner, Brian Schmidt, Gustav Tammann, Allan Sandage, Mike Pierce, John Blakeslee, George Jacoby, Robin Ciardullo, Sandra Faber, Donald Lynden-Bell, Gary Wegner, David Burstein, Alan Dressler, Roberto Terlevich, Roger Davies and Gerard de Vaucouleurs.

6. Subsequent Observations and Future Prospects

A number of important observations have been made since the publication of the Key Project results. The Supernova and H_0 for the Equation of State (SHOES) project has carried out a differential analysis of the supernova hosts NGC 4536, 4639, 3982, 3370, 3021 and 1309 with respect to NGC 4258. This eliminates uncertainties such as photometric transformations and crucially, the LMC distance. The maser distance to NGC 4258 is assumed instead. In this way, Riess *et al.* (2009) obtain $H_0 = 74.2 \pm 3.6$ km s^{-1} Mpc^{-1}. They also find a value of w in the equation of state w = P/ρ = 1.12 $\pm$ 0.12.

Most importantly, the Hubble Constant has been deduced from the position of the first acoustic peak in the small scale anisotropy of the cosmic microwave background. Given the sound horizon on the surface of last scattering (143 $\pm$ 4 Mpc) and the angular size of the first acoustic peak (0.601 $\pm$ 0.005), one obtains an angular diameter distance for the surface of last scattering of 13.7 $\pm$ 0.4 Gpc. Assuming $\Omega_M = 0.3$ and $\Omega_\Lambda = 0.7$, this yields a value of H_0 of 70 km s^{-1} Mpc^{-1}. Solving for all the cosmological model parameters, Komatsu *et al.* (2009) find $H_0 = 70.5 \pm 1.3$ km s^{-1} Mpc^{-1}.

Progress will continue in the classical distance scale of the Key Project too. The NASA mission SIM-Lite is expected to yield a rotational parallax from velocities and proper motions for the galaxies M31 and M33 to 1% accuracy. This will provide a definitive Cepheid period-luminosity relation.

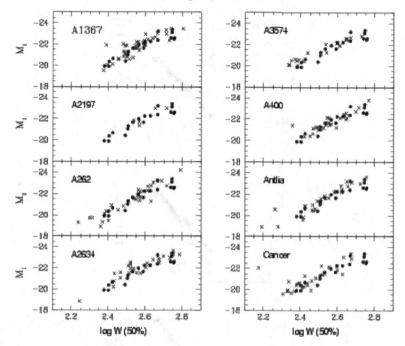

Figure 19. Tully-Fisher distances to clusters of galaxies yield the Hubble Constant.

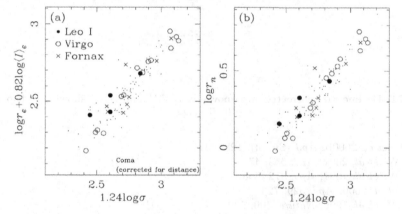

Figure 20. Calibration of the fundamental plane (left) and the D_n, σ relation (right). This is superposed on the Coma cluster (dots).

Acknowledgements

We wish to thank NASA for designating this a Key Project for the Hubble Space Telescope. We wish to thank the team for making it a success. Finally, we wish to thank the Gruber Foundation for recognizing this work.

References

Benedict, G. *et al.* 2007, *AJ*, 133, 1810
Ciardullo, R. & Jacoby, G. 1992, *ApJ*, 388, 268
Chiosi, C., Wood, P., & Capitanio, N. 1993, *ApJ* (Suppl), 86, 541
Ferrarese, L. *et al.* 2000a, *ApJ*, 529, 745

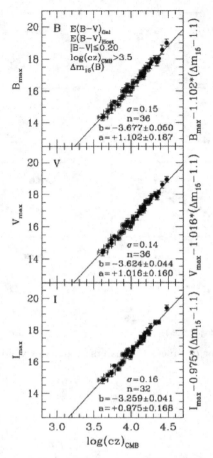

Figure 21. Decline rate corrected supernovae in B,V,I bandpasses showing a dispersion of 0.16 mag.

Ferrarese, L. *et al.* 2000b, *ApJ* (Suppl), 128, 431

Freedman, W. *et al.* 2001, *ApJ*, 553, 47

Freedman, W. *et al.* 1991, *ApJ*, 372, 455

Gibson, B. *et al.* 2000, *ApJ*, 529, 723

Herrnstein, R. *et al.* 1999, *Nature*, 400, 539

Hubble, E. 1929, *Proc. N.A.S.*, 15, 168

Kelson, D. *et al.* 2000, *ApJ* 529, 768

Kennicutt, R. *et al.* 1998, *ApJ*, 498, 181

Komatsu, E. *et al.* 1998, *ApJ* (Suppl), 180, 330

Leavitt, H. 1908, *Ann. Harvard CO*, 60, 87

Leavitt, H. & Pickering, E. 1912, *Harvard CO Circ.*, 173, 1

Macri, L. *et al.* 2006, *ApJ*, 652, 1133

Madore, B. & Freedman, W. 1991, *PASP*, 103, 933

Madore, B. & Freedman, W. 2002, *SPIE*, 4847, 156

Madore, B. & Freedman, W. 2005, *ApJ*, 630, 1054

Mould, J. & Sakai, S. 2008, *ApJ*, 686, 75

Mould, J. & Sakai, S. 2009a, *ApJ*, 694, 1331

Mould, J. & Sakai, S. 2009b, *ApJ*, 697, 996

Phelps, R. *et al.* 1998, *ApJ*, 500, 763

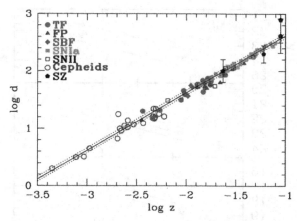

Figure 22. Redshifts and distances for Cepheids, Tully-Fisher clusters, surface brightness fluctuations, the fundamental plane, supernovae of type Ia and type II and clusters with Sunyaev-Zeldovich distances.

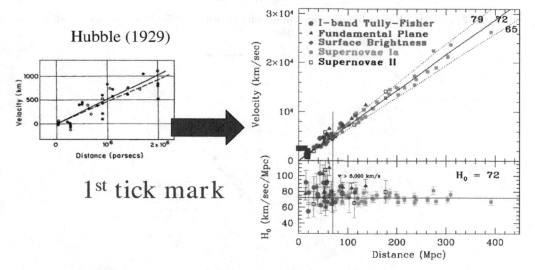

Figure 23. The combined distance indicators yield $H_0 = 72$ km s^{-1} Mpc^{-1}.

Riess, A. *et al.* 2009, *ApJ*, 699, 539

Saha, A. *et al.* 1999, *ApJ*, 522, 802

Sakai, S. *et al.* 1997, *ApJ*, 478, 49

Sakai, S. *et al.* 2000, *ApJ*, 529, 698

Sandage, A. *et al.* 2004, *A&A*, 424, 43

Scowcroft, V. *et al.* 2009, *MNRAS*, 396, 1287

Secker, J. & Harris, W. 1993, *AJ*, 105, 1358

Stetson, P. 1994, *PASP*, 106, 250

Tonry, J. *et al.* 2001, *ApJ*, 546, 681

Turner *et al.* 1998, *ApJ*, 505, 207

Zaritsky, D., Kennicutt, R., & Huchra, J. 1994, *ApJ*, 420, 87

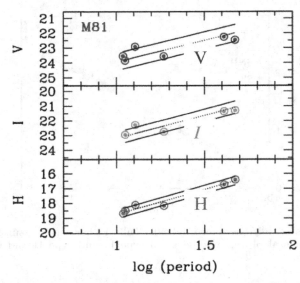

Figure 24. A NICMOS H band period luminosity relation for M81 is compared with Key Project results.

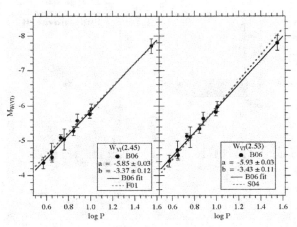

Figure 25. HST parallaxes for Galactic Cepheids yield the period-luminosity relation shown by the filled circles. The Freedman *et al.* (2001) PL relation is shown on the left and the Sandage *et al.* (2004) PL relation is shown on the right. The former is a better fit for the periods in excess of 10 days which were used in the Key Project.

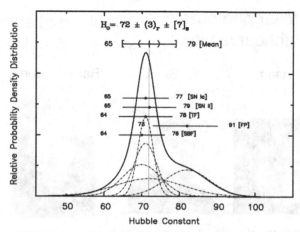

Figure 26. The probability distribution emerging from the Key Project error analysis.

Figure 27. The team photographed at the Aspen Center for Physics in 1999. From left to right: Brad Gibson, Peter Stetson, Shaun Hughes, Laura Ferrarese, Lucas Macri, Dan Kelson, Wendy Freedman, John Graham, Rob Kennicutt, John Huchra, Kim Sebo, Jeremy Mould, Shoko Sakai, Garth Illingworth and Nancy Silbermann

stellar photometry   Cepheid observing
light curve fitting   strategy & calibration

Peter Stetson   Abi Saha   Barry Madore

Figure 28. Special roles were played by three team members.

Shoko Sakai   Laura Ferrarese   Brad Gibson   Dan Kelson

Tully-Fisher relation SBF   Type Ia supernovae Dn–σ relation

Figure 29. Final papers on particular secondary distance indicators were written by four team members.

Highlights of Astronomy, Volume 15
XXVIIth IAU General Assembly, August 2009
Ian F. Corbett, ed.

The Legacies of Galileo

Franco Pacini

Arcetri Astrophysical Observatory, University of Florence, Italy
email: pacini@arcetri.astro.it

1. Introduction

The scientific community is celebrating in 2009 the International Year of Astronomy. The timing coincides with the 400th anniversary of the first astronomical use of a telescope, when Galileo's observations demonstrated that the Earth is not alone in the Universe. One can hardly think of a more important event in the history of mankind.

Figure 1. The poster announcing the open-house of the Arcetri Observatory in 2007 for children

In 2009 we also celebrate the 40th anniversary of the first human landing on the Moon and the 400th anniversary of Kepler's laws. For the occasion, I was asked, several months ago, by the IAU General Secretary to give a Discourse about Galileo during our Assembly in Rio. I am honoured by the request but I am also very concerned, since I am not an historian of Science. The only justification for my talk could be that I have had the fortune of working for about 30 years at the Arcetri Observatory, just outside Florence. This Observatory was erected in 1872 as a memorial to Galileo, in a location where the natural beauty combines with the importance of the historical heritage.

2. The Astronomical Discoveries of Galileo

It is generally agreed that Galileo did not invent the telescope. Similar instruments had existed for some time. Some of them had been used for astronomical viewing but the quality of the lenses was poor and did not lead to significant results. Galileo set for

Figure 2. Galileo Galilei

Figure 3. A view of Arcetri Observatory

himself the goal of constructing better lenses and using them for building an instrument at the cutting edge of technology.

Indeed, one of the most important legacies of Galileo is the connection of technological progress with the advancement in pure knowledge. Scientific progress in astronomy and in other areas of science has always been dependent on this connection. Over the centuries, the big steps in science (astronomy in particular) have generally been associated with innovative instrumentation. Perhaps the efforts made by Galileo to obtain a sharp view of the lunar surface can be considered the equivalent of modern advances in interferometry and adaptive optics.

When Galileo turned his telescope towards the sky, he realized that the moon is another world, with mountains, craters, plains. The similitude between lunar and terrestrial features suggests that the Earth and the Moon are made of similar materials. However, it took a long time (centuries) for astronomers to establish the identity of cosmic matter and physical laws across the whole Universe. This has been one of the main achievements of modern astrophysics.

Following the observations of the Moon (Winter 1609), Galileo made a series of additional discoveries which can be listed as follows: In both cases the observations provided support for the Copernican view of the Planetary System.

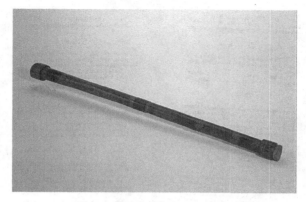

Figure 4. Galileo's telescope

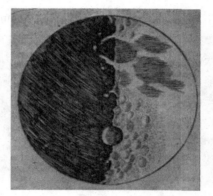

Figure 5. Half Moon as observed by Galileo

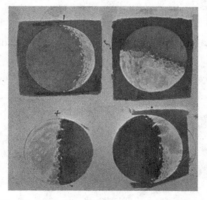

Figure 6. Phases of the Moon as observed by Galileo

Observations showed that Saturn had an elongated shape, indicating the presence of external material, the famous rings.

He saw that the number of stars visible with a telescope exceeds by far those visible with the naked eye. In addition, the Milky Way can be resolved into a very large number of stars, too many and too distant for seeing them individually with the naked eye.

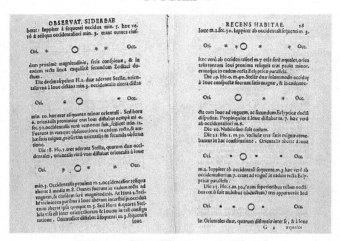

Figure 7. The 4 main satellites of Jupiter(January 1610)

Figure 8. The phases of the planet Venus(October 1610)

Figure 9. Saturn with its rings (July 1610)

The first results of Galileo's observations, made over a few months during the winter 1609-1610 were published in early 1610 in the Sidereus Nuncius. We can note that, at that time, the speed of publication was certainly much faster than at present!

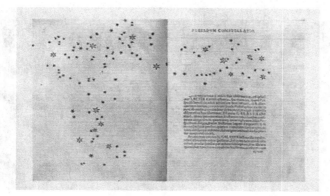

Figure 10. The Milky Way observed as individual stars

Figure 11. Sidereus Nuncius

3. Galileo in Arcetri

Galileo spent the last part of his life in Arcetri (1631 until 1642) in the Villa Gioiello. At that time he had already revolutionized astronomy, thanks to the observations carried out in Padova and Venice around 1610.

"Il Gioiello" was the typical residence of a land owner, overviewing a landscape rich of olive trees, vineyards, fruits. Roughly 400 years later, from the Observatory, we can admire a huge old oaktree and we can imagine Galileo resting under the shadow of this tree after a walk in the fields.

Galileo rented Il Gioiello in 1631, following a suggestion of his daughter Suor Maria Celeste who was a nun in a convent nearby. The signature of the rental contract took place in September 1631. The contract describes the details of the Villa which are nearly identical to those which one can presently find in the property.

When moving inside the Villa it is hard to resist a strong emotional feeling. Unfortunately, Galileo was not able to enjoy for a long time the quiet of Arcetri and the pleasure of being very close to his daughter. Indeed, in 1633, he was tried by the Church because

Figure 12. "Il Gioiello"

Figure 13. The Entrance

Figure 14. The internal Courtyard – now

he had published one of his fundamental works "Dialogo dei massimi sistemi". The Church did not accept Galileo's arguments and ordered that his work be included in the list of forbidden books. From a general point of view it is not surprising that Galileo's

Figure 15. The internal Courtyard – then

Figure 16. A view from the Observatory

Figure 17. Galileo's daughter, Suor Maria Celeste

discoveries led to conflicts with theology and philosophy: the Church could hardly tolerate such disagreements, roughly at the time of the Protestant Reformation.

However, because of his abjuration Galileo was spared the destiny of Giordano Bruno who, a few years earlier (January 1600), had been burned for heresy in a square of Rome

Figure 18. Galileo abjuration

(Campo dei Fiori). Bruno had supported the view that the Earth is not the center of the Universe and the existence of a multiplicity of worlds.

While in Arcetri, Galileo received important visitors such as the poet John Milton and his students Torricelli and Viviani. In 1638 he became blind and died in Arcetri on January 8, 1642.

Figure 19. Giordano Bruno in Campo dei Fiori in Rome

4. Galileo and the Scientific Method

Though Galileo is famous especially for his astronomical discoveries, he is also generally recognized as one of the main founders of modern science. There are many examples which support this claim. The most important is probably his rejection of the principle of authority ("ipse dixit") and the adoption of the experimental method. In many occasions Galileo stressed the importance of simplified experiments, evidencing the basic aspects of a physical phenomenon, clarifying the details. Some of his results involve what we call "conceptual experiments" (gedanken experiment). In addition to the famous discussion about the motion of a body falling inside a ship, another "conceptual experiment" was the attempt to measure the speed of light. Unfortunately light moves too fast and Galileo did not have the technology necessary to measure its speed.

On the occasion of the Year of Astronomy, the Arcetri Observatory, in collaboration with the Museum for the History of Science (both in Florence), reinvestigated the properties of Galileo's telescopes and evaluated their performance with modern techniques. Francesco Palla and his colleagues at the Arcetri Observatory have used the Sidereus Nuncius as a sort of guide for the observations. This has made it possible to compare modern observations with Galileo's drawings. The agreement is very good. It is impressive to look at a portion of the sky and to find the same stars which were catalogued by the great scientist. Furthermore, the image of Saturn clearly shows the elongated nature due to the presence of the rings.

Figure 20. The Historical Arcetri Telescope "Amici"

Figure 21. Saturn view with a Galileoscope

5. Conclusions

The life of Galileo was marked by many dramatic events, generally related to conflicts with the Church. There can be no doubt that the discoveries of Galileo are a milestone in the cultural history of mankind, one of the most important legacies, at the border between Science, Philosophy, Religion. In addition, Galileo was well aware of the importance of science in society.

He stated clearly that scientists should write in the common language of the people, rather than in Latin, because everybody should be informed and understand the new discoveries. This attitude is fundamental even in today's world, in a society dominated by Science and Technology. In many countries, especially in Italy, the Church was able to keep some control over the development of science. In other countries, Galileo's discoveries had a broader impact. One can perhaps argue that Galileo's discoveries also opened the way to science fiction and to more exchanges between art and science.

Several famous paintings (such as the one by Creti) were inspired by astronomy.

Figure 22. Jupiter in a painting by Creti

An alternative example is Paul Fontanelle who, in France, published an amusing dialogue narrating the speculations of a Philosopher with a young, pretty Lady, engaged in a variety of discussions about the possibility of extraterrestrial life.

Some years ago, the National Academy of Science of the United States stated that, at the present epoch, the number and the importance of astronomical discoveries nearly matches what happened at the time of Galileo.

The Year of Astronomy was conceived to give to the citizens of the world a special feeling for our science, following Seneca's statement *"If the stars were visible from just one place on Earth, people would never stop from travelling on that place in order to see them"*

Children are particularly fascinated by the sky. Unfortunately, in many parts of the world, stars are hidden by the light of the bombs falling from the sky. **Let's hope that the stars will soon be back.**

Figure 23. The poster announcing the open-house of the Arcetri Observatory in 2008 for children

Highlights of Astronomy, Volume 15
XXVIIth IAU General Assembly, August 2009
Ian F. Corbett, ed.

© International Astronomical Union 2010
doi:10.1017/S1743921310008161

Water on Planets

James F Bell III

Dept. of Astronomy, Cornell University, 402 Space Sciences Bldg., Ithaca NY 14853 USA
email: jfb8@cornell.edu

Abstract. Water is an abundant molecule in the Cosmos. It has exploitable and unique spectroscopic and physical properties and has been found to be ubiquitous in places that we would expect in the standard model of solar system formation and nebular condensation: beyond the snow line in outer solar system planets, moons, asteroids, and comets. However, water is also an important constituent of planetary bodies (dominating at least one of their surfaces) in the inner solar system, likely indicating significant mixing between inner and outer solar system reservoirs of water during planetary accretion and the early history of the solar system. Water has played a critical role in the differential evolution of the terrestrial planets Venus, Earth, and Mars, and the concept of the "habitable zone" where liquid water could be stable on an Earth-like planet provides a starting point for assessing the habitability of worlds in our solar system and beyond. Examples of potentially habitable environments outside this zone in our own solar system warn us that this concept should only be a guide, however-important exceptions will no doubt occur. Recent discoveries of past liquid water and abundant present subsurface ice on Mars, of water reservoirs in unexpected places like the poles of Mercury and the Moon and the subsurface of Enceladus, of water in circumstellar disks and in the atmospheres of extrasolar planets, and the expectation of the discovery of water on Earth-like worlds in the habitable zones around other stars make this an exciting time in the study of water on planets both in our own solar system, and beyond.

Keywords. water, planets, satellites, asteroids, comets, extrasolar planets, habitable zone

1. Introduction

The topic of "water on planets" covers an expansive, literally astronomical, range of objects and scientific realms. Aspects of physics, chemistry, astronomy, geology, and even biology all come into play when considering the origin and influence of water on the history and evolution of planetary (and satellite, asteroidal, cometary, and even dust grain) surfaces, interiors, and atmospheres. Outstanding reviews of this topic have been presented recently by Chambers, 2004, Drake, 2005, Encrenaz, 2008, Owen, 2008, and Albarede, 2009.

Any review of this broad topic is almost certain to be biased by the reviewer's own background, experience, and perception of the role of water in their own specific scientific studies. My own bias is to approach this topic with the perspective of a geologist. That is, on one level, to think about water on planets in a forensic sense: what clues can we observe on planets or other solar system (or extrasolar) objects that can reveal the role, if any, that water plays or has played in the history of those bodies? On another level, a geologic approach to the study of water also includes a built-in need to understand not only the specific and quantitative local observations, but also the bigger-picture, qualitative context of water's influence on classes of planetary objects and/or regions of planetary and interplanetary space. Such a perspective lends itself to the classic "What, Where, When, and Why" approach to studying this topic. Perhaps this approach is a bit cliché, but for as broad a topic as water on planets it can provide a useful framework within which to begin more specific, focused studies.

Another bias that I tip-toed around above has to do with the definition of "planet." While the IAU has recently adopted a controversial new, dynamically-based definition (IAU, 2006), my own personal working "definition" focuses on the physical and/or geophysical properties of each object itself, rather than on its particular location or orbit. For example, if an object is large enough for its own gravity to have given it a rounded shape, for internal processes such as differentiation to have occurred, for internal gravitational and/or radioactive heat sources to have enabled some form of volcanism and possibly even volatile release to form an atmosphere-these kinds of characteristics define, for me, an object that I would call a "planet." Such a definition is perhaps not so uncommon among astronomers and planetary scientists (e.g., PSI, 2006; Sykes, 2009). More importantly, it enables a more thorough, complete review and understanding of the role and influence of water on a much wider sampling of solar system bodies than a narrow definition of "planets" would otherwise allow. Why, for example, should we consider Mercury a planet and not the comparable or larger (and more water-rich) bodies Callisto, Titan, and Ganymede? By this definition, our solar system has some 35 to 40 "planets" and likely even more yet to be discovered in the Kuiper Belt.

Having made these various confessions of biases, let us proceed to review and survey the exciting and topical study of water on planets. This review begins with an overview focused on the question of **What** is water? Specifically, what are the properties of the water molecule that make it a rich object of study as well as special from the standpoint of planetary and chemical/biologic processes. I then discuss the question of **Where** do we find water (in all its forms)? Thinking about this question in our own solar system as well as extrasolar planetary and protoplanetary systems will naturally lead to the introduction of key concepts like the snow line and the habitable zone. Next I digress a bit to the specific questions of **When** (and where) has water been liquid, and for how long. The digression is worth considering because the presence and stability of liquid water is a key requirement for assessing the potential habitability of worlds-especially Earth-like worlds. Finally, I conclude by wandering a bit into more philosophical territory and considering the **Why**. That is, I attempt to explain why, in both the scientific and sociological senses, we as astronomers, planetary scientists, educators, and general students of the Cosmos, must care about and aggressively pursue the study of water on planets.

2. Consider the H_2O molecule...

Hydrogen and Oxygen are the #1 and #3 most abundant elements in the Universe, and so it should not be surprising that H_2O is an abundant and ubiquitous molecule. H_2O has a net dipole moment, meaning that it is endowed with interesting physical and chemical properties as well as some important exploitable properties from the perspective of remote sensing. For example, water is extremely spectroscopically active, exhibiting a strong and diagnostic fundamental mid-IR bending mode absorption as well as a family of overtone and combination bands in the near-IR. In the solid phase, these bands exhibit detectable variations in band strength, center, and shape that can be used to provide diagnostic information on the abundance, crystallinity, and temperature of water ice on planetary bodies, as well as the presence of impurities or exotic structural forms like clathrates (e.g., Clark, 1981; Warren, 1982; Grundy et al., 1999; de Bergh et al., 2008). In addition, the density of water's solid phase (0.92 g/cm^3) is less than the density of the liquid phase (1.00 g/cm^3). This is an unusual property among most molecules that might seem otherwise esoteric if not for the important fact that water ice floating on liquid water acts as a thermal insulator (a so-called "solid-state greenhouse"; e.g., Matson and Brown, 1989) and thus allows liquid water to remain stable under sub-freezing temperatures.

Other aspects of water's physical- and phase-related properties are also important from the standpoint of considering its history and influence on planetary bodies. For example, the phase diagram of water reveals that the boundaries between solid, liquid, and gas phases span the range of typical terrestrial surface conditions, as well as current and potential past Martian surface conditions (see §4). Thus, phase changes and their resulting volumetric and/or enthalpic implications are expected to play important roles in the energy balance and the nature of geologic processes on planetary surfaces and interiors where water is a significant component. Increasing the salinity of water also has an important effect, lowering the freezing point significantly for saline solutions (like Earth's oceans) and thus potentially significantly expanding the temperature range over which water can remain liquid. We will come back to some of the key the implications of that particular behavior in §3 and §4. Finally, another property of water that is critical in the study of its history and influence on planetary bodies is that it is an excellent and ubiquitous solvent except for the so-called "CHNOPS"-based (Carbon, Hydrogen, Nitrogen, Oxygen, Phosphorus, and Sulfur) organic molecules. The fact that most of the organic molecules that are key to complex biochemical reactions and life as we know it do not dissolve in water provides a convenient medium for the transport of energy to and removal of byproducts from these molecules as part of the chemistry of life. We will return to this topic more in §5.

3. Where do we find water?

3.1. The Snow Line

Before embarking on a survey of where we find water in our solar system and beyond, it is instructive to ask, based on our current understanding of planetary formation and solar system evolution, where should we find water in a solar system? A key concept here is that of the "snow line." In the currently-accepted model of solar system formation that begins with a hot, rotating cloud of gas and dust that collapses to a warm accreting debris disk, the snow line is the distance from the central star beyond which a particular volatile molecular component condenses as ice as the disk cools (e.g., Cassen, 1994; Lecar *et al.*, 2006). Based on modeling and observations of the heliocentric distance for the onset of cometary activity, the snow line for H_2O in the solar system today is around 2.0-2.5 AU (e.g., at $160K \leqslant T \leqslant 200K$ approximately; see review in Encrenaz, 2008). The water snow line was likely to have been farther out (perhaps near 4-5 AU, near where proto-Jupiter formed) in the early solar system (e.g., Stevenson and Lunine, 1988; Encrenaz, 2008), and of course there are different, similarly more distant "snow" lines for molecular disk volatiles like CH_4, NH_3, CO_2, and others that have lower condensation temperatures based on their saturation vapor pressure behavior.

The fact that water condenses at such a significantly higher temperature than other common nebula volatiles means that water is the first volatile phase to condense at a particular heliocentric distance. Combined with the ubiquity of water in general, this implies that water ice is likely to be the dominant volatile in a typical solar nebula, and that the snow line marks an important boundary in a typical solar system, beyond which significantly greater abundances of solids (ice) are available for planetary accretion and growth (e.g., Stevenson and Lunine, 1988; Encrenaz, 2008). The concept of a hard "line" beyond which water condenses is likely to be unrealistic, however, as local, synoptic-scale variations in nebular conditions (e.g., pressure, temperature, density, composition, grain size) almost certainly would have resulted in the water ice condensation occurring within a more fuzzy zone rather than at a specific, narrow boundary. Indeed, Podolak, 2009 has called for the use of the term "ice stability region" rather than "snow line" to acknowledge the complexities of the physics and chemistry occurring within condensing stellar and protostellar disks.

3.2. Where do we expect to find water in our solar system?

3.2.1. Giant Planets.

If the snow line (or ice stability region) concept is correct, then we would certainly expect to detect large abundances of water in the interiors and satellites of the giant planets of the outer solar system. Indeed, tropospheric water vapor has been detected from telescopic (e.g., KAO, ISO) and spacecraft (e.g., Voyager, Galileo) infrared spectroscopic observations of Jupiter and Saturn (see Encrenaz, 2008 for a review), which represent more than 91% of the planetary mass in our solar system. Interestingly, however, the measured relative abundances are lower than would be predicted from the saturation vapor pressure data alone, suggesting that other processes, perhaps such as cloud condensation, are influencing the ability of spectroscopic observations to accurately estimate the water inventories of the giant planets. Galileo probe measurements of water vapor in the Jovian atmosphere (e.g., Wong *et al.*, 2004) confirmed in general the dryer-than-expected nature of the upper troposphere, but the probe's entry into what appears to have been an anomalous "hot spot" complicates the extrapolation of the probe's measurements to the troposphere (and interior) overall.

The next largest reservoir of planetary water should be within Uranus and Neptune (7% of the planetary mass in our solar system). However, there are no data available on the abundance of tropospheric (0.1 –1.0 bar) or deeper water on these planets, despite models that predict tropospheric condensation (e.g., Atreya, 1986). ISO $\sim$ 40 m infrared emission spectra do provide evidence for stratospheric (10^{-3} to 0.1 bar) water in all four giant planets, a rather surprising observation that was used to infer the existence of significant external sources of water to these planet's atmospheres (e.g., Feuchtgruber *et al.*, 1997; Lellouch *et al.*, 2002). Such a postulate should probably not be surprising, given what we now know may be relatively common impacts of volatile-rich comets with the giant planets, witnessed first so dramatically with the impact of split comet Shoemaker-Levy 9 with Jupiter in the summer of 1994, and most recently with the apparent impact (not directly observed) of a previously unknown comet with Jupiter in the summer of 2009. Given the large uncertainties in impactor flux and composition, it is probably not possible to accurately determine whether these kinds of small body impacts could explain the observed stratospheric abundances of water in the giant planets. Nonetheless, they are clearly a potentially important source of external volatiles, for the giant planets and, as we shall discuss below, for the terrestrial planets as well.

Given the dearth of actual measurements, modeling studies continue to provide the best way to estimate the water inventory, composition, and other aspects of the interiors of the giant planets. Canonical giant planet interior models predict relatively small (e.g., 10% RJ) refractory cores surrounded by ices mixed throughout the deep atmosphere (e.g., Pollack *et al.*, 1996). More recent models, employing new measurement constraints from telescopic and Galileo mission probe and gravity data, hypothesize that the ices are not particularly well-mixed but are instead sequestered deep in the interior, in a sort of "ice mantle" just above the core (e.g., Militzer *et al.*, 2008). Indeed, models such as these are being used to advocate that Uranus and Neptune may actually be "ice giants" rather than gas giants (e.g., Lunine, 1993; Boss, 2002). Because of their lower-mass cores and larger heliocentric distances during accretion, the ratio of nebular gas to solid icy planetesimal mass accreted during the formation of Uranus and Neptune was likely much lower than that accreted during the formation of Jupiter and Saturn.

While the giant planets are certainly major reservoirs of water in our solar system (as are, likely, giant planets in other solar systems), additional measurements are clearly needed to more fully understand the O/H ratio and the absolute and relative abundances and distribution of water in their atmospheres and deep interiors. Significant new insights could come, for example, from new radio frequency observations of the strong H_2O rotational transition near 22 GHz to be obtained by the NASA Juno mission, set to orbit Jupiter beginning in 2016. In the meantime, researchers will continue to use computational methods to incorporate new and existing telescopic and spacecraft measurements and new laboratory data on the behavior of molecular species and ices at high temperatures and pressures into an improved understanding of water in the giant planets.

3.2.2. Outer Solar System Satellites and Rings.

Telescopic and spacecraft spectroscopic observations have shown that all outer solar system satellites except for tidally-heated and devolatilized Io have (or are dominated by) water ice on their surfaces (e.g., Clark *et al.*, 1986). Bulk density information is available for most of these objects from close spacecraft flyby determinations of their masses; the relatively low density values (except for Io) suggest that all of these objects have significantly high water:rock ratios in their interiors. Observations of the ring systems of Saturn, Uranus, and Neptune also reveal the presence of water ice, mixed to varying degrees with silicate dust and low-albedo organic solids (e.g., Sicardy, 2005). Overall, the satellites and rings of the outer solar system represent a significant reservoir of water and is consistent with formation of these bodies beyond the snow line, in a nebular environment dominated by ice-rich planetesimals.

3.2.3. Main Belt and Outer Solar System Asteroids, Comets, and Trans-Neptunian Objects.

3.2.3.1. Main Belt Asteroids.

Early "mapping" of the composition of the main asteroid belt using reflectance spectroscopy and a small number of spacecraft flyby encounters revealed an overall gradient in composition with heliocentric distance that appears consistent with the standard model of solar system formation: rocky and metallic ("igneous") inner solar system bodies transitioning to rocky/icy ("metamorphic"), and then ice-dominated bodies ("primitive") farther out. (e.g., Jones *et al.*, 1990; Gaffey *et al.*, 2002). While the known population of small bodies has grown by orders of magnitude recently (currently $N \sim 5 \times 10^5$) because of new surveys and instruments (Minor Planet Center, 2010), and many new dynamical complexities among asteroid populations have been revealed (e.g., Bottke *et al.*, 2002), the overall trend in composition remains an important observational constraint. For example, while the C-type main belt asteroids (which show spectroscopic evidence of hydration as well as similarities to carbonaceous chondrite meteorites) are found throughout the main belt, their distribution peaks in the outer main belt beyond the snow line (~ 2.6–3.5 AU; e.g., Nelson *et al.*, 1993; Rivkin *et al.*, 2002a).

3.2.3.2. Outer Main Belt and Trojan Asteroids.

The physical and compositional characteristics of outer main belt asteroids (dominated by D-class objects) and "special" outer solar system asteroid populations like Jupiter's Trojan asteroid clouds (N >4000; most near 5 AU; Dotto et al., 2008) and the Centaurs (N~ 250; most with perihelia between Saturn and Neptune; Barucci et al., 2008) are still much less well known compared to inner main belt and near-Earth asteroids. Many members of this population appear to have relatively featureless, red-sloped visible to near-IR spectra consistent with refractory and/or mixed refractory-organic (macromolecular hydrocarbons) surface compositions (e.g., Luu et al., 1994; Dotto et al., 2008; Barucci et al., 2008), in some ways similar to the few available spectra of comet nuclei (e.g., Jewitt and Luu, 1990; see also §3.2.3.3). That is, the lack of spectroscopic evidence for ices, organics, and other volatiles in D-type and Trojan asteroids also occurs for comet nuclei, whose bulk compositions are known to be icy but masked by a thin, dark, refractory mantling layer. A few Centaurs show evidence of water and other ices in their surface spectra (e.g., (5145) Pholus: Cruikshank et al., 1998), and/or intermittent comet-like coma activity and release of water and other volatiles (e.g., (2060) Chiron: Luu and Jewitt, 1990; Luu et al., 1990; Barucci et al., 2008). The relatively small population of known Centaurs is still poorly characterized, however. Whether or not D-type and Trojan asteroids represent a significant reservoir of water is an unresolved question that could require future up-close spacecraft missions to resolve. Given the dynamical importance of the Trojans, in particular, as "trapped" examples of potentially once more distant objects scattered into the inner solar system by the migration of the giant planets (e.g., Tsiganis et al., 2005; Morbidelli et al., 2005), understanding their composition could provide significant insights about the role of external delivery of water to Earth and the other terrestrial planets (see also §4.1).

3.2.3.3. Comets.

Cometary comae reveal prodigious evidence for water vapor and water ice, and modeling of the available data on water and dust abundances suggests water mass fractions around 80% for typical comet nuclei (e.g., Festou et al., 2004; Encrenaz, 2008). Until very recently, however, water ice had not been directly measured on comet nuclei; as mentioned above, their surfaces were instead observed to be apparently mantled by a dark, reddish layer of refractories and/or macromolecular hydrocarbons. The only direct measurement of comet nucleus water ice comes from the Deep Impact mission's 2005 impact excavation of water ice from the subsurface of comet 9P/Tempel 1 (e.g., Sunshine et al., 2006), which provided support for the hypothesis that the dark, refractory mantles of comet nuclei (and potentially other dark, reddish, primitive objects like D-type and Trojan asteroids) are relatively thin. Four comets (1P/Halley, C/1996 B2 Hyakutake, C/1995 O1 Hale-Bopp, 8P/Tuttle) have also had their D/H ratio in water reported in spectra of their comae (Horner et al., 2007; Villanueva et al., 2009). In all cases D/H was measured to be about 3 times that of Earth's standard mean ocean water D/H value, suggesting that comets and perhaps other external sources may not be the origin of the water on Earth or the other terrestrial planets (see §4.1).

3.2.3.4. Trans-Neptunian (Kuiper Belt) Objects.

The Kuiper Belt extends from ~ 30 to ~ 55 AU. Water ice has been identified in the spectra of many of the more than 1100 known Trans-Neptunian objects, such as Pluto's moon Charon, (50000) Quaoar, (90482) Orcus, 1996 TO66, (136108) Haumea, (10199) Chariklo and others (e.g., see recent review in Barucci *et al.*, 2008). Pluto itself has been found to exhibit a minor water ice component, but its spectrum is dominated by other ices (e.g., CH_4, N_2, CO, CO_2; e.g., Olkin *et al.*, 2007). Spectroscopic observations of these faint objects are challenging, and so characterization of the surface compositions of these objects is far from complete. Nonetheless, even the very incomplete census of this population so far shows it to be a reservoir rich in volatile ices, including water.

3.3. Is there water in places where we might not have expected it?

The common wisdom, according to the standard model of solar system formation, is that only rocky and metallic bodies should condense and accrete into planets inside a solar system's snow line. The expectation, then, is that the terrestrial planets in our solar system should be essentially dry objects. In a bulk sense, the densities of the terrestrial planets confirm this expectation. However, these objects are not-or perhaps were not-completely water-free. Indeed, even the relatively small fractional abundances of water on these planetary bodies has important implications for planetary evolution and, of course, life.

3.3.1. Airless Inner Solar System Bodies.

Mercury and the Moon are rocky/metallic and intrinsically dry planets. However, theoretical modeling and some radar and orbital remote sensing observations have revealed that water ice and other volatiles could potentially exist within permanently-shadowed regions near their poles (e.g., Harmon *et al.*, 1994) because of these bodies' relatively low orbital inclinations. Confirmation that the radar-bright regions near Mercury's poles are water ice deposits is a major measurement objective of the MESSENGER mission, set to enter Mercury orbit in 2011. NASA's recent LCROSS mission impacted an Atlas rocket upper stage into one of these permanently-shadowed regions near the Moon's south pole. The resulting ejecta plume was observed to contain water vapor, OH, and other volatiles, presumably deposited by cometary impacts (e.g., Colaprete *et al.*, 2009). Although the observed abundances were very small, the detection confirms the hypothesis that volatiles like water can exist stably in special places even on the harsh lunar surface, and thus provide samples of (potentially ancient) cometary materials for scientific study, as well as potential resources for future human explorers. Spectroscopic observations of Phobos, Deimos (e.g., Rivkin *et al.*, 2002b), and some of the members of the Aten, Apollo, and Amor near-Earth asteroid populations (N $\sim$ 6700 at present) have not yet revealed any specific evidence of water ice or other volatiles among any of these small inner solar system bodies, although many of these objects are C-type or other spectroscopic classes that could be expected to contain some volatile component (e.g., Binzel *et al.*, 2002).

3.3.2. Venus.

Because of its extremely high temperatures, the surface and (likely) interior of Venus are dry, and the planet's atmosphere contains only a minor trace amount of water vapor. However, the measured D/H ratio in the atmosphere is some 150 times higher than that of the Earth's oceans, strongly suggesting that there has been a substantial amount of water that has escaped over the planet's history (e.g., Donahue and Russell, 1997). This raises many questions: how did the water escape (slowly? catastrophically?), and what was the origin and abundance of the putative early reservoir of water on Venus?

Answering these questions and solving the puzzle of water of Venus will likely require future (challenging) orbital, landed, and perhaps even sample return missions.

3.3.3. Mars.

Mars is the most Earth-like of the other terrestrial planets, and it is the only other one with significant quantities of water stably existing on its surface-primarily as ice in the polar caps, but also as a small amount of atmospheric water vapor. Recent orbital and surface lander measurements have also revealed significant quantities of ice in the shallow high-latitude subsurface (e.g., Boynton *et al.*, 2008; Feldman *et al.*, 2008; Smith *et al.*, 2009). Given its distance closer to (but still well within) the snow line and lower density, Mars might be expected to contain a higher bulk water component than the other terrestrial planets. However, modeling, geochemical data from Martian meteorites and from surface and orbital measurements suggest that the Martian mantle is likely to be dry (e.g., McSween, 1994; Lunine *et al.*, 2003; McSween *et al.*, 2009). The planet's relatively small core and thus higher ratio of rocky to metallic materials would thus presumably explain the lower density compared to the other terrestrial planets. The Martian surface today is a cold (mean temp $\sim$ 215 K), low pressure (5–10 mbar), and hyperarid environment (Baker, 2001). However, there is ample evidence that liquid water once existed stably on the Martian surface (and likely, subsurface), and thus that the temperature and atmospheric pressure were likely higher. The evidence includes geologic landforms like river channels and deltas that are preserved in some of the planet's most ancient (3–4 Ga) terrains; large regions of (ancient) heavily-eroded terrain suggesting the past action of precipitation and surface runoff; and a variety of hydrated minerals, including clays, sulfates, evaporite salts, and certain iron oxides, that can only have formed by the alteration of precursor minerals in an aqueous solution (see detailed reviews in, e.g., Carr, 1996; Bell, 2008). The discovery that Mars once had liquid water raises intriguing possibilities about the potential emergence of life on the Red Planet. However, it also raises perplexing questions that are the focus of much of the ongoing spacecraft exploration of Mars: What was the source of early water on Mars? Where did the water go: subsurface? slow escape to space? catastrophic loss via impacts? And – from the standpoint of habitability – how long was the environment able to support liquid water stably on the surface?

3.3.4. Earth.

We take the presence of water on our home planet for granted because more than 70% of Earth's surface area is covered by water (equal to a global surface layer $\sim$ 3 km deep, or about 0.02% of Earth's mass). However, from a cosmochemical perspective, the bulk composition of the Earth is estimated to be only <0.1 wt.% water (e.g., Drake and Righter, 2002). Earth's thin layer of surface water cloaks what is truly a dry world. Where then did Earth's surface water come from? We return to this question in §4.1 below. Despite its small relative abundance, water has dramatic effects on our planet's geology, geochemistry, climate, and biology. For example, the oceans sequester enormous quantities (tens of bars or more) of CO_2 in solution, removing a potential source of additional greenhouse warming from the atmosphere (e.g., Walker *et al.*, 1981). Water is also thought to play a major role in the lubrication of Earth's several dozen lithospheric plates, enabling the subduction that eventually leads to melting and recycling of crustal rocks and some of the sequestered atmospheric CO_2 (e.g., Ragenauer-Lieb *et al.*, 2001). Water is responsible for erosion of surface landforms and alteration of minerals to oxidized and/or hydrated phases. And as discussed in §2 above and §5 below, liquid water is also a critical component of life on this planet.

4. The Special Appeal of Liquid Water

4.1. Origin of Earth's Water?

Because of the critical role of water in the past and present geologic and biologic evolution of our planet and probably other Earth-like planets, as well as the expectation from classical models of solar system formation that the Earth should not have formed with a significant inventory of water, there is a special interest in understanding the enigma of our planet's inventory of water. As outlined in many recent studies and reviews (e.g., Morbidelli *et al.*, 2000; Drake and Righter, 2002; Lunine *et al.*, 2003; Drake, 2005; Owen, 2008; Albarede, 2009; Holland *et al.*, 2009), the primary hypotheses being explored for Earth's water are split between those that invoke endogenic sources, with water originating from the population of planetesimals accreting in near-Earth space to form our planet, versus those that invoke exogenic sources, with water having been brought in from bodies farther out in the solar system, likely beyond the snow line. Endogenic hypotheses posit that the early Earth intrinsically formed with a significant internal component of water that, through volcanic and/or impact degassing over time, is the major source of our planet's surface water. Exogenic hypotheses invoke large early impacts of the proto-Earth with volatile-bearing planetesimals or the addition of a "late veneer" (post Moon-forming event) of water from comet and asteroid impacts. Endogenic and exogenic hypotheses have strengths and weaknesses based on available data and current modeling of solar system formation. For example, the concept of Earth accreting from "wet" planetesimals would appear to require that volatile-rich source material should have been part of the early accretion of the planet, even though most of that material may have only been available beyond the solar system's snow line (which may have been out near 4–5 AU early in the history of the solar system; e.g., Stevenson and Lunine, 1988). There are issues with the potential sources of externally-delivered water as well. As mentioned above, the measured D/H ratio in comets is inconsistent with the value in Earth's oceans, arguing against a cometary source. However, there are still important questions about whether the small number of available cometary D/H measurements truly sample the population's values; whether the D/H ratio of Earth's oceans could have changed over time due to dilution, weathering, or other processes; and whether the assumed bulk D/H ratio of the Earth overall-and indeed the bulk abundance of water in the Earth-is truly known (e.g., Abe *et al.*, 2000; Drake and Righter, 2002; Smyth *et al.*, 2006). There are also unresolved questions about the ability of near-Earth and main belt planetesimals and asteroids-primitive materials in the source region of the accreting Earth-to explain Earth's water inventory. The parent bodies of ordinary chondrite meteorites could contain from 0.1 to as much as 9 wt.% H_2O by some estimates (e.g., Zinner, 1988). The higher ranges of these estimated abundances could be adequate to justify Earth's water inventory as intrinsic rather than extrinsic, and the ordinary chondrites also have D/H ratios comparable to that of Earth's oceans (e.g., Drake and Righter, 2002). However, the issue of whether the ordinary chondrites as opposed to other primitive source materials such as the carbonaceous chondrites represent the building block materials for terrestrial planet accretion remains controversial (e.g., Morbidelli *et al.*, 2000). Recently it has been proposed that outer main belt and outer solar system asteroids (large compared to comet nuclei, and hydrated because of their formation beyond the snow line) could have been significant sources of Earth's water during and shortly after accretion. More specifically, the hypothesized migration of the giant planets during the early history of the solar system has been invoked as a source of substantial dynamical mixing/scattering of small bodies from the outer solar system to the inner solar system, and as a potential explanation for the late heavy bombardment of the Earth-Moon system that is preserved in the lunar impact record (e.g., Gomes *et al.*, 2005; Tsiganis *et al.*, 2005; Morbidelli *et al.*, 2005). In this so-called "Nice model" of solar system evolution, the orbits of many small outer solar system objects could have been dramatically altered when Jupiter and Saturn passed through a 2:1 mean-motion resonance around $\sim$ 4 Ga, scattering relatively large, volatile-rich bodies into the inner solar system (as well as the Kuiper Belt) and providing potential large-impact sources of water for the terrestrial planets. Interestingly, it may be possible to test the Nice model by confirming the volatile-rich compositions of Jupiter's Trojan asteroids-a large population of small bodies that has been trapped in Jupiter's Lagrange points since the end of the proposed period of giant planet migration (e.g., Dotto *et al.*, 2008). It could be argued that invoking cometary and/or asteroidal impacts to deliver Earth's water relies on ad hoc or stochastic events or assumptions. However, the apparent importance of stochastic impact events in the evolution of the solar system is now more appreciated than ever, with serious impact-related hypotheses being explored to explain, for example, the origin of the Moon, the late heavy bombardment, and the unusual spin properties of Uranus and Venus (e.g., Korycansky *et al.*, 1990; Canup and Asphaug, 2001; Gomes *et al.*, 2005; Davies, 2008). While the origin of Earth's water is still a highly debated and exciting topic of study, it is also of course entirely possible that a combination of both endogenic and exogenic sources could explain our planet's current surface water inventory.

4.2. Habitable Zones and Water on Other Planets

The habitable zone is a key concept or metric that enables astronomers to think more broadly about the concept of water on any planets-whether in our solar system or others. Specifically, the habitable zone is defined as the heliocentric distance from a star within which conditions would be favorable for life as we know it to exist (e.g., Hart, 1979). Even more specifically, the definition provides a guide to assess habitability (the potential to develop and sustain life) by providing an estimate of the planetary region where liquid water could be stable on a planet with Earth-like surface conditions. The formal definition of the habitable zone defines the mid-line of the zone at a heliocentric distance in AU of $(L_{star}/L_{sun})^{0.5}$, where L is the stellar or solar luminosity. The inner and outer boundaries of the habitable zone are typically assumed to range from about 95% to 135% of the midline, based mostly on analogy with our own solar system. However, the boundaries are widely recognized to be fuzzy, and the entire definition itself is understood to be significantly biased by the particular characteristics of our own solar system-as yet the only place where a habitable environment is known to exist. While the concept of the habitable zone is a useful guide and starting point for understanding the role of water in influencing a planet's potential biologic and even geologic evolution, even in our own solar system there are clearly exceptions to this "rule." Prominent among these exceptions are Jupiter's moon Europa and Saturn's moon Enceladus, both of which are well outside the classical habitable zone but both of which exhibit compelling evidence for liquid water just below their icy crusts. Evidence for a deep ocean on Europa comes from Voyager and Galileo mission geologic evidence of thin icy crustal plates floating and moving over a liquid layer, salty mineral deposits extruded onto the surface, and gravity and magnetic field measurements that suggest a salty, conducting liquid subsurface layer (e.g., Carr *et al.*, 1998; McCord *et al.*, 1999; Kivelson *et al.*, 2000). The surprising evidence for liquid water under the icy crust of tiny Enceladus (R $\sim$ 250 km) comes from Cassini mission observations of plumes of water ice emanating from warmer fractures near the satellite's south pole (e.g., Porco *et al.*, 2006). In the case of Europa, tidal heating from Jupiter provides the energy source that would prevent a subsurface ocean from freezing solid; the energy source powering Enceladus's geysers is presently unknown, but possibilities include tidal heating from Saturn and other satellites, and/or radioactive heating from rocky materials deep within the satellite. Beyond our solar system, the study of extrasolar planets has expanded rapidly over the past decade, with more than 400 extrasolar planets currently known, and more being announced frequently (see, e.g., http://exoplanet.eu/). Even more exciting is that the first planets are now being discovered within the habitable zones around other nearby, Sun-like stars (e.g., Selsis *et al.*, 2007; von Bloh *et al.*, 2007). Water vapor and water ice are now being detected in circumstellar disks and even the atmospheres of some exoplanets (e.g., Watson *et al.*, 2007; Tinetti *et al.*, 2007). The study of water in exoplanets and of the nature of extrasolar habitable zones is just beginning. It's exciting to realize that astronomers and planetary scientists are on the cusp of some significant new discoveries including, soon, the expected discovery of the first truly Earth-like planets within the habitable zones of other stars.

5. Implications: Why care about Water on Planets?

Our own solar system provides a warning that we must interpret concepts like the habitable zone with care: special environments outside the classically-defined habitable zone could still satisfy the constraints of habitability for life as we know it (and, of course, who knows about life as we don't know it?). An additional caveat is that a star's habitable zone should be expected to evolve with time as the star evolves. For example, the Sun's habitable zone has moved outwards as the solar luminosity has increased over time (e.g., Kasting and Catling, 2003). This implies that Venus was likely to have been solidly within the habitable zone early in the history of the solar system when the Sun was fainter, and that Mars will move solidly into the habitable zone in the far future as solar luminosity increases. If the Sun continues to evolve as expected, the inner edge of the solar system's habitable zone will eventually sweep past the Earth (in perhaps $\sim$ 1 Gy), vaporizing the oceans, releasing trapped CO_2 into the atmosphere, and potentially rendering our planet as inhospitable as Venus is today. While prognosticating about a planet's distant future is fraught with assumptions and uncertainties, the implications of this particular prediction are so drastic that it can't help but to compel us to learn more about stellar and planetary evolution, and about the role that water and other volatiles play in the evolution of a planet's climate.

Studying the triad of Venus, Earth, and Mars has yielded particularly important and illuminating insights about the critical role of the physical state of water in influencing the evolution of a terrestrial planet. In particular, water appears to have played a major role in the divergent evolutions of these three planets. In the standard model of solar system formation, all three planets would have formed with approximately the same starting (primarily rocky/metallic) bulk composition, and with generally thin and reducing (H_2, CH_4, etc.) atmospheres because of their relatively poor ability to retain nebular gas and their location in the relatively volatile-poor inner solar system. Differentiation, outgassing, and additional late heavy bombardment and subsequent accretion of volatiles would have resulted in the buildup of CO_2, H_2O, SO_2, N_2, etc. in their atmospheres over time, while H_2 would have slowly been lost due to escape and solar wind breakdown. At the heliocentric distance of Venus, high temperatures kept the water in the vapor phase, while at Mars's distance the water would primarily have existed as ice (even if the past atmosphere had been thick enough to allow liquid water to occasionally be stable on the surface). Earth's surface conditions were, of course, just right for liquid water to be stable, leading to the (early) formation of deep oceans. Earth's oceans provide a sink for atmospheric greenhouse gases like CO_2 and SO_2 (and of course, water vapor), either directly dissolved in water or precipitated as carbonate and sulfate rocks and stored in crustal reservoirs. The lack of oceans on Venus (currently, at least) means that the CO_2 was not sequestered and thus became the source of a massive amount of runaway greenhouse warming. And while there is good evidence for the past presence of liquid water on Mars, the lack of evidence for extensive, global-scale carbonate deposits (despite detailed searches from orbital and landed missions) argues against the planet ever having had long-lived, large-scale oceans. While CO_2 still dominates the Martian atmosphere, it is currently in too small an abundance to provide any significant greenhouse warming. If Mars still harbors significant amounts of water, it is likely locked up as polar and subsurface ice.

The ongoing discovery of the role of water (and CO_2) in the surfaces, atmospheres, and climates of Venus and Mars provides global-scale context and insight about the short- and long-term influences of these volatiles on our home planet. Indeed, NASA and other space agencies have recently adopted a "follow the water" theme for the exploration of Mars and other destinations in our solar system as well as the study of extrasolar planets. As we have seen from previous telescopic and spacecraft observations as well as laboratory and modeling studies, to follow the water focuses us on understanding key aspects of solar system formation, planetary evolution, and planetary atmospheric, surface, and interior processes all under a single theme that is both scientifically rewarding as well as interesting and exciting to the general public. Ultimately, to search for water is to search for ourselves.

Acknowledgements

I am grateful to Martha Haynes and other colleagues on the IAU Executive Committee for inviting me to present this review as an Invited Discourse at the IAU's General Assembly XXVII in Rio de Janeiro in August, 2009. I am also extremely grateful to Jonathan Lunine, Amara Graps, Dale Cruikshank, and Jamie Lloyd for providing background information, preprints, and general collegial support for my research into this exciting and timely topic. My research on the past history and present inventory of water on specific planets and other planetary bodies has been supported over the years by grants and contracts from the NASA Planetary Geology, Planetary Astronomy, and Mars Data Analysis Programs, from the Jet Propulsion Laboratory/Caltech, and from the Space Telescope Science Institute.

References

Abe, Y., E. Ohtani, T. Okuchi, K. Righter, & M. J. Drake *"Water in the early Earth", in "Origin of the Earth and Moon", edited by R. M. Canup and K. Righter, Univ. of Ariz. Press, Tucson, 2000* pp. 413–433

Albarède, F. *Volatile accretion history of the terrestrial planets and dynamic implications, Nature,* 461, 1227–1233, 2009.

Atreya, S. K., *Atmospheres and Ionospheres of the Outer Planets and Their Satellites, 234 pp., Springer-Verlag, Berlin,* 1986.

Baker, V. R., *Water and the martian landscape, Nature,* 412, 228–236, 2001.

Barucci, M. A., *"Composition and Surface Properties of Transneptunian Objects and Centaurs"* in *The Solar System Beyond Neptune*, M. A. Barucci, H. Boehnhardt, D. P. Cruikshank, and A. Morbidelli (eds.), University of Arizona Press, Tucson, pp. 143–160, 2008.

Bell III, J. F. (Editor) *The Martian Surface: Composition, Mineralogy, and Physical Properties*, Cambridge University Press (ISBN-13: 9780521866989), Cambridge, 688 pp., 2008.

Binzel, R. P., D. F. Lupishko, M. DiMartino, R. J. Whiteley, & G. J. Hahn, *"Physical Properties of Near-Earth Asteroids"*, in *Asteroids III* (W.F. Bottke Jr. et al., eds.), pp. 255–271, University of Arizona Press., 2002.

Boss, A. P., *Formation of gas and ice giant planets*, Earth Planet. Sci. Lett., 202, 513–523, 2002.

Bottke, W. F., A. Cellino, P. Paolicchi, & R. P. Binzel (editors), *Asteroids III*, Univ. Arizona Press, 785 pp.

Boynton, W. V., G. J. Taylor, S. Karunatillake, R. C. Reedy, & J. M. Keller, *"Elemental abundances determined via the Mars Odyssey GRS"*, Chapter 5 in *"The Martian Surface: Composition, Mineralogy, and Physical Properties"* (J.F. Bell III, editor), Cambridge Univ. Press, 105–124, 2008.

Canup, R. M. & E. Asphaug, *Origin of the Moon in a giant impact near the end of the Earth's formation*, Nature, 412, 708–712, 2001.

Carr, M. H., *Water on Mars*, Oxford Univ. Press, 248 pp., 1996.

Carr, M. H., Belton, M. J. S., Chapman, C. R., Davies, M. E., Geissler, P., Greenberg, R., McEwen, A. S., Tufts, B. R., Greeley, R., Sullivan, R., Head, J. W., Pappalardo, R. T., Klaasen, K. P., Johnson, T. V., Kaufman, J., Senske, D., Moore, J., Neukum, G., Schubert, G., Burns, J. A., Thomas, P., & Veverka, J., *Evidence for a subsurface ocean on Europa*, Nature, 391, 363–365, 1998.

Cassen, P., *Utilitarian models of the solar nebula*, Icarus 112, 405–429, 1994.

Chambers, J. E., *Planetary accretion in the inner Solar System*, Earth Planet. Sci. Lett., 223, 241–252, 2004.

Clark, R. N., *Water frost and ice: The near-infrared spectral reflectance 0.65-2.5 m*, J Geophys. Res., 86, 3087–3096, 1981.

Clark R. N., Fanale F. P., & Gaffey M. J., *"Surface composition of natural satellites"*, in *"Satellites"*, ed. J.A. Burns and M.S. Matthews, pp. 437-491. Tucson: Univ. Arizona Press, 1986.

Colaprete, A., Briggs, G., Ennico, K., Wooden, D., Heldmann, J., Sollitt, L., Asphaug, E., Korycansky, D., Schultz, P., Christensen, A., Galal, K., Bart, G. D., & the LCROSS Team, *An Overview of the Lunar Crater Observation and Sensing Satellite (LCROSS) Mission Results from Swing-by and Impact*, Lunar Exploration Analysis Group, November 16–19, Houston, Texas. LPI Contribution No. 1515, p. 11, 2009.

Cruikshank, D. P., T. L. Roush, M. J. Bartholomew, T. R. Geballe, Y. J. Pendleton, S. White, J. F. Bell III, J. K. Davies, T. C. Owen, C. deBergh, D. Tholen, M. P. Bernstein, R. H. Brown, K. A. Tryka, & C. M. Dalle Ore, *The composition of Centaur 5145 Pholus*, Icarus, 135, 389–407, 1998.

Davies, J. H., *Did a mega-collision dry Venus' interior?* Earth Planet. Sci. Lett., 268, 376–383, 2008.

de Bergh, C., B. Schmitt, L. V. Moroz, E. Quirico, & D. P. Cruikshank, *"Laboratory Data on Ices, Refractory Carbonaceous Materials, and Minerals Relevant to Transneptunian Objects and Centaurs"*, in *"The Solar System Beyond Neptune"*, M. A. Barucci, H. Boehnhardt, D. P. Cruikshank, and A. Morbidelli (eds.), University of Arizona Press, Tucson, pp. 483–506, 2008.

Donahue, T. M. & C. T. Russell, *"The Venus Atmosphere and Ionosphere and their Interactions with the Solar Wind: An Overview,"* in *"Venus II"*, Univ. Arizona Press, pp. 3–31, 1997.

Dotto, E., J. P. Emery, M. A. Barucci, A. Morbidelli, & D. P. Cruikshank, *"De Troianis: The Trojans in the planetary system,"* in *"The Solar System Beyond Neptune"* (Barucci, Boehnhardt, Cruikshank, Morbidelli, Eds.), pp. 383-396, University of Arizona Press, Tucson, 2008.

Drake, M. J., *Origin of water in the terrestrial planets*, Met. & Plan. Sci., 40, 519–527, 2005.

Drake, M. J. & Righter K., *Determining the composition of the Earth*, Nature, 416, 39–46, 2002.

Encrenaz, T., *Water in the solar system*, Ann. Rev. Astron. Astrophys., 46, 57–87, 2008.

Feldman, W. C., M. C. Mellon, O. Gasnault, S. Maurice, & T. H. Prettyman, *"Volatiles on Mars: Scientific results from the Mars Odyssey Neutron Spectrometer"*, Chapter 6 in *"The*

Martian Surface: Composition, Mineralogy, and Physical Properties (J.F. Bell III, editor)", Cambridge Univ. Press, 125–152, 2008.

Festou, M. C., H. U. Keller, & H. A. Weaver (editors), *Comets II, Univ. Arizona Press*, 745 pp., 2004.

Feuchtgruber, H., E. Lellouch, T. de Graauw, B. Bézard, T. Encrenaz, & M. Griffin, *External supply of oxygen to the atmospheres of the giant planets, Nature*, 389, 159–162, 1997.

Gaffey, M. J., E. A. Cloutis, M. S. Kelley, & K. L. Reed, *"Mineralogy of Asteroids", in Asteroids III (W.F. Bottke Jr. et al., eds.)*, pp. 653–667. University of Arizona Press., 2002.

Gomes, R., H. F. Levison, K. Tsiganis, & A. Morbidelli, *Origin of the cataclysmic late heavy bombardment period of the terrestrial planets, Nature 435*, 466–469, 2005.

Grundy, W. M., M. W. Buie, J. A. Stansberry, J. R. Spencer, & B. Schmitt, *Near-Infrared Spectra of Icy Outer Solar System Surfaces: Remote Determination of H2O Ice Temperatures, Icarus*, 142, 536–549, 1999.

Harmon, J. K., Slade, M. A., Vélez, R. A., Crespo, A., Dryer, M. J., & Johnson, J. M., *Radar mapping of Mercury's polar anomalies, Nature*, 369, 213–215, 1994.

Hart, M. H., *Habitable zones about main sequence stars, Icarus*, 37, 351–357, 1979.

Holland, G., M. Cassidy, & C. J. Ballantine, *Meteorite Kr in Earth's mantle suggests a late accretionary source for the atmosphere, Science*, 326, 1522–1525, 2009.

Horner, J., O. Mousis, & F. Hersant, *Constraints on the Formation Regions of Comets from their D:H Ratios, in Earth, Moon, Planets*, 100, 43–56, 2007.

International Astronomical Union, *"IAU 2006 General Assembly: Resolutions 5 and 6", http://www.iau.org/static/resolutions/Resolution_GA26-5-6.pdf*, Aug. 24, 2006.

Jewitt, D. C. & J. X. Luu, *CCD spectra of asteroids II: The Trojans as spectral analogs of cometary nuclei, Astron. J.*, 100, 933–944, 1990.

Jones, T. D., L. A. Lebofsky, J. S. Lewis, & M. S. Marley, *The composition and origin of the C, P, and D asteroids – Water as a tracer of thermal evolution in the outer belt, Icarus*, 88, 172–192, 1990.

Kasting, J. & D. Catling, *Evolution of a Habitable Planet, Ann. Rev. Astron. Astrophys.*, 41, 429–463, 2003.

Kivelson, M. G., K. K. Khurana, C. T. Russell, M. Volwerk, R. J. Walker, & C. Zimmer, *Galileo Magnetometer measurements: A stronger case for a subsurface ocean at Europa, Science*, 289, 1340–1343, 2000.

Korycansky, D. G., P. Bodenheimer, P. Cassen, & J. B. Pollack, *One-dimensional calculations of a large impact on Uranus, Icarus* 84, 528–541, 1990.

Lecar, M., M. Podolak, D. Sasselov, E. Chiang, *On the location of the snow line in a protoplanetary disk, Ap. J.*, 640, 1115–1118, 2006.

Lellouch, E., B. Bézard, J. I. Moses, G. R. Davis, P. Drossart, H. Feuchtgruber, E. A. Bergin, R. Moreno, & T. Encrenaz, *The Origin of Water Vapor and Carbon Dioxide in Jupiter's Stratosphere, Icarus*, 159, 112–131, 2002.

Lunine, J. I., *The atmospheres of Uranus and Neptune, Ann. Rev. Astron. Astrophys.*, 31, 217–263, 1993.

Lunine, J. I., J. Chambers, A. Morbidelli, & L. A. Leshin, *The origin of water on Mars, Icarus*, 165, 1–8, 2003.

Luu, J. X. & D. C. Jewitt, *Cometary activity in 2060 Chiron, Astron. J.*, 100, 913–932, 1990.

Luu, J. X., D. C. Jewitt, & E. Cloutis, *Near-infrared spectroscopy of primitive Solar System objects. Icarus* 109, 133–144, 1994.

Luu, J. X., D. C. Jewitt, & C. Trujillo, *Water Ice in 2060 Chiron and Its Implications for Centaurs and Kuiper Belt Objects, Ap. J.*, 531, L151–L154, 1990.

Matson, D. L. & R. H. Brown, *Solid-state greenhouses and their implications for icy satellites, Icarus*, 77, 67–81, 1989.

McCord, T. B., G. B. Hansen, D. L. Matson, T. V. Johnson, J. K. Crowley, F. P. Fanale, R. W. Carlson, W. D. Smythe, P. D. Martin, C. A. Hibbitts, J. C. Granahan, & A. Ocampo, *Hydrated salt minerals on Europa's surface from the Galileo near-infrared mapping spectrometer (NIMS) investigation, J. Geophys. Res.*, 104, 11827–11852, 1999.

McSween H. Y., *What we have learned about Mars from SNC meteorites, Meteoritics*, 29, 757–779, 1994.

McSween, H. Y., G. J. Taylor, & M. B. Wyatt, *Elemental composition of the Martian crust,* Science, 324, 736–739, 2009.

Militzer, B., W. B. Hubbard, J. Vorberger, I Tamblyn, & S. A. Bonev, *A Massive Core in Jupiter Predicted from First-Principles Simulations, Ap. J.,* 688, L45–L48, 2008.

Minor Planet Center, *IAU/Smithsonian Astrophysical Observatory, lists of minor planets:* http://www.cfa.harvard.edu/iau/lists/Lists.html, 2010.

Morbidelli, A., Chambers, J., Lunine, J. I., Petit, J. M., Robert, F., Valsecchi, G. B., & Cyr, K. E., *Source regions and timescales for the delivery of water on Earth, Meteor. Planet. Sci.,* 35, 1309–1320, 2000.

Morbidelli, A., H. F. Levison, K. Tsiganis, & R. Gomes, *Chaotic capture of Jupiter's Trojan asteroids in the early Solar System, Nature,* 435, 462–465, 2005.

Nelson, M. L., D. T. Britt, & L. A. Lebofsky, *Review of asteroid compositions, in Resources of Near-Earth Space, Univ. Arizona Press,* pp. 493–522, 1993.

Olkin, C. B., Young, E. F., Young, L. A., Grundy, W., Schmitt, B., Tokunaga, A., Owen, T., Roush, T., & Terada, H., *Pluto's Spectrum from 1.0 to 4.2 ?m: Implications for Surface Properties, Astron. J.,* 133, 420–431, 2007.

Owen, T., *The contributions of comets to planets, atmospheres, and life: Insights from Cassini-Huygens, Galileo, Giotto, and inner planet missions, Space Sci. Rev.,* 138, 301–316, 2008.

Planetary Science Institute, *"Petition Protesting the IAU Planet Definition,"* http://www.ipetitions.com/petition/planetprotest/, 2006.

Podolak, M., *The location of the snow line in protostellar disks,* Invited talk at IAU Symposium 263: Icy Bodies in the Solar System, Rio de Janeiro, Aug. 2009. arXiv preprint http://arxiv.org/abs/0911.4803v1.

Pollack, J. B., O. Hubickjy, P. Bodenheimer, J. J. Lissauer, M. Podolak, & Y. Greenzweig, *Formation of the Giant Planets by Concurrent Accretion of Solids and Gas, Icarus,* 124, 62–85.

Porco, C. C., P. Helfenstein, P. C. Thomas, A. P. Ingersoll, J. Wisdom, R. West, G. Neukum, T. Denk, R. Wagner, T. Roatsch, S. Kieffer, E. Turtle, A. McEwen, T. V. Johnson, J. Rathbun, J. Veverka, D. Wilson, J. Perry, J. Spitale, A. Brahic, J. A. Burns, A. D. Del Genio, L. Dones, C. D. Murray, & S. Squyres, *Cassini Observes the Active South Pole of Enceladus, Science,* 311, 1393–1401, 2006.

Ragenauer-Lieb, K., Yuen, D., & Branlund, J., *The initiation of subduction: criticalilty by addition of water? Science,* 294, p. 578–580, 2001.

Rivkin, A. S., E. S. Howell, F. Vilas, & L. A. Lebofsky, *"Hydrated Minerals on Asteroids: The Astronomical Perspective,"* in Asteroids III (W.F. Bottke Jr. et al., eds.), pp. 235–253, University of Arizona Press., 2002.

Rivkin, A. S., R. H. Brown, D. E. Trilling, J. F. Bell III, & J. H. Plassmann, *Near-infrared spectrophotometry of Phobos and Deimos, Icarus,* 156, 64–75, 2002.

Selsis, F., J. F. Kasting, B. Levrard, J. Paillet, I. Ribas, & X. Delfosse, *Habitable planets around the star Gliese 581 Astron. Astrophys.,* 476, 1373–1387, 2007.

Sicardy, B., *Dynamics and Composition of Rings, Space Sci. Rev.,* 116, 457–470, 2005.

Smith, P. H. and 35 others, *H2O at the Phoenix landing site, Science,* 325, 58–61, 2009.

Smyth, J. R., D. J. Frost, F. Nestola, C. M. Holl, & G. Bromiley, *Olivine hydration in the deep upper mantle: Effects of temperature and silica activity, Geophys. Res. Lett.,* 33, L15301, 2006.

Stevenson, D. J. & J. I. Lunine, *Rapid formation of Jupiter by diffuse redistribution of water vapor in the solar nebula, Icarus,* 75, 146–155, 1988.

Sunshine, J., A'Hearn, M. F.; Groussin, O.; Li, J.-Y.; Belton, M. J. S.; Delamere, W. A.; Kissel, J.; Klaasen, K. P.; McFadden, L. A.; Meech, K. J.; Melosh, H. J.; Schultz, P. H.; Thomas, P. C.; Veverka, J.; Yeomans, D. K.; Busko, I. C.; Desnoyer, M.; Farnham, T. L.; Feaga, L. M.; Hampton, D. L.; Lindler, D. J.; Lisse, C. M.; & Wellnitz, D. D., *Exposed Water Ice Deposits on the Surface of Comet 9P/Tempel 1, Science,* 311, 1453–1455, 2006.

Sykes, M., *Classifying Planets from a Geophysical Perspective,American Astronomical Society, AAS Meeting #214, #237.06; Bull. Amer. Astron. Soc.,* Vol. 41, p. 740, 2009.

Tinetti, G., A. Vidal-Madjar, M. Liang, J. P. Beaulieu, Y. Yung, S. Carey, R. J. Barber, J. Tennyson, I. Ribas, N. Allard, G. E. Ballester, D. K. Sing, & F. Selsis, *Water vapour in the atmosphere of a transiting extrasolar planet, Nature,* 448, 169–171, 2007.

Tsiganis, K., R. Gomes, A. Morbidelli, & H. F. Levison, *Origin of the orbital architecture of the giant planets of the Solar System, Nature*, 435, 459–461, 2005.

Villanueva, G. L., M. J. Mumma, B. P. Bonev, M. A. DiSanti, E. L. Gibb, H. Bhnhardt, & M. Lippi, *A Sensitive Search for Deuterated Water in Comet 8P/Tuttle, Ap. J. Lett.*, 690, L5–L9, 2009.

von Bloh, W., C. Bounama, M. Cuntz, & S. Franck, *The habitability of super-Earths in Gliese 581, Astron. Astrophys.*, 476, 1365–1371, 2007.

Walker, J. C. B., Hays, P. B., & Kasting, J. F., *A negative feedback mechanism for the long term stabilization of the Earth's surface temperature, J. Geophys. Res.*, 86, 9776–9782, 1981.

Warren S. G., *Optical properties of snow. Rev. Geophys.* 20: 67–89, 1982.

Watson, D. M., C. J. Bohac, C. Hull, W. J. Forrest, E. Furlan, J. Najita, N. Calvet, P. d'Alessio, L. Hartmann, B. Sargent, J. D. Green, K.H. Kim & J. R. Houck, *The development of a protoplanetary disk from its natal envelope, Nature*, 448, 1026–1028, 2007.

Wong, M. H., P. R. Mahaffy, S. K. Atreya, H. B. Niemann, & T. C. Owen, *Updated Galileo probe mass spectrometer measurements of carbon, oxygen, nitrogen, and sulfur on Jupiter, Icarus*, 171, 153–170, 2004.

Zinner, E., *"Interstellar cloud material in meteorites", in "Meteorites and the Early Solar System" (ed. Kerridge, J.)* Univ. Arizona Press, Tucson, pp. 956–983, 1988.

Highlights of Astronomy, Volume 15
XXVIIth IAU General Assembly, August 2009
Ian F. Corbett, ed.

ID 3 Evolution of Structure in the Universe

Simon D. M. White

Max-Planck-Institut für Astrophysik, Garching-bei-München, 85748 Germany
email: swhite@mpa-garching.mpg.de

Abstract. Recent studies of the Cosmic Microwave Background have provided us with a high quality image of the Universe when it was only 380,000 years old. At that time it was a near-uniform mixture of hydrogen, helium, dark matter and radiation, with no galaxies, no stars, no planets and no people, indeed no atomic nuclei heavier than Lithium. Under the action of gravity, the weak fluctuations observed in the microwave sky evolved into the extraordinarliy complex structure of our present Universe. I will show how supercomputer simulations can be used to demonstrate that such evolution does indeed reproduce the observed properties of today's galaxies and large-scale structures, thus confirming the extraordinary assumptions of the current structure formation paradigm. Only a quarter of the energy density of the present Universe is in gravitating matter; only a sixth of this matter is made of atoms or other known particles; only 5 percent of this baryonic material is currently inside galaxies. Most of today's Universe is in the form of Dark Energy; most of the gravitating matter is Dark Matter; and most of the baryons remain unseen in intergalactic space. The properties of the fluctuations measured in the microwave sky suggest that they originated very close to the Big Bang as quantum fluctuations of the vacuum itself. Everything has formed from nothing.

Keywords. cosmic microwave background, dark matter, dark energy, evolution, gravity, galaxies, simulations, structure of Universe

---

Text of presentation not available.

Highlights of Astronomy, Volume 15
XXVIIth IAU General Assembly, August 2009
Ian F. Corbett, ed.

© International Astronomical Union 2010
doi:10.1017/S1743921310008185

Do Low Luminosity Stars Matter?

María Teresa Ruiz

Department of Astronomy, University of Chile, Casilla 36-D, Santiago, Chile
email: mtruiz@das.uchile.cl

Abstract. Historically, low luminosity stars have attracted very little attention, in part because they are difficult to see except with large telescopes, however, by neglecting to study them we are leaving out the vast majority of stars in the Universe. Low mass stars evolve very slowly, it takes them trillions of years to burn their hydrogen, after which, they just turn into a He white dwarf, without ever going through the red giant phase. This lack of observable evolution partly explains the lack of interest in them. The search for the "missing mass" in the galactic plane turned things around and during the 60s and 70s the search for large M/L objects placed M-dwarfs and cool WDs among objects of astrophysical interest. New fields of astronomical research, like BDs and exoplanets appeared as spin-offs from efforts to find the "missing mass". The search for halo white dwarfs, believed to be responsible for the observed microlensing events, is pursued by several groups. The progress in these last few years has been tremendous, here I present highlights some of the great successes in the field and point to some of the still unsolved issues.

Keywords. stars, white dwarfs, low mass, brown dwarfs, luminosity function

1. Introduction

Which stars are considered "low-luminosity" stars, is a definition that evolves with time, as more powerful telescopes and detectors explore new frontiers into the "dim universe". Here, I will discuss stars with a visual luminosity $L \leqslant 0.01\ L_\odot$, corresponding to an absolute visual magnitude $M_V \sim 10$.

Figure 1 (from (Monet *et al.* (1992))) presents the Hertzprung-Russell (HR) Diagram of nearby stars with distances determined from trigonometric parallaxes. All stars in figure 1 are low-luminosity stars, the band that extends from $M_V \sim 10$ and $V - I \sim 1.6$ down to $M_V \sim 20$ and $V - I > 4$, are low mass hydrogen burning red dwarfs, that define the lower Main Sequence (MS) and cool sub-dwarfs (slightly to the left of the MS). The end of this sequence is defined by brown-dwarfs (BD) extending all the way to planet mass objects, which lie beyond this diagram. To the left of the MS in figure 1 is the white dwarf's (WD) cooling sequence.

Historically, low luminosity stars have received very little attention, in part because they are not only faint but also very stable. MS low mass stars burn their nuclear fuel at a very slow rate, therefore, they look the same for many Hubble times. Actually they have not had time to evolve like more massive stars do, the universe is not old enough for that.

On the other hand, WDs slowly evolve by cooling down, but in this case, it is only recently that we have begun to appreciate and understand the physics behind their evolution and its relevance to astrophysics in general.

Low luminosity stars constitute more than 80% of all stars in the universe, consisting of a variety of objects, like M-dwarfs, brown-dwarfs, white dwarfs, each with something to contribute to the advance of astronomical knowledge.

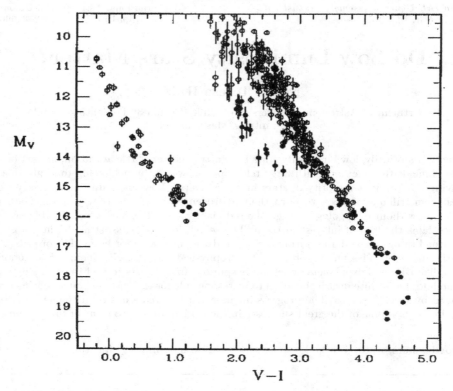

Figure 1. Hertzsprung-Russell diagram of nearby stars with distances determined from trigonometric parallaxes, showing the lower main sequence and the white dwarf's cooling sequence. From (Monet *et al.* (1992)).

2. Missing Mass Problem

One of the first bursts of interest on the study of low-luminosity stars came from the so called "missing mass problem" in the galactic disk. Analyzing the motions of stars in the direction perpendicular to the galactic plane the astronomer, Jan Oort (Oort (1965)) found that the observed motions were produced by a gravitational force that required a local mass density of $\rho_0 \sim 0.15\ M_\odot\ pc^{-3}$. At that time, the contribution to the local density from stars was $\rho_{stars} = 0.06\ M_\odot\ pc^{-3}$, and from gas and dust $\rho_{g+d} = 0.03\ M_\odot\ pc^{-3}$, therefore the total mass in the galactic disk that could be accounted for was not enough to explain the stellar motions studied by Oort (Oort (1965); Reid & Hawley (2005)).

During the 60s and 70s the "missing mass problem" attracted a lot of attention towards the search for low luminosity stars. Suddenly M dwarfs became "objects of interest", due to their well known large mass to light ratios ($M/L > 10$). This prompted several efforts to determine the stellar luminosity function (LF) aimed at having a better defined faint end, where M-dwarfs make their contribution. With the LF, and knowing the relation between mass and luminosity, for stars of different masses, one can obtain the stellar mass function (MF). The stellar mass density can then be derived by integrating the MF.

In the 70s an intense debate took place regarding the LF determined by different authors (Sanduleak (1965); Gliese (1972); Schmidt (1974); Luyten (1968); Luyten (1974); Jones (1973); Faber *et al.* (1976); Weistrop (1972); Weistrop (1976)). In order to estimate stellar luminosities (needed for the LF) of M-dwarfs, included in the different samples, it was necessary to know their distances, which were derived using spectroscopic and photometric parallaxes.

For a while, the LF determined by Weistrop (1972), Sanduleak (1965) and Gliese (1972), among other authors, implied a large density of M-dwarfs, enough to account for all Oort's missing mass. On the other hand, Luyten (Luyten (1968)) and previously van Rhijn (van Rhijn (1936)), had obtained a stellar LF indicating a local mass density of M-dwarfs that was a factor 5 to 10 lower.

The issue was discussed and settled during the IAU General Assembly of 1976. The conclusion was that the LFs that implied high densities of M-dwarfs were wrong. Discrepancies originated in the distance's determinations. Inaccurate photometry and spectral classifications, systematically under estimated distances, placing stars closer to us, thus occupying a smaller volume and implying a much larger density of stars.

The problem with the conflicting LFs was solved but we were left with the missing mass problem yet unsolved.

In spite of the progress in observational tools available to astronomy that follow that meeting in 1976, for more than two decades, Oort's missing mass was still not found.

More recent estimates (Chabrier (2001); Chabrier (2002); Robin (2001)), indicate that the local mass density considering the contribution by disk and halo (local) stars, brown dwarfs and that of gas and dust, amounts to: $\rho_{stars+bd+g+d} = 0.075 - 0.095 \ M_\odot \ pc^{-3}$.

On the other hand, the contribution by disk white dwarfs and neutron stars (Harris *et al.* (2006); Ruiz & Bergeron (2001); Holberg *et al.* (2002); Perna *et al.* (2006)) to the local density of matter is estimated to be : $\rho_{wd+ns} \sim 3.6 \times 10^{-3} \ M_\odot \ pc^{-3}$.

Adding up the contribution of all relevant low luminosity objects, mentioned above, plus that of gas and dust the total observed local mass density is : $\rho_{total} = 0.079 - 0.099 M_\odot pc^{-3}$. Therefore, decades after the meeting at the 1976 General Assembly and much effort from many groups, the missing mass had not been found.

The answer to this puzzle came from a very different approach (Creze *et al.* (1998); Bienayme (1999); Bienayme *et al.* (1999)), it came from a new analysis of stellar motions perpendicular to the galactic plane, this time using a very well defined sample of stars with distances obtained from Hipparcos Catalogs (ESA (1992); ESA (1997)). The remarkable result of these investigations was that the local dynamical density (needed to explain the observed stellar motions) was only $\rho_0 = 0.076 \pm 0.015 \ M_\odot \ pc^{-3}$, a value well below that found by Oort (Oort (1965)).

The present situation is that there is no missing mass in the local galactic disk, leaving no room for any disk dark matter component (Bienayme *et al.* (1999)). The problem was solved and this time apparently for good.

3. Renewed Interest in Low-luminosity Stars

In spite of the fact that efforts to find the missing mass among low-luminosity stars failed to do so, they were crucial to open up other lines of research on low luminosity stellar objects.

Microlensing experiments like MACHO (Alcock *et al.* (1997); Alcock *et al.* (2000)), EROS (Afonso *et al.* (2000); Lasserre *et al.* (2000)) and OGLE (Udalski *et al.* (1998)), found that the events detected towards the galactic bulge and the LMC were consistent with the lensing objects being small (in size) stars, with a mass of $\sim 0.6M_\odot$, a very good match to a WD star. These faint objects could account for about 20 % to 30 % of the dark matter in the galactic halo, in this case the prime candidates are faint Halo WDs.

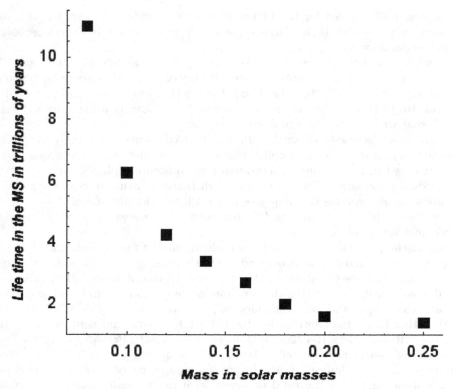

Figure 2. The evolution of a low mass star proceeds at a very slow rate, in time scales of trillions of years, (Laughlin *et al.* (1997)).

The field of brown dwarf's (BD) research and exoplanets were also spin-offs from the search for missing mass. The first BDs were identified in 1997 (Ruiz *et al.* (1997); Kirkpatrick *et al.* (1997)), and the first exoplanets in 1995 (Mayor & Queloz (1995)). Today, thanks to wide-deep surveys (like 2MASS, UKIDSS, SDSS, CFHTLS, DENIS, etc.) hundreds of BSs and exoplanets have been found in the solar neighborhood and in nearby stellar clusters.

Another source of interest in low mass stars research came from theoretical modeling of their evolution, interiors and atmospheres (Bergeron *et al.* (1995); Chabrier *et al.* (2000); Allard *et al.* (1997); Burrows *et al.* (1997); Marley *et al.* (2002); Baraffe *et al.* (1998)). Models today are quite mature, they rapidly evolve as new observations and the required physics become available. Every time a new type of object has been found models were almost ready to reveal their physical characteristics.

4. M-dwarfs

M-dwarfs are the lowest mass stars that populate the main sequence (see figure 1), they have masses between $0.5M_\odot$ to $0.08M_\odot$. Below $\sim 0.08M_\odot$ the star cannot sustain stable hydrogen (H) nuclear burning in their core, in which case they are called brown dwarfs. Temperatures of low mass M-dwarfs range from $T_{eff} \sim 3600K$ and $T_{eff} \sim 2400K$, and their visual luminosities from $L \sim 0.01L_\odot$ and $L \sim 10^{-5}L_\odot$.

The evolution of a low mass star not only proceeds at a very slow rate, in time scales of trillions of years, see figure 2 (taken from (Laughlin *et al.* (1997))), but it is somewhat different from that of more massive stars. A star with a mass $M < 0.2M_\odot$ remains fully convective over most of its life, mixing its helium (He) core with the surface H in a very efficient way. As a result almost all the H in an M-dwarf star is converted to He. The total amount of H burned by a $0.2M_\odot$ star is similar to that burned by a star like the Sun, in which only matter located in the central core participates in the nuclear process that converts H in He.

The end of an M-dwarf life, after more than 70 Hubble times, comes when the star runs out of H, they will not proceed to burn He as higher mass stars do. It will become a He white dwarf without ever going through the red giant phase (Laughlin *et al.* (1997); Chabrier & Baraffe (2000)).

Many Hubble times from now, when most stars will be dead, cold and invisible (white dwarfs, neutron stars or lack holes), these low mass stars will still be shining, and for a brief period of a few Gyrs, will have a surface temperature similar to that of the present day Sun. Maybe that would be the last chance for life in the Universe, as Laughlin *et al.* (1997) point out.

5. Brown Dwarfs

Below $\sim 0.08M_\odot$ stars cannot sustain stable hydrogen nuclear burning, they only have a brief initial period of deuterium (D) fusion (Chabrier & Baraffe (2000)). Later, they evolve cooling down supported by the pressure of their electron degenerate interiors. In this case the object is called a "brown dwarf" (BD) (Tarter (1974)).

After the first discovery of a handful of BDs, today thanks to large IR surveys, many hundreds of them are known. At the same time mature models (Chabrier *et al.* (2000); Burrows *et al.* (1997); Burrows *et al.* (2001)), are assigning physical meanings to their observed spectral characteristics (Geballe *et al.* (2002)).

As soon as the number of BDs known became large enough, they were classified according to their spectral features, which, thanks to the existence of reasonably good models were interpreted in terms of their physical parameters like temperature, metallicity, gravity etc.

The predominating features in the visual-IR spectra of M-dwarfs are TiO, CaOH, VO, H_2O, CO bands. On the other hand, L-dwarf spectra show the presence of dust clouds in their atmospheres, features of hydrides like FeH, CrH, MgH, CaH, and alkali metals like K, Na, Cs. Their effective temperatures range (spectral type L0 to L9.5) from $T_{eff} \sim 2400\ K$ and $T_{eff} \sim 1400\ K$.

Temperatures of T-dwarfs range from $T_{eff} \sim 1400\ K$ down to $T_{eff} \sim 600\ K$. In T-dwarfs the dust has completely precipitated to the bottom of the atmosphere, the most visible spectral features are CH_4, H_2O, NaI and KI.

The letter "Y" was reserved for yet a possible "new" spectral type, in case BDs with temperatures below 600 K were found, they were also expected to show the presence of NH_3 in their atmospheres (bad choice of letter, try to tell when somebody is talking about a "white dwarf" or a "Y dwarf"). It was a good idea though, because recently several Y-dwarfs have been found (Leggett *et al.* (2009); Burningham *et al.* (2008); Delorme *et al.* (2009)), with masses from 5 to 50 Jupiter masses, $T_{eff} \sim 550K$ and signs of NH_3 in their spectra. Y-dwarfs definitively enter the realm of exoplanets. This is even more clear in figure 3 (from (Burningham *et al.* (2008))), here the dark line spectrum is that of Jupiter, those bellow correspond to different Y-dwarfs, it is clear that the overall spectral shape of Y-dwarfs and Jupiter is remarkably similar.

The issue of whether low mass BDs and giant planets are the same thing is subject to some debate, although, the prevailing scenario suggests that they are not. BDs are believed to form the same way stars do, that is by the collapse of a dense core in a giant

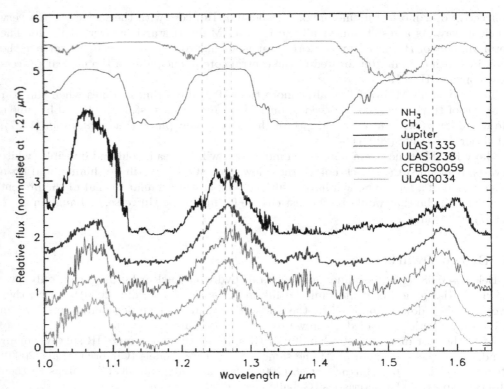

Figure 3. The close resemblance of the overall spectrum of the planet Jupiter (in solid black) and that of four Y-dwarfs, is clear from this figure taken from (Burningham *et al.* (2008)).

molecular cloud, therefore, its metallicity should reflect that of the parent cloud. On the other hand planets form from a debris disk around a star, the debris should have a higher metal content than the progenitor molecular cloud. This fact is actually observed in the solar system where Jupiter and Saturn have higher metallicities than the Sun. Other differences, consequence of their different origins are expected, like a large difference in their atmospheric pressures.

One other similarity between BDs and giant planets comes from the relation between their mass and radius. As can be seen in figure 4 (from (Chabrier *et al.* (2009))), for low mass M-dwarfs the radius is proportional to the mass while, for BDs and giant planets, radius seems to be rather insensitive to the mass, with all of them having a radius close to that of the planet Jupiter. The fits to the data in figure 4 correspond to models with different age and metallicity combinations (Chabrier *et al.* (2009)).

The difference between the mass-radius relation for stars and that for BDs and giant planets arise from the mechanisms supporting these objects against gravity. In the case of stars, this is the ideal-gas pressure, while for BDs and giant planets is the pressure from partially degenerate electrons.

6. Stellar Luminosity and Mass Functions

As a result of the search for "missing mass", a significant advance in the determination of the local stellar Luminosity and Mass Functions took place (Bochanski *et al.* (2009); Reid & Gizis (1997); Cruz *et al.* (2007)), thanks to large surveys with well defined distance

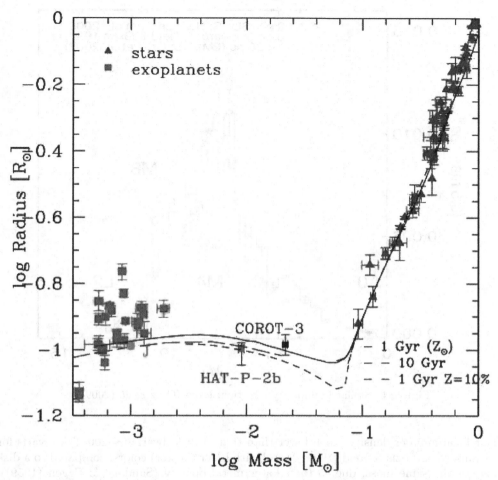

Figure 4. Mass-Radius relation for low mass stars, BDs and giant planets. The planet Jupiter is indicated with a solid dot and the letter J. Figure taken from (Chabrier *et al.* (2009)).

limits. Figure 5 (from (Cruz *et al.* (2009))), presents the local stellar LF determined by three different groups based on surveys that cover different volumes (8 pc with Hipparcos, 20 pc with 2MASS (for BDs) and 2000 pc with SDSS). Given the difference in samples, the LFs derived by these studies match pretty well, showing a decline below spectral type $\sim$ M4. The secondary peak shown in the BD's region of the LF, seems to be real (Cruz *et al.* (2007)), although this awaits confirmation from a larger volume sample of BDs.

The mass function (MF) for stars with masses $0.1 M_\odot < M < 0.8 M_\odot$ is shown in figure 6 (from (Bochanski *et al.* (2009))). Compared with previous determinations (Kroupa (2002); Miller & Scalo (1979)), the MF determined by Bochanski *et al.* (2009) presents a decline in the contribution of stars with masses below $0.27 M_\odot$, thus discarding the old idea that there could be huge numbers of low mass stars making an important contribution to the local baryonic dark matter.

6.1. *Subdwarfs Stars and BDs.*

So far we have discussed nearby disk stars, with ages of the order of a few Gyrs. However, in the solar vicinity there is also an older population of faint cool subdwarfs and BDs.

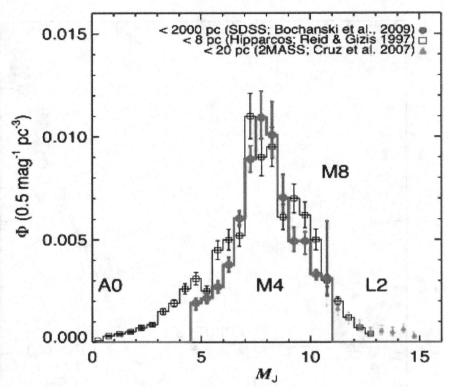

Figure 5. Stellar Luminosity Function. From (Cruz *et al.* (2009)).

Their local number density much lower than that of disk dwarfs (~ 200 disk dwarfs for each subdwarf (Gizis & Reid (1999))), and with bluer (hotter) colors, compared to a disk dwarf of the same mass, due to their lower metal opacity (Sandage & Eggen (1959)), make these cool subdwarfs hard to find. Until recently, no more than about 100 low mass cool subdwarfs were known (Gizis *et al.* (1997); Rojo & Ruiz (2003); Lepine *et al.* (2007)).

This situation has changed thanks to large area deep-surveys. For example, Lépine *et al.* (2009) discovered numerous local extreme-subdwarfs (esd) and ultra-subdwarfs (usd) from SDSS photometry (Lepine (2009)), using the color-color diagram (r-i) as a function of (g-r) (SLOAN filters), in which dwarfs of different metallicities clearly separate from each other. Subdwarfs (sd) have metallicities [Fe/H] ~ -1.2, while esd metallicity is [Fe/H] ~ -2.0 and that of usd is [Fe/H] < -2.0.

Accurate metallicities for low-mass, cool stars are difficult to obtain, at least in the optical region where absorption bands are saturated (Rojas-Ayala & Lloyd (2009); Lepine *et al.* (2007)), therefore this classification of sd, esd and usd is quite useful (and the best there is for now) when studying the older low-mass population.

Figure 7 (from (Lepine *et al.* (2007))), presents the differences in spectral features between dwarfs, subdwarfs and extreme-subdwarfs. The displayed wavelength range includes TiO bands and a CaH band. Low metallicity atmospheres, with less metals available, tend to produce hydrides (like CaH) instead of oxides (like TiO) which require two metals. In figure 7, the strength of the TiO band is maximum in solar metallicity dwarf stars (M-dwarf) and disappears towards esd type stars, while CaH does the opposite.

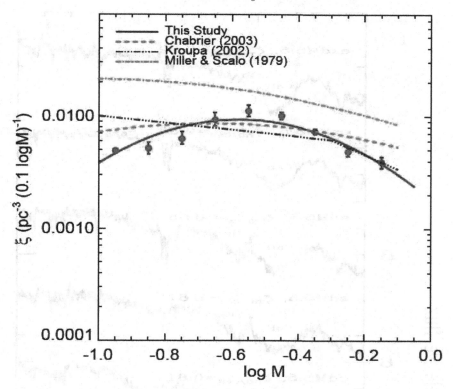

Figure 6. Stellar Mass Function. The fit is a log-normal distribution with a characteristic mass of $0.27 \pm 0.01 M_\odot$. Figure taken from (Bochanski *et al.* (2009)).

One of the latest additions to the subdwarf family has been the discovery of a few low metallicity BDs. IR surveys have revealed the existence of several L and T subdwarfs (Lepine & Scholz (2008); Burgasser *et al.* (2009); Burgasser *et al.* (2003); Cushing *et al.* (2009); Reiners & Basri (2006); Delorme *et al.* (2009)).

The field of low-mass stars, BDs and giant planets is rapidly advancing taking advantage of wide-deep surveys. We should expect much better constrained LFs, particularly at its faint end. The issue regarding differences between giant planets and BDs should be better understood, and models will evolve to be able to explain details regarding the physics of BDs (and giant planets), like for example, the relation between age-rotation-activity that in BDs seems to be different from that of stars. We are yet to find the oldest subdwarfs with no metals (they should be there).

6.2. *Cool White Dwarfs*

White Dwarfs (WD) are the solid remains of stars with main sequence masses below $\sim 8\ M_{odot}$. The typical WD mass is $\sim 0.6\ M_{odot}$ with an upper mass limit of$\sim 1.4\ M_{odot}$ and a radius of $\sim 0.01\ R_{odot}$. WDs are supported against gravity by a fully degenerate interior, in this case the radius is inversely proportional to the mass. More massive WDs are smaller and thus less luminous, than lower mass WDs.

WDs evolve by cooling therefore, in principle, if we could find the coolest WD and could determine its age (main sequence age + cooling age), then one could directly measure the age of that particular stellar system. The potential of WDs as "cosmic clocks" had long been recognized (Schmidt (1959); Winget *et al.* (1987); Wood (1992)). The disk

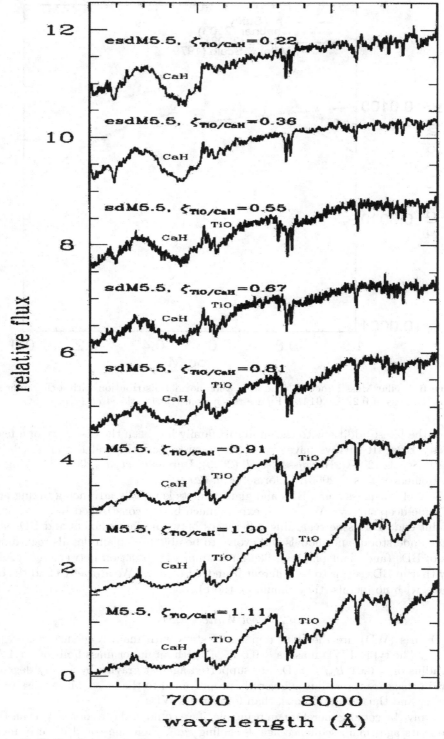

Figure 7. Stellar features in the spectra of disk dwarfs, subdwarfs and extreme-subdwarfs. Figure taken from (Lepine *et al.* (2007)).

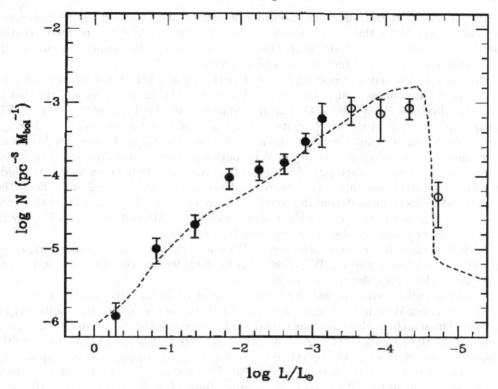

Figure 8. White Dwarf Luminosity Function (model fit by Winget *et al.* (1987)). Figure taken from (Liebert *et al.* (1988)).

WD' luminosity function in figure 8 (from (Liebert *et al.* (1988))) shows a rise towards lower luminosities (L), which is expected given that WDs evolve by cooling and the timescales strongly increase at lower temperatures, therefore the LF has a maximum at low luminosities ($L \sim 10^{-4} L_\odot$). Beyond the maximum, the LF drops abruptly, this has been interpreted as due to the finite age of the galactic disk (Winget *et al.* (1987)). The first cooling models by Winget *et al.* (1987) derived an age of 9.3 Gyr for the galactic disk (later this value was modified by improved models).

For a while this very optimistic view regarding the use of WDs as cosmic clocks prevailed (Fontaine *et al.* (2001)).

The WD LF also contains information about the star forming History of a system (Noh & Scalo (1990); Isern *et al.* (1998); Rowell *et al.* (2009); Hernanz *et al.* (1994)), bumps and inflections reveal changes in the star formation process.

Today, the optimistic view is out, the realization that a "very cool WD is not necessarily old" did it. The age of a WD strongly depends on several parameters, like mass, core composition, and atmospheric composition (Fontaine & Brassard (2005)). Most WDs have a dense C/O core of degenerate matter with a stratified atmosphere of pure H, pure He or a mixed H-He composition. In some cases traces of metals (C, Ca) are also present.

Core composition is important, for instance, early models with pure C cores (Fontaine *et al.* (2001)), cool down slowly due to an increased heat capacity, suggesting older ages than those obtained assuming a mixed C/O core, while Ne/O cores (massive WDs) and He cores (low mass WDs) cool down at different rates. Therefore in order to obtain the age of a WD, its core composition needs to be known.

Atmospheric composition is also important, a pure He atmosphere is more transparent and cools down faster than a H atmosphere. A mixed atmosphere will be more complicated, will depend on the H/He ratio. Therefore, the atmosphere composition also needs to be known in order to determine the age of a WD.

Another relation that is important to calculate the age of a WD, is the relation between the initial to final mass, that is the mass of the star in the main sequence (MS) that is the progenitor of a given mass WD. This information is needed because the age of a WD is the sum of its age in the MS plus its cooling age. For very massive WDs the MS age can be neglected compared with its cooling age, the opposite is the case for low mass WDs, however for the great majority of WDs both ages are relevant. Recent work in the determination of the semi-empirical initial to final mass function shows that it depends on metallicity and more importantly there is a large scatter which is real and reflects the fact that stars loose mass during the AGB phase, in a process that seems to be stocastic in nature. That is to say, stars with the same mass in the MS can produce a WDs with a wide range of masses (Catalan (2008); Salaris *et al.* (2009)).

Work towards a better understanding of WDs and its evolution is under way, these problems that arose regarding WD's ages will be dealt with once some missing physics becomes available in order to improve the models.

One outstanding problem that still remains unsolved is the little success of various groups in finding WDs belonging to the galactic halo (Oppenheimer (2001); Lodieu *et al.* (2009)). The search, in this case, was motivated by the results of the microlensing experiments (MACHO (Alcock *et al.* (1997); Alcock *et al.* (2000)), EROS (Afonso *et al.* (2000); Lasserre *et al.* (2000)) and OGLE (Udalski *et al.* (1998)), that suggested that up to 30% of the Halo baryonic dark matter could be in the form of WDs. So far, the observed density of matter in halo WDs has only an upper limit of $\rho_{halo\ wd} < 4 \times 10^{-5}\ M_\odot\ pc^{-3}$ (Harris *et al.* (2006)). Where are the halo WDs ? Are they much fainter than we think ? these are important questions that still need to be answered.

7. Conclusions

Low luminosity stars, almost by chance, have become "hot topics" in modern astronomy, and new fields of research have been born, like BDs and exoplanet research and the halo baryonic dark matter issue (suggested by the microlensing experiments).

Dim stars are illuminating a range of important topics in astronomy. For example, low mass stars with their long lives have remained un-evolved since their birth, they constitute ideal objects to study the Initial Mass Function in different environments. WD's LF can reveal the history of star formation. Baryonic dark matter in the Halo and Disk, exoplanet formation mechanism, age of stellar system, these are among the many astrophysical areas that low luminosity star's research is making an important contribution to.

In summary, to answer the question in the title of this presentation: **yes, low luminosity stars do matter !**

Acknowledgements

Thanks to the organizers, for the opportunity to show how important low luminosity stars are. Thanks to my friends and colleges Pierre Bergeron, Sandy Leggett, Andreas Reisseneger, José Garrido, Avril Day-Jones and René Méndez for the use of their unpublished data and for bringing to my attention publications that help me prepare this talk. Thanks to Neill Reid and Suzanne Hawley for their wonderful book that was of great help too. This work received support from CATA (BASAL) and FONDAP projects (CONICYT).

References

Alcock, C. *et al.* (The MACHO Collaboration) 1997, *ApJ* 486, 697

Alcock, C. *et al.* 2000, *ApJ* 542, 281

Afonso, C. *et al.* 2000, *ApJ* 532, 340

Baraffe, I., Chabrier, G., Allard, F., & Hauschild, PH. 1998, *A&A* 337, 403

Bergeron, P., Saumon, D., & Wesemael, F. 1995, *ApJ* 443, 764

Bienaymé, O. 1999, *A&A* 341, 86

Bienaymé, O., Crézé, M., Chereul, E., & Pichon, C. 2006, *ASP Conference Series* vol. 182, p. 301

Bochanski, J. J., Hawley, S. L., Reid, I. N., Covey, K. R., West, A. A., Golimowski, D. A., & Ivezić, Z. 2009, *Cool Stars, Stellar Systems and the Sun : 15 Cambridge Workshop* ed. E. Stempels p. 977

Burgasser, A. J., Lépine, S., Lodieu, N., Scholz, R.-D., Delorme, P., Jao, W.-C., Swift, B. J., & Cushing, M. C. 2009, *Cool Stars, Stellar Systems and the Sun : 15 Cambridge Workshop* ed. E. Stempels p. 242

Burgasser, A. J. *et al.* 2003, *ApJ* 592, 1186

Burningham, B., *et al.* 2008, *MNRAS* 391, 320

Burrows, A., Hubbard, W. B., Lunine, J. I., & Liebert, J. 2001, *Rev. of Modern Physics* 73, 719

Burrows, A., Marley, M., Hubbard, W. B., Lunine, J. I., Gillot, T., Saumon, D., Freedman, R., Sudarsky, D., & Sharp, C. 1997, *ApJ* 491, 856

Catalán, S., Isern, J., García-Berro, E., & Ribas, I. 2008, *MNRAS* 387, 1693

Chabrier, G., Baraffe, I., Leconte, J., Gallardo, J., & Barman, T. 2009, *Cool Stars, Stellar Systems and the Sun : 15 Cambridge Workshop* ed. E. Stempels p. 102

Chabrier, G. 2002, *ApJ* 567, 304

Chabrier, G. 2001, *ApJ* 554, 1274

Chabrier, G., Baraffe, I., Allard, F., & Hauschildt, P. H. 2000, *ApJ* 542, 464

Chabrier, G. & Baraffe, I. 2000, *Ann. Rev. Astron. Astrophys.* 38, 337

Crézé, M., Chereul, E., Bienaymé, O., & Pichon, C. 1998, *A&A* 329, 920

Cruz K. L. *et al.* 2009, *Astro2010: The Astronomy and Astrophysics Decadal Survey, Science White Papers, no. 60*

Cruz, K. L. *et al.* 2007, *AJ* 133, 439

Cushing, M. C., Looper, D., Burgasser, A. J., Kirkpatrick, J. D., Faherty, J., Cruz, K. L., Sweet, A., & Robyn, E. 2009, *ApJ* 696, 986

Delorme, P., Delfosse, X., Albert, L., Artigau, E., Forveille, T., Reylé, C., Allard, F., Homeier, D., & Robin, A. 2008, *Cool Stars, Stellar Systems and the Sun : 15 Cambridge Workshop* ed. E. Stempels p. 513

ESA 1997, *Hopparcos Catalog* ESA SP-1200

ESA 1992, *Hopparcos Input Catalog* ESA SP-1136

Faber, S. M., Burstein, D., Tinsley, B. M., & King, I. R. 1976, *AJ* 81, 45

Fontaine, G. & Brassard, P. 2005, *14th European Workshop on White Dwarfs, ASP Conference Series* eds. D. Koester and S. Moehler, p. 334

Fontaine, G., Brassard, P., & Bergeron, P. 2001, *PASP* 113, 409

Geballe, *et al.* 2002, *ApJ* 564, 466

Gizis, J. E. & Reid, I. N. 1999, *AJ* 117, 508

Gizis, J. E., Scholz, R. D., Irwin, M., & Jahreiss, H. 1997, *MNRAS* 292, 41

Gliese, W. 1972, *QJRAS* 13, 138

Harris, H. C. *et al.* 2006, *AJ* 131, 571

Hernanz, M., García-Berro, E., Isern, J., Mochkovitch, R., Segretain, L., & Chabrier, G. 1994, *ApJ* 434, 652

Holberg, J. B., Oswalt, T. D., & Sion, E. M. 2002, *ApJ* 571, 512

Isern, J., García-Berro, E, Hernanz, M., Mochkovitch, R., & Torres, S. 1998, *ApJ* 503, 239

Jones, D. H. P. 1973, *MNRAS* 161, 19

Kirkpartrick, J. D., Beichman, C. A., & Skrutskie, M. F. 1997, *ApJ* 476, 311

Kroupa, P. 2002, *Science* 295, 82

Laughlin, G., Bodenheimer, P., & Adams, F. C. 1997, *ApJ* 482, 420

Lasserre, T. *et al.* 2000, *A&A* 355, L39

Leggett, S. K., Burningham, B., Cushing, M. C., Marley, M. S., Pinfield, D. J., Saumon, D., Smart, R. L., & Warren, S. J. 2009, *Cool Stars, Stellar Systems and the Sun : 15 Cambridge Workshop* ed. E. Stempels p. 541

Lépine, S. 2009, *Cool Stars, Stellar Systems and the Sun : 15 Cambridge Workshop* ed. E. Stempels p. 545

Lépine, S. & Scholz, R-D 2008, *ApJ* 681, L33

Lépine, S., Rich, R. M., & Shara, M. M. 2007, *ApJ* 669, 1235

Liebert, J. Dahn, C. C., & Monet, D. G. 1988, *ApJ* 332, 891

Lodieu, N., Leggett, S. K., Bergeron, P., & Nitta, A. 2009, *ApJ* 692, 1506

Luyten, W. J. 1974, *Proper Motion Survey with the 48-inch Schmidt Telescope, XLVI*

Luyten, W. J. 1968, *MNRAS* 139, 221

Marley, M. S., Seager, S., Saumon, D., Lodders, K., Ackerman, A. S., Freedman, R. S., & Xiaohui, F. 2002, *ApJ* 568, 335

Mayor, M. & Queloz, D. 1995, *Nature* 378, 355

Miller, G. E. & Scalo, J. M. 1979, *ApJS* 41, 513

Monet, D. G., Dahn, C. C., Vrba, F. J., Harris, H. C., Pier, J. R., Luginbuhl, C. B., & Ables, H. D. 1992, *AJ* 103, 638

Noh, H.-R. & Scalo, J. M. 1990, *ApJ* 352, 605

Oort, J. H. 1965, *Galactic Structure*, ed. M. Schmidt, (University of Chicago Press)

Oppenheimer, B. R., Hambly, N. C., Digby, A. P., Hodkin, S. T., & Saumon, D. 2001, *Science* 292, 698

Perna, R., Narayan, R., Rybicki, G., Stella, L., & Treves, A. 2006, *ApJ* 594, 936

Reid, I. N. & Hawley, S. L. 2005, *New Light on Dark Stars* (Springer) p. 349

Reid, I. N. & Gizis, J. E. 1997, *AJ* 113, 2246

Reiners, A. & Basri, G. 2006, *AJ* 131, 1806

Robin, A. C., Reylé, C., & Crézé, M. 2001, *ASP Conference Series* eds. S. Deiters, B. Fuchs, A. Just, R. Spurzem & R. Wielen vol. 228, p. 553

Rojas-Ayala, B. & Lloyd, J. P. 2009, *Cool Stars, Stellar Systems and the Sun : 15 Cambridge Workshop* ed. E. Stempels p. 776

Rojo, P. M. & Ruiz, M. T. 2003, *AJ* 126, 353

Rowell, N. R., Hmably, N. C., & Bergeron, P. 2009, *Journal of Physics: Conference Series* 172, Issue 1

Ruiz, M. T. & Bergeron, P. 2001, *ApJ* 558, 761

Ruiz, M. T., Leggett, S. K., & Allard, F. 1997, *ApJ* 491, L107

Salaris, M., Serenelli, A., Weiss, A., & Miller-Bertolami, M. 2009, *ApJ* 692, 1013

Sanduleak, N. 1965, *PhD Thesis*, Case Institute of Technology

Sandage, A. R & Eggen, O. J. 1959, *MNRAS* 119, 278

Schmidt, M. 1974, *Highlights of Astronomy* ed. G. Contopoulos (Reidel) vol. 3, p. 450,

Tarter, J. C. 1974, *PhD Thesis*, University of Californiaat Berkeley

Udalsky, A. *et al.* 1998, *Acta Astron.* 48, 431

van Rhijn, P. J. 1936, *Pub. KartyenAstr. Lab. Grningen* No. 47

Weistrop, D. W. 1976, *ApJ* 204, 113

Weistrop, D. W. 1972, *AJ* 77, 849

Winget, D. E., Hansen, C. J., Liebert, J., Van Horn, H. M., Fontaine, G., Nather, R. E., Kepler, S. O., & Lamb, D. Q. 1987, *ApJ* 315, L77

Wood, M. A. 1992, *ApJ* 386, 539

Highlights of Astronomy, Volume 15
XXVIIth IAU General Assembly, August 2009
Ian F. Corbett, ed.
© International Astronomical Union 2010
doi:10.1017/S1743921310008197

JD1 – Dark Matter in Early-type Galaxies: Overview

Léon V. E. Koopmans[1] & Tommaso Treu[2,3]

[1]Kapteyn Astronomical Institute, University of Groningen, P.O. Box 800, 9700AV, Groningen, The Netherlands email: koopmans@astro.rug.nl
[2]Department of Physics, University of California, Santa Barbara, CA 93106-9530, USA; email: tt@physics.ucsb.edu
[3]Packard Fellow

Abstract. We summarize the motivations and main conclusions of the joint discussion "Dark Matter in Early-type Galaxies".

Keywords. Galaxies: elliptical and lenticular, cD – Galaxies: formation – Galaxies: evolution

1. Motivations

It is now commonly accepted that galaxies form inside deep gravitational potential wells dominated by dark matter halos. The existence of cold dark matter (CDM) is a fundamental cornerstone of our standard cosmological model.

However, our understanding of dark matter halos is far from satisfactory, particularly as far as early-type galaxies are concerned. In fact, the very existence of dark matter halos around early-type galaxies is still in question. Although they have been detected around early-type galaxies for over a decade (e.g. Saglia, Bertin & Stiavelli 1992; Franx, van Gorkom & de Zeeuw 1994), there have been claims that they are not always required (Romanowsky *et al.* 2003).

Improving our understanding of early-type galaxies and their dark matter halos is essential to test galaxies formation models. First, in the standard hierarchical model of galaxy formation, early-type galaxies are believed to result from major mergers of smaller galaxies (and halos). Therefore they provide the ultimate test of the merging hypothesis. Second, as their central regions are baryon dominated while their outer regions are believed to be dark matter dominated, they represent an ideal testing ground for the poorly understood interactions between baryons and dark matter. Third, additional information about the interplay between dark matter and baryons is provided by the still unexplained tight and non-trivial correlations between observables, known as scaling relations or scaling laws (e.g., the Fundamental Plane; Djorgovski & Davis 1987; Dressler *et al.* 1987). Fourth, being the most massive galaxies in the universe, their dark matter halos are in principle detectable out to high-redshift, thus enabling direct evolutionary studies. Finally, as the most massive galaxies in the universe they are also expected to host the most massive satellites and therefore are an ideal testing ground for the abundant satellite population predicted by CDM simulations and undetected in the local group (e.g., Kravtsov 2009).

Two sets of obstacles need to be overcome to improve our understanding of dark matter in early-type galaxies. From an observational point of view these systems typically lack the optical emission lines and diffuse gas that is so effective in tracing the dark matter halos of spiral galaxies. Therefore, alternative mass tracers must be found and exploited. From a modeling point of view, the triaxiality of early-type galaxies introduces fundamental degeneracies in interpreting projected observables in terms of intrinsic threedimensional properties (e.g. the so-called "mass-anisotropy degeneracy").

Substantial progress has been achieved on all these issues in the past few years using a combination of new technologies (e.g. integral field spectrographs on large telescopes and sensitive and high resolution X-ray satellites), new techniques (e.g. strong and weak gravitational lensing) as well as new theoretical insights (e.g. improved understanding of the role of active galactic nuclei). The goal of this meeting is to review and discuss recent progress and identify key questions for the future. In particular the following topics were identified by the organizing committee (chaired by the two authors of this summary and composed of Luca Ciotti, Wyn Evans, Ortwin Gerhard, Dan Maoz, Priyamvada Natarajan, Takaya Ohashi, and Silvia Pellegrini):

(*a*) Stellar and dark matter density profiles.
(*b*) Stellar and dark matter substructure.
(*c*) Empirical scaling relations.
(*d*) Formation mechanisms.
(*e*) Cosmic evolution.
(*f*) Observational and modeling techniques.
(*g*) Ongoing and future surveys.

2. Meeting Conclusions

Highlights of the individual contributions to the meeting are given in this volume. In this section we present a short summary of the panel discussion that concluded the meeting. The questions presented to the panel were:

(*a*) *Is there unambiguous evidence for dark matter in early-type galaxies?* The general consensus of the meeting was that observational evidence for dark matter around early-type galaxies has improved dramatically, especially with measurements currently extending to many effective radii using dynamical tracers, X-ray observations that probe the regions where stellar mass and traditionally the dark-matter halo starts, and also weak lensing which probes much further out. However, the picture is more complicated than imagined a decade ago and it now seems that the fraction of dark matter inside a fixed radius (e.g. the effective radius) might be a strong function of the mass of the galaxy, increasing both for masses larger *and* smaller than those corresponding to a luminosity of a few L_*.

(*b*) *Are modified gravity theories competitive with dark matter?* Here the jury is still out, although our understanding and observational tests have dramatically improved over the last few years (e.g. Bradač *et al.* 2006; Clowe *et al.* 2006). The development of relativistic theories of MOND has lead to new tests through gravitational lensing and stellar dynamics. Also N-body simulations that can be tested against observations are now performed in alternative gravities.

(*c*) *What methods/techniques or combination thereof are most effective at answering specific questions related to the topics of this meeting?* The consensus is that there is not a single "magic" technique or method that can answer all the questions about dark matter unambiguously, but that different methods have to be combined and compared.

(*d*) *Are there recognizable trends in dark matter properties with respect to early-type galaxies properties such as stellar mass or velocity dispersion?* Dramatic progress has been made over the last decade, showing that not all galaxies have identical distributions of dark matter, as predicted by pure dark-matter simulations. The DM content seems to be a strong function of the mass of the galaxy. Recent simulations seem to suggest that this trend is due to feedback from active galactic nuclei for the high-mass galaxies and possibly due to supernovae feedback for the lower-mass galaxies.

(*e*) *Is there dark matter substructure around early-type galaxies?* This is still very much an open question. Whereas tremendous progress has been made in the study of satellites around the Milky Way and the Local Group galaxies, far less, if anything at all is known about substructure around early-type galaxies. Here gravitational lensing appears to be a very promising method, which allows one through a number of observables (flux-ratios, astrometry, time-delays) to probe small scale fluctuations in the potential/mass-distribution of the lens galaxies. Both flux-ratio anomaly and gravitational-imaging methods have now claimed the discovery of substructure around galaxies, but considerable more research in this area is still essential.

(*f*) *What do the mass profiles of early-type galaxies look like? Are universal profiles a good fit to the data?* The answer to this question might again be a function of galaxy stellar mass. Whereas for the more massive ($> L_*$) elliptical galaxies typically isothermal-like total density profiles are found, lower-mass elliptical might still be consistent with the absence of a massive dark matter halo in their inner regions and a density profile that is steeper than isothermal. Likewise the dark matter density profiles may be close to NFW or steeper for massive early-type galaxies, and possibly flatter at dwarf galaxies and cluster scales, although much more work is needed to establish robust trends.

(*g*) *Why are there such tight scaling relations if star formation depends on micro-physics while galaxy dynamics seems dominated by gravitational (DM) macro physics?* Whereas progress has been made on explaining these tight relations and the importance of dark matter for example in the tilt of the Fundamental Plane, many other relations exist as well, including the black-hole to spheroid mass relation, the mass-metallicity relation. Although it seems clear that these relations must be tightly coupled to the formation history of early-type galaxies, theory is still not fully capable of explaining them from first principles. The existence of such tight relations implies that the micro-physics of star formation and black-hole physics somehow know about the large scale physics of galaxy formation and assembly (where gravity dominates). This suggest a strong feedback mechanism between the active nucleus and the interstellar medium. However, precisely how the AGN couples to the ISM and regulates star formation is not clear yet.

(*h*) *What open questions are "most interesting" for future studies?* As often progress may come from the most unexpected directions. However, what seem critical in the future to further our understanding, is to couple physics on small scales, be it dark-matter physics or the physics of star formation and AGN to events that occur on large macroscopic scales (i.e. galaxy assembly, etc). This requires covering an enormous dynamics range in scales and masses.

(*i*) *What tools (techniques/telescopes/instruments/surveys) will we need to answer these questions?* Major new facilities are coming online (e.g. ALMA) or will come online in the coming decade (e.g. TMT/ELT, LSST, SKA, JWST, etc) that all will have a tremendous impact on all the questions posed above. The coming decade promises to be transformational in the study of galaxies as testing ground of astrophysics and cosmology. Early-type galaxies in particular, because of their relative "simplicity" compared to other galaxies (e.g. spirals, irregulars), are expected to play a major role in improving our understanding of (i) structure formation in the early Universe (ETGs form at the highest-density peaks of the universe), (ii) super-massive black-hole formation and physics, (iii) dark-matter halos and mass-substructure, and (iv) the physics of dark-matter itself.

3. Conclusions

We are now in a golden age of studying the structure, formation and evolution of galaxies. In particular massive ETGs provide a test bed of the physics of galaxy formation, dark matter, dynamics, gravity, etc. Whereas enormous progress has been made over the last decade in these fields, we expect that the coming decades will be perhaps even more exiting. With new facilities coming online we will move from studying single ETGs to large ensembles spanning a wide range in masses, redshifts, and other properties. Dramatic improvements in theory and numerical simulations to include ever more sophisticated physics will be needed to interpret and understand the ever growing body of observational evidence. The challenge will be finding ways to implement and control all the relevant fundamental physics, tying together the small and large scales.

Acknowledgement

We wish to thank the members of the scientific organizing commitee, the speakers, and participants for making this meeting a success. We thank the IAU for their help and support of this meeting, and the city of Rio de Janeiro for providing a beautiful setting and warmXC hospitality during the meeting. The authors would like NASA for supporting HST programs #10174, #10587, #10886, #10494, #10798, #11202. L.V.E.K. is supported by an NWO-VIDI program subsidy (project number 639.042.505). T.T. acknowledges support from the NSF through CAREER award NSF-0642621 and by the Packard Foundation through a Packard Fellowship.

References

Bolton, A. S. et al. 2008, ApJ, 684, 248
Bolton, A. S. et al. 2007 ApJL, 665, L105
Bradač, M., et al. 2006, ApJ, 652, 937
Clowe, D., Bradač, M., Gonzalez, A. H., Markevitch, M., Randall, S. W., Jones, C., & Zaritsky, D. 2006, ApJL, 648, L109
Djorgovski, S. & Davis, M. 1987, ApJ, 313, 59
Ciotti, L., Lanzoni, B., & Renzini, A. 1996, MNRAS, 282, 1
Dressler, A., et al. 1987, ApJ, 313, 42
Franx, M., van Gorkom, J. H., & de Zeeuw, T. 1994, ApJ, 436, 642
Kravtsov, A. V. 2009, arXiv:0906.3295
Nipoti, C., Treu, T., Auger, M. W., & Bolton, A. S. 2009, ApJL, 706, L86
Romanowsky, A. J., Douglas, N. G., Arnaboldi, M., Kuijken, K., Merrifield, M. R., Napolitano, N. R., Capaccioli, M., & Freeman, K. C. 2003, Science, 301, 1696
Saglia, R. P., Bertin, G., & Stiavelli, M. 1992, ApJ, 384, 433
Treu, T. et al. 2009, ApJ, submitted, arXiv:0911.3392

Highlights of Astronomy, Volume 15
XXVIIth IAU General Assembly, August 2009
Ian F. Corbett, ed.

© International Astronomical Union 2010
doi:10.1017/S1743921310008203

Dark Matter and Elliptical Galaxy Dynamics

Ortwin Gerhard

Max-Planck-Institut für Ex. Physik, Giessenbachstrasse 1, D-85748 Garching, Germany
email: gerhard@mpe.mpg.de

The halos of elliptical galaxies, through their orbit and angular momentum distribution, contain important information about the formation and evolution of these systems.

There have been several recent advances in obtaining stellar kinematics at faint surface brightness levels. Planetary nebulae have been shown to trace stellar kinematics, and the velocity fields obtained with them point to more frequent kinematic misalignments and more complicated angular momentum properties in the outer halos than indicated by kinematics within the effective radius R_e (Coccato *et al.* 2009). In M87, the outer edge of the stellar halo and the transition to the unbound intracluster light has been reached at 150 kpc, where the halo velocity dispersion is below 100 km/s (Doherty *et al.* 2009). New techniques based on slitlets (Proctor *et al.* 2009) and on combining IFU data (Weijmans *et al.* 2009) have pushed the limits of absorption line spectroscopy to 3–4 R_e.

Made-to-measure adaptive particle modeling techniques have become competitive with the familiar Schwarzschild method (De Lorenzi *et al.* 2007), and first modeling results have been obtained for two intermediate luminosity ellipticals, NGC 3379 and NGC 4697 (De Lorenzi *et al.* 2008, 2009). The results show a strong mass-anisotropy-shape degeneracy in these systems with falling velocity dispersion profiles. As a consequence, these galaxies need not be devoid of dark matter, as has sometimes been suggested, but may simply be part of the lower concentration population of halos within the scatter predicted by dark matter simulations. If so, their orbit structure must be strongly radially anisotropic, which is not unexpected (Dekel *et al.* 2005).

The circular velocity curves of especially high-mass ellipticals are approximately flat within $2R_e$, but the recent study by Thomas *et al.* (2007) has shown some variety. Within R_e ~10–40% of the matter is dark, within $5R_e$ dark matter contributes ~50–80% of the mass. Thomas *et al.* (2009) have dynamically measured dark matter mean densities (DMMD) within the central $2R_e$ for a sample of Coma cluster ellipticals, and have related these to the assembly redshifts of ellipticals relative to those of spirals (updating Gerhard *et al.* 2001). They find the highest DMMDs in lower-luminosity ellipticals, that at the same luminosity (baryonic mass), the DMMD's of ellipticals are ~ 7 (~ 13) times higher than those of spirals, pointing to an assembly redshift $z = 2–3$.

References

Coccato, L., Gerhard, O., Arnaboldi, M., & Das, P. *et al.* 2009, *MNRAS*, 217
Dekel, A., Stoehr, F., Mamon, G. A., Cox, T. J., *et al.* 2005, *Nature*, 437, 707
De Lorenzi, F., Debattista, V. P., Gerhard, O., & Sambhus, N. 2007, *MNRAS*, 376, 71
De Lorenzi, F., Gerhard, O., Saglia, R. P., Sambhus, N., *et al.* 2008, *MNRAS*, 385, 1729
De Lorenzi, F., Gerhard, O., Coccato, L., Arnaboldi, M., *et al.* 2009, *MNRAS*, 395, 76
Doherty, M., Arnaboldi, M., Das, P., Gerhard, O., *et al.* 2009, *A&A*, 502, 771
Gerhard, O., Kronawitter, A., Saglia, R. P., & Bender, R., 2001, *AJ*, 121, 1936
Proctor, R. N., Forbes, D. A., Romanowsky, A. J., Brodie, J. P., *et al.* 2009, *MNRAS*, 398, 91
Thomas, J., Saglia, R. P., Bender, R., Thomas, D., *et al.* 2007, *MNRAS*, 382, 657
Thomas, J., Saglia R. P., Bender R., Thomas D., *et al.* 2009, *ApJ*, 691, 770
Weijmans, A.-M., Cappellari, M., Bacon, R., de Zeeuw, P. T., *et al.* 2009, *MNRAS*, 000, 000

Highlights of Astronomy, Volume 15
XXVIIth IAU General Assembly, August 2009
Ian F. Corbett, ed.

© International Astronomical Union 2010
doi:10.1017/S1743921310008215

JD1-The Planetary Nebulae and the Dynamics of NGC 1399

Emily McNeil[1], Magda Arnaboldi[2], Ortwin Gerhard[3], Kenneth Freeman[1], Payel Das[3] and Lodovico Coccato[3]

[1]Research School of Astronomy and Astrophysics, Australian National University,
Canberra, ACT 2601, Australia
email: emcneil@mso.anu.edu.au

[2]European Southern Observatory, Karl-Schwarzschild-Strasse 2
D-85748 Garching bei München, Germany

[3]Max-Planck-Institut für Extraterrestrische Physik, Giessenbachstrasse
85748 Garching bei München, Germany

Dynamical models of galaxies are limited by the paucity of kinematic data at large radii. Beyond the feasible limit of integrated-light spectroscopy, we rely on discrete tracers such as planetary nebulae and globular clusters. We describe a large ($\sim$200) sample of planetary nebula (PN) velocities in the outer regions of the cD elliptical NGC 1399. These data were obtained with a counter-dispersed slitless-spectroscopy technique which traces the kinematics to about 60 kpc (McNeil *et al.*, 2009).

The PNe complete the velocity and velocity dispersion profiles at these previously unattainable radii and trace the matter distribution typical of a central cluster galaxy. The results show a heterogenous population in the Fornax cluster core composed of subcomponents from NGC 1399, NGC 1404 and a mysterious low-velocity population. The kinematics are described by a low amplitude of rotation and a more gently rising velocity dispersion profile than previously thought (Arnaboldi *et al.*, 1994).

We compared these observations to spherical Jeans models for an anisotropy range of $-1 < \beta < 1$. The PN kinematics are consistent with a spherical model only if the mass is at the lower limit of the range permitted by X-rays.

Acknowledgements

E. McNeil gratefully acknowledges the support for her participation by ASA and IAU Travel Grants and Mt Stromlo Observatory funds.

References

Arnaboldi, M., Freeman, K. C., Hui, X., Capaccioli, M., & Ford, H. 1994, *The Messenger* 76, 40

McNeil, E. K., Arnaboldi, M., Freeman, K. C., Gerhard, O. E., Coccato, L., & Das, P. 2009, in prep

Highlights of Astronomy, Volume 15
XXVIIth IAU General Assembly, August 2009
Ian F. Corbett, ed.

© International Astronomical Union 2010
doi:10.1017/S1743921310008227

Probing the 2-D kinematic structure of early-type galaxies out to 3 effective radii

Robert N. Proctor[1], Duncan A. Forbes[1], Aaron J. Romanowsky[2], Jean P. Brodie[2], Jay Strader[3], Max Spolaor[1], J. Trevor Mendel[1] and Lee Spitler[1]

[1] Centre for Astrophysics & Supercomputing, Swinburne University, Hawthorn VIC 3122, Australia

[2] UCO/Lick Observatory, University of California, Santa Cruz, CA 95064, USA

[3] Harvard-Smithsonian Centre for Astrophysics, 60 Garden St., Cambridge, MA 02138, USA
email: rproctor@astro.iag.usp.br

We detail an innovative new technique for measuring the 2-D velocity moments (rotation velocity, velocity dispersion and Gauss-Hermite coefficients h_3 and h_4) using spectra from Keck DEIMOS multi-object spectroscopic observations. The data are used to reconstruct 2-D rotation velocity maps.

Here we present data for two nearby early-type galaxies to ~ 3 effective radii. From these data 2D kinematic maps are constructed. Three other galaxies have been similarly analysed (see 2009, MNRAS, 398, 91). We provide significant insights into the global kinematic structure of these galaxies, and, in some cases, challenge the accepted morphological classification. We show that between 1–3 effective radii the velocity dispersion declines very slowly, if at all, in all five galaxies. For the two galaxies with velocity dispersion profiles available from planetary nebulae data we find very good agreement with our stellar profiles. We find a variety of rotation profiles beyond 1 effective radius, i.e remain constant, decrease *and* increase with radius. These results are of particular importance to studies which attempt to classify galaxies by their kinematic structure within one effective radius, such as the recent definition of fast- and slow- rotator classes by the SAURON project. Our data suggests that the rotator class may change when larger galacto-centric radii are probed. This has important implications for dynamical modelling of early-type galaxies. The data from this study are available on-line.

Highlights of Astronomy, Volume 15
XXVIIth IAU General Assembly, August 2009
Ian F. Corbett, ed.
© International Astronomical Union 2010
doi:10.1017/S1743921310008239

Kinematic properties of early type galaxy halos using planetary nebulae

L. Coccato[1], O. Gerhard[1], M. Arnaboldi[2,3], P. Das[1], N. G. Douglas[4], K. Kuijken[5], M. R. Merrifield[6], N. R. Napolitano[7], E. Noordermeer[6], A. J. Romanowsky[8,9], M. Capaccioli[10,11], A. Cortesi[6], F. De Lorenzi[1] and K. C. Freeman[12]

[1]Max-Plank-Institut für Extraterrestrische Physik, Giessenbachstraße, D-85741 Garching bei München, Germany (E-mail: lcoccato@mpe.mpg.de); [2]European Southern Observatory,Karl-Schwarzschild-Straße 2, D-85748 Garching bei München, Germany; [3]INAF, Osservatorio Astronomico di Pino Torinese, I-10025 Pino Torinese, Italy; [4]Kapteyn Astronomical Institute, Postbus 800, 9700 AV Groningen, The Netherlands; [5]Leiden Observatory, Leiden University, PO Box 9513, 2300RA Leiden,The Netherlands; [6]School of Physics and Astronomy, University of Nottingham, University Park, Nottingham NG7 2RD, UK; [7]INAF-Observatory of Capodimonte, Salita Moiariello, 16, 80131, Naples, Italy; [8]UCO/Lick Observatory, University of California, Santa Cruz, CA 95064, USA; [9]Departamento de Fisica, Universidad de Concepción, Casilla 160-C, Concepción, Chile; [10]Dipartimento di Scienze Fisiche, Universitá Federico II, Via Cinthia,80126, Naples, Italy; [11]INAF – VSTceN, Salita Moiariello, 16, 80131, Naples, Italy; [12]Research School of Astronomy & Astrophysics, ANU, Canberra, Australia.

We studied the kinematics of the halo of a sample of 16 early type galaxies out to 5-10 effective radii using Planetary Nebulae (PNe) as kinematic tracers (Coccato et al. 2009).

We show that PNe are reliable tracers for the mean stellar population distribution and kinematics. In fact we found: i) a good agreement between the PNe number density distribution and the stellar surface brightness in the region where the two data sets overlap; ii) a good agreement between PNe and absorption-line kinematics. Therefore, the combination of photometry, absorption-line and PNe kinematics can be used to derive the entire dynamical picture of the galaxy and its halo. This is useful for the derivation of the luminous and dark matter profiles (e.g. De Lorenzi et al. 2008; 2009; Napolitano et al. 2009).

Moreover, in our analysis we found i) that the mean rms velocity profiles fall into two groups, with part of the galaxies characterized by slowly decreasing profiles and the remainder having steeply falling profiles; ii) a larger variety of velocity dispersion radial profiles; iii) that twists and misalignments in the velocity fields are more frequent at large radii, including some fast rotator galaxies; iv) that outer haloes are characterized by more complex radial profiles of the specific angular momentum-related λ_R parameter than observed within 1 R_e; v) that many objects are more rotationally dominated at large radii than in their central parts; and vi) that the halo kinematics are correlated with other galaxy properties, such as total B-band and X-ray luminosity, isophotal shape, total stellar mass, V/σ, and α parameter, with a clear separation between fast and slow rotators.

References

Coccato, L., Gerhard, O., Arnaboldi, M., et al. 2009, MNRAS, 394, 1249
de Lorenzi, F., Gerhard, O., Coccato, L., et al. 2009, MNRAS, 395, 76
de Lorenzi, F., Gerhard, O., Saglia, R., et al. 2008, MNRAS, 385, 1729
Napolitano N. R., Romanowsky, A. J., Coccato, L., et al. 2009, MNRAS, 393, 329

Highlights of Astronomy, Volume 15
XXVIIth IAU General Assembly, August 2009
Ian F. Corbett, ed.

© International Astronomical Union 2010
doi:10.1017/S1743921310008240

Modeling dark haloes in early-type galaxies: stellar kinematics at large radii

Anne-Marie Weijmans[1,2]

[1] Sterrewacht Leiden, Leiden University, Postbus 9513, 2300 RA Leiden, the Netherlands
email: weijmans@strw.leidenuniv.nl

[2] European Southern Observatory, Karl-Schwarzschild-Str 2, 85748 Garching, Germany

We developed a new method to obtain absorption line spectra of early-type galaxies at large radii, using integral-field spectrography (IFS). By using the spectrograph as a 'photon-collector' and adding the signal of many individual spaxels together in one spectrum, we obtain sufficient signal-to-noise to measure both stellar kinematics and line strengths at large radii. These can be used to determine the properties of the dark matter halo, as well as the stellar halo population.

We applied this method to the early-type galaxies NGC 3379 and NGC 821, using the SAURON IFS (Weijmans *et al.* 2009). We sampled the major axis of these galaxies with the IFS. The resulting stellar kinematics agree with data from long-slit spectrography and planetary nebulae. We construct mass models using the triaxial Schwarzschild code developed by van den Bosch *et al.* (2008) and find that inclusion of a dark halo is necessary for both systems to explain the observed kinematics. At least 30–50% of the total matter within 4 R_e is dark.

We also observed the early-type galaxy NGC 2549 with the PPak IFS, but instead of sampling the major axis, we obtained a full mosaic out to almost 5 R_e (Weijmans 2009). We find that NGC 2549 rotates in a disc-like structure over the total observed field. For this galaxy we also confirm the presence of a dark halo, with a dark matter fraction of at least 63% within 5 R_e.

References·

van den Bosch, R. C. E., van de Ven, G., Verolme, E. K., Cappellari, M., & de Zeeuw, P. T., 2008, *MNRAS*, 385, 647
Weijmans, A., 2009, PhD thesis, Leiden University
Weijmans, A., *et al.*, 2009, *MNRAS*, in press (arXiv0908.0018)

Highlights of Astronomy, Volume 15
XXVIIth IAU General Assembly, August 2009
Ian F. Corbett, ed.

© International Astronomical Union 2010
doi:10.1017/S1743921310008252

Luminous and dark matter in early-type lens galaxies

C. Grillo

LMU, Scheinerstr. 1, D-81679 München, Germany
email: cgrillo@usm.lmu.de

In the past few years gravitational lensing has allowed astrophysicists to make great progress in the understanding of the internal structure of early-type galaxies. By taking advantage of accurate photometric and spectroscopic measurements, the luminous and dark matter content of lens galaxies can in principle be disentangled (e.g., Grillo *et al.* 2008, 2009). SDSS J1538+5817 is an extraordinary strong lensing system composed of an elliptical galaxy and two equally-distant sources located, respectively, at redshifts 0.143 and 0.531 (Grillo *et al.*, submitted to ApJ). The sources are lensed into two and four images with an almost complete Einstein ring, covering a rather large region on the lens plane. By using HST/ACS and WFPC2 imaging and NOT/ALFOSC spectroscopy, we have investigated the lens total mass distribution within one effective radius. Then, we have fitted the SDSS multicolor photometry of the galaxy with composite stellar population models to obtain its luminous mass. By combining lensing and photometric measurements, we have estimated the lens mass in terms of luminous and dark matter components and studied the global properties of the dark matter halo. The exceptional lensing configuration of this system has allowed us to conclude that the galaxy dark matter density distribution is shallower and more diffused than the luminous one and the former starts exceeding the latter at a distance of approximately 1.5 times the effective radius. Extending these results to a larger number of lenses would help us to decipher the processes that rule galaxy formation and evolution in the LCDM scenario.

References

Grillo, C. *et al.* 2008, A&A, 486, 45
Grillo, C., Gobat, R., Lombardi, M., & Rosati, P. 2009, A&A, 501, 461

Highlights of Astronomy, Volume 15
XXVIIth IAU General Assembly, August 2009
Ian F. Corbett, ed.

© International Astronomical Union 2010
doi:10.1017/S1743921310008264

Weak lensing constraints on the dark matter haloes of early type galaxies

Priyamvada Natarajan[1,2]

[1]Radcliffe Institute for Advanced Study, 8 Garden Street, Cambridge, MA 02138, USA
[2]Dept. of Astronomy, Yale University, 260 Whitney Avenue, New Haven, CT 06520, USA
email: priyamvada.natarajan@yale.edu

Abstract. Constraints on the masses of dark matter haloes associated with galaxies in the field and in clusters have been successfully obtained used galaxy-galaxy lensing techniques. Weak lensing thus provides important information on the mass distribution in galaxies at large radii in a statistical fashion. Current work suggests that all galaxies possess extended dark matter haloes, however those that host galaxies in clusters are less extended and less massive.

Keywords. dark matter, galaxies

---

1. Introduction

Gravitational lensing has emerged as one of the most powerful techniques to map mass distributions on a range of scales: galaxies, clusters and beyond. The distortion in the shapes of background galaxies viewed through fore-ground mass distributions is independent of the dynamical state of the lens, therefore, unlike other methods for mass estimation there are fewer biases in lensing determinations. Here, we focus on the statistical mass estimates of field and cluster early-type galaxies obtained using weak gravitational lensing. Galaxy-galaxy lensing, the preferential alignment of faint background galaxies around bright foreground galaxies, was originally proposed as a method to constrain the masses and spatial extents of field galaxies (Brainerd, Blandford & Smail 1996), but has since been very successfully extended and applied inside clusters (Natarajan & Kneib 1996; Natarajan et al. 1998; 2002a). Constraints from this technique are statistical and probe the mass profiles at large radii, typically $r >> 0.1r_{vir}$. One advantage of probing this scale in galaxies is that uncertainties due to baryonic effects do not dominate, unlike measurements on smaller scales.

2. Galaxy-galaxy lensing constraints on field galaxies

Early results on galaxy-galaxy lensing identified a signal associated with massive haloes around typical early-type field galaxies, extending to beyond 100 kpc† (e.g. Brainerd, Blandford & Smail 1996; Hudson et al. 1998; Hoekstra et al. 2004). In particular, Hoekstra et al. (2004) report the detection of finite truncation radii based on the imaging data from the Red-Sequence Cluster Survey. Using a truncated isothermal sphere to model the mass in galaxy haloes, they find a best-fit central velocity dispersion for an L^* galaxy of $\sigma = 136 \pm 5$ kms^{-1} (68% confidence limits) and a truncation radius of 185 ± 30 kpc. However,

---

† We adopt $h = H_o/100 \, \mathrm{km\,s^{-1}\,Mpc^{-1}} = 0.7$ and $\Omega_\Lambda = 0.7$.

the most exciting results on the masses, density profiles and the galaxy-mass correlation function at the present time are from the SDSS (Sloan Digital Sky Survey). Mandelbaum and collaborators (Mandelbaum *et al.* 2006; Hirata *et al.* 2007) find that mass traces light on these large scales as $M/L \sim L^{0.5}$; the mass of a typical dark matter (DM) halo that hosts an early-type L* galaxy is $1.5 \times 10^{12} h^{-1} M_{\odot}$; more massive DM haloes host brighter galaxies and that the masses of haloes that host late-types are a factor of 2–4 lower. Fitting parametric models they find that both NFW and Einasto profiles offer good fits over the range $0.2 r_{vir} < r \, 1 h^{-1}$ Mpc; and the inferred concentration parameter c_{vir} for NFW profiles is consistent with expectations of a power-law distribution with halo mass from cosmological simulations.

3. Galaxy-galaxy lensing constraints on cluster galaxies

Isolating the galaxy-galaxy lensing signal in clusters has been a fruitful enterprise with HST (Hubble Space Telescope) quality data. Constraints have been obtained on the host haloes for fiducial early-types in ~ 10 clusters. DM haloes in the cluster environment appear to be severely truncated compared to their field counterparts. Although there is some variation amongst clusters, the typical mass of a DM halo that hosts an L* galaxy in clusters is a factor of 2-5 lower than an equivalent field galaxy. In addition, to the fact that tidal truncation in the violent cluster environment appears to strip infalling DM haloes, a strong radial trend with projected cluster-centric distance has also been detected for the cluster Cl 0024+16 (Natarajan *et al.* 2009). Utilizing mosaic-ed HST data extending well beyond the virial radius for the cluster Cl 0024+16, we find that fiducial DM haloes are more massive the further away they are from the cluster center, as expected from dynamics and crossing times (Treu *et al.* 2003). A first detection of the DM halo associated with late-type galaxies is also reported in this cluster. The DM halo hosting a late-type appears to be more severely tidally stripped than the equivalent halo hosting an early-type in the same cluster environment. Comparing with high resolution cosmological simulations, it is found that the observationally detected halo sizes and masses are consistent within the errors with expectations in ΛCDM paradigm (Natarajan & Springel 2004; Natarajan, De Lucia & Springel 2007).

4. Conclusions

Current weak lensing observations and analysis provide evidence for the existence of extended DM haloes around most galaxies. The DM haloes associated with early-type galaxies both in the field and in clusters are more massive than the haloes hosting equivalent luminosity late-types. These results offer deeper insights into the interplay between baryons and dark matter as a function of environment. Future large surveys are likely to help disentangle this relationship further.

References

Brainerd, T., Blandford, R., & Smail, I., 1996, *ApJ*, 466, 623
Hirata, C., *et al.*, 2007, *MNRAS*, 381, 1197
Hoekstra, H., Yee, H. K., & Gladders, M., 2004, *ApJ*, 606, 67
Hudson, M. J., Gwyn, S. D. J., Dahle, H., & Kaiser, N., 1998, *ApJ*, 503, 531
Mandelbaum, R., Seljak, U., Kauffmann, G., Hirata, C., & Brinkmann, J.,, 2006, *MNRAS*, 368, 715

Natarajan, P. & Kneib, J.-P. 1997, *MNRAS*, 287, 833

Natarajan, P., Kneib, J.-P., Smail, I., & Ellis, R. S. 1998, *ApJ*, 499, 600

Natarajan, P., Kneib, J.-P., & Smail, I., 2002a, *ApJ*, 580, L11

Natarajan, P. & Springel, V., 2004, *ApJ Lett.*, 617, 13

Natarajan, P., De Lucia, G., & Springel, V., 2007, *MNRAS*, 376, 180

Natarajan, P., *et al.*, 2009, *ApJ*, 693, 970

Treu, T., *et al.*, 2003, *ApJ*, 591, 53

Highlights of Astronomy, Volume 15
XXVIIth IAU General Assembly, August 2009
Ian F. Corbett, ed.
© International Astronomical Union 2010
doi:10.1017/S1743921310008276

Dark-Matter in Galaxies from Gravitational Lensing and Stellar Dynamics Studies

L. V. E. Koopmans

Kapteyn Astronomical Institute, University of Groningen, P.O. Box 800, 9700AV, Groningen,
The Netherlands; email: koopmans@astro.rug.nl

Abstract. Strong gravitational lensing and stellar dynamics provide two complementary methods in the study of the mass distribution of dark matter in galaxies out to redshift of unity. They are particularly powerful in the determination of the total mass and the density profile of mass early-type galaxies on kpc to tens of kpc scales, and also reveal the presence of mass-substructure on sub-kpc scale. I will shortly discuss these topics in this review.

Keywords. Galaxies: elliptical and lenticular, cD – Galaxies: formation – Galaxies: evolution

In recent years, the combined study of early-type galaxies using stellar dynamics and gravitational lensing has born exiting and unique results on the density profile of early type galaxies to redshift of unity, showing that massive early-type galaxies are close to homologous, have total density profile close to isothermal and do not appreciably evolve in their ensemble properties (for example Treu & Koopmans (2004), Koopmans *et al.* (2009)). Neither through lensing alone, nor through dynamics would these results have been possible to obtain. This has become even more exciting in recent years through the discovery of nearly 100 new strong galaxy-scale lens systems in the SLACS survey, for example Bolton *et al.* (2008) and development of powerful modeling techniques that self-consistently combine these two methods, using HST imaging and IFU/Longslit spectroscopic data Barnabè & Koopmans (2007), Czoske *et al.* (2008), Barnabè *et al.* (2009).

A second exiting new development is the *imaging* of mass-substructure on sub-kpc scales in these galaxies using extended arcs and Einstein rings Vegetti & Koopmans (2009). Also this is now possible thanks to the typical extended and highly-magnified nature of the lensed images in SLACS systems. Because multiple images allow one to separate the source structure from the lens-potential structure, this provides a unique method to probe unseen structure in lenses from dwarf satellites to general deviations from simple symmetries. Recently, the first detection of a high M/L-ratio dwarf satellite was reported using this method. Detection (even non-detections) promise to be a unique avanue to study the low-mass end of the dwarf-satellite mass function and possibly even probe the nature of dark matter – Vegetti & Koopmans (2009).

References

Barnabè, M. & Koopmans, L. V. E. 2007, ApJ, 666, 726
Barnabè, *et al.*, 2009, MNRAS, 399, 21
Bolton, A. S., *et al.* 2008, ApJ, 682, 964
Czoske, O., *et al.* 2008, MNRAS, 384, 987
Koopmans, L. V. E., *et al.* 2009, ApJL, 703, L51
Treu, T. & Koopmans, L. V. E. 2004, ApJ, 611, 739
Vegetti, S. & Koopmans, L. V. E. 2009, MNRAS, 1456
Vegetti, S., Koopmans, L. V. E., Bolton, A., Treu, T., & Gavazzi, R. 2009, arXiv:0910.0760
Vegetti, S. & Koopmans, L. V. E. 2009, MNRAS, 392, 945

Highlights of Astronomy, Volume 15
XXVIIth IAU General Assembly, August 2009
Ian F. Corbett, ed.

Density profile and line-of-sight mass contamination of SLACS gravitational lenses

C. C. Guimarães and Laerte Sodré Jr.

Departamento de Astronomia, IAG, Universidade de São Paulo

We use data from 58 strong lensing events surveyed by the Sloan Lens ACS Survey to estimate the projected galaxy mass inside their Einstein radii by two independent methods: stellar dynamics and strong gravitational lensing. We perform a joint analysis of both estimates examining the galaxy-lens density profile (that we approximate by a power law), the anisotropy of the velocity distribution (represented by an effective constant parameter), and a possible line-of-sigh (l.o.s.) mass contamination (which is suggested by various independent works in the literature). For each model, a likelihood analysis is performed to find the parameters that produce the best agreement between the dynamical and lensing masses, and the parameter confidence levels. The Bayesian evidence is calculated to allow a comparison among the models. We find a degeneracy among the slope of the density profile, the anisotropy parameter and the l.o.s. mass contamination. For a density profile close to isothermal, a l.o.s. mass contamination of the order of a few percent is possible, being less probable with larger anisotropy.

References

Guimarães, A. C. C. & Sodré, L. J. 2009, arXiv:0904.4381

Highlights of Astronomy, Volume 15
XXVIIth IAU General Assembly, August 2009
Ian F. Corbett, ed.

Dark Matter Determinations from *Chandra* Observations of Quadruply Lensed Quasars

David Pooley[1], Jeffrey A. Blackburne[2], Saul Rappaport[2] and Paul L. Schechter[2]

[1]Department of Astronomy, University of Wisconsin, Madison, WI 53706, USA
email: dave@astro.wisc.edu

[2]Kavli Institute for Astrophysics and Space Research, Massachusetts Institute of Technology, Cambridge, MA 02139, USA

We present *Chandra* X-ray observations of 14 quadruply lensed quasars. The X-ray data reveal flux ratio anomalies which are more extreme than those seen at optical wavelengths, confirming the microlensing origin of the anomalies originally seen in the optical data. The reduction of the anomalies in the optical, as compared to X-ray, indicates that the sizes of the optical emitting regions of the quasars must be about 1/3 the size of the projected Einstein radii of the microlensing stars. The X-ray emitting regions are essentially point sources and therefore give a microlensing signal unencumbered by source size considerations. For each lensing galaxy, we determine the most likely ratio of smooth material (dark matter) to clumpy material (stars) to explain the X-ray flux ratios. The ensemble of *Chandra*-observed quads indicates that the amount of matter projected along the lines of sight to the images (at radial distances of several kpc from the centers of the lensing galaxies) is ~90% smooth dark matter and ~10% stars.

Highlights of Astronomy, Volume 15
XXVIIth IAU General Assembly, August 2009
Ian F. Corbett, ed.

© International Astronomical Union 2010
doi:10.1017/S1743921310008306

Microlensing Diagnosis in Lensed Quasars

V. Motta[1] E. Mediavilla[2], E. E. Falco[3] and J. A. Muñoz[4]

[1] Departamento de Fisica y Astronomia, Universidad de Valparaiso, Avda. Gran Bretaña 1111, Valparaiso, Chile
email: **vmotta@dfa.uv.cl**

[2] Instituto de Astrofisica de Canarias, Universidad de La Laguna, V´ia Láctea s/n, La Laguna, Tenerife, España
email: **emg@iac.es**

[3] Harvard-Smithsonian Center for Astrophysics, Whipple Observatory, Amado, Arizona, USA
email: **falco@cfa.harvard.edu**

[4] Departament d'Astronomia i Astrofisica, Universitat de València, Burjassot, Valencia, España
email: **jmunoz@uv.es**

Simple gravitational lens models usually suffice to reproduce the positions of lensed quasar images, but they have problems to reproduce their optical flux ratios. The so-called flux ratio anomalies are thought to be produced by small-scale structure in the lens galaxies (microlensing).

Since the sizes of the emitting regions in each quasar depend on the observed wavelength, microlensing will yield a wavelength-dependent magnification: the continuum emitting region is magnified but not the narrow-line region. For a pair of images, the ratio of their emission line fluxes represents the baseline of no microlensing. Relative to this, the ratio of their continua yields the difference in microlensing magnification. We demonstrate the method, presenting the results obtained from MMT, VLT and HST spectroscopy for Q0957+561 and HE1104-1805. In these cases we found the microlensing magnification is wavelength-dependent (chromatic). A statistically significant sample of microlensing measurements using spectra will allow us to estimate the fraction of mass in the lens galaxy that is composed of compact objects (α).

We conducted a preliminary study, with microlensing measurements from spectroscopy we collected from the literature for 29 image pairs. The histogram of observed microlensing events peaks below 0.6 mag. The likelihood of the microlensing measurements using frequency distributions obtained from simulated microlensing magnification maps is explained by a low value of α ($< 10\%$). The results will improve once a homogenous and statistically significant sample of microlensing measurements is assembled.

Highlights of Astronomy, Volume 15
XXVIIth IAU General Assembly, August 2009
Ian F. Corbett, ed.

Chemo-dynamical evolution of dwarf galaxies: from flat to cuspy dark matter density profiles

S. Pasetto[1,2], E. K. Grebel[1], P. Berczik[1] and R. Spurzem[1]

[1] Astronomisches Rechen-Institut, Zentrum für Astronomie der Universität Heidelberg, Germany; [2] Max-Planck-Institut für Astronomie, Heidelberg, Germany;
email: spasetto@ari.uni-heidelberg.de

A model of an isolated dwarf spherical galaxy (dSph) is considered in its chemo-dynamical evolution with time. The system is composed by 3 γ-model density profiles: gas, stellar and dark matter, and it is realized in a spherical symmetric equilibrium configuration. The total masses used in our simulations are covering the dwarf galaxies mass range. The stability of this configuration is first tested for the system evolving under the gravity effect alone and then evolved taking into account for the most relevant stellar astrophysical processes implemented with a Smooth Particle Hydrodynamic approach. The two different kinds of evolution are compared. The dark matter evolves naturally from a centrally cuspy density profile into a flatter one within a timescale of several Gyr. The effect manifests itself naturally, without any tuned initial conditions, as soon as few standard criteria on star formation are assumedand the SN feedback on the ISM has been adopted the prescription in (Cioffi & Shull 1991) and (Bradamante *et al.* 1998). This result is expected to be a possible natural explanation for the discrepancy between observations that want flatter dark matter profiles (e.g. de Block 2005), and N-body simulations that predict cuspy dark matter profiles (Navarro *et al.* 1997). Chemical considerations are presented as a tool to follow with observational parameters the theory predictions.

References

Pasetto, E. K. Grebel, P. Berczik, & R. Spurzem, 2009 submitted A&A and references therein

Highlights of Astronomy, Volume 15
XXVIIth IAU General Assembly, August 2009
Ian F. Corbett, ed.

Modeling mass independent of anisotropy: A comparison between Andromeda and Milky Way satellites

J. Wolf.

Center for Cosmology Physics & Astronomy Frederick Reines Hall University of California
Irvine, CA 92697-4575, US; email: wolfj@uci.edu

Mass profile determinations for dispersion supported galaxies from line-of-sight velocities are subject to large uncertainties associated with the unknown stellar velocity anisotropy. We demonstrate both analytically and with available kinematic data (for systems spanning eight decades in luminosity) that the mass-anisotropy degeneracy is effectively eliminated at a characteristic radius that is close to the 3D deprojected half-light radius of the stars. This allows a simple, yet accurate formula to describe the half-light dark matter masses of all hot systems, including dwarf spheroidal galaxies (dSphs), based on directly observable parameters: $M_{1/2} = 4\,\sigma_{\rm LOS}^2\,R_{\rm half}\,/\,G$, where $R_{\rm half}$ is the 2D projected half-light radius and $\sigma_{\rm LOS}$ is the luminosity-weighted square of the line-of-sight velocity dispersion. The fact that masses are well-constrained within a characteristic stellar radius has allowed our group to perform systematic, accurate mass determinations for Milky Way dSphs and to conclude that they all have a common mass scale of approximately 10^7 $M_\odot$ within 300 pc of their centers. We extend this work to the satellite population of Andromeda using Keck/DEIMOS spectroscopy of individual stars. We find that the Andromeda dSphs are also consistent with sharing a common mass, but that it is offset from the scale of the Milky Way dSphs by a factor of ~ 2.

Highlights of Astronomy, Volume 15
XXVIIth IAU General Assembly, August 2009
Ian F. Corbett, ed.

Empirical Scaling Relations

Tommaso Treu

Department of Physics, University of California, Santa Barbara, CA 93106-9530, USA
email: tt@physics.ucsb.edu

Abstract. Early-type galaxies do not come in any shape, form, and color. Many of their observable properties obey tight correlations, also known as empirical scaling relations. The correlations are non-trivial, in the sense that they cannot be explained by simple physical or dimensional arguments. A subset of the empirical scaling relations connects baryonic observables with quantities that depend on the total gravitational potential of the galaxies, and thus on their dark matter content. These correlations are a fundamental testbed for our understanding of the formation and evolution of early-type galaxies, and, more in general, of the physical processes that determine the interplay between baryons and dark matter at galactic scales.

Keywords. Galaxies: elliptical and lenticular, cD - Galaxies: formation - Galaxies: evolution

---

The fundamental plane Djorgovski & Davis (1987), Dressler *et al.* (1987), the more fundamental plane Bolton *et al.* (2007) – together with their projections and their extension including stellar mass estimators – connect their baryonic content to their dark matter content. One key feature of these scaling relations is the so-called "tilt" of the fundamental plane (e.g. Ciotti *et al.* (1996)), i.e. the fact that M/L increases with galaxy stellar mass. The "tilt" could be the result of several competing effects, the main suspects being systematic changes in internal dynamical structure with galaxy size and stellar population changes. Recent studies (see also Cappellari, this meeting), indicate that the trends are due in some part to an increase in stellar mass to light ratio with stellar mass, resulting from the now well-established mass metallicity and mass age correlations (also known as downsizing). However, these are not sufficient, and a systematic increase in dark matter content (or stellar IMF) with stellar mass appears necessary to full reproduce the "tilt" Bolton *et al.* (2008), Treu *et al.* (2009).

These trends are particularly remarkable if one considers the tight scatter of the scaling relations. In fact, their tightness is an extremely powerful constraint on galaxy evolution models. Nipoti *et al.* (2009) showed that although dry mergers could be invoked to explain the apparent growth in size of early-type galaxies since $z \sim 2$, they tend to destroy the observed correlations. Therefore, the progenitors of present day early-type galaxies cannot have formed from galaxies obeying the same scaling relations. Likewise, early-type galaxies are unlikely to have accreted more than half of their current mass by dry mergers.

References

Bolton, A. S. *et al.* 2008, *ApJ*, 684, 248
Bolton, A. S. *et al.* 2007 *ApJL*, 665, L105
Djorgovski, S. & Davis, M. 1987, *ApJ*, 313, 59
Ciotti, L., Lanzoni, B., & Renzini, A. 1996, *MNRAS*, 282, 1
Dressler, A., *et al.* 1987, *ApJ*, 313, 42
Nipoti, C., Treu, T., Auger, M. W., & Bolton, A. S. 2009, *ApJL*, 706, L86
Treu, T. *et al.* 2009, *ApJ*, submitted, arXiv:0911.3392

Highlights of Astronomy, Volume 15
XXVIIth IAU General Assembly, August 2009
Ian F. Corbett, ed.

Scaling relations in early-type galaxies from integral-field stellar kinematics

M. Cappellari,[1] N. Scott,[1] K. Alatalo,[2] L. Blitz,[2] M. Bois,[3] F. Bournaud,[4] M. Bureau,[1] R. L. Davies,[1] T. A. Davis,[1] P. T. de Zeeuw,[5,14] E. Emsellem,[5,3] J. Falcon-Barroso,[6] S. Khochfar,[7] D. Krajnovic,[1] H. Kuntschner,[5] P.-Y. Lablanche,[3] R. M. McDermid,[8] R. Morganti,[9] T. Naab,[10] M. Sarzi,[11] P. Serra,[9] R. C. E. van den Bosch,[12] G. van de Ven,[13] A. Weijmans[14] and L. M. Young[15]

[1]University of Oxford, UK; [2]University of California, Berkeley, USA; [3]Université de Lyon, France; [4]Université Paris Diderot, France; [5]ESO, Garching, Germany; [6]IAC, La Laguna, Spain; [7]MPI for Extraterrestrial Physics, Garching, Germany; [8]Gemini Observatory, Hilo, USA; [9]ASTRON, Dwingeloo, The Netherlands; [10]Universitäts-Sternwarte München, Germany; [11]University of Hertfordshire, Hatfield, UK; [12]The University of Texas, Austin, USA; [13]IAS, Princeton, USA; [14]Leiden University, The Netherlands; [15]New Mexico Tech, Socorro, USA

Early-type galaxies (ETGs) satisfy a now classic scaling relation $R_e \propto \sigma_e^{1.2} I_e^{-0.8}$, the Fundamental Plane (FP; Djorgovski & Davis 1987; Dressler et al. 1987), between their size, stellar velocity dispersion and mean surface brightness. A significant effort has been devoted in the past twenty years to try to understand why the coefficients of the relation are not the ones predicted by the virial theorem $R_e \propto \sigma_e^2 I_e^{-1}$.

Recent studies, using independent approaches from either (i) detailed dynamical models or (ii) strong galaxy lensing, point to a genuine variation of the mass-to-light ratio M/L in galaxies as the reason for nearly all the observed 'tilt' in the FP (e.g. Cappellari et al. 2006; Bolton et al. 2008). However these studies are limited by a small and biased sample or are restricted to only the most massive ETGs respectively.

We overcome both limitations by modeling the stellar dynamics, using axisymmetric Jeans anisotropic models (JAM; Cappellari 2008), for the K-band selected, volume-limited ATLAS^{3D} sample of 263 nearby ETGs, spanning a large range of masses and with $60 < \sigma_e < 350$ km s^{-1}. A key for the project is the availability for all galaxies of high-quality integral-field kinematics observed with the SAURON spectrograph and detailed Multi-Gaussian Expansion (Emsellem et al. 1994) models of the photometry.

We confirm the genuine M/L variation and construct both the FP and the More FP (MFP; Bolton et al. 2007) for the ATLAS^{3D} sample, relating the mean surface density $\Sigma_e \equiv I_e \times (M/L)_{\rm JAM}$, σ_e and R_e. Our MFP produces a relation as tight as the FP over the full mass range. We compare the global $(M/L)_{\rm JAM}$ variation among galaxies with predictions from two-SSP stellar population models and find that variations of both dark matter (or IMF) and population are required to explain the observations.

References

Bolton, A. S., et al. 2008, ApJ, 684, 248
Bolton, A. S., Burles, S., Treu, T., Koopmans, L. V. E., & Moustakas, L. A. 2007, ApJ, 665, L105
Cappellari, M., et al. 2006, MNRAS, 366, 1126
Cappellari, M. 2008, MNRAS, 390, 71
Djorgovski, S. & Davis, M. 1987, ApJ, 313, 59
Dressler, A., et al. 1987, ApJ, 313, 42
Emsellem, E., Monnet, G., & Bacon, R. 1994, A&A, 285, 723

Highlights of Astronomy, Volume 15
XXVIIth IAU General Assembly, August 2009
Ian F. Corbett, ed.

Lenticular vs spiral galaxies: dark matter content and the Tully-Fisher relation

M. Bureau, M. J. Williams and M. Cappellari

University of Oxford, UK

We provide observational constraints on disk galaxy evolution for a sample of 28 local edge-on early-type (S0–Sb) disk galaxies. We do this in two ways: (i) we use simple dynamical modelling techniques to constrain their stellar and dark matter content (Williams et al. 2009) and (ii) we compare the zero points of the Tully-Fisher relations (TFRs; Tully & Fisher 1977) of the spirals and S0s.

For each galaxy, we model the stellar mass distribution under the assumptions of axisymmetry and a constant stellar mass-to-light ratio (M/L), and include a NFW halo. We then use an axisymmetric Jeans modelling technique assuming constant orbital anisotropy (Cappellari 2008). In this way we derive a model-dependent but purely dynamical estimate of the stellar (M/L), free from uncertainties due to the initial mass function and late phases of stellar evolution. We find a median K_S-band stellar (M/L) of 1.09 with a small rms scatter of 0.31. Dark matter typically comprises 15% of the mass within the effective radius R_e and 50% within the optical radius R_{25}.

There is a small but significant difference between the zero points of the spiral and S0 TFRs. For a given circular velocity, spirals are brighter than S0s by 0.5 mag at K_S-band, an offset smaller than previous results (e.g. Bedregal et al. 2006). We argue that our determination is to be preferred because it is free from the possible bias introduced by the comparison of rotational velocities derived from global emission line widths to stellar kinematics.

References

Bedregal, A. G., Aragón-Salamanca, A., & Merrifield, M. R., 2006, *MNRAS*, 373, 1125
Cappellari, M., 2008, *MNRAS*, 390, 71
Tully, R. B. & Fisher, J. R., 1977, *A&A*, 54, 661
Williams, M. J., Bureau, M., & Cappellari, M., 2009, *MNRAS*, accepted

Highlights of Astronomy, Volume 15
XXVIIth IAU General Assembly, August 2009
Ian F. Corbett, ed.

Scaling relations of early-type galaxies at $1 < z < 2$

Paolo Saracco[1], Marcella Longhetti[1] and Adriana Gargiulo[1]

[1]INAF – Osservatorio Astronomico di Brera, via Brera 28, 20121 Milano, Italy
email: name.surname@brera.inaf.it

We studied the scaling properties of a sample of 65 ETGs at $1 < z < 2$ with spectro-scopic confirmation of their redshift and spectral type. The sample collects proprietary (Longhetti *et al.* 2007) and archival HST data and it is composed of 30 ETGs with HST-NICMOS observations (see Saracco *et al.* 2009) and of 35 ETGs from the GOODS-South field covered by HST-ACS observations. The whole sample is covered also by ground-based optical and near-IR observations while complementary mid-IR data (Spitzer or AKARI) are available for 45 galaxies. The study of the Kormendy, the size-luminosity and the size-mass relations of these ETGs shows that a large fraction ($\sim 50\%$) of them follows the local relations. These 'normal' ETGs are not smaller (denser) than their lo-cal counterparts with comparable stellar mass, luminosity and surface brightness and no size evolution is required for them. On the contrary, the remaining half of the sample is composed of very compact ETGs with sizes (densities) 2.5-3 (15-30) times smaller (higher) than the other ETGs and than local ETGs. Thus, not all the high-z ETGs are superdense and, consequently, only some of them must experience size evolution showing that the evolutionary path of ETGs at $0 < z < 2$ is not univocal. We also find that the stellar population of normal ETGs formed at $1.5 < z_{form} < 3$ while it formed at $2 < z_{form} < 9$ in compact ETGs. This suggests that different histories of mass assembly must take place at high-z to produce both the normal and the superdense ETGs seen at $1 < z < 2$ (Saracco *et al.* 2010).

References

Longhetti, M., Saracco, P., Severgnini, P., *et al.* 2007 *MNRAS* 374, 614
Saracco, P., Longhetti, M., & Andreon, S. 2009 *MNRAS* 392, 718
Saracco, P., Longhetti, M., & Gargiulo, A. 2010 *MNRAS* submitted

Highlights of Astronomy, Volume 15
XXVIIth IAU General Assembly, August 2009
Ian F. Corbett, ed.

Scaling relations of early-type galaxies in the 6dF Galaxy Survey

C. Magoulas[1], M. Colless[2], H. Jones[2], J. Mould[1]
and C. Springob[2]

[1] School of Physics, University of Melbourne, Australia
[2] Anglo-Australian Observatory, Epping, Australia

Over 10,000 early-type galaxies from the 6dF Galaxy Survey (6dFGS) (Jones, D. H. *et al.* (2009), Jones *et al.* (2004)) have been used to determine the Fundamental Plane at optical and near-infrared wavelengths. We find that a maximum likelihood fit to an explicit three-dimensional Gaussian model for the distribution of galaxies in size, surface brightness and velocity dispersion can precisely account for selection effects, censoring and observational errors, leading to precise and unbiased parameters for the Fundamental Plane and its intrinsic scatter.

The 6dFGS is the largest NIR-selected sample in the local universe, covering a wide range of environments and masses, making it ideal for investigating the influence of dark matter on the Fundamental Plane. Taking advantage of the large and homogeneous nature of the 6dFGS sample we have explored the environmental dependence of the Fundamental Plane. We divided the sample into three sub-samples defined by a measure of richness derived from a group catalogue generated using a percolation-based algorithm. The maximum likelihood fitting algorithm was then used to obtain a robust and accurate Fundamental Plane for each individual richness sub-sample. The resulting fitted offsets and slopes are consistent with no significant difference between the Fundamental Plane of the galaxies in the richest clusters and isolated field galaxies. This has important implications for the peculiar velocities of the galaxies in the 6dFGS sample, as the measured peculiar velocities are dependent on Fundamental Plane distances to these groups and clusters.

References

Jones, D. H. *et al.* 2009, *astro-ph/0903.5451*
Jones, D. H *et al.* 2004, *MNRAS* 791, 355

Highlights of Astronomy, Volume 15
XXVIIth IAU General Assembly, August 2009
Ian F. Corbett, ed.

Galaxy formation from dry and hydro simulations

Luca Ciotti

Department of Astronomy, University of Bologna, via Ranzani 1, 40127 Bologna, Italy
email: luca.ciotti@unibo.it

Abstract. The effects of dry and wet merging on the Scaling Laws (SLs) of elliptical galaxies (Es) are discussed. It is found that the SLs, possibly established at high redshift by the fast collapse of gas-rich and clumpy stellar distributions in preexisting dark matter halos following the cosmological SLs, are compatible with a (small) number of galaxy mergers at lower redshift.

Keywords. Galaxies: elliptical and lenticular, cD – Galaxies: formation – Galaxies: evolution

---

The main results obtained in a series of papers (Ciotti & van Albada 2001; Nipoti, Londrillo & Ciotti 2003; Lanzoni et al. 2004; Ciotti, Lanzoni & Volonteri 2007; see also Ciotti 2009), are presented. It is found that 1) Parabolic dry merging in a population of low mass spheroids leads to massive Es that fail the Faber-Jackson (FJ) and Kormendy relations, being characterized by low velocity dispersion and very large effective radii. Parabolic wet merging in the same population of progenitors leads to Es in better agreement with the observed SLs, as long as enough gas for dissipation is available. 2) The edge-on structure of the Fundamental Plane (FP) is surprisingly preserved. Therefore, the FJ and Kormendy relations, despite their larger scatter, are stronger tests for merging than the edge-on FP. 3) Parabolic dry or wet merging of Es following the observed SLs over the full mass range preserve the Kormendy, FJ, and edge-on FP relations. Thus, massive Es cannot be formed by parabolic merging of low mass spheroidal galaxies, even in presence of substantial gas dissipation, but their SLs, once established by galaxy formation, are robust against merging. 4) Dark matter halos obtained from cosmological simulations define a FJ, a Kormendy, and a FP-like relation, as expected from the spherical collapse model for virialized systems. 5) Numerical simulations of cold dissipationless collapse in pre-existing dark matter halos can reproduce Sersic profiles remarkably similar to those observed, over a large radial range. Note that cold dissipationless collapse is a process which is expected to dominate the late stages of an initially dissipative process. Thus the SLs of Es, possibly established at high redshift by the fast collapse of gas rich and clumpy stellar distributions in pre-existing dark matter halos (following the cosmological SLs), can persist even in the presence of a moderate number of dry or wet mergings. Then monolithic-like collapse at early times and subsequent merging could just represent the different phases of galaxy formation (collapse) and evolution (merging, in addition to the aging of the stellar population and related phenomena).

References

Ciotti, L. 2009, *La Rivista del Nuovo Cimento* 32, n.1, 1
Ciotti, L. & van Albada, T. S. 2001, *ApJ* 552, L13
Ciotti, L., Lanzoni, B., & Volonteri, M. 2007, *ApJ* 658, 65
Lanzoni, B., Ciotti, L., Cappi, A., & Tormen, G. 2004, *ApJ* 600, 640
Nipoti, C., Londrillo, P., & Ciotti, L. 2003, *MNRAS* 342, 501

Highlights of Astronomy, Volume 15
XXVIIth IAU General Assembly, August 2009
Ian F. Corbett, ed.

© International Astronomical Union 2010
doi:10.1017/S1743921310008392

The formation of cD Halos: the case of NGC 3311 in the Hydra I Cluster

Ventimiglia G.[1], Gerhard O.[1] and Arnaboldi M.[2]

[1] Max-Planck-Institut fuer Ex. Physik, Giessenbachstrasse D-85748, Garching b. Muenchen
[2] European Organisation for Astronomical Research in the Southern Hemisphere,
Karl-Schwarzschild-Strasse 2, D-85748 Garching b. Muenchen

We have studied the core of the Hydra I cluster, around its central cD galaxy, NGC 3311. We have analyzed the kinematics a sample of 60 intracluster planetary nebulae (PNs), detected using the multi-slit imaging spectroscopy technique (MSIS, Gerhard *et al.*, 2005, Ventimiglia *et al.*, 2008). PNs are good tracers of light (Coccato *et al.*, 2009) and the MSIS allows to measure their velocities and positions at the same time. The histogram of the PN radial velocities presents several discrete components. We are comparing this result with ΛCDM hydro-dynamical simulations and other data in order to interpret it in the framework of the formation of extended halos around cD galaxies. V band photometric data around NGC 3311 have revealed the presence of an excess of light in the North-East part of the galaxy, which is spatially coincident with most of the PNs contributing to the reddest peak in the PNs LOSVD. We have measured, using Long-Slit data, the velocity of HCC26, a dwarf (DW) galaxy in the middle of the excess of light. The reddest peak in the PNs LOSVD is consistent both with the velocity of HCC 26 and of several other DWs in the same region. We are investigating the possibility that the light in excess has been stripped from these galaxies and now incorporated into the halo of NGC 3311.

References

Coccato, L., *et al.* 2009, *MNRAS* 394, 1249C
Gerhard, O., *et al.* 2005, *ApJ* 621, L93
Ventimiglia, G., Arnaboldi, M., & Gerhard, O. 2008, *AN* 329, 1057

Highlights of Astronomy, Volume 15
XXVIIth IAU General Assembly, August 2009
Ian F. Corbett, ed.

© International Astronomical Union 2010
doi:10.1017/S1743921310008409

Red Galaxies Growing in Dark Matter Halos

Michael J. I. Brown[1] and the Boötes Field Collaborations

[1] School of Physics, Monash University, Clayton, Victoria 3800, Australia
email: Michael.Brown@sci.monash.edu.au

Abstract. To understand the slow growth of massive galaxies at $z < 1$, we have modeled how these galaxies populate dark matter halos. The models are constrained with the observed luminosity function and clustering of $z < 1$ red galaxies. In the most massive halos, much of the stellar mass resides within multiple satellite galaxies rather than a single central galaxy. Consequently, massive galaxies grow slowly within rapidly growing dark matter halos.

Keywords. galaxies: elliptical and lenticular, cD, galaxies: evolution, (cosmology:) dark matter

The most massive dark matter halos are predicted to grow rapidly via mergers at $z < 1$, and there has thus been an expectation that massive galaxies will also grow rapidly via mergers. Contrary to this expectation, recent observations show that massive galaxies grow slowly at $z < 1$. To understand why this is the case, we have constrained models of how galaxies populate dark matter halos using the observed luminosity function and clustering of $z < 1$ red galaxies (White *et al.* 2007, Brown *et al.* 2008). We find that the relationship between red galaxy stellar mass and host halo mass does not evolve significantly, although the masses of individual galaxies and halos are evolving. In the most massive dark matter halos, much of the stellar mass resides within multiple satellite galaxies and diffuse intra-cluster light, rather than a single central galaxy. We also find that the stellar masses of the most massive galaxies are proportional to halo mass to the power of a third. Consequently, the stellar masses of the largest galaxies grow slowly, even though they reside within rapidly growing dark matter halos.

References

Brown, M. J. I., Zheng, Z., White, M., Dey, A., Jannuzi, B. T., Benson, A. J., Brand, K., Brodwin, M., & Croton, D. J. 2008, *ApJ* 682, 937
White, M., Zheng, Z., Brown, M. J. I., Dey, A., & Jannuzi, B. T. 2007, *ApJ (Letters)* 655, 69

Highlights of Astronomy, Volume 15
XXVIIth IAU General Assembly, August 2009
Ian F. Corbett, ed.

© International Astronomical Union 2010
doi:10.1017/S1743921310008410

The growth of the red-sequence in clusters since ≃ 1

Roberto P. Muñoz[1,2], L. F. Barrientos[1], B. P. Koester[3],
D. G. Gilbank[4], M. D. Gladders[3] and H. K. C. Yee[5]

[1]Departamento de Astronomia y Astrofisica, Pontificia Universidad Católica de Chile, Av.
Vicuña Mackenna 4860, Casilla 306, Santiago 22, Chile; email: rmunoz@astro.puc.cl;
[2]Departamento de Fisica y Astronomia, Universidad de Valparaiso,; [3]Department of
Astronomy and Astrophysics, University of Chicago,; [4]Department of Physics and Astronomy,
University of Waterloo,; [5]Department of Astronomy and Astrophysics, University of Toronto

Abstract. We use deep nIR imaging of 15 galaxy clusters at $z \simeq 1$ to study the build-up of the red-sequence in rich clusters since the Universe was half its present age. We measured, for the first time, the luminous-to-faint ratio of red-sequence galaxies at z=1 from a large ensemble of clusters, and found an increase of 100% in the ratio of luminous-to-faint red-sequence galaxies from z=0.45 to 1.0. The measured change in this ratio as function of redshift is well-reproduced by a simple evolutionary model developed in this work, that consists in an early truncation of the star formation for bright cluster galaxies and a delayed truncation for faint cluster galaxies.

Keywords. galaxies: clusters: general, galaxies: evolution, galaxies: formation

---

Galaxy clusters are gravitationally bounded structures in the Universe which are inhabited by several thousands of galaxies and filled by a hot X-ray emitting gas. The central megaparsec of clusters is dominated by early type galaxies, which are observed to obey tight empirical scaling relations as the Fundamental plane (FP) and the Color-magnitud relation (CMR). In this work, we present deep J_s and K_s-band imaging of 15 galaxy clusters at $z \simeq 1$, which were discovered in the Red-Sequence Cluster Survey (RCS-1) (Gladders & Yee (2005)) and followed up using the VLT/ISAAC instrument. We built the K_s − band luminosity function (LF) and the color-magnitude diagram (CMD) for the combined cluster sample at z=1 through the application of the B+Z method developed by Muñoz, Padilla & Barrientos (2009). We found that our K_s − band LF is well described by a Schechter function with $K_s^* = 18.82$ and $\alpha = -0.42$, and that the CMD shows a qualitatively deficit of red-sequence (RS) galaxies with $M_V > -20$. We computed the ratio between the number of luminous and faint RS galaxies (L/F ratio) within the magnitude limits defined by De Lucia *et al.* (2007) and found the value of 1.07 ± 0.28 at $z = 1$. We concluded that our value of K_s^* at $z = 1$ favors a passive evolution model with formation redshift $z_f = 3$, and that the increase of the L/F ratio towards higher redshifts can be explained if progenitors of present-day $M_V > -20$ early-type galaxies have undergone a recent burst of star formation at $z = 1$.

References

Gladders, M. D. & Yee, H. K. C. 2005, *ApJS* 157, 1

De Lucia, G., Poggianti, B. M., Aragón-Salamanca, A., White, S. D. M., Zaritsky, D., Clowe, D., Halliday, C., Jablonka, P., von der Linden, A., Milvang-Jensen, B., Pelló, R., Rudnick, G., Saglia, R. P., & Simard, L. 2007, *MNRAS* 374, 809

Muñoz, R. P., Padilla, N. D., & Barrientos, L. F. 2009, *MNRAS* 392, 655

Highlights of Astronomy, Volume 15
XXVIIth IAU General Assembly, August 2009
Ian F. Corbett, ed.

Dark matter and X-ray halo in early-type galaxies and clusters of galaxies

Takaya Ohashi

Department of Physics, Tokyo Metropolitan University, Hachioji, Tokyo 192-0397, Japan
email: ohashi@phys.metro-u.ac.jp

X-ray observations reveal extended halos around early-type galaxies which enable us to trace the dark matter distribution around the galaxies (see Mathews and Brighenti 2003 for a review). X-ray luminosities, L_X of massive early-type galaxies are $10^{40} - 10^{42}$ erg s^{-1} in 0.3–2 keV. The correlation plot between L_X and B-band luminosity L_B shows a large scatter in the sense that L_X varies by 2 orders of magnitudes for the same L_B, in the brightest end ($\log L_B \gtrsim 10.5$). The amount of the X-ray hot gas in early-type galaxies is typically a few % of the stellar mass, in contrast to clusters of galaxies which hold ~ 5 times more massive gas than stars. Matsushita (2001) showed that X-ray luminous galaxies are characterized by extended X-ray halo with a few tens of r_e, similar to the scale of galaxy groups, so the presence of group-size potentials would be strongly linked with the problem of large L_X scatter.

Fukazawa *et al.* (2006) carried out extensive X-ray study of 53 early-type galaxies. They showed that gravitational mass profiles can be divided into 2 components. The inner component ($r < r_e$) follows the stellar distribution without significant need of dark matter ($M/L \leqslant 10$), while the outer component, consistent with the NFW profile, indicates the dark matter dominance with M/L ratio exceeding 100 at around $10r_e$ for bright galaxies. In a dark-matter dominated cluster A1060, the gravitational potential profile inferred from the X-ray data follows $r^{-1.5}$, i.e. the so-called modified NFW profile, rather than r^{-1}.

X-ray observations have revealed dynamical features in the X-ray halo from a number of galaxies. X-ray morphlogies in Chandra image suggest fast gas motion for NGC 4636 and NGC 1404. Hard X-ray emission and X-ray cavities are observed in some groups (e.g. HCG 62), with the latter indicating that non-thermal pressure is overcoming the hot-gas pressure. Therefore, non-thermal processes are causing significant effects in many galaxies. On the other hand, suppression of Fe XVII resonance lines seen in the centers of several galaxies, including NGC 4636, indicates that the gas is not turbulent. Further knowledge about the nature of non-thermal components in galaxies and clusters are important in understanding the evolution of these systems.

X-ray astronomy mission ASTRO-H is planned for launch in 2014 under a major collaboration between Japan and US. This mission will carry microcalorimeters with energy resolution better than 7 eV (about 20 times better than the CCD resolution), with 36 pixels covering a field of view of $3' \times 3'$. Direct observations of hot-gas motion with $\Delta v \sim 100$ km s^{-1} from galaxies and clusters using Doppler shifts of emission lines will be carried out for the first time. Including the capabilities of hard X-ray detectors (multi-layer coated mirrors and Compton cameras), ASTRO-H will be a very powerful mission for the study of non-thermal processes in extended objects.

Keywords. X-rays: galaxies, X-rays: ISM, galaxies: halos

References

Fukazawa, Y., Botoya-Nonesa, J. G., Pu, J., Ohto, A., & Kawano, N. 2006, *ApJ* 636, 698
Matsushita, K. 2001, *ApJ* 547, 693
Mathews, W. G. & Brighenti, F. 2003, *ARAA* 41, 191

Highlights of Astronomy, Volume 15
XXVIIth IAU General Assembly, August 2009
Ian F. Corbett, ed.

X-rays and dynamics

Silvia Pellegrini

Department of Astronomy, University of Bologna, via Ranzani 1, 40127 Bologna, Italy
email: silvia.pellegrini@unibo.it

Abstract. The hot X-ray emitting interstellar medium of early type galaxies can be used in principle as a total mass tracer and a tool to determine the stellar orbital anisotropy, based on the hypothesis of hydrostatic equilibrium for it. Here the effects that deviations from equilibrium have on both estimates are shown, and a comparison is made with cases for which accurate optical and X-ray information are available.

Keywords. galaxies: elliptical and lenticular, cD – galaxies: fundamental parameters – galaxies: halos – galaxies: ISM – galaxies: kinematics and dynamics – galaxies: structure

The mass distribution and amount and the internal kinematics of an early type galaxy can be derived from X-ray observations of its hot gas properties (density and temperature profiles) and optical observations of the motions of stellar objects (stars, planetary nebulae and globular clusters; e.g., Ciotti & Pellegrini 2004). Recently, a lot of progress in data quality and analysis methods has allowed to exploit X-ray and optical information extending over similar radial ranges, and out to 10-15 effective radii (e.g., Humphrey *et al.* 2006, Coccato *et al.* 2009). One can then investigate whether there is agreement or contradiction between the results from the X-ray and optical analysis used independently, to establish the robustness of each method taken separately. For example, the mass-anisotropy degeneracy may affect optical estimates, while hydrostatic equilibrium (a necessary condition to exploit the X-ray information) may not hold due to bulk flows, or may require pressure terms other than thermal and unaccounted for. A general result concerning possible deviations of X-ray estimates from true values, that is independent of assumptions about the underlying mass profile or the internal orbital distribution, is the following: the deviations are in the sense of mass underestimates accompanied by orbital anisotropy overestimates, and viceversa. From hydrodynamical models of the hot gas flow evolution for representative galaxies one derives mass underestimates for infalling gas and overestimates for outflows (Pellegrini & Ciotti 2006). This could explain some observed discrepancies between X-ray and optical estimates of dynamical quantities, as have been noted for example within one effective radius for galaxies with detailed optical and X-ray information (e.g., NGC4472, NGC4649 and NGC1407; Ciotti & Pellegrini 2004, Hwang *et al.* 2008, Romanowsky *et al.* 2009).

References

Ciotti, L. & Pellegrini, S. 2004, *MNRAS* 350, 609
Coccato, L. *et al.* 2009, *MNRAS* 394, 1249
Humphrey, P. J. *et al.* 2006, *ApJ* 646, 899
Hwang, H. S. *et al.* 2008, *ApJ* 674, 869
Pellegrini, S. & Ciotti, L. 2006, *MNRAS* 370, 1797
Romanowsky, A. *et al.* 2009, *AJ* 137, 4956

Highlights of Astronomy, Volume 15
XXVIIth IAU General Assembly, August 2009
Ian F. Corbett, ed.

© International Astronomical Union 2010
doi:10.1017/S1743921310008446

The outer haloes of massive, elliptical galaxies

Payel Das[1], Ortwin Gerhard[1], Flavio de Lorenzi[1], Emily McNeil[2], Eugene Churazov[3] and Lodovico Coccato[1]

[1]MPI for Extraterrestrial Physics, Giessenbachstrasse, 85748, Garching, Germany email: pdas@mpe.mpg.de
[2]Research School of Astronomy and Astrophysics, Australian National University, Canberra, ACT 2601, Australia
[3]MPI for Astrophysics, Karl-Schwarzschild-Str. 1, 85741, Garching, Germany

The outer haloes of massive elliptical galaxies are dark-matter dominated regions where stellar orbits have longer dynamical timescales than the central regions and therefore better preserve their formation history. Dynamical models out to large radii suffer from a degeneracy between mass and orbital structure, as the outer kinematics are unable to resolve higher moments of the line-of-sight velocity distribution. We mitigate this degeneracy for a sample of quiescent, massive, nearby ellipticals by determining their mass distributions independently using a non-parametric method on X-ray observations of the surrounding hot interstellar medium. We then create dynamical models using photometric and kinematic constraints consisting of integral-eld, long-slit and planetary nebulae (PNe) data extending to ~50 kpc. The rst two galaxies of our sample, NGC 5846 and NGC 1399, were found to have very shallow pro jected light distributions with a power law index of ~1.5 and a dark matter content of 70–80% at 50 kpc. Spherical Jeans models of the data show that, in the outer haloes of both galaxies, the pro jected velocity dispersions are almost inde- pendent of the anisotropy and that the PNe prefer the lower end of the range of mass distributions consistent with the X-ray data. Using the N-body code NMAGIC, we cre- ated axisymmetric models of NGC 5846 using the individual PNe radial velocities in a likelihood method and found them to be more constraining than the binned velocity disper- sions. Characterising the orbital structure in terms of spherically averaged proles of the velocity dispersions we nd $\sigma_\psi > \sigma_r > \sigma_\theta$.

Acknowledgements
P. Das was supported by the DFG Cluster of Excellence.

Highlights of Astronomy, Volume 15
XXVIIth IAU General Assembly, August 2009
Ian F. Corbett, ed.

Gravitational potential and X-ray luminosities of early-type galaxies observed with XMM-Newton and Chandra

Ryo Nagino and Kyoko Matsushita

Tokyo University of Science
email: j1207705@ed.kagu.tus.ac.jp; Present address: 1-3 Kagurazaka, Shinjyuku-ku, Tokyo, Japan
matusita@rs.kagu.tus.ac.jp Present address: 1-3 Kagurazaka, Shinjyuku-ku, Tokyo, Japan

We study the dark matter content in early-type galaxies and investigate whether X-ray luminosities of early-type galaxies are determined by the surrounding gravitational potential. We derived gravitational mass profiles of 22 early-type galaxies observed with XMM-Newton and Chandra. Sixteen galaxies show constant or decreasing radial temperature profiles, and their X-ray luminosities are consistent with kinematical energy input from stellar mass loss. The temperature profiles of the other 6 galaxies increase with radius, and their X-ray luminosities are significantly higher. The integrated mass-to-light ratio of each galaxy is constant at that of stars within 0.5–$1 r_e$, and increases with radius. The scatter of the central mass-to-light ratio of galaxies was less in K-band light. At $3r_e$, the integrated mass-to-light ratios of galaxies with flat or decreasing temperature profiles are twice the value at $0.5 r_e$, where the stellar mass dominates, and at $6 r_e$, these increase to three times the value at $0.5 r_e$. This feature should reflect common dark and stellar mass distributions in early-type galaxies: Within $3 r_e$, the mass of dark matter is similar to the stellar mass, while within $6 r_e$, the former is larger than the latter by two-fold. In contrast, X-ray luminous galaxies have higher gravitational mass in the outer regions than X-ray faint galaxies. We describe these X-ray luminous galaxies as the central objects of large potential structures; the presence or absence of this potential is the main source of the large scatter in the X-ray luminosity.

These results are published in A&A, 501, 157, 2009.

JD1 - Poster Session

Abstract. During the Joint Discussion #1 a significant number of exciting new results were presented in the form of a poster. Below we list the title and authors of the posters.

(*a*) M. D. Suran, *LCDM hydrodynamical cosmological simulations*

(*b*) I. A. Lacerna, *Spatial Correlations of Halo Assembly*

(*c*) P. M de Novais, *Merging pairs of galaxies in the SDSS*

(*d*) R. Salinas, *Kinematics of the field elliptical NGC 7507 – A galaxy with little dark matter?*

(*e*) E. Iodice, *Dark Matter Content in the Polar Disk Galaxy NGC4650A*

(*f*) T. Verdugo, *The whole picture of a galaxy group: Combining Strong Lensing, Weak lensing, Dynamics and N-body simulations in SL2SJ02140-0532*

(*g*) R. Gonzalez, *Galaxy properties within DM halos and Large scale structure*

(*h*) J. A. Magana, *Structure Formation with phi^2 Dark Matter*

(*i*) P. da Cunha Ferreira, *Predicting the length-to-width ratio on gravitational arcs*

(*j*) V. E. Timofeev, *Observation of ionization jerk in the ionization chamber ASK-1*

(*k*) A. D. Ernest, *Gravitational Eigenstates in the Cosmos: The Answer to Dark Matter?*

(*l*) M. A. Dantas, *Current lookback time-redshift bounds on dark energy*

(*m*) G. B. Caminha, *Fraction of arcs in galaxy clusters: redshift evolution and the importance of magnification*

(*n*) A. S. Iribarrem, *Radial statistics of galaxy number counts at high redshifts*

(*o*) R. R. Rosa, s *Characterizing Extreme Event Dynamics in Galaxy-Sized Dark Matter Haloes*

(*p*) V. A. P. Martin, *Log Slit Spectroscopy and broad-band photometry of the peculiar galaxy ESO 287-G40*

(*q*) D. Bettoni, *The Core Fundamental Plane of low redshift radio galaxies*

(*r*) S. Bryan, *Orbits in Dark Matter Haloes*

(*s*) F. E. M Costa, *Current constraints on dark matter-dark energy interaction*

(*t*) L. Marassi, *Mass Functions in an Homogeneous Dark Energy Model*

(*u*) M. D'Onofrio, *Comparing the FP of early-type galaxies in the V and K bands*

(*v*) E. R. Carrasco *Disentangling the monster. Gemini/GMOS observations of a massive galaxy in the core of Abell 3827*

(w) E. S. Rykoff, *The Origin and Evolution of Fossil Groups*

(x) G. Caminha, *Cross section for arc formation in the perturbative approach*

Highlights of Astronomy, Volume 15
XXVIIth IAU General Assembly, August 2009
Ian F. Corbett, ed.

© International Astronomical Union 2010
doi:10.1017/S174392131000846X

JD2 - Diffuse Light in Galaxy Clusters

Magda Arnaboldi[1] and Ortwin Gerhard[2]

[1] European Southern Observatory, Karl Schwarzschild-Str 2, 85748 Garching, Germany
email: marnabol@eso.org

[2] Max Planck Institute for Extraterr. Physics, Giessenbachstrasse, 85748 Garching, Germany
email: gerhard@mpe.mpg.de

Abstract. Diffuse intracluster light (ICL) has now been observed in nearby and in intermediate redshift clusters. Individual intracluster stars have been detected in the Virgo and Coma clusters and the first color-magnitude diagram and velocity measurements have been obtained. Recent studies show that the ICL contains of the order of 10% and perhaps up to 30% of the stellar mass in the cluster, but in the cores of some dense and rich clusters like Coma, the local ICL fraction can be high as 40%-50%. What can we learn from the ICL about the formation of galaxy clusters and the evolution of cluster galaxies? How and when did the ICL form? What is the connection to the central brightest cluster galaxy? Cosmological N-body and hydrodynamical simulations are beginning to make predictions for the kinematics and origin of the ICL. The ICL traces the evolution of baryonic substructures in dense environments and can thus be used to constrain some aspects of cosmological simulations that are most uncertain, such as the modeling of star formation and the mass distribution of the baryonic component in galaxies.

Keywords. (cosmology:) large-scale structure of universe; galaxies: clusters: general; galaxies: evolution; galaxies: interactions; galaxies: structure; galaxies: kinematics and dynamics; galaxies: star clusters; (ISM:) planetary nebulae: general

1. Introduction

The Joint Discussion dedicated to the study of diffuse light in clusters took place on the 6th and 7th of August, 2009 during the XXVIIth IAU General Assembly in Rio de Janeiro. It was the first scientific meeting on this subject. This Joint Discussion provided a forum to confront observational evidence and theoretical predictions, and to identify future directions for understanding the origin and implications of this new component of galaxy clusters.

The Joint Discussion included four main sessions covering the distribution of diffuse light in clusters and groups, the kinematics of intracluster stars, the intracluster stellar populations, and the predictions of cosmological simulations for the evolution of galaxies in clusters and groups, and for the formation of the intracluster light. 14 invited plus 10 oral talks, and 14 poster papers contributed to an intense scientific exchange and set the stage for a lively scientific discussion, which concluded the workshop.

In what follows, we provide a brief summary and some selected references for the talks in the four sessions, and end with a summary of the discussion which took place on August 7th , 2009.

Figure 1. This deep image of the Virgo Cluster obtained by Chris Mihos and colleagues using the Burrell Schmidt telescope shows the diffuse light between the galaxies belonging to the cluster. North is up, east to the left. The dark spots indicate where bright foreground stars were removed from the image. M87 is the largest galaxy in the picture (lower left). From ESO PR19/2009, based on Mihos *et al.* (2005).

2. Distribution of diffuse light in cluster and groups

2.1. *Clusters at $z = 0$*

2.1.1. *Diffuse light in the Virgo cluster – C. Mihos*

Thanks to the specially adapted Burrel Schmidt telescope, the study of intracluster light in the Virgo cluster reached for the first time a photometric accuracy of significantly less than 1% of night sky emission over an area of many square degrees. Deep V ($\mu_V = 28.5$ mag/arcsec^2) and B band ($\mu_B = 29.0$ mag/arcsec^2) photometry with calibrated, quantitative photometric solutions and a well understood error model were obtained. This work has revealed faint surface brightness features over a multitude of angular scales, from narrow streams to extended diffuse halos. The B-V colour of the ICL is similar to that in the outer halos of ellipticals, except for some streamers, and there is a rough correlation with the spatial distribution of intracluster planetary nebulae. References: Mihos *et al.* (2005, 2009).

2.1.2. *Diffuse light in the Coma cluster - C. Adami*

Diffuse light in the Coma cluster is documented since the work of Zwicky in 1951. Photographic photometry in the 1970s revealed a large elongated diffuse component in the core of the cluster, around the two BCG galaxies. Several small-scale streamers and plumes were discovered in the 1990s with CCD photometry. The CFHT multi-color Coma survey shows that the diffuse light is found either in the centre of the Coma cluster or along the in-fall directions towards nearby large scale structures. The diffuse light in the cluster core was found to be distributed on different scales, as quantified by a wavelet analysis. References: Thuan & Kormendy (1977), Adami *et al.* (2005a, 2005b).

2.1.3. *Testing dark matter with the "star pile" in Abell 545: the faintest cD or the brightest ICL? - R. Salinas*

The study of surface brightness distribution of the star pile in Abell 545 shows that it can be described as the halo of a cD without the high surface brightness central galaxy. The light distribution in Abell 545 can thus be considered as independent evidence for intracluster light as a separate stellar component from the BCG halo. The spectroscopic follow-up shows that its spectrum is consistent with that of an old stellar population. References: Salinas *et al.* (2007).

2.2. *Groups at z = 0: compact and fossil groups*

2.2.1. *Diffuse light and intra-group star-forming regions in compact groups of galaxies - C. Mendes de Oliveira*

Compact groups are high density regions in the universe where the morphology of both the stars and the HI gas in galaxies is often disturbed, providing evidence for ongoing interactions. An evolutionary sequence for compact groups can be traced by the fraction of light in the diffuse component on the group scale, the intragroup light (IGL), which increases with the degree of interactions, reaching up to 30–40% in groups with many ongoing interactions. Groups with the highest fraction of IGL light are also those with the highest fraction of early type galaxies. The colour of the IGL in most cases is similar to the colours in the outskirts of the member galaxies. Star formation occurs in the intragroup medium when the compact group contains stripped HI and is in an intermediate to advanced stage of interaction. References: Da Rocha *et al.* (2008), de Mello *et al.* (2008), Torres-Flores *et al.* (2009).

2.2.2. *Wavelet analysis of diffuse intra-group optical light in compact groups of galaxies - C. Da Rocha*

The wavelet analysis of diffuse optical intra-group light in compact groups of galaxies can provide reliable measurements of the intragroup light on different scales. The diffuse light distributions in HCG 79 and HCG 51 are illustrated as prototype cases. References: Da Rocha *et al.* (2008).

2.3. *Clusters at z = 0.3*

2.3.1. *Intracluster light in moderate redshift clusters - J.J. Feldmeier*

The review of photometric measurements of ICL fractions in clusters highlights the challenges posed by such measurements and their intrinsic limitations. The ICL is difficult to measure, being at best at the level of 1% of the night sky. Critical steps in the data reduction are the sky subtraction and flat fielding, which have to be precise to 0.1%. Severe obstacles are the effects of the large scale PSF and scattered light, and the separation of ICL from BCGs, cD halos and other galaxies. Fractions of the ICL are determined from photometric measurements, integrating the light outside the assumed outer radii of individual galaxies over the region confined within an isophote with surface brightness threshold defined by the depth of the observations. Such measurements for the ICL fractions are in the range from 10 to 45% - ICL is common in galaxy clusters. References: Feldmeier *et al.* (2004a), Gonzalez *et al.* (2007), Krick and Bernstein (2007).

2.3.2. *Statistical properties of the intracluster light at z 0.25 - S. Zibetti*

To quantify the amount of ICL and determine trends with cluster properties, key issues are statistics and depth. These can be addressed by stacking many shallow images, combining them into deeper ones. This procedure was adopted for a sample of 683 clusters with BCGs from the SDSS sample. Average images were obtained after masking all detectable sources (foreground stars and galaxies), centering and scaling; they provide average properties of the diffuse optical light. The results for the integrated light fractions within 500 kpc cluster radius are BCG:ICL:galaxies = 21.9:10.9:67.2, with a robust estimate of the fraction (ICL+BCG)/(total light) being about 30%. The color of ICL is similar to the color of galaxy light in the cluster, and the radial profile of the ICL is more centrally concentrated than that of the light in cluster galaxies. References: Zibetti *et al.* (2005), Pierini *et al.* (2008).

2.3.3. *The diffuse intergalactic light in intermediate redshift clusters : RXJ0054-2823 - J. Melnick*

An "S"- shaped arc, bluer than the cD galaxies and the ICL, has been detected in the intermediate redshift cluster RXJ0054-2823 using the consecutive differential image technique. There are no emission lines associated with this arc and its redshift is consistent with the cluster redshift, i.e. it is not a lensed background galaxy image. The arc is probably formed by two spirals caught in the act of being tidally crunched by the three giant galaxies in the cluster center. References: Melnick *et al.* (1999), Faure *et al.* (2007).

3. Kinematics of Intracluster stars – individual stars and absorption line spectroscopy

3.1. *Properties of intracluster starlight – A. Zabludoff*

The intracluster stars are hard to count: yet, they are significant in the understanding of the origin of baryon vs. dark matter distributions and the enrichment of stars and gas in clusters. ICL is here detected in terms of an additional de Vaucouleurs profile required to fit the surface brightness profile of the central galaxy at large radii. A two-dimensional fit with a single profile is often poor at large radii, and fails both for the ellipticity and position angle profiles. The BCG and the outer component are aligned within 10 deg about 40% of the time; in the rest of the cases the misalignment is large. The extrapolated light of the outer component dominates the total and its color is similar to that of an old population. The increase of the velocity dispersion measured in several clusters suggests that it responds to the cluster potential. References: Kelson *et al.* (2002), Zaritsky *et al.* (2006), Gonzalez *et al.* (2007).

3.2. *Intracluster planetary nebulae and globular clusters*

Discrete objects like planetary nebulae (PNe) are excellent tracers for measuring the line-of-sight (LOS) kinematics of intracluster stars: PNe occur during the final phase of solar type stars, their number density distribution follows light, and their nebular shell re-emits more than 10% of the UV light from the stellar core in one optical emission line, [OIII]5007Å. This bright emission line makes it possible to detect and measure the velocity of such stars even in regions where the total surface brightness is too faint for absorption line spectroscopy. By surveying large areas in clusters, a suitable number of PNe can be detected and mean radial velocities and velocity dispersions can be determined.

Such investigations have been carried out in a number of nearby clusters and this section covers the recent observational results in this field. Similar studies are possible using globular clusters, assuming these also trace the distribution of stars. Studies based on globular clusters are reported in Section 4.

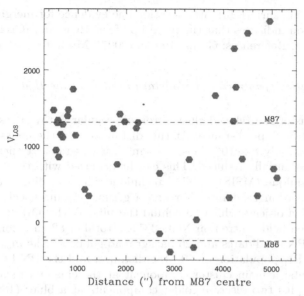

Figure 2. Distribution of line-of-sight velocity versus projected distance from the center of M87 for all spectroscopically confirmed PNs in the 6 FLAMES fields in the Virgo cluster core. From Doherty *et al.* (2009).

3.2.1. *Kinematics of diffuse light in the Virgo cluster core from planetary Nebulae - M. Arnaboldi*

The kinematics of the ICL in the Virgo cluster was studied using the PNe identified in narrow band surveys. The FLAMES spectrograph at the VLT was used for the spectroscopic campaign, with a total of 6 VLT-Flames pointings observed. From the positions and velocities of the detected PNe a projected phase-space diagram was built which illustrates the different dynamical components along the LOS in the Virgo cluster core, see Fig. 2: these are the extended halo of M87 and the ICL component. Within $R < 3600'' = 260$ kpc from M87, the ICL component is all at negative velocities relative to M87, and it is not phase-mixed. Part of this component is at velocities consistent with the idea that light of the M86 sub-group is being tidally stripped by the more massive M87 component, while the two galaxies approach each other along the LOS. By contrast, in a more distant field at $R = 4500''$, the full velocity width of the Virgo cluster is seen. References: Arnaboldi *et al.* (1996, 2004), Doherty *et al.* (2009).

3.2.2. *Dynamics of cluster cores and brightest galaxies from planetary nebulae velocities - O. Gerhard*

Numerical simulations predict that isolated galaxies acquire extended radially anisotropic halos built from accretion of smaller galaxies; the level of predicted anisotropy is consistent with results from a few nearby ellipticals studied so far. In simulated BCG galaxies, additional high-velocity components from disrupted or stripped galaxies may be superposed on the central galaxy. Observationally, the outer halos of the BCGs at the centers of the Fornax, Virgo, Coma and Hydra I clusters are in fact harbouring substructure associated with disrupted satellites. In all cases the data indicate that galaxy halos and ICL are discrete components; the former do not blend continuosly into the latter, and there is no evidence for a continuous increase of the galaxy velocity dispersion to cluster values, but rather a colder BCG halo is superposed with a hotter ICL component at radially decreasing surface brightness ratio. The evidence for merging (in Coma) and accretion/disruption indicates that the build-up of the BCG and ICL is an on-going and long-lasting process. References: Gerhard *et al.* (2007), Murante *et al.* (2007), de Lorenzi *et al.* (2009).

3.2.3. *The kinematics of intracluster planetary nebulae in the Hydra I cluster – G. Ventimiglia*

The Hydra I cluster is a relaxed cluster from its regular X-ray emission, in the Southern hemisphere at 50 Mpc distance. At this distance, even the emission line flux from the brightest PN is only 8×10^{-18} erg s^{-1} cm^{-2}; thus to detect these objects the sky noise must be substantially reduced. This can be achieved with the "multi slit imaging spectroscopy" technique (MSIS), a blind technique which combines the use of a mask with parallel slits, a narrow band filter and a grism, yielding spectra a few tens of Å wide for all emission objects which lie behind the slits. With FORS2 and the VLT, in a 6.8 arcmin^2 squared field centred on NGC 3311, a total of 82 emission line objects were identified: 56 are PN candidates and 26 background galaxies. The m_{5007} magnitudes are consistent with a PN population at 50 Mpc distance, and the PN LOSVD shows three velocity components: a main cluster component at the cluster's redshift and expected $\sigma = 600$ km s^{-1}, plus two discrete colder components at a bluer (1800 km s^{-1}) and a redder (5000 km s^{-1}) velocity, providing evidence for unmixed components in the NGC 3311 halo. References: Gerhard *et al.* (2005), Arnaboldi *et al.* (2007), Ventimiglia *et al.* (2008).

3.2.4. *Planetary nebulae in NGC 1399: the kinematics of a cD halo - E. McNeil*

A counter dispersed imaging spectroscopy study with a mosaic of 5 pointings has been carried out, covering the bright central parts and extended halo of the Fornax cD galaxy NGC 1399. These observations deliver a sample of PN positions, measured magnitudes and velocities, which can be used to construct a 2D velocity field and study the kinematics of the extended halo. In this sample, 146 PNe associated with NGC 1399 are detected, 23 PNe associated with NGC 1404, and 6 unassigned. From the projected phase space diagram v_{LOS} vs. radius, 12 PNe are identified which are associated with a new low velocity component superposed onto NGC 1399. The velocity dispersion profile at large radii in NGC 1399 is consistent with a flat profile at 250 km s^{-1}; the rise to Fornax cluster dispersion is not yet reached at these radii. References: Arnaboldi *et al.* (1994), Saglia *et al.* (2000), Mc Neil *et al.* (in preparation).

4. Intracluster stellar populations – [Fe/H] and age distribution

4.1. *Planetary nebulae as tracers of stellar populations – R. Ciardullo*

PNe can be powerful probes of the chemical history of stellar populations. However, a full analysis of the α-element metallicity distribution function based on the nebular line ratios and the detection of the temperature-sensitive [OIII]4363Å line is not yet feasible for intracluster studies. Still it is possible to search for evidence of a metal-rich intracluster population with PN deep spectroscopy, but such spectroscopic data must reach ~ 20 times fainter than [OIII] $\lambda5007$ fluxes, and the next generation of telescope may be crucial for this. PN counts (normalized to underlying luminosity) do probe the stellar population, but we need a better theoretical understanding of how these objects come to form. If the blue straggler hypothesis is correct, then we may soon be able to use PNe to measure/constrain the age of an old stellar population. References: Feldmeier *et al.* (2004b), Ciardullo *et al.* (2005), Buzzoni *et al.* (2006).

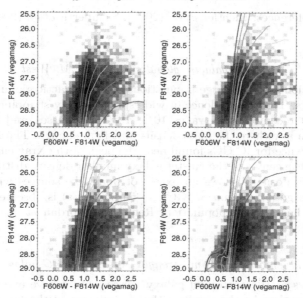

Figure 3. Results from the VICs project. Observed colour magnitude diagram of the VICS Virgo field, with a subset of the Girardi et al. (2002) isochrones for the HST ACS filters superposed. In each panel, the isochrones represent stars with metallicities of Z = 0.0001, 0.0004, 0.001, 0.0025, 0.004, 0.008, and 0.019, with redder isochrones corresponding to higher values of Z. Each panel displays a different age: log(age) = 10.1 (upper left; the contribution of the AGB was removed from these isochrones), 9.65 (upper right), 9.0 (lower left), and 8.5 (lower right). From Williams *et al.* (2007).

4.2. *The intracluster red giant star population in the Virgo cluster – P. Durrell*

Intracluster red giant branch (RGB) stars are the most numerous visible component of ICL; with RGB stars, we can determine the metallicities and constrain the ages of the stellar population(s) (latter with AGB) and relate these to the galactic origin of diffuse stellar light. The IC-RGB stars in Virgo can be studied only with very deep imaging, e.g., HST, which implies a limited field of view. The study of the VICS field yielded over 5000 IC-RGB stars whose color magnitude diagram shows a clear RGB populated by ∼ 5300 IC-RGB and AGB stars, above a small background contamination, see Fig. 3. These intracluster RGB stars are mostly old and metal-poor in the surveyed field, and they are not well-mixed even on small scales. These results are consistent with stripping of stars from a wide variety of galaxies, from dwarfs to the outer regions of ellipticals and spirals. References: Durrell *et al.* (2002), Williams *et al.* (2007).

4.3. *Intracluster globular clusters and ultra-compact dwarfs – M. Hilker*

Galaxy clusters are the places of highest galaxy density in the local universe, their cores are dominated by giant ellipticals with an extended surface brightness envelope (cD halo). Globular clusters are excellent tracers of diffuse stellar populations due to their abundant numbers and brightness. The central cluster galaxies possess extremely rich globular cluster systems (∼ 10.000 GCs) with an equal amount of red (metal-rich) and blue (metal-poor) GCs. The predominant GC population in intra-cluster regions are the blue (metal-poor) GCs which can be found out to large cluster-centric radii. Also very massive star clusters/ultra-compact dwarf galaxies (UCDs, $10^6 - 10^8 M_\odot$) are found in the intra-cluster space. Some of these might have had their origin as nuclei of now disrupted dwarf galaxies. References: Richtler *et al.* (2004), Schuberth *et al.* (2009), Misgeld *et al.* (2009).

4.4. *Intergalactic globular clusters - M. West*

Numerical simulations suggest large populations of intergalactic globulars could exist in rich galaxy clusters. There is evidence of wide-spread galaxy stripping and destruction. The Jeans mass at recombination was $\sim 10^5 - 10^6$ solar masses, and therefore globular clusters could form anywhere the mass density was high enough. The Abell cluster 1185 provides clear evidence for IC GCs found near the peak of the Xray emission which does not coincide with the center of the cluster galaxy number density distribution. The deep ACS image at this position shows an excess of point-like sources with respect to the control fields. A large population of IGCs ($\sim 50,000$ GCs) is observed in Coma, while the Virgo cluster has a small population. References: Jordan *et al.* (2009), Takamiya *et al.* (2009).

4.5. *Stellar kinematics and line strength indices in BCG halos - L. Coccato*

The bright galaxies at the centers of galaxy clusters often have extended halos whose presence is considered to be the result of the co-evolution of the central galaxy and its cluster environment. Information on the time of formation of the different components comes from the study of their stellar populations, which may provide both age and metallicity profiles. Long slit absorption line spectroscopy of the two BCG galaxies in the Coma cluster, NGC 4889 and NGC 4874, was performed at the Subaru 8 m telescope, with the FOCAS spectrograph. The integrated light spectra had adequate S/N so that absorption line indices for NGC 4889 could be determined out to 65 kpc. The results show that within $1.2R_e$ the stars of NGC 4889 must have formed over a very short time scale (< 1 Gyr), with little subsequent merging, as shown by the large α−enhancement and the measured steep metallicity gradient. The stars in the halo were formed on a longer time scale, as shown by the lower α/Fe values. This is compatible with formation in smaller systems which were subsequently accreted onto the halo of the BCG. References: Coccato *et al.* (in preparation).

5. Cosmological simulations of cluster and group formation/Origin of diffuse light

5.1. *Cluster and group formation in ΛCDM – S. White*

Starting from cosmological ΛCDM initial conditions, numerical simulations of the formation of large scale structure can now predict and reproduce the average mass profiles and properties of clusters and groups in the mass range 10^{12} M$_\odot$-10^{15} M$_\odot$. Table 1 shows some parameters for the Millenium-II simulation. Together with semi-analytic models the simulations can be used to study the stellar mass function of galaxies in groups and clusters down to 10^7 M$_\odot$, which is found to have similar shape as in the field except for the BCGs. Assuming the ICL is due to tidally stripped and disrupted galaxies, based on galaxy orbits and sizes and stripping of the DM halo, the simulations can be used to predict the fraction of cluster light in the ICL. Considerable scatter in the fractions of ICL relative to ICL + BCG (within $\sim 40 \pm 30\%$) and relative to the total cluster light (within $\sim 15 \pm 15\%$) is found, due to variations in assembly/stripping history. References: Springel, Frenk and White (2006), Hilbert and White (2009), Guo *et al.* (2009).

Table 1. The Millennium-II simulation: particle number and mass, size of simulated volume, resolution (softening), and fraction of particles in lumps at z = 0.

N_{part}	m_{part} $h^{-1} M_\odot$	L $h^{-1} Mpc$	ϵ h^{-1} kpc	$F_{halo}(z = 0)$
2160^3	6.9×10^6	100	1	0.60

5.2. *Galaxy evolution in clusters and groups – L. Mayer*

The physical mechanisms for galaxy transformation at work in cluster environments can be grouped into two categories: tidal interactions, and ICM-galaxy interactions. The latter can take place in several flavours: ram pressure stripping of cold ISM in disks; triggered star formation and strangulation; stripping of gaseous halos around galaxies. These processes help to build up the morphology density relation, i.e., that early-type galaxies (E,S0,dE,dS0) dominate in cluster cores, and explain the high number density of early-type dwarf galaxies at the centers of nearby groups and clusters, the HI deficiency of galaxies in cluster centers, and the steepening of the faint end of the luminosity function relative to the field. These transformations were studied with simulations replacing dark halos in dark matter only simulations at the time of infall with multi-component late type galaxy models. Cosmological simulations of groups show that mergers drive morphology towards elliptical systems, and that at large $R > 5R_e$ a diffuse stellar halo or intragroup light component is formed which may comprise $\sim 20\%$ of the group light. References: Mastropietro *et al.* (2005), Mayer *et al.* (2006).

5.3. *Galaxy properties in clusters: dependence on the environment – H. Muriel*

This work investigated the dependence of several galaxy properties on the environment and cluster identification techniques. Clusters of galaxies were selected from two catalogues based on the SDSS: the ROSAT-SDSS Galaxy Cluster Survey, and the MaxBCG Catalogue. Galaxies in X-ray and MaxBCG selected clusters show similar size-luminosity relations. The Faber-Jackson relation for early-type galaxies in clusters is also the same for X-ray selected and MaxBCG clusters. BCGs, non-BCG-early type galaxies in clusters and field early-type galaxies show different size-luminosity relations and have different dynamical properties. Using several criteria to classify galaxies into morphological types, the well known morphological segregation can be reproduced. These results can be related to the different processes that affect the evolution of galaxies in various environments. References: Coenda and Muriel (2009).

5.4. *Diffuse stellar component in cosmological simulations – K. Dolag*

Cosmological hydro-dynamical simulations of zoomed-in clusters have been performed with up to 4×10^6 particles in R_{vir}, radiative cooling/heating by UV background (quasars), a sub-grid two-phase model for star-formation, thermal and kinetic feedback by supernovae; and evolving stellar populations (SNIa/SNII). In these simulations it is possible to separate galaxies, the BCG, and the ICL, using substructure finding algorithms and the full phase-space information of the stellar particles. The ICL surface brightness profile is found to be steeper than the average galaxy light from cluster galaxies averaged in annuli. From the analysis of the merger trees in these cluster simulations, no preferred formation time for the ICL is found, and because it is cumulative, the largest contribution is released at $z < 1$. Most of the ICL comes from merging processes, either with the BCG or prior to infall into the cluster; other contributions are significant only outside $0.5R_{vir}$. Mergers liberate up to 30% of stars in both galaxies. Related with the finite resolution of

Figure 4. The distribution of the dark matter (DM, left), and of the stars (center), for a simulated cluster, and the distribution of stars in a high-resolution re-simulation of the same cluster (right), all at redshift $z = 0$. The frames are $6h^{-1}$ Mpc on a side, corresponding to $\approx 2R_{\mathrm{vir}}$. They show density maps generated with different logarithmic colour scales for DM and stars. The galaxy and diffuse components in each panel can clearly be seen. From Murante *et al.* (2007).

the simulations, a sizable contribution comes from dissoved galaxies, but whether this is a robust estimate cannot be firmly established at this stage. References: Murante *et al.* (2004, 2007), Dolag *et al.* (2009).

5.5. *Extended ionized and molecular gas emission in galaxy clusters – R. Oonk*

New, deep integral-field spectroscopy and imaging of the extended molecular and ionized gas distributions within the central regions of several galaxy clusters is presented, obtained with the VLT and HST. These observations show the existence of gas surrounding the BCG, extending at least up to 20 kpc from the nucleus. The H2 to HII line ratios are very high and are different from typical AGN and starburst ratios. To date no single source of excitation has successfully explained all line ratios over the entirety of the observed gas distribution. Various line diagnostics are used to constrain the properties of the observed gas and discuss its origin and fate. The question is open whether this is a birthplace of some of the stars in the ICL. References: van Weeren *et al.* (2009).

6. Summary of the discussion

The discussion at the end of the workshop was introduced by M. Arnaboldi with a quote from Uson *et al.* (1991): "Whether this diffuse light is called the cD envelope or diffuse intergalactic light is a matter of semantics; it is a diffuse component which is distributed with elliptical symmetry about the center of the cluster potential". A lively discussion then developed about the progress achieved since the time of this paper, with the main contributors being M. Arnaboldi, R. Ciardullo, P. Durrell, J. Feldmeier, C. Mihos, S. White, and A. Zabludoff. In the following, we list the main topics of discussion, and the specific conclusions and open issues that emerged for each of them. Because this list reflects only our understanding of what was being said, we apologize in advance for any errors or omissions.

ENVELOPE OF BRIGHTEST CLUSTER GALAXY VERSUS ICL:

- Sometimes the ICL is not aligned with BCG isophotes, so from this fact alone one would conclude that cD halo and ICL are different components.

- In some clusters like Coma and Abell 1185 an extended ICL component embeds more than one BCG. In Abell 545, a bright diffuse light component is seen with several galaxies embedded, but no cD. The ICL must therefore be a component distinct from a galaxy halo.

- There is kinematic evidence from PN and GC velocity distributions in the centers of nearby galaxy clusters that ICL and BCG halos are distinct physical components.

- There is a clear need for more spectroscopic data, to disentangle the contribution of halo and cluster component.

HOW TO MEASURE THE FRACTION OF ICL?

- The fraction of ICL is an observationally ill-defined quantity, either depending on arbitrary surface brightness thresholds in photometric studies, or lacking the full phase-space information to ascertain whether a star is bound to the central galaxy or not. A comparison with simulations is therefore difficult.

- Thus it would be useful if simulations were analysed to produce surface brightness maps. This would facilitate the comparison with wide field photometry measurements. One would then adopt a surface brightness threshold and make comparisons between the observed and simulated ICL plus BCG envelope fractions.

- A useful measurement to quantify the dynamical status of the ICL would be the fraction of homogenous versus irregular light or, more generally, the power spectrum over spatial scales.

TO WHAT REDSHIFT CAN WE GO WITH ICL STUDIES?

- Since much of the merging of massive galaxies occurs from z=1 to z=0, it would be interesting to reach z=0.3 with studies of ICL morphology, photometry and kinematics. Redshifts higher than $z = 0.3$ may be out of reach because of the strong surface brightness dimming effect scaling $\propto (1 + z)^{-4}$ with redshift z.

- The deep data sets obtained in weak lensing surveys may be useful for the determination of ICL morphology and photometry in intermediate redshift clusters.

METALLICITY AND COLOUR AS OBSERVATIONAL CONSTRAINTS ON HOW ICL IS MADE:

- In the VICS field in Virgo the intracluster stars are mostly old and mostly metal poor with a large metallicity spread; this is consistent with stripping of stars from a wide variety of galaxies, including the halos (but not the inner regions) of ellipticals.

- Most of the ICL in moderate redshift clusters, that is the more nearly homogeneous part, has the same color as the early-type galaxies in the cluster core.

- Blue colors are measured most often for elongated stream-like features.

- It would be very important to use cosmological simulations to predict the metallicity distribution and colours of ICL stars, and to compare this with present data.

RESULTS FROM SIMULATIONS ON THE ORIGIN OF ICL:

- In cosmological simulations of cluster formation, most of the ICl comes from the halos of evolved galaxies. Does this need to be reconciled with the results from higher-resolution simulations of galaxy evolution in dense environments?

- The importance of groups in generating ICL must be emphasized, through creating loosely bound stars in group interactions ("pre-processing") which are later unbound by the gravitational potential of the cluster into which they fall.

• Groups are generally important because of their environmental effects on galaxy evolution, and thus indirectly, the ICL.

THE HIGH-MASS END OF THE GALAXY LUMINOSITY FUNCTION:

• Is there still a problem caused by the evolution of the high mass end of the galaxy mass function from redshift $z = 1$ to $z = 0$?

• With semi-analytical models one can adjust the parameters for the AGN feedback, and recover agreement with observations. In the case of hydro-dynamical simulations, there is still a problem, and the dispersion of stars in the cluster volume as ICL would help.

FUTURE PROSPECTS:

• Currently we see a lot of new data on intracluster globular clusters from the Virgo/Coma legacy surveys. It will require lots of spectroscopic time to do the kinematic follow-up observations but the result will definitely be worth it.

• Resolved stellar populations in the Virgo and Fornax cluster ICL are obvious targets for the E-ELT, JWST, because crowding is not a problem at these low surface brightness levels.

CONCLUDING REMARK: Jorge Melnick summarized the meeting by stating that he was very pleased to see the subject of ICL to be active and strong, with extensive developments in deep surface brightness measurements, kinematics, stellar population studies, and comparisons with cosmological simulations. A very interesting meeting, much new science, and hope for more in Beijing!

7. Poster Papers

• ABS N. 123: Eduardo Cypriano *et al.* – Shrinking of cluster ellipticals: a tidal stripping explanation and implications for the intracluster light
• ABS N. 210: Marcelo Bryrro Ribeiro – Differential density statistics of the galaxy distribution and the luminosity function
• ABS N.450: Margarita Rosado *et al.* – Diffuse light in the Seyfert's Sextet
• ABS N.855: Tiberio Borges Vale *et al.* – Environmental effects on the structure of galaxy discs
• ABS N.992: Cristina Furlanetto *et al.* – Detection of gravitational arcs in galaxy clusters
• ABS N.1877: Walter Augusto Santos *et al.* – Photometric redshifts for SDSS galaxies using locally weighted regression.
• ABS N.1951: Julio Saucedo *et al.* – A study of the galaxy population in the region of A1781
• ABS N. 1962: Yasuhiro Hashimoto *et al.* – Multi-wavelength study of cluster morphology and its implications on the scaling relations, mass estimate, large scale structure, and evolution of galaxies.
• ABS N. 2076: Steven Michael Crawford *et al.* – The evolution of cluster galaxies and the diffuse intracluster light

• ABS N. 2341: Simon Nicholas Kemp *et al.* – From cDs to diffuse structures: faint light in galaxies using Schmidt and CCD data

• ABS N. 2728: Mangala Sharma *et al.* – Tracing galaxy group dynamical histories through diffuse intergalactic light.

• ABS N. 2947: Sadanori Okamura *et al.* – Observation of intracluster diffuse light in the Coma cluster

• ABS N. 2990: Roderik Overzier – Examining the Spiderweb: forming a BCG and its intracluster light at z = 2

• ABS N. 3002: Nieves Castro-Rodriguez *et al.* – Intracluster light in the Virgo cluster: large scale distribution.

Acknowledgements

We would like to acknowledge the support from the IAU, and the hard work and enthusiasm of the participants, which made this IAU Joint Discussion #2 on Diffuse Light in Galaxy Clusters both possible and stimulating.

References

Adami, C., Biviano, A., Durret, F., & Mazure, A. 2005a, *A&A*, 443, 17

Adami, C., *et al.* 2005b, *A&A*, 429, 39

Arnaboldi, M., Freeman, K. C., Hui, X., Capaccioli, M., & Ford, H. 1994, The Messenger, 76, 40

Arnaboldi, M., *et al.* 1996, *ApJ*, 472, 145

Arnaboldi, M., Gerhard, O., Aguerri, J. A. L., Freeman, K. C., Napolitano, N. R., Okamura, S., & Yasuda, N. 2004, *ApJ*, 614, L33

Arnaboldi, M., Gerhard, O., Okamura, S., Kashikawa, N., Yasuda, N., & Freeman, K. C. 2007, *PASJ*, 59, 419

Buzzoni, A., Arnaboldi, M., & Corradi, R. L. M. 2006, *MNRAS*, 368, 877

Ciardullo, R., Sigurdsson, S., Feldmeier, J. J., & Jacoby, G. H. 2005, *ApJ*, 629, 499

Coenda, V. & Muriel, H. 2009, *A&A*, 504, 347

Da Rocha, C., Ziegler, B. L., & Mendes de Oliveira, C. 2008, *MNRAS*, 388, 1433

de Lorenzi, F., *et al.* 2009, *MNRAS*, 395, 76

de Mello, D. F., Torres-Flores, S., & Mendes de Oliveira, C. 2008, *AJ*, 135, 319

Doherty, M., *et al.* 2009, *A&A*, 502, 771

Dolag, K., Murante, G., & Borgani, S. 2009, arXiv:0911.1129

Durrell, P. R., Ciardullo, R., Feldmeier, J. J., Jacoby, G. H., & Sigurdsson, S. 2002, *ApJ*, 570, 119

Faure, C., Giraud, E., Melnick, J., Quintana, H., Selman, F., & Wambsganss, J. 2007, *A&A*, 463, 833

Feldmeier, J. J., Mihos, J. C., Morrison, H. L., Harding, P., Kaib, N., & Dubinski, J. 2004a, *ApJ*, 609, 617

Feldmeier, J. J., Ciardullo, R., Jacoby, G. H., & Durrell, P. R. 2004b, *ApJ*, 615, 196

Gerhard, O., Arnaboldi, M., Freeman, K. C., *et al.* 2005, *ApJ*, 621, L93

Gerhard, O., Arnaboldi, M., Freeman, K. C., Okamura, S., Kashikawa, N., & Yasuda, N. 2007, *A&A*, 468, 815

Girardi, L., Bertelli, G., Bressan, A., Chiosi, C., Groenewegen, M. A. T., Marigo, P., Salasnich, B., & Weiss, A. 2002, *A&A*, 391, 195

Gonzalez, A. H., Zaritsky, D., & Zabludoff, A. I. 2007, *ApJ*, 666, 147

Guo, Q., White, S., Li, C., & Boylan-Kolchin, M. 2009, arXiv:0909.4305

Hilbert, S. & White, S. D. M. 2009, arXiv:0907.4371

Jordán, A., *et al.* 2009, *ApJS*, 180, 54

Kelson, D. D., Zabludoff, A. I., Williams, K. A., et al. 2002, ApJ, 576, 720

Krick, J. E. & Bernstein, R. A. 2007, AJ, 134, 466

Mayer, L., Mastropietro, C., Wadsley, J., Stadel, J., & Moore, B. 2006, MNRAS, 369, 1021

Mastropietro, C., Moore, B., Mayer, L., Debattista, V. P., Piffaretti, R., & Stadel, J. 2005, MNRAS, 364, 607

Melnick, J., Selman, F., & Quintana, H. 1999, PASP, 111, 1444

Mihos, J. C., Harding, P., Feldmeier, J., & Morrison, H. 2005, ApJ, 631, L41

Mihos, J. C., Janowiecki, S., Feldmeier, J. J., Harding, P., & Morrison, H. 2009, ApJ, 698, 1879

Misgeld, I., Hilker, M., & Mieske, S. 2009, A&A, 496, 683

Murante, G., et al. 2004, ApJ, 607, L83

Murante, G., Giovalli, M., Gerhard, O., Arnaboldi, M., Borgani, S., & Dolag, K. 2007, MNRAS, 377

Pierini, D., Zibetti, S., Braglia, F., Böhringer, H., Finoguenov, A., Lynam, P. D., & Zhang, Y.-Y. 2008, A&A, 483, 727

Richtler, T., Dirsch, B., Gebhardt, K., et al. 2004, AJ, 127, 2094

Saglia, R. P., Kronawitter, A., Gerhard, O., & Bender, R. 2000, AJ, 119, 153

Salinas, R., Richtler, T., Romanowsky, A. J., West, M. J., & Schuberth, Y. 2007, A&A, 475, 507

Schuberth, Y., Richtler, T., Hilker, M., Dirsch, B., Bassino, L. P., Romanowsky, A. J., & Infante, L. 2009, arXiv:0911.0420

Springel, V., Frenk, C. S., & White, S. D. M. 2006, Nature, 440, 1137

Takamiya, M., West, M., Côté, P., Jordán, A., Peng, E., & Ferrarese, L. 2009, Globular Clusters - Guides to Galaxies, 361

Thuan, T. H. & Kormendy, J. 1977, PASP, 89, 466

Torres-Flores, S., Mendes de Oliveira, C., de Mello, D. F., Amram, P., Plana, H., Epinat, B., & Iglesias-Páramo, J. 2009, A&A, 507, 723

Uson, J. M., Boughn, S. P., & Kuhn, J. R. 1991, ApJ, 369, 46

van Weeren, R. J., Intema, H. T., Oonk, J. B. R., Rottgering, H. J. A., & Clarke, T. E. 2009, arXiv:0910.4967

Ventimiglia, G., Arnaboldi, M., & Gerhard, O. 2008, Astronomische Nachrichten, 329, 1057

Zaritsky, D., Gonzalez, A. H., & Zabludoff, A. I. 2006, ApJ, 638, 725

Zibetti, S., White, S. D. M., Schneider, D. P., & Brinkmann, J. 2005, MNRAS, 358, 949

Williams, B. F., et al. 2007, ApJ, 656, 756

Highlights of Astronomy, Volume 15
XXVIIth IAU General Assembly, August 2009
Ian F Corbett

JD3 – Neutron Stars: Timing in Extreme Environments

Tomaso M. Belloni[1], Mariano Méndez[2] Chengmin Zhang[3]

[1] INAF – Osservatorio Astronomico di Brera, Via E. Bianchi 46, I-23807 Merate, Italy
email: tomaso.belloni@brera.inaf.it

[2] Kapteyn Astronomical Institute, University of Groningen, P.O. Box 800, 9700 AV Groningen,
The Netherlands
email: mariano@astro.rug.nl

[3] National Astronomical Observatories, Chinese Academy of Sciences, Beijing, China
email: zhangcm@bao.ac.cn

Abstract. The space-time around Neutron Stars is indeed an extreme environment. Whether they are in accreting binary systems, isolated or in non-accreting binaries (perhaps with another Neutron Star), Neutron Stars provide a window onto physical processes not accessible by other means. In particular, the study of their time variability: pulsations, quasi-periodic oscillations, thermonuclear X-ray bursts, flares and giant-flares, pulsar glitches and pulse period variations, constitutes a valuable instrument to unveil those very same physical processes. Here we briefly summarize the most important results presented at Joint Discussion 3 of the XXVII IAU General Assembly.

Keywords. stars: neutron

1. Introduction

Joint Discussion 3 (JD03) was focused on the astronomical systems harboring neutron stars, from isolated and binary radio pulsars to magnetars and accreting X-ray binaries. These systems constitute a unique tool for the study of matter under extreme conditions. Exploring the gravitational wave emission, testing General Relativity in the strong-field regime and the determination of the equation of state of nuclear matter are the major topics.

With the launch of high-energy missions such as XMM-Newton, Chandra, INTEGRAL and Suzaku for energy spectra and RossiXTE for fast time variability, we have made considerable progress in understanding the extreme environment close to compact objects. As a perspective, the future instrumentation for timing analysis, from radio observations to high-energy emissions, have been addressed in this Joint Discussion.

In the three half-day meeting of JD03, a total of 31 oral presentations were given, including six invited, seven solicited and 18 contributed, and about 50 posters were displayed. Four of the invited presentations, which were intended as broad reviews of the subject, are included in this volume. Here, we would like to summarize some of the interesting topics and contributions to the Joint Discussion.

2. Broad iron lines in bright accreting Neutron-Star binaries

A broad iron line, interpreted as a relativistically broadened line from neutral or mildly-ionized iron has been discovered in the X-ray emission of the accreting millisecond X-ray pulsar SAX J1808.4–3658 as observed by XMM-Newton in 2008 (Papitto *et al.* 2009, see Fig. 1, left panel). Relativistic models were fitted to the line: an inner radius of the

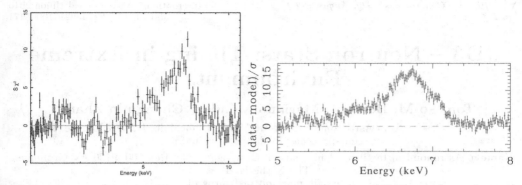

Figure 1. Left: Residuals obtained removing the broad line from the best fit model for the 2008 XMM-Newton observation of SAX J1808.4-3658 (from Papitto *et al.* 2009). Right: similar plot for 4U 1705-44 (from Di Salvo *et al.* 2009).

accretion disk of around 9 gravitational radii was found. For a 1.5 $M_\odot$ neutron star, this corresponds to a radius inside the co-rotation radius of the system, which is expected given that pulsations were also observed. A similar line was recently detected by the same group from 4U 1705–44 (Di Salvo *et al.* 2009, see Fig. 1, right panel). Similarly broadened lines were reported by other groups in the recent years: Serpens X-1 (Bhattacharyya & Strohmayer 2007; Cackett *et al.* 2008), 4U 1820–30 and GX 349+2 (Cackett *et al.* 2008) and 4U 1636–53 (Pandel *et al.* 2008), as well as from independent observations of SAX J1808.4–3658 (Cackett *et al.* 2009).

3. The binary pulsar PSR J0737–3039A/B

PSR J0737–3039A/B is a binary system consisting of two radio pulsar with a ∼2.5 hr orbit (Burgay *et al.* 2003; Lyne *et al.* 2004; Kramer & Stairs 2008; Archibald *et al.* 2009). The two pulsars have a period of 23 ms (pulsar A) and 2.8 s (pulsar B). The system is observed almost edge-on, causing eclipses. This system is providing the most stringent constraints on General Relativity in a relatively strong regime, as it yields independent tests for a number of post-keplerian parameters. As time goes on, the measurements become more precise and the limits more stringent. At present, all interval limits on the parameters contain the values expected from General Relativity.

4. γ-ray pulsars

An exciting result from the Fermi γ-ray satellite is the major increase in the discovery of γ-ray pulsars, some of which discovered in γ-rays (see Abdo *et al.* 2009). A graphical summary of the pulsar detections by Fermi can be seen in Fig. 2. The detection of radio-quiet pulsars is particularly important as it shows that γ-rays are emitted in a wider beam than radio emission, favoring models where the high-energy photons are produced in the magnetosphere and not on the polar caps of the neutron star.

5. From accreting binaries to millisecond pulsars

Evolutionary models indicate that neutron-star accreting low-mass X-ray binaries (LMXB) are the progenitors of millisecond radio pulsars, accretion being responsible for their spin-up to fast periods. Indeed, ten years ago the first X-ray millisecond pulsar was discovered in an LMXB, and now a dozen such systems are known.Recently, a

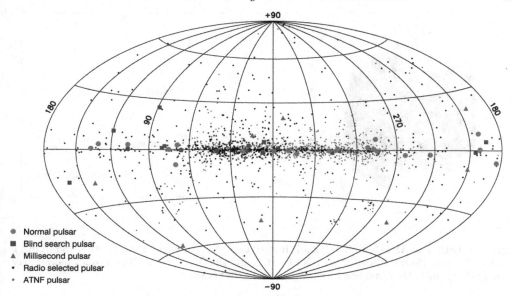

Figure 2. Galactic map of the Fermi pulsar detections (from Abdo *et al.* 2009)

millisecond radio pulsar in a circular binary system was discovered (Archibald *et al.* 2009). While many systems like this are known, optical spectra of this system obtained a decade ago show optical properties consistent with an accreting system: blue continuum, double-peaked emission lines and fast flickering. These properties are not observed today, when the optical spectrum of the companion is consistent with a solar-type star. This strongly suggests that the radio pulsar, PSR J1023+0038, within a decade turned on after an accretion phase stopped, providing compelling evidence for a direct link between these two classes of systems.

6. Future missions I: ASTROSAT

The near future for X-ray timing relies on the Indian mission ASTROSAT. This satellite, due to be launched in late 2010, will feature a number of astronomical instruments. In addition to an UV telescope, a soft X-ray telescope, a high-energy CdZnTe high-energy detector and an All-Sky Monitor, a large high-pressure collimated proportional counter will be available (see Fig. 3, left panel). This instrument, the LAXPC, will cover the energy range 3–100 keV with an effective area of $\sim$6000 cm^2 at 5 keV, comparable to that of the RXTE/PCA, but much higher above 10 keV (see Fig. 3, right panel). This instrument will take over the heritage of the RXTE/PCA and, thanks to its larger area at high energies, will open a new window on fast-timing phenomena from accreting binaries.

7. Future missions II: the Advanced X-ray Timing Array (AXTAR)

While ASTROSAT is expected to be operative in the near future, AXTAR is still at the stage of mission concept. It is a dedicated satellite for sub-millisecond timing of bright Galactic sources (Chakrabarty *et al.* 2008, see Fig. 4). Its main instrument is collimated Silicon detector covering the energy range 2–50 keV with a large collecting area (8 square meters). Such an instrument would provide a much larger number of photons, allowing to explore in much more detail the known signals and to discover weaker ones. The

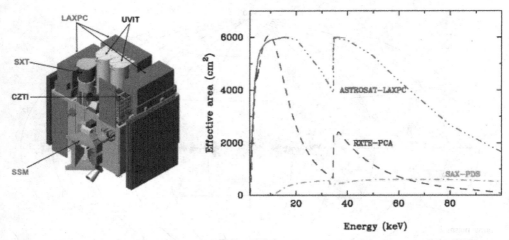

Figure 3. Left: The ASTROSAT payload. Right: The effective area curves for the ASTROSAT LAXPC in comparison to those of the RXTE/PCA and the BeppoSAX/PDS. From Chakrabarty *et al.* 2008. From `http://meghnad.iucaa.ernet.in/ãstrosat/`.

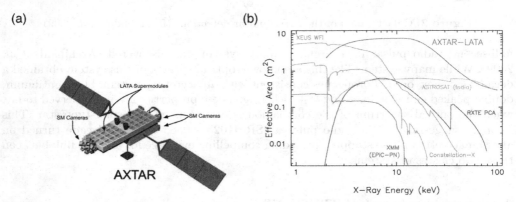

Figure 4. Left: A sketch of the AXTAR mission concept. Right: The effective area curves for AXTAR in comparison to those of other timing-capable instruments. From Chakrabarty *et al.* 2008.

main goals of the mission are to obtain measurements of general-relativistic effects in the strong field regime and to put constraint on the equation of state of neutron matter. In addition to the main large instrument with large collecting area, an all-sky monitor will ensure that a good dense monitoring of the bright X-ray sky is available for triggers and for following the long-term evolution of selected sources.

8. Future missions III: IXO

The largest endeavor of the X-ray community at the moment is IXO, the International X-Ray Observatory. This mission is currently under review by NASA, ESA and JAXA, and if accepted would fly sometime after 2017.

In its current design, IXO will carry 5 instruments on board: A Wide-Field Imager with intermediate (CCD-like) spectral resolution, a Narrow-Field Imager with high spectral resolution (using micro-calorimeters), a High-Time Resolution Spectrograph (HTRS), an X-ray polarimeter and a Grating spectrograph. The mission will cover the $0.10 - 12 keV$

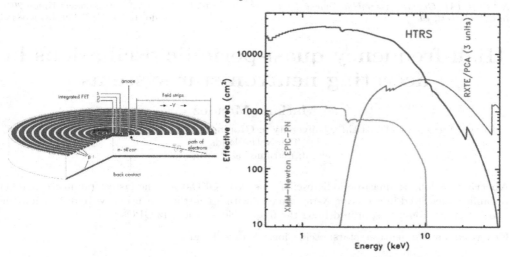

Figure 5. A schematic representation of one of the silicon-drift detectors that will be used in the HTRS (left), and the effective area of the combination of HTRS and IXO telescope in comparison with XMM-Newton and RossiXTE (right).

energy band, with a possible extension up to 40 keV. The current design anticipates 3 m^2 at 1.25 keV, 0.65 m^2 at 6 keV, and 150 cm^2 at 30 keV.

The main instrument on board IXO for timing is the HTRS (Barret *et al.* 2008). This instrument will consist of an array of silicon-drift detectors to collect the X-rays focused by the mirrors, with a spectral resolution of ~ 150 eV, a time resolution of 10 μs, and a count rate capability of 10^6 counts/s (equivalent to about 5 Crabs) with less than 10% deadtime (see Fig. 5). The main science goals are to study strong gravity around neutron stars and black holes, and to constrain the neutron-star equation of state.

Acknowledgements

We are grateful to the IAU for selecting the proposal to hold this Joint Discussion at the General Assembly, and for their help in the organization of the event in Rio de Janeiro. We are also grateful to the participants for their excellent contributions.

References

Barret, D., *et al.* 2008, *Proc. SPIE*, 7011, 70110E
Abdo, A. A. *et al.* 2009, *ApJ* submitted (arXiv:0910.1608v1)
Achibald, A. M., Stairs, I. H., Ransom, S. M. *et al.* 2009, *Science*, 324, 1411
Bhattacharyya, S. & Strohmayer, T. E. 2007, *ApJ* 664, L103
Burgay, M., D'Amico, N., Possenti, A. *et al.* 2003, *Nature*, 531, 426
Cackett, E. M., Miller, J. M., Bhattacharyya, S. *et al.*, 2008, *ApJ* 674, 415
Cackett, E. M., Altamirano, D., Patruno, A., *et al.* 2009, *ApJ* 694, L21
Chakrabarty, D., Ray, P. S., Strohmayer, T. E. *et al.* 2008, AIP Conference Proceedings, 1068, 227
Di Salvo, T., D'Aí, A., Iaria, R. *et al.* 2009,*MNRAS* 398, 2022
Kramer, M. & Stairs, I. H. 2008, *ARA&A*, 46, 541
Lyne, A. G., Burgay, M., Kramer, M. *et al.* 2004, *Science*, 303, 1153
Pandel, D., Kaaret, P., & Corbel, S. 2008, *ApJ* in press (arXiv:0808.2214)
Papitto, A., Di Salvo, T., D'Aí, A. *et al.* 2009,*ApJ* 495, L39

Highlights of Astronomy, Volume 15
XXVIIth IAU General Assembly, August 2009
Ian F. Corbett, ed.
ⓒ International Astronomical Union 2010
doi:10.1017/S1743921310008483

High-frequency quasi-periodic oscillations in accreting neutron-star systems

Mariano Méndez

Kapteyn Astronomical Institute, University of Groningen, P.O. Box 800, 9700 AV Groningen,
The Netherlands
email: mariano@astro.rug.nl

Abstract. Kilohertz quasi-periodic oscillations (kHz QPOs) are the fastest (almost coherent) variability measured in accreting X-ray binaries with a neutron-star primary. Here I review the rôle of the neutron-star spin in driving the frequencies of the kHz QPOs.

Keywords. stars: neutron, stars: oscillations, X-rays: binaries

1. Introduction

Variability is the hallmark of accreting low-mass X-ray binaries (LMXBs), binary systems containing a neutron-star (NS) or black-hole (BH) primary accreting from a normal star via an accretion disc. (In the rest of the paper I will discuss the case of LMXBs with a NS primary, although several, but not all, phenomena mentioned applies also to LMXBs with a BH primary.) Variability in LMXBs covers a broad range of time scales, ranging from several years to milliseconds, and most likely reflects different dynamical aspects of all components in these objects.

The variability over the longest time scales is due to changes in the mass flow onto the primary. It has been recently proposed in Koze & Charles 2009 that on time scales of years, the driving mechanism is probably activity in the secondary, similar to the activity seen in the Sun. While in this case the amplitude of the variability is generally mild, enhanced X-ray emission may sometimes trigger instabilities in the accretion disc (see, e.g., Cannizzo 1993 for a review), drastically increasing the rate at which mass flows onto the primary, giving rise to a so-called outburst. These outbursts are events in which the X-ray emission of the system increases by large factors within a week or so, remains high for weeks to months, and decays back to low levels within a few days to weeks.

On time scales of years to hours, we see variability due to orbital dynamics of the binary system and the accretion disc. Typical orbital motion of the binary is of the order of a few hours to a few days, while precession of the accretion disc around the primary, or radiation-induced warping of the disc, occurs on time scales of several weeks to months (Whitehurst & King 1991, Ogilvie & Dubus 2001). Closer to the primary, on time scales of a few tens of seconds to milliseconds, inhomogeneities in the accretion disc produce quasi-periodic variability in the X-ray light curves of these systems, the so-called quasi-periodic oscillations (QPOs; van der Klis 1989, van der Klis 2006). In NS LMXBs, where there is a surface stopping the accreted matter, variability also reflects instabilities in the burning of the accreted material. In the most extreme cases, unstable burning of the accreted material produces the so-called X-ray bursts, characterized by a sudden flux increase (typical rise times of less than a second) by factors of ten to hundred (Lewin *et al.* 1993). When the burning finishes, the neutron-star surface cools down and the flux decays back to pre-burst values over a few tens of seconds, and in some extraordinary cases (the so-called super bursts) over few hours to a day (e.g., Cornelisse *et al.* 2002).

Also at (or close to) the surface of the NS coherent pulsations, seen both from persistent (non-burst) emission and X-ray bursts, reflect the rotation of the neutron star itself (Wijnands *et al.* 2003).

Among all these phenomena, the ones with the shortest time scale (or the highest frequency) are the so-called kilohertz QPOs (kHz QPOs) seen in NS LMXBs. (Similar QPOs have been observed in systems containing a BH primary, at frequencies that are 5 to 10 times lower than in NS systems. The origin of the QPOs in both classes of systems may well be the same.) Although there is still no undisputed model for these oscillations, the time scale involved is tantalizingly close to the crossing time near the surface of the neutron star, therefore understanding these QPOs may bring us closer to probing the highly curved geometry of space time in the vicinity of a neutron star, or to unveil the internal constitution of a neutron star itself.

KHz QPOs were discovered almost immediately after the launch of the *Rossi X-ray Timing Explorer* (*RXTE*; Bradt *et al.* 1993). The phenomenology of these QPOs is rather complex. Here I will not aim at providing a complete picture of these QPOs (for that see for instance van der Klis 2006), but I will concentrate in a few basic aspects that have guided our attempts to try and understand this phenomenon.

2. Neutron-star spin and kHz QPOs

The kHz QPOs are relatively narrow peaks that often appear in pairs in the power density spectrum of NS LMXBs, at frequencies ν_1 and $\nu_2 > \nu_1$ that change with time. These QPOs are thought to reflect motion of matter at the inner edge of an accretion disc around the neutron star (Miller *et al.* 1998).

Burst oscillations are short-lived ($\tau \lesssim 10$s), almost coherent pulsations seen at the rise and tail of X-ray bursts in NS LMXBs. The frequency of these oscillations, ν_b, increases in the tail of the bursts to an asymptotic value that is consistent with being the same in bursts separated by more than a year time (Strohmayer *et al.* 1998). This, and the fact that in the accretion-powered millisecond X-ray pulsar (AMP) SAX J1808.4–3658 burst oscillations appear at the same frequency as the pulsations seen during persistent (non-burst) intervals (Wijnands *et al.* 2003), indicates that the frequency of these burst oscillations is equal to the spin frequency of the NS, ν_s.

The idea that the spin of the neutron star is directly involved in the mechanism that produces the kHz QPOs is deeply entrenched in the community studying this phenomenon. This idea stems from the first detection of kHz QPOs and burst oscillations in the same source, the LMXB 4U 1728–34, very early on in the *RXTE* mission. While in different observations the kHz QPOs appeared at different frequencies, ν_1 in the range $\sim 600 - 800$ Hz, and ν_2 in the range $\sim 500 - 1100$ Hz, the frequency difference of the QPOs, when both were present simultaneously, was consistent with being constant, $\Delta\nu = \nu_2 - \nu_1 \approx 363$ Hz, and also consistent with the oscillations seen during bursts in this source at $\nu_b = 363$ Hz (Strohmayer *et al.* 1996). This suggested that a beat mechanism with the neutron star spin was responsible for the kHz QPOs. Further results on other sources appeared to confirm this picture. A detailed model, the sonic-point model(Miller *et al.* 1998), explained the observed relation between the kHz QPOs and the neutron star spin in terms of a beat between material orbiting at the inner edge of the disc with the Keplerian frequency at that radius, and the spin of the NS.

As soon as kHz QPOs were discovered in 4U 1636–53 (van der Klis *et al.* 1996) with a frequency difference of $\Delta\nu = 272 \pm 11$ Hz, and burst oscillations at a frequency $\nu_b = 581$ Hz (Zhang *et al.* 1996), it became apparent that in this source $\Delta\nu$ was inconsistent with being equal to ν_b, but it was close to $\nu_b/2$. This would have been the end of the

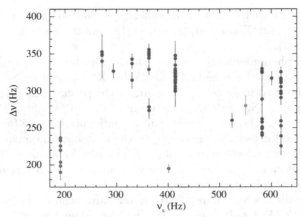

Figure 1. The frequency difference between the two simultaneous kHz QPOs, $\Delta\nu$ vs. the spin frequency, ν_s, for 13 neutron-star LMXBs (see Méndez & Belloni 2007 for references to the original papers where these data were published). The blue symbols are systems for which the spin was deduced from burst oscillations. The red points are the AMPs SAX J1808.4–3658 and XTE J1807–294. The green point is the transient accreting millisecond X-ray pulsar Aql X-1.

sonic-point model, unless in 4U 1636–53 the 581 Hz frequency seen during X-ray bursts was the second harmonic of the NS spin frequency, $\nu_b = 2 \times \nu_s$, with $\nu_s = 290.5$ Hz, e.g. if the pulsed radiation came from two antipodal poles on the NS. Although searches for a signal at half the burst oscillations frequency, the putative spin frequency of the neutron star, in the power spectrum of the bursts in 4U 1636–53 yielded no positive result (Strohmayer & Markwardt (2002)), this option remained viable.

When kHz QPOs and burst oscillations were discovered in more sources, it became apparent that there was a systematic trend in how $\Delta\nu$ and ν_b were related: For sources for which $\nu_b \lesssim 400$ Hz, $\Delta\nu \simeq \nu_b$, whereas for sources for which $\nu_b \gtrsim 400$ Hz, $\Delta\nu \simeq \nu_b/2$. These two groups of sources were then called "slow" and "fast" rotators, respectively (Muno et al. 2001). But this apparent relation is not without problems. For instance, it is now clear that in several sources $\Delta\nu$ is not constant, but changes in a very definite manner as the QPO frequencies change (e.g., Méndez et al. 1999). More importantly, in the AMP SAX J1808.4–3658, the frequency of burst oscillations is equal to the NS spin frequency (Chakrabarty et al. 2003), while two simultaneous kHz QPOs with a frequency separation $\Delta\nu \simeq \nu_s/2$ were detected (Wijnands et al. 2003). This result would then also hold for other sources in which $\Delta\nu \simeq \nu_b/2$. The sonic-point model could not explain this.

A testimony of how deep the idea of a direct link between kHz QPOs and neutron star spin is the fact that despite all this blows, beat-frequency ideas were revived adding small changes to the models in order to account for each new result that seemed to challenge the idea (see, for instance, Lamb & Miller 2003).

Recently, Méndez & Belloni (2007) suggested that the data are in fact consistent with a simpler picture in which the frequency separation between kHz QPOs, $\Delta\nu$, is independent of ν_s, with $\langle\Delta\nu\rangle$ being more or less constant across sources. In this interpretation the division between fast and slow rotators is artificial, and is in fact guided by the idea that the NS spin must be directly related to the mechanism that produces the QPOs. In Figure 1 I show $\Delta\nu$ as a function of ν_s for all sources with kHz QPOs and pulsations or burst oscillations (see Méndez & Belloni 2007 for the data and references). Except for the two AMPs, SAX J1808.4–3658 and XTE J1807–294, the data are consistent with $\Delta\nu$ being constant, or slightly decreasing with ν_s. (The apparent trend of $\Delta\nu$ decreasing with

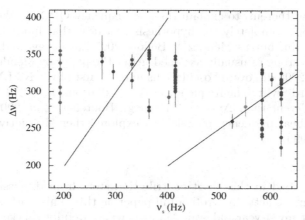

Figure 2. The same as Figure 1, but $\Delta\nu$ of the AMPs SAX J1808.4–3658 and XTE J1807–294 has been multiplied by the factor 1.5 (van Straaten *et al.* 2005, Linares *et al.* 2005). The two lines represent the relation $\Delta\nu = \nu_s$ for $\nu_s \leqslant 400$ Hz, and $\Delta\nu = \nu_s/2$ for $\nu_s \geqslant 400$ Hz, advocated by variants of the beat-frequency model.

ν_s could also be due to the fact that some sources have not been sampled well enough to show a range of $\Delta\nu$ values.)

Several authors have shown that there is a correlation between the frequency of the kHz QPOs and that of other low-frequency QPOs. More specifically, when plotted vs. the frequency of the upper kHz QPO, the frequency of the lower kHz QPO as well as the frequency of all low-frequency QPOs follow individual correlations that are consistent with being the same in at least eleven sources (see Altamirano *et al.* 2007 for an overview of these correlations). The AMPs SAX J1808.4–3568 and XTE J1807–214 show similar relations between the frequencies of the low-frequency QPOs and ν_2, but the relations are displaced with respect to those of the other sources (van Straaten *et al.* 2005; Linares *et al.* 2005). The displacement is best described as a multiplication of ν_2 by a factor close to 1.5. While this factor applied to ν_2 works for the low-frequency QPOs vs. ν_2 correlations, it does not work for the correlation between ν_1 and ν_2. Interestingly, van Straaten *et al.* (2005) and Linares *et al.* (2005) noticed that the ν_1 vs. ν_2 correlation in the AMPs and in the other sources would be reconciled if they also multiplied ν_1 by *the same factor* that they used to describe the displacement of the ν_2 vs. low-frequency QPO correlations. (Notice that ν_1 was not used to derive that factor.) The nature of this shift is unclear, but taken at face value, a multiplicative factor applied both to ν_1 and ν_2 implies that the frequency difference $\Delta\nu = \nu_2 - \nu_1$ must also be multiplied by this factor. In Figure 2 I show the same plot as in Figure 1, but here I have multiplied $\Delta\nu$ of the AMPs SAX J1808.4–3658 and XTE J1807–294 by the factor close to 1.5. For completeness, I also plot the relation predicted by beat-frequency models that advocate the idea that $\Delta\nu = \nu_s$ and $\Delta\nu = \nu_s/2$ below and above a certain spin period (in this case I took 400 Hz as the dividing spin).

3. Conclusions

I have briefly described the idea put forward by Méndez & Belloni (2007), that the frequency difference of the kHz QPOs, $\Delta\nu$, in neutron-star low-mass X-ray binaries is probably unrelated to the spin frequency, ν_s, of the neutron star. This is in contradiction to beat-frequency models that have been proposed to explain the kHz QPOs. Notice that beat-frequency mechanisms have been originally advanced in the 1980s (see van der Klis

1989 and references therein) to explain the low-frequency QPOs in these same systems. One cannot rule out completely the hypothesis of a beat that links the kHz QPOs and the neutron-star spin, but as Méndez & Belloni 2007 have suggested, the evidence for that link is not as strong as usually assumed. After the paper by Méndez & Belloni 2007, Strohmayer *et al.* 2008 discovered oscillations in one burst of the NS LMXB 4U 0614+09 with BAT, on board Swift. The frequency of the oscillations, $\nu_s = 414.7$ Hz, or half that, is significantly different than $\Delta\nu = 320$ Hz (e.g., Boutelier *et al.* 2009). Probably it is time to abandon beat-frequency models and explore other alternatives in more detail (Stella & Vietri 1999).

Acknowledgements

The topics described in this short contribution stem from discussions with Tomaso Belloni, with whom we actually wrote a full paper on this. This work was finished under the supervision of my 4-year old son, Manuel, who carefully checked that there were enough uppercase "A's" in the text. The rest of the text is the author's own responsibility.

References

Altamirano, D., van der Klis, M., Klein-Wolt, M., Méndez, M., van Straaten, S., Jonker, P. G., Lewin, W. H. G., & Homan, J. 2008, *ApJ*, 685, 436

Boutelier M., Barret D., & Miller M. C. 2009, *MNRAS*, 399, 1901

Bradt H. V., Rothschild R. E., & Swank J. H. 1993, *A&AS*, 97, 355

Cannizzo, J. K. 1993, in: Wheeler, J., (ed.), *Accretion Disks in Compact Stellar Systems*, World Scientific, Singapore, p. 6.

Chakrabarty D., Morgan E. H., Muno M. P., Galloway D. K., Wijnands R., van der Klis M., & Markwardt C. B. 2003, *Natur*, 424, 42

Cornelisse R., *et al.* 2002, *A&A*, 392, 885

Koze, M. M. & Charles, P. A. 2009, *MNRAS*, in press (arXiv:0911.4188)

Lamb F. K. & Miller M. C. 2003, *ApJ*, submitted, preprint, arXiv:astro-ph/0308179v1

Lewin, W. H. G., van Paradijs, J., & Taam, R. 1993, in W. H. G. Lewin, J. van Paradijs, & E. P. J. van den Heuvel, (eds.), *X-ray Binaries*, Cambridge Univ. Press, p. 175

Linares, M., van der Klis, M., Altamirano, D., & Markwardt, C. B. 2005, *ApJ*, 634, 1250

Méndez, M., van der Klis, M., Wijnands, R., Ford, E. C., & van Paradijs, J. 1999, *ApJ*, 511, L49

Méndez, M. & Belloni, T. 2007, *MNRAS*, 381, 790

Miller, M. C., Lamb, F. K., & Psaltis, D. 1998, *ApJ*, 508, 791

Muno M. P., Chakrabarty D., Galloway D. K., & Savov P. 2001, *ApJ*, 553, L157

Ogilvie G. I. & Dubus G. 2001, *MNRAS*, 320, 485

Stella, L. & Vietri, M. 1999, *Phys. Rev. Lett*, 82, 17

Strohmayer T. E. & Markwardt C. B. 2002, *ApJ*, 577, 337

Strohmayer T. E., Markwardt C. B., & Kuulkers E. 2008, *ApJ*, 672, L37

Strohmayer T. E., Zhang W., Swank J. H., & Lapidus I. 1998, *ApJ*, 503, L147

Strohmayer T. E., Zhang W., Swank J. H., Smale A., Titarchuk L., Day C., & Lee U. 1996, *ApJ*, 469, L9

van der Klis, M. 1989, *ARA&A*, 27, 517

van der Klis, M. 2006, in: Lewin W., van der Klis M., (eds.), *Compact Stellar X-ray Sources*, Cambridge Univ. Press, Cambridge, p. 46

van der Klis, M., van Paradijs, J., Lewin, W. H. G., Lamb, F. K., Vaughan B., Kuulkers E., & Augusteijn, T. 1996, *IAUC*, 6428, 2

van Straaten, S., van der Klis, M., & Wijnands, R. 2005, *ApJ*, 619, 455

Whitehurst R. & King A. R. 1991, *MNRAS*, 249, 25

Wijnands, R., van der Klis, M., Homan, J., Chakrabarty, D., Markwardt, C. B., & Morgan, E. H. 2003, *Natur*, 424, 44

Zhang, W., Lapidus, I., Swank, J. H., White, N. E., & Titarchuk, L. 1996, *IAUC*, 6541, 1

Highlights of Astronomy, Volume 15
XXVIIth IAU General Assembly, August 2009
Ian F. Corbett, ed.

© International Astronomical Union 2010
doi:10.1017/S1743921310008495

Accreting millisecond X-ray pulsars: recent developments

Rudy Wijnands

Astronomical Institute 'Anton Pannekoek', University of Amsterdam, Postbus 94249, 1090 GE
Amsterdam, the Netherlands
email: r.a.d.wijnands@uva.nl

Abstract. It is now more than eleven years since the discovery of the first accreting millisecond
X-ray pulsar. Since then, eleven additional systems have been found, two of them during the
last year. Here I briefly discuss the most recent developments with respect to these systems.

Keywords. pulsars:X-ray – stars:neutron – X-rays:binaries

1. Introduction

Since the discovery of the first accreting millisecond X-ray pulsar (AMXP) in 1998
(Wijnands & van der Klis 1998), eleven additional systems have been found (see Table 1
and Section 2 for the two most recently discovered AMXPs). These systems are important
because they are the evolutionary link between millisecond radio pulsars and accreting
neutron stars in low-mass X-ray binaries†. In addition, by studying the pulsations in the
AMXPs, one hopes to learn about the physics of accretion onto a magnetized object and
about the properties of neutron stars. Currently all AMXPs are transients and one must
wait for the sources to go into outburst in order to study them and their pulsations.
Here I present some highlights of the most recent developments related to the study of
AMXPs.

2. The two most recently discovered AMXPs

In the months after the IAU General assembly, two new AMXPs were discovered. The
source IGR J17511–3057 was discovered by INTEGRAL in early September 2009 (Baldovin *et al.* 2009) and soon afterwards it was found (using RXTE; Markwardt *et al.* 2009)
to be an AMXP. The neutron star in this system has a spin frequency of ∼245 Hz and
the orbital period of the system is ∼ 208 minutes (Markwardt *et al.* 2009; see also Table
1). In addition burst oscillations were observed (Watts *et al.* 2009) and a near infra-red
counterpart was discovered (Torres *et al.* 2009a,b). None of this makes this source exceptional; it would seem to be just a 'run-of-the-mill' AMXP. The second, recently discovered
AMXP does exhibit some atypical behavior. Heinke *et al.* (2009) reported the discovery of a new X-ray transient in the gobular cluster NGC 6440 (the second one in this
cluster). An RXTE observation on this cluster showed that this transient is an AMXP.
The neutron star in this system has a spin period of ∼205 Hz and the orbital period is
∼57 min (Altamirano *et al.* 2009). These properties are not different from those of other
known AMXPs. However, it was found (Heink *et al.* 2009) that this source has only very
short outbursts (of order days) and the peak X-ray luminosity during outburst is only

---

† Recently, a different type of 'link' between millisecond radio pulsars and neutron-star low–
mass X-ray binaries has been found (Archibald *et al.* 2009) opening up an additional, very
interesting window leading to understanding the relationship between the two types of sources.

Table 1. The currently known AMXPs with their basic characteristics and when the pulsations were first seen

Name	spin (Hz) (Hz)	binary period (min) (minutes)	$M_{c,min}$ ($M_\odot$)	When discovered
SAX J1808.4–3658	401	120	0.043	April 1998
XTE J1751—30 5	435	42	0.014	April 2002
XTE J0929–314	185	44	0.083	April 2002
XTE J1807–294	191	40	0.0066	February 2003
XTE J1814–338	314	258	0.17	June 2003
IGR J00291+5934	599	150	0.039	December 2004
HETE J1900.1–2455	377	84	0.016	June 2005
Swift J1756.9–2508	182	54	0.007	June 2007
Aql X-1	550	1140	?	August 2007
SAX J1748.9–2021	442	526	0.1	August 2007
NGC 6440 X-2	206	57	0.0067	August 2009
IGR J17511–3057	245	208	0.0011	September 2009

$\sim 10^{36}$ erg s^{-1} which makes it rather difficult to study in outburst. However, soon after it was discovered it was found that the source appears to be in outbursts roughly every month. This makes it the X-ray transient with the shorted known recurrence time (Heinke *et al.* 2009). The reason for this short recurrence time is not understood. Currently, monitoring campaigns using RXTE and Swift have begun to observe more outbursts of this source in order to study the evolution of its pulsations over time. Its frequent outbursts might give us unique insight into magnetic accretion if enough high-time resolution data during outburst can be collected. The discovery of this source also raises the question if many similar systems are missed, especially if they have similar short and weak outbursts but which are less frequent.

With twelve AMXPs now known, the neutron-star spin distribution can be studied in low-mass X-ray binaries. By including also the neutron-star spin measurements obtained from nearly coherent oscillations during type-I X-ray bursts (the so-called burst oscillations) the sample of sources now reaches about thirty. It was noted with a smaller sample (e.g., Chakrabarty *et al.* 2003 and references to that paper) that the neutron-star spin distribution in low-mass X-ray binaries (and also in the millisecond radio pulsars) appears to cut-off at around 600 to 700 Hz, with no neutron stars spinning faster than that† despite the fact that the break-up frequency of neutron stars is significantly larger. By using the full list of spin frequencies currently available, this conclusion still holds (see, e.g., Fig.1 which shows the sample complete until the summer of 2009 but does not include the most recent discoveries which are still consistent with this general picture). Currently, it is unclear what is the underlying physics. The most exciting explanation (although probably not the correct one) proposes that the spinning up of the neutron star in low-mass X-ray binaries is halted by the emission of gravitational waves from the stars (see Chakrabarty 2005 and Chakrabarty *et al.* 2003 and the papers referring to these two papers for a discussion of the various models).

---

† One system has a rather marginal indication of a burst-oscillation frequency (and thus a neutron-star spin) larger than 1000 Hz, but this result is highly debated and needs to be confirmed. I will not discuss this system further.

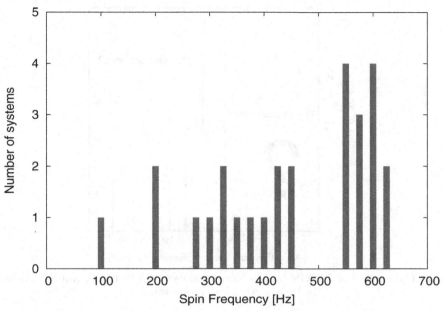

Figure 1. The number of AMXPs and burst oscillation sources (as of summer 2009) versus the observed frequency of the neutron stars (as inferred from the frequency of the pulsations or the burst oscillations). Figure made by Alessandro Patruno for the purpose of my presentation at the meeting.

3. Timing noise

One of the main reasons to study AXMPs is to determine the reaction of the neutron star to the accretion of matter. The usual way to investigate this is to search for changes in the pulse frequency (i.e., to determine a spin-frequency derivative or set upper limits on any). However, the measurement of such a frequency derivative has been severely hampered by an unmodeled component in the pulse arrival times which has been called 'timing noise' (named after a similar phenomenon, albeit on much longer time scales, seen in several radio pulsars). The presence of this timing noise has raised serious concerns about the validity of any frequency derivative reported in the literature. Recently, it has been shown that it is possible to attribute most of this timing noise to a phase offset of the pulsations that varies in correlation or anti-correlation with the X-ray flux (Patruno *et al.* 2009). If confirmed, this correlation could bias the measured spin frequencies and could artifically introduce a frequency derivative. Therefore it is very important to study all AMXPs (and hopefully also new ones) in detail to either confirm or reject this correlation. A confirmation would seriously inhibit our ability to measure any true spin period changes.. Interestingly, theoretically, these effects can rather easily be explained by assuming a wandering hot spot whose location depends on the accretion rate or by a systematic variations of the pulse profile (Patruno *et al.* 2009; Lamb *et al.* 2009, Poutanen *et al.* 2009).

4. Why only so few accreting millisecond X-ray puslars

Despite decades of observations, it is still unclear why only a dozen of the more than 150 known neutron-stars low-mass X-ray binaries exhibit pulsations. Aside from the pulsations, the other properties of the AMXPs look remarkably similar to those observed

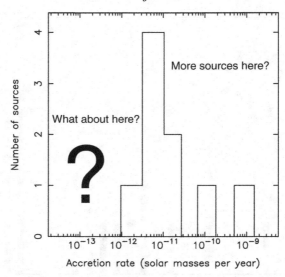

Figure 2. The number of AMXPs (as of summer 2009 and not including Aql X-1) versus their time averaged accretion rate.

from the non-pulsating sources, indicating that the difference between the two classess should be subtle. One possiblity is that the neutron stars in the non-pulsating systems have no or only a very weak magnetic field or that their magnetic axis is aligned with their spin axis. However, if this idea is correct, then those systems cannot be the progenitors of the millisecond radio pulsars unless a mechanism is available which would increase the magnetic field or misalign it with the rotation axis after the accretion has halted. Alternatively, it has been proposed that in the non-pulsating systems the pulsations (which are intrinsically there) are scatterred in a Comptonizing medium (e.g., Titarchuk *et al.* 2002) but it has been suggested that this cannot work for all systems (e.g., Göğüş *et al.* 2007; but see Titarchuk *et al.* 2007).

Based on the rather low time-averaged accretion rate ($\langle \dot{M} \rangle$) of the first AMXP discovered (SAX J1808.4–3658), Cumming *et al.* (2001) suggested that if $\langle \dot{M} \rangle$ is sufficiently high, the neutron star magnetic field might be buried by the accreted matter and does not have time to dissipate through the accreted material. However, for SAX J1808.4–3658 $\langle \dot{M} \rangle$ is sufficiently low that the magnetic field dissipation can indeed happen, giving it a magnetic field still strong enough to disturb the flow of the accreted matter. Intriguinly, nearly all of the AMXPs discovered after this (see Figure 2) have indeed rather low $\langle \dot{M} \rangle$ which would support this idea. The only outlayers are the so-called intermittent AMXPs which do not always exhibit pulsations. Those systems have high time averaged accretion rates and the intermittency of their pulsations might be related to that. Still, things are not fully clear and the burying model does have problems explaining those systems (see the discussions in Galloway *et al.* 2007, Casella *et al.* 2008, and Altamirano *et al.* 2008).

In order to make progress in our understanding of the reason (or reasons) why only such a small percentage of the neutron-star low-mass X-ray binaries exhibit pulsations, more AMXPs need to be discovered. When looking at Figure 2 (adapted from Wijnands 2008a), two strategies can be envisaged to try to enlarge the sample of AMXPs. Firstly, one should investigate the brighter transients with typically high $\langle \dot{M} \rangle$ ($> 10^{-10}$ $M_\odot$ yr^{-1}) in detail for possible short lived episodes during which they migh exhibit pulsations. Some searches have been performed already using the RXTE archive but so far no new AMXP has been found. However, such searches still need to be fully optimized and several

AMXPs might still be lurking in the RXTE archive. Secondly, if indeed most AMXPs have low $\langle \dot{M} \rangle$, one might assume (see Wijnands 2008a and references therein) that those neutron-star low-mass X-ray binaries with a very low $\langle \dot{M} \rangle$ ($< 10^{-11}$ $M_\odot$ yr^{-1}) might also exhibit coherent pulsations when in outburst. The fact that NGC 6440 X-2 might have a $\langle \dot{M} \rangle$ as low as $\sim 10^{-12} M_\odot$ yr^{-1} (Heinke *et al.* 2009) would support that idea. However, those systems are difficult to find. RXTE might not be sensitive enough to detect the pulsations leaving only XMM-Newton to be used (see discussion in Wijnands 2008b). Therefore, the discovery of those systems and the search for possible pulsations is quite challenging but could yield very important results which would help us to understand why only so few systems pulsate.

Acknowledgements

I would like to thank Alessandro Patranski for the help in preparing Figure 1.

References

Archibald, A. M., *et al.* 2009, Science, 324, 1411

Altamirano, D., Casella, P., Patruno, A., Wijnands, R., & van der Klis, M. 2008, ApJ, 674, L45

Altamirano, D., *et al.* 2009, arXiv:0911.0435

Baldovin, C., *et al.* 2009, The Astronomer's Telegram, 2196, 1

Casella, P., Altamirano, D., Patruno, A., Wijnands, R., & van der Klis, M. 2008, ApJ, 674, L41

Chakrabarty, D. 2005, Binary Radio Pulsars, 328, 279

Chakrabarty, D., Morgan, E. H., Muno, M. P., Galloway, D. K., Wijnands, R., van der Klis, M., & Markwardt, C. B. 2003, Nature, 424, 42

Cumming, A., Zweibel, E., & Bildsten, L. 2001, ApJ, 557, 958

Galloway, D. K., Morgan, E. H., Krauss, M. I., Kaaret, P., & Chakrabarty, D. 2007, Apj, 654, L73

Göğüş, E., Alpar, M. A., & Gilfanov, M. 2007, ApJ, 659, 580

Heinke, C. O., *et al.* 2009, arXiv:0911.0444

Lamb, F. K., Boutloukos, S., Van Wassenhove, S., Chamberlain, R. T., Lo, K. H., Clare, A., Yu, W., & Miller, M. C. 2009, ApJ, 706, 417

Markwardt, C. B., Altamirano, D., Swank, J. H., Strohmayer, T. E., Linares, M., & Pereira, D. 2009, The Astronomer's Telegram, 2197, 1

Patruno, A., Wijnands, R., & van der Klis, M. 2009, ApJ, 698, L60

Poutanen, J., Ibragimov, A., & Annala, M. 2009, ApJ, 706, L129

Titarchuk, L., Cui, W., & Wood, K. 2002, ApJ, 576, L49

Titarchuk, L., Kuznetsov, S., & Shaposhnikov, N. 2007, ApJ, 667, 404

Torres, M. A. P., Jonker, P. G., Steeghs, D., Simon, J. D., & Gutowski, G. 2009a, The Astronomer's Telegram, 2216, 1

Torres, M. A. P., Jonker, P. G., Steeghs, D., Damjanov, I., Caris, E., & Glazebrook, K. 2009b, The Astronomer's Telegram, 2233, 1

Watts, A. L., Altamirano, D., Markwardt, C. B., & ., Strohmayer, T. E. S. 2009, The Astronomer's Telegram, 2199, 1

Wijnands, R. 2008a, A Population Explosion: The Nature & Evolution of X-ray Binaries in Diverse Environments, 1010, 382

Wijnands, R. 2008b, American Institute of Physics Conference Series, 1068, 231

Wijnands, R., & van der Klis, M. 1998, Nature, 394, 344

Highlights of Astronomy, Volume 15
XXVIIth IAU General Assembly, August 2009
Ian F. Corbett

© International Astronomical Union 2010
doi:10.1017/S1743921310008501

Gamma-ray emission from pulsar/massive-star binaries

Gustavo E. Romero[1,2,†]

[1]Instituto Argentino de Radioastronomía (IAR, CCT La Plata, CONICET), C.C.5, (1894)
Villa Elisa, Buenos Aires, Argentina
email: romero@iar-conicet.gov.ar

[2]Facultad de Ciencias Astronómicas y Geofísicas, Universidad Nacional de La Plata, Paseo del
Bosque s/n, 1900 La Plata, Argentina
email: romero@fcaglp.unlp.edu.ar

Abstract. I present a review of the main phenomenological properties at high energies related to massive gamma-ray binaries and I discuss some aspects of pulsar models for these objects.

Keywords. Radiation mechanisms: non-thermal, stars: neutron, stars: early-type, stars: winds, outflows, gamma rays: observations, gamma rays: theory.

---

1. Introduction

Many X-ray binaries are formed by a pulsar (rapidly rotating neutron star) and a massive, hot star. In most of these systems the soft X-ray emission is originated in an accretion disk, which is truncated at some distance of the neutron star by the magnetic pressure. The matter is then channeled onto the surface of the compact object through closed magnetic field lines and it impacts onto the magnetic poles of the star, producing hard X-rays by Bremsstrahlung. It has been speculated for a long time that such systems might also produce non-thermal radiation by various mechanisms. For instance, Maraschi & Treves (1981) suggested that the interaction between the relativistic pulsar wind with the stellar wind of a Be star might result in efficient particle (electron) acceleration up to relativistic energies, with the consequent output of inverse Compton (IC) radiation in the gamma-ray domain. Such a model was later developed in more detail by Tavani *et al.* (1994) and Tavani & Arons (1997) for the pulsar/Be system PSR B1259-63.

A different type of model was proposed in the mid 1980s for Cygnus X-3, thought at that time to be a high-energy gamma-ray source. For instance, Berezinskii *et al.* (1986) considered the irradiation of the stellar atmosphere by relativistic protons produced by the pulsar. The result would be a gamma-ray and neutrino source.

Finally, at the end of the same decade, Cheng & Ruderman (1989) showed that in accreting systems, when the accretion disk rotates faster than the neutron star, a gap is open in the magnetosphere of the pulsar, and a strong potential drop is established in a charge-depleted region. Protons from the star might be accelerated in this gap along closed field lines in such a way that they would impact onto the accretion disk. This model was applied to the classic pulsar/Be binary A0535+262 by Romero *et al.* (2001), and further developed by Orellana *et al.* (2007).

In recent years several high-mass gamma-ray binaries have been detected by imaging atmospheric Cherenkov telescopes (IACTs), confirming some of these early predictions and posing new theoretical problems. In the next section I describe the main characteristics of the detected systems.

† Member of CONICET

2. Known gamma-ray binaries

High-energy emission has been detected from four confirmed massive binaries by IACTs and there is a fifth source that is almost surely a binary, but additional information must yet be gathered, especially about the orbital parameters, before the full confirmation of this latter source. With the exception of Cygnus X-1, which is a well-known black hole candidate, the spectral energy distributions (SEDs) of these systems are rather similar, with non-thermal radio emission, a power-law, featureless spectrum at X-rays, soft gamma-ray spectra, and strong variability, modulated by the orbital period (see Dubus 2006 and Skilton *et al.* 2009 for SEDs).

2.1. *PSR B1259-63*

PSR B1259-63 / SS 2883 is a binary system containing a B2Ve donor star and a 47.7 ms radio pulsar orbiting it every 3.4 years, in a very eccentric orbit with $e = 0.87$. No radio pulses are observed when the neutron star (NS) is behind the circumstellar disk (an effect due to free-free absorption). Very high-energy (VHE) gamma-rays are detected when the NS is close to the periastron or crosses the stellar disk (Aharonian *et al.* 2005a).

The VHE spectrum can be fitted with a power-law of soft index ($\Gamma \sim -2.7 \pm 0.2$†) and explained by IC up-scattering of stellar photons. The light curve shows significant variability and a puzzling behavior, that can be reproduced using variable adiabatic and IC losses, along with the changing geometry (Khangulyan *et al.* 2007). The energetic electrons are produced at the termination shock of the pulsar wind. The cold relativistic wind should also cool by IC interactions with the stellar field. Khangulyan *et al.* (2007) presents detailed predictions for *Fermi* satellite that can be used to determine the Lorentz factor of the cold wind.

PSR B1259-63 has been detected as a variable radio source (e.g. Johnston *et al.* 2005). Pulsed radio emission is most of the time at a level of a few mJy. Transient unpulsed optically thin ($\alpha = -0.6$‡) radio emission up to 50 mJy is measured close to periastron passage. No VLBI images of the radio source, located in the southern hemisphere, have been published.

2.2. *LS I +61 303*

LS I +61 303 is a binary system containing a B0.5Ve donor star and a compact object of unknown mass (upper limit of $\sim 5 \ M_\odot$) orbiting it every 26.5 days, in an eccentric orbit with $e \sim 0.5 - 0.7$. No radio pulses are observed . VHE gamma-rays are detected after periastron (Albert *et al.* 2006, Acciari *et al.* 2009). The flux is variable. There are marginal detections at phases 0.2-0.4, close to periastron (located at phase 0.23). The maximum flux is detected at phases $0.5 - 0.7$, with values of $\sim16\%$ of the Crab Nebula flux. The strong orbital modulation shows that the emission is produced, and affected, by the interplay between the two stars. The source has also been detected by *Fermi* satellite, with a strong, periodic flux (Abdo *et al.* 2009). The maximum, contrary to what happens at higher energies, occurs immediately after the periastron passage. The spectrum shows a cutoff at ~ 6 GeV, indicating that the relation with the TeV emission is complex.

At radio wavelengths, LSI +61 303 is highly variable, periodic, and has been resolved. Dhawan *et al.* (2006) obtained a series of VLBA images, spaced 3 days apart, and covering the entire orbit. The images show extended radio structures, with the phase near periastron having indeed an apparent elongation away from the primary star. Since no macroscopic relativistic velocities are measured, the authors identify this elongation as

† $F_\gamma \propto E^\Gamma$.
‡ $F_\nu \propto E^\nu$.

"a pulsar wind nebula shaped by the anisotropic environment, not a jet". However, the situation is far from clear (see Romero *et al.* 2007).

2.3. *HESS J0632+057*

HESS J0632+057 is a high-energy point-like source coincident with a B0pe star and a variable X-ray source (Hinton *et al.* 2009). The broadband spectrum looks like that of LS I +61 303, but no gamma-ray variability has been detected so far (see, nonetheless Acciari *et al.* 2009). The energetics, at the estimated distance of 1.5 kpc, requires a power of $dE/dt \sim 10^{34-35}$ erg/s, similar to other gamma-ray binaries. Recently, Skilton *et al.* (2009) have detected the source as a non-thermal radio emitter. *Fermi* observations can be used to detect variability and possible periodic behavior, as in LS I +61 303.

2.4. *LS 5039*

LS 5039 is a binary system containing an O6.5V((f)) donor and a compact object of unknown mass (upper limit of $\sim 5\ M_{\odot}$, Casares *et al.* 2005) orbiting it every 3.9 days, in a slightly eccentric orbit with $e \sim 0.35$. No radio pulses are observed. Persistent jet-like features have been reported several times with no puzzling behavior (same jet direction) up to now, suggesting a microquasar nature (Paredes *et al.* 2000). There are, however, no traces of accretion disk and no typical accretion variability. At X-rays the emission is a power-law, likely of synchrotron nature (e.g. Khangulyan *et al.* 2008). The variable TeV emission is modulated with the orbital period of the binary system (Aharonian *et al.* 2005b). The flux maximum occurs at the inferior conjunction of the compact object (Aharonian *et al.* 2006).

2.5. *Cygnus X-1*

Cygnus X-1 is a massive X-ray binary containing an O9.7Iab donor star and an accreting black hole of at least 10 solar masses. The orbital period is 5.6 days, and the orbit, circular. Persistent radio emission, sometimes resolved in a jet-like feature, reveals its microquasar nature. It is a system that seems to be intrinsically different from the previous ones.

3. Models

3.1. *Colliding wind models*

In colliding wind models, the compact object is a pulsar with a relativistic wind. This wind interacts with the stellar wind, producing a shock at the termination of the wind and particles can be accelerated there. Modulation naturally results from the changing conditions along the orbit. These type of models can reproduce the spectrum and light curve of PSR B1259-63 (Khangulyan *et al.* 2007), and they have been applied to LSI +61 303 and LS 5039 as well (Dubus 2006). However, is far from clear whether in the case of these binaries the observed radio morphology, the SEDs and the time variability can be accommodated within such a framework (see Romero et al 2007, Bosch-Ramon *et al.* 2008). The main test for colliding wind models would be the detection of the features produced by the interaction of the relativistic monoenergetic pulsar wind with the stellar field.

3.2. *Models with accretion*

In models with accretion the non-thermal emission is produced in a collimated outflow. Such an outflow might be unstable due to the complex interaction with the ambient medium. Both leptonic and hadronic jet models have been proposed for LS 5039 and LS I +61 303 (e.g. Romero *et al.* 2005, Bosch-Ramon *et al.* 2006, Paredes *et al.* 2006, Dermer &

Boettcher 2006). These binaries present an X-ray behavior different from what is observed in classic accreting black holes and NS. One possibility is that the accretion regime is dissipationless (Bogovalov & Kelner 2005) and the angular momentum is removed by the outflow. It is usually thought that accreting models in high-mass binaries require black holes, but NS with low magnetic field can produce powerful jets (e.g. Sguera *et al.* 2009). The magnetic field can decay in a few million years because of the contamination of the NS's crust by impurities contained in the accreting inflow. If the magnetic field is low, the dissipationless accretion disk can approach to the star to form a magnetic tower that would eject the inflow. The detection of weak quasi-periodic oscillations at X-rays can indicate the presence of a radiatively inefficient disk in these objects. Jets interacting with clouds might also explain rapid flares in gamma-rays, something that is difficult to account for in an extended radiative region as that expected in a colliding wind scenario.

4. Conclusions

Neutron stars in massive binary systems seem, under some specific conditions, to be able to produce non-thermal radiation up to very high energies. The relativistic particles can either be accelerated in the termination shock of a relativistic cold wind or in outflows where shocks might also mediate a process of diffusive acceleration. Opacity constraints and geometry make the non-thermal radiation of these systems periodic over a wide range of energies. The models briefly discussed in this review should be refined in the light of forthcoming multiwavelength observations in order to get a coherent picture for each source.

Acknowledgements

This work has been supported by the Argentine Agency ANPCyT (PICT-2007-00848, BID 1728/OC-AR), CONICET, and the Ministerio de Educación y Ciencia (Spain) under grant AYA 2007-68034-C03-01, FEDER funds.

References

Abdo, A. A. (Fermi coll.) 2009, *ApJ (Lett.)*, in press (arXiv:0907.4307)
Acciari, V. A. (VERITAS coll.) 2009, *ApJ*, 700, 1034
Albert, J. *et al.* (MAGIC coll.) 2006, *Science* 312, 1771
Aharonian, F. A. *et al.* (HESS col.) 2005a, *A&A* 442, 1
Aharonian, F. A. *et al.* (HESS col.) 2005b, *Science* 309, 746
Aharonian, F. A. *et al.* (HESS col.) 2006, *A&A* 460, 743
Berezinskii, V. S., Castagnoli, C. and Galeotti, P. M. 1986, *ApJ* 301, 235
Bogovalov, S. V. & Kelner, S. R. 2005, *Astron. Rep* 49, 57
Bosch-Ramon, V., Paredes, J. M., Romero, G. E., & Ribó, M. 2006, *A&A* 459, L25
Bosch-Ramon, V., Khangulyan, D., & Aharonian, F. A. 2008, *A&A* 489, L21
Casares, J., *et al.*, 2005, *MNRAS* 364, 899
Cheng, K. S. & Ruderman, M. 1989, *ApJ* 337, L77
Dermer, C. D. & Boettcher, M. 2006, *ApJ* 643, 1081
Dhawan, V., Mioduszewski, A., & Rupen, M., 2006, in Proc. of the VI Microquasar Workshop, Como-2006
Dubus, W., 2006, *A&A* 456, 801
Khangulyan, D., Hnatic, S., Aharonian, F.A., & Bogovalov, S. 2007, *MNRAS* 380, 320
Khangulyan, D., Aharonian, F. A., & Bosch-Ramon, V. 2008, *MNRAS* 383, 467
Johnston, S., Ball, L., Wang, N., & Manchester, R. N. 2005, *MNRAS* 358, 1069
Maraschi, L. & Treves, A. 1981, *MNRAS* 194, 1

Orellana, M., Romero, G. E., Pellizza, L. J., & S. Vidrih 2007, *A&A* 465, 703

Paredes, J. M., Martí, J., Ribó, M., & Massi, M., 2000, *Science* 288, 2340

Romero, G. E., *et al.* 2001, *A&A* 376, 599

Romero, G. E., Christiansen, H. R., & Orellana, M., 2005, *ApJ* 632, 1093

Romero, G. E., Okazaki, A. T., Orellana, M., & Owocki, S. P. 2007, *A&A* 474, 15

Sguera, V., *et al.* 2009, *ApJ* 697, 1194

Skilton, J. L., *et al.* 2009, *MNRAS*, in press (arXiv:0906.3411)

Tavani, M. & Arons, J. 1997, *ApJ* 477, 439

Tavani, M., Arons, J. & Kaspi, V. 1994, *ApJ (Lett.)* 433, L37

Highlights of Astronomy, Volume 15
XXVIIth IAU General Assembly, August 2009
Ian F. Corbett

© International Astronomical Union 2010
doi:10.1017/S1743921310008513

Probing fundamental physics with pulsars

Duncan R. Lorimer and Maura A. McLaughlin

Department of Physics, 210 Hodges Hall, Morgantown, WV 26506, USA
email: duncan.lorimer@mail.wvu.edu and maura.mclaughlin@mail.wvu.edu

Abstract. Pulsars provide a wealth of information about General Relativity, the equation of state of superdense matter, relativistic particle acceleration in high magnetic fields, the Galaxy's interstellar medium and magnetic field, stellar and binary evolution, celestial mechanics, planetary physics and even cosmology. The wide variety of physical applications currently being investigated through studies of radio pulsars rely on: (i) finding interesting objects to study via large-scale and targeted surveys; (ii) high-precision timing measurements which exploit their remarkable clock-like stability. We review current surveys and the principles of pulsar timing and highlight progress made in the rotating radio transients, intermittent pulsars, tests of relativity, understanding pulsar evolution, measuring neutron star masses and the pulsar timing array.

Keywords. stars: neutron, gravitation, equation of state

1. Pulsar surveys

A wide variety of successful pulsar surveys are being carried out using most of the major radio telescopes. We summarize the current status and future prospects here.

Parkes The 1.4-GHz multibeam receiver has discovered over 800 pulsars in the past decade (see Lyne 2008). Reanalyses have resulted in a new class of pulsars (McLaughlin *et al.* 2006; see 'Rotating radio transients') and continue to find new pulsars using new techniques (Keith *et al.* 2009). New data acquisition systems provide significantly improved sensitivity to millisecond pulsars (MSPs) and are being used to resurvey the sky.

Effelsberg A similar data acquisition system is now being installed on the seven-beam receiver to provide a complementary survey of the Northern sky.

Arecibo Since 2004, Arecibo has been surveying the Galactic plane with a 1400 MHz seven-beam receiver, and currently 50 pulsars have been discovered (Cordes *et al.* 2006). Among these new discoveries is a double neutron star binary (Lorimer *et al.* 2006) an eccentric binary MSP (Champion *et al.* 2008) and a number of RRATs (Deneva *et al.* 2009). Several hundred pulsars are expected from this survey over the next five years.

Green Bank A 350 MHz receiver has enabled surveys of the Northern Galactic plane (Hessels *et al.* 2008) and a drift scan survey of over 10,000 square degrees (Boyles *et al.* 2008), resulting in a combined 60 new pulsars so far, including five MSPs. The number of pulsars found in these surveys should at least double in the next two years.

Giant Metrewave Radio Telescope Surveys at 610 MHz at low (Joshi *et al.* 2009) and intermediate (Bhattacharya, private communication) latitudes have been successful, but are plagued with severe radio-frequency interference (RFI). A new survey at 327 MHz will cover 1600 square degrees in a better RFI environment, and is expected to detect roughly 250 pulsars and 30 RRATs (McLaughlin, private communication).

2. Principles of pulsar timing

Once a new pulsar is found, it is observed at least once or twice per month over the course of a year. During each observation, pulses from the neutron star traverse the interstellar medium before being received at the radio telescope, where they are dedispersed and added in phase to form a mean pulse profile. The time-of-arrival (TOA) is defined as the time of some fiducial point on the integrated profile. Since the profile has a stable form at any given observing frequency, the TOA can be accurately determined by cross-correlation of the observed profile with a "template" profile obtained from the addition of profiles from many observations at a particular observing frequency.

The TOAs are first corrected to the solar system barycenter. Following the accumulation of a number of TOAs, a simple model is usually sufficient to fit the TOAs during the time span of the observations and to predict the arrival times of subsequent pulses. The model is a Taylor expansion of the rotational frequency $\Omega = 2\pi/P$ about a model value Ω_0 at some reference epoch $\mathcal{T}_0$. Based on this model, and using initial estimates of the position, dispersion measure and pulse period, a "timing residual" is calculated for each TOA as the difference between the observed and predicted pulse phases.

Ideally, the residuals should have a zero mean and be free from systematic trends. To reach this point, the model needs to be refined in a bootstrap fashion. Early residuals show a number of trends indicating an error in one or more of the parameters, or a parameter not yet added to the model. For further details, see Lorimer & Kramer (2005).

3. Highlights from the past few years

3.1. Rotating radio transients

The discovery of the rotating radio transients (RRATs) in a reanalysis of the Parkes multibeam survey data (McLaughlin et al. 2006) demonstrates the wealth of new sources awaiting detection. The radio emission from RRATs is typically only visible for < 1 s per day making them extremely difficult to study. Their detection was made possible by searching for dispersed radio bursts (Cordes & McLaughlin 2003), which often do not show up in conventional Fourier-transform based searches (Lorimer & Kramer 2005).

Since the initial discovery, a significant effort has gone in to searching for and characterizing more RRATs. Nearly 30 are known (Hessels et al. 2008, Deneva et al. 2009, Keane et al. 2009) but only seven have timing solutions, with four of these only recently achieved (McLaughlin et al. 2009). It is clear (Figure 1a), that the RRATs exhibit varied spin-down properties. Recently, Lyne et al. (2009) reported the detection of two glitches in RRAT J1819−1458. While these events are similar in magnitude to the glitches seen in young pulsars and magnetars, they are accompanied by a long-term *decrease* in the spin-down rate, suggesting that it previously occupied the phase space populated by the magnetars. Further observations could confirm this "exhausted magnetar" hypothesis.

McLaughlin et al. (2009) find that the probabilities that the periods and magnetic fields of RRATs and normal pulsars are drawn from the same parent distributions are small ($< 10^{-3}$), with the RRATs having longer periods and higher magnetic fields. This effect appears to be real and not due to a bias against short period objects. The other spin-down derived parameters of normal pulsars and RRATs are consistent.

3.2. Intermittent pulsars

Another new class of radio pulsars, reviewed by Kramer at this meeting, are the intermittent pulsars. The prototype, PSR B1931+24 (Kramer et al. 2006a), shows a quasi-periodic on/off cycle in which the spin-down rate increases by $\sim 50\%$ when the pulsar

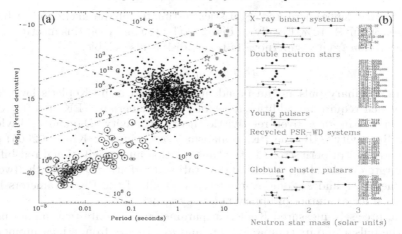

Figure 1. (a) $P - \dot{P}$ diagram of the neutron star population, with radio pulsars as dots, RRATs as open stars, soft gamma-ray repeaters as filled diamonds and anomalous X-ray pulsars as filled squares. Binaries are marked by circles, with ellipticity equal to that of the orbit and J1903+0327 marked in gray. (b) Distribution of neutron star masses as inferred from timing observations of binary pulsars and X-ray binary systems.

is on. A spectral analysis reveals a persistent periodicity that slowly varies with time in the range of 30–40 days. The pulsar switches off in less than 10 s, a timescale too small for precession and indicative of a relaxation-oscillation of unknown nature. The dramatic change in spin-down rate points to a large increase in the magnetospheric particle outflow when the pulsar switches on. The changes allow us to estimate the current density, which coincides with the predictions by Goldreich & Julian (1969) for pulsar magnetospheres and proves, for the first time, that the pulsar wind plays a substantial role in spin down.

Since PSR B1931+24 is only visible for 20% of the time, we estimate that there should be at least 5 times as many similar objects. Timing archives should be carefully mined to find pulsars with similar characteristics. A number of examples have been found, one of which, PSR J1832+0029, switched off in 2004 after 270 days of consistent detections and switched on again in 2006. Due to this apparent long timescale, we have not yet fully sampled the on/off cycles. RW find the implied magnetospheric charge density for PSR J1832+0029 to be almost four times higher than the Goldreich-Julian density! This cannot be easily reconciled by any of the current models (the debris disks of Cordes & Shannon 2006; the death valley scenario of Zhang *et al.* 2007 or the accretion model of Rea *et al.* 2008) and highlights the need for further work.

3.3. *An eccentric millisecond pulsar*

The 2.15-ms pulsar J1903+0327, found in the Arecibo multibeam survey (Champion *et al.* 2009), is distinct from all other Galactic MSPs in that its 95-day orbit has an eccentricity of 0.43! Optical observations show a possible counterpart which is consistent with a 1 $M_\odot$ star. While similar systems have been observed in globular clusters, presumably a result of exchange interactions, the standard hypothesis in which MSPs are "recycled" via the accretion of mass in a low-mass X-ray binary system cannot account for J1903+0327. A triple system now appears to be ruled out due to the lack of any change in orbital eccentricity in the system (Gopakumar *et al.* 2009). It is possible that the binary system was produced in an exchange interaction in a globular cluster and subsequently ejected, or the cluster has since disrupted. Statistical estimates of the likelihood of both these channels are roughly 1–10%. Finally, as discussed by Liu and Li at this meeting, a scenario involving accretion from a supernova fall-back disk is also viable.

3.4. *Tests of relativity*

Although most binary pulsars can be adequately timed using Kepler's laws, there are a number which require an additional set of "post-Keplerian" (PK) parameters which have a distinct form for a given relativistic theory of gravity (Damour & Taylor 1992). In General Relativity (GR) the PK formalism (see Lorimer & Kramer 2005) gives the relativistic advance of periastron, $\dot{\omega}$, the time dilation and gravitational redshift parameter, γ, the rate of orbital decay due to gravitational radiation, $\dot{P}_{\rm b}$, and the two Shapiro delay parameters, r and s. Some combinations, or all, of the PK parameters have now been measured for a number of binary pulsar systems.

Given the precisely measured Keplerian parameters, the only two unknowns are the masses of the pulsar and its companion, $m_{\rm p}$ and $m_{\rm c}$. Hence, from a measurement of just two PK parameters one can solve for the two masses. From the Keplerian mass function one can then find the orbital inclination angle i. If three (or more) PK parameters are measured, the system is "overdetermined" and can be used to test GR (or any other theory of gravity) by comparing the third PK parameter with the predicted value based on the masses determined from the other two.

As discussed by Kramer's contribution in JD06, currently the best binary pulsar system for strong-field GR tests is the double pulsar J0737−3039. In this system, with two independent pulsar clocks, five PK parameters of the 22-ms pulsar "A" have been measured as well as two additional constraints from the mass function and projected semi-major axis of the 2.7-s pulsar "B". The measurement of the projected semi-major axes gives a mass ratio $R = 1.071 \pm 0.001$. The mass ratio measurement is unique to the double pulsar and rests on the assumption that momentum is conserved. The observations of $\dot{\omega}$ and R yield the masses of A and B as $m_{\rm A} = 1.3381 \pm 0.007\,M_\odot$ and $m_{\rm B} = 1.2489 \pm 0.0007\,M_\odot$. From these values, the expected values of γ, $\dot{P}_{\rm b}$, r and s may be calculated and compared with the observed values. These four tests of GR all agree with the theory to within the uncertainties. Currently the tightest constraint is the Shapiro delay parameter s where the observed value is in agreement with GR at the 0.05% level (Kramer *et al.* 2006b).

Another unique feature of the double pulsar system is the interaction between the two pulsars. The signal from A is eclipsed for 30 s each orbit by the magnetosphere of B (Lyne *et al.* 2004) and the radio pulses from B are modulated by the relativistic wind from A during one phases of the orbit (McLaughlin *et al.* 2004). These provide unique insights into plasma physics. By modeling of the change in eclipse profiles of A over four years, Breton *et al.* (2008) fit a simple model to determine the precession of B's spin axis about the orbital angular momentum vector. This measurement agrees, within the 13% measurement uncertainty, with to the GR prediction.

3.5. *Massive neutron stars*

Multiple PK parameters measured for a number of binary pulsars provide precise constraints on neutron star masses (Thorsett & Chakrabarty 1999). As shown in Figure 1b (from Lorimer 2008), the young pulsars and the double neutron star binaries are consistent with, or just below, the canonical $1.4\,M_\odot$, but the MSPs in binary systems have, on average, significantly larger masses.

Several eccentric binary systems in globular clusters have their masses constrained from measurements of the relativistic advance of periastron and the Keplerian mass function. In these cases, the condition $\sin i < 1$ sets a lower limit on the companion mass

$m_c > (f_{\mathrm{mass}} M^2)^{1/3}$ and an upper limit on the pulsar mass. Probability density functions for both m_p and m_c can also be estimated in a statistical sense by *assuming* a random distribution of orbital inclinations. An example is the eccentric binary MSP in M5 (Freire *et al.* 2008) where the nominal pulsar mass is 2.08 ± 0.19 M$_\odot$. If these can be confirmed by the measurement of other relativistic parameters, these supermassive neutron stars will have important constraints on the equation of state of superdense matter.

Currently the largest measurement of a radio pulsar mass is the eccentric MSP binary J1903+0327 (Champion et al. 2008). Recent timing measurements of the relativistic periastron advance and Shapiro delay in this system by Freire *et al.* (2009) yield a mass of 1.67 ± 0.01 M$_\odot$. When placed on the mass–radius diagram for neutron stars (Lattimer & Prakash 2007) this pulsar appears to be incompatible with at least four equations of state. Optical measurements of the companion are required to rule out classical contributions to $\dot{\omega}$, and further timing measurements are required to verify this result.

3.6. *Pulsar timing and gravitational wave detection*

As discussed by Andersson at this meeting, the direct detection of GWs is one of the foremost goals of modern physics. Many cosmological models predict that the Universe is presently filled with an ultra low-frequency (nHz) stochastic gravitational wave (GW) background produced during the big bang era (Peebles 1993). A significant component (Jaffe & Backer 2003) is the gravitational radiation from massive black hole mergers due to Galaxy collisions at a redshift $z \sim 1$. Pulsars can be used as natural GW detectors of this background (Sazhin 1978; Detweiler 1979). pulsar acts as a reference clock, sending out regular signals which are monitored by an observer on the Earth over some time-scale T. Passing GWs perturb the local spacetime and cause a change in the observed rotational frequency of the pulsar. For regular pulsar timing observations with typical TOA uncertainties of ϵ_{TOA}, this detector would be sensitive to GWs with amplitudes $h \gtrsim \epsilon_{\mathrm{TOA}}/T$ and frequencies $f \sim 1/T$ (Bertotti *et al.* 1983; Blandford *et al.* 1984).

A natural extension of this concept is a "timing array" of a number of pulsars distributed over the sky (Hellings & Downs 1983), allowing cross-correlation of the residuals for pairs of pulsars (Foster & Backer 1990). It should therefore be possible to separate the timing noise of each pulsar from the common signature of the quadrupolar GW background from the effects of clock errors (which have a monopolar signature) and solar system ephemeris errors (which have dipolar signature).

As reviewed by Manchester in JD06, three main groups are collaborating to form an international pulsar timing array. The European Pulsar Timing Array consists of four radio telescopes which will be combined to produce a 300-m class telescope. In Australia, the Parkes Pulsar Timing Array uses the 64-m Parkes telescope to time Southern pulsars. In North America, the Green Bank and Arecibo telescopes are used in the NANOGrav collaboration. The best existing limits (Jenet *et al.* 2006) constrain the merger rate of supermassive black hole binaries at high redshift, investigate inflationary parameters and place limits on the tension of currently proposed cosmic string scenarios.

Based on current expectations of the likely strength of the GWB, a detection requires timing of 20 MSPs with 100 ns residuals for a period of five years (Jenet *et al.* 2005). In general, for residuals δt, data span T, and number of pulsars N, the sensitivity scales roughly as $\delta t^2/(NT^4)$ (Kaspi et al. 1994). Of the ~ 30 MSPs that are regularly timed internationally, five have achieved a timing precision of 100 ns or less. With improved algorithms and more sensitive observations, such precision may soon be achieved for 5–10 more. A key goal of the ongoing surveys is to find more MSPs to add to the array. This should be accomplished by the various surveys outlined in Section 1, and GW detection may be achievable within the next 5–10 years.

Acknowledgements

We thank the AAS for travel support. Our research is supported by the National Science Foundation, the Research Corporation for Scientific Advancement, West Virginia EPSCoR and the Alfred P. Sloan foundation.

References

Bertotti, B., Carr, B. J., & Rees, M. J. 1983, MNRAS, 203, 945
Blandford, R. D., Narayan, R., & Romani, R. W. 1984, J. Astrophys. Astr., 5, 369
Boyles, J., et al. 2008, BAAS, 40, 208
Breton, R., et al. 2008, Science, 321, 104
Champion, D. J., et al. 2008, Science, 324, 1411
Cordes, J. M. et al. 2006, ApJ, 637, 446
Cordes, J. M. & McLaughlin, M. A. 2003, ApJ, 596, 1142
Cordes, J. M. & Shannon, R. 2006, ApJ, 682, 1152
Damour, T. & Taylor, J. H. 1992, Phys. Rev. D, 45, 1840
Deneva, J.., et al. 2009, ApJ, in press (arXiv0811.2532)
Detweiler, S. 1979, ApJ, 234, 1100
Foster, R. S. & Backer, D. C. 1990, ApJ, 361, 300
Freire, P., Wolszczan, A., van den Berg, M., & Hessels, J. 2008, ApJ, 679, 1433
Freire, P. 2009, in proceedings of "Neutron Stars and Gamma-Ray Bursts" meeting
    (arXiv0907.3219)
Goldreich, P., Julian, 1969, ApJ, 157, 869
Gopakumar, A., Magchi, M., & Ray, A. 2009, MNRAS, in press (arXiv0908.0974)
Hellings, R. W. & Downs, G. S. 1983, ApJ, 265, L39
Hessels, J. et al., 2008, AIP Conference Proceedings, 983, 613
Jaffe, A. H. & Backer, D. C. 2003, ApJ, 583, 616
Jenet, F. A. et al. 2005, ApJ, 625, 123
Jenet, F. A. et al. 2006, ApJ, 653, 1571
Joshi, B. et al. 2009, MNRAS, in press (arXiv0906.0228)
Kaspi, V. M., Taylor, J. H., & Ryba, M. 1994, ApJ, 428, 713
Keane, E. F. 2009, MNRAS, submitted
Keith, M. et al. 2009, MNRAS, 395, 837
Kramer, M. et al. 2006a, Science, 312, 549
Kramer, M. et al. 2006b, Science, 314, 97
Lattimer, J. & Prakash, M. 2007, Physics Reports, 442, 109
Lorimer, D. R. 2008, Living Rev. Relativity, 11
Lorimer, D. R. & Kramer, M. 2005, Handbook of Pulsar Astronomy (Cambridge University
    Press)
Lorimer, D. R. et al. 2006, ApJ, 640, 428
Lyne, A. G. et al. 2004, Science, 303, 1153
Lyne, A. G. 2008, AIP Conference Proceedings, 983, 561
Lyne, A. G. et al. 2009, MNRAS, submitted.
McLaughlin, M. A. et al. 2004, ApJ, 616, L131
McLaughlin, M. A. et al. 2006, Nature, 439, 817
McLaughlin, M. A. 2009, in "Neutron Stars and Pulsars" W. Becker, ed, Astrophys & Space
    Science Library, 357, 41
McLaughlin, M. A. et al. 2009, MNRAS, in press.
Peebles, P. J. E. 1993, Principles of Physical Cosmology (New Jersey: Princeton)
Rea, N., et al., 2008, MNRAS, 391, 663
Sazhin, M. V. 1978, Sov. Astron., 22, 36
Stairs, I. H., Thorsett, S. E., & Arzoumanian, Z. 2004, Phys. Rev. Lett., 93, 141101
Thorsett, S. E. & Chakrabarty, D. 1999, ApJ, 512, 288
Zhang, B., Gil, J., & Dyks, J. 2007, MNRAS, 374, 1103

Highlights of Astronomy, Volume 15
XXVIIth IAU General Assembly, August 2009
Ian F. Corbett

© International Astronomical Union 2010
doi:10.1017/S1743921310008525

Gravitational waves from neutron stars

Nils Andersson

School of Mathematics, University of Southampton, Southampton SO17 1BJ, UK
email: na@maths.soton.ac.uk

Abstract. In this presentation, I will outline some of the different ways that neutron stars can generate gravitational waves, discuss recent improvements in modeling the relevant scenarios in the context of improving detector sensitivity, and show how observations are beginning to put interesting "constraints" on our theoretical models.

Keywords. stars: neutron, gravitational waves, stars: oscillations, pulsars: general

1. Context

Neutron stars are cosmic laboratories of exotic and exciting physics. With a mass of more than that of the Sun compressed inside a radius of about 10 kilometers, their density reaches beyond that of nuclear saturation. In essence, our understanding of these extreme circumstances requires physics that cannot be tested in terrestrial laboratories. Instead, we must try to use astrophysical observations to constrain our various theoretical models. We already have a wealth of data from radio, X-ray and gamma-ray observations, providing evidence of an incredibly rich phenomenology. We have learned that neutron stars appear in many different guises, from radio pulsars and magnetars to accreting millisecond pulsars, radio transients and intermittent pulsars. Our models for these systems remain rather basic, despite 40 years of attempts to understand the pulsar emission mechanism, glitches, accreting systems etcetera.

In the next few years we expect "gravitational-wave astronomy" to become reality. This is an exciting prospect, because gravitational-wave (GW) observations should be able to probe several key aspects of neutron star physics. Neutron stars can radiate in a number of ways. Relevant scenarios include; inspiraling binaries, rotating deformed stars, oscillations and instabilities. Modeling these different scenarios is, however, not that easy since our understanding of neutron stars relies on physics that is far from well known. To make progress we must combine supranuclear physics (the elusive equation of state) with magnetohydrodynamics, the crust elasticity, a description of superfluids/superconductors and potentially exotic phases of matter like a deconfined quark-gluon plasma or hyperonic matter. Moreover, in order to be quantitatively accurate, all models have to account for relativistic gravity.

In the last few years the first generation of large-scale interferometric GW detectors (LIGO, GEO600 and VIRGO) have reached design sensitivity in a broad frequency window (Abbott *et al* (2009)). A full years worth of high quality data was taken during the LIGO S5 run. Even though there has not yet been a detection, the experiment has already provided interesting information. At the time of writing, the LIGO detectors are running in an enhanced configuration (with sensitivity improved by about a factor of 2 compared to the S5 run). In the next five year period, they will be upgraded using more advanced technology. Once this upgrade is complete, around 2015, the second generation of ground-based detectors will reach the level of sensitivity where the first detection can be expected. Meanwhile, the discussion of third generation (3G) detectors has begun in earnest with the EU funded Einstein Telescope (ET) design study. The aim for 3G detectors is to improve the broadband sensitivity by (roughly) another order of magnitude.

2. Binary inspiral

The late stage of inspiral of a binary system provides an excellent GW source (Sathyaprakash & Schutz (2009)). As the binary orbit shrinks due to the energy lost to radiation, the GW amplitude rises and the frequency increases as well. This inspiral chirp is advantageous for the observer in many ways. First of all, it is well modeled by post-Newtonian methods and does not depend (much) on the actual physics of the compact objects involved. In fact, much of the signal is adequately described by a point-mass approximation. A key fact that makes binary systems attractive is that the amplitude of the signal is "calibrated" by the two masses (from observations one would expect to be able to infer the individual masses, the spin rates of the objects and the distance to the source). The only uncertainty concerns the event rate for inspirals in a given volume of space. Given this, it is natural to discuss the detectability of these systems in terms of the "horizon" distance d_h at which a given binary signal would be observable with a given detector. Let us assume that detection requires a signal-to-noise ratio of 8, and focus on equal mass neutron star binaries (each star has mass 1.4 $M_\odot$). For such systems, d_h would be 30 Mpc for LIGO S5. In this volume of space one would expect one event every 25-400 yrs. Advanced LIGO should improve this to d_h = 30 Mpc, and could see 2-40 events per year, while ET may reach d_h = 3 Gpc and could potentially observe thousands of events.

From these estimates we, first of all, see why it would have been surprising if a binary signal had been found in the S5 data. Given even the most optimistic rate estimate from population synthesis models, these events would be rare in the currently observable volume of space. The situation changes considerably with Advanced LIGO. Based on our current understanding, one would expect neutron star binaries to be seen once the detectors reach this level of sensitivity. However, it is also clear that if the most pessimistic rate estimates are correct, then we will not be able to gather a statistically significant sample of signals. Most likely, we will need detectors like ET to study populations.

3G detectors will likely also be required if we want to study the final stages of inspiral, including the actual merger. This is a very interesting phase of the evolution given that the merger will lead to the formation of a hot compact remnant with violent dynamics. It may also trigger a long gamma-ray burst. Most of this dynamics radiates at relatively high frequencies. Tidal disruption occurs above 600 Hz or so and the oscillations of the remnant could lead to a signal at several kHz. However, the merger signal should be rich in information. In particular, the ringdown should tell us directly whether a massive neutron star or a black hole was formed. Roughly, if the inspiral phase is observable with Advanced LIGO then ET should be able to detect the merger. In other words, the development of 3G detectors is essential if we want to study these events.

3. Rotating deformed neutron stars

GWs are generated by asymmetric dynamics. The source could be violent, like a supernova or a binary merger, or slowly evolving, like the binary inspiral. Asymmetries, either in the crust or the magnetic field, are expected to slowly leak rotational energy away from a spinning neutron star. Such sources would be the GW analogue of radio pulsars, radiating at twice the spin frequency. On the one hand, rotating neutron stars will emit low amplitude GWs, but on the other hand, they radiate continuously. Moreover, we have many potential target sources with known frequency and position. This means that observers can carry out a targeted search for known radio and X-ray pulsars.

A key question concerns what level of asymmetry one would expect a neutron star to have. This is a complicated problem, where the answer depends not only on the properties of the star, but also on it's evolutionary history. So far, modeling has mainly focused on establishing what the largest possible neutron star "mountain" would be (Haskell(2008)). Expressing the result in terms of a (quadrupole) ellipticity, theory suggests that $\epsilon < 2 \times 10^{-5} (u_{\text{break}}/0.1)$. Recent molecular dynamics simulations suggest that the breaking strain $u_{\text{break}} \approx 0.1$, much larger than had been anticipated. In comparison, terrestrial materials have $u_{\text{break}} \approx 10^{-4} - 10^{-2}$, so the neutron crust would seem to be super-strong!

Observations of targeted radio pulsars are already providing interesting results. The strongest constraint set by LIGO is $\epsilon < 7 \times 10^{-7}$ for J2124-3358 (based on 1 month of S3/4 data). Thus we know that this relatively fast spinning pulsar is far from maximally deformed. An observational milestone was later reached when S5 data was used to beat Crab pulsar "spin-down limit" by a factor of 4 or so. This result shows that GW emission does not dominate the spin-down of these systems. Although this was already "known" from the pulsar's braking index, it is clear that the observations are beginning to produce astrophysically relevant results.

It is quite easy to estimate how these results are likely to improve in the future since the effective amplitude of a periodic signal increases as the square root of the observation time. In the case of J2124-3358 one would expect analysis of the S5 data (with a factor 2 improved sensitivity, and a full year of data) to improve the constraint to $\epsilon < 10^{-7}$. Advanced LIGO, with an order of magnitude better sensitivity, but still a one year integration, should reach $\epsilon < 10^{-8}$, and ET may push the limit as far as $\epsilon < 10^{-9}$. At this point, the deformation of the star would be constrained to the micron level. One would probably expect a signal to be detected before this level is reached, but we do not know this for sure. The main issue concerns the generation mechanism for deformations. Why should the neutron star be deformed in the first place? This is an urgent problem that needs to be addressed by theorists. As far as evolutionary scenarios are concerned, accreting neutron stars in low-mass X-ray binaries have attracted the most attention (Watts & Krishnan (2009)). This is natural for a number of reasons. First of all, the currently observed spin-distribution in these systems seems consistent with the presence of a mechanism that halts the spin-up due to accretion well before the neutron star reaches the break-up limit. GW emission could provide a balancing torque. The required deformation is certainly smaller than the allowed upper limit, and in an accreting system it would be quite natural for asymmetries to develop. However, accreting systems are very messy, and we do not understand the various torques very well.

4. Oscillations and instabilities

In principle, a promising strategy for constraining neutron star physics involves observing the various modes of oscillation - Andersson (2003). Neutron stars have rich oscillation spectra, with different families of modes more or less directly associated with different core physics. The fundamental f-mode (which should be the most efficient GW emitter) scales with average density, while the pressure p-modes depend on the sound speed and the gravity g-modes are sensitive to thermal/composition gradients. In a rotating star, the inertial r-modes are restored by the Coriolis force. The r-modes are particularly interesting because they may be driven unstable by the emission of GWs.

The r-mode instability window depends on a balance between GW driving and various dissipation mechanisms. This provides a sensitive probe of the core physics. Because of this, the r-mode has been studied in a variety of contexts in the last decade. We now know that important issues concern the damping due to the vortex mediated mutual friction in a superfluid, the boundary layer at the crust-core interface and exotic bulk viscosity due to the presence of hyperons or deconfined quarks in the deep neutron star core. These problems are all very challenging. In addition, we need to be model the GW signal from an unstable r-mode. This is also difficult because, even though the r-mode growth phase is adequately described by linear theory, nonlinear effects soon become important. Detailed studies show that the instability saturates at a low amplitude due to coupling to other inertial modes. The subsequent evolution is very complex.

To model a truly realistic oscillating neutron star may be difficult, but the potential reward is considerable. This is clear from recent results for the quasiperiodic oscillations seen in the tails of magnetar flares. These oscillations have been interpreted as torsional oscillations of the neutron star crust. If this is correct, then we are already doing neutron star asteroseismology! The current models are perhaps not that precise, but they should motivate us to improve our understanding of the key physics (like the interior magnetic field and the dynamical coupling between the crust and the core). The magnetar events may also generate GWs. Having said that, LIGO found no signal from the 27/12 2004 event in SGR 1806-20. This is perhaps not surprising because pure crust oscillations do not generate strong GWs (due to the low density involved). The situation would change if the dense core were involved in the oscillation, but it is not yet clear to what extent this is the case.

5. Summary and future challenges

GW astronomy promises to provide insights into the "dark side" of the Universe. Because of their high density, neutron stars are ideal GW sources and we hope to be able to probe the extreme physics of their interiors. The potential for this is clear, in particular with 3G detectors like the ET. However, in order to detect the signals and extract as much information as possible, we need to improve our theoretical models considerably.

For binary inspirals, we need to work out when finite size effects begin to affect the evolution. We need to consider tidal resonances and compressibility in detail and ask to what extent they affect the late stages of inspiral. For hot young remnants, resulting from binary mergers or core collapse, we need to refine our large scale numerical simulations. The simulations must use "realistic" equations of state, and consider composition, heat/neutrino cooling and magnetic fields with as few "cheats" as possible. In parallel, we need improve our current understanding of neutron star oscillations and instabilities. This effort should aim at accounting for as much of the interior physics as possible. Finally, we need a clearer phenomenological understanding of pulsar glitches, accreting neutron stars, magnetar flares etcetera. These are ambitious targets, but there is no reason why we should not make good progress in the next few years. Eventually, future observations (gravitational and electromagnetic) will undoubtedly help us understand many of the aspects of neutron star physics that seem mysterious to us today.

References

Abbott, B. P., *et al* 2009, *Rep. Prog. Phys.* 72, 076901
Andersson, N. 2003, *Class. Quantum Grav.* 20, R105
Haskell, B. 2008, *Class. Quantum Grav.* 25, 114049
Sathyaprakash, B. S. & Schutz, B. F. 2009, *Liv. Rev. Relativity*, vol. 12, no. 2
Watts, A. L. & Krishnan, B. 2009 *Adv. Sp. Res.* 43, 1049

Highlights of Astronomy, Volume 15
XXVIIth IAU General Assembly, August 2009
Ian F. Corbett, ed.

© International Astronomical Union 2010
doi:10.1017/S1743921310008537

JD4 – Progress in Understanding the Physics of Ap and Related Stars

Preface

Chemically peculiar (CP) stars exhibit, simultaneously, a wide variety of physical phenomena, including diffusion, convection, magnetism, and pulsation. Thus, progress in the understanding of these objects requires the input of researchers from a variety of research fields within stellar astrophysics. The General Assembly of the IAU, in Rio de Janeiro, provided an excellent opportunity to discuss challenging new results faced in CP star research and improve the exchange of information and cooperation with experts of neighbouring scientific fields.

The Joint Discussion, entitled "Progress in understanding the physics of Ap and related stars", of 1.5 days in length, was organized by the Working Group on Ap and related stars. It comprised three sessions, each with a number of review talks along with short oral contributions. During the Joint Discussion a variety of important topics were discussed, such as the optimal exploitation of current and future instrument capabilities in the context of CP star research, the modelling of the complex atmospheres of peculiar stars and a look at CP and related phenomena in the context of Stellar Evolution. Reviews of the science highlights presented during each of the Joint Discussion's sessions are presented in these proceedings.

One important aspect of the scientific programme was the discussion that followed each session. During these sessions, the participants addressed questions such as "To what degree may complementary techniques contribute to the understanding of the CP phenomenon?", "Atomic data: what do we need as input? How confidently can we derive Global parameters of CP stars?" and "What is the key to the diversity of CP phenomena? The roles of magnetic fields, rotation, diffusion, birth environment, and binarity.". We are very grateful to Laurent Eyer, Glenn Wahlgren and Nikolai Piskunov for organizing these most fruitful discussion sessions. Similarly, we would like to thank all of those that contributed to the organization of this Joint Discussion, and all speakers and poster presenters that made the Joint Discussion possible.

Margarida S. Cunha, Chair SOC
December, 2009

Highlights of Astronomy Volume 15
XXVIIth IAU General Assembly, August 2009
Ian F. Corbett, ed.

© International Astronomical Union 2010
doi:10.1017/S1743921310008549

Pushing the limit of instrument capabilities

Denis V. Shulyak[1], Werner W. Weiss[2], Gautier Mathys[3], Laurent Eyer[4], Alexander F. Kholtygin[5], Oleg Kochukhov[6], Pierre North[7], Sergey N. Fabrika[8] & Tatyana E. Burlakova[8]

[1]Institute of Astrophysics, Georg-August-University,
Friedrich-Hund-Platz 1, D-37077 Göttingen, Germany
email: denis@phys.uni-goettingen.de

[2]Institut für Astronomie, Universität Wien,
Türkenschanzstraße 17, 1180 Wien, Austria
email: werner.weiss@univie.ac.at

[3]European Southern Observatory,
Casilla 19001, Santiago 19, Chile
email: gmathys@eso.org

[4]Observatoire de Genève,
51 ch. des Maillettes, CH-1290 Sauverny, Switzerland
email: Laurent.Eyer@unige.ch

[5]Astronomical Institute, St. Petersburg State University,
Universitetskii pr. 28, St. Petersburg, 198504, Russia
email: afkholtygin@gmail.com

[6]Department of Physics and Astronomy, Uppsala University,
Box 515, 751 20, Uppsala, Sweden
email: Oleg.Kochukhov@fysast.uu.se

[7]Ecole Polytechnique Fédérale de Lausanne,
1015 Lausanne, Switzerland
email: pierre.north@epfl.ch

[8]Special Astrophysical Observatory, Russian Academy of Sciences,
Nizhnii Arkhyz, Karachai Cherkess Republic, 369167, Russia

Abstract. Chemically Peculiar (CP) stars have been the subject of systematic research for more than 50 years. With the discovery of pulsation of some of the cool CP stars, and the availability of advanced spectropolarimetric instrumentation and high signal-to-noise, high resolution spectroscopy, a new era of CP star research emerged about 20 years ago. Together with the success in ground-based observations, new space projects are developed that will greatly benefit future investigations of these unique objects. In this contribution we will give an overview of some interesting results obtained recently from ground-based observations and discuss the future outstanding Gaia space mission and its impact on CP star research.

Keywords. stars: chemically peculiar, stars: atmospheres, stars: variables: roAp, stars: oscillations, space vehicles: instruments, Gaia mission

---

1. Brief overview on CP stars

Back in 1897, more than one century ago, the first peculiar stars were found in the course of the Henry Draper Memorial classification work at Harvard by Antonia Maury and Annie Cannon. Maury used the designation "peculiar" for the first time to describe spectral features in the remarks to the spectrum of α^2 CVn, Pickering & Maury, 1897, marking a first attempt for two-dimensional classification system considering the strength and the width of the spectral lines.

In 1974 Preston proposed the division of main-sequence CP stars into four groups according to their spectroscopic characteristics: Preston, 1974: CP1 (Am/Fm stars), CP2 (Si, SrCrEu stars), CP3 (HgMn stars), CP4 (He-weak stars). More detailed spectroscopic consideration of CP stars required introduction of new subtypes of CP stars, such as He-rich and λ Boo stars. CP2 stars, including Bp/Ap, host strong surface magnetic fields (Lantz, 1993) that are likely stable on large time-intervals (North, 1983).

Abundance peculiarities were measured using the curve-of-growth method based on simple assumptions about formation of absorption lines (models of Schuster-Schwarzschild, Unsöld, Milne-Eddington). Since that time and with development of new high-resolution, high signal-to-noise CCD based spectrometers, big progress has been made in abundance analysis of CP stars, revealing the presence of vertical, Ryabchikova *et al.*, 2006, Ryabchikova *et al.*, 2005, Ryabchikova *et al.*, 2002, Kochukhov *et al.*, 2006, and horizontal, Khokhlova *et al.*, 1997, Kochukhov *et al.*, 2004, Lüftinger *et al.*, 2009, elements separation in their atmospheres caused by the processes of microscopic particle diffusion, Michaud, 1970.

The discovery of strong stellar surface magnetic fields Babcock, 1958 opened a new research field in astrophysics – stellar magnetism. A 200 Gauss accuracy of the magnetic field detection usually obtained with photographic plates has increased to ≈ 1 Gauss with modern spectropolarimetry and new techniques (such as Least Square Deconvolution, or LSD, for example) - Donati *et al.*, 1997, Wade *et al.*, 2000.

With the discovery by D. Kurtz (Kurtz, 1978) of a 12 min pulsation period in HD 101065 a subgroup of the cool CP stars, the so-called rapidly oscillating Ap (roAp) stars, became extremely promising targets for asteroseismology, a most powerful tool for testing theories of stellar structure. Driving of the oscillations results from a subtle energy balance depending directly on the interaction between the magnetic field, convection, pulsations, and atomic diffusion. Amazing insights in the 3-D structure of stellar atmospheres became available; see, for example, Kochukhov, 2006, Freyhammer *et al.*, 2009.

2. Selected results from recent ground-based observations of CP stars

Current ground-based observations of CP stars reveal many interesting findings that require new modelling approaches and interpretations.

Accurate high-resolution observations are needed to understand the properties and origin of so-called hyper-velocity stars (HVS). These are B-type stars with peculiar galactic rest frame velocity and enhanced α elements and normal (solar) Fe abundances in their atmospheres. The origin of these stars is not well understood: one hypothesis is that they originate from the dynamical interaction of binary stars with the supermassive black hole in the Galactic Centre (GC), which accelerates one component of the binary to beyond the Galactic escape velocity. So far, however, no HVS has been unambiguously related to a GC origin. Determination of the place of ejection of a HVS requires the determination of space motion (accurate proper motions) and chemical composition, however, the later is hard to use to constrain their origin; see Przybilla *et al.*, 2008. On the other hand, if GC origin can be proved by such future astrometric missions like Gaia (see below in this contribution) then it can bring important constraints on magnetic fields in GC region and on the formation and evolution of CP stars in general.

An important issue is raised by rotational braking observed recently for magnetic Bp star HD 37776. It was shown that this star increased its rotational period by 17.7s over the past 31 years (Mikulášek *et al.*, 2008). This can not be explained by light-time effect caused, for example, by the presence of a secondary companion which is not observed in radial velocity measurements. Also, the hypothesis of free-body precession due to magnetic distortion is incompatible with light curve shapes unchanged in 31 years. The plausible scenario left is a continuous momentum loss due to magnetic braking that requires magnetically confined stellar wind, which naturally can be present in HD 37776.

Interferometry is becoming a very powerful tool in modern astrophysics since it allows for direct and model independent measurement of sizes of stellar objects. Recently, there was a report of the first detailed interferometric study of a roAp star α Cir (Bruntt *et al.*, 2008). The authors used observations from the Sydney University Stellar Interferometer (SUSI) and additional data from visual and UV observations calibrated to absolute units to accurately derive $T_{\rm eff}$ and $\log(g)$ of the star. Even thought the interferometry can provide us with accurate radii of stars, the accuracy of $T_{\rm eff}$ determination is still limited by the incompleteness of observations in all spectral ranges (that would give a value of the bolometric flux) and model atmospheres used; see Kochukhov *et al.*, 2009. Application of interferometric techniques to another Ap star β CrB is reported in this conference (JD04).

Interesting results were obtained for the well known Ap star ϵ UMa which appeared to host a brown dwarf companion. This star shows substantial variations of radial velocities measured from different spectral lines (Woszczyk & Jasinski, 1980) which allowed to derive the parameters of the system and to infer the mass of the secondary component of $M_2 = 14.7 M_{\rm Jupiter}$ (Sokolov, 2008).

There have been an extensive discussions over past years about the LiI 6708Å line which is usually used for the determination of Li abundance in magnetic Ap stars. However, due to unknown and yet unconfirmed blending with possible lines of rare-earth elements (REE) it was not definitely clear if the abundance determination suffer from systematic uncertainties or not. Recently Kochukhov, 2008 confirmed Li identification in magnetic Ap stars using detailed calculations of Zeeman pattern in Paschen-Back regime. No correlation between Li and REE line strengths was found thus ruling out the suspicion that the observed feature is due to an unidentified REE line. However the origin of Li still has to be explained theoretically by diffusion calculations and/or abundance spots on stellar surface.

High-resolution, high signal-to-noise observations allow to study not only element stratification in atmospheres of CP stars with great detail, but also the separation of different isotopes of a given element as demonstrated by Cowley *et al.*, 2009 who studied ^{40}Ca/^{48}Ca isotopic anomaly in atmosphere of selected Ap stars. This was also investigated in Ryabchikova *et al.*, 2008 who analysed calcium isotopes stratification profiles in three stars 10 Aql, HR 1217 and HD 122970 concluding that the heavy isotope concentrated towards the higher layers. Interestingly, they found no correlation in ^{48}Ca excess in atmospheres of roAp and noAp stars with the magnetic field strength.

New results of searching for the line profile variability (LPV) in the spectra of OB stars have been recently reported based on observations made with the 1.8-m telescope of Korean Bohyunsan Optical Astronomical Observatory and 6-m telescopes of Special Astrophysical Observatory, Russia. For all program stars regular and often coherent LPV was reported for all spectral lines, as shown in Fig. 1 (right panel) for the B1 supergiant ρ Leo. This coherence is connected with the presence of the stellar magnetic field. The moderate dipole magnetic field of ρ Leo with the polar field strength $B_p \approx 250$ G was detected by Kholtygin *et al.*, 2007.

Together with the regular LPV the numerous local details of line profiles in spectra of a few program stars are detected and are connected with the formation and destruction of the small-scale structures (clumps or clouds) in the stellar wind. The evidence that numerous clumps exist in the winds of the O6 star λ Ori A and O9.5 star δ Ori A was

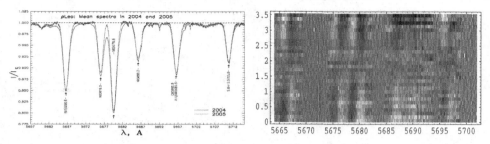

Figure 1. Left Panel: Mean spectra of ρ Leo in the spectral region $5667 - 5715\,\text{Å}$ in 2004 and 2005. **Right Panel**: *dynamical* spectra of LPV in the same region.

found. The *dynamical wavelet spectra* of LPV technique is used by Kholtygin, 2008 to determine the distribution of the line fluxes for the clump ensemble in the winds of these stars.

Many new interesting results have been obtained in the past couple of years from ground-based observations of CP stars, that lead to refinements of "traditional" analyses. Based on observations of ever increasing quality we see effects that had not been previously detected. Definitely, "new" observational techniques (such as interferometry) have a large potential. Additional observational constraints will force us to re-think some of our "well-established" views on the nature, origin and evolution of CP stars bringing bridges between them and some topics of present-day astrophysics (galactic center, lowest-mass stars, etc.).

3. Gaia space mission and CP stars

In parallel to the progress in ground-based observations, modern space missions open a number of possibilities for CP stars' research not only in the solar neighbourhood, but also much further beyond. Scheduled to be launched in 2012, Gaia is one of the cornerstone astrometric missions of the European Space Agency (ESA) which will allow the study of the general properties and characteristics of a huge sample of stars of all spectral types. Gaia will measure position, distances, space motions, radial velocities and fundamental parameters of about 1 billion(!) objects. This scanning satellite will observe every single object (galaxies, quasars, solar system objects, etc.) to 20th magnitude, thus providing important scientific constraints for nearly all fields of modern astrophysics. With the expected astrometric accuracy of about $7\mu as$ at 10th magnitude, Gaia's precision is more than 100 times higher than those of previously developed missions like Tycho and Hipparcos. Gaia has two telescopes with two viewing directions separated by $106.5°$: each telescope has a SiC primary mirror $1.45 \times 0.5\,\text{m}^2$ and 35 m focal length. The key idea of Gaia design is that the images from both telescopes are combined in one focal plane which hosts a number of CCD detectors with the total size of 4500×1966 pixels. Further detailed information can be retrieved from the official Gaia website (http://www.rssd.esa.int/index.php?project=GAIA&page=index).

Although Gaia is primarily an astrometric mission, it will also provide us with broadband spectrophotometric data in a wide spectral range. This is done by two photometers called Blue Photometer (BP) and Red Photometer (RP) with the working wavelength ranges of $330 - 680$ nm and $640 - 1000$ nm respectively. In addition, Gaia will be armed with a spectrograph for the radial velocity measurements (RVS) which operates in a narrow spectral range of $847 - 874$ nm (around CaII infrared triplet) with the resolution $R = 11500$. Every detected object will automatically pass through BP/RP and then RVS CCD's, and this opens a wide range of possibilities for stellar studies. Indeed, RP/BP

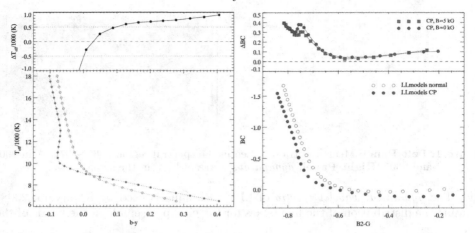

Figure 2. *Left panel*: theoretical T_{eff} as a function of Strömgren $b - y$ index for normal (open circles) and CP (filled circles) stars. *Right panel*: same but for bolometric correction BC as a function of $B_2 - G$ Genéva index. Upper subplot shows difference in BC of non-magnetic CP and 5 kG magnetic CP models with respect to models computed with solar abundances.

photometry and RVS spectroscopy will ideally allow the determination of such important parameters as T_{eff}, $\log(g)$, and metallicity based on specific calibrations and sophisticated algorithms that are presently developed: the final catalog of Gaia observations is scheduled for 2020, after five years of operation and three years of data reduction phase.

Among a billion objects, there will be a certain fraction of CP stars observed. Indeed, depending on spectral classes, $10 - 30\%$ of main-sequence B-F stars are CP, and at A0 all slowly rotating stars are CP. The determination of the position of these stars, their parameters, binarity, class of peculiarity etc., would greatly benefit our understanding of CP phenomena in general and increase the number of potentially interesting objects for detailed research in particular. However, we are faced a difficulty that for stars with moderate or strong peculiarities the standard temperature indicators are inadequate and blind application of the usual photometric calibrations may lead up to $500 - 1000$ K errors in T_{eff} determination; see, for example, Shulyak *et al.*, 2008. Generally, there are two reasons why the energy distributions of CP stars differ from those of normal: inhomogeneous horizontal and vertical elements distribution as well as the presence of strong surface magnetic fields. All these modify atmospheric structure and thus lead to abnormal photometric parameters observed for CP stars. As an example, the impact of peculiar opacity on some photometric parameters is shown in Fig. 2, which illustrates theoretically predicted behaviour of some parameters. Theoretical models were computed taking into account characteristic temperature behaviour of abundances of CP stars as derived, for example, in Ryabchikova *et al.*, 2004.

Thus, in order to distinguish between normal and CP stars as observed by Gaia, as well as to derive the type of peculiarity for individual objects, a parametrization of CP stars is needed. This, in turn, requires dedicated model atmospheres to predict their observed parameters. In our investigations we employ the LLMODELS stellar model atmosphere code to compute extensive libraries of high resolution stellar fluxes that are now used for preliminary analysis of Gaia simulated data. LLMODELS is 1-D, LTE model atmosphere code (Shulyak *et al.*, 2004) that treats the bound-bound opacity by direct, line-by-line spectrum synthesis with the fine frequency spacing ($10^5 - 10^6$ points, 10^7 in parallel mode) thus resolving individual spectral lines. The code does not use any precalculated opacity tables and no assumptions are made about the depth-dependence of the line

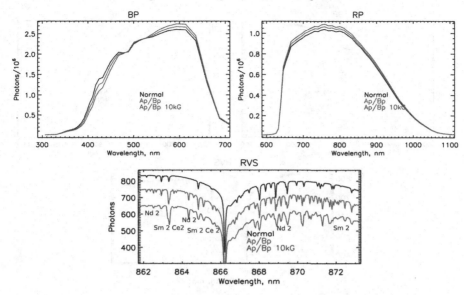

Figure 3. Gaia simulated BP/RP and RVS data for $T_{eff} = 8000$ K, $\log(g) = 4.0$ normal and magnetic and non-magnetic Ap stars.

absorption coefficient thus providing a high dynamical range in opacity calculation. This makes it possible to account for the effects of individual non-solar abundance and inhomogeneous vertical distribution of elements. It can also compute models with detailed treatment of anomalous Zeeman splitting, Kochukhov *et al.*, 2005, polarized radiative transfer in all four Stokes parameters Khan & Shulyak, 2006a, Khan & Shulyak, 2006b, and magnetohydrostatic equilibrium taken into account, Shulyak *et al.*, 2007, Shulyak *et al.*, 2009a. Chemical abundances and stratification are provided as input parameters for the LLMODELS code and kept constant in the model atmosphere calculation process. This allow us to explore the changes in model structure due to stratification that were extracted directly from observations without modelling the processes that could be responsible for the observed inhomogeneities. Such an empirical modelling can be applied to any CP star for which accurate spectroscopic observations exist; see Kochukhov *et al.*, 2009, Shulyak *et al.*, 2009b and talk JD04-i:5 of this meeting.

Having a suitable code for computing CP star spectra, it is now possible to verify how Gaia sees CP stars. To simulate Gaia data we used the Gaia Object Generator (GOG), a tool originally designed to obtain catalogue data and main database data (including mission final data) for the Gaia satellite. User source input specifications were used, allowing us directly to feed GOG with high resolution fluxes obtained by LLMODELS. Calibration and spectral noise were ignored, mostly because their final models are not yet strictly defined. The model grid of CP stars was computed taking into account characteristic chemistry and effective temperatures of the following types of CP stars: Am/Fm, λBoo, Ap, HgMn, He-weak and He-rich.

As an example, Fig. 3 illustrates the result of simulation of BP, RP, and RVS spectra for a normal $T_{eff} = 8000$ K, $\log(g) = 4.0$ star with solar composition, and two Ap stars: non-magnetic, and magnetic with assumed 10 kG surface magnetic field. The energy redistribution due to peculiar abundances and magnetic field is clearly seen, as are abnormally strong lines of REE in the RVS spectrum in case of Ap stars. Thus, there is a chance to use both low resolution spectrophotometry and high resolution spectroscopy

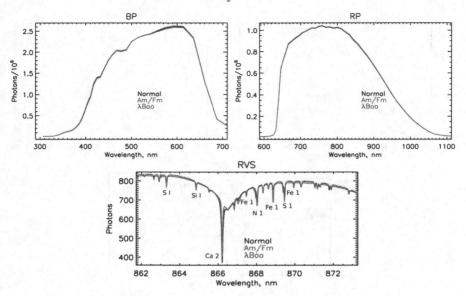

Figure 4. Same as in Fig. 3 but for Am/Fm and λ Boo star models.

to verify the type of peculiarity. The example of Am/Fm and λ Boo models is presented in Fig. 4.

It follows from our analysis that using BP/RP spectra it is only possible to distinguish between Ap and normal stars. Energy distributions of Am/Fm, λBoo, HgMn, He-weak and He-rich stars as seen by Gaia are hardly different from those of normal stars with the same $T_{\rm eff}$ and $\log(g)$. However, in RVS spectra different types of CP stars are well visible due to the presence of certain spectral features (like deep FeI and CaII lines in case of Am/Fm stars, REE elements in case of Ap stars, MnI/II features in case of HgMn stars, etc.).

Thus, using the BP/RP spectra can most likely help to see CP stars with only strong peculiarities causing substantial energy redistribution from UV to IR and thus easily detectable. On the other hand, analysis of line strength indices or relative parameters (like equivalent widths) in RVS spectra can be applied for all type of CP stars. In addition, after a standard procedure of Gaia's data processing (which will derive fundamental parameters of all observed stars), it will be possible (based on RVS spectra) to derive extended astrophysical parameters like magnetic field strength, signatures of element stratification, starspots, etc.

In the framework of this analysis we carry out calculations of extensive grids of normal and selected CP groups' model atmospheres and fluxes, as well as development, testing and implementation of modified stellar parametrization algorithms as applied to CP stars research.

During its five years' life time, Gaia will observe every object from 40 to 250 times depending upon their individual positions on the sky. The Gaia time sampling is quite irregular, with gaps of typically a month. Still, the probability to recover the periods of strictly periodic signals is high, and this opens a possibility to study a variability of Ap and related stars. The variability amplitude of magnetic CP stars depends on wavelength and is up to ≈ 0.1 mag. For stars with $T_{\rm eff} > 10\,000$ K the amplitude is lower for longer wavelength (decreasing amplitude from U to V), and with $T_{\rm eff} < 10\,000$ K variability in different filters may be in correlation and thus applying too wide filters may result in very weak or no observed variations at all. The characteristic periods are from 0.5 day to decades, and are generally stable.

The preliminary estimate of detected variable Ap stars with Gaia is about 40 000 (lower limit) and depends upon the Galaxy model used. This is achieved by the high photometric precision of Gaia in G-band (350 − 1100 nm) which is 20 mmag at $G = 20$ and ≈ 1 mmag at $10 < G < 14$. In addition, the application of two separate BP and RP photometers is a good point as it allows to study variability due to energy redistribution from UV to visual and IR.

On the other hand, the detection of roAp variability is a challenging task for Gaia since the characteristic amplitudes are below 0.01 mag and periods are between 6 and 20 minutes, however, there is a possibility to use data from every single CCD of Gaia detector separetely, thus allowing for finer time resolution which is critical for roAp stars. For instance, Mary, 2006 studied a model of roAp star HR 3831 for which they were able to recover three periods assuming stable multiperiodic sinusoidal signal with 16 frequencies (without noise). Later, Varadi *et al.*, 2009 investigated ZZ Ceti stars resulting in 65% of recovery of a period (multi-periodic nonlinear stable signal with 7 frequencies, with noise at $G = 18$ mag). Thus, the goal of using Gaia data in the light of variability research is to detect and classify correctly roAp with some period(s) and amplitude(s) characteristics.

Acknowledgements

DS would like to acknowledge the support received from the Deutsche Forschungsgemeinschaft (DFG) Research Grant RE1664/7-1 and IAU GA travel grant.

Personal thanks from DS to the GOG WEB administration team.

References

Babcock, H. W. 1958, *ApJS* 3, 141

Bruntt, H., North, J. R., Cunha, M., Brandão, I. M., Elkin, V. G., Kurtz, D. W., Davis, J., Bedding, T. R., Jacob, A. P., Owens, S. M., Robertson, J. G., Tango, W. J., Gameiro, J. F., Ireland, M. J., & Tuthill, P. G. 2008, *MNRAS* 386, 2039

Cowley, C. R., Hubrig, S., & González, J. F. 2009, *MNRAS* 396, 485

Donati, J.-F., Semel, M., Carter, B. D., Rees, D. E., & Collier Cameron, A. 1997, *MNRAS* 291, 658

Freyhammer, L. M., Kurtz, D. W., Elkin, V. G., Mathys, G., Savanov, I., Zima, W., Shibahashi, H., & Sekiguchi, K. 2009, *MNRAS* 396, 325

Khan, S. A. & Shulyak, D. V. 2006, *A&A* 454, 933

Khan, S. & Shulyak, D. 2006, *A&A* 448, 1153

Khokhlova, V. L., Vasilchenko, D. V., Stepanov, V. V., & Tsymbal, V. V. 1997, *Astron. Lett.* 23, 465

Kholtygin, A. F., Chountonov G. A., Fabrika, S.N. *et al.* 2007, in *Physics of Magnetic Stars*, Proc. Intern. Conf. held in the Special Astrophysical Observatory, Russia, August 28–31, 2006, Eds. I. I. Romanyuk and D. O. Kudryavtsev, p. 262

Kholtygin, A. F. 2008, in *Clumping in Hot Star Winds*, W.-R. Hamann, A. Feldmeier & L. M. Oskinova, eds., Potsdam: Univ.-Verl., p. 167

Kochukhov, O., Shulyak, D., Ryabchikova, T. 2009, *A&A* 499, 851

Kochukhov, O. 2008, *A&A* 483, 557

Kochukhov, O., Tsymbal, V., Ryabchikova, T., Makaganyk, V., & Bagnulo, S. 2006, *A&A* 460, 831

Kochukhov, O. 2006, *A&A* 446, 1051

Kochukhov, O., Khan, S., & Shulyak, D. 2005, *A&A* 433, 671

Kochukhov, O., Drake, N. A., Piskunov, N., & de la Reza, R. 2004, *A&A* 424, 935

Kurtz, D. W. 1978, *IBVS*, 1436, 1

Lanz, T. 1993, *ASP-CS* 44, 60

Lüftinger, T., Fröhlich, H.-E., Petit, P., Aurière, M., Weiss, W., Nesvacil, N., Gruberbauer, M., Shulyak, D., Alecian, E., Catala, C., Donati., J.-F., Kochukhov, O., & Roudier, T. 2009, *A&A*, submitted

Michaud, G. 1970, *ApJ* 160, 641

Mikulášek, Z., Krtička, J., Henry, G. W., Zverko, J., Žižǎovský, J., Bohlender, D., Romanyuk, I. I., Janík, J., Božić, H., Korčáková, D., Zejda, M., Iliev, I. K., Škoda, P., Šlechta, M., Gráf, T., Netolický, M., & Ceniga, M. 2008, *A&A* 485, 585

Mary, D. L. 2006, *A&A* 452, 715

North P. 1993, *ASP-CS* 44, 577

Pickering, E. C. & Maury, A. C., 1897, *Annals of Harvard College Observatory* 28, 1

Preston, G. W. 1974, *ARAA* 12, 257

Przybilla, N., Nieva, M. F., Tillich, A., Heber, U., Butler, K., & Brown, W. R. 2008, *A&A* 488, 51

Ryabchikova, T., Kochukhov, O., & Bagnulo, S. 2008, *A&A* 480, 811

Ryabchikova, T., Ryabtsev, A., Kochukhov, O., & Bagnulo, S. 2006, *A&A* 456, 329

Ryabchikova, T., Leone, F., & Kochukhov, O. 2005, *A&A* 438, 973

Ryabchikova, T., Nesvacil, N., Weiss, W. W., Kochukhov, O., & Stütz, Ch. 2004, *A&A* 423, 705

Ryabchikova, T., Piskunov, N., Kochukhov, O. *et al.* 2002, *A&A* 384, 545

Shulyak, D., Kochukhov, O., Valyavin, G., Lee, B.-C., Galazutdinov, G., Kim, K.-M., Han, I., & Burlakova, T. 2009, *A&A*, submitted

Shulyak, D., Ryabchikova, T., Mashonkina, L., & Kochukhov, O. 2009, *A&A* 499, 879

Shulyak, D., Kochukhov, O., & Khan, S. 2008, *A&A* 487, 689

Shulyak, D., Valyavin, G., Kochukhov, O., Lee, B.-C., Galazutdinov, G., Kim, K.-M., Han, I., Burlakova, T., Tsymbal, V., & Lyashko, D. 2007, *A&A* 464, 1089

Shulyak, D., Tsymbal, V., Ryabchikova, T., Stütz Ch., & Weiss, W. W. 2004, *A&A* 428, 993

Sokolov, N. A. 2008, *MNRAS* 385, 1

Varadi, M., Eyer, L., Jordan, S., Mowlavi, N., & Koester, D. 2009, in *Stellar Pulsation: challenges for theory and observation*, AIP Conference Series, vol 1170, 330

Wade, G. A., Donati, J.-F., Landstreet, J. D., & Shorlin, S. L. S. 2000, *MNRAS* 313, 823

Woszczyk, A. & Jasinski, M. 1980, *AcA* 30, 331

Highlights of Astronomy, Volume 15
XXVIIth IAU General Assembly, August 2009
Ian F. Corbett, ed.
© International Astronomical Union 2010
doi:10.1017/S1743921310008550

A 3-D look into the atmosphere?

G. A. Wade[1], L. Mashonkina[2], T. Ryabchikova[2], J. Krticka [3], J.
Silvester[1], O. Kochukhov[4], J. C. Sousa[5], T. Nomura[6], G. M.
Wahlgren[7], M. Gruberbauer[8], S. Hubrig [9], M. Briquet[10], K. Yüce[11],
N. Drake[12], N. Nunez [13], R. O. Gray[14], J. Ziznovsky[15], C. Cowley[16],
A. Shavrina[17], M. Dworetsky[18], O. Pintado[19], N. Polosukhina[20] &
L. Cidale[21]

[1] Royal Military College, Canada

[2] Institute of Astronomy, Russian Academy of Sciences, Russia

[3] Masaryk University, Czech Republic

[4] Uppsala University, Sweden

[5] CAUP-Centro de Astrofisica da Universidade do Porto, Portugal

[6] University of Tokyo, Japan

[7] NASA-GSFC/CUA, USA

[8] St Mary's University, Canada

[9] Astrophysikalisches Institut Potsdam, Germany

[10] Instituut voor Sterrenkunde, Katholieke Universiteit Leuven, Belgium

[11] Ankara University, Turkey

[12] Sobolev Astronomical Institute of St. Petersburg State University, Russia

[13] Casleo-UNSJ, Argentina

[14] Appalachian State University, USA

[15] Astronomical Institute of the Slovak Academy of Sciences, Slovakia

[16] Dep. of Astronomy, University of Michigan, USA

[17] Main Astronomical Observatory, National Academy of Sciences of Ukraine

[18] University College London, United Kingdom

[19] Instituto Superior de Correlación Geológica-CONICET, Argentina

[20] Crimean Astrophysical Observatory, Ukraine

[21] Faculdad de Ciencias Astronomicas y Geofisicas, Argentina

Abstract. The atmospheres of chemically peculiar stars can be highly structured in both the horizontal and vertical dimensions. While most prevalent in the magnetic stars, these structures can also exist in non-magnetic stars. In addition to providing an important window to understanding the physical processes at play in these complex atmospheres, they can also be exploited to study stellar pulsations. This article reviews contributions to the session "A 3D look into the atmosphere" of the Joint Discussion "Progress in understanding the physics of Ap and related stars". It is divided into 3 sections: "Magnetic field and surface structures", "Pulsations in the atmospheres of roAp stars/inversions", and "Spectral synthesis/atmospheric models".

Keywords. stars: chemically peculiar, stars: atmospheres, stars: variables: roAp stars, stars: magnetic fields

---

1. Magnetic field and surface structures

Magnetic field. Following their discovery by Babcock (1947), strong, organised magnetic fields have been found to exist in a small fraction (e.g. Shorlin *et al.* 2002, Power *et al.* 2008) of main sequence intermediate-mass A and B stars (the "Ap/Bp stars"). Historically, these fields have been modeled as "oblique rotators": strong, long-lived dipoles or aligned low-order multipoles, frozen into the stellar plasma, inclined relative to the stellar rotation axis, and rotating rigidly with the star. These properties, along with a lack of any strong, positive correlation of magnetic field strength with rotation or mass, has led most investigators to suppose that these fields are "fossils": the slowly-decaying remnants of magnetic field accumulated from the ISM during star formation, or generated by a pre-main sequence dynamo. Recent studies of these fields have yielded important new information about their detailed properties. Detailed mapping using the Magnetic Doppler Imaging method (Kochukhov & Piskunov 2002) has revealed the presence of complex structures in the photospheric magnetic fields of 53 Cam (Kochukhov *et al.* 2004a) and α^2 CVn (Kochukhov & Wade, in preparation), the origin of which is unknown. Silvester (this conference, JD04-o:4) reported on a continuation of this work using new measurements acquired using the ESPaDOnS and Narval instruments at the Canada-France-Hawaii Telescope and Télescope Bernard Lyot, respectively. Aurière *et al.* (2007) obtained magnetic field measurements of $\sim$ 30 Ap/Bp stars with weak or previously-undetected magnetic fields, demonstrating that each of them hosted a magnetic field stronger than about 300 G. They interpreted this apparent field strength threshold as a "critical field strength", below which magnetic fields are unstable to rotational distortion. Lignières *et al.* (2009) reported the detection of a circular polarisation signature in the mean Least-Squares Deconvolved spectral line of Vega. Interpreting this as evidence of a weak magnetic field, they proposed that Vega may be the first of a new class of magnetic A-type stars - perhaps those stars in which the initial fossil fields were weaker than the "critical field strength", and have hence decayed.

Meanwhile, intensive investigations of pre-main sequence Herbig Ae/Be stars has yielded the discovery of magnetic stars at this early phase of evolution which exhibit field properties very similar to those of Ap stars (e.g. Alecian *et al.* 2008, 2009, also Alecian's presentation at this conference, JD04-o:11), further supporting the fossil field hypothesis. At the other end of the main sequence, Bagnulo *et al.* (2006) and Landstreet *et al.* (2007, 2008) used samples of Ap stars observed in open clusters to demonstrate a dramatic decrease with age of the ensemble-averaged magnetic field strength of Ap stars - a phenomenon that does not appear to be fully explainable by stellar expansion during main sequence evolution. Finally, recent detections of variable Zeeman signatures in mean spectral lines of intermediate-mass red giant stars (e.g. Aurière *et al.* 2008, 2009) have allowed a first insight into the post-main sequence evolution of Ap stars, and potentially the origin of the magnetic fields observed in some white dwarfs.

At this conference, posters by Hubrig *et al.* (JD04-p:4) and Drake *et al.* (JD04-p:9) discussed the magnetic properties of some individual stars with strong magnetic fields.

Surface structures. The presence of strong lateral chemical nonuniformities in the atmospheres of magnetic Ap/Bp stars has long been known. Since the pioneering work in the 1970s and 80s (see Khokhlova (1974), and works by Piskunov, Rice, Hatzes) to develop techniques to "map" these structures by modeling the resultant variations of rotationally-broadened line profiles, the technique of "Doppler Imaging" has evolved significantly. Investigations such as that of Kochukhov *et al.* (2004b) of the (magnetic) rapidly oscillating Ap (roAp) star HR 3831 report maps of the magnetic field and the abundances of many chemical species, demonstrating a tremendous diversity of abundance patterns and suggestive relationships with the magnetic field and rotational geometry. Kochukhov (this conference, JD04-o:3) reported results of 3-dimensional mapping of the atmospheric chemical structures of the Ap star θ Aur: maps of not only the surface distribution of Fe, but also the vertical stratification of this element. These authors

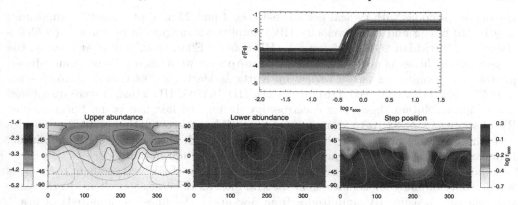

Figure 1. 3-D Doppler imaging reconstruction of the Fe distribution in the atmosphere of magnetic Ap star θ Aur (HD 40312). The vertical abundance profile for each surface element is given by a step function, characterized by the Fe abundance in the upper and lower atmospheric layers as well as the position of transition region. The bottom panel shows reconstructed maps for these three parameters. The upper panel displays all vertical abundance profiles plotted on top of each other and color-coded according to the upper abundance.

discovered that while the abundance of Fe in the lower atmosphere is not strongly correlated with the presence of spots, the abundance in the upper atmosphere and the position of the lower/upper atmosphere transition depth are. Their results are illustrated in Fig. 1. Krticka (this conference, JD04-o:1) demonstrated for the Bp star HD 7224 that the photometric variability could be acceptably explained by the opacity variations introduced by the nonuniform distributions of Si and Fe in its atmosphere. A poster by Shavrina *et al.* (JD04p:21) examined the presence of Li spots on the roAp star HD 12098, while a poster of Polosukhina *et al.* (JD04-p:27) discusses a connection of the Li spots on the surfaces of Ap stars with the magnetic field configuration.

A remarkable decade-old discovery by Ryabchikova *et al.* (1999) that has now been confirmed in several stars (Adelman *et al.* 2002, Kochukhov *et al.* 2005, Hubrig *et al.* 2006) is the presence of nonuniform distributions of Hg and other elements in atmospheres of some relatively rapidly-rotating HgMn stars. Sensitive searches for magnetic fields in these objects (e.g. Wade *et al.* 2006) have yielded no detections at the level of a few G, suggesting that the origin of these spots is not a magnetic field, but rather some other dynamical effect (e.g. rotation, and/or presence within a close binary). At this conference, posters by Nunez *et al.* (JD04-p:12) and Briquet *et al.* (JD04-p:5) focused on this phenomenon.

2. Pulsations in the atmospheres of roAp stars / inversions

An explicit review on pulsations in the atmospheres of chemically peculiar magnetic stars is presented recently by Kochukhov in 2009, and references therein. Here we briefly review the main characteristics of the rapidly oscillating Ap (roAp) stars.

General characteristics of roAp stars. A group of the roAp stars represents the coolest magnetic Ap stars and occupies the central part of the main sequence band in the effective temperature range 6300 - 8100 K. On the HR diagram roAp stars overlap with other groups of pulsating stars - δ Sct-type and solar-like pulsators - however the nature of the pulsations and their characteristics are different. roAp stars pulsate in high-overtone, low-degree p-modes with typical periods between 4 and 22 min, photometric amplitudes of 0.2 – 10 mmag and radial velocity (RV) amplitudes ranging from 3 m s^{-1} (β CrB - Hatzes & Mkrtichian 2004) to 5 km s^{-1} (HD 99563 - Elkin *et al.* 2005). At present, the roAp group includes 38 members. Thirty-two roAp stars were discovered by ground-based photometric monitoring (see a review by Kurtz & Martinez 2000), and six roAp stars (HD 75445, HD 115226, HD 116114, β CrB, HD 154708, HD 218994) were discovered using high-resolution time-series spectroscopy during the last five years. Spectroscopic RV observations of 22 roAp stars suggest a possible anti-correlation of RV amplitudes with the magnetic field strength: the stronger the field, the lower the RV amplitude. However, statistics are sparse, and this result requires confirmation.

Atmospheric pulsations. An outstanding characteristic of RV variations in roAp stars is a diversity of pulsation signatures in spectral lines of different atoms/ions. In the same star one can measure RV amplitudes from few m s^{-1} (Fe lines) to hundreds of m s^{-1} (rare-earth elements lines – REE, see Ryabchikova *et al.* 2007). This phenomenon was explained as a propagation of the acoustic pulsation wave in a chemically-stratified stellar atmosphere (Ryabchikova *et al.* 2002), where stratification results from chemical diffusion in a globally-stable atmosphere. In most roAp stars for which detailed line-by-line pulsation analysis has been performed, pulsations have a standing- or running wave character with an increase of RV amplitudes towards the outer layers. In two stars, 33 Lib (Mkrtichian *et al.* 2003) and 10 Aql (Sachkov *et al.* 2008) a definite presence of the pulsation node at the position of REE line formation layers at log $\tau_{5000} = -4$ is observed with the opposite direction of the phase jumps. It looks like in 33 Lib we have pulsational inversion: instead of the outwardly-running wave observed in most other stars, an inwardly-propagating wave exists. According to Sousa & Cunha (2008, JD04_o:7) this behaviour as well as the appearance of the nodes may be interpreted as the superposition of magnetic standing and acoustic running waves in the atmospheres of magnetic stars, mimicking an inwardly-propagating wave, node layers, etc., that depends on the magnetic field geometry and on the observer's perspective. Khomenko & Kochukhov (2009) came to the same conclusion in their 2D magneto-hydrodynamical simulations of magneto-acoustic pulsations in realistic atmospheres of roAp stars. They considered the propagation of pulsation waves with initial amplitude 100 m s^{-1}, excited by a radial pulsation mode below the visible surface of the realistic atmospheric model ($T_{\rm eff} = 7750$ K, log g= 4.0) with a constant inclined magnetic field. MHD simulations confirmed the theoretical results by Saio (2005) about the absence of pulsational magnetic field variations above 1 G. On the other hand, MHD simulations showed significant pulsational variations of the thermodynamic parameters temperature and density at different layers of the stellar atmosphere, in particular at the photospheric base where the density inversion takes place.

In some roAp stars the frequency spectrum of photometric variations differs from that measured from spectroscopic variations. Kurtz *et al.* (2006) found a 'new type' of upper atmospheric pulsations. In some stars such as 33 Lib and HD 134214, new frequencies - not observed in photometry – appear in the RV variations of Pr III-Nd III lines. These lines are formed in the upper atmospheric layers; hence, new frequencies are either amplified or even excited in these layers. Extensive photometric + spectroscopic studies are required to understand this phenomenon.

Nomura & Shibahashi (JD04_o:14) presented first results on numerical simulations of line profile variations in roAp stars. Their simplified model could reasonably explain the observed bisector variations in the Hα core in all roAp stars (an increase of RV amplitudes with the depth of the core), while it failed to explain the observed core-to-wing bisector variations in Nd III lines. Hα core bisector variations agree with the general picture of

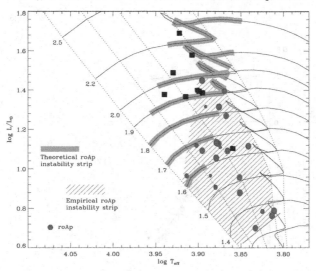

Figure 2. Theoretical and observed IS for roAp stars. Positions of the roAp stars are shown by filled circles with size proportional to parallax value, while filled squares indicate positions of the nine luminous Ap stars in which pulsations were not found.

acoustic wave propagation in the stellar atmosphere – an increase of RV towards the upper layers. Contrary to the Hα core, in most of stars we observe a decrease of the RV amplitude with the depth within the Nd III line profile.

Evolutionary status of roAp stars. The current theory of pulsations in roAp stars proposes that they are excited by a κ-mechanism acting in the hydrogen ionization zone. A globally-organized magnetic field effectively suppresses convection and makes excitation possible (Balmforth *et al.* 2001). Abundance gradients, in particular helium settling in the stable atmospheres due to diffusion, also favour the hydrogen κ-mechanism (Vauclair & Théado 2004). However, first calculations of the instability strip (IS) of roAp stars (Cunha 2002) showed that the red border is shifted to higher luminosities than is revealed by observations. Fig. 2 taken from Kochukhov (JD04_p:2) presents the position of the roAp stars with known Hipparcos parallaxes on the HR diagram. The theoretical IS is taken from the recent calculations by Théado *et al.* (2009) who considered the effect of metallicity on the pulsational excitation in roAp stars. These calculation do not differ significantly from Cunha's calculations. We also plot nine luminous Ap stars in which pulsations are predicted theoretically, but were not found in spectroscopy (Freyhammer *et al.* 2008). It is evident that the theoretical and observed IS are shifted both in luminosity and in effective temperature relative each other, which means that the currently accepted excitation mechanism model needs to be improved.

Modelling the pulsations in individual stars. Pulsation modelling of two rather rapidly rotating roAp stars HD 83368 (HR 3831 – Kochukhov 2004, 2006) and HD 99563 (Freyhammer *et al.* 2009) based on spectroscopic time-series observations obtained over the rotational period shows a single mode which is a distorted oblique dipole mode. First indirect imaging of nonradial pulsations in the roAp star HD 83368 by Kochukhov (2004) directly demonstrates an alignment of pulsations with the axis of the dipolar-like global magnetic field, thus proving the oblique pulsator model proposed by Kurtz (1982) to explain rotational modulation of the photometric pulsation amplitudes in roAp stars.

Frequency modelling of slowly rotating roAp stars are based on long-time photometric ground-based (HD 24712 – WET campaign) or space-based (γ Equ, 10 Aql – MOST;

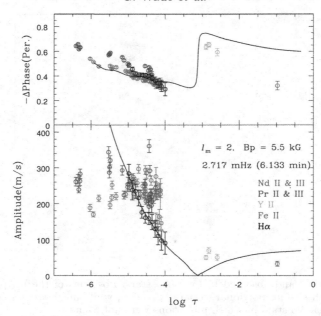

Figure 3. Theoretical (black line) and observed (open circles) pulsational RV phase and amplitude distributions in the atmosphere of the roAp star HD 24712.

α Cir – WIRE) observations. Only for HD 101065 (Przybylski's star) frequency modelling was based on long-duration spectroscopic monitoring. The frequency pattern in these stars is richer and can usually be represented by a combination of odd-even modes with ℓ=0-3 (see Gruberbauer *et al.* 2008 for γ Equ; Huber *et al.* 2008 for 10 Aql; Bruntt *et al.* 2009 for α Cir; and Mkrtichian *et al.* 2008 for HD 101065). HD 24712 is the first star where not only the frequency pattern derived by Kurtz *et al.* (2005) in WET campaign, but also RV amplitude and phase distributions for the two highest amplitude frequencies in the atmosphere were modelled (see Saio *et al.* JD11_p:16). An improved model atmosphere of HD 24712 which takes into account chemical anomalies as well as abundance stratification for a few of the most important elements – Si, Ca, Cr, Fe (Shulyak *et al.* 2009, see also next Section) – was used in the pulsation modelling. Magnetic fields are required to fit the observed frequencies in all models, with magnetic perturbations usually 2-3 times larger than actually observed. The only exception is HD 24712, where the polar intensity of the magnetic dipole, $B_p \approx 5.5$ kG, needed to represent the observed pulsation data, is roughly consistent with the observed value of 4.2 kG. Fig. 3 taken from Saio *et al.* (JD11_p:16) shows a comparison between the observed and theoretical distributions of the pulsational RV amplitudes and phases in the atmosphere of HD 24712 for the main highest amplitude frequency.

Formally, the modelling is successful, however for all models of HD 101065 and HD 24712 no high-order p-modes appropriate to the observed oscillations are excited. This is also the case for few modes in the γ Equ model. These negative results agree with the predicted roAp IS positions for these stars. These stars are too cool for p-modes to be excited by the κ-mechanism in the hydrogen ionization zone. The temperature inversion in the atmosphere of HD 24712 (Shulyak *et al.* 2009) seems too small to help excite high-order p-modes.

3. Spectral synthesis / atmospheric models

In this section, we highlight recent results on modelling of the atmospheres and spectra of chemically peculiar (CP) stars with non-uniform vertical distributions of elements, and describe advances in searching for and investigating exotic chemical elements in CP stars.

Self-consistent diffusion model atmospheres. The microscopic diffusion process is believed to be the main cause of chemical stratification in stellar atmospheres. LeBlanc *et al.* (2009) performed self-consistent calculations of atmospheric structure and element diffusion, yielding the abundance gradients for 39 elements (H-Ga, Kr-Nb, Ba, and La). No competing physical processes, including mass loss, convection, and turbulence, were considered. In the model with $T_{\rm eff} = 8\,000$ K and log g= 4, diffusion produces a highly non-uniform distribution of every included element with a lower abundance in the upper layers, above $\log \tau_{5000} = -1$, and an abundance increase in deep layers, $-1 \leqslant \log \tau_{5000} < 0$. The change in temperature due to stratification is small in the continuum and line-forming region, up to 5%. LeBlanc *et al.* (2009) found that the effect of diffusion depends strongly on effective temperature. In the hotter model with $T_{\rm eff} = 12\,000$ K, the abundance gradient for each included element is small and the calculated abundance exceeds the initial one throughout the atmosphere. Depending on the treatment of the redistribution of momentum among the various ions of an element, the temperature in the continuum and line-forming region can be up to 10% higher compared to that in a chemically homogeneous model with 10× enhanced abundances of metals. The accuracy of the self-consistent diffusion models is defined by the accuracy and completeness of atomic data, in particular transition probabilities and photoionization cross-sections for a variety of elements. For most heavy elements beyond the iron group, the required data is either incomplete or missing.

The self-consistent diffusion model atmospheres are only just beginning to be utilized in the analysis of CP stars. The theoretical results support observational findings about the abundance jumps for the Fe group and lighter metals in cool Ap stars. For chromium and iron in cool Ap stars, Ryabchikova (2008) highlighted an agreement between the predicted and empirical abundance profiles for the position and the magnitude of the abundance jumps. At this conference, Ryabchikova *et al.* (JD04_p:13) showed much smaller abundance gradients for the hotter star HD 170973 ($T_{\rm eff} = 11\,200$ K) in agreement with the prediction of the diffusion theory.

For $T_{\rm eff} = 8\,000$ K and log g = 4, LeBlanc *et al.* (2009) calculated a self-consistent diffusion model atmosphere with a horizontal magnetic field of 2 kG and found that diffusive motions of the ionized species are efficiently impeded in the upper atmosphere and metals are pushed outwards. In the layers around $\log \tau_{5000} = -3$, the temperature increases by approximately 10% compared to that in the chemically homogeneous model.

Alecian & Stift (2007) calculated the stratification of some metals (Mg, Si, Ca, Ti, and Fe) in the models with fixed atmospheric structure and magnetic fields of up to 10 kG and various inclination. They found that (i) the differences in abundance profiles between the non-magnetic and magnetic with vertical field atmospheres are small, (ii) the role of magnetic field is small for deep layers, below $\log \tau_{5000} = -1$, independently of the field inclination, and (iii) the horizontal field can be very efficient in helping the accumulation of elements in the upper atmosphere.

Non-local thermodynamic equilibrium (NLTE) line formation for Pr II-III and Nd II-III in A-type stars. The rare-earth elements play a key role in understanding the phenomenon of Ap stars. Classic LTE analysis finds a strong violation of the ionization equilibrium between the first and second ions of the REE in the atmosphere of roAp stars suggesting a non-uniform vertical distribution of these elements in the atmosphere.

Contrary to the iron group and lighter metals, the REE are, probably, accumulated in the uppermost atmospheric layers, where the LTE assumption is not valid. The difficulties of the statistical equilibrium calculations for the REE are due to incompleteness of the system of known energy levels and poorly known photoionization and collisional data. Mashonkina *et al.* (2005, 2009) constructed comprehensive model atoms for Nd II-III and Pr II-III using not only measured but also calculated energy levels. The calculations for model atmospheres with uniform distributions of elements showed that NLTE acts in the right direction and tends to reduce the abundance difference between the first and second ions, but fails to remove this disparity completely. For example, for the benchmark roAp star HD 24712 ($T_{eff} = 7250$ K), the difference in LTE abundances derived from the Pr III and Pr II lines amounted 2.1 dex, and only a 0.2 dex smaller difference was obtained in the NLTE case. The left panel of Fig. 4 illustrates the NLTE effects on the Pr II–III lines in HD 24712 when the distribution of praseodymium is assumed to be uniform with [Pr/H] = 3.

Mashonkina *et al.* (2005, 2009) then checked various abundance profiles, independently for praseodymium and neodymium, trying to fit the observed equivalent widths of the lines of both ionization stages in HD 24712 with a single abundance distribution. It was found that both elements are concentrated in the uppermost atmospheric layers, above $\log \tau_{5000} = -4$, with abundance enhancements of 4.5 dex for neodymium and 5 dex for praseodymium compared to the corresponding solar abundances. For comparison, the praseodymium and neodymium enriched layer has to be located above $\log \tau_{5000} = -8$ when stratification analysis is performed under LTE conditions. In the stratified atmosphere, the calculated NLTE effects are very large (Fig. 4) because the line formation occurs in the high atmosphere, where collisions are rare and inefficient in establishing thermodynamic equilibrium. The lines of Pr II and Nd II are strongly weakened compared to their LTE strengths, and the opposite effect is found for Pr III and Nd III. The NLTE abundance corrections may reach +1.2 dex for the first ions and −0.7 dex for the second ions.

Stratified model atmospheres with empirical abundance profiles. Kochukhov *et al.* (2009) and Shulyak *et al.* (2009) suggested the method of self-consistent analysis, where the individual abundances and chemical stratification derived for a given star are then used to recalculate its model atmosphere structure. The entire process of determining stellar parameters, chemical abundances and stratification and calculation of the model atmosphere has to be repeated until convergence is achieved. Analysis of two roAp stars with similar stellar parameters, HD 24712 ($T_{eff}/\log g = 7250$ K/4.1, Shulyak *et al.* 2009) and

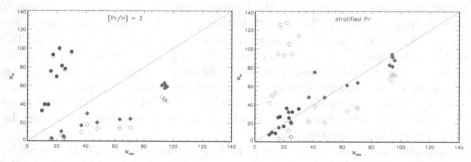

Figure 4. Observed equivalent widths of the Pr II (circles) and Pr III (diamonds) lines in HD 24712 compared with the theoretical non-LTE (filled symbols) and LTE (open symbols) equivalent widths for the uniform Pr distribution with [Pr/H] = 3 (left panel) and the stratified Pr distribution (right panel).

α Cir (7500 K/4.1, Kochukhov *et al.* 2009), showed that the calculated atmospheric structure depends on the individual abundance profiles. For α Cir, four elements, Si, Ca, Cr and Fe, were allowed to have non-uniform vertical distributions. It was found that all of them are concentrated in the lower atmosphere and this leads to a backwarming effect. The corresponding reduction of effective temperature derived from the Balmer line wings amounted to 200 K.

For HD 24712, stratification of the same four elements and, in addition, that of Sr, Ba, Pr, and Nd, was taken into account. The abundance profiles for praseodymium and neodymium were derived from the NLTE analysis of the Pr II-III and Nd II-III lines. The backwarming effect was found to be weaker in HD 24712 compared to that in α Cir due to lower abundances of chromium and iron. The accumulation of praseodymium and neodymium above log $\tau_{5000} = -4$ caused a temperature increase of up to 600 K in these same layers. It is worth noting that the changes in temperature structure due to stratification of praseodymium and neodymium did not affect the formation of the Balmer line wings and the visible continuum and did not result in any change of the stellar parameters of HD 24712.

For both stars, the differences between initial and final abundance profiles were found to be within the error bars of the determination of stratification. It should be stressed that the stellar parameter range addressed in the studies of Kochukhov *et al.* (2009) and Shulyak *et al.* (2009) is very narrow, and the validity of their findings should not be extended out of this range.

Exotic elements in CP stars. Progress in extending our knowledge of spectrum anomalies was made thanks to laboratory measurements of atomic data, calculations of atomic structures, and high-resolution spectral observations. At this conference, Wahlgren *et al.* (JD04_o:9) found nearly all lanthanides and also Pt, Au, Hg, and Bi in the ultra-violet spectrum of the spectrum transition star HR 3383. Cowley *et al.* (JD04_p:17) first investigated the spectra of Nb II, Re II, and Os II in the spectrum of the star HD 65949. Dworetsky & Maher-McWilliams (JD04_p:20) determined the gold abundance in nine HgMn stars. Cowley *et al.* (JD04-p:22, see also Cowley *et al.* MNRAS, 396, 485, 2009) gave evidence for isotope separation of calcium in the atmosphere of some CP stars from measurements of differences in the wavelengths of the cores and the wings of the infrared triplet of Ca II. Ryabchikova *et al.* (JD04_p:8) considered an effect of the completeness of the REE line data on the atmospheric structure calculations and spectral synthesis of the extreme peculiar star Przybylski's star, and concluded that the lack of these data may lead to mistaken identification of actinides in cool REE-rich stars.

Acknowledgements.

ML, ND, and TR acknowledge partial support from the International Astronomical Union for participation at the IAU XXVII General Assembly. ML and TR also thank the Russian Foundation for Basic Research (grants 09-02-08-244-z, 09-02-00002-a, 08-02-00469-a) and the Russian Federal Agency on Science and Innovation (No. 02.740.11.0247). GAW is supported by the Natural Science and Engineering Council of Canada (NSERC) and the Department of National Defence (Canada). JS is supported in part by NSERC Discovery Grants held by GAW and by David Hanes (Queen's University). KY's participation was supported by an IAU Travel Grant and by Ankara University. OK is a Royal Swedish Academy of Sciences Research Fellow supported by a grant from the Knut and Alice Wallenberg Foundation.

References

Adelman S. J., Gulliver A., Kochukhov O., & Ryabchikova T. 2002, *ApJ* 575, 449

Alecian E., Catala C., Wade G. A., Donati J.-F.., *et al.* 2008, *MNRAS* 385, 391

Alecian E., Wade G. A., Catala C., Bagnulo S., *et al.* 2009, *MNRAS* 400, 354

Alecian, G. & Stift, M. J. 2007, *A&A*, 475, 659

Aurière M., Wade G. A., Silvester J., Lignières F. *et al.*, 2007, *A&A* 475, 1053

Aurière M., Konstantinova-Antova R., Petit P., Charbonnel C., *et al.*2008, *A&A* 491, 499

Aurière M., Wade G. A., Konstantinova-Antova R., Charbonnel C., *et al.* 2009, *A&A* 504, 231

Babcock, H. W. 1947, *ApJ* 105, 105

Bagnulo S., Landstreet J. D., Mason E., & Andretta V. 2006, *A&A* 450, 777

Balmforth, N. J., Cunha, M. S., Dolez, N., *et al.* 2001, *MNRAS*, 323, 362

Bruntt, H., Kurtz, D. W., & Cunha, M. S. 2009, *MNRAS*, 396, 1189

Cunha, M. S. 2002, *MNRAS*, 333, 47

Elkin, V. G., Kurtz, D. W., & Mathys, G. 2005, *MNRAS*, 364, 864

Freyhammer, L. M., Kurtz, D. W., Cunha, M. S., *et al.* 2008, *MNRAS*, 385, 1402

Freyhammer, L. M., Kurtz, D. W., Elkin, V.G., *et al.* 2009, *MNRAS*, 396, 325

Gruberbauer, M., Saio, H., Huber, D., *et al.* 2008, *A&A*, 483, 239

Hatzes, A.P. & Mkrtichian, D. E. 2004,*MNRAS*, 351, 663

Huber, D., Saio, H., Gruberbauer, M., *et al.* 2008, *A&A*, 480, 223

Hubrig, S., Gonzalez, J. F., Savanov, I., & Schöller, M. 2006, *MNRAS*, 379, 1953

Khokhlova V. 1974, Nauchn. Inf., Vyp., 33, 11

Khomenko, E. & Kochukhov, O. 2009, *ApJ*, 704, 1218

Kochukhov, O., Piskunov, N., Sachkov, M., & Kudryavtsev, D. 2005, *A&A* 439, 1093

Kochukhov, O. & Piskunov, N. 2002, *A&A*, 388, 868

Kochukhov, O. 2004, *ApJ*, 615, L49

Kochukhov, O., Bagnulo S., Wade G.A., Sangalli L., *et al.* 2004a, *A&A* 414, 613

Kochukhov, O., Drake, N. A., Piskunov, N., & de la Reza, R. 2004b, *A&A* 424, 935

Kochukhov, O. 2006, *A&A*, 446, 1051

Kochukhov, O. 2009, *CoAst*, 159, 61

Kochukhov, O., Shulyak, D., & Ryabchikova, T. A. 2009, *A&A* 499, 851

Kurtz D. W. 1982, *MNRAS* 200, 807

Kurtz, D. W. & Martinez, P. 2000, *Baltic Astronomy* 9, 253

Kurtz, D. W. Cameron, C., Cunha, M. S. *et al.* 2005, *MNRAS* 358, 651

Kurtz, D. W. Elkin, V. G., & Mathys, G. 2006, *MNRAS* 370, 1274

LeBlanc, F., Monin, D., Hui-Bon-Hoa, A., & Hauschildt, P. H. 2009, *A&A,* 495, 937

Landstreet J. D., Bagnulo S., Andretta V., Fossati L., *et al.* 2007, *A&A* 470, 685

Landstreet J. D., Silaj J., Andretta V., Bagnulo S., *et al.* 2008, *A&A* 481, 465

Lignières F., Petit P., Bohm T., & Aurière M. 2009, *A&A* 500, 41

Mashonkina, L., Ryabchikova, T. A., & Ryabtsev, A. N. 2005, *A&A* 441, 309

Mashonkina, L., Ryabchikova, T. A., Ryabtsev, A. N., & Kildiyarova, R. 2009, *A&A* 495, 297

Mkrtichian, D. E., Hatzes, A. P., & Kanaan, A. 2003, *MNRAS* 345, 781

Mkrtichian, D. E., Hatzes A. P., Saio, H., & Shobbrook, R. R. 2008, *A&A* 490, 1109

Power J., Wade G. A., Aurière M., Silvester J., & Hanes D. 2008, *Cont. Ast. Obs. Skalnaté Pleso* 38, 443

Ryabchikova, T. A., Malanushenko, V. P., & Adelman, S. J. 1999, *A&A* 351, 963

Ryabchikova, T., Piskunov, N., Kochukhov, O., *et al.* 2002, *A&A* 384, 545

Ryabchikova, T., Sachkov, M., Kochukhov, O., & Lyashko, D. 2007, *A&A* 473, 907

Ryabchikova, T. 2008, *Contr. Astron. Obs. Skalnaté Pleso* 38, 257

Sachkov, M., Kochukhov, O., Ryabchikova, T., *et al.* 2008 *MNRAS* 389, 903

Saio, H. 2005, *MNRAS*, 360, 1022

Shorlin S. L. S., Wade G. A., Donati J. F, Landstreet J. D., *et al.* 2002, *A&A* 392, 637

Shulyak, D., Ryabchikova, T. A., Mashonkina, L., & Kochukhov, O. 2009, *A&A* 499,879

Sousa J. C., & Cunha M. S. 2008, *Contrib. Astron. Obser. Skalnaté Pleso* 38, 453

Théado, S., Dupret, M.-A., Noels, A., & Ferguson, J. W. 2009, *A&A* 493, 159

Vauclair, S., & Théado, S. 2004, *A&A* 425, 179

Wade, G. A.; Aurire, M.; Bagnulo, S.; Donati, J.-F. *et al.* 2006, *A&A* 451, 293

Highlights of Astronomy, Volume 15
XXVIIth IAU General Assembly, August 2009
Ian F. Corbett, ed.
© International Astronomical Union 2010
doi:10.1017/S1743921310008562

CP and related phenomena in the context of Stellar Evolution

J. Braithwaite[1,2], T. Akgün[3], E. Alecian[4], A. F. Kholtygin[5], J. D. Landstreet[6,7], S. Mathis[8], G. Michaud[9], J. Portnoy[10], G. Alecian[4], V. D. Bychkov[11], L. V. Bychkova[11] N. Drake[5], S. N. Fabrika[11], A. Reisenegger[3], R. Steinitz[12] M. Vick[9]

[1] Canadian Institute for Theoretical Astrophysics, Toronto, Canada

[2] Argelander Institut für Astronomie, Bonn, Germany; jonathan@astro.uni-bonn.de

[3] Departamento de Astronomía y Astrofísica, Pontificia Universidad Católica de Chile, Santiago, Chile; akgun@astro.puc.cl

[4] Observatoire de Paris, LESIA, 5 place Jules Janssen, 92190 Meudon, France evelyne.alecian@obspm.fr

[5] Department of Astronomy, Saint-Petersburg State University, Saint-Petersburg, Russia; afkholtygin@gmail.com

[6] University of Western Ontario, 1151 Richmond Street, London ON N6A 3K7, Canada

[7] Armagh Observatory, Northern Ireland

[8] Laboratoire AIM, CEA/DSM-CNRS-Université Paris Diderot, IRFU/SAp Centre de Saclay, F-91191 Gif-sur-Yvette, France; stephane.mathis@cea.fr

[9] Département de Physique, Université de Montréal, Montréal, PQ, H3C 3J7, Canada; michaudg@astro.umontreal.ca

[10] Sami Shamoon College of Engineering, Israel jacovp@sce.ac.il

[11] Special Astrophysical Observatory, Russia

[12] Physics Dept., Ben Gurion University of the Negev, Israel

Abstract. We review the interaction in intermediate and high mass stars between their evolution and magnetic and chemical properties. We describe the theory of Ap-star 'fossil' fields, before touching on the expected secular diffusive processes which give rise to evolution of the field. We then present recent results from a spectropolarimetric survey of Herbig Ae/Be stars, showing that magnetic fields of the kind seen on the main-sequence already exist during the pre-main sequence phase, in agreement with fossil field theory, and that the origin of the slow rotation of Ap/Bp stars also lies early in the pre-main sequence evolution; we also present results confirming a lack of stars with fields below a few hundred gauss. We then seek which macroscopic motions compete with atomic diffusion in determining the surface abundances of AmFm stars. While turbulent transport and mass loss, in competition with atomic diffusion, are both able to explain observed surface abundances, the interior abundance distribution is different enough to potentially lead to a test using asterosismology. Finally we review progress on the turbulence-driving and mixing processes in stellar radiative zones.

Keywords. stars: magnetic fields – stars: chemically peculiar – stars: evolution – (*magnetohydrodynamics*) MHD – stars: rotation – stars: intermediate mass – stars: pre-main sequence

1. The interesting lives of middle main sequence stars

Observations reveal that around 5–10% of intermediate-mass main-sequence (MS) stars ($1.6 - 8\,M_\odot$) have magnetic fields of strength 300 G - 30 kG (see Donati & Landstreet 2009 for a recent review of the so-called 'chemically peculiar' (CP), or Ap and Bp stars).

The fields are large-scale and apparently static, i.e. they do not evolve at all over periods of decades, and their host stars rotate much more slowly than normal A stars.

It is worthwhile to look at magnetic middle MS stars in the broader context of stellar evolution. Stars form by gravitational collapse on a timescale of a few 0.1 Myr, set by the free-fall time. At a size of a few 100 AU, angular momentum leads to slower growth through quasi-static disk accretion. The proto-star shrinks towards the MS as a T Tau star, and later, if massive enough, as a Herbig AeBe star. During early stages of collapse much magnetic flux may be retained, or lost via ambipolar diffusion, but it is not clear how to retain flux whilst the star is almost fully convective, or how the field of some Herbig AeBe stars might be related to the earlier T Tau dynamo activity phase.

On the MS, surface chemistry changes occur as a result of gravitational and radiatively driven atomic diffusion, which must compete with convection, meridional circulation and turbulent diffusion, mass loss from the surface, the magnetic field (especially in the upper atmosphere and beyond), and possibly even accretion. The complications of this multi-dimensional competition probably explain why some stars with fossil fields are marked by very distinctive chemical peculiarities (hence 'CP' stars) while other, hotter magnetic stars are not, but the details of the effects are still far from clear. However, the way in which different effects predominate with changes in stellar parameters such as mass, age and rotation rate make these stars excellent laboratories for the study of the various important internal physical processes.

Magnetic fields have been known in peculiar B stars up to about 8 $M_\odot$ for decades, but are now being found in a (still small) fraction of stars well above this mass, mostly without chemical peculiarities (e.g. Bouret *et al.* 2008 and refs. therein). These fields have significant effects on the strong stellar winds, and in extreme cases wind gas is trapped in closed field loops (Landstreet & Borra 1978; Babel & Montmerle 1997); this effect has been beautifully modelled in detail by Townsend *et al.* (2007).

The host stars of these fossil fields straddle the boundary between low mass stars which have greatly increased luminosity as giants, and high-mass stars that evolve at almost constant luminosity as supergiants. Note, that during the MS phase, R increases by a factor of 2, and $T_{\rm eff}$ decreases by 30%. The magnetic fields also pose major puzzles. Why are no fields, dynamo or fossil, found in MS F3 – F5 stars? Why do fields in MS Ap stars appear to decay on a time scale short compared to the Cowling time but still a significant fraction of the MS lifetime (Landstreet *et al.* 2008)? How and when do strong magnetic fields influence rotation rates? What happens to fields as the host stars evolve to the giant phase – can the fossil field anchor to stable layers deep in the star and persist in spite of the deep convection? Can fossil fields survive the giant phases to the end, to become the fossil fields of white dwarfs, or do these have a different origin, perhaps in a close binary system?

2. Fossil fields in Ap stars and degenerate stars

Over the 60 years since magnetic fields were discovered in Ap stars the subject has accompanied the development of magnetohydrodynamics, as theoretical explanations have been sought to explain the properties of the observed 'oblique rotators' – large-scale, static fields. Two theories have been proposed. In *core-dynamo theory* the field is generated in the core by differential rotation and convection, as is known to work in the solar envelope, and rises through the radiative zone to the surface – and herein lies a severe drawback (another being the lack of positive correlation between field strength and rotation speed): There are diffusive mechanisms to cause this rise, but the timescales are apparently too long. According to *fossil theory*, what we see is a remnant from either the original molecular cloud or a protostellar dynamo – when the star settles down, the field relaxes into a stable equilibrium, which then evolves only via the same very slow diffusive processes over timescales longer than the MS lifetime.

Theoretical work on fossil fields has focussed on demonstrating that such stable equilibria are in fact possible. This is problematic, as it can be shown that the simplest magnetic field configurations are always unstable (Goossens *et al.* 1981). Generally studies have looked only at axisymmetric fields: Tayler (1973) showed that all purely toroidal fields are prone to interchange (axisymmetric) and kink (non-axisymmetric) instabilities, and Markey & Tayler (1973) and Wright (1973) showed that purely poloidal fields are also unstable. Therefore it was concluded that a stable axisymmetric field (if it exists) must indeed have both poloidal and toroidal components. Now, such a field can be expressed as the sum of a poloidal and a toroidal component, each completely described by a single scalar function. The Lorentz force for an axisymmetric field cannot have an azimuthal (ϕ) component, since no counterpart exists in the hydrostatic fluid forces that can balance it. From this it can be shown that the two functions defining the poloidal and toroidal parts must be related. One consequence of this is that the toroidal field is entirely confined within the poloidal field lines that close inside the star. Simulations have now demonstrated that in stably stratified stars, random initial magnetic fields can evolve into roughly axisymmetric configurations with both poloidal and toroidal components, which then remain stable over a diffusive timescale (Braithwaite & Spruit 2004). It would be desirable to provide an analytic proof for the existence of such fields, and to understand how the poloidal and toroidal components help to stabilise each other. Using the energy principle developed by Bernstein *et al.* (1958), we hope to be able to show that instabilities present for toroidal fields can be eliminated by adding a poloidal component, and vice versa (Akgün & Reisenegger, in prep.) This could also allow us to determine the range of the ratio of the poloidal and toroidal components. Some progress towards that end (using a mixture of analytic and numerical methods) has been made by Braithwaite (2009), where it is found that the toroidal component can be either much stronger than, or of comparable strength to, but *not* much weaker than, the poloidal component. Meanwhile, non-axisymmetric equilibria have been found in simulations (Braithwaite 2008). Even in the most strongly magnetised stars, the Lorentz force is still typically a million times weaker than hydrostatic forces due to pressure and gravity. Therefore, a small perturbation in the background equilibrium is sufficient to balance the Lorentz force. In radiative envelopes of massive stars and in degenerate stellar interiors, matter is non-barotropic and stably stratified, which hinders radial displacements – this allows a wider range of magnetic field structures than found in barotropic fluids (Reisenegger 2009). In Braithwaite *et al.* (in prep.) we show that a star composed of matter with a barotropic equation of state is unlikely to retain a fossil field even if equilibria are available, since magnetised regions rise buoyantly and are lost through the surface on a timescale comparable to the reconnection timescale. We also show that analogous equilibria can exist outside of stars, for instance inside radio bubbles inflated by AGN – see Fig. 1.

It is informative to look at the process of relaxation of an arbitrary field into a stable equilibrium. Equilibria have special topological properities, such as the zero-gravity case where equilibrium requires that $\nabla P = (1/4\pi)(\nabla \times \mathbf{B}) \times \mathbf{B}$ and, since the Lorentz force is perpendicular to the field, the field lines reside in isobaric magnetic surfaces. Relaxation therefore requires topological change, i.e. magnetic reconnection. Magnetic helicity $H \equiv \int \mathbf{B} \cdot \mathbf{A}\, dV$ where $\mathbf{B} = \nabla \times \mathbf{A}$ is approximately conserved during relaxation, enabling us to predict the energy (and field strength) of the final equilibrium if we know the helicity of the initial turbulent field, since the energy and helicity of the equilibrium are related

Figure 1. Field lines of two magnetohydrostatic equilibria. The equilibrium on the left is axisymmetric, and can exist in a star (giving a dipolar field on the surface) as well as inside a non-gravitating bubble. On the right, a more complex non-axisymmetric equilibrium, which can only exist in a star. [Blue shading represents surface of star.]

by the (known) characteristic scale-length of the equilibrium. Unfortunately, we have few constraints on how much helicity a protostellar dynamo is likely to produce (or destroy).

As mentioned above, there are a few mechanisms which cause a fossil field to evolve and which could also bring a core-dynamo field upwards to the surface. First, finite conductivity, which 'spreads out' the magnetic energy spatially; the timescale is $\sim 10^{10}$ yrs in a MS star. Second, buoyancy: imagine a magnetised region surrounded by a less or non-magnetised medium – since the field provides pressure without mass, the magnetised region must have a lower temperature in order to have the same total pressure and density as its surroundings. Heat diffuses inwards, buoyany rise on a timescale $\tau_{\mathrm{buoy}} \sim \beta \tau_{\mathrm{K-H}}$ where β is the ratio of thermal to magnetic pressure (realistically $> 10^6$) and $\tau_{\mathrm{K-H}}$ is the thermal timescale of the star so that $\tau_{\mathrm{buoy}} \gtrsim 10^{12}$ yrs. In neutron stars there is an equivalent mechanism where the role of temperature is replaced by electron fraction Y_e. Finally, meridional circulation: a star in solid-body rotation cannot satisfy both the momentum equation and the heat equation everywhere (the *von Zeipel paradox*), and some additional circulation is required (von Zeipel 1924, Zahn 1992, Decressin *et al.* 2009 and refs. therein.) This circulation takes place on the so-called Eddington-Sweet timescale $\tau_{\mathrm{E-S}} \sim \tau_{\mathrm{K-H}}(P/P_{\mathrm{breakup}})^2$ where $\tau_{\mathrm{K-H}}$ is the stellar thermal timescale. This is just possibly relevant for the fastest rotating Ap stars.

In summary, we expect diffusive evolution to be negligible on the MS. In contrast, in neutron stars we expect the diffusive timescales to be rather shorter, in agreement with observational evidence of field decay. Some questions remain: why do not all A stars have these strong fields, and why is there such a large range in field strengths amongst the Ap stars – at what stage during formation is the Ap destiny of the star determined?

3. Magnetism and Rotation in the Herbig Ae/Be stars

Fossil field theory (see §2) predicts that fields are present throughout the MS and also earlier, in the pre-main sequence (pre-MS) phase. To explain the slow rotation of Ap/Bp stars, magnetic braking during the pre-MS has been proposed.

To test both hypotheses, E. Alecian and collaborators performed a high-resolution spectropolarimetric survey of over 100 of the pre-MS progenitors of the A and B stars, the Herbig Ae/Be stars, discovering 7 magnetic stars among 128 observed (HD 200775, HD 72106, V380 Ori, HD 190073, NGC 6611 601, NGC 2244 201, NGC 2264 83), implying that 5% of the Herbig Ae/Be stars are magnetic, as predicted by the fossil theory. Furthermore they performed a monitoring of the magnetic stars and were able to fit the temporal variations of their Stokes V profiles using the oblique rotator model (see an example for V380 Ori in Fig. 2). They conclude that these stars host large-scale organised magnetic fields with dipole strengths between 300 G and 3 kG, as predicted by the fossil theory (e.g. Alecian *et al.* 2008, Alecian *et al.* 2009, Wade *et al.*, in prep.)

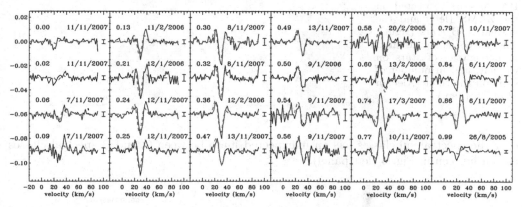

Figure 2. LSD *V* profiles of V380 Ori superimposed by the best oblique rotator model (dashed red line). The numbers close to the profiles are the rotation phase, and the small bars on the right of the profiles are the mean error bars in *V*. The profiles are sorted by increasing rotational phase, and the date of observation is indicated next to each profile (Alecian *et al.* 2009).

Then they measured the rotation velocities projected on the line of sight ($v \sin i$) of all the stars of their sample, and they first compared the $v \sin i$ distributions of the magnetic to the non-magnetic stars (Fig. 3). They find that the magnetic stars have been braked more than the non-magnetic ones, and that the braking must occur very early during the pre-MS evolution. Finally, they compared the $v \sin i$ distribution of the normal (i.e. non-magnetic and non-binary) HAeBe stars projected onto the zero-age main-sequence (ZAMS) to the normal A/B stars on the MS (Fig. 4). They find that the two distributions are very similar and conclude that the normal HAeBe stars are expected to evolve towards the ZAMS with constant angular momentum (Alecian *et al.*, in prep.)

Figure 3. $v \sin i$ histograms of the magnetic (left) and the non-magnetic (right) HAeBe stars. The y-axes are graduated in percentage of stars on the left, and in number of stars on the right (Alecian *et al.*, in prep.).

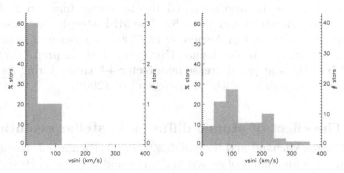

4. Magnetic fields of O and B stars: measurements, statistics and evolution

Recent measurements of the magnetic fields of early-type stars have been collected in a new catalogue of the magnetic fields of OBA stars (Bychkov *et al.* 2009). Based on the data from the catalogue together with the newest data, the statistical properties of an ensemble of the magnetic fields of OB stars were investigated. The *rms* longitudinal magnetic field $\mathcal{B}$ was used as a statistical measure of the mean field. This statistical experiment showed that this value depends weakly on the distribution of the random times of observations and on the structure of the field. The mean magnetic field averaged

Figure 4. $v \sin i$ histograms of the normal HAeBe stars projected on the ZAMS (left) and the normal A/B stars (right). The y-axes are graduated in percentage of stars on the left, and in number of stars on the right (Alecian *et al.*, in prep.).

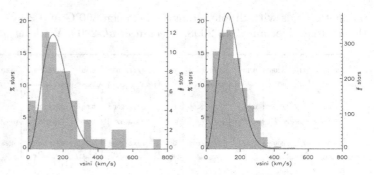

over the spectral subclasses has an unexpected jump between O and B spectral classes as can be seen in Fig. 5 (left panel).

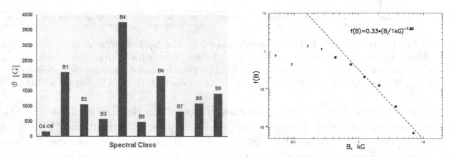

Figure 5. Left Panel: Mean magnetic fields of OB stars for different spectral subclasses. **Right Panel**: Magnetic field function for B stars.

We have calculated the differential magnetic field function (MFF) $F(\mathcal{B})$ for B stars, which is defined such that $F(\mathcal{B})d\mathcal{B}$ is the probability that a *rms* longitudinal mean field $\mathcal{B}$ lies in the interval $(\mathcal{B}, \mathcal{B} + d\mathcal{B})$ (right panel of Fig. 5). It is found that the MFF for $\mathcal{B} > 400\,\mathrm{G}$ can be approximated by the power function $F(B) = A(B/1kG)^{-\gamma}$ with parameters $A = 0.33$ and $\gamma = 1.82$. The MFF steeply decreases for $\mathcal{B} < 400\,\mathrm{G}$ – a result compatible with that of Aurière *et al.* (2007) who suggest that weak fields are destroyed by instabilities. It is also found (Kholtygin *et al.*, in prep.) that the *rms* mean magnetic field $\mathcal{B}$ of the star can decrease by a factor 4-5 times during its evolution from ZAMS to TAMS in accordance with Landstreet *et al.* (2008).

5. The effect of atomic diffusion in stellar evolution

Georges Michaud and his collaborators (O. Richard, J. Richer and M. Vick) are using abundance anomalies observed in AmFm and Horizontal Branch (HB) stars to constrain the hydrodynamics of stellar interiors. Their basic code includes all standard equations of stellar evolution to which they add 56 coupled differential equations to introduce atomic diffusion of all species included in the OPAL Rosseland opacities currently used to model stellar interiors in addition to Lithium, Berillium and Boron, which were calculated by the Montreal group. They use the original spectra that were used by OPAL to calculate their opacities. The dominant term in the transport equations involves the difference between gravitational and radiative accelerations. It is constantly recalculated for every mass shell during the evolution along with the Rosseland opacity so that the calculations take all composition changes consistently into account. For more detail see Richer *et al.* (1998), Turcotte *et al.*(1998) and Richard *et al.* (2001).

Results indicate that atomic diffusion driven by radiative accelerations can explain the abundance anomalies observed in AmFm and hot HB stars. In the absence of any competing hydrodynamical process, atomic diffusion actually leads to larger anomalies than observed. Two competing processes have now been investigated: turbulence mixing a given external mass, and mass-loss at a constant rate. Observations of Sirius A and of the Hyades star 68 Tau were found to be explained equally well by either of the two models. In the case of Sirius, 12 out of the 16 elements have their surface abundances well reproduced either by a mixed outer mass of about 10^{-5} solar mass or by a mass-loss rate of 10^{-13} solar mass per year. In the case of 68 Tau, 13 of the 14 elements are obtained within error bars. The mass-loss rate implies that after 10^8 years the matter at the surface largely comes from about 10^{-5} solar mass below the surface which is also the mixed mass in the model with turbulence. That is where the important competition between gravitational and radiative accelerations occurs in both models. Observation of pre-MS abundance anomalies would favor mass loss as a competing mechanism. For a detailed description see Richer *et al.* (2000) for the turbulence model and Vick *et al.* (2009, in prep.) for the mass loss model.

It is emphasized that there are large differences in the abundances determined by different observers and that a critical analysis of the abundance results is urgently needed to improve the situation and permit more rigorous constraints on the models.

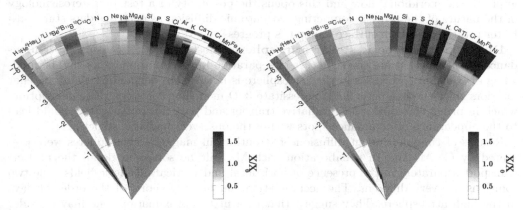

Figure 6. Concentration variations for all calculated species in two 2.50 solar mass models at 500 Myr; **Left Panel** with mass loss ($10^{-13} M_\odot$/yr) as the competing process and **Right Panel** with turbulence as the competing process. An overabundance by a factor of 1.7 or more appears black in both cases while an underabundance by a factor of 0.5 or less appears white. The radial scale is linear in r. Horizontal lines indicate the mass of the spherical shell outside a certain radius (Δm) labeled by $\log(\Delta m/M)$. As is easily seen by comparing the two panels, the interior abundances are quite different in the outer 20 % of the radius of the star but, yet, both models have the same surface abundances which also agree reasonnably well with the anomalies observed on some AmFm stars. See Vick *et al.* (2009, in prep.) for details and a similar figure at a different age. The anomalies caused by atomic diffusion affect the outer 30 % of the radius while nuclear reactions modify abundances of lighter elements within the inner 50 % of the radius leaving only a small buffer where the original abundances of the star have not been modified.

However, even if the surface abundances obtained taking either mass loss or turbulence into account are very similar, the interior composition is very different − see Fig. 6. The difference comes from the different nature of the flux in the two models. In the model with mass loss, the flux is advective and for each species approximately given by $\rho\, c(r,t)(v_D + v_W)$ where ρ is the plasma density, $c(r,t)$ is the local concentration of a given species at time t, v_D is the diffusion velocity and v_W the wind velocity. When the radiative acceleration is much larger than gravity, the v_D is positive and adds up to v_W so that $c(r,t)$ must be smaller for a given mass loss rate. In those regions of the envelope where radiative acceleration is largest on a given species, the concentration of that species is smallest in a mass loss solution. Heavy metals then tend to be underabundant between the surface convection zone and 10^{-5} solar mass below the surface whereas they are overabundant there in the model with turbulence. Helium also turns out to be distributed differently in the two models. This opens the interesting possibility of distinguishing between the two models by asterosismology. The outer $10^{-5} M_\odot$, although a small mass, occupies approximately 20 % of the radius of the star and this is where the driving often occurs most efficiently for seismic waves. The two models should have significantly different asterosismic signatures. These have not yet been calculated in detail.

On the HB, overabundances of Fe by a factor of order 50 are seen in all stars hotter than 11,000 K in the cluster M15 except for one star which has a relatively large $v \sin i$. This can be understood as the consequence of atomic diffusion driven by radiative accelerations as discussed in Michaud *et al.* (2008). The appearance of abundance anomalies in HB stars, AmFm stars and HgMn stars is explained by their slow rotation if one uses a simple meridional circulation model which involves no adjustable parameters, as is discussed in Quievy *et al.*(2009). Since this model uses advective flows, it is tempting to assume that the concentration variations seen in the advective flow of the mass loss model would also apply to the meridional flow and this opens the possibility of a test by asterosismology of the nature of the process competing with atomic diffusion as one considers stars with faster rotation in which surface anomalies progressively disappear.

Diffusion in magnetic stellar atmospheres Most magnetic CP stars show abundance variations that are the signature of separation in the atmosphere itself. The modelling of transport processes in the atmosphere is much more demanding than in stellar interiors since the magnetic fields necessitate 3–D modelling for a complete description, which in principle requires 3–D radiative transfer and particle transport. It should lead to the appearance of abundance spots within the observed magnetic structures.

Results of time dependent diffusion of strontium in magnetic atmospheres were presented by G. Alecian. In collaboration with M. Stift he solved in detail the particle transport separately in the presence of horizontal and vertical magnetic fields. The two solutions are very different. The accumulation can be very short, of the order of days in the high atmosphere. They suggest that as a first approximation, one may consider transport in the vertical direction only, since the distances are shorter; the spots are then created by the effect of the magnetic field on the vertical diffusion velocity. These results are very encouraging but many steps are still required to obtain a map of abundance anomalies at the surface of magnetic CP stars.

On the other hand, it was suggested by J. Portnoy and R. Steinitz using qualitative arguments that magnetic field gradients could lead to an additional term in the particle transport equation, perhaps explaining the over-abundance of heavier elements, and that dielectronic recombinations could be potentially important for the ionization equilibria in magnetic atmospheres.

6. Towards a coherent picture of internal transport in CP stars

The study of asteroseismology and powerful ground-based instrumentation dedicated to stellar physics is developing strongly (CoRoT, KEPLER, ESPaDOnS, etc.) generating tight constraints on the internal structure, dynamics, and magnetism of stars. For this reason stellar models are needed that include transport processes from the birth of stars to their death. A coherent picture of the dynamics of stellar radiative zones, where non-standard chemical mixing occurs, is thus required (cf. Zahn 1992, Meynet & Maeder 2000). A complex transport, *the rotational transport*, which involves several mechanisms, takes place in these regions.

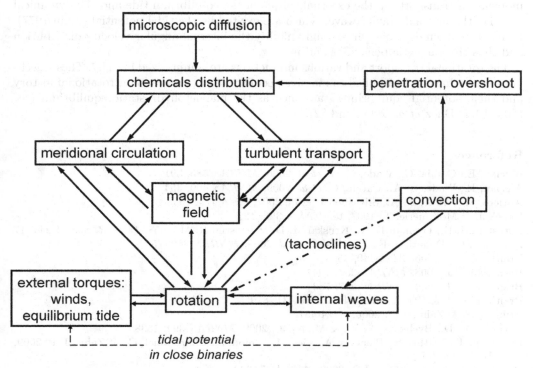

Figure 7. The complete rotational transport in stellar radiation zone.

First, rotation, structural adjustments and angular momentum losses at the surface induce 'baroclinic' or 'meridional circulation' (see §2; Maeder & Zahn 1998, Mathis & Zahn 2004, 2005) which acts simultaneously to transport angular momentum, chemicals and magnetic field by advection as well as inducing differential rotation. Next, differential rotation induces hydrodynamical turbulence via shear, baroclinic, multidiffusive and other instabilities, which of course in turn damps the differential rotation – as happens in the terrestrial atmosphere (Zahn 1983; Mathis *et al.* 2004 and refs. therein). Then, (differential) rotation interacts with fossil magnetic fields. In this case, the mean secular torque of the Lorentz force (Gough & McIntyre 1998, Mathis & Zahn 2005, Brun & Zahn 2006) and magnetohydrodynamical instabilities such as Tayler and multidiffusive magnetic instabilities (Tayler 1973, Spruit 1999, Menou *et al.* 2004) can modify the transport of angular momentum and chemicals. Eventually, a dynamo in stably-stratified stellar radiation zone is possible due to the non-linear interaction of instability-velocity field and instability-magnetic field but its detection in numerical simulations is still under exploration (Spruit 2002, Braithwaite 2006, Zahn *et al.* 2007).

Finally, internal gravity waves are excited at the boundaries with convective zones. They propagate through radiative regions where they extract or deposit angular momentum at the location where they are damped, leading to a modification of the angular velocity profile and consequently of the chemical distribution (Goldreich & Nicholson

1989; Talon & Charbonnel 2005). Because of their frequencies, they can be strongly modified by both the Coriolis acceleration (Mathis *et al.* 2008, Mathis 2009) and the Lorentz force (Kumar *et al.* 1999).

If the star is one component of a close binary system, tidal processes modify the angular momentum transport via the external torque of the equilibrium tide and the dynamical tide, i.e. the internal gravity waves which are excited by the tidal potential (Zahn 1977). Like external winds, tides can strongly modify the internal angular velocity distribution and thus the magnetic behaviour and mixing.

The rotational transport and various interactions are summarized in Fig.7. These mechanisms now have to be applied to massive peculiar stars, to study their rotational history and their potential equilibrium state such as the magnetohydrostatic equilibrium (see Moss 1977, Duez *et al.* 2009, and §2).

References

Alecian, E., Catala, C., Wade, G. A. *et al.* 2008, *MNRAS* 385, 391
Alecian, E., Wade, G. A., Catala, C. *et al.* 2009, *MNRAS* 400, 354
Aurière, M., Wade, G. A., Silvester, J. *et al.* 2007, *A&A* 475, 1053
Babel, J. & Montmerle, T. 1997, textitA&A 323, 121
Bernstein, I. B., Frieman E. A., Kruskal M. D. & Kulsrud R. M. 1958, *Proc. R. Soc. A* 244, 17
Bouret, J.-C., Donati, J.-F., Martins, F. *et al.* 2008, *MNRAS* 389, 75
Braithwaite, J. 2006, *A&A* 449, 451
Braithwaite, J. 2008, *MNRAS* 386, 1947
Braithwaite, J. 2009, *MNRAS* 397, 763
Braithwaite, J. & Spruit, H. 2004, *Nature* 431, 819
Brun, A. S. & Zahn, J.-P. 2006, *A&A* 457, 665
Bychkov, V. D., Bychkova, L. V., & Madej, J., 2009, *MNRAS* 394, 1338
Decressin, T., Mathis, S., Palacios, A., Siess, L., Talon, S., Charbonnel, C., & Zahn, J.-P. 2009, *A&A* 495, 271
Donati, J.-F. & Landstreet, J.D. 2009, *ARAA* 47, 333.
Duez, V., Mathis, S., & Turck-Chièze, S. 2009, *MNRAS*, in press (arXiv:0911.0788)
Goldreich, P. & Nicholson, P. D. 1989, *ApJ* 342, 1079
Goossens, M., Biront, D., & Tayler, R. J. 1981, *Ap&SS* 75, 521
Gough, D. O. & McIntyre, M. E. 1998, *Nature* 394, 567
Kumar, P., Talon, S., & Zahn, J.-P. 1999, *ApJ* 520, 859
Landstreet, J. D. & Borra, E. F. 1978, *ApJ* 224, L5
Landstreet, J.D., Silaj, J., Andretta V. *et al.* 2008 *A&A* 481, 465
Maeder, A. & Zahn, J.-P. 1998, *A&A* 334, 1000
Markey, P. & Tayler, R. J. 1973, *MNRAS* 163, 77
Mathis, S. 2009, *A&A* 506, 811
Mathis, S., Palacios, A., & Zahn, J.-P. 2004, *A&A* 425, 243
Mathis, S. & Zahn, J.-P. 2004, *A&A* 425, 229
Mathis, S. & Zahn, J.-P. 2005, *A&A* 440, 653
Menou, K., Balbus, S. A., & Spruit, H. C. 2004, *ApJ* 607, 564
Meynet, G. & Maeder, A. 2000, *A&A* 361, 101
Michaud, G., Richer, J., & Richard, O. 2008, *ApJ* 675, 1223
Moss, D. 1977, *MNRAS* 178, 51
Quievy, D., Charbonneau, P., Michaud, G., & Richer, J. 2009, *A&A* 500, 1163
Reisenegger, A. 2009, *A&A* 499, 557
Richard, O., Michaud, G., & Richer, J. 2001, *ApJ* 558, 377
Richer, J., Michaud, G., Rogers, F., *et al.* 1998, *ApJ* 492, 833
Richer, J., Michaud, G., & Turcotte, S.: 2000, *ApJ* 529, 338
Spruit, H. C. 1999, *A&A* 349, 189

Spruit, H. C. 2002, *A&A* 381, 923

Talon, S. & Charbonnel, C. 2005, *A&A* 440, 981

Tayler, R.J. 1973, *MNRAS* 161, 365

Townsend, R.H.D., Owocki, S.P. & Ud Doula, A. 2007, *MNRAS* 382, 139; see also http://www.astro.wisc.edu/~townsend/static.php?ref=rrm-movies

Turcotte, S., Richer, J., Michaud, G., Iglesias, C., & Rogers, F. 1998, *ApJ* 504, 539

Wright, G. A. E. 1973, *MNRAS* 162, 339

Zahn, J.-P. 1977, *A&A* 57, 383

Zahn, J.-P. 1983, *Saas-Fee Advanced Course 13, Astrophysical Processes in Upper Main Sequence Stars, eds. B. Hauck and A. Maeder, publisher: Geneva Observatory* 253

Zahn, J.-P. 1992, *A&A* 265, 115

Zahn, J.-P., Brun, A. S., & Mathis, S. 2007, *A&A* 474, 145

von Zeipel, H. 1924, *MNRAS* 84, 665

Highlights of Astronomy, Volume 15
XXVIIth IAU General Assembly, August 2009
Ian F. Corbett, ed.

© International Astronomical Union 2010
doi:10.1017/S1743921310008574

Joint Discussion 5
Modelling the Milky Way in the Era of Gaia

James J. Binney, Editor

Department of Theoretical Physics, University of Oxford, UK
email:binney@thphys.ox.ac.uk

Abstract. The body of photometric and astrometric data on stars in the Galaxy has been growing very fast in recent years (Hipparcos/Tycho, OGLE-3, 2-Mass, DENIS, UCAC2, SDSS, RAVE, Pan Starrs, Hermes, ...) and in two years ESA will launch the Gaia satellite, which will measure astrometric data of unprecedented precision for a billion stars. On account of our position within the Galaxy and the complex observational biases that are built into most catalogues, dynamical models of the Galaxy are a prerequisite full exploitation of these catalogues. On account of the enormous detail in which we can observe the Galaxy, models of great sophistication are required. Moreover, in addition to models we require algorithms for observing them with the same errors and biases as occur in real observational programs, and statistical algorithms for determining the extent to which a model is compatible with a given body of data.

JD5 reviewed the status of our knowledge of the Galaxy, the different ways in which we could model the Galaxy, and what will be required to extract our science goals from the data that will be on hand when the Gaia Catalogue becomes available.

Keywords. Galaxy: stellar content, Galaxy:evolution, Galaxy: dynamics

SOC Members: Luis A.C. Aguilar, Herwig B. Dejonghe, Kenneth C. Freeman, Ortwin Gerhard, Naoteru Gouda, Amina Helmi, Jacques R. D. Lépine, Alice C. Quillen, Annie C. R. Robin, Natalia Y. Sotnikova.

Highlights of Astronomy, Volume 15
XXVIIth IAU General Assembly, August 2009
Ian F. Corbett, ed.

The challenge raised by Gaia

Annie C. Robin

Observatoire de Besançon, Institut Utinam, Université de Franche-Comté, France
email: annie.robin@obs-besancon.fr

Abstract. Gaia will perform an unprecedented high quality survey of the Milky Way. Distances, 3D kinematics, ages and abundances will be obtained, giving access to the overall mass distribution and to the Galactic potential. Gaia data analysis will involve a high level of complexity requiring new and efficient multivariate data analysis methods, improved modelling of the stellar populations and dynamical approaches to the interpretation of the data in terms of the chemical and dynamical evolution of the Galaxy.

Keywords. Galaxy: stellar content, Galaxy: evolution, Galaxy: dynamics

---

The Gaia instruments will perform accurate photometry and astrometry up to magnitude 20 and spectroscopy to 16th magnitude. The astrometric accuracy is expected to be at the level of 10, 20, 100 μas for stars at $G = 10, 15, 20$ resp. (Brown, 2008). This astrometric accuracy will permit measurement of parallactic distances up to the Galactic center at the level of 10%. The photometry will be accurate at the level of the 0.2 mmag at G=15 and 2.5 millimag at G = 20. From the RVS, the radial velocity errors will be better than $1 \, \mathrm{km \, s^{-1}}$ for bright and cool stars, and at the end of the mission $30 \, \mathrm{km \, s^{-1}}$ for stars at $V = 16$. Gaia will also provide astrophysical parameters from the BP/RP, such as, for a star at $G = 15$, effective temperature at the level of 1 to 5%, gravity to $0.1 - 0.4$ dex, metallicity to better than 0.2 dex. Extinction will be measured on hot stars from RVS with an accuracy of $0.05 - 0.1$ mag. Thus, Gaia provides full characterization for populations to $G = 15$ (accurate distance, age, abundances, 3D velocities). Consequently, ages will be determined from the astrophysical parameters and stellar evolution models, as well as relative ages from the elemental abundances. This will enable us to trace the chemo-dynamical evolution of different populations. For example one expects to have about:

- 1.5 billion stars ($\sim 0.5\%$ of the stellar content of the Galaxy) with photometry, parallaxes and proper motions. Among them, about 9×10^8 stars belonging to the thin disc, 4.3×10^8 to the thick disc, 2.1×10^7 to the spheroid and 1.7×10^8 to the bulge.
- 200 million stars with spectroscopy ($G < 16$) (astrophysical parameters T_{eff}, $\log g$, [Fe/H], radial velocity)
- 6 million stars with elemental abundances ($G < 12$)
- Variabilities and binarity.

All populations in the Galaxy will be surveyed. Although a limited number of stars truly in the bulge will be measured with the spectrograph due to extinction and crowding, (Robin *et al*, 2005), most stars in the bulge region will have their parallaxes and proper motions measured.

In the meantime, complementary surveys are planned that will enhanced the Gaia outputs, among them LSST, RAVE, PanStarrs and JASMINE. All these surveys will efficiently complement Gaia, giving better accuracy on radial velocities for fainter stars, exploring deeper fields, furnishing denser light curves for variables, revealing dusty regions from the near infrared, etc.

Having data sets for billions of objects, covering large numbers of multi-epoch observables, the data analysis will be a real challenge. The experience with recent large scale surveys will be of some help but previous surveys have never reached this level of complexity. The question is: how to turn Gaia data into a clear understanding of Galactic dynamics and evolution?

A promising path is to use modelling as a tool for analysis, interpretation and confrontation between the data and scenarios of formation and evolution of the Milky Way. Various modelling options are pursued. Since Galactic evolution leaves traces both in the stellar kinematics and the abundances, both aspects have to be taken into account. For this reason, the stellar population-synthesis approach will be valuable, allowing us to compare scenarios of Galaxy formation and evolution with the Gaia data set by simulating catalogues with the same observables and comparable accuracies. The Besançon Galaxy model is such a project (Robin *et al*, 2003), constrained by already existing surveys like GSC2, DENIS, 2MASS, SDSS, etc., from multi-wavelength data (from X to infrared) and multivariate catalogues (photometric, astrometric and spectroscopic). It produces realistic simulations of the stellar content of the Galaxy with characteristics in agreement with our present knowledge of Galactic evolution and taking into account interstellar extinction. It is already used in Gaia preparation and is planned to be exploited for the data analysis.

Complementary to the synthetic approach, dynamical models allow the reconstruction of the Galactic potential from the space distribution and the kinematics of numerous stars or selected tracers. Several dynamical approaches are being pursued, like the Schwarzshild approach, the torus method (Dehnen & Binney 1996), N-body simulations or adaptive N-body techniques such as the "Made-to-Measure" scheme (Syer & Tremaine 1996), or quadratic programming (Dejonghe, 1989). The drawback is that the data sets are biased with parameters generally not included in the dynamical analysis – the first bias being the limiting magnitude but more complex biases also exist, such as those generated by the fact that the accuracy depends on magnitudes, colours and position on the sky, and the bias coming from interstellar extinction. As a consequence, these dynamical modelling approaches will benefit from being coupled with the synthetic approach.

Using such tools for the interpretation of Gaia data will require us to develop efficient methods of multivariate data analysis and model fitting, like genetic algorithms, or Markov Chain Monte Carlo. Inverse methods will be attempted but are not easy to handle with the large parameter space furnished by Gaia.

A huge challenge is then raised by Gaia interpretation. One can expect that no simplistic model will straightforwardly fit the data. Complemented by other surveys, the Gaia catalogue will encompass any model view. Even though imperfect, the modelling will still be useful to i) understand the imperfections of our knowledge, ii) help to interpret the findings, iii) test physical scenarios of Galaxy formation and evolution, iv) describe the dynamics of the system, in relation with the dark matter content, and v) place the scenario of formation in the cosmological context and constraints cosmological models.

References

Brown, A. G. A. 2008. AIP Conf. Series 1082, 209-215.

Dejonghe, H., 1989, *ApJ*, 343, 113.

Dehnen, W., Binney, J., 1996, ASP Conference Series, Vol. 92, pp. 393.

Robin, A. C., Reylé, C., Derrière, S., & Picaud, S. 2003, *A&A*, 409, 523

Robin, A. C., Reylé, C., Picaud, S., Schultheis, M. 2005. *A&A*430, 129.

Syer, D., Tremaine, S., 1996, *MNRAS*282, 223.

Highlights of Astronomy, Volume 15
XXVIIth IAU General Assembly, August 2009
Ian F. Corbett, ed.
© International Astronomical Union 2010
doi:10.1017/S1743921310008598

Dynamics and history of the Milky Way

Amina Helmi

Kapteyn Astronomical Institute, University of Groningen, The Netherlands
email: ahelmi@astro.rug.nl

Abstract. The structure and dynamics of the Galaxy contain information about both its current workings and its assembly history. I review our understanding of the dynamics of the disk and stellar halo, and sketch how these may be used to unravel how our Galaxy formed.

Keywords. Galaxy: dynamics, evolution

1. Dynamics of the disk(s)

The structure and kinematics of the Galactic components constrain the mass distribution and history of the Galaxy. For example the vertical dynamics of the thin disk:

• Puts limits on the distribution of mass in the disk, most of which is accounted for by the stars (Kuijken & Gilmore 1991; Holmberg & Flynn 2004). Their contribution to the circular velocity (which itself still has large uncertainties, McMillan & Binney 2009) is roughly half of that required and thus a rounder (dark) component is needed. Also the tilt of the velocity ellipsoid rules out very flattened oblate halos (Siebert *et al.* 2008).

• Its coldness has been used to constrain the amount of recent merger events (Toth & Ostriker 1992). Small dark satellites, which are so abundant in CDM simulations (Springel et al. 2008), do not induce much heating (Font et al. 2001), but mergers of 10-20% mass ratio can significantly increase the velocity dispersion of the stars.

Stewart *et al.* (2008) have found that 70% of dark-matter halos similar to that of the Milky Way ($\sim 10^{12} \, M_\odot$) have experienced a merger with an object of $\sim 10^{11} \, M_\odot$ (i.e. mass comparable to the thin disk's) in the last 10 Gyr. This would therefore be a plausible origin for the thick disk (Kazanzidis *et al.* 2008; Villalobos & Helmi 2009), also if we take into account the age distribution of its stars (Bensby & Feltzing 2009). However, this merger rate does not account for possible environmental dependences – the Local Group is in a low density region of the Universe, which must imply a smaller chance of encounters. Note as well that this class of mergers are less damaging for gas-rich disks, which is the relevant case for those lookback times (Hopkins *et al.* 2008).

Other models, besides the minor-merger scenario, have also been proposed for the formation of the thick disk. Abadi et al. (2003) suggest that it may result purely from the accretion of satellites on low inclination orbits, while Brook *et al.* (2004) find that a thick component might form early on during gas rich mergers (a different but also gaseous formation scenario has been put forward by Bournaud *et al.*, 2008). Recently Schönrich & Binney (2009) have proposed that the thick disk is composed by stars which have migrated radially via resonant mechanisms from the inner thin disk.

The dynamics of thick-disk stars encode which of these mechanisms has been dominant in the formation of this component. Recently, Sales *et al.* (2009) have shown that the eccentricity distribution is a particularly powerful discriminant. In all scenarios where the majority of the stars are formed in-situ (minor merger, gas rich mergers or migration), the distribution has a prominent peak at low eccentricity. On the other hand, when the whole disk is built by accretion, the eccentricity distribution is predicted to be flatter, reflecting the range of orbital eccentricities of satellites found in cosmological simulations.

2. Dynamics of the stellar halo

The dynamics of halo stars, and especially of those in streams, can be used to constrain: i) the total mass of the Galaxy and its spatial distribution (e.g. density and shape of the dark matter halo); and ii) the merger history of the Galaxy, as accreted objects will often deposit their debris in this component.

Models of the Sgr streams have yielded conflicting results favouring spherical, oblate or slightly prolate shapes for the Galactic dark halo depending on the set of observations used (Helmi 2004; Johnston *et al.* 2005; however see Law *et al.*(2009) who suggest it may be triaxial). Narrow streams are arguably better-suited to derive the gravitational potential in the region probed by their orbits (e.g. Eyre & Binney 2009). Koposov *et al.*(2009) have modelled GD-1 and been able to constrain the circular velocity at the Sun to be $\sim 224 \pm 13\,\mathrm{km\,s^{-1}}$, and the shape of the potential (including disk and halo) to have a global flattening $q_\phi \sim 0.87$.

Very high-resolution cosmological CDM simulations in combination with semi-analytic models of galaxy formation may now be used to make detailed predictions on the properties of the stellar halo (and in particular the accreted component, De Lucia & Helmi 2008; Cooper *et al.*, in prep.). The most recent such studies show good agreement with observations, revealing the presence of broad streams such as those from Sgr (typically originated in massive recently objects), and very narrow features, akin the Orphan Stream (Belokurov *et al.*, 2007). Furthermore, in these simulations the very chaotic build up characteristic of the hierarchical structure formation paradigm endows the stellar halo near the Sun with much kinematic substructure (Helmi *et al.*, in prep).

References

Abadi, M. G., Navarro, J. F., Steinmetz, M., & Eke, V. R. 2003, *ApJ* 597, 21
Belokurov, V., *et al.* 2007, *ApJ*, 658, 337
Bensby, T., & Feltzing, S. 2009, *arXiv:0908.3807*
Bournaud, F., Elmegreen, B. G., & Elmegreen, D. M. 2007, *ApJ* 670, 237
Brook, C. B., Kawata, D., Gibson, B. K., & Freeman, K. C. 2004, *ApJ* 612, 894
De Lucia, G., & Helmi, A. 2008, *MNRAS*, 391, 14
Eyre A. & Binney, J., 2009, arXiv:0907.0360
Font, A. S., Navarro, J. F., Stadel, J., & Quinn, T. 2001, *ApJ*, 563, L1
Helmi, A. 2004, *ApJ*, 610, L97
Holmberg, J. & Flynn, C. 2004, *MNRAS* 352, 440
Hopkins, P. F., Hernquist, L., Cox, T. J., Younger, J. D., & Besla, G. 2008, *ApJ* 688, 757
Johnston, K. V., Law, D. R., & Majewski, S. R. 2005, *ApJ*, 619, 800
Kazantzidis, S., *et al.* 2008, *ApJ* 688, 254
Koposov, S. E., Rix, H.-W., & Hogg, D. W. 2009, *arXiv:0907.1085*
Kuijken, K. & Gilmore, G. 1991, *ApJ* 367, L9
Law, D. R., Majewski, S. R., & Johnston, K. V. 2009, *ApJ*, 703, L67
McMillan, P. J. & Binney, J. J. 2009, *arXiv:0907.4685*
Sales, L. *et al.* 2009, *arXiv:0909.3858* (*MNRAS* in press)
Schönrich, R. & Binney, J. 2009, *MNRAS* 1255
Siebert, A., *et al.* 2008, *MNRAS* 391, 793
Springel, V., *et al.* 2008, *MNRAS* 391, 1685
Stewart, K. R. *et al.* 2008, *ApJ* 683, 597
Toth, G. & Ostriker, J. P. 1992, *ApJ* 389, 5
Villalobos, Á. & Helmi, A. 2009, *arXiv:0902.1624* (*MNRAS* in press)

Highlights of Astronomy, Volume 15
XXVIIth IAU General Assembly, August 2009
Ian F. Corbett, ed.

© International Astronomical Union 2010
doi:10.1017/S1743921310008604

Non-equilibrium Dynamical Processes in the Galaxy

Alice. C Quillen and Ivan Minchev

Dept. of Physics and Astronomy, University of Rochester,
Rochester, NY 14627, USA

Université de Strasbourg, CNRS, Observatoire Astronomique
11 Rue de l'Université, 67000 Strasbourg, FRANCE

Abstract. Dynamical models have often necessarily assumed that the Galaxy is nearly steady state or dynamically relaxed. However observed structure in the stellar metallicity, spatial and velocity distributions imply that heating, mixing and radial migration has taken place. Better comprehension of non-equilibrium processes will allow us to not only better understand the current structure of the galaxy but its past evolution.

During a Hubble time the Milky Way disk at the radius of the Sun has only had time to rotate 40 or 50 times. There is little time for dynamical relaxation. As larger and more precise surveys are conducted we expect even more structure to be revealed in the stellar abundance and phase space distributions. As there is little time for relaxation, structure in the phase space distribution depends on the evolution of the Galaxy.

Resonances with a bar or spiral structure cause stars to move in non-circular orbits. Libration times in Lindblad resonances can be long so evolution could take place in the non-adiabatic limit. Minchev *et al.* (2009b) proposed that the division between the Pleiades and moving groups in the solar neighborhood is associated with librations in the 2:1 Lindblad resonance with the Galactic bar (see Figure 1). These oscillations are also seen as long lived R1/R2 asymmetric ring structures in test particle simulations of bar growth (Bagley *et al.* 2009).

If pattern speeds vary, then either particles are trapped into resonance or heated as they cross the resonance. When particles are trapped into resonance their eccentricity depends on the total pattern speed change after capture. When particles cross the resonance their eccentricity can be predicted from the resonance strength and order. Bagley *et al.* (2009) suggested that the morphology of ring structures associated with a bar depends on bar pattern speed variations since growth. Peanut shaped bulges can also be modeled with a resonant trapping model (Quillen 2002). When there is more than one perturbation, chaotic heating occurs in resonances (Quillen 2003, Minchev & Quillen 2006).

Since resonances are often narrow, they can be used to place tight constraints on their pattern speed. Their location in the galaxy could be used to measure the pattern speed of a distant spiral pattern with a deep radial velocity survey (Minchev & Quillen 2007). Resonances occur when the sum of integer multiple of a star's orbital frequencies is equal to an integer multiple of a perturbing frequency such as a planet's mean motion or the pattern speed of a bar. In the Galaxy the orbital period is estimated from the tangential or v velocity component at a particular location. Thus resonances can be located on a u, v plane velocity distribution. Divisions between streams can be used to estimate bar or spiral pattern speeds (Dehnen 1999, Quillen & Minchev 2005). By matching both the velocity distribution in the solar neighborhood and simulated Oort

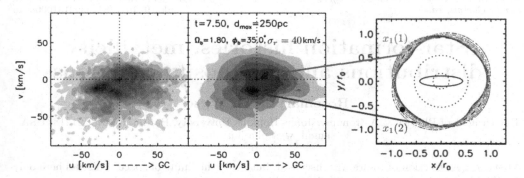

Figure 1. Orbits associated with the Bar. On the left we show the Hipparcos stellar velocity distribution (Dehnen 1998). The middle panel shows a model distribution by Minchev *et al.* (2009b). On the right we show orbits associated with the Sirius and Pleiades moving groups.

function measurements (that depend on velocity gradients) Minchev *et al.* (2007) placed an even tighter constraint on the bar pattern speed.

Differences between orbital frequencies cause the velocity and spatial distribution of stars in a narrow region in phase space to spread. This process is called "phase wrapping" and can be used to estimate the time since a merger occurred (Gomez & Helmi 2009). Uneven distributions in phase space could also be caused by large scale perturbations to the disk. The timescale for them to wrap places constraints on the time since perturbation (Minchev *et al.* 2009). These scenarios are proposed explanations for high velocity streams in the thick disk of the Galaxy. Both mergers (Quillen *et al.* 2009) and resonances (Sellwood & Binney 2002) cause radial migration. Future work can better explore the relation between structure in the phase space and abundance distributions.

In summary, dynamical structures and events leave signatures in the stellar distributions. Precise measurements can be made as observations and associated models become more comprehensive. Unveiling the current and past structure and evolution of the Galaxy will be increasingly exciting in the coming decade.

References

Bagley, M., Minchev, I., & Quillen, A. C. 2009, *MNRAS*, 395, 537
Dehnen, W. 1998, *AJ*, 115, 2384
Dehnen, W. 1999, *ApJ*, 524, L34
Gomez, F. & Helmi, A., *MNRAS* in press, 2009, arXiv0904.1377
Minchev, I. & Quillen, A. C. 2008, *MNRAS*, 386, 1579
Minchev, I. & Quillen, A. C. 2006, *MNRAS*, 368, 623
Minchev, I., Boily, C., Siebert, A., & Bienayme, O. 2009, submitted to *MNRAS*, arXiv0909.3516
Minchev, I., Quillen, A. C., Williams, M., Freeman, K. C., Nordhaus, J., Siebert, A., & Bienaymé, O. 2009, *MNRAS*, 396, L56
Minchev, I., Nordhaus, J., & Quillen, A. C. 2007, *ApJ*, 664, L31
Quillen, A. C., Minchev, I., Bland-Hawthorn, J., & Haywood, M. 2009, *MNRAS*, 397, 1599
Quillen, A. C., & Minchev, I. 2005, *AJ*, 130, 576
Quillen, A. C. 2003, *AJ*, 125, 785
Quillen, A. C. 2002, *AJ*, 124, 722
Sellwood, J. A. & Binney, J. J. 2002, *MNRAS*, 336, 785

Highlights of Astronomy, Volume 15
XXVIIth IAU General Assembly, August 2009
Ian F. Corbett, ed.

© International Astronomical Union 2010
doi:10.1017/S1743921310008616

Star-formation histories, metallicity distributions and luminosity functions

Rosemary F. G. Wyse

Department of Physics & Astronomy, Johns Hopkins University, Baltimore, MD 21218, USA
email: wysepha.jhu.edu

Abstract. A selection of topics was discussed, but given the limits of space, I discuss here only the IMF in any depth, with only a brief supplementary comment on the bulge.

1. The Stellar Initial Mass Function

The stellar Initial Mass Function is a fundamental aspect of star formation. Variations in the IMF are predicted in many theories, particularly those that invoke the Jeans mass, and associated dependences on cooling rates and pressure. The environment of the Galactic Center is rather different from the solar neighbourhood, and the dense, massive, young star clusters there provide an interesting test of the variation, or not, of the massive-star IMF. Indeed, early star-count observations had suggested a flatter IMF in the Arches star cluster, some 25 pc in projection from the Galactic Center and with age $\lesssim 3$ Myr, than in the solar neighbourhood (Figer *et al.* 1999). However, the most recent data, obtained with the NACO Adaptive Optics imager on the VLT, are entirely consistent with a standard, Salpeter (1955), slope to the IMF above $10\,M_\odot$ (Espinoza, Selman & Melnick 2009), *modulo* a level of mass segregation in the inner regions.

The level of variation in generations of massive stars that have since exploded as core-collapse supernovae can be constrained through analysis of the elemental abundances in low-mass stars that they pre-enriched. In general, if the massive-star IMF is biased towards more massive stars, the predicted ratio of alpha-elements to iron will be higher, due to the dependence of the nucleosynthetic yields on progenitor mass, within the range of core-collapse progenitors, $\sim 10\,M_\odot$ to $\sim 50\,M_\odot$ (e.g. Kobayashi *et al.* 2006). Stars formed early in the history of self-enrichment of a system will show signatures of only core-collapse supernovae, and when these stars can be identified, they show remarkably little scatter in most elemental abundance ratios, even down to very low levels of [Fe/H] (e.g. Cayrel *et al.* 2004). This constancy, within estimated errors, implies not only an invariant massive-star IMF, but also efficient mixing so that IMF-averaged yields are achieved (e.g. Wyse & Gilmore 1992; François *et al.* 2004). It should be noted that the conclusion of the invariance of the IMF from the lack of scatter is not dependent on the details of the supernova models, nor on a chemical evolution model, but is a robust conclusion. Of course, determination of what the IMF is from the value of the elemental ratios would be sensitive to models.

These conclusions hold for stars in diverse environments such as the old, metal-rich bulge (e.g. Fulbright, McWilliam & Rich 2007) and metal-poor halo. Even in the metal-poor dwarf spheroidals, where the bulk of the member stars show lower values of [α/Fe] (e.g. Venn *et al.* 2004), consistent with their broad internal spread of ages and likely self-enrichment by Type Ia supernovae (see Unavane, Wyse & Gilmore 1996), the most metal-poor – and presumably oldest – stars show the same enhanced ratio as in the field halo (e.g. Koch *et al.* 2008).

Extending this analysis to larger samples of extremely metal-poor stars is possible through exploitation of the database from the RAVE moderate-resolution spectroscopic survey of bright stars ($I < 12$; see Steinmetz *et al.* 2006 for an introduction to the survey), for which 4–8m-class telescopes suffice for follow-up high-resolution data. Preliminary results (Fulbright, Wyse, Grebel *et al.*, in prep) are very encouraging. Extremely metal poor stars, with [Fe/H] < -3 dex, have also been identified in dwarf spheroidal galaxies (e.g. Kirby *et al.* 2008; Norris *et al.* 2008), predominantly in the 'ultra-faint' systems discovered by analysis of the imaging data of the Sloan Digital Sky Survey (e.g. Belokurov *et al.* 2006, 2007).

As noted below, the ages of the bulge and halo stars are equal to look-back times corresponding to redshifts of 2 and above, implying constancy of the massive star IMF back to these early stages of the Universe, in a wide range of physical conditions.

The low-mass luminosity function can be determined in most systems by straightforward star counts, and again the constancy or otherwise of the IMF can be tested in systems that probe a wide range of physical parameters: the high-metallicity, old, dense stellar bulge; the low-metallicity, old, diffuse, dark-matter dominated dwarf spheroidals; the low-metallicity, old, dense Globular star clusters; and lastly the young disk star and open clusters. Again, all is consistent with an invariant low-mass IMF (e.g. Wyse 2005).

In terms of modelling the Galaxy, the IMF should be held fixed.

2. The Bulge

The conclusion that the bulge consists of exclusively old stars – at least those regions probed by low-reddening windows – is not new, but has been given considerably more weight by a recent HST-based analysis by Clarkson *et al.* (2008) in the Sgr low-extinction window. These authors used multi-epoch observations to identify foreground disk stars, through their proper motions. After removal of these younger stars, the colour-magnitude diagram remaining is that of an old, metal-rich population. However, much more data are required to map the stellar population in the bulge, and understand how it connects to the stellar halo, and to the inner thin and thick disks.

References

Belokurov, V. *et al.* 2006, *ApJ*, 647, L111; 2007, *ApJ*, 654, 897
Clarkson, W. *et al.* 2008, *ApJ*, 684, 1110
Figer, D. F. *et al.* 1999, *ApJ*, 525, 750
Cayrel, R. *et al.* 2004, *A&A*, 416, 1117
Espinoza, P., Selman, F. J., & Melnick, J. 2009, *A&A*, 501, 563
Figer, D. F. *et al.* 1999, *ApJ*, 525, 750
François, P. *et al.* 2004, *A&A*, 421, 613
Fulbright, J., McWilliam, A., & Rich, R.M. 2007, *ApJ*, 661, 1152
Kirby, E. *et al.* 2008, *ApJ*, 685, L43
Kobayashi, C. *et al.* 2006, *ApJ*, 653, 1145
Koch, A. *et al.* 2008, *AJ*, 135, 1580
Nissen, P. *et al.* 1994, *A&A*, 285, 440
Norris, J. E. *et al.* 2008, *ApJ*, 689, L113
Salpeter, E. E. 1955, *ApJ*, 121, 161
Steinmetz, M. *et al.* 2006, *AJ*, 132, 1645
Unavane, M., Wyse, R. F. G., & Gilmore, G. 1996, *MNRAS*, 278, 727
Venn, K. *et al.* 2004, *AJ*, 128, 1177
Wyse, R. F. G. 2005, in *The IMF 50 years later*, eds Corbelli *et al.*, (Springer) p. 201
Wyse, R. F. G. & Gilmore, G. 1992, *AJ*, 104, 144

Highlights of Astronomy, Volume 15
XXVIIth IAU General Assembly, August 2009
Ian F. Corbett, ed.

© International Astronomical Union 2010
doi:10.1017/S1743921310008628

Structure and evolution of the Milky Way: the interstellar medium perspective

François Boulanger

Institut d'Astrophysique Spatiale
email: francois.boulanger@ias.u-psud.fr

Abstract. The Herschel and Planck satellites have started imaging the sky at far-IR to mm wavelengths with an unprecedented combination of sky and spectral coverage, angular resolution, and sensitivity, thus opening the last window of the electromagnetic spectrum on the Galaxy. Dedicated observing programs on Herschel and the Planck all-sky survey will provide the first complete view at cold dust across the Galaxy, opening new perspectives on the structure and dynamical evolution of the Milky Way relevant to Gaia. The analysis and modelling of these observations will contribute to our understanding of two key questions: how do stars form from interstellar matter? how are the interstellar medium and the magnetic field dynamically coupled? The comparison with Gaia observations will contribute to build a 3D model of the Galactic extinction taking into account dust evolution between ISM components

Keywords. Interstellar medium, Star formation, Magnetic field, Interstellar dust

---

1. Introduction

The Gaia mission will produce an extraordinary stellar database and we have to consider how the data may be used to build a dynamical model of the Milky Way that will unravel its past history. To understand the evolution of the Galaxy, we need a model that describes physically how star formation proceeds from the chemical and thermo-dynamical evolution of interstellar matter. This is a vast field of research. At the meeting I highlighted the prospect of imminent advances made possible by the successful launch of the Herschel and Planck satellites. Here I focus on two main topics related to star formation: the inventory of the cold interstellar medium across the galaxy, and the structure of the Galactic magnetic field and its coupling to interstellar matter.

2. Cold interstellar matter across the Galaxy

Herschel and Planck surveys will provide the small-scale (down to the detection of individual pre-stellar cores) and global views of the distribution of cold interstellar matter in the Galaxy. Both missions will image the far-IR emission from large ($> 10\,\mathrm{nm}$) grains that account for the bulk of the dust mass. The spectral coverage will allow us to determine the dust temperature and thereby infer dust column densities and masses. With an empirical determination of the dust-to-gas mass ratio, the infrared brightness becomes a tracer of interstellar gas. It complements usual interstellar matter tracers, like the HI and CO line emission, in a unique way because the dust emission is independent of the chemical composition and physical conditions of the gas. For the first time we will have access to a complete inventory of cold interstellar matter in the Galaxy. This step forward opens several key perspectives.

The dust temperature may be used to identify the cold infrared emission from dense condensations within molecular clouds. In doing this, we will quantify the mass and distribution of matter that is presently susceptible to collapse into stars. The data will be sensitive enough to look for dust emission from High Velocity Clouds, the Magellanic Stream and the outer disk. This will help us to quantify the mass-inflow rate available to sustain star formation over the past history of the Galaxy. The dust seen in emission by Herschel and Planck is the dust that makes the extinction at optical and near-IR wavelengths. The data analysis will involve the characterisation of dust evolutionary processes that may account for the observed variations in the optical and near-IR dust extinction curve. The outcome of these dust studies will need to be taken into account to build the 3D model of extinction for Gaia.

3. The Galactic magnetic field

The Galactic magnetic field and cosmic-rays are tied to the interstellar gas. Their dynamical coupling is a prime facet of interstellar-medium physics, and many questions remain quantitatively open due to the paucity of data on the small-scale structure of the magnetic field. Planck will map the polarisation of the dust and of the synchrotron emission. The two emissions provide complementary perspectives on the structure of the Galactic magnetic field. Dust grains, unlike relativistic electrons, are coupled to the interstellar gas. Thus, the dust polarisation traces the magnetic field within the thin Galactic disk where matter is concentrated and within interstellar clouds, while the synchrotron polarisation probes the field over the whole volume of the Galaxy up to its halo. The novelty of upcoming observations ensures major progress in our understanding of the magnetised Galactic interstellar medium.

Polarisation of the dust emission results from the presence in the ISM of elongated grains with a preferred orientation. Several alignment mechanisms have been proposed. They are expected to work through interstellar clouds even if their efficiency may depend on the gas density and radiation field. It is thought to be generally true that the magnetic field acts on elongated grains so they spin with their long axes perpendicular to the field. In this case, the direction of polarisation in emission is perpendicular to $B_\perp$ the magnetic field component in the plane of sky. The degree of polarisation depends on dust properties (e.g. which grains are aligned), the efficiency of the alignment mechanism and the structure of the magnetic field within the beam.

Measurement by Planck of the polarisation of the thermal emission from aligned grains provide an unprecedented means to map continuously the orientation of the magnetic field within the ISM, from diffuse clouds to dense molecular gas. The Galactic magnetic field is commonly described as a vector sum of a regular and a random component. A first goal will be to complement existing models of the regular component. To fully describe the ordered field, we will face two open questions: (i) what is the impact of nearby bubbles powered by massive star associations on the field structure? (ii) how is the field within the thin Galactic disk, where gas and star formation is concentrated, connected to the thicker disk and the Galactic halo? The turbulent component results from the dynamical interaction between the field and interstellar turbulence. The data will also allow us to study the geometry of the magnetic field in relation to the density structure and kinematics of interstellar clouds derived from dust and gas maps. The degree of randomness in the magnetic field orientation may be combined with Doppler measurements of the turbulent gas velocity to measure the magnetic field intensity and quantify the dependence on gas density. These investigations will test theoretical and numerical studies, stressing the importance of the magnetic field in the dynamical evolution of the interstellar medium and the regulation of the star-formation efficiency.

Highlights of Astronomy, Volume 15
XXVIIth IAU General Assembly, August 2009   © International Astronomical Union 2010
Ian F. Corbett, ed.   doi:10.1017/S1743921310008641

The Milky Way Halo and the First Stars: New Frontiers in Galactic Archaeology

Timothy C. Beers[1,2], Jason Tumlinson[3], Brian O'Shea[1,2], Carolyn Peruta[1,2] Daniela Carollo[4,5]

[1] Department of Physics & Astronomy, Michigan State University,
email: beers@pa.msu.edu

[2] Joint Institute for Nuclear Astrophysics

[3] Space Telescope Science Institute

[4] Research School of Astronomy & Astrophysics, ANU, Australia

[5] INAF, Osservatorio Astronomico di Torino

Abstract. We discuss plans for a new joint effort between observers and theorists to understand the formation of the Milky Way halo back to the first epochs of chemical evolution. New models based on high-resolution N-body simulations coupled to simple models of Galactic chemical evolution show that surviving stars from the epoch of the first galaxies remain in the Milky Way today and should bear the nucleosynthetic imprint of the first stars. We investigate the key physical influences on the formation of stars in the first galaxies and how they appear today, including the relationship between cosmic reionization and surviving Milky Way stars. These models also provide a physically motivated picture of the formation of the Milky Ways "outer halo," which has been identified from recent large samples of stars from SDSS. The next steps are to use these models to guide rigorous gas simulations of Milky Way formation, including its disk, and to gradually build up the fully detailed theoretical "Virtual Galaxy" that is demanded by the coming generation of massive Galactic stellar surveys.

Keywords. astronomical data bases: surveys, Galaxy: halo, structure, methods: data analysis, n-body simulations, stars: abundances

The explosion of detailed astrometric, photometric, and spectroscopic data for stars in the Milky Way (and Local Group galaxies) that is coming upon us both now (e.g., SDSS/SEGUE, RAVE), and in the near future (PanSTARRS, SkyMapper, Gaia, LSST, SIM Lite) will fundamentally change our vision of galaxy formation and evolution. Full exploitation of this wealth of new information requires the development of sophisticated numerical models capable of producing testable predictions against which the observations can be compared. We have initiated one such effort, foreshadowed by the work of Tumlinson (2006), and continued by Tumlinson (2009). The new models predict the locations, kinematics, and chemistry of the stars in the Galactic halos, with different assumptions concerning the nature of the first stars and their effects on subsequent stellar generations. We are also exploring new methods for visualizing both the predictions and the existing databases. The first efforts will compare expectations with observed differences in the inner- and outer-halo populations, e.g., as in Carollo et al. (2007).

References

Carollo, D., et al. 2007, Nature, 450, 1020
Tumlinson, J. 2006, ApJ, 641, 1
Tumlinson, J. 2009, ApJ, submitted

Highlights of Astronomy, Volume 15
XXVIIth IAU General Assembly, August 2009
Ian F. Corbett, ed.

© International Astronomical Union 2010
doi:10.1017/S1743921310008653

Physics and Structure of the Galactic disc(s)

Ralph A. Schönrich

Max Planck Institute for Astrophysics,
Garching, D-85741, Garching, Germany
email: rasch@mpa-garching.mpg.de

Abstract. The model of Schönrich & Binney (2009) offers new ways to understand the chemo-kinematic structure of the solar neighbourhood in the light of radial mixing. The combination of chemical information with rich kinematic data reveals a still hardly explored abundance of interconnections and structures from which we can learn about both the physics and history of our Galaxy. Large upcoming datasets can be used to improve estimates of central parameters, to shed light on the Galaxy's history and to explore the unexpected way of understanding the well-known division of the Galactic disc yielded by the new model.

Keywords. Galaxy: structure, Galaxy: evolution, Galaxy: abundances, Galaxy: kinematics and dynamics

---

Radial mixing has been a vastly neglected process in modelling galactic discs until it was shown to be crucial for disc evolution from a theoretical point of view by Sellwood & Binney (2002). Schönrich & Binney (2009) demonstrated that under regular assumptions about star formation and disc structure, a perfect fit to the metallicity distribution of the Geneva-Copenhagen Survey (Nordström et al. 2004, Holmberg *et al.* 2007) was possible if radial mixing was allowed for, and that no acceptable fit could be achieved without it. This new model of chemical evolution gives a completely different history of the disc compared to classical approaches. Since a large range of different galactocentric radii contribute their populations to local datasets, there is no need for the local star-forming interstellar medium to have followed the observed density ridges in the $[\alpha/Fe]$, $[Fe/H]$ plane. Among the most important differences between classical modelling without radial mixing and models including radial migration are the predicted correlations between chemistry and kinematics along the thin-disc ridge line. In larger datasets containing $[\alpha/Fe]$ ratios, such as that of Borkova & Marsakov (2005), thin-disc stars can be picked out from stars belonging to the thick disc and halo by selecting for low $[\alpha/Fe]$ and dropping obvious halo stars. In this subsample one finds a highly significant increase of rotational velocity along the thin-disc ridge line moving from higher to lower metallicities, in nice concordance with the SB09 model. In the classical models the metal-poor stars should be the oldest thin-disc stars, so the increase in rotational velocity should be associated with an increase in velocity dispersion. The data do not show a significant increase towards lower metallicities, so the classical view is in conflict with the data.

References

Borkova, T. V. & Marsakov, V. A. 2005 *ARep*, 49, 405
Holmberg, J., Nordström, B., & Andersen, J. 2007 *A&A*, 475, 519
Nordström, B., Mayor, M., Andersen, J., Holmberg, J., Pont, F., Jörgensen, B.R., Olsen, E. H., Udry, S., & Mowlawi, N. 2004 *A&A*, 418, 989
Schönrich, R. & Binney, J 2009, *MNRAS*, 396, 203
Sellwood, J. A. & Binney, J. 2002, *MNRAS*, 336, 785

Highlights of Astronomy, Volume 15
XXVIIth IAU General Assembly, August 2009
Ian F. Corbett, ed.

© International Astronomical Union 2010
doi:10.1017/S1743921310008665

Gas flows within the Galaxy

Francoise Combes

LERMA, Observatoire de Paris, 61 Av. de l'Observatoire, F-75014, Paris
email: francoise.combes@obspm.fr

Abstract. In the recent years, more and more sophisticated models have been proposed for the gas distribution and kinematics in the Milky Way, taking into account the main bar, but also the possible nuclear bar, with the same or different pattern-speed. I review the success and problems encountered by the models, in particular in view of the new discovery of a symmetrical far-side counterpart of the 3 kpc arm. The inner part, dominated by the bar, and the outer parts, dominated by the spiral arms, can be observed from a virtual solar position, and the errors coming from kinematical distances are evaluated. The appearance of four arms could be due to a deprojection bias.

Keywords. Galaxy, dynamics, interstellar medium, bars, kinematics

1. Introduction

Reconstruction of the Galaxy's spiral arms is difficult given our internal perspective. Distances of the various features or objects are derived through a kinematical model, with near-far ambiguities inside the solar circle. One of the most successful models is that from Georgelin & Georgelin (1976) of four tightly-wound arms, traced by OB associations, optical or radio HII regions and molecular clouds. The best tracer of the Galaxy structure is the gas, atomic (HI, Liszt & Burton 1980) and molecular (CO surveys, Dame *et al.* 2001), because of its low velocity dispersion, and its confinement to the plane.

Position-velocity (P-V) diagrams are particularly instructive, revealing the high-velocity ($\Delta V = 560 \, \mathrm{km \, s^{-1}}$) Central Molecular Zone (CMZ) near zero longitude, with a molecular ring, connecting arm, 3 kpc arm, etc. The existence of a bar has long been suspected from non-circular motions towards the center, and has been directly confirmed by COBE and 2MASS (e.g. Lopez-Corredoira *et al.* 2005). Near-infrared images show clearly the peanut bulge, which is thought to be formed through vertical resonance with the bar (e.g. Combes *et al.* 1990). The CMZ has a peculiar parallelogram shape in the P-V diagram (Bally *et al.* 1988), that has been first interpreted in terms of cusped x_1 and almost circular x_2 periodic orbits, and associated gas flows, by Binney *et al.* (1991). Then Fux (1999) carried out fully self-consistent N-body and hydrodynamical simulations of stars and gas to form a barred spiral, and fit the Milky Way. He succeeded remarkably to reproduce the HI and CO P-V diagrams with a bar pattern speed of about $40 \, \mathrm{km \, s^{-1} \, kpc^{-1}}$, implying corotation at 5 kpc, and an ILR producing the x_2 orbit inside 1 kpc. The spiral structure has essentially 2 arms starting at the end of the bar.

2. More recent developments

Both external galaxies and simulations frequently show evidence for several pattern speeds in the disk and a spiral is expected to rotate slower than the bar (Sellwood & Sparke 1988). For example, by modelling gas flow in a fixed potential Bissantz *et al.* (2003) conclude that in the Galaxy $\Omega_{\mathrm{p}} = 60 \, \mathrm{km \, s^{-1} \, kpc^{-1}}$ and $20 \, \mathrm{km \, s^{-1} \, kpc^{-1}}$ for the bar and spiral, respectively. Amores et al (2009) notice that there is a ring gap in the

HI distribution at about 8.3 kpc, outside the solar circle at 7.5 kpc, and propose that it corresponds to the corotation (CR) of the spiral, which will then have a pattern speed of $\sim 25\,\mathrm{km\,s^{-1}\,kpc^{-1}}$. Simulations of gas in a barred spiral do not show ring gaps at CR, but depopulated regions at the corresponding Lagrangian points, which could correspond to the observations; there is no gap in the azimuthally averaged HI and CO gas surface density.

In the 2MASS star counts, a nuclear bar has been found by Alard (2001), and there is a corresponding CO nuclear bar (Sawada *et al.* 2004). New simulations of gas flow in a two-bar models have been done by Rodriguez-Fernandez & Combes (2008), who find a best fit when the two bars are nearly perpendicular, and the bar-spiral pattern is about $35\,\mathrm{km\,s^{-1}\,kpc^{-1}}$ (similar to Fux, 1999). The model shows the far-side twin of the 3 kpc arm, which has just been discovered in the CO P-V diagram (Dame & Thaddeus 2008). It reproduces also the connecting arm (characteristic leading dust lanes along the bar). No evidence is found of lopsidedness in the stellar potential, and the CO lopsidedness must be a purely gaseous phenomenon.

3. Remaining problems

New reconstruction of the spiral structure in the galactic plane have been attempted in the HI gas (Levine *et al.* 2006), and in the CO gas (Nakanishi & Sofue 2006, Englmaier *et al.* 2009). The best fit could be two arms, starting at the end of the bar, with a pitch angle of $12°$, although four arms are still possible. Pohl *et al.* (2008) have tried novel deprojections, by simulating the gas flow with SPH in a bar potential, and obtaining distances with a kinematic model derived from the non-circular velocity field obtained. A test of the procedure with a 2 arms+bar fiducial model, with only one pattern speed, retrieves after deprojection a four arms spiral. These recent efforts demonstrate further the difficulty of disentangling distances and dynamical effects. It is still possible that several patterns exist in the Galaxy. Other prominent features have not yet been interpreted, such as the warp or tilt of the nuclear structure, or its lopsidedness.

References

Alard, C., 2001, *A&A*, 379, L44
Amores, E. B., Lépine, J., & Mishurov, Y., 2009, MNRAS, in press (astro-ph:0907.4822)
Bally, J., Stark, A. A., Wilson, R. W., & Henkel, C., 1988, *ApJ*, 324, 223
Binney, J. , Gerhard, O. E., Stark, A. A., Bally, J., & Uchida, K. I., 2001, *MNRAS*, 252, 210
Bissantz, N., Englmaier, P., & Gerhard, O., 2003, *MNRAS*, 340, 949
Combes, F., Debbasch, F., Friedli, D., & Pfenniger, D., 1990, *A&A*, 233, 82
Dame, T., M., Hartmann, D., & Thaddeus, P., 2001, *ApJ*, 547, 792
Dame, T., M. & Thaddeus, P., 2008, *ApJ*, 683, L143
Englmaier, P., Pohl, M., & Bissantz, N., 2009, *Mem Societa Astron Italiana* (astro-ph:0812.3491)
Fux, R., 1999, *A&A*, 345, 787
Georgelin, Y. M. & Georgelin, Y. P., 1976, *A&A*49, 57
Levine, E. S., Blitz, L., & Heiles, C., 2006, *Science*, 312, 1773
Liszt, H. & Burton, B., 1980, *ApJ*, 236, 779
Lopez-Corredoira, M., Cabrera-Lavers, A., & Gerhard, O., 2005, *A&A*, 439, 107
Nakanishi, H. & Sofue, Y., 2006, *PASJ*, 58, 847
Pohl, M., Englmaier, P., & Bissantz, N., 2008, *ApJ*, 677, 283
Rodriguez-Fernandez, N. & Combes, F., 2008, *A&A*,
Sawada, T., Hasegawa, T., Handa, T. & Cohen, R. J., 2004, *MNRAS*, 349, 1167
Sellwood, J. A. & Sparke, L. S., 1988, *MNRAS*, 231, 25p

Highlights of Astronomy, Volume 15
XXVIIth IAU General Assembly, August 2009
Ian F. Corbett, ed.
© International Astronomical Union 2010
doi:10.1017/S1743921310008677

Mapping the Milky Way with SDSS, Gaia and LSST

Željko Ivezić (for the LSST Collaboration)

Department of Astronomy, University of Washington, Seattle, WA 98155, USA
email: ivezic@astro.washington.edu

Abstract. We summarize recent work on the Milky Way "tomography" with SDSS and use these results to illustrate what further breakthroughs can be expect from Gaia and the Large Synoptic Survey Telescope (LSST). LSST is the most ambitious ground-based survey currently planned in the visible band. Mapping of the Milky Way is one of the four main science and design drivers. The main $20\,000\,\mathrm{deg}^2$. survey area will be imaged about 1000 times in six bands (*ugrizy*) during the anticipated 10 years of operations, with the first light expected in 2015. Due to Gaia's superb astrometric and photometric accuracy, and LSST's significantly deeper data, the two surveys are highly complementary: Gaia will map the Milky Way's disk with unprecedented detail, and LSST will extend this map all the way to the halo's edge.

Keywords. Surveys, atlases, catalogs, astronomical data bases, astrometry, photometry

1. The Milky Way Tomography with SDSS

With the SDSS data set, we are offered for the first time an opportunity to examine in situ the thin/thick disk and disk/halo boundaries over a large solid angle, using millions of stars. In a three-paper series, Jurić *et al.* (2008), Ivezić *et al.* (2008a) and Bond *et al.* (2009) have employed a set of photometric parallax relations, enabled by accurate SDSS multi-color measurements, to estimate the distances to tens of millions of main-sequence stars. Photometric metallicity estimates based on the $u - g$ colors are also available for about six million F/G stars, and proper motions based on a comparison of SDSS and the Palomar Observatory Sky Survey positions are available for about 20 million stars.

With these distances, accurate to $\sim$10%, the stellar distribution in the multi-dimensional phase space can be mapped and analyzed without any additional assumptions. The adopted analytic models and a computer code (*galfast*†) that summarize these results, can be used to generate mock catalogs for arbitrary depths and photometric systems (including kinematic quantities). They also enable searches for substructure by subtracting the smooth background distributions. Indeed, a lot of substructure is seen in the data in all projections of the parameter space (spatial distributions, kinematics, metallicity distribution).

The extension of observations for numerous main-sequence stars to distances up to $\sim 10\,\mathrm{kpc}$ represents a significant observational advance, and delivers powerful new constraints on the dynamical structure of the Galaxy. For example, most stars observed by the Hipparcos survey are within $\sim$100 pc (Dehnen & Binney 1998). In less than two decades, the observational material for such in situ mapping with main-sequence stars has progressed from first pioneering studies based on only a few hundred objects (Majewski 1993), to over a thousand objects (Chiba & Beers 2000), to the massive SDSS data set.

These new quantitative results enable fairly robust predictions for the performance of new surveys, such as Gaia and LSST (Eyer *et al.*, in prep.). Due to Gaia's superb

---

† See http://hybrid.mwscience.net.

astrometric and photometric measurements, and LSST's significantly deeper data, the two surveys will be highly complementary: Gaia will map the spatial, metallicity and kinematic distributions of stars in the Milky Way's disk with unprecedented detail, and LSST will extend these maps all the way to the halo's edge, and will obtain large local samples of intrinsically faint sources such as L, T and white dwarfs. We briefly describe LSST in the next section.

2. Brief Overview of LSST

LSST will be a large, wide-field ground-based system designed to obtain multiple images covering the sky that is visible from Cerro Pachón in Northern Chile. The LSST design is driven by four main science themes: constraining dark energy and dark matter, taking an inventory of the Solar System, exploring the transient optical sky, and mapping the Milky Way. The current baseline design, which envisages an 8.4 m (6.7 m effective) primary mirror, a 9.6 deg^2 field of view, and a 3,200 Megapixel camera, will allow about 10,000 square degrees of sky to be covered using pairs of 15-second exposures in two photometric bands every three nights on average. The system is designed to yield high image quality as well as superb astrometric and photometric accuracy. The survey area will include 30,000 deg^2 with $\delta < +34.5°$, and will be imaged multiple times in six bands, *ugrizy*, covering the wavelength range 320–1050 nm. About 90% of the observing time will be devoted to a deep-wide-fast survey mode which will observe a 20 000 deg^2 region about 1000 times in the six bands during the anticipated 10 years of operations. These data will result in databases including 10 billion galaxies and a similar number of stars, and will serve the majority of science programs. The remaining 10% of the observing time will be allocated to special programs such as Very Deep and Very Fast time domain surveys. More information about LSST can be obtained from www.lsst.org and Ivezić *et al.* (2008b).

Each 30-sec observation will be about 2 mag deeper than SDSS imaging, and the repeated observations will enable proper-motion and trigonometric parallax measurements to $r = 24.5$, about 4-5 mag fainter limit than to be delivered by Gaia, and the coadded LSST map will reach $r = 27.5$. Due to Gaia's superb astrometric and photometric accuracy, and LSST's significantly deeper data, the two surveys will be highly complementary. As shown by Eyer *et al.*, in the range $19 < r < 20$ Gaia's and LSST errors are fairly similar (within a factor of ~ 2). Towards brighter magnitudes, Gaia's error significantly decrease (by about a factor of 10 already at $r \sim 14$), and towards fainter magnitudes, LSST will smoothly extend the Gaia's error vs. magnitude curves by over 4 mag.

References

Bond, N. A., Ivezić, Ž., Sesar, B., et al. 2009, submitted to *ApJ* (arXiv:0909.0013)
Chiba, M. & Beers, T. C. 2000, *AJ*, 119, 2843
Dehnen, W. & Binney, J. J. 1998, *MNRAS*, 298, 387
Ivezić, Ž., Sesar, B., Jurić, M., *et al.* 2008a, *ApJ*, 684, 287
Ivezić, Ž., Tyson, J. A., Allsman, R., *et al.* 2008b, arXiv:0805.2366
Jurić, M., Ivezić, Ž., Brooks, A., *et al.* 2008, *ApJ*, 673, 864
Majewski, S. R. 1993, Annual Review of *A&A*, 31, 575

Highlights of Astronomy, Volume 15
XXVIIth IAU General Assembly, August 2009
Ian F. Corbett, ed.

© International Astronomical Union 2010
doi:10.1017/S1743921310008689

Schwarzschild Models for the Galaxy

Julio Chanamé

Carnegie Institution of Washington, Department of Terrestrial Magnetism
5241 Broad Branch Rd., Washington DC 20015, USA
email: jchaname@dtm.ciw.edu

Abstract. Schwarzschild's orbit-superposition technique is the most developed and well-tested method available for constraining the detailed mass distributions of equilibrium stellar systems. Here I provide a very short overview of the method and its existing implementations, and briefly discuss their viability as a tool for modeling the Galaxy using Gaia data.

1. Introduction

Models are used to relate observations to theoretical constructs such as the phase-space distribution function. A number of simplifying assumptions are required to make the modeling process tractable and these assumptions can have profound implications for the inferred mass distribution. A delicate balance must be struck between the available observations and the complexity of the models fitted to them.

The Gaia mission will require modeling techniques that are capable of handling huge numbers of measurements, while taking full advantage of the high precision of the data. Existing implementations of the orbit-superposition method already fulfil some of these requirements but do have limitations. For example, while an assumption regarding the geometry of the gravitational potential is inescapable, Schwarzschild models can be built that are completely free from assumptions regarding the detailed orbital structure, which generally have the largest impact on the derived mass distribution.

1.1. *Schwarzschild's technique*

Given an orbit library for an assumed gravitational potential, the orbit-superposition technique (Schwarzschild 1979) finds the linear sum of those orbits that best reproduces the available observations. The success of the method relies on two aspects: (1) that the stellar system can be safely considered to be in equilibrium, and (2) that the orbit library is sufficiently comprehensive. If these two conditions are satisfied, the method is very general and free from most assumptions. Even the required assumption of a given geometry for the gravitational potential is in practice removed by the iterative nature of the technique, which calls for the construction of Schwarzschild models for an entire grid of potentials, with the final model the one that best fits the data. Of course a Schwarzschild model only provides a snapshot of the current dynamical state of the system, and the question of stability must be addressed by other means.

Schwarzschild models have been successfully used to constrain the dark-matter halos of galaxies (e.g., Rix *et al.* 1997, Thomas *et al.* 2005), to weigh supermassive black holes at the centers of both galaxies (e.g., van der Marel *et al.* 1998, Gebhardt *et al.* 2000, 2001) and globular clusters (Gebhardt, Rich, & Ho 2005). They have also been used to study the dynamics of star clusters (e.g., van de Ven *et al.* 2006). More relevant to the subject at hand, orbit-superposition models have also been used to study the dynamics of the Galactic bulge (Zhao 1996, Häfner *et al.* 2000). Existing implementations of the Schwarzschild method are usually classified/labeled according to the geometry of the stellar systems they can be applied to (spherical, axisymmetric, triaxial), and to the type of dataset they are designed to handle (continuous or discrete; see Chanamé, Kleyna, & van der Marel 2008 for a review).

Points of weakness or controversy regarding Schwarzschild modeling include: the non-uniqueness of the initial conditions used to generate orbits; the amount of smoothing or regularization of the solution that is applied and its impact on the final results; possible over-interpretation of χ^2 plots and indeterminacy of best solution; how to deal with incomplete positional sampling; and large computational costs.

2. Applicability of Schwarzschild's method to Galactic surveys

We can consider the Galaxy to be composed of two kinds of structure: (1) a smoothly distributed and old Galaxy in steady-state equilibrium, and (2) a perturbed, inhomogeneous Galaxy that changes over relatively short timescales and is not in dynamical equilibrium. While the classical Galactic structures of bulge, disk(s), and halo belong to the first category, shorter-lived structures such as tidal streams, spiral arms, and disk warps, all fall into the second one. Only the background, steady-state Galaxy is susceptible to Schwarzschild modeling. Fortunately, most of the Galaxy's mass lies in steady-state structures, so a Schwarzschild model should provide a useful first approximation to the data. However, even though the non-equilibrium mass fraction is small, this component is expected to hold clues to the history of the Galaxy, and means must be found to model it too.

Modeling data from current surveys such as SDSS and RAVE will prepare us for modeling the vastly superior Gaia data. Clever arguments such as those in Smith, Evans, & An (2009) can only benefit the applicability of Schwarzschild's technique by narrowing down the range of possible shapes and geometries of the underlying gravitational potential, and could even shed light on the optimal choice of initial conditions for orbit integration.

Acknowledgements

I thank the American Astronomical Society and the International Astronomical Union for travel grants. I acknowledge support from NASA through Hubble Fellowship grant HF-01216.01-A, awarded by the Space Telescope Science Institute, which is operated by the AURA, Inc., under NASA contract NAS5-26555.

References

Chanamé, J., Kleyna, J., & van der Marel, R. 2008, *ApJ* 682,841
Gebhardt, K., Richstone, D., Kormendy, J., *et al.* 2000, *AJ* 119,1157
Gebhardt, K., Lauer, T., Kormendy, J., *et al.* 2001, *AJ* 122,2469
Gebhardt, K., Rich, M., & Ho, L. 2005, *ApJ* 634,1093
Häfner, R., Evans, N. W., Dehnen, W., & Binney, J. 2000, *MNRAS* 314,433
Rix, H.-W., *et al.*, 1997, *ApJ* 488,702
Schwarzschild, M. 1979, *ApJ* 232,236
Smith, M. C., Wyn Evans, N., & An, J. H. 2009, *ApJ* 698,1110
Thomas, J., *et al.*, 2005, *MNRAS* 360,1355
van de Ven, G., van den Bosch, R. C. E., Verolme, E. K., & de Zeeuw, P. T. 2006, *A&A* 445,513
van der Marel, R. P., Cretton, N., de Zeeuw, P. T., & Rix, H.-W. 1998, *ApJ* 493,613
Zhao, H. S. 1996, *MNRAS* 283,149

Highlights of Astronomy, Volume 15
XXVIIth IAU General Assembly, August 2009
Ian F. Corbett, ed.

© International Astronomical Union 2010
doi:10.1017/S1743921310008690

Connecting Moving Groups to the Bar and Spiral Arms of the Milky Way

T. Antoja[1], O. Valenzuela[2], F. Figueras[1], B. Pichardo[2] & E. Moreno[2]

[1]DAM and IEEC-UB, ICC-Universitat de Barcelona, Spain, email: tantoja@am.ub.es

[2]Instituto de Astronomia de la UNAM, Mexico, email: octavio@astroscu.unam.mx

Abstract. We use test-particle orbit integration with a realistic Milky Way (MW) potential to study the effect of the resonances of the Galactic bar and spiral arms on the velocity distribution of the Solar Neighbourhood and other positions of the disk. Our results show that spiral arms create abundant kinematic substructure and crowd stars into the region of the Hercules moving group in the velocity plane. Bar resonances can contribute to the origin of low-angular momentum moving groups like Arcturus. Particles in the predicted dark disk of the MW should be affected by the same resonances as stars, triggering dark-matter moving groups in the disk. Finally, we evaluate how this study will be advanced by upcoming Gaia data.

Keywords. Galaxy: disk, kinematics and dynamics, structure, solar neighborhood, dark matter

---

The MW potential and initial conditions (cold, intermediate and hot disks) of our simulations are described in Antoja *et al.* (2009). Next figures show: a) left: the region of Hercules ($V = -40\,\mathrm{km\,s^{-1}}$) is crowded by the the spiral arms applied to the cold disk, b) middle: the bar with the hot disk creates groups at low angular momentum ($V = -100\,\mathrm{km\,s^{-1}}$) with long integration time, c) right: effects of a model with spiral arms and bar. We propose that a good fit between the observed velocity field – see Antoja *et al.* (2008) – and the simulations requires the combined model under IC1, IC2 and IC3, where the central and low angular momentum moving groups would appear simultaneously.

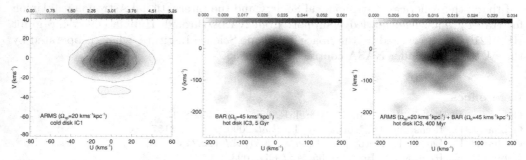

Gaia will revolutionize our knowledge of the Galactic disk. Accuracies in UVW velocities are computed using the Gaia Universe Model Snapshot (GUMS), based on the Besançon Galaxy Model, and the current estimations of the Gaia errors. We find that we will be able to perform robust statistical analysis of the velocity distribution (with accuracies better than $2\,\mathrm{km\,s^{-1}}$ in all components UVW) up to $\sim 3\,\mathrm{kpc}$ from the Sun.

References

Antoja, T., Figueras, F., Fernández, F., *et al.* 2008, *A&A*, 490, 135

Antoja, T., Valenzuela, O., Pichardo, B. *et al.* 2009, *ApJL*, 700, 78

Highlights of Astronomy, Volume 15
XXVIIth IAU General Assembly, August 2009
Ian F. Corbett, ed.

Cosmological simulations of the Milky Way

Lucio Mayer

Institute for Theoretical Physics, University of Zurich, Wintherthurestrasse 190, Zurich (CH)

Abstract. Recent simulations of forming low-mass galaxies suggests a strategy for obtaining realistic models of galaxies like the Milky-Way.

Cosmological simulations of galaxy formation are powerful tools for confronting the ΛCDM model with observational datasets. The increase in mass and spatial resolution and the improvement of sub-grid algorithms for star formation and feedback processes have recently resulted in simulated galaxies with realistic disk size and angular momentum content (Mayer et al. 2008; Governato et al. 2009a). However, simulated galaxies hosted in halos with masses $\sim 10^{12} M_\odot$ exhibit prominent bulges and structural parameters reminiscent of Sa spirals rather than of Sb/Sc galaxies. Surface densities at the solar radius are larger than that of the Milky Way (MW) by factors of a few and the more massive bulge produces a steeper rotation curve compared to that of the MW (Read et al. 2009). At halo masses $M_{\rm vir} > 2 \times 10^{12} M_\odot$ the predominance of hot-mode gas accretion counters the presence of a prominent star forming disk at $z = 0$, producing earlier-type objects resembling S0 galaxies (Brooks et al. 2009) and supporting recent estimates based on RAVE that yield $M_{\rm vir} \sim 10^{12} M_\odot$ (Smith et al. 2007).

Yet the solution to forming a realistic MW analog could be at hand. Recently we have performed galaxy-formation simulations for mass scales $< 10^{11} M_\odot$. By sampling these low-mass galaxies with several millions of particles, we achieve a mass resolution better than $10^3 M_\odot$ in the baryons, thus resolving individual molecular clouds. Star formation can now be tied to gas at molecular cloud densities ($\rho > 100$ cm^{-3}). A realistic, inhomogeneous interstellar medium is obtained that results naturally in stronger supernova outflows than when the standard star formation threshold ($\rho = 0.1$ cm^{-3}) is adopted. Such outflows efficiently remove the low-angular momentum baryonic material from the central region, suppressing the formation of a bulge and producing an object with a slowly rising rotation curve in very close agreement with observed dwarf galaxies (Governato et al. 2009b; see also Ceverino & Klypin 2009). We argue that comparable resolution of MW-sized galaxies will yield rotation curves and bulge-to-disk ratios appropriate for Sb-Sc spirals at $z = 0$. This requires increasing the number of particles employed by more than an order of magnitude.

References

Brooks, A. M., Governato, F., Quinn, T., Brook, C. B., & Wadsley, J., 2009, *ApJ*, 694, 396
Ceverino, D. & Klypin, A., 2009, *ApJ*, 895, 292
Governato, F, Brook, C. B., Brooks, A. M., Mayer, L., Willman, B., Jonsson, P., Stilp, A. M., Pope, L., Christensen, C., Wadsley, J., & Quinn, T., 2009a, *MNRAS*, 398, 312
Governato, F, Brook, C. B., Brooks, A. M., Mayer, L., Willman, B., Jonsson, P., Stilp, A. M., Pope, L., Christensen, C., Wadsley, J., & Quinn, T., 2009b, *Nature* in press (arXiv:0911.2237)
Mayer, L., Governato, F., & Kaufmann, T., 2008, *Adv. Sci. Lett.*, 1, 7
Read, J., Mayer, L., Brooks, A.M., Governato, F., & Lake, G., 2009, *MNRAS*, 397, 44
Smith, A., et al., 2007, *MNRAS*, 379, 755

Highlights of Astronomy, Volume 15
XXVIIth IAU General Assembly, August 2009
Ian F. Corbett, ed.
© International Astronomical Union 2010
doi:10.1017/S1743921310008719

Modelling the Galaxy with orbital tori

James Binney

Rudolf Peierls Centre for Theoretical Physics, Keble Road, Oxford OX1 3NP, UK
email: binney@thphys.ox.ac.uk

Abstract. The principles and advantages of torus modelling are explained.

Keywords. Galaxy: dynamics, Galaxy: evolution

1. What is torus modelling?

Torus modelling (McMillan & Binney 2008 and references therein) is a modification of Schwarzschild modelling in which orbits are numerically constructed as three-dimensional surfaces in six-dimensional phase space rather than as time sequences. These surfaces are topologically equivalent to 3-tori: if we identify each point of the floor of a room with the point of the ceiling that is vertically above it, and similarly identify corresponding points on the front and back walls, and on the right and left walls, the room becomes a 3-torus. The "angle variables" $\theta_1, \theta_2, \theta_3$ are Cartesian coordinates for position in a room that has been thus made into a 3-torus: for example, as θ_3 increments from 0 to 2π, the point moves vertically from floor to ceiling. Remarkably, as a star orbits through the Galaxy, its angle variables increase linearly in time, $\theta_i(t) = \theta_i(0) + \Omega_i t$, so its point moves in a straight line through its room-like orbital torus. Unless the frequencies Ω_i are rationally related ($\Omega_1 : \Omega_2 : \Omega_3 = n_1 : n_2 : n_3$ for integer n_i) the star eventually comes arbitrarily close to every point of its torus.

The natural labels for a torus are the three actions $J_i = (2\pi)^{-1} \oint_{\gamma_i} \mathbf{v} \cdot \mathbf{dx}$, where γ_i is the path on which θ_i increments from 0 to 2π with the other two angle variables held constant. Indeed the set of six coordinates (J_i, θ_i) are canonical coordinates for phase space. In particular the Poisson bracket of any two angle variables vanishes, $[\theta_i, \theta_j] = 0$, so tori are null in the sense that the Poincaré invariant of any part of a torus vanishes ($\int \mathbf{dv} \cdot \mathbf{dx} = 0$).

The idea of torus-modelling is to use software that returns an orbit in terms of $\mathbf{x}(\boldsymbol{\theta})$ and $\mathbf{v}(\boldsymbol{\theta})$ rather than $\mathbf{x}(t)$ and $\mathbf{v}(t)$ as a Runge–Kutta integrator does. Since in this picture the orbit is identified with the torus, it is labelled by the actions $\mathbf{J}$. The following benefits flow from expressing orbits in this way.

• Given a point in space $\mathbf{x}$ we can readily find the values of $\boldsymbol{\theta}$ at which the star reaches that point and read off the velocities $\mathbf{v}(\boldsymbol{\theta})$ with which the star passes through that point. By contrast, if we are given $\mathbf{x}(t)$, we will in general search in vain for a time when the star is precisely at the given point and will have to settle for times when it is near. It will not be clear whether we have found times that give approximations to all the possible values of $\mathbf{v}$ at the given point.

• When we integrate orbits in time, the orbit is characterised by its initial conditions. The same orbit corresponds to infinitely many different initial conditions, so it is not clear how to do a systematic survey of phase space to obtain a representative sample of orbits. This difficulty does not arise in torus modelling because the actions are essentially unique labels of orbits and the orbits with actions in the range $d^3\mathbf{J}$ occupy a volume $(2\pi)^3 d^3\mathbf{J}$ in 6-d phase space.

• Actions are adiabatic invariants so we can relate orbits in different potentials: orbits with the same actions will deform into each other if we slowly deform the potential from that of one model to that of another. No such identification is possible if orbits are characterised by their initial conditions. On account of adiabatic invariance, the distribution function (DF) of a system such as a globular cluster is invariant as, for example, loss of gas modifies the cluster's density distribution and gravitational potential. Together with the previous item, adiabatic invariance make it straightforward to specify a galaxy model uniquely and to compare the orbital structures of models that have slightly differing potentials.

• Analytic formulae for $\mathbf{x}(\boldsymbol{\theta})$ and $\mathbf{v}(\boldsymbol{\theta})$ can be specified using much less data than are required to specify $\mathbf{x}(t)$ and $\mathbf{v}(t)$. Consequently tori greatly simplify manipulation of orbit libraries. Moreover, expressions for infinitely many tori can be obtained by interpolating between data for numerically obtained tori.

• There is a simple, intuitive connection between the functional form $f(\mathbf{J})$ of the DF and the real-space properties of the system it specifies (Binney & Tremaine 2008 §4.6). Moreover, given $f(\mathbf{J})$ there is a stable scheme for evaluating the self-consistent gravitational potential, which is not always the case when a DF of the form $f(E, \ldots)$ is specified.

• The orbit-averaged Fokker-Planck equation takes an exceptionally simple form in action space (Binney & Tremaine 2008 §7.4.2), which should facilitate modelling of secular evolution.

• A set of orbital tori specify an integrable Hamiltonian which is very close to the true Hamiltonian. Perturbation theory works wonderfully well when this Hamiltonian is used as the point of reference (Kaasalainen 1994).

2. How do we obtain tori?

Analytic potentials (principally the potentials of the isochrone sphere and the multidimensional harmonic oscillator) provide analytic tori $\mathbf{x}(\boldsymbol{\theta})$, etc. We choose such a potential and refer to its structures as "toy" ones. A canonical transformation is used to map the toy torus with given actions $\mathbf{J}'$ into the target phase space. The image torus is guaranteed to be null but in general it will not lie within a hypersurface on which the target Hamiltonian is constant, as an orbital torus must. We numerically adjust the coefficients that define the canonical transformation so as to minimise the rms variation ΔH in the target Hamiltonian over the image torus. Once ΔH is small enough, the target torus provides an excellent approximation to an orbital torus.

The canonical transformation $(\mathbf{J}, \boldsymbol{\theta}) \rightarrow (\mathbf{J}', \boldsymbol{\theta}')$ is specified by its generating function

$$S(\mathbf{J}', \boldsymbol{\theta}) = \mathbf{J}' \cdot \boldsymbol{\theta} + \sum_{\mathbf{n}} S_{\mathbf{n}}(\mathbf{J}') e^{i \mathbf{n} \cdot \boldsymbol{\theta}},$$

where the sum is in principle over all vectors with integer components. The first term on the right generates the identity transformation, and the machine has to choose the coefficients $S_{\mathbf{n}}(\mathbf{J}')$ which are characteristic of the given orbit. Typically, a good approximation to an orbital torus can be obtained with a few tens of non-zero $S_{\mathbf{n}}$.

References

Binney, J. & Tremaine, S., 2008, *Galactic Dynamics*, Princeton University Press, Princeton
Kaasalainen, M., 1994, *MNRAS*, 268, 1041
McMillan, P. & Binney, J., 2008, *MNRAS*, 390, 429

Highlights of Astronomy, Volume 15
XXVIIth IAU General Assembly, August 2009
Ian F. Corbett, ed.
© International Astronomical Union 2010
doi:10.1017/S1743921310008720

In What Detail Can We Represent the Milky Way in a Conventional N-Body Model?

Victor P. Debattista

Jeremiah Horrocks Institute, University of Central Lancashire, Preston, PR1 2HE, UK
email: vpdebattista@uclan.ac.uk

Abstract. After a brief review of past N-body models of the Milky Way, I consider some of the difficulties that are inherent in the N-body approach to modelling any disk galaxy.

1. Past N-Body Models of the Milky Way

Modern efforts at modelling the Milky Way (MW) take into account that it is barred. The first large effort in this regard was carried out by Fux (1997), who ran simulations of initially axisymmetric disk+bulge+halo systems. He then compared regularly spaced outputs with a de-reddened *COBE* K-band bulge map. He scaled the kinematics of his model by requiring that the velocity dispersion matches that of Baade's window. A number of models provided reasonable fits to the data. His best fit bar angle to the sun-center line, $\psi_{\rm bar} = 28° \pm 7°$. Later he added SPH gas to his simulations and was able to fit a number of features in the CO gas $l - V$ diagram Fux (1999). He argued that the high-velocity connecting arm is due to the shocked gas within the near side of the bar. The gas distribution is also sensitive to the pattern speed of the bar, and he constrained this parameter to $\Omega_b \sim 50\,{\rm km\,s}^{-1}\,{\rm kpc}^{-1}$.

Widrow *et al.*, (2008) modelled the Milky Way by matching observations (including the rotation curve, local force field, Oort's constants, local and bulge velocity dispersions, surface density and total mass within 100 kpc) to a suite of *axisymmetric* models. N-body models of these then all produced bars. The time of bar formation depended on the Toomre Q and X parameters; in all cases $\Omega_{\rm p}$ started declining after the bar formed and a dynamically young bar is required if $\Omega_{\rm p} = 50\,{\rm km\,s}^{-1}\,{\rm kpc}^{-1}$. A problem with this approach is, however, that bar formation leads to the model departing from the observations.

N-body models are also very useful for testing models constructing using other methods. Zhao (1996) tested his Schwarzschild model of the MW bulge using N-body simulations, finding that its shape and mass distribution is stable.

2. Fundamental Limitations

Modelling the MW, or any other disk galaxy, by N-body simulations is complicated by a number effects. Foremost, disk simulations in which a bar forms are subject to considerable stochasticity. Sellwood & Debattista (2009) show that disk simulations which differ only in the seed of the random number generator used to set up the disk particles evolve quite differently. They identified a number of sources of stochasticity, including multiple disk modes, swing-amplified noise, variations in the onset and strength of bending instabilities, metastability due to upward fluctuations in $\Omega_{\rm p}$ (Sellwood & Debattista 2006), and intrinsic chaos. Stochasticity is weaker when the halo is very massive, but is never absent. Such stochasticity makes it hard to improve N-body models by iterating runs with varying parameters.

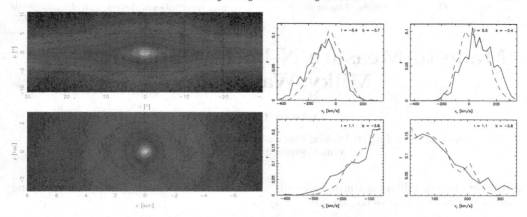

Figure 1. N-body model (left hand panels) showing the rather poor fit to the MW density. Right panels: kinematics for this model (solid lines) compared with observations (dashed lines).

Modeling is also complicated by radial migration of stars caused by transient spirals (Sellwood & Binney 2002). Roškar *et al.* (2008) show that this migration leads to significant mixing of stellar populations so that the age distribution of stars at any given radius does not reflect the star formation history at that radius. In the solar neighborhood, Roškar *et al.* (2008) estimate that as much as half the stars formed elsewhere. Since the incidence of spirals is chaotic, matching the stellar populations in simulations requires a certain degree of luck.

A third difficulty with modeling the MW is the somewhat weak constraints that kinematics of the bulge region impose on models. In order to demonstrate this, in Figure 1 I present an arbitrary disk-galaxy simulation, scaling its velocities to produce a rotation velocity of 220 km/s. The density distribution is a rather poor match to models of the MW [e.g. Bissantz & Gerhard (2002), López-Corredoira *et al.* (2005)]. However, comparing the kinematics of particles selected to lie in the bulge using selection functions that match those in observations of Rangwala *et al.* (2009) results in distributions of stellar velocities that are not substantially different from those observed.

Acknowledgements

I would like to thank the organizers for inviting me to give this review. The author acknowledges support through travel grant 2009/R1 from The Royal Society.

References

Bissantz, N. & Gerhard, O. 2002, *MNRAS*, 330, 591
Fux, R. 1997, *A&A*, 327, 983
Fux, R. 1999, *A&A*, 345, 787
López-Corredoira, M., Cabrera-Lavers, A., & Gerhard, O. E. 2005, *A&A*, 439, 107
Rangwala, N., Williams, T. B., & Stanek, K. Z. 2009, *ApJ*, 691, 1387
Roškar, R., Debattista, V. P., Quinn, T. R., Stinson, G. S. & Wadsley, J. 2008, *ApJL*, 684, L79
Sellwood, J. A. & Binney, J. J. 2002, *MNRAS*, 336, 785
Sellwood, J. A. & Debattista, V. P. 2009, *MNRAS*, 398, 1279
Sellwood, J. A. & Debattista, V. P. 2006, *ApJ*, 639, 868
Widrow, L. M., Pym, B., & Dubinski, J. 2008, *ApJ*, 679, 1239
Zhao, H. S. 1996, *MNRAS*, 283, 149

Highlights of Astronomy, Volume 15
XXVIIth IAU General Assembly, August 2009
Ian F. Corbett, ed.

© International Astronomical Union 2010
doi:10.1017/S1743921310008732

Made-to-Measure N-body Modeling of the Milky Way Galaxy

Ortwin Gerhard

Max-Planck-Institut für Ex. Physik, Giessenbachstrasse 1, D-85748 Garching, Germany
email: gerhard@mpe.mpg.de

Abstract. In this talk a brief introduction is given to made-to-measure particle methods and their potential use for modeling the Milky Way Galaxy.

1. Made-to-Measure N-body models

The term "Made-to-Measure" (M2M) for a dynamical model reproducing a set of observational data for a galaxy was coined by Syer & Tremaine (1996) (hereafter ST). M2M models can be based on distribution functions, moments, orbits, or particles, but in particular ST described an algorithm for constructing N-body equilibrium models, based on the idea of adjusting the masses or weights of the particles until the system agreed with a prescribed density distribution. The central part of their algorithm is the "force-of-change" (FOC) equation through which the particle weights are adjusted according to the mismatch of model and target density observables. Other important ingredients include a time smoothing to reduce particle noise and an entropy term to reduce large fluctuations in the weights.

As a first practical application of this method Bissantz, Debattista & Gerhard (2004) constructed a dynamical model for the Milky Way's barred bulge and disk, constraining the projected density map. However, the ST algorithm is not well-suited for mixed density and kinematic observables, and it does not allow a proper treatment of observational errors. A modified χ^2M2M algorithm which resolved both these problems was introduced by De Lorenzi et al. (2007) (DL+07) who also demonstrated the potential of the method by constructing particle models for spherical, axisymmetric, triaxial and rotating target systems. Their implementation NMAGIC has since been used by De Lorenzi et al. (2008,2009) (DL+08, DL+09) to construct dynamical models of elliptical galaxies based on photometry, Sauron, slit and planetary-nebula velocity data, and thus to explore the distribution of dark matter in these galaxies. For modeling the discrete velocities, De Lorenzi et al. (2008) introduced a likelihood scheme into the FOC equation, and they also added a separate equation for adjusting the mass-to-light ratio of the stellar system simultaneously with the observables.

A modified M2M method was introduced by Dehnen (2009). Rather than time-averaging the moments of the N-body system as ST and DL+07, he considered time-averaging the merit function, resulting in a second-order equation for adjusting the particle weights. This contains a linear damping term that acts to maximize the merit function. Writing the equations in dimensionless time (in units of orbital period) also made it possible to achieve a uniform adjustment rate per orbital time for the particle weights, despite the large range of orbital time-scales inherent in an N-body system. Dehnen (2009) used this algorithm to construct triaxial equilibrium target systems.

2. Comparisons with Schwarzschild and direct N-body methods

M2M particle models include aspects of both Schwarzschild orbit superposition methods and direct N-body simulations. When the gravitational potential of the target system is held fixed, the process of finding a distribution of particle weights to match a prescribed density field or set of kinematic observables for a galaxy, is closely related to the process of finding a distribution of orbital weights in Schwarzschild's method to fit the same constraints. Conversely, when the adjustment of particle weights is switched off and the potential is allowed to evolve, M2M particle codes reduce to N-body methods. E.g., in the work of DL+07 several equilibria are found in the first limit, while the stability of the final galaxy models in DL+08 and DL+09 is tested in the second limit. Of course, the strength of the M2M particle methods is to self-consistently evolve the particle weights and their potential simultaneously while approaching the target data (see applications in DL+08). This is perhaps the greatest advantage over Schwarzschild and N-body techniques. However, this is most useful when the particles trace the mass. Because there is no direct connection between observables and the global gravitational potential, external dark-matter potentials have still to be explored one by one. Compared to Schwarzschild and N-body techniques, M2M modeling is still in an early stage, and improvements of both the techniques and the efficiency of the method are needed.

3. M2M modeling for the Milky Way

The goal of dynamical models for the Milky Way is to uncover the fossil record of its assembly history, as described by, for example, the orbital distribution of stars of different metallicities or other population parameters. Because of the enormous detail expected from future observations, particularly from Gaia, it is likely that interesting (sub)structure in the models would be visible in the data. Thus techniques with a minimum of simplfying assumptions such as M2M may be well-suited to understand what these data will be telling us.

M2M work to date has been confined to the inner Galaxy; see Bissantz, Debattista & Gerhard (2004) and Debattista *et al.* (in preparation). These models have successfully reproduced as diverse observables as densities, radial velocity histograms and microlensing event time-scale distribution, but a comprehensive bulge model is still pending.

Modeling the nearby Galactic disk has not been tried yet. This has the problem that only a small fraction of the particles from any such model will be seen in magnitude limited surveys; see e.g., Brown *et al.* (2005). However, this may actually not be such a large problem for M2M methods such as NMAGIC. The time averaging of the observables allows suppressing the particle noise in the model observables by a factor 10-100, which eases the comparison with solar neighbourhood data.

In conclusion, made-to-measure modeling for the Milky Way has little history, but a lot of promise, and so there is a lot of work to do!

References

Bissantz, N., Debattista, V. P., & Gerhard, O. 2004, *ApJ*, 601, L155
Brown, A. G. A., Velázquez, H. M., & Aguilar, L. A. 2005, *MNRAS*, 359, 1287
Dehnen, W. 2009, *MNRAS*, 395, 1079
De Lorenzi, F., Debattista, V. P., Gerhard, O., & Sambhus, N. 2007, *MNRAS*, 376, 71
De Lorenzi, F., Gerhard, O., Saglia, R. P., & Sambhus, N., *et al.* 2008, *MNRAS*, 385, 1729
De Lorenzi, F., Gerhard, O., Coccato, L., & Arnaboldi, M., *et al.* 2009, *MNRAS*, 395, 76
Syer, D., Tremaine, S. 1996, *MNRAS*, 282, 223

Highlights of Astronomy, Volume 15
XXVIIth IAU General Assembly, August 2009
Ian F. Corbett, ed.

How to build a 3D extinction model of the Galaxy

J. R. D. Lépine[1] & E. B. Amôres[1]

[1]Instituto de Astronomia, Geofísica e Ciências Atmosféricas, Universidade de São Paulo,
Cidade Universitária, São Paulo, SP, Brazil; email: jacques@astro.iag.usp.br

Abstract. We show that anomalous extinction (deviations from the traditionally adopted $R_V = A_V/E(B-V) = 3.1$ introduces large uncertainties in the distances of stars, for distances larger than 1-2 kpc. We argue that for such distances and for directions close to the galactic plane, the use of extinction models based on the gas distribution in the Galaxy is safer, for the moment, than the use of extinction maps.

Keywords. interstellar extinction, galactic structure

1. Introduction

It is usually believed that all that we need to derive the photometric distance of a star, taking into account extinction, is the color excess $E(B - V)$. In reality the existence of anomalous extinction seriously challenges this belief. Besides this, there are two opposite views concerning the possibility of corrections for interstellar extinction based on models. One is that the dust distribution is clumpy and random. Consequently, it does not make any sense to produce models; only empirical extinction maps or tables are useful. The other idea is that the proportionality between gas and dust column densities is well established. Simple models of the gas distribution can be constructed, based on HI and CO surveys, so that extinction is predictable to some extent. Amôres & Lépine (2005) made an extensive comparison of an extinction model based on the gas distribution with a large sample of stars with known extinctions (the sample of Neckel & Klare 1980). The comparison proved that the extinction is largely predictable. Interestingly, Neckel & Klare themselves produced extinction maps based on their sample of stars, which constitutes an opportunity for a comparison between the two approaches.

2. The problem of anomalous extinction, and discussion

Neckel & Klare (1980) in a famous work, determined the spectral type and color excess for more than 7000 O and B stars situated near the galactic plane. They plotted the interstellar extinction A_V as a function of distance for many directions, and obtained some unexpected results. In many cases, one can observe in those plots a low extinction up to a distance of the order of 1 kpc (see eg. the direction $l = 128$, $b = 0$), followed by a step in the extinction up to $A_V = 2$-3, and then A_V remains constant up to about 5 kpc, the maximum distance investigated. It is not surprising to see steps in A_V, as they are explained by the presence of dense clouds along the line-of-sight. What is surprising is that there are paths of many kpc without steps, as if all the clouds were close to the Sun. Another surprising result, noted by Neckel & Klare, is that the clouds situated close to the Sun are bigger than the more distant ones.

The explanation for the two unexpected results is that the distances of the more distant stars (and clouds) are overestimated. The clouds contain dense cores in which R_V

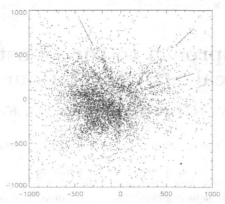

Figure 1. OB stars in the plane of the Galaxy. The distances from the Sun are indicated in parsecs, the centre of the Galaxy is outside the figure, at $(0, -7500)$. The line segments point to examples of directions of alignments of stars or "fingers of God" caused by anomalous extinction.

can be much larger than the classical value 3.1. As a consequence, the extinction is underestimated and the distances are overestimated. The existence of anomalous extinction is not a new result, we only call attention to the fact that the anomalous extinction is so widespread that it affects strongly our understanding of the local structures.

We present in Figure 1 the photometric distances of the sample of OB stars of the solar neighborhood taken from the Hipparcos catalog. We first computed the distances based on the absolute magnitudes expected from their spectral class, with the extinction estimated using $R_V = 3.1$. We also computed the distances based on the infrared H magnitudes taken from 2Mass and the relation $A_H = 0.18E(V - H)$ derived from the extinction relations by Koornneef 1983. The distances using V and H bands are not different, the ones based on H magnitudes are shown in Figure 1. Since the extinction in band H is smaller than in V, we would expect that the anomalous extinction would also be smaller. However, it can be seen in the figure that elongated structures like fingers of God are present. The elongated structures cannot be the result of any kind of calibration errors. They are the result of an unpredictable scattering in the values of R_V. The distance of the first interstellar cloud found along a line-of sight is correct, since the stars are still unaffected by extinction, but the next clouds have distances overestimated. Extinction maps, to be useful for a 3D description of the extinction in the Galaxy, would have to give the distances of all the steps in A_V, which is not possible at the moment.

How to correct for anomalous extinction? Of course when the data from Gaia become available we will be able to place real distances to the molecular clouds which are responsible for steps in the extinction, up to much larger distances. Meanwhile, it is possible that the approach of Fitzpatrick & Massa (2009) which describes R_V in terms of two color indices could be a good solution.

Acknowledgements

We thank Leticia Vaz and Nayara Torres for the compilation of the catalog of OB stars.

References

Amôres, E. B. & Lépine, J. R. D. 2005, *AJ* 130, 659
Fitzpatrick, E. L. & Massa, D., 2009, *ApJ* 699, 1209
Koornneef, J. 1983, *A&A* 128, 84
Neckel, Th. & Klare, G. 1980, *A&AS* 42, 251

Highlights of Astronomy, Volume 15
XXVIIth IAU General Assembly, August 2009
Ian F. Corbett, ed.

© International Astronomical Union 2010
doi:10.1017/S1743921310008756

A New Approach to the Construction of Dynamical Structure of our Galaxy

H. Ueda[1], N. Gouda[2], T. Yano[2], H. Koyama[3] and M. Sakagami[4]

[1]Faculty of Education and Human Studies, Akita University, Tegata-gakuen, Akita 010-0852, Japan
email: ueda@ipc.akita-u.ac.jp
[2]National Astronomical Observatory, Mitaka, Tokyo 181-8588, Japan
email:naoteru.gouda@nao.ac.jp, yano.t@nao.ac.jp
[3]Graduate School of Human and Environmental Studies, Kyoto University, Kyoto 606-8501, Japan
email: sakagami@grav.mbox.media.kyoto-u.ac.jp
[4]Department of Physics, Waseda University, Shinjuku, Tokyo 169-8555, Japan
email: koyama@gravity.phys.waseda.ac.jp

Information about positions and velocities of stars that will be gained in the era of GAIA is crucial for determining dynamical structure in our Galaxy. The distribution function of all component objects in our Galaxy is fundamental for describing its dynamics. However, only the distribution function of observable stars is obtained from space astrometry observations, and it is therefore necessary to develop theoretical studies of how to construct the distribution function of all matter including dark matter and unobservable stars using astrometric data of observable stars. This procedure falls into three categories.

1) Torus Fitting: In the first step, we find action variables $\mathbf{J}$ in any gravitational potential (Hamiltonian) whose system is almost integrable and considered to be a representation of our Galaxy. Concretely, we find the coordinate transformation between action variables and the Cartesian coordinates $(\mathbf{x}, \mathbf{p})$ in a model of our Galaxy. Although the action-angle variables $(\mathbf{J}, \theta)$ provide the most compact representation of a regular orbit, it is impossible to get analytically the coordinate transformation $(\mathbf{J}, \theta) \Leftrightarrow (\mathbf{x}, \mathbf{p})$ in an arbitrary potential. We therefore propose a new approach that numerically obtains action variables $\mathbf{J}$ and coordinate transformations $(\mathbf{J}, \theta) \Leftrightarrow (\mathbf{x}, \mathbf{p})$. It is called a "Torus Fitting method" and is an alternative of the Torus Construction method proposed by Binney and his collaborators.

2) Determination of a theoretical distribution function by M2M: In the second place, we construct the distribution function $f_{\mathrm{matter}}(J_1, J_2, J_3)$ of all matter in the system, using action variables from the first step. We do this using the made-to-measure algorithm for constructing an N-body realization of an equilibrium stellar system (please refer to Yano's presentation for details).

3) Determination of a model of our Galaxy from the distribution function of observed stars, $f_{\mathrm{obs}}(\mathbf{x}, \mathbf{p})$: Finally, we have to determine a model of the gravitational potential (Hamiltonian) of our Galaxy. Hence we need to construct the distribution function of all matter including dark matter and unobservable stars using astrometric data that include the error in distance measurements and selection biases. In this procedure a technique that utilizes Hermite polynomials is used.

From the above procedure we obtain the distribution function of all matter using space astrometry observations.

Highlights of Astronomy, Volume 15
XXVIIth IAU General Assembly, August 2009
Ian F. Corbett, ed.

© International Astronomical Union 2010
doi:10.1017/S1743921310008768

An iterative method for constructing stellar systems models: how far does it work?

Natalia Ya. Sotnikova and Sergei A. Rodionov

Saint-Petersburg State University, Saint-Petersburg, Russia
email: nsot@astro.spbu.ru

Abstract. We present a new method for constructing equilibrium phase models for stellar systems. Applications of the iterative method include both modelling of observational data and the construction of initial condition for N-body simulations.

Keywords. stellar dynamics methods: N-body simulations galaxies: kinematics and dynamics galaxies: structure

---

The aim of the iterative method (IM) is to construct an equilibrium N-body model with prescribed parameters, or constraints. Setting a given mass distribution and almost arbitrary velocities of particles, we start the iterative procedure, by letting the system go through a sequence of self-consistent evolutionary steps of short duration (iterations). At the end of each step, and before the new step is started, we transfer the new velocity distribution from a bit evolved system to a system with the initial density distribution. At this stage we need to correct individual particle velocities in accordance with imposed kinematic constraints (see details in Rodionov *et al.* 2009). We stop iterations when the velocity distribution ceases to change, which implies that the system has reached equilibrium.

We managed to construct equilibrium systems of various types – from spherical to triaxial, from one-component to multi-component, from isotropic to anisotropic (Rodionov & Sotnikova (2006); Sotnikova & Rodionov (2008); Rodionov *et al.* (2009)). Successful reconstruction of the distribution function of a model disc galaxy from its line-of-sight kinematics encouraged us to use the IM to derive the 3D kinematics of edge-on galaxies from observational data. Now we have all IR photometric parameters (for a bulge and a disc) of an edge-on galaxy NGC 4111 and obtained stellar LSVD of this galaxy at the 6-m telescope. Preliminary interpretation of kinematic and photometric observations of this galaxy in terms of its 3D structure and 3D velocity distribution showed that the IM may be very powerful method to reconstruct phase-space models of real galaxies (Sotnikova *et al.*, 2010, in preparation).

Acknowledgements

This work was supported by the Russian Foundation for Basic Research (grant 09-02-00968) and by a grant from President of the RF for support of Leading Scientific Schools (grant NSh-1318.2008.02).

References

Rodionov, S. A. & Sotnikova, N. Ya. 2006, *Astron. Rep.* 50, 983
Sotnikova N. Ya. & Rodionov, S. A. 2008, *Astron. Lett.* 34, 664
Rodionov, S. A., Athanassoula, E., & Sotnikova, N. Ya. 2009, *MNRAS* 392, 904

Highlights of Astronomy, Volume 15
XXVIIth IAU General Assembly, August 2009
Ian F. Corbett, ed.

© International Astronomical Union 2010
doi:10.1017/S174392131000877X

JD6 -Time and Astronomy
Preface and Posters

Pascale Defraigne

Royal Observatory Belgium, BE 1180 Brussels, Belgium

Responsibility for the definition of time scales left the astronomical community some 40 years ago when, in 1967, the second became defined by an atomic transition in the International System of units SI and when International Atomic Time (TAI) was defined as the primary international time scale in 1971. Atomic time is now 10^7 times more stable than the universal time UT defined by the observed Earth rotation rate and some 10^4 times more stable than the planetary orbital motions that were used to define time until 1967.

But time still interacts with astronomy in many ways: as the independent variable for the description of all dynamical systems, its stability allows one to study these systems and their perturbations. Time is therefore of major importance for astronomers, with time scales based on the SI second for practical applications and coordinate time scales for theoretical developments. Precise timing of the rotational and orbital periods of pulsars has the potential to contribute to the long-term stability of TAI, thereby returning some aspects of time keeping to astronomy. Furthermore, since observational techniques rely on the measurement of the time of propagation of electromagnetic signals, astronomy provides an important testing ground for relativity.

Those were the starting ideas of the Joint Discussion "Time and Astronomy" which gathered more than sixty participants during the IAU General Assembly 2009, in Rio de Janeiro. These proceedings provide a summary of all the subjects discussed during the meeting.

Supporting Commission: 31: Time

Co-supporting Commissions:
4: Ephemerides, 7: Celestial Mechanics and Dynamical Astronomy, 8: Astrometry, 19: Rotation of the Earth, 52 : Relativity in Fundamental Astronomy, 40 : Radio Astronomy.

Members of the Scientific Organizing Committee:
Pascale Defraigne (Belgium, chair), Aleksander Brzezinski (Poland, co-chair), Gambis Daniel (France), Sergei Klioner (Germany), Yury Ilyasov (Russia), Michael Kramer (UK), Richard Manchester (Australia), Demetrios Matsakis (US), Rendong Nan (China), Gerard Petit (France).

List of Posters:
Inspection for Secular Variation of Earth's Rotation Inferred from Chinese Ancient Astronomical Records,
  Y. Li
The difference of the Earth dynamical flattening (H) between precession observation and earth model can be reduced by $\frac{2}{3}$
  C. L. Huang, Y. Liu

Post-Newtonian Reference Frames for Advanced Theory of the Lunar Motion and a New Generation of Lunar Laser Ranging
  S. Kopeikin, Y. Xie
Status and future of the cooperative San Juan SLR Station
  R. Podest, L. Weidong, E. Actis, et al.
High-resolution Earth rotation parameters from the continuous VLBI campaign CONT08
  M. Schindelegger, J. Boehm, S. Englich, H. Schuh
Russian astronomical yearbooks in editions and program systems
  N. Glebova, G. Kosmodamianskiy, M. Lukashova, I. Netsvetaev, G. Netsvetaeva, E. Parijskaja, M. Sveshnikov, V. Skripnichenko
Design and Analysis of Nanosecond Time Synchronization System of Each Station of Each Chain of Chinese LORAN-C
  H. J. Ma, Y. H. Hu, J. F. Wu, Z. M. He
Two-Way Satellite Time and Frequency Transfer Experiment via IGSO Satellite
  X. Yang, F. Wu, Z. Li
Recent activities at NICT Space-Time Standards Group
  M. Hosokawa, K. Imamura, T. Iwama, S. Hama, J. Amagai, Y. Hanado, R. Ichikawa, Y. Koyama
Research on the technology of Common-view based on the Chinese Area Positioning System
  J. F. Wu, Y. H. Hu, X. Lu, H. J. Ma, L. Hou, Z. M. He, J. Wang
Aircraft High Dynamic Two-Way Time Synchronization Technique research
  H. J. Ma, Y. H. Hu, J. F. Wu, J. G. Wang
Studies on Algorithm of Autonomous Time Synchronization for Constellations
  X. Lu, F. Huang, H. Wu, Y. Bian
Heterodyne Digital Method in Precision Frequency Measurement
  X. Li
The precise timing of millisecond pulsars as a possible method of the study of globular clusters and Galaxy structures
  T. Larchenkova
Pulsar survey with the telescope FAST
  R. Nan
Pulsar Timing package TIMAPR upgrade
  O. Doroshenko
Pulsar timing as a tool for gravitational waves detection
  M. Sazhin
A New Algorithm for Detecting Gravitational Waves Using Pulsars
  R. Shannon
Planetary ephemerides and pulsar timing
  A. Romanoff
Pulsar Time Scale study by ensembles of simulated models
  N. Korotkova, Y. Ilyasov, M. Pshirkov

Highlights of Astronomy, Volume 15
XXVIIth IAU General Assembly, August 2009
Ian F. Corbett, ed.

Monitoring UT1 from astro-geodetic techniques at the EOP Center of the IERS

D. Gambis and C. Bizouard

Observatoire de Paris, 61 avenue de l'Observatoire, Paris, France

Abstract. Monitoring the Earth rotation is essential in various domains linked to reference frames firstly with applications in orbit determination, space geodesy or Astronomy. Secondly for geophysical studies where are involved mass motions within the different external fluid layers, atmosphere, hydrosphere, core and mantle of the earth, this on time scales ranging from a few hours to decades. The Earth Orientation Centre of the IERS is continuously monitoring the earth orientation variations from results derived from the various astro-geodetic techniques. It has in particular the task of deriving an optimal combined series of UT1 which is now based mainly on Very Long Baseline Interferometry (VLBI) with some contribution of LOD derived from GPS. We give here a brief summary concerning the contribution of the various techniques to UT1 and in aprticular how the use of LOD derived from GPS can improve the combination. More details are available in Gambis (2004) and Bizouard and Gambis (2009) and the website http://hpiers.obspm.fr/eop-pc/

Keywords. Universal Time UT1, Earth Rotation

1. UT1 determination: monitoring and accuracy

Earth Orientation Parameters describe the orientation of the earth with respect to a non rotating reference frame. One of the parameter, Universal Time UT1 represents the rotation of the Earth around its axis. Until the 1970's, UT1 was exclusively monitored by astrometric techniques based on optical instruments like photozenithal tubes, PZT, meridian refractors and astrolabes. In the 1970's the emergence of Lunar Laser Ranging (LLR) allowed to determine UT0. In 1985 its accuracy was in the range of 0.400 ms. Meanwhile VLBI technique was emerging and determined UT1 with an accuracy at least ten times better than LLR (Table 1).

Table 1. Contribution of astro-geodetic technique to the determination of UT1 and LOD

Technique	Since	EOP	Time resolution	Accuracy
ASTROMETRY	1899	UT1	5 days	1 ms
LLR	1969	UT0	1 day	0.4 ms
SLR	1976	LOD	3 days	200 μs
VLBI	1981	UT1 Standard	3-4 days	5 μs
	1981	UT1 Intensive	1 day	15 μs
	1981	LOD	3-4 days	15 μs
GPS	1993	LOD	1 day	10 μs

2. Contribution of LOD derived from GPS estimates in UT1 series, method of combined smoothing

VLBI, as the only technique referring to a non rotating celestial reference frame is the main contributor of UT1. Alternatively, satellites techniques, like GPS which are realizing their celestial frame through the orbit determination do not allow determining an accurate UT1; because of mis-modelling of various perturbations, the orbit is affected by long-term systematic variations. GPS techniques can nevertheless determine daily LOD estimates of which values are slightly biased. LOD(GPS) can be used for UT1 computation when calibrated by UT1 derived from VLBI using the so-called method of "Combined Smoothing" Vondrak *et al.*, 1999; Vondrak, 2000. Figure 1 shows the comparison of various combined UT1 series with the external Final Bulletin A series. It appears that the contribution of GPS LOD either by the direct integration of LOD(GPS) or when applying the Combined Smoothing leads to an small improvement of a few μs in the WRMS compared to the solution which does not incorporate any LOD(GPS) data. It is also striking that the contribution of intensive session is only a few μs.

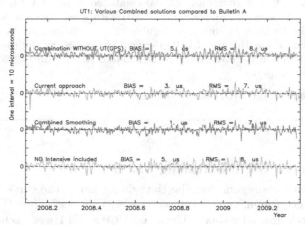

Figure 1. Comparisons of different UT1 combined series using or not UT(GPS). It appears that the contribution of GPS LOD either by the direct integration of LOD(GPS) referred as Current approach, or when applying the Combined Smoothing leads to an small improvement in the WRMS compared to the usual solution. We can remark that the inclusion of Intensive sessions does not significantly improve the final combined UT1 solution.

References

Bizouard, C. & Gambis, D., 2009, the combined solution C04 for Earth Orientation Parameters, recent improvements, Springer Verlag series, Series International Association of Geodesy Symposia, Vol. 134, Hermann (Ed.), pp. 265–270

Gambis, D., 2004, Monitoring Earth orientation using space-geodetic techniques: state-of-the-art and prospective, J. of Geodesy, Volume 78, Issue 4–5, pp. 295–303, doi 10.1007/s00190-0040394-1.

Vondrak, J. & Gambis, D., 1999, Accuracy of Earth orientation parameters obtained by different techniques in different frequency windows, in: Soffel, M., Capitaine, N. (eds.) JSR 1999 Systmes de rfrence spatio-temporels and IX. Lohrmann Colloquium, Observatoire de Paris, 206–213.

Vondrak, J. & Cepek, A., 2000, Combined smoothing method and its use in combining Earth orientation parameters measured by space techniques, Astron. Astrophys. Suppl. Ser. 147, 347–359

Highlights of Astronomy, Volume 15
XXVIIth IAU General Assembly, August 2009
Ian F. Corbett, ed.

Instability of the celestial reference frame and effect on UT1

Vladimir E. Zharov

Department of Physics, Moscow State University, Moscow, Russia
email: zharov@sai.msu.ru

Abstract. It was shown that the ICRF radio sources including the defining sources have significant apparent motion that leads to rotation of the ICRF. This rotation is transformed to secular variations of EOP that is decreased or removed if motion of sources is took into account.

Keywords. ICRF, Earth orientation parameters

VLBI is currently the only method available for measuring of the Universal Time (UT). Rotation of the Earth is described as motion of the Earth's axis of figure relative to the International Celestial Reference Frame (ICRF) that is defined by the precise coordinates of extragalactic radio sources. The rotational stability of the frame is based on the assumption that the sources have no proper motion and it means that there is no global rotation of the universe. But analysis of time series of coordinates of the ICRF radio sources shows that many of them including the defining sources have significant apparent motion, Zharov *et al.*, 2009. It is explained by motion of an emission region that is called by the ICRF source inside the jet of a quasar.

Software ARIADNA (Zharov, 2009) was used for estimation of the Earth orientation parameters (EOP) for period 1984–2008. The first solution was obtained for accepted catalog of the ICRF sources (Ma *et. al*, 1998). The second solution was obtained for improved catalog: positions of the sources were corrected and velocities of them were added. It was shown that rotation of the ICRF is due to the motions of sources. The effect of the source apparent motion has an impact on the determination of the EOP. From the first solution we have that drift of x_p-coordinate of pole is $-10 \pm 1\,\mu$as/year, y_p-coordinate of pole is $+3 \pm 1\,\mu$as/year, UT1 is $+0.15 \pm 0.05\,\mu$s/year, of nutation angles $\Delta\psi \sin\varepsilon$ and $\Delta\varepsilon$ are -3.7 ± 1.5 and $-7.5 \pm 0.7\,\mu$as/year. For second solution next drifts of the EOP were calculated : $-6 \pm 1\,\mu$as/year for x_p, $+0.5 \pm 1\,\mu$as/year for y_p, $-0.01 \pm 0.05\,\mu$s/year for $UT1$ and -3.2 ± 1.5, $-6.2 \pm 1.5\,\mu$as/year for nutation angles. There is decrease of secular variations of all EOP, specially of UT1.

Conclusions of work are: rotation of the ICRF is transformed to secular variations of EOP; catalog of sources must contain both their coordinates and apparent motion terms.

This work was supported by the RFBR grants 08-02-00971 and 07-01-00126.

References

Zharov, V. E., Sazhin, M. V., Sementsov, V. N., Kuimov, K. V., & Sazhina, O. S. 2009, *Astr. Rep.*, 53, 579
Zharov, V. E. 2009, submitted to *Astr. Rep.*
Ma, C., Arias, E. F., Eubanks, T. M., *et al.*, 1998, *AJ*, 116, 516

Highlights of Astronomy, Volume 15
XXVIIth IAU General Assembly, August 2009
Ian F. Corbett, ed.

© International Astronomical Union 2010
doi:10.1017/S174392131000880X

UT1 and relativistic theory of Earth rotation

S. A. Klioner

Lohrmann Observatory, Dresden Technical University, 01062 Dresden, Germany

Abstract. The influence of post-Newtonian effects in the rotation of Earth on UT1 are investigated and found to be negligible.

Keywords. relativity, time, reference systems

Using the numerical code for the post-Newtonian theory of Earth rotation (Klioner *et al.* (2008), Klioner *et al.* (2009)) it is not difficult to investigate the influence of the relativistic effects on the angular velocity of the Earth that is directly related to UT1. Two numerical integrations for the whole validity range of DE403 has been performed. First integration was purely Newtonian (all relativistic effects have been switched off). For the second integration several dynamical relativistic corrections were switched on: (1) full post-Newtonian torque was used; (2) relativistic time transformation between TT (argument of integration) and TDB (argument of the ephemeris) were applied; (3) relativistic scaling of parameters were applied. Using the results of both integrations, the angular velocity of proper rotation of the Earth was computed. The difference of the angular velocities obtained in two integration is shown on Fig. 1. One can see that, contrarily to claims of some authors, the influence of relativity on the LOD and therefore on UT1 is fully negligible and remains well below 1 μs. The effects of geodetic precession, being of kinematical nature, are another potential source of the variation of the LOD and will be discussed elsewhere.

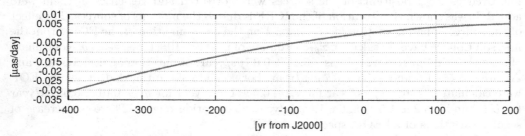

Figure 1. Difference between angular velocities of the Earth obtained in fully post-Newtonian model and in purely Newtonian one.

References

Klioner, S. A., Soffel, M., & Le Poncin-Lafitte, Chr., 2008, "Towards the relativistic theory of precession and nutation", In: The Celestial Reference Frame for the Future (Proc. of Journées 2007), N. Capitaine (ed.), Paris Observatory, Paris, 139

Klioner, S. A., Gerlach, E., & Soffel, M., 2009, Relativistic aspects of rotational motion of celestial bodies, In: Relativity in Fundamental Astronomy, Proc. of the IAU Symposium 261, S. Klioner, K. Seidelmann, M. Soffel (eds.) Cambridge University Press, Cambridge, in press

Highlights of Astronomy, Volume 15
XXVIIth IAU General Assembly, August 2009
Ian F. Corbett, ed.

© International Astronomical Union 2010
doi:10.1017/S1743921310008811

New solution of Earth Orientation Parameters in 20th century

Jan Vondrák, Cyril Ron and Vojtěch Štefka

Astronomical Institute, Acad. Sci. Czech Rep., Boční II, CZ-14131 Prague
email: vondrak@ig.cas.cz, ron@ig.cas.cz, stefka@ig.cas.cz

Abstract. We present a new solution of Earth Orientation Parameters, based on optical astrometry and catalog EOC-4.

Keywords. astrometry, reference systems, time

Recently we derived a new star catalog EOC-4 that contains not only the mean positions and linear proper motions, but also periodic changes, due to orbital motions, for double and multiple systems. The catalog, resulting from the combination of ARIHIP, TYCHO-2, Hipparcos and PPM with ground-based optical observations of latitude and universal time variations, contains 4418 stars that were observed in optical programs of monitoring Earth orientation during the 20th century; 600 of them are detected to have significant periodic components.

This catalog is now used, as a basic celestial frame, to obtain the Earth Orientation Parameters from optical astrometric observations of latitude/universal time/altitude in the interval 1899.7-1992.0. About 4.5 million individual observations from 47 different instruments (10 PZT's, 7 photoelectric transit instruments, 16 visual zenith telescopes and 14 instruments for equal altitude observations), located at 33 observatories, are used. Polar motion (x, y) is determined in 5-day steps for the whole interval studied, Universal time (UT1-TAI) covers the interval 1956.0-1992.0 (i.e., when the atomic time scale was available) also in 5-day step, and celestial pole offsets (dX, dY) with respect to most recent IAU2000 nutation and IAU2006 precession are modeled by second-order polynomials of time. As a by-product, rheological parameter $\Lambda = 1+k-l$ (combination of Love and Shida numbers k, l), governing the tidal motion of local verticals, is computed for each instrument. The average standard error of one observation is $\sigma_o = 0.184''$, slightly smaller than the one of our preceding solution with catalog EOC-3.

Apart from the dominant annual and Chandler wobbles in polar motion, it is demonstrated that Markowitz wobble (with period of about 28 years) is also present, in addition to a longer 78-year period. Spectral analysis of length-of-day changes, calculated from the differences between universal and atomic time scales UT1-TAI, confirms the tidal variations with periods 13.66, 27.56, 182 and 365 days, and also the variations caused by the atmosphere, with periods equal to 0.5, 1 and 6 years. Decadal variations with periods around 30 years are also visible. The solution will serve for further long-term Earth rotation studies.

Acknowledgements

This study was made possible thanks to the grants LC506 awarded by the Ministry of Education of the Czech Republic, and 205/08/0908 awarded by the Grant Agency of the Czech Republic.

Highlights of Astronomy, Volume 15
XXVIIth IAU General Assembly, August 2009
Ian F. Corbett, ed.

© International Astronomical Union 2010
doi:10.1017/S1743921310008823

Implementation of the IAU 2000 definition of UT1 in astronomy

Nicole Capitaine[1] and Patrick T. Wallace[2]

[1]SYRTE, Observatoire de Paris, CNRS, UPMC, 61, Av. l'Observatoire, 75014, Paris, France
email: nicole.capitaine@obspm.fr

[2]Space Science & Technology Dept., STFC/RAL, Didcot, Oxon, OX11 0QX, UK
email: patrick.wallace@stfc.ac.uk

Abstract. We recall the concepts and nomenclature associated with the IAU 2000 definition of UT1 as function of the Earth rotation angle (ERA). We comment on the complications that arise when UT1 is regarded as both an angle and a time scale. We review the IAU 2006 expressions for the position of the celestial intermediate origin (CIO) and the equation of the origins, and the associated CIO and equinox based procedures for the celestial-to-terrestrial transformation.

Keywords. astrometry, reference systems, time, Earth

According to the IAU 2000/2006 resolutions, the concepts, nomenclature and numerical expressions associated with the IAU 2000 definition of UT1 as function of the Earth rotation angle (ERA) have replaced the pre-2003 concepts, nomenclature and numerical expressions associated with Greenwich sidereal time (GST).

ERA is the angle measured along the equator of the Celestial Intermediate Pole (CIP) between the Terrestrial Intermediate Origin (TIO) and the Celestial Intermediate Origin (CIO), and increases as the Earth rotates. Due to the kinematical definition of the CIO, its position as function of time depends only on the CIP motion, while the position of the equinox depends on both the equator and ecliptic motions. UT1 is defined by its conventionally adopted linear relation to the ERA. The equation of the origins (i.e. the distance from the CIO to the equinox along the CIP equator), EO, is the link between the CIO based and equinox based expressions. The ERA based expression for GST (=ERA−EO) allows a clear distinction between UT1 regarded as a time determined by the rotation of the Earth and Terrestrial time (TT), which were merged in the pre-2003 GST expression.

The IAU 2006 expressions for the precession quantities and GST have been provided by Capitaine *et al.* (2003a), while the IAU 2000/2006 expressions for locating the CIP (i.e X, Y) and the CIO (i.e the s quantity) can be found in Capitaine & Wallace (2006). The largest change in the IAU 2006 expression for s w.r.t. the IAU 2000 one (see Capitaine *et al.* 2003b) is below 3 μas, while the corresponding change in the EO (and hence GST) is more than 5 mas. The associated CIO and equinox based precession-nutation procedures provided in Wallace & Capitaine (2006) have been the basis for the SOFA (Standards Of Fundamental Astronomy) and IERS implementations (IERS, 2009).

References

Capitaine, N., Wallace, P. T., & Chapront, J. 2003a, *A&A* 412, 567–586
Capitaine, N., Wallace, P. T., & McCarthy, D. D. 2003b, *A&A* 406, 1135–1149
Capitaine, N. & Wallace, P. T. 2006, *A&A* 450, 855–872
IERS Conventions update, 2009, `http://tai.bipm.org/iers/convupdt/convupdt_c5.html`
Wallace, P. T. & Capitaine, N. 2006, *A&A* 459, 981–985

Highlights of Astronomy, Volume 15
XXVIIth IAU General Assembly, August 2009
Ian F. Corbett, ed.

Impact of Geophysical Fluids on UT1

Richard S. Gross

Jet Propulsion Laboratory, California Institute of Technology, Pasadena, CA, 91109 USA
email: Richard.Gross@jpl.nasa.gov

Abstract. Geophysical fluids have a major impact on the Earth's rotation. Tidal variations within the oceans are the predominant cause of subdaily length-of-day (lod) variations while those within the solid body of the Earth are a major source of longer period variations; tidal dissipation within the solid Earth and oceans cause a secular change in lod. Fluctuations of the atmospheric winds are the predominant cause of nontidal lod variations on sub-decadal time scales while decadal variations are caused by interactions between the fluid core and mantle.

Keywords. Earth, time

1. Introduction

The Earth's rotation changes on all observable time scales, from subdaily to decadal and longer, reflecting the wide variety of processes causing it to change. The changes in the Earth's rotation that are observed and the models that have been developed to explain them are reviewed here.

2. Passive influence

Changes in the Earth's rotation are usually studied by applying the principle of conservation of angular momentum. Under this principle, and in the absence of external torques, the rotation of the solid Earth changes when angular momentum is transferred between it and the fluid regions with which it is in contact. But by modifying the Earth's inertia the fluid regions of the Earth influence the Earth's rotation even when they are at rest and not exchanging angular momentum with the solid Earth. If the Earth is assumed to be axisymmetric and if the fluid core is assumed to be inviscid and otherwise decoupled from the mantle, then the core cannot respond to axial changes in the rotation of the mantle. The greatest principal moment of inertia of the Earth appropriate for studying changes in the axial rotation of the Earth is then not that of the entire Earth including core C but is instead that of just the Earth's crust and mantle C_m. The presence of a passive fluid core thus has about an 11% effect on modeled lod variations.

3. Tidal variations

Tidal forces due to the gravitational attraction of the Sun, Moon, and planets deform the solid and fluid regions of the Earth causing the Earth's inertia tensor to change and thus causing the Earth's rotation to change. Tidal dissipation within the oceans and solid Earth causes the Earth's angular velocity and hence rotational angular momentum to decrease. By analyzing lunar and solar eclipse, lunar occultation, optical astrometric, and space-geodetic observations Morrison & Stephenson (2001) found that the length-of-day has increased at an average rate of $+1.8 \pm 0.1$ millisecond/century (ms/cy) during the past 2700 years. Since the observed tidal acceleration of the Moon implies that the length-of-day should be changing by $+2.3$ ms/cy due to tidal dissipation, some other mechanism must be acting to change the length-of-day by -0.5 ms/cy, the most likely candidate being glacial isostatic adjustment. The isostatic adjustment of the solid Earth in response to the decreasing load on it following the last deglaciation causes the figure of the Earth to change, and hence causes the length-of-day to change.

Of the elastic solid, inelastic solid, and dynamic ocean responses of the Earth to tidal forces in the long-period tidal band, the elastic response of the solid Earth has the greatest effect on lod, with inelastic effects being only a few percent of the elastic, and with the effect of dynamic ocean tides being most important at the fortnightly tidal frequency. Fortunately, accurate models for the effect of elastic solid, inelastic solid, and dynamic ocean tides on lod are currently available (Gross 2009).

But in the diurnal and semidiurnal tidal bands the ocean tides have the greatest effect on lod. In fact, comparisons of observations with models show the dominant role that ocean tides play in causing subdaily UT1 and lod variations, with as much as 90% of the observed UT1 variance being explained by diurnal and semidiurnal ocean tides. Apart from errors in observations and models, the small difference that remains is probably due to nontidal atmospheric and oceanic effects (Chao *et al.* 1996).

4. Nontidal variations

Since 1830 the length-of-day has been observed to vary by up to 7 ms on decadal time scales. Such large variations are too great to be caused by atmospheric fluctuations. Even if the atmosphere's motion were to cease entirely, the resulting change in lod would be only about 3 ms. So decadal variations in lod are thought to be primarily caused by core-mantle interactions. Under reasonable physical assumptions, surface observations of the secular variation of the magnetic field can be used to infer the motion of the fluid at the top of the core. Additional reasonable physical assumptions allow the angular momentum of the entire core to be computed from the inferred motion of the fluid at the top of the core. In fact, remarkably good agreement between the core angular momentum modeled in this manner and lod observations on decadal time scales is found, at least since 1900 (Ponsar *et al.* 2003).

Numerous studies have shown that atmospheric winds are the dominant cause of nontidal lod variations on time scales of a few days to a few years. For example, on intraseasonal time scales, the effect of atmospheric winds explains some 86% of the observed variance. Including the effect of surface pressure fluctuations increases the observed variance explained to 90% and additionally including oceanic effects further increases it to 92%. Similar results are obtained on seasonal and interannual time scales (Gross *et al.* 2004).

Acknowledgements

The work described in this paper was performed at the Jet Propulsion Laboratory, California Institute of Technology, under contract with the National Aeronautics and Space Administration.

References

Chao, B. F., Ray, R. D., Gipson, J. M., Egbert, G. D., & Ma, C. 1996, *J. Geophys. Res.* 101, 20151
Gross, R. S. 2009, *J. Geodyn* in press
Gross, R. S., Fukumori, I., Menemenlis, D., & Gegout, P. 2004, *J. Geophys. Res.* 109, B01406
Morrison, L. V. & Stephenson, F. R. 2001, *J. Geodyn.* 32, 247
Ponsar, S., Dehant, V., Holme, R., Jault D., Pais, A., & Van Hoolst, T. 2003, in: V. Dehant, K. C. Creager, S. Karato, & S. Zatman (eds.), *Earth's Core: Dynamics, Structure, Rotation*, Geodynamics Series (Washington DC: Amer. Geophys. Union), vol. 31, p. 251

Highlights of Astronomy, Volume 15
XXVIIth IAU General Assembly, August 2009
Ian F. Corbett, ed.

© International Astronomical Union 2010
doi:10.1017/S1743921310008847

Short-term tidal variations in UT1: compliance between modelling and observation

Sigrid Englich, Harald Schuh & Robert Weber

Institute of Geodesy and Geophysics, Vienna University of Technology,
Gusshausstrasse 27-29, 1040 Vienna, Austria
email: `sigrid.englich@tuwien.ac.at`

Abstract. The Earth rotation rate and consequently universal time (UT1) and length of day (LOD) are periodically affected by solid Earth tides and oceanic tides. Solid Earth tides induce changes with periods from around 5 days to 18.6 years, with the largest amplitudes occurring at fortnightly, monthly, semi-annual and annual periods, and at 18.6 years. The principal variations caused by oceanic tides have diurnal and semi-diurnal periods. For the investigation of the tidal effects with periods of up to 35 days, UT1 series are estimated from VLBI observation data of the time interval 1984–2008. The amplitudes and phases of the terms of interest are calculated and the results for diurnal and sub-diurnal periods are compared and evaluated with tidal variations derived from a GNSS-based LOD time series of 8 months. The observed tidal signals are finally compared to the predicted tidal variations according to recent geophysical models.

Keywords. UT1, tidal variations, VLBI, GNSS

1. Diurnal and sub-diurnal variations

We estimated the amplitudes of 70 ocean tidal terms with diurnal and semi-diurnal periods from VLBI/GNSS-observed UT1/LOD variations. The comparison with UT1 amplitudes of the conventional model (IERS Conventions (2004)), showed deviations in both tidal bands of up to 2.5 μs. The residuals in the semi-diurnal tidal band could be reduced almost to zero by accounting for the effect of the lunisolar torque on the triaxial Earth (Brzezinski & Capitaine (2002)), to which the space geodetic techniques are sensitive.

2. Variations up to 35 days

The zonal tidal terms were derived from VLBI-based UT1 time series, from which the atmospheric influence and sub-diurnal variations were removed and all signal with periods over 35 days was filtered. All terms were calculated w.r.t. the IERS conventional model as well as w.r.t. a combination of different models following Gross (2009). The comparison with the IERS model revealed large discrepancies (>40 μs) in the fortnigthly terms, whereas the combined model accounts for most of the fortnightly tidal signal.

References

Brzezinski, A. & Capitaine, N. 2002, *Proceedings of the Journees 2001*, p. 51–58
Gross, R. S. 2009, *Journal of Geodynamics*, to be published in 2009
IERS Conventions 2003, in: D. D. McCarthy & G. Petit (eds.), *IERS Technical Note 32*, 2004

Highlights of Astronomy, Volume 15
XXVIIth IAU General Assembly, August 2009
Ian F. Corbett, ed.

Determination of UT1 by VLBI

Harald Schuh[1], Johannes Boehm[1], Sigrid Englich[1] & Axel Nothnagel[2]

[1] Institute of Geodesy and Geophysics, Vienna University of Technology,
Gusshausstrasse 27-29, 1040 Vienna, Austria
email: harald.schuh@tuwien.ac.at

[2] Institute of Geodesy and Geoinformation, University of Bonn,
Nussallee 17, 53115 Bonn, Germany
email: nothnagel@uni-bonn.de

Abstract. Very Long Baseline Interferometry (VLBI) is the only space geodetic technique which is capable of estimating the Earth's phase of rotation, expressed as Universal Time UT1, over time scales of a few days or longer. Satellite-observing techniques like the Global Navigation Satellite Systems (GNSS) are suffering from the fact that Earth rotation is indistinguishable from a rotation of the satellite orbit nodes, which requires the imposition of special procedures to extract UT1 or length of day information. Whereas 24 hour VLBI network sessions are carried out at about three days per week, the hour-long one-baseline intensive sessions ('Intensives') are observed from Monday to Friday (INT1) on the baseline Wettzell (Germany) to Kokee Park (Hawaii, U.S.A.), and from Saturday to Sunday on the baseline Tsukuba (Japan) to Wettzell (INT2). Additionally, INT3 sessions are carried out on Mondays between Wettzell, Tsukuba, and Ny-Alesund (Norway), and ultra-rapid e-Intensives between E! urope and Japan also include the baseline Metsähovi (Finland) to Kashima (Japan). The Intensives have been set up to determine daily estimates of UT1 and to be used for UT1 predictions. Because of the short duration and the limited number of stations the observations can nowadays be e-transferred to the correlators, or to a node close to the correlator, and the estimates of UT1 are available shortly after the last observation thus allowing the results to be used for prediction purposes.

Keywords. UT1, VLBI, intensive sessions

1. Outline

The sensitivity of a VLBI baseline or network to UT1 is proportional to the East-West extend of the configuration. At present UT1 can be determined with an accuracy of 6-7 μs from regular 24 hour sessions and 10-15 μs from intensive sessions. We reported about state-of-the-art measurements of UT1 in different baseline or network configurations and showed examples of current investigations on the following topics:

- UT1 provided by the IVS
- 24 hours sessions, intensive sessions, e-VLBI
- Impact of errors in nutation on dUT1 from intensive sessions
- Impact of a priori troposphere gradients on dUT1 from intensive sessions
- Comparison of geodetic and atmospheric excitation with sub-diurnal periods

2. Conclusions

VLBI provides a unique capability to directly access the rotation phase of the Earth. As VLBI is the only technique to determine UT1 with high precision it is difficult to judge the accuracy of the obtained results.

Highlights of Astronomy, Volume 15
XXVIIth IAU General Assembly, August 2009
Ian F. Corbett, ed.

© International Astronomical Union 2010
doi:10.1017/S1743921310008860

Semidiurnal signal in UT1 due to the influence of tidal gravitation on the triaxial structure of the Earth

Aleksander Brzeziński[1,2] and Nicole Capitaine[3]

[1] Warsaw University of Technology, Faculty of Geodesy and Cartography, Warsaw, Poland
[2] Space Research Centre, Polish Academy of Sciences, Warsaw, Poland
[3] SYRTE, Observatoire de Paris, CNRS, UPMC, Paris, France

Abstract. The axial component of Earth rotation, which is conventionally expressed by Universal Time (UT1), contains small physical signals with diurnal and subdiurnal periods. This part of the spectrum is dominated by the tidal effects which are regular and predictable. The largest components express the influence of the gravitationally forced ocean tides with diurnal and semidiurnal periods and amplitudes up to 0.02 milliseconds (ms) in UT1 corresponding to an angular displacement of 0.30 milliarcseconds (mas); see Table 8.3 of the IERS Conventions (IERS, 2003). There are also smaller subdiurnal components (amplitudes up to 0.03 mas), designated as "spin libration" by Chao *et al.* (1991), due to direct influence of the tidal gravitation on those features of the Earth's density distribution which are expressed by the non-zonal terms of the geopotential. These components are not included in the models recommended by the IERS Conventions, in contrast to the corresponding effect in polar motion (ibid., Table 5.1).

  Here we consider in detail the subdiurnal libration in UT1. We derive an analytical solution for the structural model of the Earth consisting of an elastic mantle and a liquid core which are not coupled to each other. The reference solution for the rigid Earth is computed by using the satellite-determined coefficients of geopotential and the recent developments of the tide generating potential (TGP). We arrived to the conclusion that the set of terms with amplitudes exceeding the truncation level of 0.005 mas consists of 11 semidiurnal harmonics due to the influence of the TGP term u_{22} on the equatorial flattening of the Earth expressed by the Stokes coefficients C_{22}, S_{22}. There is an excellent agreement between our estimates for the rigid Earth and the amplitudes derived by Wünsch (1991). The only important difference is the term with the tidal code ν_2, which seems to be overlooked in the development of Wünsch. Our amplitudes computed for an elastic Earth with liquid core appear to be in reasonable agreement with those derived by Chao *et al.* (1991), but the latter model was not complete. The estimated effect is superimposed on the ocean tide influences having the same frequencies but 9 to 11 times larger amplitudes. Nevertheless, its maximum peak-to-peak size is about 0.105 mas, hence definitely above the current uncertainty of UT1 determinations. Comparison with the corresponding model of prograde diurnal polar motion associated with the Earth's triaxiality (IERS Conventions, Table 5.1) shows that: 1) the two effects are of similar size, 2) there is consistency between the underlying dynamical models, parameters employed, etc. In conclusion, we recommend adding the model developed here to the set of procedures provided by the IERS Conventions.

Keywords. Earth rotation, universal time, tidal gravitation

References

Brzeziński, A. & Capitaine, N. 2002, in: N. Capitaine (ed.), *Proc. Journees 2001*, 51, Obs. de Paris
Chao, B. F., Dong, D. N., Liu, H. S., & Herring T. A. 1991, *Geophys. Res. Letters* 18, 2007
IERS 2003, D. McCarthy & G. Petit (eds.), *IERS Technical Note* 32, Verlag des Bundesamts für Kartographie und Geodäsie, Frankfurt am Main
Wünsch, J. 1991, *Astron. Nachr.* 312, 321

Highlights of Astronomy, Volume 15
XXVIIth IAU General Assembly, August 2009
Ian F. Corbett, ed.

Current Status and Future Directions of the IERS RS/PC Predictions of UT1

W. Wooden, B. Luzum and N. Stamatakos

U.S. Naval Observatory, 3450 Massachusetts Ave, Washington, DC, USA
email: brian.luzum@usno.navy.mil

1. IERS RS/PC

The International Earth Rotation and Reference Systems Service (IERS) Rapid Service/Prediction Center (RS/PC) produces daily and weekly EOP combination and prediction solutions. The daily solutions are produced after 1700 UTC while the weekly EOP solutions are produced Thursday after 1700 UTC. These solutions include data from Atmospheric Angular Momentum (AAM) analysis and forecasts, Global Positioning System (GPS) solutions, Satellite Laser Ranging (SLR) solutions, and Very Long Baseline Interferometry (VLBI) solutions. The solutions are sent to roughly 700 people by e-mail per week and are picked up in roughly 40000 ftp downloads per month.

2. IERS RS/PC Changes

Several new or modified data sets have been incorporated into the IERS RS/PC operations. Starting in October 2007, the AAM forecasts lengths were extended from 5 days to 7.5 days. This change led to a roughly 15% improvement in UT1−UTC predictions out to 10 days in the future (Stamatakos et al. (2008)). In March 2009, a new version of the USNO UTGPS solution was incorporated into the IERS RS/PC solution that makes use of more satellites and more sophisticated orbital modeling. Starting in roughly September 2007, electronically transferred (e-transfer) VLBI intensives became available with the creation of the 'Int3' intensives. These have the potential of being available in less than 12 hours after observations. These quick-turnaround solutions appear to halve the error of the most recent combination data point in UT1−UTC and reduce the 1-day prediction error by roughly 25% (Luzum & Nothnagel (2009)).

3. Conclusions

Recent improvements in the IERS RS/PC predictions have been due to improvements in the input data series. This leads to some interesting questions. With improving data latency and increased frequency of solutions, how frequently should EOP solutions be updated? Also, what prediction length should the algorithms be optimized for? If combination and prediction solutions are made with increasing frequency, then predictions algorithms should be optimized for relatively short-term (less than 10 day) predictions.

References

Luzum, B. & Nothnagel, A. 2009, submitted to Geophys. Res. Lett.
Stamatakos, N., Luzum, B., Stetzler, B., Wooden, W., & Schultz, E. 2008, accepted in Proc. Journées Système de Référence Spatio-Temporels, Dresden, 22–24 Sept. 2008.

Highlights of Astronomy, Volume 15
XXVIIth IAU General Assembly, August 2009
Ian F. Corbett, ed.

Time Ephemeris and General Relativistic Scale Factor

Toshio Fukushima

National Astronomical Observatory of Japan,
181-8588, Mitaka, Tokyo, Japan
email: Toshio.Fukushima@nao.ac.jp

Abstract. Time ephemeris is the location-independent part of the transformation formula relating two time coordinates such as TCB and TCG (Fukushima 2009). It is computed from the corresponding (space) ephemerides providing the relative motion of two spatial coordinate origins such as the motion of geocenter relative to the solar system barycenter. The time ephemerides are inevitably needed in conducting precise four dimensional coordinate transformations among various spacetime coordinate systems such as the GCRS and BCRS (Soffel *et al.* 2003). Also, by means of the time average operation, they are used in determining the information on scale conversion between the pair of coordinate systems, especially the difference of the general relativistic scale factor from unity such as L_C. In 1995, we presented the first numerically-integrated time ephemeris, TE245, from JPL's planetary ephemeris DE245 (Fukushima 1995). It gave an estimate of L_C as $1.4808268457(10) \times 10^{-8}$, which was incorrect by around 2×10^{-16}. This was caused by taking the wrong sign of the post-Newtonian contribution in the final summation. Four years later, we updated TE245 to TE405 associated with DE405 (Irwin and Fukushima 1999). This time the renewed vale of L_C is $1.48082686741(200) \times 10^{-8}$ Another four years later, by using a precise technique of time average, we improved the estimate of Newtonian part of L_C for TE405 as $1.4808268559(6) \times 10^{-8}$ (Harada and Fukushima 2003). This leads to the value of L_C as $L_C = 1.48082686732(110) \times 10^{-8}$. If we combine this with the constant defining the mean rate of TCG-TT, $L_G = 6.969290134 \times 10^{-10}$ (IAU 2001), we estimate the numerical value of another general relativistic scale factor $L_B = 1.55051976763(110) \times 10^{-8}$, which has the meaning of the mean rate of TCB-TT. The main reasons of the uncertainties are the truncation effect in time average and the uncertainty of asteroids' perturbation. As a compact realization of the time ephemeris, we prepared HF2002, a Fortran routine to compute approximate harmonic series of TE405 with the RMS error of 0.446 ns for the period 1600 to 2200 (Harada and Fukushima 2003). It is included in the IERS Convention 2003 (McCarthy and Petit 2003) and available from the IERS web site; http://tai.bipm.org/iers/conv2003/conv2003_c10.html.

Keywords. general relativity, ephemerides, reference systems, time

---

References

Fukushima, T., 1995, Astron. Astrophys., 294, 895

Fukushima, T., 2009, in Proc. IAU Symp. 261, to appear

Harada, W. & Fukushima, T., 2003, Astron. J., 126, 2557

International Astronomical Union, 2001, in Proc. 24th General Assembly Manchester 2000, Trans. of IAU XXIVB, IAU, Paris

Irwin, A. W. & Fukushima, T., 1999, Astron. Astrophys., 348, 642

McCarthy, D. D. & Petit G., 2003, IERS Convention (2003), IERS Tech. Note 32, Obs. Paris

Soffel, M., Klioner, S. A., Petit, G., Wolf, P., Kopeikin, S. M., Bretagnon, P., Brumberg, V. A., Capitaine, N., Damour, T., Fukushima, T., Guinot, B., Huang, T.-Y., Lindegren, L., Ma, C., Nordtvedt, K., Ries, J. C., Seidelmann, P. K., Vokrouhlicky, D., Will, C. M., & Xu, C., 2003, Astron. J., 126, 2687

Highlights of Astronomy, Volume 15
XXVIIth IAU General Assembly, August 2009
Ian F. Corbett, ed.

© International Astronomical Union 2010
doi:10.1017/S1743921310008896

Atomic time scales TAI and TT(BIPM): present performances and prospects

Gérard Petit

Bureau International des Poids et Mesures
92312 Sèvres Cedex, France
email: gpetit@bipm.org

Abstract. We review the stability and accuracy achieved by the reference atomic time scales TAI and TT(BIPM). We show that they presently are at the level of a few 10^{-16} in relative value, based on the performance of primary standards, of the ensemble time scale and of the time transfer techniques. We consider how the 1×10^{-16} value could be reached or superseded and which are the present limitations to attain this goal.

Keywords. Time, Reference systems

1. Introduction

International Atomic Time TAI gets its stability from some 350 atomic clocks worldwide that generate the free atomic scale EAL and its accuracy from a small number of primary frequency standards (PFS) which are used to steer the EAL frequency. Terrestrial Time is a coordinate time in the geocentric reference system defined by the International Astronomical Union. TAI provides one realization of TT but it is not optimal because of operational constraints. The BIPM therefore computes in deferred time another realization, TT(BIPM), which is based on a weighted average of the evaluations of TAI frequency by the PFS. A new version is computed each January, the latest available being TT(BIPM08) available at ftp://tai.bipm.org/TFG/TT(BIPM).

We review the stability and accuracy achieved by the reference atomic time scales TAI and TT(BIPM). We show that they presently are at the level of a few 10^{-16} in relative value, based on the performance of primary standards, of the ensemble time scale and of the time transfer techniques. We consider how the 1×10^{-16} value could be reached or superseded and which are the present limitations to attain this goal.

2. Achieving sub-10^{-15} accuracy

The stability of atomic time scales, the PFS accuracy and the capabilities of frequency transfer all achieve about a few parts in 10^{16} in relative frequency stability.

The stability and accuracy of time scales has been extensively studied by Petit (2007). The 1-month stability is that of EAL, at $3 - 4 \times 10^{-16}$. The estimated accuracy of TT(BIPM) over recent yers is close to 5×10^{-16} or below. This is due to the ever increasing number of Cs fountain evaluations (about 170 since 1999, 45 of them in 2008), and to the improved accuracy of each evaluation. Using TT(BIPM08) as a reference shows that EAL is affected by a significant drift and that the long-term instability of TAI is between 1×10^{-15} and 1×10^{-15}, a factor two or three worse than TT(BIPM).

The time transfer techniques presently used in TAI are Global Positioning System (GPS) code measurements in a mode called all in view, which produces best results when using dual-frequency receivers, and two-way time transfer (TW) using telecommunication satellites, and the implementation of GPS phase and code measurements in TAI is under way (Petit 2009). It has been shown, see e.g. Bauch *et al.* (2006), that these techniques are capable of a frequency transfer uncertainty of 1×10^{-15} in one day (GPS code and phase, TW at short distance) or in a few days (GPS code only, TW at long distance). For TAI, where the optimal stability is searched at 1-month averaging, the contribution of frequency transfer to the instability is expected to be a few parts in 10^{16}.

3. Reaching 1×10^{-16} and beyond

¿From numerous recent publications, see e.g. in parts II, IV and V of Maleki (2009), it is clear that some frequency standards have reached a level where all systematic effects may be estimated with an uncertainty at 1×10^{-16} or below, and the number of potential candidates in this group is expected to increase continuously. Therefore PFS will have the capacity to reach the level of 1×10^{-16} and to surpass it.

Progresses in the ensemble scale EAL have been due to an increase in the number of clocks and to changes in the algorithm. Some progress can still be expected but it is unlikely that this can lead to 1-month instability of 1×10^{-16}, as this would require a 10-fold increase in the number of clocks used. We can expect a new generation of stable clocks, with a stability performance in the low 10^{-16} at 1-month averaging time and seamless continuous operation for long periods. Possible candidates may be emerging, see e.g. pp 308 and 321 in Maleki (2009).

The techniques presently used for time transfer may progress, thanks to new or modernized navigation systems, or by using more data points for TW, or by mixing the two techniques (Jiang & Petit 2009). Ultimately, however, these techniques will find their limitations. Several new techniques have emerged that should be able to to provide frequency transfer uncertainty in the low 10^{-17} region and possibly below. One is based on a microwave link to a low Earth orbit payload with one uplink and two downlink signals (Seidel *et al.* 2008) and another one (T2L2) on a two-way optical link to a low Earth orbit payload (Samain *et al.* 2008). Both promise a frequency uncertainty below 1×10^{-16} at 1-day averaging, if the clock of the space payload is stable enough. Another technique, even more performant although limited in spatial extension, transfers a stable laser frequency over a standard fiber link (Lopez *et al.* 2008).

4. Conclusions

We have shown that the present performance of the reference atomic time scales TAI and TT(BIPM) is in the region of a few parts in 10^{16} in stability and accuracy. Improving to 1×10^{-16} and below is not out of reach but may require the development of a new generation of commercial atomic clocks or a new approach to the problem of elaborating an ensemble time scale, or both.

References

Bauch, A., Achkar, J., Bize, S. *et al.* 2006, *Metrologia*, 43, 109–120

Jiang, Z. & Petit, G. 2009, *Metrologia*, 46-3, 305–314

Lopez, X., Amy-Klein, A., & Daussy, C. 2008, *Eur. Phys. J.*, D 48, 35–41

Maleki L. (ed.) 2009, *Proc 7th Symp. Freq. Standards and Metrology* (World Scientific), 308–313

Petit, G. 2007, *Proc. 21st EFTF conference*, 391–394

Petit, G. 2009, *Proc. EFTF-IFCS joint conference*, 116–119

Samain, E., Weick, J., Vrancken, P. *et al.* 2008, *Int. J. Mod. Phys.*, D 17-7, 1043

Seidel, A., Hess, M. P., Kehrer, J. *et al.* 2008, *Proc. 22nd EFTF conference*

Highlights of Astronomy, Volume 15
XXVIIth IAU General Assembly, August 2009
Ian F. Corbett, ed.

© International Astronomical Union 2010
doi:10.1017/S1743921310008902

The use of Galileo signals for time transfer metrology

Pierre Uhrich and Philip Tuckey

LNE-SYRTE, Observatoire de Paris, LNE, CNRS, UPMC,
61 avenue de l'Observatoire, 75014 Paris, France

Abstract. We briefly summarise the planned Galileo navigation system signals and their potential application in time and frequency metrology.

Keywords. time

Galileo is the global navigation satellite system (GNSS) currently being developed by the European Union, the European Space Agency and partners. Two experimental satellites are currently in operation, the in orbit validation phase involving 4 new satellites is planned to commence in 2010 and full operational capability is planned for the end of 2013. As with other GNSSs, in particular the widely-used GPS, Galileo signals will be applicable for time and frequency comparisons between distant clocks.

Galileo will provide 4 position, navigation and timing (PNT) services: Open Service, Safety of Life, Commercial Service and Public Regulated Service, as well as a search and rescue support (i.e. communications) service. The PNT services will be carried by 6 groups of signals: E5A and E5B (on either side of 1191.795 MHz), E6P and E6C (1278.75 MHz), L1P and L1F (1575.42 MHz). The E5A and E5B signals may alternatively be considered together as a single, wideband signal, called E5. The ranging signals on E5A, E5B and L1F will be unencrypted while the data channel on E5A will be unencrypted and those on E5B and L1F partially encrypted. The E6C, E6P and L1P signals will be encrypted, for commercial and governmental applications respectively.

Thus Galileo is planned to provide 4 freely-accessible single-frequency signals suitable for time and frequency comparisons using the common-view, code-based method: E5A, E5B, E5 and L1F. These signals allow for 6 dual-frequency pairs, of which those between E5 and L1 are certainly the most interesting: E5A-L1F, E5B-L1F, E5-L1F, the others being: E5A-E5B, E5A-E5, E5-E5B, each of which may be used to construct an ionosphere-free combination analogous to the well-known "P3" combination of GPS signals. The All in View and Precise Point Positioning (PPP) methods, which play an important role in time and frequency comparisons, will also be applicable to the Galileo signals, subject to the International GNSS Service (IGS) providing Galileo products similar to its current GPS products.

We note that the "CGGTTS" data exchange format, used for common view time and frequency comparisons, already includes identifiers for Galileo satellites but still needs to be extended with identifiers for the different possible Galileo signals and combinations.

We mention also that the Galileo signals allow the possibility of constructing triple-frequency combinations which eliminate both the ionospheric and the geometrical contributions, thus providing an additional tool for studying and mitigating various error terms such as code delays, multi-path, receiver noise and phase ambiguities.

Finally we note that allowing national time metrology laboratories access to the commercial and governmental Galileo signals would certainly improve measurement capabilities, as well as providing valuable feedback to the Galileo system itself.

Highlights of Astronomy, Volume 15
XXVIIth IAU General Assembly, August 2009
Ian F. Corbett, ed.

Impact of new frequency standards on the international time scales

E. Felicitas Arias[1,2] and Gianna Panfilo[1]

[1]International Bureau of Weights and Measures,
Sèvres, France
email: `farias@bipm.org`

[2]Associated astronomer at the Paris Observatory,
Paris, France

Abstract. The reference time scales maintained at the International Bureau of Weights and Measures (BIPM) are constructed with data from industrial clocks and primary frequency standards operated in national metrology laboratories and observatories world-wide distributed. Clocks are compared making use of techniques of time transfer between remote sites. The algorithm of calculation relies on clock weighting and clock frequency prediction. We briefly present hereafter the influence of some clocks on the scales, as well as the possibilities for improvement.

Keywords. atomic time scales, frequency accuracy and stability, atomic standards

---

1. Summary

The statistical treatment of industrial clock data (frequency prediction, weighting procedure) is designed to assure the high frequency stability of the scale on an averaging interval of 30 days (BIH 1974), (Guinot & Thomas 1988), (Audoin & Guinot 2001). The introduction of measurements of primary frequency standards, together with a strategy for steering the frequency of the scale contribute to improving its frequency accuracy.

International atomic time (TAI) and Coordinated Universal Time (UTC) are calculated from monthly blocks of clock data and measurements of primary frequency standards over one year. The resulting atomic time scale has long-term instabilities that make in unsuitable for applications such as timing of millisecond pulsars. Another time scale, TT(BIPMYY) (Petit2003), (Petit 2008), is calculated every year with a different algorithm that makes use of all existing data from primary frequency standards, and the resulting scale is not affected by the instabilities of TAI.

In the last five years more caesium fountains, which realize at best the second of the international system of units (SI) started contributing data to the BIPM. The statistical uncertainty of these realizations of the second is of order 10^{-16} (BIPM 2009). The frequency steering tries to compensate, in a monthly process and without degrading the stability of the scale, a drift of the frequency observed between the fountains and the beam standards. When compared to TT(BIPM), the frequency of TAI shows a drift of 4 parts in 10^{16}/month. The origin of this drift is not completely known, but studies under development seem to indicate that an incomplete model of frequency prediction of clocks in the algorithm of calculation of TAI could be partially responsible for this effect. A linear frequency prediction algorithm is applied to both, caesium industrial beams and hydrogen masers; these later represent 12% of the clock weight in TAI. While the linear model is well adapted to caesium standards, it fails to represent the frequency drift of hydrogen masers. Tests using a quadratic frequency prediction for hydrogen masers show that this model represents better the hydrogen maser's drift, and indicate that about

20% of the drift could be provoked by the inappropriate frequency model (Panfilo & Arias 2009).

While the caesium provides the primary realization of the SI second, other radiations in the microwave and optical frequencies have been recommended as secondary representations of the second, and they are called to provide in the near future primary realizations (Gill & Riehle 2006). These frequency standards have already proved to be one order of magnitude more accurate than the caesium fountains, and for their utilisation it will be necessary to make use of highly accurate frequency transfer techniques.

References

Audoin, C & Guinot, B. 2001, *The measurement of time*, Cambridge University Press

Bureau International de l'Heure 1974, *BIH Annual Report for 1973*, Observatoire de Paris

Bureau International des Poids et Mesures 2009, *BIPM Annual Report on Time Activities for 2008*, 26

Gill, P. & Rihele, F. 2006, *Proc.* 20th *EFTF*, 282

Guinot, B. & Thomas, C. 1988, *Annual Report of the BIPM Time Section*, 1, D1-D22

Panfilo, G. & Arias, E. F. 2009, *Special issue of UFFC*, submitted

Petit, G. 2003, *Proc.* 35th *PTTI*, 307

Petit, G. 2008, *Proc.* 7th *Symposium on Freqency standards and metrology*, in press

Highlights of Astronomy, Volume 15
XXVIIth IAU General Assembly, August 2009
Ian F. Corbett, ed.

© International Astronomical Union 2010
doi:10.1017/S1743921310008926

VLBI Measurements for Frequency Transfer

Hiroshi Takiguchi, Yasuhiro Koyama, Ryuichi Ichikawa, Tadahiro Gotoh, Atsutoshi Ishii, Thomas Hobiger, and Mizuhiko Hosokawa

National Institute of Information and Communications Technlogy, Japan
email: htaki@nict.go.jp

Abstract. We carried out the intercomparison experiments between VLBI, GPS and DMTD to show the VLBI can measure the right time difference. We produced the artificial change using by line stretcher. At the artificial change part, VLBI and DMTD show good agreement, less than 10ps. The quantity and sense of VLBI results match well with DMTD results. Consequently, the geodetic VLBI technique can measure the right time difference.

Keywords. time, VLBI, time and frequency transfer.

As one of the new frequency transfer technique to compare the next high stable frequency standards, we proposed the geodetic VLBI technique. Previously, we evaluated the ability of VLBI frequency transfer by comparison with GPS carrier phase frequency transfer using data from IVS and IGS. The results of the VLBI showed that the stability follows a $1/\tau$ law closely and it surpassed the stability of atomic fountain at about 10^3 seconds. And that showed the stability has reached about 2×10^{-11} at 1 sec. These results showed that geodetic VLBI technique has the potential for precise frequency transfer.

Furthermore, to show the capability of VLBI, we carried out the intercomparison experiments between VLBI, GPS and DMTD at Kashima34m-Kashima11m baseline. We inserted two line stretchers at the route of the reference signal from Hydrogen maser to Kashima11m antenna to change time delay artificially. Usually, geodetic VLBI observe the multiple sources that covered uniformly the sky alternately. And we usually estimate clock, atmosphere and station coordinates at the time of analysis. However in this experiment, we observed only one source, and we estimated only clock parameters. We employed CALC/SOLVE and NRC's PPP to analyze VLBI and GPS respectively.

The result of intercomparison between GPS and DMTD shows that there are the parts of good agreement. But at the artificial change parts, GPS show the opposite sense and large difference from DMTD. The result of intercomparison between VLBI and DMTD shows that, at the artificial change parts, VLBI and DMTD show good agreement. The quantity and sense of VLBI results match well with DMTD. For example, at one of the artificial change part, the difference between GPS and DMTD is 26ps. On the other hand, the difference between VLBI and DMTD is only 7ps. Almost all the differences between VLBI and DMTD at the artificial change parts are less than about 10ps. At other parts, VLBI shows good agreement less than 50ps for short time range, but larger difference for longer time range due to the effect of atmospheric variation. The result of our experiment clearly shows that the geodetic VLBI technique can measure the right time difference.

Acknowledgements

We would like to acknowledge IVS and IGS for the products. We are grateful that GSFC and NRC provided the analysis software.

Highlights of Astronomy, Volume 15
XXVIIth IAU General Assembly, August 2009
Ian F. Corbett, ed.

© International Astronomical Union 2010
doi:10.1017/S1743921310008938

Stability of pulsar rotational and orbital periods

Sergei Kopeikin

Department of Physics and Astronomy, University of Missouri, Columbia, MO 65211, USA
email: kopeikins@missouri.edu

Abstract. Millisecond and binary pulsars are the most stable astronomical standards of frequency. They can be applied to solving a number of problems in astronomy and time-keeping metrology including the search for a stochastic gravitational wave background in the early universe, testing general relativity, and establishing a new time-scale. The full exploration of pulsar properties requires that proper unbiased estimates of spin and orbital parameters of the pulsar be obtained. These estimates depend essentially on the random noise components present in pulsar timing residuals. The instrumental white noise has predictable statistical properties and makes no harm for interpretation of timing observations, while the astrophysical/geophyeical low-frequency noise corrupts them, thus, reducing the quality of tests of general relativity and decreasing the stability of the pulsar time scale.

Keywords. standards, time, pulsars: general

Timing observations of single and, especially, binary millisecond pulsars are widely recognized as extremely important for progression in a number of branches of modern astronomy and time-keeping metrology. In particular, the implication of pulsar timing for testing general relativity in the strong-field regime [1] and setting up the upper limit on the energy density of the stochastic gravitational wave background in the early universe [2] are among the most stimulating on-going research activities. Another important aspect of pulsar timing is metrology of time on long time intervals providing a time-scale based on the high-stable rotation of millisecond pulsars around its own axis [3, 4] and/or center of mass of a binary system [5, 6].

The accuracy of pulsar timing observations is now approaching 10 nanoseconds. Such high precision requires construction of adequate data processing software taking proper account of the relevant physical effects that can contribute to the variations of timing residuals [7]. Usually, the procedure for estimating pulsar parameters is based on the premise that white noise is a dominant source of instability in times of pulse's arrival. However, long-term monitoring of pulsars certainly reveals the presence of a non-white component of the noise having the external origin [8]. This noise is called *red* as it has a spectrum that diverges at zero frequency. The lower the timing activity of the pulsar, the further toward low frequencies one must look in order to detect the *red* noise in the timing residuals. The *red* noise also makes the residuals correlated on long time intervals.

Developing a computer code accounting for effects of the *red* noise on timing residuals and estimates of parameters is worthwhile. It requires further improving the model of the *red* noise 9 and the methods of estimation of the stability of the pulsar rotational/orbital phase [10, 11]. The present state-of-the art statistical analysis of pulsar timing data in the presence of a *red* noise has not yet reached the required level of completeness and a more elaborate technique has to be invented.

Analysis of the stability of pulsar's rotational and orbital phase is more informative in time rather than in frequency domain. This is because any noise contains both stationary and non-stationary parts, while spectral analysis of noise in frequency domain is adequate only if the noise is stationary. A non-stationary part of noise affects observed values of pulsar's parameters, increases their variances, gives rise to larger correlations and makes some of the parameters biased [10].

Another problem relates to the ergodicity of the overall timing process. There is only one, available to us, realization of the pulsar timing series. It means that one can not access the statistical ensemble to process the data. Instead, we have to apply time integration to evaluate the parameters. If the ergodicity is violated the true value of the parameters will be biased, thus, affecting our ability to maintain the pulsar time scale and to test general relativity.

Third problem is related to the mathematical structure of the estimators of pulsar's parameters. The estimators are integrals in the frequency domain with a kernel being the spectrum of the noise. Spectrum of the *red* noise is divergent, and so are the estimators. We do not know what is the best mathematical technique for making the red-noise estimators unbiased and convergent. Definite progress in this area was achieved within the theory of distributions [11].

Continuous monitoring of pulsar rotational/orbital phase is not sufficient for establishing a pulsar time scale – an ensemble of pulsars, called a pulsar timing array (PTA), is necessary [12]. Currently, two PTAs are continuously monitored by Australian [13] and Russian [14] pulsar groups. A PTA can be also used as a gravitational-wave telescope that is sensitive to radiation at nanohertz frequencies, and for improving the planetary ephemeris [15].

Acknowledgements

I acknowledge the 2009 summer fellowship support of the Research Council of the University of Missouri-Columbia.

References

[1] Kramer M. & Wex N. 2009, Class. Quant. Grav., 26, 073001
[2] Hobbs G. B., Bailes M., Bhat N. D. R., Burke-Spolaor S., Champion D. J., Coles W., Hotan A., Jenet F., Kedziora-Chudczer L., Khoo J., Lee K. J., Lommen A., Manchester R. N., Reynolds J., Sarkissian J., van Straten W., To S., Verbiest J. P. W., Yardley D., & You X. P. 2009, Publ. Astron. Soc. of Australia, 26, 103
[3] Ilyasov Yu. P., Kuzmin A. D., Shabanova T. V., & Shitov Yu. P., 1989, Proc. Lebedev Phys. Inst., Moscow, 199, 149
[4] Ilyasov Y. P., Kopeikin S. M., Sazhin M. V., & Zharov V. E. 2007, Highlights of Astronomy, 14, 479
[5] Petit G. & Tavella P., 1996, Astron. Astrophys., 308, 290
[6] Ilyasov Y. P., Kopeikin S. M., & Rodin A. E. 1998, Astron. Lett., 24, 228
[7] Edwards R. T., Hobbs G. B., & Manchester R. N. 2006, Mon. Not. Roy. Astron. Soc., 372, 1549
[8] Hobbs G., Lyne A., & Kramer M. 2006, Chinese J. Astron. Astrophys. Suppl., 6, 169
[9] Kopeikin S. M. 1997, Mon. Not. Roy. Astron. Soc., 288, 129
[10] Kopeikin S. M. 1999, Mon. Not. Roy. Astron. Soc., 305, 563
[11] Kopeikin S. M. & Potapov V. A. 2004, Mon. Not. Roy. Astron. Soc., 355, 395
[12] Foster R. S. & Backer D. C. 1990, Astrophys. J., 361, 300
[13] Manchester, R. N. 2008, AIP Conference Proceedings, 983, 584
[14] Ilyasov, Y. P. 2006, Chinese J. Astron. Astrophys. Suppl., 6, 148
[15] Hobbs, G. 2009, IAU Symposium #261. Relativity in Fundamental Astronomy: Dynamics, Reference Frames, and Data Analysis 27 April - 1 May 2009 Virginia Beach, VA, USA, #11.03; BAAS, 41, 1103

Highlights of Astronomy, Volume 15
XXVIIth IAU General Assembly, August 2009   © International Astronomical Union 2010
Ian F. Corbett, ed.   doi:10.1017/S174392131000894X

Pulsar glitches detected at Urumqi

N. Wang

40-5, South Beijing Road, Urumqi, Xinjiang 830011, China
email: na.wang@uao.ac.cn

Abstract. We present an analysis of glitches detected at Urumqi from 2000 to 2008. Statistics based on all known glitches, for example post glitch behavior were investigated.

Keywords. pusar:general – time – data analysis

Pulsar glitches are probably result from angular momentum transfer within the neutron star. They result in a sudden fractional increase in the rotational frequency ν of the pulsar with a magnitude in the range $10^{-10} < \Delta\nu_g/\nu < 510^{-6}$ (Wang *et al.*, 2000; Krawczyk *et al.*, 2003; Janssen & Stappers, 2006). In total, our work revealed 29 glitches in 19 young pulsars (Yuan *et al.*, 2009), including slow glitches in three pulsars and tiny glitches in eight pulsars. Post-glitch relaxation differ greatly from pulsar to pulsar, with some showing linear increase in $\dot{\nu}$ after the glitch. Analysis of the whole sample of known glitches show that fractional glitch amplitudes are correlated with characteristic age with a peak at about 10^5 years, but there is a spread of two or three orders of magnitude at all ages. Glitch activity is positively correlated with spin-down rate, again with a wide spread of values. For individual pulsars with many glitches, the time till the next glitch is generally proportional to the fractional glitch amplitude (Fig. 1).

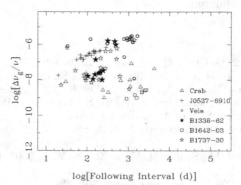

Figure 1. Fractional glitch amplitude versus length of the following interglitch interval.

Acknowledgements

This work was supported by the Knowledge Innovation Program of CAS, No. KJCX2-YW-T09, NSFC project No. 10673021, and 973 program 2009CB824800.

References

Wang, N., *et al.* 2000, *MNRAS* 317, 843
Krawczyk, A. *et al.* 2003, *MNRAS* 340,1087
Janssen, G. H. & Stappers, B. W. 2006, *AA* 457, 611
Yuan, J. P., *et al.* 2009, *MNRAS*. In preparation.

Highlights of Astronomy, Volume 15
XXVIIth IAU General Assembly, August 2009
Ian F. Corbett, ed.

© International Astronomical Union 2010
doi:10.1017/S1743921310008951

Pulsar timing array

Alexander E. Rodin

Pushchino Radio Astronomy Observatory of the Lebedev Physical Institute,
Pushchino, Moscow region, 142290, Russia
email: rodin@prao.ru

Abstract. Simultaneous timing of several pulsars distributed over the sky, so called Pulsar Timing Array (PTA), is used for a variety of metrological and astronomical applications. Three examples of PTA application are presented: link between celestial reference frames, ensemble pulsar time scale and detection of gravitational waves.

Keywords. pulsar timing, reference systems, time, gravitational waves

---

Pulsar timing coordinates, which relates with the reference frame based on the planetary ephemeris (e.g. DE200, DE405), when combining with the pulsar VLBI positions based on the coordinates of the extragalactic radio sources (ICRF) gives possibility to link the reference frames. The rotation angles between DE200 and ICRF were calculated on the basis of pulsar VLBI observations (Rodin & Sekido 2002): $A_x = -4 \pm 2$, $A_y = -13 \pm 3$, $A_z = -17 \pm 5$ mas.

The algorithm of the ensemble pulsar time PT_{ens} based on the optimal Wiener filtration method has been proposed in the paper (Rodin 2008). The method has been applied to the timing data of millisecond pulsars PSR J0613-0200, J1640+2224, J1643-1224, J1713+0747, B1855+09, B1937+21 and J2145-750. Fractional instability of TT–PT_{ens} is equal to $\sigma_z = (0.8 \pm 1.9) \cdot 10^{-15}$. A new limit of the energy density of the gravitational wave background based on σ_z (TT–PT_{ens}) was calculated to be equal to $\Omega_g h^2 \sim 3 \cdot 10^{-9}$.

Detection of gravitational waves with PTA is based on analysis of the angular correlation function (two-point correlation) of the post-fit pulsar timing residuals (Jenet *et al.* 2005). Direct use of pulsar time of arrivals (TOAs) is complicated by the fact that they contain contribution of different kinds of noise, which has different amplitudes and begins to dominate at various time intervals. Pulsar timing data were expanded into components

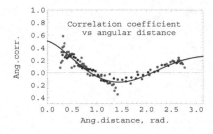

by model independent Caterpillar-SSA method (Golyandina *et al.* 2001). Two-point correlation was calculated for each component. Some components showed distinct angular correlation function (see fig.) which agrees well with the theoretically predicted one: correlation coefficient between theoretical and experimental angular correlation functions is equal to $\rho = 0.94 \pm 0.08$.

References

Golyandina, N., Nekrutkin, V., & Zhigljavsky, A., Analysis of Time Series Structure SSA and
     Related Techniques. CHAPMAN & HALL/CRC, 2001

Jenet, F., Hobbs, G., Lee, K. J., & Manchester, R. N., 2005, *ApJ*, 625L, 123

Rodin, A. E. & Sekido, M. 2002, in E. Ros, R. W. Porcas, A. P. Lobanov, & J. A. Zensus (eds.),
     Proc. 6th European VLBI Network Symposium (Bonn, Germany), p. 247

Rodin, A. E., 2008, *MNRAS*, 387, 1583–1588

Highlights of Astronomy, Volume 15
XXVIIth IAU General Assembly, August 2009
Ian F. Corbett, ed.

© International Astronomical Union 2010
doi:10.1017/S1743921310008963

Search for ultra-long gravitational waves in pulsars' rotational parameters

Maxim S. Pshirkov

Pushchino Radio Astronomy Observatory, ASC LPI, Pushchino, 142290, Russia
email: pshirkov@prao.ru

Abstract. A method is suggested to explore the gravitational wave background (GWB) in the frequency range from 10^{-12} to 10^{-8} Hz. That method is based on the precise measurements of pulsars' rotational parameters: the influence of the gravitational waves (GW) in the range will affect them and therefore some conclusions about energy density of the GWB can be made using analysis of the derivatives of pulsars' rotational frequency. The calculated values of the second derivative from a number of pulsars limit the density of GWB Ω_{gw} as follows: $\Omega_{gw} h^2 < 10^{-6}$. Also, the time series of the frequency ν of different pulsars in pulsar array can be cross-correlated pairwise in the same manner as in anomalous residuals analysis thus providing the possibility of GWB detection in ultra-low frequency range.

Keywords. gravitational waves, (stars:) pulsars: general, methods: data analysis

---

Propagation of pulsar signal in space-time perturbed by a stochastic gravitational wave field results in apparent deviations of pulsar rotational frequency; also gravitational waves affect the derivatives of pulsars rotational frequency – these values absorb the effect caused by gravitational waves with ultra long wavelengths ($\sim (cD)^{-1} < f_{gw} < T_{obs}^{-1}$).

Precise pulsar timing allows us to measure second derivative of rotational frequency for a number of pulsars. It is plausible to regard the calculated value of $\ddot{\nu}$ as caused by unknown factors that are intrinsic or extrinsic to pulsar. For detailed calculations see Pshirkov (2009). We can write down the final result:

$$\Omega_{gw} < \frac{T_{obs}^2}{2\pi^2 H_0^2} \left(\frac{\ddot{\nu}_{obs}}{\nu_0}\right)^2 \tag{0.1}$$

It is instructive to make some numerical estimates using the data for PSR B1937+21:

$$\Omega_{gw} h^2 < 10^{-6} \tag{0.2}$$

Also, the time series of the frequency ν from different pulsars in pulsar array can be cross-correlated pairwise. Comparison of angular dependence of obtained functions with known sample of GWB induced correlation similarly to usual pulsar timing (see Jenet et al. (2005)) residual analysis makes possible to detect GWB in the frequency range of our interest. That method can probe GWB down to energy density $\Omega_{gw} h^2 < 10^{-7}$.

Acknowledgements

I would like to thank the support by RFBR grants no. 09-02-00922 and 07-02-01034.

References

Jenet, F. A., Hobbs, G. B., Lee K. J., & Manchester R. N. 2005, *ApJ*, 625, L123
Pshirkov, M. S., *MNRAS*, DOI: 10.1111/j.1365-2966.2009.15221.x

Highlights of Astronomy, Volume 15
XXVIIth IAU General Assembly, August 2009
Ian F. Corbett, ed.

© International Astronomical Union 2010
doi:10.1017/S1743921310008975

Impact of Pulsar Giant Pulses on Distant Clocks Comparison

Yury P. Ilyasov,[1] Vladimir E. Zharov[2] and M. Sekido[3]

[1]Pushchino Radio Astronomical Observatory of P.N. Lebedev Physical Institute of the Russian academy of science, Pushchino, Moscow region, Russia
email: `ilyasov@prao.ru`

[2]Department of Physics, Moscow State University, Moscow, Russia
email: `zharov@sai.msu.ru`

[3] National Institute of Information and Communications Technology
Kashima, Japan
email: `sekido@nict.go.jp`

Abstract. New method of precise clocks comparison based on observation and registration of giant pulses of the millisecond pulsars is discussed. It was shown that expected accuracy of comparison is about 0.2–2 ns and depends on uncertainty of delay in the Earth ionosphere and troposphere.

Keywords. Pulsars, time transfer, giant pulses

Now it is well-known that several pulsars occasionally generate several times per hour huge superfine short pulses, which flux density is about million times more than an average level. The most famous pulsars, having such signatures, are the millisecond pulsar B0531+21, relatively young pulsar in the Crab nebulae, and the other such as J0218+4232, J1939+2134, J1959+2048. Their radio emission features are studied now in detail but real mechanism of such huge generation power is not well understood yet (Hankins & Eilek 2007, Popov *et al.* 2008). Giant pulses (GP) being as shorter as part of nanosecond and received several times per hour could be used for distant clocks precise comparison (Ilyasov *et al.* 1991, Ilyasov *et al.* 2007). Now VLBI technique is under the study to apply for time transfer within accuracy about 0.01 – 0.1 ns. But it can be done after correlation procedure of VLBI data, which takes several hours or even days. A new method of distant clock comparison could be done promptly only after communication between clock stations in the first step without any correlation procedure. When GP is about one ns duration a precision time transfer could be comparable with VLBI technique. The fast wide bandwidth facilities should be used to detect GP. In this new method the same limiting accuracy factors of VLBI should be taken into consideration. Depending on frequency of observations clock comparison can be done with accuracy of about 0.2 – 2 ns. This method is much more cheap than Two-Way Satellite Time Transfer (TWSTT) and can be used not only for ground-base clocks but for the clocks in space.

This work was supported by the RFBR grants 06-02-16816, 09-02-00922.

References

Hankins, T. H. & Eilek, J. A. 2007, *ApJ*, 670, 693
Popov, M., Soglasnov, A., *et al.* 2008, *J.Astr*, 85, 1
Ilyasov, Yu., Kuzmin, A., *et al.* 1991, Technique of clocks synchronization by pulsars signals, Patent (Soviet Union) No 1669301
Ilyasov, Yu, Oreshko, V., *et al.* 2007, Proc. of the IAA RAS (St. Petersburg), 17, 128

Highlights of Astronomy, Volume 15
XXVIIth IAU General Assembly, August 2009
Ian F. Corbett, ed.

The Parkes Pulsar Timing Array Project

R. N. Manchester

Australia Telescope National Facility, CSIRO,
PO Box 76, Epping, NSW, Australia
email: dick.manchester@csiro.au

Abstract. The Parkes Pulsar Timing Array project is timing 20 millisecond pulsars with the aims of detecting gravitational waves, establishing a time scale based on pulsar periods and improving solar-system ephemerides.

Keywords. pulsars: general — time — reference systems — gravitational waves

The Parkes Pulsar Timing Array (PPTA) project is timing 20 millisecond pulsars using the Parkes 64-m radio telescope with observations at three frequencies, 685 MHz, 1.4 GHz and 3 GHz, at 2 – 3 weekly intervals. Principal collaborators in the project are based at Swinburne University of Technology, the University of Texas at Brownsville, the University of California, San Diego, and at the CSIRO Australia Telescope National Facility. The main goals of the project are to detect gravitational waves, to establish "pulsar time" as a long-term standard of time and to improve the currently used solar-system ephemerides. Data acquisition commenced in mid-2004. Since then, data recording and analysis systems have been improved so that we are now reaching our goal of 100 ns rms timing residuals on several pulsars. Currently, nearly half of the sample has rms residuals of less than 500 ns and almost all are less than 2 μs. More complete descriptions of the project are given by Manchester (2008) and Hobbs et al. (2009).

Early results from the project (Jenet et al. 2006) put a limit of $\sim 2 \times 10^{-8}$ on the energy density of a graviational-wave background in the Galaxy relative to the closure density of the Universe. They also limited the tension of cosmic strings in the early Universe and the equation of state in the inflation era. Saito & Yokoyama (2009) also show that PTA observations put strong limits on the formation of intermediate-mass black holes at the end of the epoch of inflation.

Pulsars have a frequency stability comparable to the best atomic clocks over intervals a year or more. By combining results from the many pulsars observed in the PPTA and other PTA projects, it should be possible to establish a pulsar timescale which has better stability than the current international timescales. With continued observations it will also be possible to improve on the best current measurements of planetary masses and possibly to detect previously unknown trans-Neptunian objects.

Acknowledgement

The Parkes telescope is part of the Australia Telescope which is funded by the Commonwealth Government for operation as a National Facility managed by CSIRO.

References

Hobbs, G. B., et al. 2009, PASP 26, 103
Jenet, F. A., et al. 2006, ApJ 653, 1571
Manchester, R. N. 2008, AIP Conf. Ser. 983, 584
Saito, R. & Yokoyama, J. 2009, Phys. Rev. Letters 983, 584

Highlights of Astronomy, Volume 15
XXVIIth IAU General Assembly, August 2009
Ian F. Corbett, ed.

© International Astronomical Union 2010
doi:10.1017/S1743921310008999

Timing of binary pulsars and the search for the low-frequency gravitational waves

Vladimir A. Potapov[1] and Sergei M. Kopeikin[2]

[1] Pushchino Radio Astronomy Observatory of P.N.Lebedev Physical Inst.,
142290, Pushchino, Russia
email: potap@prao.ru

[2] Department of Physics & Astronomy University of Missouri-Columbia
223 Physics Bldg. Columbia, Missouri 65211 USA
email: kopeikins@missouri.edu

Abstract. Millisecond and binary pulsars are the most stable natural standards of astronomical time giving us a unique opportunity to search for gravitational waves (GW) and to test General Relativity. GWs from violent events in early Universe and from the ensemble of galactic and extragalactic objects perturb propagation of radio pulses from a pulsar to observer bringing about stochastic fluctuations in the times of arrival of the pulses (TOA). If one observes the pulsar over a sufficiently long time span, the fluctuations will be registered as a low-frequency, correlated noise affecting the timing residuals in the frequency range $10^{-12} \div 10^{-7}$ Hz. This work demonstrates how the standard procedure of processing of the pulsar timing data can bias the estimate of the upper limit on the density of the GW background (GWB).

Keywords. gravitational waves, binary pulsars, data analysis

We have analyzed the statistical method proposed in Jennet *et al.* (2005) (hereinafter referred to as D05) for possible detecting of GWB with Pulsar Timing Array (PTA) consisting of binary pulsars. To this end we have used the analytic formalism of our joint paper (Kopeikin and Potapov, 2004) that determines dependence of TOA residuals and pulsar's parameters on characteristics of timing noise induced by the GWB.

We have found that the procedure of fitting of the binary pulsar parameters increases the significance of the GWB detection as compared with that given in D05, in proportion to the number of orbital revolutions, $N_{orb} = T/P_{orb}$, where T is the span of observation and P_{orb} is the orbital period of the binary pulsar. The spectral sensitivity of PTA at the frequencies close to the orbital frequencies of binary pulsars is inversely proportional to the number of PTA pulsars.

We calculated the GWB detection significance for PTA with the "target parameters" of Parkes PTA (250 TOAs, 20 pulsars, TOA residuals standard deviation 100 ns, $T = 5$ yrs.). It was shown that for the long period binary pulsars ($P_{orb} \approx 1/2$ yr.) the estimated upper limit for the energy density of GWB is about two times larger as compared with the estimate obtained by the method from D05. These two estimates asymptotically converge as the number N_{orb} increases, and became practically equal after $N_{orb} > 30$. We conclude that the method of D05 can be used without restrictions for any PTA that consists of binary pulsars with rather short orbital periods ($\simeq$ a few days).

(This work is supported by RFBR under grant No 09-02-00922).

References

F. A. Jennet, G. B. Hobbs, K. J. Lee, & R. N. Manchester 1985, *Ap.J.*, **625**, L123–L126
S. M. Kopeikin & V. A. potapov 2004, *Mon.Not.R.Astron.Soc.* **355**, 395–412

Highlights of Astronomy, Volume 15
XXVIIth IAU General Assembly, August 2009   © International Astronomical Union 2010
Ian F. Corbett, ed.   doi:10.1017/S1743921310009002

JD7 -Astrophysical Outflows and Associated Accretion Phenomena

Elisabete M. de Gouveia Dal Pino[1] and Alex C. Raga[2]

[1] Universidade de São Paulo,Cidade Universitria, BR São Paulo, SP 05508-900, Brazil
[2] Instituto de Ciencias Nucleares, Universidad Nacional Autónoma de México, México

Preface

Highly collimated supersonic jets and outflows are very frequent in several astrophysical environments. They are seen in young stellar objects (YSOs), proto-planetary nebulae, compact objects (like galactic black holes or microquasars, and X-ray binary stars), active galactic nuclei, and are also possibly associated to gamma-ray bursts (GRBs) and to ultra-high energy cosmic rays sources (UHECRs). Despite their different physical scales, all these outflow classes have strong morphological similarities, but questions such as - what physics do they share? - or - can we find a universal mechanism of acceleration and collimation that operates in all classes? - remain matters of debate. The most accepted mechanism for their origin relies on a rotating accretion disk threaded by perpendicular large-scale magnetic fields and, though most of the systems producing jets contain an accretion disk around the central source, the real role that rotation and magnetic fields play in these processes is still not fully understood, nor are the highly non-linear physical processes connected to these jet-disk systems in the large parameter space involved.

Recent years have been marked by reviving worldwide interest in the studies of magnetic disk accretion and outflows both because of their potential relation to several astrophysical phenomena that are not well understood yet like GRBs and UHECRs, and also because of the fast advance in multi-dimensional computational modelling of these systems. Also, the large improvement in the observations with the building of new instrumentation and techniques now allow to resolve the base of some outflows very near the sources. The combination of multi-wavelength data has also given new insights into outflow phenomena. We thus felt that the time was ripe to revive the discussion of the origin of the astrophysical jets and their effects on the astrophysical environments. Particularly, holding it during the IAU General Assembly in Rio de Janeiro has attracted worldwide experts from a broad range of research fields.

The main scientific goal of this one-and-a-half-day Joint Discussion on Astrophysical Outflows and Associated Accretion Phenomena was to focus on understanding the driving mechanisms of jets from proto-stars (including their possible crucial link with star formation) to microquasars and AGNs. We have tried to emphasize their differences/similarities; the basic physics of the accretion-jet process in magnetized disks, including the transport of angular momentum and the development of reconnection and turbulent dynamo; cooling/heating processes, instabilities, shock structures and particle acceleration mechanisms in the jets; the impact of the jets on energy balance and turbulence feeding in the astrophysical environments; and the potential association of jet-accretion phenomena with GRBs and UHECRs.

The strategy was to try to create a powerful interdisciplinary synergy between the observational, theoretical and computational modelling domains of expertise. We had more than 110 participants from 28 countries, out of which 17 gave invited talks, 4 oral contributions and 88 poster presentations. All the sessions ended with discussion panels of the related topics and poster contributions. The meeting was concluded with a Summary Panel that highlighted what we considered the most relevant questions for continuing study. These open questions can be summarized as follows:

• To what extent are the magnetic fields advected from the environment and/or generated continuously inside the accretion disk through a dynamo process?

• What is the role of turbulence on angular momentum transport, dynamo action and large scale magnetic field generation within accretion disks?

• What is the Prandtl number regime in accretion disks over different scales (from YSO to relativistic sources)?

• What is the role of magnetic reconnection and coronal magnetic activity in the disk/coronal/jet launching process?

• Is there a connection between QPOs, general jet variability, and ejection phenomena with magnetic reconnection in the inner disk/corona boundary?

• What is the nature of the coupling between source magnetosphere and disk/corona magnetic field lines?

• Is jet collimation/recollimation of magnetic nature only?

• How relevant is steady jet production modelling?

• What is the Jet/accretion disk and star formation connection from low to high mass stars?

• What is the origin of molecular jets (prompt entrainment or intrinsic)?

• What will the new generation instruments (SKA, ALMA, etc.) be able to elucidate about the jet launching mechanism, rotation, etc?

We hope that we might be able to answer many of these questions in the near future.

Elisabete M. de Gouveia Dal Pino and Alex C. Raga, co-chairs of the SOC of the Joint Discussion on Astrophysical Outflows and Associated Accretion Phenomena, Rio de Janeiro, Brazil, August 2009

Highlights of Astronomy, Volume 15
XXVIIth IAU General Assembly, August 2009
Ian F. Corbett, ed.

Numerical Simulations of MHD Accretion Disks

J. F. Hawley

Department of Astronomy, University of Virginia, Charlottesville, VA 22904, USA
jh8h@virginia.edu

Abstract. Numerical simulations play an increasingly important role in investigating accretion disks and associated phenomena such as jets. This paper provides a few examples of recent results that have been obtained with simulations, both local or global.

Keywords. accretion, accretion disks, magnetohydrodynamics: MHD

The most energetic phenomena in the universe are systems powered by gravity through accretion, specifically accretion disks surrounding compact stars and black holes. Accretion theory has largely been based primarily on a one-dimensional time-steady model consisting an optically thick, vertically-thin, Keplerian disk with an unknown, parameterized internal stress. While analytic models of this type provide considerable insight, their limitations are now well-known, and the observational data demand moving beyond this standard. Numerical simulations provide another means for investigating accretion disks.

The governing equations of accretion are those of compressible magnetohydrodynamics (MHD). Magnetic fields render a differentially rotating fluid unstable to the magnetorotational instability - MRI; Balbus & Hawley (1991) - and the resulting turbulence accounts for the internal stress that drives accretion. These processes have been studied using local "shearing box" simulations - Hawley, Gammie & Balbus (1995). The shearing box is a Cartesian domain representing a small piece of the disk, assumed to be rotating with the local disk angular velocity, Ω. The equations include differential rotation, Coriolis force and the tidal potential. Local simulations have shown how the MRI produces angular momentum transfer and the degree to which the MRI stress acts like a Shakura-Sunyaev "α viscosity," where $\tau_{r\phi} = \alpha P$ (Balbus & Papaloizou (1999)). MRI-driven turbulence generates substantial local stress, typically with $\alpha \sim 0.01$. The orbital energy released by that stress is locally dissipated by the turbulence into heat in an eddy turnover time (Simon, Hawley, & Beckwith (2009)). Shearing box simulations that combine MHD with flux limited diffusion (Hirose, Krolik, & Blaes (2009)) find that while the stress is roughly proportional to the total pressure (including both gas and radiation), the stress *determines* the pressure, not other way around. Increases in the stress lead to stronger turbulence, more turbulent dissipation, and more heat. A consequence of this is that radiation pressure-supported accretion disks are thermally stable, in contrast to expectations based on assuming a strict α viscosity. Because of its fundamentally magnetic nature, the stress is directly proportional to the magnetic energy which is itself determined by the balance between MRI driving (at a range of scales) and turbulent dissipation. At the present time, however, we don't have a way to predict the magnetic field energy as a local function of the state of the disk.

The recognition that MHD turbulence provides the stress in disks leads to new questions. For example, because magnetic stresses don't necessarily diminish in the plunging region inside the inner most stable circular orbit (ISCO) around a black hole, the overall efficiency and luminosity of disks can be greater than predicted by models that assume "zero torque" at the ISCO. Also, the magnetic field embedded in the accretion flow can be brought into a rotating black hole's ergosphere, possibly to form a relativistic jet. Issues such as these must be investigated through global simulations of disks around rotating black holes.

De Villiers, Hawley & Krolik (2003) carried out some of the first long-term, three-dimensional MHD global disk simulations for a variety of black hole spins. These simulations, and others that followed, have shown that significant stress can continue at and inside the ISCO and all the way to the horizon for rotating black holes. The most significant unanswered question is what factors (e.g., disk thickness, thermal pressure, magnetic topology) determine the level of the stress at the ISCO. Shafee *et al.* (2008) recently simulated a thin disk around a Schwarzschild hole and suggested that the ISCO stress cuts off sharply when the disk is thin. On the other hand, Noble, Krolik, & Hawley (2009) find additional ISCO stress even for a relatively thin disk. They further considered the observational implications of the ISCO stress using simple emission and absorption models coupled with relativistic ray tracing. Work continues on this important issue.

Simulations have produced new insights into jet production. Under the right circumstances, reasonably long-lived large-scale field can be brought down and attached to the black hole. Field loops originally contained entirely within the accretion flow can expand dramatically within a cone around the rotation axis, successfully supporting jets carrying substantial Poynting flux to large distance (Hawley & Krolik (2006)) if the black hole is rotating. However, the longevity and strength of these large-scale fields—and their associated jets—depend critically on the topology of those loops (Beckwith, Hawley & Krolik (2008)). A strong, long-lived jet seems to require a net vertical field in the disk midplane whose sign remains consistent for at least an inner-disk inflow time.

In conclusion, although the time- and length-scales involved make accretion disk simulations challenging, they have revealed new details about time-dependent properties of disks, magnetic disk dynamos, jet launching mechanisms, and the dynamical properties of systems other than the standard thin disk. As the capabilities of computational hardware increase, and the development of advanced numerical codes continues, our theoretical understanding of accretion physics should substantially improve.

Acknowledgements

This work was supported by NASA grant NNX09AD14G. I thank collaborators Kris Beckwith, Jean-Pierre De Villiers, Andrew Hamilton, Julian Krolik, Scott Noble, Jeremy Schnittman, and Jacob Simon. Computational resources were supplied by the TeraGrid, supported by the National Science Foundation.

References

Balbus, S. A. & Hawley, J. F. 1991, *ApJ*, 376, 214
Balbus, S. A. & Papaloizou, J. C. B. 1999, *ApJ*, 521, 650
Beckwith, K., Hawley, J. F., & Krolik, J. H. 2008, *ApJ*, 678, 1180
De Villiers, J.-P., Hawley, J. F., & Krolik, J. H. 2003, *ApJ*, 599, 1238
Hawley, J. F, Gammie, C. F., & Balbus S. A. 1995, *ApJ*, 440, 742
Hawley, J. F. & Krolik, J. H. 2006, *ApJ*, 641, 103
Hirose, S., Krolik, J. H., & Blaes, O. 2009, *ApJ*, 691, 16
Noble, S. C., Krolik, J. H., & Hawley, J. F. 2009, *ApJ*, 692, 411
Shafee, R. *et al.* 2008, *ApJ*, 687, L25
Simon, J. B., Hawley, J. F., & Beckwith, K. 2009, *ApJ*, 690, 974

Highlights of Astronomy, Volume 15
XXVIIth IAU General Assembly, August 2009
Ian F. Corbett, ed.

© International Astronomical Union 2010
doi:10.1017/S1743921310009026

Accretion Disks around low-mass Stars unveiled by the new Generation of cm to submm Arrays

Anne Dutrey

Laboratoire d'Astrophysique de Bordeaux, Observatoire Aquitain des Science de l'Univers,
Université de Bordeaux 1, 2 rue de l'observatoire, BP 89, F-33271 Floirac Cedex, France
email: Dutrey@obs.u-bordeaux1.fr

Abstract. In the context of accretion disks, I briefly discuss the impact of three major forthcoming radio facilities: e-VLA, ALMA and SKA. These arrays are complementary by their frequency range and angular resolution. Around nearby low-mass stars, they will likely provide the first insights in the inner gas and dust disks (radius < 10-30 AU) in the area where planet formation should occur but would also allow the first investigations of the star, jet and disk connections.

Keywords. accretion disks - stars: formation - planetary systems: protoplanetary disks

---

Investigating the physics and chemistry of the dust and gas orbiting around low-mass stars is a key issue in order to understand how young planetary systems form and evolve from the residual of the molecular dense core which has formed the star: the so-called proto-planetary disk. Young disks are mostly composed of H_2 which is difficult to detect and most observations are based on other tracers.

The cm to submm wavelength domain is well suited for this purpose because 1) the dust has a moderate opacity, allowing to trace the disk mass content and 2) there are many rotational lines of simple molecules such as CO, CN, HCN, H_2CO, etc... which can probe the gas content. Other major issues, linked to the processes leading to star and planet formation are the understanding of the jet physics and its connections to both the central star and the disk. In this field, the cm wavelengths are well suited to trace free-free emission from jets or the non-thermal processes emanating from young magnetically active stars and interacting with the inner disk. Finally, due to the small angular size of disks, only large interferometers have the angular resolution needed to resolve out the inner part of disk where planets should form (10 AU $\simeq$ 0.07" at 150 pc). In the radio domain, thanks to their sensitivity and imaging capabilities, three arrays will have a major impact in this astrophysical field: e-VLA, ALMA and SKA.

e-VLA is the "Expanded Very Large Array" located in Socorro, New Mexico, USA. The VLA has been gradually improved by adding new receivers and a new correlator. In particular, the array will be operated from 1 to 50 GHz. The full upgrade will be completed in 2012. More information can be found on the NRAO site at: http://www.aoc.nrao.edu/evla/.

ALMA, the "Atacama Large Millimeter Array", is currently in construction in the Atacama desert in Chile. This will be an array of 66 antennas (54 12-m dishes for the main array and 12 7-m dishes for the compact array ACA) working between 30 GHz (this bandwidth is not yet decided) and 900 GHz on baselines as long as 15 km. The array will gradually enter in regular operation starting in 2011. More information can be found at http://almaobservatory.org.

SKA is the "Square Kilometer Array", a gigantic network of radio antennas working from 70 MHz up to 30 GHz (but the high frequency bandwidth, above 10 GHz, is still an option). Several sites are under investigation. The array is expected to be in operation around 2020. More information can be found at http://www.skatelescope.org/

In the course of understanding how stars and planets form, there are several domains which will benefit from the high sensitivity, high angular resolution images provided by these new interferometers. In this context, here is a non exhaustive list where these new instruments would provide invaluable information:

(a) Nowadays, star-disk interactions linked to stellar magnetic loops (Massi *et al.* (2006)) are mostly observed in close (spectroscopic) binary systems which present variable emissions from a few cm up to the mm range Salter *et al.* (2008). Thanks to their sensitivity and angular resolution, e-VLA and SKA would likely change our views in this domain by better constraining the magnetic activity of PMS TTauri stars.

(b) Ionized jets will be soon mapped by the e-VLA and more accurately by SKA ten year later (Rodriguez *et al.* (1994)). These high quality images would allow to study not only the jet in itself but would partially unveil its connections with the disk and the star.

(c) Dust disks, including planet forming regions (radius < 50 AU), will be imaged with ALMA, e-VLA and SKA (if the band at 20-30 GHz is implemented). This will permit to trace gaps potentially due to planet-disk interactions (Wolf and D'Angelo (2005)) but also provide better constrains on grain size distribution, dust mass estimate and its evolution with disk age. Observations around 7mm are of prime importance not only to characterize grain growth but also to disentangle with a partial contamination of the spectral energy distribution (SED) by an ionized jet (Rodman *et al.* (2006)).

(d) Mapping of gas disks will mostly remain the domain of ALMA (Dutrey *et al.* (2007)) through the observations of many molecular lines from 3 up to 0.3mm. However, only e-VLA and SKA would detect NH_3 around 25 GHz, a key molecule to understand the nitrogen chemistry and which is not yet observed in gas disks. Moreover, SKA and maybe e-VLA should be able to detect the HI line at 21 cm, tracing the neutral gas at the disk surface (Kamp *et al.* (2007)), a particularly interesting information for the disk dissipation.

(e) Last but not least, polarimetric measurements provided from both dust (Girart *et al.* (2008)) and line (through CN or C_2H Zeeman splitting or CO with the Goldreich-Kylafis effect, for example) emissions should provide estimates of the magnetic field in the inner disk at scale around 10 AU. For this purpose, sensitivity is an issue even for ALMA, particularly for line observations.

References

Dutrey, A., Guilloteau, S., & Ho, P. T. P. 2007, Proc. Proto-stars and Planets V, (Univ. of Arizona Press, Tucson), 951, p.495-506

Girart, J. M., Rao, R., & Marrone, D. P. 2008, *Ap&SS* 313, 87-90

Kamp, I., Freundling, W., & Jayaram, N. C. 2007, *ApJ* 660, 469-478

Massi, M., Forbrich, J., Menten, K. M., Torricelli-Ciamponi, G., Neidhfer, J., Leurini, S., & Bertoldi, F. 2006, *A&A* 453, 959-964

Rodriguez, L. F., Canto, J., Torrelles, J. M., Gomez, J. F., Anglada, G., & Ho, P. T. P. 1994, *ApJ* (Letters) 427, 103-106

Rodmann, J., Henning, Th., Chandler, C. J., Mundy, L. G., & Wilner, D. J. 2006, *A&A* 446, 211-221

Salter, D. M., Hogerheijde, M. R., & Blake, G. A. 2008, *A&A* (Letters) 492, 21-24

Wolf, S. & D'Angelo, G. 2005 *ApJ* 619, 1114-1122

Highlights of Astronomy, Volume 15
XXVIIth IAU General Assembly, August 2009
Ian F. Corbett, ed.

Primeval Jets from Young Stars

T. P. Ray

Dublin Institute for Advanced Studies, 31 Fitzwilliam Place, Dublin 2, Ireland
email: tr@cp.dias.ie

Abstract. The jet phenomenon lasts at least a million years for young, solar-like, stars and it occurs during a wide variety of young stellar object (YSO) phases. This includes the period when the source is highly embedded (Class 0) to when it becomes optically visible for the first time as a classical T Tauri star (Class II). Here I briefly discuss some of the properties of jets from the youngest objects.

Keywords. Protostar, YSO, jet, accretion, outflow, Herbig-Haro object, interstellar medium

1. Introduction

Low mass stars like the Sun are thought to form from so-called pre-stellar cores, cold ($\approx$10K) substructures within molecular clouds that have no internal heating source but are instead warmed by the interstellar radiation field (Kirk et al. 2007). Most of the mass of the young star is then built up during the highly embedded Class 0 phase when powerful outflows accompany high accretion rates. This is illustrated, for example, by the mm source at the centre of the dramatic Herbig-Haro (HH) 211 outflow (Lee et al. 2009). Despite the extent and power of this outflow, the core mass of the YSO is less than that of a brown dwarf. As accretion declines, the protostar enters the Class I phase characterised by rising flux at infrared wavelengths while the source remains optically invisible. Such obscuration is caused by the presence of a circumstellar disk in combination with a dusty envelope. In the Class II phase the photosphere of the newborn star is directly visible for the first time: it is now a classical T Tauri star or a Herbig Ae/Be star if it has a higher mass. The Class 0, I and II phases are thought to last about 10^5, 5.10^5 and 2.10^6 years respectively (Evans et al. 2009).

Distinguishing between these various evolutionary phases can however be difficult. Modelling the observed spectral energy distribution (SED) relies on a combination of photospheric (stellar), disk and envelope emission in which multiple scattering and re-processing of photons is important. Moreover the envelope may be partially cleared out, i.e. contain a cavity as a result of the outflow, and the disk may be flared (Robitaille et al. 2006). Add to this a random inclination angle of the disk/outflow axis to our line of sight and immediately it is seen that what might be regarded as a Class I source if viewed from a particular angle may look like a Class II source when seen from another. An added complication in constructing an SED is non-steady accretion and thus variable luminosity (Evans et al. 2009).

Some of these ambiguities may be removed by using the properties of associated outflows. For example, through a combination of proper motion studies and spectroscopy, the actual 3-D direction of a jet may be established (e.g. Caratti o Garatti et al. 2009). Assuming this is along the disk axis, the inclination angle of the disk to the line of sight may be found. Jets are also fossil records of how active their source has been in the past: with proper modelling, they can at least give us clues on the variation in activity level over thousands of years (i.e. over dynamical timescales). In particular, if outflow rates (as measured through the jet) are a proxy for accretion and accretion rates decline with time, then jets may help us distinguish between the various evolutionary phases.

In what way do the youngest (primeval) jets from Class 0 sources differ from those of more evolved protostars? In the past observations of such outflows were largely confined to studying low energy molecular transitions, e.g., the CO J=1–0 line. Such transitions trace slowly moving ambient gas, i.e. gas propelled forward by the much more highly collimated underlying atomic jet. Instead if we wish to observe the latter, at velocities close to the gravitational escape velocity from the star ($\approx 100 \mathrm{kms}^{-1}$), we have to study the outflow in the mid-infrared using for example the Infrared Spectrograph (IRS) on Spitzer. The reason for this is that the standard permitted and forbidden atomic lines in the optical and near-infrared (e.g., [SII]$\lambda\lambda$6716,6731 and [FeII]1.64μm) are too heavily obscured by ambient dust. Spitzer studies of outflows from Class 0 sources not only show that an underlying atomic component is present but that the mass loss rate in such jets is much higher than those from less embedded sources of equivalent mass (e.g., Dionatos *et al.* 2009).

Recently it has been found that many jets from young stars contain significant amounts of dust (Podio *et al.* 2009)). The presence of dust is not measured directly but is instead inferred from line ratios of refractory to non-refractory species. While depletion values are not as high as in the ISM, the gas in a jet has not attained solar abundances, at least close to the source. The dust itself appears to be intrinsic rather than entrained from the jet's surroundings, as there is very little dynamical evidence for entrainment. Moreover the dust seems to be only gradually destroyed via the mild internal jet shocks, as expected from grain models (Jones *et al.* 1994). We have only started to observe abundance ratios in jets from Class 0 sources. Again the problem is that the jet itself is relatively embedded and so the normal optical or even near-infrared lines cannot be used. We thus have to revert to mid-infrared wavelengths and initial observations also suggest large amounts of dust in such flows (Dionatos *et al.* 2009).

In the future we can look forward to studying the youngest and most embedded jets using telescopes like ALMA and the Mid-Infrared Instrument (MIRI) on the James Webb Space Telescope (JWST). In particular the MIRI IFU, given its good spatial and spectral resolution, will be particularly useful for studying the dynamics, chemical evolution, and basic properties of atomic jets from Class 0 sources.

Acknowledgements

TPR would like to thank the organisers for an excellent Joint Discussion and to acknowledge support from Science Foundation Ireland (Grant 07/RFP/PHYF790).

References

Caratti o Garatti, A., Eislöffel, J., Froebrich, D., Nisini, B., Giannini, T., & Calzoletti, L. 2009, *A&A*, 502, 579

Dionatos, O., Nisini, B., Garcia Lopez, R., Giannini, T., Davis, C. J., Smith, M. D., Ray, T. P., & DeLuca, M. 2009, *ApJ*, 692, 1

Evans, N. J., *et al.* 2009, *ApJS*, 181, 321

Jones, A. P., Tielens, A. G. G. M., Hollenbach, D. J., & McKee, C. F. 1994, *ApJ*, 433, 797

Kirk, J. M., Ward-Thompson, D., & André, P. 2007, *MNRAS*, 375, 843

Lee, C.-F., Hirano, N., Palau, A., Ho, P. T. P., Bourke, T. L., Zhang, Q., & Shang, H. 2009, *ApJ*, 699, 1584

Podio, L., Medves, S., Bacciotti, F., Eislöffel, J., & Ray, T. P. 2009, arXiv:0907.3842

Robitaille, T. P., Whitney, B. A., Indebetouw, R., Wood, K., & Denzmore, P. 2006, *ApJS*, 167, 256

Highlights of Astronomy, Volume 15
XXVIIth IAU General Assembly, August 2009
Ian F. Corbett, ed.

QPO-jet relation in X-ray binaries

Tomaso M. Belloni

INAF–Osservatorio Astronomico di Brera, Via E. Bianchi 46, I-23807 Merate, Italy
email: tomaso.belloni@brera.inaf.it

Abstract. In the past years, a clear picture of the evolution of outbursts of black-hole X-ray binaries has emerged. While the X-ray properties can be classified into our distinct states, based on spectral and timing properties, the observations in the radio band have shown strong links between accretion and ejection properties. Here I briefly outline the association between X-ray timing and jet properties.

Keywords. accretion, accretion disks, X-rays: binaries, stars: outflows

---

1. Fast time variability

The fast time variability observed in the X-ray emission from Black-Hole Binaries (BHB) can be extremely strong and complex. It is clearly connected to the spectral evolution throughout their outbursts, which can be described through the use of Hardness-Intensity Diagrams (HID; see Belloni 2009 and references therein). A total fractional ms variability of ∼40% is a major "disturbance" of the accretion flow that can hardly be ignored when trying to understand its properties. Concentrating on the most basic properties, we can identify two categories: *loud* states (LHS and HIMS in Belloni (2009)), characterized by strong flat-top noise components in the power spectra, with total fractional variability 10-40%, and *quiet* states (SIMS and HSS), with less variability in the form of a power law component. Quasi-Periodic Oscillations (QPO) are observed in all states, with a complex phenomenology. However, the HIMS-SIMS transition is very abrupt and involves the interplay between two very different "flavors" of QPO. This transition can be marked in a HID with a it QPO line (see Fig. 1). At the same time, the high-energy part of the X-ray spectrum undergoes abrupt changes through the transition (see Motta *et al.* 2009).

2. Jet ejection

The radio properties of BHB display an evident connection with the X-ray states and transitions (see e.g. Fender 2006). A relation with the states evolution was presented by Fender, Belloni & Gallo (2004) on the basis of four well-studied systems. At its basis, the unified picture of disk-jet coupling presented there identifies two regions of the HID: the hard region where a steady, compact and mildly relativistic jet is observed, and the soft region where there is no evidence of nuclear emission from the binary (see Fig. 1). The transition between these two regions marks the ejection of a fast relativistic jet, observed as a bright radio flare or, when imaged, as a superluminal jet. The position of this transition was dubbed "jet line".

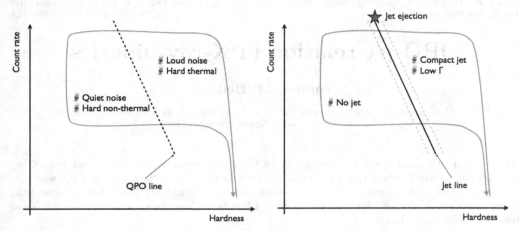

Figure 1. Left: schematic HID with the two regions identified through time variability and the 'QPO line' between them. Right: same HID, with the radio regions and the jet line'. The two fundamental lines do not coincide.

3. How do they connect?

Fender, Belloni & Gallo (2004) identified the jet line with the QPO line. This identification would lead to the attractive conclusion that the plasma responsible for the noise would be the one ejected inform of a jet. However, Fender, Homan & Belloni (2009) recently reported, on the basis of a larger sample of sources, that the two lines do not always coincide, but are close. There seems not to be a direct causal connection, as sometimes one precedes the other and vice versa. However, the close association suggests that both ejection and change in timing properties are the outcome of a complex physical transition that takes place on a longer time scale.

4. Discussion

The scheme outlined above, based on the HID and fast timing properties can be extended to neutron-star systems and even to white-dwarf binaries (see Belloni 2009; Körding *et al.* 2008; Fender 2009; Tudose *et al.* 2009). It is clear that its properties are intimately connected to the spectral state and to the characteristics of the jet. An important key is the study of the correlated variability at optical wavelengths (see e.g. Kanbach *et al.* (2001), Ganhdi *et al.* (2001)), which can shed light on this connection.

References

Belloni, T. M. 2009, in: T. M. Belloni (ed.), *The Jet Paradigm: from Microquasars to Quasars*, Lecture Notes in Physics (Heidelberg: Springer), in press (arXiv:0909.2474)
Fender, R. 2006, in: Lewin, W. H. G., & van der Klis, M. (eds.), *Compact stellar X-ray sources*, Cambridge Astrophysics Series, No. 39, Cambridge University Press, p. 381
Fender, R. P. 2009, in: T. M. Belloni (ed.), *The Jet Paradigm: from Microquasars to Quasars*, Lecture Notes in Physics (Heidelberg: Springer), in press (arXiv:0909.2572)
Fender, R. P., Belloni, T., & Gallo, E. 2004, *MNRAS*, 355, 1105
Kanbach, G., *et al.*, 2004, *Nature*, 324, 23
Gandhi, P., *et al.*, 2008, *MNRAS*, 390, L29
Körding, E., *et al.*, 2004, *Science*, 320, 1318
Motta, S., Belloni, T., & Homan, J. 2009, *MNRAS* in press (arXiv:0908.2451)
Tudose, V., *et al.*, 2009, *MNRAS* in press (arXiv:0909.3604)

Highlights of Astronomy, Volume 15
XXVIIth IAU General Assembly, August 2009
Ian F. Corbett, ed.

© International Astronomical Union 2010
doi:10.1017/S1743921310009051

Outflows in nearby AGNs

Thaisa Storchi-Bergmann

Instituto de Física - UFRGS, Campus do Vale, CP 15051, Porto Alegre, RS, Brasil
email: thaisa@ufrgs.br

Abstract. I report results of kinematic studies of the Narrow-Line Region (NLR) of nearby Active Galactic Nuclei (AGN) from integral field spectroscopy (IFS) obtained with the Gemini Telescopes, including mass outflow rates and corresponding kinetic power. The IFS has allowed the construction of velocity channel maps which provide a better coverage of the gas kinematics and do not support the presence of acceleration up to hundred parsec scales in the NLR as found in previous studies based solely on centroid velocity maps.

Keywords. galaxies: active, galaxies: jets, galaxies: nuclei, galaxies: kinematics and dynamics, galaxies: ISM, infrared: galaxies

---

Outflows are ubiquitous among AGN (Veilleux *et al.* 2005) and provide the feedback which regulates the M-σ relation (Di Matteo *et al.* 2005). The quantification of the mass and energy deposited by these outflows is thus a fundamental ingredient for the understanding of galaxy evolution. With the goal of mapping and quantifying the feedback from nuclear outflows we have been observing the nuclear region of nearby AGN using Integral Field Units (IFU) at the Gemini telescopes. In the optical, we have used the Gemini Multi-Object Spectrograph (GMOS) IFU, while in the near-infrared (near-IR), we have used the Near-Infrared Integral Field Spectrograph (NIFS) with the adaptive optics module ALTAIR. In the optical, the angular resolution is limited by the atmospheric seeing (typically 0.6 arcsec), while in the near-IR, the adaptive optics provides a much better angular resolution (≈ 0.1 arcsec).

The main results of our studies are:

• Ionized gas emission is not restricted to the outflowing region, indicating that the obscuring dusty structure Riffel *et al.* (2009) – the torus of the Unified Model (Antonucci & Miller 1985) – is clumpy (Elitzur 2006), allowing the escape of radiation along its equatorial plane. This radiation then ionizes gas from the galaxy plane.

• While the stellar kinematics is dominated by rotation in the galaxy plane (Barbosa *et al.* 2006), the gas kinematics shows both rotation and outflow. Rotation is observed mainly in the ionized gas from the galaxy plane, while the outflow is observed in an elongated or biconical structure usually following the radio structure.

• Different phases of the gas show different kinematics: while in the ionized gas the outflowing component is usually dominating, in the molecular gas (H_2) the dominant component is rotation, revealing that the H_2 gas is more settled in the galaxy plane, being also observed in inflow (Riffel *et al.* 2008).

• In many cases there is a close association between the radio structure and the gas kinematics, such as flux enhancements at the location of radio knots and increased velocity dispersion surrounding these knots (Barbosa *et al.* 2009; Riffel *et al.* 2006).

• Centroid velocity maps usually show highest velocities (typically ~ 200–300 kms) away from the nucleus, suggesting acceleration along the NLR up to ~ 100 pc from the nucleus. Our data suggest that this result may be due to projection effects which combine emission from the outflow with that from the galaxy plane. As a result, the centroid velocities show a shift of the brightest component from that in the galactic disk at zero

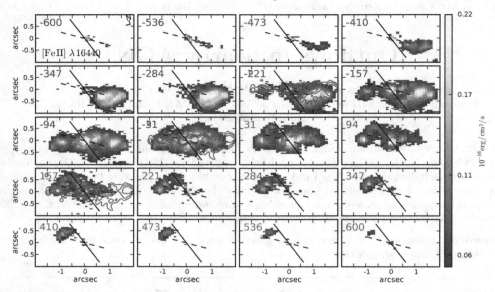

Figure 1. Channel maps obtained by integrating the flux within velocity bins of 63 kms^{-1} along the [Fe II]$\lambda\,1.64\mu$m emission-line profile from the NLR of NGC 4151 (Storchi-Bergmann *et al.* 2010). The numbers in the upper left corner of each panel are the central velocity of the bin, in kms^{-1} relative to systemic. The continuous line shows the orientation of the galaxy major axis and the dashed line shows the orientation of the bi-cone which characterize the NLR emission-line flux distributions in this galaxy. Contours are from a radio MERLIN image.

velocity – which dominates near the nucleus, to the higher velocity outflowing component – which dominates outwards.

• Channel maps, on the other hand, usually show the highest velocities close to the nucleus, as illustrated in Fig.1 and do not support acceleration along the NLR. Our data suggest that the NLR clouds are accelerated very close to the nucleus (within ≈ 10 pc), and then the flow moves at essentially constant velocity (Storchi-Bergmann *et al.* 2010).

• Mass outflow rates range from 10^{-2} to $1\,M_\odot$ corresponding to ~ 10–200 times the accretion rate to the AGN, implying that the origin of the outflow is gas from the galaxy plane being pushed by a radio jet or accretion disk wind. The kinetic power of the outflow is usually $\leqslant 10^{-3}$ times the bolometric luminosity, which is small compared with the value of ~ 5 derived for LINERs, as presented by Nemmen *et al.* (2010) at this Conference.

References

Antonucci, R. R. & Miller, J. S. 1985, *ApJ* 297, 621
Barbosa, F. K. B. *et al.* 2006, *MNRAS* 371, 170
Barbosa, F. K. B. *et al.* 2009, *MNRAS*, 396, 2
Di Matteo, T., Springel, V., & Hernquist, L. 2005, *Nature* 433, 604
Elitzur, M. & Shlosman, I. 2006, ApJ, 648, L101
Nemmen, R., Storchi-Bergmann T., & Eracleous, M. 2010, *in preparation*
Riffel R. A., Storchi-Bergmann T., Winge C., & Barbosa, F. K. B. 2006, *MNRAS* 373, 2
Riffel, R. A. *et al.* 2008, *MNRAS* 385, 1129
Riffel, R. A., Storchi-Bergmann T., & McGregor, P. J. 2009, *ApJ* 698, 1767
Storchi-Bergmann, T. *et al.* 2010, *MNRAS, in press*
Veilleux S., Cecil G., & Bland-Hawthorn J. 2005, *ARA&A* 43, 769

Highlights of Astronomy, Volume 15
XXVIIth IAU General Assembly, August 2009
Ian F. Corbett, ed.

© International Astronomical Union 2010
doi:10.1017/S1743921310009063

On the role of magnetic reconnection in jet/accretion disk systems

Elisabete M. de Gouveia Dal Pino[1], Pamela Piovezan[1,2], Luis Kadowaki[1], Grzegorz Kowal[1] and Alex Lazarian[3]

[1] IAG, Universidade de São Paulo, Rua do Matão 1226, São Paulo 05508-900, Brazil
dalpino@astro.iag.usp.br

[2] Karl-Schwarzschild-Str. 1, Postfach 1317, D-85741 Garching, GermanyMPA, Garching, Germany

[3] Astronomy Department, University of Wisconsin, Madison, WI, USA

Abstract. The most accepted model for jet production is based on the magneto-centrifugal acceleration out off an accretion disk that surrounds the central source (Blandford & Payne, 1982). This scenario, however, does not explain, e.g., the quasi-periodic ejection phenomena often observed in different astrophysical jet classes. de Gouveia Dal Pino & Lazarian (2005) (hereafter GDPL) have proposed that the large scale superluminal ejections observed in microquasars during radio flare events could be produced by violent magnetic reconnection (MR) episodes. Here, we extend this model to other accretion disk systems, namely: active galactic nuclei (AGNs) and young stellar objects (YSOs), and also discuss its hole on jet heating and particle acceleration.

Keywords. accretion disks, acceleration of particles, magnetic fields.

MR IN MICROQUASARS AND AGNS: A violent MR process between the magnetic field lines of the accretion disk and those that are anchored into the black hole may occur when a large scale magnetic field is established by turbulent dynamo in the inner disk region with a ratio between the gas+radiation and the magnetic pressures $\beta \leqslant 1$. During this process, substantial angular momentum is removed from the disk by the wind generated by the magnetic flux, which increases the disk mass accretion rate to a value near the Eddington limit. After the reconnection, the partial destruction of the magnetic flux in the inner disk will make it to return to a less magnetized condition with most of the energy being dissipated locally within the disk instead of in the outflow. The magnetic power released by MR (see Figure 1) is able to heat the coronal/disk gas and accelerate the plasma to relativistic velocities through a diffusive first-order Fermi-like process within the reconnection site that will produce intermittent relativistic ejections or plasmons (GDPL). The resulting power-law electron distribution is compatible with the synchrotron radio spectrum observed during the outbursts of these sources. We are presently testing this acceleration mechanism with fully 3D numerical simulations. The diagram of the magnetic energy rate released by violent reconnection as a function of the black hole (BH) mass spanning 10^9 orders of magnitude (Figure 1) shows that the magnetic reconnection power is more than sufficient to explain the observed radio luminosities of the outbursts, from microquasars to low luminous AGNs (LINERs and Seyfert galaxies). This result is consistent with recently found empirical relation that correlates the observed radio emission from microquasars and radio quiet AGNs to that of magnetically active stars (Laor & Behar 2008; Soker & Vrtilek 2009), suggesting that it is mainly due to magnetic activity in the coronae and therefore, is nearly independent of

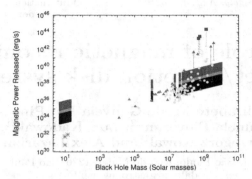

Figure 1. Magnetic power due to violent reconnection versus the BH mass for both micro-quasars and AGNs. The stars represent the observed radio luminosities for three microquasars. The circles, triangles and squares are observed radio luminosities of jets at parsec scales from LINERS, Seyfert galaxies, and luminous AGNs, respectively. The thick bars correspond to the calculated magnetic reconnection power and encompass a fiducial parameter space (see de Gouveia Dal Pino, Piovezan & Kadowaki, 2009, for details).

the intrinsic physics of the central source and the accretion disk. The correlation found in Figure 1 does not hold for radio-loud AGNs, possibly because their surroundings are much denser and then "mask" the emission due to coronal magnetic activity. In this case, particle re-acceleration behind shocks further out in the jet launching region will be probably the main responsible for the radio emission. The violent MR could also be responsible for the transition from the so called hard steep power-law state (SPLS) to the soft SPLS in microquasars (Remillard & McClintock, 2006).

MR IN YSOS: The observed flares in x-rays are often attributed to magnetic activity at the stellar corona. However, some COUP (Chandra Orion Ultra-deep Project) sources have revealed strong flares that were related to peculiar gigantic magnetic loops linking the magnetosphere of the central star with the inner region of the accretion disk. It has been argued that this x-ray emission could be due to magnetic reconnection in these gigantic loops (Favata et. al, 2005). We have extended the MR scenario described above to these sources and found that a similar magnetic configuration can be reached that could possibly produce the observed x-ray flares in most of the sources and provide the heating at the jet launching base if violent magnetic reconnection events occur with episodic, very short duration accretion rates $\sim 100 - 1000$ times larger than the typical mean accretion rates expected for more evolved (T Tauri) YSOs.

Acknowledgements

The authors acknowledge partial support from FAPESP and CNPq.

References

Blandford, R. D. & Payne, D. G. 1982, *MNRAS*, 199, 883
de Gouveia Dal Pino, E. M. & Lazarian, A. 2005, *A&A*, 441, 845
de Gouveia Dal Pino, E. M., Piovezan, P., & Kadowaki, L. 2009, *A&A* submitted
Favata, F., Flaccomio, E., Reale, F., Micela, G., Sciortino, S., Shang, H., Stassun, K. G., & Feigelson, E. D. 2005, *ApJS*, 160, 469
Laor, A. & Behar, E., 2008, *MNRAS*, 390, 847
Remillard, R. A. & McClintock, J. E. 2006, *ARA&A*, 44, 49
Soker, N. & Dil Vrtilek, S., [2009arXiv0904.0681S]

Highlights of Astronomy, Volume 15
XXVIIth IAU General Assembly, August 2009
Ian F. Corbett, ed.

© International Astronomical Union 2010
doi:10.1017/S1743921310009075

Are jets rotating at the launching?

Noam Soker

Department of Physics, Technion−Israel Institute of Technology, Haifa 32000 Israel;
soker@physics.technion.ac.il.

Abstract. I argue that the Doppler shift asymmetries observed in some young stellar object (YSO) jets result from the interaction of the jets with the circumstellar gas, rather than from jets' rotation. The jets do rotate, but at a velocity much below claimed values. During the meeting I carefully examined new claims, and found problems with the claimed jets' rotation. I will challenge any future observation that will claim to detect jet rotation in YSOs that requires the jets (and not a wind) to be launched from radii much larger than the accreting stellar radius. I conclude that the most likely jets' launching mechanism involves a very efficient dynamo in the inner part of the accretion disk, with a jets launching mechanism that is similar to solar flares (coronal mass ejection).

Keywords. ISM: jets and outflows, accretion disks

1. Introduction

In previous papers (Soker 2005, 2007a) I proposed that the interaction of the jets with a twisted-tilted (wrapped) accretion disk can form the asymmetry in the jets' line of sight velocity profiles as observed in some YSOs (e.g. Bacciotti *et al.* 2002). The claim that the observations of asymmetric Doppler shifts do not support jet rotation in YSOs was strengthened by the numerical simulations of Cerqueira *et al.* (2006). They assumed a precessing jet whose ejection velocity changes periodically with a period equals to the precession period. Practically, the dependance of the jets expansion velocity on direction around the symmetry axis leads to the same effect as the model of Soker (2005). Whereas in Soker (2005) the physical process behind this velocity profile is an interaction with the material in the jets surroundings, Cerqueira *et al.* (2006) give no justification for the periodic variation of the jets ejection speed. As far as comparison with observation is considered, it is hard to distinguish between the model of jet interaction with its surrounding (Soker 2005), and the periodic jets speed of Cerqueira *et al.* (2006).

2. Problems with claimed jets' rotation

To demonstrate the problems with the argued jet rotation, I will examine two new claims.

After the publication of my earlier papers Zapata *et al.* (2009) argued for a rotating molecular jet in Ori-S6. I find four problems with this case. More detail can be found in my presentation at the meeting:
http://iaujd-outflows.blogspot.com/2008/10/scientific-program.html
(1) In some regions the red and blue shifted components overlap. This is against expectation if the red-blue shifted components are due to jets' rotation. (2) In some regions the blue and red shifted components are disconnected. As each jet is one entity, this is against expectation if the Doppler shifts are due to jets' rotation. (3) Using the rotation interpretation at the edge of 30″ across the disk, gives a jets foot-point of 300 AU. This is larger than the size of the accretion disk given by the same authors for this object.

(4) The ring that supposedly feeds the accretion disk and the jets, rotates in opposite sense to that of the claimed jets' rotation. As Zapata *et al.* write: "The sense of rotation of the circumbinary ring is nearly opposite to that of jet and outflow, and the jet leaves the system under an angle of 45° with the ring plane." I note that a tilted jet can lead to the asymmetric red-blue shift, as in the model I proposed in 2005.

During the meeting, I was challenged to account for a very recent claim of a possible jets' rotation in HH 211 (Lee *et al.* 2009, 2007). I find two problems with the tentative claimed jets' rotation (I elaborate on these points in the appendix in the astro-ph version of this paper). (1) The blue and red components exchange sides. Namely, the velocity plots do not give a clear sense of asymmetry, and hence no unique sense of rotation. The same effect is seen in the velocity maps of HH 212 (Lee *et al.* 2008). (2) The accretion disk cannot supply the required anergy and angular momentum if the rotation is real.

My conclusion is that these types of observations give peaks in emission that show different Doppler shifts. By pure fluctuations, these might mimic rotation in some places. In same cases the sense of the fluctuations will give rotation in the same sense as that of the accretion disk. In other cases the sense will be in an opposite direction to that of the disk, and in some cases just zero rotation will be deduced. The inferred rotation is due to fluctuations that by chance can mimic rotation.

3. The launching mechanism

The talks and discussions during the meeting strengthened my view that the launching mechanism involves reconnection of magnetic field lines. Reconnection can occur between the stellar and the disk magnetic fields (e.g., de Gouveia dal Pino & Lazarian 2005; de Gouveia dal Pino *et al.* 2009), or reconnection of the disk magnetic field (Soker 2007b). Laor & Behar (2008) show that the ratio of radio luminosity to X-ray luminosity has similar values in magnetically active stars and in many accreting objects, up to radio quiet quasars. Based on this correlation I prefer the following conclusion (Soker 2007b; Soker & Vrtilek 2009): There is a very efficient dynamo in the inner part of the accretion disk, with a jets launching mechanism that is similar to solar flares (coronal mass ejection).

References

Bacciotti, F., Ray, T. P., Mundt, R., Eislöffel, J., & Solf, Jo. *ApJ*, 576, 222 (B2002)
Cerqueira, A. H., Velazquez, P. F., Raga, A. C., Vasconcelos, M. J., & de Colle, F. 2006, *A&A*, 448, 231
de Gouveia dal Pino, E. M. & Lazarian, A. 2005, *A&A*, 441, 845
de Gouveia Dal Pino, E. M., Piovezan, P., Kadowaki, L., Kowal, G., & Lazarian, A., this proceedings (IUA JD 7 at the XXVIIth IAU General Assembly)
Laor, A. & Behar, E. 2008, *MNRAS*, 390, 847
Lee, C.-F., *et al.* 2007, *ApJ*, 670, 1188
Lee, C.-F., *et al.* 2008, *ApJ*, 685, 1026
Lee, C.-F., *et al.* 2009. *ApJ*, 699, 1584
Soker, N. 2005, *A&A*, 435, 125
Soker, N. 2007a, astro-ph/0703474
Soker, N. 2007b, IAUS, 243, 195 (arXiv:0706.4241).
Soker, N. & Vrtilek, S. D. 2009, arXiv:0904.0681
Zapata, L. A., Schmid-Burgk, J., Muders, D., Schilke, P., Menten, K., & Guesten, R. 2009, *A&A* in press

Highlights of Astronomy, Volume 15
XXVIIth IAU General Assembly, August 2009
Ian F. Corbett, ed.

© International Astronomical Union 2010
doi:10.1017/S1743921310009087

On the correlation of the highest energy cosmic rays with AGNs

Vitor de Souza[1] and Peter L. s Biermman[2]

[1]Instituto de Física de São Carlos, Universidade de São Paulo, Brazil
email: vitor@ifsc.usp.br

[2]MPI for Radioastronomy, D-53121 Bonn, Germany
email: plbiermann@mpifr-bonn.mpg.de, Also at: Univ. of Bonn, Univ. of Alabama, Tuscaloosa,
AL, Univ. of Alabama at Huntsville, and at KIT Karlsruhe

Abstract. In this paper we briefly discuss the present status of the cosmic ray astrophysics under
the light of the new data from the Pierre Auger Observatory. The measured energy spectrum
is used to test the scenario of production in nearby radio galaxies. Within this framework the
AGN correlation would require that most of the cosmic rays are heavy nuclei and are widely
scattered by intergalactic magnetic fields.

Keywords. Ultra-high energetic particles, AGN correlation, energy spectrum.

1. Introduction

The production mechanism of particles with energy above 10^{20} eV has been a puzzle
since 1963 when the first measurement was done. Several models have been proposed but
the lack of experimental constraints allowed a wide range of possibilities. This panorama
has recently started to change specially after the publication of a correlation of the arrival direction of these particles with the position of AGNs done by the Pierre Auger
Collaboration [Auger Collaboration (2007)]. Since then, new scenarios have been considered. Among all candidates AGN jets [Benford & Protheroe (2008)] and powerful radio
galaxies [Rachen & Biermann (1993)] are the most favored by present data [Allard &
Protheroe (2009)].

Generally speaking, models based on sources cosmologically distributed tends to: a)
generate an abundance of light particles, b) requires low intergalactic magnetic fields to
describe the AGN correlation and c) describes the flux suppression above 4×10^{19} eV as
the interaction of the particles with the CMB [Berezinsky *et al.* (2006)]. On the other
hand, models based on local powerful sources tends to: a) generate a high flux of heavy
particles, b) requires high intergalactic magnetic fields to describe the AGN correlation
and c) describes the flux suppression above 4×10^{19} eV as the limit of the sources.

In this short article we discuss the possibility of accelerating cosmic rays in nearby
radio galaxies and compare the resulting energy spectrum with the one measured by the
Pierre Auger Observatory. Figure 1 shows the good agreement between this model and
the data. Since none of these sources in our cosmic neighborhood accelerates protons
to such high energy, according to arguments derived from the jet energetics [R. V. E.
Lovelace (1976)], heavy nuclei seem to be required. Such a heavy composition would also
be favored by the recent data of the Auger Observatory [Auger Collaboration (2009)].

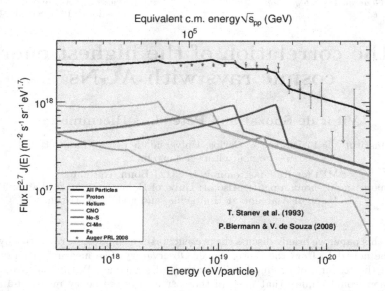

Figure 1. Energy spectrum of cosmic rays produced in radio galaxies as devised originally by Stanev *et al.* (1993) and energy shifted according to P.L.Biermann *et al.* (2009) compared to the spectrum measured by the Pierre Auger Observatory [Auger Collaboration (2008)].

2. Model Discussion

This model is based on the original argument presented by Biermann & Strittmatter (1987). In this model, a galaxy merger is considered though all its stage. Considering the standard evolution of the merging, going from the initial starburst to the spinflip of the more massive black hole, one is able to predict the formation of jets. The starburst provides through its wind-SNe the seed particles for the one-step acceleration in the jet. Applying the results of Y.-A. Gallant & A. Achterberg (1999) we can push the energy particles of the previous model up by a factor of Γ^2, leading to ultra high energy cosmic rays (UHECR).

Radio interferometry of this source have shown that in the spine of this jets Lorentz factor can reach 50 [P.L.Biermann *et al.* (2009)]. Therefore, the spectrum with all its heavy abundances is shifted straight to ultra high energy, and the fit actually is consistent with the Auger data. In this case, the most prominent source of UHECR would be Cen A.

Acknowledgements

V. de Souza is supported by FAPESP (2008/04259-0). Support for work with P. Biermann has come from the AUGER membership and theory grant 05 CU 5PD 1/2 via DESY/BMBF and VIHKOS.

References

Stanev, T., Biermann, P. L. & Gaisser, T. K. 1993, *Astron. and Astroph.* 274, 902.
Allard, D. & Protheroe, R. - *arXiv:0902.4538*.
Benford, G. & Protheroe, R. 2008, *MNRAS* 383, 663.
Rachen, J. P. & Biermann, P. L. 1993, *Astron. Astrophys.* 272,161.
The Pierre Auger Collaboration 2007 *Science* 318,938-943.
Berezinsky, V. *et al.* 2006 *Phys. Rev. D* 74,043005.
R. V. E. Lovelace 1976 *Nature* 262,649.
Biermann, P. L. & Strittmatter, P. A. 1987 *ApJ* 322,643.

The Pierre Auger Collaboration 2009, M. Unger, SOCoR workshop, Trondheim; V. de Souza, "Window on the Universe", Blois, France; J. Bellido, 31st ICRC, Lodz, Poland.

Y.-A. Gallant, & A. Achterberg *Month. Not. Roy. Astr. Soc.* 305,L6 - L10

P. L. Biermann *et al.* - *arXiv:0904.1507.*

The Pierre Auger Collaboration 2008 *Phys. Rev. Letters* 101,061101.

Highlights of Astronomy, Volume 15
XXVIIth IAU General Assembly, August 2009
Ian F. Corbett, ed.

High Resolution 3D Relativistic MHD Simulations of Jets

A. Ferrari[1], A. Mignone[1], P. Rossi[2], G. Bodo[2] and S. Massaglia[1]

[1]Dipartimento di Fisica Generale dell'Università di Torino, Italy, E-mail: ferrari@ph.unito.it
[2]INAF Osservatorio Astronomico di Torino, Italy

Abstract. We performed high-resolution three dimensional numerical simulations of relativistic MHD jets carrying an initially toroidal magnetic field responsible for the process of jet acceleration and collimation. We find that in the 3D case the toroidal field gives rise to strong current driven kink instabilities leading to jet wiggling. However, it appears to be able to maintain an highly relativistic spine along its full length.

Keywords. Jets, numerical magneto-hydrodynamics, relativity, instabilities

In the most widely accepted models of relativistic jets from AGNs magnetic fields with footpoints attached to a spinning black hole or accretion disc are the necessary element for jet acceleration and collimation (e.g. see Narayan *et al.* 2007). In particular, it is likely that a toroidal component prevails at large scales as the poloidal component decays as $B_p \propto 1/R_j^2$ while the toroidal $B_\phi \propto 1/R_j$, where R_j is the jet radius that increases from the inner galactic core to the radio jet lobes.

However, although jets must be Poynting-dominated at the origin, observational data suggest that they become kinetically dominated beyond about 1,000 gravitational radii from the central acceleration region, powered by a super-massive black-hole of $\sim 10^{8-10}$ $M_\odot$ (Sikora *et al.* 2005).

Configurations in which toroidal magnetic fields dominate are known to be violently unstable to the $m = 1$ kink instability according to the Kruskal-Shafranov criterion $|B_\phi/B_p| > 2\pi R_j/z$. The mechanism by which the toroidal field responsible for the jet acceleration can be dissipated may be related to the above mentioned instability (Giannios & Spruit 2006). At the same time relativistic jets produced in proximity of rapidly spinning black-holes with dipolar magnetic fields expand rapidly and can possibly survive up to large radii without showing appreciable disruption (McKinney & Blandford 2009).

Here we report on the first high-resolution 3D relativistic MHD simulations of the propagation of a magnetized jet injected in a uniform un-magnetized external medium. From the variety of possible magnetic field configurations, we consider the case of jets initially carrying a purely toroidal magnetic field, consistently with the previous discussion. Numerical simulations are carried out using the PLUTO code for astrophysical gasdynamics (Mignone *et al.* 2007). The chosen configuration employs a second order Godunov type scheme with the recently developed HLLD Riemann solver (Mignone *et al.* 2009) and the mixed hyperbolic/parabolic divergence cleaning to control the solenoidal constraint. The best overall resolution in our calculations corresponds to $640 \times 1600 \times 640$ computational zones with the radius of the beam spanning 20 zones.

The computational domain is initially filled with an un-magnetized medium at rest, with uniform density ρ_a and gas pressure p_a. The jet flows is injected through a cylindrical nozzle with constant density ρ_j and longitudinal velocity component given in terms of the Lorentz factor γ_j.

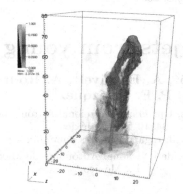

The typical structure of the jet when it has reached a length of 70 radii is given in the figure that shows the volume rendering of the tracer distribution.

The presence of a toroidal component of the field is known to drive current-driven kink instabilities which in our case are responsible for jet wiggling and beam deflection off the main longitudinal axis. The wandering of the jet head, induced by kink instability effects, may create multiple sites where the jet impacts on the external medium forming strong shocks. This behavior originates hot spots observed as in several radio jets. The asymmetry of the backflow is another feature that replicates the observational appearance of several objects.

An additional effect of the toroidal field component is a shielding of the inner jet core from any interaction with the surroundings and therefore from a loss of momentum. Correspondingly the jet remains highly relativistic all along its length. Quasi periodic shocks, formed by the combination of several effects like the interaction with the cocoon and the toroidal field pinching, can be observed all along the jet extension. A spine *plus* sheath layer structure with shocks is consistent with recent observations by the AGILE and FERMI missions and the TeV ground arrays (Donnarumma *et al.* 2009, Acciari *et al.* 2009) that jets at sub-kpc scale originate correlated variability at radio frequencies and in X and gamma rays.

Acknowledgements

This work was supported by research contracts with INAF and CINECA.

References

Acciari, V. A., Aliu, E., Arlen, T., *et al.* 2009, Science, 325, 444

Donnarumma, I., Vittorini, V., Vercellone, S., *et al.* 2008, ApJL, 691, L13

Giannios, D. & Spruit, H. C., 2006, A&A, 450, 887

McKinney, J. C. & Blandford, R. D., 2009, MNRAS, 394, L126

Mignone, A., Bodo, G., Massaglia, S., Matsakos, T., Tesileanu, O., Zanni, C., & Ferrari, A. 2007, ApJS, 170, 228

Mignone, A., Ugliano, M., & Bodo, G. 2009, MNRAS, 393, 1141

Narayan, R., McKinney, J. C., & Farmer, A.J. 2007, MNRAS, 375, 548

Sikora, M., Begelman, M. C., Madejski, G. M., & Lasota, J.-P., 2005 ApJ, 625, 72

Highlights of Astronomy, Volume 15
XXVIIth IAU General Assembly, August 2009
Ian F. Corbett, ed.

© International Astronomical Union 2010
doi:10.1017/S1743921310009105

Variable jets from young stars

A. C. Raga[1], J. Cantó[2], A. Esquivel[1], A. Rodríguez-González[1], P. F. Velázquez[1]

[1] Instituto de Ciencias Nucleares, Universidad Nacional Autónoma de México, Ap. 70-543, 04510 D. F., México
email: raga@nucleares.unam.mx

[2] Instituto de Astronomía, Universidad Nacional Autónoma de México, Ap. 70-468, 04510 D. F., México

Abstract. In this paper we discuss the fact that the observed "accelerations" (i. e., higher velocities at larger distances from the source) observed along some Herbig-Haro (HH) jets directly imply that the ejection velocity has to be time-dependent. Even though discussed in the early literature of the subject, this is an often forgotten fact.

Keywords. ISM: kinematics and dynamics – ISM: jets and outflows – ISM: Herbig-Haro objects – stars: winds, outflows

1. Introduction

There are several different possibilities for interpreting the knots found along HH jets. The two best studied ones are:

• the identification of knots with shock cells produced by Kelvin-Helmholtz instabilities at the outer boundary of the jet beam (see Micono *et al.* 1998),

• modelling the knots as "internal working surfaces" produced by an ejection velocity variability (see Raga *et al.* 1990).

The most straightforward evidence favouring (in some objects) the second of these models is the existence of jet/counterjet systems with good knot-to-knot symmetries in the two outflow lobes (examples of this are HH 111, see Gredel & Reipurth 1994 and HH 34, see García López *et al.* 2008).

In this paper, we concentrate on a second kind of evidence favouring the "internal working surface" scenario: the presence of "accelerating segments" along some jet beams (good examples of this being again HH 111, see Raga *et al.* 2002 and HH 34, see Heathcote & Reipurth 1992).

2. Accelerations

Let us assume that we have a steady, one-dimensional gasdynamic jet. In the absence of gravity and radiative losses, the flow velocity u along the outflow axis then obeys Bernoulli's theorem, which we write in the form:

$$u^2 = u_0^2 \left[1 + \frac{1}{\gamma - 1} \left(\frac{1}{M_0^2} - \frac{c_s^2}{u_0^2} \right) \right], \qquad (2.1)$$

where c_s is the sound speed at the position in which the flow velocity is u, and $M_0 = u_0/c_{s,0}$ is the Mach number, u_0 the flow velocity and $c_{s,0}$ the sound speed at the injection point. For a monoatomic gas, the specific heat ratio would be $\gamma = 5/3$.

¿From equation (2.1), we see that at the injection point (in which $c_s = c_{s,0}$) we correctly obtain $u = u_0$. Now, if the flow expands laterally as it flows along the jet axis, the resulting adiabatic expansion leads to a drop in the sound speed, and to a corresponding increase in the flow velocity (see equation 2.1). The maximum possible velocity that can be attained is:

$$u_{max} = u_0 \sqrt{1 + \frac{1}{(\gamma - 1)M_0^2}}.$$ (2.2)

For a HH jet, observations indicate that the injection Mach number is in the $M_0 \sim 10 \rightarrow 30$ range. Therefore, we would expect to see only very small variations (of at most $\sim 1\%$) in the flow velocity as one moves along the jet beam.

However, there are many observations of quite substantial (of the order of 100 % of the injection velocity) drops and "accelerations" in the flow velocity along HH jets. The drops in velocity in principle could be the result of (steady or non-steady) shocks within the jet beam. For the accelerations there is no possible mechanism other than a time-variability in the injection velocity.

In the discussion above, we have only considered a gasdynamic, non-radiative flow in the absence of an external gravitational field. The presence of gravity (e. g., from the stellar source or from the cloud core in which it is embedded) or of radiative losses will reduce the possible acceleration derived from equation (2.1). The presence of a magnetic field within the jet beam would also provide only small possible accelerations, provided that the Alfvénic Mach number at the injection point is large.

Therefore, all of the observed "accelerations" along HH jets are direct evidence for the existence of a time-variability in the ejection velocity. The theory of jets from variable sources is therefore a required ingredient for modelling many of the observed HH jets, and might also be relevant for modelling other HH objects in which clear accelerations have not yet been observed.

3. Conclusions

In this paper, we have discussed the fact that the highly supersonic "accelerations" (i. e., larger flow velocities at increasing distances from the outflow source) observed in some HH jets directly imply that the ejection velocity has to be time-dependent. This is a result of the fact that a high (gasdynamic and Alfvénic) Mach number jet has very little thermal and/or magnetic energy to convert into extra kinetic energy (i. e., to produce an outwards increase of the jet velocity). Therefore, jets with observed "accelerations" are direct evidence for the existence of a variable ejection.

Acknowledgements

This work was supported by the CONACyT grants 61547 and 101356.

References

García López, R., Nisini, B., Giannini, T., Eislöffel, J., Bacciotti, F., & Podio, L. 1008, A&A, 487, 1019
Gredel, R. & Reipurth, B. 1994, ApJ 407, L19
Heathcote, S. & Reipurth, B. 1992, AJ, 104, 2193
Micono, M., Massaglia, S., Bodo, G., Rossi, P., & Ferrari, A. 1998, A&A 333, 1001
Raga, A. C., Noriega-Crespo, A., Reipurth, B., Garnavich, P. M., Heathcote, S., Böhm, K. H., & Curiel, S. 2002, ApJ, 565, L29
Raga, A. C., Cantó, J., Binette, L., & Calvet, N. 1990, ApJ, 364, 601

Highlights of Astronomy, Volume 15
XXVIIth IAU General Assembly, August 2009
Ian F. Corbett, ed.

© International Astronomical Union 2010
doi:10.1017/S1743921310009117

Radio and Millimeter Observations of YSOs

Luis F. Rodríguez

Centro de Radioastronomía y Astrofísica, UNAM, Apartado Postal 3-72 (Xangari), 58089
Morelia, Michoacán, México
email: l.rodriguez@crya.unam.mx

Abstract. The growth of a forming star takes place by accretion from the surrounding dense medium, facilitated by a circumstellar disk. But at the same time the forming system produces collimated outflows of gas that remove excess angular momentum and magnetic flux. A key question in the field is whether we can extend the disk/jet model to stars of the largest masses or if other ingredients are present. In this note, I review recent observations from the Very Large Array and the Submillimeter Array, of IRAS 16547-4247, a massive young stellar object that exhibits evidence for both a disk and an outflow.

Keywords. stars: formation, ISM: individual (IRAS 16547-4247)

1. Introduction

Our present understanding of star formation is primarily based on observations of the relatively abundant low-mass stars. The theoretical framework for star formation (Shu *et al.* 1987, see also McKee & Ostriker 2007) has been successful in explaining the processes that occur in the formation of these low mass stars, processes that are inferred from multiwavelength observations (e.g., Lada 1991, Evans 1999). Key ingredients in this scenario are the presence of a central protostar accreting from a circumstellar disk that is surrounded by an infalling envelope of dust and gas, as well as the presence of ionized jets and molecular outflows that carry out angular momentum and mechanical energy from the accretion disk into its surroundings.

The applicability of this paradigm to the formation of massive stars remains unproven. It is possible that massive stars are formed by processes that are radically different from those that produce low-mass stars, such as by the merging of lower mass protostars (Bonnell, Bate, & Zinnecker 1998). The role of the coalescence and accretion processes in the assembling of a massive star is still under debate. If massive O stars are formed by accretion we expect that disks and jets will be present in their earliest stages of evolution. On the other hand, if they are formed through coalescence of lower-mass stars then neither disks nor jets are expected since they would be disrupted during the merging process. For a recent review on the competing ideas to explain massive star formation see Zinnecker & Yorke (2007).

2. The case of IRAS 16547-4247

2.1. *SMA Observations*

The main result of these observations, described in detail in Franco-Hernández *et al.* (2009) is the detection of strong SO_2 emission that arises only from the brightest 1.3 mm dust continuum component. The line emission shows a velocity gradient (see Fig. 1) that, if modeled as a Keplerian ring, give a mass of $\sim 20\ M_\odot$, consistent with the mass derived from the VLA observations of H_2O maser. However, the data can also be fitted with a two-component model that gives a much smaller Keplerian mass.

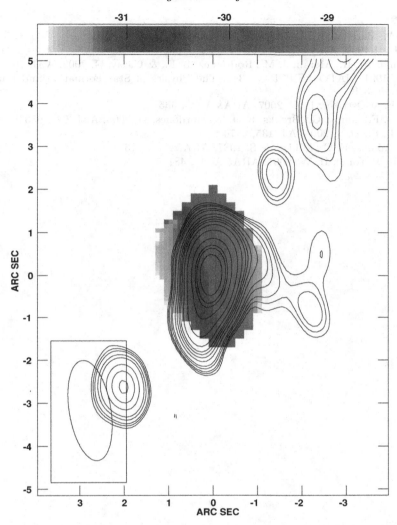

Figure 1. This image shows the first moment map of the 226.300 GHz SO_2 transition in color and the radio continuum emission at 3.6 cm with contours (Rodríguez *et al.* 2005). The color bar at the top shows the color coding for the LSR velocity of the gas in km s^{-1}. The synthesized beam for the 3.6 cm is shown in the bottom left corner of the panel. Contour levels for the 3.6 cm emission are -5, 5, 8, 10, 15, 20, 40, 60, 80, 100, 140, 180 times the rms noise level of 30 μJy beam^{-1}. The 3.6 cm source at the lower left corner of the image most probably traces an independent star and is not associated with the jet (Rodríguez *et al.* 2005). The synthesized beam for the 226.3 GHz data is 2."3 × 0."8; PA = 11°.

2.2. *VLA Observations*

In this case, the main results have been two. This first is the detailed imaging of the 3.6 cm emission from the associated thermal jet (Rodríguez *et al.* 2008), that indicate that the brightest components of the lobes show evidence of precession, at a rate of 0°.08 yr^{-1} clockwise in the plane of the sky. The second result is the detection of a group of water masers that show a velocity gradient that, if interpreted as arising in a Keplerian ring, implies a mass of ~30 $M_\odot$ for the central star(s) (Franco-Hernández *et al.* 2009). This mass is consistent with that determined at larger scales with the SO_2 emission.

References

Bonnell, I. A., Bate, M. R., & Zinnecker, H. 1998, MNRAS, 298, 93

Evans, N. J., II 1999, ARA&A, 37, 311

Franco-Hernández, R., Moran, J. M., Rodríguez, L. F., & Garay, G. 2009, ApJ, 701, 974

Lada, C. J. 1991, NATO ASIC Proc. 342: The Physics of Star Formation and Early Stellar Evolution, 329

McKee, C. F. & Ostriker, E. C. 2007, ARA&A, 45, 565

Rodríguez, L. F., Garay, G., Brooks, K. J., & Mardones, D. 2005, ApJ, 626, 953

Rodríguez, L. F., et al. 2008, AJ, 135, 2370

Shu, F. H., Adams, F. C., & Lizano, S. 1987, ARA&A, 25, 23

Zinnecker, H. & Yorke, H. W. 2007, ARA&A, 45, 481

Highlights of Astronomy, Volume 15
XXVIIth IAU General Assembly, August 2009
Ian F. Corbett, ed.

The ejection-accretion connection in young stars: Testing MHD disk winds

S. Cabrit[1], J. Ferreira[2], C. Dougados[2] and P. Garcia[3,2]

[1]LERMA, Observatoire de Paris, UMR 8112 du CNRS, 61 Av. de l'Observatoire, F-75014
Paris.   email: sylvie.cabrit@obspm.fr

[2]Laboratoire d'Astrophysique de l'Observatoire de Grenoble, UMR 5521 du CNRS, F-38041
Grenoble Cedex.   email: ferreira@obs.ujf-grenoble.fr, dougados@obs.ujf-grenoble.fr

[3]Universidade do Porto, Faculdade de Engenharia, Laboratório de Sistemas, Instrumentação e
Modelação em Ciências e Tecnologias do Ambiente e do Espaço, P-4200-465 Porto, Portugal
email: pgarcia@fe.up.pt

Abstract. Jets are ubiquitous in young accreting stars at all evolutionary stages, from deeply
embedded protostars aged less than 0.1Myr to optically revealed 10Myr old T Tauri stars. The
similar jet collimation at all ages is shown to require an effective magnetic collimation within
the inner disk regions (inside 20 AU). This fact, and the high ejection to accretion ratio $\simeq 10\%$,
appear to favor the presence of MHD disk winds. Ejection out to > 0.1 AU could explain the
velocity drop and rotation signatures across the jets, and their dust and molecular content.

Keywords. accretion disks, MHD, stars: pre–main-sequence, ISM: jets and outflows

---

The sub-arcsecond resolution achieved in recent years from the ground and in space has
revealed unprecedented details on the inner regions of young stellar jets. The improved
jet mass-fluxes confirm the correlation between ejection and accretion rates found by Hartigan *et al.* (1995), but with a higher ratio $\simeq 10\%$ if accretion shock models are adopted
(Cabrit 2007). A similar ratio is obtained in protostars (Antoniucci *et al.* 2008, Lee *et al.*
2008). Collimating such a high mass-flux on observed scales seems only achievable with
MHD disk winds:

1. The jet collimation issue and MHD disk winds

The high-velocity gas in T Tauri stars appears collimated into a narrow cone of about
20 AU in diameter at 50 AU from the source (Ray *et al.* 2007); the same jet widths
are seen in the youngest protostars, arguing against collimation by the infalling envelope
(Cabrit *et al.* 2007). The pressure of the disk atmosphere, including photo-evaporation,
is also insufficient (Cabrit 2007). Therefore, *magnetic* jet collimation seems required.

Three different MHD ejection sites may contribute to jets, probably at the same time:
stellar winds, magnetospheric winds, and disk winds (Ferreira *et al.* 2006). However,
winds from the stellar surface meet strong problems in both ejecting and collimating
such a high mass-flux (Cranmer 2009, Bogovalov & Tsinganos 2001). A higher mass-flux
may be achieved in a sporadic "reconnexion wind" along the current sheet in the distorted
stellar magnetosphere or a slower inner disk wind† (Zanni 2009, Romanova *et al.* 2009).
However the ejection occurs at a 45° angle, therefore a confining disk field would still
be required, of strength $\simeq 20$mG $\sqrt{\dot{M}_{\rm acc}/10^{-7}M_\odot {\rm yr}^{-1}}$ at a radius of 20 AU (Cabrit

---

† Both differ from the "X-wind" proposed by Shu *et al.* 1995 in that they are not locked at
corotation, and are mostly powered by magnetic torsion, and/or reconnexion of initially closed
field-lines (see also de Gouveia Dal Pino *et al.*, this volume)

2007). Such a field could be passively advected in (Shu *et al.* 2007) or provided by a centrifugal MHD disk wind launched from the inner few AUs; the strong B_ϕ component of the latter is particularly efficient at confining inner winds (Meliani *et al.* 2006, Fendt 2009, Matsakos *et al.* 2009) and predicted jet widths agree with observations (Garcia *et al.* 2001, Ray *et al.* 2007).

2. Other suggestive indications of MHD disk winds

Subarcsecond studies of atomic T Tauri jets also reveal a drop in velocity away from the jet axis, and possible rotation signatures, that could be explained by MHD disk winds (Pesenti *et al.* 2004, Ray *et al.* 2007). The rotation values impose a maximum launch radius of 0.1–3 AU, and a moderate magnetic lever arm $\lambda \leqslant 13$ yielding jet speeds < 400 km/s and an ejection to accretion ratio $\simeq 4\%$–13% in the observed range, if the MHD wind dominates the angular momentum extraction from the disk (Ferreira *et al.* 2006). However, T Tauri jet rotation signatures are still tentative due to possible contamination by shock asymmetries (e.g. Soker 2005). MHD disk winds also seem promising to explain the presence of dust and molecules in/around atomic jets (Podio *et al.* 2006, Panoglou *et al.* 2009)†. The zone of MHD wind launching would be limited by the available B-field, or by X-ray ionization of the disk surface. Infrared interferometry and ALMA observations will be decisive to further constrain the disk wind extent in young stars.

References

Antoniucci, S. , Nisini, B., Giannini, T., & Lorenzetti, D. 2008, A&A 479, 503

Bogovalov, S. & Tsinganos, K. 2001, MNRAS 325, 249

Cabrit, S. 2007, in *MHD jets and winds from young stars*, Lecture Notes in Physics (Spinger-Verlag Berlin Heidelberg), vol. 723, p. 21

Cabrit, S., *et al.* 2007, A&A 468, L29

Casse, F. & Ferreira, J. 2000, A&A 361, 1178

de Gouveia Dal Pino, E. M., Piovezan, P., & Kadowaki, L. (this volume)

Cranmer, S. R. 2009, ApJ 706, 824

Fendt, C. 2009, ApJ 692, 346

Ferreira, J., Dougados, C., & Cabrit, S. 2006, A&A 453, 785

Garcia, P., Cabrit, S., Ferreira, J., & Binette, L. 2001, A&A 377, 609

Hartigan, P., Edwards, S., & Gandhour, L. 1995, ApJ 452, 736

Lee, C.-F., *et al.* 2008, ApJ 685, 1026

Matsakos, T., Massaglia, S., Trussoni, E., *et al.* 2009, A&A 502, 217

Meliani, Z., Casse, F., & Sauty, C. 2006, A&A 460, 1

Panoglou, D., Cabrit, S., Pineau des Forêts, G., *et al.* 2009, submitted

Pesenti, N., *et al.* 2004, A&A 416, L9

Podio, L., *et al.* 2006, A&A 456, 189

Ray, T. P. *et al.* 2007, In *Protostars & Planets V*, B. Reipurth, D. Jewitt, and K. Keil (eds.), (University of Arizona Press, Tucson), pp. 231–244

Romanova, M. M.; Ustyugova, G. V.; Koldoba, A. V.; & Lovelace, R. V. E. 2009, MNRAS 399.1802

Shu, F. H., Najita, J., Ostriker, E. C., & Shang, H. 1995, ApJ 455, L155

Shu, F. H., Galli, D., Lizano, S., Glassgold, A. E., & Diamond, P. 2007, ApJ 665, 535

Soker, N. 2005, A&A 435, 125

Zanni, C.; Ferrari, A.; Rosner, R.; Bodo, G.; & Massaglia, S. 2007, A&A 469, 811

Zanni, C. 2009, in *Magnetic Fields in the Universe II*, RMxAC, 36, 284

† Note that the small launch radii < 0.1 AU inferred in molecular jets with SMA (Lee *et al.* 2008) are very uncertain since the jet widths are unresolved, and rotation signatures are strongly suppressed in this case (Pesenti *et al.* 2004). ALMA will be crucial to settle this issue

Highlights of Astronomy, Volume 15
XXVIIth IAU General Assembly, August 2009
Ian F. Corbett, ed.

© International Astronomical Union 2010
doi:10.1017/S1743921310009130

A New Look at Optical and X-ray Emission in SDSS/XMM-Newton Quasars

Monica C. B. Young[1,2]

[1] Harvard-Smithsonian Center for Astrophysics, 60 Garden St., Cambridge, MA 02138, USA
email: myoung@head.cfa.harvard.edu

[2] Department of Astronomy, Boston University, 725 Commonwealth Ave. Boston, MA 02215, USA

Abstract. We develop a new approach to the well-studied anti-correlation between the optical-to-X-ray spectral index, α_{ox}, and the monochromatic optical luminosity, l_{opt}. By cross-correlating the SDSS DR5 quasar catalog with the XMM-Newton archive, we create a sample of 327 quasars with both optical and X-ray spectra, allowing α_{ox} to be defined at arbitrary frequencies, rather than the standard 2500 Å and 2 keV. We find that while the choice of optical wavelength does not strongly influence the $\alpha_{ox} - l_{opt}$ relation, the slope of the relation flattens significantly with X-ray energy. This result suggests a change in the efficiency of X-ray photon production, where the efficiency of low energy X-ray production depends more strongly on the seed (optical/UV) photon supply. We discuss implications for line-driven wind models.

Keywords. quasars: general, galaxies: active, accretion, accretion disks

1. Introduction

A quasar's optical to X-ray continuum ionizes the characteristic broad lines and is responsible for any line-driven outflows from the accretion disk (Proga *et al.* 2000). Yet much of this continuum is never visible due to absorption in our own galaxy, so it must be parameterized by an imaginary power-law, traditionally defined by endpoints at 2500 Angstroms and 2 keV. This power-law has been observed to steepen with optical luminosity (e.g., Avni & Tananbaum 1982; Steffen *et al.* 2006). At face value, this relation suggests that efficiency of X-ray production decreases with luminosity, but theory cannot explain why this should be the case. We characterize the dependence of α_{ox} on luminosity by defining the relation at different frequencies than those traditionally used. The results have important implications for line-driven disk winds.

2. Measuring α_{ox} in the SDSS/XMM-Newton Quasar Survey

We have cross-correlated the DR5 Sloan Digital Sky Survey (SDSS) with *XMM-Newton* archival observations to obtain 792 X-ray observations of SDSS quasars. This gives the largest sample of optically selected quasars with both optical and X-ray spectra (473 with X-ray S/N > 6). We exclude broad absorption line and radio-loud quasars, as well as spectra with significant absorption or bad fits, and then we calculate α_{ox} with the standard formula: $\alpha_{ox} = \frac{log(f_x/f_o)}{log(\nu_x/\nu_o)}$ where f_x is the monochromatic flux at 1, 1.5, 2, 4, 7, or 10 keV and f_o is the monochromatic flux at 1500, 2500, or 5000 Å.

For each source with X-ray S/N > 6, we measure the monochromatic flux at 1, 1.5, 2, 4, 7 and 10 keV after fitting an unabsorbed power-law. We then interpolate a power-law through the SDSS photometry to get the luminosity at 1500, 2500, and 5000 Å.

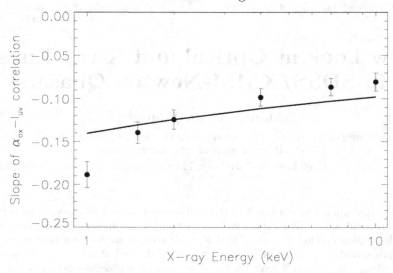

Figure 1. The slope of the $\alpha_{ox} - l_{opt}$ relation with respect to the X-ray energy. The optical wavelength is held constant at 2500 Å. The filled circles show the measured values, and the solid line shows the values predicted by the baseline effect.

3. Results

The slope of the $\alpha_{ox} - l_{opt}$ relation flattens and the dispersion tightens as the X-ray energy used to define α_{ox} increases. The effect of changing X-ray energy is much stronger than that of changing optical wavelength.

The definition of α_{ox} depends on the baseline over which it is defined, so stretching the baseline (for example, moving from 1 to 10 keV) can artificially flatten the $\alpha_{ox} - l_{opt}$ slope. However, Figure 1 shows that while changing the X-ray baseline clearly has an influence, the slope for 1 keV is significantly steeper than it would be if the intrinsic correlation were constant. The slopes at 4, 7 and 10 keV are also flatter than predicted by the baseline effect at the 1σ level. Since the baseline effect is arbitrarily normalized to the slope at 2 keV, the main result of subtracting the baseline effect is not to assess the significance of the slope at a given energy, but rather to show that the *intrinsic* slope of the $\alpha_{ox} - l_{opt}$ relation is flattening as X-ray energy increases.

4. Significance for Line-Driven Winds

The preliminary results of a population study suggest that selection effects cannot reproduce these results, so the $\alpha_{ox} - l_{opt}$ relation is likely intrinsic to quasars. This has important implications for line-driven disk winds (Murray & Chiang 1995; Proga *et al.* 2000), where UV disk photons drive a wind via radiation pressure on spectral lines. Line-driven winds depend on UV luminosity, which determines column density, and on X-ray luminosity, which determines ionization state. The traditional $\alpha_{ox} - l_{opt}$ relation predicts that at high optical luminosities, more driving radiation (high l_{opt}) and less ionizing radiation (steep α_{ox}) should encourage a strong, dense wind. The results of this study suggest that this effect is stronger than previously realized due to the significantly steeper $\alpha_{ox} - l_{opt}$ relation at 1 keV. Wind models should take this dependence into account when modeling the balance between driving radiation and overionization.

References

Avni Y. & Tananbaum H. 1982, ApJ, 262, L17

Murray N., Chiang J., Grossman S. A., & Voit G. M. 1995, ApJ 451, 498

Proga D., Stone J. M., & Kallman T. R. 2000, ApJ, 543, 686

Steffen A., Strateva I., Brandt W., Alexander D., Koekemoer A., Lehmer B., Schneider D., & Vignali C. 2006, AJ, 131, 2826

Highlights of Astronomy, Volume 15
XXVIIth IAU General Assembly, August 2009
Ian F. Corbett, ed.

© International Astronomical Union 2010
doi:10.1017/S1743921310009142

Jets from Compact X-ray Sources

Nick D. Kylafis[1,2]

[1] Foundation for Research and Technology - Hellas, Institute of Electronic Structure and Laser, 711 10 Heraklion, Crete, Greece

[2] University of Crete, Physics Department, 710 03 Heraklion, Crete, Greece
email: kylafis@physics.uoc.gr

Abstract. Jets have been observed from both neutron stars and black holes in binary X-ray sources. The neutron star jets are typically 30 times weaker than the black-hole ones. Thus, the second have been studied more extensively. Contrary to common belief, jets from compact X-ray sources are not simply "fireworks" that emit radio waves. I will demonstrate that they play a central role in the observed phenomena in both neutron star and black-hole systems. In particular, for black-hole jets, a simple jet model can explain the very stringent correlations that have been found between the power-law X-ray spectrum and a) the time lag between hard and soft X-rays and b) the characteristic frequencies observed in the power spectra. Up to now, no other model has even attempted to explain these correlations. I will present the weaknesses of the model and the improvements that need to be done to it.

Keywords. X-ray sources, black holes, neutron stars, jets

1. Short summary of presentation

It is generally perceived in our community that jets from compact X-ray sources (neutron stars and black holes) play no other role in the observed phenomena than to exhibit emission of radio waves and ejection of matter that were to fall on the compact object. To explain the X-ray spectrum, most models assume that there is a hot, static, spherical (or nearly spherical) corona near the compact object. Up to now, no physical mechanism has been proposed that creates and sustains such a static corona. Worse than this, it is not questioned whether such a structure can actually exist.

On the other hand, nobody doubts that a jet is always present when the X-ray source is in the so called hard spectral state. Furthermore, nobody doubts that the observd radio spectrum implies the existence of energetic electrons in the jet. The question then naturally arises: Why should one invoke a corona to explain with comptonization the hard X-ray spectrum, when the jet itself can do the job?

In a series of papers (Reig *et al.* (2003); Giannios *et al.* (2004); Giannios (2005); Kylafis *et al.* (2008)), we demostrated that a simple jet model can explain:

a) The entire spectrum from radio to hard X-rays.

b) The time lags between hard and soft X-rays and their correlation with Fourier frequency.

c) The narrowing of the autocorrelation function with photon energy or equivalently the increase of the rms amplitude of variability with photon energy.

d) The correlation of the hard X-ray power law index with the time lags.

e) The correlation of the hard X-ray power law index with the characteristic Fourier frequencies in the power spectrum.

Furthermore, we predicted and confirmed that the radio emission should correlate positively with the hard X-ray power law index.

Our model does not reproduce the reflected spectrum from the accretion disk. This is because we assumed, for simplicity, that the velocity of the jet is constant from its base to the top. This is, of course, unnatural. The ejected matter starts with low bulk velocity and it is gradually accelerated to its terminal velocity. We will improve our model and will report our results in the near future.

Acknowledgement: This research has been supported in part by European Union grants 39965 and 206469.

References

Giannios, D., Kylafis, N. D., & Psaltis D. 2004, *A&A*, 425, 163

Giannios, D. 2005, *A&A*, 437, 1007

Kylafis, N. D., Papadakis, I. E., Reig, P., Giannios, D., & Pooley, G. G. 2008, *A&A*, 489, 481

Reig, P., Kylafis, N. D., & Giannios, D. 2003, *A&A*, 403, L15

Highlights of Astronomy, Volume 15
XXVIIth IAU General Assembly, August 2009
Ian F. Corbett, ed.

© International Astronomical Union 2010
doi:10.1017/S1743921310009154

JD8 - Hot Interstellar Matter in Elliptical Galaxies

Dong-Woo Kim[1] and Silvia Pellegrini[2]

[1]Smithsonian Astrophysical Observatory, 60 Garden Street, Cambridge, MA 02138, USA
email: kim@cfa.harvard.edu
[2]Department of Astronomy, University of Bologna, via Ranzani 1, 40127 Bologna, Italy
email: silvia.pellegrini@unibo.it

Preface

The physical properties of the hot interstellar matter in elliptical galaxies are directly related with the formation and evolution of elliptical galaxies via star formation episodes, environmental effects such as stripping, infall, and mergers, and growth of super-massive black holes. The recent successful Chandra and XMM-Newton X-ray space missions have provided a large amount of high spatial/spectral resolution observational data on the hot ISM in elliptical galaxies. At the same time, theoretical studies with numerical simulations and analytical modeling of the dynamical and chemical evolution of elliptical galaxies have made a significant progress and start to predict various observable quantities.

As an example of rich sub-structures in the hot ISM, Chandra X-ray (blue & white) and optical (grey & white) images of a few elliptical galaxies are shown below (taken from http://chandra.harvard.edu/photo/2006/galaxies; Credit X-ray: NASA/CXC/U. Ohio/T. Statler & S. Diehl; Optical: DSS). In contrast to the stellar optical light which is smoothly distributed as known for elliptical galaxies, the shapes of the massive clouds of X-ray emitting gas reveal complex structures, indicating that a powerful source of energy must be pushing the hot gas around and stirring it up.

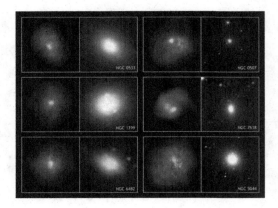

This Joint Discussion between "Galaxies" and "ISM" divisions/commissions was organized during the IAU General Assembly to bring together both observers and theorists in the field. Throughout this JD, we have discussed recent results on the hot interstellar matter in elliptical galaxies to identify important, but unsolved problems for further investigations with special emphasis on the spectral and spatial properties of the hot ISM and the comparison with the state-of-the-art theoretical models. During the one and a half day session, 23 speakers, including 4 reviewers, presented their recent achievements.

We thank all the speakers for their excellent presentations, 3 session chairs for timely managing talks and Q/A, the SOC members for their help in various organizational issues and the local organizers for their supports throughout the preparation.

Dong-Woo Kim and Silvia Pellegrini, SOC co-chairs,
Cambridge and Bologna, November 30, 2009
Scientific Organizing Committee

Françoise Combes (France),
Sofia A. Cora (Argentina),
Giuseppina Fabbiano (USA),
Alexis Finoguenov (Germany),
Brad K. Gibson (UK),
Nimisha G. Kantharia (India),
Dong-Woo Kim (USA, co-Chair),
Chiaki Kobayashi (Japan),
Cludia L. Mendes de Oliveira (Brasil),
Silvia Pellegrini (Italy, co-Chair),
Elaine M. Sadler (Australia),
Craig L. Sarazin (USA),
Thomas S. Statler (USA),
Ginevra Trinchieri (Italy)

Highlights of Astronomy, Volume 15
XXVIIth IAU General Assembly, August 2009   © International Astronomical Union 2010
Ian F. Corbett, ed.   doi:10.1017/S1743921310009166

The hot ISM of early-type galaxies

G. Fabbiano

Harvard-Smithsonian Center for Astrophysics, 60 Garden St., Cambridge MA 02138, USA
email: gfabbiano@cfa.harvard.edu

Abstract. This talks reviews the history of the discovery of the hot ISM in elliptical galaxies, and the ensuing debate on the suitability of X-ray observations of these galaxies for mass measurements. How much of the X-ray emission is truly from a hot ISM, and is this ISM in hydrostatic equilibrium? While the debate went on, a deeper understanding on the evolution of the halos was generated. High resolution *Chandra* observations are providing an answer.

Hot extended halos trapped in cluster potentials were discovered early in the history of X-ray astronomy (Kellogg & Murray 1974). It was soon realized that these halos provided a means for measuring the mass of the associated self-gravitating body, leading to huge amounts of dark matter (e.g. in the central Virgo Cluster galaxy M87, Mathews 1978; later confirmed with the first X-ray imaging telescope, the *Einstein X-ray Observatory*, Fabricant, Lecar & Gorenstein 1980). With *Einstein*, hot halos were also discovered in several early-type galaxies (Forman *et al.* 1979), leading to the realization that if the X-ray emission of E and S0 galaxies were dominated by hot halos in hydrostatic equilibrium, these observations could provide a ready means for measuring their masses (Forman, Jones & Tucker 1985). However, this assumption took a considerable leap of faith, because with the quality of these first X-ray images one could not disentangle a truly diffuse gaseous emission from the unresolved contribution of populations of low-mass X-ray binaries (LMXBs). Moreover, even if the emission was largely thermal, it could not be proved from the data that the hot halo was indeed in equilibrium (e.g., Trinchieri & Fabbiano 1985, hereafter TF85; see review Fabbiano 1989, hereafter·F89). This debate is in part still ongoing and has contributed to a deeper understanding of the evolution of early-type galaxies, and their stellar and gaseous components.

The X-ray (L_X) and B-band luminosity (L_B) are correlated, with a large scatter (the $L_X - L_B$ diagram; TF85; Forman *et al.* 1985). The interpretation of this diagram has been central to the 'halo' debate (see other talks in this meeting). Is the $L_X - L_B$ diagram the expression of halo evolution and physics, or is it 'biased' by the contribution of unresolved LMXB populations? TF85 first raised the LMXB problem, based on a comparison with the bulge of M31 and the integrated emission properties of bulge-dominated galaxies, which have L_X/L_B ratios consistent with those of 'X-ray faint' ellipticals. This hypothesis was confirmed by the spectral characteristics of the X-ray emission: harder X-ray emission was found in X-ray faint galaxies with *Einstein* and *ROSAT* (e.g., Kim, Fabbiano & Trinchieri 1992), and the CCD spectra of ASCA found the signature of a hard LMXB emission also in halo-dominated galaxies (Matsushita *et al.* 1994). With the sub-arcsecond telescope of the *Chandra X-ray Observatory*, populations of LMXBs are now obvious in the images of elliptical galaxies, and in some cases account for the bulk of the X-ray emission (e.g. NGC 3379; Brassington *et al.* 2008).

There is sufficient material from stellar outgassing to account for the hot halos. These halos may be further heated by SNIa and by gravity (if they slowly accrete to the center via cooling flows; see F89). AGN feedback may also be an important energy source (e.g. Tabor & Binney 1993). Given all this ready energy input, can we be sure that the halos are in hydrostatic equilibrium? Or are they in an outflow or wind state? Indeed, hydrodynamical modeling of the $L_X - L_B$ diagram suggested that winds, partial outflows and cooling flows could naturally explain the placement of galaxies in this diagram (Ciotti et al. 1991). Now, with *Chandra* we have set stringent limits on the amount of diffuse emission in some X-ray faint galaxies, demonstrating the presence of winds. In NGC 3379, for example, we find residual evidence of non-stellar diffuse emission in the 0.7-1.5 keV band, which is well reproduced by independent hydrodynamical modeling of the hot halo (Trinchieri, Pellegrini et al. 2008).

Optical indicators of large galaxy potentials and primordial merging are correlated with the presence of large X-ray halos (e.g., Eskridge, Fabbiano & Kim 1995; see Kormendy et al. 2009), suggesting that these halos are gravity-held. These hot halos are likely to be experiencing cooling flows, because the denser central regions would cool faster than the outer shells. There is an entire literature on cooling flows (see F89), which I will not explore here, except to say that there are two key observational diagnostics: (1) central colder gas; (2) central star formation. Neither have been detected to the amount expected. What stops the cooling flows? Radio-emitting AGN are frequently found in elliptical galaxies, and a first inkling of the interplay between these nuclear sources and the hot halos was suggested by the *Einstein* results (Fabbiano, Gioia & Trinchieri 1989). *Chandra* observations convincingly prove that AGN feedback is at play (e.g., Finoguenov et al. 2008). The evolution of the hot halos is the result of a tug of war between the pull of gravity and the push of feedback from stellar evolution (SNIa) and AGNs.

The above discussion demonstrates that caution must be used when approaching X-ray based mass measurements. In the case of 'dominant' extended hot halos hydrostatic equilibrium is probably a good approximation. However, for less X-ray luminous galaxies the halo may be far from equilibrium, and careful analysis of high resolution X-ray data (*Chandra*) and hydrodynamical modeling is required to understand the physical state of the halo.

GF would like to acknowledge support by NASA Contract NAS8-39073 (CXC).

References

Brassington, N. et al. 2008, *ApJS* 179, 142
Ciotti, L. et al. 1991, *ApJ* 376, 380
Eskridge, P. B., Fabbiano, G., & Kim, D.-W. 1995, *ApJ* 442, 523
Fabbiano, G. 1989, *ARAA* 27, 87 (F89)
Fabbiano, G., Gioia, I. M., & Trinchieri, G. 1989, *ApJ* 347, 127
Fabricant, D., Lecar, M., & Gorenstein, P. 1980, *ApJ* 241, 552
Finoguenov, A. et al. 2008, *ApJ* 686, 911
Forman, W. et al. 1979, *ApJ* 234, L27
Forman, W., Jones, C., & Tucker, W. 1985, *ApJ* 293, 102
Kellogg, E. & Murray, S. 1974, *ApJ* 193, L57
Kim, D.-W., Fabbiano, G., & Trinchieri, G. 1992, *ApJ* 393, 134
Kormendy, J. et al. 2009, *ApJS* 182, 216
Mathews, W. G. 1978, *ApJ* 219, 413
Matsushita, K. et al. 1994, *ApJ* 436, L41
Tabor, G. & Binney, J. 1993, *MNRAS* 263, 323
Trinchieri, G. & Fabbiano, G. 1985, *ApJ* 296, 447 (TF85)
Trinchieri, G., Pellegrini, S. et al. 2008, *ApJ* 688, 1000

Highlights of Astronomy, Volume 15
XXVIIth IAU General Assembly, August 2009
Ian F. Corbett, ed.

© International Astronomical Union 2010
doi:10.1017/S1743921310009178

XMM-Newton observations of elliptical galaxies in the local universe

Ginevra Trinchieri

INAF-Osservatorio Astronomico di Brera, Via Brera 28, Milano, Italy
email: ginevra.trinchieri@brera.inaf.it

XMM-Newton is well suited to the study of the X-ray properties of early-type galaxies: the wide energy band allows a characterization of the different components of the X-ray emission in galaxies, separating the gas from the compact source component through their spectral characteristics, and identifying low-luminosity absorbed AGNs; the large field of view allows a proper understanding of the large scale emission, and the separation between the galaxy and the surrounding group. Nonetheless, in spite of the much improved understanding of the X-ray characteristics of this class of sources, much of the original questions on the global X-ray properties of early-type galaxies remain. One in particular: how can we predict how much gas is there in any given galaxy? We have learned that the individual sources are tightly linked to the stellar component, both field stars and relative frequency of globular clusters. We have also learned that the central group galaxies, brighter and more extended, might represent a specific class of early-type galaxies, rather than the population as a whole. Yet we have not learned how to predict, from the stellar properties, how much hot gas a galaxy will have. Even a well selected class of sources, namely early type galaxies in isolation, where we can exclude the influence of the environment, appear to retain different amounts of the hot ISM produced by the stellar population, and display a wide range of L_x for their gaseous component for a relative narrow range of L_b, or mass [measured through L_K], as shown by Fig. 1.

Keywords. galaxies: halos; elliptical and lenticular; intergalactic medium

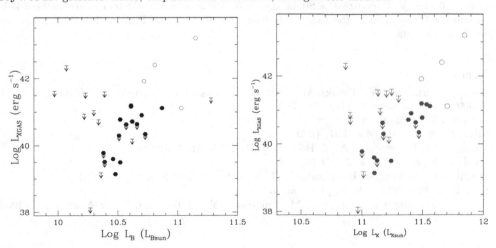

Figure 1. Comparison between the hot gas X-ray luminosity and B-band (left) or K-band (right) luminosity for isolated galaxies. "Fossil groups", namely isolated galaxies that have been found associated to a large, group-size halos, are identified by open circles. Arrows identify upperlimits to the total emission, or gas emission when associated with a symbol.

Highlights of Astronomy, Volume 15
XXVIIth IAU General Assembly, August 2009
Ian F. Corbett, ed.

Suzaku observations of early-type galaxies

K. Matsushita[1], Y. Fukazawa[2], K. Hayashi[2], S. Konami[1], R. Nagino[1], T. Ohashi[3], Y. Tawara[4] and M. Tozuka[1]

[1] Department of Physics, Tokyo University of Science, 1-3 Kagurazaka, Shinjyuku-ku, Tokyo 162-8601, Japan, email: matusita@rs.kagu.tus.ac.jp

[2] Department of Physical Science, Hiroshima University, 1-3-1 Kagamiyama, Higashi-Hiroshima, Hiroshima 739-8526, Japan

[3] Department of Physics, Tokyo Metropolitan University, 1-1 Minami-Osawa, Hachioji, Tokyo 192-0397, Japan

[4] Department of Physics, Nagoya University, Nagoya, Japan

The metal abundances in the hot X-ray emitting interstellar medium (ISM) of early-type galaxies give us important information about the present metal supply into the ISM through supernovae (SNe) Ia and stellar mass loss. In addition, O and Mg abundances should reflect the stellar metallicity and enable us to directly look into the formation history of these galaxies. The XIS instrument onboard the Suzaku satellite has an improved line spread function due to a very small low-pulse-height tail below 1 keV coupled with a very low background.

We derived abundance pattern of O, Ne, Mg and Fe of ISM of four elliptical galaxies, NGC 720 (Tawara *et al.* 2007), NGC 1399 (Matsushita *et al.* 2007a), NGC 1404 (Matsushita *et al.* 2007a), NGC 4636 (Hayashi *et al.* 2009), and two S0 galaxies, NGC 1316 (Konami *et al.* 2009), and NGC 4382 (Nagino *et al.* 2009) observed with Suzaku, and compared with those of two cD galaxies, M 87 (Matsushita *et al.* 2003), and NGC 4696 (Matsushita *et al.* 2007b) observed with XMM.

The Fe abundances of the ISM of these galaxies are about 0.5–1 solar, indicating that the present SN Ia rate is low. The O, Ne and Mg abundances are consistent with stellar metallicity. The abundance patterns of the ellipticals and cDs, and NGC 1316 are not so different from the solar pattern, using new solar abundance by Loddars (2003), and consistent with a mixture of SNe Ia and metal-poor Galactic stars. These galaxies are giant galaxies and their ISM temperatures are 0.6~1 keV. In contrast, a S0 galaxy, NGC 4382, with an ISM temperature of 0.3 keV, has a smaller O/Fe ratio in the ISM. This result means that the ISM in this galaxy contains more SNIa products.

References

Hayashi, K., Fukazawa, Y., Tozuka, M., Nishino, S., Matsushita, K., Takei, Y., & Arnaud, K. A. 2009, *PASJ* in press, arXiv:0907.2283

Konami, S., in preparation

Lodders, K. *et al.* 2003, *ApJ* 591, 1220

Matsushita, K., Finoguenov, A., & Böhringer, H. 2003, *A&A* 401, 443

Matsushita, K. *et al.* 2007a, *PASJ* 59, 327

Matsushita, K., Böhringer, H., Takahashi, I., & Ikebe, Y. 2007b, *A&A* 462, 953

Nagino, R. & Matsushita, K. in preparation

Tawara, Y., Matsumoto, C., Tozuka, M., Fukazawa, Y., Matsushita, K., & Anabuki, N. 2008, *PASJ* 60, 307

Highlights of Astronomy, Volume 15
XXVIIth IAU General Assembly, August 2009
Ian F. Corbett, ed.

© International Astronomical Union 2010
doi:10.1017/S1743921310009191

Hot gas morpoloy, thermal structure, and the AGN connection in elliptical galaxies

Thomas S. Statler[1] and Steven Diehl[1,2]

[1]Astrophysical Institute, Ohio University, Athens, OH 45701, USA
email: statler@ohio.edu

[2]Los Alamos National Laboratory, Los Alamos, NM 87545, USA

Abstract. Recent studies of the hot interstellar medium in normal elliptical galaxies have shown that (1) the gas is only approximately hydrostatic; (2) morphological disturbances are corrleated with radio and X-ray signatures of AGN; and (3) temperature gradients in the main bodies of the galaxies are correlated with nuclear activity but not with environment. An X-ray Gas Fundamental Plane (XGFP), unrelated to the stellar fundamental plane, links the global gas properties in a relation whose origin is not yet understood.

Keywords. cooling flows; galaxies: elliptical and lenticular, cD; galaxies: ISM—X-rays: galaxies

1. Introduction

Current questions regarding the X-ray gas in elliptical galaxies concern modes of heating, coupling to central AGN, and even whether the gas is hydrostatic. This last is nothing new; even in the *Einstein* era Trinchieri *et al.* (1986) noticed the asymmetry of NGC 4472, and argued that the gas is not hydrostatic at large radii.

The early years of *Chandra* saw several in-depth studies of single, interesting, usually luminous, objects that quickly became "poster children" for AGN-driven feedback. Jones *et al.* (2002) argued that features in NGC 4636 are shocks created by intermittently-fueled AGN. Finoguenov & Jones (2001) observed a striking H-shaped structure in NGC 4374, correlated with, and likely created by, expanding radio lobes.

In one of the few early studies of an X-ray faint system, Sarazin *et al.* (2001) demonstrated that low-mass X-ray binaries (LMXBs) contribute much of the total emission from NGC 4697. After removing the resolved sources, a diffuse component remained, apparently flatter than the starlight; however these authors emphasized that the gas morphology was uncertain because of the unknown contribution from unresolved LMXBs.

2. Hot ISM Properties of Normal Ellipticals

Diehl & Statler (2007) (hereafter DS1) demonstrate that the emission from gas *alone* can be isolated by taking a suitably chosen scaled difference of hard-band and soft-band images. DS1 produce gas-only images for 54 normal ellipticals observed by *Chandra*, and obtain isophotal ellipticity and position angle profiles for each system as well as an "asymmetry index" measuring the deviation from a smooth elliptical model.

The angular resolution of *Chandra* allows measurement of gas ellipticities inside the optical effective radius (R_e), where elliptical galaxies are stellar-mass dominated, offering a chance to test the assumption of hydrostatic equilibrium. DS1 find (their Fig. 6) that, contrary to this assumption, there is no correlation between optical and X-ray ellipticities. After extensive modeling, they conclude that the gas may be hydrostatic enough for radial mass profiles to be extracted, but it is not, in general, sufficiently hydrostatic for the *shape* of the gravitating mass to be inferred—asattempted by, e.g., Buote & Canizares (1994) and Buote *et al.* (2002).

The origin of morphological disturbances in the gas is addressed by Diehl & Statler (2008a). They find that the asymmetry index η is strongly correlated with the NVSS 20 cm luminosity and the X-ray luminosity of the central point source—signatures of AGN—and with the gas temperature gradient beyond $2R_e$—an indicator of environment (their Fig. 11). Since η measures small-scale asymmetries as well as large-scale lopsidedness, DS2 conclude that gas morphology is influenced in comparable measure by ram pressure in cluster environments and by AGN.

Unlike in galaxy clusters, gas temperatures in normal ellipticals do not always increase outward. Fukazawa et al. (2006) argue that positive radial temperature gradients indicate a transition to a hotter intragroup or intracluster medium. Humphrey et al. (2006) suggest that the gradient distribution is actually bimodal. Diehl & Statler (2008b) show that there is no evidence for bimodality, and that the gradients inside $2R_e$ are not affected by environment (as measured by the number density of neighbor galaxies), whereas the gradients outside $2R_e$ are (their Fig. 5). They suggest that the change from positive to negative gradients reflects a change in the role or mode of AGN heating.

3. X-Ray Gas Fundamental Plane

The global properties of the stellar distribution in elliptical galaxies follow the well known Fundamental Plane: a 2-dimensional locus in the space of effective radius, surface brightness, and velocity dispersion that is primarily a consequence of the virial theorem. Diehl & Statler (2005) find that, if one takes the analogous properties for the X-ray gas alone—half-light radius, X-ray surface brightness, and temperature, elliptical galaxies again delineate a plane, albeit a totally different one. The X-Ray Gas Fundamental Plane (XGFP) is not a consequence of the virial theorem, but is as tight as the stellar Fundamental Plane. Its origin has yet to be explained.

Acknowledgements

Support for this work was provided by the National Aeronautics and Space Administration (NASA) through Chandra Awards G01-2094X and AR3-4011X, and by National Science Foundation grants AST0407152 and AST0708284.

References

Buote, D. A. & Canizares, C. R. 1994, ApJ 427, 86
Buote, D. A., Jeltema, T. E., Canizares, C. R., & Garmire, G. P. 2002, ApJ 577, 183
Diehl, S. & Statler, T. S. 2005, ApJ 633, L21
Diehl, S. & Statler, T. S. 2007, ApJ 668, 150
Diehl, S. & Statler, T. S. 2008a, ApJ 680, 897
Diehl, S. & Statler, T. S. 2008b, ApJ 687, 986
Finoguenov, A. & Jones, C. 2001, ApJ 547, L107
Fukazawa, Y., Botoya-Nonesa, J. G., Pu, J., Ohto, A., & Kawano, N. 2006, ApJ 636, 698
Humphrey, P. J., Buote, D. A., Gastaldello, F., Zappacosta, L., Bullock, J. S., Brighenti, F., & Mathews, W. 2006, ApJ 646, 899
Jones, C., Forman, W., Vikhlinin, A., Markevitch, M., David, L., Warmflash, A., Murray S., & Nulsen, P. E. J. 2002, ApJ 567, L115
Sarazin, C. L., Irwin, J. A., & Bregman, J. N. 2001, ApJ 556, 533
Trinchieri, G., Fabbiano, G., & Canizares, C. D. 1986, ApJ 310, 637

Highlights of Astronomy, Volume 15
XXVIIth IAU General Assembly, August 2009
Ian F. Corbett, ed.

© International Astronomical Union 2010
doi:10.1017/S1743921310009208

Hot gas flows on global and nuclear galactic scales

Silvia Pellegrini[1]

[1]Department of Astronomy, University of Bologna, via Ranzani 1, 40127 Bologna, Italy
email: silvia.pellegrini@unibo.it

Abstract. One of the most significant observational improvements allowed by the high quality *Chandra* data of galaxies is the measurement of the nuclear luminosities down to low values, and of the hot ISM properties down to very low gas contents. I present here some recent developements concerning the possibility of accreting and outflowing gas, based on modeling results that take into account the role of a central supermassive black hole (MBH).

Keywords. galaxies: elliptical and lenticular, cD – galaxies: evolution – galaxies: ISM – X-rays: galaxies – X-rays: ISM

Recently, the interaction of the energy output from a central MBH with the interstellar medium (ISM) has been studied with high resolution 1D hydrodynamical simulations, including a detailed treatment of the radiative energy output and its transfer within the ISM, and of the mechanical energy output from AGN winds and jets (Ciotti, Ostriker & Proga 2009). We briefly report here the observational properties of this class of models in the X-ray band at the present epoch from a preliminary investigation (see also Pellegrini, Ciotti & Ostriker 2009).

For a bright ($L_B = 5 \times 10^{10} L_{B,\odot}$) isolated galaxy and standard assumptions concerning the stellar mass loss and SNIa rate, and the stellar and dark mass profiles, the radiative and AGN winds feedback lead to a hot ISM luminosity on the lower end of the large range observed. An external medium, as that of the outer regions of the Virgo cluster, can increase this luminosity by a factor of a few. It also causes (brief) nuclear outbursts to repeat until present epoch. The gas is inflowing within ~ 100 pc, producing a mass accretion rate on the MBH within the low radiative efficiency ADAF regime ($\dot{M}/\dot{M}_{Edd} \sim 10^{-4}$); the nuclear luminosity though ($L_{bol,nuc} \sim$ few$\times 10^{41}$ erg s^{-1} at the present epoch) is quite higher than typically observed (Pellegrini 2005a,b). Mechanical feedback from a jet can lower $L_{bol,nuc}$ to values common for early type galaxies of the local universe ($L_{bol,nuc}/L_{Edd} \sim 10^{-7} - 10^{-8}$). It also makes the nuclear outbursts less strong and rarer. The brightness profile of the hot gas can have a variety of shapes, depending on the recent flow history, the external medium and the jet heating. The temperature profile has a negative gradient, the external medium and the jet both make this profile flatter.

References

Ciotti, L., Ostriker, J. P., & Proga, D. 2009, *ApJ* 699, 89
Pellegrini, S. 2005a, *MNRAS* 364, 169
Pellegrini, S. 2005b, *ApJ* 624, 155
Pellegrini, S., Ciotti, L., & Ostriker, J. P. 2009, *Adv. Sp. Res.* 44, 340

Highlights of Astronomy, Volume 15
XXVIIth IAU General Assembly, August 2009
Ian F. Corbett, ed.

© International Astronomical Union 2010
doi:10.1017/S174392131000921X

Hyperfine structure radio lines from hot ISM in elliptical galaxies

Dmitrijs Docenko[1] and Rashid A. Sunyaev[2,3]

[1]Institute of Astronomy, University of Latvia, Rainis blvd. 19, Riga, LV-1586, Latvia
email: dima@latnet.lv

[2]Max Planck Institute for Astrophysics, Karl-Schwarzschild-Str. 1, Postfach 1317, 85741 Garching, Germany

[3]Space Research Institute, Russian Academy of Sciences, Profsoyuznaya 84/32, 117997 Moscow, Russia

Abstract. Hyperfine structure (HFS) line of ^{14}N VII ion with rest frequency of $\nu = 53.04$ GHz should be detectable from the interstellar medium in some of the densest and coolest cores of elliptical galaxies at redshifts exceeding 0.15 or so.

Keywords. galaxies: elliptical, galaxies: ISM, radio lines: ISM

Hyperfine structure (HFS) lines of highly-charged ions may open a new window in observations of hot plasmas, as first discussed by Sunyaev & Churazov (1984). Some of the relevant isotopes and ions are abundant at temperatures around $10^5 - 10^7$ K, characteristic of the hot interstellar medium (ISM) in elliptical galaxies, as well as many other types of astrophysical objects. Observations of these lines might complement soft X-ray observations with micro-calorimeters, but, in contrast to soft X-rays, they are not attenuated by the Galactic ISM and Earth atmosphere (except for the ^{14}N VII line), and allow to study observed target bulk and turbulent motions with much higher spectroscopic and angular resolution, provided that the radio telescope has sufficient sensitivity.

We estimate feasibility of HFS emission and absorption line observations from this astrophysical source type using simple theoretical estimates of spectral line absorption cross-section and emissivity (see Sunyaev & Docenko (2007), Docenko & Sunyaev (2007a) and Docenko & Sunyaev (2007b) for more details).

These estimates show that the most promising HFS line to be detected on modern instruments is one of the ^{14}N VII (rest wavelength $\lambda = 5.652$ mm) at redshifts $z > 0.15$. Using planned radio telescopes and interferometers it appears possible to observe also HFS lines of several other Mg, Si and Fe ions, including ^{57}Fe XXII line at 1.05 cm.

Acknowledgements

D.D. gratefully acknowledges the IAU travel grant.

References

Docenko, D. & Sunyaev, R. A. 2007a, in: H. Böhringer, G. W. Pratt, A. Finoguenov, & P. Schuecker (eds.), *Heating versus Cooling in Galaxies and Clusters of Galaxies*, ESO Astrophysics Symposia (Berlin, Heidelberg: Springer), p. 333

Docenko, D. & Sunyaev, R. A. 2007b, in: *From Planets to Dark Energy: the Modern Radio Universe*, Published online at SISSA, *Proceedings of Science*, p.90

Sunyaev, R. A. & Churazov, E. M. 1984, *Soviet Astron. Lett.* 10, 201

Sunyaev, R. A. & Docenko, D. O. 2007, *Astron. Lett.* 33, 67

Highlights of Astronomy, Volume 15
XXVIIth IAU General Assembly, August 2009
Ian F. Corbett, ed.

© International Astronomical Union 2010
doi:10.1017/S1743921310009221

Cool gas in brightest cluster galaxies

J. B. R. Oonk[1], W. Jaffe[1], M. N. Bremer[2] and N. Hatch[1]

[1]Leiden Observatory, Leiden University, Leiden, The Netherlands
email: oonk@strw.leidenuniv.nl

[2]H.H. Wills Physics Laboratory, University of Bristol, Bristol, United Kingdom

Abstract. Gas in galaxy clusters requires re-heating. We study the re-heating of the cool gas phases. Ionized and molecular gas is traced out to 20 kpc and found to be strongly coupled. The observed line emission may in part be explained by excitation due to hot, young stars.

Keywords. galaxies: galaxies: cooling flows, cD, clusters: general

---

Relaxed clusters of galaxies contain large quantities of gas at a variety of temperatures within their cores. To avoid catastrophic cooling, this gas needs to be re-heated. The details of this re-heating process are currently not understood.

To study the distribution and condition of the cool gas, HII and H_2, we performed deep K-band IFU observations of two clusters Abell 2597 and Sersic 159-03 with SINFONI on the VLT. These observations enable us, for the first time, to map this gas in the cores of these clusters. The distribution of the gas is filamentary and correlates well with the X-ray emission. Ionized and molecular gas is found to co-exist in both intensity and dynamics and is detected out to 20 kpc from the nucleus of the brightest cluster galaxy.

The gas near the nucleus partakes in rotation and shows a sharp increase in its velocity dispersion along the current radio axis, indicating that the gas here is stirred up by AGN outflows. Deep 5 GHz radio observations furthermore show that non-thermal plasma is spread throughout the cluster core on short timescales, $t \lesssim 10^7$ yr.

The H_2 lines are everywhere well fit by a single temperature LTE model, implying that this gas is warm, $T \sim 2300$ K. The total warm H_2 mass is $\sim 10^5$ $M_\odot$ and the total HII mass is $\sim 10^7$ $M_\odot$. The ratio of the Pa α to H_2 lines indicate a source of UV excitation rich in EUV to FUV photons (Jaffe *et al.* 2005 and Oonk *et al.* in prep.).

To study the source of excitation for the cool gas we performed deep FUV imaging with ACS-SBC on the HST and U imaging with FORS on the VLT for two clusters, Abell 2597 and Abell 2204. FUV and U continuum emission is found to exist in clumps and filaments out to 25 kpc from the nucleus of the brightest cluster galaxy. Comparing the FUV/U ratio to a black body curve we find a temperature $T_{BB} \sim 50000$ K for the young stars in cores of these clusters. Preliminary analysis suggests that the UV emission can account for the ionization seen in the $H\alpha$ emission, but detailed analysis of optical line ratios indicates the need for additional heating mechanisms (Voit *et al.* 1997).

Acknowledgements

Observations for this project were accomplished with the ESO Very Large Telescope-SINFONI/FORS and NASA/ESA Hubble Space Telescope-ACS.

References

Jaffe, W., Bremer, M. N., & Baker, K. 2005, *MNRAS* 360, 748
Voit, G. M. & Donahue, M. 1997, *ApJ* 486, 242

Highlights of Astronomy, Volume 15
XXVIIth IAU General Assembly, August 2009
Ian F. Corbett, ed.

© International Astronomical Union 2010
doi:10.1017/S1743921310009233

A multi-instrument comparison of the hot ISM in elliptical galaxies

Joel N. Bregman

Department of Astronomy, University of Michigan,
Ann Arbor, Michigan, USA
email: `jbregman@umich.edu`

Abstract. Consistent metallicities are now obtained in X-ray bright galaxies, using the Chandra ACIS and XMM PN, MOS and RGS detectors. With two temperature models, the Fe metallicity of the gas is typically solar, similar to Mg, Si, and S, but the O abundance is about half solar. These values are in conflict with models, which predict a metallicity 3-5 times higher and a Fe to O ratio near unity. This suggests that a significant fraction of metals are not becoming mixed into the hot galactic atmosphere.

Keywords. X-rays: galaxies, galaxies: ISM

We appear to be reaching an observational consensus for the metallicity of the hot gas. The metals enter the hot gas through mass loss from AGB stars plus the metals from Type Ia supernovae, which should lead to a hot ambient medium with a metallicity of 3-5 times the solar value of Fe. Supernovae should heat the gas to temperatures above the velocity dispersion temperature, an observational feature that has been known for some time. However, measuring the metallicity has been problematic and has depended upon whether one adopts one or two temperature components for the hot plasma. We used four instruments to determine the temperature for the same X-ray bright galaxies: the Chandra ACIS-S; the XMM-Newton Epic (PN and MOS); and the XMM-Newton Reflection Grating Spectrometer (RGS).

As shown in Ji *et al.* (2009, *ApJ*, 696, 2252), there is good consistency in the fits for NGC 4649 from all instruments. The abundances for Mg, Si, S, and Fe are about the same, with a value near 1.4 times the solar value, using the metallicity calibration of Grevesse and Sauval (1998). The O and Ne values are about 0.7 solar, about half that of the heavier elements. There is a discrepancy between the Chandra ACIS-S and XMM-Newton values for O, with the Chandra abundance being lower. For nine bright galaxies, we fit two-temperature models and find that he median abundances, relative to solar are 0.86 (Fe), 0.79 (Si), 0.81 (Mg), and 0.44 (O). No galaxy has a metallicity exceeding about 1.5 solar in Fe. The 2:1 ratio of Fe to O, seen in all objects with good data, has been reported previously. This distribution of metal enrichment is consistent with 65-85% of the metal enrichment coming from Type Ia SNe.

The results are in conflict with basic expectations. The metallicity is low by at least a factor of three, and this problem worsens for the lower L_X galaxies. This conflict might be removed if somehow supernovae ejecta did not mix effectively. In this case, one would expect the abundance ratios to be like that of the stars, but the O/Fe and O/Mg ratios are half that of the stars. Furthermore, there is no correlation between the stellar Fe abundance and the hot gas Fe abundance, which would be expected if the hot gas metallicity were dominated by the AGB mass loss phase. Perhaps some of this can be resolved with models that restrict the mixing of metals, and there are some presentations at this meeting that address these issues.

Highlights of Astronomy, Volume 15
XXVIIth IAU General Assembly, August 2009
Ian F. Corbett, ed.

© International Astronomical Union 2010
doi:10.1017/S1743921310009245

Abundance ratios in stars vs. hot gas in elliptical galaxies

Antonio Pipino[1]

[1] Dept. of Physics & Astronomy, University of Southern California, 90089 Los Angeles, USA
email: pipino@usc.edu

Abstract. I present predictions from a chemical evolution model for a self-consistent study of optical (i.e., stellar) and X-ray (i.e., gas) properties of present-day elliptical galaxies. Detailed cooling and heating processes in the interstellar medium are taken into account and allow a reliable modelling of the SN-driven galactic wind. The model simultaneously reproduces the mass-metallicity, colour-magnitude, L_X - L_B and L_X - T relations, and the observed trend of [Mg/Fe] with σ. The "iron discrepancy" can be solved by taking into account the dust presence.

Keywords. galaxies: elliptical and lenticular, cD - galaxies: abundances - X-rays: ISM

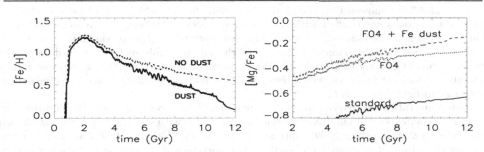

Figure 1. Predicted abundance ratios as a function of time by different models (see text).

Monolithic collapse models featuring a SN-driven galactic wind (Pipino *et al.* 2008) are shown to reproduce the largest number of observables in the optical spectrum of elliptical galaxies (e.g. Pipino *et al.* 2005, P05). Here, I made use of the P05 chemical evolution code in order to present preliminary attempts to overcome long lasting problems such as the discrepancy between the expected high Fe abundance in the post-wind phase and the observed one, as well as to explain the observed abundance ratio pattern (see Bregman, Humphrey, this Conference).

In particular, in Calura *et al.* (2008) we showed that the most recent estimates of the diffuse dust in ellipticals is enough to hide a suitable amount of Fe and reduce the gas phase abundance to the required solar value (Fig. 1, left panel). The empirical yields by François *et al.* (2004), instead, may make the predicted [Mg/Fe] closer to the observed solar value (Fig. 1, right panel). The same yields may explain why the observations hint to an under-solar [O/Mg] ratio.

I acknowledge support from the AAS and the IAU travel grants.

References

Calura, F., Pipino, A., & Matteucci, F., 2008, *A&A* 479, 669
François, P., Matteucci, F., Cayrel, R., Spite, M., Spite, F., & Chiappini, C., 2004, *A&A* 421, 613
Pipino, A., Kawata, D., Gibson, B. K., & Matteucci, F., 2005, *A&A* 434, 553
Pipino, A., D'Ercole, A., & Matteucci, F., 2008, *A&A* 484, 679

Highlights of Astronomy, Volume 15
XXVIIth IAU General Assembly, August 2009
Ian F. Corbett, ed.

© International Astronomical Union 2010
doi:10.1017/S1743921310009257

Numerical simulations of elliptical galaxies

Chiaki Kobayashi[1]

[1] The Australian National University, Mt. Stromlo Observatory, Cotter Rd., Weston ACT
2611, email: chiaki@mso.anu.edu.au

For the formation of elliptical galaxies, two scenarios, monolithic collapse vs. major merger, have been debated. We simulate the formation and chemodynamical evolution of 128 Es from the CDM initial fluctuations, using the GRAPE-SPH code that include star formation, supernovae feedback, and chemical enrichment. In our CDM-based scenario, galaxies form through the successive merging of subgalaxies with various masses.

The metallicity gradient gives one of the most stringent constraints on the galaxy formation. Kobayashi (2004) showed that gradients are destroyed by mergers and are not enough regenerated by induced star formation. In observations, there is a significant scatter (Kobayashi & Arimoto 1999), which is reproduced with our simulations (Fig. 1a). The scatter stems from the difference in the merging histories. Galaxies that form monolithically have steeper gradients, while galaxies that undergo major mergers have shallower gradients. For less-massive galaxies, Spolaor *et al.* (2009) showed that the gradients become flatter, which cannot be reproduced.

This scenario does not conflict with the observed scaling relations. The mass-metallicity relation (Fig. 1b) and the fundamental plane are also reproduced (Kobayashi 2005). An intrinsic scatter exists along the fundamental plane, and the origin of this scatter lies in differences in merging history. Galaxies that undergo major mergers tend to have larger effective radii r_e and fainter surface brightnesses, which result in larger masses, smaller surface brightnesses, and larger mass-to-light ratios M/L.

Mass-to-light ratios and baryon fractions (Fig. 1c) will be important constraints on the origin of elliptical galaxies. Our simulations are roughly consistent with the available observations with X-ray (Nagino & Matsushita 2009), optical measurements (Gerhard *et al.* 2001), and gravitational lensing (Treu *et al.* 2006). The stellar M/L correlates with the stellar mass, which can be the origin of the tilt of the fundamental plane. The total M/L does not correlate with the stellar M/L, which is contrary to the optical observations. The total M/L are almost constant as $\sim 18-28$ in $2r_e$ for $M_* \gtrsim 10^{10} M_\odot$, which is consistent with the X-ray observations, and increases upto ~ 100 for dwarf ellipticals. Thus, the baryon fraction is larger for massive galaxies.

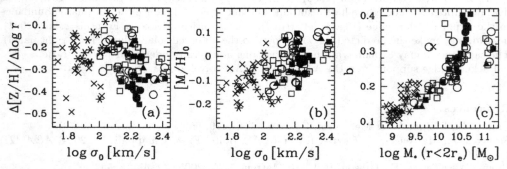

Figure 1. (a) Metallicity gradients and (b) central metallicities against central velocity dispersions for monolithic-like (filled symbols), major merger (open symbols), and dwarf ellipticals (crosses). (c) Baryon fractions against stellar mass within $2r_e$ of the simulated galaxies.

Highlights of Astronomy, Volume 15
XXVIIth IAU General Assembly, August 2009
Ian F. Corbett, ed.

© International Astronomical Union 2010
doi:10.1017/S1743921310009269

Metal abundances in the hot ISM of early-type galaxies

Philip J. Humphrey[1] and David A. Buote[1]

[1]Department of Physics & Astronomy, University of California, Irvine,
4129 Frederick Reines Hall, Irvine, CA 92697, USA
email: phumphre@uci.edu

Understanding the process of metal enrichment is one of the key problems for our picture of structure formation and evolution, in which early-type galaxies are a crucial ingredient. X-ray observations provide a powerful tool for measuring the metal distributions in their hot ISM, which is shaped by their entire history of star-formation, evolution and feedback. In Fig 1 (left panel), we summarize the results of a Chandra survey of metals in early-type galaxies, supplemented with Suzaku data (Humphrey & Buote 2006, P. Humphrey *et al.*, in prep.). Chandra is particularly suited to this study, as it enables temperature gradients and X-ray point sources to be resolved, mitigating two important sources of bias (e.g., Buote & Fabian 1998; Fabbiano *et al.* 1994). We found on average that the ISM is at least as metal-rich as the stars, and we did not find the problematical, highly sub-solar, abundances historically reported. The abundance ratios of O, Ne, Mg, Si and S with respect to Fe are similar to the centres of massive groups and clusters, suggesting homology in the enrichment process over a wide mass range. Finally, using high-quality Suzaku data, we were able to resolve, for the first time in a galaxy-scale ($\lesssim 10^{13} M_\odot$) object, a radial abundance gradient similar to those seen in some bright galaxy groups (Fig. 1, right panel).

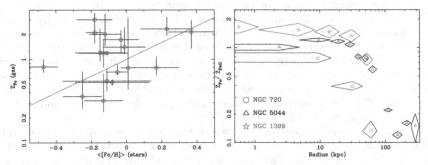

Figure 1. Left: Comparison of the ISM and stellar abundances for a sample of early-type galaxies (Humphrey & Buote 2006, P. Humphrey *et al.*, in prep.). The solid line denotes "y = x". **Right:** Radial abundance gradients for the galaxy groups NGC 1399 and NGC 5044 (Buote *et al.* 2003, 2004) and the isolated elliptical galaxy NGC 720 (P. Humphrey *et al.*, in prep.).

References

Buote, D. A. & Fabian, A. C. 1998, *MNRAS* 296, 977

Buote, D. A. *et al.* 2003, *ApJ* 595, 151

Buote, D. A. Brighenti, F., & Mathews, W. G. 2004, *ApJL* 607, L91

Fabbiano, G., Kim, D.-W., & Trinchieri, G. 1994, *ApJ* 429, 94

Humphrey, P. J. & Buote, D. A. 2006, *ApJ* 639, 136

Highlights of Astronomy, Volume 15
XXVIIth IAU General Assembly, August 2009
Ian F. Corbett, ed.

© International Astronomical Union 2010
doi:10.1017/S1743921310009270

Confronting feedback simulations with observations of hot gas in elliptical galaxies

Q. Daniel Wang[1]

[1]Department of Astronomy, University of Massachusetts,
Amherst, MA 01003, USA email: wqd@astro.umass.edu

Keywords. X-ray, elliptical galaxies, hot gas, galaxy evolution

Elliptical galaxies comprise primarily old stars, which collectively generate a long-lasting feedback via stellar mass-loss and Type Ia SNe. This feedback can be traced by X-ray-emitting hot gas in and around such galaxies, in which little cool gas is typically present. However, the X-ray-inferred mass, energy, and metal abundance of the hot gas are often found to be far less than what are expected from the feedback, particularly in so-called low L_X/L_B ellipticals. This "missing" stellar feedback is presumably lost in galaxy-wide outflows, which can play an essential role in galaxy evolution (e.g., explaining the observed color bi-modality of galaxies). We are developing a model that can be used to properly interpret the X-ray data and to extract key information about the dynamics of the feedback and its interplay with galactic environment.

First, we have constructed a 1-D model of the stellar feedback in the context of galaxy formation and evolution. The feedback is assumed to consist of two primary phases: 1) an initial burst during the bulge formation and 2) a subsequent long-lasting mass and energy injection from stellar winds and Type Ia SNe of low-mass stars. An outward blastwave is initiated by the burst and is maintained and enhanced by the long-lasting stellar feedback. This blastwave can heat the surrounding medium not only in the galactic halo, but also in regions beyond the virial radius. As a result, the smooth accretion of hot gas can be completely stopped. The long-lasting feedback can form a galactic bulge wind, which is reverse-shocked at a large radius, and can later evolve into a subsonic quasi-stable outflow as the energy injection decreases with time. The two phases of the feedback thus re-enforce each-other's impact on the gas dynamics. Present-day elliptical galaxies with significant amounts of hot gas are most likely in subsonic outflow states. The exact properties of such an outflow depend on the galaxy formation history and environment. This dependence and variance may explain the large dispersion in the L_X/L_B ratios of elliptical galaxies.

Second, to quantitatively compare the simulations with X-ray observations, we have conducted various 3-D hydrodynamical simulations with the adaptive mesh refinement code FLASH to investigate the physical properties of hot gas in and around elliptical galaxies. We have developed an embedding scheme of individual supernova remnant seeds, which allows us to examine, for the first time, the effect of sporadic SNe on the density, temperature, and iron ejecta distribution of the hot gas as well as the resultant X-ray morphology and spectrum. We find that the SNe produce a wind/outflow with highly filamentary density structures and patchy ejecta. Compared with a 1-D spherical wind model, the non-uniformity of simulated gas density, temperature, and metallicity substantially alters the spectral shape and increases the diffuse X-ray luminosity. The differential emission measure as a function of temperature of the simulated gas exhibits a log-normal distribution, with a peak value much lower than that of the corresponding 1D model. The bulk of the X-ray emission comes from the relatively low temperature and low abundance gas shells associated with SN blastwaves. SN ejecta are not well mixed with the ambient medium and typically remain very hot in the central region. Driven by the buoyancy, the iron-rich gas on average moves substantially faster than the medium and only gradually mixes with it on the way out. As a result, apparent increasing temperature and metal abundance with off-center distance can arise in the region, mimicking what have been observed in elliptical galaxies. These results, at least partly, account for the apparent lack of evidence for iron enrichment in the soft X-ray-emitting gas in galactic bulges and intermediate-mass elliptical galaxies.

Highlights of Astronomy, Volume 15
XXVIIth IAU General Assembly, August 2009   © International Astronomical Union 2010
Ian F. Corbett, ed.   doi:10.1017/S1743921310009282

Hot ISM in young elliptical galaxies

Dong-Woo Kim[1]

[1]Smithsonian Astrophysical Observatory, 60 Garden Street, Cambridge, MA 02138, USA
email: kim@cfa.harvard.edu

Abstract. Using Chandra and XMM-Newton X-ray observations of young, post-merger elliptical galaxies, we present X-ray characteristics of age-related observational results, by comparing
with typical old elliptical galaxies in terms of X-ray properties of their low-mass X-ray binaries
(LMXBs) and hot interstellar matter (ISM).

Keywords. galaxies: elliptical and lenticular, cD, X-rays: galaxies, X-rays: ISM

Stellar age in an elliptical galaxy has been measured and there are now a handful of elliptical
galaxies with intermediate stellar age (< 5 Gyr) which could link between Antennae-like merging
systems and typical old elliptical galaxies (e.g,. Thomas *et al.* 2005). We find that young elliptical
galaxies host more luminous ($> 5 \times 10^{38}$ erg s^{-1}) LMXBs which are often distributed in a
non-uniform way, indicating a possible connection with recent mergers (Kim *et al.* 2009, in
preparation). Young elliptical galaxies tend to have a small amount of hot gas, with Lx(gas)
comparable to or smaller than Lx(LMXB). By comparison, old elliptical galaxies have a wide
range of Lx(gas) and some of them (e.g., group or cluster dominant galaxies) have Lx(gas) 10
or 100 times higher than Lx(LMXB).

One of the key observables to address the age effect is the abundance ratio of Fe to α elements,
because of different production yields of SNe Ia and II. In the merger scenario of young elliptical
galaxies, we would expect an enrichment of α-elements from SN II, associated with the recent
star formation. Based on the SN yields, [Si/Fe] would be close to 0.5 if SNe Ia are dominating,
while the ratio would be close to 2.5 if SNe II are dominating. Typical old giant elliptical galaxies
have [Si/Fe] close to solar (e.g., Kim & Fabbiano 2004), indicating that the ejecta from both SNe
II (from the early star formation) and SNe Ia (continuously added later) are well mixed. With
added contribution from SNe II in young elliptical galaxies, we expect [Si/Fe] to be between 1
and 2.5 solar. By carefully selecting young and old samples (e.g, similar σ^* to avoid different
$[\alpha/Fe]^*$; weak AGN; deep Chandra observations to effectively remove LMXBs) and by rigorously
analyzing the data (extracting X-rays only from r < 30" to avoid a background problem; do
not arbitrarily tying different elements), we find that [Si/Fe] is close to 1.5-2 solar in young E's
(NGC 720 and NGC 3923), while [Si/Fe] is very close to solar in old E's (e.g, N4472, N4649).
Given the limited sample and related uncertainties, our result is consistent with the expectation.

Finally, we note that while for X-ray faint galaxies the absolute abundances of individual
elements are uncertain and controversial, the abundance ratio is better constrained because of
strong correlations between errors of different elements.

References

Kim, D.-W. & Fabbiano, G. 2004, *ApJ* 613, 933
Thomas, D. Maraston, C., Bender, R., & de Oliveira, C. M. 2005, *ApJ* 621, 673

Highlights of Astronomy, Volume 15
XXVIIth IAU General Assembly, August 2009
Ian F. Corbett, ed.

© International Astronomical Union 2010
doi:10.1017/S1743921310009294

Suzaku observation of the metallicity in the interstellar medium of NGC 1316

S. Konami[1,2], K. Matsushita, K. Sato, R. Nagino, N. Isobe, M. S. Tashiro, H. Seta, K. Matsuta, T. Tamagawa and K. Makishima

[1] Department of Physics, Tokyo University of Science, 1-3 Kagurazaka, Shinjuku-ku, Tokyo 162-8601

[2] Cosmic Radiation Laboratory, the Institute of Physical and Chemical Research, 2-1 Hirosawa, Wako, Saitama 351-0198
email: konami@crab.riken.jp

Metal abundances of the hot X-ray emitting interstellar medium (ISM) include important information to understand the history of star formation and evolution of galaxies. The metals are mainly synthesized by Type Ia (SNe Ia) and stellar mass loss in elliptical galaxies. The productions of stellar mass loss reflect stellar metallicity. SNe Ia mainly product Fe. Therefore, the abundance pattern of ISM can play key role to investigate the metal enrichment history.

S0 galaxies, which are intermediate between ellipticals and spirals, are similar to elliptical galaxies in their optical spectra. Nevertheless, S0's in the cluster environment may have been changed from spiral galaxies when falling into clusters. X-ray observations are expected to provide a clue to this issue, because we can measure the metals accumulated over the Hubble time.

We investigated metal abundances of the hot ISM in the nearby S0 galaxy, NGC1316 (Fornax A), in the Fornax group. According to Goudfrooij *et al.* (2001), several galaxies may have merged into this galaxy 3 Gyr ago. We used a 43.9 ks of archival data achieved with the Suzaku satellite, because the XIS instrument of Suzaku is particularly suitable for the spectroscopy of extended X-ray emission with a low surface brightness.

The derived abundance pattern of O, Ne, Mg, Si against Fe of the ISM is close to that of new solar abundance of Lodders (2003). Furthermore, the abundance pattern of NGC 1316 is consistent with those of elliptical galaxies measured with Suzaku. This result indicates that the total amount of present stellar mass loss and SN Ia is similar between NGC 1316 and elliptical galaxies. In contrast, the abundance pattern of the ISM in NGC 4382, a S0 galaxy with an ISM temperature of 0.3 keV, is different from that of NGC 1316 (Nagino *et al.* in prep). These difference may be attributed to morphology types or system masses. We need more sample of S0 and elliptical galaxies to investigate the difference.

References

Goudfrooij, P., Mack, J., Kissler-Patig, M., Meylan, G., & Minniti, D. 2001a, *MNRAS* 322, 643
Hayashi, K., Fukazawa, Y., Tozuka, M., Nishino, S., Matsushita, K., Takei, Y., & Arnaud, K. A. 2009, arXiv:0907.2283
Lodders, K. 2003, *ApJ* 591, 1220
Matsushita, K. *et al.* 2007, *PASJ* 59, 327
Tawara, Y., Matsumoto, C., Tozuka, M., Fukazawa, Y., Matsushita, K., & Anabuki, N. 2008, *PASJ* 60, 307

Highlights of Astronomy, Volume 15
XXVIIth IAU General Assembly, August 2009
Ian F. Corbett, ed.
© International Astronomical Union 2010
doi:10.1017/S1743921310009300

Fossil groups of galaxies: Are they groups? Are they fossils?

Renato de Alencar Dupke[1,2], Eric Miller[3], Claudia Mendes de Oliveira[4], Laerte Sodre Jr[4], Eli Rykoff[5], Raimundo Lopes de Oliveira[4] and Rob Proctor[4]

[1]University of Michigan; [2]ON/MCT; [3]MIT Kavli Institute; [4]IAG/USP; [5]UCSB
email: rdupke@umich.edu

Abstract. Fossil groups present a puzzle to current theories of structure formation. Despite the low number of bright galaxies, their high velocity dispersions and high T_X indicate cluster-like potential wells. Measured concentration parameters seem very high indicating early formation epochs in contradiction with the observed lack of large and well defined cooling cores. There are very few fossil groups with good quality X-ray data and their idiosyncrasies may enhance these apparent contradictions. The standard explanation for their formation suggests that bright galaxies within half the virial radii of these systems were wiped out by cannibalism forming the central galaxy. Since dry mergers, typically invoked to explain the formation of the central galaxies, are not expected to change the IGM energetics significantly, thus not preventing the formation of cooling cores, we investigate the scenario where recent gaseous (wet) mergers formed the central galaxy injecting energy and changing the chemistry of the IGM in fossil groups. We show a test for this scenario using fossil groups with enough X-ray flux in the *Chandra X-ray Observatory* archive by looking at individual metal abundance ratio distributions near the core. Secondary SN II powered winds would tend to erase the dominance of SN IA ejecta in the core of these systems and would help to erase previously existing cold cores. Strong SN II-powered galactic winds resulting from galaxy merging would be trapped by their deep potential wells reducing the central enhancement of SN Ia/SN II iron mass fraction ratio. The results indicate that there is a decrement in the ratio of SN Ia to SN II iron mass fraction in the central regions of the systems analyzed, varying from $99\pm1\%$ in the outer regions to $85\pm2\%$ within the cooling radius (Figure 1) and would inject enough energy into the IGM preventing central gas cooling. The results are consistent with a scenario of later formation epoch for fossil groups, as they are defined, when compared to galaxy clusters and normal groups.

Keywords. galaxies: clusters: general, X-rays: galaxies: clusters, surveys

Raimundo Lopes de Oliveira thanks FAPESP for the financial support.

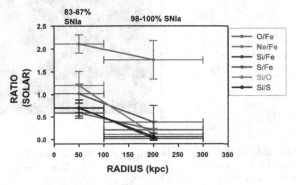

Figure 1.

Highlights of Astronomy, Volume 15
XXVIIth IAU General Assembly, August 2009
Ian F. Corbett, ed.

© International Astronomical Union 2010
doi:10.1017/S1743921310009312

A large X-ray sample of fossil groups

Eric D. Miller[1], Eli Rykoff[2], Renato de Alencar Dupke[3,4], Claudia Mendes de Oliveira[5], Timothy McKay[4] and Benjamin Koester[6]

[1]MIT Kavli Institute, [2]UCSB, [3]ON/MCT, [4]U. Michigan, [5]USP, [6]U. Chicago
email: milleric@space.mit.edu

Abstract. We present Chandra snapshot observations of the first large X-ray sample of optically identified fossil groups. For 9 of 14 candidate groups, we are able to determine the X-ray luminosity and temperature, which span a range typical of large ellipticals to rich groups of galaxies. We discuss these initial results in the context of group IGM and central galaxy ISM evolution, and we also describe plans for a deep X-ray follow-up program.

Keywords. galaxies: clusters: general, X-rays: galaxies: clusters, surveys

---

Fossil groups (FGs) are systems dominated by a single, giant elliptical galaxy, yet their X-ray emission indicates a deeper cluster-scale gravitational potential. They are thought to be old, undisturbed galaxy groups, however these systems may be younger or more active than previously thought (see Dupke *et al.* in these proceedings). These results are complicated by the small number of FGs with deep X-ray data.

To address this, we have constructed a sample of 15 FG candidates from the maxBCG cluster catalog (Koester *et al.* 2007), using the criteria $0.09 < z < 0.15$, $L_{BCG} > 9 \times 10^{11} L_\odot$, and $\Delta i > 2.0$ between the BCG and second ranked galaxy within $R_{200}/2$ (see Figure 1). We have obtained 5–10 ksec *Chandra* snapshot observations of 14 targets, and we detect diffuse X-ray emission from 11 of them at $> 90\%$ confidence, measuring T_X for 9 of these. One detection is shown in Figure 2. The measured L_X and T_X are similar to what is expected for groups of galaxies. Deep follow-up with *XMM* is necessary to measure T_X profiles, surface brightness profiles, concentration, and abundances, thereby constraining the formation mechanism of these peculiar but numerous systems.

References

Koester, B. P. 2007, *ApJ*, 660, 239

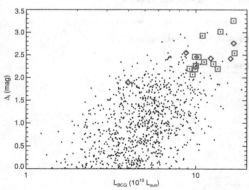

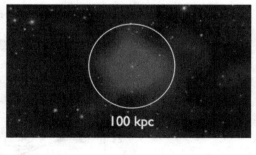

Figure 1. Magnitude differential vs. BCG luminosity for all maxBCG clusters with $9 < N_{200} < 25$; open squares identify the 15 FG candidates. Diamonds show known FGs.

Figure 2. SDSS J0856+0553, a $z = 0.09$ FG candidate. The X-ray image is plotted over the SDSS g, r, i composite image.

Highlights of Astronomy, Volume 15
XXVIIth IAU General Assembly, August 2009
Ian F. Corbett, ed.

Feedback and environmental effects in elliptical galaxies

Craig L. Sarazin[1]

[1]Department of Astronomy, University of Virginia, P.O. Box 400325, Charlottesville, VA
22904-4325, USA
email: sarazin@virginia.edu

Abstract. The role of the environment of an elliptical galaxy on its hot interstellar gas is discussed. In general, the X-ray halos of early-type galaxies tend to be smaller and fainter in denser environments, with the exception of group-central galaxies. X-ray observations show many examples of nearby galaxies which are undergoing gas stripping. On the other hand, most bright galaxies in clusters do manage to retain small coronae of X-ray emission. Recent theoretical and observational results on the role of feedback from AGN at the centers of elliptical galaxies on their interstellar gas are reviewed. X-ray observations show many examples of X-ray holes in the central regions of brightest-cluster galaxies; in many cases, the X-ray holes are filled with radio lobes. Similar radio bubbles are seen in groups and individual early-type galaxies. "Ghost bubbles" are often seen at larger radii in clusters and galaxies; these bubbles are faint in high radio frequencies, and are believed to be old radio bubbles which have risen buoyantly in the hot gas. Low frequency radio observations show that many of the ghost bubbles have radio emission; in general, these long wavelength observations show that radio sources are much larger and involve greater energies than had been previously thought. The radio bubbles can be used to estimate the total energy output of the radio jets. The total energies deposited by radio jets exceed the losses from the gas due to radiative cooling, indicating that radio sources are energetically capable of heating the cooling core gas and preventing rapid cooling.

Keywords. galaxies: clusters: general — cooling flows — galaxies: elliptical and lenticular, cD — galaxies: halos — galaxies: ISM — X-rays: galaxies — X-rays: galaxies: clusters

---

1. Environmental Effects on X-ray Emission

Given the wide range in X-ray luminosities of early-type galaxies of a given optical luminosity, the question naturally arises as to whether part of this dispersion might be due to the effects of environment on their X-ray emission. White & Sarazin (1991) suggested that elliptical galaxies in dense environments were fainter than those in sparse regions. In this and most subsequent studies, the local density was characterized by the projected galaxy density in the region around the target galaxy. Brown & Bregman (2000) argued that early-type galaxies in dense regions were more luminous; their sample included a number of group-center ellipticals. Some studies, mainly using ROSAT data, found no correlation e.g., O'Sullivan et al. 2001. More recent Chandra studies of groups and clusters (e.g, Jeltema et al. 2008; Sun et al. (2007)) seem to confirm a general anti-correlation of local density and X-ray luminosity. Observations of several galaxies in nearby clusters show evidence for ram pressure stripping; examples include M86 (Randall et al. 2008) and NGC 4552 (Machacek et al. 2006). However, despite the efficiency of stripping, most bright early-type galaxies in clusters do retain small coronae (Sun et al. (2007)).

2. AGN Feedback in Early-Type Galaxies

In recent years, evidence has been found for a coupling between supermassive black holes (SMBHs) in the centers of galaxies, and their galaxy hosts. First, the masses of the SMBHs are proportional to the bulge mass of the host, suggesting that star formation and SMBH accretion are connected. Second, the luminosity function of galaxies falls below that expected for dark matter halos at high masses in a way that can be understood if AGN suppress star formation in massive galaxies. Finally, less gas cools to low temperatures at the centers of cool core clusters, groups, and individual ellipticals than expected unless something heats the gas, and AGNs are the leading candidates. In cool core clusters, X-ray deficits ("radio bubbles") have been found at the locations of the lobes of the radio sources associated with the brightest cluster galaxies (BCGs; Fabian *et al.* 2006; Blanton *et al.* 2001); similar radio bubbles are seen in groups and individual elliptical galaxies. "Ghost bubbles"' at larger radii are also seen which lack high frequency radio emission. These are thought to be older radio bubble which have risen buoyantly through the hot gas. Recent low frequency radio observation often show that the ghost bubbles are filled with long wavelength radio emission. Recent examples include Abell 2597 (Clarke *et al.* 2005) and Abell 262 (Clarke *et al.* 2009).

Radio bubbles in clusters and ellipticals allow the determination of the total kinetic energy output of the radio jets. The energy injected by the jets must at least equal the work needed to inflate the bubbles plus the internal energy associated with their pressure plus any energy in shocks in the surrounding medium. In general, these energies greatly exceed the synchrotron radio emission from the AGN, and the energies in the radio lobes are much greater than the values expected from equipartition. These estimates indicate that radio jets in cluster BCGs and ellipticals deposit enough energy to balance radiative cooling in these systems (Dunn & Fabian 2006; Rafferty *et al.* 2006). The mechanisms by which the radio sources heat the X-ray gas are uncertain, although heating by sound waves and weak shocks is a possibility. Recent observations of Abell 2052 (Blanton *et al.* 2009) show X-ray ripples which are similar to the features seen in the Perseus cluster.

Acknowledgements

I thank Liz Blanton, Tracy Clarke, Judith Croston, and Ming Sun for comments and help with the figures in the talk. This work was supported by NASA Chandra grants GO7-8078X, GO7-8081A, GO8-9085X, GO8-9083X, NASA Herschel grant RSA1373266, and NASA HST grant HST-GO-10597.03-A.

References

Blanton, E. L., Sarazin, C. L., McNamara, B. R., & Wise, M. W. 2001, *ApJ* 558, L15
Blanton, E. L. *et al.* 2009, *ApJ* 697, L95
Brown, B. A. & Bregman, J. N. 2000, *ApJ* 539, 592
Clarke, T. E. *et al.* 2009 *ApJ* 625, 748
Clarke, T. E. *et al.* 2009 *ApJ* 697, 481
Dunn, R. J. H. & Fabian, A. C. 2006, *MNRAS* 373, 959
Fabian, A. C. *et al.* 2006, *MNRAS* 366, 417
Jeltema, T. E., Binder, B., & Mulchaey, J. S. 2008, *ApJ* 679, 1162
Machacek, M., Nulsen, P. E. J., Jones, C., & Forman, W. R. 2006 *ApJ* 648, 947
O'Sullivan, E., Forbes, D. A., & Ponman, T. J. 2001, *MNRAS* 328, 461
Randall, S. *et al.* 2008, *ApJ* 688, 208
Rafferty, D., McNamara, B. R., Nulsen, P. E. J., & Wise, M. W. 2006, *ApJ* 652, 216
Sun, M., Jones, C., Forman, W., Vikhlinin, A., Donahue, M., & Voit, M. 2007, *ApJ* 657, 197
White, R. E. III & Sarazin, C. L. 1991, *ApJ* 367, 476

Highlights of Astronomy, Volume 15
XXVIIth IAU General Assembly, August 2009
Ian F. Corbett, ed.

Scaling properties of the hot gas in early-type galaxies

Trevor J. Ponman[1] and Ewan J. O'Sullivan[1,2]

[1] School of Physics & Astronomy, University of Birmingham, Birmingham B15 2TT, UK.
[2] Harvard-Smithsonian Center for Astrophysics, 60 Garden St, Cambridge, MA02138, USA
email: tjp@star.sr.bham.ac.uk

Abstract. We examine the processes responsible for the large scatter in the X-ray/optical luminosity scaling relation of early-type galaxies.

1. The L_X-$L_{\rm opt}$ relation in early-type galaxies

Optically brighter galaxies tend to have higher X-ray luminosity. The largest sample yet presented, by Ellis & O'Sullivan (2006), is shown in Fig. 1. It is apparent that: (a) there is a factor $\sim$100 scatter in L_X around the mean relation; (b) the relation steepens at high L_B, where most galaxies lie well above the discrete source line – their X-ray emission dominated by a diffuse hot gas component; (c) many of these luminous galaxies are the bright central members of groups or clusters (BGGs or BCGs). Here we briefly examine a number of possible causes for the large scatter apparent in this relation.

2. Sources of scatter in L_X-$L_{\rm opt}$

Star formation history – Recent star formation can considerably raise the blue luminosity of a galaxy, at a given stellar mass. However, Ellis & O'Sullivan (2006) showed that replacing L_B by K-band luminosity in Fig.1, results in a very similar scatter.

X-ray data quality – The X-ray data used by Ellis & O'Sullivan (2006) came from ROSAT. However, smaller samples of higher quality datasets from Chandra and XMM (e.g. Diehl & Statler (2007), Sun (2009)) show similar scatter in the relation.

Galaxy age – It has been recognised for some years (e.g. O'Sullivan *et al.* (2001)) that spectroscopically young ellipticals have low X-ray luminosity. The compilation of Sansom *et al.* (2006) shows a decline in L_X/L_B for 1-2 Gyr after a substantial burst of star formation (e.g. following a galaxy merger), presumably due to hot gas being dispersed by the energetic galaxy winds known to be associated with starbursts. However, galaxies with larger spectroscopic ages show no age trend, and a wide scatter in L_X/L_B.

Galaxy environment – Ellis & O'Sullivan (2006) found no clear environmental trend in L_X/L_B, apart from the high values apparent in BGGs and BCGs. However, recent studies with Chandra, which is better able to resolve galaxy emission, show that L_X/L_B is suppressed by a factor $\sim$3-4 in non-central galaxies within both clusters (Sun *et al.* (2007)) and groups (Jeltema *et al.* (2008)). This presumably results from stripping of hot gas within dense environments. The properties of *isolated* early-type galaxies provide a probe of whether environmental effects can account for the bulk of the scatter in L_X/L_B. Memola *et al.* (2009) assembled L_X-L_B data for 27 isolated ellipticals, and found a good deal of scatter. However, excluding probable 'fossil groups' (Ponman *et al.* (1994)) and upper limits, the remaining sample is small. Preliminary results from a Chandra study of isolated ellipticals currently underway indicate (Humphrey private communication) that significant L_X/L_B scatter is still present, but may be limited to a factor $\sim$10.

Halo mass – Mathews *et al.* (2006) suggested that the primary factor driving L_X is the mass of the dark halo surrounding a galaxy. For a sample of ellipticals (mostly BGGs) they found that L_X/L_K increased with X-ray derived virial mass. This is consistent with the earlier result, that

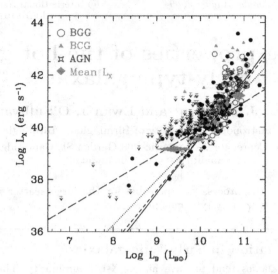

Figure 1. L_X-L_B plot for 401 early-type galaxies from Ellis & O'Sullivan (2006). The long dashed line is the discrete source contribution estimated by Kim & Fabbiano (2004), other lines show fits to various subsets of the sample, and the grey diamonds represent the mean L_X in a series of L_B bins.

BGGs have L_X/L_B ratios much higher than other elliptical galaxies. The study of NGC 4555 by O'Sullivan & Ponman (2004) shows that isolated ellipticals can also possess substantial dark matter halos.

Feedback – For a given dark halo mass, the gas content can be affected by feedback processes. Injection of energy raises the entropy, resulting in low density and X-ray luminosity. The low L_X/L_B ratio seen in ellipticals with recent star formation is probably an example of this. AGN feedback may have similar effects, if it can couple effectively to the surrounding gas. Sun (2009) finds that all radio-bright elliptical galaxies contain hot gas coronae – either compact or extensive. He postulates that AGN jets can couple to the extended hot halos found in groups or clusters, but not to compact gas cores.

References

Diehl S. & Statler T. S. 2007, *ApJ* 668, 150
Ellis S. J. & O'Sullivan E. J. 2006, *MNRAS* 367, 627
Jeltema T. E. *et al.* 2008, *ApJ* 679, 1162
Kim D.-W. & Fabbiano G. 2004, *ApJ* 611, 846
Mathews W. G. *et al.* 2006, *ApJ* 652, L17
Memola E. *et al.* 2009, *A&A* 497, 359
O'Sullivan E. J., Forbes D. A., & Ponman T. J. 2001, *MNRAS* 324, 420
O'Sullivan E. J. & Ponman T. J. 2004, *MNRAS* 354, 935
Ponman T. J. *et al.* 1994, *Nature* 369, 462
Sansom A. *et al.* 2006, *MNRAS* 370, 1541
Sun M. *et al.* 2007, *ApJ* 657, 197
Sun M. 2009, *arXiv:0904.2006*

Highlights of Astronomy, Volume 15
XXVIIth IAU General Assembly, August 2009
Ian F. Corbett, ed.

© International Astronomical Union 2010
doi:10.1017/S1743921310009348

AGN feedback in numerical simulations

Luca Ciotti

Department of Astronomy, University of Bologna, via Ranzani 1, 40127, Bologna, Italy
email: luca.ciotti@unibo.it

Abstract. The passively evolving stellar population in elliptical galaxies (Es) provides a continuous source of fuel for accretion on the central supermassive black hole (SMBH), which is 1) extended over the entire galaxy life (but declining with cosmic time), 2) linearly proportional to the stellar mass of the host spheroid, 3) summing up to a total gas mass that is > 100 times larger than the currently observed SMBH masses, 4) available independently of merging events. The main results of numerical simulations of Es with central SMBH, in which a physically based implementation of radiative and mechanical feedback effects is considered, are presented.

Keywords. X-rays: ISM - Galaxies: cooling flows - Galaxies: active

---

In a series of papers (Ciotti & Ostriker 2007; Ciotti, Ostriker & Proga 2009; Pellegrini, Ciotti & Ostriker 2009; Shin, Ciotti & Ostriker 2009; Jiang *et al.* 2009; see also Ciotti 2009) we study, with a high-resolution 1-D hydrodynamical code, the evolution of the ISM in Es under the action of SNIa heating, thermalization of the stellar mass losses, and feedback from the central SMBH. The cooling and heating functions include photoionization and Compton effects, radiation pressure is evaluated by solving the transport equation, mechanical feedback is produced by a physically based luminosity-dependent nuclear wind and jet, and star formation is also allowed. The recycled gas from the aging stars of the galaxy cools and collapses towards the center, a star-burst occurs and the central SMBH is fed. The energy output from the central SMBH pushes matter out, the accretion rate drops precipitously and the expanding matter drives shocks into the ISM. Then the resulting hot bubble ultimately cools and the consequent infall leads to renewed accretion; the cycle repeats, with the galaxy being seen alternately as an AGN/starburst for a small fraction of the time and as a "normal" elliptical hosting an incipient cooling catastrophe for much longer intervals. No steady flow appears to be possible for Eddington ratios above $\simeq 0.01$: whenever the luminosity is significantly above this limit both the accretion and the output luminosity are in burst mode. Strong intermittencies are expected at early times, while at low redshift the models are characterized by smooth, very sub-Eddington mass accretion rates punctuated by rare outbursts. One of the general consequences of our exploration is the fact that the recycled gas from dying stars can induce substantial QSO activity, even in the absence of external phenomena such as galaxy merging, while accretion feedback can be strong enough to solve the "cooling-flow" problem and to maintain the mass of the SMBH on the observed range of values.

References

Ciotti, L. 2009, *La Rivista del Nuovo Cimento* 32, n.1, 1
Ciotti, L. & Ostriker, J. P. 2007, *ApJ* 665, 1038
Ciotti, L., Ostriker, J. P., & Proga, D. 2009, *ApJ* 699, 89
Pellegrini, S., Ciotti, L., & Ostriker, J. P. 2009, *Advances in Space Research* 44, 340
Shin, M.-S., Ostriker, J. P., & Ciotti, L. 2009, *arXiv0905.4294*
Jiang, Y. F., Ciotti, L., Ostriker, J. P., & Spitkovsky, A. 2009, *arXiv0904.4918*

Highlights of Astronomy, Volume 15
XXVIIth IAU General Assembly, August 2009
Ian F. Corbett, ed.

© International Astronomical Union 2010
doi:10.1017/S174392131000935X

Effects of environment on the properties of cluster galaxies via ram pressure stripping

T. E. Tecce[1], S. A. Cora[2], P. B. Tissera[1] and M. G. Abadi[3]

[1] Instituto de Astronomía y Física del Espacio, CONICET-UBA, CC 67 Suc. 28, C1428ZAA Buenos Aires, Argentina.

[2] Facultad de Ciencias Astronómicas y Geofísicas, UNLP and Instituto de Astrofísica de La Plata, CONICET, Paseo del Bosque S/N, B1900FWA La Plata, Argentina.

[3] Observatorio Astronómico, UNC, Laprida 854, X5000BGR Córdoba, Argentina.
email: tomas@iafe.uba.ar

Abstract. We study the effect of ram pressure stripping (RPS) on the colours, cold gas content and star formation of galaxies in clusters, using a combination of N-Body/SPH simulations of galaxy clusters and a semi-analytic model of galaxy formation that includes the effect of RPS.

Keywords. Galaxies: clusters: general, galaxies: evolution, intergalactic medium

---

In the local universe, galaxy clusters have larger fractions of red, early-type galaxies than the field (Baldry *et al.* 2006) and star formation is suppressed in denser environments (Kauffmann *et al.* 2004). This could be caused by removal of cold gas from galaxies via ram pressure stripping (RPS) by the intracluster medium (ICM) (Gunn & Gott 1972).

We have studied the influence of RPS on galaxy properties using a hybrid model that combines non-radiative N-Body/SPH cosmological simulations of clusters with masses in the range $10^{14} - 10^{15} h^{-1}$ M$_\odot$ (Dolag *et al.* 2005) and a semi-analytic model of galaxy formation (Lagos *et al.* 2008) in which we implement the RPS process. We use the information provided by the gas particles in the simulations to obtain the properties of the ICM and the velocities of galaxies relative to it. This results in a more self-consistent method than previous works, which relied on analytical approximations.

We have run two sets of models, with and without RPS, with all the parameters involved in other physical processes remaining unchanged. Our results show that RPS is more important in the more massive haloes, but becomes significant only at $z \lesssim 0.5$. In the RPS model, star formation activity is strongly suppressed in dwarf galaxies (M$_* < 10^{10} h^{-1}$ M$_\odot$), which lose most of their ISM and become red and passive on a shorter timescale. RPS could then be responsible for the loss of gas in cluster dE galaxies (Boselli *et al.* 2008). Mean colours of the red and blue sequences do not change significantly, but the number of faint red galaxies increases in the RPS model for $z < 1$. The fraction of gas poor galaxies in regions around clusters increase from ~5% to ~20%, indicating that RPS acts to some degree in group environments prior to cluster infall.

References

Baldry, I. K. *et al.* 2006, *MNRAS* 373, 469
Boselli, A. *et al.* 2008, *ApJ* 674, 742
Dolag, K. *et al.* 2005, *MNRAS* 364, 753
Gunn, J. E. & Gott, J. R. 1972, *ApJ* 176, 1
Kauffmann, G. *et al.* 2004, *MNRAS* 353, 713
Lagos, C. D. P., Cora, S. A., & Padilla, N. D. 2008, *MNRAS* 388, 587

Highlights of Astronomy, Volume 15
XXVIIth IAU General Assembly, August 2009
Ian F. Corbett, ed.

AGN feedback in details and in surveys

Alexis Finoguenov[1,2]

[1] Max-Planck-Institute for Extraterrestrial Physics (MPE), Giessenbach Str., Garching,
D-85748, Germany
[2] University of Maryland, Baltimore County, 1000 Hilltop Circle, Baltimore, MD 21250, USA
email: alexis@mpe.mpg.de

Abstract. I will review the new details on AGN feedback revealed by recent Chandra observations of M84 and present its theoretical modeling. Using the results of COSMOS survey I will present the direct measurement of halo occupation statistics for radio galaxies.

Highlights of Astronomy, Volume 15
XXVIIth IAU General Assembly, August 2009
Ian F. Corbett, ed.

© International Astronomical Union 2010
doi:10.1017/S1743921310009373

Warm ionized ISM in the bulge of Andromeda galaxy

Marat Gilfanov[1] and Akos Bogdan[1]

[1] Max-Planck-Institute for Astrophysics, Karl-Schwarzschild-Str. 1, D-85748 Garching
email: mgilfanov@mpa-garching.mpg.de

Abstract. We demonstrate that unresolved X-ray emission from the bulge of M31 is composed of at least three different components: (i) Broad-band emission from a large number of faint sources – mainly accreting white dwarfs and active binaries, associated with the old stellar population, similar to the Galactic Ridge X-ray emission of the Milky Way. (ii) Soft emission from ionized gas with temperature of about ~ 300 eV and mass of $\sim 4 \times 10^6$ M$_\odot$. The gas distribution is significantly elongated along the minor axis of the galaxy suggesting that it may be outflowing in the direction perpendicular to the galactic disk. The shadows cast on the gas by spiral arms and the 10-kpc star-forming ring confirm large off-plane extent of the gas. (iii) Hard unresolved emission from spiral arms, most likely associated with protostars and young stellar objects located in the star-forming regions.

Highlights of Astronomy, Volume 15
XXVIIth IAU General Assembly, August 2009
Ian F. Corbett, ed.

Constraints on turbulent pressure in the X-ray halos of giant elliptical galaxies from resonant scattering

Norbert Werner[1], Irina Zhuravleva[2], Eugene Churazov[2,3], Aurora Simionescu[4], Steve W. Allen[1], William Forman[5], Christine Jones[5] and Jelle Kaastra[6,7]

[1] Stanford University
email: norbertw@stanford.edu

[2] Max Planck Institute for Astrophysics

[3] Space Research Institute, Moscow

[4] Max Planck Institute for Extraterrestrial Physics

[5] Harvard-Smithsonian Centre for Astrophysics

[6] SRON Netherlands Institute for Space Research, Utrecht, the Netherlands

[7] Universiteit Utrecht

Abstract. The dense cores of X-ray emitting gaseous halos of large elliptical galaxies with temperatures below about 0.8 keV show two prominent Fe XVII emission features, which provide a sensitive diagnostic tool to measure the effects of resonant scattering. We present here high-resolution spectra of five bright nearby elliptical galaxies, obtained with the Reflection Grating Spectrometers (RGS) on the XMM-Newton satellite. The spectra for the cores of four of the galaxies show the Fe XVII line at 15.01 Angstrom being suppressed by resonant scattering. The data for NGC 4636 in particular allow the effects of resonant scattering to be studied in detail. Using deprojected density and temperature profiles for this galaxy obtained with the Chandra satellite, we model the radial intensity profiles of the strongest resonance lines, accounting for the effects of resonant scattering, for different values of the characteristic turbulent velocity. Comparing the model to the data, we find that the isotropic turbulent velocities on spatial scales smaller than about 1 kpc are less than 100 km/s and the turbulent pressure support in the galaxy core is smaller than 5% of the thermal pressure at the 90% confidence level, and less than 20% at 99% confidence. Neglecting the effects of resonant scattering in spectral fitting of the inner 2 kpc core of NGC 4636 will lead to underestimates of the chemical abundances of Fe and O by about 10-20%.

JD8 Posters

My god, It's full of stars: Separating stars and gas in X-ray observations of elliptical galaxies
Bram S. Boroson, Dong-Woo Kim, and Giuseppina Fabbiano

Towards a larger sample of fossil groups and their luminosity functions
Raimundo Lopes de Oliveira, Claudia Mendes de Oliveira, Renato Dupke, Laerte Sodre Jr., Eduardo Cypriano, Eleazar Rodrigo Carrasco, and Daiana Ribeiro Bortoletto

The mass distribution of massive, X-ray luminous, elliptical galaxies
Payel Das, Ortwin Gerhard, Eugene Churazov, Alexis Finoguenov, Hans Boehringer, Flavio de Lorenzi, Emily McNeil, Roberto P. Saglia, and Lodovico Coccato

Highlights of Astronomy, Volume 15
XXVIIth IAU General Assembly, August 2009
Ian F. Corbett, ed.

© International Astronomical Union 2010
doi:10.1017/S1743921310009397

IAU Joint Discussion 9: Are the Fundamental Constants Varying in Space-time?

Paolo Molaro[1] and Elisabeth Vangioni[2]

[1] INAF –Osservatorio Astronomico di Trieste, Via Tiepolo 11, I-34131 Trieste, Italy

[2] Institut d' Astrophysique de Paris, 98bis, bd Arago, F-75014 Paris, France

The Joint Discussion on the variability of fundamental constants within the IAU GA 2009 was organized with the coordination of the IAU Division VIII Galaxies & the Universe and the support of IAU Commissions N. 47 Cosmology, N. 52 Relativity in Fundamental Astronomy, N.40 Radio Astronomy and N. 30 Radial velocities. The JD provided a timely opportunity to confront different points of view on this rather exciting subject.

The fundamental constants determine the scales of natural phenomena and many theories beyond the standard model of physics predict their variation. Such a possibility occupies quite a prominent place in cosmology since the existence of scalar fields, of the kind invoked to explain the universal acceleration, coupling with electromagnetic field or ordinary matter might lead to dynamic constants. The study of the behaviour of these quantities throughout the history of the universe is an effective way to probe fundamental physics and this is why on investigation of this subject is highly recommended in the Science Vision Document, in the ESA-ESO Working Group report on Fundamental Cosmology, and is one of the science drivers for the next facilities such as the E-ELT

High precision frequency measurements with atomic clocks in the laboratories established the fine structure constant to 17 significant figures. However, the variation is likely non linear in time and only astronomical observations can tell how they have been varying back in time. The Oklo natural reactor allows us to go back in time by about 2 Gyrs and meteorites to the birth of the solar system, QSO absorption systems bring us to about 80% of the age of the Universe while CMB and BBN to even earlier epochs. In 2001, observations of spectral lines in distant QSOs brought the first hints that the fine structure constant might change its value over time, with a variation of few parts per million. However, the subject is presently controversial with some other evidence which suggests null variation. The most active researchers in this area attended the Joint Discussion and discussed the latest developments. The issue was lively debated but a clear solution remains still to be found.

We wish to thank all participants to the meeting for their enthusiasm and for sharing their most recent results. We wish to thank the Scientific Organizing Committee for its advice and help in organizing the scientific program. The SOC was composed of

John D. Barrow, Francoise Combes, Thomas Dent, Sandro D'Odorico, Victor Flambaum, Sergei Levshakov, Carlos Martins, Michael Murphy, Cedric Ledoux, Keith Olive, Patrick Petitjean, Dieter Reimers, Raghunathan Srianand, Jean-Philippe Uzan, John Webb.

The Brazilian Astronomical society provided an efficient local organization and is also warmly thanked.

Highlights of Astronomy, Volume 15
XXVIIth IAU General Assembly, August 2009
Ian F. Corbett, ed.

© International Astronomical Union 2010
doi:10.1017/S1743921310009403

Variable Constants - A Theoretical Overview

Keith A. Olive

William I. FIne Theoretical Physics Institute, University of Minnesota, Minneapolis MN,
55455 USA

Abstract. In many theories of unified interactions, there are additional degrees of freedom which may allow for the variation of the fundamental constants of nature. I will review the motivation for such variations, and describe the theoretical relations between variations of gauge and Yukawa couplings.

Keywords. cosmology: theory, nuclear reactions, nucleosynthesis, abundances

Variations of fundamental constants are certainly possible within the context of unified theories of particle interactions. Indeed, in the context of string theories, the presence of a dilaton and other moduli fields almost guarantee that at some level gauge and Yukawa coupling constants are dynamical. Whether or not, these fields are fixed at or near the Planck scale (rendering our constants constant over effectively all of the history of the Universe) is unknown. If not, then there is the interesting possibility that the value of these constants varied over cosmological timescales (Olive (2009)).

There are a number of important astrophysical and terrestrial constraints on the fine-structure constant that must be respected. The most primordial of the limits comes from big bang nucleosynthesis (BBN) which tests for variations back to a cosmological redshift as high as $\sim 10^{10}$. The relatively good agreement between theory and observation for ^{4}He in BBN allows one to set a limit of $|\Delta\alpha/\alpha| < 0.04$ using $|\Delta Y/Y| < 4\%$ ($\Delta Y/Y$ scales with $\Delta\alpha/\alpha$). Since this limit is applied over the age of the Universe, we obtain a limit on the rate of change $|\dot{\alpha}/\alpha| < 3 \times 10^{-12}$ yr^{-1} over the last 13.7 Gyr. In the context of unified or string-inspired theories, this limit is significantly stronger and improves by about two orders of magnitude.

Very strong constraints on the variation of α can be obtained from the Oklo natural reactor which operated in Gabon approximately two billion years ago. The site has a rich uranium deposit which is naturally enriched in ^{237}U at the level of about 3.7%. The observed isotopic abundance distribution at Oklo can be related to the cross section for neutron capture on ^{149}Sm. The key isotopic ratio is that of ^{149}Sm/^{147}Sm which is 2% at the Oklo site relative to the common terrestrial value of about 90%, indicating strongly that ^{149}Sm was depleted by a thermal neutron source. Assuming that the energy difference is due to the α-dependence of the Coulomb energy alone, a limit

$$-0.56 < \Delta\alpha/\alpha \times 10^7 < 0.66 \qquad (0.1)$$

can be obtained. However, if all fundamental couplings are allowed to vary interdependently, a much more stringent limit $|\Delta\alpha/\alpha| < (1 - 5) \times 10^{-10}$ may be obtained.

There are also reported and disputed measurements of variations. These will be what make the session and these proceedings particularly interesting.

References

Olive, K. A., proceedings of JD 9 IAU Symposium, Rio 2009, to appear in Mem. S.A. It.. nd Vangioni, E., 2009 *in preparation.*

Highlights of Astronomy, Volume 15
XXVIIth IAU General Assembly, August 2009
Ian F. Corbett, ed.

© International Astronomical Union 2010
doi:10.1017/S1743921310009415

Thermodynamics in Variable Speed of Light Theories

Juan Racker[1], Paolo Sisterna[2] and Hector Vucetich[2]

[1] CONICET, Centro Atómico Bariloche, Avenida Bustillo 9500 (8400) Argentina

[2] Facultad de Ciencias Astronómicas y Geofísicas, Universidad Nacional de La PLata, Paseo del Bosque S/N (1900) La Plata, Argentina

Variable speed of light theories (VSL) are interesting because they could solve several cosmological puzzles. In this work we study the thermodynamics and Newtonian limit of the varying speed of light theory developed by J. Magueijo (Magueijo 2000). In the covariant and locally Lorentz invariant VSL proposed by Magueijo c is a dimensionless dynamical scalar field $c = c_0 e^{\psi}$, where c_0 is a constant. The matter and gravitational lagrangians are multiplied by the factors $e^{b\psi}$ and $e^{a\psi}$ respectively.

Among other phenomena the energy density and the total energy of a body in hydrostatic equilibrium may vary if c is not constant. We obtain a lagrangian for the perfect fluid following Schutz 1970. After obtaining a modified first law of thermodynamics we derive a general prescription on how to modify any equation of state up to first order in ψ.

In the Newtonian limit of this VSL theory, we show that the hydrostatic equilibrium equation is equivalent to a Newton's constant G varying equation. This equation plus an equation of state and boundary conditions determine the radius of a planet. The presence of ψ in these equations causes time variations of planetary radii. For Mercury its radius R hasn't changed more than 1 kilometer in the last 3.9×10^9 years. We find that $\left(\frac{11}{3}q - b - \frac{10}{3}\right)\dot{\psi}(t) = -\frac{1}{\delta}\frac{\dot{R}}{R} \simeq 0 \pm 5 \times 10^{-12}\mathrm{y}^{-1}$ for $\frac{\Delta R}{R} = 0 \pm 0.0004$, where ΔR corresponds to a time interval approximately equal to 3.5×10^9 years. This result can be combined with bounds for $\dot{\alpha}/\alpha$ that have been obtained using atomic clocks $\frac{\dot{\alpha}}{\alpha} = (4.2 \pm 6.9) \times 10^{-15}\mathrm{y}^{-1}$. Assuming $b = 0$ we obtain $\frac{\dot{c}}{c} = \dot{\psi} = 0 \pm 2 \times 10^{-12}\mathrm{y}^{-1}$.

White dwarfs are excellent objects to test any energy injection from a scalar field given their low luminosity and their extremely high heat conductivity. Most of them are adequately described by Newtonian physics and a polytrope type equation of state (EOS). We obtain for the dependence of the stellar internal energy $E \propto \exp\psi f(q, b, \gamma)$ where $f(q, b, \gamma) = \frac{13}{3}q - \frac{13}{3}b - \frac{14}{3}$ for white dwarfs with $\gamma = 5/3$. We assume that all the energy injected by the field ψ is radiated away, so the luminosity induced by the ψ field is $L_{\psi} = -\dot{E} = -f(\gamma, q, b)E\dot{\psi}$.

Using again the upper bound for the present time variation of α, bounding L_{ψ} by the observed luminosity of the white dwarf and assuming $b = 0$, Stein 2015B provides the strongest bound. We obtain $\frac{\dot{c}}{c} = \dot{\psi} = 0 \pm 1.4 \times 10^{-13}$. Comparing our results we see that white dwarf physics provides the strongest constraints on the VSL theory near the present epoch. Combining both bounds we obtain the b independent bound $\dot{\psi} = 0 \pm 2.2 \times 10^{-12}\mathrm{y}^{-1}$.

References

J. Magueijo. Phys. Rev. D, 62:103521, 2000.

B. F. Schutz. Phys. Rev. D, 2:2762–2773, 1970.

Highlights of Astronomy, Volume 15
XXVIIth IAU General Assembly, August 2009
Ian F. Corbett, ed.

© International Astronomical Union 2010
doi:10.1017/S1743921310009427

Accelerating universe and the time-dependent fine-structure constant

Yasunori Fujii

Advanced Research Institute for Science and Engineering, Waseda University, Tokyo, 169-8555, Japan; email: fujii@e07.itscom.net

I start with assuming a gravitational scalar field as the dark-energy supposed to be responsible for the accelerating universe. Also from the point of view of unification, a scalar field implies a time-variability of certain "constants" in Nature. In this context I once derived a relation for the time-variability of the fine-structure constant α: $\Delta\alpha/\alpha = \zeta\mathcal{Z}(\alpha/\pi)\Delta\sigma$, where ζ and $\mathcal{Z}$ are the constants of the order one, while σ on the right-hand side is the scalar field in action in the accelerating universe. I use the reduced Planckian units with $c = \hbar = M_{\rm P}(= (8\pi G)^{-1/2}) = 1$. I then compared the dynamics of the accelerating universe, on one hand, and $\Delta\alpha/\alpha$ derived from the analyses of QSO absorption lines, Oklo phenomenon, also different atomic clocks in the laboratories, on the other hand. I am here going to discuss the theoretical background of the relation, based on the scalar-tensor theory invented first by Jordan in 1955.

An important issue of this theory is the presence of different conformal frames, connected by the conformal transformations to each other. Of particular significance are the Jordan frame with the nonminimal coupling term expressed by the scalar field ϕ, so with the time-dependent gravitational constant, and the Einstein frame with the standard Einstein-Hilbert term, with the re-expressed scalar field σ. Brans and Dicke proposed an added assumption on the decoupling of ϕ from the matter Lagrangian to save the idea of Weak Equivalence Principle (WEP), to be called the BD model.

As it turns out, choosing this model is closely connected with the question which of the conformal frames is "physical." I point out that the required criterion is the constancy of matter particle masses, which provide with fundamental units of microscopic time and length, to be used for the astronomical measurements. Notice that these masses are constant or variable depending on the choice of the frame. The argument is affected, however, crucially by the presence of Λ leaving no room to define the physical frame. To be blamed is the BD model, which is replaced by what is called the scale-invariant model to identify the Einstein frame with the physical frame, at least at the classical level. Quantum effects introduce some deviations resulting in a small amount of WEP violation, implemented by the calculation in terms of quantum anomalies. This is the way the relation mentioned at the beginning is derived.

Finally I add a comment on the cosmological solutions I used. The observed acceleration is fitted by $\Lambda_{\rm eff}$ as small as 10^{-120} in the Planckian units. This small number has symbolized a fine-tuning problem. Remarkably, however, the obtained attractor solution in the Einstein frame does include $\Lambda_{\rm eff} \sim t_{*0}^{-2}$, where the asterisk in the subscript implies the cosmic time defined in the Einstein frame, and the present time $\sim 1.4 \times 10^{10}$y corresponds to $\sim 10^{60}$ in units of the Planck time. In other words, I no longer need a fine-tuning process; $\Lambda_{\rm eff}$ is small only because we are old! This is a major success of the scalar-tensor theory extended by including Λ, unparalleled by any other phenomenological approaches.

Highlights of Astronomy, Volume 15
XXVIIth IAU General Assembly, August 2009
Ian F. Corbett, ed.

© International Astronomical Union 2010
doi:10.1017/S1743921310009439

Reconstructing the evolution of dark energy with variations of fundamental parameters

N. J. Nunes[1], T. Dent[2], C. J. A. P. Martins[3] and G. Robbers[4]

[1]Institut für Theoretische Physik, Philosophenweg 16, 69120 Heidelberg, Germany
[2]School of Physics and Astronomy, Cardiff University, The Parade, Cardiff CF24 3AA, U.K.
[3]Centro de Astrofísica, Universidade do Porto, Rua das Estrelas, 4150-762 Porto, Portugal,
and DAMTP, University of Cambridge, Wilberforce Road, Cambridge CB3 0WA, U.K.
[4]Max-Planck-Institut für Astrophysik, Karl-Schwarzschild-Straße 1, D-85748 Garching bei
München, Germany

A popular candidate of dark energy, currently driving an accelerated expansion of the universe, is a slowly rolling scalar field or quintessence. A scalar field, however, must couple with other sources of matter. Consequently, its dynamical evolution can result in extra interactions between standard particles, which are mediated by the field, and to a variation in the fundamental parameters. Curiously, it has been reported that observations of a number of quasar absorption lines suggest that the fine structure constant was smaller in the past, at redshifts in the range $z = 1 - 3$ (Murphy et al. (2003), Murphy et al. (2004), but see also Srianand et al. (2007)). Could this indeed be the signature of a slowly evolving scalar field?

In this work we investigated how information can be obtained on the nature of dark energy from observational detection of (or constraints on) the variation of the fine structure constant and the proton to electron mass ratio. The reconstruction procedure is described with the purpose of forecasting the accuracy of proposed future spectrographs: ESPRESSO for VLT and CODEX for the E-ELT (Nunes & Lidsey (2004), Avelino et al. (2006), Avelino (2009)).

We discussed two parametrizations for the variation of alpha that satisfy the most stringent atomic clock constrains (Rosenband et al. (2008)) and that can also accommodate a large variation at redshift larger than unity. These parametrizations involve a sharp, recent transition in the dynamics of alpha, as well as non-trivial features in the shape of the scalar potential and the evolution of the equation of state parameter. Our results highlight the need for an independent confirmation of the quasar measurements.

NJN is supported by Deutsche Forschungsgemeinschaft, project TRR33 and thanks the organizers of the IAU and the JD9 discussion section for a stimulating meeting. The work of C.M. is funded by a Ciência2007 Research Contract, supported by FSE and POPH-QREN funds.

References

Avelino, P. P., Martins, C. J. A. P., Nunes, N. J., & Olive, K. A. 2006, Phys. Rev. D, 74, 083508
Avelino, P. P. 2009, Phys. Rev. D., 79, 083516
Murphy, M. T., Webb, J. K., & Flambaum, V. V. 2003, MNRAS, 345, 609
Murphy, M. T., Flambaum, V. V., Webb, J. K., & et al. 2004, Astrophysics, Clocks and Fundamental Constants, 648, 131
Nunes, N. J., & Lidsey, J. E. 2004, Phys. Rev. D., 69, 123511
Rosenband, T., et al. 2008, Science, 319, 1808
Srianand, R., Chand, H., Petitjean, P., & Aracil, B. 2007, Physical Review Letters, 99, 239002

Highlights of Astronomy, Volume 15
XXVIIth IAU General Assembly, August 2009
Ian F. Corbett, ed.

Searching for space-time variation of the fine structure constant using QSO spectra: overview and future prospects

J. C. Berengut[1], V. A. Dzuba[1], V. V. Flambaum[1], J. A. King[1], M. G. Kozlov[2], M. T. Murphy[3] and J. K. Webb[1]

[1] School of Physics, University of New South Wales, Sydney, NSW 2052, Australia

[2] Petersburg Nuclear Physics Institute, Gatchina, 188300, Russia

[3] Centre for Astrophysics and Supercomputing, Swinburne University of Technology, Victoria 3122, Australia

Current theories that seek to unify gravity with the other fundamental interactions suggest that spatial and temporal variation of fundamental constants is a possibility, or even a necessity, in an expanding Universe. Several studies have tried to probe the values of constants at earlier stages in the evolution of the Universe, using tools such as big-bang nucleosynthesis, the Oklo natural nuclear reactor, quasar absorption spectra, and atomic clocks (see, e.g. Flambaum & Berengut (2009)).

Comparison of atomic transition frequencies on Earth and in quasar absorption spectra can be used to measure variation of the fine-structure constant $\alpha = e^2/\hbar c$ over the last 10 billion years or so. The "many-multiplet" method Dzuba (1999) improves sensitivity to variation in α by more than an order of magnitude compared to the old "alkali-doublet" method. Study of three independent samples of data, containing 143 absorption systems over the redshift range $0.2 < z_{\text{abs}} < 4.2$, gives a variation of $\Delta\alpha/\alpha = (-0.57 \pm 0.11) \times 10^{-5}$ Murphy et al (2004). Our method and calculations have been used by other groups to analyse different data sets from different telescopes Srianand et al (2004); their results indicate no variation of α. However a re-analysis of the same data, which included 23 absorption systems from VLT/UVES, increased the error previously reported by a factor of six (Murphy et al (2007), Murphy et al (2008)).

The studies can be improved by utilising more atomic transitions that are seen in quasar spectra, however in many cases this is hampered by a lack of accurate laboratory frequency measurements. We have provided a compilation of transitions of importance to the search for α variation. They are E1 transitions from the ground state in several different atoms and ions, with wavelengths ranging from around $900 - 6000$ Å, and require an accuracy of better than 10^{-4} Å. We also discuss isotope shift measurements that are needed in order to resolve systematic effects in the study.

Researchers who are interested in performing these measurements should contact the authors directly.

References

Dzuba, V. A., Flambaum, V. V., & Webb, J. K. 1999, Phys. Rev. Lett., 82, 888

Flambaum, V. V. & Berengut, J. C. 2009, Int. J. Mod. Phys. A, 24, 3342

Murphy, M. T., Flambaum, V. V., Webb, J. K., et al. 2004, Lect. Notes Phys., 648, 131

Murphy, M. T., Webb, J. K., & Flambaum, V. V. 2007, Phys. Rev. Lett., 99, 239001

Murphy, M. T., Webb, J. K., & Flambaum, V. V. 2008, Mon. Not. R. Astron. Soc., 384, 1053

Srianand, R., Chand, H., Petitjean, P., & Aracil, B. 2004, Phys. Rev. Lett., 92, 121302

Highlights of Astronomy, Volume 15
XXVIIth IAU General Assembly, August 2009
Ian F. Corbett, ed.

© International Astronomical Union 2010
doi:10.1017/S1743921310009452

The effects of coupling variations on BBN

Keith A. Olive

William I. FIne Theoretical Physics Institute, University of Minnesota, Minneapolis MN,
55455 USA

Abstract. The effect of variations of the fundamental nuclear parameters on big-bang nucleosynthesis are modeled and discussed in detail taking into account the interrelations between the fundamental parameters arising in unified theories. Considering only ^{4}He, strong constraints on the variation of the neutron lifetime, neutron-proton mass difference are set. We show that a variation of the deuterium binding energy is able to reconcile the ^{7}Li abundance deduced from the WMAP analysis with its spectroscopically determined value while maintaining concordance with D and ^{4}He.

Keywords. cosmology: theory, nuclear reactions, nucleosynthesis, abundances

---

Big bang nucleosynthesis (BBN) is one of the most sensitive available probes of physics beyond the standard model. The concordance between the observation-based determinations of the light element abundances of D, ^{3}He, ^{4}He, and ^{7}Li, and their theoretically predicted abundances reflects the overall success of the standard big bang cosmology. Many departures from the standard model are likely to upset this agreement, and are tightly constrained.

It has also become generally accepted that there is a problem concerning the abundance of ^{7}Li. WMAP has accurately fixed the value of the baryon-to-photon ratio, $\eta = (6.23 \pm 0.17) \times 10^{-10}$ corresponding to $\Omega_B h^2 = 0.02273 \pm 0.00062$. At that value, the predicted abundance of ^{7}Li is approximately 4 times the observationally determined value, Cyburt *et al.* (2008). Several attempts at explaining this discrepancy by adjusting some of the key nuclear rates proved unsuccessful.

In unified theories of particle interactions, one generally expects that a change in the fine structure constant would directly imply a change in other gauge couplings, as well as and perhaps more importantly, variations in the QCD scale Λ_{QCD}. In addition, one might expect variations in the Yukawa couplings and Higgs vev as well. These relations can then be implemented in a BBN calculation. The resulting limit, Coc *et al.* (2007), can be expressed in terms of a limit on the variation in α

$$-3.2 \times 10^{-5} < \frac{\Delta h}{h} < 4.2 \times 10^{-5}. \qquad (0.2)$$

References

Coc, A., Nunes, N. J., Olive, K. A., Uzan, J.-P., & Vangioni, E. 2007, Phys. Rev. D, 76, 023511
Cyburt, R. H., Fields, B. D., & Olive, K. A. 2008, Journal of Cosmology and Astro-Particle Physics, 11, 12

Highlights of Astronomy, Volume 15
XXVIIth IAU General Assembly, August 2009
Ian F. Corbett, ed.

© International Astronomical Union 2010
doi:10.1017/S1743921310009464

Constraints on the variations of fundamental couplings by stellar models

A. Coc[1], S. Ekström[2], P. Descouvemont[3] and E. Vangioni[4]

[1] Centre de Spectrométrie Nucléaire et de Spectrométrie de Masse (CSNSM), UMR 8609, CNRS/IN2P3 and Université Paris Sud 11, Bâtiment 104, 91405 Orsay Campus, France

[2] Geneva Observatory, University of Geneva, Maillettes 51, 1290 Sauverny, Switzerland

[3] Physique Nucléaire Théorique et Physique Mathématique, C.P. 229, Université Libre de Bruxelles (ULB), B-1050 Brussels, Belgium

[4] Institut dAstrophysique de Paris, UMR-7095 du CNRS, Université Pierre et Marie Curie, 98 bis bd Arago, 75014 Paris, France

Abstract. The effect of variations of the fundamental constants on the thermonuclear rate of the triple alpha reaction, $^4\mathrm{He}(\alpha\alpha, \gamma)^{12}\mathrm{C}$, that bridges the gap between $^4\mathrm{He}$ and $^{12}\mathrm{C}$ is investigated. We follow the evolution of 15 and 60 $M_\odot$ zero metallicity star models, up to the end of core helium burning. They are assumed to be representative of the first, Population III stars undergoing a very peculiar evolution due to the absence of initial CNO elements (zero metallicity). The calculated oxygen and carbon abundances resulting from helium burning can then be used to constrain the variations of the fundamental constants.

Keywords. cosmology: theory, Pop III stars, nuclear reactions, nucleosynthesis, abundances

---

The equivalence principle is a cornerstone of metric theories of gravitation and in particular of General Relativity. It follows that by testing the constancy of fundamental constants one actually performs a test of General Relativity, that can be extended on astrophysical and cosmological scales. We consider here (see details in Coc *et al.* 2009, and Ekström *et al.* 2009), the very first generation of stars which are thought to have been formed a few 10^8 years after the big bang, at a redshift of $z \sim 10 - 15$, and with *zero initial metallicity*. The synthesis of complex elements in stars (mainly the possibility of the 3α-reaction at the origin of the procuction of $^{12}\mathrm{C}$) sets constraints on the values of the fine structure and strong coupling constants. This reaction is indeed very sensitive to the position of a resonance corresponding to the "Hoyle state" in $^{12}\mathrm{C}$. In order to analyze its variation with the nuclear interaction, we have used a nuclear microscopic cluster model and performed model calculations of a 15 and 60 $M_\odot$ zero metallicity star until to the end of core helium burning. We deduce the limits on the variation of the nuclear interaction to insure that the C/O ratio be of the order of unity. Depending on models, this can be releated to limits on the variations of the fine structure constant.

References

Coc, A., Ekström, S., Descouvemont, P., Meynet, G., Olive, K., Uzan, J.-Ph., & Vangioni, E., proceedings of JD 9 IAU Symposium, Rio 2009, to appear in Mem. S.A. It..

Ekström, S., Coc, A., Descouvemont, P., Meynet, G., Olive, K., Uzan, J.-Ph., & Vangioni, E., 2009 *in preparation*.

Highlights of Astronomy, Volume 15
XXVIIth IAU General Assembly, August 2009
Ian F. Corbett, ed.

WMAP 5-year constraints on α and m_e

Claudia G. Scóccola[1,2], Susana J. Landau[3] and Héctor Vucetich[1]

[1] Facultad de Ciencias Astronómicas y Geofísicas, Universidad Nacional de La Plata, Paseo del Bosque S/N 1900 La Plata, Argentina

[2] Max-Planck-Institut fur Astrophysik, Karl-Schwarzschild Str. 1, D-85741 Garching, Germany
email: scoccola@MPA-Garching.mpg.de

[3] Departamento de Física, FCEyN, Universidad de Buenos Aires
Ciudad Universitaria - Pab. 1, 1428 Buenos Aires, Argentina
Member of the Carrera del Investigador Científico y Tecnológico CONICET
email: slandau@df.uba.ar

Abstract. We have studied the role of fundamental constants in an updated recombination scenario. We focus on the time variation of the fine structure constant α, and the electron mass m_e in the early Universe. In the last years, helium recombination has been studied in great detail revealing the importance of taking new physical processes into account in the calculation of the recombination history. The equations to solve the detailed recombination scenario can be found for example in Wong *et al.* 2008. In the equation for helium recombination, a term which accounts for the semi-forbidden transition $2^3\mathrm{p}{-}1^1\mathrm{s}$ is added. Furthermore, the continuum opacity of HI is taken into account by a modification in the escape probability of the photons that excite helium atoms, with the fitting formulae proposed Kholupenko *et al* 2007. We have analized the dependences of the quantities involved in the detailed recombination scenario on α and m_e. We have performed a statistical analysis with COSMOMC to constrain the variation of α and m_e at the time of neutral hydrogen formation. The observational set used for the analysis was data from the WMAP 5-year temperature and temperature-polarization power spectrum and other CMB experiments such as CBI, ACBAR and BOOMERANG and the power spectrum of the 2dFGRS. Considering the joint variation of α and m_e we obtain the following bounds: $-0.011 < \frac{\Delta\alpha}{\alpha_0} < 0.019$ and $-0.068 < \frac{\Delta m_e}{(m_e)_0} < 0.030$ (68% c.l.). When considering only the variation of one fundamental constant we obtain: $-0.010 < \frac{\Delta\alpha}{\alpha_0} < 0.008$ and $-0.04 < \frac{\Delta m_e}{(m_e)_0} < 0.02$ (68% c.l.). We compare these results with the ones presented in Landau *et al* 2008, which were obtained in the standard recombination scenario and using WMAP 3 year release data. The constraints are tighter in the current analysis, which is an expectable fact since we are working with more accurate data from WMAP. The bounds obtained are consistent with null variation, for both α and m_e, but in the present analysis, the 68% confidence limits on the variation of both constants have changed. In the case of α, the present limit is more consistent with null variation than the previous one, while in the case of m_e the single parameters limits have moved toward lower values. To study the origin of this difference, we have performed another statistical analysis, namely the analysis of the standard recombination scenario together with WMAP5 data, the other CMB data sets and the 2dFGRS power spectrum. We see that the change in the obtained results is due to the new WMAP data set, and not to the new recombination scenario. The obtained results for the cosmological parameters are in agreement within 1σ with the ones obtained by the WMAP collaboration, without considering variation of fundamental constants.

Keywords. cosmic microwave background, early universe, cosmological parameters

Highlights of Astronomy, Volume 15
XXVIIth IAU General Assembly, August 2009
Ian F. Corbett, ed.

© International Astronomical Union 2010
doi:10.1017/S1743921310009488

Will cosmic acceleration last forever?

J. S. Alcaniz

Observatório Nacional, Rio de Janeiro - RJ, 20921-400, Brasil
email: alcaniz@on.br

Although the transition from an initially decelerated to a late-time accelerating cosmic expansion is becoming observationally established, the duration of the accelerating phase depends on the cosmological scenario and, several models, which includes our standard one, imply an eternal acceleration or even an accelerating expansion until the onset of a future cosmic singularity. In this regard, an interesting theoretical question arises if one tries to reconcile the standard description of the current cosmic acceleration with the only candidate for a consistent quantum theory of gravity we have today, i.e., Superstring theory.

As is well known, in the standard cosmological scenario, after radiation and matter dominance, the Universe asymptotically enters a de Sitter phase with the scale factor $a(t)$ growing exponentially, which results in an eternal cosmic acceleration. In such a background, the cosmological event horizon

$$\Delta = \int_{t_0}^{\infty} \frac{dt}{a(t)} \quad \rightarrow \quad \text{converges,} \tag{0.3}$$

and this is particularly troublesome for the formulation of String/M theory because local observers inside their horizon are not able to isolate particles to be scattered, which implies that a conventional S-matrix cannot be built (see e.g. Fischler et al. (2001), Hellerman et al. (2001)). This dark energy/String theory conflict, therefore, leaves us with the formidable task of either finding alternatives to the conventional S-matrix or constructing a model for the Universe that predicts the possibility of a transient acceleration phenomenon.

In fact, this latter possibility can be achieved in the context of the so-called *thawing* scalar field models in which a new deceleration period will take place in the future†. Examples of transient cosmic acceleration can also be found in brane-world cosmologies (Sahni and Shtanov (2003)), as well as in models of coupled quintessence (interacting dark matter/dark energy), as recently discussed by Costa and Alcaniz (2009).

Another interesting example of transient cosmic acceleration was recently discussed by Alcaniz et al. (2009). In this scenario, the field potential provides an equation-of-state (EoS) parameter of the type

$$w(a) = -1 + \text{const.}(a^{\kappa} + a^{-\kappa})^2 , \tag{0.4}$$

in which the competition between the double scale factor terms gives rise to a *hybrid* behavior: it is freezing over all the past cosmic evolution, is approaching the value -1 today (in agreement with current obsevational limits on w), will become thawing in the near future and will behave as such over the entire future evolution of the Universe. This behavior clearly leads to a transient acceleration phase, as shown in Fig. 1. Note that

---

† *Thawing* models describe a scalar field whose the equation-of-state parameter increases from $w \sim -1$, as it rolls down toward the minimum of its potential, whereas *freezing* scenarios describe an initially $w > -1$ EoS decreasing to more negative values (Caldwell and Linder (2005))

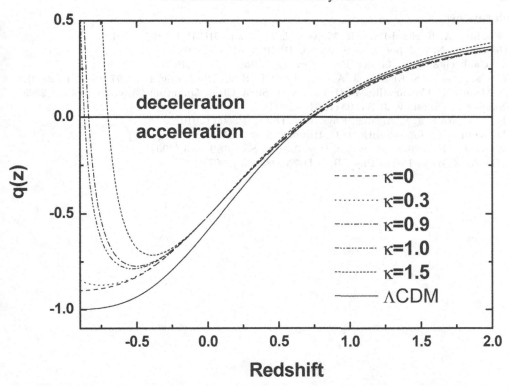

Figure 1. Cosmic deceleration/acceleration history for some selected values of κ. For comparison, the ΛCDM evolution is also shown.

similarly to thawing models, for hybrid scenarios the cosmological event horizon

$$\Delta = \int_{t_0}^{\infty} \frac{dt}{a(t)} \quad \rightarrow \quad \text{diverges}, \tag{0.5}$$

thereby allowing the construction of a conventional S-matrix describing particle interactions within the String/M-theory frameworks.

Finally, an important aspect worth emphasizing at this point is that, although thawing and hybrid models may provide a possible way to reconcile the observed acceleration of the Universe with theoretical constraints from String/M theories, they are distinguishable in what concerns the past cosmic evolution. In this regard, some recent analyses using current data from SNe Ia, LSS and CMB have explored possible variations in the $w - a$ plane and indicated a slight preference for a freezing behavior over the thawing one (Krauss *et al.* (2007), Zunckel and Trotta (2007)). For instance, Huterer and Peiris (2007) uses the Monte Carlo reconstruction formalism to scan a wide range of possibilities for $w(a)$ and find that $\sim 74\%$ are for freezing whereas only $\sim 0.05\%$ are for thawing. Similar conclusions are also obtained by Zunckel and Trotta (2007) by using the so-called maximum entropy method, where the HST/GOODS SNe Ia data showed $\simeq 1\sigma$ level preference for $w > -1$ at $z \sim 0.5$ with a drift towards $w > -1$ at higher redshifts. If such a preference for freezing EoS persists even after a systematically more homogeneous and statistically more powerful data sets become available, the combination of theoretical and observational constraints may give rise to an interesting case for hybrid models.

References

W. Fischler, A. Kashani-Poor, R. McNees, & S. Paban, JHEP **3**, 0107 (2001)

S. Hellerman, N. Kaloper, & L. Susskind, JHEP **3**, 0106, (2001)

R. R. Caldwell & E. V. Linder, Phys. Rev. Lett. **95**, 141301 (2005).

F. C. Carvalho, J. S. Alcaniz, J. A. S. Lima, & R. Silva, Phys. Rev. Lett. **97**, 081301 (2006).

J. S. Alcaniz, F. C. Carvalho, Z.-H. Zhu, & R. Silva, Class. Quantum Grav. **26** 105023 (2009).

V. Sahni & Y. Shtanov, JCAP 0311, 014 (2003).

F. E. M. Costa & J. S. Alcaniz, Phys. Rev. D **81**, 043506 (2010).

L. M. Krauss, K. Jones-Smith, & D. Huterer, New J. Phys. **9**, 141 (2007).

C. Zunckel & R. Trotta, Mon. Not. Roy. Astron. Soc. **380**, 865 (2007).

D. Huterer & H. V. Peiris, Phys. Rev. D **75**, 083503 (2007).

Highlights of Astronomy, Volume 15
XXVIIth IAU General Assembly, August 2009
Ian F. Corbett, ed.

© International Astronomical Union 2010
doi:10.1017/S174392131000949X

SNIa, white dwarfs and the variation of the gravitational constant

J. Isern[1,2], E. Garcia–Berro[1,3] and P. Lorén–Aguilar[1,3]

[1]Institut d'Estudis Espacials de Catalunya (IEEC), Edifici Nexus, c/ Gran Capità 2, E-08034 Barcelona, Spain

[2]Institut de Ciències de l'Espai (CSIC), Torre C-5 Parell, Facultat de Ciències, Campus UAB, E-08193 Bellaterra, Spain
email: isern@ieec.cat

[3]Departament de Fisica Aplicada, Universitat Politècnica de Catalunya, c/Esteve Terrades, 5, E-08860 Castelldefels, Spain
email: garcia@fa.upc.edu; email: loren@fa.upc.edu

Abstract. The critical role that the gravitational constant, G, plays in the Theory of General Relativity and the possibility, introduced by theories that unify gravity with other interactions, that G could vary in time and space have bursted the interest in detecting such variations or, at least, to bound them as tightly as possible.

White dwarfs represent the last evolutionary stage of stars with masses smaller than $10 \pm 2\,M_\odot$. Since their mechanical structure is sustained by the pressure of degenerate electrons, they do not radiate nuclear energy and their evolution is just a simple gravothermal cooling process. On the other hand, white dwarfs in close binary systems can accrete matter from the companion, experiencing nova outbursts, or in some cases they can also reach Chandrasekhar's mass and explode as a Type Ia supernova (SNIa). Since the cooling of single white dwarfs and the properties of SNIa strongly depend on the precise value of G and on its possible secular variation, white dwarfs can be used to constrain such hypothetical variations.

When white dwarfs are cool enough, their luminosity is entirely of gravothermal origin. Any variation of G modifies the energy balance of their interiors and, consequently, also modifies their luminosity. Formally, the influence of a secular variation of G can be expressed as $L = -\dot{B} + \Omega(\dot{G}/G)$ where $B = U + \Omega$ is the total binding energy, U is the total internal energy and Ω is the gravitational energy. Thus, if $\dot{G} \neq 0$ the luminosity is modified and the characteristic cooling time is different from that obtained in the case in which $\dot{G} = 0$. Detecting such variations can be done using the luminosity function of white dwarfs, which is defined as the number of white dwarfs of a given luminosity per unit of magnitude interval. It increases monotonically with the magnitude and displays a sharp cut-off due to the finite age of the Galaxy. The position of this cut-off is sensitive to the age of the Galaxy and to the value of $\dot{G}$ and, thus, it can be used to obtain a bound. White dwarfs also display g-mode pulsations driven by the κ-mechanism and the period of pulsation experiences a secular drift of $\dot{P}/P \simeq -a\dot{T}/T + b\dot{R}/R$, where a and b are model-dependent constants of the order of unity. Since both $\dot{T}$ and $\dot{R}$ depend on $\dot{G}$, the measure of $\dot{P}$ can also provide useful constraints.

Type Ia supernovae are thought to be the result of the thermonuclear explosion of a carbon-oxygen white dwarf with a mass near Chandrasekhar's limit in a close binary system. The peak luminosity of SNIa is proportional to the mass of nickel synthesized which can be considered as a fixed fraction of the mass of the exploding star, $M_{\rm Ni} \propto M_{\rm Ch} = (\hbar c)^{3/2}/m_{\rm p} G^{3/2}$. Therefore the properties of this peak with redshift can be used to test the variation of G with cosmic ages.

Although the bounds obtained in these ways have been currently superseeded by other more accurate methods, when the ongoing surveys searching for SNIa and white dwarfs will be completed, the expected bounds will be as tight as $\sim 10^{-13}$ yr^{-1}.

Keywords. Gravitation, Stars: white dwarfs, supernovae

Highlights of Astronomy, Volume 15
XXVIIth IAU General Assembly, August 2009
Ian F. Corbett, ed.

© International Astronomical Union 2010
doi:10.1017/S1743921310009506

21 cm radiation: A new probe of fundamental physics

Rishi Khatri[1] and Benjamin D. Wandelt[2]

[1,2]Department of Astronomy, University of Illinois, 1002 W. Green, Urbana, IL 61801, USA
[1]rkhatri2@illinois.edu [2]bwandelt@illinois.edu

Abstract. New low frequency radio telescopes currently being built open up the possibility of observing the 21 cm radiation from redshifts $200 > z > 30$, also known as the dark ages, see Furlanetto, Oh, & Briggs(2006) for a review. At these high redshifts, Cosmic Microwave Background (CMB) radiation is absorbed by neutral hydrogen at its 21 cm hyperfine transition. This redshifted 21 cm signal thus carries information about the state of the early Universe and can be used to test fundamental physics. The 21 cm radiation probes a volume of the early Universe on kpc scales in contrast with CMB which probes a surface (of some finite thickness) on Mpc scales. Thus there is many orders of more information available, in principle, from the 21 cm observations of dark ages. We have studied the constraints these observations can put on the variation of fundamental constants (Khatri & Wandelt(2007)). Since the 21 cm signal depends on atomic physics it is very sensitive to the variations in the fine structure constant and can place constraints comparable to or better than the other astrophysical experiments ($\Delta\alpha/\alpha = < 10^{-5}$) as shown in Figure 1. Making such observations will require radio telescopes of collecting area $10 - 10^6$ km^2 compared to ~ 1 km^2 of current telescopes, for example LOFAR. We should also

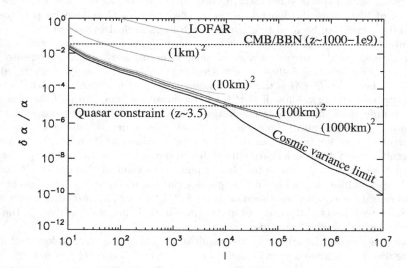

Figure 1. 21 cm constraints on α.

expect similar sensitivity to the electron to proton mass ratio. One of the challenges in observing this 21 cm cosmological signal is the presence of the synchrotron foregrounds which is many orders of magnitude larger than the cosmological signal but the two can be separated because of their different statistical nature (Zaldarriaga, Furlanetto, & Hernquist(2004)). Terrestrial EM interference from radio/TV etc. and Earth's ionosphere poses problems for telescopes on ground which may be solved by going to the Moon and there are proposals for doing so, one of which is the Dark Ages Lunar Interferometer (DALI). In conclusion 21 cm cosmology promises a large wealth of data and provides the only way to observe the redshift range between recombination and reionization.

Keywords. Cosmology: early Universe, Radio lines: general, Cosmology:diffuse radiation

Acknowledgements

Rishi Khatri thanks IAU and American Astronomical Society for travel support.

References

Furlanetto, S. R., Oh, S. P., & Briggs, F. H. 2006, *Physics Reports* 433, 181
Khatri, R. & Wandelt, B. D. 2007, *Phys. Rev. Lett.* 98, 111301
Zaldarriaga, M., Furlanetto, S. R., & Hernquist, L. 2004, *ApJ* 608, 622

Highlights of Astronomy, Volume 15
XXVIIth IAU General Assembly, August 2009
Ian F. Corbett, ed.
© International Astronomical Union 2010
doi:10.1017/S1743921310009518

Constraining a possible time-variation of the gravitational constant through "gravitochemical heating" of neutron stars

Andreas Reisenegger[1], Paula Jofré[2] and Rodrigo Fernández[3]

[1] Depto. de Astronomía y Astrofísica, Pontificia Universidad Católica de Chile,
Santiago, Chile
email: areisene@astro.puc.cl

[2] Max-Planck-Institut für Astrophysik, Garching, Germany

[3] Department of Astronomy and Astrophysics, University of Toronto, Toronto, Canada
Present address: Institute for Advanced Study, Princeton, USA

Abstract. A hypothetical time-variation of the gravitational constant G would produce *gravitochemical heating* of old ($\gtrsim 10^{7-8}$ yr) neutron stars. It could produce detectable thermal ultraviolet emission from such stars for changes as small as $|\dot{G}/G| \sim 10^{-12}$ yr^{-1}, comparable to the best existing upper limits from other methods.

Keywords. Stars: Neutron, Dense matter, Gravitation, Relativity, Pulsars: general, Pulsars: individual (PSR J0437−4715), Ultraviolet: stars

---

If the gravitational constant G is not strictly constant, as hypothesized by many authors since Dirac (1937), neutron stars would expand or contract, so the mix of degenerate fermions in their interiors would slightly depart from beta equilibrium. This induces non-equilibrium beta processes (direct or modified Urca reactions), such as the neutron beta decay ($n \rightarrow p + e + \bar{\nu}$) and its inverse ($p + e \rightarrow n + \nu$), which release energy that is invested partly in neutrino emission and partly in heating the star. After $\sim 10^{7-8}$ yr, the star arrives at a stationary state in which the temperature remains nearly constant, as the forcing through the change of G is balanced by the ongoing reactions. Using the surface temperature ($\sim 10^5$ K) of the nearest millisecond pulsar, PSR J0437−4715, inferred from ultraviolet observations with the Hubble Space Telescope (Kargaltsev *et al.* 2004), we have estimated two upper limits for this variation (Jofré *et al.* 2006): (1) $|\dot{G}/G| < 2 \times 10^{-10}$ yr^{-1}, if direct Urca reactions are allowed, and (2) $|\dot{G}/G| < 4 \times 10^{-12}$ yr^{-1}, considering only modified Urca reactions. The latter is among the most restrictive obtained by other methods (e. g., Reisenegger 2007). Further study of neutron star physics and additional observations of old neutron stars should be able to refine these constraints.

Acknowledgements

This work was supported by FONDECYT Regular Grant 1060644, the FONDAP Center for Astrophysics, and the Basal Funding Project PFB-06/2007.

References

Dirac, P. 1937, *Nature*, 139, 323
Jofré, P., Reisenegger, A., & Fernández, R. 2006, *Phys. Rev. Lett.*, 97, 131102
Kargaltsev, O., Pavlov, G. G., & Romani, R. 2004, *ApJ*, 602, 327
Reisenegger, A., Fernández, R., & Jofré, P., 2006, *Ap. Sp. Sci.*, 308, 413

Highlights of Astronomy, Volume 15
XXVIIth IAU General Assembly, August 2009
Ian F. Corbett, ed.

© International Astronomical Union 2010
doi:10.1017/S174392131000952X

Keck constraints on a varying fine-structure constant: wavelength calibration errors

Michael T. Murphy[1], John K. Webb[2] and Victor V. Flambaum[2]

[1] Centre for Astrophysics and Supercomputing, Swinburne University of Technology,
Melbourne, Victoria 3122, Australia
email: mmurphy@swin.edu.au

[2] School of Physics, University of New South Wales, Sydney NSW 2052, Australia

Abstract. The Keck telescope's High Resolution Spectrograph (HIRES) has previously provided evidence for a smaller fine-structure constant, α, compared to the current laboratory value, in a sample of 143 quasar absorption systems: $\Delta\alpha/\alpha = (-0.57 \pm 0.11) \times 10^{-5}$. The analysis was based on a variety of metal-ion transitions which, if α varies, experience different relative velocity shifts. This result is yet to be robustly contradicted, or confirmed, by measurements on other telescopes and spectrographs; it remains crucial to do so. It is also important to consider new possible instrumental systematic effects which may explain the Keck/HIRES results. Griest et al. (2009) recently identified distortions in the echelle order wavelength scales of HIRES with typical amplitudes $\pm 250\,\mathrm{m\,s^{-1}}$. Here we investigate the effect such distortions may have had on the Keck/HIRES varying α results. Using a simple model of these intra-order distortions, we demonstrate that they cause a random effect on $\Delta\alpha/\alpha$ from absorber to absorber because the systems are at different redshifts, placing the relevant absorption lines at different positions in different echelle orders. The typical magnitude of the effect on $\Delta\alpha/\alpha$ is $\sim 0.4 \times 10^{-5}$ for individual absorbers which, compared to the median error on $\Delta\alpha/\alpha$ in the sample, $\sim 1.9 \times 10^{-5}$, is relatively small. Consequently, the weighted mean value changes by less than 0.05×10^{-5} if the corrections we calculate are applied. Unsurprisingly, with corrections this small, we do not find direct evidence that applying them is actually warranted. Nevertheless, we urge caution, particularly for analyses aiming to achieve high precision $\Delta\alpha/\alpha$ measurements on individual systems or small samples, that a much more detailed understanding of such intra-order distortions and their dependence on observational parameters is important if they are to be avoided or modelled reliably.

Keywords. Instrumentation: spectrographs – Techniques: spectroscopic – Cosmology: observations - Quasars: absorption lines – Line: profiles

References

Griest, K., Whitmore, J. B., Wolfe, A. M., Prochaska, J. X., Howk, J. C., & Marcy, G. W. 2009, *ApJ*, submitted, arXiv:0904.4725v1

Highlights of Astronomy, Volume 15
XXVIIth IAU General Assembly, August 2009
Ian F. Corbett, ed.

Spatial and temporal variations of fundamental constants

S. A. Levshakov[1], I. I. Agafonova[1], P. Molaro[2] and D. Reimers[3]

[1] Ioffe Physical-Technical Institute, St. Petersburg, Russia
email: lev@astro.ioffe.rssi.ru, ira@astro.ioffe.rssi.ru

[2] INAF – Osservatorio Astronomico di Trieste, Trieste, Italy
email: molaro@oats.inaf.it

[3] Hamburger Sternwarte, Hamburg, Germany
email: st2e101@hs.uni-hamburg.de

Abstract. Spatial and temporal variations in the electron-to-proton mass ratio, μ, and in the fine-structure constant, α, are not present in the Standard Model of particle physics but they arise quite naturally in grant unification theories, multidimensional theories and in general when a coupling of light scalar fields to baryonic matter is considered. The light scalar fields are usually attributed to a negative pressure substance permeating the entire visible Universe and known as dark energy. This substance is thought to be responsible for a cosmic acceleration at low redshifts, $z < 1$. A strong dependence of μ and α on the ambient matter density is predicted by chameleon-like scalar field models. Calculations of atomic and molecular spectra show that different transitions have different sensitivities to changes in fundamental constants. Thus, measuring the relative line positions, ΔV, between such transitions one can probe the hypothetical variability of physical constants. In particular, interstellar molecular clouds can be used to test the matter density dependence of μ, since gas density in these clouds is ~ 15 orders of magnitude lower than that in terrestrial environment. We use the best quality radio spectra of the inversion transition of NH_3 $(J, K) = (1, 1)$ and rotational transitions of other molecules to estimate the radial velocity offsets, $\Delta V \equiv V_{\rm rot} - V_{\rm inv}$. The obtained value of ΔV shows a statistically significant positive shift of $23 \pm 4_{\rm stat} \pm 3_{\rm sys}$ m s^{-1} (1σ). Being interpreted in terms of the electron-to-proton mass ratio variation, this gives $\Delta\mu/\mu = (22 \pm 4_{\rm stat} \pm 3_{\rm sys}) \times 10^{-9}$. A strong constraint on variation of the quantity $F = \alpha^2/\mu$ in the Milky Way is found from comparison of the fine-structure transition $J = 1 - 0$ in atomic carbon C I with the low-J rotational lines in carbon monoxide ^{13}CO arising in the interstellar molecular clouds: $|\Delta F/F| < 3 \times 10^{-7}$. This yields $|\Delta\alpha/\alpha| < 1.5 \times 10^{-7}$ at $z = 0$. Since extragalactic absorbers have gas densities similar to those in the ISM, the values of $|\Delta\alpha/\alpha|$ and $|\Delta\mu/\mu|$ at high-z are expected to be at the same level as estimated in the Milky Way providing no temporal dependence of α and μ is present. We re-analyzed and reviewed the available optical spectra of quasars to probe $\Delta\alpha/\alpha$ from intervening absorbers. The Fe I system at $z = 0.45$ towards HE 0000–2340 provides one of the best opportunities for precise measurements of $\Delta\alpha/\alpha$ at low redshift. The current estimate is $\Delta\alpha/\alpha = (7 \pm 7) \times 10^{-6}$. With the updated sensitivity coefficients for the Fe II lines we re-analyzed the $z = 1.84$ system from the high-resolution UVES/VLT spectrum of Q 1101–264 ($FWHM = 3.8$ km s^{-1}) and found $\Delta\alpha/\alpha = (4.0 \pm 2.8) \times 10^{-6}$. The most accurate upper limit on cosmological variability of α is obtained from the Fe II system at $z = 1.15$ towards the bright quasar HE 0515–4414 ($V = 14.9$): $\Delta\alpha/\alpha = (-0.12 \pm 1.79) \times 10^{-6}$, or $|\Delta\alpha/\alpha| < 2 \times 10^{-6}$. The limit of 2×10^{-6} corresponds to the utmost accuracy which can be reached with available to date optical facilities.

Keywords. line: profiles – techniques: radial velocities – ISM: molecules – quasars: absorption lines – cosmology: observations

Highlights of Astronomy, Volume 15
XXVIIth IAU General Assembly, August 2009
Ian F. Corbett, ed.
© International Astronomical Union 2010
doi:10.1017/S1743921310009543

Searching for places where to test the variations of fundamental constants

P. Petitjean[1], P. Noterdaeme[2], R. Srianand[2], C. Ledoux[3], A. Ivanchik[4] and N. Gupta[5]

[1]Institut d'Astrophysique de Paris, Université Paris 6 & CNRS, UMR7095, 98bis Boulevard Arago, 75014, Paris, France, emailpetitjean@iap.fr,
[2]IUCAA, Post Bag 4, Ganesh Khind, Pune 411 007, India,
[3]European Southern Observatory, Alonso de Córdova 3107, Casilla 19001, Vitacura, Chile
[4]Ioffe Physical-Technical Institute of RAS, 194021 Saint-Petersburg, Russia,
[5]Australia Telescope National Facility, CSIRO, Epping, NSW 1710, Australia

Abstract. It has been realised in the last few years that strong constraints on the time-variations of dimensionless fundamental constants of physics can be derived at any redshift from QSO absorption line systems. Variations of the fine structure constant, α, the proton-to-electron mass ratio, μ, or the combination, $x = \alpha^2 g_{\rm p}/\mu$, where $g_{\rm p}$ is the proton gyromagnetic factor, have been constrained. However, for the latter two constants, the number of lines of sight where these measurements can be performed is limited. In particular the number of known molecular and 21 cm absorbers is small. Our group has started several surveys to search for these systems. Here is a summary of some of the characteristics of these absorbers that can be used to find these systems.

Keywords. Galaxies: ISM - quasars: absorption lines - physics: fundamental constants

---

The search for any variation of fundamental constants from high redshift quasar spectra. has been given tremendous interest recently with the advent of 10 m class telescopes. Since the amount of observing time required by the study of one absorption system is quite large (typically 10 to 20 hours of 10 m class telescope per quasar), the systems have to be selected carefully. In particular for μ and x, the small number of suitable systems may prevent the development of the field. We therefore have embarqued on several surveys to find these systems.

We searched for molecular hydrogen at high redshift with the VLT (Ledoux *et al.* 2003, Noterdaeme *et al.* 2008) in 77 DLAs/strong sub-DLAs (Damped Lyman-α systems), with $\log N(\text{H \textsc{i}}) \geqslant 20$ and $z_{\rm abs} > 1.8$. H_2 is detected in thirteen of the systems with molecular fractions as low as $f \simeq 5 \times 10^{-7}$ and up to $f \simeq 0.1$, with $f = 2N(\text{H}_2)/(2N(\text{H}_2) + N(\text{H \textsc{i}}))$ in the redshift range $1.8 < z_{\rm abs} \leqslant 4.2$.

For 21 cm absorbers at intermediate and high redshifts, we selected strong Mg \textsc{ii} systems ($W_{\rm r} > 1$ Å) from the Sloan Digital Sky Survey in the redshift range suitable for a follow-up with the Giant Meterwave Telescope (GMRT), $1.10 < z_{\rm abs} < 1.45$. We detected 9 new 21 cm absorption systems (Gupta *et al.* 2009). This is by far the largest number of 21-cm detections from any single survey.

References

Gupta, N., Srianand, R., Petitjean, P., Noterdaeme, P., & Saikia, D. J. 2009, MNRAS, 398, 201
Ledoux C., Petitjean, P., & Srianand, R. 2003, MNRAS, 346, 209
Noterdaeme, P., Ledoux, C., Petitjean, P., & Srianand, R. 2008, A&A, 481, 327
Srianand, R., Chand, H., Petitjean, P., & Aracil, B. 2004, PRL, 92.121302

Highlights of Astronomy, Volume 15
XXVIIth IAU General Assembly, August 2009
Ian F. Corbett, ed.

© International Astronomical Union 2010
doi:10.1017/S1743921310009555

Markov Chain Monte Carlo methods applied to measuring the fine structure constant from quasar spectroscopy

Julian King[1], Daniel Mortlock[2], John Webb[1] and Michael Murphy[3]

[1]Department of Astronomy, University of New South Wales,
Sydney, New South Wales, 2052, Australia
email: jking.phys@gmail.com

[2]Astrophysics Group, Blackett Laboratory, Prince Consort Road, Imperial College London,
London SW7 2AZ, U.K.

[3]Centre for Astrophysics and Supercomputing, Swinburne University of Technology,
Victoria, 3122, Australia

Abstract. Recent attempts to constrain cosmological variation in the fine structure constant, α, using quasar absorption lines have yielded two statistical samples which initially appear to be inconsistent. One of these samples was subsequently demonstrated to not pass consistency tests; it appears that the optimisation algorithm used to fit the model to the spectra failed. Nevertheless, the results of the other hinge on the robustness of the spectral fitting program VPFIT, which has been tested through simulation but not through direct exploration of the likelihood function. We present the application of Markov Chain Monte Carlo (MCMC) methods to this problem, and demonstrate that VPFIT produces similar values and uncertainties for $\Delta\alpha/\alpha$, the fractional change in the fine structure constant, as our MCMC algorithm, and thus that VPFIT is reliable.

Keywords. Atomic processes – Methods: numerical – Methods: statistical – Quasars: absorption lines – Quasars: individual: LBQS 2206−1958 – Quasars: individual: LBQS 0013−0029 – Quasars: individual: Q 0551−366 – Cosmology: observations

Recent years have seen sustained interest in attempting to determine the value of the fine structure constant, α in the early universe using quasar absorption lines. Differing atomic/ionic transitions have different sensitivities to α, and thus by fitting Voigt profiles to the observed profiles of quasar absorption systems, we can mesure $\Delta\alpha/\alpha = (\alpha_z - \alpha_0)/\alpha_0$, where α_z is the value of α at redshift z and α_0 is the laboratory value. Murphy *et al.* (2004) have found that $\Delta\alpha/\alpha = (-0.57\pm0.11) \times 10^{-5}$ from 143 quasar absorption systems, whereas Chand *et al.* (2004) reported $\Delta\alpha/\alpha = (-0.06 \pm 0.06) \times 10^{-5}$ from 23 measurements. However, Murphy *et al.* (2008) demonstrate that the analysis of Chand *et al.* suffers significant flaws which render both the estimate and statistical precision for $\Delta\alpha/\alpha$ of Chand *et al.* unreliable.

We apply Markov Chain Monte Carlo (MCMC) methods to investigate the $z = 1.018$, $z = 2.029$ and $z = 1.748$ absorption systems toward quasars LBQS 2206−1958, LBQS 0013−0029 and Q 0551−366 respectively. In particular, we confirm for these cases that the spectral fitting program VPFIT, used by Murphy *et al.* (2004), produces good estimates of $\Delta\alpha/\alpha$ and appropriate statistical uncertainties, and thus we regard the results of Murphy *et al.* (2004) as robust.

References

Chand, H., Srianand, R., Petitjean, P., & Aracil, B. 2004, *A&A*, 417, 853

Murphy, M. T., Flambaum, V. V., Webb, J. K., *et al.* 2004, in: S. G. Karshenboim & E. Peik (eds.), *Astrophysics, Clocks and Fundamental Constants*, Lecture Notes in Physics, Berlin Springer Verlag, Vol. 648 (Berlin: Springer Verlag), p. 131

Murphy, M. T., Webb, J. K., & Flambaum, V. V. 2008, *MNRAS*, 384, 1053

Highlights of Astronomy, Volume 15
XXVIIth IAU General Assembly, August 2009
Ian F. Corbett, ed.

Observational Determinations of the Proton to Electron Mass Ratio in the Early Universe

Rodger I. Thompson[1]

[1]Steward Observatory, University of Arizona
Tucson, AZ 85721
USA
email: `rit@email.arizona.edu`

Abstract. The values of the fundamental physical constants determine the nature of our universe from the height of mountains on earth to the evolution of the universe over its history. One of these constants is $\mu = M_P/M_e$ the ratio of the proton to electron mass. Astronomical observations provide a determination of this ratio in the early universe through observations of molecular absorption and emission lines in distant objects. Observations of molecular hydrogen in distant damped Lyman Alpha clouds provide a measurement of μ at a time when the universe was only 20% of its present age. To date there is no evidence for a change in μ at the level of 1 part in 10^5. This limit produces an observational constraint on quintessence theories for the evolution of the universe and Super Symmetric theories of elementary particles.

Keywords. Cosmology: observations

---

1. Overview

Speculation on the time stability of the fundamental constants extends back at least to the middle of the previous century. Observational constraints on the time variation of fundamental constants has centered primarily on the fine structure constant α and the ratio of the proton to electron mass μ. This review concentrates on observational constraints on μ. The most recent observations are given by Reinhold *et al.* (2006), Ubachs *et al.* (2007), Wendt and Reimers (2008), King *et al.* (2009) and Thompson *et al.* (2009). Although the first of these studies, Reinhold *et al.* (2006) and Ubachs *et al.* (2007) indicated a possible change in the value of μ the subsequent studies found no change in μ at the 10^{-5} level using the same data.

The constant value of μ at $\Delta\mu/\mu \leqslant 10^{-5}$ for look back times of 11 gigayears puts significant constraints on high energy and cosmological physics. With larger telescopes and new instrumentation, improvement by a factor of 10 should be possible in the next 5 years. This require stricts attention to the sources of systematic errors. This makes the measurement of fundamental constants in the early universe a low cost and powerful tool for the study of cosmology and high energy physics.

References

King, J. A., Webb, J. K., Murphy, M. T., & Carswell, R. F. 2009, Phys. Rev. Lett., 101, 251304
Reinhold, E. *et al.* 2006, Phys. Rev. Lett., 96, 151101
Thompson, R.I. *et al.* 2009 ApJ, 703, 1648.
Ubachs, W., Buning, R., Eikema, K. S. E., & Reinhold, E. 2007, Jour. Mol. Spec., 241, 155
Wendt, M. & Reimers, D. 2008, Eur. Phys. J. ST, 163, 197

Highlights of Astronomy, Volume 15
XXVIIth IAU General Assembly, August 2009
Ian F. Corbett, ed.
© International Astronomical Union 2010
doi:10.1017/S1743921310009579

Cosmological observations to shed light on possible variations

M. Wendt[1], D. Reimers [1] and P. Molaro[2]

[1] Hamburger Sternwarte, Universität Hamburg, Gojenbergsweg 112, 21029 Hamburg, Germany

[2] INAF –Osservatorio Astronomico di Trieste, Via Tiepolo 11, I-34143 Trieste, Italy

Cosmology contributes a good deal to the investigation of variation of fundamental physical constants. High resolution data is available and allows for detailed analysis over cosmological distances and a multitude of methods were developed. The raised demand for precision requires a deep understanding of the limiting errors involved. The achievable accuracy is under debate and current observing proposals max out the capabilities of today's technology. The question for self-consistency in data analysis and effective techniques to handle unknown systematic errors is of increasing importance. This work is motivated by numerous findings of different groups that partially are in disagreement witch each other. A large part of these discrepancies reflects the different methods of handling systematic errors. Evidently systematics are not yet under control or fully understood. We try to emphasize the importance to take these errors, namely calibration issues, into account and put forward some measures adapted to the problem. Alternative approaches for some of the steps involved are introduced.

Two independent observations of the Quasar Q0347-383 are compared to illustrate the problem of apparent precision when relying on error estimates of fitting algorithms alone. We show that the confidence to have accomplished the best numerical fit to a given data set is not very strongly correlated to the level of accuracy reached in terms of the real physical parameters. Numerous fitting algorithms used today are tested in detail for their ability to give best results for certain data. Recent works showed a trend to convert the level of statistical robustness of a fit into the resulting total error of the analysis. This approach lead to seemingly positive detections of variation in the past due to underestimated systematic errors in the reduced data itself. This work offers a test to verify the reproducibility of an analysis and it also emphasizes the plain need for better wavelength calibration in the future. Some of the approaches presented are suitable to handle absorption spectra up to that point.

In addition some procedures to deal with uncertainties in the calculated sensitivity coefficients for molecular hydrogen are highlighted. Unlike the transition frequencies for the Lyman and Werner bands, the calculated sensitivity coefficients cannot be verified in laboratories. Complex non-adiabatic effects have to be taken into account. At the present state the influence of the estimated errors in the sensitivity coefficients are negligible. However, they influence the error estimate of the analysis in principle. An issue of increasing importance for future data analysis with significantly lower uncertainties in wavelength calibration.

Highlights of Astronomy, Volume 15
XXVIIth IAU General Assembly, August 2009
Ian F. Corbett, ed.

© International Astronomical Union 2010
doi:10.1017/S1743921310009580

New limit on a varying proton-to-electron mass ratio from high-resolution optical quasar spectra

A. L. Malec[1], R. Buning[2], M. T. Murphy[1], N. Milutinovic[3],
S. L. Ellison[3], J. X. Prochaska[4], L. Kaper[2,5], J. Tumlinson[6],
R. F. Carswell[7] and W. Ubachs[2]

[1] Centre for Astrophysics and Supercomputing, Swinburne University of Technology,
Melbourne, Victoria 3122, Australia
email: amalec@swin.edu.au

[2] Laser Centre, VU University, De Boelelaan 1081, 1081 HV Amsterdam, The Netherlands

[3] Department of Physics and Astronomy, University of Victoria, Victoria, BC, V8P 1A1,
Canada

[4] University of California Observatories – Lick Observatory, University of California, Santa
Cruz, CA 95064

[5] Astronomical Institute Anton Pannekoek, Universiteit van Amsterdam, 1098 SJ Amsterdam,
The Netherlands

[6] Yale Center for Astronomy and Astrophysics, Department of Physics, New Haven, CT 06520,
USA

[7] Institute of Astronomy, University of Cambridge, Madingley Road, Cambridge, CB3 0HA, UK

Abstract. Molecular transitions recently discovered at redshift $z_{abs} = 2.059$ toward the bright background quasar J2123−0050 are analysed to limit cosmological variation in the proton-to-electron mass ratio, $\mu \equiv m_p/m_e$. Observed with the Keck telescope, the optical spectrum has the highest resolving power and largest number (86) of H_2 transitions in such analyses so far. Also, (7) HD transitions are used for the first time to constrain μ-variation. These factors, and an analysis employing the fewest possible free parameters, strongly constrain μ's relative deviation from the current laboratory value: $\Delta\mu/\mu = (+5.6 \pm 5.5_{stat} \pm 2.7_{sys}) \times 10^{-6}$. This is the first Keck result to complement recent constraints from three systems at $z_{abs} > 2.5$ observed with the Very Large Telescope.

Keywords. line: profiles – techniques: spectroscopic – methods: data analysis – quasars: absorption lines

Highlights of Astronomy, Volume 15
XXVIIth IAU General Assembly, August 2009
Ian F. Corbett, ed.

Radio measurements of constant variation, and perspectives with ALMA

Françoise Combes

LERMA, Observatoire de Paris, 61 Av. de l'Observatoire, F-75014, Paris
email: francoise.combes@obspm.fr

Abstract. In the radio domain, absorption lines in front of quasars of CO, HI, OH, HCO^+, HCN, up to NH_3 and CII are providing interesting constraints on fundamental constant variation (α and μ). With more absorbing systems, and a wider redshift range, they could be more competitive than optical studies. This could come with ALMA, with more than one order of magnitude in sensitivity.

Up to now, at intermediate and high redshift, between $z = 0.25$ to $z = 0.89$, only four absorption lines systems have been detected in the millimeter range and a fifth system at 0.765, at the OH-18cm lines (Kanekar et al. 2005). Out of these 5 systems, 3 are intervening lensing galaxies (and the background quasar is multiply imaged), and 2 correspond to an absorption of the host (PKS1413+135, B3-1504+377, for an overview see Combes & Wiklind 1996; Wiklind & Combes 1994 to 1998).

A global comparison of all molecular lines observed with the HI-21cm absorption lines in PKS1413 and B0218 systems, the two narrowest line systems, have given quite stringent constraints on $y = \alpha^2 g_p \mu$, $\Delta y/y = (-0.20 \pm 0.44) \, 10^{-5}$ and $\Delta y/y = (-0.16 \pm 0.54) \, 10^{-5}$ respectively (Murphy et al. 2001). The precision is comparable to the MM method (Murphy et al. 2003), with a limited number of absorbing systems.

The high sensitivity if the NH_3 inversion lines to variation in the μ ratio (Flambaum & Kozlov 2007) was used by Henkel et al. (2009) in a recent multi-line study of PKS1830 at $z \sim 0.9$, and Murphy et al. (2008) for B0218 at $z \sim 0.7$. They find a limit of $\Delta\mu/\mu < 1.4 \, 10^{-6}$ and $\Delta\mu/\mu < 1.8 \, 10^{-6}$ respectively.

Clearly, the radio method suffers from the rarity of the objects, and the fact that they have not yet been discovered at high redshift. The main caveats are that the lines compared come from different molecules, which might have intrinsic velocity offsets, due to several reasons, chemistry, excitation, temperature, density etc... When very different frequencies are compared (HI to CO for instance), the background continuum source has different sizes, and the absorbing medium is not the same. Also, the continuum source varies in both intensity and shape, and the compared lines are not always observed simultaneously. Only large statistics could smooth the errors down.

On the positive side, the radio domain is favoured by the high spectral resolution and the very narrow lines due to cold gas, the exquisite precision of the frequency calibration, and the well-known rest frequencies. Also, the sensitivity of the line position to the variation of constants is much higher (by a factor 100 for NH_3).

Fortunately, the sensitivity of ALMA will be able to detect many more continuum sources, to search for absorption lines, and at larger redshifts. ALMA will have a much wider bandwidth, allowing the search of absorption, even if not previously detected in the optical or HI-21cm. The redshift will be obtained directly in the millimeter.

It is possible to predict the number of continuum sources that can be selected as targets for absorption searches with ALMA. The density of flat-spectrum quasars has been shown to follow the same curve as optical quasars, a curve peaking at $z \sim 2$, very similar to the star formation history (Wall et al. 2005). They are still of significant density at $z \sim 3$.

Keywords. Galaxies: ISM – Quasars: absorption lines – Cosmology: observations

Highlights of Astronomy, Volume 15
XXVIIth IAU General Assembly, August 2009
Ian F. Corbett, ed.

© International Astronomical Union 2010
doi:10.1017/S1743921310009609

Probing fundamental constant evolution with redshifted radio lines

Nissim Kanekar[1]

[1] National Radio Astronomy Observatory, 1003 Lopezville Road, Socorro, NM 87801, USA;
e-mail: nkanekar@nrao.edu

Abstract. I report new results from radio studies of fundamental constant evolution, using comparisons between (1) conjugate satellite OH lines, and (2) inversion and rotational lines.

Keywords. Atomic processes, line:profiles, galaxies: high-redshift, radio lines: galaxies

---

Comparisons between the redshifts of radio lines from a cosmologically-distant object provide an interesting probe of changes in fundamental constants like the fine structure constant α, the proton-electron mass ratio μ, and the proton g-factor g_p (see Kanekar (2008) for a review). The best present radio techniques are based on comparisons between "conjugate" satellite OH 18cm lines, and between inversion and rotational lines; this article reports new results based on these two techniques.

The satellite OH 18cm lines are said to be "conjugate", when the lines have the same shapes, but with one line in absorption and the other in emission; this implies that the lines arise in the same gas, making them excellent probes of changes in α, μ and g_p (Kanekar, Chengalur, & Ghosh (2004)). We have obtained deep WSRT and Arecibo spectra in the conjugate satellite OH 18cm lines from the $z \sim 0.25$ source, B1413+135. A cross-correlation analysis finds that the line profiles are conjugate within the errors, but with a velocity offset of $\Delta V = (-0.23 \pm 0.09)$ km s^{-1}. This corresponds to $[\Delta G/G] = (-1.18 \pm 0.45) \times 10^{-5}$, where $G \equiv g_p \left[\mu\alpha^2\right]^{1.85}$, tantalizing ($\sim 2.6\sigma$) evidence for a change in α, μ and/or g_p over a lookback time of 2.9 Gyrs.

Comparisons between the redshifts of ammonia (NH$_3$) inversion lines and rotational lines have a high sensitivity to changes in $\mu \equiv m_p/m_e$ - Flambaum & Kozlov (2007). We have obtained deep GBT spectra in the optically-thin NH$_3$ 1-1 and CS 1-0 lines from the $z \sim 0.685$ absorber towards B0218+357. A joint fit to these lines finds a velocity offset of $\Delta V = (-0.51 \pm 0.36)$ km s^{-1}, yielding $[\Delta \mu/\mu] = (-4.9 \pm 3.5) \times 10^{-7}$. This limit, $[\Delta \mu/\mu] < 7 \times 10^{-7}$, is the best present constraint on fractional changes in μ.

The main drawback of the radio methods is the paucity of radio absorbers at high redshifts, $z > 1$. In the near future, new telescopes like ALMA and the EVLA will allow "blind" absorption surveys in the strong mm-wave rotational lines, yielding large samples of high-z molecular absorbers. Radio techniques are thus likely to play an important role in studies of fundamental constant evolution over the next decade.

Acknowledgements

I thank J. N. Chengalur and T. Ghosh for permission to present our OH results here.

References

Flambaum, V. V. & Kozlov, M. G. 2007, Phys. Rev. Lett., 98, 240801
Kanekar, N. 2008, Mod. Phys. Lett. A, 23, 2711
Kanekar, N., Chengalur, J. N., & Ghosh, T. 2004, Phys. Rev. Lett., 93, 051302

Highlights of Astronomy, Volume 15
XXVIIth IAU General Assembly, August 2009
Ian F. Corbett, ed.

© International Astronomical Union 2010
doi:10.1017/S1743921310009610

Sensitivity of microwave and FIR spectra to variation of fundamental constants

M. G. Kozlov[1] A. V. Lapinov[2] and S. A. Levshakov[3]

[1] Petersburg Nuclear Physics Institute, Gatchina, 188300, Russia
email: mgk@mf1309.spb.edu

[2] Institute of Applied Physics, Ulyanova Str. 46, Nizhni Novgorod, 603950, Russia

[3] Ioffe Physical-Technical Institute, Politekhnicheskaya Str. 26, St. Petersburg, 194021, Russia

Microwave and far infrared (FIR) spectra of atoms and molecules are in general more sensitive to the variation of the fundamental constants than optical spectra. For example, FIR transitions between levels of the ground state multiplet ^3P_J of Carbon-like ions are sensitive to α-variation, (Levshakov *et al.* (2008)). Moreover, sensitivities of the transitions (1 - 0) and (2 - 1) are different, (Kozlov *et al.* (2008)). This allows to study α-variation by comparing apparent redshifts for these two transitions of the same ion and significantly reduce systematic errors from the Doppler noise.

Molecular microwave lines are typically sensitive to the variation of the electron-to-proton mass ratio $\mu = m_e/m_p$. In some cases, such as the inversion spectrum of ammonia, this sensitivity is additionally enhanced compared to rotational and vibrational transitions, van Veldhoven *et al.* (2004). This makes ammonia very promising candidate for the studies of μ-variation, Flambaum & Kozlov (2007). In partly deuterated ammonia inversion and rotational transitions are mixed and sensitivities of microwave lines to variation of μ are significantly varying, Kozlov, Lapinov, & Levshakov (2009).

Another case, where enhancement takes place, corresponds to the Λ-doublet spectra of diatomic radicals OH and CH. These transitions are sensitive to both α- and μ-variation. Most interestingly, the coupling of the electronic angular momentum to the molecular axis for these light molecules depends on the rotational quantum number J. As a result, the sensitivity coefficients to the variation of α and μ also strongly depend on J. Therefore, one can compare apparent redshifts of Λ-doublets of different rotational levels. For some rotational quantum numbers the dimensionless sensitivity coefficients become very large (~ 1000) providing significant increase in sensitivity to variation of fundamental constants, Kozlov (2009).

Acknowledgements

This research is partly supported by the RFBR grants 08-02-00460, 09-02-00352, and 09-02-12223.

References

Flambaum, V. V. & Kozlov, M. G. 2007, Phys. Rev. Lett., 98, 240801
Kozlov, M. G. 2009, Phys. Rev. A, 80, 022118
Kozlov, M. G., Lapinov, A. V., & Levshakov, S. A. 2009, arXiv: 0908.2983
Kozlov, M. G., Porsev, S. G., Levshakov, S. A., Reimers, D., & Molaro, P. 2008, Phys. Rev. A, 77, 032119
Levshakov, S. A., Reimers, D., Kozlov, M. G., Porsev, S. G., & Molaro, P. 2008, A&A, 479, 719
van Veldhoven, J., Küpper, J., Bethlem, H. L., *et al.* 2004, Eur. Phys. J. D, 31, 337

Highlights of Astronomy, Volume 15
XXVIIth IAU General Assembly, August 2009
Ian F. Corbett, ed.

© International Astronomical Union 2010
doi:10.1017/S1743921310009622

GAME – A small mission concept for high-precision astrometric test of General Relativity

A. Vecchiato[1], Mario Gai[1], Paolo Donati[2,1], Roberto Morbidelli[1], Mario G. Lattanzi[1] and Mariateresa Crosta[1]

[1]Istituto Nazionale di Astrofisica – Osservatorio Astronomico di Torino,
Via Osservatorio 20, I-10025 Pino Torinese (TO), Italy
email: vecchiato@oato.inaf.it

[2]Università di Torino – Dipartimento di Fisica,
via Giuria 1, I-10125 Torino, Italy

Abstract. GAME (Gamma Astrometric Measurement Experiment) is a concept for a small mission whose main goal is to measure from space the γ parameter of the Parameterized Post-Newtonian formalism, Will (2001)) A satellite, looking as close as possible to the Solar limb, measures the gravitational bending of light in a way similar to that followed by past experiments from the ground during solar eclipses. In the cited formalism, deviations of the γ parameter from unity are interpreted as deviations from the predictions of General Relativity which are foreseen by several competing theories of gravity. In the present theoretical scenario, such deviations are expected to appear in the range between 10^{-5} and 10^{-7}. The most stringent experimental constraints available up to now are those of the Cassini mission, that gives $1 - \gamma \lesssim 10^{-5}$ Bertotti *et al.* (2003), while future space missions are expected to reach the 10^{-7} level of accuracy. (Vecchiato *et al.* (2003), Turyshev *et al.* (2004), Ni (2008))

Preliminary simulations have shown that the expected final accuracy of GAME can reach the 10^{-7} level, or better if the mission profile can be extended to fit a larger budget Vecchiato *et al.* (2009),Gai *et al.* (2009). This work, which has recently been extended to better assess the mission performances, has confirmed the previous results and has given indications on how further improve various aspects of the mission profile.

Moreover, thanks to its flexible observation strategy, GAME is also able to target other interesting scientific goals in the realm of General Relativity, as well as in those involving observations of selected extrasolar systems in the brown dwarf and planetary regime.

Keywords. Relativity – gravitation – astrometry – telescopes – methods: numerical – instrumentation: miscellaneous

References

Bertotti B., Iess L., & Tortora P., Sep. 2003, Nature (London), 425, 374
Gai M., Vecchiato A., Lattanzi M. G., Ligori S., & Loreggia D., Sep. 2009, Adv. Space Res., 44, 588
Ni W. T., 2008, International Journal of Modern Physics D, 17, 921
Turyshev S. G., Shao M., & Nordtvedt K. Jr., May 2004, Astronomische Nachrichten, 325, 267
Vecchiato A., Lattanzi M. G., Bucciarelli B., *et al.*, Feb. 2003, Astron. Astrophys., 399, 337
Vecchiato A., Gai M., Lattanzi M. G., Crosta M., & Sozzetti A., Sep. 2009, Adv. Space Res., 44, 579
Will C. M., 2001, Living Rev. Relativity, 2, [Online article]: cited on 14 October 2010,
http://www.livingreviews.org/Articles/Volume4/2001-4will/

Highlights of Astronomy, Volume 15
XXVIIth IAU General Assembly, August 2009
Ian F. Corbett, ed.
© International Astronomical Union 2010
doi:10.1017/S1743921310009634

VLT and E-ELT spectrographs & fundamental-constants

Paolo Molaro[1]

[1]INAF –Osservatorio Astronomico di Trieste, Via Tiepolo 11, I-34131 Trieste, Italy

The fundamental dimensionless physical constants cannot be predicted by theory but can only be measured experimentally. And so it is of their possible variation where there are several theoretical predictions but unfortunately with little theoretical guidance on the expected rate of change. The role of fundamental constants in the representation of nature as well as the implications of their variability for the Equivalence Principle and cosmology have been highlighted in many contributions at this conference (cfr K. Olive and J.P Uzan, these proceedings). Measuring the variability of the fine structure constant α or the electron-to-proton ratio μ by means of absorption lines implies the measurement of a tiny variation of the position of one or a few lines with regard to other lines which are taken as reference. For the fine structure constant the relation between its change and the doppler velocity shift is:

$$\frac{\Delta\alpha}{\alpha} \approx \frac{\Delta v}{2c}\Delta Q \tag{0.1}$$

where Q are the coefficients which describe the sensitivity of the wavelength to α. The Q values are theoretically computed and typical values are of order ≈ 0.02. Given these sensitivities in order to reveal changes of one part per million in α we need to be able to detect relative shifts of about 20 m s^{-1}. A similar relation holds for the line position the Lyman and Werner lines of molecular hydrogen and the sensitivity coefficients of to a change of μ are reflected in the K coefficients.

The precision with which the wavelength of a spectral transition can be determined from an absorption line depends on the signal-to-noise of the spectrum, the intrinsic width of the absorption line, the spectrograph resolution and pixel size. The wavelength error decreases with signal-to-noise with the decreasing intrinsic line width and with increasing spectrograph resolving power until the intrinsic line width is resolved. A convenient expression for the wavelength error of a gaussian line:

$$\sigma_0 = \frac{1}{(2\pi \ln 2)^{1/4}} \frac{1}{S/N}\sqrt{\Delta_{\text{pixel}}\text{FWHM}}.$$

However, in addition to photon noise there are the errors of instrumental origin and those coming from the wavelength calibration. It is rather questionable if these can be below 50 m/s, thus becoming the dominant source of uncertainty for measures of $\frac{\Delta\alpha}{\alpha}$ with the ambition to reach the 1 ppm level accuracy. An ideal instrument to probe fundamental constants such as the fine structure constant and the electron-to-proton mass ratio by means of absorption lines in QSOs spectra is a spectrograph which combine high throughput, high resolution and high stability and is compulsory attached to a telescope with a large photon collecting area. The ESPRESSO proposal for the incoherent combined VLT focus, and CODEX for the E-ELT, keep these recipes and, although not being optimized for this purpose, hold the promise to improve the present limits by about two orders of magnitude. Thus either these physical constants are varying within this range or they would likely escape astronomical detection.

Highlights of Astronomy, Volume 15
XXVIIth IAU General Assembly, August 2009
Ian F. Corbett, ed.

The breaking of the Equivalence Principle in theories with varying α

Lucila Kraiselburd[1,2] and Héctor Vucetich[1]

[1] Facultad de Ciencias Astronómicas y Geofísicas, Universidad Nacional de La Plata, Argentina

email: vucetich@fcaglp.fcaglp.unlp.edu.ar

[2] CONICET, Argentina
email: lulikrai@gmail.com

Abstract. The Standard Model and General Relativity provide a good description of phenomena at low energy. These theories, which agree very well with the experiment, contain a set of parameters called "fundamental constants", that are assumed invariant under changes in location and reference system. However, their possible variation has been studied since Dirac made the large numbers hypothesis (LNH). Moreover, unified field theory and extra dimensions theories such as Kaluza-Klein or Superstring theories, state not only the variation of these constants, but also the simultaneity of the variations.

The *Eötvös* effect is one of the most sensitive indicators of changes in fundamental constants. Bekenstein (2002) showed that in his theory, using a classical static particle model of matter, there is no *Eötvös* effect and therefore met the Universality of Free Fall and the Principle of Equivalence.

We present different results than those obtained by Bekenstein, Kraiselburd, Vucetich (2009). Modifying his theory, taking more realistic models of matter and using the model $TH\epsilon\mu$ techniques (Ligtman-Lee (1975) and Haugan (1979), not used before to analyze this model), very small but measurable effects have been found.

***

References

L. Kraiselburd & H. Vucetich. "Violation of the Weak Equivalence Principle in Bekenstein's theory". arXiv:0902.4146, 2009.

Jacob D. Bekenstein. "Fine-structure constant variability, equivalence principle and cosmology". *Phys. Rev.*, D66:123514, 2002.

M. P. Haugan. "Energy conservation and the principle of equivalence". *Ann. Phys.*, 118:156, 1979.

A. P. Lightman & D. L. Lee. "Restricted proof that the Weak Equivalence Principle implies the Einstein Equivalence Principle". *Phys. Rev. D*, 8:364, 1973.

Highlights of Astronomy, Volume 15
XXVIIth IAU General Assembly, August 2009
Ian F. Corbett, ed.

© International Astronomical Union 2010
doi:10.1017/S1743921310009658

Constraining Dark Matter-Dark Energy Interaction with Gas Mass Fraction in Galaxy Clusters

R. S. Gonçalves[1], J. S. Alcaniz[1] A. Dev[2] and D. Jain[3]

[1] Observatório Nacional,
Rio de Janeiro - RJ, 20921-400, Brasil

[2] Deen Dayal Upadhyaya College,
University of Delhi, Delhi - 110015, India

[3] Miranda House,
University of Delhi, Delhi - 110007, India

Abstract. The recent observational evidence for the current cosmic acceleration have stimulated renewed interest in alternative cosmologies, such as scenarios with interaction in the dark sector (dark matter and dark energy). In general, such models contain an unknown negative-pressure dark component coupled with the pressureless dark matter and/or with the baryons that results in an evolution for the Universe rather different from the one predicted by the standard ΛCDM model. In this work we test the observational viability of such scenarios by using the most recent galaxy cluster gas mass fraction versus redshift data (42 X-ray luminous, dynamically relaxed galaxy clusters spanning the redshift range $0.063 < z < 1.063$), Allen *et al.* (2008), to place bounds on the parameter ϵ that characterizes the dark matter/dark energy coupling. The resulting are consistent with, and typically as constraining as, those derived from other cosmological data. Although a time-independent cosmological constant (ΛCDM model) is a good fit to these galaxy cluster data, an interacting energy component cannot yet be ruled out.

Keywords. Cosmology – Cosmological Parameters – Coupled Quintessence – Distance Scale – Galaxy Clusters.

References

Allen, S. W. *et al.*, MNRAS, 383 (2008)

Highlights of Astronomy, Volume 15
XXVIIth IAU General Assembly, August 2009
Ian F. Corbett, ed.

Cosmological models and the brightness profile of distant galaxies

I. Olivares-Salaverri[1] and Marcelo B. Ribeiro[2]

[1] Observatório do Valongo, Universidade Federal do Rio de Janeiro, Brazil; iker@astro.ufrj.br

[2] Instituto de Física, Universidade Federal do Rio de Janeiro, Brazil; mbr@if.ufrj.br

This work aims to determine the feasibility of an assumed cosmological model by means of a detailed analysis of the brightness profiles of distant galaxies. Starting from the theory of Ellis & Perry (1979) connecting the angular diameter distance obtained from a relativistic cosmological model and the detailed photometry of galaxies, we assume the presently most accepted cosmology with $\Lambda \neq 0$ and seek to predict the brightness profile of a galaxy in a given redshift z. To do so, we have to make assumptions concerning the galactic brightness structure and evolution, assuming a scenario where the specific emitted surface brightness B_{e,ν_e} can be characterized as, $B_{e,\nu_e}(r,z) = B_0(z)J(\nu_e,z)f[r(z)/a(z)]$. Here r is the intrinsic galactic radius, ν_e the emitted frequency, $B_0(z)$ the central surface brightness, $J(\nu_e,z)$ the spectral energy distribution (SED), $f[r(z)/a(z)]$ characterizes the shape of the surface profile distribution and $a(z)$ is the scaling radius. The dependence on z is due to the galactic evolution. As spacetime curvature affects the received surface brightness, the reciprocity theorem (Ellis 1971) allows us to predict the theoretical received surface brightness. So, we are able to compare the theoretical surface brightness with its equivalent observational data already available for high redshift galaxies in order to test the consistency of the assumed cosmological model. The function $f[r(z)/a(z)]$ is represented in the literature by various different shapes, like the Hubble, Hubble-Oemler and Abell-Mihalas single parameter profiles, characterizing the galactic surface brightness quite well when the disk or bulge dependence is dominant. Sérsic and core-Sérsic profiles use two or more parameters and reproduce the galactic profile almost exactly (Trujillo et al. 2004). If we consider all wavelengths, the theory tells us that the total intensity is equal to the surface brightness, so the chosen bandwidth should include most of the SED. In order to analyze only the effect of the cosmological model in the surface brightness and minimize evolutionary effects, we assume that there exists a homogeneous class of objects, whose properties are similar in all redshifts, allowing us to carry out comparisons at different values of z. Studying the parameters that affect the galactic evolution, as well as in others geometrical tests, we will be able to infer some possible galaxy evolution which could reproduce a theoretical surface brightness profile, in order to compare with the observational data and reach conclusions about the observational feasibility of the underlying cosmological model.

Keywords. galaxies: structure, galaxies: evolution, cosmology: observations.

References

Abell, G. O & Mihalas, D. M 1966, *AJ* 71, 7

Binney, J. & Merrifield, M. 1999, *Galactic Astronomy*

Ellis G. F. R. 1971, in: Sachs, R. K. (ed.) Proc. Int. School of Physics "Enrico Fermi": pp. 104-182. Academic Press, New York (1971). Reprinted: *Gen. Relativ. Grav. (2009)* 41, 581

Ellis, G. F. R. & Perry, J. J. 1979, *Mon. Not. R. astr. Soc. (1979)* 187, 357

Trujillo, I., Peter Erwin, Asensio Ramos, A., & Graham, A. W. 2004, *AJ* 127, 1917

Highlights of Astronomy, Volume 15
XXVIIth IAU General Assembly, August 2009
Ian F. Corbett, ed.

© International Astronomical Union 2010
doi:10.1017/S1743921310009671

Calibration issues in $\delta\alpha/\alpha$

Miriam Centurión[1], Paolo Molaro[1] and Sergei Levshakov [2]

[1]INAF –Osservatorio Astronomico di Trieste, Via Tiepolo 11, I-34131 Trieste, Italy

[2]Ioffe Institute Politekhnicheskaya, Str. 26, 194021St. Petersburg, Russia

Laser Comb Wavelength calibration shows ThAr one locally unreliable with deviations up to $100\ \mathrm{m\,s^{-1}}$ (or $\delta\alpha/\alpha \approx 7\cdot10^{-6}$ for a Fe II-Mg II pair) while delivering an overall $1\ \mathrm{m\,s^{-1}}$ accuracy. Comparison of line shifts of the 5 Fe II lines with identical sensitivity to $\delta\alpha/\alpha$ offers a clean way to test local wavelength calibration errors of whatever origin.

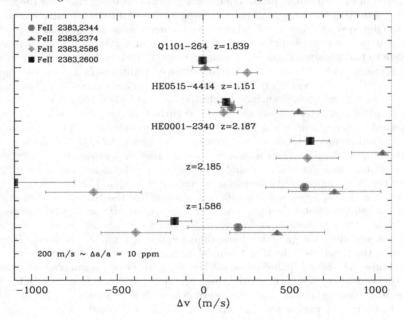

We analyzed 5 absorption systems, towards 3 QSOs. The results are shown in the Fig. Some lines are aligned within $20\ \mathrm{m\,s^{-1}}$, but others reveal large deviations reaching $200\ \mathrm{m\,s^{-1}}$ or higher (or $\delta\alpha/\alpha \geqslant 10^{-5}$). The origin of these deviations is not clearly identified These results suggest that extreme care is needed before drawing conclusions from $\delta\alpha/\alpha$ analysis based on one or only few lines.

Highlights of Astronomy, Volume 15
XXVIIth IAU General Assembly, August 2009
Ian F. Corbett, ed.

© International Astronomical Union 2010
doi:10.1017/S1743921310009683

Joint Discussion 10: 3D views on cool stellar atmospheres – theory meets observation

Organizing committee:

Carlos Allende Prieto (UK), Martin Asplund (Germany), Mats Carlsson (Norway), Márcio Catelan (Chile), Kwing Lam Chan (China), Dainis Dravins (Sweden), Hans-G. Ludwig (Chair, France), K.N. Nagendra (India), Åke Nordlund (Denmark), Natalia Shchukina (Ukraine), T. Sivarani (USA), and Matthias Steffen (Germany)

Preface

Much of what we know about the chemical composition of the Universe actually stems from the chemical composition of stars, which is often deciphered from the spectra emerging from their atmospheres. Cool, low-mass and long-living stars allow to study the evolution of the Universe's chemistry from a time shortly after the big bang until today. The observation and interpretation of stellar spectra is a classical field in astronomy but is still undergoing vivid developments. The enormous increase in available computational resources opened-up possibilities which led to a revolution in the degree of realism to which modelers can mimic Nature. High-resolution, high-stability, high-efficiency spectrographs are now routinely providing stellar spectra whose full information content can only be exploited if a very much refined description of a stellar atmosphere is at hand.

This situation motivated Commission 36 *Theory of Stellar Atmospheres* to organize an exchange of latest views on the modelling of atmospheres of cool stars, and their inherent complexities related to multi-dimensional hydrodynamics and magnetic fields. The idea materialized in the form of this Joint Discussion. Being a Joint Discussion ample time could be allocated to discussions, and indeed many lively disputes revolved around the diverse topics presented by the speakers and illustrated in posters. It is fair to say that they touched upon practically all aspects of stellar atmospheres – theoretical as well as observational with emphasis on their multi-dimensional nature.

Proceedings of the Joint Discussion were published in full in a dedicated volume of the Memorie della Società Astronomica Italiana. (Vol. 80, no. 3, 2009). The reprinted abstracts in this volume are intended to summarize the vivid exchange of views; they hopefully serve as a happy reminder to the participants of the *3D Views on Cool Stellar Atmospheres: Theory Meets Observation*, and provide some flavor of the event to those who have missed it.

We wish to thank the IAU for promoting this Joint Discussion. We are grateful to all the participants who have made it a lively meeting, and in particular to the colleagues who chaired the different sessions. Special thanks are due to the Sociedade Astronômica Brasileira who took care of the organization of the IAU General Assembly and welcomed us at Rio de Janeiro.

K. N. Nagendra, P. Bonifacio, H.-G. Ludwig, editors

Hydrodynamics and radiative transfer of 3D model atmospheres: current status, limitations, and how to make headway

M. Carlsson[1,2]

[1] *Institute of Theoretical Astrophysics, University of Oslo, P.O. Box 1029 Blindern, N-0315 Oslo, Norway*
[2] *Center of Mathematics for Applications, University of Oslo, P.O. Box 1052 Blindern, N-0316 Oslo, Norway*

Abstract. 3D MHD models are important tools for advancing our understanding of stellar atmospheres. A major computational challenge is the treatment of radiative transfer; both to get a realistic treatment of the energy transfer in the 3D modelling and for the diagnostic problem of calculating the emergent spectrum in more detail from such models. The current status, limitations and future directions of 3D MHD atmospheric modelling and the treatment of radiative transfer are here discussed.

Highest-resolution spectroscopy

D. Dravins
Lund Observatory, Box 43, SE-22100 Lund, Sweden

Abstract. 3-D models of stellar atmospheres predict spectral-line shapes with asymmetries and wavelength shifts, but the confrontation with observations is limited by blends, lack of suitable lines, imprecise laboratory wavelengths, and instrumental imperfections. Limits can be pushed by averaging many similar lines, thus averaging small random blends and wavelength errors. In non-solar cases, any detailed verification of 3-D hydrodynamics requires spectra of resolutions $R = \lambda/\Delta\lambda \approx 300,000$, soon to become available. An issue is the optical interface of high-resolution spectrometers to [very] large telescopes with their [very] large image scales, possibly requiring adaptive optics. The next observational frontier may be spectroscopy across spatially resolved stellar disks, utilizing optical interferometers and extremely large telescopes.

Granulation across the HR diagram

I. Ramírez[1], C. Allende Prieto[2], D. L. Lambert[3], L. Koesterke[3,4] and M. Asplund[1]

[1] *Max Planck Institute for Astrophysics, Postfach 1317, 85741 Garching, Germany*
[2] *Mullard Space Science Laboratory, University College London, UK*
[3] *McDonald Observatory and Dept. of Astronomy, University of Texas at Austin, USA*
[4] *Texas Advanced Computing Center, University of Texas, USA*

Abstract. We have obtained ultra-high quality spectra ($R = 180,000$; $S/N > 300$) with unprecedented wavelength coverage (4400 to 7400 Å) for a number of stars covering most of the HR diagram in order to test the predictions of models of stellar surface convection. Line bisectors and core wavelength shifts are both measured and modeled, allowing us to validate and/or reveal the limitations of state-of-the-art hydrodynamic model atmospheres of different stellar parameters. We show the status of our project and preliminary results.

Accounting for convective blue-shifts in the determination of absolute stellar radial velocities

C. Allende Prieto[1], L. Koesterke[2], I. Ramírez[3], H.-G Ludwig[4] and M. Asplund[3]

[1] *Mullard Space Science Laboratory, University College London, UK*
[2] *Texas Advanced Computing Center, The University of Texas at Austin, USA*
[3] *Max Plank Institute for Astrophysics, Garching, Germany*
[4] *CIFIST, GEPI, Observatoire de Paris, CNRS, Université Paris Diderot, France*

Abstract. For late-type non-active stars, gravitational redshifts and convective blueshifts are the main source of biases in the determination of radial velocities. If ignored, these effects can introduce systematic errors of the order of ~ 0.5 km s^{-1}. We demonstrate that three-dimensional hydrodynamical simulations of solar surface convection can be used to predict the convective blue-shifts of weak spectral lines in solar-like stars to ~ 0.070 km s^{-1}. Using accurate trigonometric parallaxes and stellar evolution models, the gravitational redshifts can be constrained with a similar uncertainty, leading to absolute radial velocities accurate to ~ 0.1 km s^{-1}.

3D radiative transfer with continuum and line scattering in low arbitrary velocity fields

A. M. Seelmann and P. H. Hausschildt

Hamburger Sternwarte, Gojenbergsweg 112, 21029 Hamburg, Germany

Abstract. With the increasing computer power of modern supercomputers, full 3D radiative transfer calculations with scattering is becoming more and more feasible. The PHOENIX/3D code is a first step towards realistic 3D model atmospheres with scattering in the line and the continuum. Numerous 3D hydro-dynamical codes exist, for example the FLASH or the CO^{5}BOLD, which can calculate 3D hydro-dynamical structures for, e.g., the Sun. This work concentrates on bringing the two approaches together in the sense of doing realistic, time-independent, radiative transfer with line scattering under considerations of the intrisitc velocity fields in a snapshot of the 3D hydro structure.

Projection methods for line radiative transfer in spherical media

L. S. Anusha and K. N. Nagendra

Indian Institute of Astrophysics, Koramangala 2nd Block, Bangalore 560 034, India

Abstract. An efficient numerical method called the Preconditioned Bi-Conjugate Gradient (Pre-BiCG) method is presented for the solution of radiative transfer equation in spherical geometry. A variant of this method called Stabilized Preconditioned Bi-Conjugate Gradient (Pre-BiCG-STAB) is also presented. These methods are based on projections on the subspaces of the n dimensional Euclidean space $\mathbb{R}^n$ called Krylov subspaces. The methods are shown to be faster in terms of convergence rate compared to the contemporary iterative methods such as Jacobi, Gauss-Seidel and Successive Over Relaxation (SOR).

The solar continuum intensity distribution

S. Wedemeyer-Böhm[1,2] and L. Rouppe van der Voort[1]

[1] *Institute of Theoretical Astrophysics, University of Oslo, P.O. Box 1029 Blindern, N-0315 Oslo, Norway*
[2] *Center of Mathematics for Applications (CMA), University of Oslo, Box 1053 Blindern, N0316 Oslo, Norway*

Abstract. For many years, there seemed to be significant differences between the continuum intensity distributions derived from observations and simulations of the solar photosphere. In order to settle the discussion on these apparent discrepancies, we present a detailed comparison between simulations and seeing-free observations that takes into account the crucial influence of instrumental image degradation. We use a set of images of quiet Sun granulation taken in the blue, green and red continuum bands of the Broadband Filter Imager of the Solar Optical Telescope (SOT) onboard Hinode. The images are deconvolved with Point Spread Functions (PSF) that account for non-ideal contributions due to instrumental stray-light and imperfections. In addition, synthetic intensity images are degraded with the corresponding PSFs. The results are compared with respect to spatial power spectra, intensity histograms, and the centre-to-limb variation of the intensity contrast. The observational findings are well matched with corresponding synthetic observables from three-dimensional radiation (magneto-)hydrodynamic simulations. We conclude that the intensity contrast of the solar continuum intensity is higher than usually derived from ground-based observations and is well reproduced by modern numerical simulations. Properly accounting for image degradation effects is of crucial importance for comparisons between observations and numerical models. It finally settles the traditionally perceived conflict between observations and simulations.

Temperature stratification in the Sun's photosphere in high horizontal resolution using Ca II H filtergrams

V. M. J. Henriques[1,2] and D. Kiselman[1]

[1] *Institute for Solar Physics, Royal Swedish Academy of Sciences, AlbaNova University Center, SE-106 91 Stockholm, Sweden*

[2] *Stockholm Observatory, Stockholm University, Sweden*

Abstract. A method to extract the temperature stratification in the Sun's photosphere using filtergrams is presented along with some high resolution results. The data was acquired with the Swedish 1-m Solar Telescope (SST) using a tunable filter in the Ca II H blue wing. Each full scan is completed in the order of seconds thus allowing for the full resolution of the SST and reasonable depth sampling to be obtained simultaneously in a shorter time than that of the evolution time scale of the photosphere. We test the quality of the method by applying it to a set of synthetic images (obtained through radiative transfer on 3D HD and MHD simulation snapshots followed by degradation) and comparing the output with the known 3D simulated atmosphere. Fine structure around bright points becomes evident in both the temperature gradient maps computed from a set of test observations and synthetic images obtained from MHD simulations.

Solar abundances and granulation effects

E. Caffau[1], H.-G. Ludwig[2,1] and M. Steffen[3]

[1] *GEPI, Observatoire de Paris, CNRS, Université Paris Diderot, Place Jules Janssen, 92190 Meudon, France*

[2] *CIFIST Marie Curie Excellence Team*

[3] *Astrophysikalisches Institut Potsdam, An der Sternwarte 16, D-14482 Potsdam, Germany*

Abstract. The solar abundances have undergone a major downward revision in the last decade, reputedly as a result of employing 3D hydrodynamical simulations to model the inhomogeneous structure of the solar photosphere. The very low oxygen abundance advocated by Asplund *et al.* (2004), A(O)=8.66, together with the downward revision of the carbon and nitrogen abundances, has created serious problems for solar models to explain the helioseismic measurements.

In an effort to contribute to the dispute we have re-derived photospheric abundances of several elements independently of previous analysis. We applied a state-of-the art 3D (CO5BOLD) hydrodynamical simulation of the solar granulation as well as different 1D model atmospheres for the line by line spectroscopic abundance determinations. The analysis is based on both standard disc-centre and disc-integrated spectral atlases; for oxygen we acquired in addition spectra at different heliocentric angles. The derived abundances are the result of equivalent width and/or line profile fitting of the available atomic lines. We discuss the different granulation effects on solar abundances and compare our results with previous investigations. According to our investigations hydrodynamical models are important in the solar abundance determination, but are not responsible for the recent downward revision in the literature of the solar metallicity.

Computation and analysis of gyrosyncrothron emission in solar flares

T. S. N. Pinto and J. E. R. Costa

Instituto Nacional de Pesquisas Espaciais – Divisão de Astrofísica, Av. dos Astronautas, 1758, CEP 12227-010, São José dos Campos - SP, Brazil

Abstract. The emission spectrum of solar flares in the range of microwave wavelengths is known to be due to the gyrosynchrotron mechanism. In this work, this emission for a few solar flares of the cycle 23 observed by the Nobeyama Radio Observatory instruments were analyzed. The information provided by the time profiles and the event images were used as the input parameters in the numerical computation of two flares' spectrum. A priori information obtained from the data was the electron spectral index, the emitting area and the angle between the magnetic field lines and the observer's line of sight. With these in hands, and following Dulk (1985) and Costa (2005) the first guesses of the magnetic field intensities were obtained. Based on a homogeneous source model where the depth is considered to be a fraction of the observed width, the emission spectra were fitted with the χ^2 criterion, with the non-thermal electron number density and the magnetic induction set as free parameters. With this simplified scenario the values obtained for one of the events were the typical, as can be seen from other flares' analysis in the literature. This flare showed an emission peak at 17 GHz resulting in an inferred magnetic induction of 663 G and an electron density of 2.6×10^6 cm^{-3}. For the second flare, the fitted spectrum showed many harmonics in frequencies higher than the peak emission (9 GHz) that resulted in a very high magnetic induction (2183 G) and a very low electron density (3.8×10^5 cm^{-3}). This analysis was a first step to model a sample of selected flares using anisotropic magnetic fields extrapolated from the photosphere using the force-free hypothesis. The final analysis will be done solving the transfer equation in this anisotropic ambient.

Testing 3D solar models against observations

T. M. D. Pereira[1,2], M. Asplund[3] and D. Kiselman[2]

[1] *Research School of Astronomy and Astrophysics, Australian National University, Cotter Rd., Weston, ACT 2611, Australia*
[2] *The Institute for Solar Physics of the Royal Swedish Academy of Sciences, AlbaNova University Center, 106 91 Stockholm, Sweden*
[3] *Max-Planck-Institut für Astrophysik, Postfach 1317, D–85741 Garching b. München, Germany*

Abstract. We present results from a series of observational tests to 3D and 1D solar models. In particular, emphasis is given to the line formation of atomic oxygen lines, used to derive the much debated solar oxygen photospheric abundance. Using high-quality observations obtained with the Swedish Solar Telescope (SST) we study the center-to-limb variation of the O I lines, testing the models and line formation (LTE and non-LTE). For the O I 777 nm triplet, the center-to-limb variation sets strong constraints in the non-LTE line formation, and is used to derive an empirical correction factor (S_H) to the classical Drawin recipe for neutral hydrogen collisions. Taking advantage of the spatially-resolved character of the SST data, an additional framework for testing the 3D model and line formation is also studied. From the tests we confirm that the employed 3D model is realistic and its predictions agree very well with the observations.

Doppler shifts in the transition region and corona

P. Zacharias[1], S. Bingert[1] and H. Peter[2]

[1] Kiepenheuer Institut für Sonnenphysik – Schöneckstrasse 6, D-79104 Freiburg, Germany
[2] Max-Planck-Institut für Sonnensystemforschung – Max-Planck-Strasse 2, D-37191 Katlenburg-Lindau, Germany

Abstract. Emission lines in the transition region and corona show persistent line shifts. It is a major challenge to understand the dynamics in the upper atmosphere and thus these line shifts, which are a signature of the mass cycle between the chromosphere and the corona. We examine EUV emission line profiles synthesized from a 3D MHD coronal model of a solar-like corona, in particular of an active region surrounded by strong chromospheric network. This allows us to investigate the physical processes leading to the line Doppler shifts, since we have access to both, the synthetic spectra and the physical parameters, i.e. magnetic field, temperature and density in the simulation box. By analyzing the evolution of the flows along field lines together with the changing magnetic structure we can investigate the mass cycle. We find evidence that loops are loaded with mass during a reconnection process, leading to upflows. After the loops disconnect from the reconnection site, they cool and drain which leads to the observed redshifts. Previous 1D loop models (neglecting the 3D nature) assumed that heating leads to evaporation and upflows followed by a cooling phase after the heating stops. The scenario modeled here is quite different, as it shows that the continuously changing three-dimensional magnetic structure is of pivotal importance to understand the mass balance between the chromosphere and the corona.

Cloud formation and dynamics in cool dwarf and hot exoplanetary atmospheres

Adam J. Burgasser

Center of Astrophysics and Space Sciences, University of California, 9500 Gilman Dr., San Diego, CA 92093, USA

Abstract. The lowest-mass stars, brown dwarfs and extrasolar planets present challenges and opportunities for understanding dynamics and cloud formation processes in low-temperature atmospheres. For brown dwarfs, the formation, variation and rapid depletion of photospheric clouds in L- and T-type dwarfs, and spectroscopic evidence for non-equilibrium chemistry associated with vertical mixing, all point to a fundamental role for dynamics in vertical abundance distributions and cloud/grain formation cycles. For exoplanets, azimuthal heat variations and the detection of stratospheric and exospheric layers indicate multi-layered, asymmetric atmospheres that may also be time-variable (particularly for systems with highly elliptical orbits). Dust and clouds may also play an important role in the thermal energy balance of exoplanets through albedo effects. For all of these cases, 3D atmosphere models are becoming an increasingly essential tool for understanding spectral and temporal properties. In this review, I summarize the observational evidence for clouds and dynamics in cool dwarf and hot exoplanetary atmospheres, outstanding problems associated with these processes, and areas where effective synergy can be achieved.

Reflectance spectra of earth-like exoplanets

M. Wagner and P. H. Hauschildt

Hamburger Sternwarte, Gojenbergsweg 112, 21029 Hamburg

Abstract. Numerical simulations on irradiated exoplanets provide spectra that contain informations about the temperature- and density structure and chemical composition of the exoplanetary atmosphere. The calculation of cool objects is challenging, because of the much more complex chemistry, i.e. the strong molecular abundances as well as the occurrence of dust formation. In order to create an object with planetary features, such as size, temperature and abundances, the stellar atmosphere code PHOENIX had to be adapted in an appropriate way. Starting with an object of Venus-like parameters in 1D (spherical setup), temperature and optical depth will be reduced to Earth-like values. But in an optically thin atmosphere, what influence might the surface texture have on the combined spectrum? An albedo module has already been embedded to serve the cases of non-angular dependence (e.g. soils, vegetation) and angular dependence, i.e. water surface. The aim is to expand the work to 3D.

Simulations of dust clouds in the atmospheres of substellar objects

B. Freytag[1,3], F. Allard[1,2], H.-G. Ludwig[3], D. Homeier[4] and M. Steffen[5]

[1] *Centre de Recherche Astrophysique de Lyon, UMR 5574: CNRS, Université de Lyon, École Normale Supérieure de Lyon, 46 allée d'Italie, F-69364 Lyon Cedex 07, France*

[2] *Institut dÁstrophysique de Paris, UMR 7095: CNRS, Université Pierre et Marie Curie-Paris 6, 98bis boulevard Arago, 75014 Paris, France*

[3] *Observatoire de Paris-Meudon, GEPI-CIFIST, F-92195 Meudon, France*

[4] *Institut für Astrophysik Göttingen, Georg-August-Universität, Friedrich-Hund-Platz 1, D-37077 Göttingen, Germany*

[5] *Astrophysikalisches Institut Potsdam, An der Sternwarte 16, D-14482 Potsdam, Germany*

Abstract. The atmospheres of brown dwarfs allow the formation of dust grains and their rain-out into deeper, invisible layers. However, observed spectra of L dwarfs can only be reproduced when static 1D models account for dust formation and its resulting greenhouse effect in the visible layers. Time-dependent hydrodynamical processes can mix up the material giving rise to complex unsteady weather phenomena on these objects. We performed radiation hydrodynamics simulations in two and three dimensions of the atmospheres of brown dwarfs with CO5BOLD, including a treatment of dust particles. We find that exponential overshoot (close to the gas convection zone), gravity waves (weak omni-present mixing), and convection within dust layers (in the thick clouds in cooler models) contribute to the atmospheric mixing, which is far from being a stationary process. The presence of dust in the atmospheres is accompanied by large temporal and spatial intensity fluctuations.

Brown dwarf parallax programs

R. L. Smart

Istituto Nazionale di Astrofisica – Osservatorio Astronomico di Torino, Strada Osservatorio 20, 10025 Pino Torinese, Italy

Abstract. Parallaxes are crucial for many brown dwarf topics from the substellar mass function to 3D atmospheric modeling. Here we discuss the current sample of brown dwarfs with parallaxes and the prospects for the near future.

Polarization: Proving ground for methods in radiative transfer

K. N. Nagendra[1], L. S. Anusha[1] and M. Sampoorna[2]

[1] *Indian Institute of Astrophysics, Koramangala, Bangalore 560 034, India*
[2] *Instituto de Astrofísica de Canarias, E-38205, La Laguna, Tenerife, Spain*

Abstract. Polarization of solar lines arises due to illumination of radiating atom by anisotropic (limb darkened/brightened) radiation. Modelling the polarized spectra of the Sun and stars requires solution of the line radiative transfer problem in which the relevant polarizing physical mechanisms are incorporated. The purpose of this paper is to describe in what different ways the polarization state of the radiation 'complicates' the numerical methods originally designed for scalar radiative transfer. We present several interesting situations involving the solution of polarized line transfer to prove our point. They are (i) Comparison of the polarized approximate lambda iteration (PALI) methods with new approaches like Bi-conjugate gradient method that is faster, (ii) Polarized Hanle scattering line radiative transfer in random magnetic fields, (iii) Difficulties encountered in incorporating polarized partial frequency redistribution (PRD) matrices in line radiative transfer codes, (iv) Technical difficulties encountered in handling polarized specific intensity vector, some components of which are sign changing, (v) Proving that scattering polarization is indeed a boundary layer phenomenon. We provide credible benchmarks in each of the above studies. We show that any new numerical methods can be tested in the best possible way, when it is extended to include polarization state of the radiation field in line scattering.

Magnetic structuring at spatially unresolved scales

J. O. Stenflo[1,2]

[1] *Institute of Astronomy, ETH Zurich, HIT J 23.6, CH-8093 Zurich*
[2] *Istituto Ricerche Solari Locarno (IRSOL), Via Patocchi, CH-6605 Locarno-Monti*

Abstract. Magneto-convection structures the Sun's magnetic field down to the magnetic diffusion scale of order 10 m, where the field ceases to be frozen-in. This is about four orders of magnitude below the current resolution limit of solar telescopes. The subpixel structuring has a dramatic effect on the derived, spatially averaged flux densities in the resolved domain, in particular on the angular distribution of the field. Thus we find that the previously reported apparent predominance of horizontal magnetic flux on the quiet Sun is an artefact of the sub-resolution structuring. Here we try to clarify how Stokes profile data may be used to explore the spatially unresolved domain. Insert text of abstract

---

Modeling the second solar spectrum

M. Sampoorna

Instituto de Astrofísica de Canarias, E-38205 La Laguna, Tenerife, Spain.

Abstract. We present an empirical approach to model the wing polarization of strong resonance lines. This procedure based on 'last scattering approximation' (LSA) was developed by Stenflo (1980) and Stenflo (1982), for coherent scattering in the laboratory frame. We generalize his empirical approach to handle partial frequency redistribution (PRD). We illustrate this approach by applying it to the Ca I 4227 Å line. The LSA approach is successful in reproducing the observed Stokes Q/I polarization, including the location of the wing polarization maxima and the minima around the Doppler core, but fails to reproduce the observed spatial variations of the far wing polarization in terms of magnetic field and PRD effects. This null result points in the direction of a non-magnetic origin, which may include local deviations from a plane-parallel stratification with an inhomogeneous solar atmosphere.

3D stellar atmospheres for stellar structure models and asteroseismology

F. Kupka[1,2]

[1] Observatoire de Paris, LESIA, CNRS UMR 8109, F-92195 Meudon, France

[2] Institute of Mathematics, University of Vienna, Nordbergstraße 15, A-1090 Vienna, Austria

Abstract. Convection is the most important physical process that determines the structure of the envelopes of cool stars. It influences the surface radiation flux and the shape of observed spectral line profiles and is responsible for both generating and damping solar-like oscillations, among others. 3D numerical simulations of stellar surface convection have developed into a powerful tool to model and analyse the physical mechanisms operating at the surface of cool stars. This review discusses the main principles of 3D stellar atmospheres used for such applications. The requirements from stellar structure and evolution theory to use them as boundary conditions are analysed as well as the capabilities of using helio- and asteroseismology to reduce modelling uncertainties and probing the consistency and accuracy of 3D stellar atmospheres as part of this process. Simulations for the solar surface made by different teams are compared and some issues concerning the uncertainties of this modelling approach are discussed.

The CIFIST 3D model atmosphere grid

H.-G. Ludwig[1,2], E. Caffau[2], M. Steffen[3], B. Freytag[1,2,4], P. Bonifacio[1,2,5] and A. Kučinskas[6,7]

[1] CIFIST – Marie Curie Excellence Team

[2] GEPI – Observatoire de Paris, CNRS, Université Paris Diderot, 92195 Meudon, France

[3] Astrophysikalisches Institut Potsdam, An der Sternwarte 16, 14482 Potsdam, Germany

[4] CRAL – UMR 5574 CNRS, Université de Lyon, École Normale Supérieure de Lyon, 46 allée d'Italie, 69364 Lyon Cedex 07, France

[5] INAF – Osservatorio Astronomico di Trieste, via Tiepolo 11, 34143 Trieste, Italy

[6] Institute of Theoretical Physics and Astronomy, Goštauto 12, Vilnius LT-01108, Lithuania

[7] Vilnius University Astronomical Observatory, Čiurlionio 29, Vilnius LT-03100, Lithuania

Abstract. Grids of stellar atmosphere models and associated synthetic spectra are numerical products which have a large impact in astronomy due to their ubiquitous application in the interpretation of radiation from individual stars and stellar populations. 3D model atmospheres are now on the verge of becoming generally available for a wide range of stellar atmospheric parameters. We report on efforts to develop a grid of 3D model atmospheres for late-type stars within the CIFIST Team at Paris Observatory. The substantial demands in computational and human labor for the model production and post-processing render this apparently mundane task a challenging logistic exercise. At the moment the CIFIST grid comprises 77 3D model atmospheres with emphasis on dwarfs of solar and sub-solar metallicities. While the model production is still ongoing, first applications are already worked upon by the CIFIST Team and collaborators.

Spatially resolving the inhomogeneous structure of the dynamical atmosphere of Betelgeuse with VLTI/AMBER

K. Ohnaka

Max-Planck-Institut für Radioastronomie, Auf dem Hügel 69, 53121 Bonn, Germany

Abstract. We present spatially resolved high-spectral resolution K-band observations of the red supergiant Betelgeuse (α Ori) using AMBER at the Very Large Telescope Interferometer (VLTI). IR long-baseline interferometry combined with spectral resolutions of 4800–12000 enables us to probe the inhomogeneous structures in the dynamical atmosphere of Betelgeuse using the CO first overtone lines near 2.3 μm. Our AMBER observations mark the highest spatial resolution (9 mas) achieved for Betelgeuse with five resolution elements over its stellar disk. The AMBER data in the CO lines reveal salient inhomogeneous structures. Particularly, the visibilities and differential/closure phases within the CO lines show that the blue and red wings of the lines originate in spatially distinct regions over the stellar disk, clearly demonstrating an inhomogeneous velocity field in the atmosphere of Betelgeuse. The AMBER observations in the CO lines can be roughly explained by a simple model, in which a patch of CO gas is moving outward or inward with velocities of 10–15 km s^{-1}, while the CO gas in the remaining region in the atmosphere is moving in the opposite direction at the same velocities. This model also suggests the presence of dense molecular layers extending to $\sim$1.4–1.5 $R_\star$ with a CO column density of $\sim 1 \times 10^{20}$ cm^{-2}. Our AMBER observations of Betelgeuse in the CO first overtone lines are the first spatially resolved study of the gas motion in a stellar atmosphere (photosphere and extended, warm molecular layers) other than the Sun and have opened a door to a better understanding of macroturbulence.

Abundance analysis of the halo giant HD 122563 with three-dimensional model stellar atmospheres

R. Collet[1], Å. Nordlund[2], M. Asplund[1], W. Hayek[3,1] and R. Trampedach[3]

[1] *Max-Planck-Institut für Astrophysik, Postfach 1317, D–85741 Garching b. München*
[2] *Niels Bohr Institute, University of Copenhagen, Juliane Maries Vej 30, DK–2100, Copenhagen, Denmark*
[3] *Research School of Astronomy and Astrophysics, Mount Stromlo Observatory, Cotter Road, Weston ACT 2611, Australia*

Abstract. We present a preliminary local thermodynamic equilibrium (LTE) abundance analysis of the template halo red giant HD122563 based on a realistic, three-dimensional (3D), time-dependent, hydrodynamical model atmosphere of the very metal-poor star. We compare the results of the 3D analysis with the abundances derived by means of a standard LTE analysis based on a classical, 1D, hydrostatic model atmosphere of the star. Due to the different upper photospheric temperature stratifications predicted by 1D and 3D models, we find large, negative, 3D−1D LTE abundance differences for low-excitation OH and Fe I lines. We also find trends with lower excitation potential in the derived Fe LTE abundances from Fe I lines, in both the 1D and 3D analyses. Such trends may be attributed to the neglected departures from LTE in the spectral line formation calculations.

3D hydrodynamical simulations of stellar photospheres with the CO^{5}BOLD code

A. Kučinskas[1,2], H.-G. Ludwig[3], E. Caffau[3] and M. Steffen[4]

[1] *Institute of Theoretical Physics and Astronomy, Goštauto 12, Vilnius LT-01108, Lithuania*
[2] *Vilnius University Astronomical Observatory, Čiurlionio 29, Vilnius LT-03100, Lithuania*
[3] *GEPI - CIFIST, Observatoire de Paris-Meudon, 5 place Jules Janssen, 92195 Meudon Cedex, France*
[4] *Astrophysikalisches Institut Potsdam, An der Sternwarte 16, D-14482 Potsdam, Germany*

Abstract. We present synthetic broad-band photometric colors of a late-type giant located close to the RGB tip ($T_{\rm eff} \approx 3640$ K, $\log g = 1.0$ and [M/H] $= 0.0$). Johnson-Cousins-Glass $BVRIJHK$

colors were obtained from the spectral energy distributions calculated using 3D hydrodynamical and 1D classical stellar atmosphere models. The differences between photometric magnitudes and colors predicted by the two types of models are significant, especially at optical wavelengths where they may reach, e.g., $\Delta V \approx 0.16$, $\Delta R \approx 0.13$ and $\Delta(V - I) \approx 0.14$, $\Delta(V - K) \approx 0.20$. Differences in the near-infrared are smaller but still non-negligible (e.g., $\Delta K \approx 0.04$). Such discrepancies may lead to noticeably different photometric parameters when these are inferred from photometry (e.g., effective temperature will change by $\Delta T_{\mathrm{eff}} \approx 60\,\mathrm{K}$ due to difference of $\Delta(V - K) \approx 0.20$).

The effective temperature scale: resolving different versions

L. Casagrande

Max Planck Institute for Astrophysics, Postfach 1317, 85741 Garching, Germany

Abstract. The effective temperature of a stellar surface is a measure of the total energy and its correct characterization plays a central role in both theory and observations. Various effective temperature scales have been proposed in literature. Despite being such a long-lived tradition and the high internal precision usually achieved, systematic differences of order 100 K among various scales are still present, thus hindering much of a progress in the field. We present an Infrared Flux Method based investigation aimed to carefully assess the sources of such discrepancies and pin down their origin. We break the impasse among different scales by using a large set of solar twins, stars which are spectroscopically and photometrically identical to the Sun, to set the zero-point of the effective temperature scale to within few degrees. Our newly calibrated, precise and accurate temperature scale applies to dwarfs and subgiants, from super solar metallicities to the most metal poor stars currently known. The effect of using spectral energy distribution computed from 3D models in the Infrared Flux Method, as well as 3D synthetic colours are also briefly outlined.

Micro- and macroturbulence derived from 3D hydrodynamical stellar atmospheres

M. Steffen[1], H.-G. Ludwig[2,3] and E. Caffau[3]

[1] *Astrophysikalisches Institut Potsdam, An der Sternwarte 16, D-14482 Potsdam, Germany*
[2] *CIFIST Marie Curie Excellence Team*
[3] *GEPI – Observatoire de Paris, CNRS, Université Paris Diderot; 92195 Meudon, France*

Abstract. The theoretical prediction of micro- and macroturbulence (ξ_{mic} and ξ_{mac}) as a function of stellar parameters can be useful for spectroscopic work based on 1D model atmospheres in cases where an empirical determination of ξ_{mic} is impossible due to a lack of suitable lines and/or macroturbulence and rotational line broadening are difficult to separate. In an effort to exploit the CIFIST 3D model atmosphere grid for deriving the theoretical dependence of ξ_{mic} and ξ_{mac} on effective temperature, gravity, and metallicity, we discuss different methods to derive ξ_{mic} from the numerical simulations, and report first results for the Sun and Procyon. In both cases the preliminary analysis indicates that the microturbulence found in the simulations is significantly lower than in the real stellar atmospheres.

3D molecular line formation in dwarf carbon-enhanced metal-poor stars

N. T. Behara[1,2], H.-G. Ludwig[1,2], P. Bonifacio[1,2,3], L. Sbordone[1,2],
J. I. González Hernández[1,2] and E. Caffau[2]
[1] CIFIST Marie Curie Excellence Team
[2] GEPI, Observatoire de Paris CNRS, Université Paris Diderot, Place Jules Janssen,
92190 Meudon, France
[3] Istituto Nazionale di Astrofisica - Osservatorio Astronomico di Trieste Via Tiepolo 11,
I-34143 Trieste, Italy

Abstract. We present a detailed analysis of the carbon and nitrogen abundances of two dwarf
carbon-enhanced metal-poor (CEMP) stars: SDSS J1349-0229 and SDSS J0912+0216. We also
report the oxygen abundance of SDSS J1349-0229. These stars are metal-poor, with [Fe/H] <
−2.5, and were selected from our ongoing survey of extremely metal-poor dwarf candidates from
the Sloan Digital Sky Survey (SDSS). The carbon, nitrogen and oxygen abundances rely on
molecular lines which form in the outer layers of the stellar atmosphere. It is known that convection in metal-poor stars induces very low temperatures which are not predicted by 'classical'
1D stellar atmospheres. To obtain the correct temperature structure, one needs full 3D hydrodynamical models. Using CO5BOLD 3D hydrodynamical model atmospheres and the Linfor3D
line formation code, molecular lines of CH, NH, OH and C_2 were computed, and 3D carbon,
nitrogen and oxygen abundances were determined. The resulting carbon abundances were compared to abundances derived using atomic C I lines in 1D LTE and NLTE. For one star, SDSS
J1349-0229, we were able to compare the 3D oxygen abundance from OH lines to O I lines in
1D LTE and NLTE. There is not a good agreement between the carbon abundances determined
from C_2 bands and from the CH band, and molecular lines do not agree with the atomic C
I lines. Although this may be partly due to uncertainties in the transition probabilities of the
molecular bands it certainly has to do with the temperature structure of the outer layers of the
adopted model atmosphere. In fact the discrepancy between C_2 and CH is in opposite directions
when using 3D and 1D models. Confronted with this inconsistency, we explore the influence of
the 3D model properties on the molecular abundance determination. In particular, the choice
of the number of opacity bins used in the model calculations and its subsequent effects on the
temperature structure and molecular line formation is discussed.

Effects of granulation on neutral copper resonance lines in metal-poor stars

P. Bonifacio[1,2,3], E. Caffau[2] and H.-G. Ludwig[1,2]
[1] CIFIST Marie Curie Excellence Team
[2] GEPI, Observatoire de Paris, CNRS, Université Paris Diderot; Place Jules Janssen,
92190 Meudon, France
[3] Istituto Nazionale di Astrofisica – Osservatorio Astronomico di Trieste, Via Tiepolo 11,
I-34143 Trieste, Italy

Abstract. We make use of three dimensional hydrodynamical simulations to investigate the
effects of granulation on the Cu I lines of Mult. 1 in the near UV, at 324.7 nm and 327.3 nm.
These lines remain strong even at very low metallicity and provide the opportunity to study the
chemical evolution of Cu in the metal-poor populations. We find very strong granulation effects
on these lines. In terms of abundances the neglect of such effects can lead to an overestimate
of the A(Cu) by as much as 0.8 dex in dwarf stars. Comparison of our computations with
stars in the metal-poor Globular Clusters NGC 6752 and NGC 6397, show that there is a
systematic discrepancy between the copper abundances derived from Mult. 2 in TO stars and
those derived in giant stars of the same cluster from the lines of Mult. 2 at at 510.5 nm and
587.2 nm. We conclude that the Cu I resonance lines are not reliable indicators of Cu abundance
and we believe that an investigations of departures from LTE is mandatory to make use of these
lines.

Monitoring mass motions of Betelgeuse's photosphere using robotic telescopes

M. Weber, T. Carroll, T. Granzer, M. Steffen and K. G. Strassmeier

Astrophysikalisches Institut Potsdam, An der Sternwarte 16, D–14482 Potsdam, Germany

Abstract. We started monitoring Betelgeuse using STELLA/SES, the STELLA échelle spectrograph fed by a robotic 1.2 m telescope on Tenerife, and the automatic photometric telescope (APT) T7 in Arizona in fall 2008. In this first observing season, we have collected 67 high resolution spectra from 390 to 900 nm at a resolution of 50,000 and a S/N between 100 and 300, and a comparable number of photometric observations in the Hα filter. In this presentation, we report on the initial findings based on this first data set: Radial velocities, effective temperature (along with surface gravity and metallicity) are automatically computed by the STELLA/SES data reduction & analysis pipeline. We compare these global measurements and the photometric brightness with velocities and temperature indicators derived from individual spectral lines, to bring these values in line with recently published observations. Furthermore we compute synthetic line profiles from state-of-the-art 3D stellar convection models, and compare the line-profiles, their shapes and positions to our observations. The final aim of the observing program is to find out if the spectral line variations can be explained using these non-magnetic convection models.

References

Asplund, M., Grevesse, N., Sauval, A. J., Allende Prieto, C., & Kiselman, D. 2004, *A&A* 417, 751

Costa, J. E. R. 2005, in: Proceedings of Magnetic Fields in the Universe, 544

Dulk, G. A. 1985, *ARAA* 23, 169

Stenflo, J. O. 1980, *A&A* 84, 68

Stenflo, J. O. 1982, *Solar Phys.* 80, 209

Monitoring mass outbursts of Betelgeuse's photosphere using robotic telescopes

Highlights of Astronomy, Volume 15
XXVIIth IAU General Assembly, August 2009
Ian F Corbett, ed.

© International Astronomical Union 2010
doi:10.1017/S1743921310009695

Detecting individual gravity modes in the Sun: Chimera or reality?

Rafael A. García*

Laboratoire AIM, CEA/DSM – CNRS - Université Paris Diderot – IRFU/SAp,
91191 Gif-sur-Yvette Cedex, France
email: rafael.garcia@cea.fr

Abstract. Over the past 15 years, our knowledge of the interior of the Sun has tremendously progressed by the use of helioseismic measurements. However, to go further in our understanding of the solar core, we need to measure gravity (g) modes. Thanks to the high quality of the Doppler-velocity signal measured by GOLF/SoHO, it has been possible to unveil the signature of the asymptotic properties of the solar g modes, thus obtaining a hint of the rotation rate in the core (García *et al.* 2007, 2008a). However, the quest for the detection of individual g modes is not yet over. In this work, we apply the latest theoretical developments to guide our research using GOLF velocity time series. In contrary to what was thought till now, we are maybe starting to identify individual low-frequency g modes...

Keywords. Sun: helioseismology, Sun: oscillations, Sun: interior, Sun: rotation

1. Observations and analysis

Gravity modes are very sensitive to the structure (e.g. Basu *et al.* 2009, García *et al.* 2008c) and the dynamics (e.g. Mathur *et al.* 2008) of the radiative zone and, in particular, to the inner core of the Sun. There have been many attempts to look for them without so far an undisputed detection of such modes (Appourchaux *et al.* 2009), although some interesting peaks and patterns have been detected with GOLF and VIRGO (e.g. Turck-Chièze *et al.* 2004, Jiménez & García 2009) with a high confidence level.

A 4500-day GOLF time series (Gabriel *et al.* 1995) starting on April 11, 1996 and calibrated into velocity (García *et al.* 2005) has been used to compute a single, full resolution power spectrum density (PSD) in spite of the different sensitivity to the visible solar disk between the blue- and the red-wing GOLF measurements (see for further details García *et al.* 1998, Ulrich *et al.* 2000). To increase the signal-to-noise ratio, we have also smoothed the PSD with a 41-nHz boxcar function as it is commonly done in asteroseismology when the signal is weak (e.g. Michel *et al.* 2008).

2. Discussion

Several of the highest peaks between 60 and 140 μHz are located around the theoretical frequencies of the dipole modes obtained by the Saclay seismic model (e.g. Mathur *et al.* 2007). Moreover, varying the splitting of the modes from 1 to 5 times the rotation rate of the radiative region, Ω_{rad}, we notice that there is a quasi complete sequence of peaks matching the model when the splitting of these modes is around 4.5 Ω_{rad} (see Fig.1). This could be the first time that individual g modes are identified in the Sun. For example, the candidate mode $\ell=1$, n$=-4$ has an amplitude of 1.8 ± 0.4 mm/s (with a signal-to-noise ratio of ~ 4) which is close to the latest theoretical predictions (Belkacem *et al.* 2009).

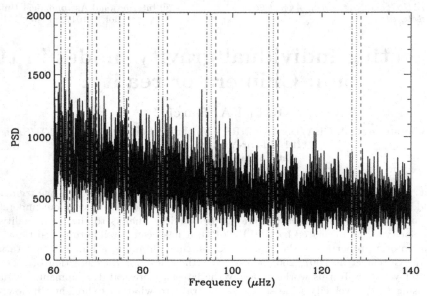

Figure 1. GOLF PSD. The dotted vertical lines are the central frequencies of the dipole g modes computed by the Saclay seismic model. The vertical dashed and dot-dashed lines are the rotational split components with a rotational splitting 4.5 times larger than in the radiative zone above 0.2 $R_\odot$.

The splittings can be estimated for each individual g-mode candidate. However, it is very difficult to obtain an accurate inference of the rotational profile of the core (García *et al.* 2008b, Mathur *et al.* 2009). The rotation rate in the core could be up to $\sim$7 Ω_{rad}, since $\sim$60% of the rotational kernel of these g modes is located in the inner core.

References

Appourchaux, T., Belkacem, K., Broomhall, A.-M. *et al.* 2009, *ARA&A*, arXiv:0910.0848
Basu, S., Chaplin, W. J., Elsworth, Y., New, R., & Serenelli, A. M. 2009, *ApJ*, 699, 1403
Belkacem, K., Samadi, R., Goupil, M. J., Dupret, M. A. 2009, *A&A* 494, 191
Gabriel, A. H., Grec, G., Charra, J. *et al.* 1995, *Sol. Phys.* 162, 61
García, R. A., Roca Cortés, T., & Régulo, C. 1998, *A&AS* 128, 389
García, R. A., Turck-Chièze, S., Boumier, P. *et al.* 2005, *A&A* 442, 385
García, R. A., Turck-Chièze, S., Jiménez-Reyes, S. J. *et al.* 2007, *Science* 316, 1591
García, R. A., Jiménez, A., Mathur, S. *et al.* 2008a, *Astron. Nachrichten* 329, 476
García, R. A., Mathur, S., Ballot, J. *et al.* 2008b, *Sol. Phys.* 251, 119
García, R. A., Mathur, S., & Ballot, J. 2008c, *Sol. Phys* 251, 135
Jiménez, A. & García, R. A. 2009, *ApJS* 184, 288
Mathur, S., Turck-Chièze, S., Couvidat, S., García, R. A. 2007, *ApJ* 668, 594
Mathur, S., Eff-Darwich, A., García, R. A.x, Turck-Chièze, S. 2008, *A&A* 484, 517
Mathur, S., García, R. A., & Eff-Darwich, A. 2009, *ASSP*, arXiv:0902.4142
Michel, E., Baglin, A., Auvergne, M. 2008, *Science* 322, 558
Turck-Chièze, S., García, R. A., Couvidat, S. *et al.* 2004, *A&A* 484, 517
Ulrich, R. K., García, R. A., Robillot, J.-M., *et al.* 2000, *A&A* 364, 799

*In collaboration with: J. Ballot, A. Eff-Darwich, R. Garrido, A. Jiménez, S. Mathis, S. Mathur, A. Moya, P. L. Pallé, C. Régulo, D. Salabert, K. Sato, J. C. Suárez and S. Turck-Chièze

Highlights of Astronomy, Volume 15
XXVIIth IAU General Assembly, August 2009
Ian F Corbett, ed.

© International Astronomical Union 2010
doi:10.1017/S1743921310009701

Dynamo action in rotating convection

Gustavo Guerrero[1,2] and Elisabete M. de Gouveia Dal Pino[2]

[1]Nordita, AlbaNova University Center,
Roslagstullsbacken 23 SE-10691, Stockholm, Sweden
email: guerrero@nodita.org

[2]Instituto de Astronomia, Geofísica e Ciências Atmosféricas (IAG), Universidade de São Paulo,
São Paulo, Brazil
email: dalpino@astro.iag.usp.br

Abstract. We present MHD numerical simulations of a rotating turbulent convection system in a 3D domain (we have used the finite volume, Goudunov type MHD code PLUTO (Mignone *et al.* 2007)). Rotating convection is the natural scenario for the study of the dynamo action which is able to generate a large scale magnetic field, like the observed in the sun. Though we have neglected in the present approach the Ω effect, due to a large scale shear, our model is appropriate to test the controversial existence of the so called α effect that arises from helical turbulence (e.g. Cattaneo & Hughes 2006, Käpylä *et al.* 2009). We start with a two-layer piecewise polytropic region in hydrostatic equilibrium (e.g. Ziegler 2002), considering one stable overshoot layer at the bottom and a convectively unstable layer at the top of the computational domain. We have allowed this hydrodynamic system to evolve up to the steady state, i.e., after about 10 turnover times (τ). Then, we introduced a seed magnetic field and let the system evolve for more $\sim 40\tau$. Our preliminary results are summarized below in Figure 2.

References

Cattaneo, F. & Hughes, D. 2006, *J. Fluid Mech.*, 553, 401

Käpylä, P., Korpi, M. & Brandenburg A. 2009, *ApJ*, 697, 1153

Mignone, A. and Bodo, G. and Massaglia, S. and Matsakos, T. and Tesileanu, O. and Zanni, C. and Ferrari, A. 2007, *ApJS*, 170, 228

Ziegler, U. 2002, *A&A*, 386, 331

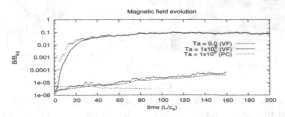

Figure 1. We show the time evolution of the mean magnetic field for models with two different rotation rates: $Ta=0$ and $Ta=10^6$. These simulations consider a vertical field (VF) boundary condition for the magnetic field (e.g., Käpylä *et al.* 2009). The case for $Ta=10^6$ with perfect conductor (PC) boundary conditions is also presented in this figure. Two significant results are evident: first, the presence of rotation allows the exponential amplification of a large scale magnetic field, which suggests the existence of a turbulent α effect operating in the convection zone, in agreement with Käpylä *et al.* (2009). Second, this result is very sensitive to the boundary conditions: for VF boundary conditions, or open boundaries, there is a growing of the large scale magnetic fields to values around 10 % of the equipartition value with the kinetic energy, while in the case with PC boundary conditions the saturation of the magnetic field occurs at the early stages of the time evolution resulting in a much smaller amplification of its strength. The evolution of the fluctuating components of the magnetic field are also shown with + symbols, for comparison.

Highlights of Astronomy, Volume 15
XXVIIth IAU General Assembly, August 2009
Ian F Corbett, ed.

© International Astronomical Union 2010
doi:10.1017/S1743921310009713

Realistic MHD numerical simulations of solar convection and oscillations in inclined magnetic field regions

Irina N. Kitiashvili[1], Alexander G. Kosovichev[2], Alan A. Wray[3] and Nagi N. Mansour[3]

[1] Center for Turbulence Research, Stanford University, Stanford, CA 94305, USA
email: irinasun@stanford.edu

[2] Hansen Experimental Physics Laboratory, Stanford University, Stanford, CA 94305, USA
email: AKosovichev@solar.stanford.edu

[3] Ames Research Center, Moffett Field, CA 94040, USA
email: Alan.A.Wray@nasa.gov and nagi.n.mansour@nasa.gov

Keywords. Sun: magnetic fields, oscillations, sunspots; convection, turbulence; methods: numerical

It is known that physical properties of solar turbulent convection and oscillations strongly depend on magnetic field. In particular, recent observations from SOHO/MDI revealed significant changes of the wave properties in inclined magnetic field regions of sunspots, which affect helioseismic inferences. We use realistic 3D radiative MHD numerical simulations to investigate solar convection and oscillations and their relationship in the presence of inclined magnetic field. In the case of highly inclined and strong 1-1.5 kG field the solar convection develops filamentary structure and high-speed flows (Fig. 1a), which provide an explanation to the Evershed effect in sunspot penumbra (Kitiashvili, *et al.* 2009).

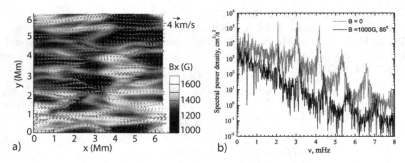

Figure 1.
a) Filamentary structure of the horizontal component of magnetic field, Bx, developed by magnetoconvection in strong magnetic field ($B_0 = 1000$ G) inclined by 85^0 towards the surface. Arrows show the velocity field. b) Oscillation power spectra of the vertical component of velocity averaged over the horizontal photospheric region for the inclined field (black) and non-magnetic region (gray) regions.

The power spectrum of large wavelength oscillations shows that compared to the non-magnetic quiet-Sun regions the amplitude of acoustic oscillations excited by turbulent magnetoconvection in sunspot penumbrae is substantially reduced (at least, by a factor 10), and that the resonant frequencies are shifted towards higher frequencies (Fig.1b). This may explain the observed frequency increase in magnetic regions.

References

Kitiashvili, I.N., Kosovichev, A.G., Wray, A.A. & Mansour, N.N. 2009, *ApJ* 700, L178

Highlights of Astronomy, Volume 15
XXVIIth IAU General Assembly, August 2009
Ian F Corbett, ed.

© International Astronomical Union 2010
doi:10.1017/S1743921310009725

How surface magnetism affects helioseismic waves

Paul S. Cally

Centre for Stellar and Planetary Astrophysics, Monash University, Clayton, Victoria, Australia 3800
email: paul.cally@sci.monash.edu.au

Abstract. It has been known for two decades that sunspots both absorb and advance the phase of solar f and p-modes. More recently, Time-Distance and other local helioseismic techniques have been used to probe active regions by exploring phase shifts which are interpreted as travel-time perturbations. Although absorption is an intrinsically magnetic effect, phase shifts may be produced by both thermal and magnetic effects (and of course flows, though these can be factored out by averaging travel times in opposite directions). We will show how these two effects alter wave phase, and conclude that phase shifts in umbrae are predominantly thermal, whilst those in highly inclined field characteristic of penumbrae are essentially magnetic. The two effects are generally not additive.

Keywords. Sun: helioseimology, Sun: magnetic fields

1. Introduction

The aim of this paper is to present a simplified view of how strong surface magnetic fields in solar active regions are expected to influence helioseismic waves incident from below, with implications for both the atmosphere above, and the internal helioseismic wave field. As first suggested by the author nearly a decade ago (Cally (2000)), magnetic field inclination is the crucial ingredient that redresses the inability of vertical magnetic field models Cally, Bogdan & Zweibel (1994) to sufficiently interact with p-modes to explain observed absorption and phase shifts. Our understanding of the process of mode transmission/conversion has advanced greatly in the last few years due to the development of several different techniques, most notably Generalized Ray Theory see Section 2. However, analysis of the phase discontinuities characteristic of reflection at caustics and of mode conversion has only been tackled this year (Cally (2009a), Cally (2009b)), and we summarize the results in Section 3.

2. Transmission and Conversion

Classical ray theory as applied to MHD waves Weinberg (1962) is based on the eikonal approximation and the resulting dispersion relation $\mathcal{D}(\omega, k_x, k_y, k_z; \mathbf{x}) = 0$ connecting the frequency ω, wave vector $\mathbf{k}$, and position $\mathbf{x}$. With ω, k_x and k_y held fixed for instance (in a horizontally invariant atmosphere), k_z changes with height z in such a way that $\mathcal{D}(\omega, k_x, k_y, k_z; z) = 0$, thus defining a relationship between k_z and z. This forms loci in the z-k_z plane which may be categorized as *fast*, *slow*, or *Alfvén*, and generally these are disjoint, meaning that solutions stay fast, slow, or Alfvén. However, at avoided crossings between these curves (mode transmission/conversion points), the eikonal approximation breaks down and a more sophisticated matching process Tracy, Kaufman & Brizard (2003) is required which reveals tunnelling between the branches. Fast/slow *transmission* occurs most readily when the *attack angle* between the wave vector $\mathbf{k}$ and the magnetic field $\mathbf{B}$ at the Alfvén/acoustic equipartition level $a = c$ is small; this represents the coupling of a predominantly acoustic wave below $a = c$ with a predominantly acoustic wave above this level, hence "transmission". At large attack angle a fast wave from below stays fast as it passes through $a = c$, but that corresponds to a change of nature from acoustic to magnetic, hence "conversion". At intermediate attack angles there is both substantial transmission and conversion: the wave splits.

If we are concerned with the part of the split wave field which returns to the solar interior as part of the "helioseismic field" of skipping acoustic waves, its path through the magnetic surface layer predominantly follows this route: (i) conversion to a magnetic (*i.e.*, fast) wave near $a = c$; (ii) refraction back downward due to the rapidly increasing Alfvén speed with height in the active region atmosphere; (iii) conversion back to acoustic as it passes downward through $a = c$ again. All this is most pronounced if the attack angle is large. Since helioseismic waves typically impinge on the surface quite steeply, this process is most favoured in highly inclined magnetic field, characteristic of sunspot penumbrae. In umbrae, where the field is more vertical and the attack angle smaller, there is mostly transmission as an acoustically dominated atmospheric wave, which may or may not reflect, depending on the relation of the wave frequency to the acoustic cutoff frequency. Even if it does though, the wave timings, both group and phase, are very different for the two scenarios.

3. Conclusions: Phase and travel-time perturbation

Local helioseismology typically identifies phase differences between skips in active regions compared with equivalent oscillations in the quiet Sun as evidence of a travel time perturbation. These may be variously identified as resulting from changes in subsurface sound speed, flows, or direct magnetic effects. Numerical modelling, both wave and ray, leads to the following conclusions regarding ray timings based on phase perturbations:

(*a*) There is a strong field inclination dichotomy;

(*b*) Thermal effects are only a good predictor of travel-time perturbations (really phase perturbations) in umbrae, where field is near vertical and waves are impinging roughly parallel to **B**;

(*c*) Travel-time perturbations in highly inclined field (penumbra) are mostly magnetic in origin;

(*d*) Magnetic effects increase strongly with $|B|$.

This travel time dichotomy may explain the observations of Couvidat & Rajaguru (2007), who notice a distinct "ring-like" structure in travel time maps in sunspot penumbrae.

References

Cally, P. S. 2000, *Solar Phys.*, 192, 395
Cally, P. S. 2006, *Phil. Trans. Roy. Soc. Lond. A* 364, 333
Cally, P. S. 2007, *AN* 328, 286
Cally, P. S. 2009a, *Solar Phys.* 254, 241
Cally, P. S. 2009b, *MNRAS* 395, 1309
Cally, P. S., Bogdan, T. J., & Zweibel, E. G. 1994, *ApJ* 437, 505
Couvidat, S. & Rajaguru, S. P. 2007, *Astrophys. J.* 661, 558
Hansen, S. C. & Cally, P. S. 2009, *SolarPhys.* 255, 193
Schunker, H. & Cally, P. S. 2006, *MNRAS*, 372, 551
Tracy, E. R., Kaufman, A. N., & Brizard, A. J. 2003, *Phys. Plasmas* 10, 2147
Weinberg, S. 1962, *Phys. Rev.*, 126, 1899

Highlights of Astronomy, Volume 15
XXVIIth IAU General Assembly, August 2009
Ian F Corbett, ed.

© International Astronomical Union 2010
doi:10.1017/S1743921310009737

Magnetic structure of sunspot under the photosphere

Elena A. Kirichek[1] and Alexandr A. Solov'ev[1]

[1]Central (Pulkovo) Observatory, Russian Academy of Sciences,
St. Petersburg, 196140 Russia
email: solov@gao.spb.ru

In recent years, the local helioseismology has become a highly effective tool for investigating subphotospheric layers of the Sun, which can yield fairly detailed distributions of the subphotospheric temperatures and large-scale plasma flows based on the spectra of the oscillations observed at the photospheric layers and the observed peculiarities of propagation of magnetoacoustic waves in this medium (Zhao *et al.* (2001), Kosovichev (2006)). Unfortunately, the effects of temperature and the magnetic field on the wave propagation speed have not yet been separated Kosovichev (2006), so that the structure of the sunspot magnetic field in deep layers, beneath the photosphere, remains a subject of purely theoretical analysis. In his analysis of some theoretical models of the subphotospheric layers of sunspots based on recent helioseismological data, Kosovichev (2006) concluded that Parker's ("spaghetti") cluster model Parker (1979) is most appropriate. In this model, the magnetic flux in the sunspot umbra is concentrated into separate, strongly compressed, vertical magnetic flux tubes that are interspaced with plasma that is almost free of magnetic field; the plasma can move between these tubes.

Our study is dedicated to the problem of the magnetic field structure in the subphotospheric layers of a sunspot regarded as a long-lived (actually steady-state) well-structured stable formation. We will demonstrate that, in the cool part of the magnetic flux tube of sunspot the Parker's model can be successfully applied, but in the region of "overheating" located, in according with helioseismological data, under the cool column of the spot, at depths from about 4 Mm, the condition of transverse (horizontal) equilibrium, i.e. the lateral balance of the total (gaseous + magnetic) pressures, cannot be satisfied in Parker's sunspot model. The magnetic flux tube of the spot must expand sharply with the depth starting from depths of about 4 Mm, so that it penetrates into deeper hot layers of the convection zone in an very expanded, diffuse form, as is assumed in the "shallow sunspot" model of Efremov *et al.* (2007), Solovev & Kirichek (2008).

We use this model to give the theoretical interpretation of the low-frequency eigen oscillations of sunspots observed both by ground-based telescopes (Pulkovo) and by space instruments (MDI/SOHO) Efremov *et al.* (2009).

References

Efremov, V. I., Parfinenko, L. D., & Solovev, A. A. 2007, *Astronomy Reports*, 51, N5, 401-410.
Efremov, V. I., Parfinenko, L. D., & Solovev., A. A. 2009, *(accepted). Solar Phys.*
Kosovichev, A. G. 2006, *Adv. Space Res.*, 38, 876
Parker, E. N. 1979, *Cosmical Magnetic Fields*, Part 1. (Claredon Press, Oxford), 608.
Solovev A. A. & Kirichek E. A. 2008, *Astrophysical Bulletin*, 63, N2, 169-180
Zhao, J., Kosovichev A. G. and Duval, T. L. 2001, *Astrophys. J.*, 557, 384

Highlights of Astronomy, Volume 15
XXVIIth IAU General Assembly, August 2009
Ian F Corbett, ed.

The future of helioseismology

Alexander G. Kosovichev

Hansen Experimental Physics Laboratory, Stanford University, Stanford, CA 94305, USA
email: AKosovichev@solar.stanford.edu

Abstract. Helioseismology has provided us with the unique knowledge of the interior structure and dynamics of the Sun, and the variations with the solar cycle. However, the basic mechanisms of solar magnetic activity, formation of sunspots and active regions are still unknown. Determining the physical properties of the solar dynamo, detecting emerging active regions and observing the subsurface dynamics of sunspots are among the most important and challenging problems. The current status and perspectives of helioseismology are briefly discussed.

Keywords. Sun: helioseismology, activity, interior magnetic fields, oscillations, sunspots

1. Outstanding problems of helioseismology

One of the most important unsolved problems of solar physics and astrophysics is understanding of the physical mechanism of the dynamo operating inside the Sun and producing 11-year activity cycles. Despite a significant progress in theoretical modeling of dynamo processes and successful laboratory experiments the basic understanding of the solar dynamo is still missing. The current theories assume that the poloidal component of magnetic field, which represents the global dipole field, is generated by helical turbulence at the bottom of the convection zone - tachocline, and that the toroidal component, which is a source of bipolar sunspot regions, is produced by stretching the poloidal field by the differential rotation (e.g. Jouve, *et al.* 2008). However, so far helioseismology observations were not able to confirm these models. If the magnetic field is generated by turbulence it is expected that when the field becomes sufficiently strong it suppresses the turbulence and affects the turbulent stresses that maintain the differential rotation. However, the helioseismic observations during a whole solar cycle were not able to detect significant variations of the rotation rate in the tachocline. They found a weak evidence of variations with a period of ~1.3 years but no changes on the 11-year scale. In addition, to explain the emergence of magnetic fields at mid- and low latitudes the strength of the toroidal magnetic field in the tachocline must be 60-100 kG. This exceeds the energy equipartition value, and requires a dynamic compression, e.g. by back reaction of updraft motions associated with emerging magnetic flux (Parker, 2009). Such motions have not been detected by helioseismology. Alternatively, the magnetic fields can be generated in the bulk of the convection zone and shaped by the subsurface shear layer (Brandenburg, 2009). The equator-ward migration of dynamo waves in this shear layer could explain the sunspot butterfly diagram. The migrating zone of sunspot formations is associated with zonal flows ("torsional oscillations"). The magnetic flux that forms sunspots tend to emerge in the shear layer between slower and faster flows. Helioseismology has established that these flows are quite deep and occupy at least the upper 30% of the convection zone. They are particularly strong at high latitudes, where they migrate towards the poles, and perhaps are related to the polar field reversals. However, the physical mechanism of these flows and their relation to the dynamo mechanism are not yet established (Howe *et al.* 2009). Synoptic observations showed that the polar field reversals correlate with the polarward transport of magnetic flux. This transport is assumed to be caused by meridional flow. Local helioseismology has detected this flow below the surface, and found that its speed significantly changes with time (e.g. Svanda, 2007). The flux-transport dynamo theories assume that after reaching the polar regions the magnetic flux is transported downward to the tachocline, and then by a reverse meridional flow towards the low-latitude regions. However, the reverse flow has not been detected. To determine the deep structure of the Sun's meridional circulation is one of the greatest challenges. Successful imaging of the tachocline structure by time-distance helioseismology provides optimism that this problem will be solved (Zhao *et al.* 2009). Implementation of modern data assimilation methods (Kitiashvili *et al.* 2008) gives a promising approach for forecasting the state of the solar dynamo and future solar activity.

Another outstanding problem of helioseismology is the detection of magnetic flux before it become visible on the surface. Local helioseismology has been successful in detecting large active regions on the far-side of the Sun, and this technique is becoming more robust (Hartlep *et al.* 2008). However, observations of emerging magnetic field below the surface are very difficult, because helioseismic signals for deep perturbations are weak, and in the upper convection zone the magnetic flux is emerging very fast, not giving enough time to accumulate a sufficient signal-to-noise ratio in the measurements. Therefore, the current efforts are focused on finding changes in the convective flow patterns associated with the emerging flux. The initial measurements have provided an evidence of upflows prior the flux emergence (Komm, *et al.* 2009). However, more systematic studies are necessary. Also, initial attempts are made to find a relationship between the evolution of subsurface sound-speed perturbations and flows and the development of the magnetic structure of active regions on surface (Kosovichev 2009).

2. Future helioseismology projects

The Helioseismic and Magnetic Imager (HMI) of board of the Solar Dynamics Observatory, scheduled for launch in 2010, will focus on investigations of the physical mechanisms and links among the solar dynamo, development of active regions and the internal dynamics leading to flares and coronal mass ejections (Kosovichev *et al.* 2007). The HMI will provide an unprecedented amount of high-resolution helioseismology data. Developing efficient and robust procedures for acoustic imaging of the deep interior is a new challenge of helioseismology. Future helioseismology projects include also currently planned out-of-ecliptic space missions, Solar Orbiter and Solar-C, which have a great potential for studying the solar dynamics in high-latitude and polar regions and for stereo-helioseismology of the tachocline and deeper interior. In addition, it is crucial to continue operating the ground-based helioseismology networks, GONG and BiSON, for monitoring the long-term behavior of our nearest star.

References

Brandenburg, A. 2009, in: *Cosmic Magnetic Fields*, Proc. IAU Symp., vol. 259, p. 159
Hartlep, T., *et al.* 2008, *ApJ*, 689, 1373
Howe, R., *et al.* 2009, *ApJ*, 701, L87
Jouve, L., *et al.* 2008, *A&A*, 483, 949
Kitiashvili, I. & Kosovichev, A. G. 2008, *ApJ*, 688, L49
Komm, R., *et al.* 2009, *Solar Phys.*, 258, 13
Kosovichev, A. G. & HMI Science Team, 2007, *Astr. Nach.*, 328, 339
Kosovichev, A. G. 2009, *Space Science Rev.*, 144, 175
Parker, E. N. 2009, *Space Science Rev.*, 144, 15
Švanda, M., *et al.* 2007, *ApJ*, 670, L69
Zhao, J., *et al.* 2009, *ApJ*, 702, 1150

Highlights of Astronomy, Volume 15
XXVIIth IAU General Assembly, August 2009
Ian F Corbett, ed.

© International Astronomical Union 2010
doi:10.1017/S1743921310009750

Numerical simulation of propagation of the MHD waves in sunspots

K. Parchevsky[1], A. Kosovichev[1], E. Khomenko[2,3], V. Olshevsky[3] and M. Collados[2]

[1] Hansen Experimental Physics Laboratory, Stanford University, Stanford, CA 94305, USA
email: kparchevsky@solar.stanford.edu

[2] Instituto de Astrofisica de Canarias, 38205 La Laguna, Tenerife, Spain
email: khomenko@iac.es

[3] Main Astronomical Observatory, NAS, 02680 Kyiv, Ukraine.

Abstract. We present results of numerical 3D simulation of propagation of MHD waves in sunspots. We used two self consistent magnetohydrostatic background models of sunspots. There are two main differences between these models: (i) the topology of the magnetic field and (ii) dependence of the horizontal profile of the sound speed on depth. The model with convex shape of the magnetic field lines near the photosphere has non-zero horizontal perturbations of the sound speed up to the depth of 7.5 Mm (deep model). In the model with concave shape of the magnetic field lines near the photosphere $\delta c/c$ is close to zero everywhere below 2 Mm (shallow model). Strong Alfven wave is generated at the wave source location in the deep model. This wave is almost unnoticeable in the shallow model. Using filtering technique we separated magnetoacoustic and magnetogravity waves. It is shown, that inside the sunspot magnetoacoustic and magnetogravity waves are not spatially separated unlike the case of the horizontally uniform background model. The sunspot causes anisotropy of the amplitude distribution along the wavefront and changes the shape of the wavefront. The amplitude of the waves is reduced inside the sunspot. This effect is stronger for the magnetogravity waves than for magnetoacoustic waves. The shape of the wavefront of the magnetogravity waves is distorted stronger as well. The deep model causes bigger anisotropy for both mgnetoacoustic and magneto gravity waves than the shallow model.

Keywords. helioseismology, magnetic fields, sunspots

We performed simulations for two types of magnetohydrostatic models of sunspots referred to as "deep" and "shallow" models with grid size $376 \times 376 \times 67$ ($\Delta x = \Delta y = 0.15$ Mm, $\Delta z_{top} = 0.05$ Mm, $\Delta z_{bottom} = 0.52$ Mm) for different strength of the magnetic field at the photospheric level (0.84 kG, 1.4 kG, and 2.2 kG).

The wave amplitude decreases when the wave enters the sunspot and restores its value when the wave passes the center of the sunspot. For the source distance of 9 Mm the wave amplitude inside the sunspot becomes bigger than outside. For the source distance of 12 Mm the wave amplitude inside the sunspot remains smaller than outside for all moments of time. For the fixed source location the wave amplitude inside the sunspot grows with the strength of the magnetic field.

For simulations with the magnetic field and multiple sources randomly distributed at the depth of 0.1 Mm the ratio of oscillation amplitudes outside and inside the sunspot increases to 4.9 ± 2 in comparison with the non-magnetic simulations 2.3 ± 0.4. The source strength was reduced inside the sunspot to simulate the suppression of the acoustic sources by the magnetic field inside sunspots. Observed ratio is 3.0-4.5 (± 0.4). This can be an evidence, that not all sources are suppressed inside the sunspot umbra.

Highlights of Astronomy, Volume 15
XXVIIth IAU General Assembly, August 2009
Ian F Corbett, ed.
© International Astronomical Union 2010
doi:10.1017/S1743921310009762

Differential Rotation of the Sun from helioseismology and magnetic field study.

Elena Gavryuseva[1] and Susanne Höfner[2]

[1] Institute for Nuclear Research of the Academy of Sciences of Russia,
Moscow, Russia
email: elena.gavryuseva@gmail.com

Abstract. Differential rotational rate of the large scale magnetic field and its temporal dependence has been evidenced at different latitudes through activity cycles 21-23. Rotational rate of the magnetic field at the latitudes above 55 degrees doesn't getting slower. Torsional waves with 11-year period are clearly seen in the magnetic field rotation. Rotation is slower where and when the magnetic field is getting stronger.

The time of the magnetic field emergence to the photosphere was estimated.

It was revealed the existence of quasi-stable over 30 years exceptionally regular and symmetric in the both hemispherest longitudinal structure. This structure is originated from the tachocline zone rotating like a rigid body with the rotation rate corresponding to the rotation in the photosphere on the 55-60 degrees of the helio-latitudes.

Keywords. Sun, magnetic field, rotation, longitudinal structure.

The measurements of the large scale solar magnetic field in the photosphere taken at Wilcox Solar Observatory since 1976 up to 2008 (http://wso.stanford.edu/synoptic.html) have been analyzed to deduce its latitudinal and longitudinal structures, its differential rotation, and their variability in time.

A longitudinal structure quasi-stable over 30 years exceptionally regular and symmetric in the both hemispheres has been found without any a priori assumption about the rotation of the Sun (Gavryuseva & Godoli (2004), Gavryuseva (2006)). This structure is originated from the tachocline zone under the convective envelope.

Comparison between the rotational rate of the large scale magnetic field and helioseismological findings of omega as a function of latitude and radius was performed.

Differential rotational rate of the magnetic field and its temporal dependence has been evidenced at different latitudes through activity cycles (Gavryuseva & Godoli (2004), Gavryuseva (2006), Gavryuseva (2007)). Rotational rate of the large scale magnetic field at the latitudes above 55 degrees doesn't getting slower. The time of emergence was estimated.

Torsional waves are clearly seen in the magnetic field rotation. Rotation is slower where and when the magnetic field is getting stronger.

Comparison between the rotation of the Sun inferred from helioseismology and the rotation of the large scale solar magnetic field permits to reveal that there are long living longitudinal structure originated from the tachocline zone rotating like a rigid body with the rotation rate corresponding to the rotation in the photosphere on the 55–60 degrees of the helio-latitudes. This longitudinal structure is clearly visible during present long interval of low solar activity. While during last cycles NN 21, 22 and 23 it was necessary to use the data set as long as 20 years about to select this regular longitudinal structure between the stochastic processes taken place in the photosphere. This study permits to combine the knowledge from helioseismology with solar activity phenomena ruled by the large scale magnetic field and to get a panoramic view on the internal structure and dynamics of the Sun.

References

Gavryuseva, E., Godoli, G. 2004, *Physics and Chemistry of the Earth*, v. 31, iss. 1–3, pp. 68-76.

Gavryuseva, E. 2006, *Solar Activity and its Magnetic Origin, Proceedings of the 233rd Symposium of the International Astronomical Union held in Cairo, Egypt, March 31 - April 4*, ed. V. Bothmer; A.A. Hady. Cambridge University Press, pp. 61–62

Gavryuseva, E. 2006, ibid, pp. 65–66

Gavryuseva, E. 2007, AIP

Gavryuseva, E. 2007, AIP

Highlights of Astronomy, Volume 15
XXVIIth IAU General Assembly, August 2009
Ian F Corbett, ed.

New advances in asteroseismology of pulsating hot subdwarf stars

S. Charpinet[1], V. Van Grootel[1], S.K. Randall[2], E.M. Green[3], G. Fontaine[4], P. Brassard[4] and P. Chayer[5]

[1] Laboratoire d'Astrophysique de Toulouse-Tarbes, Université de Toulouse, CNRS, 14 av. E. Belin, 31400 Toulouse, France
email: scharpin@ast.obs-mip.fr

[2] ESO, Karl-Schwarzschild-Str. 2, 85748 Garching bei München, Germany

[3] Steward Observatory, University of Arizona, 933 North Cherry Ave., Tucson, AZ 85721

[4] Université de Montréal, C.P. 6128, Succ. Centre-Ville, Montréal, QC H3C 3J7, Canada

[5] Space Telescope Science Institute, 3700 San Martin Drive, Baltimore, MD 21218, USA

Abstract. Hot subdwarf stars (of the sdB and sdO type) host three known classes of nonradial pulsators. Two of them feature short period ($P \sim 60 - 600$ s) accoustic mode oscillations, while the third group is characterized by slow g-mode deformations with periods of $\sim 1 - 2$h. These pulsations offer favorable grounds to infer some of the internal properties of these objects through asteroseismology. This has been exploited for the rapid p-mode sdB pulsators and the present contribution reviews some of the recent advances in this field. The long period g-mode pulsators, whose vibrations probe much deeper inside the star, are also of high interest. With the advent of space observations using CoRoT and KEPLER, the asteroseismology of these slower oscillators will also become a possibility, and likely contribute to significant breakthroughs in our understanding of these hot and compact stars.

Keywords. stars:subdwarfs; stars:oscillations; stars:interiors; stars:rotation

Quantitative asteroseismic studies of hot B subdwarf stars have been effective for several years, following the pioneering work of Brassard et al. (2001). The rapid, p-mode sdB pulsators, also known as the EC14026 or V361 Hya stars, proved to be particularly suitable for that purpose. These stars pulsate in low-order, low-degree accoustic modes with periods typically in the range $80 - 600$ s. For 12 such stars (out of 42 known), asteroseismic solutions could be derived providing accurate determinations of the main structural parameters: the mass, H-rich envelope mass, surface gravity, effective temperature, radius, absolute magnitude, and distance. For two objects, Feige 48 (Van Grootel *et al.*, 2008) and PG 1336–018 (Charpinet *et al.* 2008), information on the internal rotation could also be obtained. Such advances open up many opportunities to study this phase of stellar evolution as well as key physical process that influence the structure and evolution of stars in general (more can be found in the review paper of Charpinet *et al.* 2009).

References

Brassard *et al.* 2001, *ApJ*, 563, 1013
Charpinet *et al.* 2008, *A&A*, 489, 377
Charpinet *et al.* 2009, in *American Institute of Physics* Proceedings, in press
van Grootel, V., Charpinet, S., Fontaine, G., & Brassard, P. 2008, *A&A*, 483, 875

Highlights of Astronomy, Volume 15
XXVIIth IAU General Assembly, August 2009
Ian F Corbett, ed.

© International Astronomical Union 2010
doi:10.1017/S1743921310009786

Data Reduction Induced Errors in the Asteroseismology of Early-Type Stars

Petr Škoda

Astronomical Institute of the Academy of Sciences, Czech Republic
email: skoda@sunstel.asu.cas.cz

In the asteroseismology of early type stars it plays an important role the extremely precise high resolution spectroscopy. Especially the determination of pulsation modes from observations requires the identification of subtle changes in line profiles of a number of chemical elements. As commonly available instruments suitable for this purpose are echelle spectrographs, there is a number of possible imperfections and even systematic errors introduced by the echelle reduction procedures. We will mention several, most critical steps, that should be carefully checked during every reduction.

The errors in wavelength calibration may be caused by behaviour of two-dimensional dispersion polynomials at the edges of the chip. Further improvements may be achieved by elimination of blended lines and careful line centering algorithms (de Cuyper and Hensberge 1998) or by introduction of a new fitting formula (Hensberge and Verschueren 1989).

One of the problematic issues is the procedure of cosmics rejection by the variance-weighted optimal extraction based on estimated cross-order profile (Horne 1986). As was shown by Verschueren *et al.* (1997), it cannot bring measurable improvement on data with high SNR, but can introduce strange artifacts. Therefore the better method is the removal of cosmics from the original 2D frame using statistic medians (Pych 2004) or Laplace transform (van Dokkum 2001). The alternative, self-consistent method of optimal extraction was also suggested (Piskunov & Valenti (2002)).

The critical step before merging of separate spectral orders into one long spectrum is the removal of echelle blaze function. It is a very difficult task, some ideas being given by Škoda & Hensberge (2003), Hensberge (2007) or Škoda *et al.* (2008). The problem is much worse in case of early type stars, where some H and He lines may be extended over more that one echelle order, and so the precise order merging is required to preserve the line profile.

References

de Cuyper, J.-P. & Hensberge, H. 1998, *A&AS*, 128, 409

Hensberge, H. & Verschueren, W. 1989, *The Messenger*, 58, 51

Hensberge, H. 2007, in *The Future of Photometric, Spectrophotometric and Polarimetric Standardization*, ASP Conf. Series, 364, 275

Horne, K. 1986, *PASP*, 98, 609

Piskunov, N. E. & Valenti, J. A. 2002, *A&A*, 385, 1095

Pych, W. 2004, *PASP*, 116, 148

Škoda, P. & Hensberge, H. 2003, in *Astronomical Data Analysis Software and Systems XII*, ASP Conf. Series, 295, 415

Škoda, P., Šurlan B & Tomić, S. 2008, *Proc. SPIE*, 7014, 196

van Dokkum, P. G. 2001, *PASP*, 113, 1420

Verschueren, W., Brown, A. G. A., Hensberge, H., David, M., Le Poole, R. S., de Geus, E. J., & de Zeeuw, P. T. 1997, *PASP*, 109, 868

Highlights of Astronomy, Volume 15
XXVIIth IAU General Assembly, August 2009
Ian F Corbett, ed.

© International Astronomical Union 2010
doi:10.1017/S1743921310009798

Times-Series Photometry & Spectroscopy of the Bright Blue Supergiant Rigel: Probing the Atmosphere and Interior of a SN II Progenitor

Edward F. Guinan[1], J. A. Eaton[2], R. Wasatonic[1], H. Stewart[1], S. G. Engle[1] and G. P. McCook[1]

[1] Department of Astronomy & Astrophysics, Villanova University, Villanova, PA 19085 USA
email: edward.guinan@villanova.edu

[2] Center of Excellence in Information Systems, Tennessee State University, 3500 John A. Merritt Blvd., Box 9501, Nashville, TN 37209 USA

As the 7th brightest star and the most luminous star in the solar neighborhood, Rigel (β Orionis) is a very intriguing object. This blue supergiant (B8 Iab; V-mag = +0.05–0.18-mag; B–V = −0.03), at a distance (from Hipparcos) of ∼240±35 pc has a $\langle M_V \rangle$ = −6.7 mag. The following physical properties were determined via spectroscopic, photometric, and interferometric studies: $L/L_\odot$ ≈ 66,000 K; T_{eff} ≈ 12,000 K; $M/M_\odot$ ≈ 17±3; $R/R_\odot$ ≈ 70; τ ≈ 3–10 Myr. Interestingly Rigel has similar physical properties with the 12th mag blue supergiant progenitor of SN 1987A: Sanduleak −69° 202a. Thus Rigel (along with its co-asterism Betelgeuse) are likely to be the nearest progenitors of a Type II supernova. Such a nearby explosion would be V ≈ $−11^{th}$ mag (similar to a quarter moon).

Intensive photometric observations were carried out using telescopes in Pennsylvania and Arizona. High resolution & high S/N spectroscopy was carried out by Eaton using the TSU 2.0 m Automatic Spectroscopic Telescope (AST) during 2008/09. Together, they show complex light and RV variations on times scales of hours, weeks, and months. Systematic RV variations of up to 10 km/s and light changes of up to 0.12-mag were found on similar timescales. Preliminary analyses of these data have been carried out for periodicities using FFT and CLEAN-est routines.

Evidence of cyclic/periodic oscillations are present in some of the datasets, in addition to stochastic variations. These observations have been carried out in preparation of continuous, ultra-high precision photometry planned with the Canadian MOST satellite for nearly one month during November/December, 2009. Our preliminary study indicates that Rigel will be an excellent target for asterioseismic studies with MOST and as well as the upcoming BRITE-Constellation Mission. The continuous ultra-high precision photometry from space expected from MOST will yield important information on the possible presence of p- and g- mode oscillations in this star. The study of these will permit Rigel's internal structure to be probed and compared to stellar interior models. After the frequencies and spacing are identified they will be compared to various modern interior models of evolved stars that match Rigel's measured physical properties.

Keywords. stars: atmospheres, early-type, evolution, fundamental parameters, individual (Rigel), oscillations, supernovae: general, variables: other

Acknowledgements

We acknowledge support for this research from NSF/RUI Grant AST05-07536 and NASA/MOST Grant NNX09AH28G. The photometry from Pennsylvania was conducted by R. Wasatonic.

Highlights of Astronomy, Volume 15
XXVIIth IAU General Assembly, August 2009
Ian F Corbett, ed.

© International Astronomical Union 2010
doi:10.1017/S1743921310009804

Asteroseismology of rapidly rotating pulsators

W.A. Dziembowski[1,2]

[1] Warsaw University Observatory, Aleje Ujazdowskie 4, 00-478 Warszawa, Poland
email: wd@astrouw.edu.pl

[2] Copernicus Astronomical Center, ul. Bartycka 18, 00-716 Earszawa Poland

Abstract. Oscillation frequencies were determined for a number of rapidly rotating main sequence stars. However, real seismic probing is still ahead of us. I review here tools that we have for modeling pulsation in rotating stars and their potential application to seismic sounding.

Keywords. stars: oscillations - stars: rotation

1. Introduction

The most important goal of asteroseismology is probing internal rotation and particularly interesting when rotation is rapid. For the upper main sequence stars, we may adopt that beginning with the equatorial velocity of 100 km/s rotation should be regarded rapid because then it matters for stellar evolution. With such a definition, most of β Cephei and δ Scuti stars are rapidly rotating pulsators. For the lower main sequence stars, even 10 km/s is a large velocity, implying enhanced magnetic activity. So far, assessments of the interior rotation rate were made only for some slowly rotating objects.

2. The tools

Interpretation of stellar oscillation spectra requires calculation of evolutionary stellar models and their oscillation properties. For rapidly rotating stars, such complete tools are not yet available.

A cubic (in rotation rate, Ω) **perturbation theory** was developed by Soufi *et al.*(1998) and Suárez *et al.* (2006) for uniform and shellular rotation, respectively. In the case of p-modes, more important than including the Ω^3 terms is taking into account effects of coupling between close modes of degree differing by 2. When such effects are included, the treatment may be adequate even somewhat above 100 km/s equatorial velocity.

The traditional approximation allows for a strict separation of radial and latitudinal eigenfunctions. It is applicable to slow modes in not too rapidly rotating stars as it ignores effects of centrifugal distortion. Townsend (2005) carried an extensive survey of such modes in main sequence B stars using his generalization of this approximation to nonadiabatic oscillations.

Limitations of the traditional approximation were discussed by Lee & Saio (1997) and Saio *et al.* (2006), who compared results obtained with this approximation and **a more accurate treatment** employing expansion in Legendre function. In their approach the distortion was treated as a linear perturbation. They showed that most of retrograde modes, which are unstable with the traditional approximation, are found with their more accurate treatment.

Accurate description of all effects of rotation was considered so far only to adiabatic oscillations. Reese *et al.* (2006) developed a method employing Chebyshev and spherical function expansion. The method was originally applied to uniformly rotating polytropic models. Results of application to differentially rotating realistic stellar models were published recently by Reese *et al.* (2009). Lovekin *et al.* (2009) solved similar problems with their finite difference 2D code.

3. Important applications

β **Cephei stars** are both most important and promising objects for probing internal rotation. They occur in the mass range, where fast rotation influences stellar evolution up to the end as a supernova explosion or a cooling white dwarf. Estimate of rotation rate in deep interior was made for several objects(see e.g. Dziembowski & Pamyatnykh, 2008). However, none of them is a rapid rotator. The main advantage of β Cep stars is instability of modes sensitive to rotation rate in different layers, including the μ-gradient zone. The main difficulty is that only some of the unstable modes are detected. Mode identification is then crucial but it is not an easy task.

Rich oscillation spectra for four **Be stars** were found with data from the *MOST* satellite (Cameron *et al.* 2008, and references herein). Two sequences of the peaks were identified with unstable high-order prograde g-modes corresponding to $\ell = m = 1$ and 2. This identification led the authors to the conclusion that all four stars rotate with the equatorial velocity close to the maximum (Keplerian) value. This is an important finding, but we should expect more information from these rich spectra. There are chances for determination the frequency spacing between consecutive modes from which we could derive constrains on the internal structure and rotation rate.

For **solar-like pulsators**, which choose high-order p-modes, even at 20 km/s equatorial rotation rate the centrifugal force has a significant impact on theoretical frequency spectra. The asymptotic equality of the ℓ and $\ell + 2$-mode frequencies implies that the mode coupling is very significant. Multiplets of close low degree modes overlap. Still, it should be possible to recover data needed to obtain some measure of the internal rotation rate.

In lower main stars, rapid rotation is always associated with an enhanced magnetic activity. The related complications of the oscillation spectra are expected large. However, prospects for learning something about subphotospheric rotation and magnetic field in young active stars is so interesting that undertaking the difficult seismic analyzes of such objects is worth the effort.

Acknowledgements

The work was supported by MNiSW grant number N N203 379636.

References

Cameron, C., Saio & H, Kusching, R. *et al.* 2008, *ApJ*, 685, 489
Dziembowski, W. A. & Pamyatnykh, A. A. 2008, *MNRAS*, 385, 2061
Lee, U. & Saio, H. 1997, *ApJ*, 491, 839
Lovekin, C. C., Deupree, R. G., & Clement, M. J. 2009, *ApJ*, , 693, 677L
Ligniéres, F. & Rieutord, M. 2006, *A&A* 455, 607
Reese, D., MacGregor, K. B., Jackson, S. *et al.* 2009, *arXiv0903.4854*
Saio, H, Cameron, C.,Kusching, R. *et al.* 2007, *ApJ*, 654, 544
Soufi, F., Goupil, M. J., & Dziembowski, W. A. 1998, *A&A* 334, 911
Suárez, J. C., Goupil, M. J. & Morel, P. 2006, *A&A* 449, 673
Townsend, R. H. D. 2005, *MNRAS*, 360, 465

Highlights of Astronomy, Volume 15
XXVIIth IAU General Assembly, August 2009
Ian F Corbett, ed.

© International Astronomical Union 2010
doi:10.1017/S1743921310009816

Magnetoacoustic oscillations in Ap stars

Margarida S. Cunha[1]

[1] Centro de Astrofísica da Universidade do Porto, mcunha@astro.up.pt

Abstract. In this paper I briefly summarize a number of robust results derived from observational and theoretical studies of roAp stars over the past three decades.

Keywords. stars: chemically peculiar, stars: magnetic fields, stars: oscillations

1. Introduction

Rapidly oscillating Ap (or roAp) stars are chemically peculiar magnetic stars that exhibit pulsations with typical periods in the range of 5 to 20 minutes. Well below the photosphere the magnetic field is unlikely to play an important role in the dynamics of the oscillations. However, in the outer layers it influences the oscillations both directly, through the action of an additional restoring force, and indirectly, through the interaction with outer convection. As a result of this, asteroseismic studies of these pulsators provide a unique opportunity to infer about the properties of photospheric and subphotospheric stellar magnetic fields and about the interaction between the magnetic field, diffusion and convection, in stars other than the sun.

2. Observational results

roAp stars were first discovered and characterized through high-speed photometry Kurtz (1982). Despite some earlier attempts aiming at the detection of pulsational radial velocities in this pulsators e.g.]matthewsetal88,libbrecht88, it was with the advent of the high-spectral resolution spectrographs that the intense spectroscopic campaigns developed, providing a wealth of new data on these stars e.g. seefor a recent review]kochukhov09. From three decades of observations of roAp stars, a number of interesting and, in some cases, intriguing properties emerged, namely:

- roAp stars are found among the coolest chemically peculiar magnetic stars, having effective temperatures approximately within the range of 6400 K − 8200 K;
- The oscillations are of high frequency, often close to, or even above, the acoustic cutoff frequency for the corresponding stellar models;
- In some multiperiodic pulsators the frequencies of the oscillations are close to equally spaced, but in a number of stars that pattern is not present, or is restricted to a particular frequency range in the oscillation spectrum;
- The analysis of high-resolution spectroscopic time series of roAp stars reveled that the structure of the pulsations in the atmospheric layers is complex, often including running components, similarly to what is observed in regions of strong magnetic field of the solar atmosphere.

3. Theoretical results

Over the past two decades, theoretical studies of roAp stars for earlier results were aimed primarily at interpreting the observed oscillation spectra or at studying pulsation stability in models of these pulsators. Comparisons between model predictions and observations has also been carried out for a limited number of stars Cunha (2001), Cunha et al.(2003)Cunha, Fernandes, & Monteiro, Gruberbauer et al., 2008, Huber et al., 2008), Bruntt et al., 2009. Some of the aspects in which progress has clearly been made, as well as some of the open questions that still remain can be summarized as follows:

• The oscillations in roAp stars are likely to be excited by the opacity mechanism acting in the region of hydrogen ionization. Nevertheless, in most cases such excitation requires the suppression of the thin outer convective region of the star by the action of the magnetic field. Moreover, the red edge predicted from models is significantly hotter than that derived from observations.

• The interaction between the magnetic field and pulsations in the outer layers of roAp stars results in energy being transferred between magnetic and acoustic waves. This provides a natural mechanism for mode reflection even for oscillations with frequencies above the acoustic cutoff. Nevertheless, part of the mode energy is expected to be dissipated by running acoustic waves in the atmosphere and running magnetic waves in the interior.

• The direct effect of the magnetic field on pulsations results in perturbations to the oscillation frequencies that in some cases are as large as tens of μHz. This provides a natural explanation to what is observed in the oscillation spectra of several multiperiodic roAp stars, in particular the significant deviations from the equally spaced pattern that would be expected in the absence of the magnetic field;

• The correct interpretation of the data acquired in recent high-resolution spectroscopic campaigns requires a detailed analysis of the eigenfunctions in the magnetic boundary layer, taking into account the horizontal and vertical structure of the oscillations, as well as the horizontal and vertical distribution of chemical elements.

4. Conclusions

This is a very exciting time for studies of magnetic pulsators. Recently, significant amounts of excellent seismic data of roAp stars have been acquired and additional high-quality data is envisaged as result of the recent launch of the NASA satellite Kepler. These observations posed new challenges to those attempting to model these pulsators. Ultimately, the correct modelling of these stars will provide a unique insight into important physical processes taking place inside the stars, such as diffusion and convection, and into the way they interplay with magnetic fields.

Acknowledgements

MSC acknowledges the support of the Portuguese MCTES and of the FSE, of the European Union, through the programme POPH.

References

Bruntt, H., Kurtz, D. W., Cunha, M. S., *et al.* 2009, MNRAS, 396, 1189
Cunha, M. S. 2001, MNRAS, 325, 373
Cunha, M. S. 2007, Communications in Asteroseismology, 150, 48
Cunha, M. S., Fernandes, J. M. M. B., & Monteiro, M. J. P. F. G. 2003, MNRAS, 343, 831
Gruberbauer, M., Saio, H., Huber, D., *et al.* 2008, A&A, 480, 223
Huber, D., Saio, H., Gruberbauer, M., *et al.* 2008, A&A, 483, 239
Kochukhov, O. 2009, Communications in Asteroseismology, 159, 61
Kurtz, D. W. 1982, MNRAS, 200, 807
Libbrecht, K. G. 1988, ApJ, 330, L51
Matthews, J. M., Wehlau, W. H., Walker, G. A. H., & Yang, S. 1988, ApJ 324, 1099
Sousa, S. G. & Cunha, M. S. 2008, MNRAS, 386, 531
Théado, S., Dupret, M.-A., Noels, A., & Ferguson, J. W. 2009, A&A, 493, 159

Highlights of Astronomy, Volume 15
XXVIIth IAU General Assembly, August 2009
Ian F Corbett, ed.

Complex asteroseismology of the B-type main sequence pulsators

Jadwiga Daszyńska-Daszkiewicz, Przemysław Walczak

Instytut Astronomiczny, Uniwersytet Wrocławski, ul. Kopernika 11, 51-622 Wrocław, Poland
emails: daszynska@astro.uni.wroc.pl, walczak@astro.uni.wroc.pl

Abstract. We present examples of an extended asteroseismic modelling in which we aim at fitting not only pulsational frequencies but also certain complex parameter related to each frequency. This kind of studies, called **complex asteroseismology**, has been successfully applied to a few main sequence B-type pulsators and provided, in particular, plausible constraints on **stellar opacities**. Here, we briefly describe our results for three early B-type stars.

Keywords. stars: early-type, pulsations; individual: θ Oph, ν Eri, γ Peg; atomic data

1. Introduction

It is now well known that pulsational frequencies are a primary seismic probe of stellar structure and evolution. A power of these seismic tools has been firstly explored for the Sun and then for various types of pulsating variables at different evolutionary stages. Important results have been obtained also for the main sequence pulsators of the B spectral type. Knowledge of the internal structure and evolution of these massive stars is of great importance for astrophysics because they form the CNO elements and are progenitors of Type II Supernovae.

Each pulsational frequency, ν, is associated with the complex, nonadiabatic parameter f, which describes the bolometric flux perturbation normalized to the radial surface displacement. The f parameter is embedded in the expression for a complex amplitude of the light variations. Thus, having multicolour photometric data, one can try to extract empirical values of f together with the mode degree, ℓ, and compare them with theoretical counterparts. A value of f is determined in subphotospheric layers which have rather weak contribution to the frequency value. Therefore, ν and f constitute the two asteroseismic tools complementary to each other and combining them in a seismic survey yields a new kind of information. In the case of the B-type pulsators, complex asteroseismology provides a critical and unique test for stellar opacities and the atomic physics.

In this short report, we summarise our results for three early B-type pulsators: θ Ophiuchi, ν Eridani and γ Pegasi.

2. The β Cep star θ Ophiuchi

θ Oph is the B2IV type star in which 7 pulsational frequencies were detected from photometry in a range from 7 to 8 c/d. Three of them appeared also in spectroscopy. Using both the OP and OPAL opacity data, we found a family of seismic models with different parameters (M, $T_{\rm eff}$, α_{ov}, Z), which reproduced two centroid frequencies: ν_3 ($\ell = 0$, p_1) and ν_6 ($\ell = 1$, p_1). In general, seismic models with lower metallicity, Z, demanded a higher core overshooting, α_{ov}. Then, we went a step further by a requirement of fitting simultaneously also the f parameter corresponding to the radial mode. A comparison of empirical and theoretical values of f pointed substantially to a preference for the OPAL tables. For more details see Daszyńska-Daszkiewicz & Walczak (2009a).

3. The hybrid β Cep/SPB pulsators: ν Eridani and γ Pegasi

ν Eri and γ Peg are the most multimodal pulsating stars of early B spectral types. Asteroseismic studies of these variables have been intensified after the recent photometric and spectroscopic multisite campaigns which led to a detection of next frequencies typical for the β Cep type as well as entirely new peaks in the SPB frequency domain.

The frequency analysis of the ν Eri data revealed 14 peaks: 12 of the β Cep type and 2 low frequency modes typical for the SPB pulsations. In our seismic analysis, we first looked for stellar models with different $(M, T_{\rm eff}, \alpha_{ov}, Z)$, which reproduced three centroid frequencies: ν_1 ($\ell = 0$, p_1), ν_4 ($\ell = 0$, g_1) and ν_6 ($\ell = 1$, p_1). Available data allowed us to determine the f parameter for eight high frequency modes and one SPB mode. Then, we could compare empirical and theoretical values of f in a wide range of pulsational frequencies, i.e., from 0.6 to 8 c/d. The obtained consistency is very encouraging and brings a great seismic potential. The value of f depends not only on mode frequency, but also on the shape of eigenfunctions. In the low frequency region, f strongly depends on the mode degree, ℓ, whereas for the high frequency modes it is independent of ℓ.

A closer comparison of empirical and theoretical values of f for the radial mode indicated again a preference for the OPAL opacities. Moreover, the OPAL seismic models had larger effective temperatures and masses which fit better observational values of $T_{\rm eff}$ and luminosity. More details can be found in Daszyńska-Daszkiewicz & Walczak (2009b).

For more than 50 years, γ Peg was considered as a monoperiodic star. The analysis of the recent MOST and ground based photometric and spectroscopic data showed up 8 frequencies of the β Cep type and 6 peaks of the SPB type. Using these observations, we were able to determine the empirical values of f for four β Cep modes and for all six SPB modes. The work on this variable is in the making and results will appear soon.

4. Conclusions

Presented examples demonstrate a great potential of **complex seismic studies** consisting in fitting simultaneously pulsational frequencies and corresponding values of the complex, nonadiabatic parameter f. In particular, complex asteroseismology of the B-type pulsators yields a valuable constraints on **stellar opacities**. Heretofore, our results indicate a preference for **the OPAL tables**.

One of the most important results is that empirical values of f are determinable also for **high order g modes**. The f parameter of the β Cep and SPB modes have a different dependence on the mode frequency and degree, ℓ. Therefore, complex asteroseismology of the **hybrid pulsators** can give much better assessment of stellar model precision.

Acknowledgements

The work was supported by Astronomical Institute of Wrocław University and by the Polish MNiSW grant N N203 379636.

References

Daszyńska-Daszkiewicz, J., Walczak, P. 2009a, *MNRAS* in press
Daszyńska-Daszkiewicz, J., Walczak, P. 2009b, in: M. Marconi, D. Cardini, M.P. Di Mauro (eds.) *Synergies between solar and stellar modelling*, Rome HELAS Workshop, *AP&SS* in press

Highlights of Astronomy, Volume 15
XXVIIth IAU General Assembly, August 2009
Ian F Corbett, ed.

Asteroseismology of massive stars with the MOST satellite

Anthony F. J. Moffat[1] and The MOST Team

[1]Département de physique, Université de Montréal,
C.P. 6128, Succursale Centre-ville, Montréal, QC, H3C 3J7, Canada
email: moffat@astro.umontreal.ca

Abstract. Since 2003 the MOST (Microvariability and Oscillations of STars) microsatellite has obtained typically a month of non-stop, minute-of-time resolution, high-precision, single-broadband optical photometry for each of a significant number of Galactic OB and WR stars. Numerous p- and g-modes were clearly detected in several OB stars, including discovery of g-modes for the first time in a blue supergiant (Saio *et al.* 2006). True rotation periods were found for some SPBe pulsators (Cameron *et al.* 2008). Many O stars are remarkably quiet. Five presumably single WR stars have been observed so far, each interesting in its own way. In particular, the cool WR stars WR123 (WN8) and WR103 (WC9d) both show mostly short-lived, multimode oscillations with most of the Fourier power occurring on a day or longer timescale (Moffat *et al.* 2008a). WR123 also revealed a fairly stable 10-hour periodicity (Lefèvre *et al.* 2005). All of these oscillations probably arise in the stellar cores. WR111 (WC5) shows no (coherent) oscillations above the detection limit of 0.05 mmag in the 10-minute period range predicted for strange-mode pulsations at a level of 2 mmag (Moffat *et al.* 2008b). WR110 (WN5-6 and a stronger-than-average X-ray source) and WR124 (WN8h, i.e. in contrast with the previously observed, hydrogen-free WR123 of otherwise similar subtype), both strongly variable with MOST, are currently being analyzed. The next target just observed (late-June to early Aug 2009) is the 30-day eclipsing binary CV Ser = WR113 (WC8d + O8-9IV). Besides stellar oscillations, we will also search for orbital-phase dependent, stochastic variability in CV Ser as wind clumps in the WR component's dense wind pass in front of the O-star.

Keywords. Space vehicles, stars: early-type, stars: oscillations (including pulsations), stars: Wolf-Rayet.

References

Cameron, C., Saio, H., Kuschnig, R., Walker, G. A. H., Matthews, J. M., Guenther, D. B., Moffat, A. F. J., Rucinski, S. M., Sasselov, D., Weiss, W. W. 2008, *ApJ*, 685, 489

Lefèvre, L., Marchenko, S. V., Moffat, A. F. J., Chené, A. N., Smith, S. R., St-Louis, N., Matthews, J. M., Kuschnig, R., Guenther, D. B., Poteet, C. A., Rucinski, S. M., Sasselov, D., Walker, G. A. H., Weiss, W. W. 2005, *ApJ*, 634, L109

Moffat, A. F. J., Marchenko, S. V., Lefèvre, L., Chené, A.-N., St-Louis, N., Zhilyaev, B. E., Aerts, C., Saio, H., Walker, G. A. H., Matthews, J. M., Kuschnig, R., Cameron, C., Rowe, J. F., Guenther, D. B., Rucinski, S. M., Sasselov, D., Weiss, W. W. 2008a, in: A. de Koter, L. J. Smith, & L. B. F. M. Waters (eds.), *ASP Conference Series* (San Francisco: Astronomical Society of the Pacific), 388, 29

Moffat, A. F. J., Marchenko, S. V., Zhilyaev, B. E., Rowe, J. F., Muntean, V., Chené, A.-N., Matthews, J. M., Kuschnig, R., Guenther, D. B., Rucinski, S. M., Sasselov, D., Walker, G. A. H., Weiss, W. W., 2008b, *ApJ*, 679, L45

Saio, H., Kuschnig, R., Gautschy, A., Cameron, C., Walker, G. A. H., Matthews, J. M., Guenther, D. B., Moffat, A. F. J., Rucinski, S. M., Sasselov, D., Weiss, W. W. 2006, *ApJ*, 650, 1111

Highlights of Astronomy, Volume 15
XXVIIth IAU General Assembly, August 2009
Ian F Corbett, ed.

© International Astronomical Union 2010
doi:10.1017/S1743921310009841

Application of the Rayleigh-Ritz variational technique for coronal loop oscillations

Narges Fathalian[1], Hossein Safari[2] and Sadollah Nasiri[1,2]

[1]Institute for Advanced Studies in Basic Sciences, P. O. Box 45195-1159, Zanjan, Iran

[2]Department of Physics, Zanjan University, P. O. Box 45195-313, Zanjan, Iran

Abstract. We studied the fast kink modes of a cylindrical model of coronal loops, in coronal conditions, stratified density and low-β plasma. The mode frequencies and profiles are calculated.

Keywords. Sun: corona, Sun:oscillations

1. Introduction

Several theoretical models have been developed to explore the nature and propagation of waves of coronal loops in 1-D models. Here, a Rayleigh-Ritz variational method developed for the oscillations of 2-D coronal loop model.

2. Overview

The MHD equations in the matrix representation reduces as (see e.g., Sobouti 1981 and Hasan & Sobouti 1987).

$$\mathbf{WZ} = \mathbf{SZE}, \tag{2.1}$$

where E is a diagonal matrix whose elements are the eigenvalues ω_l and $\mathbf{Z}$ is the matrix of the expansion coefficients. The elements of $\mathbf{W}$ and $\mathbf{S}$ are determined from the equations of motions. We adapt a Rayleigh-Ritz procedure and approximate the linear series in Eq. (2.1) by a finite number of terms, say n. The matrix blocks S_{ls} and $W_{ls}; l, s = 1, 2, 3$ become $n \times n$ matrices.

3. Results

Our numerical results show that: a) for unstratified loop, as expected, ω_n is proportional to its mode number (i.e., $\omega_n \approx n\omega_1$). b) for stratified loops, all the fundamental, first, second, and the third overtone kink ($m = 1$) frequencies increases with increasing stratified density scale height. For typical loop lengths, $100 - 400$Mm, the density scale heights fall in the range of $13 - 108$Mm. These results are in agreement with Verwichte et al. (2004), McEwan et al. (2006), and Safari et al. (2007). The application of variational method in real model of coronal loop are implementing in our group.

References

Hasan, S. S. & Sobouti, Y. 1987, *MNRAS*, 228, 427

McEwan, M., Donnelly, G. R., Díaz, A. J., & Roberts, B. 2006, *A&A*, 460, 893

Safari, H., Nasiri, S., & Sobouti Y. 2007, *A&A*, 470, 1111

Sobouti, Y. 1981, *A&A*, 100, 319

Verwichte, E., Nakariakov, V. M., Ofman, L., & DeLuca, E. E. 2004, *Sol. Phys.*, 223, 77

Highlights of Astronomy, Volume 15
XXVIIth IAU General Assembly, August 2009   © International Astronomical Union 2010
Ian F Corbett, ed.   doi:10.1017/S1743921310009853

Testing new models of M dwarfs

A. Baran[1,†], S.D. Kawaler[2]
and J. Krzesinski[1]

[1] Cracow Pedagogical University, Krakow, Poland
email: sfbaran@cyf-kr.edu.pl

[2] Iowa State University Ames, USA,
†Present address:Iowa State University, Ames, USA

Abstract. We present our new observing project searching for pulsations in M dwarfs.

1. M dwarfs

Young low-mass main sequence stars are the dominant population in our Galaxy ($>70\%$ of all stars). Precise modelling of mechanical and thermal properties of the stars is very important from very wide range of physical and astrophysical reasons: from understanding fundamental physical problems to astrophysical and cosmological implications. Low mass M dwarfs have radius $0.1 \sim 0.7 R_{Sun}$ Effective temperature is $< 5\,kK$ and $\log g = 3.5 \div 5.5$. Spectroscopically, they are characterized by strong and wide molecules absorption lines (TiO, VO, H_2O and CO). An understanding of how the stars are born is very crucial aspect to learn the complete theory of formation of M dwarfs. Recent calculations (SK) revealed that young low-mass main sequence stars can show stellar oscillations driven by ϵ mechanism. The mechanism is based on instability of energy production in the chemical reactions so it must work close to the center of these stars. If the perturbations have enough time to grow in amplitude they can cause periodic change in stellar brightness with the period connected to the dynamical time scale.

2. Goals of the project

(a) to search for stellar oscillations in M dwarfs
(b) to obtain light curves of those M dwarfs which are components of eclipsing systems
(c) to detect other types of variability, caused by rotation or chromospheric activity
(d) to detect planetary transits
(e) to make all data obtained in this project publically available.

To achieve our goals we plan to perform photometry of a sample of early type M dwarfs with masses between 0.4 and 0.6 M_{Sun}. First observations have been performed at Mt. Suhora Observatory and at IRSF in SAAO. Our sample is limited to cool stars only so infrared telescopes are particularly desirable for this project.

To publish all data obtained a database on PHP and MySQL server have been already prepared. It provides information about stars, useful details for observers as well as an interactive tool for making plots of light curve and the Fourier transform. The database is accessible by any web browser at 149.156.24.35/~andy/mdwarfs.

Acknowledgements

This work is supported by grant N N203 379736 kindly provided by MNiSW.

Highlights of Astronomy, Volume 15
XXVIIth IAU General Assembly, August 2009
Ian F Corbett, ed.

© International Astronomical Union 2010
doi:10.1017/S1743921310009865

Testing the hot-flasher scenario with asteroseismological tools. First Results

M. M. Miller Bertolami[1,2], A. H. Córsico[1,2] and L. G. Althaus[1,2]

[1]Facultad de Ciencias Astronómicas y Geofísicas, UNLP, B1900FWA, La Plata, Argentina

[2]Instituto Astrofísica La Plata, CCT-La Plata, UNLP-CONICET
email: mmiller@fcaglp.unlp.edu.ar

Abstract. A core helium flash after the departure from the red giant branch (i.e. "hot-flasher scenario") offers one of the most promising explanations for the origin of He-sdO stars. Recently, Miller Bertolami *et al.* (2008) have shown that many surface properties of H-deficient sdO stars (the He-sdO stars) could be explained through this scenario if chemical diffusion is taken into account. In this context the He-sdO stars formed during a hot-flasher event would transform into H-rich hot-sdB stars (33000-38000 K) as a consequence of diffusion of the remaining H towards the surface of the star. Thus, some hot sdBs might be the descendants of He-sdO stars that have previously burnt most of their H-content and, thus, a very thin H envelope should be expected (10^{-9} to 10^{-10} $M_\odot$, see Miller Bertolami *et al.* 2008 for details). Interestingly enough, the location of these sdB stars in the $\log g - T_{\rm eff}$ diagram should overlap with the domain of the rapidly pulsating (p-mode) EC 14026 stars. This fact opens the interesting possibility of employing asteroseismology to investigate the existence of hot-sdB stars characterized by such very thin H envelopes. In this preliminary investigation, we explore the sensitivity of the acoustic pulsation spectrum of EC 14026 stars to the thickness of the H envelope.

The pulsation analysis presented in this work was performed with the help of the adiabatic radial and nonradial pulsation code employed by our group in numerous asteroseismological studies of white dwarfs and pre-white dwarfs (see Córsico *et al.* 2008 and references therein). The stellar models adopted in the present study where extracted from the 0.48150 $M_\odot$ (Z= 0.001) sequence of Miller Bertolami *et al.* (2008). Given the exploratory nature of this work, in order to analyze the effects of different thicknesses of the H-rich envelope on the pulsation spectrum of sdB stars we have artificially added a H-rich envelope in the outermost layers of our initial model. We have considered four different thicknesses of the H envelope ($\log M_{\rm env}/M_\odot \sim -4, -6, -8, -10$) in addition to the self-consistent model (which lacks a H envelope). Next, we have pursued the evolution of the five sequences during the evolution on the HB. For each stellar model, we have computed the radial and nonradial p-modes with periods longer than 20 seconds, thus comfortably covering the observed period range of EC 14026 stars (80 - 400 sec).

Our results show that the cycle of trapping is markedly smaller for the case of thick H-envelope models ($M_{\rm env} > 10^{-7}$ $M_\odot$) than for the thin H-envelope models ($M_{\rm env} < 10^{-7}$ $M_\odot$). We find that sdB stars with very thin H envelopes ($M_{\rm env} < 10^{-7}$ $M_\odot$) would not display almost any kind of trapping features in their frequency distribution at the observed range in EC 14026 stars ($\nu < 13$ mHz). Consequently their frequency spectra should be significantly different from that of normal sdB stars. We plan to explore in future works to which extent the shape of the chemical transitions (here adopted as simple gaussian profiles) affects the mode-trapping features.

Keywords. stars: evolution, stars: horizontal-branch, stars: subdwarfs, stars: oscillations.

References

Miller Bertolami, M., M., Althaus, L. G., Unglaub, K. & Weiss, A 2008, A&A, 491, 253

Córsico, A. H., Althaus, L. G., Kepler, S. O., Costa, J. E. S., & Miller Bertolami, M. M. 2008, A&A, 478, 869

Highlights of Astronomy, Volume 15
XXVIIth IAU General Assembly, August 2009
Ian F Corbett, ed.

Hot DQ white dwarfs: a pulsational test of the mixing scenario for their formation

A. D. Romero[1,2], A. H. Córsico[1,2], L. G. Althaus[1,2] & E. García-Berro[3,4]

[1]Facultad de Ciencias Astronómicas y Geofísicas, Universidad Nacional de La Plata, Argentina

[2]Instituto de Astrofísica La Plata, IALP, CONICET-UNLP

[3]Departament de Fisica Aplicada, Escola Politécnica Superior de Castelldefels, Universitat Politécnica de Catalunya

[4]Institut d'Estudis Espacials de Catalunya
email: aromero@fcaglp.unlp.edu.ar

Hot DQ white dwarfs constitute a new class of white dwarf stars, uncovered recently within the framework of SDSS project. There exist nine of them, out of a total of several thousands white dwarfs spectroscopically identified. Recently, three hot DQ white dwarfs have been reported to exhibit photometric variability with periods compatible with pulsation g-modes. In this contribution, we presented the results of a non-adiabatic pulsation analysis of the recently discovered carbon-rich hot DQ white dwarf stars. Our study relies on the full evolutionary models of hot DQ white dwarfs recently developed by Althaus et al. (2009), that consistently cover the whole evolution from the born-again stage to the white dwarf cooling track. Specifically, we performed a stability analysis on white dwarf models from stages before the blue edge of the DBV instability strip ($T_{\rm eff} \approx 30\,000$ K) until the domain of the hot DQ white dwarfs ($18\,000 - 24\,000$ K), including the transition DB→hot DQ white dwarf. We explore evolutionary models with $M_* = 0.585 M_\odot$ and $M_* = 0.87 M_\odot$, and two values of thickness of the He-rich envelope ($M_{\rm He} = 2 \times 10^{-7} M_*$ and $M_{\rm He} = 10^{-8} M_*$).

We found that at epochs in which the models have He-dominated atmospheres, they exhibit g-mode pulsations typical of DBV stars. When the white dwarf models become carbon-dominated atmospheres, they continue being pulsationally unstable with similar characteristics than DB models, showing overstable g-modes primarily driven through the κ-mechanism due to the partial ionization of C. The blue edge of DQVs is hotter for less massive models and for thinner He envelopes. In particular, the instability domains of DBVs and DQVs are clearly separated in the $T_{\rm eff} - \Pi$ plane for the case of models with $M_{\rm He} = 10^{-8} M_*$, but the transition is continuous in the case in which $M_{\rm He} = 2 \times 10^{-7} M_*$. The periods detected in SDSS J142625.70+575218.4 and SDSS J220029.08-074121.5 are well within our theoretical ranges of excited periods, but the period at $\Pi \sim 1052$ s corresponding to SDSS J234843.300-94245.3 is long in excess, and is not well accounted for by our models. Finally, our work demonstrate that the diffusive/convective mixing scenario not only is able to nicely explain the origin of hot DQ white dwarfs, but it also accounts for the variability of these stars. We caution that the results of the present work could somewhat change if the frozen convection approximation were relaxed in our stability analysis. This calculations provide strong support to the convective-mixing picture for the formation of hot DQs. We found theoretical evidence that a fraction of pulsating DB white dwarfs —those characterized by thin He-rich envelopes— would become pulsating DQ white dwarfs.

References

Althaus, L. G., Garcia-Berro, E., Córsico, et al., 2009, ApJL, 693, L23

Highlights of Astronomy, Volume 15
XXVIIth IAU General Assembly, August 2009
Ian F. Corbett, ed.

© International Astronomical Union 2010
doi:10.1017/S1743921310009889

JD 12: The First Galaxies - Theoretical Predictions and Observational Clues

No contributions were received from this Joint Discussion.

Highlights of Astronomy, Volume 15
XXVIIth IAU General Assembly, August 2009
Ian F. Corbett, ed.

© International Astronomical Union 2010
doi:10.1017/S1743921310009890

JD13 – Eta Carinae in the Context of the Most Massive Stars

Theodore R. Gull[1] and Augusto Damineli[2]

[1]Laboratory for Extraterrestial Planets and Stellar Astrophysics, Code 667, NASA/GSFC, Greenbelt, MD, 20771, USA, email: Theodore.R.Gull@nasa.gov
[2]IAGSP, Universidade de Sao Paulo, Rua do Matao 1226, Sao Paulo, 05508-900, Brazil, email: damineli@astro.iag.usp.br

Eta Car, with its historical outbursts, visible ejecta and massive, variable winds, continues to challenge both observers and modelers. In just the past five years over 100 papers have been published on this fascinating object. We now know it to be a massive binary system with a 5.54-year period. In January 2009, η Car underwent one of its periodic low-states, associated with periastron passage of the two massive stars. This event was monitored by an intensive multi-wavelength campaign ranging from γ-rays to radio. A large amount of data was collected to test a number of evolving models including 3-D models of the massive interacting winds. August 2009 was an excellent time for observers and theorists to come together and review the accumulated studies, as have occurred in four meetings since 1998 devoted to Eta Car. Indeed, η Car behaved both predictably and unpredictably during this most recent periastron, spurring timely discussions.

Coincidently, WR140 also passed through periastron in early 2009. It, too, is a intensively studied massive interacting binary. Comparison of its properties, as well as the properties of other massive stars, with those of Eta Car is very instructive. These well-known examples of evolved massive binary systems provide many clues as to the fate of the most massive stars.

What are the effects of the interacting winds, of individual stellar rotation, and of the circumstellar material on what we see as hypernovae/supernovae? We hope to learn.

Topics discussed in this 1.5 day Joint Discussion were:

η Car: the 2009.0 event: Monitoring campaigns in X-rays, optical, radio, interferometry
WR140 and HD5980: similarities and differences to η Car
LBVs and Eta Carinae: What is the relationship?
Massive binary systems, wind interactions and 3-D modeling
Shapes of the Homunculus & Little Homunculus: what do we learn about mass ejection?
Massive stars: the connection to supernovae, hypernovae and gamma ray bursters
Where do we go from here? (future directions)

The Science Organizing Committee:

Co-chairs: Augusto Damineli (Brazil) & Theodore R. Gull (USA). Members: D. John Hillier (USA), Gloria Koenigsberger (Mexico), Georges Meynet (Switzerland), Nidia I. Morrell (Chile), Atsuo T. Okazaki (Japan), Stanley P. Owocki (USA), Andy M.T. Pollock (Spain), Nathan Smith (USA), Christiaan L. Sterken (Belgium), Nicole St Louis (Canada), Karel A. van der Hucht (Netherlands), Roberto Viotti (Italy) and Gerd Weigelt (Germany)

Website for talks and posters:
http://astrophysics.gsfc.nasa.gov/research/etacar/IAUJD.html

Figure 1. Professors Sveneric Johansson and Vladelin Letokhov discussing the stimulated emission properties of the ionized ejecta surrounding η Car. Both researchers passed away this past year. Their interest in atomic spectroscopy and enthusiasm was infectious to all.

1. Oral Presentations

1.1. *Dedication to Prof. Sveneric Johansson*
Henrik Hartman

Professor Sveneric Johansson is remembered for his important contributions to the knowledge on atomic data, focusing on the iron group elements in general and singly ionized iron, Fe II, in particular. His work includes term analysis of several important ions, and measurements of atomic parameters for astrophysicaly important elements. His thorough knowledge of atomic structure also allowed major contributions to the analysis of complex astronomical spectra and atomic photo processes. Sveneric is greatly missed as an ingenious scientist, positive colleague and a great friend.

Sveneric received his PhD from Lund University in 1978 under the supervision of Professor Bengt Edlén, on the subject of term analysis of Fe II (the spectrum of Fe$^+$). This work continued to be his main research topic for more than 35 years. Sveneric led classical atomic spectroscopy into a new era of measurements with crucial astronomical applications. He spent a sabbatical year at NASA's Goddard Space Flight Center with Dave Leckrone during 1987-1988, starting up a collaboration for the upcoming Hubble Space Telescope (HST) mission and the χ Lupi pathfinder project. The high resolution spectrographs onboard HST challenged existing laboratory atomic data bases. Sveneric foresaw the need of high-accuracy ultraviolet data and directed, together with Ulf Litzén, the Lund University spectroscopy laboratory to measure wavelengths, isotopic shifts and line structures needed to interpret astronomical observations. Spectroscopic investigations included iron, yttrium, mercury, boron, gold, ruthenium, nickel, thallium, platinum, and zirconium.

The high cosmic abundance of iron makes Fe II lines abundant in a variety of astronomical objects. For quantitative analyses the intrinsic strength of the spectral lines need to be known. In 2001 Sveneric founded the Atomic Astrophysics group at Lund University and organized the FERRUM project, an international collaboration on oscillator strengths for iron group elements. The aim of this project is to present a fully evaluated and consistent set of values, experimental and theoretical, that can be used for astronomical analyses.

Throughout his career Sveneric also analyzed complex astronomical emission line spectra, and was especially interested in atomic photo processes. Together with Professor Vladilen Letokhov he identified and developed the idea of stimulated emission (LASER) in gas condensations close to the massive star Eta Carinae. From the strange behavior observed and ionization structure of the high ionization lines, they derived the concept of resonance-enhanced two-photon ionization (RETPI) of Ne and Ar atoms as an explanation for the production of these ions.

In addition, it is with great sadness, that we learnt of Dr. Vladelen Letokhov's passing during 2009. He is greatly missed by colleagues and friends all over the world. During his productive career he published nearly 900 articles and 16 monographs. Sveneric's and Vladilen's work on photo processes culminated in their book 'Astrophysical Lasers' (Oxford Press, 2009) published earlier this year.

1.2. The historical background on η Car
D. John Hillier

Eta Carinae, a spectacular object, is one of the most luminous stars in the galaxy, and exhibits a wide range of interesting phenomena with implications for many areas of astrophysics. In this presentation we provide a brief summary of key discoveries and an introduction to some jargon associated with η Car.

Eta Carinae, a spectacular object, is one of the most luminous stars in the galaxy, and exhibits a wide range of interesting phenomena with implications for many areas of astrophysics. In this presentation we provide a brief summary of key discoveries and an introduction to some jargon associated with η Carinae.

In the 1840's η Carinae underwent a giant outburst and ejected a nebula which we call the Homunculus. The event was so impressive that η Carinae was classified as a peculiar SN. With the onset of dust formation, it suffered a dramatic drop in brightness by $\sim$6 magnitudes (e.g., van Genderen *et al.* 1984, Space Sci. Rev., 39, 317). In the early 1890's η Carinae underwent a smaller outburst ejecting the Little Homunculus nebula (discovered with the HST; Ishibashi *et al.* 2003, AJ, 125, 3222).

The Homunculus is a bipolar nebula whose axis is tilted at about 41° to our line of sight. H_2 emission and dust is confined to a thin outer layer, while [Fe II] & [Ni II] emission lines originate inside this shell (Smith *et al.* 2006, ApJ, 644, 1151). From infrared observations the mass of the Homunculus is inferred to exceed $10 M_\odot$ (Smith *et al*, 2003, AJ, 125, 1458), and is possibly as large as $20 M_\odot$ (Smith *et al.* 2007, ApJ, 655, 911). In contrast, the mass of the Little Homunculus is $\sim 0.1 M_\odot$ (Smith 2005, MNRAS, 357, 1330).

S-condensation ejecta (a condensation to the south of the Homunculus) are N enhanced and CO depleted, consistent with the influence of CNO processing (Davidson *et al.* 1982, ApJ, 254, L47). A similar abundance pattern is seen in the star (Hillier *et al.* 2001, ApJ, 553, 837). As η Carinae is located in a region of massive star formation (Walborn *et al.* 1977, ApJ, 211, 181), it is inferred that it is a young, but evolved, massive star.

Speckle observations showed that η Carinae is composed of 4 'star-like' objects (Weigelt *et al.* 1986, A&A, 163, L5). Subsequent HST observations revealed that the brightest of these is truly star-like, while the remaining 3 are small nebula (the Weigelt blobs) which emit the narrow permitted and forbidden lines that are prominent in ground-based spectra (Davidson *et al.* 1995, AJ, 109, 1784); they are prominent because the primary star suffers additional extinction ($\sim$5 magnitudes in 1997; Hillier *et al.* 2001, ApJ, 553, 837).

The discovery of a 5.5 year variability cycle (Damineli 1996, ApJ, 460, L49) led to the realization that η Carinae is a binary system (Damineli *et al.* 1997, New Astr., 2,

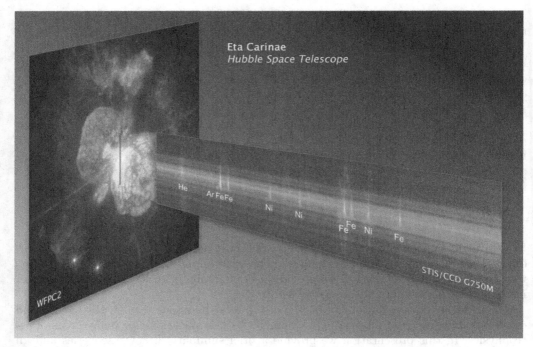

Figure 2. An example of what is so intriguing about η Car: the extended wind and ejecta. A 0.1″-wide slit of the Hubble Space Telescope Imaging Spectrograph samples the extended structure surrounding η Car as imaged by Hubble Space Telescope. Continuum and broad line emission at the center of the spectrum originate from the extended interacting winds. Narrow forbidden emission lines shifted with velocities up to 500 km s^{-1} come from the interior of the Homunculus, thrown out in the 1840s. An estimated 10 to 20 $M_\odot$ was ejected during the Great Eruption as well as up to 0.5 $M_\odot$ in the lesser eruption of the 1890s. How did the ejecting star survive and what clues does this provide us on the late stages of massive stellar evolution? (Image courtesy of NASA and STScI)

107). A wide range of phenomena, including infrared (Whitelock *et al.* 2004, MNRAS, 352, 447), X-ray (Ishibashi *et al.* 1999, ApJ, 524, 983; Corcoran 2005, AJ, 129, 2018), radio (Duncan *et al.* 1999, ASP Conf. Ser. 179, 54), and line variability (Damineli *et al.* 2008, MNRAS, 386, 2330) indicate that we are dealing with a binary system with a large orbital eccentricity ($\epsilon \sim 0.9$).

The spectrum of the primary is similar to the P Cygni star HDE 316285. Modeling places a lower limit of $60R_\odot$ on the radius of the central star, although with a re-interpretation of the He I emission lines a larger radius ($\sim 240R_\odot$) is now preferred. Because of the very dense wind we observe the wind — not the "normal" photosphere of the star ($\dot{M} \sim 10^{-3}$ $M_\odot$/yr; Hillier *et al.* 2001, ApJ, 553, 837).

UV spectra reveal multiple systems of narrow absorption lines arising from neutral and singly ionized metals, and from H_2 (Gull *et al.* 2005, ApJ, 620, 442). The two dominant systems are associated with the Little Homunculus and the Homunculus, with other systems thought to be related to structures arising from the periodic interaction between the winds of the primary and secondary stars.

HST observations show that the central star has brightened – by over a factor of 3 since the first HST observations (Martin *et al.* 2004, AJ, 127, 2352). This is presumably due to a reduction in extinction, since spectra of the star, and the Weigelt blobs, have not shown dramatic changes. Variability observations show that spectral changes occur throughout the 5.5 year cycle. This provides additional evidence for binarity; the variability most likely arises from illumination effects of the Weigelt blobs as the secondary star (believed to be responsible for ionizing the Weigelt blobs) moves in its orbit.

HST observations show that the broad He I emission lines most likely originate in the neighborhood of the wind-wind interface, and are not excited by the primary star. They exhibit complex radial velocity and profile variations which are broadly consistent with those expected in a binary system (Nielsen *et al.* 2007, ApJ, 660, 669).

1.3. *The 2009 monitoring campaign*

1.3.1. *The X-ray light curve*
Michael F. Corcoran & Kenji Hamaguchi

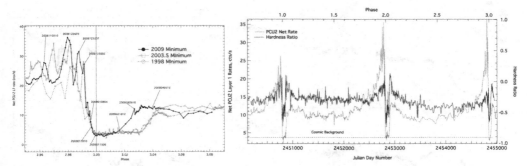

Figure 3. Left: Overplot of Eta Car's 3 X-ray minima observed by RXTE (in the 2–10 keV band). The 2009 minimum showed an abrupt recovery compared to the two earlier minima. Right: RXTE hardness ratio compared to the PCU2 net rate. All three minima show a marked increase in hardness towards the end of the X-ray minima through flux recovery.

X-ray photometry in the 2–10 keV band of the the supermassive binary star Eta Car has been measured with the Rossi X-ray Timing Explorer from 1996–2009 (see Fig. 1). The ingress to X-ray minimum is consistent with a period of 2024 days. The 2009 X-ray minimum began on January 16 2009 and showed an unexpectedly abrupt recovery starting after 12 Feb 2009. This is about one month earlier than the flux recovery in the two earlier minima (in 2003.5 and 1998). This recovery roughly corresponds in phase to the "shallow minimum" of Hamaguchi *et al* (2007 ApJ 663, 522), and suggests that for the most recent cycle the "shallow minimum" was very shallow indeed, or did not occur at all. Figure 1 also shows the hardness ratio measured by RXTE compared to the RXTE fluxes. The X-ray colors become harder about half-way through all three minima and continue until flux recovery. The behavior of the fluxes and X-ray colors for the most recent X-ray minimum (which corresponds to the time of periastron passage of an unseen companion star) suggests a significant change in the inner wind of Eta Car and might suggest that the star is entering a new unstable phase of variable mass loss.

1.3.2. *Optical photometry of the 2009.0 event of η Car*
Eduardo Fernández-Lajús, Cecilia Fariña, Juan P. Calderón, Martän A. Schwartz, Nicolás E. Salerno, Carolina von Essen, Andrea F. Torres, Federico N. Giudici, Federico A. Bareilles, M. Cecilia Scalia & Cintia S. Peri

During the last "event" that ocurred in 2009.0, η Car was the target of several observing programs. Through our optical photometric monitoring campaign, we recorded in detail

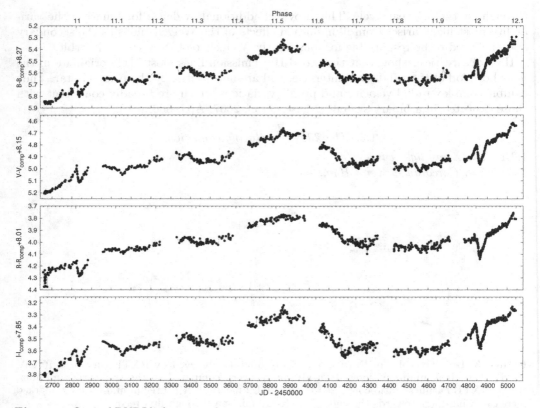

Figure 4. Optical BVRI light curves from monitoring η Car by the La Plata Observatory. While the fluxes are trending brighter, most noticeable are the broad bump associable with apastron and the two narrow drops associated with the 2003.5 and 2009.0 periastron events.

the behavior of the associated "eclipse-like" event, which happened fairly on schedule. In this work we present the resulting $UBVRI$ and Hα light curves, and a new determination of the present period length.

Our ground-based photometry was performed from the beginning of the 2009 observing season of η Car, using two telescopes at La Plata Observatory and Complejo Astronómico El Leoncito, both located in Argentina. The $UBVRI$ and Hα light curves obtained display once more an "eclipse-like" appearance. This feature is preceded by an ascending branch which peaks a maximum one month later. A sudden drop of 0.15 - 0.26 mag (depending on the band) reaches a minimum nearly simultaneously in the six bands. Then, the recovery phase starts and the brightness increases steeply up to the end of the season. The color indices show some particularities during the event, specially a blueing peak in $V - R$. Although the general trend of this event is quite similar to that of the 2003.5, there are some differences, specially the deeper dips of the minima and the high increasing rate after the "eclipse-like" feature. Our long term photometry shows some evidence of systematic brightenings of the central region (r < 3") relative to the complete "Homunculus" (r < 12") occurring just after each of these last two events.

Our results provided more observational evidence on the periodic origin of the events occurring at η Car, in accordance with the proposed binary nature of this object.

1.3.3. *VLTI/AMBER interferometry and VLT/CRIRES spectroscopy of η Car across the 2009.0 spectroscopic event*

Gerd Weigelt, José H. Groh, Thomas Driebe, Karl-Heinz Hofmann, Stefan Kraus, Dieter Schertl, P. Bristol, Augusto Damineli, Theodore Gull, Henrik Hartman, Florian Kerber, Florentin Millour, Koji Murakawa & Krister E. Nielsen

η Car's 2009.0 spectroscopic event provided a unique opportunity to study the changes of η Car's primary wind and wind-wind interaction region. The goals of VLTI/AMBER observations in 2008 and 2009 were to study the wavelength-dependent shape of η Car's aspherical stellar wind and wind-wind interaction region across the 2009.0 spectroscopic event. We carried out a large number of VLTI/AMBER observations with spectral resolution of 12000 in April 2008, January 2009, March 2009, and April 2009. We measured that the size of the wind did not significantly change at the wavelength of the Brγ 2.16 μm line during our event observations from Jan 1 to 8. However, during the event, the size of the HeI 2.06 μm emitting region collapsed from 17 mas (continuum-subtracted 50% encircled energy diameter before the event) to only 6 mas during the event. Therefore, we found strong evidence for the collapse of the wind-wind interaction zone during periastron passage.

In addition, we obtained near-IR long-slit spectroscopy of η Car with very high spatial (0.2″) and spectral ($R = 100\,000$) resolution using VLT/CRIRES. These unique data provided definitive evidence that high-velocity material, up to ~ -1900 km s^{-1}, was present in the wind region of Eta Car during the 2009.0 periastron passage. A broad, high-velocity absorption is seen in He I λ10833 only in the spectrum of 2008 Dec 26 – 2009 January 07, which strongly suggests a connection with the periastron passage, since a brief appearance of high-velocity material was also detected during previous periastron passages. We suggest that the high-velocity absorption is either formed directly in the wind of the companion star or, most likely, is due to shocked, high-velocity material from the wind-wind collision zone.

1.3.4. *HeII λ4686 in η Car: The Data and Modeling:*

Augusto Damineli, Mairan Teodoro, João E. Steiner, Nidia I. Morrell, Rodolfo H. Barbá, G. Sollivela, Roberto C. Gamen, R. Eduardo Fernández-Lajús, Federico Gonzalez, C. A. O. Torres, José Groh, Luciano Fraga, C.B. Pereira , Marcelo Borges Fernandes, M. I. Zevallos & Peter McGregor

The intrinsic emission of He II is quite repeatable from cycle to cycle. The He II λ4686 line flux rises by a factor of $\approx$10 in the 2 months preceding phase zero. There are two local maxima in the month preceding the minimum and a secondary maximum $\approx$50 days after phase zero. The rising before phase zero resembles that seen in X-rays, but with remarkable differences. The He II line flux increases by a factor of $\approx$10 as compared to only a few times in X-ray emission. Both light curves collapse before phase zero, but the collapse of He II is shifted by 16.5 days relative to the X-ray collapse. The minimum in He II is reached a week after phase zero. Since the X-ray variability is measured in the range 2–10 keV, and comes mainly from the vertex of the wind-wind shock cone, it is probably not common to the He II emitting region, which comes from gas at lower temperature. The He II line indicates a high luminosity source in the system, but it is not clear where it comes from. One possible source is the collision of the secondary star wind, since the SED derived from Parkin *et al.* 2009 (MNRAS 394, 1758) models indicates the presence of 10 times more He$^+$ ionizing photons than those passing through this atomic transition. Recombination of the shocked secondary wind is not the only source for the He$^+$ ionizing photons. As the shock cone migrates deep in the wind of the primary star,

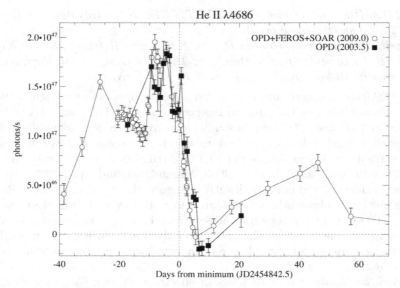

Figure 5. Line flux (photons per second) in the He II λ4686 spectral line along cycles #11 (2003.5) and #12 (2009.0).

a huge amount of hard photons are free to escape and ionize the inner walls of the wind-wind collision zone.

1.4. *3-D Modeling and Application*

1.4.1. *3-D models of the colliding winds in η Car*
Julian M. Pittard, E. Ross Parkin, Michael F. Corcoran, Kenji Hamaguchi & Ian R. Stevens

A 5.5 yr periodicity is now firmly established for η Car, with variations seen at radio, sub-mm, infrared, optical, and X-ray energies (Duncan:2003, Abraham:2005, Corcoran:2005, Damineli:2008). The overwhelming consensus is that this emission is regulated by the presence of an (unseen) companion, with the emission either originating in the wind-wind collision region between the stars (e.g., as for the X-rays, see (Pittard:2002), or being influenced by its presence and the low-density cavity which the wind of the companion star bores into the dense wind of the LBV primary (e.g., as for the radio emission).

The X-ray emission from η Car is believed to originate in the hot plasma created by the high-speed wind of the companion star shocking against the denser LBV wind (e.g., Pittard:1998, Pittard:2002). We present a recent analysis of the *RXTE* X-ray lightcurve, using a 3-D model with spatially and energy dependent X-ray emission (Parkin:2009). The model fails to obtain a good match to the data through the minimum and over-predicts the hardness of *XMM-Newton* spectra (Hamaguchi:2007). We find that the pre-shock speed of the companion wind must substantially decrease around periastron passage, and that this reduction lasts for longer than expected post-periastron. This implies that the companion wind no longer shocks at high speed against the LBV wind at this time. We speculate that this is either because the wind-wind collision region deforms into a multitude of oblique, radiative shocks, or the LBV wind completely overwhelms it and accretes onto the companion star (Soker:2005). We conclude by presenting 3-D hydrodynamical models of the colliding winds, noting several interesting features as the stars swing through periastron passage.

1.4.2. *3-D Numerical Simulations of Colliding Winds in η Car & WR140*
Atsuo T. Okazaki

We report on the results from 3-D SPH (Smoothed Particle Hydrodynamics) simulations of colliding winds in the supermassive binary η Car and the proto-typical Wolf-Rayet binary WR 140. For simplicity, both winds are assumed to be either isothermal or adiabatic, and a simplied radiation force on the wind particles is adopted. Our simulations show that in η Car the lower density, faster wind from the secondary carves out a spiral cavity in the higher density, slower wind from the primary, whereas in WR 140 it is the lower density, primary (an O4-5V star) wind that carves out a spiral cavity in the denser wind from the secondary (a WC9 star). Because of their very high orbital eccentricities, both systems show a similar, asymmetric shape of interaction surface: the cavity is very thin on the periastron side, whereas it occupies a large volume on the apastron side. A closer look, however, reveals differences caused by the differences in the wind momentum ratio and the speed of the slower wind: the interaction cone is wider and the spiral structure is more tightly wound in η Car than in WR 140. These differences are likely to affect the observational appearances of these interesting binaries.

1.4.3. *Precession and Nutation in η Car*
Zulema Abraham & Diego Falceta-Gonçalves

Although the overall shape of the X-ray light curve of η Car can be explained by the high eccentricity of the binary orbit, other features, like the asymmetry near periastron passage and the short quasi-periodic oscillations seen at those epochs, have not yet been accounted for. We explain these features assuming that the rotation axis of η Car is not perpendicular to the orbital plane of the binary system. As a consequence, the companion star will face η Car on the orbital plane at different latitudes for different orbital phases and, since both the mass loss rate and the wind velocity are latitude dependent, they would produce the observed asymmetries in the X-ray flux. We were able to reproduce the main features of the X-ray light curve assuming that the rotation axis of η Car forms an angle of 29 degrees with the axis of the binary orbit. We also explained the short quasi-periodic oscillations by assuming nutation of the rotation axis, with amplitude of about 5 degrees and period of about 22 days. The nutation parameters, as well as the precession of the apsis, with a period of about 274 years, are consistent with what is expected from the torques induced by the companion star.

1.4.4. *Accretion onto the Companion of η Car*
Amit Kashi & Noam Soker

The Accretion Model was introduced to explain observations along the entire orbit, mainly those close around the spectroscopic event. We use the standard parameters of the system and show that near periastron the secondary is very likely to accrete mass from the slow dense wind blown by the primary. The condition for accretion (that the accretion radius is large) lasts for several weeks. The exact duration of the accretion phase is sensitive to the winds' properties that can vary from cycle to cycle.

We find that: (1) The secondary accretes $\sim 2 \times 10^{-6} M_\odot \mathrm{yr}^{-1}$ close to periastron. (2) This mass possesses enough angular momentum to form a geometrically thick accretion belt, around the secondary. (3) The viscous time is too long for the establishment of equilibrium, and the belt must dissipate as its mass is blown in the re-established secondary wind. This process requires about half a year, which we identify with the recovery phase of η Car from the spectroscopic event.

We attribute the early exit in the 2009 event to the primary wind that we assume was somewhat faster and of lower mass loss rate than during the two previous X-ray minima. This results in a much lower mass accretion rate during the X-ray minimum, and consequently faster recovery of the secondary wind and the conical shell.

Mass transfer is an important process in the evolution of close massive star binaries. The high luminosity and ejected mass of many eruptive events can be explained by mass transfer, e.g., the Great Eruption of η Car.

1.4.5. *The outer interacting winds of η Car revealed by HST/STIS*
Theodore R. Gull (presented by Michael F. Corcoran)

High spatial resolution (0.1″) with moderate spectral resolution has been applied to mapping the extended wind structure of η Car. Emission lines of [Ne III], [Ar III]. [Fe III], [S III] and [N II] show an extended outer structure associable with the extended wind interaction regions. [Fe II] reveals the structure of the primary wind. We followed the spectro-images of these lines from the 1998.0 through the 2003.5 minima, finding changes in structure and velocity as the two massive winds, originating from a highly eccentric massive binary, interact.

Comparison of the forbidden line emission spatial structures to 3-D models (see Gull *et al.*, 2009, MNRAS 396, 1308) shows 1) that the He I and H I, consistent with the observations of Weigelt *et al* (2007, A&A 474, 87), originate deep within the 0.1″ limit of HST angular resolution, 2) that the broad [Ne III], [Fe III], [Ar III], [S III] and [N II] profiles are blue-shifted relative to the broad H I, Fe II and [Fe II] profiles. Moreover, the spatial distributions of the high excitation, forbidden emissions are oriented in a NE to SW distribution in the form of arcuate velocity loops that evolve in strength and spatial location across the broad high state of the binary system.

Based upon the 3-D SPH models of Okazaki (see above), the forbidden high excitation emissions originate from compressed structures in the outer regions of the interacting winds that flowed out in the previous cycle. FUV radiation is channeled by the spiral cavity carved out by the lesser wind of η Car B, the less massive, but hotter companion, with a spectral distribution of a mid O-star. As η Car B, in the highly eccentric orbit, spends the majority of the orbit near apastron, the blue-shifted, spatial distributions of the high excitation, forbidden emission, and the excitation of the blue-shifted Weigelt condensations, demonstrate that apastron is on the near side of η Car A with periastron passing on the far side. Moreover, because of the high eccentricity of the binary system, the outer, hot, low density cavity is spirally-shifted in the orbital plane by about 45 to 60° relative to the orbital major axis, known from the X-ray curve to be tilted at 45° from the sky. Combining this information leads to placement of the orbital plane close to, if not in, the plane defined by the skirt of the Homunculus, whose planar axis is aligned to the axis of symmetry of the bipolar Homunculus and Little Homunculus.

Continued mapping of the spatial distribution provides the potential to map portions of the interacting winds as they distort throughout the 5.5 year period.

1.5. *Mass loss in single and binary massive stars:*
1.5.1. *What causes the X-ray flares in Eta Carinae?*
Anthony F. J. Moffat & Michael F. Corcoran

We examine the rapid variations in X-ray brightness ("flares"), plausibly assumed to arise in the hard X-ray emitting wind-wind collision zone (WWCZ) between the two stars in eta Car, as seen during the past three orbital cycles by RXTE. The observed flares tend to be shorter in duration and more frequent as periastron is approached (see the figure), although the largest flares tend to be roughly constant in strength at all phases. Among the plausible scenarios (1. the largest of multi-scale stochastic wind clumps from the LBV component entering and compressing the hard X-ray emitting

WWCZ, 2. large-scale corotating interacting regions (CIR) in the LBV wind sweeping across the WWCZ, or 3. instabilities intrinsic to the WWCZ), the first one appears to be most consistent with the observations. This requires homologously expanding clumps as they propagate outward in the LBV wind and a turbulence-like power-law distribution of clumps, decreasing in number towards larger sizes, as seen in Wolf-Rayet winds.

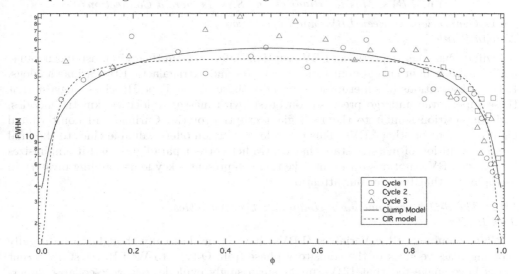

Figure 6. Full width half maximum (in days) for the identified flares vs. orbital phase. Green symbols are from cycle 1, blue symbols cycle 2, and red symbols cycle 3. The smooth curves are the best-fit models: clump model, long-dashed line; CIR model, short dashed line.

1.5.2. *Revealing the mechanism of the Deep X-ray Minimum of η Car*
Kenji Hamaguchi, Michael F. Corcoran & the η Car 2009 campaign observational Team

The multi-wavelength observing campaign of the colliding wind binary system, η Car, targeted at its periastron passage in 2003 presented a detailed view of the flux and spectral variations of the X-ray minimum phase. The X-ray spectra showed a strange Fe K line profile, without significantly varying the hard band slope above 7 keV. The result, combined with 3-D modeling studies, suggests that the X-ray minimum originates from either an eclipse of most of the emission by a porous absorber or a large change of the plasma emissivity.

The key to solve this problem would be in the deep X-ray minimum phase when X-ray emission from the central point source plunges. We therefore launched another focussed observing campaign of η Car with the *Chandra*, *XMM-Newton* and *Suzaku* observatories during the periastron passage in early 2009. Five *Chandra* spectra taken during the deep minimum revealed an underlying non-variable X-ray component from the central point source. With similar X-ray characteristics, it would be the Central Constant Emission (CCE) component discovered in 2003. Instead, the 2009 data showed it has a very hot plasma of $kT \sim 4-6$ keV. The other, variable component, probably originating in the wind-wind collision (WWC), decreased from the hard energy band above ~ 4 keV around the onset of the deep minimum and recovered only in the hard band at the end. These phenomena are consistent with a picture that the hottest plasma at the WWC convex was hidden behind an optically thick absorber first and cooler plasmas in the WWC tail followed: i.e., the deep minimum would be driven by an X-ray eclipse. On the other hand, *Suzaku* did not find any extremely embedded X-ray source ($N_{\rm H} \lesssim 10^{25}$ cm^{-2}) in spectra above 10 keV during the X-ray minimum; *XMM-Newton* spectra showed strong deformation in the iron K line as in the last cycle; the X-ray minimum recovered earlier in 2009 without significant N_H change from the 2003 cycle. These results suggest that the WWC plasma activity significantly changed during the X-ray minimum.

1.6. *LBVs, Massive Binaries and SNs: Is there a Connection?*

1.6.1. *Connections between LBVs and Supernovae*
Nathan Smith

I will discuss the properties of LBV eruptions inferred from their circumstellar nebulae and from their light curves in historical examples and extragalactic Eta Carinae analogs. Recent observations of supernovae, especially those of the Type IIn class, suggest that these supernovae undergo precursor outbursts with masses, velocities, kinetic energies, and composition similar to the 1843 giant eruption of Eta Carinae and non-terminal giant eruptions of other LBVs. This possible connection offers valuable clues to the final pre-SN evolution of massive stars that contradict current paradigms, and it emphasizes that giant LBV eruptions (or events like them) represent a key long-standing mystery in astrophysics that begs for our attention.

1.6.2. *The S-Dor phenomenon in Luminous Blue Variables*
José H. Groh

While Luminous Blue Variables (LBVs) have been classically thought to be rapidly evolving massive stars in the transitory phase from O-type to Wolf-Rayet stars, recent works have suggested that LBVs might surprisingly explode as a core-collapse super-nova. Such a striking result highlights that the evolution of massive stars through the LBV phase is far from being understood. LBVs exhibit photometric, spectroscopic, and polarimetric variability on timescales from days to decades, probably caused by different physical mechanisms.

I presented the latest results on the long-term S Dor-type variability of LBVs, in particular regarding changes in bolometric luminosity, the Humphreys-Davidson limit, and the role of rotation. The S Dor-type variability characterized by irregular visual magnitude changes on timescales of decades, with a typical amplitude of $\Delta V \simeq 1 - 2$ mag, and corresponding changes in effective temperature and hydrostatic radius. During visual minimum, the star is typically hot, while at visual maximum, a cooler effective temperature is obtained. How the S Dor-type variability relates to the powerful giant eruptions is not clear, although it could be possible that a relatively large amount of stellar mass, which is not ejected from the star, is taking part in the S Dor-type variability. This would suggest that the S Dor-type variability is a failed giant eruption.

At least for AG Car, a significant reduction ($\sim$50%) in the inferred bolometric luminosity from visual minimum to maximum has been determined, and a high rotational velocity has been obtained during minimum. I will present evidence that fast rotation is typical in Galactic LBVs that show S-Dor type variability, and will discuss how these recent results put strong constraints on the progenitor, current evolutionary stage, and fate of LBVs.

1.6.3. *Pulsational instability in massive stars: implications for SN and LBV progenitors*
Matteo Cantiello & Sung-Chui Yoon

Most massive stars experience a pulsational instability induced by κ-mechanism, when the surface temperature sufficiently decreases. The amplitude of pulsations grows very fast, and may result in very high mass loss rates. We propose a new scenario for massive star evolution based on our new calculations of this pulsational instability, where the initial mass of SNe progenitors increases according to the order: SN IIp-- > SN IIn-- >SN IIL-- >SN IIb-- >SN Ib/c. Moreover, the pulsation appears strong in the early core He-burning stage for M $\geqslant$40M_o, and may lead to the formation of LBVs. We also argue that stellar eruptions like SN 2008S may be related to this instability.

1.6.4. *Hydrodynamical Models of Type II-P SN Light Curves*
Melina C. Bersten, Omar Benvenuto, & Mario Hamuy

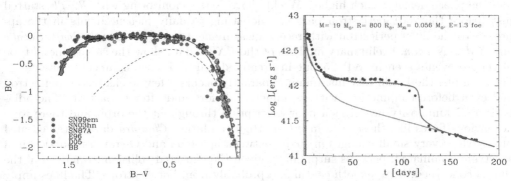

Figure 7. Hydrodynamical models for TypeII-P SN Light Curves. Left: Bolometric correction versus B-V. Right: Light curve for SNeII-P

We present computations of bolometric light curves (LC) of type II plateau supernovae (SNe II-P) obtained using a newly developed, one-dimensional Lagrangian hydrodynamic code with flux-limited radiation diffusion. We derive a calibration for bolometric corrections (BC) from *BVI* photometry (see figure 7, left) with the goal of comparing our models with a large database of high-quality *BVI* light curves of SNe II-P. The typical scatter of our calibration is 0.1 mag. As a first step, in our comparison we have determined the physical parameters (mass, radius and energy) of two very well observed supernovae, SN 1999em (see figure 7, right) and SN 1987A. Despite the simplifications used in our code we obtain a remarkably good agreement with the observations and the parameters derived are in excellent concordance with previous studies of these objects.

1.7. *Massive Binaries and η Car: What is the Relationship?*

1.7.1. *WR140 & WR25 in X-ray relation to η Car*
Andrew M. Pollock & Michael F. Corcoran

WR 25 (WN6ha+O) and WR 140 (WC7+O5) are both X-ray bright binaries of long period and high eccentricity, whose individual stellar and wind and collective binary parameters are much better known than those of η Car. Observations at different orbital phases thus show how X-rays are produced by colliding winds under physical and geometrical conditions that are quite well defined at any one time but which vary considerably around the orbit. As WR 25 is 7' from η Car, there are more observations than would otherwise be the case, a few of which during the 2003 *XMM–Newton* campaign led to the recognition of brightness and absorption variations that were soon shown to coincide with a periastron passage of the 208-day $e \approx 0.6$ optical radial velocity orbit discovered by Gamen *et al.* 2006, *A&A* 460, 777. Their orbit was used in early 2008 to plan a month-long daily ToO campaign with the soft X-ray XRT instrument aboard the *Swift* GRB Observatory. As well as the relatively shallow eclipse by the extended Wolf-Rayet wind, a sudden overall decrease between quadrature and conjunction is most obviously interpreted as a stellar eclipse by the WN6ha primary, thought to be one of the most massive stars in the Galaxy. Repeatability is good within the relatively modest statistical limits of the few dozen measurements available, spread unevenly over several cycles. The luminosity increases monotonically between apastron and periastron from the surface that provides the backdrop for the eclipses.

Observing conditions for WR 140 are more favourable. It has an orbit well-established by Marchenko *et al.* 2003, *ApJ*, 596, 1295, of longer 7.94-year period and higher $e \approx 0.881$ eccentricity. It is also a brighter X-ray source. As a result, measurements are more precise and the phase density much higher. Weekly hard X-ray monitoring with *RXTE* started just before the 2001 periastron passage, increasing to daily measurements in the approach to the 2009 periastron with recent measurements also made with *Swift*, *Suzaku* and *XMM–Newton*. Preliminary analysis of the *RXTE* data show the same general type of eclipse events seen in WR 25 but in greater detail and with significant differences. For example, the luminosity maximum apparently occurs a few weeks before periastron and even before conjunction. with asymmetries before and after periastron. The adiabatic $1/D$ luminosity law gives a poor description throughout the orbit and there were no obvious flares like those seen in η Car. High resolution *Chandra* data obtained at 4 phases show very small changes in shape between apastron and O-star conjunction in a spectrum dominated, perhaps surprisingly given the expected collisionless nature of the shocks concerned, by a smooth continuum probably from hot electrons. The lines imply complete mixing of shocked material from both winds. Details of the velocity profiles are more difficult to understand, especially the absence of the highest velocity blue-shifted material near periastron.

1.7.2. *The Erupting Wolf-Rayet System HD 5980 in the SMC: A (Missing) Link in Massive Stellar Evolution or a Freak?*
Rodolfo H. Barbá

The Wolf-Rayet eclipsing binary system HD 5980 in the Small Magellanic Cloud has shown a peculiar behaviour along the past years. In 1994 the star developed an unpredicted eruption and changed its spectrum from WN-type to one resembling those of Luminous Blue variables (LBV). In this presentation, I will review observational aspects of this unique system, emphasizing those similarities and differences with extreme LBV objects like η Car. I will briefly describe a century of photometric and spectroscopic records of the star, and depict a new analysis of the spectroscopic data obtained during the outburst phase, and the present WN-E stage. Also, I will discuss the different scenarios proposed to explain the LBV-like behaviour (rapid rotators, tidal interactions, single star evolution).

1.7.3. *The Extragalactic η Car Analogs*
Schuyler D. Van Dyk

Powerful eruptions of massive stars, such as η Car are often referred to as "supernova (SN) impostors," because some observational aspects can mimic the appearance of a true SN. During the Great Eruption during the 1800's of η Car, the star greatly exceeded the Eddington limit, with its bolometric luminosity increasing by ~ 2 mag. The total luminous output of such an eruption ($\sim 10^{49.7}$ erg) can rival that of a SN, to such a degree that some impostors initially are assigned designations as SNe, even in modern extragalactic SN searches. A number of extragalactic SN impostors are known, such as SNe 1954J, 1961V, 1997bs, 1999bw, 2000ch, 2001ac, 2002kg, 2003gm, NGC 2363-V1, etc. I will present here the latest results for those that can be considered η Car analogs. Not all impostors are as powerful as η Car, and aretherefore not considered true analogs to

η Car; some cases are more like the "classical" LBVs (e.g., S Dor), where the bolometric luminosity remains constant during an eruption, as the star's envelope expands or its wind becomes optically thick, and the apparent temperature cools to $\sim$8000 K. Like η Car, the precursor star for each analog is expected to survive the eruption and return to relative quiescence. Some have had eruption survivors identified (SNe 1954J, 1961V), using the *Hubble Space Telescope*, some have seemingly "vanished" after outburst (SNe 1997bs, 1999bw), and one (SN 2000ch) continues in outburst after almost a decade. Only one (SN 1999bw) has shown evidence for dust emission, based on *Spitzer Space Telescope* observations, and the emission has apparently faded from detection. Studying the characteristics of the analogs provides us with a greater understanding of η Car itself and of the evolution of very massive stars.

1.8. *Summary and Discussion*
Nidia I. Morrell, Michael F. Corcoran, Anthony F.Moffat & Julian Pittard

After a brief brainstorm session, Mike Corcoran, Tony Moffat, Nidia Morrell and Julian Pittard came up with the following list of questions and highlights, which served as a basis for a half-hour open discussion on future studies of η Car:

How to better constrain the orbital and wind parameters of both stars in η Car?

What is its future evolution?

What caused the Great Eruption? Which star erupted?

What is the nature of the companion star? (Very urgent!)

Whats the connection between WRs, LBVs and supernovae?

How to explain the strictly cyclic, bizarre behavior of the He II 4686 emission, which emerges only within several months of periastron passage?

What is the role of a companion star in driving the formation, evolution and instabilities of η Car and other binary LBVs?

Does dust form in η Car?

Does η Car pulsate?

2. POSTERS:

2.1. *A full cycle 7 mm light-curve of η Car*
Zulema Abraham, Pedro P. Beaklini & Carlo Miceli

It is now well established that the light curve of Eta Carinae has a periodic behavior at all wavelengths, from mm waves to X-rays. These light curves are characterized by the presence of a sharp dip, with duration that depends on wavelength, being longer at X-rays. At mm wavelengths, the dip was detected during the last four cycles, but only during the 2003.5 minimum the light curve was obtained with daily resolution. At that epoch, the 7 mm light curve, obtained with the Itapetinga radiotelescope, in Atibaia, Brazil, followed the X-ray decaying behavior but showed a strong peak, not seen at other wavelengths, before reaching the minimum. This peak was attributed to free-free emission of the 107 K optically thick gas located at the wind-wind collision contact surface. Here, we report the 7 mm light curve of the complete 2003-2009 cycle, including the 2003.5 and 2009.0 minima, both obtained with daily resolution. We show for the rst time that: (a) the duration of the minima are the same at 7 mm and at X-rays; (b) The peak at 7 mm seen after the minimum is 2003.5 appeared again in 2009.0, with the same phase, duration and shape; (c) two other strong peaks were observed before the 2009.0 minimum, coincident with the peaks observed at X-rays, which supports the previous assumption that they are formed at the wind-wind shock interface.

2.2. *The multiple zero-age main-sequence O star Herschel 36*

Julia I. Arias, Rodolfo H. Barbá, Roberto C. Gamen, Nidia I. Morrell, Jesús Maíz Apellániz, Emilio J. Alfaro, Nolan R. Walborn, Alfredo Sota, Christian M. Bidin

We present a study of the zero-age main-sequence O star Herschel 36 in M8, based on high-resolution optical spectroscopic observations spanning six years. This object is denitely a multiple system. We propose a picture of a close massive binary and a companion of spectral type O, most probably in wide orbit about each other. The components of the close pair are identied as O9 V and B0.5 V. The orbital solution for this binary is characterized by a period of 1.5415±0.00001 days. With a spectral type O7.5 V, the third body is the most luminous component of the system. It also presents radial velocity variations with short (a few days) and long (hundreds of days) timescales, although no accurate temporal pattern can be discerned from the available data. Some possible hypotheses to explain the variability are briey addressed and further observations are suggested.

2.3. *Spatially extended wind emission in the massive binary systems VV Cep & KQ Pup*

Wendy Hagen Bauer, Theodore R. Gull, Philip Bennett & Jahanara Ahmad

VV Cep and KQ Pup are binary systems consisting of M supergiant primaries with B main-sequence companions which orbit within the extensive M supergiant winds. VV Cep undergoes total eclipses and was observed with the HST/STIS Spectrograph at several epochs which spanned total eclipse through "chromospheric eclipse" as lines from ions like Fe I weakened and disappeared through first quadrature. KQ Pup comes close to eclipsing its hot companion and was observed to be in chromospheric eclipse (showing weak absorption from Fe I in the M supergiants chromosphere) by STIS in October 1999. Two-dimensional reprocessing of the STIS echelle spectra has revealed spatially extended emission in all observations of these two systems. Emission arising from gas thought to be associated with the hot component shows spatial extension consistent with the STIS spatial point spread function. The spatially extended flux seen outside total eclipse arises from emission in transitions expected to be observed from the winds of cool supergiants. VV Cep was observed at enough epochs to map out radial velocity structure within the wind. It is consistent with model predictions for wind flow in a binary system in which the wind outflow is comparable with the M supergiants orbital velocity. Spatially resolved wind and wind interaction structures of these two stars and of η Car reinforce the need for imaging spectroscopy and added capabilities of integral field units for mapping these complex interacting binary systems.

2.4. *Abundances and depletion of iron-peak elements in the Strontium filament of η Car*

Manuel A. Bautista, Henrik Hartman, Marcio Meléndez, Theodore R. Gull, Katharina Lodders & Mariela Martinez

We carried out a systematic study of elemental abundances in the Strontium Filament, a peculiar metal-ionized structure located in the skirt plane of the Homunculus, ejecta surrounding η Car. To this end we interpret the emission spectrum of neutral C and singly ionized Al, Sc, Ti, Cr, Mn, Fe, Ni, and Sr using multilevel non-LTE models for each ion. The atomic data for most of these ions is limited and of varying quality, so we carried out ab initio calculations of radiative transition rates and electron impact excitation rate coefcients for each of these ions. The observed spectrum is consistent with an electron density $\approx 10^7 \text{cm}^{-3}$ and a temperature between 6000 and 7000 K. The observed spectra are consistent with large enhancements in the gas phase Sr/Ni, Sc/Ni, and Ti/Ni abundance ratios relative to solar values. Yet, the abundance ratios Cr/Ni, Mn/Ni, and Fe/Ni are roughly solar. We explore various scenarios of elemental depletion in the context of nitrogen-rich chemistry, given that the stellar ejecta has enriched nitrogen at the expense of greatly depleted oxygen and carbon due to mixing in the >60 $M_\odot$ star. Finally, we discuss the implications of these findings for the generation of dust during the evolution of supermassive stars from main sequence to pre-supernova stage.

2.5. *A fast ray tracing disk model for 10μ interferometric data fitting:*
First application on the B[e] star CPD57 2874
Philippe Bendjoya, Giles Niccolini, & Amando D. de Souza

We present here a parametric dust disk model (P2DM) developed to fit interferometric observations in a much faster computing time than the classical Monte Carlo Modeling Approach. P2DM combined with a Levenberg-Markward minimisation algorithm allows us to derive both crucial physical and geometrical parameters. This model is restricted to wavelengths around and above 10 microns (no gas, no scattering) making it useful for VLTI-MIDI (and future MATISSE) observations and implies that more elaborate modelling is necessary to get a deeper understanding of the physical processes responsible of the observed disks. Neverthelss, this fast and physical model is useful for exploring the physical parameter phase space and to provide starting values for more powerful models. We present the model and its applica- tion to the supergiant B[e] CPD -57 2874 star observed with VLTI-MIDI.

2.6. *A search for relics of interstellar bubbles originated by LBV progenitors:*
Cristina E. Cappa, Silvina Cichowolski, Javier Vasquez & J. R. Rizzo

The strong stellar winds of massive O stars sweep up and compress the surrounding gas creating interstellar bubbles in their environs. In this modified environment, massive stars evolve into Luminous Blue Variables (LBVs), which are the immediate progenitors of WR stars. Using the Canadian Galactic Plane Survey (CGPS) and Southern Galactic Plane Survey (SGPS) we searched for HI interstellar bubbles associable with O-type progenitors of a number of galactic LBVs and LBV candidates. We found HI cavities and shells that probably originated from the massive progenitors of P Cygni, G79.29+0.46, AG Carinae, and He3-519.

2.7. *Massive binaries and rotational mixing*
Selma E. de Mink, Marreo Cantiello, Norbert Langer & Onno R. Pols

In massive stars fast rotation is the cause of efficient internal mixing, which leads to the transport of hydrogen burning products from the core to the stellar envelope. This results in hot and overluminous stars, which stay compact as they gradually evolve into massive helium stars (e.g. Yoon & Langer, 2005). While non-rotating stars in close binaries experience severe mass loss as soon as their radius exceeds the Roche lobe radius, fast-rotating stars, which are efficiently mixed, stay compact and can avoid the onset of mass transfer.

This can occur in wide binaries (orbital periods much larger than about 10 days) where the rotation rate of the stars is not affected by tides during the main sequence evolution. Alternatively, this can occur in massive binaries with orbital periods smaller than 3 days. Tides force the stars to rotation rates high enough to trigger efficient mixing (De Mink *et al.* 2008, 2009). This type of evolution leads naturally to the formation of compact Wolf-Rayet binaries and is potentially interesting as an explanation for the formation of massive black hole binaries such as M33 X-7 and IC10 X-1.

2.8. *MHD numerical simulations of wind-wind collisions in massive binary systems*
Diego Falceta-Gonçalves & Zulema Abraham

In past years, several massive binary systems have been studied in details at both radio and X-rays wavelengths, revealing a whole new physics present in such systems. Large emission intensities from thermal and non-thermal sources showed us that most of the radiation in these wavelengths originates at the wind-wind collision region. OB and WR stars present supersonic and massive winds that, when under collision, emit largely in X-rays and radio due to the free-free radiation, as well as in radio due to synchrotron emission. However, in the latter case, magnetic fields play an important role on the emission distribution. Astrophysicists have been modeling free-free and synchrotron emission from massive binary systems based on purely hydrodynamical simulations and ad hoc assumptions regarding the distribution of magnetic energy and the field geometry in order to study the non-thermal source. In this work we provide a number of the first MHD numerical simulations of wind-wind collision in massive binary systems. We study the free-free emission, characterizing its dependence on the stellar and orbital parameters. We also study, self consistently, the evolution of the magnetic field at the shock interfaces, obtaining also the synchrotron energy distribution integrated along different lines of sight.

2.9. On the peculiar variations of two southern B[e] stars
Marclo B. Fernandes, Michaela Kraus, Olivier Chesneau, Jiri Kubát, Amando D. de Souza, Francisco X. de Araujo, Philippe Stee & Anthony Meilland

In this work, we present the peculiar variations shown by two B[e] stars, namely the SMC supergiant LHA115-S23 and the galactic unclassied object HD50138, mainly based on high resolution optical spectroscopic data. The spectra of LHA115-S23 revealed the disappearance of photospheric He I absorption lines in a period of only 11 years. Due to this, the star has changed its MK classication from B8I to A1Ib, becoming the first A[e] star identified. Concerning HD50138, the brightest known B[e] star, based on our data, taken with a difference of 8 years, it is possible to see the presence of strong spectral variations, probably associated with an outburst, which took place prior 2007. A detailed spectroscopic description, the projected rotational velocities, the modeling of their spectral energy distributions, and the discussion about the possible nature and circumstellar scenarios for these two curious B[e] stars are provided.

2.10. Interferometric analysis of peculiar stars with the B[e] phenomenon
Marcio B. Fernandes, Olivier Chesneau, Denis Mourard, Michaela Kraus, Philippe Stee, Armando D. de Souza, Anthony Meilland, Florentin Millour & Samer Kanaan

Stars that present the B[e] phenomenon are known to form a very heterogeneous group. This group is composed by objects in different evolutionary stages, like high- and low-mass evolved stars, intermediate-mass pre-main sequence stars and symbiotic objects. However, more than 50% of the confirmed B[e] stars have unknown evolutionary stages, being called as unclassified B[e] stars. The main problem is the absence of reliable physical parameters and of knowledge of their circumstellar geometries. Based on this, high-angular resolution interferometry is certainly an important tool to answer several questions concerning the nature of these stars, including a possible evolutionary link between B[e] supergiants and LBV stars, like η Car. In this work, we present the results related to a sample of objects, namely HD50138, HD45677, HD62623 and MWC361 based on observations using VLTI/MIDI, VLTI/AMBER and CHARA/VEGA.

2.11. Numerical models for 19th century outbursts of η Car
Ricardo F. González Domínguez

We present new results of two-dimensional hydrodynamical simulations of the eruptive events of the 1840s (the great) and the 1890s(the minor) eruptions suffered by the massive star, η Car. The two bipolar nebulae commonly known as the Homunculus (H) and the Little Homunculus (LH) were formed from the interaction of these eruptive events with the underlying stellar wind. We assume a colliding wind scenario to explain the shape and the kinematics of both Homunculi. Adopting a more realistic parametrization of the phases of the wind, we show that the LH is formed at the end of the 1890s eruption when the post-outburst η Car wind collides with the eruptive flow, rather than at the beginning (as claimed in previous works; González *et al.* 2004a, 2004b). The regions at the edge of the LH become Rayleigh-Taylor unstable and develop filamentary structuring that shows some resemblance with the observed spatial structures in the polar caps of the inner Homunculus (Smith 2005). We also find the formation of some tenuous equatorial, high-speed features.

2.12. *Discovery of a new WNL star in Cygnus with Spitzer*

Vasilii Gvaramadze, Sergei Fabrika, Wolf-Rainer Hamann, Olga Sholukhova, Azamat F. Valeev, Vitaly P. Goranskij, Anatol M. Cherepashchuk, Dominik J. Bomans & Lidia M. Oskinova

We report the serendipitous discovery of an infrared ring nebula in Cygnus using the archival data from the Cygnus-X Spitzer Legacy Survey and present the results of study of its central point source. The optical counterpart to this source was identied by Dolidze (1971) as a possible Wolf-Rayet star. Our follow-up spectoscopic observations with the Russian 6-m telescope confirmed the Wolf-Rayet nature of this object and showed that it belongs to the WN8-9h subtype.

2.13. *VLT-CRIRES observations of η Car's Weigelt blobs & Strontium Filament*

Henrik Hartman, José Groh, Thedore R. Gull, Hans U. Kaufl, Florian Kerber, Vladilen Letokhov & Krister E. Nielsen

We have taken Very Large Telescope-CRIRES observations of η Car, focused on the Weigelt condensations (WC) and the Strontium Filament (SrF). These are nebular regions, in the close vicinity to Eta Car, with complex emission line spectra. The two regions show, however, strikingly different physical conditions and abundances. The WC are driven by far-UV radiation from the hot companion (Eta Car B). The radiation is internally redistributed to hydrogen emission which enables exotic atomic photo processes, such as Resonance Enhanced Two-Photon Ionization (RETPI) and stimulated emission (LASER). The lines proposed for the stimulated emission are the 1.68 and 1.74 mm transitions from the c4F7/2 level in Fe II (i.e. the spectrum of Fe^+).

The Strontium Filament received its name from the initial discovery of [Sr II], lines from singly-ionized strontium. Modeling of the emission spectrum has revealed strange abundances (see separate poster by Bautista *et al.* at this meeting), and spectral lines with complex line profiles. The main emission component is consistent with a creation of the ejecta in the 1890s.

We present a preliminary analysis of the ejecta in the NIR, using high spectral (R= 90,000) and spatial resolution ($\approx$0.3") spectra obtained with CRIRES in April 2007. The data allow us to study the individual ejecta in detail, at a spectroscopic phase where the effects due to η Car B's periastron passage is negligible.

We all acknowledge the tremendous contributions by Sveneric Johansson and Vladilen Letokhov to the field of plasma physics, the understanding of the physical processes in the WC, and the final contribution with their book *Astrophysical Lasers* (Oxford, 2009).

2.14. *Radiative transfer Modeling of rotational modulations in the B supergiant HD 64760*
Alex Lobel & Ronny Blomme

We develop parameterized models for the large-scale structured wind of the blue super-giant, HD 64760 (B0.5 Ib), based on best fits to Rotational Modulations and Discrete Absorption Components (DACs) observed with IUE in Si IV λ1400. The fit procedure employs the WIND3D code with non-LTE radiative transfer (RT) in 3-D. We parameterize the density structure of the input models in wind regions (we term "Rotational Modulation Regions" or RMRs) that produce Rotational Modulations, and calculate the corresponding radial velocity field from CAK-theory for radiatively-driven rotating winds. We find that the Rotational Modulations are caused by a regular pattern of radial density enhancements that are almost linearly shaped across the equatorial wind of HD 64760. Unlike the Co-rotating Interaction Regions (CIRs) that warp around the star and cause DACs, the RMRs do not spread out with increasing distance from the star. The detailed RT fits show that the RMRs in HD 64760 have maximum density enhancements of ∼17 % above the surrounding smooth wind density, about two times smaller than hydrodynamic models for CIRs. Parameterized modelling of Rotational Modulations reveals that nearly linear-shaped (or 'spoke-like') wind regions co-exist with more curved CIRs in the equatorial plane of this fast rotating B-supergiant. We present a preliminary hydrodynamic model computed with Zeus3D for the RMRs, based on mechanical wave excitation at the stellar surface of HD 64760.

2.15. *Parameterized structured wind modelling of massive hot stars with Wind3D*
Alex Lobel & Jesús A. Toalá

We develop a new and advanced computer code for modelling the physical conditions and detailed spatial structure of the extended winds of massive stars with three-dimensional (3-D) non-LTE radiation transport calculations of important diagnostic spectral lines. The WIND3D radiative transfer code is optimized for parallel processing of advanced input models that adequately parameterize large-scale wind structures observed in these stars. Parameterized 3-D input models for Wind3D offer crucial advantages for high-performance transfer computations over ab-initio hydrodynamic input models. The acceleration of the input model calculations permits us to investigate and model a much broader range of physical (3-D) wind conditions with Wind3D. We apply the new parameterization procedure to the equatorial wind-density structure of Co-rotating Interaction Regions (CIRs) and calculate the wind velocity-structure from CAK-theory for radiatively-driven rotating winds. We use the parameterized CIR models in WIND3D to compute the detailed evolution of Discrete Absorption Components (DACs) in Si IV UV resonance lines. The new method is very flexible and efficient for constraining physical properties of extended 3-D CIR wind structures (observed at various inclination angles) from best fits to DACs in massive hot stars. We compare the results with an accurate hydrodynamical model for the DACs of B0.5 Ib-supergiant HD 64760, and apply it to best fit the detailed DAC evolution observed with *IUE* in B0 Iab/Ib-supergiant HD 164402.

2.16. *3D modeling of eclipse-Like events in η Car*
Thomas I. Madura, Theodore R. Gull, Atsuo Okazaki & Stanley Owocki

We discuss recent efforts to apply 3D Smoothed Particle Hydrodynamics (SPH) simulations to model the binary wind collision in η Car, with emphasis on reproducing BVRI photometric variations observed from La Plata Observatory. Photometric dips occurring concurrently with X-ray minima seen with RXTE provide further evidence for binarity in the system. We investigate the role of the unseen secondary star, focusing on two effects: 1) an occultation of the secondary by the slower, extended optically thick primary wind; and 2)a Bore-Hole effect, wherein the fast wind from the secondary carves a cavity in the dense primary wind, allowing increased escape of radiation from the hotter/deeper layers of the pri mary s extended photosphere. Such models may provide clues on how/where light is escaping the system, the directional illumination of distant material (e.g., the Homunculus, the Little Homunculus, the purple haze, Weigelt blobs,etc.) and the parameters/orientation of the binary orbit.

2.17. *The Other Very Massive Stars in the Carina Nebula as observed with HST*
Jesús Maíz Apellániz, Nolan R. Walborn, Nidia I. Morrell, Ed P. Nelan & Virpi S.
Niemela

We have used HST/ACS+FGS and ground-based data to study 10 WNha, O2-4 supergiant, and O3.5 main-sequence stars in the Carina Nebula. HD 93129 Aa+Ab is the most massive known astrometric binary. Its motion is currently being followed with STIS spectroscopic observations planned for the fall of 2009. Previously unknown resolved components are detected: an ~8 M$_\odot$ star for HD 93162 (=WR 25) and two ~1 M$_\odot$ stars for Tr 16-244. Overall, at least 8 of the 11 most massive stars in the Carina Nebula are members of multiple systems. The NUV-to-NIR photometry has been processed with the new version (v3.1) of the CHORIZOS code using Geneva isochrones with ages of 1.0 Ma and 1.8 Ma. Most stars in our sample are found to have visual total extinctions between 1.0 and 2.2 mag but HD 93162 and Tr 16-244 are more extinguished. The ratio of total to selective extinction R_{5495} is found to vary between 3.0 and 4.5 and is positively correlated with the total extinction. For a fixed age for the full sample, the Trumpler 14 stars are underluminous for their spectral types, hence implying a small age ($\lesssim1$ Ma) for the cluster. HD 93250 is overluminous for its spectral type, a possible indication of an undetected (by spectroscopic, interferometric, or imaging methods) massive companion. The three WRs (22, 24, and 25) and HD 93129 Aa have evolutionary (initial) masses above 90 M$_\odot$, i.e. values comparable to that of η Car.

2.18. *The High Angular Resolution Multiplicity of Massive Stars*
Brian D. Mason, William I. Hartkopf, Douglas R. Gies, Theo A. ten Brummelaar, Nils
H. Turner, Chris D. Farrington & Todd J. Henry

Conducted on NOAO 4-m telescopes in 1994, the first speckle survey of O stars (Mason *et al.* 1998) had success far in excess of our expectations. In addition to the frequently cited multiplicity analysis, many of the new systems which were first resolved in this paper are of significant astrophysical importance. Now, some ten years after the original survey, we have re-examined all systems analyzed before. Improvements in detector technology allowed for detection of companions missed before as well as systems which may have been closer than the resolution limit in 1994. Also, we made a first high-resolution inspection of the additional O stars in the recent Galactic O Star Catalog of Maíz-Apellániz & Walborn (2004). In these analyses we resolved four binaries not detected in 1994 due to the enhanced detection capability of our current system or kinematic changes in their relative separation. We also recovered four pairs, confirming their original detection. In the new sample, stars are generally more distant and fainter, decreasing the chance of detection. Despite this, eight pairs were detected for the first time.

In addition to many known pairs observed for testing, evaluation and detection characterization, we also investigated several additional samples of interesting objects, including accessible Galactic WR stars from the contemporaneous speckle survey of Hartkopf *et al.* (1999), massive, hot stars with separations which would indicate their applicability for mass determinations (for fully detached O stars masses are presently known for only twelve pairs), and additional datasets of nearby red, white, sub and G dwarf stars to investigate other astrophysical phenomena. In these observations, in addition to those enumerated above we resolved seventeen pairs for the first time.

Massive stars have also been a important observing program for the CHARA Array. Preliminary results from Separated Fringe Packet solutions of interferometric binaries are also presented.

2.19. Far-IR Spectroscopic Imaging of the ISM around η Car
Hiroshi Matsuo, Takaaki Arai, Tom Nitta & Aya Kosaka

To study interstellar material around η Car, we have performed far-infrared imaging spectroscopic observations using a Fourier transform spectrometer onboard the Japanese infrared satellite AKARI. We have obtained images of C II, N II, and O III covering the 15 arcmin $\times$10 arcmin area centered at η Car. The O III and C II lines were found widespread, but peaked toward Carinae nebulae, which gives an indication of interaction of ejecta and molecular clouds. The N II line is weak and only partially observed around η Car. Comparison with ionized hydrogen and non-thermal emission at millimeter-wave O III emission is coincident with ionized region while C II emission is peaked at different positions but similar to the position angle of the Homunculus nebulae, which may indicate that we are observing interactions of old ejecta with molecular clouds.

2.20. Stellar forensics with SNe & GRBs:
Deciphering the size & metallicity of their massive progenitors
Maryam Modjaz

Massive stars die violently. Their explosive demise gives rise to brilliant fireworks that constitute supernovae and long GRBs, and that are seen over cosmological distances. By interpreting their emission and probing their environment, we get insights into the size, make-up, mass loss history and metallicity of their massive progenitor stars that are situated at extragalactic distances.

I will present extensive X-ray, optical and NIR data on SN 2008D which was discovered serendipitously with the NASA Swift satellite via its X-ray emission from shock breakout. It is a supernova of Type Ib, that is, a core-collapse supernova whose massive stellar progenitor had been been stripped of most, if not all, of its outermost hydrogen layer, but had retained its next-inner helium layer, before explosion. I will discuss the signicance of this supernova, the derived size of its Wolf-Rayet progenitor, what it tells us about the explosive demise of massive stars, and its implications for the supernova-GRB connection. Furthermore, I will present observational results that confirm low metallicity as a key player in determining whether some massive stars die as GRB-SN or as an ordinary SN without a GRB. I show that the oxygen abundances at the SN-GRB sites are systematically lower than those found near ordinary broad-lined SN Ic, at a cut-off value of $0.3-0.5$ Z_{solar}.

2.21. Rapid Spectrophotometric Changes in R127 and Reversal of the Decline
Nidia I. Morrell, Roberto C. Gamen, Nolan R. Walborn, Rodolfo H. Barbá, Katrien Uytterhoeven, Artemio Herrero, Christiopher Evans, Ian Howarth & Nathan Smith

R127, the famous Luminous Blue Variable in the Large Magellanic Cloud, was found in the peculiar early-B state and fainter in January 2008, suggesting that the major outburst which started sometime between 1978 and 1980 was drawing to a close, and that the star would presumably continue to fadeand move to earlier spectral types until reaching its quiescent Ofpe/WN9 state. Archival data showed that the main spectral transformation from the peculiar A-type state at maximum started between 2005 and 2007, and that it was in close concordance with features in the light curve. However, subsequent observations during 2008 and early 2009 have shown that the spectrum of R127 is now returning to a cooler, lower excitation state, while the photometry shows a new brightening of the star. A speculative 7-year cycle during the decline bears further investigation. The curious behavior of R127 provides an opportunity to gain further insight into the rapid transitional stages in the late evolution of very massive stars.

2.22. *Interaction of LBVs with their molecular outskirts*
J. Ricardo Rizzo, Javier Vasquez, Aina Palau, Francisco M. Jimenez-Esteban &
Cristina Cappa

LBV stars disturb the ISM by their mass-loss events and high UV flux. Here we show the distribution of CO and isotopomers around some LBV stars, including η Car. Detected shells and high density regions point to the presence of shocked regions and PDRs.

2.23. *The Luminous Blue Variable Stars in M33:*
the Extended Hot Phase of Romano's Star (GR 290)
Corinne Rossi, Vito Francesco Polcaro, Silvia Galleti, Roberto Gualandi, Laura Norci
& Roberto F. Viotti

Romano's Star (GR290) is an LBV in M33. Recently, the star underwent a dramatic decrease in the visual, that was accompanied by a marked increase of the spectral line excitation. Presently, GR290 appears to be in the hottest phase ever observed in an LBV. More than 100 emission lines have been identified in the 3100−10000Å range covered by the WHT spectra, including the hydrogen Balmer and Paschen series, He I and He II, C III, N II-III, Si III-IV, and many forbidden lines of [O III], [N II], [S III], [Ar III] and [Fe III]. Many lines, especially the He I triplets, show a P Cygni profile with an E-A radial velocity difference of about 400 km/s. The 2008 spectrum appears quite similar to that of a typical WN8-9 star. During 2003−2009 GR290 varied between the WN11−WN8 spectral types, with the hottest spectrum corresponding to a fainter visual magnitude. This temperature-visual luminosity anticorrelation suggests variation at constant Mbol. GR290 might just present the key evidence that will help to bridge the LBV and WNL evolutionary phases.

2.24. *X-Ray modeling of η Car and WR140 from hydrodynamic simulations*
Christopher Russell, Michael F. Corcoran, Atsuo Okazaki, Thomas I.Madura
& Stanley Owocki

The colliding wind binary (CWB) systems η Car and WR140 provide unique laboratories for X-ray astrophysics. Their wind-wind collisions produce hard X-rays, which have been monitored extensively by several X-ray telescopes, such as RXTE and Chandra. To interpret these X-ray light curves and spectra, we apply 3D hydrodynamic simulations of the wind-wind collision using both smoothed particle hydrodynamics (SPH) and nite difference methods. We nd isothermal simulations that account for the absorption of X-rays from an assumed point source of X-ray emission at the apex of the wind-collision shock cone can closely match the RXTE light curves of both η Car and WR140. We are now applying simulations with self-consistent energy balance and extended X-ray emission to model the observed X-ray spectra. We present these results and discuss efforts to understand the earlier recovery of η Car's RXTE light curve from the 2009 minimum.

2.25. Accretion onto the secondary of η Car during the spectroscopic event
Noam Soker & Amit Kashi

We show that near periastron passage the shocked primary wind becomes gravitationally bound to the secondary star. This results in accretion flow onto the secondary star that almost shuts down the secondary wind. The accretion process is the mechanism of the deep X-ray minimum. Not only in the present η Car, but also during the great eruption, accretion played a key role.

2.26. New massive, eclipsing, double-lined spectroscopic binaries Cyg OB2-17 & NGC 346-13
V. E. Stroud, J. S. Clark, I. Negueruela, D. J. Lennon & C. J. Evans

Massive, eclipsing, double-lined spectroscopic binaries are not common but necessary to understand the evolution of massive stars as they are the only direct way to determine the masses of OB stars and therefore obtain mass-luminosity functions. They are also the progenitors of energetic phenomena such as X-ray binaries and γ-ray bursts. We discuss results from photometric and spectroscopic studies of two binary systems: Cyg OB2-B17 which is a semidetached binary located in the Cyg OB2 association and comprised of 2 O supergiants; and NGC 346-13 which is a system located in the Small Magellanic Cloud and comprised of a semi-evolved B1 star and a hotter, optically fainter secondary, suggesting mass transfer in the system.

2.27. Monte Carlo radiative transfer in stellar wind
Brankica Šurlan & Jiří Kubát

As a first step towards solution of the radiative transfer equation in clumped stellar wind we started to develop a code for the formal solution of the radiative transfer equation for given velocity, temperature, and density stratification. Wind structure was taken from a model calculated using a NLTE code by Krtička & Kubát (2004, A&A 417, 1003). Wind opacity consists of line scattering under Sobolev approximation and of the electron scattering. As our first preliminary results we plot the P Cygni profile of the line obtained from our calculation. This work has been supported by grants 205/08/0003 and 205/08/H005 (GA ČR).

2.28. Gamma-ray observations of the η Car region
Marco Tavani, Sabina Sabatini, Roberto Viotti, Michael F. Corcoran, Elena Pian & the AGILE TEAM

We present the results of extensive observations by the gamma-ray AGILE satellite of the Galactic region hosting the Carina nebula and the colliding wind binary η Car. The AGILE gamma-ray satellite monitored the η Car region in several occasions during the period 2007 July to 2009 January. AGILE detects a gamma-ray source consistent with the position of η Car. The average gamma-ray flux above 100 MeV integrated over the pre-periastron period 2007 July - 2008 October is $F = (37 +/- 5) \times 10^{-8}$ ph/cm^2/sec corresponding to an average gamma-ray luminosity of $L = 3.4 \times 10^{34}$ erg/sec for a distance of 2.3 kpc. AGILE also detected a remarkable 2-day gamma-ray flaring episode of on 11-13 October 2008, most likely caused by a colliding wind transient particle acceleration episode. The pre-periastron gamma-ray emission appears to be erratic, and is possibly related to transient acceleration and radiation episodes in the strongly variable colliding wind shocks in the system. Our results provide the long sought first detection above 100 MeV of a colliding wind binary, and have important theoretical implications.

2.29. *Long-term variability of η Car*
Mairan Teodoro

During the last 50 years, η Car has increased its brigthness at variable rates. For instance, the central source presented V=8 from 1910 to 1940, when it suddenly increased its brightness by 1 magnitude in a few years. Since then, the brightness has increased almost linearly with time at a rate of approximately 0.03 mag per year. However, after the spectroscopic event of 1997.9, the rate increased to 0.2 mag per year and remained so until mid-2006, when a drop in the brightness of the central source was observed (almost 30 per cent in less than one year!). In this work we present the results of our study on the long-term variability of the central source of Eta Car, showing that, while the central source is getting brighter, the equivalent width of the lines are getting weaker from cycle to cycle. Besides, our results indicate that at least in the last 4 events, the behaviour of the high- and intermediary-excitation lines near the spectroscopic event have not changed signicantly.

2.30. *η Car around the 2009 periastron - a new view with X-shooter*
Christina Thöne, Theodore R. Gull, Guido Chincarini, Elena Pian, Henrik Hartman, Sandro D'Odorico & Lex Kapor

We observed the η Car binary system with the X-shooter spectrograph at the VLT during commissioning phase that spanned the latest periastron event of the system on Jan. 11 2009. X-shooter covers the whole spectral range from the UV (3000Å to the IR (2.5 μm) simultaneously with medium resolution ($R = \lambda/\delta\lambda = 4000 - 9000$). Two long slits were placed on the Homunculus skirt radially extending out from the star in opposite directions at three different epochs in January (510 d after periastron), March and June. At visible wavelengths, the Strontium Filament was sampled with three sub-slits of the 1.8" ×4" Integral Field Unit (IFU) in January. The shape of the Balmer lines in the opposite slit positions can give us information about the orientation of the orbit of the secondary star. The absence of PCygni absorption on the south-west slit indicates that the secondary enters from the south-western side ionizing the wind material causing the absorption in the north-east slit. The X-ray emission, which disappears during periastron due to the collapse of the shock front of the winds, recovered surpisingly early in 2009. High ionization lines were still not visible again in the data of the March run while they are still visible in the outer regions of the radial slits in January since those regions had not yet seen the shut off of the FUV radiation due to the light travel time.

2.31. *INTEGRAL observations of η Car*
Roland Walter & Jean-Christophe Leyder

If relativistic particle acceleration takes place in colliding-wind binaries, then hard X-rays and γ-rays are expected through inverse Compton scattering of the copious UV radiation field. The *INTEGRAL* satellite provided hard X-ray images of the Carina region with a much higher spatial resolution than previously available. Based on observations taken far from periastron, a bright source was detected at the position of η Car up to 100 keV. Two additional nearby hard X-ray sources could also be resolved. This is the first unambiguous detection of η Car at hard X-rays. There is no other X-ray source in the hard X-ray error circle, bright enough to match the hard X-ray flux.

The average hard X-ray emission of η Car in the 22-100 keV energy range is very hard (with a photon index $\Gamma \approx 1$) and its luminosity (7×10^{33}erg/s) is in agreement with the predictions of inverse Compton models and corresponds to about 0.1% of the energy available in thewind collision.

New *INTEGRAL* observations were taken during the 2009 periastron passage, and the first results are presented. Only a 5-σ upper-limit could be derived. This is consistent with a lower fraction of very energetic particles during periastron than outside. This could perhaps be linked with electron cooling by the extreme radiation field.

2.32. *BRITE-Constellation*
Werner W. Weiss, Anthony F. Moffat & the BRITE-Constellation Team

BRITE-Constellation, a project developed since 2003 by researchers at Canadian and Austrian Universities presently consists of *UniBRITE* and *BRITE-Austria/TUG-SAT1*, which are two 20 cm cube nanosatellites. Each will host a 30 mm aperture telescope with a CCD camera equipped with either a red (550 to 700 nm) or a blue (390 to 460 nm) lter, to perform high-precision two-color photometry of the brightest stars in the sky for up to several years. Depending on the orbit and the position of the *BRITE* targets the photometry can be obtained contiguously during many orbits for many months, with gaps during individual orbits, or only for certain periods of the year.

The primary science goals are studies of luminous stars in our neighbourhood, representing objects which dominate the ecology of our Universe, and of evolved stars to probe the future development of our Sun.

A launch of *UniBRITE* and *BRITE-Austri* in 2009 is envisioned and an expansion proposal of the *BRITE-Constellation* by two additional spacecraft of the same construction is currently under review in Canada.

Highlights of Astronomy, Volume 15
XXVIIth IAU General Assembly, August 2009
Ian F. Corbett, ed.
© International Astronomical Union 2010
doi:10.1017/S1743921310009907

JD14 – Examining the PDR-molecular cloud interface at mm and IR wavelengths

R. Simon, V. Ossenkopf, M. Röllig and J. Stutzki

I. Physikalisches Institut, Universität zu Köln, Germany
email: simonr@ph1.uni-koeln.de, ossk@ph1.uni-koeln.de, roellig@ph1.uni-koeln.de,
stutzki@ph1.uni-koeln.de

Abstract. Much progress has been made in recent years towards a better understanding of the physical and chemical processes in Photo-dissociation/Photon-dominated Regions (PDRs), both observationally and in terms of detailed physical and chemical modelling. This article highlights some of the problems and new opportunities observers and modellers are facing.

Keywords. astrochemistry, ISM: clouds, ISM: evolution, ISM: kinematics and dynamics, ISM: structure, radio lines: ISM

1. Introduction

PDRs are regions of the interstellar medium (ISM), where the physical and chemical conditions and the structure of the clouds are dominated by the local photon density leading to photo-dissociation and photoelectric heating. As part of the cosmic cycle of matter, PDRs play a key role in the cloud/star-formation interaction and ISM evolution. They cover a wide range in UV field strength, density, geometry, and metallicity, and due to the filamentary and clumpy nature of clouds, PDRs are found everywhere. Understanding the origin of PDR emission in our Galaxy is crucial for the interpretation of observations in external galaxies, where individual PDRs cannot be resolved.

In PDRs, Far-UV (FUV) photons from young stars control the gas heating (photo-electric effect) and chemistry. The PDR gas cools predominantly in the Far-infrared (FIR) through dust continuum, but also spectral line emission (up to a few percent of the total cooling). Attenuation of the FUV radiation from the cloud edges to their more shielded interiors leads to gradients in the temperature and the chemistry. Chemical models predict an observationally confirmed layered structure with typical scales of a few 0.01 pc ($10''$ at Galactic distances). Details of this stratification can only be resolved in Galactic edge-on clouds where the spatial resolution is high.

2. Observations and modelling

PDRs are complex in various aspects (geometry, excitation, dynamics, clumpiness) and deriving physical parameters from observations is challenging due to the degeneracy of parameters such as abundance and excitation temperature. Physical modelling of the source structure, temperature, density, and UV field gradients together with the chemical modelling is therefore important. The Horsehead and Orion Bar PDRs (Goicoechea et al. (2009), van der Wiel et al. (2009)) are prime examples of extensively studied PDRs where such knowledge from observations is being incorporated in the models.

While current PDR models provide reasonable fits to many observed abundances and the general layering in the PDR, several details in the interplay between chemistry and dynamics or radiation transfer are still poorly understood. In some cases, this leads to orders of magnitude differences for predictions from different models and between models and observations. Hence, an important aspect in the study of PDRs is the intercomparison of the different models in benchmark studies (Röllig *et al.* (2007)) to identify strengths and weaknesses, trigger improvements, and ensure qualitative and quantitative agreement of the models in at least simple cases.

Two recent observational studies towards the Orion Bar and Horsehaed PDRs illustrate the discrepancies between observations and model predictions. First, the SO abundance predicted from single, large clump models of the Orion Bar (van der Wiel *et al.* (2009)) is much higher than observed, implying very high sulfur depletion, whereas CS and HCS$^+$ observations in the Horsehead (Goicoechea *et al.* (2006)) are compatible with low sulfur depletion. Second, the abundances of radicals such as HCO and small hydrocarbons are observed much higher in the Horsehead than predicted by steady state models (Pety *et al.* (2007), Gerin *et al.* (2009)). Whether such findings are due to peculiarities of the sources, abundant carriers or important processes missing in the models (neutral-neutral reactions (e.g., with oxygen), grain surface chemistry, photo-desorption of ices, and photo-erosion of grains) will require improvements in modelling and new observations alike, which are being pusued by the different research groups as knowledge progresses.

On the modelling side, such improvements include updates of chemical reaction rates, dissociative channels, and branching ratios where possible, inclusion of fractionation, H_2 formation and excitation, dust modelling (to cover arbitrary dust distributions and surface reactions), the impact of PAHs (evaporation and charged grains), sophisticated radiative transfer (non-local, non-LTE), influence of X-rays, dynamics and kinematics (turbulence, shocks, photo-evaporation, advection), time dependence, geometry (detailed modelling of 3D source structure), and clumpiness.

New observing opportunities with Herschel, STO, SOFIA, and ALMA, largely complementary to, e.g., IRAM 30 m, PdBI, JCMT, NANTEN2, APEX, Spitzer, will provide high spatial and spectral resolution observations with the goal to obtain a complete inventory of main cooling lines of warm and dense gas: mid-J CO, [C I], [C II], [O I], water, as well as N II and others to disentangle contributions to C II from various phases (warm ionized medium, H II regions). Herschel for the first time offers the unique opportunity to observe light hydrides (e.g., water, OH, CH, CH$^+$, NH), which are key species at the knots of the chemical networks and building blocks for larger molecules. In this context, the two Herschel guaranteed time key programs WADI and HEXOS are dedicated to the study of prototype PDRs (and shocks) to cover a large area in parameter space (WADI) and the full spectral range between 480 and 5000 GHz in the Orion Bar (HEXOS).

Building on a wealth of data already present in the literature, lots of preparatory observations for the two key programs are planned and well underway, including spectral line and continuum observations at the highest frequencies accessible with ground based telescopes and Spitzer spectroscopy. All these observations will enable a calibration of the chemical models and ultimately lead to a deeper understanding of PDRs and their role in the evolution of molecular clouds.

References

Gerin, M. *et al.* 2009, *A&A* 494, 977
Goicoechea, J. R. *et al.* 2009, *A&A* 498, 771
Goicoechea, J. R. *et al.* 2007, *A&A* 456, 565
Pety *et al.* 2007, *Molecules in Space and Laboratory* J. L. Lemaire, F. Combes (eds)
Röllig, M. *et al.* 2007, *A&A* 467, 187
van der Wiel, M. H. D. *et al.* 2009, *A&A* 498, 161

Highlights of Astronomy, Volume 15
XXVIIth IAU General Assembly, August 2009
Ian F. Corbett, ed.

© International Astronomical Union 2010
doi:10.1017/S1743921310009919

Warm molecular gas in the M17 SW nebula

J. P. Pérez-Beaupuits[1], M. Spaans[1], M. Hogerheijde[2] and R. Güsten[3]

[1] Kapteyn Astronomical Institute, PO Box 800, 9700 AV, Groningen, The Netherlands

[2] Leiden Observatory, Leiden University, PO Box 9513, 2300 RA, Leiden, The Netherlands

[3] Max-Planck-Institut für Radioastronomie, Auf dem Hügel 69, 53121 Bonn, Germany
email: jp@astro.rug.nl

Abstract. High resolution maps of the ^{12}CO $J = 6 \rightarrow 5$ line and the [C I] $^3P_2 \rightarrow {}^3P_1$ (370 μm) fine-structure transition in the Galactic nebula M17 SW are presented. The maps were obtained using the dual color multiple pixel receiver CHAMP$^+$ on the APEX† telescope.

Keywords. galactic: ISM, galactic: individual: M17 SW, molecules: ^{12}CO , atoms: [C I]

Observations of mid-J molecular lines are used to trace the warm (50 to few hundred K) and dense gas ($n(H_2) > 10^5$ cm^{-3}) across the interface region of the M17 SW nebula. Figure 1 shows the transition between the ionization front, traced by the 21 cm emission (Brogan & Troland, 2001, ApJ, 560, 821) , the atomic gas traced by the [C I] $^3P_2 \rightarrow {}^3P_1$ transition and the molecular gas traced by the ^{12}CO $J = 6 \rightarrow 5$. The warm gas extends up to a distance of ~ 2.2 pc from the M17 SW ridge. The structure and distribution of the [C I] map indicate that its emission arises from an interclump medium with densities of the order of 10^3 cm^{-3}. The warmest gas is located along the ridge of the molecular cloud, close to the ionization front. The peak emissions of the ^{12}CO $J = 6 \rightarrow 5$ line and [C I] are ~ 850 K km^{-1} s and ~ 280 K km^{-1} s , respectively. These maps, along with the ^{13}CO $J = 6 \rightarrow 5$ and ^{12}CO $J = 7 \rightarrow 6$ lines, also observed with CHAMP$^+$, are reported in Pérez-Beaupuits *et al.* (2009, A&A, *accepted*, arXiv:0910.4937v2).

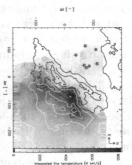

Figure 1. Grey scale map of the ^{12}CO $J = 6 \rightarrow 5$ line in M17 SW, with $9.4'' \times 9.4''$ resolution. The *black* contour lines correspond to the 21 cm continuum emission reported by Brogan & Troland (2001) with $10'' \times 7''$ resolution. The *white* contour lines correspond to the $^3P_2 \rightarrow {}^3P_1$ 370 μm fine-structure transition of [C I] ($9.4'' \times 9.4''$ resolution). The contour levels are the 25%, 50%, 75% and 90% of the peak emissions. The *open stars* indicates the O and B ionizing stars.

Acknowledgements

We are grateful to the MPfIR team and the APEX staff for their help and support during and after the observations. We are grateful to C. Brogan for providing the 21 cm map, and to A. Baryshev and W. Boland for their contribution in building CHAMP$^+$. Construction of CHAMP$^+$ is a collaboration between the Max-Planck- Institut für Radioastronomie Bonn, SRON Groningen, the Netherlands Research School for Astronomy (NOVA), and the Kavli Institute of Nanoscience at Delft University of Technology, with support from the Netherlands Organization for Scientific Research (NWO) grant 600.063.310.10.

† This publication is based on data acquired with the Atacama Pathfinder Experiment (APEX). APEX is a collaboration between the Max-Planck-Institut fur Radioastronomie, the European Southern Observatory, and the Onsala Space Observatory.

Highlights of Astronomy, Volume 15
XXVIIth IAU General Assembly, August 2009
Ian F. Corbett, ed.

© International Astronomical Union 2010
doi:10.1017/S1743921310009920

Spitzer reveals what's behind Orion's Bar

Robert H. Rubin[1,2,3]

[1]NASA Ames Research Center, M.S. 245-6, Moffett Field, CA 94035-1000, USA
email: rubin@cygnus.arc.nasa.gov

[2] Orion Enterprises

[3] Kavli Institute for Astronomy & Astrophysics, Peking University

Abstract. We present *Spitzer* Space Telescope observations of 11 regions in the Orion Nebula all southeast of the Bright Bar. Our Cycle 5 program obtained deep spectra with both the IRS short-high (SH) and long-high (LH) modules with aperture grid patterns chosen to very closely match the same area in the nebula. Previous IR missions observed only the inner few arcmin (the 'Huygens' region). The extreme sensitivity of *Spitzer* in the 10-37 μm spectral range permitted us to measure many lines of interest to much larger distances from the exciting star θ^1 Ori C.

Keywords. ISM: abundances, H II regions, individual (Orion Nebula)

1. Synopsis

Orion is the benchmark for studies of the interstellar medium, particularly for elemental abundances. *Spitzer* observations provide a unique perspective on the neon and sulfur abundances. The Ne/H abundance ratio is especially well determined, with a preliminary value of 1.12×10^{-4}.

There are corresponding new ground-based spectra taken by Co-I Bob O'Dell. We find that the electron density vs. distance from θ^1 Ori C for [S III] (*Spitzer*) and [S II] (CTIO) decreases in a very similar distribution from $\sim$1000 to $\sim$100 cm^{-3}. The optical data are used to estimate the electron temperature, optical extinction, and the S$^+$ fraction. From our preliminary results we estimate the Ne/S abundance ratio by number by observing the dominant ionization states of Ne (Ne$^+$, Ne^{++}) and S (S^{++}, S^{3+}) with *Spitzer*. The optical data are used to correct our *Spitzer*-derived Ne/S ratio for S$^+$ which is not observed with *Spitzer*. The median value (excluding all three outermost 'Veil' positions) adjusted for the optical S$^+$/S^{++} ratio is Ne/S = 14.5.

A dramatic find is the presence of high-ionization Ne^{++} all the way to the outer optical boundary $\sim$12 arcmin from θ^1 Ori C. This IR result is robust, whereas the optical evidence from observation of high-ionization (e.g. O^{++}) at the outer optical boundary suffers uncertainty because of the scattering of emission from the much brighter inner Huygens region.

The *Spitzer* spectra are consistent with the Bright Bar being a high-density 'localized escarpment' in the larger Orion Nebula picture. Hard ionizing photons reach most solid angles well SE of the Bright Bar.

Our spectra are the deepest ever taken in these outer regions of Orion over the 10-37 μm range. Tracking the changes in ionization structure via the line emission to larger distances provides much more leverage for understanding the far less studied outer regions. The outer Veil is likely a new H II region-PDR interface.

Acknowledgements

Support is from Spitzer Space Telescope Cycle 5 program 50082. This research is the combined effort of C.R. O'Dell, G. Ferland, J. Simpson, S. Colgan, & I. McNabb.

Highlights of Astronomy, Volume 15
XXVIIth IAU General Assembly, August 2009
Ian F. Corbett, ed.
© International Astronomical Union 2010
doi:10.1017/S1743921310009932

Atomic carbon in an infrared dark cloud

Volker Ossenkopf[1,2,3]**, Chris W. Ormel**[3,4]**, Robert Simon**[1]**, Kefeng Sun and Jürgen Stutzki**[1]

[1]I. Physikalisches Institut der Universität zu Köln, Zülpicher Straße 77, 50937 Köln, Germany;
[2]SRON Netherlands Institute for Space Research, P.O. Box 800, 9700 AV Groningen,
Netherlands; [3]Kapteyn Astronomical Institute, University of Groningen, P.O. Box 800, 9700
AV Groningen, Netherlands; [4]Max-Planck-Institut für Astronomie, Königstuhl 17, 69117
Heidelberg, Germany

Infrared dark clouds (IRDCs) are potential sites of massive star formation, dark in the near-infrared, but in many cases already with indications of active star-formation from far-infrared and submm observations. They are an ideal test bed to study the role of internal and external heating on the structure of the molecular cloud material.

We report the first spectroscopic mapping of an atomic carbon line in an IRDC, G48.65-0.29, using the 2×4 pixel SMART receiver of the KOSMA observatory. Comparing the intensities with the ^{13}CO 1-0 data from the BU-FCRAO Galactic Ring Survey (Jackson *et al.* (2006)) shows a relatively constant line ratio, ranging from about 0.2 to 0.5.

When comparing the influence of the irradiation by embedded sources and by the external Galactic UV field for the production of atomic carbon, we find that most [C I] emission can be attributed to externally illuminated surfaces. A comparable contribution from internal star-formation to the atomic carbon production is only found for the Southern region hosting one of the most evolved embedded sources.

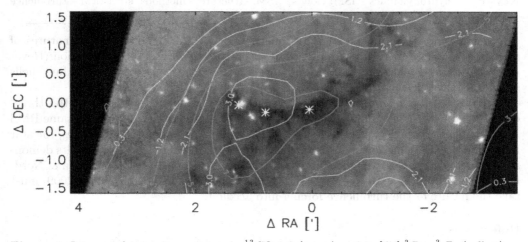

Figure 1. Integrated intensity contours in ^{13}CO 1-0 (green) and in [C I] $^3P_1 - ^3P_0$ (yellow) on top of a Spitzer IRAC false-colour image, composed of the 8μm, 5.8μm, and 3.6μm channels. The contours are labelled by antenna temperatures integrated from 30 to 41 km/s. Known submm cores (Ormel *et al.* (2005)) are indicated by stars.

References

Ormel, C. W., Shipman, R. F., Ossenkopf, V., & Helmich, F. P. 2005, A&A, 439, 613
Jackson, J. M., Rathborne, J. M., Shah, R. Y., Simon, R. *et al.* 2006, ApJS 163, 145

Highlights of Astronomy, Volume 15
XXVIIth IAU General Assembly, August 2009
Ian F. Corbett, ed.

© International Astronomical Union 2010
doi:10.1017/S1743921310009944

Solenoidal versus compressive turbulence forcing

C. Federrath[1], J. Duval[2], R. S. Klessen[1], W. Schmidt[3] and M.-M. Mac Low[4]

[1]Zentrum für Astronomie der Universität Heidelberg, Institut für Theoretische Astrophysik, Albert-Ueberle-Str. 2, D-69120 Heidelberg, Germany

[2]Astronomy Department at Boston University, 725 Commonwealth Avenue, Boston, MA 02215, USA

[3]Institut für Astrophysik Göttingen, Friedrich-Hund-Platz 1, 37077 Göttingen, Germany

[4]Department of Astrophysics, American Museum of Natural History, Central Park West at 79th Street, New York, NY 10024-5192, USA

Abstract. We analyze the statistics and star formation rate obtained in high-resolution numerical experiments of forced supersonic turbulence, and compare with observations. We concentrate on a systematic comparison of solenoidal (divergence-free) and compressive (curl-free) forcing (Federrath *et al.* 2009 a,b), which are two limiting cases of turbulence driving. Our results show that for the same RMS Mach number, compressive forcing produces a three times larger standard deviation of the density probability distribution. When self-gravity is included in the models, the star formation rate is more than one order of magnitude higher for compressive forcing than for solenoidal forcing.

Keywords. hydrodynamics – ISM: clouds – ISM: structure – methods: statistical – turbulence

Observational data indicate that turbulence in the ISM exhibits both signatures of solenoidal and compressive forcing, depending on the region under consideration (Heyer *et al.* 2006; Hily-Blant *et al.* 2008; Goodman *et al.* 2009). In particular, expanding shells show statistical characteristics similar to compressively driven turbulence.

Compressive forcing produces a three times larger standard deviation of the three-dimensional and the column density probability distributions (PDFs) for the same RMS Mach number (Federrath *et al.* 2008). When self-gravity is added to the models, the star formation rate is about 25 times higher for compressive forcing. These two results demonstrate that star formation models based on the turbulent density PDF (Padoan & Nordlund 2002; Krumholz & McKee 2005; Elmegreen 2008; Hennebelle & Chabrier 2009) must take the nature of the turbulence forcing into account.

References

Elmegreen, B. G. 2008, *ApJ*, 672, 1006
Federrath, C., Duval, J., Klessen, R. S., Schmidt, W., & Mac Low, M.-M. 2009a, *arXiv:0905.1060*
Federrath, C., Klessen, R. S., & Schmidt, W. 2008, *ApJ*, 688, L79
Federrath, C., Klessen, R. S., & Schmidt, W. 2009b, *ApJ*, 692, 364
Goodman, A. A., Pineda, J. E., & Schnee, S. L. 2009, *ApJ*, 692, 91
Hennebelle, P. & Chabrier, G. 2009, *ApJ*, 702, 1428
Heyer, M. H., Williams, J. P., & Brunt, C. M. 2006, *ApJ*, 643, 956
Hily-Blant, P., Falgarone, E., & Pety, J. 2008, *A&A*, 481, 367
Krumholz, M. R. & McKee, C. F. 2005, *ApJ*, 630, 250
Padoan, P., & Nordlund, Å. 2002, *ApJ*, 576, 870

Highlights of Astronomy, Volume 15
XXVIIth IAU General Assembly, August 2009
Ian F. Corbett, ed.

Simulating the chemistry and dynamics of molecular clouds

S. C. O. Glover[1], C. Federrath[1], M.-M. Mac Low[2] and R. S. Klessen[1]

[1] Institut für Theoretische Astrophysik, Zentrum für Astronomie der Universität Heidelberg,
Albert-Ueberle-Strasse 2, 69120 Heidelberg, Germany
email: sglover@ita.uni-heidelberg.de

[2] Department of Astronomy, American Museum of Natural History, Central Park West at 79th
Street, New York, NY 10024, USA

Abstract. We have performed high-resolution three-dimensional simulations of turbulent interstellar gas that for the first time self-consistently follow its coupled thermal, chemical and dynamical evolution. Our simulations have allowed us to quantify the formation timescales for the most important molecules found in giant molecular clouds (H_2, CO), as well as their spatial distribution within the clouds. Our results are consistent with models in which molecular clouds form quickly, within 1–2 turbulent crossing times, and emphasize the crucial role of density inhomogeneities in determining the chemical structure of the clouds.

Keywords. astrochemistry – molecular processes – methods: numerical – ISM: clouds – ISM: molecules

One of the main difficulties involved in the numerical modelling of the formation and evolution of giant molecular clouds (GMCs) is the fact that their chemical, thermal and dynamical evolution is strongly coupled. Prior attempts to model GMCs have typically dealt with this problem with ignoring one or more aspects of this coupling, e.g. by assuming that the gas remains isothermal in hydrodynamical simulations. As a first step towards producing more realistic cloud models, we have developed a lightweight treatment of cloud chemistry and cooling that has allowed us to perform high-resolution three-dimensional simulations of turbulent interstellar gas that accurately model the gas chemistry, cooling, and the hydrodynamics. Full details of this treatment and of the simulations that we have performed can be found in Glover *et al.* (2009). In this contribution, we restrict ourselves to highlighting a few of our most important results.

We find that CO formation occurs rapidly in dense, turbulent gas. Most of the CO in our simulations forms within the first 1–2 turbulent crossing times, on a similar, but slightly slower timescale to the H_2. These short chemical timescales suggest that the limiting timescale in the formation of a GMC is not the time required to convert H to H_2 and C^+ to CO, but rather the time required to assemble the cloud material from the low density ISM. We also find that the CO abundances produced in our simulations are highly inhomogeneous, and are not well correlated with either the local gas density or with the effective visual extinction. Instead, it appears to be a combination of these quantities that best predicts the resulting CO abundance. This has important consequences for attempts to infer the structure of real GMCs from observations of the CO emission.

References

Glover, S. C. O., Federrath, C., Mac Low, M.-M., & Klessen, R. S. 2009, *MNRAS*, in press; arXiv:0907.4081

Highlights of Astronomy, Volume 15
XXVIIth IAU General Assembly, August 2009
Ian F. Corbett, ed.

© International Astronomical Union 2010
doi:10.1017/S1743921310009968

Constraining How Star Formation Proceeds: Surveys in the Sub-mm and FIR

Doug Johnstone[1,2]

[1] National Research Council Canada, Herzberg Institute of Astrophysics,
5071 West Saanich Rd, Victoria, BC, V9E 2E7, Canada
email: doug.johnstone@nnrc-cnrc.gc.ca

[2] Department of Physics & Astronomy, University of Victoria, Victoria, BC, V8P 1A1, Canada

Abstract. Coordinated multi-wavelength surveys of molecular clouds are providing strong constraints on the physical conditions within low-mass star-forming regions. In this manner, Perseus and Ophiuchus have been exceptional laboratories for testing the earliest phases of star formation. Highlights of these results are: (1) dense cores form only in high column density regions, (2) dense cores contain only a few percent of the cloud mass, (3) the mass distribution of the dense cores is similar to the IMF, (4) the more massive cores are most likely to contain embedded protostars, and (5) the kinematics of the dense cores and the bulk gas show significant coupling.

1. Core Locations Within Molecular Clouds

Large samples of pre- and proto-stellar cores within nearby molecular clouds have been investigated primarily through (sub)millimeter continuum observations with single-dish telescopes, which preferentially reveal small and dense regions of dust emission and therefore are ideal for locating the cores. Two such surveys, (Johnstone *et al.* (2004) and Kirk *et al.* (2006)), taken with SCUBA at the JCMT, have covered significant portions of Ophiuchus and Perseus respectively.

Comparison between the core locations and the bulk material in the clouds, as measured by near IR extinction [see Lombardi & Alves (2001) for the technique], reveals that the cores are biased toward the highest column density regions (measured in A_v of extinction) within their individual clouds. Despite the fact that most of the mass of the molecular cloud is found at low column density, few cores are found in these extended regions. In Ophiuchus the cores are almost exclusively found above $A_v = 15$ Johnstone *et al.* (2004) while the mean A_v is 4. In Perseus the cores are almost exclusively found above $A_v = 5$ Kirk *et al.* (2006) while the mean A_v is only 2. Interestingly, the Perseus cores appear slightly offset from the column density peaks within the cloud, perhaps evidence for triggering events playing an important role in producing cores and protostars.

The mass of the cloud locked up within the cores is minimal, as might be expected given the low star formation efficiency expected for clouds. For both Perseus and Ophiuchus, the total mass in cores is only a few percent of the entire cloud mass. Even in the higher column density zones, the cores only account for ten to twenty percent of the cloud mass Johnstone *et al.* (2004), Kirk *et al.* (2006). This is an important point to recognize: the cores appear to be a separate, insignificant, component of the cloud and their physical properties should be considered distinct from those of the bulk cloud.

2. Core Properties

Considering the cores as a distinct sample reveals that the core mass function has a similar appearance to that of stars (see Ward-Thompson *et al.* (2007) and references therein); the majority of cores and the majority of the mass in the cores is found in the lower mass cores - like for stars. Whether this relationship is evidence for the masses of stars being pre-selected at the core stage is debatable but the distribution does provide a starting point for comparison between observations and simulations Hatchell & Fuller (2008), Swift & Williams (2008).

Collating mid-IR observations of star-forming regions taken by *Spitzer* Evans *et al.* (2003) with the submillimeter core maps allows for a robust determination of which cores harbor proto-stellar sources Jørgensen *et al.* (2007), Jørgensen *et al.* (2008). Detailed analysis shows that the proto-stellar cores are more massive and more likely to be highly concentrated than the starless cores, as might be expected if the massive cores are more likely to collapse under their own weight. As well, mid-IR detections, indicating a heating source within the proto-stellar cores, are strongly biased toward the center of each submillimeter core, implying that the proto-star is kinematically coupled to the core Jørgensen *et al.* (2007).

3. Core Kinematics

The cores themselves appear kinematically coupled to the bulk cloud, as noted by Kirk *et al.* (2007) in an analysis of N_2H^+ (1–0) and $C^{18}O$ (2–1) line profiles. The N_2H^+ line widths are dominated by thermal motions, especially for the starless core sample [this is also seen in a complementary NH_3 (1,1) dataset Rosolowsky *et al.* (2008)]. The $C^{18}O$ gas, however, preferentially probes the bulk gas surrounding the core and reveals a much wider line width, consistent with the notion that non-thermal motions dominate on larger scales within molecular clouds. Interestingly, the line centroids of the $C^{18}O$ and N_2H^+ observations are typically offset by less than the sound speed, suggesting that the non-thermal motions in the cloud do not lead to harsh conditions at the core boundaries. These observations are hard to reconcile with simplified simulations of turbulence alone, requiring magnetic fields or other softening agents on small scales Kirk *et al.* (2009).

The launch of *Herschel* and the commissioning of SCUBA2 at the JCMT in 2009 ensure that there will be ever more impressive far infrared and (sub)millimeter maps of nearby star-forming regions by 2011 when ALMA begins Early Science.

Acknowledgements

DJ thanks his collaborators on these projects: H. Kirk, J. Di Francesco, M. Tafalla, J. Jørgensen, S. Basu and all the members of the COMPLETE Team. DJ is supported by a Natural Sciences and Engineering Research Council Discovery Grant.

References

Evans, N. J., II *et al.* 2003, PASP, 115, 965
Hatchell, J. & Fuller, G. A. 2009, A&A, 482, 855
Johnstone, D., Di Francesco, J., & Kirk, H. 2004, ApJL, 611, 45
Jørgensen, J. K., Johnstone, D., Kirk, H., & Myers, P. C. 2007, ApJ, 656, 293
Jørgensen, J. K., Johnstone, D., Kirk, H., Myers, P. C. *et al.* 2008, ApJ, 683, 822
Kirk, H., Johnstone, D., & Di Francesco, J. 2006, ApJ, 646, 1009
Kirk, H., Johnstone, D., & Tafalla, M. 2007, ApJ, 668, 1042
Kirk, H., Johnstone, D., & Basu, S. 2009, ApJ, 699, 1433
Lombardi, M. & Alves, J. 2001, A&A, 377, 1023
Rosolowsky, E. W., Pineda, J. E., Foster, J. B., Borkin, M. A. *et al.* 2008, ApJS, 175, 509
Swift, J. & Williams, J. 2008, ApJ, 679, 552
Ward-Thompson, D., André, P., Crutcher, R. *et al.* 2007, Protostars and Planets V, 33

Highlights of Astronomy, Volume 15
XXVIIth IAU General Assembly, August 2009   © International Astronomical Union 2010
Ian F. Corbett, ed.   doi:10.1017/S174392131000997X

Testing PDR models against ISO fine structure line data for extragalactic sources

M. Vasta[1], M. J. Barlow[1], S. Viti[1], J. A. Yates[1] and T. A. Bell[2]

[1]Department of Physics and Astronomy, University College London, Gower Street, London
WC1E 6BT, UK. email: mv@star.ucl.ac.uk

[2]Caltech, Department of Physics, MC 320- 47, Pasadena, CA 91125, USA

Abstract. Studies of our own Galaxy and observations of external galaxies have suggested that stellar ultraviolet radiation can ionize vast volumes of a galaxy and that far-ultraviolet radiation impinging on neutral cloud surfaces is responsible for a large fraction of the observed far-infrared (FIR) spectral line emission that cools the gas (Crawford & al. (1985)). Fine structure (FS) emission lines can be used as tracers of nebular conditions such as density, excitation and ionization. By virtue of their different excitation potentials and critical densities, FS emission lines provide an insight into the energetics and chemical composition of the regions from which they originate. The far infrared [C II]158 μm, [O I]145 μm and [O I]63 μm fine structure emission lines obtained with the Infrared Space Observatory (ISO) from 35 extragalactic sources are examined to investigate the chemical abundances and large scales physical properties of these sources. Line fluxes are compared with a grid of PDR models previously computed using the UCL_PDR code. We overplotted our model predictions against flux ratios from the [C II]158 μm and [O I]63 μm and 145 μm ISO LWS fluxes. In this section we will only discuss the sensitivity of the ratios to changes in the input parameters. We find that the average radiation field G_0 is $60-8 \times 10^2$ and the average density n_H $10^4 - 9 \times 10^4$ cm^{-3}. While ionised carbon, because of its ionisation potential, can be found in both neutral gas and ionised gas clouds, species such as ionised nitrogen [N II], with ionisation potential of 14.53 eV, can arise only from H II regions. The 11 sources that have detections of both [C II] 158 μm and [N II] 122 μm have mean and median [C II]$_{158}$/[N II]$_{122}$ flux ratios of 10.2 and 5.9 respectively. A H II region [C II]$_{158}$/[N II]$_{122}$ ratio of 1.6 implies that H II region contribute only 16% (mean case) and 27% (median case) of the overall [C II] 158 μm flux that is observed. We used the above predicted H II region [C II]$_{158}$/[N II]$_{122}$ ratio of 1.6 along with the observed [N II] 122 μm fluxes, to correct the observed [C II] 158 μm flux of these 11 sources for HII region contributions. We estimate that 10-60% of the [C II] is excited in ionised regions. When accounting for the contribution to the [CII] 158 μm by H II regions we found that our models fitted better the observations. We modeled the oxygen emission line profile emitted from an ensemble of PDRs and found a clear [O I] 63 μm self-absorbed profile. We estimate that approximately 20-70% of the [O I] 63 μm intensity may be suppressed through oxygen self-absorption depending on the physical parameters of the PDR regions. This work has been submitted for publication to MNRAS, Vasta et al. (2009).

Keywords. FIR fine structure emission lines, PDRs, Extragalactic Sources, ISO LWS, Oxygen self-absorption.

References

Crawford, M. K. Genzel, R. Townes, C. H., & Watson, D. M 1980, ApJ755, 771

Highlights of Astronomy, Volume 15
XXVIIth IAU General Assembly, August 2009
Ian F. Corbett, ed.

The Phase Structure of the ISM in Galaxies

Mark G. Wolfire[1]

[1] Astronomy Department, University of Maryland, College Park, MD 20742, USA
email: mwolfire@astro.umd.edu

Abstract. Diffuse gas in the Galaxy is observed to exist as cold ($T \sim 100$ K) neutral atomic gas (CNM) and warm neutral atomic ($T \sim 8000$ K) gas (WNM). In addition to these "thermal" phases, gas can also exist as warm ($T \sim 8000$ K) ionized gas, cold ($T \sim 10$ K) molecular gas and in warm ($T \sim 100 - 500$ K) interface regions or Photodissociation Regions (PDRs) on the surfaces of molecular clouds. The same chemical and thermal processes that dominate in the PDRs associated with molecular clouds are also at work in the diffuse neutral gas. Two additional "phases" are gas associated with GMCs that has H_2 but no or little CO, and short lived or transient phases such as shocks, shears, and turbulence. I will first review the different gas phases in the Galaxy, their physical conditions and their dominant cooling lines. I will also discuss the observations and theoretical modeling in support of turbulence versus thermal instability as the driving force in producing the "thermal" gas phase distributions. Rough estimates for the distribution of phases in the Galaxy and the origin of the dominant emission lines has been conducted by previous telescopes (e.g., COBE, BICE) but with low velocity and low spectral resolution. The distribution and mass of the various gas phases is important for sorting out the role of SN in setting ISM pressures and in driving ISM turbulence. In addition, understanding the Galactic phase distribution is important in interpreting observations of extragalactic systems in which beams encompass several emission components. I will review the potential for future observations by e.g., STO, SOFIA, and Herschel to detect and separate phases in Galactic and extragalactic systems.

Keywords. ISM: clouds, ISM: general, ISM: HII regions, galaxies: ISM

1. Introduction

What are some of the big ISM questions that might be answered by looking at phases? In this short written contribution I focus on the WNM and CNM phases. Which of the various ISM components dominates the [C II] emission and where? The warm ($T \sim 8000$ K) neutral gas (WNM) produces faint [C II] and [O I] and the cold ($T \sim 100$ K) neutral gas (CNM) produces mainly [C II]. The [C II] line mainly, but also with [O I], [N II], [C I], and CO are the dominant coolants of the ISM. Knowing the luminosity of the cooling lines, gives the energy input into the gas from either radiative heating or from turbulence or shocks. What is the ISM engine that drives gas into various phases? Part of answering this is to determine the distribution of CNM and WNM as well as the thermal pressure $P_{\rm th}$. The distributions have implications for constraining the role of SN in setting pressures, the volume filling factor of hot gas, and shock processing of material.

2. WNM and CNM Emissivity and Thermal Pressure

We can estimate the emissivity in the WNM and CNM from phase diagrams ((Wolfire, et al. 2003)). In a plot of thermal pressure, $P_{\rm th}/k$ versus density n, and for an isobaric multiphase medium, the points of thermal stability correspond to the WNM and CNM phases. At $P_{\rm th} > P_{\rm max}$ gas is converted to the cold phase, but at $P_{\rm th} < P_{\rm min}$ the gas is converted to the warm phase. The cooling rate in the cold gas is a factor of 10 times larger than the cooling in the warm gas. So for comparable column densities, the [C II] emission from the cold gas is always brighter, and one can map out the CNM clouds in emission. Previously, CNM clouds were only seen in absorption using H I or UV absorption spectroscopy.

With this information one can 1)estimate the CNM contribution to the [C II] emission. 2) Along with H I get the WNM/CNM mass distribution as a function of position and test global models of the ISM. If SNRs dominate the topology of the ISM you expect more CNM than WNM. Most of the volume is filled with a hot ISM. On the other hand if SNRs are small and contained by magnetic fields then most of the volume is filled by WNM. 3)Along with H I the C II gives the thermal pressure in CNM clouds. Kulkarni & Heiles (1987) suggested that this was a better way to get the thermal pressure than C I, or C II in absorption. This is because for a given C II cooling rate/H the pressure changes by only a factor of two for a wide range in temperatures.

Comparing thermal pressure in arm and interarm regions we can asses the role of thermal pressure in driving gas into cold phases which eventually leads to star formation. As expected for thermal instability (TI), does the the CNM/WNM rise as the thermal pressure goes up? If there is considerable WNM at $P_{\text{th}} > P_{\max}$ or CNM at $P_{\text{th}} \ll P_{\min}$ then dynamical processes are likely to dominate with dynamical times much less than the cooling times. Experiments of this kind could be carried out with STO, SOFIA, and Herschel observations of Galactic and extragalactic sources.

The question of phases is not without controversy. A great review article by Vázquez-Semadeni (2009) is entitled "Are there phases in the ISM". Much of the controversy centers on Heiles & Troland (2003) which has turned into a bit of an urban legend. The legend is that 50% of the gas mass is in thermally unstable temperatures and TI plays little role in creating CNM and WNM gas. It appears that most everyone misses that there are two distributions plotted in their Fig. 2. The distribution in temperatures for the in-plane gas shows $\sim 75\%$ of the warm gas within the 7000-9000 K range exactly as expected for TI. Only %25 of the gas is outside this range, and when the CNM is included only $\sim 15\%$ of the gas mass is at thermally unstable temperatures. The out of plane distribution looks nothing like the in-plane distribution with much of the gas outside of 7000-9000 K. I would conclude that the in-plane gas is dominated by TI, while the out of plane is dominated by dynamical processes. Numerical simulations give mixed results showing either no or weak TI (e.g., Gazol et $al.$ (2001)) or significant TI (e.g., Koyama & Ostriker (2009)). The results depend on the model resolution, heating rates, cooling rates, and type and amplitude of the turbulence (Gazol et $al.$ 2005).

Acknowledgements

M. G. W. was supported in part by a NASA LTSA grant #NNG05G64G.

References

Gazol, A., Vázquez-Semadeni, E., & Kim, J. 2005, ApJ, 630, 911
Gazol, A., Vázquez-Semadeni, E., Sánchez-Salcedo, F. J., & Scalo, J. 2001, $ApJL$, 557, L121
Heiles, C. & Troland, T. H. 2003, ApJ, 586, 1067
Koyama, H. & Ostriker, E. C. 2009, ApJ, 693, 1316
Kulkarni, S. R. & Heiles, C. 1987, Interstellar Processes, 134, 87
Vázquez-Semadeni, E. 2009, in: M. de Avillez (ed.), The $Role$ of $Disk$-$Halo$ $Interaction$ in $Galaxy$ $Evolution$, EAS Publication Series (Paris: EDP), 2009 in press
Wolfire, M. G., McKee, C. F., Hollenbach, D., & Tielens, A. G. G. M. 2003, ApJ 587, 278

Highlights of Astronomy, Volume 15
XXVIIth IAU General Assembly, August 2009
Ian F. Corbett, ed.

© International Astronomical Union 2010
doi:10.1017/S1743921310009993

Diagnosing the ISM in star-forming regions

Willem A. Baan[1], Edo Loenen[1,2] and Marco Spaans[2]

[1]ASTRON, Dwingeloo, The Netherlands [2]Kapteyn Institute, Groningen, The Netherlands

Abstract. A report on studies using the observed line ratios of high-density molecular tracers to diagnose the physics and chemistry of the ISM in star-formation environments.

Keywords. Stars: formation – ISM: molecules – Radio lines: ISM – Galaxies: starburst

---

Molecular line emissions may be used for the diagnostics of physical and chemical conditions of the Interstellar Medium (ISM) and for understanding the dominant heating processes. Molecular species react differently to environmental influences but the ensemble of their line ratios provides a most sensitive probe of the physics and chemistry. A multi-molecule multi-transition interpretation serves to estimate densities, temperatures, radiation fields, and chemical abundances of the emission regions. The observed line ratios (of HCN, HNC, HCO^+, CN, and CS) will be interpreted using an extensive physical/chemical modeling network, which includes large numbers of molecular species and reactions as well as the radiation transfer. The objective of these studies is to establish a benchmark for modeling the ISM variations in star-formation regions in resolved Galactic sources, in partially resolved nearby galaxies, and in unresolved distant sources.

Isolated Galactic star-formation regions (Loenen *et al.* 2008) can be consistently diagnosed using observed line ratios as a distance-dependent mixed signature of a PDR and a warm molecular envelope. A single CS transition already shows the Sulpher depletion, while the HCO^+ reveals a variation in the CR flux with galacto-centric distance.

The integrated ISM in starburst nuclei observed over a large range of FIR luminosities shows distinct excitation regimes relating to the mean density and the nature of heating source (Loenen *et al.* 2008). The HCO^+/HCN and HCO^+/HNC ratios are sensitive to density. The HNC/HCN ratio discriminates between heating by X-rays (XDRs), UV-photons (young PDRs), and added mechanical heating due to star-formation feedback.

Molecular emissions of high-density tracer molecules also provides insights into the time-evolution of a starburst and the nuclear ISM. The extragalactic data displays clear relations between molecular emissions of high-density tracer molecules from the starburst nucleus and the FIR luminosity. This behavior and the observed variation in the line ratios can be explained with a simple model for the evolution of the starburst and of the depletion and destruction of the high-density ISM during the outburst (Baan *et al.* 2008). In this picture, the observed ratio of the (nuclear) HCN over the (extended) CO(1-0) emissions may serve as an evolutionary time scale. Modeling results suggest that the observed line ratios vary with time (during the evolution of the starburst) as a function of mean nuclear density and of variation of the heating sources (Baan *et al.* 2009).

References

Baan, W. A., Henkel, C, Loenen, A. F, Baudry, A., & Wiklind, T 2008, A&A 477, 747
Baan, W. A., Loenen, A. F., & Spaans, M. 2009, A&A submitted
Loenen, A. F., Spaans, M., Baan, W. A., & Meijerink, R. 2008, A&A 488, L5.
Loenen, A. F., Baan, W. A., & Spaans, M 2009, A&A submitted

Highlights of Astronomy, Volume 15
XXVIIth IAU General Assembly, August 2009
Ian F. Corbett, ed.

© International Astronomical Union 2010
doi:10.1017/S1743921310010008

Search for High-Extinction Regions in the Small Magellanic Cloud

M.-Y. Lee[1], S. Stanimirović[1], J. Ott[2], J. Th. van Loon[3], A. D. Bolatto[4], P. A. Jones[5,6], M. R. Cunningham[5], K. E. Devine[1] and J. M. Oliveira[3]

[1]Department of Astronomy, University of Wisconsin-Madison, USA
email: lee@astro.wisc.edu [2]National Radio Astronomy Observatory, [3]Astrophysics Group,
Keele University, [4]Department of Astronomy, University of Maryland, [5]School of Physics,
University of New South Wales, [6]Departamento de Astronomia, Universidad de Chile

We have applied the unsharp-masking technique to the 24 μm image of the SMC, obtained with the *Spitzer*, to search for high-extinction regions. Fifty-five candidate regions of high-extincion (namely high-contrast regions, HCRs) have been identified from the decremental contrast image. HCRs have a size of $8 - 14$ pc and a peak contrast at 24 μm of $2 - 2.5\%$. To constrain physical properties of HCRs, we have performed observations of NH_3, N_2H^+, HNC, HCO^+, and HCN toward one of the HCRs, HCR LIRS36–east, using the ATCA and the Mopra telescope. No molecular line emission detected, but upper limits to column densities of molecular species suggest that HCRs are moderately dense with $n \sim 10^3$ cm^{-3}. Two interesting properties of HCRs are shown below.

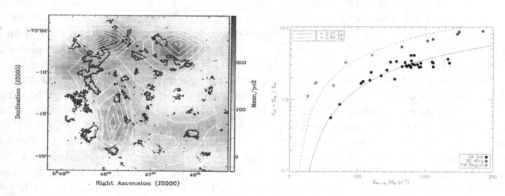

Figure 1. (Left) Selected HCRs in the southwest bar of the SMC overlaid in black contours on the H_2 surface density image from Leroy *et al.* (2007). CO(1–0) emission is overlaid in white contours. HCRs are located in the outskirts of CO clouds but in regions with significant amount of H_2. For details, see Lee *et al.* (2009). (Right) Molecular fraction (R_{H_2}) as a function of total gas surface density (Σ_{HI+H_2}) for HCRs. Based on the similar amounts of H I and H_2 surface densities, HCRs most likely represent the regions where atomic-to-molecular hydrogen transition occurs in the SMC. Their molecular fraction is in agreement with the theoretical prediction from Krumholz *et al.* (2009) for metallicity of $1/4 \sim 1/2$ M$_\odot$.

References

Krumholz, M. R., McKee, C. F., & Tumlinson, J. 2009, *ApJ*, 693, 216
Lee *et al.*, 2009, *AJ*, 138, 1101
Leroy *et al.*, 2007, *ApJ*, 658, 1027

Highlights of Astronomy, Volume 15
XXVIIth IAU General Assembly, August 2009
Ian F. Corbett, ed.

Chemical tracers of dense gas in extragalactic environments

Serena Viti[1] and Estelle Bayet[1]

[1]Dept. Physics and Astronomy, University College London, Gower Street, WC1E 6BT, UK
email: sv@star.ucl.ac.uk

Abstract. Within the context of ALMA (and future missions such as SPICA) we present some recent observational and theoretical work on molecular line emissions from extragalactic environments.

Keywords. galaxy: ISM; stars: formation; ISM: molecules; astrochemistry

Extragalactic molecular line emission is an ensemble of PDRs and dense (star forming?) gas where the spatial and temporal effects are 'diluted' in the beam. Several molecular species (CO, HCN, HNC, CS, HCO$^+$...) are observable, especially in nearby galaxies. The question that we address is: can we use molecules as probes of the physical and chemical conditions of galaxies?

We have performed extensive theoretical (Bayet *et al.* 2008a, Bayet *et al.* 2009a) and observational (Bayet *et al.* 2008b, Bayet *et al.* 2009b) studies of the chemistry of star forming as well as photodominated regions representing different types of galaxies. Our theoretical studies provide us with trends in the chemistry in response to changes in the physical conditions of galaxies. A list of molecular tracers of star forming as well as PDR gas in different categories of galaxies can be found in our papers (see above references), but a particularly useful result to report here is the comparison of the molecular tracers predicted by dense star-forming core models with those derived from PDR models. For example while SO$_2$ and H$_2$S should both be delectable in star forming gas in starburst galaxies, they are inappropriate tracers of PDRs. By contrast HCO$^+$ is predicted to be undetectable in dense star forming cores while enhanced in PDR-dominated galaxies. For high redshift sources, H$_2$CS and H$_2$CO are both predicted to be inadequate molecular tracers of PDRs, while both are good signatures of dense star-forming cores.

Observationally we have carried out an extensive survey of CS (predicted to be a high density tracer) in order to better determine the properties of the very dense star forming gas over a large range of physical conditions (Bayet *et al.* 2009b) and an highlight of our work is that high J CS lines, in particular the 7-6, seem to be tracing a gas with a minimum density of 10^7 cm^{-3}, typical of 'hot cores' and hence of high mass star formation.

References

Bayet, E., Viti, S., Williams, D. A., & Rawlings, J. M. C. 2008, *ApJ* 676, 978
Bayet, E., Lintott, C., Viti, S., Martin-Pintado, J., Martin, S., Williams, D. A., & Rawlings, J. M. C. 2008, *ApJL* 685, L35
Bayet, E., Viti, S., Williams, D. A., & Rawlings, J. M. C., Bell, T. 2009, *ApJ* 696, 1466
Bayet, E., Aladro, R., Martin, S., Viti, S., & Martin-Pintado, J. 2009, *ApJ* in press

Highlights of Astronomy, Volume 15
XXVIIth IAU General Assembly, August 2009
Ian F. Corbett, ed.

© International Astronomical Union 2010
doi:10.1017/S1743921310010021

AKARI Far-Infrared View of Nearby Galaxies

H. Kaneda[1], T. Suzuki[2], T. Onaka[3], I. Takase[3], M. Yamagishi[1], D. Ishihara[1] and I. Sakon[3]

[1] Graduate School of Science, Nagoya University, Chikusa-ku, Nagoya 464-8602, Japan
email: kaneda@u.phys.nagoya-u.ac.jp

[2] National Astronomical Observatory of Japan, Mitaka, Tokyo 181-8588, Japan

[3] Graduate School of Science, The University of Tokyo, Bunkyo-ku, Tokyo 113-0033, Japan

Abstract. We have observed 57 nearby galaxies in the far-infrared with the Far-Infrared Surveyor on *AKARI* to study the properties of dust in various environments.

Keywords. ISM: dust, extinction — infrared: galaxies — galaxies: ISM

AKARI far-infrared (IR) observations of 57 nearby galaxies have been carried out in part of the *AKARI* mission program "ISM in our Galaxy and Nearby Galaxies" (Kaneda *et al.* 2009a). The Far-IR Surveyor (FIS) on *AKARI* has 4 photometric bands at the wavelengths of 65, 90, 140, and 160 μm, which are of great use to accurately determine spatial variations in the properties of dust. For face-on spiral galaxies such as M 101 and M 81, we spectrally decompose dust emission into warm and cool components and spatially resolve each component, whereby we obtain physical insight into relationship between star formation rates and ISM densities (e.g. Suzuki *et al.* 2007). Large dynamic ranges of signal detection of the FIS provided by a special read-out mode is another advantage; we clearly confirm the presence of far-IR dust in the halo of the edge-on starburst galaxy NGC 253 without saturation problems at its very bright nucleus (Kaneda *et al.* 2009b). For NGC 1316, we find that far-IR emission is extending along the mid-IR jet-like structures, getting softer away from the galactic nucleus (Fig.1). Spatial information on vertical structures as well as face-on ones of galaxies is important in view of material circulation in a galaxy. Our *AKARI* data will be complementary to both existing and upcoming far-IR data of nearby galaxies such as those from *Spitzer* and *Herschel*.

AKARI is a JAXA project with the participation of ESA.

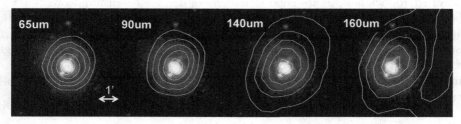

Figure 1. *AKARI* far-IR 4-band contour maps of NGC 1316, overlaid on the *AKARI* 11 μm image. Contour levels are logarithmically drawn from 10 to 80 % of the peak brightness.

References

Kaneda, H., Koo, B.-C., Onaka, T., & Takahashi, H. 2009a, *Adv.Sp.Res.* 44, 1038
Kaneda, H., Yamagishi, M., Suzuki, T., & Onaka, T. 2009b, *ApJ* 698, L125
Suzuki, T. *et al.* 2007, *PASJ* 59, S473

Highlights of Astronomy, Volume 15
XXVIIth IAU General Assembly, August 2009
Ian F. Corbett, ed.

The interplay of dense gas and stars in M33

Carsten Kramer[1], Christof Buchbender[1], Guillermo Quintana-Lacaci[1], Jonathan Braine[2], Pierre Gratier[2] and Erik Rosolowsky[3]

[1] Instituto Radioastronomia Milimetrica (IRAM), Av. Divina Pastora 7, E-18012 Granada, Spain
email: kramer@iram.es

[2] Observatoire Bordeaux, LAB - UMR 5804, 2, rue de l'Observatoire, BP 89 -33271 Floirac Cedex - France

[3] University of British Columbia, 3333 University Way, Kelowna, BC V1V 1V7, Canada

Abstract. We are studying the interplay of star formation and its 'fuel', the molecular gas (diffuse and dense) at selected positions along the major axis of M33. We have observed the ground-state transitions of HCN, HCO$^+$, and ^{13}CO using the IRAM 30m telescope. These data will complement existing CO, HI, Spitzer, and radio continuum maps. Furthermore, these data will be complemented by far-infrared maps of [CII], H$_2$O, [OI], [NII], and the dust continuum taken with Herschel in the open time key project HERM33ES.

M33 is a spiral galaxy located at a distance of 840 kpc. Observations of small scale structures in M33 do not suffer from distance ambiguities like galactic observations do. The IRAM-30m beam at 89 GHz is 27″, which corresponds to only 92 pc. M33 is seen almost face-on allowing to study individual cloud complexes.

It is the nearest late-type galaxy, is chemically and dynamically young, roughly 10 times less massive than the Milky Way and with an overall metallicity subsolar by a factor 2-3. With a well-defined metallicity gradient (Magrini *et al.* 2007, Rosolowsky & Simon 2008), M33 is a particularly interesting object, filling in the metallicity gap between the Milky Way and the Magellanic Clouds. Studying metallicity effects, e.g. on the CO-to-H$_2$ conversion factor, is also of importance for interpreting the emission of objects at high redshifts.

Due to their large dipole moments, HCN and HCO$^+$ trace the dense gas ($n > 10^4$ cm^{-3}) in galaxies. And as stars condense out of the densest material, these molecules may be good tracers of the star formation rate (SFR). The role of HCN as tracer of the SFR was recently discussed by Wu *et al.* (2005), combining Galactic data with data of nearby galaxies and starbursts. They show that HCN is a much better tracer of star formation than CO (see also Gao & Solomon 2004). A similar conclusion should hold for HCO$^+$. Previous studies of the HCO$^+$/HCN ratio find values around 1, e.g. 1.1 in M31 (Brouillet *et al.* 2005) or 0.5 - 1.6 in a sample of galactic nuclei, depending only slightly on environment (Krips *et al.* 2008).

Using the IRAM 30m telescope, we detected HCO$^+$(1-0) and HCN(1-0) at several positions along the major axis of M33 out to 4 kpc. We find rather large HCO$^+$/HCN ratios of integrated intensities of $\sim$ 10 in the central regions. In addition we find indications of a rise of HCN intensities with radius, while the SFR also drops with distance. More data are currently observed to secure this finding. The large HCO$^+$/HCN ratios in the center may reflect the low N/H abundance of M33. It is lower than the N/H solar value (Asplund *et al.* 2005) by a factor 5 or more (Magrini *et al.* 2007, 2009, priv. com.). We will use models of the chemical and physical structure of photon dominated regions (Roellig *et al.* 2007) to confirm this speculation.

Highlights of Astronomy, Volume 15
XXVIIth IAU General Assembly, August 2009
Ian F. Corbett, ed.

© International Astronomical Union 2010
doi:10.1017/S1743921310010045

NRO Legacy Project: Survey of Giant Molecular Clouds in M33

N. Kuno[1,2], T. Tosaki[3], S. Onodera[1], K. Muraoka[4], H. Kaneko[2],
T. Sawada[5], K. Nakanishi[2,5], S. Komugi[6], Y. Tamura[7], K. Kohno[7],
R. Kawabe[1], N. Arimoto[2,5] and S. Okamoto[7]

[1]Nobeyama Radio Observatory, National Astronomical Observatory of Japan, [2]Graduate
University for Advanced Studies (SOKENDAI), Japan, [3]Joetsu University of Education,
Japan, [4]Osaka Prefecture University, Japan, [5]National Astronomical Observatory of Japan,
[6]Insititute of Space and Astronautical Science, Japan Aerospace Exploration Agency, Japan,
[7]University of Tokyo, Japan

As the Nobeyama Radio Observatory Legacy Project: Survey of Giant Molecular
Clouds in M33, we have been mapping M33 in CO(1-0) with the multi-beam receiver
BEARS equipped on the 45-m telescope using the OTF mapping technique since 2007.
The purpose of this project is to investigate the physical properties of GMCs and understand the evolutionary process from GMC formation to star formation in GMCs by comparing with various data such as CO(3-2), 1.1 mm continuum obtained with ASTE10m
telescope at Atacama and the optical data obtained with SUBARU. We identified 87
GMCs using the first year data of CO(1-0) and observed 28 GMCs among them in
CO(3-2) with ASTE (Onodera 2009, PhD thesis, University of Tokyo). From the comparison of these lines, it was shown that the CO(3-2)/CO(1-0) ratio increases with star
forming activity in the GMCs. Furthermore, we found that more massive GMCs tend to
have higher CO(3-2)/CO(1-0) ratio. Since the ratio is thought to be an indicator of the
fraction of warm and dense molecular gas, our results imply that the fraction of warm
and dense gas increases with GMC mass. Especially, since the ratio in the GMCs with
low star forming activity is in the range where the ratio depends mainly on the density,
we speculate that dense gas fraction increases with GMC mass.

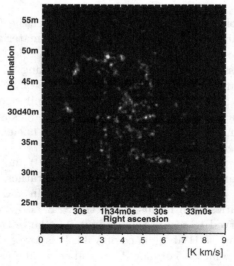

Figure 1. CO(1-0) integrated intensity map of M33.

Highlights of Astronomy, Volume 15
XXVIIth IAU General Assembly, August 2009
Ian F. Corbett, ed.

© International Astronomical Union 2010
doi:10.1017/S1743921310010057

The radio-infrared correlation in galaxies

F. S. Tabatabaei, R. Beck and E. Berkhuijsen

Max-Planck-Institut für Radioastronomie, Auf dem Hügel 69, 53121 Bonn, Germany
email: tabataba@mpifr-bonn.mpg.de

The radio–infrared correlation holds within galaxies down to scales of about 50 pc (Hughes et al. 2006, Tabatabaei et al. 2007a). It was explained as a direct and linear relationship between star formation and IR emission. However, one fact making the IR-star formation linkage less obvious is that the IR emission consists of at least two emission components, cold dust and warm dust. The cold dust emission may not be directly linked to the young stellar population. Furthermore, understanding the origin of the radio–IR correlation requires to discriminate between the two main components of the radio continuum emission, free-free and synchrotron emission. Although cosmic ray electrons originates also from the star forming regions (supernovae remnants; final episodes of massive stars), the synchrotron–IR correlation may not be as tight as thermal–IR correlation locally, as a result of convection and diffusion of the cosmic ray electrons from their place of birth. The magnetic field distribution may further modify the correlation.

We present a multi-scale study of the correlation of IR with both the thermal and non-thermal (synchrotron) components of the radio continuum emission from the nearby galaxies M33 and M31. Using the Spitzer MIPS IR data at $70 \, \mu m$ and $160 \, \mu m$, we derive extinction maps which are used to correct Hα and to separate the radio thermal and nonthermal components at 20 cm, following Tabatabaei et al. (2007b). Our scale-by-scale analysis in M33 using wavelet functions showed that the correlation is almost perfect on all scales between the radio thermal emission and the warm dust emission. The thermal radio–cold dust in M31 is not as tight as that in M33 on scales smaller than 2 kpc, indicating different dust heating sources: young massive stars in M33 but the interstellar radiation field in M31. The synchrotron–IR correlation is better in M33 than in M31 locally, however, it is better in M31 than in M33 globally. Tabatabaei et al. (2008) showed that M33 is dominated by strong turbulent magnetic field in star forming regions. In M31, however, Fletcher et al. (2004) found a stronger large scale regular magnetic fields than the small scale turbulent field. Therefore, the different scale-dependency of the synchrotron–IR correlation in M33 and M31 can be explained by their different magnetic fields and hence the propagation of cosmic ray electrons. Regarding that the star formation rate per area in M33 is about 10 times larger than that in M31 (Tabatabaei et al.. in prep.), we conclude that the magnetic fields and cosmic rays enhanced in star forming regions can cause a good correlation between the synchrotron and IR emission locally. This phenomena should be visible in late-type galaxies with on-going star formation and pronounced in starburst galaxies.

References

Fletcher, A., Berkhuijsen, E. M., Beck, R., & Shukurov, A. 2004, A&A 414, 53
Hughes, A., Wong, T., Ekers, R. et al. 2006, MNRAS 370, 363
Tabatabaei, F. S., Beck, R., Krause, M. et al. 2007a, A&A 466, 509
Tabatabaei, F. S., Beck, R., Krügel, E. et al. 2007b, A&A 475, 133
Tabatabaei, F. S., Krause, M., Fletcher, A., & Beck, R. 2008, A&A 490, 1005

Highlights of Astronomy, Volume 15
XXVIIth IAU General Assembly, August 2009
Ian F. Corbett, ed.

© International Astronomical Union 2010
doi:10.1017/S1743921310010069

Molecular lines studies at redshift greater than 1

Françoise Combes

LERMA, Observatoire de Paris, 61 Av. de l'Observatoire, F-75014, Paris
email: francoise.combes@obspm.fr

Abstract. Observations of CO molecules in the millimetrer domain at high redshift (larger than 1), have provided interesting informations about star formation efficiency, and its evolution with redshift. Due to the difficulty of the detections, selection effects are important. The detection if often due to gravitational amplification. Objects selected by their (far)infrared flux, are in general associated to ULIRGS, mergers with starburst in the nuclear regions. Quasars have been selected as powerful optical sources, and have been found to be associated to starbursts, rich in gas. The gas fraction appears to be much higher at redshift greater than 1. Quasars allow to probe the end of the reionisation period, and the relation between bulge and black hole mass. However these selection bias could have led us to miss some gaseous galaxies, with low-efficiency of star formation, such as the more quiescent objects selected by their BzK colors at $z = 1.5$ or 2.

Keywords. Galaxy, molecules, high redshift, star formation, quasars

1. Introduction

Up to 2009, there is about 50 high redshift sources detected in the CO lines. This domain of research has grown quickly, from the first discovery in 1992, of the faint IRAS source F10214+4724 at $= 2.3$ (Brown & van den Bout 1992, Solomon *et al.* 1992). Most of the sources have a redshift between $z = 2$ and 3. Recently, Iono *et al.* (2009) observed 30 of these sources homogeneously in CO(3-2) with SMA, and found a very good correlation between CO and FIR luminosities, even for QSO. This means that, although the contribution of the AGN to the FIR luminosity increases with power, the starburst always dominate.

The advantage to detect the CO lines, is to obtain the efficiency of star formation (ratio of FIR luminosity to the gas content), which evolves quickly with redshift, and when resolved, study the kinematics and mass at high z: for example, the source SMM J2399-0136 at $z = 2.808$, has been spatially resolved with the IRAM interferometer, and a hint of rotation curve has been found at $z = 2.808$ (Genzel *et al.* 2003). Taking into account the lens amplification factor of 2.5, the H_2 mass derived is $6~10^{10}~M_\odot$, and the dynamical mass is $3.0~10^{11}/\sin^2 i~M_\odot$, uncertain, since the inclination of the object is not well determined.

2. Recent results

Thanks to strong gravitational amplification, it was possible to detect CO line emission towards 6 quasars at z larger than 4, and determine their surprising properties.

The most distant quasar detected, SDSS J1148+5251 at $z = 6.4$, is a unique object, at the end of the epoch of reionization. The Gunn-Peterson trough is detected in HI absorption in its spectrum (Fan *et al.* 2003, White *et al.* 2003). In this powerful starburst, of surface density 1000 $M_\odot$/year /kpc^2, an impressive list of molecules has been detected, and even ionised carbon CII (158μm redshifted at 1mm), cf Walter *et al.* (2009). Surprisingly the HCN molecule is not detected, while HCN is the better tracer of star formation, well correlated with FIR (Gao & Solomon 2004).

One of the common features of the z > 4 quasars, resolved in the CO(2-1) line with the VLA, is that they are gas-rich mergers of galaxies, with complex morphologies, and molecular extent of about 5kpc, much larger than local ULIRGs (Riechers *et al.* 2008a,b). Once corrected for their strong amplification, their H_2 mass is up to 10^{11} $M_\odot$, a significant fraction of the dynamical mass. although the latter is ill-defined, given the perturbed shapes. Their black hole masses, derived assuming their AGN radiate at Eddington luminosity, or from the nuclear emission lines, appear an order of magnitude higher than expected from the Magorian relation. But uncertainties are large.

Their star formation surface density is saturating around 1000 $M_\odot$/year /kpc^2, as for Eddington limited star formation, i.e. dust opacity limited gas surface density. The average H_2 column density reaches 10^{24} cm^{-2}, and the volumic density 10^4 cm^{-3}.

Objects at more moderate SFR are not detected, unless strongly amplified. Three Lyman-break galaxies (LBG) at z~ 3, such as the Cosmic eye (Coppin *et al.* 2007) with a magnification of ~ 30, also satisfy the same FIR-CO luminosity relation. Their star formation rate is of the order of 50 $M_\odot$/year, with a starburst time-scale of 40Myr, they are the high-z analog of local LIRGs.

The most efficient star forming objects appear to be the Submillimeter Galaxies (SMG), which have been selected by their FIR luminosity, redshifted in the mm. They are more efficient than ULIRGs, and are very compact starbursts, of radius lower than 1kpc, or less. But all detected objects at high z are not so efficient. Recently, the BzK galaxies, selected for their red colors, have been detected with much more CO emission than expected, at z = 1.5-2 (Daddi *et al.* 2008). These are also ULIRGs, but with a much larger molecular content, and a time-scale to consume their gas of ~ 2 Gyr. They are extended 10kpc disks, and their CO excitation is low, peaking at CO(3-2) like the Milky Way. Another star forming BzK galaxy however was not detected by Hatsujade *et al.* (2009), and certainly, they have a wide range of properties. Due to their low excitation, it might be appropriate to use a higher CO-to-H_2 conversion ratio than for ULIRGs (Dannerbauer *et al.* 2009).

SMG are much more concentrated, more compact, they are expected to be major mergers remnants, with low angular momentum, and precursors of elliptical galaxies (Bouche *et al.* 2007). Do they actually trace massive haloes? In the GOODS-N field, a cluster of radio galaxies and SMG at z = 1.99 has been studied, it is the strongest known association of SMG (Chapman *et al.* 2008), with an overdensity of 10. SMG appear also to be associated to filaments traced by Lyman alpha emitters (LAE). In SSA22, a protocluster region at z = 3.1, traces a filament, where SMG have been detected at 1.1mm with AzTEC on ASTE (Tamura *et al.* 2009).

SMG are also sometimes associated to another type of objects, Lyman-alpha blobs (LAB, Geach *et al.* 2007), huge ionised gas nebulae, excited by a central starburst or AGN. In local clusters at z = 0.4-1, such as CL0024+16, the LIRGs detected by Spitzer at 24μm can be detected in CO (Geach *et al.* 2009).

The fact that at high redshift, most detected objects are quasars, allows to study the AGN-starbutst association, and possible AGN feedback on CO emission. For instance, APM08279+5255 at z = 3.9 is a lensed QSO, (amplification factor 50), and one of the brightest object in the sky; it has been observed with mm and cm telescopes, the CO lines from CO(1-0) to CO(11-10) are detected. Recent 0.3" resolution CO(1-0) mapping with the VLA (Riechers *et al.* 2009), reveals that the emission is not extended, as previously thought. The amplification factor could also be lower. The CO line emission is co-spatial with optical/NIR and X-rays, and very compact. The best model shows that the CO is in a circumnuclear disk of 550 pc radius, inclined by 25°, with a gas mass of 1.3×10^{11} $M_\odot$. There is no hint of the influence of the AGN feedback.

3. Perspectives

About 50 systems are presently detected in molecular lines at high redshift. Given the different properties of the various categories of objects, it becomes obvious that results are dominated by selection effects. ULIRGs have a very efficient star formation, their molecular gas has a compact distribution, and is highly excited. But BzK objects have much more extended gas, with normal star formation efficiency and consumption time-scales.

Present results are also strongly biased by lensing magnification.

One of the main robust results, is that quasars and starbursts are intimately linked, and the AGN activity does not seem to quench star formation, at these redshifts.

References

Bouche N., Cresci, G., Davies, R. *et al.* 2007, ApJ 671, 303
Brown R. L. & van den Bout P. A. 1992, ApJ 397, L19
Chapman, S. C., Neri, R., Bertoldi, F. *et al.* 2008, ApJ 689, 889
Coppin, K. E. K., Swinbank, A. M., Neri, R. *et al.* 2007, ApJ 665, 936
Daddi E., Dannerbauer, H., Elbaz, D. *et al.* 2008, ApJ 673, L21
Dannerbauer, H., Daddi, E., Riechers, D. A. *et al.* 2009, ApJ 698, 178
Fan, X., Strauss, M. A., Schneider, D.P. *et al.* 2003, AJ 125, 1649
Gao Y., & Solomon P. 2004, ApJ 606, 271
Geach, J. E., Smail, I., Chapman, S. C. *et al.* 2007, ApJ 655, L9
Geach, J. E., Smail, I., Coppin, K. *et al.* 2009, MNRAS 395, L62
Genzel, R., Baker, A. J., Tacconi, L. *et al.* 2003, ApJ 584, 633
Hatsukade, B., Iono, D., Motohara, K. *et al.* 2009, PASP 61, 487
Iono, D., Wilson, C. D., Yun, M. S. *et al.* 2009, ApJ 695, 1537
Riechers, D. A., Walter, F., Brewer B. *et al.* 2008a, ApJ 686, 851
Riechers, D. A., Walter, F., Carilli, C. *et al.* 2008b, ApJ 686, L9
Riechers, D. A., Walter, F., Carilli, C. *et al.* 2009, ApJ 690, 463
Solomon P., Radford, S. J. E., & Downes, D. 1992, Nature 356, 318
Tamura, Y., Kohno, K., Nakanishi, K. *et al.* 2009, Nature 459, 61
Walter F., Riechers, D., Cox, P. *et al.* 2009, Nature 457, 699
White, R. L., Becker, R. H., Fan, X., & Strauss, M. A. 2003, AJ 126, 1

Highlights of Astronomy, Volume 15
XXVIIth IAU General Assembly, August 2009
Ian F. Corbett, ed.

© International Astronomical Union 2010
doi:10.1017/S1743921310010070

Vibrationally Excited HC$_3$N in NGC 4418

F. Costagliola[1] and S. Aalto[1]

[1]Department of Radio and Space Science, Chalmers University of Technology,
Onsala Space Observatory, SE-439 92 Onsala, Sweden
email: francesco.costagliola@chalmers.se, saalto@chalmers.se

Keywords. galaxies: evolution, galaxies: individual: NGC 4418, galaxies: starburst, galaxies: active, radio lines: ISM, ISM: molecules

Luminous infrared galaxies (LIRGs) emit most of their radiation in the infrared region of the spectrum in the form of dust thermal continuum, with typical luminosities of $L_{IR} > 10^{10}$ L$_\odot$. The central power source responsible for the total energy output is deeply buried in the dusty central regions of these objects and its origin still unclear. Recent studies by Spoon *et al.* (2007) and Aalto *et al.* (2007) suggest that some LIRGs might represent early obscured stages of active galaxies, either AGNs or starbursts, and thus play a fundamental role in galaxy formation and evolution.

NGC 4418 is an almost edge-on Sa-type LIRG with deep mid-infrared silicate absorption features suggesting that the inner region is enshrouded by large masses of warm (85 K) dust (Evans *et al.* (2003)). *This object has one of the highest luminosities of HC_3N $J = 10–9$ emission (compared to HCN $J = 1–0$) found for an external galaxy and a tentative detection of a vibrational HC_3N line was reported by Aalto et al. (2007).*

Here we report the first confirmed extragalactic detections of vibrationally excited HC$_3$N in the LIRG NGC 4418. The observations were carried out in December 2007 and August 2008 with the IRAM 30m telescope on Pico Veleta, Spain. We detected the HC$_3$N rotational transitions $J = 10–9$, 16–15, 17–16, 24–23, 25–24, 28-27, 30-29. For the $J = 10–9$, 17–16, 25–24 and 28-27 lines, we also detected rotational transitions of the $v_7 = 1$ vibrationally excited levels. For the $v_6 = 1$ lines we generally have upper limits, the only clear detection being in the $J = 25–24$ band.

When observed lines are plotted on a population diagram, they show three main temperature components. The $v = 0$ rotational levels clearly show two temperature components at 20 K (for $E_u < 100$ K) and 529 K (for larger E_u). The $v_7 = 1$ transitions have an excitation temperature of 80 K that is consistent with the dust temperature of 85 K found by Evans *et al.* (2003) for the inner 0.5″. This rotational temperature can either be due to the IR radiation field dominating the excitation, or it can reflect the gas kinetic temperature in the case that gas and dust are in thermal equilibrium. The low temperature, 20 K, component, could be coming from more extended gas, further away from the nuclear warm dust. If we compare the populations of different rotational levels with the same J, we can get the vibrational temperature, that describes the excitation of the vibrational modes. This can be done for the $J = 25–24$ band, for which we have both $v_6 = 1$ and $v_7 = 1$ lines. The resulting vibrational temperature is 500 K, comparable to the value fitted to the high-J levels of the $v = 0$ state. The bending modes v_6 and v_7 have both critical densities greater than 10^8 cm^{-3} and are thus most likely radiatively excited. The derived vibrational temperature then reflects the temperature of a radiation field. Assuming that our 500 K component represents the temperature of the IR continuum, the required source size θ for the emitting region can be estimated by the relation $L_{IR} \propto \theta^2 T^4$, that gives $\theta = 0.01''$, that corresponds to a linear diameter of 1.45 pc. This is an upper limit to the size of the emitting region, in the case of it being responsible for all the observed IR flux. Our interpretation of the HC$_3$N excitation then leads to an extremely compact radiation source in the core of NGC 4418. From the population diagram, the hot component has a column density of 8.3×10^{15} cm^{-2}, that can be assumed to be a lower limit to the total HC$_3$N column. The HC$_3$N excitation and abundances seem similar to those found for hot cores in Sgr B2 in the Galactic

Centre (de Vicente *et al.* (2000)). That the hot (500 K) component could be associated with a buried AGN cannot be excluded, but the large HC_3N abundances may prove to be difficult to reconcile with an AGN.

Acknowledgements

We thank the staff at the IRAM 30m telescope for their kind help and support during our observations. Furthermore, we would like to thank the IRAM PC for their generous allocation of time for this project. This research was supported by the EU Framework 6 Marie Curie Early Stage Training programme under contract number MEST-CT-2005-19669 "ESTRELA" and by the European Community Framework Programme 7, Advanced Radio Astronomy in Europe, grant agreement no. 227290, "RadioNet".

References

Spoon, H. W. W., Marshall, J. A., Houck, J. R., Elitzur, M., Hao, L., Armus, L., Brandl, B. R., & Charmandaris, V. 2007, *ApJ* 654, L49–L52

Aalto, S., Monje, R., & Martín, S. 2007, *A&A* 475, 479–485

Evans, A. S., Becklin, E. E., Scoville, N. Z., Neugebauer, G., Soifer, B. T., Matthews, K., Ressler, M., Werner, M., & Rieke, M. 2003, *AJ* 125, 2341–2347

de Vicente, P., Martín-Pintado, J., Neri, R., & Colom, P., 2000 *A&A* 361, 1058–1072

Highlights of Astronomy, Volume 15
XXVIIth IAU General Assembly, August 2009
Ian F. Corbett, ed.

ⓒ International Astronomical Union 2010
doi:10.1017/S1743921310010082

Mid-IR Spectroscopy of Submm Galaxies: Extended Star Formation in High-z Galaxies

K. Menéndez-Delmestre[1], A. W. Blain[2], I. Smail[3], D. M. Alexander[4], S. C. Chapman[5], L. Armus[6], D. Frayer[6], R. J. Ivison[7,8] and H. Teplitz[6]

[1] NSF Astronomy & Astrophysics Postdoctoral Fellow; Carnegie Observatories, Pasadena, CA
email: kmd@obs.carnegiescience.edu [2] Caltech, Pasadena, CA
[3] Computational Cosmology, Durham, UK [4] Physics, Durham, UK
[5] Inst. of Astronomy, Cambridge, UK [6] SSC, Pasadena, CA [7] Astronomy Technology Centre, Edinburgh, UK [8] Inst. for Astronomy, Edinburgh, UK

Abstract. Ultra-luminous infrared galaxies (ULIRGs; L $> 10^{12}$ L$_\odot$) are quite rare in the local universe, but seem to dominate the co-moving energy density at $z > 2$. Many are optically-faint, dust-obscured galaxies that have been identified only relatively recently by the detection of their thermal dust emission redshifted into the sub-mm wavelengths. These submm galaxies (SMGs) have been shown to be a massive objects (M$_\star \sim 10^{11}$ M$_\odot$) undergoing intense star-formation(SFRs $\sim 10^2 - 10^3$ M$_\odot$ yr^{-1}) and the likely progenitors of massive ellipticals today. However, the AGN contribution to the far-IR luminosity had for years remained a caveat to these results. We used the *Spitzer* Infrared Spectrograph (IRS) to investigate the energetics of 24 radio-identified and spectroscopically-confirmed SMGs in the redshift range of $0.6 < z < 3.2$. We find emission from Polycyclic Aromatic Hydrocarbons (PAHs) – which are associated with intense star-formation activity – in $> 80\%$ of our sample and find that the median mid-IR spectrum is well described by a starburst component with an additional power-law continuum representing $< 32\%$ AGN contribution to the far-IR luminosity. We also find evidence for a more extended distribution of warm dust in SMGs compared to the more compact nuclear bursts in local ULIRGs and starbursts, suggesting that SMGs are not simple high-redshift analogs of local ULIRGs or nuclear starbursts, but have star formation which resembles that seen in less-extreme star-forming environments at $z \sim 0$.

Keywords. infrared: galaxies — galaxies: starburst — galaxies: AGN — galaxies: submillimeter

---

Deep X-ray studies suggest that $> 28 - 50\%$ of SMGs host an obscured AGN (Alexander *et al.* 2005). Although X-rays provide one of the most direct routes to estimate the luminosities of AGN, under high column densities such as found in SMGs, hard X-ray photons may be completely absorbed. The mid-IR provides an indirect insight to the dust-enshrouded nature of SMGs. We present the largest sample of SMGs observed in the mid-IR with *Spitzer* IRS (see Menéndez-Delmestre *et al.* 2009, hereafter MD09, for full discussion). Our main results are:

1. We find that $> 80\%$ of the SMGs in our sample display luminous PAH features and only four SMGs have continuum-dominated spectra (weak or absent PAH features). This indicates that, although some diversity exists within the population, SMGs are in general a population dominated by intense star-forming activity.

2. The composite SMG spectrum is best fit by the combination of a starburst template and an additional power-law continuum representing a maximum AGN contribution of $< 32\%$ to the far-IR luminosity in SMGs (see Fig. 1).

3. We quantify the strength of the silicate absorption feature and find that the distribution in $\tau_{9.7\mu m}$ for SMGs falls below that of local ULIRGs and the most obscured low-redshift nuclear starburst-dominated galaxies. This suggests that SMGs have lower dust obscuration to their mid-IR continuum emitting regions than these local samples.

4. Comparison of 7.7/11.3 PAH flux ratios suggests that SMGs host similar radiation envi-

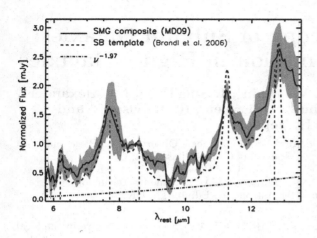

Figure 1. The SMG composite spectrum (solid line) is dominated by strong PAH emission features (vertical dashed lines). We fit this composite with a local starburst template (dashed spectrum) and derive an upper limit to the contribution from an AGN by making the conservative assumption that the additional continuum emission (dot-dashed line) arises solely from an obscured AGN. As described in MD09, we estimate that AGN activity contributes < 32% to the far-IR luminosity in typical SMGs. We emphasize that this is a strong upper limit for the SMG population, since a fraction of the red continuum in this composite spectrum likely arises from dust emission heated by optically-thick HII regions.

ronments to local starbursts. However, the 7.7/6.2 PAH ratio is lower in SMGs than in local nuclear starbursts. A stronger 6.2-μm PAH emission relative to the 7.7μm PAH feature may be attributed to lower extinction by ice (at 6μm) along the line of sight to SMGs

We conclude that the detailed mid-IR spectral properties of SMGs show several differences to local ULIRGs and nuclear starbursts. The differences in Si-absorption strengths and 7.7/6.2 PAH ratios can be most easily explained by a difference in the extinction to the mid-IR continuum and line emitting regions of these galaxies, with the SMGs showing systematically lower extinction. Considering that SMGs are very dusty objects, a lower extinction suggests that the warm dust is distributed over spatial scales significantly more extended in SMGs than found in local ULIRGs and nuclear starbursts.

References

Alexander, D. M., Bauer, F. E., Chapman, S., Smail, I., Blain, A., Brandt, W. N., & Ivison, R. 2005, *ApJ*, 632, 736
Brandl, B. R. *et al.* 2006, *ApJ*, 653, 1129
Chapman, S., Blain, A., Smail, I., & Ivison, R. 2005, *ApJ*, 622, 772
Krabbe, A., Böker, T., & Maiolino, R. 2001, *ApJ*, 557, 626
Menéndez-Delmestre, K. *et al.* 2009, *ApJ*, 699, 667 (MD09)

Highlights of Astronomy, Volume 15
XXVIIth IAU General Assembly, August 2009
Ian F. Corbett, ed.

© International Astronomical Union 2010
doi:10.1017/S1743921310010094

Dust production in supernovae: the case of Kepler's SNR

E. M. Reynoso[1]†, H. L. Gomez[2] and L. Dunne[3]

[1]IAFE (CONICET-UBA) and Physics Dept. (FCEyN, UBA), Buenos Aires, Argentina
[2]School of Physics and Astronomy, Cardiff University, The Parade, Wales, UK
[3]School of Physics and Astronomy, University of Nottingham, UK

Abstract. We compare the submillimetre (submm) emission with the H I and CO distribution towards Kepler's supernova remnant (SNR), and conclude that 0.1 to 1.2 $M_\odot$ of dust originates from Kepler. Such rates are sufficient to explain the origin of dust in high redshift galaxies.

Keywords. dust, supernova remnants, individual: Kepler's SNR, submillimetre

1. Results and conclusions

The role of supernovae as significant sources of dust is still unresolved. To explore this subject, we analyze the submm emission towards Kepler's SNR - Morgan *et al.* 2003 - in connection with the 3-dimensional picture of the interstellar medium (ISM) around this source as inferred from H I and CO line data, with the aim of differentiating the amount of dust corresponding to the ISM or to the remnant. To ensure that the submm emission structure is not affected by the observing technique employed, simulations of chopping over different possible emission distributions were carried out.

We find a ring-like structure of submm emission coincident with the remnant's shell, which cannot be explained by a synchrotron source, chopping on-off nearby interstellar structures, swept-up interstellar material, nor by line-of-sight contamination. Indeed, there is no correlation between the submm peaks and the H I or CO emission, while the molecular gas is found to be diffuse and optically thin and has a negligible contribution to the submm emission towards the SNR. After subtracting a synchrotron component using a spectral index map, we estimate a dust mass of 0.1 or 1.2 $M_\odot$ depending on the absorption coefficient (Gomez *et al.* 2009). This result agrees with the range of estimates for Cas A (from IR and submm data: Krause *et al.* 2004, Rho *et al.* 2008, Dunne *et al.* 2009) and implies that supernovae can still be considered as significant sources of dust.

Acknowledgements

E. M. R. is partially supported by grants PIP-CONICET 114-200801-00428 and UBACyT X482. H.L.G. would like to acknowledge the support of LCOGT.

References

Dunne L. *et al.*, 2009, MNRAS, 394, 1307
Gomez, H. L., Dunne, L., Ivison, R. J., Reynoso, E. M., Thompson, M. A., Sibthorpe, B., Eales, S. A., DeLaney, T. M., Maddox, S., & Isaak, K. 2009, *MNRAS* 397, 1621
Krause O., *et al.*, 2004, Nature, 432, 596
Morgan H. L., Dunne L., Eales S., Ivison R. J., & Edmunds M. G., 2003, ApJ, 597, L33
Rho J., *et al.*, 2008, ApJ, 673, 271

† Member of the Carrera del Investigador Cient´ıfico, CONICET, Argentina.

Highlights of Astronomy, Volume 15
XXVIIth IAU General Assembly, August 2009
Ian F. Corbett, ed.

© International Astronomical Union 2010
doi:10.1017/S1743921310010100

Detecting the first quasars with ALMA

Dominik R. G. Schleicher[1,2,3], Marco Spaans[4] and Ralf S. Klessen[1]

[1] Zentrum für Astronomie der Universität Heidelberg, Institut für Theoretische Astrophysik, Albert-Ueberle-Str. 2, 69120 Heidelberg, Germany. E-Mail: dschleic@ita.uni-heidelberg.de

[2] Leiden Observatory, P.O. Box 9513, NL-2300 RA Leiden, the Netherlands

[2] ESO Garching, Karl-Schwarzschild-Str. 2, 85748 Garching bei München, Germany

[4] Kapteyn Astronomical Institute, University of Groningen, P.O. Box 800, 9700 AV, Groningen, the Netherlands

Abstract. We show that ALMA is the first telescope that can probe the (dust) obscured central region of quasars at $z > 5$ with a maximum resolution of ~ 30 pc employing the 18 km baseline.

Keywords. astrochemistry - telescopes - galaxies: active - high-redshift - ISM

We explore the possibility to detect the first quasars with ALMA - Schleicher, Spaans & Klessen(2009). For this purpose, we adopt the Seyfert 2 galaxy NGC 1068 as a reference system and calculate the expected fluxes if this galaxy were placed at high redshift. This choice seems justified due to the absence of any indication for an evolution in metallicity in high-redshift quasars, and is conservative due to the moderate column densities in NGC 1068, leading to correspondingly moderate fluxes.

On large scales, the dominant excitation mechanism are soft UV-photons produced by the starburst, leading to strong emission in the [CII] 158 μm and the [OI] 63 μm line. Due to its unprecedented angular resolution, ALMA can probe the central 200 pc of high-redshift quasars. Calculations with the X-ray dominated region code of Meijerink & Spaans(2005) show that strong emission in the high-J CO lines can be expected in these regions (see Fig. 1). As these lines are redshifted into the ALMA bands, this may provide the first opportunity to detect these lines. Flux estimates for various fine-structure lines are provided in the paper.

References

Meijerink, R. & Spaans, M., 2005, A&A, 436, 397
Schleicher, D. R. G., Spaans, M., & Klessen, R. S., 2009, submitted to A&A

† Member of the Carrera del Investigador Cient´ıfico, CONICET, Argentina.

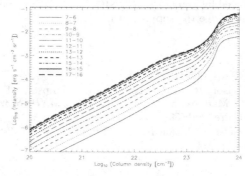

Figure 1. The expected intensities in the high-J CO lines as a function of the cloud column density in the central region of a system like NGC 1068.

Highlights of Astronomy, Volume 15
XXVIIth IAU General Assembly, August 2009
Ian F. Corbett, ed.

© International Astronomical Union 2010
doi:10.1017/S1743921310010112

JD15 – Magnetic Fields in Diffuse Media

Elisabete M. de Gouveia Dal Pino and Alex Lazarian, Eds.

Preface

Most of the baryonic matter in the Universe is permeated by magnetic fields which affect many, if not most, of astrophysical phenomena both, in compact sources and in diffuse gas.

Recent years have been marked by a worldwide surge of interest in the astrophysical magnetic fields, their origin, and their influence on the formation and evolution of astrophysical objects (stars, galaxies, cooling flows). This growing interest is in part due to the fact that it has become possible to trace magnetic fields in molecular clouds, over vast extensions of the Milky Way and to study extragalactic magnetic fields, including fields in clusters of galaxies. With the combination of various techniques, such as Zeeman and Faraday rotation measurements with synchrotron and aligned grain polarimetry, it is now possible to undertake quantitative observational studies of magnetic fields, the results of which can be compared with high resolution dynamo and MHD turbulence simulations. This brings the field to a new stage and, at the same time, calls for addressing fundamental questions, such as the correspondence of basic processes in astrophysical media with computer simulations, and our real understanding of the processes that we rely to infer magnetic fields from observations.

We thus felt that it was time to invite researchers of magnetic fields to debate the vital questions related to the origin of astrophysical magnetic fields in diffuse gas, their effects on transport processes in the interstellar medium of spiral galaxies and in the intracluster medium in order to get better insight into key astrophysical processes like star formation, acceleration of cosmic rays, and transfer of matter and energy between the diffuse and dense gas.

This two-day Joint Discussion on "Magnetic Fields in Diffuse Media" held at the IAU General Assembly, in Rio de Janeiro, provided an excellent forum for this timely undertaking. Because of the cross-disciplinary nature of the subject it was able to catch the attention and interest of a wide range of astrophysicists, including specialists in space and Solar physics.

The meeting counted with about 100 participants from 28 countries including 20 invited contributions, 6 oral contributions and 63 poster presentations. All the sessions ended with discussion panels of the related topics. We have also awarded prizes for the five most outstanding posters which were nominated by an anonymous competent referee. In the closure, a Summary Panel highlighted what might be the most relevant questions for continuing the study in the field:

• What is the role of turbulent magnetic reconnection and ambipolar diffusion on star formation triggering?

• What is the role of MHD turbulence on clouds formation?

• What is the dominant mechanism in the galactic dynamo: is it driven by turbulence and/or magnetic helicity?

• To what extent is the kinetic-MHD turbulence dominant in the intergalactic medium?

• What is the nature of the cosmic ray and MHD turbulence coupling in the intergalactic and intracluster medium?

• What is the role of supernovae, AGNs and mergers on turbulence and magnetic field feeding at the cores and outskirts of galaxy clusters?

• What will the coming new generation of instruments (SKA Pathfinders, LSST, SOFIA, PLANCK, LOFAR, ALMA, SKA) unveil about large scale cosmic field origin and structure?

Elisabete M. de Gouveia Dal Pino and Alex Lazarian, co-chairs of the SOC of the Joint Discussion on Magnetic Fields in Diffuse Media,
Rio de Janeiro, Brazil, August 2009

Highlights of Astronomy, Volume 15
XXVIIth IAU General Assembly, August 2009
Ian F. Corbett, ed.

Zeeman Splitting in the Diffuse ISM

Carl Heiles

Astronomy Department, University of California, Berkeley CA 94720-3422, USA
email: heiles@astro.berkeley.edu

Abstract. Many Zeeman-spitting measurements in the diffuse Galactic Interstellar Medium have been made of the 21-cm line in both absorption and emission. Typical field strength is about 6 μG, with enhancement in shocked regions; the magnetic, turbulent, and cosmic-ray pressures are comparable and considerably larger than the thermal gas pressure. For PhotoDissociation Regions, Carbon recombination lines show intriguing results for single-dish measurements. OH Megamasers in Ultra-Luminous Infrared Galaxies show easily detectable fields whose strength ranges up to at least 20 mG. Upper limits for several damped Ly-α systems range down to a few μG. The z=0.692 system against 3C286 was reported in the literature to have a large field strength, but this result is wrong.

Keywords. ISM: magnetic fields, galaxies: magnetic fields, masers, polarization

1. Historical Introduction

Galactic ISM Studies in the 21-cm line. Zeeman splitting of the Interstellar Medium (ISM) was discovered for the 21-cm line in absorption against the powerful continuum source Cas A, and a few other sources, by Verschuur (1969). No further work was done in absorption for more than thirty years, when Heiles and Troland (a series culminating in 2005) surveyed some 70 sources using the Arecibo telescope for their "Millennium Survey". Meanwhile, work concentrated on mapping the field using the 21-cm line in emission. This resulted in a catalog of 429 positions (Heiles, 1988; Heiles, 1989; Heiles, 1994; Myers *et al.* 1995) observed with the Hat Creek 85-foot telescope (which was destroyed by a windstorm in winter 1993), plus a few more by Verschuur (1989). Meanwhile, Verschuur (1995a, 1995b) severely criticized all this work, claiming that all emission results were spurious owing to polarized sidelobes. This criticism was put to rest by analysis of the North Celestial Pole data (Heiles 1996), the analysis of his criticism by Heiles (1998), and the observations and analyses by Robishaw (2008).

Nevertheless, polarized beam structure plagues measurements of Zeeman splitting in 21-cm line emission, even for the Green Bank Telescope with its clear aperture and supposedly minimal sidelobes (Robishaw and Heiles 2009).

Galactic and Extragalactic Studies of Small Sources. Polarized sidelobes are irrelevant for small sources, including OH masers and star forming regions (see Crutcher, this conference). Observations have been performed of three additional classes of small source: PhotoDissociation Regions (PDRs; often called Photon-Dominated Regions), which are the highly compressed regions lying between HII regions and the un-ionized ambient medium; OH Megamasers (OHMs) at moderate redshifts in Ultra-Luminous Infrared Galaxies (ULIRGs); and Damped Lyman-α Absorption (DLA) line systems. PDRs yield intriguing results that justify the time and effort involved in high-resolution mapping of the Zeeman splitting, for example with the VLA. OHMs have turned into a field with many interesting and strong results, currently being pursued with vigor. DLAs have only negative results, despite a recent report of an amazingly strong field towards 3C286 (Wolfe *et al.* 2008) which has subsequently been shown to be wrong (see below).

2. Some Details and Results

Magnetism is a Major Force in the Diffuse ISM. The Millennium Survey of Zeeman splitting of the 21-cm line in absorption showed that the Cold Neutral Medium has a well-defined median field strength of about 6 μG (Heiles & Troland 2005). This makes it comparable to the volume-average Galactic field strength (Ferrière 2001) and means it dominates the gas thermal pressure by a factor of typically about 3 and is comparable to the turbulent pressure and the cosmic-ray pressure. The emission results are consistent with this, and in particular show field amplification in morphologically interesting regions such as shocks (Heiles 1989).

Zeeman Splitting is Probably Mappable in some PDRs. In his thesis, Robishaw has Zeeman splitting results on the Carbon Recombination Line emission in several regions, obtained with the GBT. The formal significance on the results is roughly 3σ, which is barely adequate for a single source, but because 4 out or 5 sources have this level of significance the probable reality of significant detections is enhanced. The regions include DR21 (920 ± 310 μG), Orion (290 ± 150 μG), M17 (2530 ± 860 μG), and DR21OH (4940 ± 2500 μG); the only source examined with a formally insignificant detection was W3, with (60 ± 110 μG), but this might simply reflect unfavorable viewing geometry.

Zeeman Splitting of OHMs in ULIRGS. OH Megamaseers are often easy to detect, at least at Arecibo; similarly, their Stokes V signals, which reveal Zeeman splitting, are often strong and easy to detect. Robishaw (2008) presents a pilot study of 6 sources, which have fields ranging up to about 20 mG. We are currently surveying about 70 OHMs—the known list in Arecibo's declination range of –1 to 39 deg – and have many detections.

Zeeman Splitting in Damped Ly-α systems. Lane & Heiles (in preparation) observed and found upper limits in 3 DLAs in 2 sources, 0738+313 (1.8 ± 2.4 μG for z=0.0912; 12.5 ± 9.0 μG for z=0.2212) and 0235+164 ($\lesssim 10$ μG, z=0.524). Wolfe *et al.* (2008) observed 3C286 and reported the amazing 84 ± 9 μG. It turns out, however, that this result is wrong. Owing to a software bug, a phase angle was not accounted for. This angle happened to be about 90°, which meant that the purported circular polarization is really linear. The absorption line is strongly linearly polarized and is of interest in itself. However, the magnetic field limit is (12.7 ± 5.8) μG, an insignificant result.

References

Ferrière, K. 2001, RevModPhyss, 73, 1031

Heiles, C. 1988, ApJ, 324, 321

Heiles, C. 1989, ApJ, 336, 808

Heiles, C. 1994, ApJ, 424, 208

Heiles, C. 1996, ApJ 466, 224

Heiles, C. 1997, ApJS 111, 245

Heiles, C. 1998, Astrophys. Lett. and Comm. 37, 85

Heiles, C. & Troland, T. H./ 2005, ApJ 624, 773

Myers, P. C., Goodman, A. A., Rüsten, R., & Heiles, C. 1995, ApJ, 442, 177

Robishaw, T. 2008, Ph.D. Thesis, University of California, Berkeley (AAT 3331778)

Robishaw, T. & Heiles, C. 2009, PASP, 121, 272

Robishaw, T., Quataert, E., & Heiles, C. 2008, ApJ, 680, 981

Verschuur, G. L. 1969, ApJ, 156, 861

Verschuur, G. L. 1989, ApJ, 339, 163

Verschuur, G. L. 1995a, ApJ, 451, 624

Verschuur, G. L. 1995b, ApJ, 451, 645

Wolfe, A. M., Jorgenson, R. A., Robishaw, T., Heiles, C., & Prochaska, J. X. 2008, Nature, 455, 638

Highlights of Astronomy, Volume 15
XXVIIth IAU General Assembly, August 2009
Ian F. Corbett, ed.

Cosmic magnetic field observations with next generation instrumentation

Rainer Beck

Max-Planck-Institut für Radioastronomie, Auf dem Hügel 69, 53121 Bonn, Germany
email: rbeck@mpifr-bonn.mpg.de

Abstract. The origin of magnetic fields in the Universe is an open problem in astrophysics and fundamental physics. Forthcoming radio telescopes will open a new era in studying cosmic magnetic fields. Low-frequency radio waves will reveal the structure of weak magnetic fields in the outer regions and halos of galaxies and in intracluster media. At higher frequencies, the EVLA and the SKA will map the structure of magnetic fields in galaxies in unprecedented detail. All-sky surveys of Faraday rotation measures (RM) towards a huge number of polarized background sources with the SKA and its pathfinders will allow us to model the structure and strength of the regular magnetic fields in the Milky Way, the interstellar medium of galaxies, in galaxy clusters and the intergalactic medium.

Keywords. Techniques: polarimetric, galaxies: clusters, galaxies: magnetic fields

1. Origin of cosmic magnetic fields

The origin of the first magnetic fields in the Universe is still a mystery (Widrow 2002). A large-scale primordial field is hard to maintain in a young galaxy because the galaxy rotates differentially, so that field lines strongly wind up and reconnect during galaxy evolution. Turbulent "seed" fields in young galaxies can originate e.g. from the Weibel instability in shocks during cosmological structure formation (Lazar *et al.* 2009), or can be injected by the first stars or jets generated by the first black holes (Rees 2005). In any case a mechanism to sustain and organize the magnetic field is required. The most promising mechanism is the dynamo (Beck *et al.* 1996) which transfers mechanical energy into magnetic energy. In young galaxies a small-scale dynamo with turbulence driven by star formation (Brandenburg & Subramanian 2005) probably amplified the seed fields from the protogalactic phase to the energy density level of turbulence within less than 10^9 yr, followed by a large-scale dynamo generating regular fields (2009).

Magnetic fields in galaxy clusters are probably seeded by AGNs and amplified in the intracluster medium by a turbulent dynamo (Ryu *et al.* 2008), giving rise to diffuse radio halos. Cluster mergers may compress the field and generate "radio relics" which are highly polarized and best observable at low frequencies.

2. Tools to measure interstellar magnetic fields

Most of what we know about galactic and intergalactic magnetic fields comes through the detection of radio waves. *Synchrotron emission* is a measure of the strength of the total magnetic field component in the sky plane. Linearly polarized synchrotron emission emerges from ordered fields, and *B*-vectors indicate the field orientation. The orientation of the polarization vectors is changed in a magnetized thermal plasma by *Faraday rotation*. Faraday rotation measures (RM) from multi-wavelength observations yield the strength and direction of the regular field component along the line of sight. *Zeeman splitting* of radio spectral lines also measures the regular field along the line of sight.

3. Prospects with future telescopes

Future radio telescopes will greatly widen the range of observable magnetic phenomena. To resolve the detailed structure of the ISM and halo fields, high-resolution, deep polarization observations at high frequencies, where Faraday effects are small, require a major increase in sensitivity for continuum observations which will be achieved by the Extended Very Large Array (EVLA) and the planned Square Kilometre Array (SKA). Polarized emission is an excellent tracer of interactions between galaxies and of cluster mergers. The SKA will also allow to measure the Zeeman effect in weak magnetic fields in the Milky Way and in nearby galaxies.

Low-frequency radio telescopes like the Low Frequency Array (LOFAR, under construction) and the low-frequency SKA array will be suitable instruments to search for synchrotron radiation from low-energy cosmic-ray electrons which suffer only moderately from energy losses, allowing to trace magnetic fields in the outer regions and halos of galaxies, in the Milky Way and in galaxy clusters (Beck 2009).

Radio spectro-polarimetric observations in many narrow frequency channels allows application of *RM Synthesis* (Brentjens & de Bruyn 2005). If the emitting medium has a relatively simple structure, *Faraday tomography* will become possible. This method is going to revolutionize radio polarization observations.

Polarized pulsars and extragalactic sources can be used to observe a grid of RM values to measure the magnetic field structure of intervening objects. A sufficiently high number density of sources requires large sensitivity and/or high survey speed. The *POSSUM* all-sky survey at 1.4 GHz with the planned Australia SKA Pathfinder (ASKAP) telescope will measure about 80 RM per square degree. The SKA will be able to detect several 1000 RM per square degree. The SKA *Magnetism Key Science Project* plans an all-sky RM grid which will contain about 10^4 RM values from pulsars with a mean spacing of $\simeq 30'$ and about 10^8 RM from compact polarized extragalactic sources at a mean spacing of just $\simeq 1.5'$ (Gaensler *et al.* 2004). This survey will be used to model the structure and strength of the magnetic fields in the Milky Way, in intervening galaxies, and in the intergalactic medium. The distance range of this method is much larger than by direct imaging of the polarized emission from the intervening galaxy. Looking back into time, the future telescopes will also shed light on the origin and evolution of cosmic magnetic fields in galaxies (Kronberg *et al.* 2008) and in clusters (Krause *et al.* 2009). The detection of RM from the intergalactic medium would allow us to probe the existence of magnetic fields in such rarified regions, measure their intensity, and investigate their origin and their relation to the structure formation in the early Universe.

References

Arshakian, T. G., Beck, R., Krause, M., & Sokoloff, D. 2009, *A&A*, 494, 21

Beck, R. 2009, *Rev. Mex. AyA* 36, 1

Beck, R., Brandenburg, A., Moss, D., Shukurov, A., & Sokoloff, D. 1996, *ARAA* 34, 155

Brandenburg, A. & Subramanian, K. 2005, *Phys. Rep.* 417, 1

Brentjens, M. A. & de Bruyn, A. G. 2005, *A&A* 441, 1217

Gaensler, B. M., Beck, R., & Feretti, L. 2004, *New Astr. Rev.* 48, 1003

Krause, M., Alexander, P., Bolton, R., *et al.* 2009, *MNRAS*, in press

Kronberg, P. P., Bernet, M. L., Miniati, F., *et al.* 2008, *ApJ* 676, 70

Lazar, M., Schlickeiser, R., Wielebinski, R., & Poedts, S. 2009, *ApJ* 693, 1133

Rees, M. J. 2005, in: R. Wielebinski, & R. Beck (eds.), *Cosmic Magnetic Fields* (Berlin: Springer), p. 1

Ryu, D., Kang, H., Cho, J., & Das, S. 2008, *Science* 320, 909

Widrow, L. M. 2002, *Rev. Mod. Phys.* 74, 775

Highlights of Astronomy, Volume 15
XXVIIth IAU General Assembly, August 2009
Ian F. Corbett, ed.

Turbulent diffusion and galactic magnetism

Axel Brandenburg & Fabio Del Sordo

NORDITA, Roslagstullsbacken 23, SE-10691 Stockholm, Sweden; and
Department of Astronomy, Stockholm University, SE-10691 Stockholm, Sweden

Abstract. Using the test-field method for nearly irrotational turbulence driven by spherical expansion waves it is shown that the turbulent magnetic diffusivity increases with magnetic Reynolds numbers. Its value levels off at several times the rms velocity of the turbulence multiplied by the typical radius of the expansion waves. This result is discussed in the context of the galactic mean-field dynamo.

Keywords. turbulence, (magnetohydrodynamics:) MHD, galaxies: magnetic fields

---

The galactic dynamo is believed to be powered by supernova-driven turbulence. This type of forcing does not directly produce vorticity; it can only be produced indirectly through oblique shocks, i.e. through the baroclinic term. The aim of this work is to assess whether vorticity is actually important for the dynamo.

The galactic magnetic field has a strong large-scale component which is generally believed to be due to a mean-field dynamo of $\alpha\Omega$ type that is governed by the equation

$$\frac{\partial \overline{B}}{\partial t} = \nabla \times (\overline{U} \times \overline{B} + \overline{EMF} - \eta\mu_0 \overline{J}), \quad \text{where} \quad \overline{emf}_i = \alpha_{ij}\overline{B}_j - \eta_{ij}\mu_0 \overline{J}_j \qquad (0.1)$$

is the mean electromotive force, $\overline{J} = \nabla \times \overline{B}/\mu_0$ is the mean current density, $\overline{U}$ is the mean flow, and μ_0 is the vacuum permeability. In order to assess its efficiency, one needs to determine the tensors α_{ij} (the "α effect") and η_{ij} (turbulent magnetic diffusivity). Note that α_{ij} is a pseudo tensor and non-vanishing diagonal components can only be constructed from a combination of polar and axial vectors, and would therefore be vanishing in the absence of stratification and rotation. The η_{ij} tensor, on the other hand, does not require this, and it should be finite even in the completely homogeneous case. This is the case considered in the present study, which is a necessary intermediate step.

For homogeneous flows η_{ij} is an isotropic tensor, which we write as $\eta_{ij} = \eta_{\mathrm{t}}\delta_{ij}$, where η_{t} is the turbulent magnetic diffusivity. A relevant concern in mean-field theory is that turbulent transport coefficients such as η_{t} must stay finite even in the limit of large values of the magnetic Reynolds number, defined here as $R_{\mathrm{m}} = u_{\mathrm{rms}}/\eta k_{\mathrm{f}}$, where u_{rms} is the rms velocity of the turbulence, and k_{f} is the wavenumber corresponding to the scale of the energy-carrying motions. Given the importance of a possible R_{m} dependence, it is necessary to perform so-called *direct* simulations, where no subgrid scale modeling is used. This implies that we must make compromises regarding the strength of the forcing and consider only subsonic flows. Following earlier work of Mee & Brandenburg (2006) we consider a flow driven by random expansion waves of radius $R = 2/k_{\mathrm{f}}$ (not to be confused with the magnetic Reynolds number R_{m}) and determine η_{t} using the test-field method of Schrinner *et al.* (2005) in the implementation of Brandenburg (2005).

The evolution of internal energy and hence entropy is not relevant to our question about turbulent transport coefficients. Therefore we consider an isothermal equation of state where the pressure p is proportional to the density ρ with $p = \rho c_{\mathrm{s}}^2$, with c_{s} being the isothermal speed of sound. We adopt a Gaussian potential forcing function f_{f} of the form $F_{\mathrm{f}}(\boldsymbol{x}, t) = \nabla\phi$ with $\phi(\boldsymbol{x}, t) = N \exp\left\{[\boldsymbol{x} - \boldsymbol{x}_{\mathrm{f}}(t)]^2/R^2\right\}$, where $\boldsymbol{x} = (x, y, z)$ is the position vector, $\boldsymbol{x}_{\mathrm{f}}(t)$ is the random forcing position, R is the radius of the Gaussian, and N is a normalization factor. We consider a time dependence of $\boldsymbol{x}_{\mathrm{f}}$ with a forcing time $\delta t_{\mathrm{force}} \approx (u_{\mathrm{rms}}k_{\mathrm{f}})^{-1}$ that defines the interval during which $\boldsymbol{x}_{\mathrm{f}}$ remains constant. We use the PENCIL CODE (http://pencil-code.googlecode.com) which is a non-conservative, high-order, finite-difference code (sixth order in space and third order in time) for solving the compressible hydrodynamic and hydromagnetic equations.

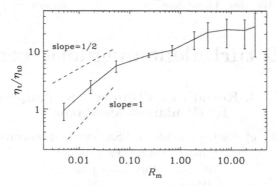

Figure 1. Dependence of η_t on $P_m = 1$.

In Fig. 1 we plot the dependence of η_t on R_m. Following earlier work of Sur *et al.* (2008) we normalize η_t by $\eta_{t0} \equiv u_{rms}/3k_f$. Note that, for low values of R_m, η_t increases proportional to R_m^n with n between 1/2 and 1. For larger value of R_m, η_t seems to levels off at a value of about 20 times η_{t0}. Expressing this in terms of u_{rms} and the typical radius R of the expansion waves, we find that $\eta_t \approx 4u_{rms}R$. Note also that η_t is always positive, in contrast to analytic predictions for irrotational turbulence using the first-order smoothing approximation (Krause & Rädler 1980).

Based on these results we can conclude that nearly irrotational turbulence is at least as efficient as vortical turbulence in diffusing mean magnetic field. Clearly, our study is still at a preliminary stage. It is important to clarify a possible dependence of our results on the microscopic magnetic Prandtl number and, in the nonlinear regime, on the magnetic field strength. Next, we need to consider the case with rotation and stratification which should then lead to an α effect, as well as turbulent pumping. This would provide an opportunity to compare with early predictions by Ferrière (1992) for this type of flows.

Acknowledgements

We acknowledge the allocation of computing resources provided by the Swedish National Allocations Committee at the Center for Parallel Computers at the Royal Institute of Technology in Stockholm and the National Supercomputer Centers in Linköping. This work was supported in part by the European Research Council under the AstroDyn Research Project 227952 and the Swedish Research Council grant 621-2007-4064.

References

Brandenburg, A. 2005, *Astron. Nachr.*, *326*, 787
Ferrière, K. 1992, *ApJ*, *389*, 286
Krause, F. & Rädler, K.-H. 1980, *Mean-field magnetohydrodynamics and dynamo theory* (Pergamon Press, Oxford)
Mee, A. J. & Brandenburg, A. 2006, *MNRAS*, *370*, 415
Schrinner, M., Rädler, K.-H., Schmitt, D., *et al.* 2005, *Astron. Nachr.*, *326*, 245
Sur, S., Brandenburg, A., & Subramanian, K. 2008, *MNRAS*, *385*, L15

Highlights of Astronomy, Volume 15
XXVIIth IAU General Assembly, August 2009
Ian F. Corbett, ed.

© International Astronomical Union 2010
doi:10.1017/S174392131001015X

2D and 3D turbulent magnetic reconnection

A. Lazarian[1], G. Kowal[2], E. Vishniac[3], K. Kulpa-Dubel[4], and K. Otmianowska-Mazur[4]

[1] University of Wisconsin-Madison, USA, `lazarian@astro.wisc.edu`
[2] University of São Paulo, Brazil
[3] McMaster University, Canada
[4] Kracow University, Poland

Abstract. A magnetic field embedded in a perfectly conducting fluid preserves its topology for all times. Although ionized astrophysical objects, like stars and galactic disks, are almost perfectly conducting, they show indications of changes in topology, *magnetic reconnection*, on dynamical time scales. Reconnection can be observed directly in the solar corona, but can also be inferred from the existence of large scale dynamo activity inside stellar interiors. Solar flares and gamma ray busts are usually associated with magnetic reconnection. Previous work has concentrated on showing how reconnection can be rapid in plasmas with very small collision rates. Here we present numerical evidence, based on three dimensional simulations, that reconnection in a turbulent fluid occurs at a speed comparable to the rms velocity of the turbulence, regardless of the value of the resistivity. In particular, this is true for turbulent pressures much weaker than the magnetic field pressure so that the magnetic field lines are only slightly bent by the turbulence. These results are consistent with the proposal by Lazarian & Vishniac (1999) that reconnection is controlled by the stochastic diffusion of magnetic field lines, which produces a broad outflow of plasma from the reconnection zone. This work implies that reconnection in a turbulent fluid typically takes place in approximately a single eddy turnover time, with broad implications for dynamo activity and particle acceleration throughout the universe. In contrast, the reconnection in 2D configurations in the presence of turbulence depends on resistivity, i.e. is slow.

Keywords. galaxies: magnetic fields — physical processes: MHD — physical processes: turbulence — methods: numerical

Reconnection in 3 dimensions versus 2 dimensions

Turbulence is ubiquitous in magnetized astrophysical fluids and it makes magnetic reconnection fast according to the model by Lazarian & Vishniac (1999, henceforth LV99). The LV99 model naturally generalizes Sweet-Parker model for the case of turbulent field lines (Fig. 1). Unlike earlier attempts to invoke turbulence into reconnection, the LV99 model provided testable predictions of how reconnection rate changes with the power and scale of the turbulence.

The testing of the model in Kowal *et al.* (2009) were found consistent with the prediction of the LV99 model. For instance, Fig. 2a shows that magnetic reconnection is independent of resistivity. Taking into account the differences of the Lundquist numbers of numerical simulations and astrophysical fluids this would not constitute a solid prove of the reconnection being fast. However, Kowal *et al.* (1999) also successfully tested the scaling of the reconnection rates predicted by LV99. This allows us to claim the success of the LV99 model of fast reconnection.

An explanatory note is also due. The LV99 model does not require turbulence being strong, i.e. it does not require strong bending of magnetic field lines by turbulence. The testing of the model in Kowal *et al.* (2009) were performed for subAlfvenic driving. At the similar circumstances, turbulence in 2D, which was the focus of the research prior to LV99 work does not show fast reconnection Fig.2b (Kulpa-Dubel *et al.* 2009). This confirms the assesment in LV99 that the turbulent reconnection is only fast in 3D.

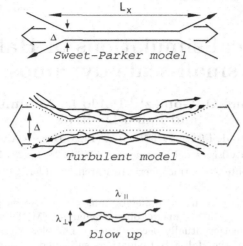

Figure 1. *Upper plot*: Sweet-Parker model of reconnection. The outflow is limited by a thin slot Δ, which is determined by Ohmic diffusivity. The other scale is an astrophysical scale $L \gg \Delta$. *Middle plot*: Reconnection of weakly stochastic magnetic field according to LV99. The model that accounts for the stochasticity of magnetic field lines. The outflow is limited by the diffusion of magnetic field lines, which depends on field line stochasticity. *Low plot*: An individual small scale reconnection region. The reconnection over small patches of magnetic field determines the local reconnection rate. The global reconnection rate is substantially larger as many independent patches come together. From Lazarian *et al.* (2004).

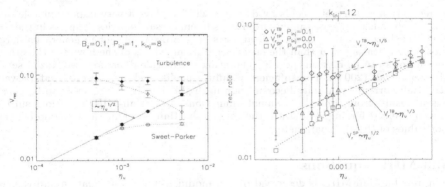

Figure 2. *Left panel*: Reconnection in 3D is fast, i.e. independent of resistivity, when turbulence is present. *Right panel*: Reconnection in 2D is slow in the presence of turbulence, i.e. it depends on resistivity. Even weak dependence of resistivity makes reconnection negligible in astrophysical consitions.

References

Kowal, G., Lazarian, A., Vishniac, E. T., & Otmianowska-Mazur, K. 2009, ApJ, 700, 63

Kulpa-Dybel, K., Kowal, G., Otmianowska-Mazur, K., Lazarian, A., & Vishniac, E. 2009, A& A, submitted, arXiv:0909.1265

Lazarian, A., & Vishniac, E. T. 1999, ApJ, 517, 700

Lazarian, A., Vishniac, E. T., & Cho, J. 2004, ApJ, 603, 180

Highlights of Astronomy, Volume 15
XXVIIth IAU General Assembly, August 2009   © International Astronomical Union 2010
Ian F. Corbett, ed.   doi:10.1017/S1743921310010161

Numerical simulations of Hall MHD small-scale dynamos

Daniel O. Gómez[1,2] Pablo D. Mininni[1,3] and Pablo Dmitruk[1]

[1]Departamento de Física, FCEN-UBA, (1428) Buenos Aires, Argentina

[2]Instituto de Astronomía y Física del Espacio, CC 67 Suc 28, (1428) Buenos Aires, Argentina

[3]National Center for Atmospheric Research, Boulder, CO 80307, USA

Abstract. Much of the progress in our understanding of dynamo mechanisms, has been made within the theoretical framework of magnetohydrodynamics (MHD). However, for sufficiently diffuse media, the Hall effect eventually becomes non-negligible. We present results from three dimensional simulations of the Hall-MHD equations subjected to random non-helical forcing. We study the role of the Hall effect in the dynamo efficiency for different values of the Hall parameter, using a pseudospectral code to achieve exponentially fast convergence.

Keywords. ISM, magnetic fields, MHD, turbulence.

1. Introduction

The generation of magnetic fields by dynamo activity plays a very important role in a wide range of astrophysical objects, ranging from stars to clusters of galaxies. The gas in these objects is characterized by turbulent flows. Helical flows have proved efficient in generating large-scale dynamos, i.e. on scales larger than the energy-containing eddies of the flow. On the other hand, non-helical flows can be instrumental in generating small-scale dynamos (see Kazantsev (1968), also Haugen *et al.* (2004)). For sufficiently diffuse media such as the one that pervades the ISM, the Hall current eventually becomes non-negligible and its effect needs to be considered. We present results from three dimensional simulations of the Hall-MHD equations subjected to random non-helical forcing. We study the role of the Hall effect in the dynamo efficiency for different values of the Hall parameter.

2. Hall-MHD equations

Incompressible Hall-MHD is described by the modified induction equation and the equation of motion, i.e. dissipative Navier-Stokes equation,

$$\frac{\partial \boldsymbol{B}}{\partial t} = \nabla \times [(\mathbf{U} - \epsilon \nabla \times \boldsymbol{B}) \times \mathbf{B}] + \eta \nabla^2 \mathbf{B} \tag{2.1}$$

$$\frac{\partial \mathbf{U}}{\partial t} = -(\mathbf{U} \cdot \nabla)\mathbf{U} + (\boldsymbol{B} \cdot \nabla)\boldsymbol{B} - \nabla \left(P + \frac{B^2}{2}\right) + \mathbf{F} + \nu \nabla^2 \mathbf{U} \tag{2.2}$$

where $\nabla \cdot \mathbf{B} = 0 = \nabla \cdot \mathbf{U}$ and $\mathbf{F}$ denotes a solenoidal and non-helical external force, delta-correlated in time. The velocity $\mathbf{U}$ and the magnetic field $\mathbf{B}$ are expressed in units of a characteristic speed U_0, ϵ measures the relative strength of the Hall effect, η is the magnetic diffusivity, and ν is the kinematic viscosity. Note that the measure of Hall effect ϵ can be written as $\epsilon = \frac{c}{\omega_{pi} L_0} \frac{U_A}{U_0}$ where L_0 is a characteristic length scale (the size of the box is $2\pi L_0$).

3. Results

We performed simulations of the HMHD equations with 256^3 spatial resolution, using a pseudospectral code - Mininni *et al.* (2005). We first generate a stationary hydrodynamic turbulence by applying a non-helical forcing. In a second stage, a random and small magnetic field is

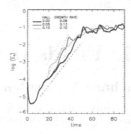

Figure 1. (a) Kinetic (thick) and magnetic (thin) dissipation rates vs. time for $\epsilon = 0.00, 0.05, 0.10$. (b) Magnetic energy vs. time, listing the corresponding growth rates.

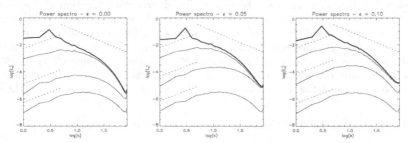

Figure 2. Total energy spectrum (thick trace) at $t = 72$. Magnetic energy spectra at $t = 18, 36, 72$ (from left to right) are also shown. The Kolmogorov and Kazantsev spectra are overlaid (dotted trace) for reference.

introduced at small scales. The viscous (thick trace) and resistive (thin) dissipation rates vs. time are shown in Fig. 1a for three runs with different values of the Hall parameter ϵ. The exponentially fast growth of magnetic energy is displayed in Fig. 1b. Even though the growth rate for moderate Hall ($\epsilon = 0.05$) is larger than from a purely MHD run (i.e. $\epsilon = 0.00$), for somewhat larger Hall ($\epsilon = 0.10$) the growth rate is reduced, confirming similar results by Mininni *et al.* (2005) for large-scale dynamos. The growth of the magnetic energy spectrum is shown in Fig. 2. The Kazantsev slope $E_k \propto k^{3/2}$ (Kazantsev (1968)) provides a reasonable approximation at small wavenumbers for all these cases, while the kinetic energy spectrum remains always close to Kolmogorov (i.e. $E_k \propto k^{-5/3}$). At saturation, the total magnetic energy reaches a sizeable fraction of the total kinetic energy (15% to 20%).

The key to understand the impact of the Hall effect on the dynamo is to study its role on the shell-to-shell energy transfer, which is discussed in Mininni *et al.* (2009) (see details in Mininni, Alexakis & Pouquet (2007)).

References

Mininni, P. D., Alexakis, A., & Pouquet, A. 2007, *J. Plasma Phys.* 73, 377
Haugen, N. E. L., Brandenburg, A., & Dobler, W. 2004, *Phys. Rev. E* 70, 016308
Mininni, P. D., Gómez, D. O., & Mahajan, S. M. 2005, *Ap.J.* 619, 1019
Kazantsev, A. P. 1968, *Sov. Phys. JETP* 26, 1031
Mininni, P. D., Dmitruk, P., & Gómez, D. O. 2009, *in prep.*

Highlights of Astronomy, Volume 15
XXVIIth IAU General Assembly, August 2009
Ian F. Corbett, ed.

Role of Magnetic Fields in Star Formation

Richard M. Crutcher

Department of Astronomy, University of Illinois, Urbana, IL 61801, USA
email: crutcher@illinois.edu

Abstract. I describe two recent projects to test star formation theory using Zeeman observations. First, using Bayesian analysis, the probability distribution function of the magnitude of the total magnetic field strength B_t and its dependence on volume density $n(H)$ were inferred from Zeeman observations of the line-of-sight strengths B_z. The result was that from one molecular cloud to another B_t ranges uniformly between values close to zero and a maximum B_0, and that B_0 scales as $n^{2/3}$. Second, observations of the *ratio* of the mass/flux (M/Φ) between the core and envelope regions of four dark clouds yielded values < 1. All of these results disagree with predictions of the strong magnetic field, ambipolar diffusion driven theory of star formation.

Keywords. ISM: magnetic fields, stars: formation, techniques: polarimetric

1. Introduction

The role of magnetic fields in star formation remains unclear. One reason is that radio Zeeman observations yield only the line-of-sight component B_z of the magnetic vector **B**. Here we very briefly describe two efforts to overcome this limitation in order to test star formation theory with observations of magnetic field strengths.

2. Bayesian analysis of Zeeman data

It is possible to infer statistical information about the total magnetic field strength B_t in a sample of interstellar clouds by making assumptions. One assumption is that the direction of **B** is random from cloud to cloud, so that the set of possible observed B_z range from zero up to the full magnitude B_t of **B**. Another assumption concerns the probability distribution function (pdf) of the magnitude of the total strength of the 3D magnetic field and its relation to the pdf of the observed B_z. Heiles & Crutcher (2005) have discussed this assumption. They considered four analytic functions to describe the pdf of B_t: a Kronecker delta function, a flat or uniform distribution, a weighted Gaussian function, and a Gaussian function. All assumed possible forms for the pdf yield mean and median values for B_z roughly equal to $0.5B_t$, so if one is only interested in inferring the approximate mean or median of B_t from a set of B_z measurements, the form of the distribution of B_t within the set of clouds does not matter very much.

However, having only approximate information about the mean or median value of B_t significantly limits our knowledge of interstellar magnetic fields. In order to overcome this limitation, Crutcher *et al.* (2010) carried out a Bayesian statistical analysis of H I, OH, and CN Zeeman surveys of diffuse and molecular clouds. Although a number of studies were carried out, here I give the results of only the most comprehensive study. The relationship between the cloud density n and the maximum magnetic field in a cloud was taken to be $B_{max}(n) = B_0$ for $n < n_0$, $B_{max}(n) = B_0 (n/n_0)^\alpha$ for $n > n_0$. Finally, the pdf of B_t was parameterized by a flat pdf with the smallest values cut off depending on the value of a parameter f; then $fB_0 < B_t < B_0$. The flat pdf over the full range 0 to B_0 is given by $f = 0$ and the delta-function pdf by $f = 1$, so this parameterization covers a wide range in possible pdf's. Results for the median values of the four parameters were: $B_0 \approx 10$ μG, $n_0 \approx 300$ cm^{-3}, $\alpha \approx 0.67$, and $f \approx 0.03$ (which is essentially the flat pdf over the full range $0 < B_t < B_0$.

3. Core/envelope mass-to-flux ratio measurements

Crutcher *et al.* (Crutcher, Hakobian & Troland (2009)) carried out OH Zeeman observations toward the envelope regions surrounding four molecular cloud cores for which detections of B_z had been achieved in the same lines, and evaluated the *ratio* of mass to magnetic flux (M/Φ) between the cloud core and envelope. This relative M/Φ measurement reduces uncertainties in previous studies, such as the angle θ between **B** and the line of sight. They calculated $\mathcal{R} \equiv [M/\Phi]_{core}/[M/\Phi]_{env} = [N_{OH}/(B_z\cos\theta)]_{core}/[N_{OH}/(B_z\cos\theta)]_{env}$. With the assumption that $\theta_{core} \approx \theta_{env}$, as predicted by idealized strong-field models, the unknown θs drop out, so $\mathcal{R}$ can be directly found from observations. The result was that for all four clouds, $\mathcal{R} < 1$. The idealized ambipolar diffusion theory of core formation requires the ratio of the central to envelope M/Φ to be approximately equal to the inverse of the original subcritical M/Φ, or $\mathcal{R} > 1$. The probability that all four of these clouds have $\mathcal{R} > 1$ is 3×10^{-7}; the results are therefore significantly in contradiction with the hypothesis that these four cores were formed by ambipolar diffusion.

Mouschovias and Tassis (unpublished) strongly criticized this result, first arguing that the apparent variation of B_z among the four positions observed for each envelope proved that the θs varied greatly due to large-scale bending of field lines. Hence, ignoring the putative variation in the measured B_z led to too small an uncertainty in $\mathcal{R}$. However, the observed variation in the B_z was at the $2 - 3\sigma$ level – that is, statistically insignificant. Second, they carried out their own analysis of the data incorporating the putative variations in $(B_z)_{env}$, which led to larger uncertainties in the $\mathcal{R}$ such that $\mathcal{R} > 1$ was not excluded. However, they employed the relationship above for $\mathcal{R}$ that assumes that $\theta_{core} \approx \theta_{env}$ so the $\cos\theta$s divide out. Use of the expression for $\mathcal{R}$ without the individual θs while assuming that the θs vary widely from position to position is internally inconsistent. For both reasons, their criticism is therefore invalid.

4. Conclusions

The Bayesian analysis found that the distribution of the B_t ranges from very small values up to a B_{max} that scales with density approximately as $n^{2/3}$. The power-law scaling is not consistent with a strong magnetic field model, but rather with one in which magnetic energy is much weaker than gravity (highly supercritical). Also, the fact that a significant population of molecular clouds must have very small magnetic field strengths implies that for many molecular clouds, magnetic fields do not dominate. The relative M/Φ experimental results were not consistent with the "idealized" strong field, ambipolar diffusion theory of star formation. The conclusion is that the role of magnetic fields in star formation is complicated and diverse, with much work remaining to be done.

References

Crutcher, R. M., Wandelt, B., Heiles, C., Falgarone, E., & Troland, T. H. 2010, *ApJ* submitted
Crutcher, R. M., Hakobian, N., & Troland, T. H. 2009, *ApJ* 692, 844
Heiles, C. & Crutcher, R. M. 2005, *Cosmic Magnetic Fields*, Lecture Notes in Physics (Heidelberg: Springer), vol. 218, p. 23

Highlights of Astronomy, Volume 15
XXVIIth IAU General Assembly, August 2009
Ian F. Corbett, ed.

© International Astronomical Union 2010
doi:10.1017/S1743921310010185

From Magnetized Cores to Protoplanetary Disks

Susana Lizano[1] and Frank H. Shu[2]

[1] CRyA, UNAM, Apdo. Postal 72-3, 58089 Morelia, Michoacán, México
email: s.lizano@crya.unam.mx

[2] Physics Department, University of California at San Diego, La Jolla, CA 92093, USA
email: fshu@physics.ucsd.edu

Abstract. We highlight several recent theoretical results that show how magnetic fields, with the magnitudes currently observed in molecular clouds, affect the structure and evolution of dense cores and protoplanetary disks to form stars and planets.

Keywords. magnetic fields, stars: formation, accretion: accretion disks

1. Introduction

Magnetic fields have been observed in molecular clouds with strengths large enough to give support against gravity. The relevant parameter is the mass-to-flux ratio, $\lambda = 2\pi G^{1/2} M/\Phi$, where G is the gravitational constant, M is the mass of the cloud, and Φ is the magnetic flux that permeates the cloud; $\lambda > 1$ is required for stability. Observations of magnetic fields at different gas densities show $-0.5 < \log(\lambda) < 0.5$ (e.g., Zeeman OH measurements, Troland & Crutcher 2008; CN measurements, Falgarone *et al.* 2008). Recently, beautiful high resolution submillimeter dust polarization maps obtained with the SMA show the hourglass shape predicted by models of magnetic fields dragged in during the gravitational collapse, both in the low mass source NGC1333 IRAS 4 (Girart *et al.* 2006), and in the massive star forming regions G31.41+0.31 (Girart *et al.* 2009) and W51 (Tang *et al.* 2009). In G31.41+0.31 there is also evidence of magnetic braking by a strong magnetic field. In W51, the polarization maps obtained with BIMA show a uniform, large-scale magnetic field while the SMA map shows the small-scale dragged field in the individual sources e2 and e8. Recently, Crutcher *et al.* (2009) measured the core and envelope mass-to-flux ratio of 4 low-mass star forming clouds and did not find the increase of λ in the cores predicted by ambipolar diffusion models. It is important to note that their observations used the Arecibo and Green Bank telescopes which have very low angular resolution of several arc minutes. In contrast, the dust polarization SMA observations mentioned above probe scales of hundreds to thousands of AU (angular scales of arc seconds) and find the predicted behavior.

2. Gravitational Collapse and Magnetic Field Dissipation

Ideal MHD models of the gravitational collapse of magnetized rotating clouds show that the magnetic field tends to acquire a "split monopole" geometry, with $B \sim a^3 t/(G^{1/2} r^2)$, where a is the sound speed and r is the radial coordinate (Galli *et al.* 2006). The magnetic field trapped in the central star becomes so large that it brakes the rotating infalling material such that the azimuthal velocity goes to to zero at the origin, i.e., no centrifugally supported disk is formed. This behavior was found in numerical models by Allen *et al.* (2003). Recent numerical simulations of Hennebelle & Fromang (2007) and Mellon & Li (2008) showed that centrifugal disks can only be formed in ideal MHD conditions for $\lambda > 20 - 80$. Since $\lambda < 4$ is observed in molecular clouds, field dissipation is a prerequisite for disk formation. Moreover, stars have $\lambda_* \sim 10^3 - 10^4$, thus, the magnetic field has to be dissipated by even larger amounts to form stars, which is the classical flux problem. Nevertheless, at densities $n > 10^9$ cm^{-3}, Ohmic dissipation is efficient. Shu *et al.* (2006) showed that the process of Ohmic dissipation occurs at scales of ten's of AU, i.e., disk size scales, and that enough flux can be lost during the gravitational collapse phase to form stars. Recently, Gonçalves *et al.* (2008) successfully applied this model to the submillimeter dust polarized emission from the low mass protostar NGC1333 IRAS 4A.

3. Magnetized Protoplanetary Disks

When an accretion disk forms during the gravitational collapse phase it will drag the magnetic field from the parent core. The disk will evolve subject to two diffusive processes: viscosity, ν, due to turbulent and magnetic stresses, that produces accretion towards the star and transfer of angular momentum outside, and resistivity, η, due to microscopic collisions and the magnetorotational instability (MRI), which allows matter to slip across field lines. Shu *et al.* (2007) studied the structure of steady state models of magnetized disks and found that their masses, sizes and magnetic field strengths are consistent with observations of disks around young stars. In these disks, the dragging of field lines by accretion is balanced by the outward field diffusion only if the ratio $\eta/\nu \sim A \ll 1$, where A is the disk aspect ratio. Moreover, the magnetic tension due to a poloidal magnetic field threading the disk will produce subkeplerian rotation. Subkeplerian rotation poses a problem to launch disk winds: they either have to be warm to overcome the potential barrier or they need a dynamically fast diffusion across the magnetic field lines. Nevertheless, such a large diffusion also produces sonic accretion speeds which imply too short a lifetime for the disks, less than 5000 yr (Shu *et al.* 2008). Another important effect of subkeplerian rotation in protoplanetary disks is that, at a given radius, an embedded protoplanet orbits with keplerian speed and thus, experiences a headwind from the slower gaseous disk. The resulting velocity mismatch results in energy loss from the orbit and inward migration (Adams et al. 2009). In particular, subkeplerian migration reduces the migration time and dominates over Type I migration for small planets (less than one Earth mass), and/or close orbits ($\leqslant 1$ AU).

References

Adams, F. C., Cai, M. J., & Lizano, S. 2009, ApJL, 702, L182
Allen, A., Li, Z.-Y., & Shu, F. H. 2003, ApJ, 599, 363
Troland, T. H. & Crutcher, R. M. 2008, ApJ, 680, 457
Crutcher, R. M., Hakobian, N., & Troland, T. H. 2009, ApJ, 692, 844
Falgarone, E., Troland, T. H., Crutcher, R. M., & Paubert, G. 2008, A & A, 487, 247
Galli, D., Lizano, S., Shu, F. H., & Allen, A. 2006, ApJ, 647, 374
Girart, J. M., Beltrán, M. T., Zhang, Q., Rao, R., & Estalella, R. 2009, Science, 324, 1408
Gonçalves, J., Galli, D., & Girart, J. M. 2008, A & A, 490, L39
Hennebelle, P. & Fromang, S. 2008, A & A, 477, 9
Mellon, R. R. & Li, Z.-Y. 2009, ApJ, 698, 922
Shu, F. H., Galli, D., Lizano, S., & Cai, M. 2006, ApJ, 647, 382
Shu, F. H., Galli, D., Lizano, S., Glassgold, A. E., & Diamond, P. H. 2007, ApJ, 665, 535
Shu, F. H., Lizano, S., Galli, D., Cai, M. J., & Mohanty, S. 2008, ApJL, 682, L121
Tang, Y.-W., Ho, P. T. P., Koch, P. M., Girart, J. M., Lai, S.-P., & Rao, R. 2009, ApJ, 700, 251

Highlights of Astronomy, Volume 15
XXVIIth IAU General Assembly, August 2009
Ian F. Corbett, ed.

© International Astronomical Union 2010
doi:10.1017/S1743921310010197

Intense velocity-shears and magnetic fields in diffuse molecular gas: from 10 pc to 5 mpc

Edith Falgarone[1] and Pierre Hily-Blant[2]

[1]LERMA/LRA, CNRS UMR 8112, Ecole Normale Supérieure & Observatoire de Paris, 24 rue Lhomond, 75005 Paris, France, email: edith.falgarone@ens.fr

[2]LAOG, CNRS UMR 5571, Université Joseph Fourier, BP 53, 38041 Grenoble, France
email: pierre.hilyblan@obs.ujf-grenoble.fr

Abstract. Regions of intense velocity-shears are identified on statistical grounds in nearby diffuse molecular gas: they form conspicuous thin (~ 0.03 pc) and parsec-long structures that do not bear the signatures of shocked gas. Several straight substructures, ~ 3 mpc thick, have been detected at different position-angles within one of them. Two exhibit the largest velocity-shears ever measured far from star forming regions, up to 780 km s^{-1} pc^{-1}. Their position-angles are found to be also those of 10-parsec striations in the $I(100\mu m)$ dust emission of the large scale environment. The **B** field projections, where available in these fields, are parallel both to the parsec- and to one of the milliparsec-scale shears. These findings put in relation the small-scale intermittent facet of the gas velocity field and the large scale structure of the magnetic fields.

Keywords. Turbulence, MHD, ISM:magnetic fields, ISM: kinematics and dynamics, ISM: structure, ISM: evolution

Long thought to be featureless and of minor importance in molecular cloud dynamics, diffuse molecular gas is now recognized to be an important fraction of molecular clouds mass, and to harbor intense dynamical activity and still unexplored small-scale structure. These have been progressively revealed by large scale sensitive maps, see Goldsmith *et al.* 2008, Hily-Blant & Falgarone 2009, observations of its molecular richness (see references in Snow & McCall 2006), its ubiquitous small-scale structure (Falgarone *et al.* 1998) and its supersonic turbulence. We propose that turbulent dissipation is one of the missing clues in understanding the main features of diffuse gas, and the processes at the origin of its condensation. In this context, shocks are expected to be the main drivers of turbulent dissipation. What are the observations telling us?

Magnetic field intensity: A Bayesian analysis of a large sample of Zeeman measurements (Crutcher *et al.*, submitted) shows that the **B** intensity in the diffuse medium is likely not to increase with density, up to $n = 300$ cm^{-3}, suggesting that either the gas flows along the field lines (Hennebelle *et al.* 2008, Nakamura & Li 2008) or that shocks do not increase the **B** field intensity because they enhance ambipolar diffusion (Li & Nakamura 2004). Both scenarios predict shocks perpendicular to **B**.

*Velocity-shears and **B** orientation:* At the parsec-scale, intense velocity-shears of 15 to 30 km s^{-1} pc^{-1} have been disclosed on statistical grounds in two translucent regions of the Polaris Flare and the Taurus cloud: they are a manifestation of the intermittency of turbulence, traced by the non-Gaussian tails of probability distribution functions of CO line velocity centroid increments (CVI) (Hily-Blant *et al.* 2008, Hily-Blant & Falgarone 2009). Their locations form thin (< 0.05 pc) and coherent structures that extend over > 1 pc (Fig. 1). In the Taurus field, this locus is parallel to the **B** field projection, a finding possibly in line with the alignment of striations in the ^{12}CO(1-0) line velocity centroids with the local **B** field projection in one Taurus edge (Heyer *et al.* 2008). These intense velocity-shears do not bear shock signatures (e.g. density enhancement).

At the milliparsec-scale, the pc-scale coherent structure in the Polaris Flare splits into eight straight substructures (field observed with the IRAM interferometer: rectangle in Fig. 1, Falgarone, Pety and Hily-Blant 2009). They are the sharp edges of extended CO-layers and some exhibit the highest velocity-shears ever measured in non-star forming clouds, up to 780 km s^{-1} pc^{-1}. Their position angles, measured within ± 10 °, are all different ($PA \sim 100°, 60°, 165°$), while

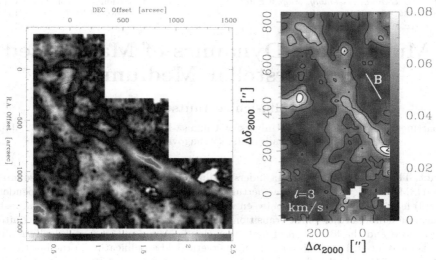

Figure 1. Extrema of ^{12}CO line CVI in the Polaris Flare *(left)* and the Taurus cloud *(right)* (from Hily-Blant *et al.* 2008, Hily-Blant & Falgarone 2009). 1 arcmin is 0.045 pc at the distance of both fields, $d = 150$ pc.

the **B** field projection measured 1° North has $PA = 108 \pm 19$ ° (Heiles 2000). Unexpectedly, the PAs of the small-scale velocity-shears are found among those of 10-parsec straight dust $I(100\mu m)$ streaks of the Polaris Flare.

The observations therefore suggest that there is a prevalence of velocity-shear and vorticity over compression in diffuse molecular gas, and that the most intense shear-layers appear to be parallel to the **B** field projections. It is remarkable that orientations of mpc-scale shears be recovered at the 10 pc-scale. These results warrant further observations both over large dynamic ranges to improve statistics and smaller scales with ALMA. They demand comparison with advanced numerical simulations.

References

Falgarone, E., Panis, J.-F., Heithausen, A., *et al.* 1998, *A&A* 331, 669
Falgarone, E., Pety, J. and Hily-Blant, P. 2009, A&A in press, arXiv0910.1766F
Goldsmith, P. F., Heyer, M., Narayanan, G., *et al.* 2008, *ApJ* 680, 428
Heiles C. 2000, *AJ* 119 923
Hennebelle, P., Banerjee, R., Vázquez-Semadeni, E., *et al.* 2008, *ApJ* 486, L43
Heyer, M., Gong, H., Ostriker, E., & Brunt, C. 2008, *ApJ* 680, 420
Hily-Blant, P. & Falgarone, E. 2007, *A&A* 469, 173
Hily-Blant, P. & Falgarone, E. 2009, *A&A* 500, L29
Hily-Blant, P., Falgarone, E., & Pety, J. 2008, *A&A* 481, 367
Li, Z.-Y. & Nakamura, F. 2004 *ApJ* 609, L83
Nakamura, F. & Li, Z.-Y. 2008, *ApJ* 687, 354
Snow, T. P. & McCall, B. J. 2006 *AARA* 44, 367

Highlights of Astronomy, Volume 15
XXVIIth IAU General Assembly, August 2009
Ian F. Corbett, ed.

© International Astronomical Union 2010
doi:10.1017/S1743921310010203

Multi-Phase Dynamics of Magnetized Interstellar Medium

Shu-ichiro Inutsuka

Department of Physics, Nagoya University, Chikusa-ku, Nagoya, Aichi 464-8602, Japan
email: inutsuka@nagoya-u.jp

Abstract. The recent progress in our understanding of the dynamics of muliti-phase interstellar medium (ISM) is reviewed. Non-linear perturbations (e.g., shock waves or time-dependent radiation field) lead to the interchange between warm phase and cold phase via thermal instability. Dynamical modelling of this phase transition dynamics is essential in describing ubiquitous turbulence in ISM and the formation of molecular clouds. A concept of magnetically multi-phase medium is introduced. Recent finding of the magnetic field amplification in the blast wave propagating in magnetized multi-phase ISM is providing a strong motivation for rapid acceleration of cosmic rays.

Our understanding on the physical processes in the transition between warm neutral medium (WNM) and cold neutral medium (CNM) is substantially increased in the last decade. The basic property of thermal instability (hereafter, TI; Field 1965) and resultant phase transition dynamics of interstellar medium (ISM) is reviewed in, e.g., Inutsuka *et al.* (2005), Hennebelle *et al.* (2007). When the ISM is swept-up by a shock wave, the gas temperature in the shock-compressed gas first increases by shock-heating but eventually decreases because of higher cooling rate in dense gas. During gas temperature is in the range from $\sim 6,000$ K to ~ 300 K, the gas is subject to thermal instability. The multi-dimensional nonlinear evolution of TI has been first studied by Koyama & Inutsuka (2002). They have found that TI generates supersonic turbulence that does not decay as long as the shock wave continues its propagation (see also, Audit & Hennebelle 2005; Hennebelle & Audit 2007; Koyama & Inutsuka 2000, 2004; Inutsuka & Koyama 2002, 2004, 2007; Inoue *et al.* 2006, 2007, 2008, 2009; Nagashima *et al.* 2005, 2006; Yamada *et al.* 2007; Heitsch *et al.* 2005; Gazol *et al.* 2005; Vazquez-Semadeni *et al.* 2006, TI in the shock propagating model is continously happening in "fresh" thermally unstable gas that is continuously provided by shock compression and heating. The motion of CNM that is supersonic with respect to the CNM sound speed but subsonic to WNM can survive shock dissipation in surrounding WNM.

Using two-dimensional two-fluid magnetohydrodynamic simulations Inoue & Inutsuka (2008, 2009) have studied the converging supersonic flows of magnetized warm neutral medium that generate a shocked slab of thermally unstable gas in which cold clouds form, and shown that the effect of magnetic field on the phase transition dynamics is crucial. They found that in the shocked slab magnetic pressure dominates thermal pressure and the thermal instability grows in the isochorically cooling, thermally unstable slab that leads to the formation of HI clouds whose number density is typically $n \lesssim 10^2 \, \mathrm{cm}^3$, even if the angle between magnetic field and converging flows is small. They also found that even if there is a large dispersion of magnetic field, evolution of the shocked slab is essentially determined by the angle between the mean magnetic field and converging flows. Thus, the direct formation of molecular clouds by piling up warm neutral medium does not seem to be a typical molecular cloud formation process, unless the direction of supersonic converging flows is biased to the orientation of mean magnetic field by some mechanism. However, when the angle is small, the H I shell generated as a result of converging flows is massive and possibly evolves to molecular clouds, provided gas in the massive H I shell is piled up again along the magnetic field line. Thus, another subsequent shock wave can again pile up the gas of the massive shell and produce a larger cloud. We thus emphasize the importance of multiple episodes of converging flows, as a typical formation process of molecular clouds. A natural consequence of this complicated history of molecular clouds is highly non-uniform spatial distribution of the ratio of magnetic pressure to gas pressure. In such configurations, large gas pressure gradients can be partially balanced by the magnetic pressure gradient, thus, maintaining high density contrast in dense clouds. More detailed analyses on the physical property of this *magnetically multi-phase medium* is clearly needed to understand the realistic evolution of ISM.

Since molecular clouds form through the contraction of HI gas, the question arises as to whether this structure is maintained in the molecular phase or not. Recently Hennebelle & Inutsuka (2006) investigated whether the warm neutral atomic hydrogen (WNM) can exist in molecular clouds. They considered the dissipation of MHD waves propagating in the WNM inside the cloud, and found that it is sufficient to allow the existence of WNM inside a molecular cloud of size $\simeq 1$ pc having pressure equal to $\simeq 10 \times P_{\rm ISM}$. This result suggests the possibility that channels of magnetized WMN may provide efficient energy injection for sustaining internal turbulence which otherwise decays in a crossing time.

Inoue, *et al.* (2009) examined magnetohydrodynamic simulations of the propagation of a strong shock wave (e.g., due to a young supernova remnant) through the interstellar two-phase medium composed of small-scale cloudlets and diffuse warm neutral medium. They have shown that the shock-compressed shell becomes turbulent owing to the preshock density inhomogeneity, and magnetic field amplification takes place in the shell. The maximum field strength is determined by the energy equi-partition condition in the post-shock region, which gives the field strength on the order of 1 mG in the case of shock velocity $\sim 10^3 {\rm kms}^{-1}$. This strongly magnetized turbulent region can be an ideal site for cosmic ray acceleration, and agrees with the spatial scale of time-dependent X-ray hot spots recently observed in young supernova remnants (Uchiyama *et al.* 2007).

References

Audit, E. & Hennebelle, P. 2005, A&A, 433, 1

Field, G. B. 1965, ApJ, 142, 531

Gazol, A. Vazquez-Semadeni, E., & Kim, J. 2005, ApJ 630, 911

Heitsch, F. *et al.* 2005, ApJ, 633, L113

Hennebelle, P. & Audit, E. 2007 A&A 465, 431

Hennebelle, P. & Inutsuka, S. 2006 ApJ 647, 404

Hennebelle, P., Mac Low, M.-M., & Vazquez-Semadeni, E. 2008 "Structure formation in the universe", Ed. G. Chabrier (Cambridge University Press)

Inoue, T. & Inutsuka, S. 2008, ApJ 687, 303 ; 2009, ApJ 704, 161

Inoue, T., Inutsuka, S., & Koyama, H. 2006, ApJ 652, 1131 ; 2007, ApJ 658, L99

Inoue, T. Yamazaki, R., & Inutsuka, S. 2009, ApJ 695, 825

Inutsuka, S., & Koyama, H. 2002, ApSS, 281, 67; 2004, RMxAC 22, 26; 2007, ASPCS, 365, 162;

Inutsuka, S., Koyama, H., & Inoue, T. 2005, AIPCP 784, 318

Koyama, H. & Inutsuka, S. 2000, ApJ, 532, 980 ; 2002, ApJ 564, L97 ; 2004, ApJ 602, L25

Nagashima, M., Koyama, H., & Inutsuka, S. 2005, MN, 361, L25 ; 2006, ApJL 652, L4

Uchiyama, Y., *et al.* 2007, Nature, 449, 576

Vazquez-Semadeni, E., *et al.* 2006, ApJ, 643, 245

Yamada, M., Koyama, H., Omukai, K., & Inutsuka, S. 2007, ApJ 657, 849

Highlights of Astronomy, Volume 15
XXVIIth IAU General Assembly, August 2009
Ian F. Corbett, ed.

© International Astronomical Union 2010
doi:10.1017/S1743921310010215

Atomic alignment: New Diagnostics of Magnetic Field in Diffuse Medium

Huirong Yan and A. Lazarian

[1] University of Arizona, 1629 E University Blvd, Tucson, 85719, USA
email: yan@lpl.arizona.edu

[2] University of Wisconsin-Madison, 475 Charter St., Madison, 53706, USA
email: alazarian@wisc.edu

Abstract. We discuss a new technique of studying magnetic fields in diffuse astrophysical media, e.g. interstellar and intergalactic gas/plasma. This technique is based on the angular momentum alignment of atoms and ions in their ground or metastable states. The alignment reveals itself in terms of the polarization of the absorbed and emitted light. The corresponding studies of magnetic fields can be performed with multiband spectropolarimetry, from UV, optical, to IR/radio. A unique feature of these studies is that they can reveal the 3D orientation of magnetic field. We mention several cases of interplanetary, circumstellar and interstellar magnetic fields for which the studies of magnetic fields using ground state atomic alignment effect are promising.

Keywords. ISM: magnetic fields, polarization, atomic processes.

1. Conditions for atomic alignment

The basic idea of the atomic alignment is quite simple. The alignment is caused by the anisotropic deposition of angular momentum from photons. In typical astrophysical situations the radiation flux is anisotropic. As the photon spin is along the direction of its propagation, we expect that atoms scattering the radiation from a light beam to be aligned (see Hawkins 1955; Varshalovich 1968). Such an alignment happens in terms of the projections of angular momentum to the direction of the incoming light. For atoms to be aligned, their ground state should have non-zero angular momentum. Therefore fine (or hyperfine) structure is necessary to enable various projection of atomic angular momentum to exist in their ground state.

Owing to the magnetic precession, the atoms with different projections of angular momentum will be mixed up. Magnetic mixing happens if the angular momentum precession rate ν_L is higher than the rate of the excitation from the ground state τ_R^{-1}, which is true for many astrophysical conditions, e.g., interplanetary medium, ISM, intergalactic medium, etc. As the result, angular momentum is redistributed among the atoms, and the alignment is altered according to the angle between the magnetic field and radiation field θ_r.

Long-lived alignable metastable states that are present for some atomic species between upper and lower states may act as proxies of ground states. Absorptions from these metastable levels can also be used as diagnostics for magnetic field therefore.

2. Range of applicability

Many species can be aligned. The corresponding lines (including both absorption and emission) can be used as the diagnostics. A number of lines with the maximum degree of polarization have been provided in YLa,b,c, Yan & Lazarian 2008b. An incomplete list of objects where effects of alignment should be accounted for arises from our studies (YLa,b,c). These include diffuse medium in the early Universe, quasars, AGNs, reflection nebulae, high and low density ISM, circumstellar regions, accretion disks and comets. One can easily add more astrophysical objects to this list. For instance, Io sodium tail can be studied the same way as sodium tail of comets (YLb).

In general, in all environment when optical pumping is fast compared with the collisional processes we expect to see effects of atomic alignment and magnetic realignment of atoms. The

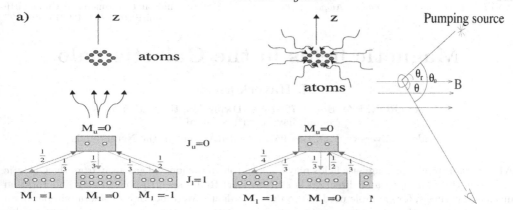

Figure 1. *Left*: A toy model to illustrate how atoms are aligned by anisotropic light. Atoms accumulate in the ground sublevel $M = 0$ as radiation removes atoms from the ground states $M = 1$ and $M = -1$; *right*: Typical astrophysical environment where the ground-state atomic alignment can happen. A pumping source deposits angular momentum to atoms in the direction of radiation and causes differential occupations on their ground states. In a magnetized medium where the Larmor precession rate ν_L is larger than the photon arrival rate τ_R^{-1}, however, atoms are realigned with respect to magnetic field. Atomic alignment is then determined by θ_r, the angle between the magnetic field and the pumping source. The polarization of scattered line also depends on the direction of line of sight, θ and θ_0. (From YLc)

wide variety of atoms with fine and hyperfine structure of levels ensures multiple ways that the information can be obtained. Comparing information obtained through different species one can get deep insights into the physics of different astrophysical objects. If the implications of atomic alignment influenced the understanding of particular features of the Solar spectrum (Landi Degl'Innocenti E. 1999, Trujillo Bueno 1999), then the studies of atomic alignment in a diffuse astrophysical media can provide much deeper and yet unforeseen changes in our understanding of a wide variety of physical processes.

As the resolution and sensitivity of telescopes increases, atomic alignment will be capable to probe the finer structure of astrophysical magnetic fields including those in the halo of accretion disks, stellar winds etc. Space-based polarimetry should provide a wide variety of species to study magnetic fields with.

References

Hawkins, W. B. 1955, Phys. Rev. 98, 478

Landi Degl'Innocenti E. 1999, in *Solar Polarization*, K. N. Nagendra, & J. O. Stenflo, eds. (Kluwer Academic Publisher), p. 61

Trujillo Bueno, J. 1999, in *Solar Polarization*, K. N. Nagendra, & J. O. Stenflo, eds. (Kluwer Academic Publisher), p. 73

Varshalovich, D. A. 1968, Astrofizika, 4, 519

Yan, H. & Lazarian, A. 2006, *ApJ*, 653, 1292 (YLa)

Yan, H. & Lazarian, A. 2007, *ApJ*, 657, 618 (YLb)

Yan, H. & Lazarian, A. 2008a, *ApJ*, 677, 1401 (YLc)

Yan, H. & Lazarian, A. 2008b, in *Magnetic Fields In The Universe II*, RevMexAA, astro-ph/0806.3703 (YLd)

Highlights of Astronomy, Volume 15
XXVIIth IAU General Assembly, August 2009
Ian F. Corbett, ed.

© International Astronomical Union 2010
doi:10.1017/S1743921310010227

Magnetic fields in the Galactic halo

M. Haverkorn[1,2]

[1] ASTRON, P.O. Box 2, 7990 AA, Dwingeloo, the Netherlands
email: haverkorn@astron.nl

[2] Leiden University, P.O. Box 9513, 2300 RA, Leiden, the Netherlands

Abstract. Interstellar magnetic fields play a major role in the ionized gas away from the Galactic disk, but their strength and direction is still unclear. Radio spectro-polarimetry and rotation measure synthesis, for example used in the Parkes Galactic Meridian Survey and a number of all (southern and northern) sky surveys, enable determination of the properties of magnetic fields in the Galactic halo and their role in the disk-halo interaction.

Keywords. magnetic fields,polarization,techniques: polarimetric,ISM: magnetic fields,ISM: general,Galaxy: halo,radio continuum: ISM

1. Introduction

As magnetic fields are frozen into ionized gas, and have comparable energy density to the (turbulent) gas and cosmic rays, they are believed to be an important part of the organization of the Galactic halo†. However, not much is known about the magnetized Galactic halo: estimates of its scale height vary widely, and there is no agreement on its parity across the Galactic plane e.g. Han *et al.* (1997), Frick *et al.* (2001). Radio spectro-polarimetry will play a large role in clarifying these issues in the near future.

2. The Parkes Galactic Meridian Survey (PGMS)

The Parkes Galactic Meridian Survey (PGMS) is a radio spectro-polarimetric survey at 2180-2420 MHz, performed with the 64m single-dish at Parkes, NSW, at about 9 arcmin resolution. The survey spans a 5° wide strip around Galactic longitude $l = 254°$, in the Galactic latitude range $0° > b > -90°$‡. The scientific goals of the PGMS are two-fold. Firstly, limits of Galactic foreground polarization, in preparation for measurements of the polarization B-mode in the CMB, are described in Carretti *et al.* (2009). Secondly, the PGMS is used to study the

---

† The term "Galactic halo" in this context means the thick disk of magnetized interstellar gas and non-thermal emission that exists away from the Galactic plane.

‡ It also includes a $10° \times 10°$ extension centered at $l = 251°$ and $b = -35°$, observed to coincide with the area observed in the BOOMERanG mission to detect Cosmic Microwave Background (CMB) polarization (Masi *et al.* (2006)).

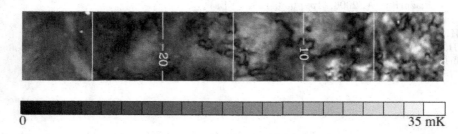

Figure 1. Polarized intensity in part of the Parkes Galactic Meridian Survey closest to the Galactic plane, i.e. in the range $251.5° < l < 256.5°$, $0° > b > -30°$.

polarization structure in the Galaxy, from the Galactic disk through the disk-halo interaction region into the halo.

Figure 1 shows polarized intensity $\sqrt{Q^2 + U^2}$ in the PGMS from the Galactic plane down to $b = -30°$. Structure in polarization, caused by differential Faraday rotation in the interstellar medium, is strong and on small scales close to the plane, and becomes weaker and varying on larger scales going away from the plane. However, degree-scale structure in polarized synchrotron radiation can still be seen on a 10 mK level all the way to the Galactic South pole (Carretti *et al.* (2009)).

Polarized intensity maps of the same region at the lower frequency of 1.4 GHz (Testori *et al.* (2008)) shows major depolarization around the Galactic plane at $b <\sim 30°$ (discussed in Wolleben *et al.* (2006)). At 2.3 GHz, however, Faraday depolarization is much less because of its strong wavelength dependence. Therefore, the PGMS enables us to study polarized emission down to much lower Galactic latitudes. Ongoing work on comparison of the two surveys will result in estimates of the strength and scale of fluctuations in rotation measure (RM) from the plane up to the Galactic halo. Also, dozens of polarized extragalactic point sources have been detected in the PGMS, allowing comparison of RM of background point sources and diffuse Galactic emission.

3. Other ongoing and future surveys

New technologies such as high spectral resolution spectro-polarimetry and new analysis methods such as RM synthesis (Burn(1966), Brentjens & de Bruyn (2005)) are enabling a great step in the study of galactic magnetism. The largest effort to realize this promise is the Global Magneto-Ionic Medium Survey (GMIMS, Wolleben *et al.* (2009)), a global project to map the diffuse polarized emission over the entire sky in six separate surveys in the approximate frequency bands of 300-800 MHz, 800-1300 MHz and 1300-1800 MHz, in both northern and southern skies. Related efforts are the S-band Polarization All-Sky Survey (S-PASS; PI Carretti) at 2.3 GHz and the Southern Twenty-cm All-sky Polarization Survey (STAPS; PI Haverkorn) at 1.4 GHz, the latter of which may form one of the GMIMS surveys. Continuous coverage of the whole sky from 300 MHz to 1.8 GHz will result in superb Faraday depth resolution in the RM synthesis analysis.

Low-frequency telescopes such as the LOw Frequency ARray (LOFAR) will expand the reach of these surveys to very low RM, i.e. very weak magnetic fields and/or old synchrotron electrons.

References

Brentjens, M. A. & de Bruyn, A. G. 2005, *A&A*, 441, 1217

Burn, B. J. 1966, *MNRAS*, 133, 67

Carretti, E., Haverkorn, M., McConnell, D., Bernardi, G., *et al.* 2009, *A&A*, submitted.

Frick, P., Stepanov, R., Shukurov, A., & Sokoloff, D. 2001, *MNRAS*, 325, 649

Han, J. L., Manchester, R. N., Berkhuijsen, E. M., & Beck, R. 1997, *A&A*, 322, 98

Masi, S., Ade, P. A. R., Bock, J. J., *et al.* 2006, *A&A*, 458, 687

Testori, J. C., Reich, W., & Reich, P. 2008, *A&A*, 484, 733

Wolleben, M., Landecker, T. L., Reich, W., & Wielebinski, R. 2006, *A&A*, 448, 411

Wolleben, M., Landecker, T. L., Carretti, E., Dickey, J. M., *et al.* 2009, *Cosmic Magnetic Fields: From Planets, to Stars and Galaxies*, IAU Symposium, vol. 259, p. 89

Highlights of Astronomy, Volume 15
XXVIIth IAU General Assembly, August 2009
Ian F. Corbett, ed.

© International Astronomical Union 2010
doi:10.1017/S1743921310010239

Large scale magnetic fields of our Galaxy

JinLin Han

National Astronomical Observatories, Chinese Academy of Sciences, Jia20, DaTun Road,
ChaoYang District, Beijing, China. email: hjl@nao.cas.cn

Abstract. Large-scale magnetic fields in the Galactic disk have been revealed by distributions of pulsar rotation measures (RMs) and Zeeman splitting data of masers in star formation regions, which have several reversals in arm and interarm regions. Magnetic fields in the Galactic halo are reflected by the antisymmetric sky distribution of RMs of extragalactic radio sources, which have azimuthal structure with reversed directions below and above the Galactic plane. Large-scale magnetic fields in the Galactic center probably have a poloidal and toroidal structure.

Keywords. ISM: magnetic fields, pulsars: general, Galaxy: structure

---

Magnetic fields of our Galaxy is fundamentally important in understanding the detected cosmic rays, star formation processes, and the origin of magnetic fields in the universe. Measuring the large-scale Galactic magnetic fields is the first and necessary step for further researches. Here I summarize the updated knowledge of large-scale magnetic fields in our Galaxy.

In the Galactic disk, the best probes for the magnetic fields are pulsars. RMs, together with dispersion measures, of pulsars near the tangential regions can directly and most efficiently to reveal the magnetic fields along the spiral arms where the line of sight is almost parallel to the large-scale fields. It has now been confirmed using the updated RM list of 1025 pulsars (Han *et al.* 2009) that magnetic fields in the spiral arms (i.e. the Norma arm, the Scutum and Crux arm, and the Sagittarius and Carina arm) are always counterclockwise in both the first and fourth quadrants, though occasionally some disordered fields appear in some segments of arms. Magnetic fields between these arms are always clockwise. Magnetic fields in arm regions and interarm regions show clear continuity in the fourth quadrant to the first quadrant. The arm-interarm field reversals are very probably related to the streaming motions of the interstellar gas (Han *et al.* 2006). We noticed that the averaged variation of RMs of extragalactic radio sources along the Galactic longitudes (Brown *et al.* 2007) are consistent with the field reversal pattern obtained from pulsar RMs (see Han 2007 for comparison).

Supposing that the large-scale magnetic fields always go along arms, pulsar RMs become less efficient to probe the magnetic fields in the regions where the lines of sight to pulsars get more perpendicular to the spiral arms. Also, distant pulsars are very faint and hard to be observed for RMs, especially in the farther half of the Galactic disk. The distribution of the median magnetic fields of OH masers in HII and star formation regions can be used as supplementary tools for large-scale magnetic fields (Han & Zhang 2007) in both the near and farther halves of the Galactic disk. Good perspective is justified by apparent coherent large-scale reversals in the distribution of magnetic fields from Zeeman splitting data of OH masers in a few kpc (see Fig. 1). More measurements of distant masers are needed to confirm this.

In the Galactic halo, large-scale magnetic fields have been qualitatively revealed by the antisymmetric RM sky of extragalactic radio sources in the inner Galaxy (Han *et al.* 1997 and 1999): Radio sources of $0° < l < 90°$ mostly have positive RMs above the Galactic plane, but negative below the plane. Radio sources of $270° < l < 360°$ mostly have positive RMs below the Galactic plane, and mostly negative above the plane. This is consistent with the large-scale toroidal fields in the halo with reversed directions above and below the Galactic plane. The antisymmetry has been recently confirmed by the RM distribution of much more radio sources (Taylor et al. 2009). However, the properties of the large-scale halo fields (e.g. field strength varies in what form and with what scale in height and radius) are not known yet.

In the Galactic center, within a few hundred pc, both poloidal and toroidal magnetic fields have been observed. See Han (2009) for a short summary and Ferriere (2009) for an extensive review, and please consult the references therein. The highly polarized non-thermal radio filaments within $1°$ from the Galactic center indicate poloidal magnetic fields in the region. The

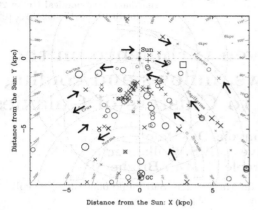

Figure 1. The distribution of the magnetic field median from Zeeman splitting measurements of OH masers (cross and circles) in 137 objects or HI or OH lines of 17 molecular clouds (plus and squares) projected onto the Galactic plane. Spiral arms and the magnetic field directions (arrows) derived from pulsar RM data are indicated. All crosses and pluses indicate the clockwise maser fields viewed from the Northern Galactic pole, and all circles and squares with counter-clockwise fields. See Han & Zhang (2007) for details.

filaments are probably illuminated flux tubes, with a field strength of about 1 mG. The diffuse radio emission of extent 400 pc indicates for a weak pervasive field of tens of μG. Polarized thermal dust emission detected in the molecular cloud zone at sub-mm wavelength is probably related to the toroidal fields parallel to the Galactic plane and complements the poloidal fields shown by the vertical filaments. It is possible that toroidal fields in the clouds are sheared from the poloidal fields, which explains the antisymmetric RM distribution of radio sources in this very central region. In a larger region, including the bulge, the large-scale magnetic fields should be closely related to the material structure, such as the bar. Observations for the magnetic fields there are very rare.

Acknowledgements

The author is supported by the National Natural Science Foundation (NNSF) of China (10821061, 10773016 and 10833003) and the National Key Basic Research Science Foundation of China (2007CB815403).

References

Brown, J. C., Haverkorn, M., Gaensler, B. M., *et al.* 2007, *ApJ*, 663, 258
Ferriere, K. 2009, A&A, in press (arXiv:0908.2037)
Han, J. L. 2007, in: *From Planets to Dark Energy: the Modern Radio Universe*, Proceedings of Science (Published online at SISSA), Pos(MRU)072
Han, J. L. 2009, IAU Symp.259, p.455 (arXiv:0901.1165)
Han, J. L., Manchester, R. N., Berkhuijsen, E. M., & Beck, R. 1997, *A&A*, 322, 98
Han, J. L., Manchester, R. N., & Qiao, G. J. 1999, *MNRAS*, 306, 371
Han, J. L., Manchester, R. N., Lyne, A. G., Qiao, G. J., & van Straten, W. 2006, *ApJ*, 642, 868
Han, J. L., van Straten, W., Manchester, R. N., & Demorest, P. B. 2009, *ApJ*, to be submitted
Han, J. L. & Zhang, J. S. 2007, *A&A*, 464, 609
Taylor, A. R., Stil, J. M., & Sunstrum, C. 2009, *ApJ*, 702, 1230

Highlights of Astronomy, Volume 15
XXVIIth IAU General Assembly, August 2009
Ian F. Corbett, ed.

© International Astronomical Union 2010
doi:10.1017/S1743921310010240

The role of SN-driven turbulence on the formation of outflows, inflows and cooling flows: from Galaxies to Clusters of Galaxies

E. M. de Gouveia Dal Pino[1],
D. Falceta-Gonçalves[2], J. S. Gallagher[3],
C. Melioli[4] A. D'Ercole[4] and F. Brighenti[4]

[1]IAG, Universidade de São Paulo, Rua do Matão 1226, São Paulo 05508-900, Brazil,
dalpino@astro.iag.usp.br

[2]NAC, Universidade Cruzeiro do Sul, Rua Galvão Bueno 868, São Paulo 01506-000, Brazil

[3]Astronomy Department, University of Wisconsin, Madison, WI, USA

[4]University of Bologna, Italy

Abstract. Star forming galaxies often exhibit hot halos with structures that resemble chimneys and fountains extending for several kpc above the galaxy. Observations indicate that they are probably produced by supernovae (SNe) which blow superbubbles that carve holes in the disk. Through these holes, high speed material is injected and expands buoyantly up to a maximum height and then returns to the disk pulled by the galaxy gravity. This circulating gas in a fountain tends to condense out forming high-velocity clouds and filaments. Starburst galaxies also show evidence that the spectacular winds that arise from their disk are fed by SNe explosions. Similarly, at galaxy cluster scales, most massive clusters exhibit rich filamentary structure of ionized gas which is distributed all around the central galaxy. We discuss here the role that SNe bubbles play in driving outflows and filamentary structures both at galaxy and galaxy-cluster scales. With the help of HD and MHD numerical simulations, we show in particular that SN-driven turbulence may play a key role at helping a central AGN halting and "isotropize" the cooling flow in the central regions of a galaxy cluster.

Keywords. supernovae, outflows, galaxies, galaxy clusters.

The role of SNe at Galactic Scales: The ejection of gas out of the disc in late-type galaxies is related to star formation and is due mainly to Type II SNe. We have recently explored the development of fountains in the Milky Way in order to understand their dynamical evolution and their influence on the redistribution of the freshly delivered metals over the disc (Melioli *et al.* 2008, 2008). We performed 3D hydrodynamical, non-equilibrium radiative cooling simulations where the whole Galaxy structure, the differential rotation and the supernova explosions generated either by a single or by multiple generations of fountains were considered. A typical fountain powered by 100 Type II SNe may eject material up to ~ 2 kpc which then collapses back mostly in the form of dense, cold clouds and filaments. To investigate the dynamical evolution of multiple generations of fountains (fueled by SNe from ~ 100 OB associations), we have considered the observed size-frequency distribution of young stellar clusters of the Galaxy. As in the case of a single fountain, multiple fountains are able to form only intermediate-velocity clouds above the disc. This indicates that the high-velocity clouds (HVCs) which have been detected at galactic altitudes higher than ~ 5 kpc are probably formed by material that has been captured from the surrounding intergalactic medium (IGM) (e.g., de Gouveia Dal Pino *et al.* 2009 and references therein). Another possibility is that they have been accelerated to higher latitudes by a cosmic-ray driven wind (e.g., Hanasz *et al.* 2009). The simulations also reveal that most of the lifted gas falls back on the disc within a radial distance $\Delta R \simeq 0.5$ kpc from the place where the fountain flow originated. This implies that the fountains do not change significantly the radial profile of the disc chemical abundance or metallicity. The simulations also allowed us to consistently calculate the feedback of the star formation on the halo gas. We found that the hot gas gains about 10% of all the SN II energy produced in the disc which suggests that the SN feedback more than compensates for the halo radiative losses. Besides, it allows a quasi steady-state disc-halo circulation after about 150 million years. We have also considered the possibility of mass infall from the IGM and its interaction with the clouds that are formed by the fountains. Though the simulations are not suitable to reproduce the slow rotational pattern that is typically observed in the haloes around disc galaxies, they indicate that the presence of an external gas infall may help to slow down the rotation of the gas in the clouds and thus the amount of angular momentum that they transfer to the coronal gas, as previously suggested in the literature.

The role of the SNe at galaxy clusters scales: Perseus (Abell 426) is commonly regarded as a prototype of the cooling core clusters of galaxies. NGC 1275, the central galaxy in the Perseus cluster, is the host of gigantic hot bipolar bubbles inflated by AGN jets observed in the radio as Perseus A. It presents a spectacular H_α-emitting nebulosity surrounding NGC 1275, with loops and filaments of gas extending to over 50 kpc. The origin of the filaments is still unknown, but probably correlates with the mechanism responsible for the giant buoyant bubbles. Motivated by the findings described in the previous session, we have performed 2.5 and 3-dimensional MHD simulations of the central regions of this cluster in which turbulent energy triggered by star formation and SNe explosions is introduced (Falceta-Gonçalves *et al.* 2009). The simulations reveal that the turbulence injected by massive stars could be responsible for the nearly isotropic distribution of filaments and loops that drag magnetic fields upward as indicated by recent observations. Weak shell-like shock fronts propagating into the ICM with velocities of 100-500 km/s are found, also resembling the observations. The isotropic outflow momentum of the turbulence slows the infall of the intracluster medium, thus limiting further starburst activity in NGC 1275. As the turbulence is subsonic over most of the simulated volume, the turbulent kinetic energy is not efficiently converted into heat and additional heating is required to suppress the cooling flow at the core of the cluster, and simulations combining the MHD turbulence with the AGN outflow can reproduce the temperature radial profile observed around NGC 1275. We conclude, therefore that, while the AGN mechanism is the main heating source, the SNe are crucial to *isotropize* the energy distribution.

Acknowledgements

EMGDP and DFG acknowledge partial support from FAPESP and CNPq.

References

de Gouveia Dal Pino, E. M., Melioli, C., D'Ercole, A., Brighenti, F. C., & Raga, A. 2009, Adv. Spa. Res., in press, arXiv:0803.3835v1

Falceta-Gonçalves, D., de Gouveia Dal Pino, E. M., Gallagher, J. S., & Lazarian, A., 2009, ApJ Lett., in press

Hanasz, M., Otmianowska-Mazur, K., Kowal, G., & Lesch, H. 2009, A&A, 498, 335

Melioli C., Brighenti, F. C., D'Ercole, A., & de Gouveia Dal Pino, E. M. 2008, MNRAS, 388, 573

Melioli C., Brighenti, F. C., D'Ercole, A., & de Gouveia Dal Pino, E. M. 2009, MNRAS, 399, 1089

Highlights of Astronomy, Volume 15
XXVIIth IAU General Assembly, August 2009
Ian F. Corbett, ed.

© International Astronomical Union 2010
doi:10.1017/S1743921310010252

Magnetic fields in dwarfs versus early-type galaxies

Krzysztof T. Chyży

Astronomical Observatory, Jagiellonian University,
ul. Orla 171, 30-244 Kraków, Poland; email: `chris@oa.uj.edu.pl`

Abstract. According to a recent systematic study of dwarf irregular galaxies the production of their magnetic fields appears to be regulated mainly by the surface density of the galactic star-formation rate. Magnetic fields in nearby dwarfs are typically weak, with the mean value of the total field strength three times smaller than in the normal spirals. Dwarfs with stronger fields reveal vivid star-forming activity, have clear signs of current or recent gravitational interactions, are more massive and evolved systems. Recently discovered strong regular fields in an early-type ringed galaxy NGC 4736 also indicates that even without spiral density waves an effective generation of strong magnetic fields is possible in any type of galaxy if only starburst characteristics are reached.

Keywords. galaxies: irregular, galaxies: dwarf, galaxies: starburst, galaxies: magnetic fields

1. Generation of magnetic fields in dwarfs

It is well known that dwarf galaxies are the most numerous species in the Universe. But they are also tiny and weak objects, rather difficult to observe, especially in the radio domain. Therefore, it is not well known whether strong magnetic fields could be easily generated in such low-mass objects. Surprisingly, in an optically bright, dwarf irregular galaxy NGC 4449, a very strong radio polarized emission was detected 2000. With the strength of the total magnetic field of about $12\,\mu G$ and of the regular component about $8\,\mu G$, it is very similar to typical massive and grand-design spiral galaxies (Beck 2005). Are such strong magnetic fields a rule or an exception among the dwarf galaxies?

The recent systematic study of radio emission and polarization in nearby dwarf galaxies indicates that their magnetic fields are predominately weak (Chyży *et al.* 2009). The mean value of the total magnetic field strength is $4\,\mu G$, three times smaller than in the normal spirals. It is found that magnetic field strength depends primarily on the surface density of the galactic star formation rate. The strongest magnetic fields in the nearby dwarfs have been detected only in highly evolved, most massive dwarfs, showing also clear effects of past interactions. Such interactions could probably trigger their star formation bursts. In such conditions of enhanced star-forming activity strong magnetic fields may effectively be produced through a small-scale dynamo process.

It seems that the high-redshift dwarfs could have a considerably higher star formation rate than the local analogs and thus could also have stronger magnetic fields or even totally regular ones (Arshakian *et al.* 2009). However, in such a case distant dwarfs would be completely different from the local ones, great majority of which show only weak magnetic fields.

2. Dwarfs versus early-type galaxies

The dependence of the magnetic field strength upon the star formation rate is observed not only for low-mass systems. It is also well visible in case of early-type galaxies. NGC 4736, for example, reveals a ringed morphology caused by the inner Lindblad resonance due to the galactic oval. The resonance causes the accumulation of gas and triggers star formation in the ring. It is the only region within the galaxy where young stars are produced. Inside and around the ring magnetic fields are exceptionally strong reaching $30\,\mu G$ (Chyży & Buta 2008). They even ignore

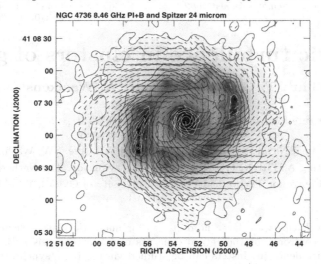

Figure 1. The radio intensity contour map of NGC 4736 at 8.46 GHz and 8.″6x8.″6 resolution with observed magnetic field vectors of the polarized intensity superimposed upon the infrared 24 µm image (from Spitzer survey of SINGS galaxies: Kennicutt *et al.* 2003). The contours are at 0.02, 0.06, 0.19, 0.30, 0.48, 0.77, 1.54 mJy/beam area.

the ring morphology and cross it at a large pitch angle forming a very coherent spiral pattern (Fig. 1).

If we compare NGC 4736 with the Sombrero galaxy, which is another early-type object, clear differences are to be seen. In the Sombrero galaxy, which produces young stars ten times slower than NGC 4736, only some weak patches of magnetized medium with strength of up to 6 µG were detected (Krause *et al.* 2006). Hence, even without spiral density waves an effective generation of strong magnetic fields is still possible in an early-type galaxy if only starburst characteristics are attained. This resembles the starbursting dwarfs, which also show strong fields, but here the starbursting ring is triggered not by gravitational interactions like in dwarfs but by the Lindblad resonance.

Acknowledgements This work was supported by the Polish Ministry of Science and Higher Education through grants: 92/N-ASTROSIM /2008/0 and 3033/B/H03/2008/35.

References

Arshakian, T. G., Beck, R., Krause, M., & Sokoloff, D. 2009, A&A 494, 21
Beck, R. 2005, In: Cosmic Magnetic Fields, eds. R. Wielebinski & R. Beck, Springer, p. 41
Chyży, K. T., Beck, R., Kohle, S., Klein, U., & Urbanik, M. 2000, A&A 355, 128
Chyży, K. T., & Buta, R. J. 2008, ApJ 677, L17
Chyży, K. T., Weżgowiec, M., Beck, R., & Bomans, D. J. 2009, A&A, to be submitted
Kennicutt, R. C., Armus, L., Bendo, G. *et al.* 2003, PASP, 115, 928
Krause, M., Wielebinski, R., & Dumke, M. 2006, A&A 448, 133

Highlights of Astronomy, Volume 15
XXVIIth IAU General Assembly, August 2009
Ian F. Corbett, ed.

© International Astronomical Union 2010
doi:10.1017/S1743921310010264

Magnetic turbulence in clusters of galaxies

T. A. Enßlin[1], T. Clarke[2], C. Vogt[3], A. Waelkens[1], and A. A. Schekochihin[4]

[1] Max-Planck-Institute for Astrophysics, Karl-Schwarzschild-Str.1, 85741 Garching, Germany.
ensslin@mpa-garching.mpg.de

[2] Interferometrics Inc., Naval Research Lab., 4555 Overlook Ave. SW, Washington, DC USA.

[3] Stichting ASTRON, P.O.Box 2,7990 AA Dwingeloo, The Netherlands.

[4] Rudolf Peierls Centre for Theoretical Physics, University of Oxford, 1 Keble Road, Oxford OX1 3NP, UK

Abstract. Galaxy clusters are large laboratories for magnetic plasma turbulence and therefore permit us to confront our theoretical concepts of magnetogenesis with detailed observations. Magnetic turbulence in clusters can be studied via the radio-synchrotron emission from the intra-cluster medium in the form of cluster radio relics and halos. The power spectrum of turbulent magnetic fields can be examined via Faraday rotation analysis of extended radio sources. In case of the Hydra A cool core, the observed magnetic spectrum can be understood in terms of a turbulence-mediated feedback loop between gas cooling and the jet activity of the central galaxy. Finally, methods to measure higher-order statistics of the magnetic field using Stokes-parameter correlations are discussed, which permit us to determine the power spectrum of the magnetic tension force. This fourth-order statistical quantity offers a way to discriminate between different magnetic turbulence scenarios and different field structures using radio polarimetric observations.

Keywords. galaxies: clusters: general, magnetic fields, turbulence

---

How can we compare turbulence models to observations?

Clusters of galaxies are excellent laboratories for studying turbulent magnetic fields via polarized radio synchrotron emission of relativistic electron populations, which can be found predominately in merging clusters, like Abell 2256 (Fig. 1). But also non-merging cluster exhibit turbulent magnetic fields with Kolmogorov-like power spectra, as an analysis of a Faraday rotation map of the cool core region of the Hydra A cluster shows (also Fig. 1). This turbulence is at least partly stirred by the buoyant motion of radio gas bubbles from the feedback of a central AGN into cooling cluster gas.

Also higher order statics of the magnetic fields may be measurable in future via radio observations. Fig. 2 shows two magnetic field scenarios with identical power spectra, one MHD simulation, and one solenoidal Gaussian random field. Fig. 3 shows their magnetic and tension force power spectra; the former being identical, whereas the latter discriminates the two scenarios. The error bars show how well this could be measured using the Stokes correlators method developed by Waelkens *et al.* (2009).

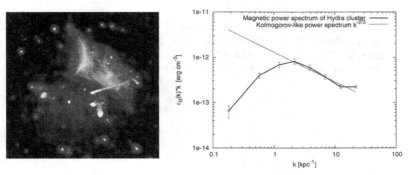

Figure 1. Left: Diffuse radio emission of Abell 2256. A Mpc-sized radio relic marks the probable location of a cluster merger shock wave (top-right structure), and a radio halo indicates the existence of turbulent magnetic fields in the cluster center (roundish central object). From Clarke & Enßlin (2006) **Right:** The magnetic power spectra of the cool core of the Hydra cluster by Vogt & Enßlin (2005).

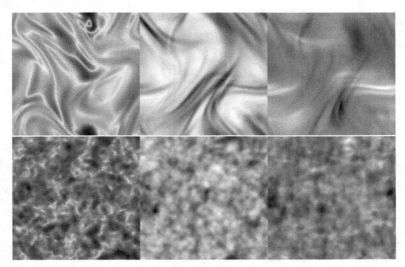

Figure 2. MHD Simulation (**top**, Schekochihin et al. (2004)) and Gaussian random (**bottom**). **Left:** Magnetic energy density. **Middle:** Total synchrotron intensity **Right:** Stokes Q synchrotron polarization.

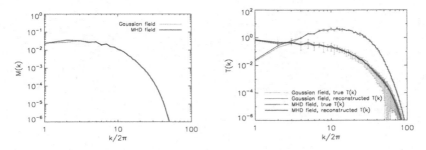

Figure 3. The magnetic power spectra of the two cases are similar (**left**), although their magnetic tension force power spectrum differ substantially, as revealed by Stokes-correlators (**right**).

References

Clarke, T. E. & Enßlin, T. A. 2006, AJ, 131, 2900
Schekochihin, A. A., Cowley, S. C., Taylor *et al.* 2004, ApJ, 612, 276
Vogt, C. & Enßlin, T. A. 2005, A&A, 434, 67
Waelkens, A., Schekochihin, A. A., & Enßlin, T. A. 2009, MNRAS398, 1970

Highlights of Astronomy, Volume 15
XXVIIth IAU General Assembly, August 2009
Ian F. Corbett, ed.
© International Astronomical Union 2010
doi:10.1017/S1743921310010276

Cosmic rays in magnetized intracluster plasma

L. Feretti[1], A. Bonafede[1], G. Giovannini[1,2], F. Govoni[3] & M. Murgia[3]

[1] Istituto di Radioastronomia INAF, Via Gobetti 101, 40129 Bologna, Italy

[2] Dipart. Astronomia, Univ. Bologna, Via Ranzani 1, 40127 Bologna, Italy

[3] Oss. Astronomico Cagliari, Loc. Poggio dei Pini, Strada 54, 09012 Capoterra (CA), Italy

1. Magnetic Fields in Clusters of Galaxies

A breakthrough in the studies of magnetic fields in clusters of galaxies has been reached in recent years from the analysis of the Rotation Measure of sources seen through the magnetized cluster medium (Govoni & Feretti 2004). The results obtained can be summarized as follows: (i) magnetic fields are present in all clusters; (ii) at the center of clusters undergoing merger activity the field strenght is around 1 μG, whereas at the center of relaxed cooling core clusters the intensity is much higher ($\sim$ 10 μG); (iii) a model involving a single magnetic field coherence scale is not suitable to describe the observational data, because of different scales of field ordering and tangling.

Assuming a magnetic field power spectrum: $|B_\kappa|^2 \propto \kappa^{-n}$ (Murgia *et al.* 2004), the range of spatial scales is found between $30 - 500$ kpc and the spectral index n is in the range $2 - 4$ (note that in A2255, Govoni *et al.* 2006 obtain a flatter index at the center, and a steeper index at the periphery, likely due to different turbulence development).

In addition, the cluster magnetic field intensity shows a radial decline linked to the thermal gas density n_e as $B \propto n_e^x$. A trend with $x = 1/2$ is expected if the B field energy scales as the thermal energy, while $x = 2/3$ if the B field results from a frozen-in field during the cluster collapse. The values of x derived so far are in the range $0.5 - 1$.

2. Diffuse Radio Emission

Magnetic fields at the μG level in the intracluster plasma are illuminated by cosmic rays, which give rise to diffuse cluster radio emission of synchrotron origin. While magnetic fields are ubiquitous in clusters, the radio emitting electrons are currently not known to be present in all clusters, although their presence is revealed in several conditions (merging and relaxed clusters), at different cluster locations (center, periphery, intermediate distance), on very different scales (100 kpc to >Mpc), and generally is related to a high degree of magnetic field ordering. Most spectacular examples of diffuse radio emission are the giant radio halos and relics detected in merging clusters. Several giant double relics, located on opposite side with respect to the cluster center, are presently known (e.g. Bonafede *et al.* 2009, van Weeren *et al.* 2009). Radio halos of smaller size have also been detected, as well as mini-halos and small relics in cooling core clusters. New halos have been detected in A851, A1213, A1351, A1995, A2034 and A2294 (Giovannini *et al.* 2009, also Giacintucci *et al.* 2009 for A1351). Among them, the cluster A1213 is remarkable because its X-ray luminosity is about 10 times weaker than that associated with clusters hosting radio halos (as derived so far).

All diffuse radio sources have in common the very steep radio spectra, implying that the radiating particles have short lifetimes, and need to be reaccelerated.

3. Radio − X-ray connection in radio halos

The properties of radio halos are linked to the properties of the host clusters (Cassano *et al.* 2006, Giovannini *et al.* 2009), in particular: a) the radio power of both small and giant halos correlates with the cluster X-ray luminosity (i.e. gas temperature and total mass); b) the radio spectra of halos are affected by the cluster temperature, being flatter in hotter clusters, and in

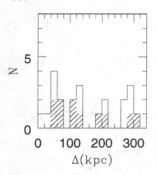 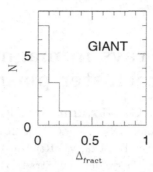 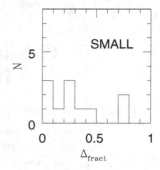

Figure 1. Left panel: Values of the offset Δ between the radio and X-ray centroids in kpc. The dashed area refers to giant halos. Middle and Right panels : Fractional offset (Δ/radio halo size) for giant halos (size $\geqslant$ 1 Mpc) and small halos (size $<$ 1 Mpc), respectively.

hotter cluster regions (Orrú *et al.* 2007); c) in a number of well-resolved clusters, a point-to-point spatial correlation is observed between the halo radio brightness and the cluster X-ray brightness (Govoni *et al.* 2001).

A step forward is to check whether the last property is common to all radio halos, i.e. whether radio halos are generally distributed as the X-ray thermal gas. Using a sample of clusters having good radio and X-ray data, we have analysed the position of the radio halo with respect to that of the X-ray gas distribution. The left panel of Fig. 1 shows that both giant and small radio halos can be significantly shifted, up to hundreds kpc, with respect to the centroid of the host cluster. To highlight radio halos with the most pronounced asymmetric distribution, we have then derived the ratio between the radio-X-ray offset and the halo size. From the middle and right panels of Fig.1, we deduce that halos can be very asymmetric with respect to the X-ray gas distribution, and this becomes more relevant when halos of smaller size are considered. A possibility is that the asymmetry in the structure originates by magnetic field fluctuations as large as hundreds of kpc, as suggested by Vacca et al. (2009) on the basis of magnetic field modeling.

References

Bonafede, A., Giovannini, G., Feretti, L., Govoni, F., & Murgia, M., 2009, A&A 494, 429
Cassano, R., Brunetti, G., & Setti, G., 2006 MNRAS 369, 1577
Giacintucci, S., Venturi, T., Cassano, R., *et al.*, 2009, ApJl in press, eprint arXiv:0909.0437
Giovannini, G., Bonafede, A., Feretti, L., *et al.*, 2009, A&A in press, eprint arXiv0909.0911
Govoni, F., Enßlin, T. A., Feretti, L., & Giovannini, G., 2001, A&A 369, 441
Govoni, F. & Feretti, L., 2004, Int. J. Mod. Phys. D, Vol., 13, 1549
Govoni, F., Murgia, M., Feretti, L., *et al.*, 2006, A&A 460, 425
Murgia, M., Govoni, F., & Feretti, L. , 2004, A&A 424, 429
Orrú, E., Murgia, M., Feretti, L., *et al.*, 2007, A&A 467, 943
Vacca V., Murgia, M., Govoni, F., *et al.*, 2009, A&A, Submitted
van Weeren, R. J., *et al.* 2009, A&A in press, eprint arXiv0908.0728

Highlights of Astronomy, Volume 15
XXVIIth IAU General Assembly, August 2009
Ian F. Corbett, ed.

© International Astronomical Union 2010
doi:10.1017/S1743921310010288

Magnetic Fields and Cosmic Rays in Galaxy Clusters

Klaus Dolag

MPI for Astrophysic, Karl-Schwarzchild-Str. 1, D-85741 Garching, Germany.

Abstract. In galaxy clusters, non-thermal components such as magnetic field and high energy particles keep a record of the processes acting since early times till now. These components play key roles by controlling transport processes inside the cluster atmosphere and beyond and therefore have to be understood in detail by means of numerical simulations. The complexity of the intra cluster medium revealed by multi-frequency observations demonstrates that a variety of physical processes are in action and must be included properly to produce accurate and realistic models. Confronting the predictions of numerical simulations with observations allows us to validate different scenarios about origin and evolution of large scale magnetic fields and to investigate their role in transport and acceleration processes of cosmic rays.

Keywords. magnetic fields, methods: numerical, galaxies: clusters: general

1. Introduction

Magnetic fields have been detected in galaxy clusters by radio observations, via the Faraday rotation signal of the magnetized cluster atmosphere towards polarized radio sources in or behind clusters (see Carilli & Taylor 2002 for a recent review) and from diffuse synchrotron emission of the cluster atmosphere (see Govoni & Feretti 2004; Ferrari *et al.* 2008, for recent reviews). However, our understanding of their origin is still very limited. Furthermore, the origin and the evolution of the population of cosmic rays within galaxy clusters are tightly connected to the dynamics of the system and to the evolution of the magnetic field. Therefore, cosmological MHD simulations are a valuable tool to investigate and distinguish different scenarios. See Dolag *et al.* 2008 for a recent review and figure 1 for a typical simulation network used to simulate cosmological, large scale structures including non-thermal effects.

2. Radio Emission

The diffuse radio emission within galaxy clusters is produced by synchrotron radiation of relativistic electrons with the cluster magnetic fields. Such diffuse emission – often refered to as giant radio haloes – is detected over regions spanning Mpc in size. One basic problem in explaining this phenomena is that the cooling time of such relativistic electrons is much shorter than their diffusion time over the region of interest. Therefore they basically have to be produced locally within the whole radio emitting region. One, often discussed mechanism to produce such relativistic electrons is the so called secondary model, where the relativistic electrons are a product by scattering of cosmic ray protons with thermal protons. Cosmic ray protons can for example be produced within accretion shocks and then advected into the cluster, or directly produced within merger shocks. Due to their larger mass compared to the cosmic ray electrons they can diffuse throughout the radio emitting region within the galaxy cluster without undergoing significant energy losses. Although the scaling between radio luminosity and temperature of the simulated clusters agree well with the observed ones, there are no indications that simulations would be able to produce the class of galaxy clusters, for which no radio emission is observed. In fact, the scatter in the predicted scaling relation is very small, as also found in previous studies (Dolag & Ensslin 2000, Miniatti *et al.* 2001, Pfrommer *et al.* 2007). Even cluster which are undergoing two major merger events lead to very elongated loops along the scaling relation which can not bridge the gap between the clusters with and without observed radio emission

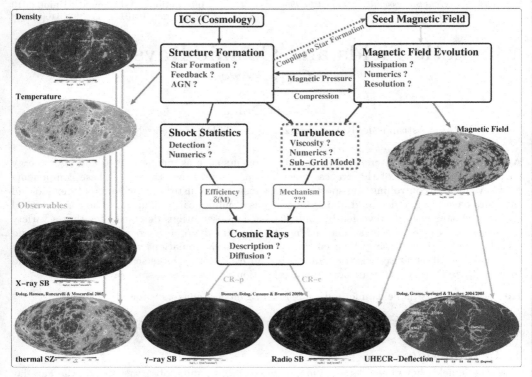

Figure 1. typical simulation network used to simulate cosmological, large scale structures including non-thermal effects.

(see Donnert *et al* 2009c). Furthermore, such models can not reproduce the observed spectrum of the radio emission for cluster like coma (see Donnert *et al.* 2009b).

3. Conclusions

The increasing amount of available radio data – both, for rotation measures as well as for diffuse radio emission – are driving our understanding of magnetic fields and cosmic rays in galaxy clusters. The improvements in the interpretations of these data over the last years are revealing a quite complex structure of the magnetic fields within galaxy clusters. Also the improvements in the numerical methods are producing more robust predictions for the magnetic field in galaxy clusters, which are helping to interpret the observations. Therefore, in the last years, a consistent picture of the magnetic fields in clusters of galaxies has been emerged from both, numerical work and observations.

Simulations of individual processes like shear flows, shock/bubble interactions or turbulence/merging events predict consistently a super-adiabatic amplification of magnetic fields within such processes. This now has been largely confirmed through direct cluster simulations within a cosmological context. It is worth mentioning that this common result is obtained by using a variety of different codes (see Dolag & Stassyszyn 2009), which are based on different numerical schemes. Within this context, various observational aspects are reproduced. Moreover, the overall amount of amplification of the magnetic field driven by the structure formation process lead to a final magnetic field strength at a level, sufficient to link models that predict magnetic field seed by various different processes with the magnetic fields observed in galaxy clusters. In fact, the imprint of structure formation onto the magnetic field within galaxy clusters is such strong, that no measurable properties of the initial magnetic seed fields remain inside galaxy clusters. Therefore the only place we can hope to still find signs of the original process of magnetization are mildly non linear regimes of structure formation like filaments (Dolag *et al.* 1999,2002/2005, Donnert *et al.* 2009a).

Such models of magnetic field in galaxy clusters allow also to constrain the origin of cosmic rays within galaxy clusters when confronted with observations of the diffuse radio emission. Although so called secondary models are able to produce sufficient radio emission, a detailed comparison shows that they fail to produce some key observational aspects. Most striking they overproduce the number of galaxy clusters which are expected to show radio emission as well as the fail to produce the observed spectral shape for the diffuse radio emission (see Dolag & Ensslin 2000, Donnert *et al.* 2009b/2009c). All this demonstrate the power of such cosmological MHD simulations to learn more about non thermal components like magnetic fields and cosmic rays within galaxy clusters and the large scale structure.

Acknowledgments

KD acknowledge supported by the DFG cluster of excellence "Origin and Structure of the Universe".

References

Carilli, C. T. & Taylor, G. B. 2002, *ARAA*, 40, 319

Clarke, T. E., Kronberg, P. P., & Boehringer, H. *ApJ*, 547, L111

Dolag, K., Bykov, A. M., & Diaferio, A. 2008, *SSRev*, 134, 311

Dolag, K., Bratelmann, M., & Lesch, H. 1999, *A&A*, 348, 351

Dolag, K., Bratelmann, M., & Lesch, H. 2002, *A&A*, 387, 383

Dolag, K. & Ensslin, T. 2000 *A&A*, 362, 151

Dolag, K., Grasso, D., Springel, V., & Tkachev, I. 2005, *Journal of Cosmology and Astro-Particle Physics*, 1, 9

Dolag, K. & Stasyszyn, F. 2009, *MNRAS*, 398, 1678

Donnert, R., Dolag K., Lesch, H., & Mueller, E. 2009, *MNRAS*, 292, 1008

Donnert, R., Dolag K., Cassano, R., & Brunetti, G., *MNRAS*, in press

Donnert, R., Dolag K., Brunetti, G., & Cassano, R., *in prep*

Ferrari, C., Govoni, F., Schindler, S., Bykov, A. M., & Rephaeli, Y. 2008, *SSRev*, 134, 93

Govoni, F. & Feretti, L. 2004, *International Journal of Modern Physics D*, 13, 1549

Miniatti, F., Johns, T. W., Kang, H., & Ryu, D. 2001, *ApJ*, 512, 233

Pfrommer, C., Ensslin, T. A., Jubelgas, M., Springel, & Dolag, K. 2007 *MNTAS*, 378, 385

Highlights of Astronomy, Volume 15
XXVIIth IAU General Assembly, August 2009
Ian F. Corbett, ed.

© International Astronomical Union 2010
doi:10.1017/S174392131001029X

Properties of MHD turbulence and its consequences for the ISM and ICM

D. Falceta-Gonçalves[1], G. Kowal[2] & A. Lazarian[3]

[1]Núcleo de Astrofísica Teórica, Universidade Cruzeiro do Sul - Rua Galvão Bueno 868, CEP
01506-000, São Paulo, Brazil
[2]Jagiellonian University, ul. Orla 171, 30-244 Kraków, Poland
[3]University of Wisconsin, Madison, 475 N. Charter St., WI 53711, USA

Abstract. It is well known that the interstellar (ISM) and intergalactic (ICM) media are
threaded by large scale magnetic fields. The understanding of its role on the dynamics of the
media is, however, still in progress. For the ISM, magnetic fields may control or, at least, play
a major role on the turbulence cascade leading to the star formation process. The ICM, on
the other hand, is assumed to be thermally dominated but still the magnetic field may play an
important role on the processes of acceleration and propagation of cosmic rays. In this work
we provide a review of the latest theoretical results on the evolution of MHD turbulence under
collisional and collisionless plasma approaches.

Keywords. hydrodynamics, turbulence, methods: numerical, ISM: magnetic fields, galaxies:
clusters: general

1. Turbulence in collisional magnetized plasmas

Turbulence is, in general, characterized by chaotic fluid motions that result in diffusion of
matter and dissipation of kinetic energy. These processes are of great importance, mostly the
understanding of how the energy cascades through different scales and gets dissipated in a
plasma. In Kolmogorov's theory, large turbulent eddies in an incompressible fluid would fragment
to form smaller cells. In this theory, it is assumed that the turbulence is homogeneous, isotropic,
scale invariant and local (i.e. waves interact with other waves with similar wavenumber k only).
Here, a single physical timescale is defined $\tau_l \sim l/\delta v_l$, the turnover time, being l the size of the
eddy and v_l its velocity. The energy transfer rate, assumed to be constant at all scales, will then
be $\dot{\epsilon} \sim \delta v_l^2/\tau_l$, and the relations $E(k) \propto k^{-5/3}$ and $\delta v_l \propto l^{1/3}$ are obtained.

The insterstellar medium (ISM), however, is known to be threaded by magnetic fields. In
this situation, magnetic perturbations play a role on the energy budget - as new wave modes
arise - of the system and change the cascading process of the turbulence. Goldreich & Sridhar
(1995) showed that if wave-wave interactions are strong in magnetized plasmas, i.e. $\tau_A = \tau_s$,
being $\tau_A = l/c_A$ with c_A the Alfven speed, the turbulence is anisotropic ($l_\parallel \propto l_\perp^{2/3}$). The cells
are elongated in the magnetic field direction. GS95 model predicted that $E(k_\perp) \propto k_\perp^{-5/3}$, i.e.
alfvenic perturbations are incompressible. Compared to the spectra of the solar wind, which
shows almost kolmogorov's cascade but small anisotropy, the answer is that the GS anisotropy
occurs at the local magnetic field reference frame. Observations, on the other hand, integrate
along the line of sight randomly oriented field lines reducing the observed anisotropy. Obsevations
of the ISM have also revealed roughly a Kolmogorov spectrum, as shown in Armostrong, Rickett
& Spangler (1995). Though it is supersonic at most scales (therefore compressible), this scaling
can be explained by alfvenic turbulence (see Cho & Lazarian 2003).

2. Turbulence in collisionless magnetized Plasmas

Magnetized and low density (weakly collisional) plasmas are known to present anisotropic
pressures with respect to the magnetic field orientation, which can last considerably long compared to the dynamical timescales of certain systems. Under certain conditions, gyrotropic plasmas give rise to new wave modes (see Passot & Sulem 2006) and instabilities depending on

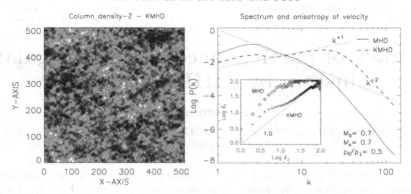

Figure 1. Column density map for the CGL-MHD model with $p_\parallel/p_\perp = 0.5$, and its spectrum of velocity compared to the collisional MHD case (Falceta-Gonçalves, Kowal & Lazarian 2009)

the pressure ratio, e.g. the firehose ($p_\parallel > p_\perp$) and mirror ($p_\perp > p_\parallel$) instabilities, which acts as additional particle accelerating mechanisms and may result in an increase in magnetic field energy. The intracluster medium (ICM) is possibly the most suitable environment for the study of gyrotropic plasma effects.

In this work we simulated the evolution of turbulence in collisionless plasmas, by solving the set of double-isothermal CGL-MHD equations.

In Fig. 1 we present the column density map and spectra of velocity obtained for the CGL-MHD model compared, for the model where $p_\parallel/p_\perp = 0.5$. The CGL-MHD instabilities are responsible for changes in the statistics of the turbulence. Here, the firehose instability is responsible for a curvature of the magnetic field lines. The curved tubes tend to slow down and trap the flowing gas. Since the growth rate is larger at small scales, we expect this effect to create more granulated maps. It is noticeable the power excess at large values of k due to the fast growth of the instabilities in these scales. The slopes of velocity and density spectra for the MHD models range between -1.7 to -2.0 at the inertial range, while CGL-MHD simulations show positive slopes ($\sim +1$) in velocity spectra, and we must point that we obtain flat density spectra ($\alpha \sim 0$).

Since ICM plasma is believed to be collisionless, we may expect these intabilities may be present. As a result, comparison with observed spectra, the increase of magnetic field energy and particle acceleration, must be studied in more details.

References

Armstrong, J. W., Rickett, B. J., & Spangler, S. R. 1995, *ApJ* 443, 209

Cho, J. & Lazarian, A. 2003, *MNRAS* 345, 325

Falceta-Gonçalves, D., Kowal, G., & Lazarian, A. 2009, *ApJ*, submitted

Goldreich, P. & Sridhar, S. 1995, *ApJ* 438, 763

Passot, T. & Sulem, P. L. 2006, *J. Geophys. Res.*, 111, 4203

Highlights of Astronomy, Volume 15
XXVIIth IAU General Assembly, August 2009
Ian F. Corbett, ed.

© International Astronomical Union 2010
doi:10.1017/S1743921310010306

Models of particle acceleration in galaxy clusters by MHD turbulence

G. Brunetti

INAF Istituto di Radioastronomia, via P. Gobetti 101, I–40129, Bologna, Italy
email: brunetti@ira.inaf.it

Abstract. Present radio data provide indirect evidence that diffuse radio emission in the central cluster regions may originate from turbulent-acceleration of relativistic particles. I was invited to discuss models of particle acceleration by MHD turbulence in clusters and in these pages I briefly touch the main points of my talk.

Keywords. acceleration of particles, radiation mechanisms: nonthermal, turbulence, galaxies: clusters: general

1. Introduction

Galaxy clusters have a key role in the hierarchy of the large scale structure as they are the largest bound objects in the Universe, where most of the gravitational energy is dissipated. The discovery of Mpc-scale diffuse synchrotron sources in a fraction of galaxy clusters demonstrates that the Inter-Galactic-Medium (IGM) is magnetised at μG level and that relativistic particles are mixed with the thermal IGM (e.g. Carilli & Taylor 2002). Synchrotron emission provides unique diagnostic of the large scale magnetic fields and relativistic particles in these environments and is a probe of the complex physical processes in the IGM (e.g. Rudnick *et al* 2009). Observations support a connection between diffuse synchrotron emission and cluster mergers, suggesting that a fraction of the energy dissipated during mergers is channelled into non-thermal components in the IGM (e.g. Enßlin *et al.* 1998; Cassano & Brunetti 2005; Pfrommer *et al.* 2006).

Giant radio halos are the most spectacular examples of non-thermal phenomena in the IGM (e.g. Ferrari *et al* 2008). These diffuse synchrotron sources are typically centered on the X-ray emitting region of galaxy clusters and extend for Mpc-scales. They can originate from secondary electrons injected during proton-proton collisions (e.g. Blasi & Colafrancesco 1999), or by relativistic electrons re-accelerated by MHD turbulence generated during cluster-cluster mergers (e.g. Brunetti *et al* 2001; Petrosian 2001).

2. Stochastic particle acceleration and evolution of radio halos

We believe that turbulence in galaxy clusters is sub–sonic but strongly super–Alfvénic and that the main source of turbulence-dissipation is the collisionless damping on thermal electrons and protons. Yet a fraction of the turbulent energy can be channelled into the re-acceleration of pre-existing relativistic particles (e.g. Brunetti & Lazarian 2007). Stochastic particle acceleration by MHD turbulence is rather inefficient in the IGM, that implies a maximum energy for the re-accelerated electrons of a few GeV, due the short electron-cooling time. This is expected to cause a steepening at higher frequencies in the spectrum of turbulence-driven radio halos that is a key-feature of the turbulent scenario and allows to provide indirect evidences in favour of this mechanism (e.g. Thierbach *et al.* 2003; Brunetti *et al.* 2008).

In this scenario the formation and evolution of giant radio halos is complex and tightly connected with the evolution of the hosting clusters (e.g. Cassano & Brunetti 2005). Present data suggest that radio halos are generated only in a fraction of massive and merging clusters and that they are "suppressed" when clusters become more dynamically relaxed, in a time-scale < 1 Gyr. This short time-scale is most likely due to the fast cooling of the emitting electrons as soon as cluster-turbulence is dissipated, although magnetic field dissipation in the emitting regions might play a role (Brunetti *et al* 2009). Calculations that consider the acceleration of relativistic particles (protons and secondary electrons due to proton-proton collisions in the IGM) by merger-driven turbulence, show that radio halos and hard X-ray emission are expected in massive galaxy clusters only in connection with merging events, while gamma ray emission, at some level, should be long-living and more common in clusters (e.g. Brunetti 2009).

3. LOFAR : unveiling radio halos in galaxy clusters

The steep spectrum of radio halos makes these sources ideal targets for observations at low radio frequencies (e.g. Enßlin & Röttgering 2002). In addition if turbulence plays a key role in accelerating relativistic particles in radio halos, very steep spectrum halos should be more common in the universe and would show up preferentially at these lower frequencies (e.g. Cassano *et al.* 2006).

The Low Frequency Array (LOFAR) promises an impressive gain of two orders of magnitude in sensitivity and angular resolution over present instruments in the frequency range 15–240 MHz, and as such will open a new observational window on the Universe. LOFAR is expected to provide a leap forward in our understanding of the origin and evolution of the relativistic matter and magnetic fields in galaxy clusters. The discovery of about 400 giant radio halos at redshift $\leqslant 0.7$ is expected from LOFAR surveys at 120 MHz, most of them with very steep spectra (Cassano *et al.* 2009).

Acknowledgements

Support through grants ASI-INAF I/088/06/0, PRIN-INAF2007 and PRIN-INAF2008 is acknowledged.

References

Blasi, P. & Colafrancesco, S., 1999, APh 12, 169
Brunetti, G., 2009, RMxAC 36, 201
Brunetti, G., Setti, G., Feretti, L., & Giovannini, G., 2001, MNRAS 320, 365
Brunetti, G. & Lazarian, A., 2007, MNRAS 378, 245
Brunetti, G., *et al.* 2008, Nature 455, 944
Brunetti, G., Cassano, R., Dolag, K., & Setti, G., 2009, A&A in press; arXiv:0909.2343
Carilli, C. L. & Taylor, G. B., 2002, ARA&A 40, 319
Cassano, R. & Brunetti, G., 2005, MNRAS 357, 1313
Cassano, R., Brunetti, G., & Setti, G., 2006, MNRAS 369, 1577
Cassano, R., Brunetti, G., Röttgering, H., & Brüggen, M., 2009, A&A in press; arXiv:0910.2025
Enßlin, T. A., Biermann, P. L., Klein, U., & Kohle, S., 1998, A&A 332, 395
Enßlin, T. & Röttgering, H., 2002, A&A 396, 83
Ferrari, C., Govoni, F., Schindler, S., Bykov, A. M., & Rephaeli, Y., 2008, SSRv 134, 93
Petrosian, V., 2001 ApJ 557, 560
Pfrommer, C., Sppringel, V., Enßlin, T. A., & Jubelgas, M., 2006, MNRAS 367, 113
Rudnick, L., *et al.* 2009, arXiv:0903.0824
Thierbach, M., Klein, U., & Wielebinski, R., 2003, A&A 397, 53

Highlights of Astronomy, Volume 15
XXVIIth IAU General Assembly, August 2009
Ian F. Corbett, ed.

NEI Modelling of the ISM - Turbulent Dissipation and Hausdorff Dimension

Miguel A. de Avillez[1] and Dieter Breitschwerdt[2]

[1] Department of Mathematics, University of Évora, 7000 Évora, Portugal
email: mavillez@galaxy.lca.uevora.pt

[2] Department of Astronomy & Astrophysics, Technical University of Berlin, D-10623 Berlin, Germany
email: breitschwerdt@astro.physik.tu-berlin.de

Abstract. High-resolution non-ideal magnetohydrodynamical simulations of the turbulent magnetized ISM, powered by supernovae types Ia and II at Galactic rate, including self-gravity and non-equilibriuim ionization (NEI), taking into account the time evolution of the ionization structure of H, He, C, N, O, Ne, Mg, Si, S and Fe, were carried out. These runs cover a wide range (from kpc to sub-parsec) of scales, providing resolution independent information on the injection scale, extended self-similarity and the fractal dmension of the most dissipative structures.

1. Introduction

In star forming disk galaxies, matter circulation between stars and the interstellar medium (ISM), in particular the energy input by supernovae, determines the dynamical and chemical evolution of the ISM, and hence of the galaxy as a whole. So far ISM models used radiative cooling assuming collisional ionization equilibrium (CIE), which is a good approximation providing the cooling time is much longer than the recombination times (e.g., Kafatos 1973). A condition verified for most of the ions for temperatures $> 10^{5.8}$ K. At lower temperatures departures from equilibrium are expected. While in CIE the ionization fractions depend only on the temperature and are sharply peaked, in NEI these same fractions depend on the dynamical and thermal history of the plasma. These departures affect the local cooling, which is a time-dependent process that controls the flow dynamics, feeding back to the thermal evolution by a change in the density and internal energy distribution, which in turn modifies the thermodynamic path of non-equilibrium cooling (Breitschwerdt & Schmutzler 1999).

2. ISM Modelling

Following Avillez & Breitschwerdt (2007) we study the evolution of the ISM in a patch of the Galaxy with $0 \leqslant (x, y) \leqslant 1$ kpc size in the Galactic plane, and $|z| \leqslant 10$ kpc perpendicular to it using non-ideal MHD equations coupled to the time-dependent calculation of the ionization structure of the plasma. The adopted model includes: (i) SNe types Ia and II occurring at the Galactic SN rate - 40% of the SNe II occur in the field (ii) gravitational field of the stellar disk, (iii) local self-gravity, (iv) heat conduction, (v) backgroung heating due to the UV photon field, (vi) mean (of 3.0μ G) and turbulent magnetic field components corresponding to a total field of 4.5 μ G, and (vii) radiative cooling calculated on the spot and using solar abundances of Asplund *et al.* (2005). The NEI calculation takes into account the time-dependent ionization structure of the 10 most important elements in nature – H, He, C, N, O, Ne, Mg, Si, S, and Fe - and includes collisional ionization by thermal electrons, autoionization, charge exchange reactions, radiative and dielectronic recombination, photoionization, comptonisation, Auger effect and ionization of H and He by photo- and Auger electrons.

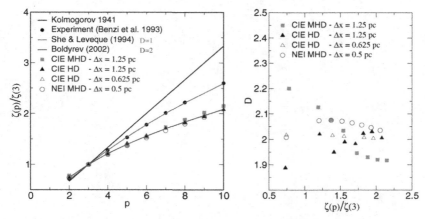

Figure 1. *Left:* The scalings $\zeta(p)/\zeta(3)$ vs. order p. Black, red and blue lines refer to Kolmogorov (1941), She-Levéque (1994) and Boldyrev (2002) models, respectively; Bullets refer to data of Benzi *et al.* (1993); Triangles (black and red) and green squares refer to CIE HD (1.25 pc and 0.625 pc resolutions) and MHD (1.25 pc resolution) runs of Avillez & Breitschwerdt (2007); Blue circles correspond to NEI MHD (0.5 pc resolution) run discussed here. *Right:* The Hausdorff dimension D of the most dissipative structures vs. $\zeta(p)/\zeta(3)$, with $p = 2, 4, ..., 10$ shown in the left panel.

3. Fractal Dimensions

We measured the velocity structure functions $\langle \delta v_l^p \rangle$ and determined the corresponding scalings $\zeta(p)/\zeta(3)$ and the Hausdorff dimension D of the most dissipative structures (Figure 1). D oscillates around 2, indicating that turbulent energy is dissipated preferentially through shocks in the HD cases and a combination of shocks and current sheets in the MHD runs. In the presence of the magnetic field (ideal and non-ideal calculations) D departures from 2 due to the anisotropy induced by the magnetic field. The NEI and CIE runs have similar fractal dimensions for the most dissipative structures and reproduce experimental and numerical results previously published.

Acknowledgements

M. A would like to thank the organizers for the invitation.

References

Asplund, M., Grevesse, N., & Sauval, A. J. 2005, *ASP Conf. Ser.* 336, 25
Avillez, M. & Breitschwerdt, D. 2007, *ApJ* (Letters) 665, 35
Benzi, R., Ciliberto, S., Tripiccione, R., Baudet, C., Massaioli, F., & Succi, S. 1993, *Phys. Rev. E* 48, R29
Boldyrev, S. 2002, *ApJ* 569, 841
Breitschwerdt, D. & Schmutzler, T. 1999. *A&A* 347, 650
Kafatos, M. 1993, *ApJ* 182, 433
Kolmogorov, A. N. 1941, *C.R. Acad. Sci. URSS* 30, 301
She, Z.-S. & Levèque, E. 1994, *Phys. Rev. Lett.* 72, 336

Highlights of Astronomy, Volume 15
XXVIIth IAU General Assembly, August 2009
Ian F. Corbett, ed.

© International Astronomical Union 2010
doi:10.1017/S174392131001032X

Whole Heliosphere Interval: Overview of JD16

David F. Webb[1],[2] Sarah E. Gibson[3] and Barbara J. Thompson[4]

[1] ISR, Boston College, Institute for Scientific Research, 140 Commonwealth Ave., Chestnut Hill, MA 02467

[2] Also at: Air Force Research Laboratory, Hanscom AFB, MA, 01731-3010, USA
email: david.webb@hanscom.af.mil

[3] HAO/NCAR, P.O. Box 3000, Boulder, CO 80301-3000, USA
email: sgibson@ucar.edu

[4] NASA/Goddard Space Flight Center, Greenbelt, MD 20771, USA
email: barbara.j.thompson@nasa.gov

Abstract. The Whole Heliosphere Interval is an international observing and modeling effort to characterize the three-dimensional interconnected solar-heliospheric-planetary system, i.e., the "heliophysical" system. WHI was part of the International Heliophysical Year, on the 50th anniversary of the International Geophysical Year, and benefited from hundreds of observatories and instruments participating in IHY activities. WHI describes the 3-D heliosphere originating from solar Carrington Rotation 2068, March 20–April 16, 2008. The focus of IAU JD16 was on analyses of observations obtained during WHI, and simulations and modeling involving those data and that period. Consideration of the WHI interval in the context of surrounding solar rotations and/or compared to last solar minimum was also encouraged. Our goal was to identify connections and commonalities between the various regions of the heliosphere.

Keywords. Sun: magnetic fields, Sun: coronal mass ejections (CMEs), interplanetary (IP) medium, (Sun:) solar-terrestrial relations

1. Characterizing the 3–D Heliosphere

JD16 began with the introductory talk of Thompson *et al.* (2009), presented by S. Gibson, which overviewed the science and status of the Whole Heliosphere Interval (WHI) effort. The overarching science goals for WHI are 1) to characterize the three-dimensional solar minimum heliosphere and 2) to connect the origins and effects of solar structure and activity through the solar wind to the Earth and other planetary systems. The challenges of incorporating a vast array of data, Sun to Earth, and comparing these to three-dimensional models, both local and global, were discussed in this talk (see figure 1) and also in the poster of Schroeder *et al.* (2009). A Sun-to-Earth mapping of WHI structures during Carrington Rotation (CR) 2068, both observed and modeled, was also presented in the poster of Gibson *et al.* (2009c).

The invited talk of Bisi *et al.* (2009a) (this issue), presented by B. Jackson, (see also Bisi *et al.* (2009b)) summarized heliospheric observations, including interplanetary scintillation (IPS) observations from the Solar-Terrestrial Environment Laboratory (STELab) in Japan, as well as in-situ solar wind observations obtained by the STEREO, ACE, and Wind spacecraft for the WHI period. The global structure of the inner heliosphere was reconstructed tomographically from IPS observations, and compared favorably to in-situ measurements. The invited talk of Manoharan (2009) (this issue) described another

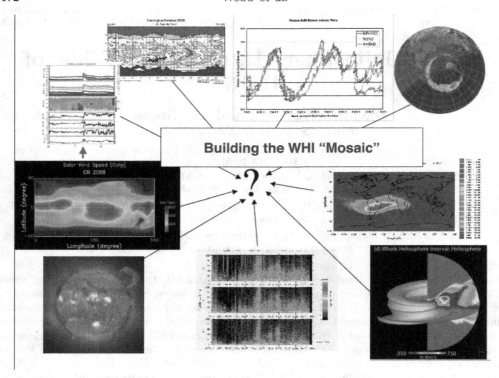

Figure 1. A fundamental and necessary challenge of WHI is combining data and models from a wide variety of sources. WHI participants have analyzed data from the solar subsurface, through the heliosphere, to the Earth's atmosphere. The above figure shows examples of the many types of data and model simulations comprising the WHI "Mosaic." Clockwise from upper right: Auroral UV emission, projected onto a map of Earth's north pole (Polar UVI), Energetic particle flux above Earth in the inner Van Allen belt (Jason-1 Doris), Modeled 3-D solar wind speed for CR 2068 (MAS Model), PC1 wave power spectrogram from Atlantic Search Coil Magnetometer at Cape Dorset, Antarctica, SOHO/EIT image of the solar chromosphere showing a giant erupting prominence, Solar wind speed for CR 2068 from IPS measurements (Ooty Radio Telescope), density, electron flux and magnetic field in a CME from IMPACT on STEREO-B, Magnetic polarity and coronal hole boundary map for CR 2068 (HelioSynoptics), and solar wind speed observed at STEREO-A, Wind, and STEREO-B, ballistically projected back to solar "origination" date.

observational analysis of the heliosphere using IPS data from the Ooty Radio Telescope, and compared the WHI period and the current solar minimum to the solar minimum of 1996. This analysis found that the band of low-speed solar wind along the equatorial belt was considerably wider during WHI than in the past minimum. The invited talk of Dal Lago *et al.* (2009) (this issue) presented cosmic ray observations from the Global Muon Detector Network (GMDN). They found that the cosmic ray gradient observed during WHI was consistent with the location of the heliospheric current sheet (HCS).

The invited talk of Riley *et al.* (2009) (this issue) presented a 3-D model of the solar corona and inner heliosphere during WHI. The MHD model included energy transport processes, estimated plasma temperature and density in the corona, and generally provided a global framework for the Sun and heliosphere during WHI. Model results were consistent with the IPS tomographic reconstruction of a wider band of variable and low-speed wind around the equator for WHI, in particular as compared to the 1996 Whole Sun Month (WSM) period. Moreover, the model showed that there was a significant

The "Active Side" of the WHI Sun: March 29, 2008

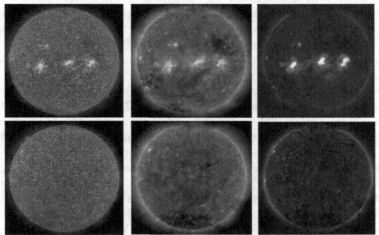

The "Quiet Side" of the WHI Sun: April 11, 2008

Figure 2. SOHO EIT 304A, 195A, and Hinode XRT soft X-ray observations of the corona during the "quiet" side of WHI (bottom) and the active side of WHI (top).

source of high-speed solar wind streams during WHI, as opposed to WSM, discussed further below.

The current solar minimum has been longer and quieter than any other observed during the Space Age, as described in an invited talk by D. Hassler. Our hope was that WHI would provide observations characteristic of this deep solar minimum, but also that the Sun would have sufficient structure and activity to make connections from Sun to Earth clear and compelling. Our hopes were realized.

For approximately the later half of the WHI period, the Sun presented a sunspot-free, deep solar minimum face (figure 2) (bottom panels). During this "quiet" side of WHI, on April 14, 2008, a NASA sounding rocket was launched from White Sands Missile Range in New Mexico, USA with a prototype EUV Variability Experiment (EVE), as described in the poster by Chamberlin & Woods (2009), presented by M. Haberreiter (see also Woods et al. (2009) and Chamberlin et al. (2009)). The main purpose of this rocket flight was to provide the fifth underflight calibration for the Solar EUV Experiment (SEE). However, of great importance to WHI and solar minimum science, this rocket flight also provided the first observations of the solar EUV irradiance at high spectral resolution during solar cycle minimum conditions. The higher spectral resolution measurements will help resolve outstanding concerns about the previous solar soft X-ray irradiance results made with broadband photometers shortward of 27 nm, and the measurements provide the most accurate reference for the 2008-2009 solar cycle minimum.

Figure 2 (top panels) shows soft X-ray and extreme ultraviolet observations of the Sun during the first half of WHI on CR 2068. Large, low-latitude coronal holes and solar active regions (ARs) are apparent. These departures from the quiet Sun seen later in the rotation led to both eruptive activity and solar wind structure that inspired many of the presentations at JD16.

Petrie & Amari (2009) presented models of the magnetic structure of the three ARs visible during WHI, i.e., NOAA 10987, 10988 and 10989, day by day as they traversed the solar disk. Daily observations from the the SOLIS Vector Spectro-magnetograph were used as boundary conditions on the nonlinear force-free magnetic field code XTRAPOL. Physical quantities such as free magnetic energy and helicity were computed with reference to the potential field of these ARs, and properties of model fields were compared with virial estimates based on the magnetograms. The interconnectivity of the the three regions was also considered in the context of global potential field models of CR 2068.

The invited talk of Leamon & McIntosh (2009a) (also Leamon & McIntosh (2009b)) demonstrated the influence of both coronal holes and ARs on the distribution of open magnetic field lines. In particular, they presented a new method of visualizing the solar photospheric magnetic field based on the "Magnetic Range of Influence" (MRoI). The MRoI is a simple realization of the magnetic environment, reflecting the distance required to balance the integrated magnetic field contained in any magnetogram pixel. Maps of MRoI allow for easy visual inspection of where sub-terrestrial field lines in a Potential Field Source Surface (PFSS) model connect to the photosphere and thus the source of Earth-directed solar wind. High MRoI can result from either a flux concentration (e.g., an AR), or from a large predominantly unipolar region (e.g., a coronal hole) which lead to very different wind conditions at 1 AU. In this study, the MRoIs during both the WHI and WSM periods were determined. In particular, the WHI period differed greatly from the WSM period in that the large, low-latitude coronal holes dominated the Sun-Earth connectivity and were the source of fast solar wind that drove responses at Earth.

2. Connecting Origins and Effects of WHI Structure and Activity

High-speed streams emerging from low-latitude coronal holes were a key characteristic of the WHI interval, and for the months both before and after WHI. Although sunspot numbers were lower than the previous minimum, long-lived regions of open magnetic flux at low latitudes meant that the Earth was periodically driven by these regions as they appeared rotation after rotation. The invited talk of Dal Lago *et al.* (2009) (this issue) described two streams during WHI that created corotating interaction regions (CIRs), with highly fluctuating magnetic fields, that had geomagnetic impact. Dense structures associated with these CIRs, as well as other transient structures, were reconstructed in the inner heliosphere from Solar Mass Ejection Imager (SMEI) data, as described in the talk of Jackson *et al.* (2009). This dense material was then compared with the CIR velocity patterns from the aforementioned IPS data. The invited talk of Maris & Maris (2009) (this issue) described the extent that the wind streams of WHI and surrounding rotations drove geomagnetic storms, and compared the high-speed streams of WHI and surrounding rotations to those seen during WSM. The talk of Gibson *et al.* (2009a) also compared the origins and impacts of high-speed streams during WHI to those of WSM, noting that the outer radiation belt observed by the GOES satellites during WHI were at levels more than 3× that of WSM, and that both the radiation belts and aurorae exhibited the periodicities of the solar wind due to the recurring high speed streams.

Most of the ∼60 CMEs or other transients during WHI occurred during the first, active half of the interval (see figure 2). A working group was organized to study these transients, both individual events and how the events fit into the global and evolutionary pattern of the solar magnetic field. The overarching goal was to trace the effects of solar structure and activity through the solar wind to Earth, other planets and spacecraft.

The discussion of activity was keynoted in an invited talk by Sterling (2009) (this issue) who gave an overview of CME observations and models. He described his examination of the lower solar atmosphere at EUV wavelengths during WHI, searching for indications of eruptive events. He used EIT data from SOHO and EUVI data from the two STEREO spacecraft, and found that, of the 3 main ARs during WHI, the most potential-like region 10989 was also the most dynamic, also confirmed by Petrie & Amari (2009).

An overview of specific projects involving data on activity during the WHI interval was given by Webb *et al.* (2009). He discussed the origins and early development of the CMEs, including data on their associated ARs and filaments and prominences, the use of multi-spectral data to study the propagation of the transients over large distances and in 3 domensions, and studies of the events in the context of the global magnetic field and the heliospheric current sheet. He discussed the use of helioseismology data from SOHO MDI (D. Haber) to determine flows as a function of depth in two areas of activity during CRs 2068 and 2069. Haber found a possible vortex area before the April 5-9 series of events, and flows before the April 26 event. These periods are being studied globally and compared with changes in H-alpha synoptic charts (with P. McIntosh).

Cremades *et al.* (2009) discussed a study to characterize all the CMEs that originated during WHI, including their source regions and ensuing IP structures, aiming toward a global description of their role in determining heliospheric conditions. They also included narrower ejective features such as "jets". Data from the STEREO and SOHO spacecraft allow the study of 3-D properties of the CMEs and multi-wavelength images from these missions and from ground-based observatories help characterize the CME source regions. The association of the Earth-directed CMEs with IP CMEs (ICMEs), including magnetic clouds, was assessed using data from the ACE spacecraft at L1. These results during WHI were then compared with those from the WSM in 1996. They found for WHI vs WSM that there were more ejecta on all size scales, less ambiguity in source identification, ejecta occurred at all position angles with jets at polar latitudes, and the solar wind was less oriented along the Parker spiral direction. Models of magnetic clouds observed during this solar cycle minimum were also discussed in the poster by Vandas & Romashets (2009).

Lisnichenko & Podladchikova (2009) described results during WHI using the continuous EUV imaging of the solar corona by the STEREO/SECCHI/EUVI and SOHO/EIT telescopes to detect key surface signatures of CMEs. These include the global EIT waves and coronal "dimming" regions. They compared the results using both visual examination of images and with a detection algorithm called NEMO (Novel EIT wave Machine Observing). They produced a catalog of WHI EIT wave and dimming events and discussed their morphology. Visually they found 10 large-scale EUV events associated with CMEs. Using SOHO data NEMO autonomously detected 40 Events, and using STEREO data detected 42 events. The WHI events had substantially smaller geometrical sizes than events observed in 1997-2005.

Energetic, fast CMEs with shocks (e.g., Type II bursts) occurred during WHI on March 25 and just after WHI on April 26, 2008 and were associated with arcades, coronal dimmings and EUV waves. The March 25 CME was a classic limb CME, and the April 26 CME originated on the disk and appeared as an ICME at STEREO-B. The March 25 event was discussed in the talks by Gopaslwamy *et al.* (2009a) (see also Gopalswamy *et al.* (2009b)) and Webb *et al.* (2009). The STEREO spacecraft were separated by about 50°, so the CME was a disk event from STEREO-B and a limb event for STEREO-A. One of the important aspects of this CME is that it was well observed by the STEREO/SECCHI inner coronagraph (COR1) when the metric type II burst was in progress, so they were able to obtain the shock height with respect to the CME.

Other details of the April 26 event were described by Jackson *et al.* (2009) and Webb *et al.* (2009). A brief IP type II suggested that this event had the lowest starting frequency ever observed, with implications for the medium through which shock propagates. At STEREO-B the Fe charge state data suggested that there was a CIR-type interface with bidirectional streaming indicating possible ejecta. SMEI density movies confirm this showing that the CME broke through the pre-existing CIR with its core passing just east and north of STEREO-B with some plasma hitting Earth.

The three WHI ARs had several transients starting on April 5 until their west limb passges several days later. A west-limb CME on April 9 was asociated with 10989 when it was over the limb and a filament to its east. The CME was slow with an erupting prominence and post-CME current sheet observed by Hinode XRT. Del Zanna (2009) reported on EUV spectral observations performed by the SOHO Coronal Diagnostic Spectrometer during the filament eruption, together with broad-band imaging observations from X-rays (Hinode/XRT) to the EUV (STEREO, TRACE). The EUV spectra were used to explain the observed signatures in the images. Surprisingly, little evidence of heating was found in these observations, including in the X-ray emission. Webb *et al.* (2009) also presented analyses of the filament, CME and current sheet from XRT observations by S. Savage and D. McKenzie, and Hinode, STEREO, and SOHO observations by Landi *et al.* (2009).

Rapid pulsations in solar flares at sub-THz, mid-IR and GHz ranges observed by the Solar Submillimeter-wave Telescope (SST) in Argentina were discussed in the talk by Kaufmann *et al.* (2009) and the poster by Cassiano *et al.* (2009). Rapid radio pulses in solar bursts have been known for many decades and are common to all events when observed with enough time resolution and sensitivity. These new observations combined with other imaging and spectral data can be used to better understand the processes in ARs that lead to flares and CMEs.

Besides the modeling of the WHI ARs by Petrie & Amari (2009) discussed earlier, two other talks dealt with modeling issues. In the talk of Stepanov, Tsap and Kopylova (2009), given by Y. Tsap, they used the ballooning instability driven by kink oscillations to model the formation of the cusp-shaped structure atop a coronal loop that can form during a CME as a result of magnetic reconnection. They compared their results to microwave and hard X-ray observations from an event on April 15, 2002. In an invited talk, Kusano *et al.* (2009) described data-driven simulations of Sun-Earth connection events which goals are to understand the physical mechanisms of the missing links in the Sun-Earth system, and to develop a physics-based model applicable for the prediction of the onset of solar eruptions.

3. Space Weather and Future Work

Kusano *et al.* (2009) also discussed the use of end-to-end modeling to predict the influence of solar eruptions on the terrestrial environment, or space weather. Their group is developing a new simulation model of the heliosphere, consisting of several sub-models of solar ARs, the corona, IP space, Earths magnetosphere, and particle acceleration. The model was applied to the geoeffective event on Dec. 13–15, 2006. The simulation was driven by Hinode vector magnetograms of the magnetic field of the flaring AR, and solar wind observations including IPS data. Future work includes better coupling of the sub-models, developing a high-accuracy interpolation between different metrics, improving the SEP model, and combining MHD and hybrid simulations. Although there were no Hinode vector field data during WHI, it was suggested that ground-based SOLIS magnetograms might be used to drive the simulation for WHI events.

In an invited talk, Raulin *et al.* (2009a) (this issue) discussed the future of the IHY and its campaigns, of which WHI was one. The IHY program took place over the 5 year period 2004–2008 and was a major international effort that involved the deployment of new instrumentation, new observations from the ground and in space, and a strong education component. Under the United Nations, instrument arrays were deployed to provide global measurements of heliophysical phenomena. The UN is planning to continue this effort aimed at understanding the impacts of Space Weather on Earth and the near-Earth environment as a new program, the International Space Weather Initiative (ISWI). In a poster Raulin *et al.* (2009b) described one of the UN IHY programs called SAVNET. This is the South America VLF Network which consists of seven tracking receivers spread over Latin America. The SAVNET array is monitoring anomalies of subionospheric propagating VLF waves that reveal changes of the electrical properties of the low ionospheric D- and E-regions. Results were shown of SAVNET's capability to monitor solar activity on short, minutes to hours, and long, solar cycle, time scales.

Schuch *et al.* (2009) and Kemmerich *et al.* (2009) discussed the use of ground-based muon detectors to measure the high energy galactic cosmic ray intensity and its modulation by the magnetic field in transients such as CMEs and CIRs. The limited observations during WHI were discussed. Schuch *et al.* (2009) proposed to combine an existing 6-country network of global muon multi-directional detectors telescopes, called the GMDN, with STEREO observations to develop a new method to track in 3-D solar-heliospheric structures. The reduction in cosmic ray counts at Earth's surface caused by ICMEs and their shocks traveling in IP space can be detected as much as ten hours in advance with the GMDN, permitting more accurate and reliable Space Weather forecasts.

Finally, Gavryuseva (2009) discussed a study of the relationships between the solar and IP magnetic fields, the solar wind characteristics near Earth and geomagnetic perturbations using Wilcox Solar Observatory observations of the large-scale solar magnetic field and OMNI solar wind data over 30 years and 3 solar cycles. The goal was to search for variations of the latitudinal and longitudinal structure of the photospheric field to make long-term predictions of solar activity leading to space weather events. The temporal dependence of the differential rotational rate was determined as a function of latitude and a quasi-stable longitudinal structure persisted over the 30-year period.

Acknowledgements

We thank the following who were either members of the SOC for JD16 or reviewed the papers in this section: D. Banerjee, M. Bisi, A. Breen, H. Cremades, N. Crosby, J. Davila, R. Forsyth, A. Galvin, K. Georgieva, J. Kozyra, I. Mann, P. Manoharan, G. Petrie, G. Poletto, K. Shibata and R. Stamper. The NASA STEREO mission consists of 4 primary suites of remote sensing and in situ instruments on two spacecraft and comprises multiple instruments developed by many worldwide groups. Hinode is a Japanese mission developed and launched by ISAS/JAXA, with NAOJ as domestic partner and NASA and STFC (UK) as international partners, and in co-operation with ESA and NSC (Norway). SOHO is a project of international collaboration between ESA and NASA. SMEI is a collaborative project of the US Air Force Research Laboratory, NASA, the University of California at San Diego, the University of Birmingham, UK, Boston College, and Boston University.

References

Bisi, M. M., Jackson, B. V., Clover, J. M., Hick, P. P., Buffington, A., & Tokumaru, M. 2009a, *Highlights of Astronomy, XXVIIth IAU General Assembly, August 2009, Ian F. Corbett, ed. (this issue)* 15

Bisi, M. M., Jackson, B. V., Buffington, A., Clover, J. M., Hick, P. P., & Tokumaru, M. 2009b, *Solar Phys.* 256, 201

Cassiano, M. M. *et al.* 2009, *XXVII IAU General Assembly Abstract Book* 391

Chamberlin, P. C. & Woods, T. N. 2009, *XXVII IAU General Assembly Abstract Book* 387

Chamberlin, P. C., Woods, T. N., Crotser, D. A., Eparvier, F. G., Hock, R. A., & Woodraska, D. L. 2009, *Geophys. Res. Lett.*, 36, 5, CiteID L05102

Cremades, H., S. Dasso, & C. H. Mandrini, 2009 *XXVII IAU General Assembly Abstract Book* 387

Dal Lago, A. *et al.* 2009, *Highlights of Astronomy, XXVIIth IAU General Assembly, August 2009, Ian F. Corbett, ed. (this issue)* 15

Del Zanna, G. 2009, *XXVII IAU General Assembly Abstract Book* 386

Gavryuseva, E. A. 2009, *XXVII IAU General Assembly Abstract Book* 387

Gibson, S. E., Kozyra, J. U., De Toma, G., Emery, B. A., Onsager, T., & Thompson, B. J. 2009a, *XXVII IAU General Assembly Abstract Book* 386

Gibson, S. E., Kozyra, J. U., De Toma, G., Emery, B. A., Onsager, T., & Thompson, B. J. 2009b, *J. Geophys. Res.* 114, A09105, doi:10.1029/2009JA014342

Gibson, S. E. *et al.* 2009c, *XXVII IAU General Assembly Abstract Book* 392

Gopalswamy, N. *et al.* 2009a, *XXVII IAU General Assembly Abstract Book* 390

Gopalswamy, N. *et al.* 2009b, *Solar Phys*, 259: 227254

Jackson, B. V. *et al.* 2009, *XXVII IAU General Assembly Abstract Book* 386

Kaufmann, P., C. G. Gimnez de Castro, E. Correia, J. E. R. Costa, J.-P. Raulin, & A. S. Vlio 2009, *XXVII IAU General Assembly Abstract Book* 391

Kemmerich, N. *et al.* 2009, *XXVII IAU General Assembly Abstract Book* 392

Kusano, K. *et al.* 2009, *XXVII IAU General Assembly Abstract Book* 390

Landi, E., J. C. Raymond, M. P. Miralles, & H. Hara 2009, *Astrophys. J.*, submitted

Leamon, R. J. & McIntosh, S. W. 2009a, *XXVII IAU General Assembly Abstract Book* 387

Leamon, R. J. & McIntosh, S. W. 2009b, *Astrophys. J. Lett* 697, L28

Lisnichenko, P. and O. Podladchikova 2009, *XXVII IAU General Assembly Abstract Book* 387

Manoharan, P. K. 2009, *Highlights of Astronomy, XXVIIth IAU General Assembly, August 2009, Ian F. Corbett, ed. (this issue)* 15

Maris, G. & Maris, O. 2009, *Highlights of Astronomy, XXVIIth IAU General Assembly, August 2009, Ian F. Corbett, ed. (this issue)* 15

Petrie, G. & Amari, T. 2009, *XXVII IAU General Assembly Abstract Book* 386

Raulin, J.-P., J. M. Davila, T. Bogdan, K. Yumoto, & J. Leibacher 2009a, *XXVII IAU General Assembly, August 2009, Ian F. Corbett, ed. (this issue)* 15

Raulin, J.-P., F. Bertoni, & H. Rivero 2009b, *XXVII IAU General Assembly Abstract Book* 389

Riley, P., Linker, J. A., & Mikic, Z. 2009, *Highlights of Astronomy, XXVIIth IAU General Assembly, August 2009, Ian F. Corbett, ed. (this issue)* 15

Schroeder, P., Thompson, B., & Gibson, S. 2009, *XXVII IAU General Assembly Abstract Book* 392

Schuch, N. J. *et al.* 2009, *XXVII IAU General Assembly Abstract Book* 388

Stepanov, A. V., Y. T. Tsap, & Y. G. Kopylova 2009, *XXVII IAU General Assembly Abstract Book* 387

Sterling, A. C. 2009, *XXVII IAU General Assembly, August 2009, Ian F. Corbett, ed. (this issue)* 15

Thompson, B. J. *et al.* 2009, *XXVII IAU General Assembly Abstract Book* 390

Vandas, M. & E. Romashets 2009, *XXVII IAU General Assembly Abstract Book* 391

Webb, D. F. *et al.* 2009, *XXVII IAU General Assembly Abstract Book* 389

Woods, T. N. *et al.* 2009, *Geophys. Res. Lett.*, 36, 1, CiteID L01101

JD16 ORGANIZING COMMITTEE

Barbara J. Thompson (Chair, USA)   Sarah E. Gibson (Co-chair, USA)
David F. Webb (Co-chair, USA)   Dipankar P.K. Banerjee (India)
Andrew R. Breen (UK)   Hebe Cremades (Argentina)
Norma B. Crosby (Belgium)   Robert J. Forsyth (UK)
Antoinette B. Galvin (USA)   Katya Y. Georgieva (Bulgaria)
Janet U. Kozyra (USA)   Ian R. Mann (Canada)
Giannina Poletto (Italy)   Kazunari Shibata (Japan)
Richard Stamper (UK)

Acknowledgements

This joint discussion was coordinated through IAU Division II (Sun and Heliosphere) and sponsored and supported by IAU Division X (Radio Astronomy), and by IAU Commissions 10 (Solar Activity), 12 (Solar Radiation and Structure) and 49 (Interplanetary Plasma and Heliosphere).

Funding support by the
International Astronomical Union,
International Heliophysical Year
and
National Aeronautics and Space Administration
is gratefully acknowledged.

Highlights of Astronomy, Volume 15
XXVIIth IAU General Assembly, August 2009   © International Astronomical Union 2010
Ian F. Corbett, ed.   doi:10.1017/S1743921310010331

A Summary of 3-D Reconstructions of the Whole Heliosphere Interval and Comparison with in-Ecliptic Solar Wind Measurements from STEREO, ACE, and Wind Instrumentation

Mario M. Bisi[1] B. V. Jackson[1], J. M. Clover[1], P. P. Hick[1,2], A. Buffington[1] and M. Tokumaru[3]

[1] Center for Astrophysics and Space Sciences, University of California, San Diego,
9500 Gilman Drive #0424, La Jolla, CA 92093-0424, USA
email: mmbisi@ucsd.edu or Mario.Bisi@gmail.com

[2] San Diego Supercomputer Center, University of California, San Diego,
9500 Gilman Drive #0505, La Jolla, CA 92093-0505, USA

[3] Solar-Terrestrial Environment Laboratory (STELab), Nagoya University,
Furo-cho, Chikusa-ku, Nagoya 464-8601, Japan

Abstract. We present a summary of results from simultaneous Solar-Terrestrial Environment Laboratory (STELab) Interplanetary Scintillation (IPS), STEREO, ACE, and Wind observations using three-dimensional reconstructions of the Whole Heliosphere Interval – Carrington rotation 2068. This is part of the world-wide IPS community's International Heliosphysical Year (IHY) collaboration. We show the global structure of the inner heliosphere and how our 3-D reconstructions compare with in-ecliptic spacecraft measurements.

Keywords. (Sun:) solar wind, (Sun:) solar-terrestrial relations

1. Introduction

Interplanetary Scintillation (IPS) is the rapid variation in radio signal from a compact distant natural source produced by turbulence/variations in density of the solar wind (e.g., Hewish, Scott, & Wills 1964). Observations of IPS allow the solar wind velocity (and an inferred value of density) to be determined over a large range of heliographic latitudes and solar elongations. Multi-antenna observations of IPS at 327 MHz used in this paper are from simultaneous recordings of the same radio source by up to four antennas separated by baselines of up to ~200 km. These allow solar wind velocity to be measured to a high degree of accuracy. Density values for the solar wind can be inferred from the "normalised scintillation level" (g-level) of IPS observations.

IPS observations using the Solar-Terrestrial Environment Laboratory (STELab) arrays, Nagoya University, Japan (Kojima & Kakinuma 1987), are used routinely for the time-dependent three-dimensional (3-D) tomographic reconstructions. These have a one-day cadence and 20° × 20° digital resolution for current STELab IPS data, but are smoothed by a Gaussian filter that interpolates temporally and spatially over regions with few data points (e.g., Jackson & Hick 2005). The resolution is predicated by the number of lines of sight available for the reconstructions. During the Whole Heliosphere Interval (WHI), there are few lines of sight (particularly those contributing to the density reconstruction); thus, there was sometimes incomplete coverage for individual days

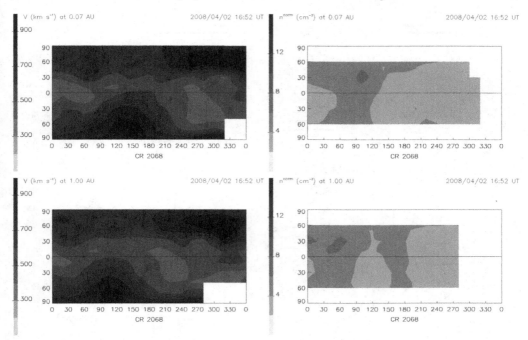

Figure 1. CR-averaged synoptic plots from the 3-D reconstructions at the "source surface" (15 R$_\odot$) (top) and at 1 AU (bottom) for velocity (left) and density (right). The "evolution" of structure can be seen from near the Sun to Earth's distance. From these, an *in situ* comparison is made with solar-wind measurements near 1 AU (Figure 2).

reconstructed. For the details of the 3-D reconstruction methods, see Bisi *et al.* (2009) and references therein.

In this paper, we summarise our 3-D reconstructions of the WHI Carrington rotation (CR) 2068 using the STELab IPS data and expand on the Bisi *et al.* (2009) paper. We concentrate on global average CR structure (through the summing and then averaging of all the available days reconstructed) and comparison with multi-point in-ecliptic *in situ* measurements.

2. Carrington-Rotation-Averaged 3-D Reconstruction Summary

Full details on the reconstructions for this period can be found in Bisi *et al.* (2009). Using the same data and time-dependent tomography here, the 3-D reconstructions for each day were averaged over the entire rotation (CR 2068). Data coverage, and thus reconstruction, was far greater in velocity than it was in g-level (density proxy) and so the CR-averaged density reconstruction has larger gaps in coverage.

Figure 1 shows the CR-averaged synoptic maps in both velocity and density at 0.07 AU (15 R$_\odot$) and 1 AU. A fairly-well defined and "tilted" streamer belt can be seen as a lower velocity region, and some small structures are visible in density. The lower velocities follow the heliospheric current sheet (HCS) mapped near the solar surface at this time. Note how structure changes from near the Sun to near the Earth, and that the time shift in solar-wind velocity structures is clearly seen as a result of the ∼radial propagation out to 1 AU. The density plots in our analysis show no dominant density structure that

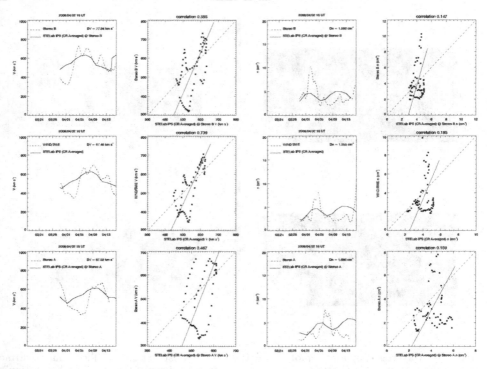

Figure 2. The lefthand plots for each of the six pairs shown compare the CR-averaged reconstructed solar wind velocity (left) and density (right) at STEREO-B, Wind, and STEREO-A
(top-to-bottom – solid line), and those measured by each respective spacecraft (dashed line).
The spacecraft data, originally hour-averaged, have been averaged over a day to match that of
the IPS time-dependent reconstruction. On the right of each pair are plots of the correlation of
the two data sets in each case; the dashed lines represent a 100% correlation and the solid lines
represent the best fit.

follows the HCS during this period, even though the results of our composite of transient
features match the density enhancements observed at 1 AU.

Figure 2 shows an extraction from the CR-averaged synoptic maps displayed in
Figure 1 at 1 AU with solar wind measurements taken from STEREO and Wind instruments. Overall, the CR-averaged velocity compares well with these measurements;
density compares well sometimes (with disagreement likely due to a limited number of
observations going into the 3-D reconstruction). For unknown reasons, the ACE density
data are very different from Wind, and the SOHO|CELIAS data seem more similar to
Wind than ACE. The extraction from the tomography agrees better with Wind density
than with ACE density.

3. Conclusions

The CR-averaged synoptic maps show good velocity structure both at the source surface and near Earth, with some structure visible in the density maps (but to a lesser
extent). The overall comparison of these CR-averaged data are good with the "ground
truth" *in situ* data, especially for velocity. Although the comparison is not as good as
those obtained with the individual day time-dependent extractions shown by Bisi *et al.*
(2009), it does verify our tomographic technique showing that we can reproduce the

synoptic velocity (and density) structure throughout a large portion of WHI with these available STELab IPS data.

Acknowledgements

UCSD authors acknowledge NASA (grant NNX08AJ116), the NSF (grants ATM 0331513, ATM0852246, ATM0925023), and AFOSR (grant FA9550-06-1-0107) for funding for these analyses. The IPS observations were carried out under the solar wind program of the Solar-Terrestrial Environment Laboratory (STEL) of Nagoya University. We also thank the Wind|SWE, SOHO|CELIAS, ACE|SWEPAM, and STEREO|PLASTIC Groups for making their data freely available on the internet.

References

Bisi, M. M., Jackson, B. V., Buffington, A., Clover, J. M., Hick, P. P., & Tokumaru, M. 2009, *Solar Phys.* 256, 201

Hewish, A., Scott, P. F., & Wills, D., 1964, *Nature* 203, 1214

Jackson, B. V. & Hick, P. P., 2005, in *Astrophys. and Space Sci. Lib.*, 314, *Solar and Space Weather Radiophysics: Current status and future developments*, ed. Gary, D. & Keller, C. U. (Kluwer Academic Publ., Dordrecht.), 355

Kojima, M. & Kakinuma, T. 1987, *J. Geophys. Res.* 92, 7269

Highlights of Astronomy, Volume 15
XXVIIth IAU General Assembly, August 2009   © International Astronomical Union 2010
Ian F. Corbett, ed.   doi:10.1017/S1743921310010343

Peculiar Current Solar-Minimum Structure of the Heliosphere

P. K. Manoharan

Radio Astronomy Centre, National Centre for Radio Astrophysics,
Tata Institute of Fundamental Research, Udhagamandalam (Ooty), 643001, India
email: mano@ncra.tifr.res.in

Abstract. In this paper, I review the results of 3-D evolution of the inner heliosphere over the solar cycle 23, based on observations of interplanetary scintillation (IPS) made at 327 MHz using the Ooty Radio Telescope. The large-scale features of solar wind speed and density turbulence of the current minimum are remarkably different from that of the previous cycle. The results on the solar wind density turbulence show that (1) the current solar minimum is experiencing a low level of coronal density turbulence, to a present value of ∼50% lower than the previous similar phase, and (2) the scattering diameter of the corona has decreased steadily after the year 2003. The results on solar wind speed are consistent with the magnetic field strength at the poles and the warping of heliospheric current sheet.

Keywords. turbulence, scattering, Sun: corona, Sun: magnetic fields, Sun: coronal mass ejections (CMEs), solar wind, solar-terrestrial relations

1. Interplanetary Scintillation

In this study, a large amount of interplanetary scintillation (IPS) data obtained from the Ooty Radio Telescope (ORT), operating at 327 MHz (Swarup *et al.* 1971), has been employed to study the 3-D evolution of the heliosphere over the period 1989–2009. The IPS observations made with the ORT can provide the velocity of the solar wind and the scintillation index (m) in the heliocentric distance range of $R \sim 10$–250 solar radii $(R_\odot)$ and at all heliographic latitudes. The value of m is a measure of electron-density turbulence in the solar wind $(m^2 \sim \int \delta N_e^2(z)\,dz)$, along the line of sight (z) to the radio source (e.g., Manoharan *et al.* 2000). The normalized scintillation index, $g = m(R)/<m(R)>$ (i.e., observed index normalized by its long-term average) enables the comparison of levels of density turbulence obtained from different sources. However, the value of g is linearly related to δN_e only in the weak-scattering region at distances at $>40\ R_\odot$. For example, an m-R profile attains the peak value at the strong-to-weak scattering transition point, which typically occurs $\sim 40\ R_\odot$ for IPS at 327 MHz (e.g., Manoharan 1993; 2006). In this study, the solar wind velocity and turbulence images have been exclusively obtained from weak-scattering data. However, the contour of constant level of turbulence in a year at different latitudes has been determined using peaks of several m-R profiles.

2. Solar Cycle 23: Three-Dimensional Solar Wind

Figure 1a shows the latitudinal distributions of solar wind speed and density turbulence (g) observed at Ooty over the solar cycle 23. These plots are similar to the well-known '*butterfly diagram*' of photospheric magnetic field intensity. They have been made by tracing backward/forward from the measurement location onto a sphere of radius ~ 100 $R_\odot$, which approximately corresponds to the mid range of distances covered in the observations utilized to generate the plots.

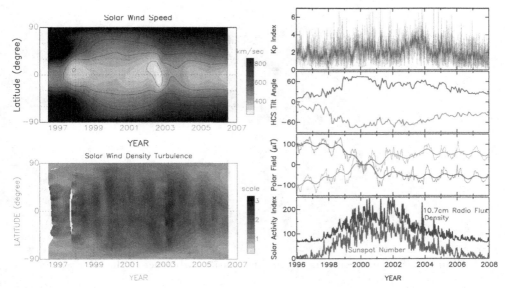

Figure 1. (a) Latitudinal distributions of solar wind speed (*left, top*) and density turbulence (*left, bottom*). The contour levels drawn on the speed gray-scale image are 350, 450, 550, and 650 kms^{-1}. (b) The stack plot (*right*) shows geo-magnetic disturbance index, Kp, intensity of polar magnetic field, tilt angle of heliospheric current sheet (HCS), and solar activities (i.e., sunspot number and solar radio flux density at 10.7 cm).

It is evident in the '*latitude-year*' speed plot that during minimum of the solar cycle, polar regions are dominated by high speed streams ($\sim$600–800 kms^{-1}) from open-field coronal holes, and low and variable flow speeds ($\leqslant$500 kms^{-1}) are observed at the low- and mid-latitude regions of the complex/closed field corona. But, there are marked differences in latitudinal extents of low- and high-speed flow regions between the current and previous minimum phases. For example during 1996–97, the low-speed flow was confined to $\sim\pm30°$ of the equatorial belt; whereas at the current minimum, it extends to a latitude range of $\sim\pm50°$. These low-speed wind widths also correlate with the latitudinal warping (tilt angle) of the heliospheric current sheet (HCS) over a small amplitude, $\sim\pm15°$, at the previous minimum and a moderate amplitude, $\sim\pm30°$, at the current minimum, respectively (Figure 1b). These results suggest (i) a near-dipole magnetic field of the Sun for the previous minimum, around the year 1997 and (ii) a never-approached dipole-field geometry during the current minimum. It is in good agreement with the result from other independent IPS speed measurements obtained from the Solar-Terrestrial Environment Laboratory (e.g., Tokumaru *et al.* 2009). Thus, the HCS tilt of the corona of the current phase tends to resemble a condition similar to that of moderate activity, but without activity!

These changes in the latitudinal extents of low-speed wind have also influenced the high-speed flows from the polar regions. In the current minimum phase, high-speed regions at the poles have remarkably shrunk towards the poles (Figure 1a, *top*). Moreover, the speed of the high-latitude ($>50°$) wind is considerably less for the current minimum than for the previous minimum. These findings nicely correlate with the polar field strength, which is $\sim$40–50% weaker for the current minimum phase. The magnetic pressure associated with the polar coronal holes seems to determine the acceleration of the high-speed wind. The weak field may be due to the fact that the polar field has not fully developed after the field reversal around the year 2000 (Figure 1b).

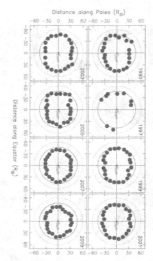

Figure 2. Shape of contours of constant density turbulence in the solar wind.

2.1. *Density-Turbulence Structures*

The drifting of density structures from high to low latitudes, seen in Figure 1a (*bottom*), is caused by the slow and gradual movement of concentrated magnetic field regions of the corona. It is likely due to the migration of small/medium-size coronal holes from polar to low latitude regions and the high-speed wind from these coronal holes inter-acting with the low-speed wind, causing compression in front of the high-speed stream. The latitudinal spread of density patterns is also consistent with the HCS tilt angle (Figure 1b), which is maximum at the time of polarity reversal of the cycle and during which a large number of coronal mass ejections (CMEs) dominate the heliosphere (e.g., Yashiro *et al.* 2004).

The density plot is also consistent with the solar wind disappearance period, around mid 1999 (low level of density turbulence), and co-rotating interaction regions (CIRs) dominating the heliosphere in the first half of 2003 (intense density turbulence along the latitudinal direction). It may be noted that during late October and early November 2003, a number of CME events prevailed in the interplanetary medium. However, the effect of a CME over the Sun-Earth distance is limited to 2–4 days after its onset and weakens with solar distance. Whereas, in the case of CIR events, the influence of each event is seen for several days, with a systematic increase in density turbulence with radial distance. Moreover, they show a latitudinal pattern. For CIR events during 2003, the latitudinal distribution and the radial evolution of the enhancement of turbulence have been observed (Manoharan 2008). The CIR-dominated period, in the first half of 2003, is in agreement with the moderate-to-severe storms observed at the Earth (Figure 1b, refer to Kp index plot). As observed in the speed plot, the latitudinal extents of low-turbulence regions at the poles also show remarkable changes between the current and previous minimum phases. The average level of turbulence at the current minimum seems to be considerably lower than that of the previous cycle.

2.2. *Scattering Diameter of the Corona*

Figure 2 displays constant $\delta N_e(R)$ plots, at different phases of the solar cycle. The plot for the year 1997 has been limited by weak-scattering observations and the last plot includes data up to May 2009. In general, a given level of turbulence is observed closer to the Sun at the poles than at the equator. However, depending on the phase of the solar cycle, the diameter of the contour can vary along the poles, but, remains nearly the same along the equator (e.g., Manoharan 1993).

The important point to note in this analysis is that after the year 2003, the overall diameter of the $\delta N_e(R)$ contour has gradually decreased with respect to the Sun's center. In other words, the same level of turbulence seems to move closer to the Sun. Thus, the radial dependence of turbulence $(C_{N_e}^2(R) \sim [\delta N_e(R)]^2$, which typically varies as $R^{-4})$ suggests that the scattering diameter of the corona has gradually shrunk towards the Sun. In other words, the scattering power $(C_{N_e}^2(R))$ remained nearly the same at all latitudes between 1989 and 2003, but decreased $\sim$50% around middle of 2009 at low-latitudes.

3. Discussion and Conclusions

The present large-scale 3-D features of solar wind speed and density turbulence are remarkably different from that of the previous cycle. In the current minimum phase, the extent of the low-speed region along the equatorial belt is considerably wider than that of the previous cycle; whereas the high-speed regions have shrunk towards the poles in contrast to the low-latitude extent of the previous cycle. The other important result of this study is that after the year 2003, the overall scattering diameter of the corona has gradually decreased with respect to the Sun's center. These results are consistent with the ecliptic and off-ecliptic studies (e.g., Lee *et al.* 2009; McComas *et al.* 2008; Smith & Balogh 2008; Tokumaru *et al.* 2009).

The weak fields observed at the poles, as well as corresponding solar wind speed and density turbulence for the current low activity, are possibly caused by the changes in the movement of large-scale fields, as the reversal of polarity progresses. It is linked to the rate of poleward and equatorward meridional flows, which transport the unbalanced magnetic flux (e.g., Sheeley 2008). Moreover, the flux-transport dynamo has predicted weak polar fields and a long solar cycle (e.g., Choudhuri *et al.* 2007).

Acknowledgements

I thank all the members of the Radio Astronomy Centre for making the Ooty Radio Telescope available for IPS observations. I also thank the National Space Science Data Center for OMNI data and the Wilcox Solar Observatory for the magnetic field data. This work is partially supported by the CAWSES-India Program, which is sponsored by ISRO.

References

Choudhuri, A. R., Chatterjee, P., & Jiang, J. 2007, Phy. Rev. Lett., 98, 131103
Lee, C. O. *et al.* 2009, Solar Physics, 256, 345
Manoharan, P. K. 1993, Solar Physics, 148, 153
Manoharan, P. K. *et al.* 2000, ApJ, 530, 1061
Manoharan, P. K. 2006, Solar Physics, 235, 345
Manoharan, P. K. 2008, in B. N. Dwivedi & U. Narain (eds.), *Physics of the Sun and its Atmosphere*, (World Scientific, Singapore), p235–266
McComas, D. J. *et al.* 2008, Geophys. Res. Lett., 35, 18103
Sheeley, Jr., N. R. 2008, ApJ, 680, 1553
Smith, E. J. & Balogh, A. 2008, Geophys. Res. Lett., 35, L22103
Swarup, G. *et al.* 1971, Nature Phys. Sci., 230, 185
Tokumaru, M. *et al.* 2009, Geophys. Res. Lett., 36, L091001
Yashiro, S. *et al.* 2004, J. Goephys. Res., 109, 7105

Highlights of Astronomy, Volume 15
XXVIIth IAU General Assembly, August 2009   © International Astronomical Union 2010
Ian F. Corbett, ed.   doi:10.1017/S1743921310010355

On Cosmic Rays, IP Structures and Geospace Consequences During WHI

A. Dal Lago[1], F. L. Guarnieri[2], M. R. da Silva[1], W. D. Gonzalez[1], C. R. Braga[1], N. J. Schuch[3], K. Munakata[4], C. Kato[4], J. W. Bieber[5], T. Kuwabara[5], M. Tokumaru[6], M. L. Duldig[7] and J. E. Humble[8]

[1] National Institute for Space Research (INPE), Sao Jose dos Campos, Brazil
email: dallago@dge.inpe.br

[2] Universidade do Vale do Paraiba (UNIVAP), Sao Jose dos Campos, Brazil

[3] Southern Regional Space Research Center (CRS/INPE), Santa Maria, Brazil

[4] Physics Department, Shinshu University, Japan

[5] Bartol Research Institute and Department of Physics and Astronomy, Univ. of Delaware, USA

[6] Solar-Terrestrial Environment Laboratory, Nagoya University, Japan

[7] Australian Antarctic Division, Tasmania, Australia

[8] School of Mathematics and Physics, University of Tasmania, Tasmania, Australia

Abstract. This work presents some observations during the period of the Whole Heliosphere Interval (WHI) of the effects of interplanetary (IP) structures on the near-Earth space using three sets of observations: magnetic field and plasma from the Advanced Composition Explorer (ACE) satellite, ground-based cosmic ray data from the Global Muon Detection Network (GMDN) and geomagnetic indices (Disturbance storm-time, Dst, and auroral electrojet index, AE). Since WHI was near minimum solar activity, high speed streams and corotating interaction regions (CIRs) were the dominant structures observed in the interplanetary space surrounding Earth. Very pronounced geomagnetic effects are shown to be correlated to CIRs, especially because they can cause the so-called High-Intensity Long-Duration Continuous AE Activity (HILDCAAs) - Tsurutani and Gonzalez (1987). At least a few high speed streams can be identified during the period of WHI. The focus here is to characterize these IP structures and their geospace consequences.

Keywords. Cosmic Rays, CIRs, AE index, WHI

1. Introduction

In this work, we will address ground cosmic ray observations from the Global Muon Detector Network (GMDN) during the Whole Heliosphere Interval (WHI) period, CR2068, from March 20–April 16, 2008 (http://ihy2007.org/WHI/), as well as geomagnetic effects correlated to CIRs observed during this period.

2. Global Muon Detection Network Observations During WHI

Since December 1992, a two-hemisphere network has acquired data using a pair of muon detectors at Nagoya (Japan) and Hobart (Australia), which have detection areas of $36m^2$ and $9m^2$, respectively. A small $4m^2$ prototype was installed in São Martinho da Serra (Brazil) in 2001 (Da Silva, *et al.* (2004)), which was upgraded in December 2005, increasing its detection area to $28m^2$ (Okazaki, *et al.* (2008)). These detectors are multidirectional, allowing the simultaneous recording of intensities in various viewing directions. In March 2006, the Global Muon Detector Network (GMDN) was completed with a new detector at Kuwait University (Kuwait), with a detection area of $9m^2$. Unlike the others, this muon detector is a hodoscope designed specifically for measuring the loss cone anisotropy (Munakata, *et al.* (2000); Leerungnavarat, *et al.* (2003)).

Figure 1 shows observations from the GMDN for the first half of the WHI period, from DOY 85 to 90. From top to bottom is shown the cosmic ray best-fit density, measured x, y, z anisotropies and calculated x, y, z cosmic ray gradient.

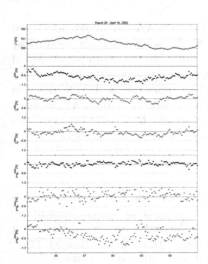

Figure 1. From top to bottom: cosmic ray best-fit density, measured x, y, z anisotropies and calculated x, y, z cosmic ray gradient.

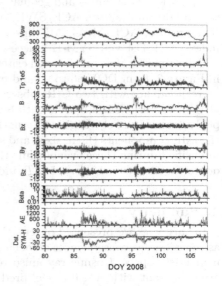

Figure 2. ACE solar wind velocity, proton number density and temperature, ACE IMF intensity, x, y and z components, plasma beta, auroral index (AE), Dst and SYM-H indeces.

From Figure 1, it is possible to see a negative x anisotropy during the period. Okazaki, *et al.* (2008), using GMDN observations, have interpreted similar anisotropies in terms of north-south cosmic ray gradients. In this particular case of Figure 1, a negative x anisotropy is consistent with the $A < 0$ epoch of the solar cycle, for which a simple drift model predicts a local cosmic ray maximum in the Heliospheric Current Sheet (HCS) (see Okazaki, *et al.* (2008) for further details). ACE observations of IMF longitude (not shown) shows that the Earth was north of the HCS. Following the methodology of Okazaki, *et al.* (2008), the last panel of Figure 1 shows the cosmic ray gradient in the negative z direction, which is consistent with the fact that the Earth was north of the HCS.

3. High Intensity, Long Duration, Continuous AE Activity Observations During WHI

Tsurutani and Gonzalez (1987) defined a class of events which they called HILDCAAs, which stands for High Intensity, Long Duration, Continuous AE Activity. Its been debated that these events are not continuous substorms, but a new form of energy deposition in the auroral ionosphere (Tsurutani, *et al.* (2004); Guarnieri (2006)). The criteria for an event to be called HILDCAAs are: (1) AE peak of, at least, 1000 nT; (2) the event must last for, at least, two days; (3) the AE values cannot decrease to less than 200 nT for more than 2 hours at a time; and (4) the event must occur outside the main phase of a geomagnetic storm. The interplanetary magnetic field during these events presents large amplitude fluctuations (Alfvén waves) instead of southward directed fields (Tsurutani and Gonzalez (1987); Gonzalez, *et al.* (1999)). HILDCAAs are found to be important in the descending and minimum phase of the solar cycle (Guarnieri (2006)), such as the WHI period. Figure 2 shows interplanetary observations from the Advanced Composition Explorer (ACE satellite) and geomagnetic indices from the WHI period (CR 2068). From top to bottom is shown ACE/SWEPAM solar wind velocity, proton number density and temperature, ACE/MAG IMF intensity, x, y and z components, plasma beta, auroral index (AE), Dst and SYM-H indeces.

Following the above mentioned criteria, two periods of HILDCAAs are clearly identified in Figure 2, from DOY 86 to 90 and from DOY 96 to 100. These two periods are in close correspondence with two CIRs observed during this period. The CIRs can be identified as the two periods in Figure 2 with high velocities (first panel), preceded by high density peaks (second panel). Highly fluctuating magnetic field components can be identified in the IMF components (fifth, sixth and seventh panels). The auroral electrojet, AE, index is intensified during these events while the disturbance storm-time, Dst, and SIM-H indices do not reach values below $-50nT$, indicating weak geomagnetic storm activity.

4. Summary

In this work we presented observations from the GMDN from the WHI period, as well as the geospace consequences associated with the interplanetary structures present in the near-Earth space. Cosmic ray anisotropies and gradients derived from the GMDN data are consistent with $A < 0$ model predictions, following the work done by Okazaki, *et al.* (2008). Two CIRs were identified during the WHI period using ACE observations. These structures caused HILDCAA events. Such events are more frequent during descending and minimum phases of the solar cycle, such as the WHI period.

Acknowledgements

The authors would like to acknowledge N. Ness (Bartol Research Institute), D. J. McComas (Southwest Research Institute), R. Lepping (NASA GSFC), K. Ogilvie (NASA GSFC), and the CDAWeb for ACE interplanetary magnetic field and plasma data; and Kyoto WDC for the Dst index, CNPq of Brazil for projects 303798/2008-4, 472031/2007-4.

References

Da Silva, M. R., Contreira, D. B., Monteiro, S., Trivedi, N. B., Munakata, K., Kuwabara, T., & Schuch, N. J. 2004, *Astrophys. Space Sci.* 290, 3-4
Gonzalez, W. D., Tsurutani, B. T., & Clua de Gonzalez, A. L. 1999, *Space Sci. Rev.* 88
Guarnieri, F. L. 2006, *In: Geophysical Monograph Series* 167
Leerungnavarat, K., Ruffolo, D., Bieber, J. W. 2003, *ApJ* 593, 587
Munakata, K., J. W. Bieber, S. I. Yasue, C. Kato, M. Koyama, S. Akahane, K. Fujimoto, Z. Fujii, J. E. Humble, & M. L. Duldig 2000, *J. Geophys. Res.* 105, 27
Okazaki, Y., Fushishita, A., Narumi, T., Kato, C., Yasue, S., Kuwabara, T., Bieber, J. W., Evenson, P., Da Silva, M. R., Dal Lago, A., Schuch, N. J., Fujii, Z., Duldig, M. L., Humble, J. E., Sabbah, I., Kota, J., & Munakata, K. 2008, *ApJ* 681, 1
Tsurutani, B. T. & Gonzalez, W. D. 1987, *Space Sci. Rev.* 35
Tsurutani, B. T., W. D. Gonzalez, F. L. Guarnieri, Y. Kamide, X.-Y. Zhou, & J. K. Arbal 2004, *J. Atmosph. Sol. Terr. Phys.* 66, 167

Highlights of Astronomy, Volume 15
XXVIIth IAU General Assembly, August 2009
Ian F. Corbett, ed.

© International Astronomical Union 2010
doi:10.1017/S1743921310010367

Global MHD Modeling of the Solar Corona and Inner Heliosphere for the Whole Heliosphere Interval

Pete Riley, Jon A. Linker and Zoran Mikic

Predictive Science, Inc.
9990 Mesa Rim Road, Suite 170, San Diego, CA 92121, USA.

Abstract. With the goal of understanding the three-dimensional structure of the solar corona and inner heliosphere during the "Whole Heliosphere Interval" (WHI), we have developed a global MHD solution for Carrington rotation (CR) 2068. Our model, which includes energy transport processes, such as coronal heating, conduction of heat parallel to the magnetic field, radiative losses, and the effects of Alfvén waves, is capable of producing significantly better estimates of the plasma temperature and density in the corona than have been possible in the past. With such a model, we can compute emission in extreme ultraviolet (EUV) and X-ray wavelengths, as well as scattering in polarized white light. Additionally, from our heliospheric solutions, we can deduce magnetic field and plasma parameters along specific spacecraft trajectories. We have made detailed comparisons of both remote solar and *in situ* observations with the model results, allowing us to: (1) Connect these disparate sets of observations; (2) Infer the global structure of the inner heliosphere; and (3) Provide support for (or against) assumptions in the MHD model, such as the empirically-based coronal heating profiles.

Keywords. Sun: corona, Sun: evolution, Sun: magnetic fields, Sun: solar wind, interplanetary medium

---

1. Introduction

Whole Heliosphere Interval (WHI), which ran from March 20 through April 16, 2008, and coincided with Carrington Rotation (CR) 2068, is providing a unique opportunity for both observers and modelers to collaborate in an effort to understand the three-dimensional structure and evolution of the solar corona and inner heliosphere. It builds on the previous Whole Sun Month (WSM) interval, which proved to be exceptionally successful. WHI occurred on the way to the current solar minimum, which has, thus far, been unique in a number of ways. For example, the polar photospheric flux is lower than the previous minimum by $\sim 40\%$ (Svalgaard and Cliver 2007) and the coronal holes are noticeably smaller Kirk *et al.* (2009). Measurements by *in situ* spacecraft show substantial differences between the current minimum and the previous three. As of late 2008, Ulysses polar observations, in particular, suggested that: (1) The interplanetary magnetic field (IMF) was $\sim 36\%$ lower than the previous minimum Smith and Balogh (2008); (2) The scaled number density was $\sim 17\%$ lower McComas *et al.* (2008); and (3) The scaled temperature was $\sim 14\%$ lower McComas *et al.* (2008). It was also determined that the bulk solar wind speed was $\sim 3\%$ lower, although this may not represent a statistically significant change. The profiles of high-speed streams upstream of Earth also seem to be unique, being stronger, longer in duration, and more recurrent than the previous minimum Gibson *et al.* (2009).

To understand the three-dimensional structure during the WHI, and, more generally, the unique features of the current solar minimum, we have undertaken a detailed

Pete Riley, Jon A. Linker & Zoran Mikic

Pete Riley , Jon A. Linker & Zoran Mikic

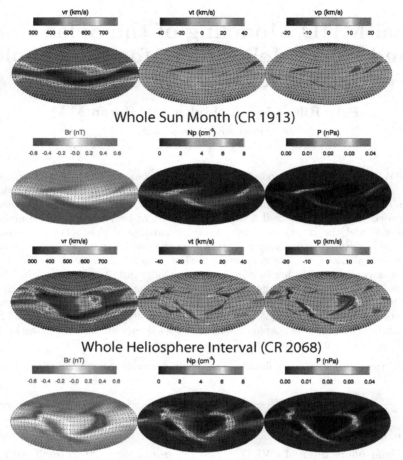

Figure 1. Mollweide projection maps of radial speed (v_r), meridional speed (v_r), azimuthal speed (v_p), radial magnetic field (B_r), number density (N_p), and thermal pressure (P) for Carrington rotation 1913 (top), corresponding to the Whole Sun Month (WSM) interval, and 2068 (bottom), corresponding to the Whole Heliosphere Interval (WHI).

investigation involving magnetohydrodynamic (MHD) modeling of the global structure of the corona and inner heliosphere, analysis of remote solar and *in situ* measurements, and interpretation and connection of the data using the simulation results. Our model results allow us to explore the physical connections between the various phenomena and synthesize these diverse observations into a coherent picture. In this brief report, we highlight one specific aspect of this study: A comparison of the large-scale three-dimensional structure of the inner heliosphere during WHI and WSM.

2. Modeling the Large-Scale Structure of the Heliosphere during WSM and WHI

MHD models have proven highly successful in interpreting and understanding a wide array of solar and heliospheric phenomena. They provide a global context for connecting diverse datasets and understanding the physical interrelationship between often dissimilar phenomena Riley *et al.* (1996, 2001a,b, 2002, 2003); Riley (2007). Our group has studied the properties of the ambient solar wind for a number of years, and found that, in general, our model can reproduce the essential large-scale features of the solar wind. While these past comparisons demonstrate the success of the MHD model, the simplified polytropic approximation used has limitations. In the current study, we have developed coupled global thermodynamic MHD simulations driven by observed photospheric magnetic fields to study the large-scale, quasi-stationary properties of the WHI and understand the differences between the current solar minimum and the previous one, as characterized by the Whole Sun Month (WSM) interval (August/September, 1996).

In Figure 1, we show the three components of speed, together with the radial magnetic field strength, number density, and thermal pressure for WSM and WHI. The differences are quite remarkable. First, the "band of solar wind variability," that is, the region of typically slower, but more variable solar wind, and roughly centered about the helio-equator, extends to significantly higher latitudes during WHI. Second, the polar speeds are essentially the same for the two minima (confirmed by Ulysses observations). Third, a significant source of fast solar wind in the ecliptic plane derives from equatorial coronal holes during WHI. Fourth, the structure of CIRs is more complex during WHI: The systematic, opposed tilts observed during the declining phase of solar cycle 22 Gosling *et al.* (1995); Riley *et al.* (1996, 2001a,b) are not nearly as well defined during WHI; the equatorial coronal holes producing more localised "U" shaped interaction regions Riley *et al.* (2003).

3. Closing Remarks

In this brief report, we have summarized one aspect of our modeling effort to support the goals of the WHI Campaign. Model results will be contributed to the WHI repository (http://ihy2007.org/WHI/obs_models.shtml) and will be made available through Predictive Science's website (http://www.predsci.com).

References

Kirk, M. S., W. D. Pesnell, C. A. Young, & S. A. Hess Webber, *ArXiv e-prints* (2009), 0901.1158.
Gibson, S. E., J. U. Kozyra, G. d. Toma, B. A. Emery, T. Onsager, & B. J. Thompson, *Geophys. Res. Lett.* **114**, A09105 (2009).
Gosling, J. T., S. J. Bame, D. J. McComas, J. L. Phillips, V. J. Pizzo, B. E. Goldstein, & M. Neugebauer, *SSR* **72**, 99 (1995).
McComas, D. J., R. W. Ebert, H. A. Elliott, B. E. Goldstein, J. T. Gosling, N. A. Schwadron, & R. M. Skoug, *Geophys. Res. Lett.* **35**, 18103–+ (2008).
Riley, P., J. T. Gosling, L. A. Weiss, & V. J. Pizzo, *J. Geophys. Res.* **101**, 24349 (1996).
Riley, P., J. A. Linker, & Z. Mikić, *J. Geophys. Res.* **106**, 15889 (2001a).
Riley, P., J. A. Linker, Z. Mikić, & R. Lionello, "MHD Modeling of the Solar Corona and Inner Heliosphere: Comparison with Observations," in *Space Weather, Geophysical Monograph Series*, edited by P. Song, H. J. Singer, and G. L. Siscoe, AGU, Washington, DC, 2001b, vol. 125, p. 159.
Riley, P., J. A. Linker, & Z. Mikić, *J. Geophys. Res.* **107**, DOI 10.1029/2001JA000299 (2002).
Riley, P., Z. Mikić, & J. A. Linker, *Ann. Geophys.* **21**, 1347 (2003).
Riley, P., J. A. Linker, Z. Mikić, D. Odstrcil, T. H. Zurbuchen, D. Lario, & R. P. Lepping, *J. Geophys. Res.* **108**, 2 (2003).
Riley, P. *Journal of Atmospheric and Terrestrial Physics* **69**, 32–42 (2007).
Smith, E. J. & A. Balogh, *Geophys. Res. Lett.* **35**, 22103–+ (2008).
Svalgaard, L. & E. W. Cliver, *Ap. J. Lett.* **661**, L203–L206 (2007).

Highlights of Astronomy, Volume 15
XXVIIth IAU General Assembly, August 2009
Ian F. Corbett, ed.

WHI High-Speed Streams at Geospace

G. Maris[1] and O. Maris[2]

[1] Institute of Geodynamics, RO-020032, Bucharest, Romania
email: gmaris@geodin.ro

[2] Institute for Space Sciences, RO-077125, Bucharest, Romania
email: maris@spacescience.ro

Abstract. The fast solar wind analysis during the Whole Heliospheric Interval (Carrington Rotation no. 2068: March 20 - April 16, 2008) is herein presented. The analysis is based on the data recorded by space experiments aboard the ACE and SOHO missions. The high-speed streams in the solar wind were determined by their main parameters: duration, maximum velocity, velocity gradient. The fast solar wind was also analyzed during the preceding and following Carrington Rotations, nos. 2067 and 2069, respectively. The main properties of the rapid streams and their solar sources were analysed in terms of the present solar minimum phase. The geospace response to the fast streams was evaluated using the geomagnetic index dynamics. A comparative analysis of the high-speed streams registered during Whole Heliospheric Interval with the ones determined during Whole Sun Month was also made.

Keywords. solar wind - high-speed streams, solar-terrestrial relations

1. Introduction

An internationally coordinated observing and modeling program, Whole Heliospheric Interval (WHI), focused on observations originating from Carrington Rotation (CR) 2068, March 20 - April 16, 2008 was developed during IHY. The interval was chosen during the prolonged minimum between solar cycles 23 and 24, one solar cycle after Whole Sun Month (WSM), August 10–September 8, 1996, in order to make a comparative analysis of solar quiet intervals near solar cycle minima. The 3-D observing views of the Sun during CR 2068 were possible aboard STEREO, SOHO, ACE missions in quadrature with Ulysses (operated at a reduced rate). Using all ground and space data, it would be possible to accomplish the main goals of the program: to characterize the 3-D solar minimum heliosphere and to trace the effects of solar structure and activity through the solar wind to the Earth and other planetary systems. This paper presents a brief overview of the high-speed streams (HSSs) in the solar wind that appeared during WHI as well as during the two adjacent Carrington Rotations, 2067 and 2069. A comparison of the HSSs registered during WHI with the ones during WSM is made. The HSS impact on the terrestrial magnetosphere, evaluated by Ap and Dst geomagnetic index dynamics and by a regression line analysis, ends the paper.

2. HSSs during WHI

The HSSs have a multitude of definitions given by different authors; the simplest is often the best: *a large increase in the solar wind velocity lasting for several days.* Some catalogues of HSSs were set up (Lindblad and Lundstedt, 1989 and the references therein: Mavromichalaki *et al.* 1998; Maris and Maris, 2008) covering solar cycles 20–23. We extended the HSS catalog to 2008, during WHI. We used the same selection procedure of the streams as Linblad and Lundsted because it allows for a more precise determination of the HSS beginning and end. So, we selected as a 'high-speed stream' a solar wind flow having $\triangle V1 \geqslant 100$ km/s that lasted for two days, where $\triangle V1$ was the difference between the smallest 3-hr mean plasma velocity for a given day (V0) and the largest 3-hr mean plasma velocity for the following day (V1). Solar wind velocity and plasma density registered by ACE/SWEPAM and SOHO/CELIAS/MTOF experiments were used during the WHI interval as well as during the two adjacent CRs to determining the fast streams. The HSSs obtained are presented in Fig. 1 and Table 1.

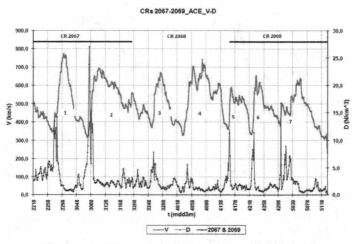

CRs 2067-2069_ACE_V-D

Figure 1. HSSs registered during CRs 2067 - 2069.

Table 1. HSSs during CRs 2067-2069 (HSS numbers correspond to numbers in Fig. 1; A* - the Bz polarity alternates during the stream; WHI HSSs in italics).

CR	No	M	D	3-h	V0	V1	dt1	Vmax	Dur	$\triangle$Vmax	IMF	Bz
2067	1	2	27	1	345.7	562.0	15	770.3	9.4	424.6	−	A*
	2	3	7	8	319.3	444.3	7	693.0	18.0	373.7	+	+
2068	*3*	*3*	*25*	*8*	*368.3*	*566.7*	*8*	*663.7*	*7.6*	*295.4*	*−*	*−*
	4	*4*	*3*	*3*	*336.0*	*501.3*	*13*	*734.3*	*12.9*	*398.3*	*+*	*+*
	5	4	16	2	359.3	590.7	6	590.7	5.9	231.4	+	+
2069	6	4	22	5	335.7	648.7	10	648.7	8.0	313.0	−	−
	7	4	30	5	387.3	499.7	4	637.0	9.5	249.7	+	+

The significance of the columns in Table 1 are: columns 1 and 2 give the numbers of CRs and streams determined (the same as in Fig.1); columns 3-5 give the start data of the streams, by: M - month; D - day; 3-h - 3-hr interval of the start day; column 6: V0 - minimum (pre-stream) 3-hr velocity of the stream (km/s); column 7: V1 - maximum 3-hr velocity in the second day of the stream; column 8: dt1 - time interval between V0 and V1 (in number of 3-hr intervals); column 9: Vmax - maximum velocity of the stream; column 10: Dur - duration of the stream, in days; column 11: $\triangle$Vmax = Vmax - V0 is maximum gradient of the plasma velocity; columns 12 and 13: the interplanetary magnetic field (IMF) polarity and the dominant polarity of Bz component during the stream, respectively. Note that 'dominant polarity' was chosen as the polarity covering more than 80% of HSS duration (Bz in GSM coordinate system, given by OMNI2 data base, http://omniweb.gsfc.nasa.gov). Solar sources of the streams no. 3 and 4 were coronal holes (CHs) observed in the solar corona at low latitudes. The complex structure of the CH, the source of stream no. 4, resulted in the complex structure of the stream. All the streams registered during CRs 2067–2069 were recurrent, as a consequence of their CH sources (Fig. 2).

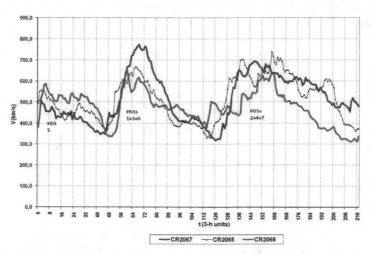

Figure 2. HSSs during CR 2067-2069, for each CR, superimposed over each other. The recurrent nature of the HSSs (1+3+6) and HSSs (2+4+7), respectively, is easily seen. HSS no. 5 (CR 2069) is superimposed over the end of HSS no. 2 (at the beginning of CR 2068).

All the streams had Vmax between 590-770 km/s, $\triangle$Vmax, between 231-424 km/s and, long durations (>6 days). Some of them had complex structure (nos. 2 and 4) determined by the successive fast plasma flows generated by the different regions of the CH sources passing over the central zone of the Sun.

HSSs determined during the WSM interval (1996) are represented in Fig. 3 and all HSS parameters during CRs 1912-1914 are given in Table 2. HSSs nos. 2-5 correspond to the WSM interval (the corresponding numbers are noted in Fig. 3). The complex structures were registered during HSSs nos. 3 and 4, similar to HSSs no. 2 and 4 during WHI (Fig. 1). The streams during CRs 1912-1914 had durations $\leqslant$ 6.4 days with only two exceptions (HSSs nos. 6 and 10). Maximum velocities (Vmax) of the four HSSs during WSM were less 619 km/s and $\triangle$Vmax $\leqslant$ 228 km/s (Table 2).

3. Geomagnetic Effects of the WHI Streams

We analyzed the geomagnetic activity during WHI following the classification of the Sunspot Index Data Center (Brussels) as: active periods, with 15 < Ap < 30; minor geomagnetic storm, 29 < Ap < 50; major geomagnetic storm, 49 < Ap < 100; and, severe geomagnetic storm, Ap $\geqslant$ 100 *(http://sidc.oma.be/educational/classification.php#geoA).* During all HSSs registered in the three CRs, the geomagnetic field presented active periods but only two minor geomagnetic storms appeared as consequences of the HSSs nos. 1 and 3. During HSS no. 3, Bz was southward and favorable to the wind energy transfer to the terrestrial magnetosphere and the correlation coefficient (r) between stream velocity and Dst index was −0.78. Although HSS no. 4 registered the second highest Vmax and $\triangle$Vmax, only sporadically, some energy of the fast stream entered the terrestrial magnetosphere because Bz was northward (positive) and, r (V;Dst) = −0.59. A good

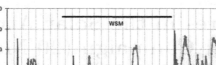

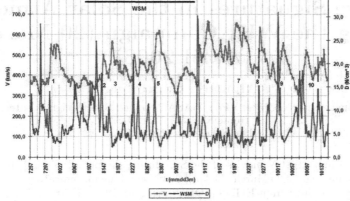

Figure 3. HSSs registered during CRs 1912-1914.

Table 2. HSSs during CRs 1912-1914 (HSS numbers correspond to numbers in Fig. 3; A* - the Bz polarity alternates during the stream; WSM HSSs in italics).

CR	No	M	D	3-h	V0	V1	dt1	Vmax	Dur	$\triangle$Vmax	IMF	Bz
	1	7	30	6	358.3	552.3	6	553.0	6.1	194.7	+	−
1912	*2*	*8*	*13*	*1*	*357.0*	*503.0*	*14*	*503.0*	*3.2*	*146.0*	*+*	*A**
	3	*8*	*16*	*3*	*388.3*	*566.3*	*8*	*566.3*	*5.6*	*178.0*	*+*	*A**
	4	8	22	3	367.3	500.0	9	500.0	6.4	132.7	+	−
1913	5	8	28	6	390.0	607.3	8	618.7	4.5	228.7	+	−
	6	9	9	6	348.0	551.3	5	665.7	9.4	317.7	+	−
	7	9	19	1	454.3	656.3	12	656.3	5.4	202.0	+	−
1914	8	9	25	1	411.3	636.7	15	636.7	4.5	225.4	+	−
	9	10	2	2	336.0	520.0	6	559.0	4.8	223.0	−	+
	10	10	8	1	337.7	525.3	13	527.0	7.4	189.3	+	−

correlation was obtained for HSS no. 6 (r = −0.75) and even for HSS no. 1 (r = −0.72) where the highest Vmax and $\triangle$Vmax disturbed the terrestrial magnetosphere although Bz < 0 occurred during < 70% of the HSS duration.

Acknowledgements

The authors express their gratitude to the referee for their constructive comments and helpful suggestions. The authors acknowledge the solar wind data obtained by courtesy of ACE/SWEPAM and SOHO/CELIAS/MTOF teams; SOHO is a mission of international cooperation between ESA and NASA. One of the authors (GM) acknowledges the IAU travel grant to IAU General Assembly 2009 in Rio de Janeiro, Brazil.

References

Lindblad, B. A. & Lundstedt, H. 1989, *Sol. Phys* 120, 145–152.
Mavromichalaki, H. & Vassilaki, A. 1998, *Sol. Phys* 183, 181–200.
Maris, O. & Maris, G. 2008, *at: http://www.spaceweather.eu/ in Cap. "Data Catalogs for SW".*

Highlights of Astronomy, Volume 15
XXVIIth IAU General Assembly, August 2009
Ian F. Corbett, ed.

Eruptive Signatures in the Solar Atmosphere During the WHI Campaign (20 March–16 April 2008)

Alphonse C. Sterling†

Space Science Office, VP62, NASA Marshall Space Flight Center,
Huntsville, AL 35812 USA
email: alphonse.sterling@nasa.gov

Abstract. We examined EUV movies of the Sun during the period of the Whole Heliospheric Interval (WHI) campaign of 20 March–16 April 2008, searching for indications of eruptive events. Our data set was obtained from EIT on SOHO, using its 195 Å filter, and from EUVI on the two STEREO satellites, using their 171 Å, 195 Å, 284 Å, and 304 Å filters. Here we present a table showing results from our preliminary search.

Keywords. Sun: activity, Sun: corona, Sun: coronal mass ejections (CMEs), Sun: flares, Sun: UV radiation

1. Introduction

Positive identification of the source regions on the solar disk of solar eruptions is, in many cases, a difficult problem. Of course there are frequently cases where there are obvious features on the solar disk, such as solar flares or strong intensity dimming regions, that occur in tandem with the launch of a coronal mass ejection (CME). There are other cases however where CMEs occur with only very weak solar-disk signatures, or even no signature at all - see, for example, (Hudson *et al.* (1998), Webb *et al.* (1998), Robbrecht *et al.* (2009)). During the Whole Heliospheric Interval (WHI) campaign of 20 March–16 April 2008, several satellites and ground-based observatories performed coordinated observations of the Sun and heliosphere, and therefore this period is an appropriate one for revisiting the connection between CMEs and their solar source regions. We have initiated such an investigation, and here we summarize some early results of an initial survey. Our focus was to identify visually what appear to be eruptive events in the lower solar atmosphere during the WHI period. This initial investigation can serve as a basis for more detailed studies to follow. Future investigations will make comparisons between our list with lists of observed CME eruptions.

2. Analysis Procedure

We first examined movies from the EUV Imaging Telescope (EIT) on the SOHO spacecraft over the WHI period for indications of eruptions, usually in the form of intensity dimmings (e.g., Hudson *et al.* (1998)), filament eruptions, or transient brightenings coupled with rapid changes in the local coronal structure (cf. Canfield *et al.* (1999)). From EIT we used images in the 195 Å filter, as it had the highest cadence during the period (~12 min). After composing an initial list of low-coronal eruptive-like signatures with

---

† Present address: JAXA/Institute of Space and Astronautical Science, Hinode Group, 3-1-1 Yoshinodai, Sagamihara, Kanagawa 229-8510, Japan

EIT 195 Å images, we supplemented the list using movies from the EUVI imagers on the STEREO spacecraft. EUVI has better spatial resolution than EIT (~ 1.6"/pixel vs. 2.6"/pixel), and higher time cadence (~ 3—6 min for EUVI). We examined EUVI images taken with their 171 Å, 195 Å, 284 Å, and 304 Å filters.

3. Results

During the WHI period, four active regions (ARs) appeared on the solar disk: NOAO ARs 10987, 10988, and 10989 appeared early in the WHI period, and 10990 began developing at the end of the WHI period. Many of the events we identified as "eruptive looking" emanated from or near the ARs. Table 1 presents an overview of our results. Seven of the events (events 3, 7, 10, 12, 15, 16, and 17) originated from an active region, while three of the events (events 4, 6, and 14), while near the limb, almost certainly had no direct connection to an active region. The connection between the remaining events and active regions was marginal or uncertain, as indicated in the Table 1 notes.

4. Discussion

We identified nearly 20 features that appeared eruptive, but not necessarily all were "ejective," i.e., an eruption that produced an ejection that left the Sun (e.g., Moore *et al.* (2001)). Some eruptions do not produce ejections into the heliosphere, sometimes even when the eruptions are accompanied by large flares (e.g., Green *et al.* (2002), Wang & Zhang (2007)), but instead the material is confined by strong overlying magnetic field. Such "confined eruptions" often will not produce a CME (e.g., Moore *et al.* (2001), Török & Kliem (2005)), or could result in a weak secondary CME or outflow (Bemporad *et al.* (2005)). To identify ejective eruptions we look for persistent dimmings, an expelled prominence, etc. Confined eruptions generally produce relatively transient dimmings or brightenings, and sometimes a filament can be seen to have its upward movement thwarted (e.g., Ji *et al.* (2003)). In marginal cases however it can be difficult to determine whether an eruption is confined or ejective, and so for several events in Table 1 we indicate the connection is uncertain. AR 10989 erupted several times, and some of these were clearly ejective. AR 10987 probably produced at least one ejective eruption (event 11), but overall ARs 10987 and 10988 displayed fewer eruptive signatures than AR 10989, according to our visual survey results.

In several cases we initially missed events in movies taken with the EIT and EUVI 195 Å filter, but could identify them readily in movies from a different EUVI filter. This could be an important consideration when attempting to identify disk sources of CMEs.

Acknowledgements

A.C.S. was supported by funding from NASA's Office of Space Science through the Solar Physics Supporting Research and Technology Program and the Sun-Earth Connection Guest Investigator Program. The author thanks D. Webb and H. Cremades for valuable discussions.

References

Bemporad, A., Sterling, A. C., Moore, R. L., & Poletto, G. 2005, *ApJ*, 635, L189
Canfield, R. C., Hudson, H. S., & McKenzie, D. E. 1999, *Geophys. Res. Lett.*, 26, 627
Green, L. M., Matthews, S. A., van Driel-Gesztelyi, L., Harra, L. K., & Culhane, J. L. 2002, *Solar Physics*, 205, 325

Hudson, H. S., Lemen, J. R., St. Cyr, O. C., Sterling, A. C., & Webb, D. F. 1998, *Geophys. Res. Lett.*, 25, 2481

Ji, H., Wang, H., Schmahl, E. J., Moon, Y.-J., & Jiang, Y. 2003, *ApJ*, 595, L135

Moore, R. L., Sterling, A. C., Hudson, H. S., & Lemen, J. R. 2001, *ApJ*, 552, 833

Robbrecht, E., Patsourakos, S., & Vourlidas, A. 2009, *ApJ*, 701, 283

Török, T., & Kliem, B. 2005, *ApJ*, 630, L97

Wang, Y., & Zhang, J. 2007, *ApJ*, 665, 1428

Webb, D. F., Cliver, E. W., Gopalswamy, N., Hudson, H. S., & St. Cyr, O. C. 1998, *Geophys. Res. Lett.*, 25, 2469

Table 1. "Eruptive-Looking" Coronal Events Over WHI Period: 20-Mar - 16 Apr 2008

Event	Date (M/DD)	UT	First Instrument[1]	Location	Comments:
1	3/22	11:45	EUVI-B 195	At E limb.	Weak coronal opening or ejection. Possible precursor to 14:05 UT event.
2	3/22	14:05	EUVI-B 195	At E limb.	Coronal opening, and ejection. Behind the limb for EIT. Probably from AR 10989 during its early stages.
3	3/25	18:34	EIT 195	AR 10989 at E limb.	Coronal opening and dimming, bright transient flare.
—	3/26	—	EUVI-B 195	—	Extensive activity between ARs 10987 and 10988, but no obvious ejections. Maybe a failed eruption visible in EUVI-B 284 from about 16:26 UT between those two ARs, perhaps producing a weak ejection (cf. Bemporad *et al.* (2005)).
4	3/27	c16:06	EUVI-B 304, etc.	NE limb in STEREO B.	Slowly-erupting prominence in 304 Å; faint in other EUVI filters. Probably ejective, but not certain.
5	3/27	17:26	EUVI-B 284	S of AR 10988.	Filament erupts, but probably not ejective. Source region rooted adjacent to but not in the AR, but AR magnetic field may have triggered onset.
6	3/29	07:47 onward.	EIT 195	Along E limb.	Large-scale flows seen in 195 Å filters and EUVI-B 171 Å; continues until next day. Slowly-evolving prominence in EUVI-B 304 Å; may eject from Sun at $\gtrsim 11$ UT.
7	3/30	05:22	EIT 195	AR 10989.	Eruptive flare, removing corona (dimming) to SE. Also well seen in EUVI-B 284.
8	3/30	>7:46	EUVI-B 304	S of ARs 10988 & 10989.	Large filament eruption, faint in non-304 filters. From EUVI-A 304 however, it may not be ejective.
9	4/01	04:26	EUVI-B 284	N of AR 10989.	Eruption with source outside of AR, but with remote connections to the AR. Prominent in 284 Å but weak in 195 Å. May be confined.
10	4/05	05:34	EIT 195	AR 10989 SE side.	Eruption with dimming. From EUVI-B 284 Å there may be a partial ejection, but uncertain; EUVI 171 Å shows dimming that suggests an ejection however.
11	4/05	15:35	EUVI-A	W limb, probably from AR 10987	Coronal opening and dimming.
12	4/05	19:34	EIT 195	AR 10989, SE side.	Same location as event at 05:34 UT on same day. Filament eruption clear in 304 Å; probably ejective, but uncertain.
13	4/09	09:58	EIT 195	W limb, probably from AR 10989.	Filament and cavity clearly erupting in EUVI-A. In EUVI-B 284 appears as very fast ejection.
14	4/11	16:06-24:00	EUVI-B 304	NE limb.	Very slow filament eruption; seen in 304 Å but not noticed in other filters. Faint and behind the limb in EIT 195 Å.
15	4/16	06:46	EUVI-B 284	AR 10990.	Eruption from AR; possibly ejective, but uncertain.
16	4/16	10:06	EUVI-B 284	AR 10990.	Relatively strong eruption from AR, likely ejective.
17	4/16	18:06	EUVI-B 284	AR 10990.	Eruption from AR, likely ejective.

Notes:

[1] Instrument in which feature was first noticed.

Highlights of Astronomy, Volume 15
XXVIIth IAU General Assembly, August 2009
Ian F. Corbett, ed.

© International Astronomical Union 2010
doi:10.1017/S1743921310010392

The Future of IHY Campaigns: Transition to the International Space Weather Initiative

Jean-Pierre Raulin[1], Joseph M. Davila[2], Thomas Bogdan[3], Kiyohumi Yumoto[4] and John Leibacher[5]

[1] CRAAM/EE, Universidade Presbiteriana Mackenzie, São Paulo, Brazil
email: raulin@craam.mackenzie.br

[2] NASA - Goddard Space Flight Center, Greenbelt, MD, USA
email: joseph.m.davila@nasa.gov

[3] NOAA - Space Weather Prediction Center, Boulder, CO, USA
email: Tom.Bogdan@noaa.gov

[4] SERC - Kyushu University, Fukuoka, Japan
email: yumoto@serc.kyushu-u.ac.jp

[5] NSO - National Solar Observatory, Tucson, AZ, USA
email: jleibacher@nso.edu

Abstract. We will present the relevant activities performed during the International Heliophysical Year (IHY) program during the 5 year period 2004 - 2008. The IHY was a major international effort that involved the deployment of new instrumentation, new observations from the ground and in space, and a strong education component. Under the United Nations Office for Outer Space program called Basic Space Science Initiative (UNBSSI), instrument arrays have been deployed to provide global measurements of heliophysical phenomena. As a result, significant scientific and educational collaborations emerged between the organizing groups and the host country teams. In view of the great successes achieved by the IHY during these years, we propose to continue the highly successful collaboration with the UN program to study the universal processes in the solar system that affect the interplanetary and terrestrial environments, and to continue to coordinate the deployment and operation of new and existing instrumentation arrays aimed at understanding the impacts of Space Weather on Earth and the near-Earth environment. To this end, we propose a new program, the International Space Weather Initiative (ISWI). The ISWI strongly complements the International Living With a Star (ILWS) program, providing more attention nationally, regionally, and internationally for the ILWS program. Based on a three-year program activity, the ISWI would provide the opportunity for scientists around the world to participate in this exciting quest to understand the effect of space disturbances on our Earth environment.

Keywords. astronomical data bases: miscellaneous, Sun: general, solar-terrestrial relations, solar system: general.

1. Introduction

The International Heliophysical Year (IHY) started in February, 2007 and was a major international effort with three main objectives: (i) to advance the understanding of fundamental heliophysical processes that govern the Sun, Earth and heliosphere, (ii) continue the tradition of international research and advancing the legacy of the International Geophysical Year (IGY) in 1957, and (iii) demonstrate the beauty, relevance and significance of space and Earth science to the world. IHY marks the 50th anniversary of IGY 1957 which was one of the most successul international science program of all times by establishing a new basis for the development of new space science and technology. IHY therefore extended the concept and frontiers of Geophysics to Heliophysics, by studying not only atmospheric and solar-terrestrial physics, but also other planets and the outer reaches of the heliosphere and how it interacts with the interstellar medium.

2. Main Achievements of IHY

IHY science objectives were reached through the study of universal processes. A major scientific activity was developed with the Coordinated Investigation Programs (CIP) when thousands of scientists worlwide from 6 regions and 71 countries interacted for a better understanding of heliophysical processes. As a result 65 CIPs from 200 international institutions in North and South America, Europe, Africa, Asia and Australia were proposed, and many regional and international workshops took place.

One of the main components of the IHY was the collaboration with the United Nations through the United Nations Basic Space Science Initiative (UNBSSI). This initiative was dedicated to the establishment of observatories and instrument arrays for increasing the understanding of space science and the viability of space science research, engineering and education in developing countries and regions not yet involved in space research. As a result IHY provided the framework for the deployment of arrays of small instruments to perform global measurements of space physics-related phenomena. The framework of the UNBSSI/IHY collaboration was defined and established in a series of international workshops hold by UNOOSA (United Nations Office for Outer Space Affairs) in cooperation with the space agencies NASA, ESA and JAXA. During these planning meetings held during the period 2005 - 2009 in the United Arab Emirates, India, Japan, Bulgaria and South Korea, scientists from developing and industrialized nations met to discuss and develop collaborative projects. Low-cost scientific instruments were provided by scientists from developed countries to scientists from developing nations, and the latter provided sites for the instruments, to maintain them and control the data quality. As a result, 14 arrays of simple and low cost instruments were installed in all continents, to provide observations of the Sun and its varying activity, the solar wind and solar wind-planetary relationship, space weather, solar variation impacts on the Earth atmosphere, cosmic rays and the limits of the heliosphere.

The last objective of the IHY was to communicate the unique scientific results of the program to the scientific community and to the public. As mentioned earlier, several workshops were organized to present and discuss the results obtained from about 65 coordinated investigation programs proposed all over the world. The educational outreach during IHY and the dissemination to a larger public were also important. Many students got their graduate and/or post-graduate degrees by being involved in the instrument arrays installation and in the data analysis and interpretation, and these will in turn be able to raise the interest and curiosity of younger people to the beauty of space physics phenomena. During the IHY period about 15 summer schools were organized in North and Latin America, the Asian-Pacific region, Europe and Africa, and the Balkan-Black Sea-Caspian region, with most of the schools exceeding 100 students.

3. International Heliophysical Year to International Space Weather Initiative

IHY 2007 which has strengthened the mark of 50 years of space exploration officially came to an end on February 2009. During these few years, this program has been dedicated to the study of the heliosphere as an integrated system composed of the Sun, the interplanetary medium, the planets and the Earth environment. At the same time that humanity is celebrating the International Year of Astronomy in 2009, many of the IHY 2007 participants hope for an amplification of their efforts.

To build on the legacy from the IHY achievments, a new UN initiative has been adopted by the UNOOSA: the International Space Weather Initiative (ISWI). The goal of the ISWI is to continue the highly successful collaboration with the UN to encourage new and existing scientific collaborations to study the universal processes in the heliosphere and their interaction with planetary environments, and to continue to coordinate the deployment and operation of new and existing instrumental arrays. Finally, ISWI will strongly complement the International Living With a Star (ILWS) program providing more attention nationally, regionally and internationally for ILWS activities.

IHY has allowed establishing and deploying many instrumental programs to study heliospheric physics, a discipline which includes solar physics as well as planetary magnetospheric and ionospheric physics. However, and because of this multi-disciplinar environment, there is a strong need for data access to and further analysis by a broader scientific community. To this end, universal data access and analysis systems are required. Similarly, virtual observatories should be of great help to address scientific questions that span disciplinary boundaries. In practice, such observatory networks should be able to facilitate science research performing through different heliospheric domains, and broaden significantly the user base.

During the new ISWI program emphasis should be given to space weather studies and their implications on human life and technology both in space and on the surface of the Earth. Understanding extreme events in our solar system is a necessary condition for safe human space travel to planets in the future. Such an effort to derive practical applications of science is also important for developing countries to get better local support for ground based observations of space phenomena and data transfer and handling. The resulting data products also need to be coordinated to allow predictive relationships that enable the forecasting of space weather to be established.

During ISWI a large effort will be made to support and encourage space science courses and curricula in academic instituitions that provide support to instrumental arrays. Similarly it will be important to develop public outreach material unique to the ISWI program, and to coordinate its distribution.

A three-year plan for ISWI has been presented, discussed and recommended by the Science and Technical Subcommittee (STSC) to the UN COPUOUS (Committee on the Peaceful Uses of Outer Space). As a result, this first plan for ISWI will be part of an upcoming UN General Assembly resolution.

4. Conclusions

The International Heliophysical Year was a major program of scientific collaboration which led to great achievments. To build upon this success, the community proposes a new initiative: the International Space Weather Initiative to provide the opportunity for scientists around the world to participate in this exciting quest to understand the effect of space disturbances on our terrestrial environment.

Acknowledgements

JPR would like to acknowledge the support of CNPq (Grant 304433/2004-7), FAPESP (Grant 2006/02979-0) and MACKPESQUISA agencies.

Highlights of Astronomy, Volume 15
XXVIIth IAU General Assembly, August 2009
Ian F. Corbett, ed.

© International Astronomical Union 2010
doi:10.1017/S1743921310010409

Special Session 1 - IR and Sub-mm Spectroscopy - a new tool for studying Stellar Evolution

Glenn Wahlgren[1], Hans Käufl[2] & Florian Kerber[2]

[1] Goddard Space Flight Center, NASA/GSFC, Catholic University of America, US Greenbelt, MD 20771, United States

[2] ESO, Karl-Schwarzschild-Strasse 2, 85748 Garching bei München, Germany

Preface

Infrared astronomy has come into its own over the last decade. Based on mature detector technology and sophisticated instrumentation it is contributing exciting science in many fields of astrophysics. Stellar evolution is a field that has long been dominated by ultraviolet and optical work, but one that has benefited from a strongly increasing contribution from the infrared (IR) and sub-millimeter (sub-mm) domains. In particular, spectroscopy in these domains holds the promise to enable important advances through quantitative analysis of individual stars and stellar systems.

All facets of stellar evolution, from star formation to advanced stages of stellar evolution will be impacted by the higher sensitivity and fidelity of the new astronomical spectral observations. Infrared and sub-mm observations, in particular, are needed for accurate quantitative analysis of: proto-stars and the formation of multiple star systems, formation of proto-planetary and debris disks, massive star properties and winds, stellar outer envelopes for mass loss in AGB stars, velocity fields for mass outflows and jets from young stars and eruptive variables, tests of stellar nucleosynthesis using post-main sequence stars, chemical composition studies for cool stars, cold gas, and high excitation, low density plasmas, stellar magnetic fields from magnetically sensitive IR spectral lines, and chemistry and kinematics of red giants and supergiants in metal-rich stellar clusters and galaxy fields.

The progress to be made in astrophysics is linked to advances in technology. In the recent past there have been at least three obvious advances in instrumental capabilities:

• a tremendous gain in sensitivity, be it the follow up of first-generation spacecraft (*IRAS* and *ISO*) by today's *Spitzer Space Telescope* or *WISE* or when comparing the first-generation single-detector CVF photometers at seeing limited 2 to 4 m class telescopes with multi-mode instruments or diffraction limited 8 to 10 m telescopes with $(2k)^2$-detectors being considered state-of-the-art,

• a gain in resolving power over the IRAS resolving power of $\frac{\lambda}{\Delta\lambda} \approx 20$. At the onset it was barely possible to discriminate solid-state features from atomic emission lines. Adequate spectral resolution was only achieved with Fourier transform spectrometers or heterodyne receivers, albeit with extremely low sensitivity. Only recently have sensitive spectrographs become available, allowing for meaningful single-line absorption spectroscopy in stellar atmospheres,

• and adaptive optics, which has allowed for diffraction limited spectroscopy relevant to investigating the circumstellar environment, is now readily available. More sophisticated techniques, such as interferometry or spectro-astrometry, routinely provide for spectroscopic information on milli-arcsec scales.

At this point, the rapid evolution in IR spectroscopy, mostly based on detector technology, has subsided and substantially larger telescopes are needed. Detailed trade-offs are required from first-generation instrumentation as to how to use the detector pixel real-estate: echelle cross-dispersion versus single-order integral-field spectroscopy. It appears that for stellar evolution spectral coverage (cross dispersion) is more important than the integral field. However, an even more basic concern is that first-generation instrumentation on Extremely Large Telescopes (ELTs) may not include spectrographs capable of resolving stellar lines. Unfortunately, no IR multi-object fiber-fed spectrographs are on the horizon, even though this concept has been proven invaluable in optical spectroscopy. In addition, although spectropolarimetry in the IR offers fundamentally better performance (higher contrast and weaker fields possible), this technique has unfortunately not yet been implemented anywhere.

An optimistic outlook on IR and sub-mm spectroscopy is justified by several developments coming together at this point in time, which will effectively turn it into a new tool that is considerably more potent than the sum of its parts.

- Evolution of technology relevant to IR and sub-mm spectroscopy is continuing to advance, resulting in evermore powerful combinations of instruments and detectors. The *Spitzer* and AKARI IR space observatories and modern high-resolution spectrographs (Phoenix, CRIRES, VISIR, TEXES) operating in the 1 to 25 micrometer region are cases in point. In the near term several large dedicated facilities on the ground (ALMA), at high altitudes (SOFIA), and in space (*Herschel Space Telescope*) will become available, which will significantly enhance the observational capabilities in terms of spatial and spectral resolution, sensitivity and access to wavelength domains.

- Large IR surveys, such as UKIDSS and the *Spitzer* Legacy Programs, or the upcoming VISTA surveys, produce extensive public data sets that can be mined for intriguing objects and follow-up observations with state-of-the-art instrumentation. For example, large IR surveys (e.g., Gould Belt) are now providing complete samples of young stellar objects in clouds down to the planetary mass regime, allowing time scales for different evolutionary phases and processes of clustered versus non-clustered star formation to be tested.

- The IR spectral range accessible today spans some four octaves; therefore, advances in calibration techniques are necessary to remove the need to rely on a single method, e.g. atmospheric features, for frequency calibration. With the vast development of sensitive high-dispersion grating spectrographs the issue of frequency calibration in IR spectroscopy became relevant. The calibration standards available today, telluric atmosphere hollow cathode lamps characterized in the IR and the NIST legacy, connecting a broad selection of molecular transitions to be used in gas-cells to the time standard, have paved a very convenient way. More sophisticated methods, etalons and frequency combs, are on the horizon, ensuring that frequency calibration never will be a limiting factor in spectrum analysis. Indeed, absolute precision traceable to the time standard will soon become the norm.

- The advances in stellar modeling have also helped to establish spectrophotometric standard stars. While in the near-IR one can select true standard stars, such as white dwarfs, in the thermal-IR regime compromises must be made due to the reduced instrument sensitivity. Good general modeling of SEDs and detailed spectra for atomic lines are now available.

- Excellent facilities in laboratory astrophysics are starting to make significant progress in establishing the fundamental data necessary to provide the required ground truth for the IR domain, covering many aspects of atomic, molecular and solid state physics. Calibration sources (discharge lamps and gas cells) traceable to laboratory standards are becoming available for the IR. These advances in calibration techniques are changing the paradigm for wavelength calibration to an approach similar to the ultraviolet-optical domain. Still, the Earth's atmosphere will remain a valuable tool for wavelength calibration in some spectral regions.

- Sophisticated modelling of stellar atmospheres and the theory of stellar evolution is providing a more complete picture with which the growing mosaic of observations and measurements can be compared. State-of-the-art database infrastructure is making all such data easily accessible on an unprecedented scale.

- As has been demonstrated at this conference, and elsewhere in the literature, the HITRAN data base combined with a line-by-line radiative transfer code now also allows the removal of effects from telluric absorption to a high level of precision, e.g. by employing a realistic treatment of pressure broadening in combination with true atmospheric profiles available from meteorological databases. Only occasional telluric standard star observations are required, hence the efficiency of ground-based spectroscopy benefits fundamentally, an important issues when it comes to ELTs, where observing time will be very expensive.

- Finally, progress is being made towards better characterizing the properties of the Earth's atmosphere and its impact on astronomical observations. Active real-time correction of seeing through adaptive optics (AO) facilities, also involving laser guide stars, will become routine operations. Especially in the era of ELTs, source confusion will be a real problem. Only AO-assisted spectroscopy will allow unambiguous studies of stellar populations beyond the closest members of the local group. This is only possible in the near-IR and thus, to prepare ELT programs for near-IR stellar spectroscopy is of paramount importance. Measurement of the water vapour content of the atmosphere above an observatory is about to become a routine practice. A stand-alone monitor of precipitable water vapour allows to assess the quality of a given night and will greatly facilitate the scheduling of IR observations in service mode.

It was our aim that this General Assembly Special Session foster collaboration between various fields, bringing together experts from theoretical and observational astrophysics, instrumentation and laboratory spectroscopy to develop an integrated approach to applying IR and sub-mm spectroscopy to the study of stellar evolution. In combination, these fields hold the key for the scientific success of the current and planned facilities. New observations will foster new thinking about old problems, reveal unexpected phenomena and lead to transformative thinking. This conference occurred at a particularly advantageous time for transferring knowledge in IR and sub-mm spectroscopy from mission to mission. Certain space missions have produced a wealth of data; *Spitzer Space Telescope* completed its cryogenic mission and entered the "warm" phase, while AKARI had ended operations after completing an all-sky survey. Others are either undergoing their on-orbit commissioning phase (*Herschel Space Telescope*) or are making advanced preparations (SOFIA). New ground-based facilities have matured to be able to present results of unprecedented quality.

This symposium engaged the laboratory data producers and users in fruitful discussions that educated the astronomers on the limitations of available data and lead to new collaborations and projects for laboratory investigations and calculations for basic data that are necessary to perform reliable quantitative analyses of the new astronomical spectra. From the perspective of atomic and molecular spectroscopy the advances in astronomical instrumentation at IR and sub-mm wavelengths poses many challenges. Despite the fact that IR spectroscopy has been an active research field for decades, the advent of recent, relatively high-resolution astronomical spectroscopy reveals stark deficiencies in the completeness and accuracy of the atomic, molecular, and solid-state data needed for both the simple tasks of line identification and chemical abundance analysis and more complex tasks of modeling stars and their environments. It is sometimes overlooked that molecular spectroscopy in the radio/sub-mm regime and near-IR spectroscopy observe the same molecules, with the only difference being that radio lines are nearly always pure rotational transitions, while IR transitions also change their vibrational state. Still, protagonists of these two world-views do not often enough make sensitivity calculations for the respective other technology, even though the cross-fertilization might be fundamental. We hope this workshop has helped to combine the two communities, so that ALMA and the 8 to 40 m class telescopes will be utilized in a coherent way.

The broad interest in this Special Session was attested to by the support from a number of Commissions (14, 34, 36, 44, 45) and Working Groups (Red Giants), in additon to the sponsoring Commission 29, representing four IAU Divisions. The conference was well attended at all sessions, and fostered lively discussions. The oral presentations were complemented by poster papers, which are listed at the end of the table of contents. The written contributions presented in *Highlights* for this Special Session are meant to convey the nature of the oral presentations, all of which were of high quality. The speakers were charged with presenting the broader picture in their subjects so that after a few days the attendees could synthesize the status of IR spectroscopy. It was clear that all aspects of stellar astronomy and its supporting technologies are still finding their way in this era of increased activity at IR and sub-mm wavelengths, which makes for an exciting time of rapid discovery and awareness.

It is a great pleasure to acknowledge the contributions of the remainder of the Organizing Committee (France Allard, Thomas Ayres, Steven Federman, Carol Grady, Bengt Gustafsson, Kenneth Hinkle, Chiyoe Koike, John Lattanzio, Gillian Nave, Livia Origlia, Peter Schilke, Jonathan Tennyson, Stepan Urban, and Ewine van Dishoeck) for their role in creating the scientific program. We would also like to thank the IAU for extending financial support to a number of our attendees, Session Chairpersons, and the Brazilian LOC for accomodating all of our needs.

Glenn M. Wahlgren, Florian Kerber, Hans Ulrich Käufl, co-chairs SOC, Washington DC,

Garching bei München, December 1, 2009

Highlights of Astronomy, Volume 15
XXVIIth IAU General Assembly, August 2009
Ian F. Corbett, ed.
© International Astronomical Union 2010
doi:10.1017/S1743921310010410

SpS1-Infrared and submillimetre-wave spectroscopy as probes of stellar evolution

Sun Kwok[1]

[1]Department of Physics, The University of Hong Kong, Hong Kong, China
email: sunkwok@hku.hk

1. Introduction

For over a hundred years, optical spectroscopy has been the main tool to study stellar structure and evolution. Photospheric spectra of the electronic transitions of atoms and ions are used to determine the temperature and elemental abundance. Beyond atomic and ionic lines, only the electronic transitions of a few simple molecules (C_2, CN, H_2O, TiO, CH, etc.) appear in the optical photospheric spectra. With the recent development of infrared and submm spectroscopy, a wide range of molecules have been observed, specially in cool atmospheres of red giants and brown dwarfs. We also realize that beyond the photosphere, a stellar system consists of chromosphere, corona, and stellar wind. Both young and evolved stars possess extensive circumstellar regions and the atoms, molecules, and solid particles in this environment radiate a wide range of lines and bands observable at infrared and submm wavelengths.

2. Atoms and ions

Many of the fine structure lines of common neutral atoms and ions lie in the infrared part of the spectrum. The ground state of neutral carbon (C I) has two fine-structure lines $^3P_2 - {}^3P_1$ (370 μm) and $^3P_1 - {}^3P_0$ (609 μm). Neutral oxygen (O I) has fine-structure lines $^3P_1 - {}^3P_2$ (63 μm) and $^3P_0 - {}^3P_1$ (145 μm). Other common fine-structure lines of ions include the $^2P_{3/2} - {}^2P_{1/2}$ line of C II at 158 μm, the $^3P_2 - {}^3P_1$ (122 μm) and $^3P_1 - {}^3P_0$ (205 μm) lines of N II, and the $^3P_2 - {}^3P_1$ (52 μm) and $^3P_1 - {}^3P_0$ (88 μm) of [O III]. Atoms with an odd number of nucleons can have hyperfine lines observable in the infrared. Examples are the 158 μm $^2P_{3/2} - {}^2P_{1/2}$ $F = 2-1$ line of [^{13}C II], the 9.0 μm $^3P_1 - {}^3P_2$ line of [Na IV] and the 3.66 μm line of [Al VI].

3. Molecules

Since the development of mm/submm receivers, the rotational transitions of over 60 molecules have been discovered in the circumstellar envelopes of evolved stars. These include inorganics (e.g., CO, SiO, SiS, NH_3, AlCl, etc.), organics (CH_4, H_2CO, CH_3CN, etc), radicals (CN, C_2H, C_3, HCO^+, etc.), chains (e.g., HCN, HC_3N, HC_5N, etc.), and rings (C_3H_2). The rotational transitions of many of the lighter molecules lie in the submillimeter range. Some examples are the $N = 2 \rightarrow 1$ $^2\Pi_{3/2} \rightarrow {}^2\Pi_{1/2}$ at 149.390 μm $(- \rightarrow +)$ and 149.091 μm $(+ \rightarrow -)$ of CH, the ground-state ortho rotational transition $1_{10} - 1_{01}$ (537.7 μm) of H_2O, the $(N, J) = 3, 3 \rightarrow 1, 2$ (615.7 μm) of O_2, the $1_{11} - 0_{00}$ (127.65 μm) line of methylene (CH_2).

Infrared spectroscopy from space has opened up the possibility of detecting the stretching and bending modes of molecules. For example, the fundamental vibrational mode of HCN consists of the stretching modes ν_1 (100 $\leftrightarrow$ 000) at 4.8 μm and ν_3 (001 $\leftrightarrow$ 000) at 3 μm and bending mode ν_2 (010 $\leftrightarrow$ 000) at 14 μm. The 13.7 μm ν_5 *cis*-bend of acetylene (C_2H_2) is commonly observed in absorption in extreme carbon stars. For CO_2, in addition to the $01^10 \leftrightarrow 00^00$ fundamental ν_2 bending mode at 14.97 μm, hot bends (030 $\leftrightarrow$ 020 at 15.40 μm, 020 $\leftrightarrow$ 010 at 16.18 μm) and combination bands (100 $\leftrightarrow$ 010 at 13.87 μm and 110 $\leftrightarrow$ 020 at 13.48 μm) have been observed. C_4H_2 at 15.9 μm, ν_{11} of C_6H_2 at 16.1 μm), ν_4 of benzene (C_6H_6) at 14.84 μm, and cyanopolyynes (ν_5 of HC_3N at 15.1 μm, ν_7 of HC_5N at 15.6 μm) have been detected in absorption in the proto-planetary nebula AFGL 618. Pure carbon chains (C_n) have no permanent electric-dipole moment and therefore have no permitted rotational transitions, but their stretching and bending modes can be observed in the infrared. For example, the stretching modes of C_3 (ν_3 at

4.90 μm) and C_5 (ν_5 at 4.62 μm) have been detected in absorption in carbon star spectra. The ν_2 $(0, 1^1, 0) \leftrightarrow (0, 0^0, 0)$ bending mode of C_3 near 158 μm can be observed with the *Herschel Space Observatory*. $J = 2 - 0$ at 28.2188 μm) and *ortho* H_2 ($S(1)$, $J = 3 - 1$ at 17.0348 μm) have been observed by the *Spitzer Space Telescope*. Since HD is no longer homonuclear and has a dipole moment, the $v = 0, J = 1 \rightarrow 0, J = 2 \rightarrow 1$, and $J = 3 \rightarrow 2$ rotational transitions at 112.072, 56.2, and 37.7 μm can be used to provide accurate determination of HD and to infer the deuterium abundance.

4. Solids

The 9.7 μm Si$-$O stretching and the 18 μm O$-$Si$-$O bending modes of amorphous silicates are the most commonly observed solid-state features in stars, with over 4000 such spectra recorded by the *IRAS Low Resolution Spectrometer*. The O$-$H stretching mode (3 μm) and bending mode (6 μm) of amorphous water ice is also widely seen. Other common forms of ice features include the 4.67 μm C$-$O stretch of CO, the 15.3 μm O$-$C$-$O bend of CO_2, the 7.7 μm C$-$H deformation mode of CH_4. A family of infrared emission features at 3.3, 6.2, 7.7, 8.6, and 11.3 μm have been identified as due to stretching and bending modes of aromatic compounds. Features at 3.4 and 6.9 μm can be attributed to the C$-$H stretching and bending modes of aliphatic units. Broad emission plateaus at 8 and 12 μm are likely to be due to the in-plane and out-of-plane bending modes of a mixture of aliphatic groups attached to aromatic rings. Amorphous carbonaceous solids with mixed sp^2/sp^3 structures are now known to be rapidly synthesized in the circumstellar environment of evolved stars.

5. Conclusions

With new observing facilities such as *Herschel*, *SOFIA*, and *ALMA*, we will have powerful tools in the infrared and submm wavelengths to probe the physical and chemical structures of stellar environments.

Acknowledgements

This work is supported by a grant from the Research Grants Council of the Hong Kong Special Administrative Region, China (Project No. HKU 7020/08P).

References

Kwok, S. 2007, *Physics and Chemistry of the Interstellar Medium*, University Science Books

Highlights of Astronomy, Volume 15
XXVIIth IAU General Assembly, August 2009
Ian F. Corbett, ed.

SpS1-Preparing for the harvest from large infrared surveys

Deborah L. Padgett[1]

[1] Spitzer Science Center, California Institute of Technology, M/C 220-6, Pasadena, California
91214, USA
email: dlp@ipac.caltech.edu

1. The legacy of IRAS

During the past decade, there has been a revolution in the availability of multi-wavelength astronomical surveys. From the Sloan Digital Sky Survey (SDSS) to the NRAO VLA Sky Survey (NVSS), astronomical research based on publicly accessible datasets is becoming standard practice in the community. Beginning with the Infrared Astronomical Satellite (IRAS) mission, infrared surveys have played a critical role in stellar astronomy by identifying cool and dusty stars worthy of spectroscopic characterization. IRAS' four photometric bands at 12, 25, 60, and 100 μm were ideal for detecting dusty circumstellar material. All-sky surveys like IRAS reveal the brightest members of each class of rare objects, optimizing their follow-up strategy. The case of debris disks around main sequence stars demonstrates this utility. IRAS detected dust disks around four nearby stars, Beta Pictoris, Fomalhaut, Epsilon Eridani, and Vega. The "Fabulous Four" remain the best studied debris disks, despite hundreds of additional examples discovered by the *SpitzerSpaceTelescope*. In the nearly 30 years since IRAS was launched, its highly reliable catalog of just 250000 sources, modest by modern standards, with arcminute scale resolution and 0.3 - 1 Jy sensitivity, has generated over 10,000 references in ADS. This is a success story by any measure.

2. Ground-based surveys

Ground-based infrared surveys have led the way in discovering and characterizing brown dwarfs. The Two Micron All Sky Survey (2MASS) mapped the sky in the near-infrared, with sensitivity limits of $J = 15.8$, $H = 15.1$, and $K = 14.3$. Together with the SDSS and DENIS surveys, 2MASS has determined the space density of ultracool dwarf stars (Reid *et al.* 2008) and T dwarfs (Metchev *et al.* 2008). The new UKIRT Infrared Deep Sky Survey (UKIDSS) is approximately five magnitudes deeper than 2MASS, but it only covers a quarter of the sky. Early results include the lowest temperature brown dwarf (550 K; Burningham *et al.* 2009). The high quality astrometry and photometry in 2MASS makes it an invaluable tool in constructing stellar spectral energy distributions and conducting proper motion studies. The 2MASS catalogs and image server can be found at IRSA (irsa.ipac.caltech.edu), which is also the archive for data from IRAS, *Spitzer*, and WISE.

3. Large galactic surveys with *Spitzer*

The *SpitzerSpaceTelescope* is an 85 cm telescope which operated cryogenically from 2003 to 2005 at wavelengths from 3.6 to 160 μm. Its warm mission at 3.6 and 4.5 μm is ongoing. Although *Spitzer* is a pointed observatory, it has conducted a number of large "Legacy" surveys of interest to stellar astronomers. Typically, these projects include enhanced data products, such as large mosaics and source catalogs, available via IRSA and the Spitzer Leopard archive tool. These include: 1) imaging surveys of the galactic plane from 3.6 to 70 μm: GLIMPSE (PI: Churchwell), MIPSGAL (PI: Carey), GLIMPSE II (PI: Churchwell), MIPSGAL II (PI: Carey), GLIMPSE 3D (PI: Benjamin), SMOG (PI: Carey). Together, these surveys map most of the Milky Way within ± 5 degrees of the plane. Typical sensitivities for these surveys are tens of microJy at the shortest IRAC bands and 1 mJy at MIPS 24 μm. 2) imaging surveys of nearby star-forming regions from 3.6 to 160 μm. These include Cores to Disks (c2d; PI: Evans), Taurus 2 (PI: Padgett), Gould's Belt (PI: Allen), and Cygnus-X (PI: Hora). These are imaging maps of individual star-forming clouds with areas of up to 45 square degrees (Taurus). The sensitivity of these maps ($\sim$10 μJy at the shortest IRAC bands) are higher than the galactic plane surveys due to longer individual exposures. 3) pointed surveys of nearby stars. These include the FEPS survey (PI: Meyer) of several hundred stars from 5 Myr to 5 Gyr and a variety of large guaranteed time and general observer programs which observed many stars within 50 pc for debris disks.

4. The Wide-field Infrared Survey Explorer (WISE)

WISE is a 40 cm telescope which will be cryogenically cooled by solid hydrogen when it is launched into a sun-synchronous orbit in early December 2009. WISE will scan the sky continuously, mapping the entire sky 1.5 times in its ten months of expected cryogenic lifetime. Its large format infrared detectors will produce simultaneous images at 3.3, 4.5, 12, and 22 μm with resolution of 6 arcseconds (12 at 22 μm). The sensitivity is expected to be 0.12 mJy @ 3.3 μm, 0.16 mJy @ 4.6 μm, 0.85 mJy @ 12 μm, and 4 mJy @ 22 μm in the minimum coverage area on the ecliptic. WISE will detect all of the warm debris disks around solar-type stars within 100 pc. It will also be the premier tool for finding nearby low temperature brown dwarfs. Predictions indicate that WISE should detect at least a handful of 300 K T dwarfs and quite likely will observe a brown dwarf system closer than Proxima Cen. Information on WISE can be found at www.astro.ucla.edu/ wright/WISE/.

References

Reid, I. N. *et al.* 2008, *AJ* 136, 1290

Metchev, S. A., Kirkpatrick, J. D., Berriman, G. B., & Looper, D. 2008, *Ap.J.* 676, 1281

Burningham,B. *et al.* 2009, *MNRAS* 395, 1237

Highlights of Astronomy, Volume 15
XXVIIth IAU General Assembly, August 2009
Ian F. Corbett, ed.

SpS1-The Spitzer atlas of stellar spectra

David R. Ardila, W. Makowiecki, S. van Dyk, I. Song, J. Stauffer, J. Rho, S. Fajardo-Acosta, D.W. Hoard and S. Wachter

[1] Spitzer Science Center, California Institute of Technology, M/C 220-6, Pasadena, California 91214, USA
email: dlp@ipac.caltech.edu

We present Spitzer Space Telescope spectra of 147 stars (R~64 - 128, $\lambda\lambda = 5 - 35$ μm, S/N~100) covering most spectral and luminosity classes within the HR diagram. The spectra are available from the NASA/IPAC Infrared Science Archive (IRSA) and from the first author's webpage (http://web.ipac.caltech.edu/staff/ardila/Atlas/). The Atlas contains spectra of 'typical' stars, which may serve to refine galactic synthesis models, study stellar atmospheres, and establish a legacy for future IR missions, such as JWST.

The targets were selected from available public archival observations, complemented with a Spitzer cycle 5 DDT proposal (pid 485, PI Ardila). The priority was to include naked photospheres observed in the complete spectral range of the low-resolution Infrared Spectrograph (IRS). The presence of excess was determined by comparing the observed spectrum to an Engelke function (Engelke 1992). We have avoided objects described as young stars, RS CVn stars, Be stars, or spectroscopic binaries in the Simbad database. Spectral types were taken from the literature. We have included Wolf-Rayet and Emission-Line stars, as well as some stars of intrinsic interest (i.e. Cepheids, blue stragglers) independently of whether or not they present excess.

The spectra provided by IRS/SSC pipeline S18.7.0 were corrected for residual calibration errors, repeated observations were averaged, and the orders were matched. The error in the overall flux level is that of the standard IRS calibration (~5%). The spectra have not been corrected for interstellar extinction. In this Atlas, most stars from bright luminosity classes show interstellar silicate absorption at 10 μm as well as reddened spectral slopes. For nearby (intrinsically faint) stars, the reddening correction is negligible.

The main sequence presents few spectral features over most of the temperature range. Lines from the Humphreys series of Hydrogen are present for some early A dwarfs. The fundamental SiO band (7.55 μm) first becomes noticeable at M0 V and remains strong until M6.5 V when it becomes confused with the H_2O band at 6.27 μm. At later spectral classes, the spectra develop strong CH_4 and NH_3 bands (Cushing et al. 2006). The SiO band is strongly gravity dependent, and is observed as early as K0 for giants (also noted by Heras et al. 2002), remaining strong until M6 III, even while the water band starts developing (around M0 III).

This Atlas inherits a rich history of efforts in classifying the mid-IR spectra of stars. In particular, larger atlases based on IRAS (e.g. Kwok et al. 1997) and ISO (e.g. Kraemer et al. 2002) data are available. The primary goal of those atlases has been to establish an intrinsic IR classification scheme, one that does not rely on the optical spectral classification. The Spitzer Atlas, on the other hand, seeks to understand the spectral sequence at mid-IR wavelengths and as such it provides complementary information to those efforts.

References

Cushing, M. C. et al. 2006, *ApJ*, 648, 614
Heras, A. M. et al. 2002, *A&A*, 394, 539
Kraemer, K. E. et al. 2002, *ApJS*, 140, 389
Kwok, S. et al. 1997, *ApJS*, 112, 557

Highlights of Astronomy, Volume 15
XXVIIth IAU General Assembly, August 2009
Ian F. Corbett, ed.

© International Astronomical Union 2010
doi:10.1017/S1743921310010446

SpS1-From molecular clouds to massive stars

Maria R. Cunningham

School of Physics, University of New South Wales
Sydney, NSW, 2052, Australia
email: maria.cunningham@unsw.edu.au

1. Introduction

The life cycles of massive stars have a major impact on the evolution of galaxies, while in turn, position in galaxy has a major impact on the efficiency and type of star formation which occurs in a molecular cloud (see e.g. Luna *et al.* 2006). However, exactly how massive stars form, on what timescales, and how they shape their environments during this active and energetic phase is poorly understood.

2. Stages of Massive Star Formation

The processes leading from the formation of the molecular phase of the interstellar medium (ISM) to the formation of a massive star or cluster may be broken into a number of stages.

Formation of a giant molecular cloud: Giant molecular clouds (GMCs) are predominantly found in the spiral arms of the Galaxy, embedded in atomic hydrogen gas (HI), forming from HI in regions of enhanced density. The causes of the density enhancements are not known, but it is likely that turbulence, gravitational instability and converging flows of atomic gas all play a part (see e.g. Hennebelle *et al.* 2008). The origin of GMCs is important to the massive star formation story. Molecular clouds are turbulent over a wide range of environments and size scales (Heyer & Brunt 2004), and turbulence likely plays a significant part in regulating star formation (see review by Mac Low & Klessen 2004). The energy to drive turbulence must be constantly replenished, or the turbulent motions will quickly die out (see e.g. Stone, Ostriker, & Gammie, 1998). While, the origin and size scales of the energy injection are unknown, there is now a consistent body of literature pointing to the likelihood that turbulence is predominantly driven at scales as large as or larger than the extent of the molecular component of the gas (Brunt, Heyer & Mac Low 2009). Hence, the turbulent cascade may start in the atomic gas.

From giant molecular cloud to dense cores: *There is a basic problem which needs to be explained by any model for massive star formation:* In the Milky Way galaxy, only one star forms each year, on average, compared to the hundreds which should be forming if the gas in molecular clouds is simply collapsing under its own weight. Some physical process must act to limit the rate of star formation, and turbulence and magnetic fields have both been suggested as possibilities. Whether stars form: 1) quickly due to triggering by turbulent motions, with disruption by turbulent disturbances then limiting star formation efficiency, or 2) slowly due to gravitational collapse regulated by magnetic fields is debated (see e.g Elmegreen & Scalo 2004; Mouschovias *et al.* 2006), although both phenomena are likely important (see e.g. Nakamura & Li 2008).

In an effort to characterise turbulence observationally, we have mapped a 0.6 × 1.2 square-degree region of the southern Galactic plane near G333, in molecular transitions near 3 mm that trace varying critical densities (e.g. ^{13}CO, CS, N_2H^+), using the Mopra radio telescope (Bains *et al.* 2006; Wong *et al* 2008; Lo *et al* 2009). We have used spatial power spectrum (Lazarian & Pogosyan 2000) and delta variance (Bensch, Stutzki & Ossenkopf 2001) analyses to follow the turbulent cascade through the molecular gas between the scales of ∼0.5 pc to 20 pc. The results are reported elsewhere (Bains *et al.* 2009; Jones *et al.* in prep) but, in summary, all molecules show the same power spectrum slope, suggesting that: 1) turbulent energy is injected at scales larger than ∼20 pc (consistent with Brunt *et al.* 2009), and 2) passes through the gas to the smallest scales examined (∼0.5 pc) without dissipation, suggesting that the turbulent cascade continues to these smaller scales, affecting the formation of dense cores, and their fragmentation into individual protostars.

From dense cores to stars: The process by which a dense core evolves into a massive star or cluster is also debated. Carolan *et al.* (2009) find that massive star formation is most likely to occur by accretion occurring directly from the gas onto the protostar, producing a massive protostellar core, in what is often called the *core* accretion model. The discovery of an isolated massive star forming core in the G333 region (Lo *et al.* 2007) supports this. This is in contrast to the recent observations and simulations of Smith, Longmore & Bonnell (2009), who find that the *competitive* accretion model seems more likely to apply. In this model, only small cores are able to form directly from the gas phase, as accretion is halted by radiation pressure once an object becomes too massive. Massive stars then form by the coalescence of these smaller cores.

References

Bains I. *et al.*, 2006, *MNRAS*, 367, 1609
Bains I. *et al.*, 2009, *PASA*, 26, 110
Bensch F., Stutzki J., & Ossenkopf V., 2001, *A&A*, 366, 636
Brunt, C. M., Heyer, M. H., & Mac Low, M.-M. 2009, *A&A*, 504, 883
Carolan P. B. *et al.*, 2009, *MNRAS*, 1353
Elmegreen, B. G. & Scalo, J. 2004, *ARAA*, 42, 211
Hennebelle, P., Banerjee, R., Vázquez-Semadeni, E., Klessen, R. S., & Audit, E. 2008, *A&A*, 486, L43
Heyer, M. H. & Brunt, C. M. 2004, *ApJ*, 615, L45
Lazarian A. & Pogosyan D., 2000, *ApJ*, 537, 720
Lo N. *et al.*, 2009, *MNRAS*, 395, 1021
Lo N., Cunningham M., Bains I., Burton M. G., & Garay G., 2007, *MNRAS*, 381, L30
Luna A., Bronfman L., Carrasco L., & May J., 2006, *ApJ*, 641, 938
Mac Low, M.-M., & Klessen, R. S. 2004, *Reviews of Modern Physics*, 76, 125
Mouschovias, T. C., Tassis, K., & Kunz, M. W. 2006, *ApJ*, 646, 1043
Nakamura, F. & Li, Z.-Y. 2008, *ApJ*, 687, 354
Smith, R. J., Longmore, S., & Bonnell, I. 2009, arXiv:0908.3910
Stone J. M., Ostriker E. C., & Gammie C. F., 1998, *ApJ*, 508, L99
Wong T. *et al.*, 2008, *MNRAS*, 386, 1069

Highlights of Astronomy, Volume 15
XXVIIth IAU General Assembly, August 2009
Ian F. Corbett, ed.

© International Astronomical Union 2010
doi:10.1017/S1743921310010458

Development of jets, outflows and HH objects

A. C. Raga[1], D. López-Cámara[1], J. Cantó[2], A. Esquivel[1],
A. Rodríguez-González[1] and P. F. Velázquez[1]

[1]Instituto de Ciencias Nucleares, Universidad Nacional Autónoma de México, Ap. 70-543,
04510 D. F., México
email: raga@nucleares.unam.mx

[2] Instituto de Astronomía, Universidad Nacional Autónoma de México, Ap. 70-468,
04510 D. F., México

1. Introduction

The entrainment of molecular material through a mixing layer along the walls of a HH jet beam has been modeled analytically (Cantó & Raga 1991; Stahler 1994) and numerically (Taylor & Raga 1995; Lim et al. 1999). However, when full radiative jet simulations are carried out, the molecular, environmental material remains within a dense shell which follows the shape of the leading bow shock. Because of this, no molecular material reaches the outer boundary of the jet beam, and therefore no "side-entrainment" of molecular gas into the fast jet beam takes place.

Therefore, if one wants to model objects in which a central, well collimated, fast molecular jet is observed (e.g., the spectacular HH 212, see Correia et al. 2009 and references therein), it is necessary to either assume that the jet flow is initially partially molecular (Völker et al. 1999) or to choose parameters such that molecules are formed within the jet beam (Raga et al. 2005).

2. Jet in a side-wind

In this paper, we propose the following possibility. Let us assume that an outflow source has a velocity of a few km s^{-1} relative to the surrounding environment (a velocity which would be consistent with the random velocity of T Tauri stars). This will produce a streaming motion of the environment relative to the jet beam, which for appropriate parameters will be able to push the far bow shock wing into contact with the jet beam. In this situation, the dense shell of environmental material (which follows the bow shock wings) will touch the body of the jet beam, therefore allowing side-entrainment of molecular material to take place.

Figure 1 shows the results obtained from a 3D simulation. In this simulation, a jet with a top-hat initial cross section with a velocity $v_j = 150$ km s^{-1}, density $n_j = 10^3$ cm^{-3}, temperature $T_j = 10^3$ K and radius $r_j = 10^{15}$ cm moves into a homogeneous environment of density $n_{env} = 200$ cm^{-3} and temperature $T_{env} = 10$ K. The environment streams past the jet source (in a direction perpendicular to the outflow axis) at a velocity $v_{env} = 5$ km s^{-1}. Both the jet and the environment are initially neutral (except for singly-ionized carbon), and a parametrized radiative energy loss is included. The ionization of hydrogen is explicitly followed.

The mid-plane density stratification of fig. 1 shows that the dense, post-bow-shock shell is indeed swept into direct contact with the jet beam. Through this shell/jet beam interface, direct entrainment of molecular gas into the jet beam can take place.

3. Conclusions

In this short paper we discuss the possibility of side-entrainment of molecular material into a HH jet as a result of a motion of the jet source through the surrounding, molecular environment. This model will be described in more detailed in a future paper.

Acknowledgements

This work was supported by the CONACyT grants 61547, 101356 and 101975.

t = 450 yr

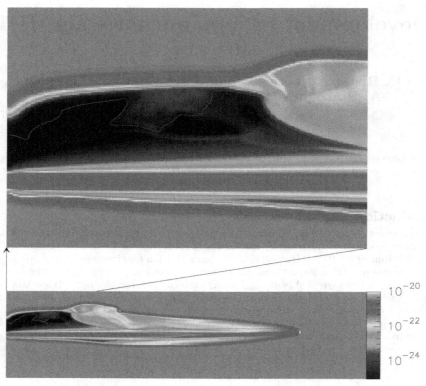

Figure 1. The density stratification (on the plain including the jet axis and the direction of incidence of the streaming environment) obtained from the numerical simulation described in the text after a 450 yr time-integration. The bottom frame has an axial extent of 2×10^{17} cm. The density (in g cm^{-3}) is shown with the look-up-table given by the bottom right bar. The single, black/grey solid line (red respectively in the online material) indicates the interface between the jet and the environmental material.

References

Cantó, J. & Raga, A. C. 1991, ApJ, 372, 646
Correia, S., Zinnecker, H., Ridgway, S. T., & McCaughrean, M. J. 2009, A&A, 505, 673
Lim, A. J., Rawlings, J. M. C., & Williams, D. A. 1999, MNRAS, 308, 1126
Raga, A. C., Williams, D. A., & Lim, A. J 2005, RMxAA, 41, 137
Taylor, S. D. & Raga, A. C. 1995, A&A, 296, 823
Stahler, S. W. 1994, ApJ, 422, 616
Völker, R., Smith, M. D., Suttner, G., & Yorke, H. W. 1999, A&A, 343, 953

Highlights of Astronomy, Volume 15
XXVIIth IAU General Assembly, August 2009
Ian F. Corbett, ed.

SpS1-Circumstellar disks & their evolution: Dust

Carol A. Grady[1]

[1]Eureka Scientific & Goddard Space Flight Center, Greenbelt, MD 20771 USA
email: Carol.A.Grady@nasa.gov

Circumstellar disks are an intrinsic part of star formation and are also where planets form, migrate, and where the materials capable of producing life-bearing worlds are produced. The most flamboyant signatures of the presence of disks are at infrared through millimeter wavelengths, where thermal emission from dust dominates the system light. The discovery and characterization of the emission from such disks has been a major activity for ground-based observatories and space missions (IRAS, ISO, MSX, AKARI, and *Spitzer*), and continues with the newest generation of infrared (IR) capabilities.

Spectral Energy Distributions (SEDs): The presence of excess light compared with the stellar photosphere not only demonstrates the presence of circumstellar material, but its shape offers insight into its spatial distribution of dust. There is a rapid decay in disk frequency over the first 5 Myr, consistent with the formation time for large asteroids, and dating of Saturn (Pascucci & Tachibana 2009). Association average SEDs indicate grain growth to 10× interstellar sizes by 1-2 Myr (Bouwman *et al.* 2008; Sargent *et al.* 2009). The frequency of *transitional* disks, those with warm dust deficits relative to the average SEDS, indicating cleared lanes or cavities, increases over 2-5 Myr (Brown, 2009).

After 8-12 Myr, disk fractional IR luminosities (F_{IRE}/F_*) are $\leqslant 10^{-3}$, and warm dust becomes rare (Moór *et al.* 2009). Overall F_{IRE}/F_* drops with age, but with a large dispersion, permitting identification of *debris* disks which have undergone recent collisional activity (Wyatt 2008). Debris disk SEDs fall into two classes: those which can be fit by discrete blackbody components, indicating material confined to particle belts or rings, and those with more continuous dust distributions (e.g. disks) (Morales *et al.* 2009). Now the SED data can be combined with locations of Jovian-mass planets to map the planetary system architecture (Chen *et al.* 2008; Reidemeister *et al.* 2009).

Solid-State Features: Spectral features due to grains with radii $\leqslant 10\mu m$ are often seen. Absorption is typical of either embedded objects, or stars which are viewed with sight lines passing near the (high inclination) disk midplane. Otherwise, emission is seen.

Ices and Polycyclic Aromatic Hydrocarbons (PAHs): Water and simple molecular ices are seen in young stellar object (YSO) and high inclination disk spectra (Boogert *et al.* 2008; Pontoppidan *et al.* 2008; Gibb *et al.* 2004). Crystalline ice was detected in emission for a few stars by ISO (e.g. Malfait *et al.* 1999), and more recently in debris disks (Chen *et al.* 2008; Lisse *et al.* 2008). Water ice has been detected in reflectance spectra for a Herbig F star disk (Honda *et al.* 2009), but has not been seen in Herbig Ae star disks (Debes *et al.* 2008). Emission in a suite of mid-IR bands identified with PAHs is typical of Herbig Ae stars (Keller *et al.* 2008), is less common in T Tauri stars (Geers *et al.* 2006) and is not seen toward low-mass embedded YSOs (Geers *et al.* 2009). The PAH emission can extend over ≈100 AU of the disk (Habart *et al.* 2006; Geers *et al.* 2007).

Dust: A wealth of silicate, oxide, and other dust features are commonly observed in the spectra of Herbig Ae, T Tauri, and some debris disks. Recent work in this area has focussed on establishing typical grain sizes, degree of silicate crystality, and chemistry for intermediate-mass, T Tauri stars, and brown dwarfs using *Spitzer* IRS data. In addition to the distinct features, broader emission components can be identified using mineral libraries based on Solar System objects. There have been efforts to compare the mineralogy of the inner disks with outer disks from integrated light spectra (Sargent *et al.* 2009). Mid-IR interferometry demonstrates that silicate crystallinity drops with radius for Herbig Ae and bright T Tauri disks (Leinert *et al.* 2004; van Boekel *et al.* 2006). Chemical inventories of dust are now available for a few debris disks (Lisse *et al.* 2008).

Transient Phenomena: Variable solid-state emission has now been reported for Herbig Ae (Sitko *et al.* 2008) and T Tauri stars (Woodward *et al.* 2004). Increased silicate crystallinity during EX Lup's outburst demonstrates that significant grain processing occurs in the inner disks of pre-main-sequence stars (Ábrahám *et al.* 2009). Debris disk transient phenomena include large F_{IRE}/F_*, crystalline water ice, and species indicating hypervelocity impacts (Chen *et al.* 2008; Lisse *et al.* 2009).

The Future: *Herschel*, SOFIA, and SPICA will study crystalline water ice features, crystalline silicates, and hydrous silicates, providing an expanded set of mineral and ice diagnostics. ALMA observations will permit verification of disk structures inferred from modeling of SEDs. JWST, as well as providing high contrast imaging, will also enable spatially-resolved spectroscopic studies of small grains. In tandem with ground-based facilities, there will be new disk identifications from AKARI and WISE data, and more synoptic studies. The disk chemical inventories will facilitate systematic studies of whether *stellar* compositional peculiarities are inherited from their molecular cloud.

References

Ábrahám, P. *et al.*, 2009, *Nature* 459, 224.

Boogert, A. *et al.*, 2008, *ApJ* 678, 985.

Bouwman, J. *et al.*, 2008, *ApJ* 683, 479.

Brown, J. *et al.*, 2009, (this proceedings).

Chen, C. *et al.*, 2009, arXiv 0906.3744.

Chen, C. *et al.*, 2008, *ApJ* 689, 539.

Debes, J. *et al.*, 2008, *ApJ* 673, L191.

Geers, V. *et al.*, 2006, *A&A* 459, 545.

Geers, V. *et al.*, 2007 *A&A* 476, 279.

Geers, V. *et al.*, 2009, *A&A* 495, 837.

Gibb, E. *et al.*, 2004, *ApJS* 151, 35.

Habart, E. *et al.*, 2006, *A&A* 449, 1067.

Honda, M. *et al.*, 2009, *ApJ (Letters)* 690, L110.

Keller, L. *et al.*, 2008 *ApJ* 684, 411.

Leinert, Ch. *et al.*, 2004, *A&A* 423, 537.

Lisse, C. *et al.*, 2008, *ApJ* 673, 1106.

Lisse, C. *et al.*, 2009, arXiv 0906.2536.

Malfait, K. *et al.*, 1999, *A&A* 345, L181.

Moór, A. *et al.*, 2009, *ApJ* 700, L25.

Morales, F. *et al.*, 2009, *ApJ* 699, 1067.

Pascucci, I. & Tachibana, S. 2009, in *Protoplanetary Dust: Astrochemical and Cosmochemical Perspectives* Cambridge Planetary Science (No. 12), editors D. Apai and D. Lauretta, (Cambridge: Cambridge University Press) (in press).

Pontoppidan, K. *et al.*, 2008, *ApJ* 678, 1005.

Reidemeister, M. *et al.*, 2009, arXiv 0905.4688

Sargent, B. *et al.*, 2009, *ApJS* 182, 477.

Sitko, M. *et al.*, 2008, *ApJ* 678, 1070.

van Boekel, R. *et al.*, 2006, *Advances in Stellar Interferometry* SPIE 6268, editors J. Monnier, M. Schöler, W. Danchi, E 13.

Woodward, C. *et al.*, 2004 *Debris Disks and the Formation of Planets*, ASP Conf. Ser. 324 editors L. Caroff, L. J. Moon, D. Backman, E. Praton (ASP: San Francisco), p. 224.

Wyatt, M. C. 2008, *ARA&A* 46, 339.

Highlights of Astronomy, Volume 15
XXVIIth IAU General Assembly, August 2009
Ian F. Corbett, ed.

© International Astronomical Union 2010
doi:10.1017/S1743921310010471

SpS1-Gas in protoplanetary disks

Miwa Goto[1]

[1] Max-Planck-Institut für Astronomie, Königstuhl 17, Heidelberg, D-69117, Germany

1. Introduction

High resolution infrared spectroscopy is the key technique to look at the inner regions of protoplanetary disks. As molecular hydrogen is an inefficient emitter, CO gas is the single most important molecular probe of the disk. The energy gaps of the vibrationally excited levels ($\Delta E > 3000$ K) and the critical density required to keep the molecules in the excited state ($n_c \sim 10^{10} \mathrm{cm}^{-3}$) match well to the physical condition of the inner regions of protoplanetary disks. In order to resolve the vibrational lines of different rotational states, a spectral resolving power of $\lambda/\Delta\lambda > 10\,000$ is necessary; or even higher ($> 30\,000 - 100\,000$), if we would like to fully resolve the gas kinematics. Scoville *et al.* (1980) provided the fundamentals of the excitation mechanisms, which is essential for the interpretation of the vibrational transitions of CO, and pioneered the study of the circumstellar environment with infrared CO lines in the observation of BN (Scoville *et al.* (1983)). The bandhead emission of CO at 2.3 μm from young stars was unambiguously attributed to the circumstellar disks by Carr (1989) and Najita *et al.* (1996), because the gas kinematics matches well to what is expected from Keplerian rotation. Since then, the gas kinematics have been extensively used to shed light on peculiar disk structures, such as the inner truncation (Brittain *et al.* 2003), the outer truncation (Najita *et al.* 2008), and the gap (van der Plas *et al.* 2008; though this is an oxygen forbidden line).

2. High angular resolution

The most notable breakthrough in CO observation in the past decade was the advent of 8-m class telescopes equipped with adaptive optics systems. The high angular resolution attained by these new devices allows us to spatially resolve the line emission of a few nearby disks, without relying on the line kinematics. This can be accomplished by either the spectroastrometric technique (Pontoppidan *et al.* 2008; van der Plas *et al.* 2009) or in a fully resolved manner as in the case of CO line emission from the Herbig Ae star HD 141569A (Goto *et al.* 2006). HD 141569A is a transition object in the class of Herbig Ae/Be stars with little infrared excess at wavelengths shorter than 10 μm. The line images of its disk show that the inner disk is apparently truncated at a radius of 11 AU, which is about half of the nominal gravitational radius of the star. The size of the inner cavity, which was too large to be accounted for by dust sublimation or magnetospheric truncation, strongly indicates that the inner disk is removed by photoevaporation (Clarke *et al.* 2001).

3. Spectroastrometry

Spectroastrometry is an alternative way to squeeze out spatial information from protoplanetary disks by measuring the centroid of an emission line for various wavelengths. In this technique, the size of the point-spread function is not the hard limit of the accuracy of the measurement anymore, but the signal-to-noise ratio of the emission lines. With an integral field spectrograph, spectroastrometry can be done even in two dimensions. An example is the spectroastrometry of Brγ of TW Hya performed by SINFONI at the VLT (Goto *et al.* 2009, submitted). The centroid of the emission line is measured to an accuracy of a hundredth of a pixel, making it possible to resolve the gas kinematics on the physical scale of the radius of the star.

4. Temporal Development

High resolution CO observations are also used to observe the transient phenomena. Pre-main sequence stars are all more or less photometrically variable. The most outstanding variables among them are the outbursts of FU Ori or EX Lup type stars, where the young stars brighten up by several magnitudes in visible light, once in a decade to a century. EX Lup started its biggest outburst on record in early 2008. The gas kinematics during the outburst were monitored at CO 4.7 μm for five months for a total of six epochs. The observations revealed that the inner disk is cut off at 0.2–0.3 AU inward and drained out, while the outer disk does not change. Toward the end of the massive accretion phase, the line profile varied rapidly with a flipping of the asymmetry on timescales of a few days to a week, suggesting that a hot spot emerged on the orbiting gas disk (Goto *et al.* 2009 in prep.).

5. Future Direction

METIS is one of the infrared instruments currently being proposed for E-ELT. It offers imaging and spectroscopic capabilities at 3–14 μm and is fully compatible with the adaptive optics system. The most notable feature of the instrument is the integral field spectroscopy at 3–5 μm. With an improvement of the angular resolution by a factor of five, this instrument will be the cornerstone to the next breakthrough in the understanding of the physical and chemical evolution of the gas in protoplanetary disks.

References

Brittain, S. D. *et al.* 2003, *ApJ*, 588, 535

Carr, J. S. 1989 *ApJ*, 345, 522

Clarke, C. J., Gendrin, A., & Sotomayor, M. 2001, *MNRAS*, 328, 485

Goto *et al.*, 2006, *ApJ*, 652, 758

Najita, J., Carr, J. S., Glassgold, A. E., Shu, F. H., & Tokunaga, A. T. 1996, *ApJ*, 462, 919

Najita, J. R., Crockett, N., & Carr, J. S. 2008, *ApJ*, 687, 1168

Pontoppidan, K. M., *et al.*, 2008, *ApJ*, 684, 1323

Scoville, N., Kleinmann, S. G., Hall, D. N. B., & Ridgway, S. T. 1983, *ApJ*, 275, 201

Scoville, N. Z., Krotkov, R., & Wang, D. 1980, *ApJ*, 240, 929

van der Plas, G. *et al.*, 2008, *A&A*, 485, 487

van der Plas, G. *et al.*, 2009, *A&A*, 500, 1137

Highlights of Astronomy, Volume 15
XXVIIth IAU General Assembly, August 2009
Ian F. Corbett, ed.

© International Astronomical Union 2010
doi:10.1017/S1743921310010483

SpS1-Dust and gas clearing in transitional disks

Joanna M. Brown[1]

[1] Max-Planck-Institut für extraterrestrische Physik, Garching bei München, Germany

Understanding how disks dissipate is essential to studies of planet formation. Infrared observations of young stars demonstrate that optically-thick circumstellar disks disappear from around half the stars in low-mass star-forming regions by an age of 3 Myr and are almost entirely absent in 10 Myr old associations (e.g. Haisch et al., 2001). Accretion ceases on the same approximate timescale (e.g. Calvet et al. 2005). The disappearence of gas and dust - planetary building material - places stringent limits on the timescales of giant planet formation. During this crucial interval, planet(esimal)s form and the remaining disk material is accreted or dispersed. Mid-infrared spectrophotometry of protoplanetary disks has revealed a small sub-class of objects in the midst of losing their disk material. These disks have spectral energy distributions (SEDs) suggestive of large inner gaps with low dust content, often interpreted as a signature of young planets. Such objects are still rare although Spitzer surveys have significantly increased the number of known transitional objects (e.g. Brown et al. 2007, D'Alessio et al., 2005). However, spectrophotometric signatures are indirect and notoriously difficult to interpret as multiple physical scenarios can result in the same SED. Recent direct imaging from millimeter interferometry has confirmed the presence of large inner holes in transitional disks, providing additional constraints and lending confidence to current SED interpretations (Brown et al. 2008, Brown et al. 2009, Andrews et al. 2009, Isella et al., 2009).

Most transitional disks are discovered through the properties of their dust emission and much less is known about the gas content. However, the state of the gas has strong implications for the method of disk evolution. Photoevaporation requires a complete absence of gas in the inner regions, while grain growth and settling can occur with high gas densities. Stellar and planetary companions may allow gas in the inner regions depending on the mass and orbital radius of the companion. Spectroscopy provides tracers of the gas throughout the disk and reveals both the physical and chemical structure. For example, high resolution 5 micron spectra reveal that CO gas is often present inside the dust holes Salyk et al., 2009, Pontoppidan et al., 2008. Prospects for the future are particularly good as Herschel, ALMA and the next generation of ground-based high resolution spectrometers will provide additional clarity on current results as well as access to a larger sample of transitional disks too faint for current instruments.

References

Andrews, S. M., Wilner, D. J., Hughes, A. M., Qi, C., & Dullemond, C. P., 2009, ApJ, 700, 1502

Brown, J. M. et al., 2007, ApJL, 664, L107

Brown, J. M., Blake, G. A., Qi, C., Dullemond, C. P., & Wilner, D. J., 2008, ApJL, 675, L109

Brown, J. M., Blake, G. A., Qi, C., Dullemond, C. P., Wilner, D. J., & Williams, J. P., 2009, ApJ, 704, 496

Calvet, N. et al., 2005, AJ, 129, 935

D'Alessio, P. et al. 2005, ApJ, 621, 461

Haisch, Jr., K. E., Lada, E. A., & Lada, C. J., 2001, ApJL, 553, L153

Isella, A., Carpenter, J. M., & Sargent, A. I., 2009, ApJ, 701, 260

Pontoppidan, K. M., Blake, G. A., van Dishoeck, E. F., Smette, A., Ireland, M. J., & Brown, J., 2008, ApJ, 684, 1323

Salyk, C., Blake, G. A., Boogert, A. C. A., & Brown, J. M., 2009, ApJ, 699, 330

Highlights of Astronomy, Volume 15
XXVIIth IAU General Assembly, August 2009
Ian F. Corbett, ed.

SpS1-Spectroscopic observations of young disk evolution with Herschel and ALMA

W. R. F. Dent

ALMA JAO, Santiago, Chile
email: wdent@alma.cl

1. Introduction

In the next few years, both *Herschel* and ALMA will be providing unique new insights into the physics and chemistry of protoplanetary disks. In particular, they will be used to study how disks evolve from massive embedded systems around young Class 0 objects, through low-mass disks around optically-visible T Tauri stars, to debris disks around stars on the main-sequence. Gas dominates the mass in the younger systems, but in debris systems there is very little - if any. How does the gas disappear, what is the effect of this on planetary formation, and what is the role of "transition" disks? I outline some of the areas where these two large facilities will contribute to these studies, focussing on the Herschel Key project, GASPS, and looking forward to the role of ALMA.

2. Herschel and GASPS

Herschel is a 3.5m telescope launched in early 2009, and currently at the L2 point, 1.5×10^6 km from Earth. It has three cryogenic instruments operating in the far-infrared. Focussing on the spectroscopic capabilities of these instruments, HIFI is a heterodyne receiver for 480-1250 and 1410-1910 GHz (de Graauw *et al.*, 2008), PACS is an instrument with a 5x5 IFU of resolution $\sim$2000, operating in the region 60-200 μm (Poglitsch *et al.*, 2008), and SPIRE includes an imaging FTS of resolution up to 1000 for the 200-700 μm region (Griffin *et al.*, 2008). Several large programs studying disks with spectroscopy have been allocated time, and it is likely that many more open-time projects will be using these instruments during the spacecraft lifetime. The current large projects include a study of the nearby "Fab-4" debris disks, HOBYS (a study of massive young stars), WISH (studying water in young star-forming regions), a HIFI spectral survey of young stellar objects, GASPS (studying gas in protoplanetary systems), DIGIT (dust, ice and gas in young systems) and HOPS (observing protostars in Orion). We focus here on GASPS, which is tailored to study the evolution of gas and dust in protoplanetary disks.

The aim of GASPS is to study several key gas tracers and dust in a wide range of visible young stellar objects. The sample of $\sim$200 targets covers a range of stellar types, disk dust mass, age, UV & X-ray flux, and Hα strength. The species to be observed are [OI] (63 and 145 μm), [CII] (157 μm) and lines of H$_2$O at 78 and 180 μm; these are predicted to be the brightest in such objects. The targets are selected from the nearest star formation regions and associations of ages from 1 to 30 Myr, including Taurus, β Pic Moving Group, η Cha, Tuc Hor and Upper Sco. They are chosen to have relatively low extinction, in order to minimise contamination from nearby cloud emission. Line strengths and ratios will be used to derive physical parameters, by comparison with an extensive model database (Woitke *et al.*, 2009). PACS will not have sufficient spectral resolution to resolve the lines; therefore, the survey will rely on comparison with these models, with dust SEDs, and with other gas species. From this, we will investigate trends over the sample, looking at what parameters affect the line strengths and ratios. GASPS is a survey, and will provide line and continuum strengths. But to derive detailed information about individual disks, both spectral and spatial resolution are necessary; this is what ALMA will provide.

3. ALMA and future studies of gas in disks

ALMA will be a 66-element interferometer, operating at wavelengths from 350 μm to 9 mm at an altitude of 5km in northern Chile. It will have movable antennas and baselines up to 10km, providing resolutions from 1 arcsec down to less than 20 mas at what is probably the prime

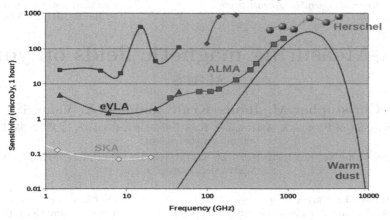

Figure 1. Sensitivity of ALMA, *Herschel* and other facilities. The upper lines (= least sensitive) refer to existing instruments, the solid curve shows a typical SED from warm dust.

observing band, 850 μm. The gain over existing interferometers will be in three main factors: resolution ($\sim$20x), sensitivity (20x more baselines) and sensitivity ($\sim$40x). Figure 1 compares the sensitivity of ALMA with other facilities.

ALMA will be ideal for studying the structure, chemistry and physics of protoplanetary and debris disks, which requires improvements in all three factors to make significant advances. In particular, chemical differentiation, turbulence, in/outflows will be seen on scale of a few AU in the nearest star-formation regions. As well, it should reveal non-axisymmetric disk structure, in both self-gravitating disks and in systems affected by young protoplanets (e.g. Narayanan *et al.*, 2006). The study of dust-rich debris disks should be significantly advanced by ALMA: detection of 1 Kuiper Belt's mass of dust, of the remnant CO and CI in young debris disks (Kamp *et al.*, 2003), and the detection of classical debris disks in Taurus will be possible. The resolution should allow study of clumps on sub-AU scales in disks, looking at structure and motions in these systems.

ALMA will be fully operational by 2013. However, an opportunity to perform early science will be available in 2011. At that time the number of antennas will be no less than 16, and baselines will be limited to 500-1000 m. The resolution will be $\sim$0.5 arcsec at 850 μm with sensitivity already significantly improved on existing facilities.

References

de Graauw *et al.*, 2008, *SPIE* 7010, 4
Griffin *et al.*, 2008, *SPIE* 7010, 6
Kamp *et al.*, 2003, *A&A* 397, 1129
Narayanan *et al.*, 2006, *ApJ* 647, 1426
Poglitsch *et al.*, 2008, *SPIE* 7010, 5
Woitke, Kamp, & Thi 2009, *A&A* 501, 383

Highlights of Astronomy, Volume 15
XXVIIth IAU General Assembly, August 2009   © International Astronomical Union 2010
Ian F. Corbett, ed.   doi:10.1017/S1743921310010501

SpS1-Measuring magnetic fields on young stars

Christopher M. Johns–Krull[1] and Jeff A. Valenti[2]

[1] Department of Physics & Astronomy, Rice University, Houston, TX 77005, USA
email: cmj@rice.edu

[2] Space Telescope Science Institute, 3700 San Martin Dr., Baltimore, MD 21210, USA
email: valenti@stsci.edu

T Tauri stars (TTSs) are young ($\sim$few Myr) late type stars that have only recently emerged from their natal molecular cloud material to become visible at optical wavelengths. It is now generally accepted that accretion of circumstellar disk material onto the surface of a TTS is controlled by a strong stellar magnetic field (e.g. see review by Bouvier *et al.* 2007). The stellar field appears critical for explaining the rotational properties of TTSs (Bouvier *et al.* 2007, Herbst *et al.* 2007) and may also play a critical role in driving the outflows seen from many of these sources (e.g. Shang *et al.* 2007, Mohanty & Shu 2008). As a result, there is a great deal of interest in measuring the magnetic field properties of TTSs (e.g. Johns–Krull 2007, Donati *et al.* 2008). In particular, disk locking theories predict that an equilibrium is established where the disk is trunctated at or close to corotation and the stellar rotation rate depends only on the (assumed) dipolar magnetic field strength, the stellar mass, radius, and the mass accretion rate in the disk (see Bouvier *et al.* 2007).

Recent efforts to measure magnetic field properties on TTSs have focused on detecting Zeeman broadening of lines in Stokes I spectra or on the detection of circular polarization in Stokes V spectra. Zeeman broadening measurements are sensitive to total magnetic flux, while polarization measurements give information on field geometry, though they often miss most of the flux (e.g. Reiners & Basri 2009). The majority of results published to date are based on Zeeman broadening measurements. Rapid rotation of young stars complicates the observations; however, the λ^2 dependence of Zeeman broadening compared to the λ^1 dependence of Doppler broadening allows this to be overcome by using lines (e.g. Ti I) in the near-IR (e.g. ~ 2 μm). Recent measurements (e.g. Johns–Krull 2007, Yang *et al.* 2008) find field strengths of the correct magnitude, but the predicted star to star variations are not observed.

Acknowledgements

C. M. J-K. would like to acknowledge partial support from the NASA Origins of Solar Systems program through grants NNG06GD85G and NNX08AH86G to Rice University.

References

Bouvier, J., Alencar, S. H. P., Harries, T. J., Johns–Krull, C. M., & Romanova, M. M. 2007, *Protostars and Planets V* 479
Donati, J.-F. *et al.*, 2008, *MNRAS* 386, 1234
Herbst, W., Eislöffel, J., Mundt, R., & Scholz, A. 2007, *Protostars and Planets V* 297
Johns–Krull, C. M. 2007, *ApJ* 664, 975
Mohanty, S. & Shu, F. H. 2008, *ApJ* 687, 1323
Reiners, A. & Basri, G. 2009, *A&A* 496, 787
Shang, H., Li, Z.-Y., & Hirano, N. 2007, *Protostars and Planets V* 261
Yang, H., Johns–Krull, C. M., & Valenti, J. A. 2009, *AJ* 136, 2286

Highlights of Astronomy, Volume 15
XXVIIth IAU General Assembly, August 2009
Ian F. Corbett, ed.

© International Astronomical Union 2010
doi:10.1017/S1743921310010513

High-resolution near-IR spectroscopy: from 4m to ELT class telescopes

E. Oliva[1] and L. Origlia[2]

[1] INAF - Osservatorio di Arcetri, taly, email: oliva@arcetri.astro.it

[2] INAF - Osservatorio di Bologna, Italy, email: livia.origlia@oabo.inaf.it

1. Introduction

High-resolution (HR) near-IR spectroscopy is opening new windows in our understanding of several hot topics of modern planet, stellar and extragalactic astrophysics, and it will have a huge impact in the JWST and ALMA era and beyond. The much reduced extinction at these wavelengths allows to pierce the dust embedding those objects which are heavily obscured in the optical. Moreover, at high redshifts several spectral features, commonly exploited when studying local galaxies, are shifted into the near-IR. However, despite its scientific potential, the field of HR IR spectroscopy and its related science is developing very slowly, because of the lack of optimized instruments with the necessary combination of spectral resolution and coverage.

High spectral resolution ($R \sim 100,000$) is crucial to properly resolve the spectral features and derive accurate chemical and kinematic information of individual stars, stellar clusters and galaxies. A wide spectral coverage in a single exposure is highly desirable for a complete screening of chemical abundances in stellar populations or for extremely accurate radial velocity measurements for extra-solar planet search, which also require high stability. A wide spectral coverage is also crucial to unveil the nature and physical properties of poorly explored objects, like e.g. very low mass dwarfs or transient objects.

2. Scientific highlights

The inner Galaxy, where dust obscuration makes optical observations virtually impossible, is the most suitable laboratory to study stellar physics, kinematics, evolution and chemical enrichment in the high metallicity domain, with a major impact in our understanding of extragalactic bulges and elliptical galaxies. Using existing 4-10m class telescopes one can study the bright, evolved stellar populations of red giants and super-giants, Rich, Origlia & Valenti (2007). However, pristine abundances, which are best recorded in much fainter main sequence stars, can only be accessed with an ELT.

Extragalactic stellar clusters trace the star formation history and evolution of their host galaxies. HR integrated spectroscopy in the near-IR offers a unique chance to characterize the chemical composition and kinematics of these objects even in extincted nuclear and/or starforming regions (Larsen *et al.* (2008)). With existing telescopes one can measure stellar clusters in the Local Group, while an ELT is mandatory to study old stellar clusters out to Virgo, and young super star clusters out to 50-100 Mpc.

HR near-IR spectroscopy is also a unique tool to constrain the pristine metal and dust formation in low density environments, as traced by Lyα absorption systems at high redshifts (Prochaska *et al.* (2007), Wolfe (2005)).

Other relevant topics in stellar astrophysics, which strongly benefit from HR IR spectroscopy are circumstellar disks, brown dwarfs, stellar magnetic fields and stellar winds. We finally mention that HR near-IR spectroscopy is also a unique tool for the characterization of planet atmospheres in the solar system, (Lellouch *et al.* (2009)), the search for exo-planets (Marcy & Butler (1998)) with habitable conditions and the physics of exo-planet atmospheres (Brown (2001), Swain, Vasishti & Tinetti (2008)).

Moving from 4-8m class telescopes to an ELT, one can pierce objects and/or spatially resolved structures, with 1 to 2 orders of magnitude better limiting sensitivities and/or at $\geqslant 5$ times larger spatial resolution and distance.

Table 1. Expected limiting Vega magnitudes of SIMPLE on the E-ELT in 2hr integration.

S/N	$I(0.90~\mu m)$	$Y(1.05~\mu m)$	$J(1.25~\mu m)$	$H(1.65~\mu m)$	$K(2.20~\mu m)$
10	20.2	20.7	20.9	21.1	20.6
50	17.6	18.1	18.3	18.6	18.4
100	16.2	16.7	16.9	17.2	17.1

3. A *simple* instrumental concept

We developed an instrument concept for a HR near-IR spectrometer virtually independent on the telescope diameter, following from a detailed study of existing HR optical and IR spectrographs. The current baseline characteristics of the spectrograph are as follows (Oliva & Origlia 2008).

- spectrograph in vacuum cooled and thermostated at cryogenic temperatures
- fixed position at the telescope
- resolving power of at least $R = 100,000$ (goal 150,000)
- complete 0.85-2.5 μm spectral coverage in a single exposure
- fixed spectral format, cross-dispersed echellogram

Cross dispersion is performed by means of prisms in double pass mode, which provide a minimum inter-order distance of about 60 pixels. The detector can be a mosaic of three 2048^2 Hawaii-II RG arrays with $18\mu m$ pixels or a mosaic of three 4096^2 arrays with $15\mu m$ pixels, a format which may soon become the standard for HgCdTe arrays. The sky projected slit width is 0.38", 0.19" and 0.038" on a 4m, 8m, 40m. On the largest telescopes the instrument needs to be assisted by adaptive optics to concentrate the light in the slit. The limiting magnitudes for point sources scale with telescope size as follows

$$m_{lim} = \text{constant} + 5.0\log_{10}(D_{tel}) + 2.5\log_{10}(SLE)$$

where D_{tel} is the telescope diameter and SLE is the fraction of light falling in the slit. A dedicated study of such an instrumental concept for the 42m E-ELT (i.e. the "SIMPLE" study) is underway. The expected limiting magnitudes are summarized in Table 1.

References

Brown, T. M. 2001, *ApJ* 553, 1006
Larsen, S. S., Origlia, L., Brodie, J., & Gallagher, J. S., 2008, *MNRAS* 383, 263
Lellouch, E., Sicardy, B., deBergh, C., Käufl, H., Kassi, S., & Campargue, A., 2009, *A&A* 495, L17
Marcy, G. W. & Butler, R. P. 1998, *ARA&A* 36, 57
Prochaska, J. X., Chen, H.-W., Dessauges-Zavadsky, M., & Bloom, J. S. 2007, *ApJ* 666, 267
Oliva, E. & Origlia, L. 2008, *SPIE* 7014, 560
Rich, R. M., Origlia, L., & Valenti, E. 2007, *ApJ* 665, 119
Swain, M. R., Vasishti, G., & Tinetti, G. 2008, *Nature* 452, 329
Wolfe, A. M., Gawiser, E., & Prochaska, J. X. 2005, *ARA&A*, 43, 861

Highlights of Astronomy, Volume 15
XXVIIth IAU General Assembly, August 2009
Ian F. Corbett, ed.

SpS1-Instrumentation for sub-millimeter spectroscopy

Luis F. Rodríguez[1]

[1] Centro de Radioastronomía y Astrofísica (UNAM), Apartado Postal 3-72 (Xangari),
58089 Morelia, Michoacán, México
email: l.rodriguez@crya.unam.mx

1. Introduction

The fields of millimeter and sub-millimeter interferometry have been developing for more than 30 years. At millimeter wavelengths the most important interferometers are the Combined Array for Research in Millimeter-Wave Astronomy (CARMA), the Plateau de Bure Interferometer (PdBI), and the Nobeyama Millimeter Array (NMA). At sub-millimeter wavelenghts, the most powerful interferometer is the SubMillimeter Array (SMA, for a detailed description, see Ho *et al.* 2004).

The important discoveries made by these interferometers laid the ground-work for a project without precedent in ground-based radio astronomy: the Atacama Large Millimeter Array (ALMA).

2. The Atacama Large Millimeter Array

ALMA is a major international collaboration between Europe, North America, East Asia, and Chile that is unique in most aspects. The interferometer will be constituted by 50 antennas of 12 m diameter plus the Atacama Compact Array (ACA), which will participate in the array with four 12 m antennas and twelve 7 m antennas. As a whole, ALMA will have its antennas distributed in a variable configuration that will spread from about 150 meters at its most compact, up to 16 km. The total collecing area of the array will be larger than 6500 m^2. The largest configuration will take advantage of the geography of the site (see Fig. 1), with the longest arms stretching between the hills of the site. The ACA will be located in a compact configuration, about 50 m in diameter, which will allow sensitive wide-field imaging and total power measurements. These configurations will provide an angular resolution from 1$.''$4 in its most compact state, up to 0$.''$015 in its most extended state.

ALMA will cover a frequency range from 31 to 950 GHz (10 to 0.3 mm). With superconducting SIS mixers as receivers, sensitivities close to the quantum limits will be achieved. Future instruments will need to have much larger collecting areas to improve on ALMA. The correlator will be able to cover bandwidths of 16 GHz with 4096 channels. Finally, the site is also unique; at an altitude of 5 km and with outstanding characteristics, the Chajnantor plains is one of the best sites in the world. A detailed review of ALMA is given by Tarenghi (2008).

3. Sub-millimeter spectroscopy

The sub-millimeter range is expected to be ideal for the study of thermal phenomena. Dust emission scales as $S_\nu \propto \nu^{2-4}$, depending on the opacity. The emission from lines will grow as $S_\nu \propto \nu^5$, and as demonstrated by the SMA, a true "forest" of molecular lines is expected. As a matter of fact, the density of detectable line emission in the sub-millimeter may turn out to be a problem, with spectral confusion being reached in some objects.

ALMA is expected to make major progress in several lines of research. Protoplanetary disks around forming stars of all masses will be resolved and their morphology and dynamics will be studied in great detail. The search for effects such as sub-Keplerian rotation and the presence of gaps and structures induced by star formation will become standard practice. Star formation is also characterized by collimated outflows of gas and ALMA will allow the investigation of these

Figure 1. Google-Earth view of the ALMA site with a possible distribution of antennas superposed. This image shows the array in its most extended configuration.

phenomena with unprecedented spatial and spectral resolution. The understanding of massive star formation is now at an impasse because the sources are relatively far and small; however, ALMA will be able to produce significant advances. Evolved cool stars are surrounded by shells of molecular gas, whose chemistry is still poorly understood. Finally, the problem of galaxy formation and starburst galaxies will undergo enormous advances with ALMA. We are in the dawn of a new era of millimeter and sub-millimeter interferometry.

Acknowledgements

I thank Paul T. P. Ho and Lars Nyman for providing me with much of the material used in the presentation of this talk. I acknowledge the support of DGAPA, UNAM, and of CONACyT (México).

References

Ho, P. T. P., Moran, J. M., & Lo, K. Y. 2004, *Ap. J.* 616, L1
Tarenghi, M. 2008, *Astrophysics and Space Science* 313, 1

Highlights of Astronomy, Volume 15
XXVIIth IAU General Assembly, August 2009
Ian F. Corbett, ed.

© International Astronomical Union 2010
doi:10.1017/S1743921310010537

SpS1-SOFIA studies of stellar evolution

R. D. Gehrz[1], E. E. Becklin[2] and T. L. Roellig[3]

[1]Department of Astronomy, School of Physics and Astronomy, 116 Church Street, S. E.,
University of Minnesota, Minneapolis, MN 55455, USA, email: gehrz@astro.umn.edu

[2]Universities Space Research Association, NASA Ames Research Center, MS 211-3, Moffett
Field, CA 9403, USA, email: ebecklin@sofia.usra.edu

[3]NASA Ames Research Center, MS 245-6, Moffett Field, CA 94035, USA,
email: Thomas.L.Roellig@nasa.gov

1. Introduction

The U.S./German Stratospheric Observatory for Infrared Astronomy (SOFIA, Figure 1) is a 2.5-meter infrared airborne telescope in a Boeing 747-SP flying in the stratosphere at altitudes as high as 45,000 feet where the atmospheric transmission averages $\geqslant 80\%$ throughout the 0.3 - 1600 μm spectral region. SOFIA's first-generation instruments include broadband imagers, moderate

Figure 1. Left: SOFIA and F/A-18 safety chase plane during its second checkout flight on May 10, 2007. Right: Ground operations tests at Palmdale in March, 2008. NASA photos.

resolution spectrographs capable of resolving broad features due to dust and large molecules, and high resolution spectrometers suitable for kinematic studies of molecular and atomic gas lines at km s^{-1} resolution. These and future instruments will enable SOFIA to make unique contributions to studies of the physics and chemistry of stellar evolution for many decades. Science flights will begin in 2010. A full operations schedule of at least 100 flights per year will begin in 2014 and will continue for 20 years. The SOFIA Guest Investigator (GI) program, open to investigators worldwide, will constitute the major portion of the SOFIA observing program.

2. Spectroscopic studies of stellar evolution with SOFIA

Stellar evolution in interstellar medium (ISM) plays a central role in the chemical evolution of the Universe (see Figure 5 in Becklin *et al.* 2007). New stars form from the ISM and produce heavy elements by main-sequence and explosive nucleosynthesis. The winds of post-main-sequence stars, novae, and supernovae inject processed material into the ISM, increasing its metallicity and therefore enriching the metal content of the next generations of stellar systems. A full understanding of the way that stars participate in this recycling requires studies of the physics and chemistry of all phases of stellar evolution from birth to death. SOFIA, with its wide wavelength coverage and high spectral resolution capabilities, is destined to play a dominant role in this field. Below, we give several specific examples of the role SOFIA will play in spectroscopic studies of various phases of stellar evolution.

3. SOFIA and regions of star formation in the ISM

SOFIA's imaging instruments will provide detailed, broad/narrow band studies of regions of star formation to identify protostellar condensations and regions where the winds of embedded red giant stars, supergiant stars, and supernovae are interacting with the ISM. Follow-up studies with SOFIA's moderate and high resolution spectrometers will probe the detailed composition of the gas and dust, the physical dynamics, and other physical properties associated with these phases of stellar evolution. In particular, SOFIA can observe the atomic fine structure lines of [O I] at 63 and 145 μm and of [C II] at 158 μm. These lines are bright in regions illuminated or shocked by the outflows of massive stars and supernova explosions. SOFIA's first light spectrometer (GREAT) will be unique in its ability to resolve these lines at the sub-km s^{-1} level and probe in detail the physical conditions and kinematics in these regions.

4. SOFIA and evolved stellar systems

As an example or SOFIA's capability to study the late stages of stellar evolution, we consider nova explosions due to thermonuclear runaways caused by matter accreted onto white dwarfs in binary systems. Novae may contribute substantial amounts of isotopes of light metals to the ISM. IR observations can be used to quantify the physical parameters of novae explosions and assess their contribution to ISM chemical abundances (Gehrz 2008). Abundances in novae ejecta can be deduced from IR dust emission features and IR forbidden emission lines. Recent IR observations have shown that some novae have ejected shells that were extremely overabundant compared to solar values in C, N, O, Ne, Mg, Al, and Si. The physical properties and mineralogy of dust produced by novae have been shown to be similar to those of the small grains in comets.

SOFIA will be a unique platform for observing novae because of its mobility and rapid deployability. First, monitoring of the temporal development of nova explosions requires observations occurring at all declinations and developing on time-scales of days, weeks, and months. Second, SOFIA's spectroscopic capabilities will enable the recording of many forbidden lines obscured by the atmosphere from ground-base observatories and heretofore unavailable to the spectrometers of other space missions. These lines are crucial for determining accurate elemental abundances in novae ejecta. SOFIA's spectrometers will provide a powerful probe of the physical conditions and kinematics of the ejected gas through studies of the atomic fine structure lines such as [O I] (63, 145 μm), [O III] (52, 88 μm), [O IV] (25.9 μm), [C II] (158 μm), [Si II] (34 μm), and [S I] (26 μm) and [S III] (18.7 μm). Numerous forbidden neon lines in obscured spectral regions are also available to SOFIA. Neon is a major product of nova explosions.

References

Becklin, E. E., Tielens, A. G. G. M., Gehrz, R. D., & Callis, H. H. S. 2007, *Proc. SPIE* 6678, 66780A-1

Gehrz, R. D. 2008, in *Classical Novae, 2nd Edition*, eds. M. F. Bode and A. Evans, Cambridge University Press: Cambridge, p. 167.

Highlights of Astronomy, Volume 15
XXVIIth IAU General Assembly, August 2009
Ian F. Corbett, ed.

© International Astronomical Union 2010
doi:10.1017/S1743921310010549

Extreme adaptive optics in the mid-IR: The METIS AO system

Remko Stuik[1], Laurent Jolissaint[1], Sarah Kendrew[1], Stefan Hippler[2] and Bernhard Brandl[1]

[1]Leiden Observatory, Leiden University, P.O. Box 9513, 2300 RA Leiden, The Netherlands
email: stuik@strw.leidenuniv.nl

[2]Max Planck Institute for Astronomy, Königstuhl 17, 69117 Heidelberg, Germany

METIS, the Mid-infrared ELT Imager and Spectrograph, is currently in its phase A study as one of the candidate first-light instruments for the European Extremely Large Telescope. METIS will feature several observational modes, ranging from diffraction limited imaging in L, M and N-bands to high-resolution Integral Field spectroscopy for the L and M-bands. METIS in its current design gives sensitivities similar to *Spitzer* in imaging and low-resolution spectroscopy and with its high-resolution spectrograph will provide unprecedented line sensitivity. The design of METIS is optimized for both galactic science cases (*e.g.* conditions in the early solar system, formation and evolution of proto-planetary disks and properties of exoplanets) and extragalactic science cases (*e.g.* the growth of Supermassive Black Holes). METIS will require a high-order adaptive optics (AO) system to meet its scientific goals, both to provide correction for atmospheric turbulence as well as reduce the impact of wind shake, leading to a residual image motion of 3 - 5 mas rms. METIS is expected to feature both an internal Single Conjugate AO system as well as an external Laser Tomography AO system. The challenges for the METIS AO system are mainly in the broad correction range, an excellent image stability required for coronagraphy and in providing a high sky coverage to be available for as many science targets as possible. An additional challenge for METIS is the need to compensate for composition turbulence, mainly in the form of fast fluctuations in water vapor concentration. Water vapor fluctuations impact the performance of METIS in several ways: Atmospheric dispersion causes a broadening of the point-spread function, both in the science channel and the wavefront channel, but can be corrected using a Atmospheric Dispersion Corrector. Variations in the water vapor composition cannot be corrected this way and are currently estimated to give a residual image motion of $\leqslant 10$ mas rms. This effect can, especially for coronagraphy, not be neglected. Chromatic optical path difference errors, caused by changes in the index of refraction along the path through the atmosphere were found to be negligible in the case of METIS due to attenuation by the outer scale (at typical values of 25 m). Chromatic anisoplanatism is the effect that the light at different wavelengths travels through a slightly different light path through the atmosphere and can be–at least partly–corrected. The last effect is composition turbulence, mainly caused by fast (> 1 Hz) fluctuations in the water vapor content. Based on data for ALMA and radiometer probes, this leads to a maximum loss in Strehl ratio between 5 and 10%. This mainly has an impact on coronagraphy and the METIS AO team is actively investigating ways to compensate also water vapor turbulence. The main challenge is currently obtaining reliable data on the distribution and magnitude of precipitable water vapor fluctuations.

Highlights of Astronomy, Volume 15
XXVIIth IAU General Assembly, August 2009
Ian F. Corbett, ed.

SpS1-Silicate dust formation around AGB stars

Hans Ulrich Käufl[1], Florian Kerber[1] and Bernhard Aringer[2,3]

[1]European Southern Observatory, Karl Schwarzschild Straße, D-85748 Garching bei München, Germany, email: hukaufl@eso.org, fkerber@eso.org

[2]Department of Astronomy, University of Vienna, Türkenschanzstraße 17, 1180 Wien, Austria
[3]Osservatorio Astronomico di Padova - INAF, Vicolo dell'Osservatorio 5, 35122 Padova, Italy
email: aringer@astro.univie.ac.at

VISIR, the VLT Imager and Spectrograph for the Mid-Infrared is a multi-mode instrument, featuring also a high resolution Echelle spectrograph with a spectral resolving power $\frac{\nu}{\Delta\nu}$ of $\approx$30 000 or 10 km s^{-1} at $\nu \approx$ 30 000 GHz ($\lambda \approx$ 10 μm). A limited long-slit mode as well as a general cross-dispersed mode are available. The Echelle grating is illuminated with a 200 mm diameter collimated beam. Cross-dispersion is achieved by a pair of grisms in the pre-slit optics. The entire frequency interval corresponding to the "10 μm-window" from 22 400 to 39 500 GHz is fully accessible, albeit sequentially. This interval contains a multitude of fundamental molecular rotational-vibrational bands such as SiO, OH$^-$, H$_2$O, NH$_3$, CH$_4$ and many other hydrocarbonates. Since its commissioning in April 2004, VISIR has been plagued by artifacts introduced from its detector. The cross-dispersed mode is especially handicapped, as it is the most demanding mode for dynamic range of illumination. Now an ambitious upgrade with a pair of newly developed 1k^2 As:Si detectors is underway, which will fully resurrect the spectroscopic mode. This will also increase the frequency interval accessible in one exposure by 240% while changing from critical to 3-pixel sampling. Even in the absence of extra spectral features this increase is quite valuable for absorption line spectroscopy, as the limiting factor in analysis often is the definition of the photospheric continuum.

While the typical width of molecular rotational-vibrational absorption lines in giant stars is more of the order of 3−5 km s^{-1} VISIR will still allow for spectroscopic studies of the molecular gas surrounding RGB and AGB stars. One obvious science project after recommissioning is to propose parallel observations with VISIR at $\nu \approx$ 37 500 GHz ($\lambda \approx$ 8 μm) and with the CRIRES spectrograph at $\nu \approx$ 75 000 GHz at three times higher resolution to measure simultaneously the SiO overtone and fundamental band, in combination with low resolution spectra, again with VISIR, sensitive to silicate dust. This could be combined with other high-resolution IR spectra to constrain the atmospheric structure even better. Even shocks, atomic hydrogen emission (*Brackett$_\alpha$* $\nu =$ 74 074 GHz) is a proven diagnostic observed in Mira-stars, can be investigated.

Silicate dust is ubiquitous in the universe. It is found in planetary systems, circumstellar disks of stars in various stages of their evolution, in molecular clouds, or in tori of AGNs. It is a well established observational fact that AGB stars are surrounded by silicate dust, while undergoing sometimes even extreme mass loss ($\dot{M}$ up to $10^{-3} M_\odot/yr$). Still, a detailed understanding of the process of dust formation, as well as the efficiency, is lacking. Thus the amount of dust produced by AGB stars, and consequently the relative importance of this source to produce a major constituent of the ISM, remains unclear. Synoptic observations of a sample of pulsating AGB stars will provide for a fundamentally new data set. Observing the dust and its parent molecule with many different transitions in combination with state-of-the-art models will allow for a new approach to settle the issue of silicate dust production.

Highlights of Astronomy, Volume 15
XXVIIth IAU General Assembly, August 2009
Ian F. Corbett, ed.

© International Astronomical Union 2010
doi:10.1017/S1743921310010562

A tool for modelling telluric spectra

Alain Smette[1], Hugues Sana[1,2] and Hannes Horst[1,3]

[1]ESO, Alonso de Cordova 3107, Casilla 19001, Vitacura Santiago, Chile
email: asmette@eso.org

[2]Universiteit van Amsterdam, Sterrenkundig Instituut 'Anton Pannekoek',
Postbus 94249, – NL-1090 GE Amsterdam, The Netherlands   email: hsana@eso.org

[3]now at EOS GmbH - Electro Optical Systems, Niederlassung Chemnitz,
Annaberger Strasse 240, D-09125 Chemnitz, Germany   email: Hannes.Horst@eos.info

1. Introduction

Accurate synthetic telluric spectra are required for efficient use of telescope time, in particular, with large telescopes and high-resolution NIR spectroscopy: (i) In the preparation of observations, are the telluric features at the same wavelength as spectroscopic features of scientific interest? Since water vapor is the molecule whose abundance varies most in the atmosphere, what values of precipitable water vapor are suitable to carry out successful observations? Are the observations of a telluric star required? Or better, can telluric features in the science spectrum be accurately represented by an appropriate synthetic spectrum? This point is also very important in the planning of telescope time, as observations of a telluric star may sometimes take longer than the one of the science target. (ii) In the analysis of the observations, how do telluric lines affect the scientifically interesting features in the observed spectrum? Is it possible to recover the useful information when telluric star observations could not be obtained, do not have sufficient SNR, or suffer from a significant change in instrumental or observing conditions?

A number of authors have already proposed synthetic telluric spectra to help analyze spectral features, e.g., Bailey, Simpson & Crisp (2007), but did not include the possibility to change the temperature and pressure profile and the column densities of the relevant molecules in the Earth atmosphere. In the following, we describe *molecfit*, a tool to provide accurate synthetic spectra for use in Exposure Time Calculators, and in determining column densities of relevant molecules in the atmosphere based on either sky emission spectra in the thermal infra-red or in transmission spectra at any wavelength long-wards of 400 nm. In its most complex form, it accurately models the telluric lines affecting a given spectrum. A systematic check of the accuracy of molecfit over the NIR range covered by CRIRES (Käufl *et al.* 2004) is being carried out.

2. Requirements

A tool aiming at accurately modelling telluric spectra requires: (a) an accurate radiative transfer code, able to take into account the most important aspects of the radiative transfer physics taking place in the Earth atmosphere; we choose RFM, the Reference Forward Model (Dudhia 2008); (b) an accurate and complete database of molecular parameters, such as HITRAN (Rothman *et al.* 2009); (c) an atmospheric profile – appropriate for the time of observations – that accurately describes the change of temperature, pressure and abundance of the different molecules with altitude: The molecular abundance profiles are provided by RFM. Temperature and pressure profiles adapted for the time and place of the observations can be obtained through http://ready.arl.noaa.gov/READYamet.php.

Adapting the code to reproduce *observed* telluric lines requires additional capabilities: *(d)* the column densities of the different molecules involved have to be adjusted to match those at the time of the observations; *(e)* the wavelength calibration provided by a data reduction tool is usually not accurate enough and needs to be also re-determined based on the telluric lines themselves; *(f)* the spectral resolution may depend on the accurate light profile at the entrance slit; in the case of CRIRES, it may fluctuate with the adaptive optics profile; its value must therefore also be adjusted by the code; *(g)* the response curve of the whole system (instrument,

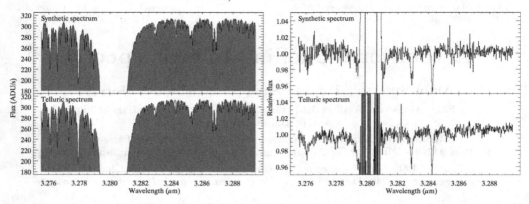

Figure 1. Comparison of correction between a telluric and a synthetic spectrum. *(Left:)* shaded (red in the online version): extract of a CRIRES spectrum of a star showing 2 intrinsic absorption features; *(solid line, top:)* synthetic spectrum obtained with molecfit; *(solid line, bottom:)* telluric star spectrum taken with an airmass difference smaller than 0.01. *(Right:)* Relative difference between the science spectrum and the synthetic spectrum *(top)* and a telluric spectrum *(bottom)*.

telescope, atmosphere) needs to be adjusted for accurate corrections; *(h)* some data points, either fixed (detector defects) or wavelength dependent (known stellar features) must not be used by the fitting algorithm.

3. Applications

This tool is already used to create the sky spectra (both emission and absorption) used by the CRIRES and E-ELT Exposure Time Calculators and to measure the amount of precipitable water vapor with CRIRES and VISIR (Lagage *et al.* 2004) in real time (cf. : http://www. eso.org/sci/facilities/paranal/sciops/CALISTA/pwv/data.html). It is currently being adapted to measure systematically the column density of other molecules, such as CO_2, CO, O_3, CH4.

As shown in Fig. 1, the quality of the correction by the synthetic spectrum can be very similar to (or better than) the one by the telluric spectrum.

References

Bailey, J., Simpson, A., & Crisp, D. 2007, PASP 119, 228
Dudhia, A. 2008 http://www.atm.ox.ac.uk/RFM/index.html
Käufl *et al.*, 2004, SPIE 5492, 1218
Lagage, P. O. *et al.* 2004, The Messenger 117, 12
Rothman, L. S., *et al.*, 2009, JQSRT 110, 9

Highlights of Astronomy, Volume 15
XXVIIth IAU General Assembly, August 2009
Ian F. Corbett, ed.

© International Astronomical Union 2010
doi:10.1017/S1743921310010574

Building-up a database of spectro-photometric standards from the UV to the NIR

J. Vernet[1], F. Kerber[1], V. Mainieri[1], T. Rauch[2], F. Saitta[1],
S. D'Odorico[1], R. Bohlin[3], V. Ivanov[4], C. Lidman[4], E. Mason[4],
A. Smette[4], J. Walsh[5], R. Fosbury[5], P. Goldoni[6], P. Groot[7],
F. Hammer[8], L. Kaper[9], M. Horrobin[10], P. Kjaergaard-Rasmussen[11]
and F. Royer[8]

[1]ESO, Karl-Schwarzschild-Str. 2, 85748 Garching bei München, Germany; [2]Institute for
Astronomy and Astrophysics, Kepler Center for Astro and Particle Physics, Eberhard Karls
University, Sand 1, 72076 Tübingen, Germany; [3]STScI, 3700 San Martin Drive, Baltimore,
MD 21218, USA; [4]ESO, Alonso de Cordova 3107, Vitacura, Santiago 19, Chile; [5]ST-ECF,
Karl-Schwarzschild-Str. 2, D-85748 Garching bei München, Germany; [6]Service
d'Astrophysique, Centre d'Etudes de Saclay, F-91190 Gif-sur-Yvette Cédex, France; [7]Radboud
Univ. Nijmegen, Postbus 9010, 6500 GL Nijmegen, The Netherlands; [8]Observatoire de Paris,
5, place Jules Janssen, F-92195 Meudon, France; [9]Astronomical Institute Anton Pannekoek,
Universiteit van Amsterdam, Kruislaan 403, 1098 SJ Amsterdam, The Netherlands;
[10]Physikalisches Institut Universität zu Köln Zülpicher Str. 77 D-50937 Köln; [11]Niels Bohr
Institute for Astronomy, Blegdamsvej 17, DK-2100 Copenhagen, Danemark

We present results of a project aimed at establishing a set of 12 spectro-photometric standards over a wide wavelength range from 320 to 2500 nm. Currently no such set of standard stars covering the near-IR is available. Our strategy is to extend the useful range of existing well-established optical flux standards (Oke 1990, Hamuy *et al.* 1992, 1994) into the near-IR by means of integral field spectroscopy with SINFONI at the VLT combined with state-of-the-art white dwarf stellar atmospheric models (TMAP, Holberg *et al.* 2008). As a solid reference, we use two primary HST standard white dwarfs GD71 and GD153 and one HST secondary standard BD+17 4708. The data were collected through an ESO "Observatory Programme" over ∼40 nights between February 2007 and September 2008.

Flux measurements were done in eight telluric absorption-free windows (one in the *J*-band, four in the *H*-band, three in the *K*-band). This careful choice of 'clean' atmospheric windows and the stability of SINFONI make it possible to achieve an accuracy of 3 to 6% depending on the wavelength band and stellar magnitude, which is well within our original goal of 10% accuracy.

While this project was originally tailored to the needs of the wide wavelength range (320-2500 nm) of X-shooter on the VLT, it will also benefit any other near-IR spectrographs, providing a huge improvement over existing flux calibration methods.

References

Oke, J. B. 1990, *A&A* 99, 1621
Hamuy, M., Walker, A. R., Suntzeff, N. B., *et al.*, 1992, *PASP* 104, 533
Hamuy, M., Walker, A. R., Suntzeff, N. B., *et al.*, 1994, *PASP* 106, 566
Holberg, J. B., Sion, E. M., Oswalt, T., *et al.*, 2008, *ApJ* 135, 1225

Highlights of Astronomy, Volume 15
XXVIIth IAU General Assembly, August 2009
Ian F. Corbett, ed.

SpS1-High-resolution infrared spectroscopy at high and low altitudes

Sarah Kendrew[1]

[1] Leiden Observatory, Leiden, Netherlands
email: kendrew@strw.leidenuniv.nl

1. Introduction

The advantages of a high altitude, dry site for ground-based astronomy at infrared (IR) wavelengths are well-known: the lower temperature and pressure associated with increased altitude reduce the emissivities of both atmosphere and telescope, and a lower atmospheric absorption improves the transmission of IR radiation. The next generation of IR instruments under development (for ELTs) will open up a new discovery space, particularly in high-resolution (HR) spectroscopy, which will not have a space-based counterpart and has proven to be a powerful tool for studying all stages of stellar evolution (e.g. Jaffe et al., 2003). I present here a summary of quantitative work into transmission-dependent aspects of HR IR spectroscopy at high and low altitudes †.

2. Findings

Modelled transmission profiles have been successfully used to improve telluric line removal in HR IR spectra (Mandell et al, 2008, Smette et al, 2009). Accurate transmission modelling is complex, requiring a detailed knowledge of vertical atmospheric profiles in T, P and molecular constituents. The combination of models and experimental data can lead to a better understanding of the interplay between these factors.

The largest transmission gains with altitude are found in spectral regions dominated by water absorption; species with high abundances in the upper atmosphere, such as O_3, show only a moderate improvement. The regions benefiting the most from high and dry sites are the 6.5 μm region of M-band, between 6.5-8 μm (N-band) and the entire Q-band (17-25 μm), which is dominated by water line absorption.

The relative velocity between Earth and a (galactic) science target can be used to Dopplershift diagnostic spectral lines from behind their telluric equivalent. This technique is commonly used for e.g. 13.7 μm C_2H_2 absorption (also for 4.7 μm CO emission), where an improvement of 25% in transmission can result from this method (depending on the target co-ordinates). In targets that push the limits of the instrumental sensitivity the observing efficiency is much improved when observing from a high-altitude site.

References

Jaffe et al., 2003, ApJ 596(2), 1053
Mandell, A. et al., 2008, ApJ 681, L25
Smette, A. et al., 2009, these proceedings

† Work presented here uses the 2008 edition of HITRAN, the Reference Forward Model (RFM), more information: http://www.atm.ox.ac.uk/RFM/index.html, and a standard tropical atmosphere profile.

Highlights of Astronomy, Volume 15
XXVIIth IAU General Assembly, August 2009
Ian F. Corbett, ed.

© International Astronomical Union 2010
doi:10.1017/S1743921310010598

Atmospheric water vapour content over La Silla Paranal Observatory

R. Querel[1], F. Kerber[2], R. Hanuschik[2], G. Lo Curto[3],
D. Naylor[1], M. Sarazin[2] and A. Smette[3]

[1] Institute for Space Imaging Science, University of Lethbridge, Lethbridge, AB, Canada

[2] European Southern Observatory, Garching, Germany

[3] European Southern Observatory, Santiago, Chile

Water vapour is the principle source of opacity at infrared wavelengths in the earth's atmosphere. Measurements of atmospheric water vapour serve two primary purposes when considering operation of an observatory: long-term monitoring of precipital water vapour (PWV) is useful for characterizing potential observatory sites, and real-time monitoring of PWV is useful for optimizing use, in particular for mid-IR observations.

In support of site testing for the European Extremely Large Telescope (E-ELT), we have used La Silla and Paranal as calibration sites to ground-truth satellite measurements of PWV. To this end, dedicated measurement campaigns have been conducted over both sites and Las Campanas Observatory (LCO) through a collaboration between the University of Lethbridge, scientists from the European Southern Observatory, and the Giant Magellan Telescope site test team at LCO.

Several independent measurement techniques were used in this study. Continuous measurements were obtained using IRMA infrared radiometers at 20 μm. PWV was also retrieved from spectra covering the wavelength range from the visible to the infrared using facility instruments (FEROS, HARPS and MIKE, UVES, CRIRES, VISIR) through fitting their data to simulated atmospheric spectra. A prototype 3-band lunar spectrophotometer, operating at $\sim$0.94 μm, was also employed.

Local meteorological data were provided by a series of radiosonde launches timed to coincide with satellite overpasses. The radiosondes provided in situ measurements of PWV, and time and location specific atmospheric profiles. Together, this multi-faceted approach has resulted in a unique data set. Integral to this analysis is a site specific atmospheric radiative transfer model (BTRAM), common to all retrieval schemes.

We also have reconstructed the PWV history over the La Silla Paranal observatory, by analysing > 2000 high-resolution echelle spectra of spectrophotometric standard stars using archival FEROS and UVES data. The extracted PWV values were compared to ENVISAT and GOES satellites to allow a statistical analysis of PWV over several years.

Acknowledgements

We would like to acknowledge Greg Tompkins for his electronics expertise; the AstroMeteorology Group at the Universidad de Valparaiso for launching the radiosondes; and Joanna Thomas-Osip and Gabriel Prieto (Las Campanas) for excellent collaboration. D.N. acknowledges support from NSERC, AIF and CFI.

References

Thomas-Osip, J., McWilliam, A., Phillips, M. M. *et al.*, 2007, PASP, 119, 697
Querel, R. R., Naylor, D. A., Thomas-Osip, J. *et al.*, 2008, SPIE, 7014E.172Q

Highlights of Astronomy, Volume 15
XXVIIth IAU General Assembly, August 2009
Ian F. Corbett, ed.

Astrophotonics and IR astronomy

J. W. OByrne[1], J. Bland-Hawthorn[1], R. Haynes[2], A. Horton[2], J. Bryant[1] and J. G. Robertson[1]

[1] School of Physics, University of Sydney, [2] Anglo-Australian Observatory

Working in collaboration with industry, the University of Sydney, the Anglo-Australian Observatory and Macquarie University are developing new 'astrophotonic' solutions to problems in astronomical instrumentation. A key first step involves overcoming the limitations imposed by multimode (MM) optical fibres that have been used by astronomers for many years to transport or reformat light from the telescope focus to an optical spectrograph. These large-core MM fibres maximise light into an astronomical instrument but at the expense of propagating many unpolarized modes. Until recently, this has deterred the use of more complex in-fibre processing of the light since this is typically limited to single-mode (SM) propagation. A MM to SM converter, known as a 'photonic lantern', was first demonstrated by Leon-Saval *et al.* (2005). If the number of transverse modes equals the number of SM fibres, and if a gradual and adiabatic transition between the MM fiber and the ensemble of SM fibres can be achieved, lossless coupling can take place in either propagation direction. Noordegraaf *et al.* (2009) demonstrated an efficient 1 x 7 photonic lantern (1 MM input and 7 SM outputs) for the first time.

The application of a photonic lantern and a SM fibre device designed to suppress emission from hydroxyl molecules (OH) in the Earth's upper atmosphere has recently been demonstrated in initial on-sky tests by Bland-Hawthorn *et al.* (2009). The OH-suppressing fibre, using a fibre Bragg grating, and a comparison fibre without a grating but also with a photonic lantern in series, were pointed at the sky. Light from the other ends of the fibres was re-imaged onto the focal plane of IRIS2 near-IR spectrograph/imager at the Anglo-Australian Telescope (AAT). The spectral region of interest was isolated with an *H*-band spectroscopic filter. The fibre Bragg gratings reduced the emission from 63 OH lines to less than 1% at a resolving power of 10,000 although the suppressed region did not reach the dark sky limit because the fibres saw a 10 degree field of view and therefore includes the light from many stars. The experiment effectively cleaned up half of the spectroscopic *H*-band at low loss (~10%). A test of a complete *H* and *J*-band OH-suppressing fibre will be carried out later in the year.

This device can be applied in any project requiring spectra at near-IR wavelengths, for example high-redshift quasars or cool stars such as L and T dwarfs. A more detailed discussion of the scientific implications is presented by Ellis & Bland-Hawthorn (2009).

Other photonics devices are also under development. In particular, an integrated photonic spectrograph (IPS) using an arrayed-waveguide grating (AWG) structure has recently had its first on-sky demonstration using IRIS2 in June 2009.

References

Bland-Hawthorn, J., Ellis, S. C., Haynes, R., & Horton, A. 2009, *AAO Newsletter* No.115, 15
Ellis, S. C & Bland-Hawthorn, J. 2008, *MNRAS* 386, 47
Leon-Saval, S. G., Birks, T. A., Bland-Hawthorn, J., & Englund, M. A. 2005, *Opt. Lett.* 19, 2545
Noordegraaf, D., Skovgaard, P. M. W., Nielsen, M. D., & Bland-Hawthorn, J. 2009, *Opt. Express* 17, 1988

Highlights of Astronomy, Volume 15
XXVIIth IAU General Assembly, August 2009
Ian F. Corbett, ed.

Atomic data for IR and sub-mm wavelengths

Gillian Nave

National Institute of Standards and Technology, Gaithersburg, MD, USA
email: gnave@nist.gov

1. Introduction

Atomic spectra in the infrared and sub-mm wavelength regions can be divided into two broad categories: electric dipole-allowed transitions, and forbidden lines due to transitions within the ground term or between low-lying levels of the same parity. Both are of potential importance in the interpretation of astrophysical spectra. Allowed transitions can provide diagnostic information for stellar photospheres, particularly for elements that are not accessible in the visible region. Electric-dipole forbidden lines are important diagnostics of low-density plasmas, such as nebulae and the interstellar medium. In order to interpret astrophysical spectra, accurate atomic data are required. This paper summarizes the techniques for measuring atomic data and lists the most important compilations and databases.

2. Techniques for measuring IR atomic data

Forbidden line wavelengths: Forbidden lines from neutral through doubly-ionized species are rarely observed in the laboratory. Although wavelengths of a few lines have been directly measured using laser spectroscopy (e.g. Cooksy, Blake, & Saykally 1986), or by analysis of astronomical spectra (e.g. Feuchtgruber, Lutz & Beintema 2001), accurate wavelengths for these lines are usually determined from optimized energy levels. The most common technique is to measure wavelengths of allowed lines in the visible and ultraviolet regions using Fourier transform spectroscopy (FTS) (e.g. Aldenius & Johansson 2007) and use these to measure the energy levels involved in the forbidden transition. This technique is limited to neutral through doubly-ionized species with resonance lines above about 140 nm. This includes most of the transition-group elements.

Forbidden line oscillator strengths: The majority of the oscillator strengths of forbidden lines in atomic spectra databases are from atomic structure calculations. However, some measurements have been made by combining lifetimes obtained from laser-excitation of ions in a storage ring with branching fractions obtained from astrophysical spectra (e.g. Hartman *et al.* 2008).

Allowed line wavelengths: Electric dipole-allowed lines in the IR region can originate from low-excitation levels, but most are due to transitions between highly-excited levels. Wavelengths for both can be measured by Fourier transform spectroscopy. Extensive lists of IR data include Ne I (Sansonetti, Blackwell & Saloman 2004), Ar I (Whaling *et al.* 1995), Ar II (Whaling *et al.*), Ti I (Forsberg 1991), Fe I (Nave *et al.* 1994), Co I (Pickering & Thorne 1996), Co II (Pickering *et al.* 1998), Ni I (Litzén *et al.* 1993), and Kr I (Sansonetti & Greene 2007). Some rare earth elements also have their principal resonance lines in this region (e.g. Ce III, Johansson & Litzén 1972).

Allowed line oscillator strengths: There have been few measurements of atomic oscillator strengths in the IR. Some of the most extensive include Fe I (O'Brian *et al.* 1991), Ti I (Blackwell-Whitehead *et al.* 2005) and Mn I (Blackwell-Whitehead *et al.* 2006). The latter two papers are part of a collaborative effort to obtain IR atomic data for cool star analysis by Lund University, Sweden, Imperial College London, UK, and NIST, USA. All have been measured by combining branching fraction measurements using Fourier transform spectroscopy with level lifetimes using laser-induced fluorescence.

IR wavelength calibration: Wavelengths of Th/Ar hollow cathode lamps suitable for calibration of IR spectrographs have been published by Kerber, Nave & Sansonetti (2008). Measurements of other elements of potential interest for spectrograph calibration are contained in a companion paper (Kerber *et al.* 2009).

G. Nave

3. Databases and compilations

The last compilation of atomic data dedicated to the IR region was by Outred (1978). Another valuable list of line identifications in the IR region is the key to identification of solar features in the ATMOS spectrum by Geller (1992). Additional databases including IR atomic data include the NIST Atomic Spectra Database, the Vienna Atomic Line Database, the Atomic linelist at Kentucky and Kurucz atomic line database. In addition to atomic data for allowed transitions, these databases also include wavelengths and calculated oscillator strengths for forbidden lines.

References

Aldenius, M. & Johansson, S. 2007, *A&A* 467, 753.

Atomic linelist at Kentucky http://www.pa.uky.edu/peter/atomic/

Blackwell-Whitehead, R. J., Xu, J. L. , Pickering, J. C., Nave, G., & Lundberg, H. 2005 *MNRAS* 361, 1281.

Blackwell-Whitehead, R. J., Lundberg, H, Nave, G., Pickering, J. C., Jones, H. R. A, Lyubchik, Y., Pavlenko, Y. V., & Viti, S. 2006 *MNRAS* 373, 1603.

Cooksy, A. L., Blake, G. A., & Saykally, R. J. 1986, *ApJ* 305, L89.

Forsberg, P 1991, *Phys. Scr* 44, 446.

Feuctgruber, H. Lutz, D., & Beintema, D. A 2001 *ApJS* 136, 221.

Geller, M., 1992 *NASA ref. publ.* 1224, vol III.

Hartman, H., Gurell, J., Lundin, P., Schef, P., Hibbert, A., Lundberg, H., Mannervik, S., Norlin, L.-O., & Royen, P. 2008, *A&A* 480, 571.

Johansson, S. & Litzén, U. 1972, *Phys. Scr.* 6, 139.

Kerber, F., Nave, G., & Sansonetti, C. J. 2008 *ApJS* 178, 374.

Kerber, F., Aldenius, M., Bristow, P, Nave, G., Ralchenko, Y., & Sansonetti, C. J. this volume.

Kurucz atomic line database http://www.cfa.harvard.edu/amp/ampdata/kurucz23/sekur.html

Litzén, U., Brault, J. W., & Thorne, A. P. 1993, *Phys. Scri.* 47, 628

Nave, G., Johansson, S., Learner, R. C. M., Thorne, A. P, & Brault, J. W. 1994, *ApJS* 94, 221

NIST Atomic Spectra Database. Ralchenko, Yu., Kramida, A. E., Reader, J. and NIST ASD Team 2008 [Online]. http://physics.nist.gov/asd3

O'Brian, T. R., Wickliffe, M. E., Lawler, J. E., Whaling, W., & Brault, J. W. 1991 *JOSA B* 8, 1185.

Outred, M. 1978 *J. Phys. Chem. Ref. data* 7, 1.

Pickering, J. C., Thorne, A. P. T. 1996, *ApJS* 107, 761.

Pickering, J. C., Raassen, A. J. J., Uyling, P. H. M., Johansson, S. 1998, *ApJS* 117, 261.

Sansonetti, C. J., Blackwell, M. M., Saloman, E. B. 2004 *J. Res. NIST* 109, 371

Sansonetti, C. J. & Green, M. B 2007, *Phys. Scr.* 75, 577.

VALD database: http://vald.astro.univie.ac.at/ vald/php/vald.php

Whaling W., Anderson W. H. C., Carle, M. T., Brault J. W., & Zarem, H. A. 1995 *J. Quant. Spectrosc. Radiat. transfer* 53, 1

Whaling W., Anderson W. H. C., Carle, M. T., Brault J. W., & Zarem, H. A. 2002, *J. Res. NIST* 107, 149.

Highlights of Astronomy, Volume 15
XXVIIth IAU General Assembly, August 2009
Ian F. Corbett, ed.

© International Astronomical Union 2010
doi:10.1017/S1743921310010628

SpS1-Laboratory spectroscopy of small molecules

Peter F. Bernath

Department of Chemistry, University of York, Heslington, York, YO10 5DD, UK
email: pfb500@york.ac.uk

This contribution focuses on the study of 'cool' sources with surface temperatures in the range of about $500 - 4000$ K. In this temperature range spectra are dominated by strong molecular absorption and the tools of modern chemical physics can be applied to compute the molecular opacities needed to simulate the observed spectral energy distributions. (See Bernath (2005) for an introduction to molecular spectroscopy including line intensities and Bernath (2009) for a recent astronomical review article.)

1. Hot H_2O, CH_4 and NH_3

The vibration-rotation and pure rotational lines of H_2O and OH appear strongly in cool oxygen-rich objects. The dominant carbon-containing molecule is CO at high temperatures and CH_4 for cooler objects, such as T-type brown dwarfs and 'hot Jupiter' extrasolar planets. For nitrogen, N_2 is the high temperature molecular form, but NH_3 appears in T dwarfs and is predicted to be the distinguishing feature for very cool brown dwarfs ('Y'-type) with surface temperatures below about 700 K.

Excellent high resolution ground-based spectra of the Sun are available from the Kitt Peak Fourier transform spectrometer (FTS) and from low earth orbit with the Advanced Composition Explorer (ACE) FTS. The spectra are dominated by the strong first overtone and fundamental vibration-rotation bands of CO. The Meinel bands (OH vibration-rotation bands) near 3 μm are also strong and the OH pure rotational lines are prominent at longer wavelengths. The vibration-rotation bands of CH and NH can be seen at 3 μm and pure rotational lines of NH are also present.

A new very high signal-to-noise solar atlas in the $750 - 4400$ cm^{-1} region has recently been prepared by Hase et al. (2009) from ACE satellite observations. Based on ACE, ATMOS and laboratory spectra, the analysis of OH has recently been improved by Bernath & Colin (2009) and similar work is underway for CH and NH.

At 5800 K, the Sun's photosphere is too hot for water to exist, but by 3900 K the concentration of OH and H_2O are equal. A large number of unassigned lines were noticed in two Kitt Peak sunspot atlases; it was suspected that these lines were due to hot water but the available laboratory data were inadequate to confirm this. Comparison of a new laboratory emission spectrum of H_2O at 1800 K with the sunspot absorption spectrum identified most of the unassigned sunspot lines as H_2O lines. Wallace et al. (1995) proved that there is 'water on the Sun'. Through variational calculations of the energy levels using a high quality ab initio potential energy surface, Polyansky et al. (1997) were able to assign most of the strong lines.

The water emission in the 500–13 000 cm^{-1} spectral region was recorded with a high-resolution FTS using an oxy-acetylene torch as a source. Work on this 3000 K spectrum of H_2O has just finished with the publication by Zobov et al. (2008) of the last paper in the series. The latest water linelist of Barber et al. (2006) contains more than 500 million lines and is recommended for simulation of spectral energy distributions of cool stars, brown dwarfs and extrasolar planets.

The situation for CH_4 is much less satisfactory than for H_2O although Nassar & Bernath (2003) have published hot emission spectra and there is considerable theoretical activity, for example by Warbier et al. (2009) A similar situation exists for NH_3 with calculations underway, for instance, by Huang et al. (2008)

We have decided to take an empirical approach to determine new linelists for hot CH_4 and NH_3. A reasonably complete list of line positions, line strengths and lower state energy levels is needed. We have therefore begun to record spectra of hot NH_3 and CH_4 over a wide range of temperatures and to calibrate the line positions and strengths using the HITRAN database of Rothman *et al.* (2009) Empirical lower state energies are determined from the temperature dependence of the line intensities.

2. Metal hydrides

The only metal hydride seen clearly in the solar photosphere spectrum is MgH, detected via the $A^2\Pi-X^2\Sigma^+$ transition. Sunspot spectra convincingly show lines of the visible and near infrared electronic transitions of AlH, MgH, CaH and FeH. Both the $A^2\Pi-X^2\Sigma^+$ and $B'^2\Sigma^+-X^2\Sigma^+$ electronic transitions of MgH can be seen in sunspot spectra. Shayesteh *et al.* (2003) completed a set of MgH observations in the infrared and visible that had led earlier in 2003 to the discovery of the linear MgH_2 molecule. From the MgH B–X observations, Shayesteh *et al.* (2007) determined all bound ground state levels with spectroscopic precision and obtained a very precise dissociation energy of 11104.7 cm^{-1}.

The most important application of the spectroscopy of metal hydrides is in the classification of L-type dwarfs as discussed by Kirkpatrick (2005). The FeH $F^4\Delta-X^4\Delta$ electronic transition near 1 μm and the $E^4\Pi-A^4\Pi$ transition near 1.58 μm are seen strongly along with the $A^6\Sigma^+-X^6\Sigma^+$ transition of CrH near 861 nm. Dulick *et al.* (2003) prepared a linelist for the F–X transition of FeH, Burrows *et al.* (2002) prepared one for CrH and Burrows *et al.* (2005) for TiH (which has not been detected yet in L-dwarfs) by extrapolating laboratory observations and using line intensities based on *ab initio* calculations. These molecular opacities are in modest agreement with observations, but high resolution comparisons indicate that improvements are needed.

References

Barber, R. J., Tennyson, J., Harris, G. J. & Tolchenov 2006, MNRAS 368, 1087

Bernath, P. F. 2005, *Spectra of Atoms and Molecules*, Second Edition, Oxford UP, New York

Bernath, P. F. 2009, Int. Rev. Phys. Chem., in press

Bernath, P. F. & Colin, R. 2009, J. Mol. Spectrosc., in press

Burrows, A., Ram, R. S., Bernath, P., Sharp, C. M. & Milsom, J. A. 2002, Astrophys. J. 577, 986

Burrows, A., *et al.* 2005, ApJ 624, 988

Dulick, M., *et al.* 2003, ApJ 594, 651

Hase, F., Wallace, L., McLeod, S. D., Harrison, J. J. & Bernath, P. F. 2009, J. Quant. Spectrosc. Rad. Transfer, in preparation

Huang, X., Schwenke, D. W., & Lee, T. J. 2008, J. Chem. Phys. 129, 214304

Kirkpatrick, J. D. 2005, Annu. Rev. Astron. Astrophys. 43, 195

Nassar, R. & Bernath, P. 2003, J. Quant. Spectrosc. Rad. Transfer, 82, 279

Polyansky, O. L., *et al.* 1997, Science 277, 346

Rothman, L. S., *et al.* 2009, J. Quant. Spectrosc. Rad. Transfer 110, 533

Shayesteh, A., *et al.* 2007, J. Phys. Chem. A 111, 12495

Shayesteh, A., Appadoo, D. R. T., Gordon, I. & Bernath, P. F. 2003, J. Chem. Phys. 119, 7785

Wallace, L., *et al.* 1995, Science 268, 1155

Warmbier, R., *et al.* 2009, A&A 495, 655

Zobov, N. F., *et al.* 2008, MNRAS 387, 1093

Highlights of Astronomy, Volume 15
XXVIIth IAU General Assembly, August 2009
Ian F. Corbett, ed.

SpS1-The evolution of brown dwarf infrared spectroscopic properties

France Allard & Isabelle Baraffe[1]

[1] Centre de Recherche Astrophysique de Lyon, UMR 5574: CNRS, Université de de Lyon, École Normale Supérieure de Lyon, 46 allée dItalie, F-69364 Lyon Cedex 07, France
email: fallard@ens-lyon.fr

Brown dwarfs (hereafter BDs) are formed, like stars, by interstellar cloud collapse, but attaining masses of less then $0.075\,M_\odot$ (Baraffe *et al.* 1998), i.e. too low core temperatures ($<3.5 \times 10^6$ K) to stabilize the nuclear burning of the hydrogen PP chain. Therefore, even the most massive BDs begin cooling after some 10^9 yrs. However, for masses above $0.06\,M_\odot$, core temperatures become hotter than the lithium burning temperature (2.4×10^6 K). All BDs above $0.013\,M_\odot$ ($13\,M_{\rm Jup}$) reach core temperatures above the 1.0×10^6 K necessary to burn deuterium from about 10^7 yrs. The IAU has adopted the definition of the planetary regime as objects having masses below the deuterium burning conditions. But BDs are likely to form well below this limit into the planetary mass regime down to some $5\,M_{\rm Jup}$. It is therefore convenient, in the absence of indices on their formation mechanisms, to call them planetary mass objects or *planemos*.

While cooling over time, their effective temperature ranges from 3000 K for the most massive or youngest BDs to below 100 K for the least massive and oldest planemos. They therefore encompass several spectral types, extending the spectral classification from M to L, T and eventually Y. Below 1500 K, they become partially electron degenerate, causing their radius to stabilize near a value of $1\,R_{\rm Jup}$. This causes the surface gravity to reach a maximum value of $\log g{=}5.4$ at 1500 K before decreasing below even 3.5, towards the planetary mass regime. Too few constraints (eclipsing binaries) are available for these faint objects in the mass-radius diagram, while the stellar regime has been successfully explained by Baraffe *et al.* (1998). For more informations on the core properties and its evolution see the review by Chabrier & Baraffe (2000).

While burning deuterium below some 10^7 yrs depending on their mass, BDs can be fully convective rapid rotators with magnetic spots. The composition of their atmosphere is then typical of M dwarfs stars, with H_2, CO and N_2 the most abundant molecules, and TiO, VO, CO and H_2O the most important opacities. Their spectra also show nearly saturated hydride bands (CaH, CAOH, AlH, MgH, SiH), which play an important role in the spectra of low-metallicity subdwarfs where double-metal molecules deplete faster then hydrides. Pressure broadening is important in these atmospheres. To distinguish M-type BDs from M dwarf stars, one uses indicators of low gravity and of low core temperatures, i.e. the lithium test or the detection of the 6708Å lithium line in BDs.

As a BD cools below 2500K, clouds begin to form in its atmosphere, and the condensation of refractory elements begins to deplete important opacities such as VO, TiO and H_2O, revealing underlying absorption bands (CrH, FeH, water vapor at 0.93 mμ) and atomic lines. The L spectral type occurs when the depletion of refractory elements begins affecting TiO and VO band strengths in the optical. The silicate clouds cause a greenhouse effect (or blanketing effect in astronomy), redistributing flux scattered off of micron-size dust grains to infrared wavelengths, beyond the peak of the spectral distribution ($\geqslant 1.2\ \mu$m). Forsterite (Mg_2SiO_4) and enstatite ($MgSiO_3$) provide the strongest dust opacities in these objects. Observations of L dwarfs with *Spitzer* have revealed a corresponding absorption feature around 9 μm (Cushing *et al.* 2006). While clouds affect greatly the atmosphere and spectroscopic properties of BDs, the core evolution and its contraction and cooling history is only negligibly affected. However, their formation history and the magnetic field generation at young age may affect BD evolution more substantially (Baraffe, Chabrier & Gallardo, 2009).

As BDs cool to below 1800 K, silicate clouds sink gradually with the convection zone well below the photosphere. The photosphere, freed of its greenhouse effect and of most of its optical molecular opacities, cools to below the local gas temperature of 1400 K. At such low temperatures, the most stable carbon-bearing molecule is methane. And below 1000 K the most stable azote molecule is NH_3. The detection of these molecules at near-infrared wavelengths is therefore the criteria for the identification of, respectively, T-type and Y-type BDs. The transparent optical wavelengths, allowing line formation into deep and dense atmospheric layers, result in the formation of quasi-molecules (with H_2). This is the case of the resonance doublets of $K I$ at 0.77 μm and $Na I D$ at 0.58 μm, which are broadened to several thousands of angstroms from the line centre. The red wing of the $K I$ doublet defines the shape of the 1.0 μm flux peak. The first Y dwarf could be CFBD0059 with a marginal detection of NH_3 at 1.5 to 1.55 μm in the H bandpass (Delorme *et al.* 2008). The 11 μm band of NH_3 has been detected in T dwarfs.

The coolest, least massive BD known is ULAS 1335 ($T_{eff} = 575$ K (Burningham *et al.* 2008). At such low temperatures, one reaches into the planemo regime and conditions for water vapor condensation. If the mean difference between BDs and planets is their formation mechanism, planets might be distinguished from BDs by an enrichment of elements similar to that measured by the Galileo probe in the atmosphere of Jupiter. However, cloud formation mechanisms in the atmosphere complicate the determination of abundances: the condensation of refractory elements onto grains, and the upwelling of molecules due to convective motions and their generated turbulences and gravity waves (Freytag *et al.* 2009), leads to non-equilibrium chemistry (Saumon *et al.* 2006).

In summary, the evolution of the spectral properties of BDs depends therefore on the details of the cloud formation mechanisms, on the hydrodynamical properties of the atmosphere, and on accurate reaction rates for the molecules involved valid for the density and temperature conditions prevailing in these atmospheres. Model atmospheres rely on detailed and complete (to high J-values) molecular line lists, dissociation energies and oscillator strengths. For the study of M dwarfs, for instance, no line list is yet available for CaOH. For the study of T and Y dwarfs more precise and complete line lists for NH_3, CH_4, H_2S, CO_2 and PH_3 are needed. Refractory indexes as a function of wavelength are also required for condensates LiCl, Na_2S, H_2O, NH_3 and NH_4SH, for example. Model atmospheres (of e.g. Allard *et al.* 2001), synthetic spectra, colors and evolution models are available via the web site (http://phoenix.ens-lyon.fr/simulator).

References

Allard, F. *et al.* 2001, *ApJ* 556, 357

Baraffe, I. *et al.* 1998, *A&A* 337, 403

Baraffe, I. *et al.* 2009, *ApJ* 702, L27

Burningham, B. *et al.* 2008, *MNRAS* 391, 320

Chabrier, G. & Baraffe, I. 2000, *ARA&A* 38, 337

Cushing, M. C. *et al.* 2006, *ApJ* 648, 614

Delorme, P. *et al.* 2008, *A&A* 482, 961

Freytag, B. *et al.* 2009, *A&A* in preparation

Saumon, D. *et al.* 2006, *ApJ* 647, 552

Highlights of Astronomy, Volume 15
XXVIIth IAU General Assembly, August 2009
Ian F. Corbett, ed.

© International Astronomical Union 2010
doi:10.1017/S1743921310010641

AKARI near-infrared spectroscopy of brown dwarfs

**Issei Yamamura[1], Takashi Tsuji[2], Toshihiko Tanabé[2]
and Tadashi Nakajima[3]**

[1]Institute of Space and Astronautical Science, JAXA, Japan
[2]Institute of Astronomy, The University of Tokyo, Japan
[3]National Astronomical Observatory, Japan

1. Introduction

Brown dwarfs (hereafter BDs) are of particular interest because of their extremely low-temperature atmospheres for comparison with atmospheres of giant planets. Aiming to obtain clues to understand the formation and disappearance of dust clouds and molecular abundances in BD photospheres, we conducted an observation programme of space-borne near-infrared spectroscopy of bright BDs with the Infrared Camera (IRC) on-board AKARI.

2. NIR spectra of BDs

Eleven BDs were observed during the cold phase of the AKARI mission (May 2006–Aug. 2007). The spectra cover the wavelength range 2.5 to 5.0 μm with a resolving power of roughly 100. Spectral types of the observed targets range from L5 to T9.

Seven targets have good quality spectra. We detect CO_2 for the first time in spectra of late-L and T-type dwarfs. We also confirm CO bands in a T9 dwarf spectrum, supporting the idea of non-equilibrium chemistry in the photosphere of late-T dwarfs.

3. Model analysis and discussion

All spectra are reasonably well fitted by the Unified Cloudy Model (UCM; Tsuji 2002, 2005). Observed CO and CO_2 bands in late-L to T dwarf spectra are much stronger than that expected from the UCM assuming thermal equilibrium chemistry. We can reproduce the depth of CO bands in T9 dwarf spectra by introducing vertical mixing effects into the model. However, the same mechanism does not work for mid-T and late-L dwarfs, since CO is rather abundant in the upper photosphere from the beginning under LTE. Alternative mechanisms need to be considered, which is a subject of future investigations.

We continue the AKARI observations in its warm phase.

Acknowledgements

This research is based on observations with AKARI, a JAXA project with the participation of ESA.

References

Tsuji, T. 2002, *ApJ* 575, 264
Tsuji, T. 2005, *ApJ* 621, 1033

Highlights of Astronomy, Volume 15
XXVIIth IAU General Assembly, August 2009
Ian F. Corbett, ed.

© International Astronomical Union 2010
doi:10.1017/S1743921310010653

Ultra cool dwarfs with companions

R. G. Kurtev[1], V. D. Ivanov[2], R. Jayawardhana[3] and J. H. Borissova[1]

[1]Department of Physics and Astronomy, Valparaiso University, Chile
email: radostin.kurtev@uv.cl; jura.borissova@uv.cl

[2]European Southern Observatory, Santiago, Chile, email: vivanov@eso.org

[3] Toronto University, Canada, email: rayjay@astro.utoronto.ca

The binary fraction in the sub-stellar regime is a topic of discussion. The lower masses of ultra cool dwarfs (UCDs) with respect to the other stars make them even more important because a measurable effect on their radial velocity (RV) or luminosities can be caused by extremely low mass companions. Some UCDs in young star forming regions are bright enough to be studied with existing high resolution instrumentation. The UCDs are intrinsically faint in the optical and the optical RV measurements are affected by "rotationally modulated inhomogeneous surface features" that can mimic a companion, while the near-infrared (NIR) RVs are less prone to them. Therefore, we decided to monitor the RV of six UCDs in the NIR. Blake *et al.* (2007) demonstrated RV measurement accuracy of 300-600 m s^{-1} in the NIR using telluric calibration.

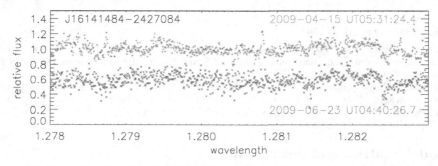

Figure 1. Spectra of J16141484-2427084 obtained at two epochs.

Here we present the first results of a search for unresolved companions (with NIR measured RV) around a sample of young low mass members of the Upper Sco star-forming region based on high resolution (R∼50000) *J*-band GEMINI(S)/PHOENIX spectra obtained during the program GS-2009A-Q-68. The spectra for two epochs of J16141484-2427084 are presented in Fig 1. Our targets were selected from the list of Slesnick *et al.* (2006). Two of them, J16141484-2427084 and J16095217-2136283, present radial velocity variations above the sensitivity threshold of about 700-900 m s^{-1}. The obtained ΔV, corrected for the Earth barycentric velocity, are 3.58±0.73 km s^{-1} and -2.61±0.93 km s^{-1} respectively. A further model analysis will be carried out to improve the precision and to search for differences in the radial velocity of other candidates.

RGK and JHB would like to acknowledge the support from FONDECYT projects #1080154 and #1080086.

References

Blake *et al.* 2007, *ApJ*, 666, 1198
Slesnick *et al.* 2006, *AJ*, 131, 3016

Highlights of Astronomy, Volume 15
XXVIIth IAU General Assembly, August 2009
Ian F. Corbett, ed.

© International Astronomical Union 2010
doi:10.1017/S1743921310010665

SpS1-Digging in the solar COmosphere with NAC

Thomas R. Ayres[1]

[1] Center for Astrophysics & Space Astronomy, University of Colorado, Boulder, CO, USA

1. Introduction

The solar "COmosphere" is an enigmatic region of cold gas (temperatures as low as $\sim$3500 K) coexisting in the low chromosphere with plasma much hotter ($\sim$7000 K). This zone probably consists of patchy clouds of cool gas, seen readily in off-limb emissions of CO 4667 nm lines, threaded by hot gas entrained in long-lived magnetic filaments as well as transient shock fronts. The COmosphere was not anticipated in classical 1D models of the solar outer atmosphere, but is quite at home in the contemporary 3D highly dynamic view, which one might call the Magnetic Complexity Zone.

COmospheres apparently are an integral part of the atmospheres of virtually all late-type stars, at least those of solar temperature and cooler, as deduced from studies of over-strong CO 4667 nm absorptions in the thermal IR. Early ideas (1980's) concerning the origin of the COmosphere focused on what has been called "molecular cooling catastrophes." Alternatively, the cold clouds might be a transient response to adiabatic cooling phases of large amplitude acoustic disturbances.

COmospheres are significant in the context of this Special Session, because they can have an impact on the general issue of CNO abundances in evolved stars, and particularly the isotopomers that are keys to exploring chemical evolution in the giant branch.

2. New Observations

The best hope for probing the physical structure of COmospheres is in observational studies of the Sun. Great progress has been made over the past two decades in this regard, in boosting spectral and spatial resolution of thermal infrared imaging. Most recently (ca. 2008), the NSO Array Camera (NAC) has been commissioned on the McMath-Pierce solar telescope at Kitt Peak, and very recently (April 2009), a new Integral Field Unit (IFU; developed at California State, Northridge) has been made available for engineering tests. The IFU is an "image slicer" that takes a $6'' \times 8''$ patch of the solar surface, and optically reorganizes it into 25 $8''$-high slitlets, separated by small gaps, and arrays these along the 100 mm entrance slit of the vertical spectrograph. In conjunction with the Horizontal IR Adaptive Optics bench (HIRAO), one now can obtain 2D spectral snapshots of the COmosphere, especially the CO off-limb emissions. In fact, during the engineering run, my colleagues and I were able to obtain off-limb images of the 2313 nm *first-overtone* lines of CO. These features are in general 100× thinner than the 4667 nm fundamental bands, and the fact that they also show the same off-limb signature implies that the 'CO clouds' must occupy a relatively large fraction of the chromospheric volume below about 1000 km. This type of work holds great promise for the future, especially when carried to the 4 m Advanced Technology Solar Telescope (ATST), slated for construction later this year, and operations toward the end of the decade.

Acknowledgements

I thank the National Science Foundation for providing support for this project.

Highlights of Astronomy, Volume 15
XXVIIth IAU General Assembly, August 2009
Ian F. Corbett, ed.

© International Astronomical Union 2010
doi:10.1017/S1743921310010677

SpS1-Infrared spectroscopy of post-AGB objects

Kenneth H. Hinkle[1], Sean D. Brittain[2] & Richard R. Joyce[1]

[1]National Optical Astronomy Observatory, Tucson, AZ, USA
(Operated by the Association of University for Research in Astronomy, Inc., under cooperative agreement with the National Science Foundation)
email: hinkle@noao.edu, joyce@noao.edu

[2]Clemson University, Clemson, SC, USA
email: sbritt@clemson.edu

Post-AGB (pAGB) objects are low to intermediate initial mass ($\leqslant 8$ $M_\odot$) objects that have terminated normal nuclear burning and as a result are undergoing rapid evolution toward the white dwarf sequence. In classical pAGB objects evolution is to hotter effective temperatures at roughly constant luminosity. However, there are also several classes of pAGB objects that have revived nuclear burning after approaching or being on the white dwarf sequence. These include objects with delayed final helium shell flashes (e.g. Sakurai's star) and white dwarf mergers (e.g. R CrB stars). Binary evolution plays a critical role in many of these systems. A group of pAGB supergiants with large infrared excesses are suspected to be binaries that have undergone common envelope evolution. Further details on many of these objects can be found in reviews, see for instance Van Winckel (2003) and Herwig (2005).

The use of infrared (IR) tools has greatly increased our understanding of pAGB evolution. The strong infrared excess of these objects makes them very bright in the infrared and easily detectable in the mid-infrared. Many objects become embedded in their AGB mass loss and are optically obscured. The reduction of the obscuration and the brightness of pAGB objects in the IR makes them ideal targets for detailed analysis using IR spectroscopy. We report on three specific groups of pAGB objects.

Post-common envelope binaries: Van Winckel *et al.* (1995) identified a class of pAGB single-lined, long-period, spectroscopic binaries. The primary star of these binaries is an early-type supergiant with very peculiar abundances. These objects typically have strong IR excesses, carbon-rich circumstellar dust, and a photospheric abundance pattern characterized by a severe deficiency of refractory elements and a near-solar abundance for volatile elements. The orbits of these objects suggest previous common envelope evolution when the primary was at the tip of the AGB. A circumbinary Keplerian rotating disk appears a common feature of these pAGB binaries (De Ruyter *et al.* 2006).

Hinkle *et al.* (2007) report on high-resolution infrared spectroscopy in the 2.3-4.6 μm region. In the prototype HR 4049 the 4.6 μm spectrum shows a rich forest of CO and H_2O emission lines. All the spectral lines observed in the 2.3-4.6 μm spectrum are circumbinary in origin and originate in oxygen rich gas in contrast to the extended circumstellar carbon-rich gas. The emission and absorption line profiles show that the circumbinary gas is located in a thin, rotating layer near the dust disk. The velocities of the disk and the spectroscopic orbit yield masses for the individual stars, for HR 4049 $M_{A\ I} \sim 0.58$ $M_\odot$ and $M_{M\ V} \sim 0.34$ $M_\odot$. Chemical processing in the disk is the likely method for producing the chemical peculiarities seen on the stellar photosphere.

Final flash objects: Final helium shell flash evolution occurs in stars that have terminated their AGB stage and are either on the constant luminosity pAGB track or just entering the early white dwarf sequence. A final helium burning thermal pulse returns the star briefly to the AGB (Herwig 2005). The final flash episode is an astronomically very brief but probably fairly common event in stellar evolution. The mass loss associated with the final flash planetary nebula stage has been proposed to explain the shape and composition of exotic multi-shell planetary nebulae.

Hinkle *et al.* (2008) undertook a detailed investigation of the 80[+] year old final flash object V605 Aql. The cloud of ejecta was imaged at high spatial resolution in both optical emission lines and infrared continuum. The He I 10830 Å spectrum was also observed. The obscuring circumstellar shell was shown to be a disk with extended structure including knots. The obscuration of V605 Aql by dust marked the emergence of the hot white dwarf remnant from an optically thick pseudo-photosphere. This white dwarf drives a 2500 km s^{-1} wind principally in one plane resulting in a circumstellar disk. Where the wind encounters the circumstellar disk He I 10830 Å emission is created and hot, $\sim$1500 K, grains are generated. Grains exit the disk at 350 K and the accompanying gas is then expanding at $\sim$140 km s^{-1}. The strong concentration of the mass loss to the disk suggests the white dwarf is now rotating rapidly. There is convincing evidence in the literature that this process is also seen in V4334 Sgr and is still going on in the old final flash objects A30 and A78.

Double degenerate mergers: The category of hydrogen-deficient luminous stars includes, in order of increasing effective temperature, the hydrogen-deficient carbon (HdC) stars, the R Coronae Borealis (RCB) stars, and the extreme helium (EHe) stars. The origins of the HdC, RCB and EHe stars have remained a puzzle for decades. Two different scenarios have been presented. In one, the double-degenerate (DD) scenario the H-deficient supergiant is formed from the merger of a He white dwarf (WD) with a carbon-oxygen WD. In the other, these H-deficient stars result from a final, pAGB helium shell flash. The remaining H-rich envelope is ingested by the He-shell and the ensuing nucleosynthesis includes large-scale conversion of H to He.

Clayton *et al.* (2005, 2007) made the remarkable discovery from medium-resolution infrared spectra that the oxygen in HdCs was primarily the isotope ^{18}O and not the usual dominant isotope ^{16}O. This work supports the origin of HdC and RCB stars via the DD scenario. Exploratory calculations by Clayton *et al.* (2007) suggested that the merger is very rapid and induces nucleosynthesis that converts ^{14}N by α-capture to ^{18}O resulting in a high abundance of ^{18}O relative to ^{16}O. García-Hernández *et al.* (2009) have followed up with a high spectral resolution of C, N, O abundances in the HdC and RCB stars and support the results of Clayton *et al.*(2005, 2007).

References

Clayton, G. C., Geballe, T. R., Herwig, F., Fryer, C., & Asplund, M. 2007, *ApJ*, 662, 1220

Clayton, G. C., Herwig, F., Geballe, T. R., Asplund, M., Tenenbaum, E. D., Engelbracht, C. W., & Gordon, K. D. 2005, *ApJ*, 623, L141

De Ruyter, S., Van Winckel, H., Maas, T., Lloyd Evans, T., Waters, L. B. F. M., & Dejonghe, H. 2006, *AA*, 448, 641

García-Hernández, D.A., Hinkle, K. H., Lambert, D.L., & Eriksson, K. 2009, *ApJ*, 696, 1733

Herwig, F. 2005, *ARA&A* 43, 435

Hinkle, K. H., Brittain, S. D., and Lambert, D. L. 2007, *ApJ*, 664, 501

Hinkle, K. H. *et al.* 2008, *AA*, 479, 817

van Winckel, H., Waelkens, C., & Waters, L. B. F. M. 1995, *AA*, 293, L25

van Winckel, H. 2003, *ARA&A* 41, 391

Highlights of Astronomy, Volume 15
XXVIIth IAU General Assembly, August 2009
Ian F. Corbett, ed.

FIR and sub-mm line observations of AGB and post-AGB nebulae

Valentín Bujarrabal[1]

[1] Observatorio Astronómico Nacional, Ap 112, E-28803 Alcalá de Henares, Spain
email: v.bujarrabal@oan.es

The existing information on asymptotic giant branch (AGB) and post-AGB objects from FIR and submm spectroscopy is still scarce. Observations from the ground are often very difficult to calibrate and show low S/N ratios. However, we expect to obtain in the near future an impressive amount of high-quality data, from the *Herschel* space telescope. ALMA is also expected to provide high-quality maps of submm lines.

AGB stars eject significant amounts of gas, which forms spherical circumstellar envelopes (CSEs) expanding at moderate velocity (e.g. Habing 1996). AGB CSEs have been systematically observed by means of mm-wave molecular lines, for instance the intense low-J rotational transitions of CO ($J = 1$–0 and $J = 2$–1). Such lines are known to probe the bulk of the circumstellar material, requiring very low excitation conditions. However, they are not efficient in discriminating gas with relatively high excitation (over ~ 100 K). So, they are not very useful to study the inner CSE, at distances smaller than $\sim 10^{16}$ cm. For such a purpose, we need observations of lines requiring intermediate excitation conditions, i.e. between about 100 K and 1000 K, in almost all cases at submm and FIR wavelengths. The processes relevant to explain mass loss actually take place in these inner layers: pulsation of shells very close to the star and dust formation at a somewhat larger distance. The study of inner shells is crucial to understand those processes, and, therefore, AGB and post-AGB evolution, which is thought to be driven by the very copious mass ejection of these stars.

Preliminary works including submm lines (mostly of CO) have been particularly useful for measuring the temperature profiles in AGB CSEs and for checking the presence of variations in the mass-loss rate in the last ~ 1000 yr (e.g. Groenewegen *et al.* 1998, Teyssier *et al.* 2006, Decin *et al.* 2007, Ramstedt *et al.* 2008). The temperature was confirmed to vary with distance to the star following approximately a potential law, $T \propto 1/r^{\alpha}$, with $\alpha \sim 1$. In general, the data are compatible with constant mass-loss rate; in one object, IRC +10011, a decrease by at least a factor 10 of the mass-loss rate took place about 800 yr ago. We expect that data in the near future will very significantly increase our knowledge on this topic. Some existing space observations have also been useful to detect water vapor emission in O-rich and even C-rich stars (see Maercker et al. 2008, 2009; Melnick *et al.* 2001, González-Alfonso *et al.* 2007). However, SWAS and ODIN data only contain the very low-excitation and opaque $1_{1,0}$–$1_{1,1}$ line, and ISO detected only opaque lines using very low spectral resolution. Therefore, the interpretation of the existing data is very preliminary and in need of more observations to refine models.

As mentioned, the end of the AGB phase is thought to be due to the huge amounts of material (most of the initial stellar mass) ejected. The process is spectacular. In less than ~ 1000 yr, the large and cold red giant, surrounded by a spherical, slowly expanding CSE, becomes a tiny and very hot blue dwarf, around which a young planetary nebula (PN) shows, in general, a clear axial symmetry and fast bipolar flows. The appearance of the central star is due to the ejection of all stellar layers over the core, which is now visible (the blue dwarf). The nebula evolution is thought to be due to shock interaction between the fossil AGB envelope and post-AGB axial jets carrying very high energy and momentum (e.g. Bujarrabal *et al.* 2001). The result is that huge amounts of material (a fraction of a solar mass) are accelerated to $50 - 100$ kms^{-1} in just a few hundred years. The physical processes at work, including both the wind interaction and the fast wind ejection, are still not well understood.

Young PNe have been well observed thanks to mm-wave molecular lines. Accurate maps of CO emission (e.g. Sánchez Contreras *et al.* 2004) reveal the density and velocity distribution of most of the nebular gas. In general, the detected gas is always in expansion; only for one object, the Red Rectangle, has a rotating disk been well mapped (Bujarrabal *et al.* 2005), and whose presence, if confirmed to be systematic in these objects, could explain the energetic post-AGB jets. Again, such observations are not efficient in discriminating warm regions, which must be studied to understand the PN shaping by shocks propagating in the massive component and the innermost rotating layers. Results from submm observations are still rare (e.g. Nakashima *et al.* 2007) and have not yielded conclusive results. We expect that future data will be crucial to improve our knowledge on the evolution of these objects.

Finally, I would like to mention interest in the study of PDRs in young PNe. PDRs appear in post-AGB evolution as an intermediate phase between the cold and molecule-rich AGB CSEs and the very hot gas in old PNe, which is fully ionized by the UV stellar field. PDRs are warm and rich in neutral or slightly-ionized plasma. Their study requires almost necessarily space telescopes. ISO detected some PDR tracers, namely fine-structure lines of O I, C II and some other atom (Castro-Carrizo *et al.* 2001, Fong *et al.* 2001). C II was particularly efficient to measure the total PDR mass, while the O I line at 63 μm was useful to check models and study the excitation conditions. Massive (close to 1 ms^{-1}) PDRs were found in some young PNe, like NGC 7027 and NGC 6302. But the lack of sensitivity and spectral resolution severely hampered these works, and systematic studies of this transitional component must still be performed.

References

Bujarrabal, V., Castro-Carrizo, A., Alcolea, J., & Sánchez Contreras, C. 2001, *A&A* 377, 868
Bujarrabal, V., Castro-Carrizo, A., Alcolea, J., & Neri, R. 2005, *A&A* 441, 1031
Castro-Carrizo, A., Bujarrabal, V., Fong, D., *et al.* 2001, *A&A* 367, 674
Decin, L., Hony, S., & de Koter, A. 2007, *A&A* 475, 233
Fong, D., Meixner, M., Castro-Carrizo, A., *et al.* 2001, *A&A* 367, 652
González-Alfonso, E., Neufeld, D.A., & Melnick, G. J. 2007, *ApJ* 669, 412
Groenewegen, M. A. T.; van der Veen, W. E. C. J., & Matthews, H. E. 1998, *A&A* 338, 491
Habing, H. J. 1996, *A&AR* 7, 97
Maercker, M., Schöier, F. L., Olofsson, H., *et al.* 2008, *A&A* 479, 779
Maercker, M., Schöier, F. L., Olofsson, H., *et al.* 2009, *A&A* 494, 243
Melnick, G. J., Neufeld, D. A., Ford, K. E. S., *et al.* 2001, *Nature* 412, 160
Nakashima, J., Fong, D., Hasegawa, T., *et al.* 2007, *AJ* 134, 2035
Ramstedt, S., Schöier, F. L., Olofsson, H., & Lundgren, A. A. 2008, *A&A* 487, 645
Sánchez Contreras, C., Bujarrabal, V., & Castro-Carrizo, A. 2004, *ApJ* 617, 1142
Teyssier, D., Hernández, R., Bujarrabal, V., *et al.* 2006, *A&A* 450, 167

Highlights of Astronomy, Volume 15
XXVIIth IAU General Assembly, August 2009
Ian F. Corbett, ed.

Spatially-resolved high-spectral resolution observations of the red supergiant Betelgeuse

K. Ohnaka[1]

[1] Max-Planck-Institut für Radioastronomie, Auf dem Hügel 69, 53121 Bonn, Germany
email: kohnaka@mpifr-bonn.mpg.de

Red supergiants (RSGs) experience slow, intensive mass loss up to 10^{-4} $M_\odot$ yr^{-1}. Despite its importance not only in stellar evolution but also in the chemical enrichment of the interstellar matter, the mass loss mechanism in RSGs is not well understood. A better understanding of the outer atmosphere of RSGs is a key to unraveling the mass-loss mechanism in these stars. High spatial resolution observations in IR molecular lines are very effective for probing the physical properties of the inhomogeneous outer atmosphere. We observed the prototypical RSG Betelgeuse (M1-2Ia–Ibe) in the CO first overtone lines with the spectro-interferometric instrument AMBER at the ESO's Very Large Telescope Interferometer (VLTI) using baselines of 16, 32, and 48 m. Details of the observations and the modeling are described in Ohnaka *et al.* (2009). The high-spectral ($R = 4800$–12000) and high-spatial resolution (9 mas) provided with AMBER allowed us to study inhomogeneities seen in the individual CO first overtone lines. Our AMBER observations represent the highest spatial resolution achieved for Betelgeuse, corresponding to five resolution elements over its stellar disk. The AMBER visibilities and closure phases in the K-band continuum can be reasonably fitted by a uniform disk with a diameter of 43.19 ± 0.03 mas or a limb-darkening disk with 43.56 ± 0.06 mas and a limb-darkening parameter of $(1.2 \pm 0.07) \times 10^{-1}$. On the other hand, our AMBER data in the CO lines reveal salient inhomogeneous structures. The visibilities and phases (closure phases, as well as differential phases representing asymmetry in lines with respect to the continuum) measured within the CO lines show that the blue and red wings originate in spatially distinct regions over the stellar disk, indicating an inhomogeneous velocity field that makes the star appear different in the blue and red wings. Our AMBER data in the CO lines can be roughly explained by a simple model, in which a patch of CO gas is moving outward or inward with velocities of 10–15 km s^{-1}, while the CO gas in the remaining region in the atmosphere is moving in the opposite direction at the same velocities. These velocities compare favorably with the macroturbulent velocities of 10–20 km s^{-1} derived by spectroscopic analyses. Also, the AMBER data are consistent with the presence of warm molecular layers (so-called MOLsphere) extending to $\sim$1.4–1.5 $R_\star$ with a CO column density of $\sim 1 \times 10^{20}$ cm^{-2}. However, the present data are insufficient to constrain the surface pattern uniquely or to reconstruct an image. Our AMBER observations of Betelgeuse are the first spatially resolved study of the macroturbulence in a stellar atmosphere (photosphere and possibly MOLsphere as well) other than the Sun. The spatially resolved CO gas motion is likely to be related to convective motion in the upper atmosphere or intermittent mass ejections in clumps or arcs.

References

Ohnaka, K. *et al.* 2009, *A&A* in press, astro-ph/0906.4729.

Highlights of Astronomy, Volume 15
XXVIIth IAU General Assembly, August 2009
Ian F. Corbett, ed.

Infrared and sub-mm observations of cataclysmic variables

A. Evans

Astrophysics Group, Keele University, Keele, Staffordshire, ST5 5BG, UK

1. Introduction

Although cataclysmic variables (CVs) come in a wide variety of shapes and sizes, the essential ingredients are a compact primary star and a Roche-lobe-filling secondary. In most cases the cool component is a main sequence dwarf, and the compact component a white dwarf (WD). Material from the cool component flows through the inner Lagrangian point via an accretion disc onto the surface of the WD; the flow near the WD is significantly affected by the strength of the magnetic field the WD may have (see Warner for a review of CVs). CVs are characterised by regular eruptions, ranging in energetics and frequency from 'dwarf novae', in which eruptions of amplitude $\sim 3 - 4$ mag in the visual occur every few days to weeks, to classical novae (CNe) in which the eruption is explosive, due to thermonuclear runaway (TNR) in material accreted on the surface of the WD (see Bode & Evans for a review of CNe).

2. CVs in the infrared

The cool component and accretion disc dominate the SED of CVs in the near infrared (IR) but in some cases an excess is seen at longer wavelengths. The most likely explanation is emission by dust in the CV system and the origin of this material is problematic. Possibilites include a circumbinary dusty disc, a remnant of the material produced during the common envelope phase of the CV progenitor, and material lost from the CV via winds, the accretion flow or ejected in outbursts (see Brinkworth).

3. Classical novae

CNe are a subset of the CVs in which the layer accreted on the WD undergoes TNR. The accreted layer is ejected explosively and some $10^{-4} M_\odot$ of material is ejected at $\sim 1000\,\mathrm{km\,s^{-1}}$; however, there is a long-standing problem in that far more material is ejected in the explosion than seems to be predicted by TNR models, with possible implications for the long-term evolution of CVs. Elemental abundances in the ejecta are highly non-solar, and CNe are expected to be a significant source of ^{13}C, ^{17}F, ^{22}Na (which decays to ^{22}Ne) and ^{26}Al in the Galaxy. A CN eruption goes through an optically thick phase, a free-free phase, a dust-forming phase, a nebular phase and a coronal phase, although not all CNe produce dust, and not all are coronal.

IR spectroscopy of CNe is key in determining abundances in the ejecta, and for characterising the dust mineralogy in dusty novae. Also, since some of the WD is mixed into the accreted layer and participates in the TNR, the nature of the WD (whether CO/ONe) is also determined by IR spectroscopy; abundances therefore shed light on the TNR and on the nature of the WD. Dust-forming CNe display 'chemical dichotomy', in which both O-rich and C-rich dusts are present simultaneously; this arises because CO formation in the ejecta does not go to saturation.

The WD progenitor may have shed some $1 M_\odot$ of material before the binary became a CV, and in some rare cases this material may be seen in the IR and sub-mm in the CN environment.

4. The recurrent nova RS Oph

The best studied recurrent nova (RN) is RS Oph, which had its sixth known eruption in 2006. The eruption in RS Oph is also due to TNR on a WD but in this case, the WD is close to the Chandrasekhar limit (and so RS Oph is a candidate Type Ia supernova), and the material ejected in the explosion runs into and shocks the wind of the red giant.

The 2006 eruption was the subject of observations from the X-ray to the radio. *Spitzer* observations revealed a rich emission line spectrum and yielded a major surprise: the presence of hot dust. Unlike CNe, in which dust condenses in the ejecta, the dust in the RS Oph environment must precede the explosion, and is likely located in a circumbinary disc (see Evans *et al.* for a review of RNe and the 2006 eruption of RS Oph).

5. Symbiotic stars

Symbiotic stars have spectra showing evidence of a hot component (e.g. by virtue of high excitation lines) and an evolved star with prominent TiO, VO bands (see Corradi *et al.*). Many of these systems also undergo nova-like eruptions, and so RS Oph is sometimes classified as a symbiotic. As well as displaying the IR emission and dust features seen in objects like RS Oph and CNe, some also display maser emission. This offers the prospects of measuring orbital motion and even parallax.

6. Conclusions

The binarity of CN systems and the kinetics of mass ejection play a major role in shaping the ejecta. Thus, a CN eruption resembles the events that occur during the late phases of evolution of binary stars but with the advantage that the process can be observed in a human lifetime. CNe are important therefore for understanding the shaping of planetary nebulae. The explosions of RS Oph resemble those of supernovae, again in fast-forward.

It is with some justification that CNe have been described as "unique laboratories in which several poorly-understood astrophysical processes (e.g. mass transfer, common envelope evolution, molecule and grain formation) may be observed in real time."

In the course of the evolution from wide binaries to CVs, a great deal of material is deposited in the interstellar medium. This, together with the mass-loss during eruptions, indicates that they contribute to the chemical evolution of their host galaxies.

Acknowledgements

I thank the IAU for financial support.

References

Bode, M. F. & Evans, A. 2008, eds, *Classical Novae*, second edition, Cambridge University Press

Brinkworth, C. S., *et al.* 2007, *ApJ* 659, 1541

Corradi, R. L. M., Mikołajewska, J., Mahoney, T. J. 2003, eds, *Symbiotic Stars Probing Stellar Evolution*, ASP Conference Series Vol. 303

Evans, A., Bode, M. F., O'Brien, T. J., Darnley, M. J. 2008, eds, *RS Ophiuchi (2006) and the Recurrent Nova Phenomenon*, ASP Conference Series Vol. 401

Warner, B. 1995, *Cataclysmic Variable Stars*, Cambridge University Press

Highlights of Astronomy, Volume 15
XXVIIth IAU General Assembly, August 2009
Ian F. Corbett, ed.

© International Astronomical Union 2010
doi:10.1017/S1743921310010719

Globular clusters in the near–infrared

E. Valenti[1,2], L. Origlia[3] and R. M. Rich[4]

[1]ESO - European Southern Observatory,
Av. Alonso de Cordova, 3107 Casilla 19001, Santiago, CHILE

[2]Pontificia Universidad Catolica de Chile, Departamento de Astronomia y Astrofisica,
Av. Vicuña Mackenna 4860, 782-0436 Macul, Santigo, CHILE

[3]INAF - Osservatorio Astronomico di Bologna,
Via Ranzani, 1, 40127 Bologna, ITALY

[4]Department of Physics and Astronomy, Math-Sciences 8979, UCLA, Los Angeles CA
90095-1562, USA

The study of Globular Cluster (GC) stellar populations (SPs) addresses fundamental astrophysical questions ranging from stellar structure, evolution and dynamics, to Galaxy formation. Indeed, they represent: *i) fossils* from the remote and violent epoch of Galaxy formation, *ii)* test particles for studying Galaxy dynamics and stellar dynamical model, and *iii)* fiducial templates for studying integrated light from distant stellar systems. In particular, *high resolution spectroscopy* of GC SPs provides abundance patterns which are crucial for understanding the formation and chemical enrichment time-scale of the host galaxy. Here the major results on Galactic GCs based on high-resolution near-infrared (near-IR) spectroscopy are briefly reviewed. Optical and IR spectroscopy are complementary tools to investigate SPs in different environments, the latter being more suitable in the case of moderately-high extinction regions ($A_V \geqslant 2$) and high metallicity.

Since the Galactic halo GCs are mainly a low-to-intermediate metallicity population located in low-extinction regions, they have been extensively studied with high-resolution optical spectrographs, to derive accurate information about their chemistry, kinematics and dynamics. However, the few available IR spectroscopic studies of Halo GCs demonstrated how this spectral region can be exploited to derive unique chemical properties, which are not easily observable in the optical (i.e. C isotopes and ^{19}F). By using high resolution (R=10,000) K-band spectra of giants in ω Cen, Smith, Terndrup & Suntzeff (2002) derived C isotope ratios from the first overtone CO bandhead at 2.3 μm. Because ω Cen is the only GC showing star-to-star iron abundance variations, they could investigate the behavior of the $^{12}C/^{13}C$ ratio over the metallicity ([Fe/H]), finding no correlation. Smith *et al.* (2005) derived, for the first time, fluorine abundances in the Halo GC M4 from high-resolution (R=50,000) K-band spectra. The authors found that ^{19}F, only measurable in the IR, shows: *i)* star-to-star abundance variations, *ii)* a correlation with O, and *iii)* an anti-correlation with Na and Al, hence suggesting pollution from stars with M> $3.5M_\odot$. Finally, Yong *et al.* (2008) used H and K-band spectra (R=50,000) of giants to perform a detailed chemical analysis of the Halo cluster NGC 6712. They found star-to-star abundance variations of C, N, O, F and Na.

Because of the high level of extinction and metal content, M giants in Bulge GCs are suitable targets for high-resolution IR spectroscopy. In this framework, with the final goal being to trace the formation and chemical enrichment time-scales of the Bulge, our group collected high-resolution (25,000⩽R⩽100,000) H and K-band spectra of giant stars (near the Red Giant Branch Tip) for a number of GCs and fields. H band spectra of cool giants shows many absorption features due to both metals (Fe I, Ca I, Si i, Mg I, Ti I, Al I, etc.) and molecules (CO, OH), which allow accurate abundance analysis over a wide range of metallicities, up to the super-solar regime (see left panel of Fig. 1). Moreover, the IR CO bandheads and OH roto-vibrational lines provide the most robust estimates of oxygen and carbon abundances and their isotopic ratios in cool stars (Melendez, Barbuy & Spite (2001)). From the data analyzed so far (seven GCs) we found low C isotopic ratios ($<^{12}C/^{13}C> \leqslant 8$), which suggests that mixing mechanisms due to cool-bottom processing are at work during the evolution along the Red Giant Branch, and as summarized in the right panel of Fig. 1, an overall [α/Fe] enhancement up to solar metallicity, which points towards an early and rapid Bulge formation scenario.

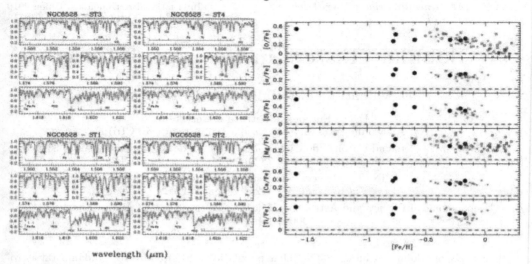

wavelength (μm)

Figure 1. *Left Panel:* Selected portions of the observed NIRSPEC echelle spectra (*dotted lines*) of four giants in NGC 6528 with the best-fitting synthetic spectrum (*solid lines*) superimposed. A few important molecular and atomic lines of slight interest are marked (from Origlia, Valenti & Rich (2005)). *Right Panel:* Plot of α-element to iron abundance ratios as a function of [Fe/H] for observed giants in Bulge clusters (*filled circles* - from Origlia, Rich & Castro (2002), Origlia & Rich (2004), Origlia, Valenti & Rich (2005), Origlia *et al.* (2005), Origlia, Valenti & Rich (2008)) and fields (*squares* - from Zoccali *et al.* (2006), Lecureur *et al.* (2007), *x-crosses* - from Rich & Origlia(2005), Rich, Origlia & Valenti (2007)). *Green dashed lines* mark the solar reference value.

References

Lecureur, A., Hill, V., Zoccali, M., *et al.* 2007, *A&A* 465, 799L
Melendez, J., Barbuy, B. & Spite, F. 2001, *ApJ* 556, 858
Origlia, L., Rich, M. R. & Castro, S. 2002, *AJ* 123, 1590
Origlia, L. & Rich, M. R. 2004, *AJ* 127, 3422
Origlia, L., Valenti, E. & Rich, M. R. 2005, *MNRAS* 356, 1276
Origlia, L., Valenti, E., Rich, M. R., Ferraro, F. R. 2005, *MNRAS* 363, 879
Origlia, L., Valenti, E. & Rich, M. R. 2008, *MNRAS* 388, 1419
Rich, R. M. & Origlia, L. 2005, *ApJ* 634, 1293
Rich, R. M., Origlia, L. & Valenti, E. 2007, *ApJ* 665, 119
Smith, V. V., Terndrup, D. M. & Suntzeff, N. B. 2002, *ApJ* 579, 832
Smith, V. V., Cunha, K., Ivans, I. I., *et al.* 2005, *ApJ* 633, 392
Yong, D., Melendez, J., Cunha, K., *et al.* 2008, *ApJ* 689, 1020
Zoccali, M., Lecureur, A., Barbuy, B., *et al.* 2006, *A&A* 457, L1

Highlights of Astronomy, Volume 15
XXVIIth IAU General Assembly, August 2009
Ian F. Corbett, ed.
© International Astronomical Union 2010
doi:10.1017/S1743921310010720

Spitzer observations of molecules and dust in evolved stars in nearby galaxies

Mikako Matsuura

Department of Physics and Astronomy, University College London, Gower Street, London
WC1E 6BT, United Kingdom; email: mikako@star.ucl.ac.uk

1. Introduction

Molecules and dust are formed in and around the asymptotic giant branch (AGB) stars and supernovae (SNe), and are ejected into the interstellar medium (ISM) through the stellar wind. The dust and gas contain elements newly synthesised in stars, thus, dying stars play an important role in the chemical enrichment of the ISM of galaxies. However, quantitative analysis of molecules and dust in these stars had been difficult beyond our Galaxy. The high sensitivity instruments on-board the *Spitzer Space Telescope* (*SST*; Werner *et al.* 2004) have enabled us to study dust and molecules in these stars in nearby galaxies. Nearby galaxies have a wide range in metallicity, thus the impact of the metallicity on dust and gas production can be studied. This study will be useful for chemical evolution of galaxies from low to high metallicity.

2. Observations and results

We present 5–35 μm spectra, obtained by the infrared spectrometer (IRS) on-board the *SST*. Targets are mainly asymptotic giant branch (AGB) stars located in five nearby galaxies within 200 kpc. The main objectives are: 1) to exploit the dust property, molecular abundance, and mass loss of AGB stars for a wide range of metallicities of the host galaxies, and 2) to investigate the contribution of mass loss from AGB stars on the galactic chemical evolution. The metallicities of these galaxies range from half of the solar metallicity (Large Magellanic Cloud) to 5 % of the solar metallicity (Sculptor dwarf spheroidal galaxy). We found the following results.

• Carbon-rich AGB stars have abundant C_2H_2 molecules in low metallicity galaxies (Matsuura *et al.* 2005, 2006, 2007; Sloan *et al.* 2009). At low metallicity, as the initial oxygen abundance is low, after carbon elements are synthesised in stars, it leads to a high carbon-to-oxygen (C/O) abundance ratio in carbon-rich AGB stars. A similar effect is found for amorphous carbon (Groenewegen *et al.* 2007).

• The SiC dust fraction is lower at lower metallicity, but this fraction suddenly increases at the end of the AGB phase (Gruendl *et al.* 2008). The dust condensation sequence (SiC, and amorphous carbon/graphite) changes at low metallicity.

• We found that the gas mass ejected from all AGB stars and SNe almost the same in the LMC (Matsuura *et al.* 2009). AGB stars are one of the major dust sources in the LMC. AGB dust at low metallicity tends to be more carbonaceous than oxidized.

References

Groenewegen M. A. T., *et al.*, 2007, *MNRAS* 376, 313
Gruendl R. A. *et al.*, 2008, *ApJ* 688, L9
Matsuura M., *et al.*, 2005, *A&A* 434, 691
Matsuura M., *et al.*, 2006, *MNRAS* 371, 415
Matsuura M., *et al.*, 2007, *MNRAS* 382, 1889
Matsuura M., *et al.*, 2009, *MNRAS* 396, 918
Sloan G. C., *et al.*, 2009, *Science* 323, 353
Werner M. W., *et al.*, 2004, *ApJS* 154, 1

Highlights of Astronomy, Volume 15
XXVIIth IAU General Assembly, August 2009
Ian F. Corbett, ed.

SpS1-POSTER PAPERS

High resolution atomic spectroscopy at Lund Observatory
R. Blackwell-Whitehead, H. Hartman, H. Nilsson
Measurement of the sulphur abundance in halo stars by using Mult. 3 at 1045 nm
E. Caffau, L. Sbordone, H. Ludwig, P. Bonifacio, M. Spite
A far-infrared AKARI-FIS survey of post-AGB stars and (proto) planetary nebluae
N. Cox, P. Garcia-Lario, A. Garcia-Hernandez, A. Manchado
PO molecular line prospects for sunspot umbrae
A.A. de Almeida, G.C. Sanzovo, R. Boczko
K and L-band spectroscopy of Be stars
A. Granada, M.L. Arias, L.S. Cidale
VLT-CRIRES Observations of Eta Carinae ejecta: Weigelt blobs and Sr filament
H. Hartman, J. Groh, T.R. Gull, H.U. Käufl, F. Kerber, V. Letokhov, K.R. Nielsen
Submillimeter spectral line observations of early stage protostellar candidates
T. Hill, M.R. Cunningham, V. Minier
Modelling of circumstellar dust around symbiotic Miras in the infrared
T. Jurkic, D. Kotnik-Karuza
Standards and reference data for calibration of infrared spectrographs
F. Kerber, M. Aldenius, P. Bristow, G. Nave, Y. Ralchenko, C.J. Sansonetti
Infrared lines in NLTE model atmospheres of hot stars
J. Kubát, B. Šurlan
CRIRES in self-aligning spectrograph mode, a fundamental improvement for quantitative stellar
spectroscopy
H.U. Käufl, J.-F. Picard, A. Smette, F. Kerber, U. Seemann
CRIRES-POP: A new high resolution near-IR spectral library
*T. Lebzelter, S. Bagnulo, T. Dall, H. Hartman, G. Hussain, U. Käufl, M. Lederer, F. Nieva,
N. Przybilla, S. Ramsay, A. Seifahrt, A. Smette, S. Uttenthaler, G.M. Wahlgren*
Dust in S stars
S. Lorenz-Martins, H.S. de Almeida
Far-infrared spectroscopic imaging of interstellar material around Eta Carinae
H. Matsuo, T. Arai, T. Nitta, A. Kosaka
Infrared integral field spectrometer observations of molecular hydrogen lines on the planetary
nebula NGC 7293
M. Matsuura
Evolution among extreme Carbon stars and post-AGBs
A.F.P. Sánchez, A.B. de Mello, S. Lorenz-Martins
CRIRES: A new era in high-resolution infrared spectroscopy
U. Seemann, H.U. Käufl, A. Reiners
A view on dead OH/IR stars
T.C.K. Silva, S. Lorenz-Martins, R. Ortiz, A.F.P. Sánchez
Heavy element abundances of cool supergiants from near-IR spectra
G.M. Wahlgren, K.G. Carpenter, R.P. Norris

Highlights of Astronomy, Volume 15
XXVIIth IAU General Assembly, August 2009
Ian F. Corbett, ed.

SpS2-The International Year of Astronomy 2009

Pedro Russo[1], Catherine Cesarsky[2] and Lars Lindberg Christensen[1]

[1]ESO, 85748 Garching bei München, Germany, [2]CEA Saclay, Bâtiment Siège, 91191 Gif sur Yvette cedex, France
email: prusso@eso.org, catherine.cesarsky@cea.fr, lars@eso.org

Part I: Introduction

IAU Welcome

The International Astronomical Union (IAU) launched 2009 as the International Year of Astronomy (IYA2009) under the theme, The Universe, Yours to Discover. IYA2009 marked the 400th anniversary of the first astronomical observation through a telescope by Galileo Galilei. It has been, and still is, a global celebration of astronomy and its contributions to society and culture, with a strong emphasis on education, public engagement and the involvement of young people, with events at national, regional and global levels throughout the whole of 2009. UNESCO endorsed IYA2009 and the United Nations proclaimed the year 2009 as the International Year of Astronomy on 20 December 2007. These proceedings aim to give a brief account of IYA2009, from its inception to the present and how its legacy will influence the future of astronomy communication on a planet-wide scale.

Astronomy is one of the oldest fundamental sciences. It continues to make a profound impact on our culture and is a powerful expression of the human intellect. Huge progress has been made in the last few decades. One hundred years ago we barely knew of the existence of our own Milky Way. Today we know that many billions of galaxies make up our Universe and that it originated approximately 13.7 billion years ago. One hundred years ago we had no means of knowing whether there were other solar systems in the Universe. Today we know of more than 400 planets around other stars in our Milky Way and we are moving towards an understanding of how life might have first appeared. One hundred years ago we studied the sky using only optical telescopes and photographic plates. Today we observe the Universe from Earth and from space, from radio waves to gamma rays, using cutting-edge technology. Media and public interest in astronomy have never been higher and major discoveries are front-page news throughout the world.

We are now in a position to reflect upon IYA2009, taking an objective view on projects that have taken place and we can see how the astronomy education and public outreach landscape is changing. There are certainly many initiatives to consider, from the twelve Cornerstone projects to the thousands of national activities that have brought together hundreds of thousands of people in many countries for astronomy-themed events. Take, for example, the Sunrise Event on New Year's Day in Busan City, South Korea, where more than 400 000 people took part. In Brazil, the 2009 Brazilian Olympiad of Astronomy and Astronautics saw more than 750,000 students participate from 32,500 schools. In Paraguay, the IYA2009 launch featured a concert with more than 1600 musicians and an audience of over 15,000.

IYA2009 was not strictly limited to Earth-bound activities. The IYA2009 logo and motto were proudly displayed on the Ariane 5 rocket that sent two frontline space observatories into space in May 2009: ESA's Herschel and Planck flagship missions. Also that month,

Figure 1. IYA2009 Opening Event, Paris, France on 14 January 2009. *Credit: IAU/José Francisco Salgado*

astronauts performed repairs and equipped the NASA/ESA Hubble Space Telescope with the latest in instrument technology. To honour IYA2009, astronaut Mike Massimino took onboard with him a replica of Galileo's telescope as well as an IYA2009 flag. So, evidently, it was a momentous event on and off our planet!

However, it is important to not just note successes, but also areas for improvement. IYA2009 should be seen as a learning experience; since it was the first time ever that such a huge network, consisting of as many as 148 nations working together on a single science communication venture, was put together, not all challenges could be met. Looking back on these will give future astronomy popularisers a head start and help them make their own projects as effective as possible. Finally, perhaps, looking back at the actions and events and the popular reaction, we will be able to truly gauge how often and how deeply IYA2009's motto, **"The Universe, Yours to Discover"**, was fulfilled during the Year.

Welcome Statement by UNESCO

Yolanda Berenguer[1], Pedro Lessa[2]
[1]Focal Point for the International Year of Astronomy (2009), UNESCO
e-mail Y.Berenguer@unesco.org
[2]Coordinator, UNESCO Rio de Janeiro Antenna Office
e-mail Pedro.Lessa@unesco.org.br

Six months have passed since the opening ceremony of the International Year of Astronomy (IYA) was held in UNESCO headquarters in Paris. At this gathering of more than 900 professional and amateur astronomers, professors, scientists and students, strong messages of support to science were expressed by ministerial representatives from Italy, France, Japan and Czech Republic. In addition, a space industry representative, Thales

Alenia, affirmed that we are living in a golden age of science and that we should continue to stimulate innovation in science and technology as it is the only way to avoid another global crisis of the kind we are experiencing at the moment.

The opening ceremony gathered eminent scientists such as Nobel Prize laureates, world-renowned astronomers, astrophysicists and writers, who presented the latest discoveries and scientific developments in astronomy and shared their views and vision of the Universe in the decades to come with great enthusiasm. Many speakers referred to the valuable contributions made by people of different cultural backgrounds and underlined the importance of international cooperation, especially in the development and implementation of projects that necessitate major observational infrastructure and capacity, both in space and on the ground.

As the United Nations lead agency for the IYA, UNESCO has ensured that all member states are aware and involved in the celebration of the Year. Highlighting the scientific, educational and cultural components, the Director-General, Mr Koichiro Matsuura, encouraged all nations, through a joint statement with the IAU President, Professor Catherine Cezarsky, to consider the IYA as an excellent opportunity to cooperate at regional and international levels, and especially in forging links between professional and amateur astronomers, to promote astronomy education in developing countries in particular and to help safeguard natural and cultural heritage sites linked to astronomical observations. In response to the call of the Director-General, numerous national committees have been created and have registered as IYA national nodes, whose main objective is to widen the reach of astronomy to non-scientists.

In close cooperation with IAU, the World Heritage Centre of UNESCO is implementing the Astronomy and World Heritage initiative, launched in 2003, which encourages the nomination of cultural properties linked to astronomy. Many State Parties to UNESCO's World Heritage Convention (1972) have designated national institutions that could identify the most representative sites and propose them for nomination on the World Heritage List.

UNESCO supports several IYA Cornerstone projects, such as the Galileoscope project. Eight sub-Saharan African countries (Ethiopia, Ghana, Kenya, Mozambique, Nigeria, Senegal, Tanzania and Uganda) will receive high quality, low cost, easy-to-assemble Galileoscopes, which will be distributed to schools through UNESCO's Associated Schools Project (ASP). The ASP is a network of educational institutions in developing countries ranging from kindergarten to teacher-training institutes that have been selected by ministries of education to carry out pilot and innovative activities. We hope to identify more countries in Africa and in other regions in the near future that will participate in the Galileoscope project.

The other UNESCO-supported Cornerstone projects are Universe Awareness, Developing Astronomy Globally, From Earth to the Universe and Dark Skies Awareness. All these focus on generating the interest, imagination, fascination and knowledge of the general public, and especially younger people, in astronomy. UNESCO sponsors stargazing events organised by the Observatoire de Paris/Meudon for the general public such as the one held in celebration of 100 Hours of Astronomy. Support will also be provided to other outreach activities such as lectures and exhibitions.

To build up the skill set of teachers, UNESCO and IAU's Commission 46 on Education Development conducted a pilot teacher-training workshop in Ecuador and Peru last June. The workshop introduced a new teaching methodology in astronomy, which was evaluated positively by the participating teachers and which has rekindled their interest in teaching this subject, once equipped with new materials and skills.

All the activities mentioned are being implemented in the framework of UNESCO's Space

Education Programme, which promotes space-related subjects and disciplines in schools and universities. As we all know, space not only brings a new dimension to science education, but also provides new knowledge and perspectives, especially in the study of the Earth, its systems and resources. Stimulating and maintaining the interest of the younger generation in science through space will hopefully lead them to take up science and engineering careers in the future.

But more important, space subjects develop the critical thinking process, problem-solving and participatory decision-making skills of individuals. These traits are fundamental and central to quality education in preparation of the next generation of scientists, future custodians of the planet Earth and future explorers of outer space.

Let me conclude by assuring you of the continuous commitment of UNESCO in science education and capacity building with a view to sustaining a safe and secure planet and peaceful exploration of outer space. I wish you a good meeting and fruitful discussions on the way forward of the International Year of Astronomy beyond 2009.

IYA2009: Behind the Scenes
Pedro Russo, Lars Lindberg Christensen, Mariana Barrosa and Lee Pullen
IAU/IYA2009 Secretariat, ESO

The recipe for an International Year
International Years have a long and varied history, from the first International Polar Year in 1882/1883, to modern equivalents such as the International Heliophysical Year in 2007/2008. Studying these previous initiatives, it became clear that a recipe of sorts involving the necessary components of a successful International Year could be concocted. First and foremost, a good idea is needed. This must be something to capture people's imagination, be relevant to society, and ideally have the potential to continue beyond the year in question. Next, it must be possible to put a strong case together in order to persuade policy makers of the value of having a year dedicated to the theme. This leads to the next two recipe points: a UN Body recommendation, leading to a UN Proclamation. Without these, official International Year status would be impossible. These are large hurdles to overcome, but the list continues. A large network, ideally already in existence and that can be built upon, is required if the initiative is to have global reach. There must be ideas for national and worldwide activities, as well as the funds to realise these. Finally, there must be genuine enthusiasm, engagement and excitement from all involved parties. Clearly, this is a demanding recipe!

IYA2009 proclamation roadmap
From the very beginning it was recognised that astronomy ticked many of the recipe boxes, and was an ideal candidate for an International Year. However, the route to official UN Proclamation was necessarily long. The following list outlines some of the key moments.

- 2002: The idea Franco Pacini reasoned that as mathematics had a Year in 2000 and Physics was set to have one in 2005, there was potential for astronomy to achieve the same level.
- 2003: IAU GA, Sydney IAU members voted unanimously in favour of Resolution B3, which recommended that 2009, the 400th anniversary of Galileo's accomplishments and the real birth of modern telescopic astronomy, be declared the Year of Astronomy.

- 2005: UNESCO endorsement Italy submitted a request to UNESCO.
- August 2005: UNESCO endorsed 2009 as the International Year of Astronomy.

- 2006: IAU GA, Prague Special session helped to reinforce plans for IYA2009 that would come to the fore in the event of a positive decision by the UN.

- 2007: Lobbying took centre stage. In order to convince the UN that a particular topic deserves International Year status, many countries must show their support.

- December 2007: UN Proclamation: **17 December 2007**, the news that astronomers had been waiting for arrived: the UN accepted this recommendation. 2009 was officially to be the International Year of Astronomy!

Strategic planning: setting up the Network
Once the UN Proclamation was confirmed, organisation could begin in earnest. The IAU was the logical choice for the central coordination role, as it is the world's largest body of professional astronomers. One problem became quickly apparent: the IAU had 64 national members, although 194 sovereign states are recognised by the UN.
A plan was devised to overcome this. Countries with professional astronomers, most often through academia, were researched. If this was not possible, or sometimes in conjunction with, professional organisations, active and visible amateur and enthusiast astronomers were identified. Neighbouring countries were also asked to support nearby nations who might be lacking in experts. Help was also requested from UNESCO delegations. Over time, a long list of astronomy experts from nations around the globe was amassed. These would later become National Nodes and Single Points of Contact.
During this research phase, successful transnational science communication and education institutions often cropped up. Great potential was seen in these lending their valuable expertise by supporting and implementing activities around the globe. The first set of Organisational Nodes had been discovered.

Strategic planning: defining the project
It is now possible to list some of the defining moments of the planning process for IYA2009, in chronological order:
2003 Establishment of the IAU Working Group, which defines IYA2009.
2005 Establishing rationale and vision.
2006 Goals and objectives.
2006–2007 Launching and devising initial projects:

(*a*) Portal to the Universe (idea presented for the first time in the C55 Business Meeting in the IAU GA in Prague)
(*b*) The Universal Times Cosmic Diary
(*c*) "*Connect people with the night sky and to help them become aware of light pollution issues*" Dark Skies Awareness
(*d*) "*Arrange a series of live webcasts over a 24-hour period from telescopes around the world.*" 100 Hours of Astronomy (Around the World in 80 Telescopes)
(*e*) "*Type of Opening: International Polar Year of World Year of Physics?*" IYA2009 Opening Ceremony
(*f*) SPoCs meeting in Garching, March 2007
(*g*) CAP2007 conference in Athens, October 2007
(*h*) Logo studies, 2007
July 2007 The IYA2009 Secretariat was established at the European Southern Observatory's headquarters in Garching, near Munich in Germany. This is to be the central hub of IYA2009 activities, coordinating during the planning, execution and evaluation phases.

Figure 2. IYA2009 The development of the logo

Figure 3. The role of the Secretariat expressed graphically

By this point a significant amount of groundwork had been completed, and a clear view of the Year was emerging. Figure 3 shows schematically how the various aspects were designed to be interconnected.

Strategic planning: funding the project
It was recognised as early as the "International Year Recipe" stage, that without adequate funding any venture on such a grand scale would be doomed from the beginning. National funding was deemed to be the responsibility of National Nodes. It was agreed that this was the only practical way of delegating over so many countries. Global funding would be used to finance the IYA2009 Secretariat, to provide operations and communication products, and to provide seed funding to the Cornerstone projects. Initial estimates placed this funding as 300 000 EUR to 1 million EUR, if major sponsors could be found. The next step was fundraising. Organisations, institutions and agencies related to astronomy, space science and the natural sciences were contacted. Many of these were from the Organisational Associates list formed earlier. Private companies were offered the opportunity to become Global Official Partners or Global Sponsors. The strategy was to

initially send direct mail and then follow up with personal calls to specific contacts and fundraisers. An elaborate Google Spreadsheet was used to keep track of proceedings. Thirty three Organisational Associates agreed to provide financial backing, along with three Global Sponsors. Unfortunately, no Global Official Partners were found, but a very respectable total of 650 000 EUR had been guaranteed.

Global Projects: Cornerstones

IYA2009 is supported by 12 Cornerstone projects. These are global programmes of activities centred on specific themes and are some of the projects that help to achieve IYA2009's main goals. Some are based on new ideas, such as From Earth To The Universe, 100 Hours of Astronomy, and the Galileoscope. Others are independent projects, including Universe Awareness and She is an Astronomer. Some consortia of organisations have joined under the same umbrella, such as the Galileo Teacher Training Program and Dark Skies Awareness. One Cornerstone, Galilean Nights, is a follow-up from an earlier initiative.

A franchising approach was taken with these Cornerstones. They have a common identity and central coordination. Key is that they have common goals, defined by IYA2009. They are also encouraged to share resources and expertise. Cornerstones are also financially independent thanks to seed funding from the IYA2009 Secretariat, of around 15 000 EUR per project.

Global Projects: Special projects and products

IYA2009 Special projects are intended to give large global projects (which satisfy the vision of IYA2009) greater international recognition and an opportunity to link with celebrations worldwide.

Criteria were established, which projects must satisfy. These include aligning with at least one of IYA2009's goals; being global in scope; being financially independent; and evidence must be given of successful implementation, in the form of human resources, funding, planning, or other relevant factors. In total, 16 projects have met these criteria. Special products are commercial products that satisfy the vision of IYA2009 to achieve greater international recognition, are an opportunity to link with celebrations worldwide, and to use the IYA2009 global network to reach out. As with the Special projects, specific criteria must be met to achieve Special product status. These include aligning with IYA2009's vision, being available globally, being adaptable to other languages, and offering a financial contribution (minimum 5000 EUR per product). During IYA2009 there were eight Special products.

Keeping the momentum

Having such a large network presents its own problems, one of which is ensuring adequate communication to maintain momentum. Several methods have been adopted to help. First is the website, which features news articles and points of interest. There are also daily updates on the web and new media outlets, such as Twitter, Facebook, and Portal to the Universe. This is in addition to weekly newsletters sent to the Single Points of Contacts, Cornerstones, Media Partners and any others who wish to be included. The Communicating Astronomy with the Public journal is published quarterly, and features IYA2009 stories, updates and best practices. Taken together, a lot of time and effort has been put into keeping the momentum of IYA2009 up over the Year.

What's next?

IYA2009 was never planned as simply a year-long series of activities; from the initial planning stage it was seen as a springboard to increased astronomy popularisation among the
public, and improved networking among professionals and enthusiasts. Now the challenge
is to realise this aim, and ensure that IYA2009's legacy lasts far into the future.

Part II: National Node Reports

IYA2009 Activities Status Report - Brazil

Tasso A. Napoleão IYA2009 National Node, Brazil e-mail tassonapoleao@gmail.com

Introduction

Starting in 2007, Brazil has built a large National Network for IYA2009, consisting of 249 Local Nodes spread all over the country. Out of that total, 25% are universities and research centres, 15% are planetariums and science museums, and 60% are amateur astronomers groups. A small staff of with 10 members (the Brazilian National Node) helped the Brazilian Single Point of Contact to coordinate all national activities.

National activities summary

The balance of 2009 first half has shown a total of 1114 IYA2009 events in the country. All were free and open to the general public. The breakdown of those activities shows that public observing sessions, combined with general astronomy talks, were the preferred kind of activity (41% of the total events number). Public star parties came close (34%), followed by workshops and conferences (14%), special planetarium sessions (6%), trade fairs and exhibitions (4%) and others (1%).

IYA2009 highlights

The most significant events in the first half of 2009 were:

IYA2009 Opening Week: Held between January 19 and 28, this was a huge celebration of astronomy in the whole country to start IYA. Some 225 public events were scheduled simultaneously in 57 Brazilian cities, with 15 000 people attending.

Astronomy in Carnival parades In February, during the most popular festival in Brazil (the Carnival), astronomy was chosen as the theme for three "Samba School" parades, in the cities of Rio de Janeiro, Brasilia and Atibaia. Millions of people watched those parades all across the country.

100 Hours of Astronomy (100HA) A huge success: with 212 public events in 64 cities, Brazil was ranked second worldwide in 100HAs total number of events. About 40 000 Brazilians took their first glance through a telescope during 100HA.

Brazilian Astronomy and Astronautics Olympiad (OBA) Major IYA2009 educational event in Brazil, OBA was held in May 2009, reaching about 860 000 students and 75 000 teachers, in 10 300 schools spread all over the country.

Milky Way Marathon A national "star-hunting" educational campaign against light pollution, held from June to September. Will be conducted every year.

Cosmic Landscapes A FETTU-type photographic exhibit, shown in more than 250 cities throughout Brazil during the second half of 2009.

SNCT (National Science and Technology Week) The official science popularisation week in Brazil, coordinated by the Brazilian Government. SNCT took place in late October and included dozens of IYA2009 events in 2009.

Year-end prospects At least 2500 public events in Brazil are expected for the whole year of 2009, with two million people participating in at least one IYA2009

event. Most of the structure, and the National Network built for IYA2009, will be maintained in activity for the subsequent years by the professional and amateur communities in order to further increase scientific awareness and to improve formal and informal science education after 2009.

International Year of Astronomy 2009 in France: Highlights and Perspectives

Chantal Levasseur-Regourd[1], Françoise Combes[2] and the French steering committee
[1]UPMC Univ. Paris 06 / LATMOS-IPSL, BP 3, 91371 Verrieres, France
e-mail aclr@latmos.opsl.fr
[2]LERMA, Observatoire de Paris, 77 avenue Denfert Rochereau, 75014 Paris, France
e-mail francoise.combes@obspm.fr

The main task of our steering committee, consisting of professional astronomers providing voluntary help for IYA2009 in France, has been to trigger the proposal of projects and ensure their coherence, visibility and coordination. Emphasis was towards observational projects coupled with astronomical animations, projects oriented towards disabled persons (e.g., translation of conferences in sign language, Braille astronomy books, accessibility of observing sites, video recordings for hospitals) and projects for younger public (with "Main à la Pâte" and "Science à l'Ecole") and universities.

The year began with the international Opening Ceremony at UNESCO in Paris, where more than 100 countries were represented in a gathering of about 1000 people, including students. Amongst other international events in France, the IAU Symposium 260 in January, the Invisible Universe colloquium in June and the Mars workshop in September may be mentioned.

French amateur astronomers efficiently coordinated observations and related astronomical activities for 100 Hours of Astronomy, which gathered more than 70 000 attendees in 50 observing sites and for Galilean Nights, for which more than 100 observing sites were open. As far as other international Cornerstones are concerned, Dark Skies Awareness was highlighted (e.g., project of the first international reserve in Europe around the Pic-du-Midi observatory), as well as Developing Astronomy Globally, with special efforts towards French overseas regions and territories. Also, we have participated in Cosmic Diary and She Is An Astronomer, have provided a French translation of the various documents prepared for the Galileoscope project, and have developed the Mutual Event of Galilean Satellites task group.

Amongst other French projects of interest for a wide public, it makes sense to mention the 100 conferences on astronomy that took place throughout the year and all over the country, the public observing sessions in conjunction with the 40th anniversary of Apollo 11 ("Nuits des Etoiles"), the numerous special planetarium shows, the special exhibitions in science museums, the exhibitions in public transport sites, the special events in conjunction with the European Heritage days in September, and the numerous astronomical activities during one whole week of November ("Fte de la Science"). Besides, an impressive variety of high quality books, DVD, movies, e-exhibitions (free downloadable), plays, concerts and stamps devoted to astronomy have appeared throughout the year. Finally, it is already obvious that all these activities will not disappear at the end of 2009. Sustainable outreach projects are already on track, e.g., websites, exhibitions, planetarium shows, and professional–amateur collaborations.

IYA2009 in Mexico
Silvia Torres-Peimbert Instituto de Astronoma,UNAM, Mexico
e-mail: silvia@astroscu.unam.mx

The celebration of the International Year of Astronomy in Mexico has been full of very interesting developments. Although professional astronomers are located in very few cities in our vast country, the associated activities have extended out to many more locations.

Star party
Global star parties near 22 archaeological and historical sites took place on 31 January. Many of the star parties were surrounded by displays, conferences, concerts and videos. These events were possible through the sponsorship of many public and private institutions, as well as the participation of amateur astronomers. This activity was very successful.

Exhibits
A large exhibit of astronomical images in Mexico City has also been of great interest to the general population. We have collected 96 astronomical images of very high quality (Tenorio-Tagle, Perez, Cruz-Gonzalez & Torres-Peimbert, private communication). This exhibit has been shown along several avenues in Mexico City, where it has been well attended. It has also been displayed in Guadalajara, Guanajuato and will be displayed at the Fair of Astronomy. A reduced version has exhibited in the cities of Cuernavaca and Queretaro and in several shopping malls in Mexico City. We have also prepared a small image exhibit, of 24 images with captions, that is being shown in many intermediate schools in Mexico City where it is accompanied by astronomical lectures. This modest exhibit is being copied by several groups around the country. Exhibits with historical instruments of the Observatorio Astronomico Nacional have been shown in Mexico City, Guanajuato and Ensenada. A series of photographs of women astronomers by the French photographer Robin Cerrutti have been displayed very prominently.

Cultural events
Several other cultural events have adopted astronomy as their theme, in particular, the Festival Cervantino adopted Galileo and the Telescope: 400 Years as the central theme for its 35th presentation. Among their initiatives was to commission twenty graphic artists to depict one aspect of Galileo and so the exhibition, Twenty Faces of Galileo, was created. An art contest for children in three different levels (kindergarten, grades 1–3 and 4–6) was organised, and the winning

pictures will be exhibited at the Fair. In addition two science fiction story contests were organised, one for youngsters under 16 and another without age limit. The winners will have their stories published. Moreover a set of astronomical radio spots lasting two to three minutes are being broadcast daily in several radio stations around the country and can be copied without charge for distribution.

Fair of Astronomy

To close the activities we are planning a Fair of Astronomy to be held in downtown Mexico City. The setup is extraordinary, and will include equipment related to physics or astronomy from the science museum of the National Autonomous University of Mexico (UNAM) and a set of exhibits of astronomical topics. We are planning multiple activities running continuously and that encompass astronomical lectures, physics workshops and short planetarium shows, as well as a representation of Galileo's workshop; a cosmic ray detector with the corresponding explanations, a puppet show with the life of Galileo, and a variety of child-oriented workshops. We plan a set of exhibits that will include 100 astronomical images; basic optical concepts and telescopes; the Solar System and the search of extraterrestrial life; historical instruments; the location of the Solar System in the Milky Way; 3D images from space explorations; Eighteen Basic Astronomical Questions; the winners of the astrophotography contest; and the winners of the children's art contest about the Universe. In addition a very singular exhibit, Twenty Faces of Galileo, is to be presented, where different artists will illustrate their view of Galileo. It is directed at the general public and will surely be of interest to many high school students.

Conclusion

In conclusion, the efforts have been very fruitful and the community very supportive. We have reached many youngsters and members of the general public.

IYA2009 in Russia: Preliminary Results

Oleg Yu. Malkov Institute of Astronomy, 48 Pyatnitskaya St., Moscow 119017, Russia
e-mail malkov@inasan.ru

The International Year of Astronomy 2009 (IYA2009) is a great event for the scientific and cultural life of all nations. Although IYA2009 activities are on several levels, the majority of IYA2009 events take place locally and nationally. The National Committee of Russian Astronomers is responsible for the IYA2009 event organisation in the Russian Federation and serves as a National Node. It establishes collaborations between professional and amateur astronomers, science centres and science communicators, teachers, lecturers and popularisers.
The Russian National Node current activities are reported here, in particular:

(*a*) Organising the All-Russia Conference "Astronomy and Society" (25–27 March, Moscow), the central event of the International Year of Astronomy 2009 in Russia;

(*b*) Organising the UNESCO congress "Astronomy and World Heritage: Across Time and Continents" (20–24 August, Kazan);

(*c*) Establishing a service at the IYA2009 site where questions from the community concerning astronomy will be answered, with participation of professional astronomers in preparing the answers (25 experts answered about 150 questions by mid-October);

(*d*) Developing pages of the websites of research institutes and university departments of astronomy that will primarily address the general community, schoolchildren, amateur astronomers and cover scientific, social, and educational activities of the corresponding institutions; creating public relations and popular astronomy structures in observatories and institutions;

(*e*) Establishing and supporting contacts with organisations of amateur astronomers, participating in the Astrofest congress of amateur and professional astronomers, defining and formulating problems where amateurs are able to contribute to professional astronomy;

(*f*) Special events for children and youths: competitions of children's astronomical drawings, with winners determined by an open ballot; visits of astronomers with lectures to schools, summer camps, orphan homes; astronomical Olympiads, conferences and team observations;

(*g*) Preparing an expert prognosis of the future of astronomy (see results in Malkov [2009], *Proceedings of Science*, in press);

(*h*) Organising the Autumn 100 Hours of Astronomy (24–27 September, with extension of this period in some institutions): about 15 000 visitors;

(*i*) Publishing a calendar for 2009, issuing a postage stamp and envelope on the occasion of IYA2009; as well as participation in some Cornerstone IYA2009 projects.

Acknowledgements: We would like to acknowledge our partners and sponsors: Novosibirsk Instrumentmaking Factory, MEADE, CELESTRON, Lenta.ru.

IYA2009 in South Africa
Kevin Govender SAAO, South Africa, e-mail kg@saao.ac.za

Activities in South Africa were coordinated by a steering committee chaired by the South African Astronomical Observatory (SAAO), with representatives from all major astronomy-related bodies including government, research, outreach and amateur organisations. In the years preceding IYA2009, the build-up activities included national stakeholder meetings; astronomy education and outreach surveys; setting up of websites and email groups; consolidation of astronomy resources; development of the "astroguide" and "astroCD" (two useful resources to assist outreach volunteers); and a series of training workshops to equip stakeholders with astronomy outreach skills.

IYA2009 in South Africa kicked off in style on New Year's Eve with a once-in-a-lifetime opportunity for the public to enjoy a full night at the SAAO in Sutherland, home to the Southern African Large Telescope (SALT). For the first time in its 37-year history in Sutherland the SAAO dedicated an entire night to the public with astronomers showing people how the telescopes worked, discussing what they do and even carrying out observations for the public. The rural town of Sutherland came to life for a week-long star party that preceded this unique night, with members of the community participating actively in the celebrations and hosting the hundreds of people who made the long trip. This event was soon followed by another mass media event — the official launch of IYA2009 in South Africa by the Minister of Science and Technology at a partial solar eclipse on 26 January. Both these events were broadcast on national television and radio and published in many newspapers.

The momentum from these events carried South Africa through many other activities including the well-attended public lecture series that was linked to a special issue of *Quest* (a popular science magazine); an astronomy-themed Scifest Africa (the biggest Science Festival in the region), as well as National Science Week (the biggest science outreach event on the South African calendar); and participation in various Cornerstones including Dark Skies, 100 Hours of Astronomy (part of which was a SALT inclusive "Around the World in 80 Telescopes" webcast), Universe Awareness, Galileo Teacher Training, From the Earth to the Universe, and Developing Astronomy Globally. The very popular "telescopes on Lion's Head" full Moon hike has established itself as a popular way of doing "extreme astronomy" and outreach training workshops with stakeholders and volunteers were held all over South Africa, as well as in four other African countries.

A grant from the Department of Science and Technology attracted people from all outreach areas in the country including science centres, planetariums, observatories and astronomy societies, all of whom are now involved in the IYA2009 celebrations. This grant also gave rise to an astronomy travelling exhibition and an equipment resource pool both available for willing volunteers to use to promote astronomy. Things are still at full pace with many activities lined up for the rest of 2009† and we still look forward to events such as the Southern African Association of Science and Technology Centres (SAASTEC) conference to be held in Sutherland; a closing star party just before the new year; and the Communicating Astronomy with the Public conference in Cape Town from 15–19 March 2010.

Situated in one of the most underdeveloped regions in the world, South Africa continues to lead the development of astronomy in Africa, with one of the driving factors being the understanding that astronomy gives us perspective. Perspective about how small we are in the Universe, yet how special the conditions are for us to be alive. Perspective, in the bigger context, about the importance (or lack thereof) of skin colour, country borders, ethnicity, religious beliefs, xenophobia and wars. Perspective about the astronomical implications of environmental degradation. Perspectives that are needed in order to make the world a better place!

IYA2009 in Japan

Norio Kaifu NAOJ, Open University of Japan, SPoC and IYA2009 committee chair of Japan
e-mail Kaifunorio@aol.com

† http://www.astronomy2009.org.za

IYA2009 has been extremely successful and well accepted by the Japanese public and media. Let me give some typical numbers for the year up to October; 2700 events officially recognised by the IYA2009 Japan Committee have been held, 500 books related to astronomy or the Universe and 1000 bookstores all through Japan have registered for the IYA2009 special book exhibition "Starry Night Book Fair", 170,000 visitors came to the IYA2009 Tokyo Exhibition, 4.4 million people reported their stargazing to the "Look up the Sky" webpage, and there have been 50 million hits on the IYA2009 Japan website.

The IYA2009 Japan Committee was established in July 2007 and made up from representatives of a variety of astronomical research, education and outreach organisations. In Japan about 200 observatories with 0.3–2-metre aperture telescopes, 300 planetariums and many science museums are run by the local authorities for the public, and their wholehearted contribution as well as the participation of educational organisations and schools made the IYA2009 Japan activities both wide in perspective and strong.

The IYA2009 Japan Committee hosted 17 Japanese projects, among them:

(*a*) IYA2009 opening event: held on 4 January in 40 science museums with 6000 participants;

(*b*) You are Galileo!: distributed thousands of telescope kits for children in Japan and overseas;

(*c*) Stars of Asia: IYA2009 Asia cooperated to publish a book of Asian myths and legends of stars;

(*d*) IYA2009 Travelling Exhibition on Astronomy: Tokyo, Sendai, Niigata, Nagoya, and Osaka;

(*e*) Look up at the Sky!: Ten million people stargazing. The target is 10% of the Japanese population;

(*f*) Starry Sky Book Fair: IYA2009 special book fair in cooperation with booksellers;

(*g*) Reproduction of Galileo's Telescopes: hands-on experience of Galileo's observations;

(*h*) 22 July, Total eclipse & safe sunglasses: caused a boom in Japan. The day was almost cloudy yet still many people enjoyed the event. The need for safe glasses was widely recognised;

(*i*) 7 July, Traditional Tanabata Light Down Campaign: the Star Festival "Tanabata";

(*j*) Tanabata Lectures: about 90 universities held IYA2009 Tanabata lectures for the public;

(*k*) Mr Galileo and his Friends: A series of cartoons about IYA2009 characters. Visit our website!

Japan also actively joined in the IYA2009 Cornerstone projects, i.e.;

(*a*) 100 Hours of Astronomy,

(*b*) Galileo Teacher Training Program,

(*c*) Dark Skies Awareness,

(*d*) Universe Awareness,

(*e*) Cosmic Diary,

(*f*) Developing Astronomy Globally,

(*g*) Galileo Telescopes,
(*h*) Galilean Nights, etc.

We will hold an IYA2009 Grand Finale for two days in December and discuss the IYA2009 Legacy.

Finally, it is already obvious that all these activities will not disappear at the end of 2009. Sustainable outreach projects are already on track, e.g., websites, exhibitions, planetarium shows, and professional–amateur collaborations.

IYA2009 in Spain

M. Villar-Martín[1], **T. Gallego**[1], **E. García**[1], **V. Martínez Pillet**[2]
[1] Instituto de Astrofísica de Andalucía-CSIC, Spain. Members of the Spanish IYA2009 node
[2] Instituto de Astrofísica de Canarias, Spain. Member of the *Comision Nacional de Astronomía*
e-mail montse@iaa.es

IYA2009 is a truly national, collaborative effort in Spain. Research centres, universities, science museums, schools, amateur astronomical societies... all are actively involved in the organisation of IYA2009 activities with a keen interest to transform the event in 2009, together with many other countries, into a worldwide astronomical celebration. IYA2009 is being promoted in Spain at the highest level by the CNA (*Comision Nacional de Astronoma*). There is a steering committee which coordinates at national level and a 16-person Working Group. CSIC (National Research Council) and the Ministry of Science and Innovation have funded the national coordination. The Spanish Astronomical Society has also actively promoted different initiatives, run by hundreds of other bodies.

Statistics: 1 January – 30 June 2009

(*a*) Spanish network for IYA2009: 150 organisations;

(*b*) Number of activities organised within the framework of IYA2009: at least 1985;

(*c*) Number of bodies that have organised one or more IYA2009 activities: at least 1000;

(*d*) Cornerstone IAU projects that Spain participates in: 9;

(*e*) Cornerstone national projects: 11.

Some IYA2009 highlights in Spain include:

(*a*) 26 March: teachers and students in 639 schools measured the Earth's radius, participating in a simple, cheap and educational initiative;

(*b*) The most important results of Spanish astronomy during the last three decades have been presented in a book using the format of brief interviews with the authors. A special contribution of the Spanish Astronomical Society for IYA2009;

(*c*) Dozens of amateur astronomical societies have shown the wonders of the night sky to thousands of people in national/international star parties through the year;

(*d*) Astronomy has been accessible to people with special needs, with several projects such as the production of a free open source software for motor disabled people and a planetarium programme for visually impaired people;

(*e*) Museums and planetariums have launched two special IYA2009 programmes that are being shown in many planetariums: *Evolution*, a tribute to Galileo and Darwin, and *Jors, Jars, Jurs and the Galigalitos*, a show for children;

(*f*) A 2009 and 2010 calendar, several TV programmes, radio podcasts and an ambitious exhibit about the role of women in astronomy have been produced;

(*g*) More than 300 activities were celebrated in Spain during 100 Hours of Astronomy.

IYA2009 in Romania: Between Education and Outreach

Magda Stavinschi[1] and Catalin Mosoia[2]
[1]IYA2009 SPoC for Romania [2] Brief Press ltd
e-mail magda_stavinschi@yahoo.fr

The International Year of Astronomy has the potential to become one of the very first steps in showing that astronomy is the locomotive of science, thanks to education and outreach activities. Worldwide celebrations of astronomy are, in Romania, part of a "Triennium" of three years of activities dedicated to the International Heliophysical Year (IHY2007), the National Year of Astronomy celebrating 100 years of the Bucharest Observatory (NYA2008) and the International Year of Astronomy (IYA2009).

All the activities represent direct examples of outreach connected with education, ideas linked with practices for making the public aware of the beauty and value of astronomy. The first step of this journey was the opening ceremony in Sibiu, well known as a cultural capital of Europe (2007). At this time an agreement was established between the National Commission of UNESCO for Romania and the authorities of Sibiu. Assisted by teachers, children launched small rockets into the sky, drawing attention to the 500th anniversary of the birth of Conrad Haas, the first person to describe a multistage rocket in writing.

Professional astronomers have participated as consultants for the science items in the news. Hand in hand with amateurs astronomers they have organised attractive national events for the public such as "She is an Astronomer". The astronomical amulet which young women wear during the year is also supporting the programme and one of the IAU Global Cornerstone projects. 100 Hours of Astronomy has been a national success. At the international level, the Romanian programme, run by the Society for Meteors and Astronomy was highly commended. Also worth mentioning is the Hands-on-Universe project chaired by the Romanian university of Craiova.

IYA2009 projects can give people involved in outreach activities a chance to show how science is best communicated through the mass media, and particularly through science journalism. We think that education (the relationships between astronomy and mathematics, physics, chemistry, biology, music, literature, history, for example), cultural heritage (IAU/UNESCO: "Astronomy & World

Heritage — Universal Treasures"), science journalism (improving the dialogue between scientists and media professionals), and outreach (leaflets, presentations, books, postal marks, workshops, such as "The Nights of the Cultural Institutes" held at the British Council, Romania, training sessions) are interconnected and might be used to spread useful astronomical information. All these have a high potential to attract young students, not only to the study of astronomy but also to science.

One of the platforms involved is the Science Newspaper, one of the IYA2009 media partners that act as a platform for teachers, scientists, and the public.

As a conclusion, we think that better science communication leads to a more efficient outreach that in turn contributes to a better education.

Part III: Organisational Node Reports

ESO's Activities for the International Year of Astronomy 2009

Lars Lindberg Christensen[1] **& Douglas Pierce-Price**[1]
[1]ESO – European Southern Observatory, Garching, Germany. e-mail: lars@eso.org

The European Southern Observatory (ESO) has played a major role in the International Year of Astronomy 2009 (IYA2009) project since planning began in 2003. ESO is hosting the IAU's IYA2009 Secretariat, which coordinates the Year globally. ESO is an Organisational Node and one of the Organisational Associates of IYA2009, and was also closely involved in the resolution submitted to the UN by Italy, which led to the UN's 62nd General Assembly proclaiming IYA2009.

ESO IYA2009 projects and activities

There has been a range of ESO-specific activities throughout 2009, from local to global in scope, and aimed at a range of levels of interest.

In Search of our Cosmic Origins is a planetarium show about ALMA, the Atacama Large Millime-ter/submillimeter Array. The show, produced by ESO and the Association of French Language Planetariums in collaboration with the Planetarium of Augsburg, is available in multiple languages and in formats for full-dome video and classical planetariums.

ESO, in collaboration with the IAU, has produced a book and movie celebrating the 400th anniversary of the telescope, *Eyes on the Skies* (Christensen & Schilling, 2009). The hardcover book is available in several languages, as is the movie on DVD and Blu-ray. Over 270 000 copies of the movie have been distributed, and it has also been broadcast numerous times on television.

The GigaGalaxy Zoom project reveals the full sky as it appears with the unaided eye from one of the darkest deserts on Earth, then zooms in on a rich region of the Milky Way using an amateur telescope, and finally uses the power of a professional telescope to reveal the details of an iconic nebula.

In terms of events, IYA2009 has been featured in ESO exhibitions throughout the year, including the global IYA2009 opening ceremony at UNESCO in Paris, the German IYA2009 opening event in Berlin, and many more. ESO will also participate in a number of activities near its headquarters in Garching, near Munich, Germany, including the Open House day on the Garching campus, planned for 24 October 2009.

Activities in Chile

ESO has organised a wide range of activities and projects for IYA2009 in Chile, our host nation. These include the distribution of a special multi-latitude Southern Hemisphere planisphere, running a series of Science Cafs, promoting a network of schools, revisiting classical science experiments such as the determination of the size of the Earth, opening a permanent astronomical exhibition at the Huanchaca Museum of the Desert in Antofagasta, and hosting Open House at Paranal, La Silla, and APEX/ALMA. ESO is also a partner in the GalileoMobile

IYA2009 Special Project, has supported the first Regional Congress "Astronomy at Schools" in the II Region of Chile, and has arranged several exhibitions of images in Chile including TWAN (The World at Night), ESO Heritage and GigaGalaxy Zoom.

IYA2009 Global Cornerstone projects at ESO

In addition to its ESO-specific activities, ESO is involved in many of the IYA2009 Global Cornerstone projects, and is playing a leading role in four of them:

Around the World in 80 Telescopes (Pierce-Price et al., 2009), part of the 100 Hours of Astronomy organised by ESO, was a record-breaking and unprecedented, live, 24-hour public webcast giving members of the public a snapshot of 80 research observatories around the world during a single 24-hour period. The webcast reached well over well over 107 000 viewers (estimated 200 000) over the 24 hours.

The Portal to the Universe (Christensen & Gay, 2008) is a global, one-stop portal for online astronomy content, for content providers, laypeople, press, educators, decision-makers and scientists. ESO, together with ESA/Hubble, is providing the portal infrastructure.

In the Cosmic Diary, professional scientists put a human face on astronomy through blogs, talking about not just the latest astronomical news, but what it is like to be an astronomer. The project is coordinated from the IYA2009 Secretariat at ESO, and 14 of our researchers are participating in the project's ESO blog.

ESO is providing the infrastructure and resources for the Galilean Nights, which will encourage people all around the world to participate in stargazing events from 22–24 October 2009.

References

(a)   Christensen, L.L. & Gay. P. 2008, *The Portal to the Universe: A comprehensive gateway to what's new in astronomy*, in ASP Conference Series 400, ISBN: 978-1-58381-672-1

(b) Christensen, L.L. & Schilling, G. 2009, *Eyes on the Skies 400 Years of Telescopic Discovery*, Wiley-VCH, ISBN 3-527-40865-7

(c) Pierce-Price et al. 2009, CAPjournal, 6, 18.

IPS Activities during IYA2009

Alexandre Cherman[1] **and Jon Elvert**[2] [1]Rio de Janeiro Planetarium, Brazil, [2]Louisiana Art & Science Museum, USA
e-mail: acherman@rio.rj.gov.br

The International Planetarium Society (IPS) is the global association of planetarium professionals. It has nearly 700 members from 35 countries around the world. They represent schools, colleges and universities, museums, and public facilities of all sizes, including both fixed and portable planetariums. The primary goal of the Society is to encourage the sharing of ideas among its members through conferences, publications and networking.

During the International Year of Astronomy, the IPS is strongly encouraging all planetariums worldwide, whether members or not, to embrace Cornerstone projects suggested by the IAU. The IPS especially supports the Galileoscope, 100 Hours of Astronomy and Globe at Night. In addition, the IPS is distributing, free to all its members, a planetarium show called *Two Small Pieces of Glass*, presenting the history of the telescope.

IYA2009 and planetarium shows

A planetarium show consists of a script, which may or may not have been previously prepared. Usually, for live shows, planetarians tend to improvise for the audience, based on their dome experience. But most of the shows are pre-recorded, and involve varying levels of pre-production.

For IYA2009, the IPS strongly encouraged all planetariums and other major content developers to produce shows related to the theme. Here is a list of shows produced in 2009 with a Galilean theme:

Probing the Heavens with Galileo, produced by Sternevent and distributed by Carl Zeiss of Germany;

Touching the Edge of the Universe, produced by the University of Applied Sciences in Kiel, Germany for the European Space Agency (ESA);

Galileo: The Power of the Telescope, produced by the Milwaukee Public Museum, Wisconsin, USA;

Galileo Skies produced by the Virginia Living Museum, Virgina, USA;

Le reve de Galileo (Galileo's Dream), produced and performed by the planetarium at the Cite des sciences et de Industrie in Paris, France. This production is a "live" performance in the planetarium by actors in costumes;

ALMA: In Search for our Cosmic Origins, produced in French and English by ESO, the French planetarium affiliate of the IPS and Mirage 3D;

Augen im All (Eyes in the Universe), produced by European Space Agency (ESA) and planetariums in Germany, Austria, and Switzerland;

Galileo Live! — A "live" performance in the planetarium of Galileo's life and discoveries produced by Canadian planetariums and nationally funded;

Evolution (Evolucion), produced by Spanish planetariums;

Pingo's Birthday (O Aniversrio do Pingo), produced by the Rio de Janeiro Planetarium and nationally funded;

Two Small Pieces of Glass, produced by IPS members: Imiloa Astronomy Center of Hawaii; Buhl Planetarium & Observatory at the Carnegie Science Center; Interstellar Studios. The show was distributed free to all IPS members worldwide. Preliminary data shows an estimated 10%+ increase in attendance.

The IPS website had primary links to:

IYA2009 Dark Skies Awareness, (particularly in connection with Earth Day on 28 March and International Astronomy Day on 2 May);

Galileoscopes, Estimated that some 25 000+ telescope units were distributed through IPS planetariums/museums resulting in increased public star parties;

100 Hours of Astronomy, planetariums worldwide participated in hosting events from 2–5 April;

Galileo Teacher Training Program, Let There Be Night.

All in all, the IPS was well represented by planetariums worldwide and seems to have satisfied the IAU ideal behind IYA2009 well.

The International Year of Astronomy 2009: Activities of the Astronomical Society of the Pacific

Bruce Partridge[1], [2], James Manning[1]
[1]Astronomical Society of the Pacific
[2]Haverford College, Pennsylvania, USA, e-mail: bpartrid@haverford.edu

The Astronomical Society of the Pacific (ASP) is a partner, along with the American Astronomical Society, of US activity to celebrate the IYA. I will describe the ASP's signature IYA2009 activities and resources, with special emphasis on those that will carry the education and outreach momentum engendered by IYA2009 beyond the calendar year 2009.

The Astronomical Society of the Pacific (ASP), founded in California 120 years ago was the first organisation of astronomers in the US. From the beginning it included both amateur and professional astronomers, and was devoted to what we would now call education and public outreach. As a consequence, the aims of IYA2009 match the ASP's mission well, that is "to increase the understanding and appreciation of astronomy by engaging scientists, educators, enthusiasts and the public to advance science and science literary". We list below some of the Society's IYA2009 activities. In some of these, the ASP is leading the effort; in others, we serve as a link to disseminate IYA2009 materials prepared by others. From the outset, the ASP has given priority to services and activities that will last beyond the calendar year 2009. Galileo's discoveries of late 1609 were important, but progress in astronomy did not stop abruptly on December 31 of that year. It would be a shame if the momentum of the IYA2009 developed this year were to dissipate.

One of our major contributions is the ASP's Discovery Guides. These are coordinated online1 resources for use by amateur astronomers in their outreach activities. They are specifically linked to NASA's IYA2009 calendar of monthly themes and objects. The Discovery Guides is one of the programmes we hope to continue after 2009.

Next in importance is our participation in the Galileo Teacher Training Program. The ASP ran a very successful training programme at its annual meeting about a month after the IAU General Assembly.

We have also developed a web-based3 directory linking users to educational resources, both those of the ASP and others, which will definitely be continued after 2009.

Next, we list a number of activities in which we supported the work of others. We helped mount a display in the San Francisco area of the From Earth to the Universe materials. We have also worked on dark skies awareness and have distributed more than 2000 Galileoscopes to our members and other astronomers in the US. The ASP was also substantially involved in a television programme called

400 Years of the Telescope; we also distributed toolkits to amateur astronomers on the design of telescopes. Finally, in the summer of 2008, the Society held a symposium on IYA2009 activities, which was published as a book3 and is a useful compendium of IYA2009 activities.

Links 1 For more information see http://www.astrosociety.org/iya/guides.html
2 http:// www.site.galileoteachers.org/
3 http://www.digitaluniverse.net/cosmicclearinghouse/
4 http://www.aspbooks.org/a/volumes/table_of_contents/?book_id=433

Part IV: IYA2009 Global Projects
Status Report for IYA2009 Special Projects
Mariana Barrosa[1,2]

[1]International Astronomical Union [2]ESO – European Southern Observatory
e-mail: mbarrosa@eso.org

IYA2009 acknowledges that the Cornerstone projects are not the only large initiatives that have contributed to the vision and goals of IYA2009. For this, a category within the global structure has been set aside for IYA2009 Special projects, intended to give large global projects that satisfy the vision of IYA2009 greater international recognition and an opportunity to link with celebrations worldwide. These projects were selected according to some very clear criteria: they should satisfy at least one of the IYA2009 goals and align with the IYA2009 vision; they should be global projects; they should be financially independent; they should demonstrate sufficient potential for successful implementation, and the bulk of the project should be implemented during 2009.
Several proposals were presented to the IYA2009 Secretariat and accepted as Special projects. Here we give an overview of the status of implementation and results of these.

(a) The World at Night (TWAN) - one people, one sky: TWAN is a bridge between art, humanity and science. Its aim is to create and exhibit a collection of stunning photographs and time-lapse videos of the world's most beautiful and historic sites against a night time backdrop of stars, planets and celestial events.

(b) 400 Years of the Telescope: The main feature of this project is the high definition documentary, *400 Years of the Telescope*, with footage from the globe's major observatories and a series of interviews by an international group of professional astronomers. But the project also includes a full-dome and traditional planetarium programme, *Two Small Pieces of Glass*, coordinated outreach programmes with educational organisations, amateur astronomy organisations, national broadcasters and planetariums, an interactive website1, the *400 Years of the Telescope* companion coffee table book and DVD, and a monthly newsletter.

(c) The mutual phenomena of the Galilean satellites of Jupiter: In this project the organisers encourage people to participate in observations of the mutual phenomena of the Galilean satellites of Jupiter, as part of an international network of observers. This network started gathering scientific data of high interest more than 20 years ago to find out more about the Galilean satellites of Jupiter: Io, Europa, Ganymede and Callisto. At the end of the campaign, all the observations will be collected and published in an international journal.

(d) *Around the World, Around the Sky*: This film takes up the title of a ten-film series on the history of astronomical observatories from antiquity until today broadcasted by Arte in 1990 - *Tours du Monde, Tours du Ciel*. This new project also deals with astronomy and observatories, but is set in the present, with a new story and a new treatment. It is a journey of exploration around the world, visiting working astronomical observatories to understand their observations and discoveries of the Universe.

(*e*) *Millions of Earths*: This is a 52-minute documentary, directed by Alain Tixier. The planets of our little Solar System are no longer the only ones on our map of the Universe. For the last thirteen years, hundreds of others - exoplanets - have been recorded. Where are they? Who records them day after day and how? *Millions of Earths* conducts a survey among different teams of researchers and finds out about the innovative techniques used to spot these new celestial bodies.

(*f*) Celebrating the 1919 Eclipse at Principe: In the spring of 1919, British astronomer Arthur Eddington travelled to the small island of Principe, to carry out what would become one of the landmark experiments of contemporary physics. His measurements confirmed Albert Einstein's theory of general relativity and kick-started our modern understanding of the Universe and how it evolved. Ninety years later, a group got together and went back to commemorate this event with the people of So Tom and Prncipe.

(*g*) The Sky - Yours to Discover: Most of the constellations we identify in the sky nowadays are based on those listed centuries ago. This project invites children and young people to gaze up at the sky and identify stars, connect stars with imaginary lines, create new constellations and original stories related to them.

(*h*) *BLAST!*: it's astrophysics Indiana Jones-style! The movie takes the viewer on a journey around the world and across the Universe to launch a revolutionary new telescope on a NASA high-altitude balloon. The movie follows this arduous scientific pursuit through several catastrophic failures in exotic locales before arriving at transcendent triumph on the desolate Antarctic ice.

(*i*) StarPeace: This project is organised by the non-profit non-governmental organisation Sky Peace and by the Astronomical Society of Iran, aimed at connecting people living on two sides of the land or sea borders of different countries by conducting joint star parties to show how the sky, being the same everywhere, could act as a bridge to join the people of the world regardless of the race, culture or nation they belong to. It is made possible by volunteer participation of active amateur astronomical groups around the globe.

(*j*) GalileoMobile: An itinerant science education project aimed at bringing IYA2009 closer to young people in South America, by fostering the will to learn through the exciting wonders of our Universe, while supplying local teachers with educational resources to sustain activities. GalileoMobile aspires to extend its impact through the production of a documentary. The trip will take place in October and November 2009, when the GalileoMobile will visit Peru, Bolivia and Chile.

(*k*) *Naming Pluto*: A 13-minute British documentary about Venetia Burney Phair, the English schoolgirl who named the planet Pluto in 1930, aged 11 years. Featuring Sir Patrick Moore, Dr. Allan Chapman, the Royal Astronomical Society and NASA, *Naming Pluto* looks back at the extraordinary human story of a young girl who made astronomical history and witnesses her long awaited reunion with planet she named, 77 years later, on the evening of her 89th birthday.

(*l*) *The Eye 3D*: In June 2009, a film crew of German 3D film experts travelled to Chile's Atacama Desert, one of the most arid places on earth, home of the VLT (Very Large Telescope) of the European Southern Observatory (ESO), to make a 3D-documentary about the most powerful optical telescope in the world.

Figure 4. IYA2009 Cornerstone Chairs present at the IAU GA. From left to right : Rick Fienberg (Galileoscope), Carolina Ödman (UNAWE), Kevin Govender (Developing Astronomy Globally), Mariana Barrosa (Cosmic Diary), Lars Lindberg Christensen (Portal to the Universe), Catherine Cesarsky (IAU IYA2009 EC WG Chair), Megan Watzke (From Earth to the Universe), Rosa Doran (Galileo Teachers Training Program), Mike Simmons (100 Hours of Astronomy), Constance Walker (Dark Skies Awareness), Pedro Russo (IYA2009 Coordinator) and Helen Walker (She is an Astronomer). Credit: IAU.

Links
http://www.400years.org

100 Hours of Astronomy
Mike Simmons[1], Douglas Pierce-Price[2] and the 100 Hours of Astronomy Task Group
[1]AWB - Astronomers Without Broders, [2]European Southern Observatory
e-mail: msimm@ucla.edu

The 100 Hours of Astronomy Cornerstone project of IYA2009 (100HA) was a worldwide event consisting of a wide range of sidewalk astronomy activities, live webcasts from research observatories and science centres, and other public outreach events during the period 2–5 April 2009.
100HA began with an opening event at the Franklin Institute in Philadelphia, USA, in combination with a major exhibition of early astronomy artefacts that included one of Galileo's two existing telescopes on loan from the Museum of the History of Science in Florence, Italy. Live internet streaming from the Franklin Institute carried the opening ceremonies, a virtual tour of the exhibition and activities with students who had built small telescopes. This led into a webcast from

science centres worldwide, organised by the Association of Science-Technology Centers, which included a variety of presentations.

An unprecedented 24 hours of live webcasts from research observatories around the world followed. Organised and hosted by ESO, Around the World in 80 Telescopes drew large online audiences throughout the 24-hour period, including not only individuals but also crowds at astronomy institutions ranging from planetariums and science centres to amateur astronomy clubs. Recordings of the segments from all observatories are available for viewing online[1].

Amateur astronomers worldwide took the spotlight on 4 April. The 24-hour Global Star Party began at thousands of locations as darkness swept across the planet. Photos from many of the star parties can be viewed in the 100HA Photo Gallery[2].

Thousands more events of various types took place worldwide as well, as organisers improvised to take advantage of the attention paid to these four special days. Other Cornerstone projects and Task Groups conducted special programmes as well. The IYA2009 Solar Physics Task Group led SunDay on 5 April, a day devoted to solar observing and educational outreach. Educational and commercial facilities that offer remotely operated telescopes participated under the banner of 100 Hours of Remote Observing by donating time. Major partners, both organisers and collaborators, are listed online[3]. The 100HA Task Group is particularly grateful to its sponsors, led by the major sponsor, Celestron, which is also a Global Sponsor of IYA2009.

Total attendance at all 100 Hours of Astronomy events cannot be accurately known since there were thousands of events conducted that were not registered on the 100HA website, but an estimated one million people - and possibly more - took part in each of the two largest components, Around the World in 80 Telescopes and the 24-Hour Global Star Party. 100 Hours of Astronomy was truly a worldwide event, as seen in media coverage from around the world that touted local events as part of the larger global effort. 100HA demonstrated the tremendous interest in astronomy among the public, and that this interest is universal. It also showed what can be accomplished with a grassroots effort of thousands of enthusiastic volunteers organised through social networks with minimal central resources. Planning is underway for a post-IYA2009 follow-up that will take advantage of the networks and excitement created by this historic event.

Links

[1] http://www.100hoursofastronomy.org/component/webcast/webcast/6
[2] http://www.100hoursofastronomy.org/photo-galleries
[3] http://www.100hoursofastronomy.org/partners

Cosmic Diary: Meet the Astronomers, See Where They Work, Know What They Know
Mariana Barrosa[1,2] & Lee Pullen[2]
[1] ESO - European Southern Observatory, e-mail: mbarrosa@eso.org
[2] IAU - International Astronomical Union

One of the Cornerstone projects of the International Year of Astronomy 2009, the Cosmic Dairy is not just about astronomy, but more what it is like to be an astronomer.

The official website1 went live on 1 January 2009. It presents a collection of blogs which aim to put a human face on astronomy. Professional scientists blog in text and images about their lives, families, friends, hobbies and interests, as well as their work, latest research findings and the challenges that face them.

Currently, the Cosmic Diary has 58 bloggers from 30 different nationalities, posting in different languages: English, Portuguese, German, Japanese, Turkish and Spanish.

The first bloggers were "recruited" in 2008 through the National Single Points of Contact. They were later joined by institutional blogs from NASA, ESA, ESO, JAXA and the Royal Observatory Greenwich (Cosmic Diary 1894).

The topics of the posts are free; bloggers write about what they want so through the Cosmic Diary we can read about science, politics, family life, travels and holidays.

In addition to the regular posts, some of the bloggers have been asked to write a feature article about a scientific topic. Throughout the year, a number of bloggers have explained their specialist fields of expertise to the public via these features. These translate cutting-edge scientific research to a wide audience, giving people unprecedented access to those at the forefront of scientific discoveries. The articles are posted online, with a new one going live every two weeks. In total we will have 26 articles by the end of 2009. The idea is to use these later to produce the Cosmic Diary Book, which will be the legacy of this project.

Some facts and figures, from 1 January 2009 to 30 July 2009:

(a) a budget of 11 000 Euros from IYA2009 funds, divided between editing and coordination (53%), web development (30%) and design (17%);

(b) 1157 posts;

(c) 116 241 Visits + RSS Feeds (number unknown) (source: Google Analytics);

(d) an average 543 visits/day + RSS Feeds (number unknown) (source: Google Analytics);

(e) visits came from 216 countries/territories (source: Google Analytics);

(f) traffic sources: 49.89% from referring sites; 28.66% from search engines; 21.44% direct traffic; 0.01% others;

(g) Cosmic Diary Facebook group with 602 members.

Link
http://www.cosmicdiary.org

Portal to the Universe
Lars Lindberg Christensen, Lars Holm Nielsen & Adam Hadhazy[1]
[1]ESO e-mail: lars@eso.org

The science of astronomy never ceases to amaze. Every day new results are published that trigger a cascade of press releases, blogs, podcasts and media coverage.

As is the nature of the internet, new providers pop up, while old ones disappear all the time. To accommodate to these oft-changing sources of astronomy information the European Southern Observatory (ESO) set up the Portal to the Universe (Christensen & Gay, 2008) – a dynamic project that serves as a global, one-stop portal for online astronomy content. The Portal is a service to laypeople, press, educators, decision-makers and even scientists. Content providers also benefit from the added visibility of their products and the increased traffic to their materials.

The portal indexes and aggregates content including news, blogs, video and audio podcasts, images, and videos. Web 2.0 collaborative tools, such as the ranking of different services according to popularity, help the user to sift constructively through the wealth of information available. A range of "widgets" (small applications) have also been developed to tap into all sorts of existing "live data", such as near-live pictures of the Sun, live positions of spacecraft and live observations from telescopes.

The Portal enables real-time access to content by aggregating (pulling) from providers of astronomy information, and distributing (pushing) this content to users. The Portal also indexes and archives, and thereby maintains a central repository of useful information. Modern internet standards such as RSS feeds and standardised metadata make it possible to tie all the suppliers of astronomy information together with a single automatically updating portal that only requires minimal human moderation. The result is a technologically advanced site that brings together strands of astronomy content from across the Internet.

In its first six months of operation, the PTTU had more than 250 000 visitors, featured more than 3500 press releases, almost 2100 podcast episodes and 21 000 blog posts.

Among the latest developments for the Portal is that we now have a real Editor-in-Chief, Adam Hadhazy, who will be taking the lead in exploiting the Portal's potential. A new section for astronomy twitter feeds has been made, as well as an intelligent algorithm to sort astronomy content from non-astronomy content. The latter turned out to be very important as much good astronomy content is produced by "mixed" channels that include social science and many other non-astronomy stories.

We encourage participation from anyone interested. We welcome content on the Portal from media outlets, bloggers, and astronomy enthusiasts. Everyone can submit astronomy and space-relevant RSS feeds for press releases, podcasts, blogs, and image archives.

ESO, together with ESA/Hubble, is providing the portal infrastructure and is partly sponsoring the project together with IAU/IYA2009.

References

Christensen, L. L. & Gay. P. 2008, *The Portal to the Universe: A comprehensive gateway to what's new in astronomy*, in ASP Conference Series 400, ISBN: 978-1-58381-672-1

IYA2009 Cornerstone She is an Astronomer

H. J. Walker[1]

[1]STFC Rutherford Appleton Laboratory, UK
e-mail: helen.walker@stfc.ac.uk

Gender equality is a fundamental principle of human rights. It is one of the UN Millennium goals to promote gender equality and empower women, and it is an IAU/UNESCO IYA2009 goal to improve gender-balanced representation of scientists at all levels. Although some countries do have a high percentage of women working as astronomers, women are significantly underrepresented at senior levels in most countries. The IAU itself finds only 13% of its members are women. The aim of the IYA2009 Cornerstone project, She is an Astronomer, is to provide information to female professional and amateur astronomers, students and those interested in the gender equality problem in science. One of the outcomes from the project was Resolution B4 on Supporting Women in Astronomy, adopted by the IAU General Assembly in Rio de Janeiro.

An objective of the project is to build a database where people can get information about the subject, ask questions and find answers. A website1 has been developed with the assistance of Quentin Stanley, Anita Heward and Emily Baldwin, using information supplied by the Task Group and the web group. The website is hosted by the Royal Astronomical Society. This is the main tool to address the aims and objectives of the project; it will grow during the year and remain as a legacy after IYA2009. The main areas where information is being gathered are:

(a) profiles of living and historic female astronomers, a largely invisible part of the astronomy community in the past;

(b) resources available to women astronomers;

(c) events taking place during the year;

(d) an area for national ambassadors of She is an Astronomer to populate with information;

(e) a forum where issues and topics can be discussed (which will start later in the year).

We are still gathering data, statistics and resources, and these are the areas where it is difficult to track down material in a form where it can be compared from country to country.

The website was launched in April and immediately astronomers began getting in touch with news and events. People from 26 countries have contacted the She is an Astronomer website. Around 90 women were invited to send in their profiles and by the end of July we had received 27 replies from 11 countries. The profiles are from women at all stages of their careers, and doing a wide variety of jobs. We have profiles for around 20 female astronomer pioneers. There are examples of different types of events which people are holding to celebrate She is an Astronomer, and the posters for IYA2009 at the IAU show that there have been a lot more events taking place, which is really great news. Some countries have very active ambassadors, such as Athena Coustenis in France and the group in Spain led by Francesca Figueras, who as part of their programme created a calendar with 12 historic female astronomers. This was such a huge success that the calendar text has been translated into English and the calendar for 2010 will be available as a download from the website (in both Spanish and English).

The profiles we received show that women have a real passion for astronomy; they love doing it. They will work hard and put in long hours. They think the situation for women is getting better, but active support is needed. One of the questions in the profile is "what recommendation would you make to young women starting their career in astronomy" and this has provided to be an amazing resource of hints and suggestions. There is advice for the individual such as do what you enjoy, keep a positive attitude, pick the one thing you excel at, get a mentor, get a life, start your family (if you want one) when you are young, energetic and flexible. For the female astronomer as part of a family, the advice is to find (and/or train) a supportive partner, to explain to your family why you love your work so they will understand and support you, and to discuss things with them and keep on discussing. We are urged to learn how the system works, join committees, get information and support, and give talks.

Link
http://www.sheisanastronomer.org

Dark Skies Awareness: An IYA2009 Cornerstone Project
Constance E. Walker[1]

[1]National Optical Astronomy Observatory, USA, e-mail: cwalker@noao.edu

The preservation of dark skies is a growing global concern, yet it is one of the easiest environmental problems people can address on local levels. For this reason, Dark Skies Awareness was created as a Cornerstone project of the International Year of Astronomy 2009. Its goal has been to raise public awareness of the impact of artificial lighting on local environments by getting people worldwide involved in a variety of programmes. The programmes provide resources on light pollution for new technologies such as a presence in Second Life, social networking and podcasts, for local thematic events at national parks and observatory open houses, for international thematic events like International Dark Skies Week and Earth Hour, for a programme in the arts like an international photo contest, for global citizen–science programmes that measure night-sky brightness worldwide, and for educational materials like a kit with a light-shielding demonstration. Dark Skies Awareness has also supported the concept of Dark Skies Communities through the IAU 2009 B5 resolution, the Starlight Reserve Concept and the International Dark Sky Places programme.

The Dark Skies Awareness programmes have been successfully implemented around the world during IYA2009. The 33 countries noted for their participation in these and other education outreach programmes on dark skies awareness are Argentina, Australia, Austria, Brazil, Canada, Chile, China, Columbia, the Czech Republic, Finland, France, Germany, Greece, Hungary, Iran, Ireland, Italy, Japan, the Former Yugoslav Republic of Macedonia, Mexico, the Netherlands, New Zealand, Poland, Portugal, Romania, the Russian Federation, Saudi Arabia,

Slovakia, Slovenia, South Africa, Turkey, the United Kingdom and the United States. In particular, these comprised half of the countries that contributed over 15 700 measurements during the two weeks of the GLOBE at Night citizen science campaign - twice the number of measurements on average from previous years.

In terms of programme sustainability beyond IYA2009, many of the resources can be downloaded (the posters, brochures, guides, etc.) and so are sustainable at least as status quo. Funding is hopeful for future podcasts, two of the star hunt programmes and the educational kits. The Second Life presence will continue for as long as the IYA2009 island is open and the social networking sites will continue at least as they are now. The Dark Skies Discovery programme in the United Kingdom has funding to continue, as does Earth Hour and the US National Park Service programme. The Earth and Sky Photo Contest is most likely a one-time event. For as long as the Dark Skies Awareness website is running, Dark Skies Communities, as well as the other programmes, will be promoted.

For information on how to become involved in any of these programmes, people are invited to visit www.darkskiesawareness.org.

IYA2009 Dark Skies Awareness was partially funded by the IYA2009 Secretariat Office, as well as from the US National Science Foundation (NSF) Astronomy Division, via the US National Optical Astronomy Observatory (NOAO). NOAO is operated by the Association of Universities for Research in Astronomy, Inc. under cooperative agreement with the NSF.

Developing Astronomy Globally

Kevin Govender[1]
[1]SAAO, South Africa, e-mail kg@saao.ac.za

The Developing Astronomy Globally (DAG) Cornerstone project was initiated in order to ensure that IYA2009 benefits those countries that do not have strong astronomical communities by stimulating astronomy in underdeveloped regions. From the outset DAG was seen as a project to kick off activities that would last well beyond 2009. As such it was aligned with the IAU's strategic plan for development entitled *Astronomy for the Developing World*.

Coordinated from South Africa, DAG started off by conducting a global astronomy survey targeting mainly underdeveloped regions. The purpose of the survey was to establish a "bottoms-up" evaluation of the state of astronomy in any given country. This data could then be used to plan development activities both for the country and for the region. Focus areas of the survey were threefold: professional, public and school level education. Participants in the survey were asked to rank their countries within each focus area and provide associated explanations and descriptions of the local situations. This survey continues to grow with a dynamic user-controlled web page for each country.

Early in the year DAG coordinated the attendance of African students at the IYA2009 opening event in Paris. This project involved selecting students from five African countries and arranging all the logistics necessary to get them to the opening ceremony. Funds were provided by UNESCO and feedback from the students was extremely positive. From there the administration of grants continued, with DAG managing a full grants process (including preparation of calls, selection of grants, payments and monitoring) for astronomy "seed funding" aimed at developing countries across the globe. An initial 12 grants were awarded to Rwanda, Kenya, the Former Yugoslav Republic of Macedonia, Nepal, Uganda, Tajikistan, Mongolia, Uruguay, Ethiopia, Nigeria, Gabon and Nicaragua (each grant was less than 1000 and carried project-specific implementation conditions). DAG has also developed, together with the IAU's Commission 46, a model for "astronomy stimulation visits" and proceeded to raise funds and rally support for a pilot programme in East Africa. This programme is envisaged to take place in Nairobi, Kenya, in November 2009. Other projects included the coordination of a committee for the selection and distribution of donated telescopes worldwide (Galileoscopes and Celestron telescopes); the consolidation and distribution of "offline" electronic resources that will be available on CDs and DVDs and be free to copy (targeted at countries or regions without abundant internet access); and the development of a new interactive website which includes dynamic country surveys and the establishment of an opportunities database for students and lecturers from developing regions.

In terms of the future of DAG, beyond IYA2009, much of what has been started will feed well into the IAU strategic plan for astronomy in the developing world. Almost all activities have served as pilot projects that have demonstrated what can be done to develop astronomy globally. The momentum of IYA2009 has been utilised to drive these pilot projects thus far, but after 2009 they will fall into the hands of the IAU's Global Development Office (which should be established soon after IYA2009) as well as the IAU's Commission 46 (Astronomy Education and Development).

The Galileo Teacher Training Program

Rosa Doran[1,2]
[1]NUCLIO Núcleo Interactivo de Astronomia [2]GHOU Global Hands-on Universe
e-mail rosa.doran@nuclio.pt

Training teachers in the use of modern tools for science education is certainly the trend of the future. Never before has the challenge of triggering student's interest in science topics and the promotion of a more dynamic and interactive classroom environment been so crucial. The vision of the Galileo Teacher Training Program (GTTP) is to be a provider of a strong network and a source for a dynamic training for educators worldwide. In 2009 more than 50 nations had GTTP representatives named in an effort that may be seen as the seeding effort of the project. Training sessions are being promoted in all corners of the planet and a strong network of educators and promoters raises a promise of a strong legacy of IYA2009.

GTTP is being built with the support of an already existing network of astronomy education promoters, the Global Hands-On Universe Association. The European branch of this group has piloted a well-structured effort in Europe, an effort awarded a silver medal by the European Commission in the scope of the

Figure 5. Galileo Teachers Training Program Session at ES/ESERO in the Netherlands, Credit: ESA/GTTP

Long Life Learning Awards in the category of Information and Communication Technologies. GTTP is being built using such experience as guidance, but facing a much bigger challenge since it also aims to embrace nations that are only very recently awakening to the new technologies.

All participants of GTTP training sessions are entitled to a participation certificate. Educators applying the learned tools in classroom are awarded a Galileo Teacher Certificate and promoters of training session a Galileo Ambassador Award. All sessions must be submitted and approved by the GTTP task group. The minimum requirement for a session is to address elementary themes and/or concepts of astronomy, use resources that address at least three types of activities, such as naked eye or small telescope observations, hands-on activities, and new technologies, including robotic telescopes and data mining.

In summary, GTTP is much more than training teachers; within the scope of the programme we also foresee the creation of a good repository of rated resources, the existence of an efficient network that will act as a 24-hour helpdesk. As an additional support to those engaging in this challenging path, we also intend to promote several thematic campaigns where GTTP teachers are invited to apply the learned resources to the study of specific themes and promotion of real scientific research in classroom.

GTTP will certainly be a strong and sustainable legacy of IYA2009, empowering educators to use astronomy as a trigger to a new paradigm in science education, based more on IBSE (Inquiry Based Science Education). GTTP can also be a powerful tool to promote global citizenship awareness and a tool to help bridge the gap between the developed and developing world. Experienced teachers may be a good link to newcomers and the network of ambassadors will certainly be built beyond all borders.

An Update on Universe Awareness
Carolina J. Ödman[1]

[1]Universe Awareness, the Netherlands,
e-mail odman@strw.leidenuniv.nl

Universe Awareness (UNAWE) has almost four years of experience enthusing young children with the scale and beauty of the Universe. UNAWE is an outreach programme with a strong social vision aiming at broadening the minds of children, awakening their curiosity in science and stimulating global citizenship. UNAWE uses the inspirational aspects of astronomy to instil a culture of peace and tolerance. Universe Awareness started in 2004 as an idea. In 2006 UNAWE obtained a grant from the Netherlands Ministry of Education, Culture and Science to develop the international programme for three years. This grant allowed for the establishment of a small international office based in Leiden, the Netherlands, and some travel. In 2007, UNAWE was chosen as one of the global Cornerstone projects of the International Year of Astronomy 2009, and this has played an important role in bringing a number of new participants into the programme.

After almost four years, UNAWE has been implemented in over 30 countries around the globe. Each country's programme is suited to its local conditions. UNAWE is implemented in schools, science museums, by development NGOs, astronomers, educators, etc. Most participants are volunteers, but in some countries national funding has been secured to hire people and produce resources. UNAWE has grown significantly thanks to the International Year of Astronomy. Some countries are running a UNAWE programme without the knowledge of the UNAWE international office. When we come across such groups, we extend a warm welcome and invitation to join the international network.

UNAWE has brought together volunteers and participants from a number of professional backgrounds. From creative artists to professional astronomers, child development specialists and journalists as well as students and teachers, the diversity of the UNAWE community is one of its greatest assets. In terms of products, UNAWE can boast a collection of educational resources that are found on its international resource website1 and distributed across the websites of the various national UNAWE programmes. UNAWE has received substantial coverage in the national and international press, for example, in *Physics Today*. In addition to this, a number of books, DVDs and other products have come out of the community's work. National UNAWE programmes, sometimes with the help of the international office, have received a number of small grants that have enabled specific projects to be carried out. As an example UNAWE was awarded an IYA2009 grant from the European Astronomical Society that enabled the translation into English of an astronomical story-book from children, originally produced in Spanish by Spanish-speaking partners from over ten countries.

Owing to its innovative approach and visible success, UNAWE has been called in as expert in various projects, for example, the UNESCO World Report on Cultural Diversity. This enables UNAWE to bring the voices of the people on the ground to high-level circles. To conclude, the success of UNAWE probably lies in its novel approach using astronomy for young children's development, its social goals, the geographic and professional diversity and the openness of its community.

Link http://www.unawe.org

From Earth to the Universe in IYA2009
Megan Watzke[1] & Kimberly Arcand[1]
[1] Chandra X-ray Center, Cambridge, MA, USA, e-mail mwatzke@cfa.harvard.edu

From Earth to the Universe (FETTU) is an image collection of astronomical objects ranging from our home planet through the galaxy to the furthest corners of the Universe. These images contain data from both telescopes on the ground and in space that observe in many different types of light, from radio to optical to X-rays and beyond.

FETTU, one of the IAU IYA2009 global Cornerstone projects, has shown the incredible appeal of astronomical images and the science they contain to the general public. This series of images, with captions now translated into dozens of languages, has appeared in nearly 70 countries and on every continent except Antarctica.

The goal of FETTU has been to engage the largest possible populations, in particular those who might not generally visit science centres or planetariums. The images were selected using a number of criteria, including the diversity of the objects, wavelengths and others. But their inherent aesthetic appeal was a major factor, because FETTU displays these images as much as art as science. Furthermore, the FETTU exhibits were largely placed in non-traditional science venues such as public parks, transportation stations, art festivals, shopping malls, libraries, etc.

By combining the beauty of these images with their placement in accessible venues, FETTU has been embraced tremendously around the world. With some 500 separate exhibitions over the course of IYA2009, it is impossible to list any decent representation of the highlights here. It can be said, however, that the presence of astronomy has been felt in countries of all sizes, regions, and politics through FETTU. This includes recent displays in halls of the Iranian Parliament, in the heart of Moscow, in a prison in Portugal, and across Bolivia.

We believe the success of FETTU demonstrates that the wonders of astronomy are universal in their appeal and remind us that no matter where we are, we all live under the same sky. We hope that the reach of FETTU can continue into 2010 and beyond and provide one model for science outreach in the future.

More information on FETTU, a project produced and directed by the Chandra X-ray Center at the Smithsonian Astrophysical Observatory: www.fromearthtotheuniverse.org

Astronomy and World Heritage: The IYA2009 Cornerstone Project

Anna P. Sidorenko[1,2] & Clive L.N. Ruggles[3,4]

[1]UNESCO World Heritage Centre, Paris, France, e-mail: A.Sidorenko@unesco.org
[2]Coordinator of the Thematic Initiative Astronomy and World Heritage
[3]Emeritus Professor of Archaeoastronomy, University of Leicester, UK
[4]Chair, IAU Working Group on Astronomy and World Heritage

Introduction

The cosmos has captivated the imagination of civilisations throughout the ages. The desire to understand or interpret what people see in the sky is often reflected in architecture, petroglyphs, urban planning and other cultural representations. These material testimonies of astronomical observations, found in all geographical regions, span all periods from prehistory to today.

UNESCO and the IAU are working together to promote collaboration in research and education as part of UNESCO's Thematic Initiative, Astronomy and World Heritage. This project creates an opportunity to evaluate and recognise the importance of astronomical heritage in terms of the enrichment of the history of humanity, the promotion of cultural diversity, and the development of international exchange.

The fact that Astronomy and World Heritage has been recognised as one of the IYA2009 Cornerstone projects reflects the fact that support from the international community is vital if we are to save cultural properties connected with astronomy from progressive deterioration and to recognise astronomical heritage by the inclusion of the most representative of these properties on the World Heritage List.

Conversely, our status as an IYA2009 Cornerstone project has given us an opportunity to develop several key projects during 2009, namely:

(a) the ICOMOS–IAU Thematic Study on the Heritage Sites of Astronomy, which will become a document of the Convention;

(b) the expansion of the Astronomy and World Heritage Timeframe currently held on the UNESCO website into a more broadly accessible database and a public forum; and

(c) the publication of a special issue of UNESCO's quarterly *World Heritage Magazine* with Astronomy as its featured theme. This is published in three languages and is distributed to States Parties and other interested organisations and individuals throughout the world. A separate article in this volume ("Astronomy and World Heritage" by Clive Ruggles) describes these activities in more detail.

The Cornerstone project has also brought about some key collaborations, most notably between the Thematic Initiative and

(a) the Ancient Skies project1, a global scientific project that is striving to collect, verify and publish available information about various human cultures, their astronomical knowledge and its representation in the sky within a single web accessible knowledge base; and

(b) the Starlight Initiative, which is working to protect the natural heritage of the dark night sky.

In the remainder of this article we focus upon the background to the Initiative itself and the complementary activities that have supported the key projects mentioned above.

The World Heritage Convention, science heritage, and the development of the Initiative

The 1972 Convention concerning the protection of cultural and natural World Heritage has provided a unique opportunity to preserve exceptional properties world-wide and to raise awareness about scientific concepts linked to these properties.

UNESCO's mission regarding World Heritage is to assist the States Parties to this Convention to safeguard sites inscribed on the World Heritage List, to support activities led by States Parties in the preservation of World Heritage, and to encourage international cooperation in heritage conservation.

In 1994, the World Heritage Committee adopted a Global Strategy whose objective is to promote activities for a representative and balanced World Heritage List, in order to fully reflect the cultural and natural diversity of heritage of outstanding universal value.

Properties with a relationship to science are amongst the least represented on the UNESCO World Heritage List and the significance of these properties, located in all the regions of the world, is not sufficiently recognised. Recognising this, and recognising also the absence of an integrated thematic approach for sites that have a symbolic or direct connection to astronomy, the UNESCO World Heritage Centre, in close consultation with its States Parties, developed, in 2005, the Thematic Initiative "Astronomy and World Heritage". Its main aim was, and remains, to provide an opportunity to identify the properties connected with astronomy, to keep their memory alive, and to preserve them from progressive deterioration, through the inscription of the most representative properties on the World Heritage List.

A principal objective of the Initiative has always been to establish a link between Science and Culture through the recognition of the scientific values of cultural sites linked to astronomy. The identification, preservation and the promotion of these properties are fields of action in the implementation of this programme.

Implementing the Initiative

In May 2007, the Executive Committee of the IAU unanimously adopted a proposal to establish an official partnership with UNESCO within the framework of the Initiative in order to ensure its effective implementation. An Implementation Strategy for the Initiative was developed jointly by UNESCO and the IAU, and this was duly examined by the World Heritage Committee at its 32nd session in Quebec, Canada, in 2008.

In signing the Memorandum of Understanding in October 2008, whose purpose was to carry out this implementation strategy, UNESCO and the IAU underlined the fundamental role that culture plays in scientific progress and, conversely, that science plays in our cultural enrichment. This is a step towards the recognition of the importance of astronomical heritage world-wide, in terms of its enrichment of the history of humanity, the promotion of cultural diversity, and the enhancement of international exchange.

The collaboration aims to share best practice, to increase the role of the World Heritage Convention, and provide an opportunity to raise public awareness — especially among the young — about astronomical heritage. This will allow us to enhance the links between science, education, culture, and communication. UNESCO and the IAU are also working together to encourage States Parties to the World Heritage Convention to actively participate in the development and implementation of the Thematic Initiative.

The significance of this collaboration lies in three essential questions:

(*a*) How can we identify astronomical sites of Outstanding Universal Value?
(*b*) How can we protect and promote them?
(*c*) What benefits can States Parties and communities draw from adopting this path?

Milestones, supporting activities, and the IYA2009
The Global Thematic Study on astronomical heritage being developed jointly by the International Council on Monuments and Sites (ICOMOS) — the Advisory Body to the World Heritage Committee that is concerned with cultural nominations — and the IAU Working Group on Astronomy and World Heritage represents the first major milestone for the Initiative. It will establish a methodological approach for the consideration of sites associated with astronomy on the basis of the World Heritage criteria, and provide support for the preparation of possible nominations for the World Heritage List.

The Working Group discussion meetings held in 2009 in Spain, Brazil, Russia and Italy have succeeded in widening the input to the Thematic Study, resulting, for example, in the inclusion of a section on Space Heritage contributed by a group of authors from the Russian Federation as well as the addition of several important case studies.

The publication of an issue of the UNESCO World Heritage Magazine devoted to astronomical and science heritage is another milestone achieved during 2009 itself.

Beyond 2009, the focus of the Initiative will shift towards common efforts to promote the identification and preservation of astronomical sites and their associated technological heritage through public awareness-raising campaigns and international projects. This is a crucial and vital step in safeguarding these sites for future generations.

In his address on the occasion of the Opening Ceremony of the IYA2009 in January 2009, the Director-General of UNESCO underlined that *"the sky belongs to everyone, and everyone has the right to enjoy the wonders it holds, to seek to discover its greatest mysteries. Astronomy brings us together, across borders, religions and beliefs; it is an instrument of peace and understanding among peoples."*

Link http://www.ancient-skies.com

Part V: Selected Activities and Programmes
Astro–Gyaan: FAQs in Astronomy

A. Sule[1] [1]Homi Bhabha Centre for Science Education, India
e-mail aniket.sule@gmail.com

As most astronomy popularisers and educators know, the public are curious about astronomical phenomena. However, most astronomical information they receive is not based on serious texts, but on media reports and press releases. The majority of the population will not make a deliberate effort to acquire knowledge which may not be directly useful in their daily lives and will be happy to gather whatever information comes their way, without paying too much attention to its authenticity. Even in the case of media reports, people's attention spans are short and they are likely to avoid long documentaries. The project Astro–Gyaan (AG) has been tailored to suit this behaviour pattern.

In Sanskrit-based languages, "Gyaan" means knowledge. In almost every public talk by astronomers, people tend to ask typical questions like "what is a black hole?", "are there aliens?", "what makes the Sun shine?" etc. If these questions are answered in a short, crisp manner aided by graphics and animation, we would succeed in quenching the thirst for knowledge of a significant majority. This is the basic premise of the AG project. The list of questions is as follows:

What makes the Sun shine?	What is meant by the expansion of the Universe?
Does solar activity affect the Earth?	Why is the sky blue?
What are comets?	Are the shapes of the constellations real?
Why can we not see a very long total solar eclipse?	What is cosmic microwave background radiation?
Why was Pluto made a dwarf planet?	What causes tides?
Are all stars like our Sun?	What causes the phases of the Moon?
How do the stars form?	Are we alone in the Universe?
Do other stars have planets like the Earth around them?	Why do astronomers probe Universe in multiple wavelengths?
What is a galaxy?	What are the next steps for Indian space science?
What is a black hole?	Which are India's major observational facilities?

What are cosmic rays?	Is astrology a science?
What are dark matter and dark energy?	How to become an astronomer?

Key features of the project are as follows:

the total number of questions was 24;

each question was answered by a different prominent Indian astronomer;

the length of each film was restricted to between 1 and 1.5 minutes;

the selection of speakers automatically included diverse representation from different research institutes in India as well as diversity in gender, geography and ethnicity.

The answers were aided by animations and graphics available in the public domain with appropriate credits. The entire cost of the project is borne by the Homi Bhabha Centre for Science Education (HBCSE), Mumbai, India. The clips are offered to the state television for free while the IPR is maintained with HBCSE. HBCSE plans to upload these clips to the web in the near future and also to distribute them on CDs to schools and colleges. The original clips are shot in the English language, but they will be translated to various regional languages for wider audiences.

We hope that a project like AG will help change people's outlook towards astronomy and create a positive image about scientists working in the field.

Astronomia.pl Portal Activity in 2008–2009

Krzysztof Czart[1] & Jan Pomierny[1]

[1] Astronomia.pl -Polish Astronomy Portal, Poland

e-mail k.czart@astronomia.pl

Astronomia.pl or Polish Astronomy Portal[1] is the main astronomical portal in Poland. During IYA2009 it was the task of the portal to manage the Polish national website for IYA2009. The portal also engaged in preparations to start the Galaxy Zoo 2 project in Poland, organised the International Conference of Young Astronomers ICYA2009, and gave patronage and media support to many astronomical events. The next big task is the Aurora Polaris project, aimed at elderly learners and blind people. Astronomia.pl also supported global activities. The main content of the website is news coverage of science, education and communication topics from astronomy, a database of articles, newsletters, discussion forum, galleries and more. The portal and its activity are described in detail in Czart & Pomierny (2006, 2008) and Czart (2005).

IYA2009 Polish website[2]: The main task of the portal during IYA2009 has been creating and maintaining the Polish national website for IYA2009. The website acts as a database of events in Poland, and it is also a source of basic information about astronomy in Poland and holds addresses of scientific and educational institutions related to astronomy.

International Conference of Young Astronomers (ICYA2009)[3]: About 150 participants from more than 30 countries attended the conference, organised by the biggest Polish universities and astronomical societies from 7–13 September in Krakow. The plan is to organise the conference each year in a different country.

Aurora Polaris[4]: Aurora Polaris is a cooperative project between institutions from Poland, the United Kingdom, Slovakia and Greece. It is funded by the Grundtvig programme of the European Union. During the project resources have been created for visually impaired people and for elderly learners. The task of the Astronomia.pl portal was to prepare podcasts about astronomical topics.

Galaxy Zoo and Galaxy Zoo 2[5]: Galaxy Zoo is a programme aimed at the general public. People were encouraged to classify galaxies. Preparation of a Polish version was a great success and attracted many people to try this task. It is the only non-English version of the project website.

Star count global activities: Astronomia.pl supported global activities during IYA2009. One example is a star counting project, for which we have prepared Polish materials and promoted them in our country. During 2008–2009 we supported GLOBE At Night 2008–2009 and The Great World Wide Star Count 2008 in this way.

References

Czart, K. & Pomierny, J. 2008, ca07.conf, 412
Czart, K. & Pomierny, J. 2006, IAUSS, 2, 62
Czart, K. 2005, EAS Publication Series, 16, 97

Links

1 The portal addresses are http://www.astronomia.pl and the separate websites http://www.kopernik.pl, http://www.planetarium.pl and http://www.astrowww.pl

2 http://www.astronomia2009.pl

3 http:// www.icya2009.org

4 http:// www.aurora-polaris.eu

5 http:// www.galaxyzoo.org

Special Session 2 IYA2009: Poster Overview
Pedro Russo[1,2]

[1]IAU/IYA2009 Secretariat, [2]ESO, email prusso@eso.org

The International Year of Astronomy 2009 (IYA2009) featured tens of thousands of events worldwide. These were organised and implemented by the many professionals, amateurs and volunteers who built IYA2009 into the most successful science education and public outreach project ever undertaken. Not even the two full weeks of the IAU General Assembly would have been enough to hear the reports from the different grassroots initiatives. Below you can find the full list of posters (and authors) presented during the Special Session, and this list gives an overview of the multiple initiatives around the world.

List of Posters

• New Astronomical Observatory in Morocco: Contribution to Research and Education (Zouhair Benkhaldoun)

- Space Sciences in the Developing Countries: Position and Potential (Babagana Abubakar)
- IYA2009 Activities in Armenia (Areg Martin Mickaelian)
- Joyful and Scientific aspects of Astrophotography (Mojtaba Taheri, Mohammad Javad Ajdadi, Sara Khalafinejad)
- IYA2009 in Cuba (Oscar Alvarez)
- Astronomical Society of Shiraz University and Similar Societies (Sara Khalafinejad, Saeed Hojjatpanah, Fatemeh Kamali)
- IYA2009 in Uruguay (Tabare Gallardo)
- OA–UNI: an Astronomical Observatory in the Peruvian Andes (Erick Meza, José Ricra, Antonio Pereyra)
- The UNI –Astronomy Group, 23 Years of Astronomical Outreach in Perú (Erick Meza, José Ricra, William Cori, Antonio Pereyra)
- The Secrets of the Birth and the Death of Galileo Galilei (Elena A. Gavryuseva)
- The Role of Langitselatan as the Media on Astronomy Education in Indonesia (Avivah Yamani, Ferry M Simatupang, Aldino Adry Baskoro, Emanuel Sungging Mumpuni)
- Ecuador Taking Part in the IYA2009 Celebration (Ericson Daniel Lopez)
- IGP's Activities for the IYA2009 (Jesús Antonio Dalmau Cam, José Kaname Ishitsuka Iba)
- IYA Activities at the University of Texas at Austin (Mary Kay Hemenway)
- 1000 Telescopes for 1000 schools in the UK (Helen Joan Walker)
- Australian Aboriginal Astronomy: Comets, Meteors, and Cosmic Impacts (Duane W. Hamacher)
- *Noche de las Estrellas*, a Massive Celebration in Mexico (S.Torres-Peimbert, J. Franco, I. Cruz-Gonzalez, B. Pichardo)
- Pakistan Celebrates IYA2009 (Malik Ghulam Murtaza)
- The Activities of Infinity — IYA2009 (Piero Galeotti)
- Sky Observation and Astronomy Teaching in the Tertiary Level in India (Rabindra Kumar Bhattacharyya)
- IYA2009 in Albania (Mimoza Hafizi)
- Education in Astronomy: Discussing Science Technology and Society (Artur Justiniano Junior)
- The Multiverse and the Mind: Exploring Cosmology's New Infinities (D. Kala Perkins)
- The You are Galileo! Telescopes (Hidehiko Agata)
- IYA2009/Peru: The UNMSM will Establish the Astronomy career as Homage to IAU (Maria Luisa Aguilar Hurtado)
- Developing Astronomy at the Brazilian Pantanal Region (Telma C. Couto da Silva, Marcos G. G. C. Lima, Celio R. Pinheiro, Denilton C. Gaio, Shozo Shiraiwa)
- Activities of the IYA2009 in Argentina (Olga I. Pintado)
- The Astronomy Club in an Isolated School in Argentina (Olga I. Pintado, Analia Juarez, Lidia Salvatierra, Carola Gomez)
- Japanese IYA2009 Activities (Kazuhiro Sekiguchi)

- Encouraging Spanish-speaking Children to Explore the Universe with UN-AWE en Español (R. M. Ros, C. Ödman)
- International Year of Astronomy 2009 Activities in India (Ranjeev Misra, Dipankar Bhattacharya)
- International Year of Astronomy 2009 Activities in Turkey (Ali Alpar)
- Non-scientists' Perception of Telescopes and the Light They Collect (Erika D. Grundstrom, Roger S. Taylor)
- The IYA09 Activities and Approaches of IIA (Prajval Shastri, Sabyasachi Chatterjee)
- The International Year of Astronomy in Minas Gerais. The Activities of the Center of Sciences Gaia: Itinerant Planetarium (Peter Leroy Faria)
- Students As Starry Messengers For IYA2009 - Puerto Rico (Carmen Pantoja, Mayra E. Lebrón)
- The International Year of Astronomy in Puerto Rico – Progress Report (Mayra E. Lebrón, Carmen Pantoja, Daniel Altschuler, José Alonso)

- IYA2009 - Puerto Rico Activities for the Visually Impaired Public (Carmen Pantoja, Gloria Maria Isidro, Mayra E. Lebrón).

Figure 6. The IAU IYA2009 EC WG present at the IAU GA in Rio de Janeiro, Brazil. From left to right: Ian Robson, Norio Kaifu, Kevin Govender, Mariana Barrosa (ex-officio), Claus Madsen, Catherine Cesarsky (Chair), Dennis Crabtree, Mary Kay Hemenway, Lars Lindberg Christensen (Secretary), Pedro Russo (ex-officio). Credit: IAU

Part VI: Legacy & Conclusions

Building on IYA2009: The IAU Strategic Plan Astronomy for the Developing World

George Miley[1,2]
[1]Leiden Observatory, Leiden, University, The Netherlands
[2]IAU Vice President, Education and Development
e-mail: miley@strw.leidenuniv.nl

Fostering astronomy in developing countries has long been regarded by the IAU as an important task. During the past two decades the IAU has conducted a range of educational activities under the auspices of Commission 46. These activities were directed mainly towards stimulating astronomy at university level. The IYA2009 and the increase in scope and size of astronomy outreach activities that it has inspired led the IAU Executive Committee to conduct a review of our educational and development. The EC charged me with leading an effort to produce a strategic plan for this area. After obtaining input from a large number of experts and stake holders, we produced a blueprint for IAU educational and development programmes during the period 2010–2020. This plan, *Astronomy for the Developing World: Building from IYA2009*, has been approved by the Executive Committee and two resolutions endorsing it have been submitted for endorsement at the closing business meeting of this General Assembly. I shall here briefly review some aspects of this plan[1].

Firstly, it shows that astronomy can play a unique role in furthering education and capacity-building throughout the world. Astronomy combines science and technology with inspiration and excitement. The South African government's 1996 policy on science and technology puts an argument for this view succinctly: *"It is important to maintain a basic science competence in flagship sciences such as physics and astronomy for cultural reasons. Not to offer them would be to take a negative view of our future, the view that we are a second-class nation, chained forever to the treadmill of feeding and clothing ourselves."*

Secondly, the plan summarises present educational activities and analyses the present state of astronomy development globally. The programme groups (PGs) of IAU Commission 46 have long conducted an impressive range of activities to further astronomy in developing countries. These include the organisation and funding of national and regional astronomy schools and visits by astronomers to developing nations. A new PG is presently being set up, directed at primary and secondary schools. Outside the formal IAU ambit, there are now also several complementary activities devoted specifically to astronomy education and outreach, including programmes for children.

Thirdly, the long-term vision of the plan is that eventually all countries should participate at some level in astronomical research and that all children throughout the world will be exposed to knowledge about astronomy and the Universe. The plan outlines specific goals for working towards this vision.

Fourthly, the "meat" of the plan is a strategy for achieving these goals, namely:

(*a*) An integrated strategic approach that includes primary, secondary and tertiary education, research and public science outreach. For each country the mix of relevant activities will be based on the future potential for research and education. Because of its relative underdevelopment, sub-Saharan Africa will receive special attention.

(*b*) Using IYA2009 as a springboard. Many of the IYA2009 global Cornerstones will be continued and supported, after the IYA2009 has finished. Examples of activities that should be continued are the Galileo teacher Training Program, UNAWE and the Galileoscope. Also, the huge network of IYA2009 contacts (e.g., SPoCs) that has been built up in IAU member states and other countries is a valuable resource that will be used for future capacity-building activities.

(*c*) Enlarging the number of active volunteers by recruiting more members and augmenting the pool of volunteers by doctoral students, postdoctoral trainees and talented non-member experts on pre-tertiary education and outreach, including amateur astronomers.

(*d*) Initiation of new activities. Major proposed new initiatives include (i) an endowed lectureship programme to provide semi-popular lectures on inspirational topics at the high-school level and (ii) long-term institute twinning between established astronomy institutes and university departments in less developed countries.

(*e*) Creation of a small Global Development Office (GDO). Mobilising a large number of volunteers and implementing new programmes needs professional co-ordination.

(*f*) Increasing regional involvement and the adoption of a bottom-up approach, with a substantial degree of decentralisation.

(*g*) Exploiting the internet and new tools, such as archives, robotic telescope networks and the Tunisian mobile science outreach "astro-bus".

Fifthly, the plan envisages a flexible implementation of the strategy, in step with available funding. The annual direct cost will be an order of magnitude larger than that of the present cost of the IAU development programme. Obtaining resources will be a huge challenge that will need action on various fronts. Several possibilities are outlined in the plan.

A vigorous fund-raising campaign will be needed, coordinated by the GDO. Although this is a difficult time for fund-raising, the funding climate is likely to improve during the decade. Before attempting to raise external funds, there are three prerequisites.

1. A credible plan. Without one that appeals to potential fund-givers, there is no chance of obtaining increased funding.

2. Management must be seen to be sufficiently professional to warrant support. Setting up the GDO is essential.

3. The IAU must show commitment and has approved an increase in funding for relevant activities from 10.

Should astronomical researchers become involved in such development activities? I suggest that the answer is a resounding yes, both for reasons of morality and expediency. Facilities needed to carry out frontier astronomical research become more expensive every year. The willingness of society to fund these magnificent machines sets an ultimate limit on what can be achieved. The decision of whether or not to construct a billion-dollar astronomical research facility is inevitably a political one. By devoting a tiny fraction of astronomical resources to global development and education, we enhance the image of astronomy as a whole and make politicians more receptive to research proposals. Mobilising astronomy in the service of global development is a cost-effective strategy.

The rationale for using astronomy to stimulate sustainable international development is clearly stated in the plan and illustrated on its front and back covers. Astronomy provides an inspirational and unique gateway to technology science and culture, three fundamental characteristics of developed nations. By mobilising large numbers of talented and creative scientists, engineers and teachers in the service of international development the plan will be a cost effective spin-off of one of the most profound adventures of our civilisation, the exploration of the Universe.

Link
[1] http://iau.org/static/education/strategicplan091001.pdf

IYA2009 Legacy and Conclusions Catherine Cesarsky

Haut-commissaire l'Energie Atomique, CEA Saclay, Btiment Sige, 91191 Gif sur Yvette cedex, France

e-mail catherine.cesarsky@cea.fr

Since its inception, the International Year of Astronomy 2009 was planned to be more than just a series of activities occurring over 12 months. It has been designed and implemented as a springboard for the popularisation of astronomy with a much longer timeframe in mind. In 2010, the IYA2009 Secretariat and external assessors will carry out a thorough evaluation of IYA2009. Lasting effects will be a key area of focus, and we already have some idea of what to anticipate. Perhaps the most impressive statistic from IYA2009 is the sheer size and scale of the astronomy network that has been created: the largest in history. 148 countries, from Afghanistan to Zimbabwe, have joined together to work toward the common goal of making astronomy accessible to all; the International Year of Astronomy 2009 truly has been international! Individuals and groups in all of these countries have been collaborating both internally and across borders on projects beneficial to us all. The relationships forged between scientists, communicators, teachers, and enthusiasts during IYA2009 should remain far into the future, and it is hoped they will only become stronger with time. Sharing resources and expertise is a win-win situation, as IYA2009 has shown.

Some of the Cornerstones will be incorporated into IAU plans. A prime example is Dark Skies Awareness, since participation in the protection of the sky is an essential duty of the IAU. Thanks to Developing Astronomy Globally and also to the general networking effort, developing nations have enjoyed increased links with astronomy groups and organisations at home and abroad. New openings and opportunities at both the professional and amateur level instigated during IYA2009 are set to continue, allowing expertise within these countries to be maximised, and helping global astronomy research and science communication. The IAU has been at the forefront of these efforts, and consolidating links between the IAU and developing nations is seen as a priority in the brand new IAU Strategic Plan for Astronomy Development. From the IYA2009 networks, we know that efficient organisation is the foundation of success. This is when having an organisation like the IAU to coordinate efforts really comes into its own. Education was a strong theme during the Year, emphasised in particular by the Galileo Teacher Training Cornerstone, and there is much potential in building on the existing efforts to extend the reach of science in general and astronomy in particular, on a world level. Thus, IYA2009 is a springboard for the enhancement of IAU educational activities as set in the Strategic Plan.

Combining increased opportunities for developing nations with improved education, the Universe Awareness project (UNAWE) tackled difficult issues head-on during IYA2009. Its aim of creating internationally an awareness of our place in the Universe and on Earth, targeted at children in underprivileged environments, has inspired many. Clearly this programme must continue in 2010 and beyond. Providing a wealth of educational material is a factor that deserves to be highlighted. During IYA2009 resources were disseminated and put to good use.

Celestron and Japanese "You are Galileo" telescopes, as well as large numbers of Galileoscopes have been donated, mainly to developing countries. The Galileoscopes, low-cost telescope kits, result from one of the IYA2009 Cornerstones, allowing educators to utilise excellent quality, but accessible tools to improve their astronomy communication. Galileoscopes will continue to be sold after 2009, but at a higher price.

Many other astronomy-related products have been developed during 2009. As an example, consider the Cosmic Diary, an IYA2009 Cornerstone. Throughout the year professional astronomers have been keeping blogs about their lives and work, allowing the public to see what life as a scientist is really like. As part of the initiative these bloggers have produced feature articles about their areas of expertise, explaining complex ideas in easy-to-understand language. These features will form the basis of a book to be published in 2010, providing both a legacy and an additional avenue of communication.

Several movies, often accompanied by books, have been produced and shown on TV. *400 years of the Telescope*, the most extensively shown, has also given rise to a widely distributed, splendid planetarium show. *Eyes on the Skies*, an IAU-produced, highly educational DVD movie celebrating the 400th anniversary of the telescope, subtitled in many languages, can look forward to an extended career in classrooms, astronomy clubs and homes, while *Tours du Monde, Tours du Ciel* flies its viewers all over the world and beyond to visit observatories at all wavelengths, including cosmic ray experiments; an unforgettable experience. *Blast!* has also been extremely popular with audiences of all kinds. Theatre productions and music have all been successful as well. All of these products provide a foundation to build upon for years to come.

Other Cornerstones, and most of the Special projects, will also survive 2009. Large steps forward have been made for the designation of astronomical sites by the UNESCO World Heritage programme. These give historical sites prominence and prestige, and help ensure that the public is aware of their importance. More work remains to be done in the coming years. Protecting and preserving our astronomical cultural heritage for future generations to appreciate must remain a priority.

Other ventures are set to continue, such as the Special project, The World At Night, which shows stunning images above landmarks worldwide. In the same vein, From Earth to the Universe, a set of astronomical images ready with captions for exhibits, which has been displayed all over the world, will be kept up-to-date and available. Another IYA2009 Cornerstone project, the Portal to the Universe, is entrenched in the area of astronomy news and is expected to expand further. Its long-term aim is to become the ultimate source of astronomical updates for the public, allowing anyone and everyone to have easy access to the latest developments in this science.

She Is An Astronomer, which promotes gender-equality, has gathered much interesting material on its website and will soon hold an international workshop in England. In conjunction with the Cosmic Diary, this Cornerstone can help to present a modern image of astronomers to the public. The stereotype of oddball figures with long beards in towering observatories is not only inaccurate, but also damaging. Helping to reshape preconceptions and expectations is notoriously difficult, but also necessary. The extent to which IYA2009 has had a positive impact in this area will only be known with time.

Last but not least, large-scale public observing programmes, following the model of the famous worldwide events, 100 Hours of Astronomy and the Galilean Nights, will continue to be organised.

In conclusion, with IYA2009 we hope to have fostered: an increased awareness by society that we are living in an extraordinary era of discoveries about the Universe; a modern image of astronomers in the eyes of the public; a clear demonstration that a career in astronomy is also for women and minorities; the creation of international networks of scientists, communicators, teachers and amateurs, which should remain in existence far beyond 2009; a wealth of educational material on astronomy, books, films, movies for television, DVDs, theatre, planetarium shows, and music related to astronomy; the inception of a new set of goals for the IAU embedded in the Strategic Plan, of a partnership between IYA2009 and UNESCO; and the birth of many vocations at the professional and amateur level. It is evident that there are many strands to astronomy popularisation. IYA2009 has put the wheels in motion, but in many ways the work is only just beginning. For the Year to have a lasting legacy, the momentum gained must be effectively utilised to keep pushing forward, breaking barriers and keeping this most dynamic of sciences at the forefront of people's imaginations.

Acknowledgements: IYA2009 would not have been such a worldwide success without the vision, drive and sense of organisation of Lars Lindberg Christensen, and the skill, enthusiasm and hard work of Pedro Russo, splendidly assisted by Mariana Barrosa and the staff writer Lee Pullen. In the first half of 2010, the IYA2009 Secretariat will prepare a comprehensive final online-only report, an Executive Summary as a printed brochure and a coffee-table book on IYA2009. This will of course require the help of all involved to send them the reports on their activity in a timely manner. This is the last favour that I am asking from the many individuals who have made this very special year surpass our expectations, and in particular from all the Chairs of Cornerstones, Special Groups and Task Groups, and our marvellous army of SPoCs (Single Points of Contact).

I also wish to thank all of our sponsors, starting with Thales Alenia Space, Celestron and History Channel, following with our long list of generous Organisational Associates. Thanks also go to our enthusiastic Media Partners.

SpS3 – Astronomy in Antarctica

Special Session 3, IAU General Assembly XXVII
Rio de Janeiro, Brazil, August 6-7 2009

Preface

This was a 2-day meeting held during the XXVIIth International Astronomical Union General Assembly in Rio de Janeiro in 2009.

Antarctica offers a range of remarkable conditions that provide asuperlative environment for observational astronomy from visible to millimetre wavebands, as well as for high energy astrophysicsexperiments. This meeting discussed the current state of Antarctic astronomy, with winter-time facilities now operating at both the South Pole and Dome C on the high plateau, and activity underway at Domes A and F. The status offacilities at these sites was reviewed at the meeting and new science results presented, including from the International Polar Year of 2007/08.

Scientific Rationale

Antarctica provides unique conditions for a wide range of astronomicalobservations. The cold, dry air above the high Antarctic plateauprovides the best ground-based conditions for many observations at thermal infrared and sub-millimetre wavelengths. The stable air, low levels of high-altitude turbulence and narrow boundary layer over thesummits of the plateau provide for superb seeing in the optical. Thecircumpolar wind provides suitable conditions for long durationballoon flights. The vast quantities of pure ice, on a stableplatform, provide unsurpassed conditions for neutrino telescopes. The high geomagnetic latitude provides unique conditions for cosmic raydetection.Over the past decade Antarctica has seen a wide range of experiments designed to exploit these conditions for a variety of astronomical observations. Extensive site testing on the plateau has establishedthe great potential for observational astronomy from optical tomillimetre wavelengths. At the South Pole there have been infrared(SPIREX), sub-millimetre (e.g. AST/RO, Viper) and several CMBR (e.g. Python, DASI, ACBAR) telescopes operating. Particle physicsexperiments, particularly cosmic ray air shower arrays (e.g. SPASE)and neutrino telescopes (AMANDA), have been developed. Coastalstations, such as McMurdo, have hosted long-duration balloon flights,such as the BOOMERANG CMBR experiment. The high plateau site of DomeC has now completed its third season of winter-time operations, withthe first site-testing experiments deployed there demonstrating superboptical seeing conditions. The first expedition ever to Dome A, the summitof the Antarctic plateau, was undertaken in 2005. Initialinvestigations have also been undertaken regarding the suitability ofDome F for future astronomical observations.At the 2,900m US Amundsen-Scott South Pole station there are now twomajor facilities for astrophysics. These are the cubic kilometrecollecting volume IceCube neutrino telescope, and the 10m South PoleTelescope, to probe dark energy through for SZ measurements of distantgalaxy clusters. For the French-Italian 3,200m Concordia Station atDome C, the 80cm IRAIT mid-IR telescope is under construction, and adesign study completed for the 2.4m optical/IR PILOT telescope. Chinaconducted the first traverse to 4,200m Dome A, the highest pointon the Antarctic plateau, and returned there in 2008 in the PANDAprogram of the International Polar Year. At the Japanese base of Fujiat the 3,800m Dome F the first investigations on the suitability ofthis site for astronomical observations have begun.International involvement in these experiments is high. Internationalcollaboration in Antarctica has been productive and effective, withboth SCAR (the Scientific Committee for Antarctic Research) and theIAU sponsoring sub-committees to foster developments in astrophysicalresearch there. This meeting will further all these objectives.

Conference Programme

The meeting took place over 1.5 days, and followed the format of thesuccessful meetings held during the IAU General Assemblies in Sydney and Prague.The first day reviewed the experiments of the past few years.Speakers from the major facilities were invited to report on theirachievements. Highlights from the science conducted in the infrared,sub-millimetre, CMBR and particle astrophysics were presented. The openingsession also provided an introduction to the field ofAntarctic astronomy for the non-specialist, as well as the role of astronomy in SCAR.The second day featured discussion on plans for the development ofthe high plateau stations and the astronomical facilities they mightprovide in the near future.

Day 1, August 6, 2009:

- Session 1: An Overview of Astronomy in Antarctica
  - Astronomy in Antarctica: an overview *by Michael Burton*
  - The SCAR 'Astronomy & Astrophysics from Antarctica' Scientific Research Program *by John Storey*
- Session 2: The South Pole
  - BICEP: a cosmic microwave background telescope at the South Pole *by Yuki Takahashi*
  - The 10m South Pole Telescope *by John Carlstrom*
  - IceCube neutrino observatory at the South Pole *by Kirill Filimonov*
  - Observing the Universe from the South Pole *by Vladimir Papitashvili*
- Session 3: Dome C
  - ARENA, a roadmap for astronomy at Concordia Station (Dome C) *by Nicolas Epchtein and given by Hans Zinnecker*
  - Future plans for Dome C *by Vincent Coudé du Foresto*
  - The LUCAS program: detecting vegetation and traces of life in the Earthshine *by Danielle Briot*
- Session 4: Dome A
  - CSTAR and future plans for Dome A *Xiangqun Cui*
  - The PLATO observatory: robotic astronomy from the Antarctic plateau *by Michael Ashley*

Day 2, August 7, 2009:

- Session 5: Dome F
  - Solar cycles and supernovae embedded in a Dome F ice core *by Yuko Motizuki*
  - Plans for Dome F *by Takashi Ichikawa*
- Session 6: Other Sites
  - Site testing activities on the Greenland Ice Cap *by Michael Andersen*
  - The Stratospheric Terahertz Observatory (STO) *by Gordon Stacey*
- Session 7: Visions for Antarctic Astronomy
  - Science for the Antarctic plateau: what should we do? *by Hans Zinnecker*
- Session 8: Business Meeting
  - Matters for Discussion
  - Election of Chair for Working Group
  - The SCAR Scientific Research Program

Most of the talks given at this meeting can be downloaded from the IAU Working Group for Antarctic Astronomy website, at URL http://www.phys.unsw.edu.au/jacara/iau.

Michael Burton, Chair SOC
Sydney, Australia, October 31, 2009

Scientific Organising Committee
Michael Burton (Chair, Australia)
Carlos Abia (Spain)
John Carlstrom (USA)
Vincent Coudé du Foresto (France)
Xiangqun Cui (China)
Sebastián Gurovich (Argentina)
Takashi Ichikawa (Japan)
James Lloyd (USA)
Mark McCaughrean (UK)
Gino Tosti (Italy)
Hans Zinnecker (Germany)

Highlights of Astronomy, Volume 15
XXVIIth IAU General Assembly, August 2009
Ian F. Corbett, ed.

Astronomy in Antarctica in 2009

Michael G. Burton

School of Physics, University of New South Wales, Sydney, NSW 2052, Australia
email: m.burton@unsw.edu.au

Abstract. This article summarises the subject matter of Special Session 3 at IAU General Assembly XXVII in Rio de Janeiro, Brazil, which took place on August 6-7, 2009. In it, we overview the state of Astronomy in Antarctica as it is in 2009. Significant astronomical activity is now taking place at four stations on the Antarctic plateau (South Pole, Domes A, C & F), as well as at the coastal station of McMurdo.

Keywords. Antarctica, Telescopes, Site Testing, Instrumentation.

---

1. Overview

As is now well known (e.g., Storey (2005)), Antarctica offers remarkable conditions for a range of astronomical observations across both the photon and the particle spectrum. This is especially so on the summits of the Antarctic plateau on account of the extremely cold, dry and stable atmosphere. The conditions enable measurements from optical to sub-millimetre wavelengths that would have greater sensitivity and/or sharper imaging quality than measurements made with equivalent facilities at temperate-latitude sites, as well as opening new windows for regular ground-based viewing in the mid-IR and THz spectral regimes. Furthermore, the pure ice provides a novel detector for the capture and detection of particles, in particular neutrinos. At coastal locations the circumpolar vortex provides opportunities for long duration balloon flights that might last several weeks. The ice flow off the plateau also concentrates meteorites that have fallen over the continent into a few locations where they can be readily collected. In the following section we briefly review some of the astronomy ventures taking place over Antarctica. International collaboration is an integral part of these activities at each of the locations. Astronomy is now also established within SCAR – the equivalent body to the IAU for Antarctic science – as a formal research program. Lack of space precludes referencing in this article; however further information on the activities mentioned below is to be found in the accompanying articles in this Journal.

2. Antarctic Stations conducting Astronomy

2.1. *South Pole: the US Amundsen-Scott Station*

The South Pole Station dates back to the IGY of 1957-58, with astronomical activity pioneered there by the late Martin Pomerantz beginning in 1979. The establishment of the "Dark Sector" in 1994 set the scene for what is a major Observatory today. Four astronomical experiments are currently funded; the AMANDA and IceCube neutrino observatories (the latter will comprise of a cubic kilometre detector 2 km beneath the ice surface), the 10m South Pole Telescope (SPT) measuring the SZ-effect in galaxy clusters to probe the equation of state for dark energy, and BICEP measuring the polarization of the cosmic microwave background emission in order to search for gravity waves from the inflation of the Universe.

2.2. *Dome C: the French / Italian Concordia Station*

Concordia Station opened for winter operation in 2005. Noted for its ice core measurements of a column nearly 1 million years in depth, the principal astronomical activity so far has been site testing and instrument characterisation. It is clear that the surface boundary layer, for instance, is much narrower than at South Pole, and the average wind speed is lower. European interest in the station has been piqued through the European Union-funded ARENA network program, which has examined possible options for the site. Highest priority of these is a 2.5m class IR-optimised telescope (PILOT/PLT), a collaboration also including Australia. Interest in the sub-mm is also high, as are the prospects for long-time series measurements, CMBR experiments, solar astronomy and, in particular, an IR interferometer for studying exo-zodiacal emission.

2.3. *Dome A: the Chinese Kunlun Station*

The first humans only visited Dome A in 2005 with a Chinese traverse to the summit of the Antarctic plateau. No humans have yet wintered over at the site. During 2009 China began construction of a new station, Kunlun. The PLATO autonomous observatory operated through the winters of 2008 & 2009, gathering the first astronomical data from the site, as well as site testing data, including on the boundary layer and the atmospheric transparency in the sub-millimetre. These have shown that the boundary layer is exceedingly narrow (less than Dome C), and that the air is exceedingly dry. The Chinese Center for Antarctic Astronomy is now examining options for ambitious facilities, including a network of 0.5m telescopes, a 4m optical/IR telescope and a 15-30m sub-mm/THz telescope.

2.4. *Dome F: the Japanese Fuji Station*

The Japanese station at Dome Fuji has been used to collect ice cores (with one fascinating astronomical result reported by Motizuki at the meeting), and has operated through an Antarctic winter. No astronomical experiments have yet been conducted, but some site testing data has been obtained. The site would appear to offer comparable qualities to Domes A and C. The site testing program is being enhanced and there are plans for both optical/IR and sub-mm/THz facilities. Pilot studies for prototype facilities in these bands have commenced.

2.5. *McMurdo: the US Long Duration Balloon Facility (LDBF)*

Over the Antarctic summer season the circumpolar vortex provides an opportunity for long duration (several weeks) flights of balloons launched from coastal locations, providing the rational behind the LDBF at McMurdo station. Several astronomical experiments have been launched from here, perhaps the most notable being the BOOMERanG experiment which determined that the Universe was flat. In 2010-11 a 0.8m diameter THz spectral imaging telescope will be launched (STO – the Stratospheric Terahertz Observatory) with the aim of mapping large scale N^+ and C^+ emission over the southern Galactic plane.

References

Storey, J. W. V. 2005, Antarctic Science, 17, 555

Highlights of Astronomy, Volume 15
XXVIIth IAU General Assembly, August 2009
Ian F. Corbett, ed.

Astronomy and Astrophysics from Antarctica: a new SCAR Scientific Research Program

J. W. V. Storey

School of Physics, University of New South Wales, Sydney NSW 2052, Australia
email: j.storey@unsw.edu.au

Abstract. In July 2008 the IAU became a union member of the ICSU body SCAR—the Scientific Committee on Antarctic Research. At the same time, SCAR initiated a Planning Group to establish a Scientific Research Program in *Astronomy and Astrophysics from Antarctica*. Broadly stated, the objectives of Astronomy and Astrophysics from Antarctica are to coordinate astronomical activities in Antarctica in a way that ensures the best possible outcomes from international investment in Antarctic astronomy, and maximizes the opportunities for productive interaction with other disciplines.

Keywords. Antarctica, the Arctic, Site Testing

1. What is SCAR?

SCAR—the Scientific Committee on Antarctic Research—was established as an ICSU body in 1957 and held its first meeting in 1958. It currently has 31 Full Members (those countries with active scientific research programme in Antarctica), 4 Associate Members (those countries without an independent research programme as yet or which are planning a research programme in the future) and, with the recent addition of the IAU, 9 Union Members (those ICSU scientific unions that have an interest in Antarctic research). Like the IAU, SCAR and its divisions hold a number of meetings and conferences throughout the year. Every two years an Open Science Conference is held, similar in size and breadth to the IAU General Assemblies. More information on SCAR is available from the SCAR web site: http://www.scar.org/.

SCAR is organised into three Standing Scientific Groups. These are the:
- Standing Scientific Group on GeoSciences
- Standing Scientific Group on Life Sciences
- Standing Scientific Group on Physical Sciences

In addition, there are five Scientific Research Programmes (SRPs) whose focus is on international scientific coordination. Currently, these are:
- Antarctic Climate Evolution (ACE)
- Subglacial Antarctic Lake Environments (SALE)
- Evolution and Biodiversity in the Antarctic (EBA)
- Antarctica and the Global Climate System (AGCS)
- Interhemispheric Conjugacy Effects in Solar-Terrestrial and Aeronomy Research (ICESTAR)

At the end of 2009, ICESTAR will no longer be an SRP, making way for *Astronomy and Astrophysics from Antarctica* (AAA) as a new Scientific Research Program.

2. Astronomy and Astrophysics from Antarctica (AAA)

AAA has set for itself the following goals. To:

(*a*) Coordinate site-testing experiments to ensure that results obtained from different sites are directly comparable and well understood,

(*b*) Build a data base of site-testing data that is accessible to all researchers,

(*c*) Increase the level of coordination and cooperation between astronomers, atmospheric physicists, space physicists and meteorologists,

(*d*) Extend existing Antarctic site-testing and feasibility studies to potential Arctic sites; for example, in Greenland and Canada,

(*e*) Define and prioritise current scientific goals,

(*f*) Create a roadmap for development of major astronomical facilities in Antarctica,

(*g*) Stimulate international cooperation on major new astronomical facilities in Antarctica.

3. Data Archiving

Section III.1.c of the Antarctic Treaty (1959) states that "Scientific observations and results from Antarctica shall be exchanged and made freely available." To assist with meeting this requirement, SCAR has established the Standing Committee on Antarctic Data Management (SCADM). SCADM helps facilitate co-operation between scientists and nations with regard to scientific data, and advises on the development of the Antarctic Data Management System. AAA will work closely with SCADM to maximise the usefulness and accessibility of data collected by astronomers in Antarctica.

4. Structure of AAA

To achieve its goals, AAA will be structured as four themes:

- Site testing, validation and data archiving.
- Arctic site testing.
- Science goals.
- Major new facilities.

Each of these themes will be managed by a working group. Each working group has a chair and a vice-chair, and a variable number of members. All IAU members, and others, are warmly encouraged to join one or more of these working groups and to contribute to the success of the AAA program.

Acknowledgements

I thank the other members of the AAA SRP Planning Group for their assistance in putting the program together: Michael Andersen, Philip Anderson, Michael Burton, Xiangqun Cui, Nicolas Epchtein, Takashi Ichikawa, Albrecht Karle, James Lloyd, Silvia Masi & Lifan Wang

References

The Antarctic Treaty, 1959. Available at http://www.ats.aq/ (Accessed September 2009).

Highlights of Astronomy, Volume 15
XXVIIth IAU General Assembly, August 2009
Ian F. Corbett, ed.
© International Astronomical Union 2010
doi:10.1017/S174392131001077X

BICEP: a cosmic microwave background polarization telescope at the South Pole

Yuki D. Takahashi for the BICEP collaboration

Department of Physics, University of California, Berkeley, CA 94720-7300, USA
email: yuki@bolo.berkeley.edu

Abstract. BICEP was a telescope designed to probe the polarization of the cosmic microwave background (CMB) for the signature of gravitational waves produced during the epoch of inflation. The instrument was developed by a team of scientists from Caltech/JPL, UC Berkeley, and UC San Diego. It was installed at the South Pole in November 2005 and the CMB observations were conducted from February to November each year with one winter-over scientist responsible for operating and maintaining the instrument. Taking advantage of the excellent atmospheric conditions at the South Pole, we mapped 2% of the sky at 100 and 150 GHz. We completed 3 years of observations from 2006 to 2008, mapping the CMB polarization anisotropy at degree angular scales with unprecedented sensitivity. In 2010, a next generation instrument, BICEP2, will be installed on the existing telescope mount for an even deeper survey.

Keywords. Cosmic Microwave Background, Instrumentation: polarimeters, Gravitational Waves.

1. Introduction

BICEP was specifically designed to search for a signature of gravitational waves from inflation by studying the polarization of the CMB. The ultimate goal is to find direct evidence for inflation through gravitational waves that would have been generated during inflation. Those gravitational waves would have resulted in polarization of the CMB with a "B-mode" pattern. The resulting B-mode polarization anisotropy has an angular power spectrum that is expected to peak at around 2° angular scales, and whose magnitude allows us to constrain the inflationary energy scale.

2. Experiment Design

Because the potential signal is expected to be under a μK rms, the design priorities for BICEP were sensitivity and systematic error control. For maximum sensitivity, we chose the South Pole site because of its highly transparent and stable atmosphere. We observed in atmospheric transmission windows near the peak of the CMB blackbody spectrum, at two frequency bands (100 and 150 GHz) as a guard against potential foreground contamination.

The BICEP telescope has a modest aperture of 25 cm, leading to a beam size of $\sim$1°, which is adequate to resolve the B-mode anisotropy at its peak. This small aperture allows for aggressive shielding of sidelobes as well as simple and effective implementation of calibration measurements. To measure CMB polarization, we difference pairs of polarization sensitive bolometers (PSBs), similar to those recently launched on the Planck space-based telescope. The telescope is a simple on-axis refractor with feedhorns coupling the radiation onto 49 pairs of PSBs.

3. Deployment and Observations

We integrated and tested the instrument from 2003 to 2005 and deployed it to the South Pole in November 2005 for 3 years of operation until December 2008. During the summer seasons, we performed careful calibration measurements to characterize the instrument and to ensure that systematic errors are subdominant to the noise level. Takahashi *et al.* (2009) describes the instrumental properties characterized, including the bolometer temporal response, PSB pair beam mismatch, far sidelobes, spectral bandpass, polarization orientations and polarization efficiency. For CMB observations, we chose the cleanest available 2% of sky where the dust emission is minimized at 150 GHz. We also observed polarized emission from the Galaxy, including that from mid-latitude dust.

4. Results

BICEP achieved mapping depth on the Galaxy similar to that expected by the Planck survey. Our maps of polarized emission from the Galactic plane are being used to calibrate the polarized response of Planck. In our CMB field, BICEP probes much deeper. Our E-mode spectrum is sample variance limited up to the first peak. BICEP has lowered the upper limits on the B-mode by an order of magnitude with only the first 2 years of data and aggressive cuts. From the analysis of these data, the tensor-to-scalar ratio has been constrained to $r < 0.73$ with 95% confidence (Chiang *et al.* 2009). Systematic errors are controlled to well below this limit, and the analysis of the full 3-year data is ongoing.

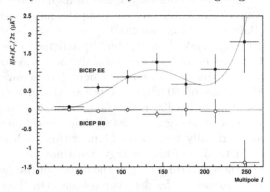

Figure 1. BICEP at the South Pole.   **Figure 2.** Two-year polarization spectra.

5. Future

Observations will continue in 2010 with BICEP2, which has a highly packed antenna-coupled transition-edge sensor bolometer array. With over 200 PSB pairs, mapping speed is expected to increase $\sim$5 fold. Extrapolating from the achieved sensitivity of BICEP, BICEP2 can expect to reach the sensitivity necessary to begin probing physically interesting ranges of amplitudes for gravitational wave signal from inflation.

We acknowledge support by NSF Grant OPP-0230438 and the US Antarctic Program.

References

Chiang, H. C., *et al.* 2009, submitted to *ApJ*, arXiv:0906.1181
Takahashi, Y. D., *et al.* 2009, submitted to *ApJ*, arXiv:0906.4069

Highlights of Astronomy, Volume 15
XXVIIth IAU General Assembly, August 2009
Ian F. Corbett, ed.

IceCube neutrino observatory at the South Pole: recent results

Kirill Filimonov for the IceCube Collaboration
http://icecube.wisc.edu

University of California, Berkeley, CA 94720-7300, USA
email: filimonov@berkeley.edu

Abstract. The IceCube neutrino observatory, the largest particle detector in the world (1 km^3), is currently being built at the South Pole. IceCube looks down through the Earth to filter out lower-energy particles and uses optical sensors embedded deep in the ultra-clean Antarctic ice to detect high energy neutrinos via Cherenkov radiation from charged particles produced in neutrino interactions. A summary of selected recent results is presented.

Keywords. Neutrinos

The upper limit on the flux of high energy neutrinos from optically thin extra-galactic sources can be estimated using the measured flux of high energy (10^{18} eV) cosmic rays [Bahcall & Waxman 2001]. To detect the diffuse flux of neutrinos of $dN/dE_\nu \lesssim 5 \cdot 10^{-8} E_\nu^{-2} \text{ GeV}^{-1}\text{cm}^{-2}\text{s}^{-1}\text{sr}^{-1}$ predicted in this way demands very large detector volumes. Both water and ice are naturally occurring optically transparent media for detection of neutrino interactions using Cherenkov radiation. When completed, IceCube will comprise 80 strings of 60 Digital Optical Modules (DOMs) each, deployed between depths of 1,450 m and 2,450 m in the clear ice below the Amundsen-Scott Station at the South Pole [Achterberg *et al.* 2006]. The typical inter-string separation is 125 m and the DOMs are vertically spaced by 17 m within each string. The detector geometry is optimized for detection of high-energy neutrinos ($E_\nu > \text{TeV}$). In addition, 6 closely spaced (72 m) strings will form a core in the center of the IceCube array, reducing the energy threshold for a subset of the detector volume. On the surface, a pair of ice filled tanks containing two DOMs are placed close to each string to form the IceTop air shower array. The surface array will be used to measure cosmic ray composition and for calibration and background studies. At the time of this conference, IceCube is taking data with 59 strings. Here we concentrate mainly on the data taken with 40 strings during 2008.

1. Moon Shadow

As the Earth travels through the interstellar medium, the Moon blocks some cosmic rays from reaching the Earth. This results in a relative deficit of muons measured by IceCube from the direction of the Moon. The resulting deficit (Moon shadow) can be used to calibrate detector angular resolution and pointing accuracy. Figure 1 shows the distribution of reconstructed muons per 1.25° square bin, relative to the position of the Moon. A 5.2σ deficit is observed for the Moon bin. From this, we can conclude that IceCube has no systematic pointing error larger than the search bin, 1.25° [Boersma *et al.* 2009].

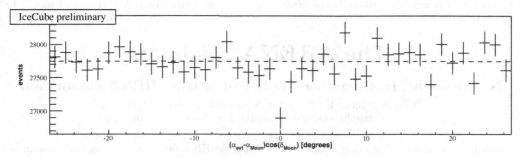

Figure 1. Number of events per 1.25° square bin, relative to the position of the Moon. The declination of the reconstructed track is within 0.625° bin from the declination of the Moon.

2. Point Source Search

One of the primary physics goals of IceCube is to identify point sources of astrophysical neutrinos. Six months of the IceCube 40-string dataset were unblinded for a point-source search. A maximum likelihood analysis, including an energy term to help separate a hard power-law spectrum from the softer backgrounds, was performed over the whole sky on a very fine grid, along with a list of 39 *a priori* source candidates. A sky map of directions of 17,777 events (6,796 up-going neutrino candidates and 10,981 high energy down-going muons – the backgrounds for the northern and southern sky searches, respectively) and their significance is presented in Figure 2. The most significant location in the all-sky

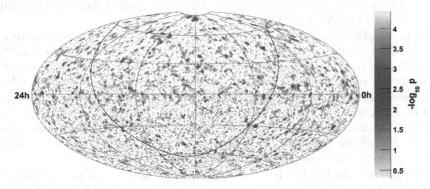

Figure 2. A sky map of the point source candidates (black dots) and their significance (color/shade intensity) in 175.5 days of IC40 data (IceCube preliminary).

search was at r.a. = 114.95°, dec = 15.35°, although an equal or greater significance shows up in 611 out of 1000 scrambled skymaps, meaning the post-trial p-value for the all-sky search is 61.1%. For the list of 39 pre-defined sources, the most significant source was PKS 1622-297 with a pre-trial p-value of 5.2%. Again from scrambling, the post-trial p-value for the source list was 61.8%. No evidence of a point source of high-energy neutrinos has been found with IceCube so far [Dumm *et al.* 2009]. Upcoming analyses of a larger dataset with the IC59 and larger configurations will yield much better sensitivity.

References

Bahcall, J., & Waxman, E., 2001, *Phys. Rev. D* **64**, 023002
Achterberg, A., *et al.* IceCube Collaboration, 2006, *Astropart. Phys* **26**, 155
Boersma, D., *et al.* IceCube Collaboration, 2009, Proc. of the 31st ICRC
Dumm, J., *et al.* IceCube Collaboration, 2009, Proc. of the 31st ICRC

Highlights of Astronomy, Volume 15
XXVIIth IAU General Assembly, August 2009
Ian F. Corbett, ed.

The ARENA roadmap

N. Epchtein[1], H. Zinnecker[2] on behalf of the ARENA consortium

[1]CNRS-Fizeau, University of Nice Sophia Antipolis, France
[2]Astrophysikalisches Institut Potsdam, Germany

Abstract. We present the main achievements of the ARENA network and a set of recommendations for the development of astronomy and astrophysics at CONCORDIA (Dome C)

Keywords. Site Testing, Instrumentation: high angular resolution, Surveys, Sun: infrared, Stars: general, Cosmology:cosmic microwave background, Infrared: stars, Galaxies: high-redshift

1. The ARENA network

ARENA (*Antarctic Research, a European network for Astrophysics*) is an initiative to draw out a roadmap for the development of Astronomy and Astrophysics at the French Italian Station Concordia at Dome C sponsored by the European Commission. It gathers some 100 scientists, engineers and polar technicians from 7 countries (6 in Europe and Australia). The activity of the network covers the 4-year period 2006-2009 encompassing the International Polar Year (IPY, 2008-9).

Dome C is one of the highest loci in Antarctica, and CONCORDIA (see Figure 1) one of the rare stations run all year round inside the continent. Created mainly to undertake the deepest drilling of the ice cap down to 3,000 m, Dome C happens to offer also compelling conditions for astronomical observations in a wide range of frequencies and techniques.

In 2005, several laboratories in Europe were struck by the exceptional seeing conditions reported by an Australian team. The atmosphere above a thin boundary layer of some 30 m is free of turbulence and, thus, the seeing exceptionally good (300 mas) above this layer. Dome C appears as an extremely appealing site for the rapid build-up of the first multispectral international observatory in Antarctica.

Realizing the potential of Dome C, a successful proposal was submitted to the EC in 2005, which had the following goals, i) aggregate and give access to the site assessment data collected so far, ii) identify the most compelling science programmes, iii) propose a few case studies of instrumental devices compliant with the polar conditions, iv) evaluate the logistics requests to set up one or several astronomical facilities, v) stimulate the interest of the public (especially during the IPY).

The ultimate goal of the network is to raise strong arguments in favour of the creation of an International Observatory in the forthcoming decade at Dome C (the so called "ARENA roadmap"). A set of some 25 specific tasks led by experts were carried out and, later on, 6 working groups were set up to prepare independently their roadmaps in their respective areas. The latter are briefly described in the following section.

2. Working Groups achievements

Wide field optical/infrared surveys. Under a grant of the Australian government UNSW/AAO carried out in 2008 a phase A study for the PILOT project, a 2.5 m class telescope that would serve as a pathfinder for future larger telescopes equipped with a suite of focal imaging instruments from the visible to the far infrared. European and

Figure 1. The Concordia station in 2007: main buildings, the site testing instruments, and the Concordiastro towers *(courtesy E. Aristidi)*

Australian astronomers came to a less ambitious project that would basically focus on the spectral range in which Antarctica brings an obvious advantage: the near thermal infrared and especially the 2.3-3.5 μm window hardly accessible from the ground. The PLT (*Polar Large telescope*) is a descoped version of PILOT. This project is considered as the most mature in its cost range and is fully supported for an immediate phase B study (2010-2013). In the meantime, IRAIT, an 80-cm IR dedicated telescope will provide rapidly (as of 2011) the first IR images and invaluable clues on the IR sky at Dome C.

Submillimetre-wave dish. Measurements of water vapour content definitely show that Dome C supersedes any other site in the THz regime and in particular in the 200 μm window. After ruling out a project to clone a 12 m ALMA antenna, configured for Antarctic conditions and installed shortly as a pathfinder, the working group eventually proposed a 25 m diameter dish (*the Antarctic Submillimetre Telescope*), envisioning a much more scientifically compelling project to exploit the 200-400 μm windows. It should rapidly enter a phase A study through a joint venture between Italy (INAF- IEE) and France (ThalesAleniaSpace, CEA, Saclay).

Optical/infrared interferometry. The ultimate goal of the interferometric community is to set up a kilometric array of optical/NIR telescopes. The Antarctic plateau with its immense flat areas and its unique atmospheric conditions would be the perfect location to install such an instrument (KEOPS project). The best pathway to this gigantic instrument is, however, still uncertain. Several pathdfinders have been proposed among which the Aladdin concept (a nulling interferometer to measure *"exozodis"*) is currently the most advanced, essentially by Observatoire de Paris, Nice and Liège Universities and AMOS. This instrument is described in more details by V. Coudé du Foresto elsewhere in this session.

Long time series. The basic advantage of polar sites is to provide long dark periods perfectly suited to the study of periodic variations of astronomical objects (sun, stars, planet transits). Several projects are underway, among them the most advanced is A–STEP, a 40 cm telescope aimed to measure planetary transits. This instrument led by Nice Observatory/UNSA is under construction at Dome C and should provide first images in 2010. Other more complex and robotic projects are proposed and supported by ARENA such as ICE-T (a twin telescope with an ultra precise photometer) at AIP/Germany, and SIAMOIS (an interferometer to measure oscillations of stars) at Paris Observatory.

Cosmic Microwave Background. Antarctic sites have a very stable atmosphere and allow long integration times of the same area of the sky. They are very appropriate to measure tiny flux variations in the millimetre wave range and Dome C is likely to be even more stable than the South Pole. A French-Italian consortium (APC University of Paris 7 and University of Roma, la Sapienza) is undertaking a project to measure the B polarization of the CMB using a bolometer and an interferometer (BRAIN/QUBIC). Dome C has an additional advantage being 15° away from the pole in latitude. For polarization experiments it is actually crucial to measure the same polarization direction with different inclinations of the axis of the polarimeter.

High angular resolution imaging solar physics. Dome C is an outstanding site for high angular resolution especially during Summer. Simultaneously, the sky is coronal and the seeing is excellent. Antarctica is suitable for very high angular resolution imaging of the solar photosphere and corona. An instrument consisting of a solar interferometric imager (AFSIIC) is proposed, consisting of 3 telescopes of 70 cm on top of a tower of about 30 m.

More details on all this are given in the ARENA website (http://arena.unice.fr) and in the Proceedings of the 3 conferences organized by ARENA (Epchtein & Candidi, 2007), (Zinnecker *et al.*, 2008), and in particular(Spinoglio & Epchtein, 2010).

3. Conclusions and Recommendations

Dome C is the only place where one can expect to undertake outstanding programmes of astronomy in Antarctica in the next decade from the visible to the submillimetre range. The excecutive committeee of ARENA is preparing a series of recommendations in conclusion of the ARENA roadmap. Among them, it is recommended to create an astronomical observatory at Dome C with a stable structure, to proceed with the internationalization of the CONCORDIA station, to pursue the site assessment, open wide access to the site qualification data, and facilitate the comparison of site characteristics, to strongly support the currently on-going instruments (IRAIT, A-STEP, QUBIC), to seek funding and support for a couple of additional small instruments (ICE-T, SIAMOIS) to start phase B studies of PLT for a first light before 2020, and phase A studies for AL-ADDIN, AFSIIC and AST. It is also highly recommended that a strong cooperation be initiated between all the countries involved in astronomical developments on the Antarctic continent in order to define a global policy.

Acknowledgements

All participants in the ARENA workprogramme are warmly thanked for their invaluable contributions and the preparation of the roadmap. We are endebted to the European Commission for funding and supporting the ARENA network under contract RICA026150 of the Sixth Framework Programme.

References

Epchtein, N. & Candidi M. (eds.), 2007, Proc. First ARENA Conference on *"Large Astronomical Infrastructure at Concordia, Prospects and Constraints for Antarctic Optical/IR Astronomy"*, EAS Publication Series Vol. 25, EDP

Zinnecker H., Epchtein, N., & Rauer R. (eds.), 2008, Proc. Second ARENA Conference on *"The Astrophysical Science Cases at Dome C"*, EAS Publication series Vol. 33, EDP

Spinoglio, L. & Epchtein, N. (eds.), 2010, Proc. Third ARENA Conference on *"An Astronomical Observatory at CONCORDIA for the Next Decade"*, EAS Publication series Vol. 40, EDP, in press

Highlights of Astronomy, Volume 15
XXVIIth IAU General Assembly, August 2009
Ian F. Corbett, ed.

The LUCAS program: detecting vegetation and traces of life in the Earthshine

Danielle Briot[1], Luc Arnold[2], Stéphane Jacquemoud[3],
Jean Schneider[4], Karim Agabi[5], Eric Aristidi[5], Jérôme Berthier[4],
Erick Bondoux[6], Zalpha Challita[6], Denis Petermann[6],
Cyprien Pouzenc[6] and Patrick Rocher[4]

[1] Observatoire de Paris-Meudon, 61 avenue de l'Observatoire, 75014 Paris
email: danielle.briot@obspm.fr

[2] Observatoire de Haute-Provence, France

[3] Université Paris-Diderot, IPGP, France

[4] Observatoire de Paris-Meudon

[5] Université Nice Sophia-Antipolis

[6] Concordia Station, Dome C, Antarctica

Abstract. The aim of the LUCAS program is to observe chlorophyll and atmospheric molecules in the Earthshine spectrum in order to prepare the detection of life in terrestrial extrasolar planets to be discovered. Actually, observations from Antarctica offer a unique possibility to study the variations of Earthshine spectrum during Earth rotation while various parts of Earth are facing the Moon. Special instrumentation for the LUCAS program was designed and put in the Concordia station in the Dome C. Observations are in progress.

Keywords. Astrobiology, Earth, Moon

1. Introduction

The detection of extrasolar planets has given a new impulse to the research of life in the Universe. In order to prepare the detection of life on Exo-Earths or SuperEarths located in the hability zone, when their images could be seen, we study the detection of terrestrial life, Earth being seen as a dot. Earthshine, which is located in the dark part of the Moon inside the crescent, corresponds to the Earth light backscattered by the non-sunlit Moon. Due to the lunar surface roughness, any place of the Earthshine reflects all the enlighted part of the Earth facing the Moon. So a spectrum of the Moon Earthshine directly gives the disk-averaged spectrum of the Earth, as could be seen extrasolar planets. In this spectrum we can see molecules in the planet atmosphere, like oxygen and ozone, which may be biologic markers, and also the chlorophyll spectrum due to the vegetation reflectivity. The chlorophyll spectrum presents a very typical sharp edge in the near-infrared, near 700 nm, the so-called Vegetation Red Edge (VRE). Since 2002, (see Arnold *et al.* (2002) and Woolf *et al.* (2002)), several studies detected the VRE from Earthshine observations (see a review of the results in Arnold 2008). Although the values of VRE are only a few percent, we found larger values when continents are facing the Moon and smaller values in case of an ocean (see Arnold *et al.* (2002) and Hamdani *et al.* (2006)). To detect the variation of the Earthshine as a function of the Earth landscapes facing the Moon, long observational times are necessary, so as to observe the rotating Earth. This is possible only in the case of observations from very high latitude places, and even more so when near the pole, during a total diurnal cycle (nycthemere).

Other information about Earthshine can be found in the same volume (Briot, 2009).

2. Dedicated Instrumentation and the State of the Art

Concordia station at the Dome C offers such a possibility. After checking the darkness of the sky during Earthshine observations, the LUCAS *(LUmière Cendrée en Antarctique par Spectroscopie)* experiment was then imagined in 2006, installed in 2007, and the first observations planned for the southern winter of 2008. A dedicated instrumentation for Earthshine spectroscopic observations was designed and built at Haute Provence and Paris-Meudon observatories. It is made off a 20-cm diameter Schmidt-Cassegrain telescope that feeds a low resolution slit spectrograph. The spectrograph, based on a 300 grooves/mm reflecting grating, has a 500 to 900 nm spectral range and a resolution of about 100 at 700 nm. The camera is a KAF402ME-based CCD detector. Tests carried out at the Haute-Provence observatory validated the instrumentation: overall optical alignment focus, and data acquisition. Due to the extreme weather conditions in Antarctica, the full instrument (telescope, spectrograph and detector) is insulated to withstand the very low temperatures that prevail in the Concordia Station. The internal temperature of the instrument is regulated thanks to a set of PT100 temperature detectors and a heater that prevent the camera shutter to cool down to negative temperatures. During the 2008 campaign, we had some problems of shutter and heat insolation. The feedback we got from the first observing campaign in 2008 was very important to detect, analyze and correct instrumental problems due to extreme temperature and extreme physical conditions. Some important instrumental improvements were carried out for the 2009 winterover campaign: new thermalization, new impermeable box, high quality (military) connections, new dome, etc. These improvements, as well as the ingenuity of the winterover observers, were very efficient and we have obtained Moon spectra, Earthshine and enlightened Moon, during each observational sequence since 2009 June solstice, that is to say during each Moon cycle since the first or second day after the New Moon, up to the First Quarter. Continuous observations times spend up to 8 hours running. Obviously, so long an observational time of the Earthshine is impossible at low or moderate latitudes.

3. Conclusion

Actually, LUCAS is the first program with spectroscopic observations at Dome C. We obtain Earth vegetation spectra during several hours (up to 8 hours) and we will detect variations during Earth rotation, as it will be possible in the future for extrasolar planets. As such, it is also a test for the design and improvement of small instrumentation, data collecting and management of observations in Concordia's extremely cold environnement.

References

Arnold, L. 2008, *Space Sci. Revs* 135, 323
Arnold, L., Gillet, S., Lardière, O., Riaud, P., & Schneider, J. 2002, *A&A* 392, 231
Briot, D. 2009, *Highlights of Astronomy* vol. 15, (this volume), Special Session 6
Hamdani, S., Arnold, L., Foellmi, C., Berthier, J., Billeres, M., Briot, D., François, P., Riaud, P., & Schneider, J. 2006, *A&A* 460, 617
Woolf, N. J., Smith, P. S., Traub, W. A., & Jucks, K. W. 2002, *ApJ* 99, 225

Highlights of Astronomy, Volume 15
XXVIIth IAU General Assembly, August 2009
Ian F. Corbett, ed.
© International Astronomical Union 2010
doi:10.1017/S1743921310010811

The PLATO observatory: robotic astronomy from the Antarctic plateau

M. C. B. Ashley[1], G. Allen[2], C. S. Bonner[1], S. G. Bradley[3], X. Cui[4],
J. R. Everett[1], L. Feng[5], X. Gong[4], S. Hengst[1], J. Hu[6], Z. Jiang[6],
C. A. Kulesa[7], J. S. Lawrence[1,8,9], Y. Li[10], D. M. Luong-Van[1],
M. J. McCaughrean[11,12], A. M. Moore[13], C. Pennypacker[14], W. Qin[10],
R. Riddle[13], Z. Shang[16], J. W. V. Storey[1], B. Sun[10], N. Suntzeff[17],
N. F. H. Tothill[11], T. Travouillon[15], C. K. Walker[7], L. Wang[5,17],
J. Yan[5], H. Yang[10], D. G. York[18], X. Yuan[4], X. Zhang[5], Z. Zhang[10],
X. Zhou[6] and Z. Zhu[5]

[1] School of Physics, University of New South Wales, NSW 2052, Australia

[2] Solar Mobility Pty Ltd, Thornleigh, NSW 2120, Australia

[3] Physics Department, University of Auckland, Auckland 1142, New Zealand

[4] Nanjing Institute of Astronomical Optics and Technology, Nanjing 210042, China

[5] Purple Mountain Observatory, Nanjing 210008, China

[6] National Astronomical Observatories, Chinese Academy of Sciences, Beijing 100012, China

[7] Steward Observatory, University of Arizona, Tucson, AZ 85721, USA

[8] Department of Physics and Engineering, Macquarie University, NSW 2109, Australia

[9] Anglo-Australian Observatory, NSW 1710, Australia

[10] Polar Research Institute of China, Pudong, Shanghai 200136, China

[11] School of Physics, University of Exeter, Exeter, EX4 4QL, UK

[12] European Space Agency, ESTEC, Postbus 299, 2200 AG Noordwijk, The Netherlands

[13] Caltech Optical Observatories, California Institute of Technology, Pasadena, CA 91125, USA

[14] Lawrence Berkeley National Laboratory and Space Sciences Laboratory, University of California, Berkeley, CA 94720, USA

[15] Thirty Meter Telescope Project, Pasadena, CA 91107, USA

[16] Tianjin Normal University, Tianjin 300074, China

[17] Physics Department, Texas A&M University, College Station, TX 77843, USA

[18] Department of Astronomy and Astrophysics and The Enrico Fermi Institute, University of Chicago, Chicago, IL 60637, USA

Abstract. PLATO is a 6 tonne completely self-contained robotic observatory that provides its own heat, electricity, and satellite communications. It was deployed to Dome A in Antarctica in January 2008 by the Chinese expedition team, and is now in its second year of operation. PLATO is operating four 14.5cm optical telescopes with 1k×1k CCDs, a wide-field sky camera with a 2k×2k CCD and Sloan g, r, i filters, a fibre-fed spectrograph to measure the UV to near-IR sky spectrum, a 0.2m terahertz telescope, two sonic radars giving 1m resolution data on the boundary layer to a height of 180m, a 15m tower, meteorological sensors, and 8 web cameras. Beginning in 2010/11 PLATO will be upgraded to support a Multi Aperture Scintillation Sensor and three AST3 0.5m schmidt telescopes, with 10k×10k CCDs and 100TB/annum data requirements.

Keywords. Site Testing, Instrumentation: miscellaneous, Atmospheric Effects, Telescopes

1. Introduction

The potential of Antarctica, in particular the high plateau in the Antarctic interior, to provide the best astronomical observing sites on the Earth's surface has been widely discussed in the literature (see, e.g., Ashley *et al.* 2004). Dome C (3260m altitude) on the plateau shows a median seeing of 0.27 arcseconds above a 30m turbulent boundary layer (Lawrence *et al.* 2004). However, some of the best potential sites, such as Dome A, do not yet have the infrastructure to support people over the winter. To explore these sites, both for site-testing and for simple astronomical experiments, it is necessary to have a reliable source of power and internet connectivity. Hence PLATO.

PLATO, short for "plateau observatory", is a self-contained astronomical observatory designed to provide 1kW of electricity in order to run experiments with no human presence for up to a year before servicing (Lawrence *et al.* 2009). Electricity is generated by a combination of solar power and diesel engines running on Jet-A1 fuel. PLATO has "on-board" supervisor computers that provide internet access via Iridium modems, and allow many aspects of the facility to be controlled. For the convenience of instrument designers PLATO provides a thermally-insulated environment inside a 10-foot shipping container that can be temperature controlled, usually at least 50°C above the ambient temperature—which can fall to below −75°C in winter.

2. PLATO Performance

2.1. *Power system*

PLATO was installed at Dome A by the Chinese expedition team during January 2008, and ran for 204 days that year, stopping due to an exhaust leak from its engines in early August. At the time of writing (2009 November 9), PLATO has been running for 301 days continuously. Details of the engine system are given by Hengst *et al.* 2009.

2.2. *Iridium communications*

With two Iridium modems, PLATO can reliably transfer $\sim$ 30MB of data per day from Antarctica. The transfer occurs over an "ssh" socket connection, and uses a custom Perl script that copes efficiently with the partial transfer of large files, while simultaneously allowing bidirectional control of PLATO via a "bash" command-line interface. Iridium also provides reliable absolute time for the PLATO instruments, accurate to ±20 ms.

2.3. *Scientific instruments*

The original PLATO instruments, and results from 2008, are described by Yang *et al.* 2009. The PreHEAT instrument showed spectacularly-high atmospheric transmission at a wavelength of 450 microns—a paper describing these results is in preparation.

During 2009, the instruments described in the abstract have been operating. All have worked well and returned data for much of the year. Three papers on CSTAR results are in preparation. Snodar (Bonner *et al.* 2009) has given excellent statistical information on the height of the atmospheric boundary layer, with 10 second or better time resolution, and 1m spatial resolution, throughout the year.

2.4. *Diagnostic information*

The data stream from PLATO includes health and status information such as bus voltages and engine temperatures. This information is available from a webpage updated every minute, with the data usually between 1 and 4 minutes old. Any anomalies with the data trigger the transmission of an SMS message to one or more mobile phones. In practice,

PLATO can operate for weeks at a time with no need for outside intervention. Such intervention is usually only necessary to change instrument parameters or to work around sub-system failures. The redundant nature of much of PLATO's design has allowed us to continue operating despite the occasional electrical and mechanical problems.

There are 8 web-cameras of various types to monitor the sky conditions and instrument icing. One of the cameras is inside the Engine Module, and can assist with diagnosing engine problems such as the exhaust leak that stopped PLATO during 2008. This camera includes a microphone, to measure the engine RPM and general health.

3. Future Plans

PLATO is serviced each Austral summer by the Chinese expedition team organized by the Polar Research Institute of China. At a minimum, servicing involves replacing the six diesel engines, changing the lubricating oil, and filling the Jet-A1 fuel tank. The opportunity is also taken to maintain and upgrade the scientific experiments.

For 2010 we are adding a sub-millimeter Fourier Transform Spectrometer, and a lunar SHABAR to measure the contribution of the boundary layer to astronomical seeing.

Beyond 2010, we will need to replace PLATO's Engine Module with a higher-power version to support three 0.5-m Antarctic Schmidt Telescopes (AST3) under construction at NIAOT. There are also plans to install a Multi Aperture Scintillation Sensor to provide data on the free atmosphere contribution to the astronomical seeing.

4. Conclusions

The PLATO concept has proven its reliability through two successful periods of operation during 2008 and 2009. In practice, the lack of on-site people during winter has not been a major impediment, and has had some benefits: e.g., the instruments have to be designed from the outset for full automation, which tends to lead to greater reliability and longer uptimes.

Acknowledgements

We thank the Polar Research Institute of China for making our productive collaboration possible, and the members of the 2008 and 2009 Chinese expedition teams for their efforts installing and servicing PLATO. We acknowledge funding from the Chinese PANDA International Polar Year project, the Chinese Academy of Science, the National Natural Science Foundation of China, the US National Science Foundation, the Australian Antarctic Division, and the Australian Research Council through the Discovery Projects and Linkage International schemes.

References

Ashley, M. C. B., Burton, M. G., Lawrence, J. S., & Storey, J. W. V. 2004, *Astron. Nachr*, No. 6-8, 619-625

Bonner, C. S., Ashley, M. C. B., Lawrence, J. S., Luong-Van, D. M., & Storey, J. W. V. 2009, *Acoustics Australia*, 37, 47–51

Hengst, S., Luong-Van, D. M., Everett, J. R., Lawrence, J. S., Ashley, M. C. B., Castel, D., & Storey, J. W. V., 2009, *Int. J.Energy Res.*, DOI: 10.1002/er.1595, in press

Lawrence, J. S., Ashley, M. C. B., Tokovinin, A., & Travouillon, T. 2004, *Nature*, 431, 278–281

Lawrence, J. S., Ashley, M. C. B., Hengst, S., Luong-Van, D. M., Storey, J. W. V., Yang, H., Zhou, X., & Zhu, Z. 2009, *Rev. Sci. Inst.*, 80, 064501-1–064501-10

Yang, H. *et al.* 2009, *PASP*, 121, 174-184

Highlights of Astronomy, Volume 15
XXVIIth IAU General Assembly, August 2009
Ian F. Corbett, ed.
© International Astronomical Union 2010
doi:10.1017/S1743921310010823

Supernovae and solar cycles
embedded in a Dome F ice core

Yuko Motizuki, Yoichi Naka and Kazuya Takahashi
for Dome F glaciological astronomy collaboration

RIKEN Nishina Center, Hirosawa 2-1, Wako, Japan
email: motizuki@riken.jp

Abstract. We have recently found signals of candidates for two historical supernovae and past solar cycles in a depth profile of nitrate ion concentrations in an ice core portion corresponding to the 10th and the 11th centuries. This ice core was drilled in 2001 at Dome Fuji (Dome F) station in Antarctica. We briefly review our findings and discuss why Dome F is appropriate for this study.

Keywords. Sun: general, Supernovae: individual (SN 1006, Crab Nebula)

1. Supernova and Solar Cycle Signals in Ice Cores

Ice cores are known to be rich in information regarding past climates, and the possibility that they record astronomical phenomena has also been discussed. Rood *et al.* (1979) were the first to suggest that nitrate ion (NO_3^-) concentration spikes observed in the depth profile of a South Pole ice core might correlate with the known historical supernovae: Tycho (AD 1572), Kepler (AD 1604) and SN 1181 (AD 1181). Their findings, however, were not supported by subsequent examinations by different groups using different ice cores (*e.g.*, Risbo *et al.* 1981; Herron 1982; Legrand & Kirchner 1990), and the results have remained controversial and confusing (Green & Stephenson 2004; Dreschhoff & Laird 2006).

Motizuki *et al.* (2009) presented a precision analysis of an ice core drilled in 2001 at Dome F station in Antarctica. It revealed highly significant three NO_3^- spikes dating from the 10th to the 11th century. Two of them were coincident with SN 1006 (AD 1006) and the Crab Nebula SN (AD 1054), within the uncertainty of their absolute dating based on known volcanic signals. They concluded that the coincidence had a confidence level much larger than 99%.

Moreover, by applying time-series analyses to the measured NO_3^- concentration variations, the authors discovered very clear evidence of an 11-year periodicity that can be explained by solar modulation. The 11-year periodicity was obtained with the 99.9 % confidence level by using the epoch-folding method, which has a clear mathematical basis. This was one of the first times that a distinct 11-year solar cycle has been observed for a period before the landmark studies of sunspots by Galileo Galilei with his telescope. See Motizuki *et al.* (2009) for details.

2. Uniqueness of the Precipitation Environment at Dome F

Dome F is located at 77.2°S, 39.4°E, and its altitude of 3,810 m is the highest point in east central Antarctica. It is natural to wonder whether Dome F site is unique enough to catch such astronomical phenomena. The crucial point here is the degree of stratospheric

Table 1. Tritium concentration in snow corresponding to the deposition in 1966 reported in Antarctica. Extracted from Table 1 of Kamiyama *et al.* (1989).

Point	Tritium, TU
DC (Dome F)	4,200
South pole [Pit A]	2,800
Dome C	700
Halley Bay	620

components contained in ice cores, because both supernovae and solar activities can affect nitrogen oxide production in the stratosphere.

The uniqueness of the precipitation environment of Dome F has been shown from ionic and tritium measurements - Kamiyama, Ageta & Fujii 1989. First, the chemical composition there differs sharply from sea salts. Second, as is shown in Table 1, at Point 'DC' (the site of Dome F) the measured tritium content deposited in 1966 in relation to nuclear weapon tests was much larger than those observed at Dome C and Halley Bay, a coastal site. The tritium content was also observed to increase rapidly in the region above 3600m, where the effects of katabatic wind and the circumpolar vortexes become small. All results indicate that most of the ions in the snow at Dome F precipitate directly from the stratosphere, not from the troposphere (see *e.g.*, Kamiyama *et al.* 1989).

3. Future Prospects

The extension of our analyses to deeper and shallower depths is in progress. Our preliminary results suggest several other historical supernova candidate spikes in the past 2,000 years. Next we are planning to analyse another fresh Dome F core with more detailed core dating. As noted above, Dome F may be an appropriate place to investigate stratospheric or astronomical information. We also encourage the examination of our results by using ice cores recovered from other sites in Antarctica.

Acknowledgements

This collaboration is organized by RIKEN, National Institute of Polar Research (NIPR), Shinshu University and National Institute for Environmental Studies (NIES) to perform ion concentration measurements of ice cores with high time resolution and to analyse the results, especially in relation to astronomical phenomena.
We would like to thank all members of Dome F glaciological astronomy collaboration. This work was supported in part by a Grant-in-Aid for Scientific Research from the Japan Society for the Promotion of Science.

References

Dreschhoff, G. A. M. & Laird, C. M. 2006, *Advances in Space Res.* 38, 1307
Green, D. A. & Stephenson, F. R. 2004, *Astroparticle Phys.* 20, 613
Herron, M. M. 1982, *J. Geophys. Res.* 87, 3052
Kamiyama, K., Ageta, Y., & Fujii, Y. 1989, *J. Geophys. Res.* 94, 18, 515
Legrand, M. R. & Kirchner, S. 1990, *J. Geophys. Res.* 95, 3493
Motizuki, Y., Takahashi, K., Makishima, K., Bamba, A., Nakai, Y., Yano, Y., Igarashi, M., Motoyama, H., Kamiyama, K., Suzuki, K., & Imamura, T. 2009, submitted to *Nature*, http://arxiv.org/abs/0902.3446
Risbo, T., Clausen, H. B., & Rasmussen, K. L. 1981, *Nature* 294, 637
Rood, R. T., Sarazin, C. L., Zeller, E. J., & Parker, B. C. 1979, *Nature* 282, 701

Highlights of Astronomy, Volume 15
XXVIIth IAU General Assembly, August 2009
Ian F. Corbett, ed.

© International Astronomical Union 2010
doi:10.1017/S1743921310010835

Future plans for astronomy at Dome Fuji

Takashi Ichikawa

Astronomical Institute, Tohoku University, Aoba, Sendai 980-8578, Japan
email: ichikawa@astr.tohoku.ac.jp

Abstract. In Antarctica the cold and dry air is expected to provide the best observing conditions on the Earth for astronomical observations from the infra-red to the sub-millimetre. To utilise these advantages of Antarctica, we have devised a plan to construct an astronomical observatory at Dome Fuji, which is located in inland Antarctica. For pilot research and site testing at Dome Fuji, we have developed 40 cm infrared and 30 cm THz telescopes, which are durable for the harsh environment of inland Antarctica. As our project for astronomical research at Dome Fuji is approved for the 3-year program of NIPR, we will start the site testing and pilot research for astronomy at Dome Fuji from 2010.

Keywords. Site Testing, Galaxies: evolution, Infrared: general, Submillimetre

1. Introduction

Antarctica is expected to provide the last windows open to space for ground-based astronomical observations. Especially, the highest regions of the Antarctic plateau above 3,000 m elevation are an attractive environment for observational astronomy. Due to the low temperature, thermal noise at infra-red wavelengths is much lower in Antarctica than other temperate sites. The dry atmosphere, with little water vapour, is more transparent from the infra-red to the sub-millimetre. At the summits of the plateau the wind speed is low and the atmosphere is stable, so that no violent storms and blizzards exist. Since the surface inversion layer is thin, we expect good seeing. As well as the South Pole, there are better sites for astronomy at several bases on the summits of the plateau, Dome C, Dome A, and Dome Fuji (also known as Dome F) (see Fig. 1). Astronomers in the world pay attention to these sites, which have extremely good conditions. In this context, a Japanese group has organized a consortium consisting of four universities (Tohoku, Tsukuba, Rikkyo, Nagoya) and two institutes (National Institute of Polar Research and National Astronomical Observatory) to promote astronomy at Dome Fuji.

2. Dome Fuji

Dome Fuji station is located at $-77°$ 19' 01"S, $39°$ 42' 12"E; 1,000 km inland on the Antarctic Continent at 3,810 m above sea level (see Fig. 1), which is the second-highest summit of the Antarctic ice sheet. It was established in 1995 by NIPR for a deep ice drilling program and atmospheric observations. The year-round average temperature is about $-54°$C, and in winter the temperature falls as low as $-80°$C. Due to this low temperature, thermal noise at infra-red wavelengths is much lower in Antarctica than other sites. Although the site is on the border of the aurora oval, this is not a drawback for infrared and THz astronomy. In the 2006 summer we carried out monitoring observations of the atmospheric turbulence in the boundary layer (up to the altitude of 1,000 m) using a SODAR, and of transparency using a 220 GHz radiometer. However, it is in winter that the superior characteristics of Antarctica appear, so that it is necessary for us to examine turbulence and transparency then.

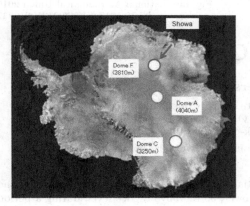

Figure 1. Dome Fuji and Syowa stations.

Figure 2. 40 cm infrared telescope at Rikubetsu, the coldest place in Japan

3. Science Goals

To enjoy the advantages in Antarctica, we are planning to construct 2m-class infrared and 10m-class THz telescopes. Thanks to the low background and high transmittance, a 2m-class telescope has the capability of 8m-class telescopes located at Mauna Kea in the near-/mid-infrared. An infrared survey observation in K-dark band at 2.4 μm will give the deepest and widest dataset for the high-z universe with reasonable cost and observation time. It will reach deeper than those of VISTA and UKIDSS by 1–2 mag. The THz telescope will target dusty galaxies at the high-z universe to study galaxy evolution in its early phase of star formation enshrouded in dust. The long polar night in winter is favourable for searching for variable objects with long periods, such as extra-solar planets in orbit in a habitable zone. The observation of molecules (e.g., CO, H_2O, CH_4) at the second eclipse will provide information on the exo-planet's atmosphere. To undertake such observations we are making near-/mid-infrared instruments for imaging and spectroscopy.

4. Future Plans

NIPR has planned the construction of a new permanent winter-over station with raised floors at Dome Fuji in the next 6-year program, because the old station was buried under snow. Astronomical facilities are also expected. Before the completion of the station, we will start astronomical site testing and pilot research (e.g., CO survey in Galactic plane, faint stellar halo of nearby galaxies, second eclipse of exoplanets) with small telescopes (Lundock & Ichikawa 2008; Murata *et al.* 2008) and site-testing equipments from 2010. In collaboration with an Australian group (Storey *et al.*), we will construct a PLATO observatory for Dome Fuji, which will enable us to make unmanned operation of the instruments before the winterover station is constructed.

Acknowledgements

This work has been supported in part by a Grant-in-Aid for Scientific Research (21244012) of the Ministry of Education, Culture, Sports, Science and Technology in Japan.

References

Lundock, R. & Ichikawa, T. 2008, SPIE, 7014, 89
Murata, C., *et al.* 2008, SPIE, 7019, 79

Highlights of Astronomy, Volume 15
XXVIIth IAU General Assembly, August 2009
Ian F. Corbett, ed.
© International Astronomical Union 2010
doi:10.1017/S1743921310010847

Site testing on the Greenland Ice Cap

M. I. Andersen[1], K. Pedersen[2] & A. N. Sørensen[2]

[1] Dark Cosmology Centre, Niels Bohr Institute, Univ. Copenhagen, Juliane Mariesvej 30,
DK-2100 Kbh Ø, Denmark
email: mia@dark-cosmology.dk

[2] Niels Bohr Institute, Univ. Copenhagen, Juliane Mariesvej 30, DK-2100 Kbh Ø, Denmark

Abstract. We present a site testing program initiated at the SUMMIT station on the Greenland Ice Cap. A DIMM was mounted in the SWISS tower, 39 m above the ice level, during a period of 3 weeks in the late Arctic summer 2008. Tracking Polaris, the DIMM obtained continuous seeing measurements. The campaign was hampered by poor weather and the measured seeing was fluctuating, suggesting that the boundary layer was very unstable. However, during short periods, the un-calibrated seeing went below 0″.5, indicating that the free atmosphere seeing above Greenland is not significantly different from what is found above the Antarctic plateau.

Keywords. Site Testing, Atmospheric Effects

1. Introduction

The Antarctic plateau has over the past 30 years been established as an observing platform of increasing importance (Indermuehle, Burton & Maddison (2005)). The key parameters which makes sites on the Antarctic plateau, such as Dome C, Dome F and Dome A, interesting, is the excellent seeing above a very thin boundary layer (Lawrence *et al.* (2004), Aristidi *et al.* (2009)), the cold environment and the extremely low precipitable water vapor. A limitation of Antarctic is that only about one third of the sky is visible. A similar site in the Northern hemisphere could complement the Antarctic plateau.

2. Greenland as an Astronomical Site

The only approximately similar site in the Northern hemisphere is the Greenland Ice Cap. Its area is 7 times smaller than Antarctica, but the altitude of summit, the highest point on the ice cap, is 3,221 m and thus almost identical to Dome-C. Because the scale height of the boundary layer under stable high pressure conditions is many orders of magnitude smaller than the extent of these ice caps, there is no reason to believe that the behavior of the boundary layer above the Greenland Ice Cap is fundamentally different from what is observed above the Antarctic plateau. There is thus reason to believe that one can get access to the free atmosphere from a tower of manageable height.

As compared to the Antarctic plateau, the Greenland Ice Cap is much more affected by external weather systems. However, Greenland has 56,000 inhabitants and an infrastructure which works year around, with daily connections between Kangerlussuaq international airport and Copenhagen on commercial airlines. There is also access by sea all year. In the summer season there is access to SUMMIT via USAF C-130 planes. In the winter season there is access with Twin-Otter planes and medical evacuations can be carried out, except for a few days a year when the temperature at SUMMIT drops below −55°C.

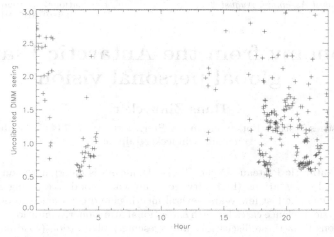

Figure 1. Uncalibrated seeing data for one day during the campaign. The measured differential image motion varies on short time scales, suggesting that the height of the boundary layer varied quite dramatically during these measurements.

3. A Tower-Mounted Seeing Monitor

The reference instrument for seeing measurement is the Differential Image Motion Monitor (DIMM) (Sarazin, M. & Roddier, F. (1990)). The challenge in characterizing sites on ice plateaus with a DIMM is that it must be located above the boundary layer, which has a completely dominant contribution to the seeing. The strategy which was identified as likely to be most successful for measuring the free atmosphere seeing above the Greenland Ice Cap at SUMMIT was therefore to place a DIMM, which can track Polaris, in the 50 m high Swiss tower. This DIMM is pointed to the celestial pole and brings the light from Polaris onto the axis of the DIMM telescope through the use of achromatic prisms. The achromatic prisms are mounted in a wheel which rotates at the siderial rate. The DIMM is foreseen to be able to operate fully autonomously.

4. First Seeing Measurements

During three weeks in August 2008, in the late Arctic summer, the DIMM was deployed in the Swiss tower, 39 m above the ice. Unfortunately the weather was not co-operative. There were clouds during most of the days and more or less continuous data could only be taken during some part of one day. The DIMM has not yet been cross-calibrated against another DIMM, so only un-calibrated seeing measurements are shown in Fig. 1. After calibration, the best seeing measurements are likely 10%-20% better, i.e. closer to 0″.4. Permanent installation of the DIMM is intended from 2010 or 2011.

Acknowledgements

We acknowledge support from the NSF Arctic programme at SUMMIT and from the Instrument Centre for Danish Astrophysics

References

Aristidi, E. *et al.* 2009 *Astron. and Astrophys.* 499, 955
Indermuehle, B., Burton, M., & Maddison, S. 2005 *PASA* 22, 73
Lawrence, J. S. *et al.* 2004 *NATURE* 431, 278
Sarazin, M. & Roddier, F. 1990 *Astron. and Astrophys.* 227, 294

Highlights of Astronomy, Volume 15
XXVIIth IAU General Assembly, August 2009
Ian F. Corbett, ed.

© International Astronomical Union 2010
doi:10.1017/S1743921310010859

Astronomy from the Antarctic Plateau: a global personal vision

Hans Zinnecker

Astrophysikalisches Institut Potsdam, An der Sternwarte 16, D-14482 Potsdam, Germany
email: hzinnecker@aip.de

Abstract. The Antarctic Plateau (Dome C, also Dome A) is emerging as an especially good sitefor astronomical observations (high, dry, cold, no wind, good free seeing above a certainboundary layer). Over the last few years, several meetings and conferences tookplace to discuss potential astrophysical science cases for such exceptional atmosphericconditions. I summarise my personal conclusions from these discussions and presenta global vision (roadmap) for Antarctic Astronomy for future optical, near-IR, thermal-IR, andfar-IR/sub-mm observations. The need for international collaboration between Europe, Australia and China is stressed.

Keywords. Antarctica, Telescopes, Site Testing, Instrumentation.

1. Introduction

The main advantages of the Antarctic Plateau ($\sim 3{,}200$ m) for astronomyare the weather conditions. At Dome C it is cold, high, dry and the sky quite stable. There are few clouds, hardly any wind and there are few aurorae. There is excellent night-time seeing in the Antarctic winter (May – August) above a ground layer of some 30 m, with a median of about ~ 0.3 arcsec in thevisible. There are also 4 hours of excellent seeing on the ground duringday-time afternoon in the Antarctic summer (Nov – Feb), especially in thethermal infrared (remember the FWHM of the seeing scales with the inverse1/5 power of the wavelength) which allows us to make infrared astronomicalobservations in Antarctica in the summer-time, i.e. when the conditionsare much less harsh than in the winter time and without building towers (cf. Zinnecker (2007)). The Italian 80 cm IRAIT telescope will be the firstto prove the concept. Of course, optical observations require winter-timewhich in turn require telescopes on towers (except long-time seriesoptical photometric observations, where the seeing is not critical).

Although Dome A and Dome F may be slightly better astronomical sites (see the contributions by X. Cui from China and T. Ichikawa from Japan, respectively, in this Special Session and Proceedings), as a matter of fact Dome C and its infrastructure, i.e. the French-Italian Concordiastation, exist today and could be used (or even expanded) for optical/infrared/sub-mm astronomy, political problems notwithstanding.

2. How best to exploit the Antarctica Plateau for astronomy now

Astronomy in Antarctica is already going on at the Amundsen-Scott South Polestation. The South Pole 10 m radio/sub-mm telescope is installed andtaking data and the BICEP 25 cm aperture is studying CMB polarisationanisotropy (J. Carlstrom; Y.D. Takahashi, these Proc.). This raises the question whether Dome Cshould also have a big radio antenna, to carry out sub-mmobservations at 450 and 350 microns, and pushing for THz observationsin the 200 microns window. This would be a very good idea consideringthe very low precipitable water vapour (PWV) values, often less than 0.25 mm. However, to be competitive a 25 m dish is needed (see the science case described by Minier (2008)) which wouldbe very expensive and would stretch the current logistics at theDome C Concordia station. In my view, this discourages going forsuch a Antarctic Submillimetre Telescope (AST) telescopeat the present time; furthermore, there is a major push for Dome Awhich is claimed to be an even better sub-mm site than Dome C andwhere our Chinese colleagues plan to build a 15 m-class sub-mm–telescope. Finally, we should probably wait to see whether the 25 m CCAT (Cornell–Caltech Atacama Telescope, now under consideration) willbe realised on a very high site (5,600 m) near Cerro Chajnantor or not.

Leaving aside sub-mm telescopes, what should we do in optical and infrared astronomy? We have mentioned the advantages of the much reduced sky background in the thermal infrared regime (3 – 5 microns) compared to more temperate sites (factor ~ 30) due to the very cold atmosphere. There are several science cases for this part of the spectrum using a 2 – 3 m-class telescope (Burton (2005)), including molecular hydrogen surveysof the inner Galaxy, cool protostar and evolved star searchesin the LMC/SMC and also Galactic star forming regions in thesouthern hemisphere (e.g. Carina, Vela, etc.). These would allbenefit greatly from the increased spatial resolution with anAntarctic 2.5 m telescope, compared to Spitzer data obtained witha 85 cm telescope (factor 3 gain in angular resolution at 4 microns). Follow-up studies for NASA's soon-to-be-launched WISE (40 cm) near-to-thermal infrared all-sky survey telescope will also be an issue. It has been questioned whether all this would represent 'killer' science as opposed to fairly unique science. Perhaps the best use of a 2.5 m telescope (small enough to be affordable now but large enough to be a pathfinder for future large structures) may be to exploit the K_{dark} window (2.4 microns, Casali (2007)) wherethe reduced sky background allows us to really go deep (much below UKIDSS and VISTA) and carryout extra-galactic near-infrared surveys down to $K_{dark} = 25$ mag. to study distant galaxy populations at high redshifts. JWST will provide competition, but only in principle and not in practice, as the FOV of JWST/NIRCAM is 3.4×3.4 arcmin. while an Antarctic survey telescope wouldhave 20×20 arcmin. or more, depending on the cost of IR-detectors.

3. Roadmap for a Dome C Astronomical Observatory

I consider a roadmap to be a temporal sequence of major milestones, with a beginning and an end. It should give realistic timescales to implement telescopes or instruments of increasing complexity. The beginning and first milestone would be to operate and obtain data with IRAIT as soon as possible (2010). The CAMISTIC bolometer cameraon IRAIT will probe the quality of the site for sub-mm/THz (particularly for 200 microns) imaging observations. The second milestone should be an IR-optimised PLT–like telescope, if necessary with ground layeradaptive optics (first light in 5–10 years, 2015–2020). This should take advantage of the good Antarctic summer seeing andstart operating in the day-time, doing 0.3 arcsec 2.5–5 microns imaging. At the same time, robotic optical telescopes like ICE–T could operatein the Antarctic winter, without towers (as the seeing is not criticalfor long-time series photometry). I would leave the lead for 2–4 m class optical telescope to our Chinese friends. The third milestone (in 10–15 years time) would be ALADDIN, the two-telescope thermal–IRinterferometer; it would then still be in time to act as an exo-zodiprecursor for DARWIN (the latter won't happen before 2025). The 4thmajor milestone (around 2030) could be a THz interferometer array (of the order of a dozen telescopes) complementing ALMA and SOFIA and being a pathfinder for a far-IRinterferometer in space (ESA's future FIRI mission). The ultimateand 5th milestone that I see on the horizon (in the 2030-2040 timeframe) would be an optical/IR kilometric array of 2 m class telescopes.

4. The need for international collaboration

Astronomy in Antarctica is expensive for any country trying toestablish and maintain a permanent base on this remote continent. For example, the total operating costs of the Concordia station are 8 million Euro per year. Still, astronomy in Antarctica ismuch cheaper than ESA missions from space. Space projects are usually donein co-operation between several nations. Similarly, it would betoo costly if every nation would want to build their own suiteof telescopes in Antarctica. We have to share our resources. This means France and Italy will *invite* other European partners to participate in astronomy from Dome C. It also means that Dome C and Dome A should each concentrate and specialise on the best possible science doable from their respective sites. While Dome C might be best suited for near-infraredand thermal-infrared observations, Dome A (at 4,200 m) might be bestfor sub-mm and THz astronomy. Collaboration would reducecost and help to exchange scientific, technical, and logistic experience. These and other issues could be discussed at an IAU symposium on Antarctic Astronomy during the nextIAU General Assembly in 2012 in Beijing/China. By then, Chinese plans for Dome A will have firmed up, while Europe and Australia may have teamed up again at Dome Cpursuing their common interests for a PLT-like telescope. The first IRAIT measurements and results at around that time would hopefullyshow that Dome C lives up to its expectation as the premier astronomicalobserving site on the planet.

Acknowledgements

I would like to thank Michael Andersen and Mark McCaughreanfor critical discussions, as well as Michael Burton for hispatience with the manuscript. Travel support from the ARENAEC network the German DAAD is gratefully acknowledged.

References

Burton, M. G., Lawrence, J. S., Ashley, M. C. B., *et al.* 2005, *PASA* 22, 199
Casali, M. 2007, in: N. Epchtein and M. Candidi (eds.), *1ˢᵗ ARENA Conference: Large Astronomical Infrastructures at Concordia*, EAS Publications Series, vol. 25, p. 201
Minier, V. *et al.* 2008, in: H. Zinnecker, N. Epchtein & H. Rauer (eds.), *2ⁿᵈ ARENA Conference: The Astrophysical Science Cases at Dome C*, EAS Publications Series, vol. 33, p. 21
Zinnecker, H., Andersen M. I., & Correia, S. 2007, in: N. Epchtein and M. Candidi (eds.), *1ˢᵗ ARENA Conference: Large Astronomical Infrastructures at Concordia*, EAS Publications Series, vol. 25, p. 183

Highlights of Astronomy, Volume 15
XXVIIth IAU General Assembly, August 2009
Ian F. Corbett, ed.

© International Astronomical Union 2010
doi:10.1017/S1743921310010860

CSTAR and future plans for Dome A

Xiangqun Cui
On behalf of the Chinese Center for Antarctic Astronomy (CCAA)

National Astronomical Observatories / Nanjing Institute of Astronomical Optics and Technology, CAS *Present address: NIAOT, 188 Bancang Street, Nanjing 210042, P. R. China*
email: xcui@niaot.ac.cn

Abstract. The first set of Chinese Antarctic telescopes at Dome A is called CSTAR. It consists of four 14.5 cm wide-field telescopes and was installed at Dome A during the traverse of 2007/2008. CSTAR successfully operated for 135 days in 2008 and for more than 200 days in 2009. This paper briefly introduces recent developments in Chinese Antarctic astronomy and their international collaborative activities. It also describes future plans for Dome A, as the building of Kunlun Station began in January of this year.

Keywords. Antarctica, Telescopes, Site Testing.

1. Introduction

In 2004, a Chinese traverse team led by Yuansheng Li, the senior researcher of the Polar Research Institute of China (PRIC), reached Dome A, from Zhongshan Station. Shortly following this event, the first time humans had ever been to Dome A, Lifan Wang & Xiangqun Cui, with the collaboration of the LAMOST project, organised a workshop on "Wide Field Survey Telescopes at Dome C/A" in Beijing between 3 to 4 June of 2005. This workshop started Chinese astronomical activities and the international collaboration at Dome A. In June 28, 2005, organised by Jun Yan, Lifan Wang and Xiangqun Cui, a further meeting on "Antarctic Astronomy at Dome A" was held at Purple Mountain Observatory (PMO) in Nanjing. In November 2005, the meeting "Wide Field Astronomy on the Antarctic Plateau" was organised by Lifan Wang & Enrico Cappellaro in Padua, Italy. Through the efforts of Jun Yan, Xiangqun Cui, Longlong Feng & Lifan Wang, the Chinese astronomical community joined the PANDA project, which was a Chinese Key international program for the IPY between 2007 to 2010. A MoU between USNW–NAOC–PRIC was signed for collaborations on Antarctic astronomical research in December 2006. This includes the international collaboration on site testing of Dome A with Texas A & M University, USA, University of New South Wales, Australia, California Institute of Technology, USA, Solar Mobility, Australia, Thirty Meter Telescope Project, USA, University of Arizona, USA and the University of Chicago, USA. In addition to these activities, the Chinese Center for Antarctic Astronomy (CCAA) was established on December 24, 2006 in PMO, including the member institutes of Purple Mountain Observatory (PMO), Nanjing Institute of Astronomical Optics & Technology (NIAOT), National Astronomical Observatories (NAOC), Polar Research Institute of China (PRIC), Institute of High Energy Physics (IHEP), Shanghai Astronomical Observatory (SAO), Nanjing University, Tianjin Normal University and the Institute of Electronic Engineering.

Since then, in early 2007 building of the Chinese Small Telescope Array (CSTAR), which consists of 4×14.5 cm wide field telescopes, began in China, and the international collaboration on site survey instruments using the PLATO laboratory started. In the

2007/2008 traverse CSTAR and PLATO were successfully installed and operated (including the site survey instruments SNODAR, GATTINI, Sonics & Pre-HEAT) by Dr. Xu Zhou and Dr. Zhenxi Zhu. In the 2008/2009 traverse Dr. Xuefei Gong successfully maintained these instruments. The Chinese Kunlun Base Station at Dome A was also established. By January 27, 2009 236 square meters of the main building had been built for Kunlun Station. The complete building will be constructed next summer, during the 2009/2010 traverse.

2. CSTAR

CSTAR consists of four 14.5 cm small telescopes with a 20 sq deg FOV (4.5 × 4.5 degrees) and 1k × 1k frame transfer CCD for each, and with g, r, i filters and unfiltered respectively. The image quality is 90% of the light energy circled in 2 pixels. A fixed observing direction of the South Pole area was adopted for these pioneering observations.

The scientific purposes of CSTAR are to: (1) measure atmospheric extinction, (2) measure sky brightness, (3) search for variable stars, (4) search for transiting exo-planets and to (5) find bright SNe, Novae or afterglows of GRBs.

After 135 days observation with 20 seconds exposure time for each image, 271,041 good frames were obtained. Some results have been obtained from the data processing will be published soon.

See references Cui, et al. (2008), Gong, et al. (2008), Kulesa, et al. (2008), Lawrence, et al. (2008), Lawrence, et al. (2008), Tothill, et al. (2008), Yang, et al. (2009) and Yuan, et al. (2008) for papers describing Chinese activities at Dome A.

3. Future Plans for Dome A

2008-2010:
- Continue site survey and astronomical observations
- Upgrade PLATO and CSTAR
- Develop and operate a Fourier Transform Spectrometer (FTS)
- Develop and operate 3 × 50cm/70cm modified Schmidt telescopes (AST3)

2011-2015:
- Continue operation of FTS and AST3
- Develop and operate a 1m Optical/Infrared telescope
- Develop and operate a 5m Sub-mm/THz telescope

2012-2020:
- Develop and operate a 4m wide field Optical/Infrared telescope
- Develop and operate a 15m Sub-mm/THz telescope

References

Cui, X. Q., et al. 2008, Proc. SPIE 7012, 70122D
Gong, X. F., et al. 2008, Proc. SPIE 7012, 701848
Kulesa, C. A, et al. 2008, Proc. SPIE 7012, 701249
Lawrence, J. S., et al. 2008, Proc. SPIE 7012, 701227
Lawrence, J. S., et al. 2008, Rev. Sci. Inst. in press
Tothill, N., et al. 2008, Proc. EAS 33, 301
Yang, H. G., et al. 2009, PASP 121, 174
Yuan, X. Y., et al. 2008, Proc. SPIE 7012, 70124G

Highlights of Astronomy, Volume 15
XXVIIth IAU General Assembly, August 2009
Ian F. Corbett, ed.

© International Astronomical Union 2010
doi:10.1017/S1743921310010872

Special Session 4: Astronomy Education between Past and Future

Jean-Pierre de Greve, editor

Department of Physics, Vrije Universiteit Brussel, BE1050 Brussels, Belgium
email: jpdgreve@vub.ac.be

Organizing committee:
Jean-Pierre de Greve (Chair, Belgium), Rajesh Kochhar (India), Edward F. Guinan (USA), : John B. Hearnshaw (New Zealand), George K. Miley (Netherlands), Ian E. Robson (UK), Rosa M. Ros (Spain), Il Seong Nha (Rep. of Korea), Malcolm G. Smith (USA), and Antonio Videra (Brasil)

Preface

The special session aims at discussing an integrated approach of the different efforts to increase and promote the teaching and learning of astronomy in the world, with emphasis on developing countries. To this end, attention will be given to research on education, specifically in the field of physics, to best practices of the use of astronomy in educational systems (specifically in developing countries), and to innovative learning initiatives other than formal education. The Special Session aims also at creating a universal perspective wherein modern (post-Copernican) astronomy will presented as an intellectual cumulus. The objective of the session is to disseminate best practices in teaching and learning activities of astronomy and to give an opportunity to learn about initiatives in different cultural and socio-economic settings. The special session also wants to give food-for-thought and proposals for reflection for an integrative approach, and for optimization processes, to enhance the interest in astronomy and its role as a trigger towards science education in the educational systems, with emphasis on the developing countries. The outcome should be a sensitization of teachers and students alike to the concept of a universal history of astronomy and creation of some reliable source material which can be used as a teaching aid in a culture-specific context. The outcome could be a set of recommendations for future integrated actions, and eventually recommendations on new initiatives, framed into the new decadal policy plan.

Jean-Pierre de Greve, Rajesh Kochhar, Edward F. Guinan, co-chairs SOC.

Highlights of Astronomy, Volume 15
XXVIIth IAU General Assembly, August 2009
Ian F. Corbett, ed.

Challenges in Astronomy Education

Jean-Pierre De Greve

Department of Physics, Vrije Universiteit Brussel, BE1050 Brussels, Belgium
email: jpdgreve@vub.ac.be

Abstract. Astronomy is an attractive subject for education. It deals with fascination of the unknown and the unreachable, yet is uses tools, concepts and insights from various fundamental sciences such as mathematics, physics, chemistry, biology. Because of this it can be well used for introducing sciences to young people and to raise their interest in further studies in that direction. It is also an interesting subject for teaching as its different aspects (observation techniques, theory, data sampling and analysis, modelling,?) offer various didactical approaches towards different levels of pupils, students and different backgrounds. And it gives great opportunities to teach and demonstrate the essence of scientific research, through tutorials and projects. In this paper we discuss some of the challenges education in general, and astronomy in particular, faces in the coming decades, given the major geophysical and technological changes that can be deducted from our present knowledge. This defines a general, but very important background in terms of educational needs at various levels, and in geographical distribution of future efforts of the astronomical community. Special emphasis will be given to creative approaches to teaching, to strategies that are successful (such as the use of tutorials with element from computer games), and to initiatives complementary to the regular educational system. The programs developed by the IAU will be briefly highlighted.

Teaching Astronomy for Development

Edward F. Guinan[1], Laurence A. Marschall [2]

[1] *Villanova University, Tampa, FL, USA*
[2] *Gettysburg College,Gettysburg, PA, USA*

Abstract. The primary goals of the IAU Commission 46 Program Teaching Astronomy for Development (TAD) are to aid in "the enhancement of a country's astronomy education and astronomical research in support of education". The IAU-TAD program continues to vigorously support astronomy education, teaching, research and outreach programs in developing countries. TAD programs supported over the last few years have included the following countries: Columbia, Iran, Kazakhstan, Mongolia, Morocco, Nepal, North Korea (DPR- Korea), the Philippines, Vietnam. Examples and outcomes of some of these programs are discussed. Also discussed are the future plans for the TAD program as well as practical information on how to apply for a TAD program for your country. Additional information about the IAU/TAD program can be provided by the TAD chairs at e-mail addresses: edward.guinan@villanova.edu and marschal@gettysburg.edu. Information on TAD programs is also available on the IAU Commission 46 website.

Astronomy Education in Thailand From Junior School To University

Boonrucksar Soonthornthum[1,2], Kam-Ching Leung[1]
[1] - *University of Nebraska, USA*
[2] - *National Astronomical Research Institute of Thailand*

Abstract. In 2001 the government introduced astronomy into the natural science school curriculum. While it was an excellent idea, there was a shortage of teachers with training in astronomy! In order to educate the teachers we needed to expand the astronomy programs at the university level. In this paper we describe the historical development of astronomy from the 1600 to present day Thailand.

The first observatory was established in 1685 during the reign of King Narai. King Rama IV calculated the time and location of the solar eclipse in 1868. Modern astronomical activity started in the 80s with several graduates returning from studying overseas. The first international conference was held in Chiang Mai in 1995. The most important astronomy education government policy statement was announced in 2001 to integrate astronomy into the school curriculum. The second important government policy on astronomical research, education and outreach was the establishment of National Astronomical Research Institute of Thailand, NARIT in 2004. The education programs in the current decade will be fully described in our paper.

Teacher Professional Development

Mary Kay Hemenway
University of Texas at Austin

Abstract. One challenge in providing professional development is that the teachers are unaware of what they really need to become more effective in teaching astronomy. Designing a professional development program for the naïve learner includes consideration of the science standards (what concepts their students may be tested upon), activities to make them aware of their own understanding of the topics, activities to engage them at a level above what their students need to know, and an activity to assist them in integrating their new knowledge. Specific examples of program planning and elements will be provided from Teacher Professional Development workshops offered at McDonald Observatory over the past eight years.

Papers on Astronomy Education in Proceedings of IAU Meetings: a review of 1988–2006

Paulo Sergio Bretones, Jorge Megid Neto
Faculdade de Educação - UNICAMP

Abstract. The IAU as the biggest international organization of professional Astronomy has been dedicated to questions on Astronomy Education in schools and for the general public, by means of the Commission 46 (Astronomy Education and Development). In the last 20 years, many works were presented on Colloquia and Special Sections of General Assemblies.

In this context, we reviewed the complete papers on Astronomy Education published in the IAU proceedings. We have focused our analyses on the following documents: a) The Teaching of Astronomy, IAU Colloquium 105 (1988); b) New Trends in Astronomy Teaching, IAU Colloquium 162 (1998); c) Astronomy for Developing countries, XXIV GA (2001); d) Teaching and Learning Astronomy, XXV GA (2005); e) Astronomy for the developing world, XXVI GA (2007); f) Innovations in Astronomy Education, XXVI GA (2008). A total of 322 papers were published in these events, excepting posters, and 283 papers were selected and analysed in this review that deal specifically with Astronomy Education.

We have selectioned the following descriptors to the classification these papers based on literature: country of the work; school level approach or public education; topics of contents in Astronomy; focus of the study; method of research; theoretical framework.

Related to the country of origin of the work, researchers from USA represents 35.6% of the total of papers, followed by United Kingdom (9.9%), India (4.9%), France (4.6%) and Canada (4.2%), among other countries.

The greatest percentage of works was related to university education (37.8%) and to public education (27.6%), what is justified by the fact that Astronomy is not a specific discipline in the curriculum in the basic education in the majority of the countries. The basic levels of school appear with very low frequency: elementary – 7.8%; secondary or middle school – 10.6%; and high school – 12.0%. There are 33.9% of works that deal on general approaches for all school levels.

Related to the focus of the study, there is a predominance to present non-school programs (28.3%) and studies of curricular discussions (26.3%). Follow the works on development and discussion of teaching materials (18.4%) and the studies of learning and teaching in astronomy education in the schools (15.5%). Other studies were related to students or teachers understanding (3.2%), teacher education (7.1%) and studies on history of Astronomy or history of Astronomy Education (4.9%).

With respect to the method of research, the great majority of the papers were classified in Research & Development (67.5%). They presented reports of education experiences of the authors or reports of the current situation in Astronomy Education in countries, typical of researchers from the hard science areas, those dedicated to present what they are doing without a more deep discussion of the theoretical framework, or even the use of scientific methodology of data gathering. Essays (20.5%), consisting of free exposition of ideas in the field of astronomy education, in almost the totality with absence of references or explicit theoretical framework follow in number. Typical methods of education research such as survey (8.8%), empirical research (2.1%), content analysis (1.4%) and review (1.1%) were not very present. In this context, studies of action research or case study were not found. Among the 283 selected papers, 87.6% did not show any explicit theoretical references in the text of the work.

With respect to the topics of contents, most studies do not deal with specific topics of Astronomy (83.4%). Much less frequent were studies of Sun-Earth-Moon System (6.7%), Stars (4.2%), Solar System (3.5%) and Sun (2.6%).

In this sense, even with the efforts to elevate Astronomy for public outreach and classrooms, these works always are oportune for the exchanges of experiences. However, from the point of view of educational research, deeper treatments that deal with epistemological questions, teaching and learning processes proper of the area of Astronomy at different school levels and theoretical framework are necessary, which could stimulate the development of scientific investigations as it happens in other areas of knowledge.

A universal history of astronomy as a teaching aid

Rajesh Kochhar

Institute of Science Education and Research, Chandigarh

Abstract. Human beings are born astronomers. Astronomy is more than a branch of modern science. It is a symbol of the collectivity and continuity of humankind's cultural heritage. This mixture of science and culture is astronomy's strength as well as dilemma. Strength, because support for astronomy transcends all boundaries; dilemma, because this support transcends science also. History is an exercise in reconstructing the past, carried out in the present with an eye on the future. Thus paradoxical as it may seem history is an instrument that converts the past into a bridge between the present and the future. More specifically, history of astronomy is an enquiry into how human perception of their cosmic environment has evolved with time. It is relatively an easy matter to discuss the history of modern astronomy as western astronomy. But if we wish to advance the cause of astronomy and see its development world-wide, we must place post-Galilean developments in a wider spatial and temporal context. The 19th century historiography consciously projected modern science (including modern astronomy) as a characteristic product of the Western civilization decoupled from and superior to its antecedents, with the implication that all material and ideological benefits arising from modern science were reserved for the West. As a reaction to this, the orientalized East has often tended to view modern science as "their" science, distance itself from its intellectual aspects, and seek to defend, protect and reinvent "our" science and the alleged (anti-science) Eastern mode of thought. This defensive mind-set works against the propagation of modern astronomy in most of the non-Western countries. There is thus need to construct a history of world astronomy that is truly universal and unselfconscious, but at the same time rigorous. The universal history of astronomy would be placed what we may call the framework of Cultural Copernicanism. Just as Copernican principle in cosmology tells that the universe does not have a preferred location or direction, Cultural Copernicanism would imply that no cultural or geographical area or ethnic or social group can be deemed to constitute a benchmark for judging and evaluating others.

Within this framework, modern astronomy should be seen as the current end of a cultural and intellectual cumulus to which different cultures and regions have contributed at different epochs. The past should not be pitted against the present, but evaluated in its own right and as leading to the present.

Astronomy Education in the Maya region, some past and future lessons

Maria Cristina Pineda de Carias

Central America Suyapa Astronomical Observatory, National Autonomous University of Honduras

Abstract. Increasing has become now the education of topics of Astronomy in the schools and colleges of Central America, particularly in Honduras, where an astronomic observatory works from the late nineties dedicated to the development of Astronomy and Astrophysics, Space Science and Archaeoastronomy, as part of the academic university activities. It worries nevertheless how, teaching and learning of fundamental topics such as the seasons of the year, or rotation and revolution of the Earth around the Sun are done, because they do not contribute fully to understand related environmental conditions. Textbooks and bibliographic information available to teachers and students, places emphasis on understanding the concepts of equinoxes and solstices as isolated terms, leaving out entirely the zenith sun passages, the apparent motions of the Sun or, the amount of solar radiation that every day we get in different dates of the year for the fact of living in the tropics. Our studies in Archaeoastronomy reveal how the apparent motion of the Sun do not pass unnoticed among the ancient Mayans, who, on the contrary, knew and used this knowledge to measure divisions of day and to mark relevant dates of the tropic year to anticipate the arrival of the rainy seasons or the time for land preparation for agricultural purposes. This paper takes a number of archaeoastronomical discoveries in the Maya region of Copan, in Honduras, as astronomical lessons from the past that should be teach nowadays, as present and future lessons among children, young students, and population, to let them to understand how the analogy between the ideas of astronomy with aspects of everyday life help to improve living conditions and preserve the environment.

Experience in Astronomy Education made by the OAVdA with the Schools in the Aosta Valley

Andrea Bernagozzi

Astronomical Observatory of the Autonomous Region of the Aosta Valley, Saint-Barthéelemy
Loc. Lignan 39, 11020 Nus (AO), ITALY
email: info@oavda.it, andrea.bernagozzi@unimi.it

Abstract. The Astronomical Observatory of the Autonomous Region of the Aosta Valley (OAVdA), in the italian Alps at the border with France and Switzerland, was opened in 2003. A formal agreement of cooperation with the italian National Institute of Astrophysics (INAF) has been established in 2006. Since then, scientific research has been the main activity in OAVdA, but we are probably among the few observatories where researchers and scientists dedicate 30% of their time to initiatives of public outreach and education.

In order to open a channel between schools and the OAVdA, we decided to emphasize the principle that the knowledge that is being studied at school now is the result of the scientific researches made by scientists 10, 100, 1000 years ago. So the knowledge that will be studied at school tomorrow is the result of the scientific researches of today, including the ones made at the OAVdA.

From this principle it follows that the astrophysicists working at the OAVdA are necessarily involved in initiatives to transfer their scientific results to students, and education activities are based on research methods and original data collected at the OAVdA.

Since 2006, thanks to the support of the local administrations and institutions, students of the schools of the Aosta Valley, 5 to 18 years old, have been exposed, at different levels, to the world of scientific research. The most successful examples are:

- The multi-year education activity *Saint-Roch Étoiles* (*Stars of Saint-Roch*, 2008–2013), where pupils from 5 to 12 years old of the Istituzione Scolastica Saint-Roch in Aosta are introduced to the world of scientific research through *meet the scientist* sessions (see the website http://www.scuole.vda.it/stroch/progetto_etoiles/frame2.htm)

- The education activity *Oltre Galileo... La Vallée guarda lontano* (*Beyond Galileo... The Aosta Valley is looking far*, 2008–2009) where 18 years old students of the Istituzione Scolastica Binel-Viglino in Pont-Saint-Martin and in Saint-Vincent faced a real problem of scientific research: the calculation of the heliocentric distance of an asteroid at the opposition from an original data set collected during the research activity at the OAVdA (see Carbognani 2008).

- The education activity *Tutti pazzi per Marte* (*There's Something About Mars*, 2008–2009) where 18 years old students of the Istituzione Scolastica Binel-Viglino in Pont-Saint-Martin and in Saint-Vincent made the unique experience of calculating Mars' distance with the parallax method during the December 2007 opposition. Actually this education activity, done in cooperation with the Astronomical Association of Southern Africa (ASSA), was a real research project. The method we developed with students and teachers resulted in a paper published on an international peer-reviewed journal (see Cenadelli *et al.* 2009).

References

Carbognani, A. 2008, *le Stelle* n. 63, p. 60
Cenadelli, D., Zeni, M., Bernagozzi, A., Calcidese, P., Ferreira, L., Hoang, C., Rijsdijk, C. 2009, *European Journal of Physics* vol. 30, p. 35

Universe Awareness: Innovations in Astronomy education and communication

Carolina Johanna Ödman, George Kildare Miley
Universe Awareness & Leiden Observatory, the Netherlands

Abstract. Universe Awareness (UNAWE) started in 2005 with the aim of inspiring young children in underprivileged environments with the scale and beauty of the universe. The goals of the programme are to broaden children's minds, awaken their curiosity in science, demonstrate the power of rational thought and stimulate tolerance and world citizenship.

The programme brings a stimulating international dimension and its aims are of an educational and social nature. The methods to achieve the vision of UNAWE are devised locally by communities in over 20 countries, leading to a dynamic network of open source education and a vast body of know-how and best practice. The success of UNAWE in diverse environments and the commitment of volunteers and partners throughout the world show that comprehensive citizen participation can reach and engage the public in ways never before applied in science communication.

The Starlight Reserve: a route towards public education in astronomy and light pollution control – the New Zealand experience

John B. Hearnshaw[1], Margaret E. Austin[2], Graeme Murray[3]

[1] - *University of Canterbury*
[2] - *UNESCO N.Z. national commission*
[3] - *Earth and Sky Ltd, Tekapo, New Zealand*

Abstract. We have produced a case study proposal for the Tekapo region surrounding Mt John Observatory in the central South Island of New Zealand to become a Starlight Reserve. The proposal has been made to UNESCO's Starlight Initiative. A Starlight Reserve would be an excellent means of promoting awareness of the night sky amongst the public, as well as an additional way of helping control light pollution in the neighbourhod of our observatory. In the 24 months since we have discussed the idea of a Starlight Reserve in New Zealand (since early 2007) there has been intense media interest in this proposal.

National Network of Public Outreach in Astronomy CIDAUNAWE

Enrique Torres
1 - Centro de Investigaciones De Astronomia - CIDA
2 - Universe Awareness

Abstract. CIDA has been carrying out, for several years now, astronomy workshops for school teachers. In 2007-2008 a complete astronomy workshop has been implemented complete with support material, with an emphasis on teaching and pedagogical techniques that will convey astronomy to children in a fun way. In addition, we are working in collaboration with the UNESCO Universe Awareness (UNAWE) program. So, there have been build a National Network of Public Outreach in Astronomy with growing impact in schools all around the country

AGA Regulations and the Future of Astronomy in South Africa

Ramotholo Sefako
SAAO, South Africa

Abstract. South Africa intends to exploit its strategic advantages of geography and infrastructure by encouraging the development of large telescopes operating at radio and optical wavelengths. In order to ensure that conditions remain optimal for doing astronomy, some wide ranging legislation, Astronomy Geographic Advantage (AGA) Act, has been formally declared. This will be discussed with the main focus on the development and implementation of the regulations that are meant to protect astronomy advantage areas against light pollution and increased dust levels (optical astronomy) and radio interferences (radio astronomy). I will also discuss what all this mean for the future of astronomy research and development in South Africa.

Popularisation of Astronomy by Organising the Second IOAA

Chatief Kunjaya[1,2], Taufiq Hidayat[1,2], Suhardja D. Wiramihardja[1,2]
[1] *Department of Astronomy, Institut Teknologi Bandung, Bandung, Indonesia*
email: kunjaya@as.itb.ac.id
[2] *Bosscha Observatory, Lembang, West Java, Indonesia*

Abstract. In 2006, Indonesia was appointed to host the second International Olympiad on Astronomy and Astrophysics (IOAA), after the first one in Thailand. The realisation of the appointment into actual event was done in August 19-28, 2008, during which the second IOAA was performed in Bandung city. The participants were high school students under 20 years old from 22 countries. The event has been held successfully with publication in printed media, internet and television. The event has attracted more students to astronomy which was indicated by significant increase of applicants to the department of Astronomy in 2008.

Acknowledgements

We would like to thank The Ministry of Education of the Republic of Indonesia for supporting the IOAA event, Institut Teknologi Bandung and Het Leids Kerkhoven-Bosscha-Fonds for providing travel support.

The COSPAR Capacity Building Programme:concepts, objectives, experience, future

Carlos Gabriel[1], Peter Willmore[2], Mariano Méndez[3], Pierre-Philippe Mathieu[4],
Joachim Vogt[5]
[1] - *ESA/ESAC, Spain*
[2] - *University of Birmingham, UK*
[3] - *University of Groningen, the Netherlands*
[4] - *ESA/ESRIN, Italy*
[5] - *Jacobs University Bremen, Germany*

Abstract. The COSPAR Capacity Building Workshops have been conceived to meet the following objectives:

• to increase knowledge and use of public archives of space data in order both to broaden the scope of research programmes in developing countries and also to ensure that scientists in those countries are aware of the full range of facilities that are available to them,

• to provide highly-practical instruction in the use of these archives and the associated publicly-available software, and

• to foster personal links between participants and experienced scientists attending the workshops to contribute to reducing the isolation often experienced by scientists in developing countries.

Since 2001 a total of nine workshops have been successfully held in different scientific areas (X-ray astronomy, Space Optical and UV Astronomy, Magnetospheric Physics, Space Oceanography and Planetary Science) in nine developing countries (Brazil, India, China, South Africa, Morocco, Romania, Uruguay, Egypt and Malaysia). Last year COSPAR started a fellowship programme to enable young scientists who have participated in a Workshop to build on skills gained there, by carrying out a 2-4 week research project at one of the several collaborating institutes.

We will discuss the modalities of the workshops, the so-far gained experience, and the future including collaborations with other institutions sharing the aim of increasing the scientific activities in developing countries.

Outreach, heritage and innovation

Julieta Norma Fierro Gossman *Instituto de Astronomia, UNAM*
Abstract. It is easier to build knowledge and develop skills when they make sense to us. Many nations have an extraordinary heritage and must make use of it to convey knowledge using elements such as national pride. In order to run outreach programs funding is fundamental. Using the media in fresh ways is a great tool to innovate ways for the general public to appreciate astronomy.

The South African National Astrophysics and Space Science Programme (NASSP)

Patricia Ann Whitelock
1 - South African Astronomical Observatory
2 - Astronomy Department University of Cape Town
Abstract. South Africa has some major astronomical facilities, in-use and on the drawing board, but a very small indigenous astronomy community. In 2002 the national observatories and the major universities combined their resources to start a unique collaborative training programme. This takes students with a BSc in physics or related discipline and prepares them to do a PhD in astrophysics. It has attracted students from all over Africa and most recently African Americans through a collaboration with the USA National Society of Black Physicists. This presentation will describe NASSP, it successes, its aspirations and its challenges.

Past, present and future of graduate astronomy education in Mexico

Christine Allen
UNAM, Mexico
Abstract. A brief historical review of the graduate astronomy program at UNAM is presented. The strengths and weaknesses of the present program are assessed and a new program, soon to be implemented, is presented. Experiences from the graduate astronomy programs at other Mexican institutions are also discussed.

Early approach to Astronomy and the digital gap.

Andrea Sanchez Saldias[1], Renzo Valentini[2], Herbert Cucurullo[2], Mariana Martinez[3]
1 - Departamento de Astronomia, Facultad de Ciencias, Uruguay

2 - Educacion Secundaria, Uruguay
3 - Observatorio Astronomico Los Molinos, Montevideo, Uruguay

Abstract. Uruguay has had the tradition of having Astronomy as a curricular discipline in High-school for the past 100 years. In 2009, together with the International Year of Astronomy, topics of Astronomy will be added to the Primary School's national curriculum, from kindergarden to 6th grade (3 to 11 year-old-children). We believe this conjunctural situation is a unique oportunity to work with teachers, give them update courses and perform follow-up activities about the success of the different enterprises, considering both the results from Montevideo (the country's capital city) and the provinces. In addition, the Plan CEIBAL (www.ceibal.edu.uy/) was implemented in Uruguay in 2007, under the motto 'one laptop per child', in an effort to attenuate the digital gap between children who could access to computes in their scholls and/or homes and those who couldn't. In 2009 every child attending to a state school possesses one laptop especially designed for Plan CEIBAL. In this work, and given the particular situations in the Education of Astronomy previously outlined, and through the corner stone Project of the IYA 2009 UNAWE-Uruguay (www.unawe.edu.uy), the multicultural origins of modern astronomy are illustrated in an international effort to enrich children's minds, awaken their curiosity in science and stimulate global citizenship and tolerance. Games, songs, hands-on activities, comics and live trades on the Internet are devised for children of age 4 and up in a coordinated manner with other UNAWE communities around the world.

UNAWE-Uruguay is aimed at the exchange of ideas and materials through the establishment of a network devoted to teacher's update and interdisciplinary workshops.

Astronomical Education in Tajikistan. Present and Future

Ibadinov Khursand

Institute of Astrophysics of Tajik Academy of Sciences, Tajikistan

Abstract. The centre of astronomy in Tajikistan is the Institute of Astrophysics of the Academy of Sciences of Tajikistan. This Institute makes a valuable contribution to science and training astronomical personnel and in astronomical education. Now in Tajikistan the reform of education continues. Now astronomy is studied in schools together with physics. It creates the certain difficulties in teaching astronomy at schools. The astronomy at schools is necessary for studying as an independent subject. At physical and mathematical faculties of universities the astronomy, but in various volumes also is studied. From 1999 it is most full astronomy and astronomical objects are studied at the Tajik National University (TNU) at specialty Astronomy. In 2008 TNU are introduced degree Baccalaureate, Specialist and Magisterial. At the Tajik State Pedagogical University in 2007 are restored training teachers in the specialty Physicist and astronomy. At universities also there are problems with teaching astronomy and astronomical subjects. At universities of Tajikistan there is no independent astronomical base (observatory), little the educational literature in a state language (Tajik). The level of knowledge of the pupils acting in universities on physical and mathematical faculty is low. Therefore it is necessary to undertake additional measures. In 2006 is restored the Small Academy of Sciences (SAS) of Tajikistan. At the Institute of Astrophysics of Tajik Academy of Sciences acts the section Astrophysics SAS for scholars. In Khujand city acted the Planetarium. In 2006 the Institute of Astrophysics, TNU and the Astronomical Society of Tajikistan, with the support IBSP of the UNESCO, on the base of Hisar astronomical observatory organized Training Methodical Center (TMC) Tajastro on astronomy. The listeners of center are students, graduate students, young scientists of Institute of Astrophysics and universities, teachers of physics and astronomy of secondary schools. It is necessary develop this TMC and organize Training-Methodical-Centre for Central Asian countries on the base Hisar astronomical observatory.

Progress of the Southeast Asia Astronomy Network (SEAAN)

Busaba Hutawarakorn Kramer[1,2], Hakim L. Malasan[3], Boonrucksar Soonthornthum[2]

1 - Max-Planck-Institute for Radioastronomy
2 - National Astronomical Research Institute of Thailand
3 - Institut Teknologi Bandung, Indonesia

Abstract. The initative of the establishment of the Southeast Asia Astronomy Network (SEAAN), was presented in the Special Session 5: Astronomy for the Developing World during the IAU XXVIth General Assembly. The aims are to establish effective mechanisms for nurturing and sharing the development and experiences in astronomy research and education among Southeast Asian countries. The first meeting of the Southeast Asia Astronomy Network (SEAAN2007) was held during the Thai National Astronomy Meeting (TNAM2007), 22-24 March 2007, hosted by NARIT and Ministry of Science and Technology, Thailand. We report here on the progress of the network and future plans.

The Brazilian Olympiad of Astronomy and Astrophysics and the International Year of Astronomy

Joao Batista Garcia Canalle[1], Jaime Fernando Villas da Rocha[2]

1 UERJ, Brazil
2 - MAST/MCT, Brazil

Abstract. The Brazilian Olympiad of Astronomy and Astrophysics (OBA in Portuguese) is the biggest program in the South hemisphere of popularization of Astronomy and Astrophysics. It begun in 1998 and in 2009 we are organizing the XII OBA. In the last year took part in the XI OBA 437.000 students belonging to all the Brazilian states and involved 5.535 schools. The final number of participants in the XII OBA will be presented in this panel. But the main point here is just to present the relations between the OBA and the International Year of Astronomy (IYA). From the federal government it was possible to get twice the normal money that we spend annually. So we prepared a huge divulgation of the XII OBA between all the schools of Brazil, but the media also helped us in these efforts, so that almost 10.000 schools begun to participate in the Brazilian Olympiad in 2009, which summed with the 15 thousand that was already taking part in the OBA, in these year we will have 25.000 schools in the event. We prepared some didactic and entertaining material to the students. We sent to each schools, for example, a nice puzzle with a beautiful image of the solar system; a stellar watch and the complete orientation on how to reproduce it in large scale and how to use it; the complete instruction how to construct a solar watch; a domino using planetary images. We initiated a program to detect meteorites using the help of students to find them. We sent to each school a small star with powder of meteorites inside it. We sent annually to all schools of the OBA some practical challenge. In this year we asked them to determine the mass of the Earth (using the mass of the Moon as unity). We asked them also to make models of spheres to represent the planets and the Sun on the same scale just to see the proportion between planets and the Sun. We will give them a huge yellow ball to represent the Sun. As in 2009 we can celebrate the first Moon landing we asked them to launch a rocket to the sky using very simple materials. We also organized a contest of draws between the students. We are buying galileoscopes to give them to each school that is taking part in the XII OBA. It is probable that almost 1.000.000 of Brazilian students will take part in this event. Finally, on last October, in Montevideo with the presence of astronomers representing many countries of South and Central America, and Mexico we founded the Latin American Astronomy and Astrophysics Olympiad. The first one will take place in Brazil October 2009.

Education of Astronomy in Argentina: a Global Vision and the study of special cases of no-formal education

Beatriz Garcia

1 - Consejo Nacional de Investigaciones Cientificas y Tecnicas, Argentina

2 - UTN - Facultad Regional Mendoza, Argentina

Abstract. With its foundations in Mathematics and Physics, Astronomy is a scientific discipline that cuts transversally all others: is possible to establish links with Biological Sciences, Social Sciences, Earth Sciences, and Technology. In Argentina, the teaching of professional Astronomy has less than a century. With two traditional schools (Cordoba and La Plata), and a recently one (San Juan), the country has only 175 professional astronomers recognized, and around 30 PhD students. Despite this small number of scientists, Astronomy in Argentina has produced remarkable discoveries. By contrast, in middle and primary levels of education, Astronomy opens a way with great difficulty, and there are only a few activities in the classroom, related to its contents. However, some experiences in the field of non-formal education, but developed by professional scientists, give results that are even surprising. This proves, in our view, there is no better diffuser of a discipline that one for which it is part of their daily activity, and ends with the myth that, in general, and with some exceptions, scientists are poor communicators. This presentation describes the current state of Education in Argentina, from a historical perspective, with a projection towards the new challenges, taking into account the capabilities and needs of the country, in terms of human resource training, and will show a series of experiences, developed in Universities and Institutions devoted to Astronomy and Astrophysics research, with a significant social impact.

Learning astronomy as a lifetime experience

Magda Stavinschi

Astronomical Institute of the Romanian Academy

Abstract. The main goal of this special session is to know how the Rate of Astronomical Discovery can be accelerated. Is it really possible? The last centuries elapsed since Galileo and especially the last decades of incredible progress of space and ground based astronomy seem to indicate a positive answer. However, there is a factor which can not only stop the speed at which the universe can be known and discovered but which can even lead to a dramatic regress through false research interpretations, by overlooking the possible implications of these discoveries for humanity and last but not least, by affecting or even destroying humanity. Naturally, it is education. As long as the young people are not systematically informed about the universe, do not know the stages when it was discovered and do not know their own astronomical history, they do not know how to preserve their astronomical heritage, to interpret the knowledge they have, not only for the progress of astronomy, but also for that of science in general. Thus, not only the rate of astronomical discovery, but even the very future of mankind might come to be questioned.

Coining Sign Language for Astronomy and Space Science Terminology

Johnny Cova S.

Centro de Investigaciones de Astronomia (CIDA), Venezuela

Abstract. Teaching science to school children with hearing deficiency and impairment can be a rewarding and valuable experience for both teacher and student, and necessary to society as a whole in order to reduce the discriminative policies in the formal educational system.

The one most important obstacle to the teaching of science to deaf or hearing impaired students is the lack of vocabulary in sign language to express the precise concepts encountered in cientific endeavor. In an ongoing collaborative project between Centro de Investigaciones de Astronomia "Francisco J. Duarte" (CIDA), Universidad de Los Andes (ULA) and Unidad Educativa Especial Bolivariana de Maturin (UEEBM) initiated in 2006, we have attempted to fill this gap by coining new signs for the terminology of astronomy and space sciences. During five two-day workshops with deaf or hearing impaired students a total of 296 concepts of mathematics, astronomy and space sciences were coined into sign language using an interactive method which we describe in the text. These new signs have been made available to the Deaf Community through the interactive web page www.cienciaensenas.org for their use and evaluation. The goal of the project is to incorporate these terms into Venezuelan Sign Language (LSV) and provide the vocabulary for the communication of science among the Deaf Community.

Bringing the Universe to a confined world: Astronomy in prison

Danielle Briot

Observatoire de Paris, 61 avenue de l'Observatoire, 75014 Paris, Fance

email: danielle.briot@obspm.fr

Abstract. Since more than a decade, some french astronomers go in prisons to talk and discuss about astronomy with the prisoners. The words "astronomy" and "prison" placed side by side represent at least two paradoxes: first, it is astonishing that people with a very distressing past and who live in very hard conditions are interested in something which corresponds to pure knowledge and without any direct utility; secondly, the aim of astronomy is to open the world in all its infinite totality whereas, in prisons, the audience is locked and confined as much as possible. However, the audience shows a strong interest, as it can be seen by the size of the audience and the numerous questions asked. Many reasons exist to go in prisons to speak and discuss about astronomy. We quote mainly: it is important for prisoners who sometimes could have a feeling that they are abandoned by the society, that professional scientists move for them; it is interesting to take advantage of a special time in the life of these people so that they meet and discuss with scientists whom they would rarely have the opportunity to meet outside; this prison time can also turned to account by the discovery of a new field of knowledge or the extension of already acquired knowledge; the discovery of a new pure and disinsterested knowledge can help to bear the prison regime which could be very hard and distressing, physically as well as mentally; as for any popularisation talk, it is the duty of astronomer to give people an account of tax money; the last but not the least, we have to not neglect the pleasure for the speaker to talk about a subject of which we are fond in front of interested people. More we give to the prisoners information to think, more they will have some tools to choose their way and become reintegrated in society after they came out of prison. We give some points of information about features of audiences: as in any lecture, the audience is very heterogeneous, however rather surprisingly a certain number of prisoners use their prison time to learn and think. We give some information about prisons in France. It is sometimes difficult to obtain authorization to enter in prison, that corresponds to a third paradox: to come into prison is complicated. The present prison overpopulation rate complicates even more the process, due to the extra work of warders implied by this overpopulation. We briefly review what is done to day in France in this domain: several initiatives exist in the Paris region concerning various prisons and a very interesting and active work is done in the South-West in France. As a conclusion, we do insist on the importance to bring culture and knowledge to some people who could feel to be rejected by the society, and by this way, to give them some tools and elements to think, to light the way, and finally to help them to take the good decisions after prison time.

Learning astronomy as a lifetime experience

Rosa Maria Ros

Applied Mathematica 4, Technical University of Catalonia

Abstract. It is well known that all of us are learning throughout our lives. This includes learning languages, learning how to use computers, learning to dance... but normally people never think that we could also include learning about astronomy. Why not? Astronomy is in the very basis of our lives and astronomers are taking giant steps forward, as a consequence it is important that our society knows more and more about astronomy and new challenges in this field.

From primary school pupils to retired people, everybody is interested in astronomy and can be attracted to this area of science if the kind of presentation is appropriate. Of course lectures and videos are a few simple ways of involving people in astronomy, but it is necessary to think of a set of different possibilities in keeping with the kind of public who will receive the presentation. A set of tales and plays to children can be more attractive than only using a film. Children love to take part in animated activities, moving their bodies and using their imagination. Competitions and challenging ways are appropriate for teenagers and university students and may be some excursions can be a good idea to get retired people involved in astronomy. A set of examples of these experiences will be taken into account in this presentation.

Astronomy for African Development: A Review of Activities

Kevin Govender

1 - South African Astronomical Observatory
2 - Southern African Large Telescope

Abstract. This presentation will review activities dealing with both the development of astronomy in Africa and the utilisation of astronomy for African development. To address the former, the progress on the development plan for astronomy in Africa will be discussed. The latter will focus mainly on the progress in the case of the SALT (Southern African Large Telescope) Collateral Benefits Programme (SCBP) which was set up to ensure African societal benefit from astronomy.

Astronomy in Equatorial Africa

Johnson Ozoemenam Urama

Department of Physics and Astronomy, University of Nigeria, Nsukka

Abstract. Africa lies astride the equator and extends almost equal distances ($\sim 37\,$deg) north and south of the equator. Equatorial Africa, here, refers to those countries in the continent that lie within about 20 degrees north and south of the equator. Nearly 80% of the countries in Africa lie in this region. Not much is known about modern astronomy here, and there is no astronomical research facility, of any type, in this region. However, their ancient architecture, folklore, myths, religion, calendar, etc. are quite rich in astronomy. One of the ways of creating awareness and interest in astronomy in this region is by unearthing the body of traditional knowledge of astronomy possessed by peoples of the different ethnic groups in equatorial Africa and to bridge the gap between this cultural astronomy and modern astronomy by providing scientific interpretation to deserving cosmogonies and ancient astronomical practices. Here, we discuss the prospects and challenges for popularizing astronomy in this region through cultural astronomy and other methods.

Ethnoastronomy: a fancy subject or a non-western epistemological breakthrough?

Luiz Carlos Jafelice

UFRN, Brazil

Abstract. Non-western views of the skies are inevitably an ethnocentric designation, not because of the "non-western" adjective, but mainly because of the assumption that cultures which are not closely related to the western one have "views of the skies", as if "sky" was in fact "something apart from earth" - and, by extension, "from life" - as westerns are convinced it is. As the separation between scientific and humanities cultures keeps widening, most astronomers and - what is the most worrying - most astronomy teachers assume that "the importance of non-western views of the skies for astronomy teaching in both developing and developed countries" are meant to supposedly supply cultural illustration, but not to introduce other significant and original epistemological contributions - as if, at the end, all world views had better reduce to the western one. We argue that the main importance of those generic "non-western views of the skies" for astronomy teaching, wherever it takes place, should be to deconstruct the conception that the scientific thought is epistemologically privileged and to contribute for an honest support to cultural and epistemological diversities. One notices, however, that pedagogical approaches usually adopted strengthen the wishful thinking of an ontological superiority of the scientific epistemology and regards all other human possibilities of knowledge construction amusing and culturally interesting, although useless in the modern technological world. Such a posture contributes: to widen the separation between those two cultures; to reinforce cultural prejudices; and to drive people away from science. That way, the science teaching programme efficiently helps to maintain everything the way it is - including all prejudice and exclusion procedures. The western thinking values the cognitive approach to knowledge matters and interprets any human psyche attribute other than reason - like affection, intuition, etc. - as a hindrance to the process of conquering the objective knowledge. Our ethnographic research shows that those other psyche attributes are the essentials of the epistemological process as a whole, particularly that concerning autochthon or traditional knowledge. The anthropological and biological approaches to the knowledge process demand a complete revision of our concept of epistemology. We present educational interventions we have developed and applied in Carnaúba dos Dantas (RN, Brazil) aiming: 1) the integration between generations - the one of older members of the community, formed by experts on traditional knowledge, the one of teachers, and the one of youngsters; and 2) the inclusion of traditional environmental knowledge - naturally including those which our culture segregates as belonging to the astronomy domain - into everyday school activities. We further present some guidelines on procedures for those interested in introducing a more humanitarian world view through the inclusion of local traditional knowledge to environmental education and generation integration, which can truly favour a comprehension and an authentic acceptance of diversities, including epistemological ones. (This work is supported by the Brazilian agency CNPq.)

Hawaii Students' Authentic Astronomy Research Projects

Mary Ann Kadooka, Michael Nassir, James Armstrong

University of Hawaii - Institute for Astronomy

Abstract. Since 2007, HI STAR, the Hawaii Student Teacher Astronomy Research, residential summer program has been striving to equip 12-17 year old students with the necessary research skills and background to conduct original research projects. Students with a passion for astronomy have been recruited through mini-workshops conducted on five out of six islands since 2005.

For one week in June, the students with their teachers have thrived on physics and astronomy lectures and image processing, photometry, and light curve lessons. They also learn to do remote observing with the 2 meter Faulkes Telescope located on Haleakala, Maui and the 16 inch DeKalb Observatory Telescope in Auburn, Indiana. The HI STAR participants then work in groups with inspiring astronomer mentors who have developed projects on comets, asteroids, galaxies, nebulae, variable stars and extrasolar planets using image data sets. Our students continue to be supported by their mentors so their projects are entered in Science Fairs the following year. Our major goal of encouraging students to have college majors in science, technology, engineering and mathematics fields is slowly being accomplished. Key to this endeavor is continuous mentoring even when they become college students. The favorable outcomes of our program are attributed to our network of enthusiastic astronomer mentors, committed teacher advisers and motivated students.

Guided School Visits to the Astronomy Museum in Brazil: new approaches

Flávia Requeijo[1], Cecília Maria Pinto do Nascimento[2], Andréa Fernandes Costa[2], Maria das Mercês Navarro Vasconcellos[3]

1 - Instituto de Geociências, Universidade Estadual de Campinas, Hawai
2 - MAST/MCT
3 - Museu da Vida

Abstract. Museums and schools are distinct learning environments, but can act together to expand their educational opportunities. In Brazil, while schools have a daily presence in our lives for at least 10 years, the museums are rarely visited. We propose that museums and schools should work together in a partnership, so that a school's visit to a museum doesn't become a single, one-time ocurrence. This work presents a study of the recently conceived guided tour "Where Do We Live?", which includes elements of Astronomy and its history, and discusses a few relationships between science and society. In order to elaborate the tour, or "trail" as we call it, a few exhibits from the museum were selected according to goals previously set by a team of educators from the Museu de Astronomia (MAST). Also, the communication pattern during the tour, which used to be rather unidirectional (guide-to-visitor), is now based on group discussions mediated by the guide. Students' reactions during the visit to the MAST's scale model of the Solar System suggest that the discussion promoted therein motivates a deep reflection on our place in the universe and on the sustainability of life on Earth. Results from questionnaires applied to teachers indicate that our suggested activities are frequently performed at the school before the visit. This, we believe, profoundly enhances the students' experience at the museum. Also, it suggests that a partnership between our museum and the schools is actually taking place, and that our planned school visit may be more than just a "leisurely field trip", but an event capable of generating reflection before and after it occurs. Thus, we emphasize Astronomy's large potencial to the construction of a more critical view of our actions and their consequences over planet Earth.

Astronomy Education and Research in Colombia: Achievements and Challenges for the Next Decade

Juan Rafael Martinez Galarza
Leiden Observatory, The Netherlands

Abstract. Despite its human, geographical and historical advantages, Colombia remains today a country that does not produce significant astronomical research. However, a continuously growing professional community, most of them part of a scientific diaspora, supported by an active amateur community, has started an ambitious program to transform Colombia into an astronomically active nation, with specific goals to achieve in the coming 10 years. I describe the current status of astronomical research and education at all levels in the country, the recent history of its astronomical society, the educational achievements (including the first undergraduate program in astronomy), and the proposed projects which will lead the future development of this science in the country. I emphasize the need for suitable astronomical infrastructure as a mean to educate the next generation of Colombian professional astronomers, and the role that the international astronomical community could play by supporting these projects.

The impact of IAU educational programs on Peruvian astronomy

Ivan Ramirez

Max Planck Institute for Astrophysics, Garching, Germany

Abstract. By year 1996 there was only one professional astronomer born in Peru, Maria Luisa Aguilar, who was also the first Peruvian IAU member. In 1997 the number of Peruvians holding a doctoral degree in astronomy was only 2. Today, the number has increased to about 10, with possibly another 10 being added to the list within the next 5 years. This exponential increase has been in part due to educational programs by the IAU. I will summarize the status of astronomers from Peru, emphasizing the impact of IAU programs on their professional careers and provide a suggestion to encourage the return of those who are currently working elsewhere.

Astronomers Without Borders: Worldwide Connections through Astronomy

Mike Simmons

Astronomers Without Borders

Abstract. Astronomers Without Borders (AWB) is a new global organizational that furthers understanding and goodwill across national and cultural boundaries using the universal appeal of astronomy, a common language spoken by all those who share an interest in the sky. A growing network of Affiliate organizations brings together clubs, magazines and other organizations involved in astronomy and space science. Forums, galleries, video conferences and more interactive technologies are used to connect participants from around the world. Projects include Sharing Telescopes and Resources (STAR), which gathers telescopes and other resources in developed countries and donates them to clubs in undeveloped countries. The World at Night (TWAN), an IYA2009 Special Project, has built a team of specialty photographers who create wide-angle images of the night sky in important natural and historic settings around the world that dramatically demonstrate the universal nature and appeal of the night sky. AWB has a lead role in organizing the IYA2009 cornerstone project and that key project's legacy will be carried on AWB through IYA2009 and beyond.

Ten years of the first mobile science program in Brazil

Horacio Dottori[1] & Basílio Santiago[1]

[1] *Departamento de Astronomia, Universidade Federal do Rio Grande do Sul, CEP: 91501-970, Porto Alegre, Brazil*

email: `santiago@if.ufrgs.br`

Abstract. Observatório Educativo Itinerante (Mobile Educational Observatory, OEI) is an educational program meant to provide background to elementary and high school teachers who want to use Astronomy to enhance their teaching skills in sciences in general, most specially Physics, Maths, Geography and Chemistry. OEI is the first mobile science program in Brazil, having started its activities in June 1999. It is therefore completing ten years on the road. During this time, it has taken Astronomy to about 1300 school teachers in more than 40 locations in the South of Brazil. The teachers courses last from 20 hours to 180 hours and are usually taken on weekends. An emphasis is given on simple experiments, which can be reproduced with little cost, and in astronomical observation. OEI also visits schools and other public places offering lectures and stargazing. The program makes frequent use of information technology, including public applets, desktop simulations, and web based educational material (www.if.ufrgs.br/oei).

Need and possibilities of astronomy teaching in school

Irma Talvikki Hannula
University of Helsinki, Finland

Abstract. The purpose of the work is to create a research-based foundation for planning the structure, content and methods of astronomy teaching in the Finnish comprehensive school. At first, a critical analysis of the significance of astronomy teaching from the point of view of the educational aims was made, in order to verify the need of it and to find significance factors, which would offer a basis for defining principles of astronomy teaching. The significance of astronomy teaching is defined to consist of all such factors through which astronomy teaching can promote the educational aims of school. Secondly, the conceptual structure of astronomy, which forms the core of teaching, was analysed in view of extracting principles for astronomy teaching. Construction of the pupil's world picture requires that teaching must follow the line of 'creation of meanings' in the development of astronomy. Thirdly, the preconditions for astronomy education were searched by a query sent to teachers, plus overviews of present textbooks, curricula and teacher training programmes, supported by own experiences from over twenty years at school, from teacher education and international work in preparation of recommendations for astronomy teaching. Finally, the principles resulting from the studies of significance and conceptual structure were gathered and concretised into a suggestion for planning guidelines of astronomy teaching.

Hands On Universe Education In China

Hongfeng Guo
National Astronomical Observatories, Chinese Academy of Sciences, Beijing 100012, China
Abstract. This presentation will give a brief introduction on the status and demands of astronomical education in primary and middle schools, the application of Hands On Universe education program in China and the recent developments of China Hands On Universe education projects. The Galileo Teacher Training Program in China will be also mentioned in this presentation.

Astronomy Teacher Training for Elementary Schools: a Multidisciplinary Approach

Gustavo de Araujo Rojas, Adilson J. A. de Oliveira, Alexandra Bujokas Siqueira
Universidade Federal de São Carlos, UFSCar, Brazil

Abstract. We report the results of a two-week teacher training course held at Universidade Federal de São Carlos, Brazil. The course focused on using artistic and Web 2.0 tools to share knowledge about Astronomy among both teachers and pupils. During the first week, the students (19 Physics teachers from Primary, Secondary and Undergraduate levels) attended a series of lectures where topics such as World Models, Earth-Sun-Moon System, Stellar Evolution, and Large-Scale Structure of the Universe were discussed. The second week consisted of a five-day workshop where four abilities were worked out:

1) how to edit a blog; 2) how to produce a webvideo using classical narrative structures and some professional screenplay techniques; 3) how to share pictures and create social bookmarking; and 4) how to play a Role Play Game themed on Astronomy.

Those tools were selected because they support the pedagogical approach of multiliteracies, which means to be able to select content about a specific theme (Astronomy in this particular experience) and shape that content in order to share it with a wider community, using different languages. After the workshops, the teacher's production was evaluated according to three categories: quality of content produced; ability to shape information about Astronomy; and motivation. The results suggest that the group was able to select meaningful topics and related trustful information, but they were not still prepared to make good use of the interactive tools. Thus, we conclude that learning about how digital media language and Web 2.0 tools work also must be taught on teacher training curriculum, in order to innovate on Astronomy teaching and learning.

GLOBE at Night and Citizen-Science: Providing an Effective Trigger for Astronomy and Light Pollution Education

Constance E. Walker[1]

[1] *National Optical Astronomy Observatory, 950 N. Cherry Ave., Tucson, AZ 85719 USA*
email: cwalker@noao.edu

Abstract. One of the most productive programs in the International Year of Astronomy's (IYA2009) Dark Skies Awareness Cornerstone Project has been GLOBE at Night. The GLOBE at Night program has endeavored to promote social awareness of the dark sky by getting the general public to measure light pollution and submit results on-line. During IYA2009 alone, over 15,700 measurements from 70 countries were contributed during the 2-week campaign period. That amount is twice the number of measurements on average from previous years. This included digital meter measurements, which were used to measure quantitatively the sky brightness in magnitude/square arcsecond. 73% of all measurements were from the U.S. (all 50 states). 900 measurements were from Chile. There were over 200 measurements each from the Czech Republic, Hungary and the United Kingdom. 13 other countries reported more than 100 measurements (Argentina, Australia, Canada, Colombia, Finland, Germany, Macedonia, Mexico, Poland, Romania, South Africa, Spain and Turkey).

The GLOBE at Night website explains clearly the simple-to-participate-in 5 step program and offers background information and interactive games on key concepts. Teacher (and Family) Guides come in 13 different languages. And to help advertise the campaign, there are downloadable postcards and flyers (in Spanish too). The report page is user-friendly and map page has data in various formats. The program has been expanded to include trainings of the general public, but especially educators in schools, museums and science centers, in unique ways. Education kits for Dark Skies Awareness have been distributed at these training workshops. The kit includes material for a light shielding demonstration, a digital "Sky Quality Meter" and "Dark Skies Ranger" activities. The activities are on how unshielded light wastes energy, how light pollution affects wildlife and how you can participate in a citizen-science star-hunt like GLOBE at Night. In addition, projects are being developed for what to do with the data once it is taken.

There were particularly spirited and creative GLOBE at Night campaigns around the world in 2009. One such "poster child" was carried out by 6500 students in northern Indiana. The students produced 3,391 GLOBE at Night measurements. To visualize the magnitudes of dark sky lost to light pollution, these students removed over 12,000 of the 35,000 stacked LEGO blocks that represented an ideal night sky across the school district.

This is a glimpse into the many accomplishments of GLOBE at Night this year and we will endeavor to continue the program beyond IYA2009. For more information on GLOBE at Night, visit `www.globeatnight.org`, and on any of the citizen-science star hunts, visit `www.darkskiesawareness.org`. This work was supported by a grant from the U.S. National Science Foundation (NSF) Astronomy Division. The U.S. National Optical Astronomy Observatory is operated by the Association of Universities for Research in Astronomy, Inc. under cooperative agreement with the U.S. NSF.

Solar eclipses and the International Year of Astronomy

Jay M. Pasachoff

Abstract. Abstract missing

The IAU Strategic Plan for Development of Astronomy

George Miley

Leiden Observatory, the Netherlands

Abstract. A short presentation of the IAU decadal strategic plan has already been given in SpS2 p 604.

Closing remarks

Jean-Pierre De Greve

Vrije Universiteit Brussel, BE1050 Brussels, Belgium

email: jpdgreve@vub.ac.be

There is no way that I could give you a resume with highlights of this meeting. For that, it was too diverse and too rich. But then, what did we learn during this meeting? Let me tell you what I learnt.

First of all, both the talks, but certainly the posters, showed how much dedicated effort astronomers give to astronomy education throughout the world. These different experiences with different local contextual settings must give me enough food for thought to improve on my own efforts. By-the-way, the posters of Special Session 4 are the only ones that can remain on display during the whole two weeks of the GA.

Secondly, the large variety, but also the large amount of astronomers involved, gives me strength to continue what I'm doing. I'm not alone. There is a whole community of astronomers working on that double objective: using astronomy education to gain interest for science and to make this world a better world.

Thirdly, both posters and presentations convinced me that we should use each others initiatives, products and competences more than we do today, because sometimes we re-invent the wheel, or spent too much time on something that could be done more efficient with input from another initiative. So, stronger networking in Astronomy education seems to me a must.

And last, as was indicated by Irma Hannula, there is not enough in-depth research on astronomy education in the educational systems.

If we want to convince ministries, administrations and the educational world as a whole of the benefits of astronomy education, we have to come up with studies that show that astronomy education does work to reach the earlier mentioned objectives: gaining interest in science, and contributing to a better world.

I very heartedly thank all participants for their contributions, posters, presentations, comments and questions. It made this a lively session. With that I close Special Session 4.

SpS4 - Posters

Adapting high-tech into simple classroom activities

Adrielli E. O. Pereira[1], Ana P. A. Benassi[1], Maria Clara Amon[1], José Osvaldo de Souza[2], Vera Jatenco-Pereira[2]

[1] *Escola Estadual Patriarca da Independência, Vinhedo, SP, Brazil* [2] *Universidade de São Paulo*

First Steps to Prepare Radioastronomers in Uzbekistan

Alisher S.Hojaev[1],[2]

Present address: Astronomy Dept., NUUz,Tashkent, 700174, Uzbekistan.
[1] *Department of Astronomy, National University of Uzbekistan, Vuzgorodok, Tashkent, 700174, Uzbekistan*
[2] *RT-70 Radioobservatory, Uzbek Academy of Sciences, Tashkent, Uzbekistan email: ash(at)astrin.uzsci.net*

Exploring the Stars: Astronomy Outreach at The University of Western Ontario

Alyssa M Gilbert, Margaret D Campbell-Brown
The University of Western Ontario, USA

Power of not knowing

Aniket Sule
Homi Bhabha Centre for Science Education, India

The Scientific Work Done By The 30 inch Reynolds Reflector At Helwan

Ashraf Ahmed Shaker
National Research Institute of Astronomy and Geophysics (NRIAG), Helwan, Cairo, Egypt

Planetario Malargüe, a bridge between knowledge, current research and scientific development withthe general audience

Beatriz Garcia[1],[2], Marisa Marañon[3], Andres Risi[3], Roberto Bandiera[3]
1 - Consejo Nacional de Investigaciones Cientificas y Tecnicas, Argentina
2 - UTN - Facultad Regional Mendoza, Argentina
3 - Planetario Malargüe

Astronomical facts in movies: rigurosity, speculation and fiction

Beatriz E. García[1],[2], Estela M. Reynoso[3],[4]
1 - Consejo Nacional de Investigaciones Cientificas y Tecnicas, Argentina
2 - UTN - Facultad Regional Mendoza, Argentina
3 - Institute for Astronomy and Space Physics, Argentina
4 - Departamento de Física, FCEN-Universidad de Buenos Aires

The use of Photometric Techniques in Teaching Science Projects

Bruna F. dos Santos[1], Cíntia P. Tomaz[1], Emerson L. Medeiros[1], Maria Clara I. Amon[1], Osvaldo de Souza[2], Jane Gregorio-Hetem[2]
[1] *Escola Estadual Patriarca da Independência, Vinhedo, SP, Brazil* [2] *Universidadede São Paulo*

Astrophysics in Burkina Faso

Claude Carignan[1,2], *Luc Turbide*[1] & *Jean Koulidiati*[2]
it[1] *Département de Physique, Université de Montréal, C. P. 6128, Succ. Centre-ville, Montréal, Qc, CANADA H3C 3J7 email: claude.carignan@umontreal.ca*
[2] *Observatoire d'Astrophysique de l'Université de Ouagadougou, UFR/SEA, BP 7021, Ouagadougou, Burkina Faso*

Astronomy Virtual Exposition at University of Brasilia's Virtual Museum

Carlos Eduardo Quintanilha[1], *Jose Leonardo Ferreira*[1], *Gilberto Lacerda Santos*[1], *Cassio Costa Laranjeiras*[2]
1 - Institute of Physics University of Brasilia, 2 - IF/UnB, Brazil

A High School science project: looking for Milky Way doubles in SDSS

Caroline Soares Mello , Danilo Macedo da Silva , Walter Augusto Santos , Laerte Sodré Jr. IAG-USP, University of São Paulo, Brazil

Itinerant Planetarium: A Cultural Activity in the Realm of Astronomy Diffusion and Popularization.

Cássio Costa Laranjeiras[1,2], *Carlos Eduardo Quintanilha*[1,3], *Marina Leite da Silveira*[1], *Robson Evangelista*[1], *Camila Jéssica Letti*[1], *Victor Souza Magalhães*[1], *Ana Carolina Salvatti*[1], *Davi Araújo Quaresma Lemos*[1], *Murilo Timo Neto*[1], *Décio Cardozo Mourão*[1], *Gabriela Cunha Possa*[1], *José Leonardo Ferreira*[1], *Rayssa Bruzaca de Andrade*[1], *Leila Lobato Graef*[1], *Adriana Eliza Correa*[1]
1 - Institute of Physics, University of Brasilia, 2 - Museum of Science and Technology of Brasilia, 3 - Brazilian Space Agency, Brazil

Building the bridge between astronomy research and education with SOFIA

Cecilia Scorza
Max Planck Institute for Astronomy, Germany

Hypathia of Alexandria

Doina George Ionescu
1 - Astronomical Institute of the Romanian Academy, Romania, 2 - Arkansas Space Center, USA

Astronomy Outreach at Outdoor Music, Movie, or Theatre Events

Donald Lubowich
Department of Physics and Astronomy, Hofstra University, USA

Motivating the public about the usefulness of scientific thinking

Enrique Vazquez-Semadeni
1 - Universidad Nacional Autonoma de Mexico, 2 - Centro de Radioastronomía y Astrofísica, Universidad Nacional Autónoma de México

PInE: an easy-to-use astronomical image processing software for the Telescópios na Escola project

Evandro Luquini[1], *Alberto Krone-Martins*[2,3], *Laerte Sodré Jr.*[2], *Osvaldo de Souza*[2]
1 - Depto. de Engenharia Elétrica, Universidade Presbiteriana Mackenzie, 2 - IAG-USP, University of São Paulo, Brazil, 3 - Observatoire de Bordeaux-France

Children of the Sun, Children of the Moon - An Exhibition on the Native

Flavia Pedroza Lima
1 - Fundação Planetário da Cidade do Rio de Janeiro
2 - Instituto de Geociências, Universidade Estadual de Campinas, Brazil

Be an Astronomer for a week-end

Gonzalo Tancredi[1,2], *Herbert Cucurullo*[1], *Sebastian Bruzzone*[1], *Santiago Roland*[1], *Raul Salvo*[1], *Mariana Martinez*[1]
1 - Observatorio Astronomico Los Molinos, MEC, Uruguay

2 - Departamento de Astronomia, Facultad de Ciencias, Uruguay

100 Hour Certificate Course in Astronomy and Astrophysics
G.S.D. Babu
M.P.Birla Institute of Fundamental Research, 43/1 Race Course Road, Bangalore 560001, India.

The recent capabilities of the 74 inch Kottamia telescope of Egypt
Hamed Abdelhamid Ismail
Zentrum fur Astronomie, Universitat Heidelberg, Germany

Modern Astronomy Education
Irina I. Kumkova, Veniamin ladimirovich Vityazev
Sobolev Astronomical Institute, Saint-Petersburg State University, Russia

Chankillo (Peru): Astronomy and society in Americas' earliest known solar observatory
Ivan Augusto Ghezzi[1,2], Clive Ruggles[3]
1 - Pontificia Universidad Catolica del Peru, 2 - Yale University, USA, 3 - University of Leicester, UK

Program Offerings in Astronomy in the Philippines
J.R.F. Torres[1] & M.B.N. Kouwenhoven[1,2,3]
[1] Department of Astronomy, Rizal Technological University, the Philippines
email: jrftorres2002@yahoo.com
[2] The Kavli Institute for Astronomy and Astrophysics, Peking University, Yi He Yuan Lu 5, Hai Dian Qu, Beijing 100871, P.R. China
email: kouwenhoven@kiaa.pku.edu.cn
[3] Department of Physics and Astronomy, University of Sheffield, Hicks Building, Hounsfield Road, Sheffield S3 7RH, UK

The Colombian Undergraduate Program in Astronomy
Jorge I. Zuluaga
Instituto de Física - FCEN, Universidad de Antioquia, Clle.67 No.53-108, Medellín, Colombia
email: jorge.zuluaga@siu.udea.edu.co

Astronomy for Beginners
José Leonardo Ferreira[1], Victor de Souza Magalhães[1], Suzana de Souza e Almeida Silva[1], Thais Carneiro Oliveira[1], Carlos Eduardo Quintanilha[1], Décio Cardoso Mourão[2]
1 - IF/UnB, 2 - FEG/UNESP, Brazil

Star Week - A Successful Campaign in Japan
Junichi Watanabe
National Astronomical Observatory of Japan, Japan

Review of Education and Popularization of Astronomy in Korea
Kang Hwan Lee
Gwacheon National Science Museum, Korea

Using a Sundial to Teach Astronomy
Karina Pavanelli Mendes da Silva[1,2], Roberto Boczko[1,2]
1 - IAG-USP, University of São Paulo, Brazil,2 - Departamento de Astronomia - IAG/USP, Brazil

Kyiv Republican Planetarium and problems of astronomical education in Ukraine
Klim I. Churyumov[1], Olena V. Dirdovs'ka[2], Nataly S. Kovalenko[2], Anna V. Melnik[1]
1 - Kyiv Shevchenko National University, 2 - Kyiv Republican Planetarium, Ukraine

Stimulating Astronomy in Developing Regions - African Focus
Lolan Naicker, Kevin Govender
SAAO, South Africa

University students and school teachers understanding Moon Phases

Maria de Fátima O. Saraiva[1], Maria Helena Steffani[2,3]
1 - Universidade Federal do Rio Grande do Sul, Departamento de Astronomia, Brazil, 2 - Planetário José Baptista Pereira, Brazil, 3 - Universidade Federal do Rio Grande do Sul, Departamento de Física, Brazil

In search of Brazilian meteorites

Maria Elizabeth Zucolotto[1], João Batista Garcia Canalle[2]
1 - Museu Nacional/UFRJ, 2 UERJ, Brazil

A public-private partnership in Astronomy Education and Outreach: Building the "Haus der Astronomie" in Heidelberg

Markus H Poessel
Max Planck Institute for Astronomy, Heidelberg, Germany

Exploring the Solar System with a Human Orrery

Melanie A. Gendre, Peter Newbury
University of British Columbia, Canada

International Astronomical Search Collaboration In China

Minchen Zhou
National Astronomical Observatories, Chinese Academy of Sciences, Beijing 100012, China

Brazil, Uruguay and Argentina joint observation of March 2009 equinox

Néstor Camino[1,2], Alejandro Gangui[3,4], Andrea Sanchez[5], Maria Helena Steffani[6,7], Maria de Fátima Saraiva[8], Thais Cortellini Abrahão[9], Luciana da Cunha Ferreira[10], Odilon Giovannini Junior[11], Luiz Carlos Gomez[12], Antônio Kanaan[13], Gentil Bruscato[12], Juan Manuel Martinez[2], Cristina Terminiello[1], Esteban Dicovskiy[14], Leonor Bonán[4,15], Javier Feu[16], Maria Iglesias[4], Nora Luján[17], Héctor Méndez[18], Hilda Suárez[15], Cynthia Quinteros[4], Herbert Cucurullo[19], Renzo Valentini[20], Zilk H. Meurer[21], L. Felipe Tamborena Barros[22], Roberta Moretti[22], Leonardo Decker[22], Graziela L. Costa[22], Anelise Audibert[22], Antonio P. Ourique[23], Fabiano Berte[23]

1 - Complejo Plaza del Cielo, Esquel (AR), 2 - Fac. de Ingeniería, UNPSJB, Esquel (AR), 3 - IAFE - CONICET (AR), 4 - CEFIEC, FCEyN - UBA (AR), 5 - Departamento de Astronomía, Universidad de la Republica, Montevideo (UY), 6 - Planetário José Baptista Pereira, UFRGS (BR), 7 - Departamento de Física, IF-UFRGS (BR), 8 - Departamento de Astronomia, IF-UFRGS (BR), 9 - Faculdade de Educação, USP (BR), 10 - Estação Ciências, Casa da Física, Manaus (BR), 11 - Departamento de Física e Química, UCS (BR), 12 - Clube de Astronomia, Colégio Militar de Porto Alegre (BR), 13 - Departamento de Física, UFSC (BR), 14 - Escuela Normal Superior N 10, Buenos Aires (AR), 15 - Escuela Normal Superior N.o 3, Buenos Aires (AR), 16 - A.P.F.A., Secretaria Gran Buenos Aires Sur (AR), 17 - Clubes de Ciencias, Ministerio de Educación, Ciudad de Buenos Aires (AR), 18 - Instituto Superior de Formación Docente N35, Monte Grande (AR), 19 - Consejo de Educación Secundaria, Observatorio Astronómico Los Molinos (UY), 20 - Consejo de Educacion Sedundaria (UY), 21 - Colégio Estadual Barro Vermelho Rio Pardo - RS (BR), 22 - IF - UFRGS (BR), 23 - Licenciatura em Matemática, UCS (BR)

Astronomy education connecting graduate and undergraduate studies and public outreach

Wilton Dias[1], Agenor Pina[1], Bruno Castilho[2], Mariângela de Oliveira-Abans[2] and Max Faúndez-Abans[2]
[1] Universidade Federal de Itajubá, Av. BPS 1303, 37500-903 Itajubá, MG, Brazil: email: newton@unifei.edu.br, wilton@unifei.edu.br, agenor@unifei.edu.br,
[2] MCT/Laboratório Nacional de Astrofísica, Rua Estados Unidos 154, 37504-364 Itajubá, MG, Brazil: email: bruno@lna.br, mabans@lna.br, mfaundez@lna.br

A review of IYA activities in Nigeria vis-à-vis the role of CBSS

Okere Bonaventure Iheanyi
Centre for Basic Space Science (CBSS), University of Nigeria, Nsukka Nigeria

The role of Internet resources in astronomical education; the experience of developing the largest astronomical site in Russia
Oleg Sergeevich Bartunov
Sternberg State Astronomical Institute of M.V. Lomonosov Moscow State University, Russia.

The use of Argus/TnE remote observations in teaching science
Osvaldo de Souza[1,2]*, Jane C. Gregorio-Hetem*[1]
1 - IAG-USP, University of São Paulo, Brazil, 2 - IF/USP, Brazil

Astronomy Communication of a new kind: Creating Ultra-sophisticated Astronomical Observatories in the Resorts for Everyone: Experiences, Challenges and Opportunities for Tomorrow
Parag Shankar Mahajani
Milkyway Citizens, India

The Teaching of Astronomy During the Last Years of Elementary School: A Proposal of Didactic Material for Teacher Support
Patrícia Amaral[1,2]*, Cássio Costa Laranjeiras*[3,1]
1 - Science Education Graduate Programme, University of Brasilia, 2 - CAPES - Coordenação de Aperfeiçoamento de Pessoal de Nível Superior, 3 - Institute of Physics, University of Brasilia

The Cultural Value of Astronomy: The Application of Sky Knowledge in the Weather Prediction and Administration of Traditional Medicine amongst The Luo Tribe of East Africa.
Paul Baki Olande and C.O. Obure
Department of physics, University of Nairobi, P.O box 30197,00100, Nairobi, Kenya

Geographic Cosmography: a new area of studies of Astronomy in the Brazilian Geography teaching
Paulo Henrique Azevedo Sobreira
Planetário da Universidade Federal de Goiás

Astronomy with an 8-inch
Rabindra Kumar Bhattacharyya
Calcutta University, India

The diffusion and education of Astronomy today: Planetaries, observatories and society. There is a consent?
Rafael Santiago Girola
Universidad Nacional de Tres de Febrero, Argentina

A Solar Station in Ica - Mutsumi Ishitsuka: A research center to improve education at the university and schools
Raul Andres Terrazas
Solar Station Ica

Astronomy Education with the NURO Telescope and the SARA Telescope in Remote Observing Mode
Ronald G. Samec[1]*, Evan R. Figg*[1]*, Reid Melton*[1]*, Christa M. Labadorf*[1]*, Robert T. McKenzie, Jr.*[1]*, Danny R. Faulkner*[2]*, Walter Van Hamme*[3]
1 - Astronomy program, Physics Department Bob Jones University, 2 - University of South Carolina, Lancaster, 3 - Department of Physics, Florida International University

The atmospheric conditions of some astronomical educational observatory sites in Uzbekistan
Sabit Ilyasov
Ulugh Beg Astronomical Institute, Astronomicheskaya 33, Tashkent 100052, Uzbekistan

Student Astronomical Society of Shiraz University and similar societies

Sara Khalafinejad , Saeed Hojjatpanah , Fatemeh Kamali
Shiraz University, Iran

Fun and interdisciplinary solar astrophysics for middle and high school students

Silvia Calbo Aroca, Cibelle Celestino Silva
Instituto de Física de São Carlos

San Luis Coelum: Observing and Learning Astronomy in Schools

Stella Maris Malaroda and San Luis Coelum Team
Universidad de La Punta, Av. Universitaria s/n, Ciudad de La Punta,
5710, San Luis, Argentina
email: `smalaroda@ulp.edu.ar`

The Planetarium of the Goiás Federal University: a life history in the Central Brazil

Suélen Alonso de Almeida[1] , Paulo Henrique Azevedo Sobreira[2]
1 - Universidade Federal de Goiás, Brazil, 2 - Planetário da Universidade Federal de Goiás, Brazil

Activities for the sky observation and its identification adapted to visually impaired people

T.P. Dominici[1] , E. Oliveira[2] , V. Sarraf[3] & F. Del Guerra[4]
[1] *Laboratório Nacional de Astrofísica (MCT/LNA), Rua Estados Unidos, 154, 37504-364, Itajubá, MG, Brazil: email: tdominici@lna.br*
[2] *Universidade do Grande ABC (UNIABC), Avenida Industrial, 3330, Santo André, SP, Brazil: email: ednilsonoliveira@ig.com.br*
[3] *Fundação Dorina Nowill, Rua Dr. Diogo de Faria, 558, Vila Clementino, São Paulo, SP, Brazil: email: vsarraf@gmail.com*
[4] *Elo3 Integração Empresarial, Rua Afonso Braz, 473, cj. 175, São Paulo, SP, Brazil: email: ferdelguerra@elo3.com.br*

Impressionist skies by MONET

*Ulf Seemann[1,2] , Present address: Fluid Mech Inc., 24 The Street, Lagos, Nigeria., Ekaterina Georgieva					Atanasova[3,4]*
& the MONET consortium[5,6,7]
[1] *European Southern Observatory, Karl-Schwarzschild-Str. 2, D-85748 Garching, Germany email: useemann@eso.org*
[2] *Institut für Astrophysik, Georg-August Universität Göttingen, Friedrich-Hund-Platz 1, D-37077 Göttingen, Germany*
[3] *Tuparev Technologies Inc, 3 Sofijski Geroj Str., 1612 Sofia, Bulgaria*
[4] *Institute of Astronomy, BAS, 72 Tzarigradsko shaussee Blvd., 1784 Sofia, Bulgaria*
[5] *McDonald Observatory at the University of Texas at Austin, 82 Mt. Locke Rd, TX 79734, USA*
[6] *South African Astronomical Observatory, 6920 Sutherland, South Africa*
The full paper is available in electronic form at URL
`http://www.astro.physik.uni-goettingen.de/~seemann/publications/`

Argentine Eratosthenes Project 2009

Victoria Bekeris[1] , Flavia Bonomo[2] , Beatriz García[3,4] , Guillermo Mattei[1] , Diego Mazzitelli[1] , Silvina Ponce Dawson[1] , Constanza Sánchez Fernández de la Vega[5] , Francisco Tamarit[6] , Jorge Zgrablich[6]
1 - Departamento de Física, Facultad de Ciencias Exactas y Naturales, Universidad de Buenos Aires, 2 - Departamento de Computación, Facultad de Ciencias Exactas y Naturales, Universidad de Buenos Aires (Argentina), 3 - UTN - Facultad Regional Mendoza, Argentina, 4 - Pierre Auger Observatory, Argentina, 5 - Departamento de Matemáticas, Facultad de Ciencias Exactas y Naturales, Universidad de Buenos Aires, 6 - Asociación Física Argentina

So equal and so different: the physics in different environments

Yamila Miguel[1,2] , Patricia Rey[2] , Claudio Quiroga[2] , Andrea Torres[3,2] , Stella Ramírez[2]

1 - Facultad de Ciencias Astronómicas y Geofísicas, Universidad de La Plata, 2 - Universidad Nacional de La Plata, Argentina, 3 - Facultad de Ciencias Astronómicas y Geofísicas, Universidad Nacional de La Plata, Argentina

Astronomy in Saudi Arabia: Future Vision of KACST

Zaki A Al-Mostafa , Abdulrahman H Maghrabi , Saad M Al-Shehre , Aymen S Kordi
Astronomy Department, King Abdulaziz City for Science and Tech. Saudi Arabia

Highlights of Astronomy, Volume 15
XXVIIth IAU General Assembly, August 2009
Ian F Corbett, ed.

SpS5: ACCELERATING THE RATE OF ASTRONOMICAL DISCOVERY

Ray P. Norris

CSIRO ATNF, PO Box 76, Epping NSW 1710, Australia
email: Ray.Norris@csiro.au

Abstract. Special Session 5 on Accelerating the Rate of Astronomical Discovery addressed a range of potential limits to progress: paradigmatic, technological, organizational, and political. It examined each issue both from modern and historical perspectives, and drew lessons to guide future progress. A number of issues were identified which may regulate the flow of discoveries, such as the balance between large strongly-focussed projects and instruments, designed to answer the most fundamental questions confronting us, and the need to maintain a creative environment with room for unorthodox thinkers and bold, high risk, projects. Also important is the need to maintain historical and cultural perspectives, and the need to engage the minds of the most brilliant young people on the planet, regardless of their background, ethnicity, gender, or geography.

Keywords. history and philosophy of astronomy, telescopes, cosmology: miscellaneous, sociology of astronomy, education

1. Introduction

The 3.5 day Special Session on "Accelerating the Rate of Astronomical Discovery" was born in a conference-bus discussion in Lithuania between Ray Norris and Clive Ruggles on what would be the most appropriate meeting to mark the International Year of Astronomy. As the public announcements regarding the 2009 IYA have emphasized, new astronomical discoveries are currently being made at an extraordinary rate, while the invention of the telescope ushered in an equally momentous "golden age of discovery 400 years ago. We concluded that this would be an optimum occasion to review the extent to which astronomers are achieving the optimal rate of astronomical discovery. Can we identify and overcome the limits to progress? What steps can be taken to accelerate the rate of expansion of astronomical knowledge? What lessons can be learnt both from the past?

The meeting addressed a range of potential limits to progress: paradigmatic, technological, organizational, and political. It examined each issue both from modern and historical perspectives, and drew lessons to guide future progress. We were fortunate in being able to attract some internationally-renowned speakers, including four IAU Presidents (past, present, and future), pulsar discoverer Jocelyn Bell Burnell, and physicist and popular science writer Lawrence Krauss.

The meeting was divided into 12 sessions, each focussing on one particular topic, and introduced by one or two invited speakers, followed by contributed papers. This brief summary cannot possibly review every one of the many excellent papers, but instead focuses on a few highlights. A full conference proceedings is published by "Proceedings of Science" (Norris & Ruggles 2010).

2. Back to the Future

The meeting was opened by the new President of the IAU, Bob Williams, who spoke on "Discovery and the Culture of Astronomy", and who pointed out that the astronomical discovery processes evolve slowly with time, and have no systematic procedure by which they are evaluated and modified. For example, time on major telescopes is allocated by a peer-review system, and yet arguably the most significant research conducted by the Hubble Space Telescope (the Hubble Deep Fields) was not awarded time through a peer-review process. Diversity is important, and should be encouraged throughout astronomy. Furthermore, we should regularly review how we conduct the business of astronomy, and recommend change where necessary.

Historian of Science David de Vorkin introduced a historical perspective by examining what has worked and what hasn't. While no absolute answers are available, a series of case studies gave us hints to what might be most productive. We learnt that the great discoveries of the past relied not only on skill and insight, but also on the willingness to take risks, the ability to form productive collaborations, and the value of diversity and versatility.

As a case study, IAU President-elect Norio Kaifu examined the phenomenal rise of astronomy in Japan. A key factor here was the presence of visionary and inspirational leaders (including, of course, Kaifu himself, although he would be too modest to say so!). It was also important to monitor carefully the developments elsewhere, and be prepared to use cutting-edge technology to jump ahead of the international state-of-play.

The status of astronomy as a science was questioned by Lawrence Krauss. He noted that, while the last few years have produced a revolution in our picture of the Universe on its largest scales, this revolution has also raised the question of whether cosmological concepts such as inflation and Dark Energy are in fact falsifiable. If not, then cosmology has arguably crossed the border from physics to metaphysics.

Magda Stavinschi pointed out the importance of teaching students not only about current astronomy, but also of our astronomical heritage. Knowledge of that history is essential to provide the context within which discoveries and theories can be evaluated, or major projects planned. Magda noted that "one cannot have a future without knowing ones past."

3. Creativity, Innovation, and Culture

Simon White warned us of the dangers of becoming too enamoured with big projects, billion-dollar telescopes, and the resulting intellectual bandwagons. While acknowledging the value of such projects, it is important that we do not become immersed in a corporate mentality in which we fail to recognise creative ideas that depart from the juggernaut of academic orthodoxy.

Some of the greatest scientific discoveries in the past, Ken Kellermann told us, have not in fact come from testing theories, but from applying technology to broadening our view of the Universe. In fact even when telescopes have been built primarily for a particular research goal, their major discoveries are often unrelated to that goal. Knowing that to be the case, we must try to build flexibility into future telescopes, and, even more importantly, to ensure our minds and research processes are sufficiently flexible to recognise the unexpected.

The discovery of pulsars is a prime example of this process, and pulsar discoverer Jocelyn Bell Burnell explained the importance of a mindset receptive to bits of scruff on a chart recorder, while less prepared minds might dismiss them as unimportant. Other key factors included an intimate knowledge of the instrument, able to distinguish the exceptional from the routine, and a stimulating research group in which creativity and risk-taking are encouraged.

Past IAU President Ron Ekers built on this theme to show how science instruments, both in astronomy and in other areas of science, grow exponentially and inexorably from "little science" to "big science". The ensuing change of culture raises the challenge maintaining creativity and flexibility in the face of the rigid and bureaucratic processes necessary to build a "Big Science" instrument. This may be rephrased as a challenge to maintain a vigorous and nimble "Small Science" community while developing the "Big Science" instruments necessary for the advance of the research field.

The tendency of our processes to quash creative thought was also noted by Catherine Cesarsky who noted that decadal plans, visions, and roadmaps are helpful, at least to foster the inception of large projects, which undoubtedly foster first-rate science. But the resulting orthodoxy can also quench originality, and amplify bandwagon effects, so that time allocation committees discriminate against original, risky, or unorthodox projects.

Eric Feigelson reviewed the importance of cross-disciplinary research, discussing the successes, challenges and opportunities. He showed that cross-disciplinary approaches have led to significant advances in astronomy, and that cross-disciplinary researchers make profound, revolutionary transformations in our thinking. If the paths of cross-disciplinary astronomy are groomed, and talented scientists tread them in a spirit of creativity, then we can envision a blossoming of astronomy in new areas.

The influence of society on astronomical discovery was discussed by George Miley. Paradigm-changing observational facilities have become so sophisticated and expensive that they require global funding. As a result, funding decisions are inevitably political, and astronomy needs to inform these decisions using broad-based arguments, such as those in the new IAU strategic plan, which outline astronomy's benefits to society. Participation in such activities is a cost-effective way for astronomers to stimulate funding for astronomical research. On the other hand, such developments tend to stifle nonconformity amongst astronomers, which may affect future astronomical discoveries.

Clive Ruggles argued that astronomy would benefit from the mutual awareness and comprehension between modern astronomy and cultural world-views such as "indigenous cosmologies", whose goals, like those of science, are to make sense of the cosmos. The necessary links require modern astronomers to understand and respect both the heritage and the non-western frameworks of thought that give rise to these cultural perspectives. One of the most obvious potential benefits could derive from common attitudes towards the natural heritage of astronomy, namely dark skies.

4. Data and Information

Our current growth is measured not only in the size of the instruments but also in the Terabytes of data, the Teraflops required to process them, and new tools for data access, enabling new modes of astronomical discovery. George Djorgovski shared his vision of new ways of accessing data, fomenting a revolution affecting every stage of the research endeavour. We must be prepared that even our fundamental scientific methodology may change, and we must expect new areas of research, such as time-domain astronomy, to become increasingly significant.

Francoise Genova explained how the tools and processes of the Virtual Observatory can enable these far-reaching changes, and will facilitate the growth of our "knowledge infrastructure".

The impact of information science on astronomy was explored by Ray Norris, who pointed out that current astronomers are largely drawn from privileged backgrounds, and exclude those proto-Einsteins who are held back by their background or unconventionality. Web 2.0 and its successors could enormously raise the intellectual resources available to astronomy if we could figure out how to build a bridge to tap into these unrecognised intellects which are undoubtedly out there, not only in developing countries but perhaps also in our own patent offices.

5. Education and Opportunity

Rajesh Kochhar noted a declining interest in basic science, and a shift away from physical sciences towards life sciences, and argued that research needed to be closely linked to education. In developing countries, the development of a tradition of training and mentoring of the finest young minds to conduct top-quality research is arguably more important than conducting the research itself.

On the other hand, Patricia Whitelock showed how South Africa is successfully using astronomy, and developing new telescopes, such as SALT and Meerkat, as a tool for national human development. In particular, the National Astrophysics and Space Science Programme is overcoming the legacy of apartheid to raise the level of science engagement and education amongst young Africans, and help them participate in the process of research and discovery.

Any doubt about the excitement that can be found in astronomy education was quashed by a memorable presentation by Julieta Fierro, who alerted us to the opportunity of raising the scale and standard of our outreach projects. As well as underlining the importance of outreach and education, and demonstrating novel ways to engage the audiences, she stressed how outreach must be taken seriously rather than being an afterthought: it must be interesting, diverse, simple, pertinent, and peer reviewed.

6. Panel Discussion and Conclusion

The meeting finished with a lively panel discussion in which some of the speakers were invited to answer questions from each other and from the floor. Prominent in the ensuing discussion was the role of large strongly-focussed projects and instruments, designed to answer the most fundamental questions confronting us. Such instruments continue to be immensely successful both in answering the questions for which they were designed, and just as importantly, discovering new phenomena by probing unexplored regions of the observational parameter space. Both these approaches are important, and new telescopes need to be designed to address both. To do so, they must be used by people who understand the instruments intimately, and are able to recognise the difference between a new discovery and an instrumental artefact.

Against the need for such focussed projects must be balanced the need to maintain a creative environment with room for unorthodox thinkers and bold, high risk, projects. We must be wary of conservative time allocation processes and funding agencies, of bandwagon effects, and the stifling of those who venture outside the box. We must also be wary of a "one-size-fits-all" approach: while the meeting emphasised the importance of cross-fertilisation and multi-wavelength approaches, we also need people who have very deep knowledge in their own particular field or technique.

Also important are the need to maintain historical and cultural perspectives, and the need to provide opportunities for disadvantaged students to participate in astronomy. We must provide universal access to information, while avoiding drowning great discoveries in a sea ofunrefereed and insignificant publications.

Perhaps the greatest area of common ground is the need to invest in the education and engagement of the scientists of the future. It is important that we train our students to understand their instruments as well as their science, to teach them to think nimbly, and not necessarily follow the well-funded bandwagons. But we must also recognise the millions of potential brilliant scientists of the future who are currently prevented from entering the arena of science either because of social or financial needs, or because, like Einstein, they don't quite fit the model of a good mainstream student. We have the technology to reach them, but we do not yet have the processes to do so.

The challenge to astronomy is to find ways to engage the minds of the most brilliant young people on the planet, regardless of their background, ethnicity, gender, or geography. By doing so we can not only accelerate astronomy, but we can generate real benefits to wider human society.

Acknowledgements

I would like to thank my SOC co-chair, Clive Ruggles, the members of the SOC, and all the speakers and poster presenters for a stimulating, provocative, and memorable meeting.

References

Norris, R. P. & Ruggles, C. L. N. (eds) 2010, Proceedings of Science, PoS(SpS5)

SpS6 - Planetary Systems as Potential Sites for Life

Preface

Special Session 6 entitled "Planetary Systems as Potential Sites for Life" was organized jointly by Commission 16 (Physical Study of Planets and Satellites), Commission 51 (Bio-Astronomy), and Commission 53 (Extrasolar Planets). It took place over two days (August 10-11) during the XXVIIth General Assembly of the IAU held in Rio de Janeiro, Brazil.

Until recently, the characterization of solar system objects, the search for extrasolar planets, and the study of the origins of life were carried out in parallel, although not totally in isolation from each other. Indeed, many bridges have produced significant advances among these disciplines in the past few decades. This situation is rapidly evolving however, and a more integrated field is emerging. More and more, national agencies are creating programs under the general heading of "Origins", with the aim of integrating investigations related to planetary formation and evolution, the search and characterization of extrasolar planetary systems, and exobiology.

Solar system studies have reached a point where several objects have been identified as possible sites for life, past or present (Mars, Europa, Enceladus, Titan). The characterization of extrasolar planetary systems is already providing us with hints of habitable worlds around other stars. Modeling and laboratory experiments are also rapidly progressing, leading the way for more and more refined observational investigations. Thus, a synergy is truly operating.

The purpose of this Special Session was to explore the many facets of this emerging multidisciplinary field, to summarize recent advances, and to bring together investigators in the various disciplines in order to foster future projects and collaboration. Some time was also devoted to the evocation of several outstanding figures in the development of this field. The session was divided into six parts: I. Sites for Life in the Solar System; II. Laboratory and Space Experiments; III. The Search for Low-Mass Extrasolar Planets; IV. Habitability of Extrasolar Planets; V. Missions and Surveys under Development; VI. Remembering Pioneers in Astrobiology.

A total of 26 oral contributions were presented, most of them invited papers, and nearly 40 posters were on display during the second week of the General Assembly. The diversity and completeness of these presentations were outstanding. The following proceedings, which are organized according to the six parts of the session, are excellent samples of the wealth of information that was shared with the numerous audience.

We sincerely thank all the speakers and participants.

Régis Courtin (Chair SOC), Alan Boss, and Michel Mayor (co-Chairs SOC)
Meudon, Washington, and Genève, November 30, 2009

The Scientific Organizing Committee was: Carlo Blanco (Italy), Alan Boss (USA), Guy Consolmagno (Vatican City), Cristiano Cosmovici (Italy), Régis Courtin (France), Pascale Ehrenfreund (The Netherlands), Leonid Ksanfomality (Russian Federation), Luisa Lara (Spain), David Latham (USA), Michel Mayor (Switzerland), Melissa McGrath (USA), Karen Meech (USA), David Morrison (USA), John Spencer (USA), Victor Tejfel (Kazakhstan), and Stephane Udry (Switzerland).

Highlights of Astronomy, Volume 15
XXVIIth IAU General Assembly, August 2009
Ian F. Corbett, ed.

Europa, Enceladus, and Titan
as possible sites for life

Régis Courtin

LESIA, CNRS & Observatoire de Paris, 92195 Meudon, France
email: regis.courtin@obspm.fr

Abstract. Despite quite distinct bulk properties, Europa, the third largest Jovian satellite (d=3138 km), and the Saturnian satellites Enceladus (d∼500 km) and Titan (d=5151 km) share a remarkable common feature which is a strong indication of the presence of liquid water at some level below the surface. The possibilities for the development of life organisms on these bodies are reviewed.

1. Europa and Enceladus

On Europa, observational evidence from the Galileo orbiter strongly suggested the presence of a water ocean beneath an ice shelf of unknown thickness. Gravity measurements showed that the outer layer has a density of 1000 kg.m^{-3} and a thickness between 80 and 170 km (Anderson *et al.* 1998). In addition, spectroscopic studies of the surface composition showed the presence of salty minerals along the prevalent cracks in the icy crust, most probably hydrated sulfate and carbonate minerals (hexahydrite, epsomite, and natron) (McCord *et al.* 1998). A possible explanation of the presence of these salty minerals is a global salty water ocean, which reaches the surface in places of recent geologic activity and leaves salts behind. However, the presence of these salts does not unambiguously require a global ocean (Chyba & Phillips 2002). Despite remaining uncertainties on its existence, one can evaluate the potential biomass in Europa's water ocean. Conservative estimates suggest that it could be limited to ∼10^{23}-10^{24} prokaryotic-analog cells, and that it would be very difficult to supply the ocean with enough radiation-produced oxygen to sustain a terrestrial-type macrofauna (Chyba & Phillips 2001a). More favorable assumptions on the availability of oxygen and/or hydrogen through the ice shell and in the ocean itself led Chyba & Hand (2001b) to estimate a steady-state biomass of ∼10^{13}-10^{15} g, compared with the terrestrial biomass of ∼10^{18} g.

On Enceladus, the discovery of the quasi-continuous ejection of plumes from the south polar region led to a "cold geyser" scenario whereby liquid water stored in high pressure pockets below the surface is vented into space in the form of vapor and fine ice particles (Porco *et al.* 2006, Matson *et al.* 2007, Waite *et al.* 2009). The presence of an alkalin ocean, with a composition dominated by NaCl, NaHCO$_3$, Na$_2$CO$_3$, and K$^+$, is strongly suggested by the analysis of the ejected icy particles (Postberg *et al.* 2009). Thus, the internal environment is thought to be favorable for aqueous catalytic chemistry, permitting the synthesis of many complex organic compounds. Following the approach used by Chyba & Phillips (2002) for Europa, Parkinson *et al.* (2008) estimated at 10^{19}-10^{20} the total number of cells that could exist in an ecosystem underneath the ice crust in the vicinity of the plume vents. They conclude that on Europa and Enceladus, the biomass per unit area could have reached similar orders of magnitude, i.e. 10^{13}–10^{14} cells.km^{-2}.

2. Titan

On Titan, indirect evidence for a deep (water) ocean comes from radar measurements (Lorenz *et al.* 2008). Even more to the point perheaps, the rich organic chemistry taking place in the atmosphere, as well as the presence of free H_2 near the surface, and the existence of a methane cycle involving rains/rivers/lakes, support the notion of the sustainability of life. Indeed, Baross *et al.* (2007) noted that Titan's environment meets the absolute requirements for life, including thermodynamic disequilibrium, an abundant carbon inventory, and a fluid environment. They go as far as to conclude that "this makes inescapable the conclusion that if life is an intrinsic property of chemical reactivity, life should exist on Titan". Specifically, Schulze-Makuch & Grinspoon (2005) have speculated on possible life using metabolic pathways involving acetylene, while McKay & Smith (2005) have considered the possibility of widespread methanogenic life in liquid methane, with hydrogenation reactions as a possible source of free energy.

3. Conclusions and future prospects

Many questions still need to be resolved regarding the sustainability of life on Europa, Enceladus, and Titan. Are their putative sub-surface water oceans real, and if so, what is the thickness of the overlying ice-shell? What is the role of cryovolcanism in surface-interior interactions? What is the amount of free energy available at the surface or in the subsurface environments?

Because of the most favorable combination of parameters for the development of life, Titan may be considered as the prime target in the search for extraterrestrial life in the solar system, with Europa (and Mars) as close alternative(s) (Shapiro & Schulze-Makuch 2009). Many more high-quality data are expected for Titan and Enceladus from the Cassini Solstice mission (2010–2017), and these will provide a more solid basis in the search for ecosystems on these bodies. On a longer timescale, dedicated international missions (EJSM/Laplace to Europa, and TSSM/TandEm to Titan/Enceladus) are under study for the 2020-2025 horizon.

References

Anderson, J. D., Schubert, G., Jacobson, R. A., Lau, E. L., Moore, W. B., & Sjogren, W. L. 1998, *Science* 281, 2019

Baross, J. A., Benner, S. A., Cody, G. D. *et al.* 2007, in: *The Limits of Organic Life in Planetary Systems* (Washington: National Academies Press)

Chyba, C. F. & Phillips, C. B 2001a, *Proc. Natl. Acad. Sci. USA* 98, 801

Chyba, C. F. & Hand, K. P. 2001b, *Orig. Life & Evol. Biosph.*

Chyba, C. F. & Phillips, C. B. 2002, *Orig. Life & Evol. Biosph.* 32, 47

Lorenz, R., Stiles, B. W., Kirk, R. L., Allison, M. D., Persi del Marmo, P., Iess, L., Lunine, J. I., Ostro, S. J., & Hensley, S. 2008, *Science* 319, 1649

Matson, D. L., Castillo, J. C., Lunine, J., & Johnson, T. V. 2007, *Icarus* 187, 569

McCord, T., Hansen, G. B., Fanale, F. P. *et al.* 1998, *Science* 280, 1242

McKay, C. P. & Smith, H. D. 2005, *Icarus* 178, 274

Parkinson, C. D., Liang, M.-C, Yung, Y. L., & Kirschivnk, J. L. 2008, *Orig. Life & Evol. Biosph.* 38, 355

Porco, C., Helfenstein, P., Thomas, P. C. *et al.* 2006, *Science* 311, 133

Postberg, F., Kempf, S., Schmidt, J., Brilliantov, N., Beinsen, A., Abel, B., Buck, U., & Srama, R. 2009, *Nature* 459, 1098

Schulze-Makuch, D. & Grinspoon, D. H. 2005, *Astrobiology* 5, 560

Shapiro, R. & Schulze-Makuch, D. 2009, *Astrobiology* 9, 335

Waite, J. H., Lewis, W. S., Magee, B. A. *et al.* 2009, *Nature* 460, 487

Highlights of Astronomy, Volume 15
XXVIIth IAU General Assembly, August 2009
Ian F. Corbett, ed.

© International Astronomical Union 2010
doi:10.1017/S1743921310010914

Photoabsorption of hydrocarbons in Titan's atmosphere

Fabíola P. Magalhães, Gerardo G. B. de Souza and Heloisa M. Boechat-Roberty

Universidade Federal do Rio de Janeiro - Observatório do Valongo, Brazil
email: fabiola02@astro.ufrj.br

Abstract. Titan, the largest satellite of the planet Saturn, has a thick atmosphere which consists of nitrogen (N_2) and methane (CH_4). In 2004, the Cassini-Huygens mission observed the occultation of two stars through the atmosphere of Titan and measured ultraviolet (UV) absorption spectra. Through these spectra it was possible to identify the molecular species contained in this environment. In the present work, we have simulated a spectrum of this atmosphere using some molecules such as CH_4, C_2H_2, C_2H_4, C_2H_6, C_4H_2, and C_6H_6. Our cross sections data were experimentally obtained using the electron energy-loss technique, where the electron energy-loss spectra, measured high incident energies and in small scattering angles, are similar to photoabsorption spectra. The comparison of our synthetic spectrum with that measured by Cassini shows that this method is very efficient for identifying molecules as well as estimating abundances.

1. Introduction

Titan's atmosphere is primarily composed of nitrogen but methane is the nextmost abundant constituent (Wilson & Atreya 2004). The interaction of solar X-rays and ultraviolet photons or energetic electrons with these constituents produce a rich and complicated organic chemistry that leads to the formation of new molecules such as hydrocarbons and nitriles. These latter are organic species containing a triple CN bond, such as HCN, which is a precursor of the nitrogenous base of DNA, the adenine molecule. In 2004, the Cassini-Huygens mission observed the occultation of two stars through the atmosphere of Titan, Shaula (λ Scorpion) and Spica (α Virgo) and obtained UV absorption spectra from 110 nm to 190 nm. This atmospheric transmission of the stellar flux can be understood by the Lambert-Beer law:

$$I(\lambda) = I_0(\lambda) \ e^{-\tau} \ \longrightarrow \ \ln \ \left(\frac{I}{I_0} \right) = -\tau(\lambda)$$

where I(λ) is the stellar flux as a function of the wavelength (λ) after being attenuated by the absorbing material in Titan's atmosphere, I_0 is the stellar flux before traversing the atmosphere and τ (λ) is the optical depth given by:

$$\tau(\lambda) = \Sigma_i \ \sigma_i(\lambda) N_i$$

where $\sigma_i(\lambda)$ is the cross section (cm^2) and N_i is the column density (cm^{-2}) for each molecular species i.

2. Methodology

We have obtained photoabsorption spectra converted from the electron energy-loss spectra for molecules such as benzene, naphthalene and anthracene (Boechat-Roberty *et al.* 2004, de Souza *et al.* 2002, Boechat-Roberty *et al.* 1997) and other species. Briefly, we have measured electron energy-loss spectra (EELS) at electron incident energy of 1000 eV and small scattering angles. In these conditions, the generalized oscillator strength (measured by EELS) tends to be the optical oscillator strength (Lassettre & Skerbele 1974). To obtain the photoabsorption cross-section, each point in the EELS was extrapolated to zero scattering angle using the universal formula (Boechat-Roberty *et al.* 2004). To compute the Titan synthetic spectrum, we used the cross-sections of the following species: CH_4, C_2H_2, C_2H_4, C_2H_6, C_4H_2, and C_6H_6. The column densities, initially adopted from Shemansky *et al.* (2005), were adjusted to achieve the best fit.

3. Results and conclusions

Figure 1 represents the synthetic spectrum obtained from the cumulative product of the photoabsorption cross-section by the respective column density of each molecule. The comparison of our result with the Cassini spectrum shows a very good agreement.

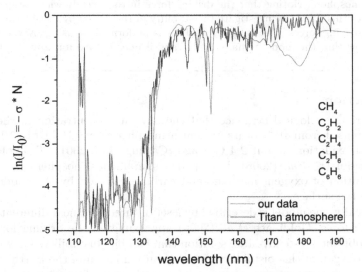

Figure 1. Comparison of our synthetic spectrum with the spectrum obtained by the Cassini spectrometer.

Acknowledgements

The authors would like to thank the financial support from Conselho Nacional de Desenvolvimento Cientfico e Tecnolgico (CNPq) and FAPERJ.

References

Boechat-Roberty, H. M., Rocco, M. L. M., Lucas, C. A., Fernandes, M. B., & de Souza, G. G. B. 1997, *J. Phys. B: At. Mol. Opt. Phys.* 30, 3369

Boechat-Roberty H. M., Rocco, M. L. M., Lucas, C. A., & de Souza, G. G. B. 2004, *J. Phys. B: At. Mol. Opt. Phys.* 37, 1467

de Souza, G. G. B., Boechat-Roberty, H. M., Rocco, M. L. M., & Lucas, C. A. 2002, *J. Electron Spec & Rel Phen.* 123, 315

Lassetre, E. N. & Skerbele, A. 1974, *Methods Exp. Phys.* B3, 868

Shemansky, D. E., Stewart, A. I. F., West, R. A., Esposito, L. W., Hallett, J. T., & Liu, X. 2005, *Science* 308, 978

Wilson, E. H. & Attreya, S. K. 2004, *J. Geophys.* 109, E06002

Highlights of Astronomy, Volume 15
XXVIIth IAU General Assembly, August 2009
Ian F. Corbett, ed.

The great oxidation of Earth's atmosphere

Zdzislaw E. Musielak[1], Manfred Cuntz[1] and Dipanjan Roy[2]

Department of Physics, University of Texas at Arlington, Arlington, TX 76019, USA
email: zmusielak@uta.edu, cuntz@uta.edu

[2]Institut des Sciences du Mouvement UMR CNRS 6233, Université de la Méditerranée, 13288
Marseille, France
email: dipanjan.roy@etumel.univmed.fr

Abstract. A simplified model of the Earth's atmosphere consisting of three nonlinear differential equations with a driving force was developed by Goldblatt *et al.* (2006). They found a steady-state solution that exhibits bistability and identified its upper value with the great oxidation of the Earth's atmosphere. Noting that the driving force in their study was a step function, it is the main goal of this paper to investigate the stability of the model by considering two different more realistic driving forces. The stability analysis is performed by using Lyapunov exponents. Our results show that the model remains stable and it does not exhibit any chaotic behavior.

1. Introduction

There is strong geological evidence that the oxygen concentration in the Earth's atmosphere increased from 10^{-5} of its present atmospheric level (PAL) to 10^{-2} PAL (the so-called great oxidation) about 2.4 Gyr ago (Catling *et al.* 2001). To explain this phenomenon, Goldblatt *et al.* (2006) developed a simplified atmospheric model in which the time evolution of oxygen, methane and carbon is driven by the sudden decline of reductants to the surface.

Mathematically, the model is described by a set of three nonlinear differential equations with a driving force. Goldblatt *et al.* (2006) approximated the driving force by a step function and obtained a steady-state solution that exhibits bistability. They identified the high value of oxygen in this bistability as the great oxidation of the Earth's atmosphere. In this paper, we consider the same mathematical model but with two different and more realistic driving forces. We investigate stability by using Lyapunov exponents. Our results show that the model is stable and does not exhibit chaos.

2. Results and discussion

The set of three nonlinear equations that describes the time evolution of oxygen, methane and carbon is discussed in detail by Goldblatt *et al.* (2006) and Cuntz *et al.* (2009), and it will not be repeated here. However, we want to emphasize that the explicit dependence on time of the driving force makes the set of equations to be non-autonomous. A standard method to perform stability analysis of such a system of equations is to investigate its Lyapunov exponents. The fact that the Lyapunov exponent method is a valuable tool for studying dynamical systems with bistability has been shown in the literature (Badzey & Mohanty 2005).

The number of Lyapunov exponents for a dynamical system is the same as the number of dependent variables representing the system. However, if the system is non-autonomous, then one additional Lyapunov exponent must be added (Wu *et al.* 2007). For the considered model, the equations describing the time evolution of oxygen and

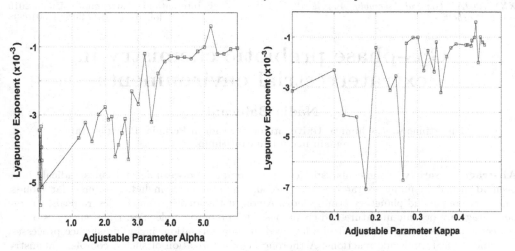

Figure 1. Lyapunov exponent for methane is plotted versus the parameters α and κ of the two driving forces used in our numerical calculations.

methane are mutually dependent and non-autonomous, which means that three Lyapunov exponents are needed to investigate the stability of this subset of two equations. Note that the equation describing the time evolution of carbon is independent and the corresponding Lyapunov exponent is zero (Cuntz et al. 2009).

To determine the stability of the model, we consider two driving forces. The so-called exponential driving force is given by $r(t) = r_0 e^{-\alpha t/t_0}$ and the logistic driving force is given by $r(t) = r_0 \kappa^{-t/t_0}$ with $t_0 = 10^7$ yrs and α and κ as parameters. The system equations are solved numerically by implementing one of these driving forces. Thereafter, the resulting three Lyapunov exponents are calculated. Since two of the exponents (oxygen and methane) are negative, and the third one corresponding to the explicit time dependency of the system is zero, the system can be classified as periodic (Hilborn 1994). The Lyapunov exponent for methane calculated for the two driving forces is plotted in Fig. 1. It shows that the exponent remains negative for all values of α and κ considered.

3. Conclusions

We studied the model of the Earth's atmosphere developed by Goldblatt et al. (2006) by using two different driving forces. Our results show that the model remains stable and it does not exhibit any chaotic behavior for the considered range of parameters. If confirmed by further studies, this result carries important implications for the biochemistry of early Earth as discussed by Kasting (2006) and others, and may be of interest for the investigation of Earth-type exoplanets as well.

References

Badzey, R. L. & Mohanty, P. 2005, *Nature* 437, 995
Catling, D. C., Zahnle, K. J., & McKay, C. P. 2001, *Science* 293, 839
Cuntz, M., Roy, D., & Musielak, Z. E. 2009, *ApJ* (Letters) 706, L178
Goldblatt, C., Lenton, T. M., & Watson, A. J. 2006, *Nature* 443, 683
Hilborn, R. C. 1994, *Chaos and Nonlinear Dynamics*, (Oxford: Oxford University Press)
Kasting, J. F. 2006, *Nature* 443, 643
Wu, X., Wang, J., Lu, J., & Iu, H. H. C. 2007, *Chaos, Solitons & Fractals* 32, 1483

Highlights of Astronomy, Volume 15
XXVIIth IAU General Assembly, August 2009
Ian F. Corbett, ed.

Gas-phase prebiotic chemistry in extraterrestrial environments

Nadia Balucani

Dipartimento di Chimica, Università degli Studi di Perugia, Perugia, Italy
email: nadia.balucani@unipg.it

Abstract. A variety of molecular species up to complex polyatomic molecules/radicals have been identified in many extraterrestrial gaseous environments, including interstellar clouds, cometary comae and planetary atmospheres. Amongst the identified molecules/radicals, a large percentage are organic in nature and encompass also prebiotic molecules. Different types of microscopic processes are believed to be involved in their formation, including surface processes, ion- and radical- molecule reactions. A thorough characterization of such a complex chemistry relies on a multi-disciplinary approach, where the observations are complemented by accurate chemical modeling. Unfortunately, a literature survey reveals that only a small percentage of the elementary reactions considered in the available models have been characterized in laboratory experiments. In this contribution, a brief overview will be given of recent experimental techniques that have allowed us to reach a better description of neutral-neutral gas-phase reactions, which might be responsible for the formation of simple prebiotic molecules.

In the sequence of steps which are believed to have led from elementary particles to the dawn of life, an important one is certainly the formation of simple prebiotic molecules from parent species abundant in the Universe. The combination of H, O, C, N, S and other atoms into molecules and their subsequent chemical evolution in extraterrestrial environments are witnessed by the identification of more than one hundred molecules in the interstellar clouds (encompassing also prebiotic molecules such as hydrogen cyanide, glycolaldehyde, formamide) and by the gas-phase chemical evolution of the atmospheres of several solar objects like Titan. The formations of organic molecules/radicals can certainly be regarded as simple processes when compared to other unknown phenomena that have led to the first living organisms. Still, the formation mechanisms of many of the observed gaseous organic molecules/radicals are not well-characterized, while a comprehension of these processes can certainly help to understand the preconditions for the birth of life (Balucani 2009).

Different types of molecular processes are believed to be involved, including radiative association and recombination, surface-induced processes, photon- or particle-induced ionization and ion-molecule reactions, photon- or particle-induced dissociation and radical-molecule reactions. The concomitance of all these phenomena and the complexity of the considered environments require a modeling approach, where all the relevant molecular processes compatible with the boundary conditions should be included with the appropriate parameters that describe them. As far as neutral-neutral reactions are concerned, provided that they are thermodynamically feasible, the relevant parameters are (i) the kinetic rate constant and (ii) the product branching ratio. Given the low temperature of interstellar clouds and several solar objects, like Titan, the kinetic rate constants should be measured at temperatures as low as T = 10-100 K.

There are two experimental approaches to cool gases and study their reactions: cryogenic cooling and expansion methods. Cryogenic cooling is limited by the saturation vapour pressure of the reactant gases, while supersonic expansion through a collimating axisymmetric, converging-diverging Laval nozzle, leads to the production of a cold and uniform supersonic flow for many gas mixtures. This is the approach used in the CRESU (*Cinétique de Réaction en Ecoulement Supersonique Uniforme*) technique, which has allowed measuring rate constants down to very low T (as low as 15 K) for many neutral-neutral reactions (Smith & Rowe 2000). From these studies experimental evidence has been obtained that some reactions, including those leading to prebiotic molecules like cyanoacetylene, are very fast, with rate constants in the gas kinetics range at very low T. Also, the CRESU results have pointed out that the extrapolation at very low T of the Arrhenius dependence of the rate constant outside the range of T investigated (commonly used in astrochemical modeling) is not warranted.

Elegant as the CRESU studies are, they do not allow determining the nature of the products and their branching ratio. Nevertheless, this piece of information is fundamental to constructing a realistic model, because the products of one reaction are the reactants of a subsequent one. In this respect, a powerful technique is the crossed molecular beam (CMB) method with mass spectrometric (MS) detection. In CMB experiments bimolecular reactions are studied under well-defined conditions. Differently to a bulk experiment, the reacting species are confined in separate beams which cross each other at a specific angle and collision energy. The products are formed at the collision center and then fly undisturbed towards the detector because of the large mean free path achieved by operating at a very low pressure. In this way, the results of well-defined molecular collisions are observed and secondary/wall collisions avoided. The use of MS detection makes the method *universal*, as any species can be ionized in the ionizer which precedes the mass filter and it is possible to determine the mass and the gross formula of all possible products of a bimolecular reaction. In recent years the CMB-MS method has been successfully applied to the study of reactions leading to prebiotic molecules. For example, several reactions producing nitriles (Balucani *et al.* 2009, Balucani & Kaiser 2001, Gu *et al.* 2009), organosulphur (Leonori *et al.* 2009) and organophosphorus (Guo *et al.* 2007) compounds have been characterized.

In conclusion, powerful experimental techniques devoted to the study of neutral-neutral reactions can significantly contribute to reach an unprecedented knowledge of the plausible gas-phase chemistry that has preceded the appearance of life.

Acknowledgements

Financial support from the Italian MIUR (Ministero Istruzione Università Ricerca) under projects PRIN (2007H9S8SW_004) is acknowledged.

References

Balucani, N. 2009, *Int. J. Mol. Sci.* 10, 2304.

Smith, I. W. M. & Rowe, B. R. 2000, *Acc. Chem. Res.* 33, 261.

Balucani, N., Bergeat, A., Cartechini, L., Volpi, G. G., & Casavecchia, P. 2009, *J. Phys. Chem. A*, in press, DOI: 10.1021/jp904302g

Balucani, N. & Kaiser, R. I. 2001, *Acc. Chem. Res.* 34, 699.

Gu, X., Kaiser, R. I., Mebel, A. M., Kislov, V. V., Klippenstein, S. J., Harding, L. B., Liang, M. C., & Yung, Y. L. 2009, *ApJ* 701, 1797.

Leonori, F., Petrucci, R., Balucani, N., Casavecchia, P., Rosi, M., Skouteris, D., Berteloite, C., Le Picard, S. D., Canosa, A., & Sims, I. R. 2009, *J. Phys. Chem. A*, in press, DOI: 10.1021/jp906299v

Guo, Y., Gu, X., Zhang, F., Sun, B. J., Tsai, M. F., Chang, A. H. H., & Kaiser, R. I. 2007, *J. Phys. Chem. A* 111, 3241.

Highlights of Astronomy, Volume 15
XXVIIth IAU General Assembly, August 2009
Ian F. Corbett, ed.

© International Astronomical Union 2010
doi:10.1017/S174392131001094X

Astrochemistry on the EXPOSE/ISS and BIOPAN/Foton experiments

H. Cottin[1], Y. Y. Guan[1], P. Coll[1], D. Coscia[1], N. Fray[1], F. Macari[1], F. Stalport[1], F. Raulin[1], C. Szopa[2], D. Chaput[3], M. Viso[4], M. Bertrand[5], A. Chabin[5], F. Westall[5] and A. Brack[5]

[1]LISA (Univ. Paris 7 & Paris 12, CNRS), av. du Gal de Gaulle, 94010 Créteil Cedex, France
email: herve.cottin@lisa.univ-paris12.fr

[2]LATMOS-IPSL (UPMC, UVSQ, CNRS), 4 place Jussieu, 75005 Paris Cedex, France
[3]CNES, Centre Spatial de Toulouse, 18 avenue Edouard Belin, 31401 Toulouse Cedex 9, France
[4]CNES, 2 place Maurice Quentin, 75039 Paris Cedex 01, France
[5]CBM (CNRS), rue Charles-Sadron, 45071 Orléans Cedex 2, France

Abstract. We describe three space experiments designed to expose to space conditions, and more specifically to solar UV radiation, selected samples of organic and mineral material.

Solar UV radiation is a major source of energy for initiating chemical evolution towards complex organic structures but it can also photodissociate the most elaborate molecules. Thus, Solar UV can erase the organic traces of past life on the surface of planets, such as Mars (Oro & Holzer 1979), destroy organic molecules present on meteorites, Barbier *et al.* 1998, influence the production of distributed sources in comets (Cottin *et al.* 2004) or initiate chemistry in Titan's atmosphere (Sagan & Thompson 1984). In the interstellar medium, the UV radiation field emitted by stars in the galaxy is also responsible for the chemical evolution and the extraordinary diversity of detected organic molecules.

To improve our knowledge of the chemical nature and evolution of organic molecules involved in extraterrestrial environments with astrobiological implications, we have developed a series of three space experiments implemented on the Russian automated capsule FOTON, or outside of the International Space Station. The goal of these experiments is to expose to space conditions, and more specifically to solar UV radiation, selected samples of organic and mineral material. The UVolution experiment was flown in September 2007, during 12 days, in the ESA BIOPAN facility set outside the FOTON M3 capsule (Demets, Schulte & Baglioni 2005). The PROCESS experiment has been installed on the ISS EXPOSE-Eutef facility from February 2008 to August 2009, and the AMINO experiment is currently on the ISS EXPOSE-R facility and will be returned to ground after 1.5 years of exposure to space conditions (Rabbow *et al.* 2009).

Most of the previous astrochemistry experiments implemented in space so far were carried out in open cells exposed to solar UV radiation (Barbier *et al.* 1998, Barbier *et al.* 2002, Boillot *et al.* 2002). In these types of experiments, solid organic samples are deposited behind a UV-transparent window and exposed to the flux of solar radiation. If the studied molecule is sensitive to energetic photons, its photodestruction can be quantified when the sample are returned to Earth for analysis. However, gaseous products resulting from photolysis are lost to space. A first use of closed cells was reported in (Ehrenfreund *et al.* 2007), while a more advanced concept is presented in (Cottin *et al.* 2008). Such sealed cells allow study of the chemical evolution in the gaseous phase, as well as investigation of heterogeneous processes, such as the degradation of solid compounds and the release of gaseous fragments.

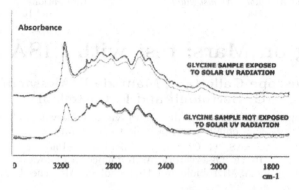

Figure 1. Infrared spectra of two open cells containing glycine before and after exposition in space during UVolution. The two top spectra show measurements for a cell exposed to solar UV (variation=-11%), while the two spectra at the bottom show measurements for a flight control cell (not exposed to solar UV, variation=0%). From the integrated exposition time of samples to UV (29 hrs), we estimate a half-lifetime of 8 days (±10%) for glycine at 1 AU.

Samples returned to Earth after the UVolution experiment have been analyzed in the laboratory. After processing of the measurements, the photochemical lifetime of the molecule at 1 AU is calculated, and can be extrapolated at other heliocentric distances and other astrophysical environments (diffuse interstellar medium, dark clouds).

Figure 1 shows an example of data processing in the case of glycine using infrared measurements. The measurements are also compared to experimental simulations performed in the laboratory on the same kind of samples with usual VUV lamps. The ratio space/lab result depends on the molecules and can be as high as a few hundreds, showing that laboratory VUV lamps are rather poor simulators of the solar flux.

The first use of closed cells has been successful. The analysis of an exposed mixture of CH_4/N_2 resulted as expected in the formation of C_2H_6 and C_2H_4 which were detected by gas chromatography. Complete results of the UVolution experiment will be published soon (Guan *et al.* 2009, in prep., Stalport *et al.* 2009, Stalport *et al.* 2009, in prep.).

References

Barbier, B., Chabin, A., Chaput, D., & Brack, A. 1998, *Planet. Space Sci.* 46, 391

Barbier, B., Henin, O., Boillot, F., Chabin, A., Chaput, D., & Brack, A. 2002, *Planet. Space Sci.* 50, 353

Boillot, F., Chabin, A., Buré, C., Venet, M., Belsky, A., Bertrand-Urbaniak, M., Delmas, A., Brack, A., & Barbier, B. 2002, *Orig. Life Evol. Biosph.* 32, 359

Cottin, H., B'enilan, Y., Gazeau, M.-C., & Raulin, F. 2004, *Icarus* 167, 397

Cottin, H., Coll, P., Coscia, D., Fray, N., Guan, Y. Y., Macari, F., Raulin, F., Rivron, C., Stalport, F., Szopa, C., Chaput, D., Viso, M., Bertrand, M., Chabin, A., Thirkell, L., Westall, F., & Brack, A. 2008, *Adv. Space Res.* 42, 2019

Demets, R., Schulte, W., & Baglioni, P. 2005, *Adv. Space Res.* 36, 311

Ehrenfreund, P., Ruiterkamp, R., Peeters, Z., Foing, B., Salama, F., & Martins, Z. 2007, *Planet. Space Sci.* 55, 383

Oro, J. & Holzer, G. 1979, *Journal of Molecular Evolution* 14, 153

Rabbow, E., Horneck, G., Rettberg, P., Schott, J.-U., Panitz, C., L'Afflitto, A., von Heise-Rotenburg, R., Willnecker, R., Baglioni, P., Hatton, J., Dettmann, J., Demets, R., & Reitz, G. 2009, *Orig. Life Evol. Biosph.*, in press (DOI 10.1007/s11084-009-9173-6)

Sagan, C. & Thompson, W. R. 1984, *Icarus* 59, 133

Stalport, F., Guan, Y. Y., Coll, P., Szopa, C., Macari, F., Raulin, F., & Cottin, H. 2009, *Astrobiology*, submitted

Highlights of Astronomy, Volume 15
XXVIIth IAU General Assembly, August 2009
Ian F. Corbett, ed.

Surviving on Mars: test with LISA simulator

Giuseppe Galletta[1,2], Maurizio D'Alessandro[3]
G. Bertoloni[4] and F. Castellani[4]

[1] Dipartimento di Astronomia, Università di Padova, vicolo Osservatorio 3, 35122 Padova, Italy
email: giuseppe.galletta@unipd.it

[2] CISAS "G. Colombo", Università di Padova
[3] INAF, Osservatorio Astronomico di Padova
[4] Dipartimento di Istologia, Microbiologia e Biotecnologie Mediche, Università di Padova

Abstract. We present the results of experiments performed in the Padua simulators of planetary environments, named LISA, used to study the limits of bacterial life on the planet Mars. The survival of *Bacillus* strains for some hours in Martian environment is briefly discussed.

1. Introduction

Mars is a typical target for exploring the possibility that life arose on a planet different from Earth. Its present surface conditions are very inhospitable, if compared with most of the terrestrial environments, but several lifeforms have shown a strong capability to adapt to very harsh conditions and to survive even for some period in circumterrestrial space. If life, similar or different from the terrestrial one, existed on Mars in the past, some life forms could have adapted to the climate changes and could have survived in some ecological niche, near the soil or deep under the surface. Moreover, while exploring Mars, we may incidentally or voluntary drop terrestrial lifeforms able to survive and to be reactivated once brought back in laboratories. Finding conditions that allow the survival of lifeforms in a Martian environment can have a double value: increasing the hope of finding extraterrestrial life and defining the limits for a terrestrial contamination of planet Mars.

2. Simulating Mars

In order to understand if some lifeforms may have survived on Mars, we conceived and built two simulators of its environments (Galletta *et al.* 2006, Galletta *et al.* 2007) where to perform researches with bacteria strains: LISA (Laboratorio Italiano Simulazione Ambienti), that allows six simultaneous experiments, and a single-experiment version of it (mini-LISA). Our LISA environmental chambers may reproduce the conditions of many Martian locations near the surface (temperature ranging from 133K to 293K, atmospheric composition with the 95% of $CO2$ at a pressure between 6 and 9 mb, strong UV radiation). Since we use a 500 liters reservoir of liquid nitrogen, refuelled once per week, experiments not longer than 25 hours may be performed inside LISA, while in mini-LISA there is theoretically no time limit.

We proceeded keeping in mind two important caveats: First, we don't have (yet) neither Martian bacteria nor Martian soils to use for the tests, and so we must use terrestrial surrogates. Second, lifeforms on a different planet may have a fully different combination of nucleic acids and aminoacids, so our conclusions could be fully wrong if applied to a "Martian life". However, they are useful for contamination- or terraforming- studies. Inside our simulators we have studied the survival of several bacterial strains belonging to

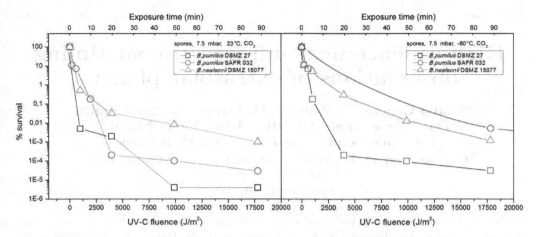

Figure 1. Survival of spores in LISA simulator at +23 and -80°C vs. exposure to UV light.

the genus *Deinococcus*, and to the endospore forming genera *Bacillus* and *Clostridium*. Cellular or endospores suspensions were layered on sterile coverslip dehydrated under sterile air flux, introduced in dedicated plates and then exposed to a pressure of 7.5 mbar of CO_2 and to ~ 4 W m^{-2} of UVC light. We simulated both "Martian summer day" (23°C) and "winter day" (-80°C) conditions for several hours.

3. Results and conclusions

In our experiments, we found that desiccation effect (water escape because of low pressure) may strongly decrease the survival of vegetative cells, but not of spores. In Martian environment, UV light appears to have the most cytocidal effect, while atmospheric gases or temperature are not relevant to the survival of cells or spores. Vegetative cells are inactivated by UV light in a few minutes, while spores may survive for hours (Fig. 1). Two of our *Bacillus* strains, *B. pumilus* and *B. Nealsonii*, have a particular capability to survive in Martian conditions without being screened by dust or other shields. As endospores suspension, they survive at least 4 hours and in some cases up to 28 hours in Martian conditions. We simulated the dust coverage happening on the real planet by blowing on the samples a very smaller quantity of grain of volcanic ash or dust of red iron oxide. Samples covered by these dust grains have shown a high percentage of survival, indicating that under the surface dust, if life was present on Mars in the past, some bacteria cell could still be present.

Acknowledgements

We would like to acknowledge the Air Liquide Italy - North East region for kindly providing us the liquid nitrogen in support of this research. This work has been funded by the University of Padua funds (ex 60%).

References

Galletta, G., Ferri, F., Fanti, G., DAlessandro, M., Bertoloni, G., Pavarin, D., Bettanini, C., Cozza, P., Pretto, P., Bianchini, G., & Debei, S. 2006, *Origins Life & Evol. Biosph.* 36, 625

Galletta, G., DAlessandro, M., Bertoloni, G., Fanti, G., Dainese, E., Pelizzo, M., Ferri, F., Pavarin, D., Bettanini, C., Bianchini, G., & Debei, S. 2007, *Mem. S.A.It.* 78, 608

Highlights of Astronomy, Volume 15
XXVIIth IAU General Assembly, August 2009
Ian F. Corbett, ed.

High cadence near-infrared transit timing observations of extrasolar planets

Claudio Cáceres[1,2], Valentin D. Ivanov[1], Dante Minniti[2]
Dominique Naef[1], Claudio Melo[1], Elena Mason[1],
Fernando Selman[1] and Grzegorz Pietrzynsky[3]

[1]European Southern Observatory [2]Pontificia Universidad Católica de Chile
[3]Universidad de Concepción

email: ccaceres@eso.org

Abstract. Currently the only technique sensitive to Earth mass planets around nearby stars (that are too close for microlensing) is the monitoring of the transit time variations of the transiting extrasolar planets. We search for additional planets in the systems of the hot-Neptune GJ-436 b, and the hot-Jupiter XO-1 b, using high cadence observations in the J and K_S bands, with the SofI and ISAAC instruments from La Silla Paranal Observatory. New high-precision transit timing measurements were used to derive new ephemeris. No statistically significant timing deviations were detected. We demonstrate that the high cadence ground based near-infrared observations are successful in constraining the mean transit time to 30 sec, and are a viable alternative to space missions.

1. Description

We present the first results from our timing study of individual transits of extrasolar planets with infrared detectors, using the *FastPhot* mode available in SofI@NTT, and ISAAC@VLT in La Silla - Paranal Observatory, which provides us with an unprecedented time resolution of 0.05-0.3 sec, and minimum "dead" time for readout (0.1%), generating a series of data cubes, with the target and one reference star in each windowed frame. The analyzed planets were the hot-Jupiter XO-1 b, and the hot-Neptune GJ-436 b.

2. Results

Figure 1 shows a light curve of planet XO-1 b with an exposure time of 0.08 sec. The total number of data points is $\sim 190,000$. In the best fitting model determination, the error calculation was developed with a Bootstrapping simulation which takes into account the presence of red noise in the data (Cáceres et al. 2009). The O-C diagram for the hot-Neptune GJ-436 b is also shown. These observations were taken on 17 May 2007, with the SofI@NTT at La Silla, in poor weather conditions. About 25,000 data points were collected using an exposure time of 0.24 sec, and for the best fitting model, we select stellar parameters from Gillon et al. (2007), and planetary parameters from Torres et al. (2007).

An individual transit timing accuracy of 30 sec is achieved. The data show some TTVs (Transit Time Variations) of up to 98 sec. However, these deviations are consistent with zero, within their respective uncertainties. Further observations with higher accuracy are necessary to better constrain the properties of these system and to address the question of whether they contains other planets.

The new ephemeris thus obtained are:

$$\text{XO-1 b:} \qquad T_C = 2453808.91682(13) + E \times 3.9415128(28) \text{HJD}$$

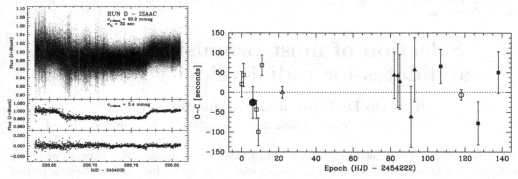

Figure 1. *Left.* A high cadence transit light curve for the transiting planet XO-1 b, taken with ISAAC@VLT the night of May 5, 2007. The best fitting model is shown, with planetary and stellar parameters taken from Holman *et al.* (2006) and McCullough *et al.* (2006) respectively. *Right.* The O-C diagram for our timing measurement of the transiting planet GJ 436b, and the data in the literature (see Cáceres *et al.* 2009 for more details).

$$\text{GJ-436 b:} \qquad T_C = 2454222.61588(12) + E \times 2.6438986(16)\,\text{HJD}$$

3. Conclusions

We achieve transiting timing accuracies of about 30 sec for individual transits. We find no significant evidence for perturbations of the orbital motion of GJ-436 b nor XO-1 b by other bodies in the system. Of course, a proper test of this hypothesis will require monitoring of multiple transits with the same or even higher accuracy. We demonstrate that the ground-based high-cadence observations of transiting extrasolar planets is an excellent technique for constraining the parameters of extrasolar planetary systems, because of the statistical significance of the obtained timing measurements. The timing precision is comparable with the space-based observations, making this method a good alternative to space missions, with their high cost and limited lifetime.

Acknowledgements

DM and CC are supported by the Basal Center for Astrophysics and Associated Technologies, and the FONDAP center for Astrophysics 15010003. This poster is also supported by Fondo ALMA-CONICYT 31070007, and the European Southern Observatory.

References

Cáceres, C., Ivanov, V. D., Minniti, D., Naef, D., Melo, C., Mason, E., Selman, F., & Pietrzynski, G. 2009, *A&A*, in press (arXiv:0905.1728)

Gillon, M., Pont, F., Demory, B.-O., Mallmann, F., Mayor, M., Mazeh, T., Queloz, D., Shporer, A., Udry, S., & Vuissoz, C. 2007, *A&A* 472, 13

Holman, M. J., Winn, J. N., Latham, D. W., O'Donovan, F. T., Charbonneau, D., Bakos, G. A., Esquerdo, G. A., Hergenrother, C., Everett, M. E., & Pál, A. 2006, *ApJ* 652, 1715

McCullough, P. R., Stys, J. E., Valenti, J. A., Johns-Krull, C. M., Janes, K. A., Heasley, J. N., Bye, B. A., Dodd, C., Fleming, S. W., Pinnick, A., Bissinger, R., Gary, B. L., Howell, P. J., & Vanmunster, T. 2006, *ApJ* 648, 1228

Highlights of Astronomy, Volume 15
XXVIIth IAU General Assembly, August 2009
Ian F. Corbett, ed.

© International Astronomical Union 2010
doi:10.1017/S1743921310010975

Selection of most promising CoRoT candidates for radial-velocity follow-up

Ronaldo Da Silva[1] and Adriana Silva-Valio[1,2]

[1]Instituto Nacional de Pesquisas Espaciais, Brazil, [2]Mackenzie University, Brazil

Abstract. We used the method of Silva & Cruz (2006), which distinguishes between planetary and stellar companions by fitting transit light curves, to select the most promising CoRoT candidates to be monitored with radial-velocity measurements. Testing this method on the light curves of confirmed CoRoT exoplanetary systems shows that the estimated radius for such planets is smaller than $2\,R_{\rm Jup}$, while for most of the light curves in which no planet has been detected, the secondary companion has an estimated radius larger than $2\,R_{\rm Jup}$. We present preliminary results concerning other light curves for which no planet has been detected yet.

The huge number of data released by the CoRoT mission makes very hard the selection of candidates to be monitored with radial-velocity measurements. Silva & Cruz (2006) proposed a method based on transit light curve fitting to distinguish between planetary and stellar companions by estimating the radius of the secondary. Here, we use the same method to select the best CoRoT candidates for planetary systems to be observed with radial velocities. After fitting a model to the observed transit, we determined the radius R_2 of the secondary companion (using the Kepler's third law and mass-radius relations), the semimajor axis, and the inclination angle of the orbital plane. We consider as good planetary candidates those objects for which we obtain $R_2 < 2\,R_{\rm Jup}$.

At present, we have analysed light curves from the first three public runs. The Table below shows examples of our results. Triangular-shaped transits are excluded by our method because they are probable caused by grazing eclipses of stars in a binary system. Apparently good candidates can also be excluded because they have have $R_2 > 2\,R_{\rm Jup}$ in spite of a planet-like transit shape. For the moment, we have a total of 22 good candidates for planetary systems. This is still ongoing work and soon we hope to be able to share a larger list of candidates to be monitored in the context of the RV follow-up.

CoRoT ID	Run	P [days]	a [AU]	i [deg]	M_1 [$M_\odot$]	R_1 [$R_\odot$]	R_2 [$R_{\rm Jup}$]	Remarks
0102811578	IRa01	1.66882	0.034	77	1.6	2.0	3.15	triangular
0101095286	LRc01	5.053	0.092	71	3.7	7.0	4.11	triangular
0102855534	IRa01	21.72	0.207	> 89	2.2	3.1	3.20	apparently good
0211660858	SRc01	8.825	0.134	> 89	2.2	3.1	4.09	apparently good
0102763847	IRa01	10.5328	0.083	88	0.7	0.7	0.80	good candidate
0101086161	LRc01	6.2125	0.068	86	1.1	1.2	0.96	good candidate

Acknowledgements

We thank the financial support from Fundação de Amparo a Pesquisa do Estado de São Paulo (FAPESP) in the form of a fellowship.

References

Silva, A. V. R. & Cruz, P. C. 2006, *ApJ* 642, 488

Highlights of Astronomy, Volume 15
XXVIIth IAU General Assembly, August 2009
Ian F. Corbett, ed.

Orbital stability of Earth-type planets in stellar binary systems

Jason Eberle, Manfred Cuntz and Zdzislaw E. Musielak

Department of Physics, University of Texas at Arlington, Arlington, TX, USA
email: wjeberle@uta.edu, cuntz@uta.edu, zmusielak@uta.edu

Abstract. An important factor in estimating the likelihood of life elsewhere in the Universe is determining the stability of a planet's orbit. A significant fraction of stars like the Sun occur in binary systems which often has a considerable effect on the stability of any planets in such a system. In an effort to determine the stability of planets in binary star systems, we conducted a numerical simulation survey of several mass ratios and initial conditions. We then estimated the stability of the planetary orbit using a method that utilizes the hodograph to determine the effective eccentricity of the planetary orbit. We found that this method can serve as an orbital stability criterion for the planet.

1. Introduction and method

Based on observational evidence, a significant number of extrasolar planets are found in binary and multiple stellar systems, see Patience *et al.* 2002, Eggenberger *et al.* 2004, which is a strong motivation to pursue detailed theoretical studies of planetary orbital stability in stellar binaries. Previous results have been obtained by, e.g., Musielak *et al.* (2005), Cuntz *et al.* (2007), and Eberle *et al.* (2008). In the following, we present a new method that relies on a differential geometrical approach based on the analysis of the curvature of the hodograph in the synodic coordinate system. We found that when orbital instability occurs, the median of the eccentricity distribution exceeds unity. This criterion can be included in detailed simulations in an automated mode allowing the identification of planetary ejection in a straightforward manner. Intuitively, this criterion also agrees with the most basic property of conic sections representing closed orbits for $e < 1$ and open orbits for $e \geqslant 1$, although it utilizes a modified definition of eccentricity (Eberle & Cuntz 2010).

2. Results and discussion

We define $\mu = M_2/M$ with $M = M_1 + M_2$, where M_1 and M_2 are the masses of the primary and secondary star, respectively. Furthermore, ρ_0 denotes the planet's relative initial distance from its host (primary) star. The mass of the planet is assumed to be very small compared to the stellar binary masses. If the planet orbits close to the primary star, the effective eccentricity remains low for the entire duration of the simulation. When the parameters are such that the zero velocity curve is nearly intersecting the L1 equilibrium point (C_1 contour, green in Fig. 1), the effective eccentricity increases. For mass ratios less than 0.1 the zero velocity contour corresponding to L2 (C_2, red in Fig. 1) is a suitable limit for instability. For initial conditions where the planet is far enough away that the zero velocity curve as already opened up at L3 (C_3 contour, black in Fig. 1), the planet is not well defined to be orbiting the primary star and quickly escapes the region of the binary stars and for most initial conditions, the planet quickly returns and rapidly

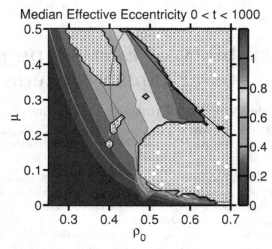

Figure 1. See text for details.

approaches one of the stars. The white regions with black crosses represent simulations that were terminated before the maximum 1000 year limit due to captures by one of the stars or ejection from the system. One unstable region is in the upper mass ratio range, roughly from $\mu = 0.35$ to 0.50 and another is from 0.00 to 0.23. In between these two regions ($\mu = 0.24$ to 0.34) is a quasi-stable region in which the effective eccentricity increases almost uniformly. The boundaries of this region remain basically the same for the duration of the simulation. Of course, we should not be surprised if the quasi-stable region were to shrink over a longer time period as 1000 orbits is not a very long time when considering long term stability. In fact, there is already a instability "island chain" slowly forming right in the middle of this quasi-stable region. The reasons for this instability "archipelago" are still under investigation.

3. Conclusions

A comparison with the zero-velocity contour of the planetary orbit shows that this criterion, which is a sufficient criterion for orbital instability, closely coincides with the opening of the zero-velocity contour at the Lagrange point L3, located to the right of the primary star, as discussed by Cuntz *et al.* (2007) and Eberle *et al.* (2008). If instability occurs, extremely large effective eccentricities are found, indicating a highly hyperbolic planetary orbit in the synodic coordinate system, a stark indication of planetary escape. Although our results have been obtained for the special case of the restricted three-body problem, we expect that it may also be possible to augment this criterion to planets in generalized stellar binary systems, and to use it concerning long-term simulations.

References

Cuntz, M., Eberle, J., & Musielak, Z. E. 2007, *ApJ*(Letters) 669, L105
Eberle, J., Cuntz, M., & Musielak, Z. E. 2008, *A&A* 489, 1329
Eberle, J. & Cuntz, M. 2010, *A&A*, submitted
Eggenberger, A., Udry, S., & Mayor, M. 2004, *A&A* 417, 353
Musielak, Z. E., Cuntz, M., Marshall, E. A., & Stuit, T. D. 2005, *A&A* 434, 355 & Erratum 2008, *A&A* 480, 573
Patience, J., White, R. J., Ghez, A. M. *et al.* 2002, *ApJ* 581, 654

Highlights of Astronomy, Volume 15
XXVIIth IAU General Assembly, August 2009
Ian F. Corbett, ed.

© International Astronomical Union 2010
doi:10.1017/S1743921310010999

Detecting planetary signals
with Bayesian methods

Samuli Kotiranta and Mikko Tuomi

Department of Physics and Astronomy, University of Turku, Turku, Finland
email: samuli.kotiranta@utu.fi

Abstract. In this paper we present an application of Bayesian model comparison to the radial velocity measurements of suspected extra-solar planetary system host star.

The Bayesian methods can be seen as an alternative to the classical statistical tools. Although these methods are generally well known, their use in astronomy has been concentrated on few specific areas where parameter space is high dimensional and analysis is difficult to perform through more conventional 'frequentist methods'. However, explicit use of Bayesian tools may reveal important new properties from simpler systems as well.

The inverse solution, of a k-planet radial velocity model $r_i(t_i) = [\sum_{j=1}^{k} K_j[\cos(\nu_j(t) + \omega_j) + e_j \cos\omega_j]]_i + \gamma_i + \epsilon_{Ii} + \epsilon_{Ji}$ was calculated with Markov chain Monte Carlo (MCMC) with Metropolis-Hastings algorithm (Hastings 1970). Here r is measured value at time t while K, ν, ω and e are the radial velocity semi-amplitude, true anomaly, longitude of pericentre and eccentricity, respectively. Parameter γ is the reference velocity. The instrument error ϵ_I and intrinsic stellar radial velocity activity ('jitter') ϵ_J are assumed to be normally distributed and jitter is kept unfixed.

The actual procedure consists of comparing k-planet model with the given data. Using this process we get marginal integrals that correspond to the probabilities of each model with k companions. When the most probable model has been found as a result, the Markov chains con be used to estimate the joint probability distribution function of the free parameters.

Because each model consist of k non-interacting planets there is a possibility that solution – even though statistically the most probable – appears to be physically unlikely. The dynamical analysis for the solution is done by taking a number of random initial value sets from the probability distributions. These were then integrated numerically with Bulirch-Stoer method (Bulirsch & Stoer 1966).

As an successful example of this procedure we found evidence of a new planet, namely HD 11506 c (Tuomi & Kotiranta 2009).

Acknowledgements

S. K. acknowledges the IAU and the Magnus Ehrnrooth Foundation for travel grants that made it possible to participate in the IAU XXVIIth General Assembly.

References

Bulirsch, R. & Stoer, J. 1966, *Numer. Math.* 8, 1
Hastings, W. 1970, *Biometrika* 57, 97
Tuomi, M. & Kotiranta, S. 2009, *A&A* 496, L13

Highlights of Astronomy, Volume 15
XXVIIth IAU General Assembly, August 2009
Ian F. Corbett, ed.

© International Astronomical Union 2010
doi:10.1017/S1743921310011002

Multiplicity study of exoplanet host stars

Markus Mugrauer, Ralph Neuhäuser
Christian Ginski and Thomas Eisenbeiss

Astrophysical Institute and University-Observatory Jena, Germany
email: markus@astro.uni-jena.de

Abstract. We present recent results of our ongoing multiplicity study of exoplanet host stars.

1. New low-mass stellar companions of exoplanet host stars

In our imaging campaign, carried out with SofI/NTT and UFTI/UKIRT, we directly detected so far several new companions of exoplanet host stars. Among them HD 3651 B the first T dwarf companion of an exoplanet host star (1, 2), HD 27442 B a white dwarf which is the secondary of the most evolved exoplanet host star system presently known (3), as well as the binary companion of HD 65216, whose B component is a low mass star, while HD 65216 C is either a massive brown dwarf or a very low-mass star (4). Recently, we identified two new low-mass stellar companions of the exoplanet host stars HD 125612 and HD 212301. The co-moving companion of HD 125612 is a wide M4 dwarf $(0.18 \, M_\odot)$, located about 4750 AU south-east of its primary. The co-moving companion of HD 212301 is a close M3 dwarf $(0.35 \, M_\odot)$, which we found at about 230 AU north-west of the exoplanet host star. The binaries HD 125612 AB and HD 212301 AB are two new members in the continuously growing list of exoplanet host star systems of which more than 40 are presently known (5).

2. Lucky-Imaging search for close companions of exoplanet host stars

We started a search for close stellar companions of exoplanet host stars at the Calar Alto Observatory in Spain, using the Lucky-Imaging (L-I) technique. The observations are carried out with the 2.2 m telescope and its L-I camera AstraLux in the I-band. We take several thousand images with integration times down to 30 ms, and choose a total integration time of about 30 min per target. After standard data-reduction, our L-I pipeline measures the Strehl-ratios of all images, and then selects only those images with the highest Strehl-ratios (selection rates from 1 to 10%). Finally, all selected images are shifted and combined. With the achieved AstraLux detection limit, beyond 1 arcsec ($\sim$40 AU of projected separation at the average distance of our targets), we are sensitive to all stellar companions (with $M > 0.08 \, M_\odot$) around our targets. Hence, close stellar companions that remain invisible in seeing limited observations are clearly detectable.

References

Mugrauer, M., Seifahrt, A., Neuhäuser, R., & Mazeh, T. 2006, *MNRAS* 373, L31
Mugrauer, M., Seifahrt, A., Neuhäuser, R., Mazeh, T., & Schmidt, T. 2007, *IAUS* 240, 638
Mugrauer, M., Neuhäuser, R., & Mazeh, T. 2007, *A&A* 469, 755
Mugrauer, M., Seifahrt, A., & Neuhäuser, R. 2007, *MNRAS* 378, 1328
Mugrauer, M. & Neuhäuser, R. 2009, *A&A* 494, 373

Highlights of Astronomy, Volume 15
XXVIIth IAU General Assembly, August 2009   © International Astronomical Union 2010
Ian F. Corbett, ed.   doi:10.1017/S1743921310011014

Modelling extrasolar planetary atmospheres

France Allard

Centre de Recherche Astrophysique de Lyon,
UMR 5574: CNRS, Université de Lyon, École Normale Supérieure de Lyon,
46 allée d'Italie, F-69364 Lyon Cedex 07, France
email: fallard@ens-lyon.fr

Abstract. The atmospheres of close-in Extrasolar Giant Planets (EGPs) experience important stellar radiation, raising the question of the heat redistribution around the planetary surface and of the importance of photochemistry effects for their spectral properties. They experience mass loss via quasi-thermal escape of their lightest elements. They rotate and experience tidal effects. Model atmospheres struggle to include even part of this complexity. Some address the dynamics of the atmospheres as a whole (3D) as subjected to rotation, or as patches of the surface (wind studies), compromising on the details of the composition and radiative/convective properties. Others solve the composition and radiative/convective properties, compromising on dynamical effects such as rotation. In this paper, we review existing model atmospheres for EGPs, and present the first high spatial resolution local (as opposed to global) 2/3D radiation hydrodynamic simulations of EGP atmospheres including dust cloud formation.

The SED of a Hot Jupiter is composed of thermal radiation left over from contraction and formation processes, stellar radiation reflected off the planetary surface, and stellar radiation absorbed by the stellar surface and reemitted redwards by thermal radiation. Planet-to-star flux contrast levels are found to be around 7 dex at the reflection peak (5000Åfor a G2 type star) and 5 dex at near-infrared wavelengths. Since this contrast is more favorable at near-IR and Infrared wavelengths, caracterisation studies of planetary candidates found by imaging use model atmospheres ignoring the effects of impinging radiation on the shape of their SED, and account for stellar irradiation by applying an achromatic correction factor. These models (in Allard *et al.* 2001) are readily available via a web simulator (http://phoenix.ens-lyon.fr/simulator), and tested for the study of brown dwarfs which cover a similar range of parameters (Teff, surface gravity) than for young planetary mass objects.

While the study of brown dwarfs can be enlightening for the study of planetary atmospheres – brown dwarf atmospheres are the site of an onset of dust cloud formation, strong rotation, and/or magnetic fields – their study thus far proceeded using only 1D static, often plane-parallel model atmospheres of various degrees of radiative transfer sophistication: Opacity Sampling vs. K-coefficient techniques, NLTE, photoionisation and photochemistry vs. Equilibrium Chemistry, stationary particle diffusion solutions, cloud models, assuming adiabaticity instead of using the Mixing Length Theory for convective mixing, etc.

For instance, Barman, Hauschildt & Allard (2001), Barman *et al.* (2002), and Barman, Hauschildt & Allard (2005) have explored the impact of dust cloud formation, NLTE and the photoionisation of sodium, and a reconstruction of the planetary surface with 1D static models, using the (at depth) entropy matching technique often used in a similar way for the study of binary stars in astrophysics. This technique allows to explore the orbital phase variations of the SED. These models led to the co-discovery of water vapor in the STIS spectro-photometric observations of HD209458 b (Barman 2007).

A further leap in terms of detailed model atmospheres was recently achieved by Freytag *et al.* (2009a), who have modeled the atmospheres of VLMs, BDs, and PMOs by local 2D and 3D high-resolution Radiation HydroDynamic (RHD) simulations, including a cloud model and dynamical molecular transport as well as impinging radiation by a parent star. These models led to a stratospheric temperature inversion needed to explain the SED of Hot Jupiters, and predict dust clouds to form easily in the atmospheres of the CoRot planets, except perhaps for CoRot-3b where dust is expected to form cirrus-like clouds in the optically transparent upper layers (Freytag *et al.* 2009b). These models suggest the formation of gravity waves, temporal intensity variability, and dynamical mixing in brown dwarf and Hot Jupiter atmospheres. Further advance on this front requires therefore the systematic and consistent account of non-equilibrium chemistry. Thus far, only some species have been explored, often using rates extrapolated to the atmospheric conditions of brown dwarfs (Saumon *et al.* 2006).

Another issue is the strong radiation field to which are subjected Hot Jupiters, and which raises, among others, the question of the redistribution of the stellar light around the planetary surface: *i*) via radiation (requiring 3D radiative transfer solutions), and *ii*) via winds due to pressure contrasts and/or rotation (requiring hydrodynamical studies). The entropy matching technique led to a day (1995K)-to-night (500K) side temperature contrast of 1500K. Iro, Bézard & Guillot (2005), using time-dependant 1D radiative transfer, have accounted for wind velocities corresponding to the expected rotational period of a gravitationally locked (with the star) exoplanet. However, the self-consistent account of rotation is a yet unsolved challenge for the understanding of brown dwarfs and planetary properties. Accounting for rotation, while neglecting the opacities and convection as a trade-off, led to a day (1700K)-to-night (300K) contrasts of the order of 600K.

A further step was achieved by Fortney *et al.* (2006) who computed phase spectra based on the results of hydrodynamical wind studies (Cooper & Showman 2006), themselves based on the radiative timescales of Iro, Bézard & Guillot (2005). These simulations led to stratospheric winds which could explain the observed Spitzer 4 to 6 μm flux excess. A review of dynamical models of Hot Jupiter atmospheres is available in Showman, Menou & Cho (2007).

References

Allard, F., Hauschildt, P. H., Alexander, D. R., Tamanai, A., & Schweitzer, A. 2001, *ApJ* 556, 357

Barman, T. 2007, *ApJ* (Letters) 661, L191

Barman, T. S., Hauschildt, P. H., & Allard, F. 2001, *ApJ* 556, 885

Barman, T. S., Hauschildt P. H., & Allard, F. 2005, *ApJ* 632, 1132

Barman, T. S., Hauschildt, P. H., Schweitzer, A., Stancil1, P. C., Baron, E., & Allard, F. 2002, *ApJ* (Letters) 569, L51

Cooper, C. S. & Showman, A. P. 2006, *ApJ* 649, 1048

Fortney, J., Cooper, C., Showman, A., Marley, M., & Freedman, R. 2006, *ApJ* 652, 746

Freytag, B., Allard, F., Ludwig, H.-G., Homeier, D., & Steffen, M. 2009a, *A&A*, in press

Freytag, B., Allard, F., Ludwig, H.-G., Homeier, D., & Steffen, M. 2009b, *Proceedings of the Conference "Molecules in the Atmospheres of Extrasolar Planets"* held at Paris Observatory, November 19, 2008.

Hauschildt, P. H., Allard, F., & Baron, E. 1999, *ApJ* 512, 377

Iro, N., Bézard, B., & Guillot, T. 2005, *A&A* 436, 719

Saumon, D., Marley, M. S., Cushing, M. C., Leggett, S. K., Roellig, T. L., Lodders, K., & Freedman, R. S. 2006, *ApJ* 647, 552

Showman, A. P., Menou, K., & Cho, J. Y. K. 2008, *ASP Conference Series* Vol. 398, 419

Highlights of Astronomy, Volume 15
XXVIIth IAU General Assembly, August 2009
Ian F. Corbett, ed.

Defining the envelope
for the search for life
in the Universe

Lynn J. Rothschild

Mail Stop 239-20, NASA Ames Research Center, Moffett Field, CA 94035-1000, USA
email: lynn.j.rothschild@nasa.gov

Abstract. The search for life in the universe relies on defining the limits for life and finding suitable conditions for its origin and evolution elsewhere. From the biological perspective, a conservative approach uses life on earth to set constraints on the environments in which life can live. Conditions for the origin of life, even on earth, cannot yet be defined with certainty. Thus, we will describe what is known about conditions for the origin of life and limits to life on earth as a template for life elsewhere, with a particular emphasis on such physical and chemical parameters as temperature, pH, salinity, desiccation and radiation. But, other life forms could exist, thus extending the theoretical possibility for life elsewhere. Yet, this potential is not limitless, and so constraints for life in the universe will be suggested.

---

1. Introduction

To find something one has to know where to look. From Earth to the edge of the observable universe is about 46.5 billion light-years, so for the moment we must narrow the search considerably to where it is possible for life to reside, while staying within our technological bounds. Life is always likely to be based on organic carbon because carbon is the fourth most common element in the universe, its chemical versatility, the discovery of organic compounds elsewhere, and the fact that we are made of it (Rothschild 2009). Thus, as a first order organic compounds must be stable and function in order for life to exist, which provides a theoretic maximum envelope for life. Because Earth is only one place that life is known thus far, the minimum envelope is derived by assessing the environmental limits for life on Earth. But what a minimum envelope, as life swarms over Earth in many niches that until recently seemed uninhabitable.

2. The extremes of life

The extreme environments and examples of organisms that inhabit them are listed in Table 1. All of these environments at some point make it difficult for organic carbon to stay intact and/or for a solvent – such as water for Earth-based life – to stay liquid (see review in Rothschild 2009). Further, each environment can add other complexities. For example, at low temperatures membranes loose their fluidity and enzymatic reactions are slowed to the point that they cannot sustain life. Radiation and oxidative damage are of particular interest as they provide physical and chemical limits to life, but also act to mutate the genetic material. To this end, our lab has conducted experiments in high ultraviolet environments from the Bolivian Altiplano to Mount Everest, and by transporting biological samples to 33 km on high altitude balloons through Stanford's BioLaunch program.

Environment	Type	Definition	Example
Temperature	hyperthermophile	growth >80° C	*Pyrolobus fumarii*-113° C strain 121
	thermophile	growth 60-80° C	
	mesophile	growth 15-60° C	*Homo sapiens*
	psychrophile	growth <15° C	*Psychrobacter*, insects
pH	alkaliophile	pH >9	OF4 (10.5); 12.8?
	acidophile	low pH loving	*Cyanidium, Ferroplasma*
Desiccation	xerophile	cryptobiotic anhydrobiotic	tardigrades
Salinity	halophile	2-5 M NaCl	*Haloarcula, Dunaliella*
Radiation		high radiation	*Deinococcus radiodurans*
Oxygen	anaerobe	cannot tolerate O_2	*Clostridium*
	miroaerophil	low levels of O_2	*Methanococcus jannaschii*
	aerophile	mid to high O_2	*Homo sapiens*
Pressure	barophile/ piezophile	pressure/ weight loving	*Shewanella* viable at 1600 MPa tardigrades
Vacuum		tolerates vacuum	tardigrades, insects, microbes, seeds
Gravity	hypo/hypergravity	<1 g / >1g	none known
Chemical	gasses, metals	tolerates high levels	CO_2 (*Cyandium cadarium*); Cu/As/ Cd/Zn (*Ferroplasma acidarmanus*)
Electricity			electric eel

Table 1. Examples of extremophiles. Adapted from Rothschild & Mancinelli (2001).

3. Could it happen again?

To assess if life could arise again and inhabit these niches, once again the biodiversity of life on Earth provides clues. When more than one organism has converged on a solution for an environmental extreme, it gives us more confidence that this evolutionary adaptation is not a one time event (Rothschild 2008). Multiple organisms have evolved to function at low and high temperature, low and high pH and so on. This suggests that there may be some universality – at least given the starting points of organic carbon and liquid water as a solvent – to the extreme environments for life.

4. Where is the field heading?

As we find more locations in our solar system and beyond that meet single variable constraints for life, attention must be paid to polyextremophiles, or organisms that can cope with multiple extreme parameters. Will this show that all possible niche space is occupied or that there are environmental combinations that, for some reason, cannot be occupied?

References

Rothschild, L. J. 2008, *Phil. Trans. Royal Soc. B* 363, 2787

Rothschild, L. J. 2009, in: C. Bertka (ed.), *Exploring the Origin, Extent and Future of Life*, (Cambridge: Cambridge University Press), p. 113

Rothschild, L. J. & Mancinelli, R. L. 2001, *Nature* 409, 1092

Highlights of Astronomy, Volume 15
XXVIIth IAU General Assembly, August 2009
Ian F. Corbett, ed.
© International Astronomical Union 2010
doi:10.1017/S1743921310011038

Earthshine observations and the detection of vegetation on extrasolar planets

Danielle Briot

Observatoire de Paris, 61 Avenue de l'Observatoire, 75014 Paris, France
email: danielle.briot@obspm.fr

Abstract. To prepare future observations of terrestrial planets and the detection of life, we search for life on the planet Earth seen as a point source. Observations of Earthshine is a convenient way to see Earth as a remote planet. The vegetation reflectance spectrum presents a sharp edge in the near infrared: the Vegetation Red Edge. Observational programs in progress are described, particularly our observations at the Concordia station in Antarctica.

1. Introduction

Earthshine (or Ashen Light) is the light from the Moon which is reflecting the illuminated part of the Earth that is facing it. The first known publication of the correct explanation of the origin of Ashen Light is in the "Sidereus Nuncius" published by Galileo Galilei in 1610. While Kepler attributed the correct interpretation to his master Maestlin (1550-1631), the oldest unpublished one is found in manuscripts of Leonardo da Vinci dated 1506–1509. Another unpublished explanation predating the "Siderius Nuncius" can be found in "Pensieri", a collection of notes written between 1578 and 1597 by Fra Paolo Sarpi, a friend of Galileo.

As early as 1912, Arcichovsky (1912) suggested looking for a signature of chlorophyll in Earthshine to calibrate for chlorophyll in the spectra of other planets. However, this approach was not technically feasible at the time and was soon forgotten, even though such observations allowed Tikhoff (1914) to deduce the blue colour of our planet due to Rayleigh scattering (more about Tikhoff can be found in Tejfel 2009, or Briot *et al.* 2004). Nearly nine decades later, it was rediscovered independently when the detection of extrasolar planets kindled new interest in the search for life. Indeed, due to the roughness of the lunar surface, any lightray in Earthshine corresponds to a reflection of the entire disk of the Earth, a situation that simulates the case of an exoplanet seen as a point source.

One can reasonably think that extrasolar terrestrial planets, or Super Exo-Earths, will be detected and extensively studied in the forthcoming decades. Therefore, we need to plan for the detection of life. Vegetation, which covers approximately 60% of the surface of the continents, plays an important role in the Earth albedo. Chlorophyll is the fundamental pigment of photosynthesis. The spectrum of chlorophyll presents a small hump in the green wavelength range, so we see grass and plants as green, but it also exhibits a much more important sharp rise in the near infrared ($\approx 700\,nm$), the so-called Vegetation Red Edge (VRE), which can hardly be confused with another spectral feature.

2. Past results and present observations

The first detections of the VRE in the Earth spectrum were obtained by Arnold *et al.* (2002) and by Woolf *et al.* (2002). The red side of the spectrum shows the presence of O_2 and H_2O absorption bands and of the VRE of chlorophyll, while the blue side clearly shows the Huggins and Chappuis ozone (O_3) absorption bands (see for example Hamdani et al. 2006). Atmospheric molecules like oxygen and ozone are possible biomarkers.

Values of the VRE collected from the literature are given in Arnold (2008). Its magnitude is only a few percent, and Arnold et al. (2002), and Hamdani et al. (2006), have shown that it is lower when an ocean is facing the Moon (1.3% for the Pacific Ocean), than when forested lands are visible (4% for Africa and Europe). At low or mid-latitudes, Earthshine can only be observed during morning or evening twilight, and only two distinct portions of the Earth surface can be facing the Moon for a given telescope. If observations are carried out at very high latitudes, Earthshine can be observed during a large fraction of the day during about 8 periods in the year, and because of Earth rotation, different terrestrial "landscapes" are facing the Moon. For this reason, we use the Concordia station, a French-Italian base for scientific research, including astronomical observations, located at the Dome C in Antarctica (lat = 75°S, lon = 123°E, altitude = 3220 m) where the polar night lasts 3 months, the mean air temperature is −50.8°C, and the lowest air temperature is −84.4°C. A dedicated instrument called LUCAS was built (Briot et al. 2009). After technical setbacks during the 2008 winterover campaign, successful observations were carried out during the three usable lunar cycles since the June 2009 solstice. Observation runs lasted up to 8 hours, which is an impossible feat at mid-latitudes.

3. Conclusions

Observations of Earthshine represent a good example of a scientific topic which was first carefully investigated, then abandoned, and finally becomes very interesting again. Some geophysical applications also exist, according to the recommandations made by the NASA Navigator Program: "continued observations of Earthshine are needed to discern diurnal, seasonal, and interannual variations". As such, LUCAS is also a test for the design and improvement of small instrumentation, data collecting and management of observations in a extremely cold environment as in the Concordia station.

Acknowledgements

Many thanks are due to all the members of the LUCAS team: J. Schneider, S. Jacquemoud, L. Arnold, J. Berthier, and P. Rocher, and specials thanks to the valiant winterover observers: K. Agabi, E. Aristidi, E. Bondoux, Z. Challita, D. Petermann, and C. Pouzenc.

References

Arcichovsky, V. M. 1912, Ann. Inst. Polytech. Don Tsar. Alexis (Novotcherkassk) 1(17), 195
Arnold, L. 2008, Space Sci. Revs 135, 323
Arnold, L., Gillet, S., Lardière, O., Riaud, P., & Schneider, J. 2002, A&A 392, 231
Briot, D., Schneider, J., & Arnold, L. 2004, Extrasolar planets: Today and Tomorrow, ASP Conference Series, Vol. 321, 2004
Briot, D., Arnold, L., Jacquemoud, S., Schneider, J., Agabi, K., Aristidi, E., Berthier, J., Bondoux, E., Challita, Z., Petermann, D., Pouzenc, C., & Rocher, P. 2009, Highlights of Astronomy Vol. 15, (this volume), Special Session 3
Hamdani, S., Arnold, L., Foellmi, C., Berthier, J., Billères, M., Briot, D., François, P., Riaud, P., & Schneider, J. 2006, A&A 460, 617
Tejfel, V. 2009, Highlights of Astronomy Vol. 15, (this volume), Special Session 6
Tikhoff, G. A. 1914, Mitteilungen der Nikolai-Hauptsternwarte zu Poulkovo, n°62, Band VI_2, 15
Woolf, N. J., Smith, P. S., Traub, W. A., & Jucks, K. W. 2002, ApJ 99, 225

Highlights of Astronomy, Volume 15
XXVIIth IAU General Assembly, August 2009
Ian F. Corbett, ed.

© International Astronomical Union 2010
doi:10.1017/S174392131001104X

The discovery of glycolaldehyde in a star forming region

Maria T. Beltràn[1], Claudio Codella[1], Serena Viti[2] Roberto Neri[3] and Riccardo Cesaroni[1]

[1] Arcetri Observatory (Italy), [2] UCL (UK), [3] IRAM (France)

Abstract. Glycolaldehyde is the simplest of the monosaccharide sugars and is directly linked to the origins of life. We report on the detection of glycolaldehyde (CH_2OHCHO) towards the hot molecular core G31.41+0.31 through observations with the IRAM PdBI (Plateau de Bure Interferomter) at 1.4, 2.1, and 2.9 mm.

The CH_2OHCHO emission comes from the hottest ($\geqslant$ 300 K) and densest ($\geqslant 2\times10^8$ cm^{-3}) region closest ($\leqslant 10^4$ AU) to the (proto)stars. The comparison of data with gas-grain chemical models of hot cores suggests for G31.41+0.31 an age of a few 10^5 yr. We have also shown that only small amounts of CO need to be processed on grains in order for existing hot core gas-grain chemical models to reproduce the observed column densities of glycolaldehyde, making surface reactions the most feasible route to its formation (Beltràn et al. 2009).

Figure 1 shows the brightness temperature scale of the CH_2OHCHO ($20_{2,18}$-$19_{3,17}$), ($14_{0,14}$-$13_{1,13}$), and ($10_{1,9}$-$9_{2,8}$) at 220463.87, 143640.94, and 103667.91 MHz, respectively, as observed towards the central position of the G31.41+0.31 hot core - see Beltràn et al. 2009 for details. Rest frequencies are pointed out by vertical bars.

- Upper panel: the glycolaldehyde line is blended with the CH3CN (12-11; K = 8) line. Two additional lines are present: (i) $^{13}CH_3CN$ ($12_6 - 11_6$; labeled by K'), and (ii) HCOOCH$_3$-A ($25_{11,15} - 26_{9,18}$) (2204445.79 MHz; Eu = 272 K) which could contain an emission contribution due to the CH_2OHCHO ($18_{4,14} - 17_{4,13}$) (220433.51 MHz; Eu = 108 K) line. The continuous line shows the fit to the group of three lines formed by the CH_2OHCHO ($20_{2,18} - 19_{3,17}$), CH$_3$CN (12-11; K = 8), and $^{13}CH_3CN$(12-11; K' = 6); the dotted lines draw the three individual Gaussian curves used for the fit.

- Middle panel: the CH_2OHCHO line is part of a spectral pattern containing also the HCCC^{13}CCN (143636.63 MHz; Eu = 183 K), C$_2$H$_3$CN ($33_{2,31} - 32_{4,28}$ (143646.50 MHz; Eu = 620 K), and C$_2$H$_5$OH ($29_{2,28} - 28_{3,26}$ (143651.78 MHz; Eu = 415 K) lines. The results of the fit as drawn as in the upper panel.

- Lower panel: besides the glycolaldehyde emission, an unidentified spectral pattern is present around 103674 MHz. The solid curve shows the fit of the isolated CH_2OHCHO line.

References

Beltràn, M., Codella, C., Viti, S., Neri, R., & Cesaroni, R. 2009, *ApJ* (Letters) 690, L93

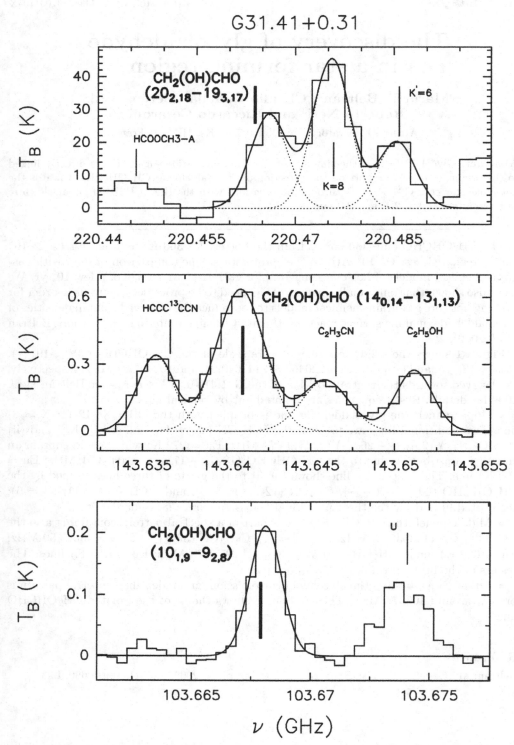

Figure 1. Beam-averaged spectra in the hot core of G31.41+0.31 (after Beltràn *et al.* 2009)

Highlights of Astronomy, Volume 15
XXVIIth IAU General Assembly, August 2009
Ian F. Corbett, ed.

© International Astronomical Union 2010
doi:10.1017/S1743921310011051

The "Living with a Red Dwarf" program: XUV radiation and plasma environments of hosted planets and impacts on habitability

Edward F. Guinan, Scott G. Engle, Trisha Mizusawa
George P. McCook, A. Wolfe and Jeffrey Coughlin

Department of Astronomy & Astrophysics, Villanova University
800 E. Lancaster Ave, Villanova, PA 19085 USA
email: edward.guinan@villanova.edu

Abstract. We describe the LWARD program on dM stars and habitability in their environment.

Red dwarf (dM) stars are overwhelmingly the most numerous stars in our Galaxy. These cool, low-luminosity, low-mass stars comprise >75% of all stars. They have very long lifetimes of 30–200+ Gyr and have nearly constant luminosities for tens of Gyrs. Because of this, red dwarfs have essentially fixed (but close-in) Habitable Zones (HZs). Determining the number of dM stars with planets and assessing planetary habitability is critically important because such studies can indicate how common life is in the universe. Due to their longevity, nearby old dM stars are obvious targets for SETI programs.

Our program – "Living with a Red Dwarf" – addresses these questions by investigating the nuclear evolution and rotation-age dependent magnetic-dynamo coronal and chromospheric XUV properties of dM stars for stellar ages ranging from ∼10 Myr to ∼12 Gyr (and corresponding rotation rates). Indeed, this study shows that, like our Sun and solar-type stars, young dM stars spin rapidly (and have robust dynamos) but lose angular momentum over time (via magnetic winds) and spin down. Also studied is how the stellar emissions (radiances as well as stellar wind fluxes) affect hosted planets and impact on the suitability for life. In this pilot study, we have selected ∼25 red dwarfs (dM0–5 stars) as proxies for red dwarfs with different ages and spectral types.

Our initial results indicate that, under certain circumstances, the more luminous dM0–3 stars can be suitable hosts for habitable planets. However, the HZs for such planets will be close to the host star (<0.5 AU) and result in tidally-locked HZ planets. Habitability of such planets may require a thick (UV shielding) atmosphere and protective magnetosphere. For instance, a super-Earth (2–10 $M_\oplus$ for sufficient gravity) with a liquid iron-nickel core may retain its atmosphere by generating strong magnetic fields, even with slow rotation - Guinan & Engle 2009, Guinan, Engle & Dewarf 2009.

Acknowledgements

This research is supported by grants from NASA/FUSE (NNX06AD38G) and NSF (AST-0507542 & AST-0507536) which we gratefully acknowledge. The "Living with a Red Dwarf" Program is on the web at: http://astronomy.villanova.edu/LWARD/

References

Guinan, E. F & Engle, S. G. 2009, *arXiv:0901.1860*
Guinan, E. F., Engle, S. G., & Dewarf, L. E. 2009, in: *Future Directions in Ultraviolet Spectroscopy*, American Institute of Physics Conf. Proc., Vol. 1135, p. 244

Highlights of Astronomy, Volume 15
XXVIIth IAU General Assembly, August 2009
Ian F. Corbett, ed.

© International Astronomical Union 2010
doi:10.1017/S1743921310011063

The dynamical architecture
and habitable zones
of the planetary system 55 Cancri

Jianghui Ji[1,2], Hiroshi Kinoshita[3], Lin Liu[4] and Guangyu Li[1,2]

[1]Purple Mountain Observatory, Chinese Academy of Sciences, Nanjing 210008, China
email: jijh@pmo.ac.cn

[2]National Astronomical Observatories, Chinese Academy of Sciences, Beijing 100012, China

[3]National Astronomical Observatory, Mitaka, Tokyo 181-8588, Japan

[4]Department of Astronomy, Nanjing University, Nanjing 210093, China

Abstract. We performed numerical simulations to study the secular orbital evolution and dynamical structure of the quintuplet planetary system 55 Cancri using the self-consistent orbital solutions of Fischer *et al.* (2008).

The nearby star 55 Cancri is of spectral type K0/G8V with a mass of $0.92 \pm 0.05\,\mathrm{M_\odot}$. Two giant planets are reported to be trapped in a 3:1 orbital resonance and Fischer *et al.* (2008) revealed a fifth planet in this system. Herein, we report on simulations of the secular evolution and dynamical structure of this system.

We show that this system can be stable for at least 10^8 yr. In addition, we extensively studied the planetary configuration of four outer companions with one terrestrial planet in the wide region 0.790–5.900 AU to examine the existence of a potential asteroidal structure and Habitable Zones (HZs). We find that there are unstable regions for the orbits about 4:1, 3:1 and 5:2 mean motion resonances (MMRs) with the outermost planet in the system, and several stable orbits can remain at 3:2 and 1:1 MMRs, a configuration which bears some resemblance with the asteroidal belt in the solar system.

From a dynamical point of view, candidate HZs for the existence of more potential terrestrial planets reside in the range between 1.0 AU and 2.3 AU for relatively low eccentricities. Moreover, our numerical simulations suggest that additional Earth-like planets can coexist with the five known planets in this system over secular timescale. This result suggests that abundant measurements and space missions (e.g. SIM Lite) should focus on this system in the future. The detailed results of this work may be found in Ji *et al.* (2009).

Acknowledgements

This work is financially supported by the National Natural Science Foundations of China (Grants 10973044, 10833001, 10573040, 10673006), NSF of Jiangsu Province, and the Foundation of Minor Planets of Purple Mountain Observatory.

References

Fischer, D. A., Marcy, G. W., Butler, R. P. *et al.* 2008, *ApJ* 675, 790
Ji, J., Kinoshita, H., Liu, L., & Li, G. 2009, *Res. Astron. & Astrophys.* 9, 703

Highlights of Astronomy, Volume 15
XXVIIth IAU General Assembly, August 2009
Ian F. Corbett, ed.

© International Astronomical Union 2010
doi:10.1017/S1743921310011075

The detectability of habitable exomoons

David M. Kipping, Stephen J. Fossey and Giammarco Campanella

University College London, Dept. of Physics & Astronomy, Gower Stree, London, N1 0XE, UK
email: d.kipping@ucl.ac.uk

Abstract. Despite the number of known exoplanets increasing on an almost weekly basis, the question as to whether exoplanets host moons remains unanswered. Exomoons could be potential seats for life, as well as improving our understanding of planetary formation and celestial mechanics. Here we summarize our findings from an investigation into how detectable habitable-zone exomoons are with Kepler-class photometry.

1. Introduction and methodology

One of the principal goals of exoplanetary science is to discover whether habitable worlds, like the Earth, are common throughout the Galaxy. Indeed, determining the fraction of habitable planets around normal stars is the primary science goal of the *Kepler Mission* (Borucki *et al.* 2008). However, it has been proposed that moons could also be potentially common seats for life (Williams *et al.* 1997) and the task of detecting these bodies should be considered as pertinent as the search for a habitable planet.

Kipping (2009a, 2009b) proposed that exomoons could be unambiguously detected around transiting planets by coupling transit timing variations (TTV) and transit duration variations (TDV). The ratio of the two amplitudes provides both the mass and period of an exomoon; factors which influence the potential habitability. In this study (Kipping *et al.* 2009), we simulate the effect for a range of possible star-planet-moon configurations and generate the expected uncertainities using the noise properties of Kepler. We are then able to evaluate the range of habitable-zone exomoons which may be detectable with Kepler, or photometry of equal quality.

2. Summary of results

- Exomoons are most easily detected around low-density exoplanets, because both the planetary transit signal and the perturbation due to an exomoon are maximal.
- Habitable-zone exomoons down to $0.2\,M_\oplus$ may be detected around the optimum targets (based on a $0.4\,M_\odot$ M-dwarf at 10pc).
- Dynamically stable $1\,M_\oplus$ exomoons may be surveyed for up to 200 pc away, consituting 25,000 stars within Kepler's field of view.

References

Borucki, W. J. *et al.* 2008, in: M. Livio, K. Sahu & J. Valenti (eds.), *A Decade of Extrasolar Planets around Normal Stars*, (Cambridge: Cambridge University Press), p. 36
Kipping, D. M. 2009a, *MNRAS* 392, 181
Kipping, D. M. 2009b, *MNRAS* 396, 1797
Kipping, D. M., Fossey, S. J., & Campanella, G. 2009, *MNRAS*, in print
Williams, D. M., Kasting, J. F., & Wade, R. A. 1997, *Nature* 385, 235

Highlights of Astronomy, Volume 15
XXVIIth IAU General Assembly, August 2009
Ian F. Corbett, ed.

© International Astronomical Union 2010
doi:10.1017/S1743921310011087

Testing models for the formation of habitable planets

Yamila Miguel and Adrián Brunini

Facultad de Ciencias Astronómicas y Geofísicas (Universidad Nacional de La Plata)
Instituto de Astrofísica de La Plata (CCT La Plata, CONICET)
email: ymiguel@fcaglp.unlp.edu.ar

Abstract. We report on a statistical study of models for the formation of habitable planets.

Although planets with masses less than $15\,M_\oplus$ have already been detected, the sample is still not large enough and we also have to rely on what we know from our own Solar System, as well as on computational models, to gain a better understanding of the planetary formation process. Our principal aim is to analyse which factors will influence the formation of these habitable planets. To this end, we developed a simple code which allows us to form a large population of planets and study them statistically, analysing which parameters favor their formation.

We consider a population of 1000 different discs based on the Minimum Mass Solar Nebula, each one with a different star mass. The formation and evolution of several planets per disc are considered, taken into account the collisions among them as a source of potential growth. The embryos grow in the oligarchic growth regime and when a core reaches the critical mass for starting the gas accretion, the envelope starts to grow with a characteristic growth time proportional to $(M_p/M_\oplus)^b$. We perform simulations with two different values of b: Model I assumes the value obtained by fitting the results of Fortier, Benvenuto & Brunini (2007) ($b = -1.91$), while Model II adopts the value used by Ida & Lin (2004) ($b = -3$). We also consider type I and II regimes of embryos' migration, where we test different values for retarding the type I migration: $C_{migI} = 1, 10, 100$.

As shown in Miguel & Brunini (2008, 2009), the distributions of mass and semi-major axis strongly depend on the gas accretion model considered. We find that the population of habitable planets is larger when Model I is considered because the gas accretion rate is slower. As a consequence, runaway gas accretion is reached by the cores at a larger mass, so there are terrestrial planets that never reach this "crossover mass" and more planets form with small masses. We also find that when type I migration is slower, there are less planets reaching the inner edge of the disc, and we end up with a major population of terrestrial planets. We also analyse the effects produced by the collisions between the embryos. We find that there are a lot of planets beyond the stability limit, which are not gravitationally bound and should not be considered. As a result, the population of habitable planets is also strongly influenced by the collisions, a very important effect that must be taken into account when we analyse terrestrial planets.

References

Fortier, A., Benvenuto, O., & Brunini, A. 2007, *A&A* 473, 311
Ida, S. & Lin, D. N. C. 2004, *ApJ* 604, 388
Miguel, Y. & Brunini, A. 2008, *MNRAS* 387, 463
Miguel, Y. & Brunini, A. 2009, *MNRAS* 392, 391

Highlights of Astronomy, Volume 15
XXVIIth IAU General Assembly, August 2009
Ian F. Corbett, ed.

© International Astronomical Union 2010
doi:10.1017/S1743921310011099

Mineralogical study of proto-planetary disks in FU Orionis stars

Emanuel J. Sainz[1,2] and Mercedes Gomez[2]

[1]Facultad de Matematica, Astronomia y Fisica, UNC, Argentina
[2]Observatorio Astronomico de Cordoba, UNC, Argentina

Abstract. We report on mid-infrared (8–13 μm) spectroscopic observations of eight FU Orionis type objects and discuss a qualitative initial mineralogical analysis.

1. Introduction

We carried out mid-infrared (8–13 μm) spectroscopic observations of eight FU Orionis type objects (FU Orionis, V1057 Cygni, V1515 Cygni, L1551 IRS5, RNO 1B/1C, OO Serpentis, PP 13S and ISO-ChaI 192) with the Gemini North and South Telescopes.

2. Results

We observe a predominance of silicates at 9.7 μm in our sample. This feature appears in emission in V1057 Cyg, V1515 Cyg and FU Ori, and in absorption in the rest of the objects. However, most of the observed FU Orionis stars show evidence of dust grains processing in the mid-IR spectra, specially in emission: the broadening of the observed profile is indicative of grain growth and of the existence of crystaline compounds such as enstatite and forsterite. The only exception corresponds to PP 13S, the spectrum of which is well fitted by a dust composition of only olivine, demonstrating little or null grain processing in this disk.

V1057 Cyg and V1515 Cyg have low signal to noise spectra (S/N~5) and thus the presence of crystalline components is not certain. In the case of L1551 IRS 5 and ISO-ChaI 192, the peak of the absorption feature does not coincide with the peak of the coefficient of absorption of the olivine. The possible cause of this displacement might be the presence of pyroxens or olivines richer in iron, instead of the proposed pure olivine composition.

In general, Class I objects show emission features, while in Class II stars these features are observed in absorption. RNO 1B, classified as Class II, with an emission spectra, is the only exception. However this may be due to the inclination of the surrounding disk with respect to the line of sight (Polomski *et al.* 2005).

References

Polomski, E. F., Woodward, C. E., Holmes, E. K., Butner, H. M., Lynch, D. K., Russell, R. W., Sitko, M. L., Wooden, D. H., Telesco, C. M., & Piña, R. 2005, *AJ* 129, 1035

Highlights of Astronomy, Volume 15
XXVIIth IAU General Assembly, August 2009
Ian F. Corbett, ed.
© International Astronomical Union 2010
doi:10.1017/S1743921310011105

Interior structure models of terrestrial exoplanets and application to CoRoT-7 b

Frank W. Wagner, Frank Sohl, Heike Rauer
Hauke Hussmann and Matthias Grott

Institute of Planetary Research, German Aerospace Center, Berlin-Adlershof, Germany
email: frank.wagner@dlr.de

Abstract. In this study, we model the internal structure of CoRoT-7 b, considered as a typical extrasolar terrestrial planet, using mass and energy balance constraints. Our results suggest that the deep interior is predominantly composed of dry silicate rock, similar to the Earth's Moon. A central iron core, if present, would be relatively small and less massive (<15 wt.% of the planet's total mass) as compared to the Earth's (core mass fraction 32.6 wt.%). Furthermore, a partly molten near-surface magma ocean could be maintained, provided surface temperatures were high enough and the rock component mainly composed of Earth-like mineral phase assemblages.

1. Introduction and methodology

Already more than a dozen low-mass ($<15\,M_\oplus$) extrasolar planets have been discovered. CoRoT-7 b is the first exoplanet among those for which the radius *and* mass have been accurately determined. In units of Earth equivalents, these are $1.68 \pm 0.09\,R_\oplus$ (Léger *et al.* 2009) and $4.8 \pm 0.8\,M_\oplus$ (Queloz *et al.* 2009), respectively. The average compressed density of CoRoT-7 b of 5.6 ± 1.3 Mg m^{-3} is thus similar to that of the Earth (5.515 Mg m^{-3}) and suggests a terrestrial-type bulk composition (Queloz *et al.* 2009). Modeling the interior structure of CoRoT-7 b is an important leap forward in understanding the origin and evolution of terrestrial-type bodies in the solar system and beyond. In the following, we present a set of four-layer interior models that have been used to infer the planet's bulk composition and physical state.

Our one-dimensional model assumes spherically symmetric and fully differentiated planets in hydrostatic and thermal equilibrium. The interior structure of these planets is obtained by solving the mass and energy balance equations in conjunction with an equation of state (EoS) for the internal density distribution (e.g. Sohl & Schubert 2007). The chosen Vinet EoS facilitates extrapolation to exceptionally high pressures (Hama & Suito 1996, Valencia *et al.* 2007). The thermal profile across the lower mantle and core is assumed to be fully adiabatic and thermal boundary layers at pressure-induced phase transitions and chemical boundaries, e.g. the core-mantle boundary, are accounted for by constant temperature jumps according to Earth-like values. A mixing length formulation (Sasaki & Nakazawa 1986, Abe 1997) is adopted to calculate the radial distribution of temperature within the upper mantle and to simulate a lithospheric structure within the uppermost part. The basic idea behind this concept is that internally generated heat is primarily transfered by vertical motion of fluid parcels which, after migrating for size-dependent characteristic length scales, will entirely loose their individuality.

2. Results and discussion

Scaling laws for terrestrial-type exoplanets are obtained and a relationship between radius R and mass M according to $R \propto M^{0.266}$ has been established within one to ten Earth masses, in accordance with previous work (e.g. Valencia *et al.* 2006, Sotin *et al.* 2007). Model planets are subdivided into an upper mantle composed of olivine and pyroxene in equal amounts, underlain by lower perovskite and post-perovskite mantle layers, and a central ϵ-iron core. The density models of CoRoT-7 b are constrained by the planet's total radius of $1.68 R_\oplus$ (Léger *et al.* 2009). We consider iron core mass fractions of 5, 15, and 65 wt.%, corresponding to planetary masses of 5.3, 5.8, and $10 M_\oplus$, respectively.

With regards to the recently reported mass range $4 - 5.6 M_\oplus$ (Queloz *et al.* 2009) for CoRoT-7 b, a Mercury-like (core mass fraction 70 wt.%) bulk composition can be safely ruled out, and, augmented by significant internal compression, only a small and less massive iron-rich core may be present at depth. It is even conceivable that an iron-rich core is lacking, turning CoRoT-7 b perhaps into a coreless, pure silicate planet (Elkins-Tanton & Seager 2008).

Moreover, we have extended our model approach by including a lithospheric model to better constrain the radial distribution of temperature within the upper mantle. Due to the close proximity of CoRoT-7 b to its primary, a high surface temperature of about 1800 K can be expected. The comparatively thin lithospheric layer on top of the upper mantle has only a marginal impact on the deeper interior. Comparing the calculated temperature profile with pressure-dependent rock melting temperatures (Gasparik 1994), we find that the uppermost part of the mantle of CoRoT-7 b could be at least partly molten, provided the mantle was composed of mineral phase assemblages of forsterite and enstatite, similar to the Earth's.

3. Conclusions

In summary, CoRoT-7 b may represent a dry, rock-rich planet predominantly composed of silicates, similar to the Earth's Moon. An iron-rich core at depth would be small and less massive or even non-existent, suggesting that CoRoT-7 b may have originated in the iron-depleted region beyond the snowline and lost its volatile mass fraction when subsequently moving toward its primary. Furthermore, albeit strongly dependent on environmental conditions and dominant mantle mineralogy, the planet might harbour a near-surface magma ocean with depth extending across the uppermost mantle.

Acknowledgement

This research is supported by the Helmholtz Alliance "Planetary Evolution and Life".

References

Abe, Y. 1997, *Phys. Earth Planet. Inter.* 100, 27
Elkins-Tanton, L. T. & Seager, S. 2008, *ApJ* 688, 628
Gasparik, T. 1994, *Mineral. Mag.* 58A, 321
Hama, J. & Suito, K. 1996, *J. Phys.: Condens. Matter* 8, 67
Léger, A., Rouan, D., Schneider, J. *et al.* 2009, *A&A* 506, 287
Queloz, D., Bouchy, F., Moutou, C. *et al.* 2009, *A&A* 506, 303
Sasaki, S. & Nakazawa, K. 1986, *J. Geophys. Res.* 91, 9231
Sohl, F. & Schubert, G. 2007, in: T. Spohn (ed.), *Treatise on Geophysics 10* (Amsterdam: Elsevier), p. 27
Sotin, C., Grasset, O., & Mocquet, A. 2007, *Icarus* 191, 337
Valencia, D., O'Connell, R. J., & Sasselov, D. D. 2006, *Icarus* 181, 545
Valencia, D., Sasselov, D. D., & O'Connell, R. J. 2007, *ApJ* 656, 545

Highlights of Astronomy, Volume 15
XXVIIth IAU General Assembly, August 2009
Ian F. Corbett, ed.
© International Astronomical Union 2010
doi:10.1017/S1743921310011117

Mars Science Laboratory (MSL) and the future missions to Mars

Michel Cabane[1] and the SAM Team[2,3,4]

[1] LATMOS-UPMC, Univ. Paris 6, 4 place Jussieu, 75005 Paris, France
email: cabane@latmos.ipsl.fr

[2] Goddard Space Flight Center, Greenbelt, MD 20771, USA

[3] Jet Propulsion Laboratory, California Institute of Technology, 4800 Oak Grove Drive,
Pasadena, CA 91109, USA

[4] LISA, Univ. Paris 7 & Paris 12, 94010 Créteil, France

Abstract. After their formation, and for almost 0.5 Gyr, Mars and Earth evolved in the same manner. On Earth, life occured very early, around 3.6 Gy ago. Then, the hypothesis that life might have occured also on Mars, and be extinct, is not unrealistic. If this is true, then complex molecules could be present, representing an early state of life-building blocks. Missions to Mars (Viking 1-2, Phoenix) have carried instruments capable to search for molecular indicators, although up to now, no positive detection has been obtained. Future missions to Mars (MSL-NASA, Exomars-ESA) will use enhanced experiments to try to end this quest.

In 1976, Viking 1 and 2 were very ambitious missions to search for life on Mars. Unfortunately, all the experiments devoted to this goal failed to reveal any organic molecules, even the most simple, Klein, 1977. Nevertheless, molecules could have been missed by Viking (Navarro-Gonzalez *et al.* 2006). In 2008, Phoenix lander performed evolved gas analysis experiments which, again, gave no result (phoenix.lpl.arizona.edu/index.php). The odd thing in these two missions is that they should have observed some of the 2 to 3.10^8 grams of organic matter that comes to Mars each year via micrometeorites (Flynn, 1995). Benner *et al.* (2000) showed that complex organic molecules may be transformed, under the effect of oxidants, into refractory molecules that cannot be analyzed by conventional means (pyrolysis, followed by gas chromatography and/or mas spectrometry), then experimental devices have to be rethought. Methane was detected recently in the atmosphere, new results (Mumma *et al.* 2009), showing CH_4 plumes containing dozens of ppbs, require a better understanding of its apparent lifetime (much less than predicted) and sources (serpentinization, H_2O-CH_4 clathrates, ...?).

In 2011, NASA will send a rover to Mars – MSL (Mars Science Laboratory) – with a 70 kg science payload to explore Mars during about 2 terrestrial years (one Martian year) and roll on about 20 km. MSL science payload consists in ten instruments (marsprogram.jpl.nasa.gov/msl/). One of them, SAM (ael.gsfc.nasa.gov/marsSAM.shtml – PI: P. Mahaffy, GSFC-NASA), will analyze atmosphere and powdered rock and soil delivered by MSL's sample acquisition and processing system. SAM consists of the association of a Gas Chromatograph (GC, six columns devoted to the detection of inorganic gases, light or heavy organics, and chirality), a Quadrupole Mass Spectrometer (MS, from 2 to 535 Da) and a tunable UV Laser Spectrometer (TLS : H_2O, CH_4, C and O, especially devoted to isotopy). Part of SAM investigations will be devoted to the study of the atmosphere (CH_4, and other trace species, noble gases, down to the ppb ; isotopy down to 10 per mil). Analysis by coupling pyrolysis at T = 1000°C and GC, MS and TLS, will allow to know what happens to powdered soil samples when heated up to 1000°C (Viking heated at about 500° C), as well for possible delivery of organic molecules inbedded into the sample, as well as of for gases issued of the decomposition of minerals that may be linked to the presence of such molecules (clays, carbonates, sulfates, were formed in the first ages of Mars and release their structural gases H_2O, CO_2, SO_2 at known temperatures). Organic molecules might be decomposed into CO_2 during heating, or fully refractory. Consequently, wet chemistry will be performed on soil samples by some of the 74 ovens of SAM. Cooking at low temperature organic molecules with a devoted reagent transforms these molecules into vaporizable ones, this is the so-called derivatization process. One may hope that, using pyrolysis and/or derivatization, one might detect down to some 10^{-12} mole of chemical species in the sample.

Mars exploration will also be performed by the European Space Agency (ESA). The July 2009 session of MEPAG foresaw an ESA-NASA association, in which NASA would simultaneously launch in 2018 its mid-size rover MRR and the ESA Exomars rover. MRR is intended to prospect Mars and cache samples to be brought to Earth by Mars Sample Return mission in 2022-24, whereas Exomars (www.esa.int/esaMI/ExoMars/) will continue the Exo-Astrobiological prospection started by MSL. Exomars, with its proposed 14 kg science payload will move on Mars surface for 1/4 Martian year. Compared to MSL, it has the advantage of drilling the soil down to 1 m ; its main differences with MSL are also the presence of a subsurface radar (WISDOM) and of a Raman-LIBS (Laser Induced Breakdown Spectroscopy) that allows an other way to detect and analyze organic molecules into a mineral mixture. In Exomars, one also finds MOMA (www.mps.mpg.de/en/projekte/exomars/moma/ – Team coordinator: F. Goesmann, MPS-Germany), which works the same way as SAM; a Laser Diode Mass Spectrometer has been added to the Pyro-Deriv-GC-MS analysis complex, that extracts molecules from their substrate. Other instruments have been proposed for future misions, among them UREY (Bada *et al.* 2008), in which a sub-critical water extractor will allow to identify organic molecules by means of capillary electrophoresis.

Orbiters, landers and rovers currently procure detailed informations that help to specify geology, geochemistry and evolution of Mars since its formation. Concerning the search for life, new developments of complex techniques, a better understanding of the chemical-geochemical processes on Mars will allow to answer the puzzling question of "life on Mars". In 2011, this will be implemented with the next vehicle on the surface of Mars – MSL with its complex instrumentation – followed by MRR and Exomars in 2018.

References

Bada, J. L., Ehrenfreund, P., Grunthaner, F., Blaney, D., Coleman, M., Farrington, A., Yen, A., Mathies, R., Amudson, R., Quinn, R., Zent, A., Ride, S., Barron, L., Botta, O., Clark, B., Glavin, D., Hofmann, B., Josset, J. L., Rettberg, P., Robert, F., & Sephton, M. 2008, *Space Science Reviews* 135, 269

Benner, S. A., Devine, K. G., Matveeva, L. N., & Powell, D. H. 2000, *PNAS* 97, 2425

Flynn, G. J. 1995, *Earth, Moon and Planets* 71, 469

Klein, H. P. 1977, *JGR* 82, 4677

Mahaffy, P. 2008, *Space Science Reviews* 135, 255

Mumma, M. J., Villanueva, G. L., Novak, R. E., Hewagama, T., Bonev, B. T., DiSanti, M. A., Mandell, A. M., & Smith, M. D. 2009, *Science* 323, 1041

Navarro-Gonzalez, R., Navarro, K., delaRosa, J., Iniguez, E., Molina, P., Miranda, L. D., Morales, P., Cienfuegos, E., Coll, P., Raulin, F., Amils, R., & McKay, C. P. 2006, *PNAS* 103, 16089

Highlights of Astronomy, Volume 15
XXVIIth IAU General Assembly, August 2009
Ian F. Corbett, ed.
© International Astronomical Union 2010
doi:10.1017/S1743921310011129

Characteristics of the Kepler target stars

Natalie M. Batalha[1], William J. Borucki[2], David G. Koch[2] Timothy M. Brown[3], Douglas A. Caldwell[4] and David W. Latham[5]

[1] Department of Physics and Astronomy, San Jose State University, San Jose, CA 95192

[2] Mailstop 244-30, NASA Ames Research Center, Moffett Field, CA 94035

[3] Las Cumbres Observatory Global Telescope Network, 6740 Cortona Dr., Ste 102, Goleta, CA

[4] SETI Institute, 515 N. Whisman Rd., Mountain View, CA 94043

[5] Harvard-Smithsonian Center for Astrophysics, 60 Garden St., Cambridge, MA 02138

Abstract. The Kepler Mission successfully launched March 6, 2009, beginning its 3.5-year mission to determine the frequency of Earth-size planets in the habitable zones of late-type stars. The brightnesses of over 100,000 stars are currently being monitored for transit events with an expected differential photometric precision of 20 ppm at V=12 for a 6.5-hour transit. The same targets will be observed continuously over the mission duration in order to broaden the detection space to orbital periods comparable to that of Earth. This paper provides an overview of the selection and prioritization criteria used to choose the stars that Kepler is observing from the > 4.5 million objects in the 100 square degree field of view. The characteristics of the Kepler targets are described as well as the implications for detectability of planets in the habitable zone smaller than $2\,R_\oplus$.

A ground-based observing campaign was initiated several years before launch to determine the apparent magnitude (Kepler passband), the surface gravity, effective temperature, and stellar radius of all stars in the Kepler field using multi-color (g,r,i,z, and D51) photometry (Borucki *et al.* 2008). This effort required multiple visits to more than 1,600 pointings. The multi-color photometry was fit to synthetic Castelli & Kurucz (2004) models (transformed to agree with observed M67 star colors) and constrained by a Bayesian prior probability distribution based on the statistics of stars in the solar neighborhood. Completeness with regards to stellar classification is at the 80% level down to 17th magnitude. The resulting stellar properties are archived in the Kepler Input Catalog which is publicly available at the Multi-Mission Archive at Space Telescope Science Institute. The derived parameters form the basis of target selection and prioritization.

Given effective temperature, surface gravity, stellar radius, expected photometric precision, and a measure of the crowding in the photometric aperture, we compute the minimum detectable planet radius, $R_{p,min}$, for every star at 1) the inner edge of the habitable zone; 2) half that distance; and 3) an orbit 5 stellar radii from the central star. The crowding metric is derived from simulated images using data from the Kepler Input Catalog and knowledge of the instrument characteristics (e.g. random noise sources, pixel response function, etc). The crowding is used to determine the dilution of the transit which translates to a larger $R_{p,min}$. The end product is a sample of $\sim 188,000$ stars brighter than 16th magnitude that fall on silicon for which a planet of radius $\leqslant 2\,R_\oplus$ is detectable (at acceptable confidence levels) over the duration of the mission. We note that stellar variability is not simulated for the purposes of target selection.

The target list is prioritized according to a weighted combination of $R_{p,min}$, apparent magnitude, and habitability (planets in the HZ are given higher priority). The list is then trimmed to meet flight segment constraints ($\leqslant 170,000$ targets and $\leqslant 5.44$ million pixels).

Mag/T_e(K)	10500	9500	8500	7500	6500	5500	4500	3500	Total
7.5	2	8	8	8	8	7	0	0	41
8.5	8	20	26	24	50	16	7	8	159
9.5	9	31	81	65	117	88	11	4	406
10.5	27	37	100	209	405	362	40	9	1189
11.5	24	58	172	396	1495	1356	157	39	3697
12.5	33	43	230	678	4148	4761	625	62	10580
13.5	34	51	170	737	9250	15841	2218	159	28460
14.5	3	0	0	0	4791	29291	4401	552	39038
15.5	7	3	0	0	4261	43132	11188	1828	60419
Total	147	251	787	2117	24525	94854	18647	2661	143989

The above Table reports the star counts (stars with $\log g \geqslant 3.5$ from amongst the 160,000 highest priority targets), binned by apparent magnitude and effective temperature. 16,011 stars ($\log g < 3.5$) are not represented in Table 1. The smaller giants make it into the target sample when the apparent magnitude is bright enough to allow for the detection of planets as small as $2\,R_\oplus$ in an orbit as close as $5\,R_*$. The targets are dominated by G-type stars on or near the Main Sequence and stars fainter than 14th magnitude. At the temperature extremes, we have $\sim 3,000$ M-type Main Sequence stars (low intrinsic brightness) and < 200 O and B-type stars. OB-type stars are rare due not only to their short MS lifetimes, but also due to the fact that the field of view was chosen to avoid young stellar populations (OB associations and star forming regions).

We compute the number of detections expected given the target star population described above assuming that every star has a terrestrial-sized planet in its habitable zone. The calculation includes a red noise contribution derived from the solar irradiance variations. It yields an expected detection rate of hundreds of planets with radii $\leqslant 2\,R_\oplus$ in the habitable zone. A cursory look at the first 30 days of science data of the 20 brightest G dwarfs in the sample suggests that stars as photometrically quiet as the Sun are as common as was expected by examining galactic populations models, the age-rotation-activity relation, and standard spin-down rates (Batalha *et al.* 2002). In particular, we identify 6 of the 20 that are indistinguishable (in terms of the stellar variability characteristics) from SOHO irradiance data of similar duration. The pre-selection of targets, made possible by pre-launch classification of the stars in the field of view, will significantly decrease the number of false positives so that the ground-based follow-up resources can be used more efficiently.

Acknowledgements

Kepler was selected as the 10th mission of the Discovery Program. Funding for this mission is provided by NASA, Science Mission Directorate.

References

Batalha, N. M., Jenkins, J., Basri, G. S., Borucki, W. J., & Koch, D. G. 2002, in: F. Favata, I.W. Roxburgh & D. Galadi (eds.), *Stellar Structure and Habitable Planet Finding*, ESA SP-485, p. 35

Borucki, W., Koch, D., Basri, G., Batalha, N., Brown, T., Caldwell, D., Christensen-Dalsgaard, J., Cochran, W., Dunham, E., Gautier, T. N., Geary, J., Gilliland, R., Jenkins, J., Kondo, Y., Latham, D., Lissauer, J. J., & Monet, D. 2008, in: Y.-S. Sun, S. Ferraz Mello & J.-L. Zhou (eds.), *Exoplanets: Detection, Formation, and Dynamics*, Proc. of IAU Symposium No 249 (Cambridge: CUP), p. 17

Castelli, F. & Kurucz, R. L. 2004, arXiv:astro-ph/0405087

Highlights of Astronomy, Volume 15
XXVIIth IAU General Assembly, August 2009 © International Astronomical Union 2010
Ian F. Corbett, ed. doi:10.1017/S1743921310011130

SIM Lite astrometric observatory
for detection of Earth-like planets

Xiaopei Pan, Michael Shao and Renaud Goullioud

Jet Propulsion Laboratory, California Institute of Technology
4800 Oak Grove Dr, Pasadena, CA 91109, USA
email: xiaopei.pan@jpl.nasa.gov, michael.shao@jpl.nasa.gov, renaud.goullioud@jpl.nasa.gov

Abstract. The SIM Lite Astrometric Observatory is an optical astrometry mission for detection of Earth-like planets and investigation of dark matter, galaxy assembly, black holes etc. SIM Lite is a minimum system that performs the full NRC recommenced SIM science programs. This paper summarize the latest status and progresse of the SIM Lite space mission.

1. SIM Lite performance

The SIM Lite mission is based on development of the SIM PlanetQuest project and has completed a ten-year technology program which was confirmed by multiple external independent reviews. SIM Lite is a cost-effective system, and has passed eight technology gates in component technology, subsystem level testing and system level testing. The SIM Lite instrument includes a 6 m Michelson Stellar Interferometer (MSI) with a 50 cm science siderostat aperture at visual bands (450–950 nm). Another 4 m MSI and a 30 cm telescope are used for the guide system. The SIM Lite spacecraft will use an Earth-trailing solar orbit with five year mission time, which can be extended to ten years. The global astrometry accuracy of the SIM Lite mission is 4 µas. The narrow angle astrometry has accuracy of 1 µas for a single measurement, and 0.04 µas for mission accuracy.

2. Detecting Earth-analogs around nearby FGK stars

The SIM Lite mission is an astrometric exoplanet search recommended by the Exoplanet Task Force (AAAC2008), and is free of the limits of astrophysical noise, which is the main limiting factor for radial velocity techniques now. SIM Lite is the only mission which can detect Earth-like planets around nearby FGK stars. Planet surveys by the SIM Lite mission are divided into three categories: the Deep Survey observes 60 stars within 10 pc for Earth analogs; the Broad Survey will search about 2000 stars for planets with mass of a few Earths and greater; and the Young Planet Survey will study about 200 stars of 1–100 Myr ages. Figure 1 compares SIM Lite with other techniques, and demonstrates SIM Lite's capabilities for rocky planet detection.

In particular SIM Lite has the unique capability to search for exoplanets around binary stars. Current explorations for exoplanets are biased to avoid binary stars because the signatures of two stellar objects confuse the tiny signals of planets. Spectroscopic binaries with separation of 11 mas or less will not be resolved by the SIM Lite instrument and can be treated as one star for planet detection. The movements caused by stellar orbital motions can be modeled and easily removed afterwards. In fact wide binary stars (a few arcsecond or more) will use the same sets of reference stars in SIM Lite observation, and can save 80% of the observation time for the second object. SIM Lite definitely is a powerful tool to discover the diversity of planetary system architectures.

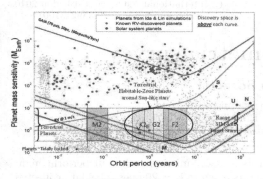
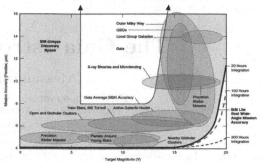

Figure 1. SIM Lite detection of Earth-analogs for nearby FGK stars

Figure 2. SIM Lite wide-angle discovery space over bright and faint objects

A NASA-directed double-blind, planet-finding capability study (DBS) was conducted in 2008. The study (Shao *et al.* 2009) confirms the predicted SIM Lite capability for detection of Earth-like planets even in complex planetary systems. It has been demonstrated that SIM Lite can detect multiple planets down to one Earth mass around the nearest 60 FGK stars, and meets the astrometric mission recommendation of the AAAC Exoplanet Task Force.

3. Precision astrometry over bright and faint objects

SIM Lite represents an entirely new measurement capability in astronomy and a revolution in astrometry that will exceed all previous measurements by a factor of 100-1,000. SIM Lite has a target-oriented instrument which has flexible integration time for bright targets ($V = -1.5\,\mathrm{mag}$) or faint targets ($V = 20\,\mathrm{mag}$). Also, SIM Lite can easily adjust its observing schedule to respond to Target of Opportunity. Unprecedented high accuracy ($4\,\mu$as) for wide-angle astrometry by SIM Lite provides the best opportunity to explore dark matter, the dynamics of galaxies, astrometric microlensing, massive black holes, and so on. The unique discovery space of SIM Lite in astrophysics is shown in Figure 2 (see details in Davidson *et al.* 2009).

4. Conclusions

SIM Lite is the only mission concept that is ready to go now, and is capable of finding one-Earth-mass planets in the habitable zone of nearby Sun-like stars. SIM Lite's unique capability for detection of exoplanets in binary star systems provides an excellent opportunity to study architectures and evolution of planets. Three survey categories of SIM Lite provide fundamental statistics of exoplanets for future missions. In addition SIM Lite will map the local dark matter, probe the Milky Way, and investigate supermassive black holes.

References

Shao, M., Zhai, C., Catanzarite, J., Loredo, T., McArthur, B., & Benedict, F. 2009, *BAAS* 41, 268

Davidson, J., Unwin, S., Edberg, S., Danner, R., & Nemati, B. (eds.) 2009, *SIM Lite Astrometric Observatory*, JPL Publication 400-1360

Highlights of Astronomy, Volume 15
XXVIIth IAU General Assembly, August 2009
Ian F. Corbett, ed.

© International Astronomical Union 2010
doi:10.1017/S1743921310011142

The Gaia astrometric survey

Alessandro Sozzetti

INAF - Osservatorio Astronomico di Torino,
Strada Osservatorio 20, 10025 Pino Torinese (TO), Italy
email: sozzetti@oato.inaf.it

Abstract. In its all-sky survey, the ESA global astrometry mission Gaia will perform high-precision astrometry and photometry for 1 billion stars down to $V = 20$ mag. The data collected in the Gaia catalogue, to be published by the end of the next decade, will likely revolutionize our understanding of many aspects of stellar and Galactic astrophysics. One of the relevant areas in which the Gaia observations will have great impact is the astrophysics of planetary systems. This summary focuses on a) the complex technical problems related to and challenges inherent in correctly modelling the signals of planetary systems present in measurements collected with a space-borne observatory poised to carry out precision astrometry at the micro-arcsecond (μas) level, and b) on the potential of Gaia μas astrometry for important contributions to the astrophysics of planetary systems.

1. Introduction

The Gaia all-sky survey, due to launch in Spring 2012, will monitor astrometrically, during its 5-yr nominal mission lifetime, all point sources (stars, asteroids, quasars, extragalactic supernovae, etc.) in the visual magnitude range $6 - 20$ mag, a huge database encompassing $\sim 10^9$ objects. Using the continuous scanning principle first adopted for Hipparcos, Gaia will determine the five basic astrometric parameters (two positional coordinates α and δ, two proper motion components μ_α and μ_δ, and the parallax ϖ) for all objects, with end-of-mission precision between 6 μas (at $V = 6$ mag) and 200 μas (at $V = 20$ mag).

Gaia astrometry, complemented by on-board spectrophotometry and (partial) radial velocity information, will have the precision necessary to quantify the early formation, and subsequent dynamical, chemical and star formation evolution of the Milky Way Galaxy. The broad range of crucial issues in astrophysics that can be addressed by the wealth of the Gaia data is summarized by Perryman *et al.* (2001). One of the relevant areas in which the Gaia observations will have great impact is the astrophysics of planetary systems (Casertano *et al.* 2008), in particular when seen as a complement to other techniques for planet detection and characterization (Sozzetti 2009).

2. Astrometric modelling of planetary systems

The problem of the correct determination of the astrometric orbits of planetary systems using Gaia data (highly non-linear orbital fitting procedures, with a large number of model parameters) will present many difficulties. For example, it will be necessary to assess the relative robustness and reliability of different procedures for orbital fits, together with a detailed understanding of the statistical properties of the uncertainties associated with the model parameters. For multiple systems, a trade-off will have to be found between accuracy in the determination of the mutual inclination angles between pairs of planetary orbits, single-measurement precision and redundancy in the number of observations with respect to the number of estimated model parameters. It will constitute a challenge to correctly identify signals with amplitude close to the measurement uncertainties, particularly in the presence of larger signals induced by other companions and/or sources of astrophysical noise of comparable magnitude. Finally, in cases of multiple-component systems where dynamical interactions are important (a situation experienced already by radial-velocity surveys), fully dynamical (Newtonian) fits involving an n-body code might have to be used to properly model the Gaia astrometric data and to ensure the short- and long-term stability of the solution (see Sozzetti 2005).

All the above issues could have a significant impact on Gaia's capability to detect and characterize planetary systems. For these reasons, within the pipeline of Coordination Unit 4 (object processing) of the Gaia Data Processing and Analysis Consortium (DPAC), in charge of the scientific processing of the Gaia data and production of the final Gaia catalogue to be released sometime in 2020, a Development Unit (DU) has been specifically devoted to the modelling of the astrometric signals produced by planetary systems. The DU is composed of several tasks, which implement multiple robust procedures for (single and multiple) astrometric orbit fitting (such as Markov Chain Monte Carlo and genetic algorithms) and the determination of the degree of dynamical stability of multiple-component systems.

3. The legacy of Gaia

Using Galaxy models, our current knowledge of exoplanet frequencies, and Gaia's estimated precision ($\sim 10~\mu$as) on bright targets ($V < 13$), Casertano *et al.* (2008) have shown how Gaia's main strength will be its ability to measure astrometrically actual masses and orbital parameters for possibly thousands of giant planets, and to determine the degree of coplanarity in possibly hundreds of multiple-planet systems. Its useful horizon for planet detection (encompassing $\sim 3 \times 10^5$ stars) extends as far as the nearest star-forming regions (e.g. Taurus at $d \simeq 140$ pc) for systems with massive giant planets ($M_p \gtrsim 2 - 3~M_J$) on $1 < a < 4$ AU orbits around solar-type hosts, and out to $d \sim 30$ pc for Saturn-mass planets with similar orbital semi-major axes around late-type stars.

In summary, Gaia holds promise for crucial contributions to many aspects of planetary systems astrophysics, in combination with present-day and future extrasolar planet search programs. For example, the Gaia data, over the next decade, will allow us to a) significantly refine of our understanding of the statistical properties of extrasolar planets, b) carry out crucial tests of theoretical models of gas giant planet formation and migration, c) achieve key improvements in our comprehension of important aspects of the formation and dynamical evolution of multiple-planet systems, d) provide important contributions to the understanding of direct detections of giant extrasolar planets, and e) collect essential supplementary information for the optimization of the target lists of future observatories aiming at the direct detection and spectroscopic characterization of terrestrial, habitable planets in the vicinity of the Sun.

References

Casertano, S., Lattanzi, M. G., Sozzetti, A., Spagna, A., Jancart, S., Morbidelli, R., Pannunzio, R., Pourbaix, D., & Queloz, D. 2008, *A&A* 482, 699
Perryman, M. A. C., de Boer, K. S., Gilmore, G., Høg, E., Lattanzi, M. G., Lindegren, L., Luri, X., Mignard, F., Pace, O., & de Zeeuw, P. T. 2001, *A&A* 369, 339
Sozzetti, A. 2005, *PASP* 117, 1021
Sozzetti, A. 2009, *EAS Publication Series*, in press (arXiv:0902.2063)

Highlights of Astronomy, Volume 15
XXVIIth IAU General Assembly, August 2009
Ian F. Corbett, ed.
© International Astronomical Union 2010
doi:10.1017/S1743921310011154

The SEE-COAST concept

Anthony Boccaletti[1], Alessandro Sozzetti[2], Jean Schneider[1] Pierre Baudoz[1], Giovanna Tinetti[3] and Daphne Stam[4]

[1]Observatoire de Paris, 5 pl. J. Janssen, 92195 Meudon, France
email: anthony.boccaletti@obspm.fr, jean.schneider@obspm.fr, pierre.baudoz@obspm.fr

[2]INAF - Osservatorio Astronomico di Torino, str. Osservatorio 20, 10025 Pino Torinese, Italy
email: sozzetti@oato.inaf.it

[3]Department of Physics and Astronomy, UCL, Gowerstreet, London WC1E6BT, UK
email: g.tinetti@ucl.ac.uk

[4]SRON, Sorbonnelaan 2, 3584 CA Utrecht, The Netherlands
email: d.m.stam@sron.nl

Abstract. The SEE COAST concept is designed with the objective to characterize extrasolar planets and possibly Super Earths via spectro-polarimetric imaging in reflected light. A space mission complementary to ground-based near IR planet finders is a first secure step towards the characterization of planets with mass and atmosphere comparable to that of the Earth. The accessibility to the Visible spectrum is unique and with important scientific returns.

1. Historic and context

Radial Velocity technique has been one of the most prolific method to identify extrasolar planets in the past decade (374 objects listed in http://exoplanet.eu). The comparison with our own Solar System already suggests a large diversity (in mass, eccentricity, semi-major axis ...). One of the next step is the study of exoplanet atmospheres and we must be also prepared for diversity. In that context, direct detection is required as we need to collect the planetary photons to perform for instance spectroscopy. But also, a few self-luminous planets have been already detected either from the ground and space using direct imaging, Kalas *et al.* 2008, Lagrange *et al.* 2009, and temperature and mass have been estimated from evolutionary models. In addition, direct imaging covers a parameter space that is complementary to Radial Velocity as it is more sensitive to long period. The ever-growing interest for Super Earths (telluric planets that are more massive than the Earth) has led us to propose a space telescope operated in the visible for spectroscopic and polarimetric analysis of these objects. SEE COAST (Super Earth Explorer - Coronagraphic Off-Axis Space Telescope) was first proposed to Cosmic Vision at ESA in 2007 but not selected. The science potential and requirements are briefly described.

2. Astrophysical requirements and instrumental concept

The Core Science program of SEE-COAST is to explore the diversity of planets especially focusing on Super Earths and mature Jovian planets as a goal (young/massive giants, brown dwarfs and debris disks are also part of the program). We anticipated that in the coming years Radial Velocity instruments will provided many targets accessible with SEE-COAST. Several tools are considered for this study :

• Spectroscopy: a spectral resolution of 40 to 80 in the visible and near IR (0.4-$1.2\,\mu$m) is required to measure the presence of several molecular species like H_2O, O_2, CH_4, CO_2.

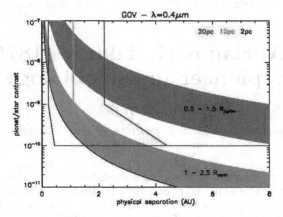

Figure 1. Star to planet contrast compared to instrumental specifications (colored curves) scaled according to wavelength and star distance (for an albedo of 0.5 and a G0V star).

If the radius is known, the albedo can be inferred from the reflected flux. Rayleigh scattering at short wavelengths gives the column density and constraints on pressure.

• Polarimetry: the polarimetric signal of jovian and telluric planets depends on wavelength and shows strong complementarities with spectroscopy to characterize planetary surfaces and clouds (Stam 2008).

• Variability: spectral and polarimetric signal are expected to vary on seasonal basis. Several visits are therefore requested to detect changes in the planetary climate.

A small 1.5m telescope in space if equipped with high contrast imaging facilities and appropriate instruments can carry out this program for nearby stars (<20 pc), nicely complementing planet finders on 8-m class telescopes for instance. It would prepare for more ambitious programs that aim to detect Earth analogs and search for trace of life. The main requirements for SEE COAST are summarized in Figure 1 which shows the theoretical contrasts in reflected light for 2 classes of planets, the giants and tellurics, compared to instrumental specifications. Telluric planets are accessible for very nearby stars within 5 AU providing small inner working angle ($2\lambda/D$) and large contrast (10^{-10}) are met. Even an Earth-like planet can be detected if any around the closest star. Spectroscopy will be feasible on giant planets within 10 AU for stars closer than 40 pc. The shortest wavelengths ($0.4\,\mu m$) are clearly more favorable for detection (contrast curves scale linearly with λ). Young planets are also preferential targets for SEE COAST. At large temperatures (>600-800 K) the thermal emission produces a significant signal at $1.2\,\mu m$ and translates into contrasts of about 10^{-9} for tellurics and 10^{-7} for giants around M stars (independently of separation). Hence, spectro-polarimetry of young objects is feasible around distant stars (50-100 pc). Instrumental aspects of the mission and technical developments are detailed in Schneider *et al.* (2009). The SEE COAST concept will be further elaborated and proposed to the next ESA Cosmic Vision call for missions.

References

Lagrange, A.-M., Gratadour, D., Chauvin, G. *et al.* 2009, *A&A* 493, L21

Kalas, P., Graham, J., Chiang, E. *et al.* 2008, *Science* 322, 1345

Stam, D. 2008, *A&A* 482, 989

Schneider, J., Boccaletti, A., Mawet, D., Baudoz, P. *et al.* 2009, *Exp. Astron.* 23, 357

Highlights of Astronomy, Volume 15
XXVIIth IAU General Assembly, August 2009
Ian F. Corbett, ed.
© International Astronomical Union 2010
doi:10.1017/S1743921310011166

Gavriil Adrianovich Tikhov (1875-1960) a pioneer in astrobiology

Victor Tejfel

Fessenkov Astrophysical Institute, Almaty, Kazakhstan
email: tejf@hotmail.com

Abstract. Astrobiology emerged as a scientific discipline in Kazakhstan more than half-a-century ago and flourished for many years under the leadership of Gavriil A. Tikhov, the oldest Pulkovian astrophysicist, member of the Academy of Sciences of the KazSSR (ASKSSR), and corresponding member of the Academy of Sciences of the USSR.

The background for the founding of this new discipline is the following: in 1941, despite the war, the government supported an expedition to Alma-Ata, organized by astronomers from Moscow and Leningrad, for the purpose of observing the total solar eclipse of 21 September 1941. At the end of the war, not all these astronomers had left Alma-Ata. Even during the war, under the initiative of academician V. G. Fessenkov, plans were carried out for an astrophysical observatory. With support from academician K. I. Satpaev, the Institute of Astronomy and Physics was created within the ASKSSR. Later, in 1950, it was divided in two: the Physical-Technical Institute, and the Astrophysical Institute, which now bears the name of its founder, V. G. Fessenkov. Tikhov also stayed in Alma-Ata and continued his own work as a member of the ASKSSR. He was well known as a researcher of Mars, one of the authors of the first Pulkovo *Course of Astrophysics*, the inventor of several optical devices for astrophysical research, and a tireless observer whose observing logbooks may still serve as a example of carefulness and accuracy.

Before describing the founding of astrobiology by Tikhov, it is worth recalling some facts from his biography and his contributions to the development of domestic astrophysics. He was born in 1875 in a small village near Minsk (Belorussia), the son of a railway employee. After 4 years of training at Moscow University, which he finished in 1897, Tikhov continued his education at the Sorbonne University in Paris. There, he was introduced to the Russian astronomer A.P. Gansky, and to J. Janssen, director of the Meudon Observatory. On his advice, Tikhov, together with French astrophysicists, made a balloon flight to observe the Leonids meteor shower in November 1899. More details about his life and activity may be found in his memoirs (Tikhov 1959), in a book by one of his disciples (Suslov 1980), or in other publications (for example, Eremeeva 1966).

Tikhov devoted his years in Alma-Ata to the question of life outside the Earth. He led the organization of the Sector of Astrobotany in the ASKSSR. Its staff was composed of astronomers, physicists and biologists who investigated the fluorescence of plants in the infrared and a number of other optical features. Along with the astrobotany research, some of the astronomers and post-graduate students carried out spectrophotometric observations of the Moon, planets, comets, asteroids, as well as variable and magnetic stars. Their primary goal was to study the optical properties of plants in various climatic conditions, including those which were thought to exist on Mars and Venus. The basic idea was the possibility for plants to accomodate to severe climatic conditions by changing their spectral reflectivity. The existence of vegetation on Mars was seriously considered at that time because observations over several decades had shown seasonal changes, such as variations in the size of the polar caps or in the colouring of the darker areas called "seas". Unlike terrestrial vegetation, however, the spectrum of Martian "seas" did not show the characteristic absorption of chlorophyll. This served as a powerful argument for opponents to the presence of life on Mars. But the spectroscopic study of plants that grow in cold climates had shown that the absorption band of chlorophyll becomes wider. This finding was the basis for Tikhov's assumption that, in very low temperature conditions, chlorophyll absorption could become so wide that it would go undetected within a narrow spectral region. The global spectral reflectivity of Martian plants would then be significantly decreased, and the absorption of solar energy would be higher.

Tikhov was a convinced opponent of geocentrism and believed that life is widespread in the universe, contrary to skeptics who thought terrestrial conditions were unique for the development of living organisms. In the 1940-50s, the possibility of life on Mars was a subject of active discussions among scientists and drew attention from a broad audience of people. At Tikhov's lectures (he never shone away from giving public lectures even in his old age), it was sometimes impossible to find room. There and then, a new scientific discipline was born which Tikhov called "Astrobiology". In 1953, under this title, he published a popular book (Tikhov 1953), while his earlier "Astrobotany" (Tikhov 1949) was published by the ASKSSR. Both of these books were based on research carried out in the Sector of Astrobotany and on earlier articles published by the ASKSSR. Results from his research activity and from that of his disciples are contained in the Proceedings of the Sector of Astrobotany and in selected works (Tikhov 1954-1960).

Although Tikhov lived most of his life before the space era, he witnessed its first achievements (satellites, automatic lunar stations). Soon after his death in 1960, the activity of the Sector of Astrobotany ceased by decision of the ASKSSR. At first, city authorities decided to keep Tikhov's house and observatory, along with its environmental grounds (the "Astrobotany Garden") as historical and cultural momentoes and to place them under state protection. It was even proposed to create an astronomical complex (planetarium, museum, public observatory), an initiative which the municipal government supported. But later, these plans were cancelled despite numerous protests in the press (Tejfel 1984, Gostev 1987, Tejfel 1987, Tulegenova 1987), and both the house and observatory were thoughtlessly torn down, including the "Astrobotany Garden".

Interest in the works of Tikhov was keenly shared by foreign scientists. In this respect, the visit which A.G. Wilson (Lovell Observatory) payed to Tikhov in 1958 is very indicative. Wilson had come especially to Alma-Ata after participating in the IAU General Assembly held in Moscow that year. During his visit, Wilson made the following statement: "America has recognized the merits of Tsiolkovskij too late; we correct this mistake by a recognition of Tikhov's works". Many years later, it is clear to us that this was not simply a diplomatic compliment.

References

Eremeeva, A. I. 1966, *Oustanding astronomers of the world* (Moscow: Kniga Publ.), p. 293

Gostev A. 1987, *The destroyed memory?* Ogni Alatau (25 Aug. 1987)

Suslov, A. K. 1980, *Gavriil Adrianovich Tikhov* (Leningrad: Nauka Press), p. 120

Tejfel, V. 1984, *Stars and memory* Ogni Alatau (12 Sept. 1984)

Tejfel, V. 1987, *Not over yet* Kazakhstanskaya Pravda (1 Jul. 1987)

Tikhov, G. A. 1949, *Astrobotany*, (Alma-Ata: Acad. Sci. KazSSR), p. 23

Tikhov, G. A. 1953, *Astrobiology*, (Moscow: Molodaya Gvardiya Publ.), p. 67

Tikhov, G. A. 1954-1960, *Basic works in 5 volumes*, (Alma-Ata: Acad. Sci. KazSSR)

Tikhov, G. A. 1959, *Sixty years at a telescope*, (Moscow: Detgiz Publ.), p. 160

Tulegenova, A. 1987, *An echo above ruins*, Kazakhstanskaya Pravda (11 Jan. 1987)

Highlights of Astronomy, Volume 15
XXVIIth IAU General Assembly, August 2009
Ian F. Corbett, ed.

Jean Heidmann (1923-2000) and SETI

Régis Courtin

LESIA, CNRS & Observatoire de Paris, 92195 Meudon, France
email: regis.courtin@obspm.fr

Abstract. Jean Heidmann (1923-2000) began his research career as a radio-astronomer in 1959 at Paris Observatory, investigating the structure of galaxies and the distance scale in the nearby universe. In the early 1980's, his scientific interest broadened to the search for extraterrestrial intelligence and he became a strong advocate of SETI, either from the ground or from space.

1. Introduction

Jean Heidmann was born May 18 1923 in Jeumont (Nord). In 1946, then a fresh graduate in engineering from Ecole Centrale, he joined the physics laboratory headed by Louis Leprince-Ringuet at Ecole Polytechnique. He was given the task of solving one of the burning questions in cosmic rays physics at the time, namely the emission of fast deutons in nuclear disintegrations generated by cosmic rays. Between 1948 and 1951, this project led him to the Wills Physics Laboratory in Bristol, where he worked under the supervision of Nevill Mott, and to the Laboratory of Nuclear Studies at Cornell University, where he worked for two years under the direction of Hans Bethe. In June 1951, he defended a PhD Thesis at Cornell, and another at Paris University in December. The members of his jury were no less than Louis de Broglie, Francis Perrin, and Louis Leprince-Ringuet, probably the top three French physicists at the time.

2. Astronomer, science editor, communicator, and advocate of SETI

Appointed "Maître de Recherches" at "Centre National de la Recherche Scientifique" (CNRS) in 1955, Heidmann continued working at Ecole Polytechnique for a few more years. In 1959, he was named "Astronome" and invited by Jean-François Denisse to join the radio-astronomy team at the Meudon branch of the Paris Observatory. The design of the Nançay Radio-Telescope was already well advanced. Starting in 1965, and through the 1970's, Heidmann and his colleagues worked with this powerful instrument to investigate the hydrogen content of galaxies as measured by the 21cm line (Chamaraux, Heidmann & Lauqué 1970, Balkowski *et al.* 1972). Using optical or UV telescopes as well, they also contributed to the multi-wavelength study of Markarian galaxies (Bottinelli *et al.* 1975), and to the determination of the Hubble constant (Heidmann 1977). Together with Italian astronomer C. Casini, he identified the so-called clumpy irregular galaxies (Casini & Heidmann 1976).

Heidmann was one of the first editors-in-chief of the European journal *Astronomy and Astrophysics* from 1973 to 1979. He was also very adept at communicating astronomy to students and to the public, either through books (Heidmann 1979, Heidmann 1980, Heidmann 1982, Heidmann 1986, Heidmann 1992, Heidmann 1998, Heidmann *et al.* 2000), numerous public conferences, or a radio program on the national network.

Towards the end of his career, his passion for SETI which started around 1982, became the main thrust behind his work. With François Biraud, and in collaboration with Jill

Tarter and Samuel Gulkis, he carried out the first search programs in Europe using the Nançay radiotelescope. One such program consisted in using rotational frequencies of pulsars to select preferred search frequencies in the 1-10 GHz Galactic radio window (Heidmann, Biraud & Tarter 1989), whereas another involved a small sample of newly discovered extrasolar planets (Biraud *et al.* 1997).

Even after his official retirement in 1993, Heidmann continued working on various aspects of SETI (Heidmann 1993, Heidmann 1994a) and was the driving force behind innovative SETI space concepts. For instance, in 1994, he proposed the establishment of a RFI-free site inside crater Saha, on the far side of the Moon (Heidmann 1994b). Until his death, he was a strong advocate of this proposal in various arenas (Heidmann 2000a, Heidmann 2000b). With Claudio Maccone, he also proposed to use the gravitational focusing effect of the Sun to detect extraterrestrial civilizations, and to study the Galactic center at high-gain (Heidmann & Maccone 1994).

His international involvement was very strong: in 1990, he was the lead-organizer of the Third International Symposium on Bioastronomy held at Val-Cenis, France (Heidmann & Klein 1991); within the International Astronautical Academy, he led the SETI Commitee and coordinated the "Lunar SETI Cosmic Study" started in 1998; within the International Astronomical Union, he was a member of the Organizing Committee of Commission 51 on Bioastronomy since 1988, and its Secretary since 1991.

Heidmann's convictions about the eventual success of SETI were so powerful that he dared propheticize that the "S" of SETI, after standing for "search", would one day mean "survey", then "study", and ultimately "sociology" of extraterrestrial intelligence.

References

Balkowski, C., Bottinelli, L., Gouguenheim, L., & Heidmann, J. 1972 *A&A* 21, 303

Biraud, F., Heidmann, J., Tarter, J., & Airieau, S. 1997 *Proc. IAU Colloquium No. 161* (Bologna: Editrice Compositori), p. 689

Bottinelli, L., Gouguenheim, L., Heidmann, J., & Duflot, R. 1975 *A&A* 41, 61

Casini, C. & Heidmann, J. 1976, *A&A*, 47, 371

Chamaraux, P., Heidmann, J., & Lauqé, R. 1970 *A&A* 8, 424

Heidmann, J. 1977 in: C. Balkowski & B.E. Westerlund (eds.), *Proc. IAU Colloquium No. 37* (Paris: CNRS), p. 487

Heidmann, J. 1979 *Au delà de la Voie Lactée, un étrange Univers* (Paris: Hachette)

Heidmann, J. 1980 *Relativistic Cosmology: an introduction* (Berlin: Springer-Verlag)

Heidmann, J. 1982 *Extragalactic Adventure* (Cambridge: Cambridge University Press)

Heidmann, J. 1986 *L'odyssée cosmique* (Paris: Denoël)

Heidmann, J. 1992 *La vie dans l'Univers* (Paris: Fayard)

Heidmann, J. 1993 in: F. Bertola & U. Curi (eds.), *The Anthropic Principle. Proc. Second Venice Conference* (Cambridge: Cambridge University Press), p. 111

Heidmann, J. 1994a *J. Brit. Interplan. Soc.* 47, 71

Heidmann, J. 1994b *Acta Astronaut.* 32, 471

Heidmann, J. 1998 *Intelligences extraterrestres* (Paris: Odile Jacob)

Heidmann, J. 2000a *Acta Astronaut.* 46, 661

Heidmann, J. 2000b *Adv. Space Res.* 26, 371

Heidmann, J., Biraud, F., & Tarter, J. 1989 *Proc. 40th Int. Astron. Congress*, Malaga, Spain

Heidmann, J. & Klein, M. J. (eds.) 1991 *Bioastronomy. The Search for Extraterrestrial Life*, Lecture Notes in Physics 390 (Berlin: Springer-Verlag)

Heidmann, J. & Maccone, C. 1994 *Acta Astronaut.* 32, 409

Heidmann, J., Vidal-Madjar, A., Reeves, H., & Prantzos, N. 2000 *Sommes-nous seuls dans l'Univers ?* (Paris: Fayard)

Highlights of Astronomy, Volume 15
XXVIIth IAU General Assembly, August 2009
Ian F. Corbett, ed.
© International Astronomical Union 2010
doi:10.1017/S174392131001118X

George W. Wetherill (1925-2006): physicist, geochemist, planetary scientist, astrobiologist

Alan P. Boss

Carnegie Institution, Washington, DC, USA
email: boss@dtm.ciw.edu

Abstract. Trained as a physicist, George W. Wetherill (1925-2006) made seminal contributions to the fields of geochemical dating, meteoritical and asteroidal science, and the theory of the formation of terrestrial planets, evolving along the way into one of the first astrobiologists.

1. Introduction

George W. Wetherill was born on 12 August 1925 in Philadelphia, Pennsylvania. He attended the University of Chicago, earning four degrees: Ph.B., S.B., S.M., and Ph.D. Upon completion of his doctorate in nuclear physics (Wetherill 1953), Wetherill became a staff member at the Carnegie Institution's Department of Terrestrial Magnetism (DTM) in Washington, D.C., where physicists had branched out into new areas of science. In 1960, he left DTM to become a professor of geophysics and geology at UCLA, also serving as chairman of the Department of Planetary and Space Science from 1968 to 1972. In 1975, Wetherill returned to DTM as Director, a position he held until 1991. He remained on the active DTM staff until retiring in 2001. George Wetherill died on 19 July 2006.

2. Research accomplishments

Wetherill can be rightfully called the father of modern theories of the formation of the Earth. Wetherill was in the vanguard of the effort to place planet formation theory on a solid basis, but his work spanned much of the disciplines of Earth and planetary science.

Wetherill and his colleagues at DTM and Carnegie's Geophysical Laboratory revolutionized the field of geochemical dating of rocks. Wetherill conceived of the concordia diagram (Wetherill 1956, 1963), which uses the decay of radioactive uranium into lead to provide accurate dates for when the rocks crystallized. Wetherill's concordia diagram was a concept that found immediate and lasting acceptance, and stands as a singular achievement in the Earth sciences. It opened up the field of geological dating for events that happened billions of years ago on the Earth and on other rocky bodies (Wetherill 1975a).

Wetherill's second major contribution dealt with the orbital evolution of asteroids and of other small bodies in the Solar System (Wetherill 1967, 1969, 1974, 1987). He was the first to show that debris kicked out from meteorite impacts on Mars could be expected to end up on Earth, as has been spectacularly verified by the dozens of Martian meteorites found to date on Earth in Antarctica and elsewhere.

Using these skills in orbital dynamics, Wetherill launched into his third major research area, planet formation modeling, in 1975. He quickly became the world's leading authority on the process by which the rocky inner planets formed through impacts between progressively larger and larger bodies (Wetherill 1975b, 1976, 1980, 1990, 1994b).

Wetherill was the first to point out that the Earth's formation involved impacts by bodies as large as Mars on the growing Earth (Wetherill 1985, 1986). One such giant impact is the now-accepted explanation for how the Earth's Moon was formed – an off-center giant impact produced a spray of hot rock that ended up in orbit around the Earth and then formed the Moon. In 1995, the first solid evidence of planetary systems orbiting other stars like our Sun was presented. Astronomers have since discovered over 350 planets in orbit around nearby stars, and recently may have found the first rocky planets. George Wetherill was way ahead of these discoveries (Wetherill 1991, 1994a, 1996). His models of terrestrial planet formation, combined with the latest discoveries, imply a rich future for those who seek to find Earth-like planets in our neighborhood of the galaxy.

3. Honors

Wetherill received a wide range of awards for his accomplishments throughout his life. He was a member of the U. S. National Academy of Sciences, a Fellow of the American Academy of Arts and Sciences, the American Geophysical Union, and the Meteoritical Society. He won the Harry H. Hess Medal of the American Geophysical Union, the Leonard Medal of the Meteoritical Society, the G. K. Gilbert Award of the Geological Society of America, the G. P. Kuiper Award of the Division of Planetary Sciences of the American Astronomical Society, and the Henry Norris Russell Lectureship of the American Astronomical Society. In 1997 Wetherill received the National Medal of Science, the nation's highest scientific award.

Acknowledgements

I thank the American Astronomical Society for a travel grant that partially supported my attendance at the Rio de Janeiro IAU General Assembly, as well as partial support from NASA Astrobiology Institute grant NCC2-1056.

References

Wetherill, G. W. 1953, *Phys. Rev.* 92, 907
Wetherill, G. W. 1956, *I. Trans. Amer. Geophys. Union* 37, 320
Wetherill, G. W. 1963, *J. Geophys. Res.* 68, 2957
Wetherill, G. W. 1967, *J. Geophys. Res.* 72, 2429
Wetherill, G. W. 1969, in: P. M. Millman (ed.), *Meteorite Research* (Dordrecht: D. Reidel), p. 579
Wetherill, G. W. 1974, *Annual Rev. Earth Planet. Sci.* 2, 303
Wetherill, G. W. 1975a, *Ann. Rev. Nucl. Sci.* 25, 283
Wetherill, G. W. 1975b, *Proc. 6th Lunar Sci. Conf.* (New York: Pergamon), p. 1539
Wetherill, G. W. 1976, *Proceedings 7th Lunar Sci. Conf.* (New York: Pergamon), p. 3245
Wetherill, G. W. 1980, *Annual Review of Astronomy and Astrophysics* 18, 77
Wetherill, G. W. 1985, *Science* 228, 877
Wetherill, G. W. 1986, in: W. K. Hartmann, R. J. Phillips & G. J.Taylor (eds.), *Origin of the Moon* (Houston: Lunar and Planetary Institute), p. 519
Wetherill, G. W. 1987, *Phil. Trans. R. Soc. Lond., Series A* 323, 323
Wetherill, G. W. 1990, *Ann. Rev. Earth. Plan. Sci.* 18, 205
Wetherill, G. W. 1991, *Science* 253, 535
Wetherill, G. W. 1994a, *Astrophysics & Space Science* 212, 23
Wetherill, G. W. 1994b, *Geochimica et Cosmochimica Acta* 58, 4513
Wetherill, G. W. 1996, *Icarus* 119, 219

Highlights of Astronomy, Volume 15
XXVIIth IAU General Assembly, August 2009
Ian F. Corbett, ed.

© International Astronomical Union 2010
doi:10.1017/S1743921310011191

Special Session 7
Young Stars, Brown Dwarfs, and
Protoplanetary Disks

Jane Gregorio-Hetem[1]
and Silvia Alencar[2]

[1] Universidade de São Paulo, IAG/USP, Brazil
email: jane@astro.iag.usp.br
[2] Departamento de Física, ICEx/UFMG, Brazil,
email: silvia@fisica.ufmg.br

Preface

In recent years our knowledge of star, brown dwarf and planet formation has progressed immensely due to new data in the IR domain (Spitzer telescope), new X-ray campaigns such as the *Chandra Orion Ultradeep Project* (COUP) and the *X-ray Emission Survey of Taurus* (XEST), with *XMM-Newton*, as well as adaptive optics results and synoptic studies of young stellar and substellar objects.

Disk accretion and jet outflows, which are intimately associated with the formation and early evolution of stars, has also provided new insights in the early stellar years. High angular resolution observations provide amazing images of protoplanetary disks around young stars at various evolutionary stages and different environments. Jet structures have also been resolved on scales of a few to a hundred AU. More spectacular results related to disk and jet structure and dynamics are already starting to appear from interferometry (VLTI, Keck/I) and improved imaging systems such as HST/ACS.

At the same time, observing programs to find planets by transit, such as the CoRoT mission, are expected to probe a new planetary mass domain and provide new results on planet formation. CoRoT Additional Programmes are also exciting opportunities to study photometric and spectroscopic variability of PMS stars that present a rich variety of types, which includes pulsation, hot and cold spots, higher modes of vibrations, stellar multiplicity, accretion processes, jets, disk debris passing in front of the star, Algol-type occultations, among many others.

Following approval by the IAU Executive Commitee, this IAU Special Session on *Young Stars, Brown Dwarfs, and Protoplanetary Disks* allowed a meeting with a broad range of subjects linked to star/planet formation and evolution, taking advantage that researchers from different, but closely related areas, attended the IAU General Assembly in 2009. The integration of talks and discussions provided excellent scientific exchange between researchers from different backgrounds to achieve significant debate about open issues in the formation and early evolution of stars, brown dwarfs, and planets.

The papers presented here, in the format of extended abstracts, give a summary of the range of topics covered in our meeting. Most of the authors have consented to make available the slides of their oral presentations at our website, where the poster contributions can also be found: http://www.fisica.ufmg.br/ss7iau09/.

The Scientific Programme benefited from the contribution of the Scientific Organising Committee (S.O.C.), suggesting names of invited speakers that fulfilled the aims of our meeting. The final choice of talk contributions was a hard task for the S.O.C., due to the high demand (77 submitted abstracts for oral contributions) to be fitted in a short schedule (2.5 days meeting). In total, the Scientific Programme counted on 9 invited talks, 25 oral presentations, and 79 poster presentations. We estimated about a hundred people attending the Special Session 7 (SpS7).

Unfortunately two of the invited speakers, Ray Jayawardhana and Isabelle Baraffe, could not come. Folowing their suggestions, their talks were presented respectively by Ilaria Pascucci and Subu Mohanty. We are greatful to them, for accepting the substitution role and for doing excellent presentations. A special thank goes to the session chairs Celso Batalha, Suzana Lizano, Hans Zinnecker, Eric Feigelson, Michael Sterzik, and Sylvie Cabrit.

The Local Organising Committee also contributed to the success of the SpS7. We greatly acknowledge Daniela Lazzaro and all the colleagues from Rio de Janeiro, who direct or indirectly helped us in the running of this special session. We also thank Beatriz Barbuy who provided financial support (CNPq Proc. No. 451483/2009) for travel expenses of some participants of the SpS7.

Scientific Organising Committee

Silvia Alencar (Brazil, Co-chair),
Nuria Calvet (USA),
Gilles Chabrier (France),
Francesca D'Antona (Italy),
Eric Feigelson (USA),
Jane Gregorio-Hetem (Brazil, Co-chair),
Sergei Lamzin (Russia),
Susana Lizano (Mexico),
Robert Mathieu (USA),
Thierry Montmerle (France),
Antonella Natta (Italy),
Bo Reipurth (USA),
Hsien Shang (Taiwan),
Michael Sterzik (Chile),
Ewine van Dishoeck (Netherlands),
Hans Zinnecker (Germany).

Highlights of Astronomy, Volume 15
XXVIIth IAU General Assembly, August 2009
Ian F. Corbett, ed.

Accretion Disks in the Sub-Stellar Realm: Properties and Evolution

Ray Jayawardhana

University of Toronto

Abstract. It is now well established that young brown dwarfs harbor accretion disks –and thus undergo a T Tauri phase– similar to their low-mass stellar counterparts. The supporting evidence includes infrared and millimeter observations of the dust component as well as optical and infrared spectra with signatures of gas accretion and outflow. Recent findings suggest that disks are common even around young planetary mass objects. The ubiquity of circum-sub-stellar disks not only hints at a common formation scenario for PMOs, brown dwarfs and stars, but also offers a new regime for investigating processes such as episodic accretion, grain growth and disk clearing.

Highlights of Astronomy, Volume 15
XXVIIth IAU General Assembly, August 2009
Ian F. Corbett, ed.
© International Astronomical Union 2010
doi:10.1017/S174392131001121X

2D mapping of ice species in molecular cores*

Jennifer A. Noble[1], H. J. Fraser[1], K. M. Pontoppidan[2], Y. Aikawa[3] and I. Sakon[4]

[1] Department of Physics, University of Strathclyde, 107 Rottenrow, Glasgow G4 0NG, Scotland
email: jennifer.noble@strath.ac.uk
[2] California Institute of Technology, Division of Geological and Planetary Sciences, MS 150-21,
Pasadena, CA 91125, U.S.A.
[3] Department of Earth and Planetary Sciences, Kobe University, Kobe 657-8501, Japan
[4] Department of Astronomy, Graduate School of Science, University of Tokyo, 7-3-1 Hongo,
Bunkyo-ku, Tokyo 113-0003, Japan

Abstract. We present data from our ice mapping program IMAPE on the AKARI satellite.
Initial results show a correlation between the abundance of $CO_{2(s)}$ and $H_2O_{(s)}$, consistent with
previous studies. We can trace abundances of molecules across a core using a single observation.

Keywords. astrochemistry, ISM: clouds, ISM: molecules, techniques: spectroscopic

1. Introduction

In dense cores, much of the molecular material is frozen on the surface of dust grains.
AKARI allows the simultaneous observation of multiple lines of site (*los*) through a core.
We observed a 1'x1' region towards 20 cores, between 2.5–5.0 μm. Data was reduced
using our own pipeline (Noble *et al.* in prep.), producing spectra for 31 *los*.

2. Results and Conclusions

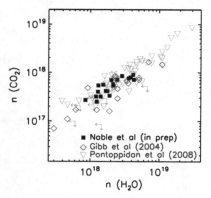

Figure 1. Correlation plot of $n(CO_2)$ vs $n(H_2O)$ for
31 *los* in the AKARI IMAPE program.

The abundance of H_2O and CO_2 was calculated for each *los* using laboratory data, and
is presented in Figure 1. Abundances agree with previous studies (as shown in Figure 1)
and a clear correlation is seen between $n(H_2O)$ and $n(CO_2)$ in the cores observed.

References

Gibb, E. L. *et al.* 2004, *ApJS*, 151, 35
Pontoppidan, K. M. *et al.* 2008, *ApJ*, 678, 1005

*Based on observations with AKARI, a JAXA project with the participation of ESA.

Highlights of Astronomy, Volume 15
XXVIIth IAU General Assembly, August 2009
Ian F. Corbett, ed.

Evolution of Young Stars and Their Disks in Serpens

Isa Oliveira[1], Bruno Merín[2], Klaus Pontoppidan[3] and Ewine van Dishoeck[1,4]

[1] Sterrewacht Leiden, Leiden University, P.O. Box 9513, 2300 RA Leiden, The Netherlands
email: oliveira@strw.leidenuniv.nl
[2] Herschel Science Center, European Space Agency (ESA),
P.O. Box 78, 28691 Villanueva de la Cañada (Madrid), Spain
[3] California Institute of Technology, Division for Geological and Planetary Sciences,
MS 150-21, Pasadena, CA 91125, USA
[4] Max-Planck Institut für Extraterrestrische Physik,
Giessenbachstrasse 1, 85748 Garching, Germany

Abstract. Unbiased, flux-limited surveys of protoplanetary disks and their parent stars currently exist for only a few clouds, primarily Taurus and IC 348, selected primarily by optical and near-IR data. Such surveys are essential to address questions of disk evolution as a function of stellar parameters such as spectral type, age, accretion activity and environment. Using the 'Cores to Disks' (c2d) Spitzer Legacy Program, we discovered a new population of young stellar objects (YSOs) in a region of only 0.8 $\deg^2$ in the Serpens Molecular Cloud. This sample contains 150 mid-IR bright ($\geqslant$ 3 mJy at 8 μm) YSOs with infrared excess, having a broad range of SED types and luminosities. Serpens is therefore a unique target region for obtaining a complete, well-defined sample of multi-wavelength observations of young stars in a possible evolutionary sequence. Compared with other clouds such as Taurus and Chamaeleon, Serpens has an exceptionally high star-formation rate (5.7×10^{-5} $M_\odot$ yr^{-1}). Follow-up complimentary observations in the optical, near- and mid-infrared (Spitzer/IRS GO3) have allowed us to characterize both the central stars and the surrounding disks. The shape and slope of the mid-infrared excess provide information on the flaring geometry of the disks. The spectral features give constraints on grain growth and mineralogy, which in turn probes heating and radial mixing. The presence of PAH features traces UV radiation, whereas Hα and Brγ are used as diagnostics of accretion. Assuming that all stars within a sufficiently small region are nearly coeval, this provides direct constraints on the importance of environment and initial conditions on disk evolution. In this meeting, we have presented our latest results on this rich populations of YSOs, as detailed in Oliveira et al. (2009, 2010). We have discussed connections between the evolution of the disks and that of their harboring stars, and the processes that determine the evolutionary sequence of protoplanetary disks.

Keywords. stars: pre–main sequence – stars: planetary systems: protoplanetary disks

References

Furlan, E. *et al.* 2006, *ApJS* 165, 568
Furlan, E. *et al.* 2008, *ApJS* 176, 184
Harvey, P. *et al.* 2007, *ApJ* 663, 1149
Kessler-Silacci, J. *et al.* 2006, *ApJ* 639, 275
Oliveira, I. *et al.* 2009, *ApJ* 691, 672; and in prep.
Oliveira, I. *et al.* in prep.
Olofsson, J. *et al.* *A&A* accepted

Highlights of Astronomy, Volume 15
XXVIIth IAU General Assembly, August 2009
Ian F. Corbett, ed.

© International Astronomical Union 2010
doi:10.1017/S1743921310011233

The End of Accretion: Transition Disks, Dissipation, and Links to Planet Formation

James Muzerolle[1]

[1]Space Telescope Science Institute, Baltimore, MD, 21218, USA
email: muzerol@stsci.edu

Abstract. I review recent studies of transitional protoplanetary disks, in which the planet forming regions are being cleared of material. The dust and accretion characteristics of these objects as a function of their environment and host star age and mass reveal important clues as to how disk structure is modified and ultimately destroyed. Clearing mechanisms such as grain growth, dynamical interactions with embedded planets or stellar companions, and photoevaporation are likely all involved, pointing to diverse disk evolutionary paths.

Keywords. accretion, accretion disks; circumstellar matter; planetary systems: protoplanetary disks; stars: pre–main-sequence

1. Introduction

Observational estimates of the lifetimes of circumstellar disks are essential for constraining planet formation theories. Studies focused on near-infrared emission probing dust in the innermost regions of disks have shown a clear decline in excess frequency with stellar age, indicating typical disk lifetimes of roughly 3 Myr, with a wide dispersion from 1-10 Myr. Meanwhile, measurements of mass accretion rates have provided complementary information on the gas content of disks, showing a general decline with stellar age consistent with expectations from models of viscous evolution. However, there is a large dispersion in accretion rates, as well as disk lifetimes, at any given stellar age that is not consistent with simple evolutionary models. The mechanisms for disk dispersal, including grain/planetesimal growth, giant planet formation, and photoevaporation, are still very much open to debate.

A key piece of the puzzle is provided by the small number of disks *in transition*. Such objects still harbor substantial amounts of primordial material but show signs of significant clearing in their inner regions. A handful of such transition disks with inner "holes" have been studied in some detail in recent years, particularly with the advent of *Spitzer* Space Telescope observations. Modeling of spectral energy distributions (SEDs) of transitional disks have inferred inner holes ranging from $\sim 3 - 24$ AU in size (e.g., Calvet *et al.* 2005). Follow-up interferometric observations of some of these objects have directly imaged these holes (e.g., Hughes *et al.* 2009; Brown *et al.* 2009).

2. Transition disk statistics

Relatively large numbers of transition disks are now being identified from *Spitzer* surveys of star forming regions and young stellar clusters. Below I list some of the important results that have been found to date on the properties of these objects.

• **3 types of transition disks:** (1) classical transition disks, sources with small or zero dust excess at $\lambda \lesssim 10$ μm indicative of an optically thick outer disk with an AU-scale inner "hole" where the dust has been rendered optically thin or completely evacuated; (2)

gapped or "pre-transitional" (Espaillat *et al.* 2007) disks, objects that exhibit a similar SED dip but more substantial short-wavelength excess indicative of an AU-scale disk gap bounded by optically thick inner and outer disk annuli; (3) "weak-excess", "evolved", or "homologously-depleted" disks, sources with small or zero short-wavelength excess and weak mid-infrared excess suggestive of a flat or optically thin disk.

• **Accretion properties:** A majority of type 1 and 2 objects show evidence for active accretion. This indicates that they still harbor substantial amounts of gas within their dust holes/gaps, and suggests that planetesimal or giant planet formation may be more likely in these cases. A much smaller fraction of type 3 objects show significant accretion activity, indicating either a lack of gas in the inner disk or very small accretion rates. When combined with the generally low disk masses inferred by Cieza *et al.* (2008), this suggests that photoevaporation may be an important mechanism for the type 3 transition.

• **Trends with stellar age:** The frequency of type 1 transition disks increases with stellar age ($\sim$ 1% of the total stellar population at 1 Myr and $\sim$ 5 − 10% at 5 Myr; Muzerolle *et al.* 2009). The trend is even stronger when type 3 objects are included; these comprise a majority of the disks still present at 5-10 Myr. The frequencies suggest that the clearing duration may be age-dependent, and at least in the case of type 3 objects may be of order the total disk lifetime (e.g., Currie *et al.* 2009).

• **Trends with stellar mass:** There is an apparent deficit of type 1 transition disks around stars with masses $M \lesssim 0.3\ M_\odot$ (Muzerolle *et al.* 2009). By contrast, the vast majority of type 3 transition disks are around M-type stars. The clearing mechanisms responsible for the different transition types thus apper to depend on the stellar mass. One possibility for the type 1 discrepancy is that giant planet formation is less common in disks around low mass stars. Conversely, the long lifetime of type 3 disks may be explained if photoevaporation is less efficient around low mass stars.

• **Stellar multiplicity:** High spatial resolution imaging has shown that at least one transition disk, the type 1 object CoKu Tau/4, is in fact circumbinary (Ireland & Kraus (2008)). However, the recent survey of Kraus *et al.* (see Kraus, this volume) has failed to find stellar companions to any other Taurus stars with transition disks. The age and stellar mass trends are also inconsistent with binarity being a significant contributor.

The properties and trends of transition disks suggest the existence of multiple evolutionary pathways (Lada *et al.* 2006; Cieza *et al.* 2007). More than one process is likely involved with the clearing of disk material. Further detailed measurements of properties such as disk masses, mass accretion rates, and hole sizes are needed to better constrain which of these mechanisms may be at work in any given object (Najita *et al.* 2007).

References

Brown, J. M., Blake, G. A., Qi, C. *et al.* 2009, *ApJ*, 704, 496

Calvet, N. *et al.* 2005, *ApJ*, 630, L185

Cieza, L., Padgett, D. L., Stapelfeldt, K. R. *et al.* 2007, *ApJ*, 667, 308

Cieza, L. A., Swift, J. J., Mathews, G. S., & Williams, J. P. 2008, *ApJ*, 686, L115

Currie, T., Lada, C. J., Plavchan, P., Robitaille, T. P., Irwin, J., & Kenyon, S. J. 2009, *ApJ*, 698, 1

Espaillat, C. *et al.* 2007, *ApJ*, 664, L111

Hughes, A. M. *et al.* 2009, ApJ, 698, 131

Ireland, M. J. & Kraus, A. L. 2008, *ApJ*, 678, L59

Lada, C. J., Muench, A. A., Luhman, K. L. *et al.* 2006, *AJ*, 131, 1574

Muzerolle, J., Allen, L., Megeath, S. T. *et al.* 2009, *ApJ*, submitted

Najita, J., Strom, S. E., & Muzerolle, J. 2007, *MNRAS*, 378, 369

Highlights of Astronomy, Volume 15
XXVIIth IAU General Assembly, August 2009
Ian F. Corbett, ed.

© International Astronomical Union 2010
doi:10.1017/S1743921310011245

Young Stellar Object Variability at IRAC Wavelengths: Clues to Star and Planet Formation

John R. Stauffer[1], Maria Morales[1] and Luisa Rebull[1]

[1]Spitzer Science Center, Caltech, Pasadena, CA 91125, USA
email: stauffer@ipac.caltech.edu

In Morales *et al.* (2009), we have recently investigated the mid-infrared (3.6 to 8.0 micron) variability of young-stellar objects (YSOs) using the IRAC camera on the Spitzer Space Telescope. Specifically, we obtained synoptic photometry of about 70 YSOs in the ~1 Myr old IC1396A globule over a 14 day period. More than half of the YSOs were detectably variable, with amplitudes up to about 0.2 magnitudes. About a third of these objects showed quasi-sinusoidal light curves with apparent periods of typically 5 to 12 days. At least two families of models can explain such light curves: (a) a Class II YSO with a photospheric hot spot which locally heats the inner circumstellar disk which is viewed from slightly above the disk plane, and (b) a YSO with a warped disk or with some other non-axisymmetric inner disk density profile, also seen with a view angle slightly above the disk plane. The two models can both yield light curve shapes and amplitudes similar to what we observe in the mid-infrared, but produce very different light curves at shorter wavelengths dominated by the stellar photosphere. Because we only had IRAC photometry for IC1396A, we were not able to discriminate between the two models for this set of data.

We have just begun a much more extensive monitoring program to obtain synoptic observations of >1500 YSOs in Orion and eleven other star formation regions. The new effort - called YSOVAR - is a Spitzer "Warm Mission Exploration Science" program allocated 550 hours of observing time. For each cluster, we will obtain about 100 epochs of IRAC photometry, normally mostly in a 40 day window of time. We will also obtain a large body of optical and near-IR photometry in order to discriminate between the different physical models. Observations of IC1396A and the inner one square degree of the Orion Nebula Cluster will be obtained during Sept.-Dec. 2009. All of the IRAC data from the program will be made publicly available within 90 days of their being pipeline processed. The optical and near-IR photomtry will also be made public, as rapidly as we can do so.

Details about YSOVAR, the clusters to be observed, and the schedule of observations can be found at http://ysovar.ipac.caltech.edu/.

References

Morales, M., Stauffer, J., Rebull, L. *et al.* 2009, *ApJ* 702, 507

Highlights of Astronomy, Volume 15
XXVIIth IAU General Assembly, August 2009
Ian F. Corbett, ed.

© International Astronomical Union 2010
doi:10.1017/S1743921310011257

Accretion dynamics and star-disk interaction in NCG 2264

Silvia H. P. Alencar[1], Jérôme Bouvier[2], Claude Catala[3], Matilde Fernandez[4], Jorge Filipe Gameiro[5], Nathalia Fonseca[1], Marcelo Medeiros Guimarães[1], Jane Gregorio Hetem[6], Pauline Telles McGinnis[1], Estelle Moraux[2] and Paula Teixeira[7]

[1]Departmento de Física, ICEx, UFMG, Belo Horizonte, MG, 31270-901, Brazil
email: silvia@fisica.ufmg.br

[2]Laboratoire d'Astrophysique de Grenoble, BP 53, F-38041 Grenoble, France

[3]Observatoire de Paris-Meudon, 61 Av. de l'Observatoire, 75014 Paris, France

[4]Instituto de Astrofisica de Andalucia, Apartado 3004, 18080 Granada, Spain

[5]Universidade do Porto, Rua das Estrelas, 4150-762 Porto, Portugal

[6]Universidade de São Paulo, IAG/USP, São Paulo, SP, 05508-090, Brazil

[7]ESO, Karl-Schwarzschild-Strasse 2, D-85748 Garching, Germany

The Corot satellite observed the young stellar cluster NGC 2264 during 23 days in March 2008. This was the first time a group of young accreting stars, classical T Tauri stars (CTTS), were followed ininterruptedly with high photometric accuracy for such a long run. Before the Corot observations, AA Tau (Bouvier *et al.* 2003, *A&A*, 409, 169 and Bouvier *et al.* 2007, *A&A*, 463, 1017) was one of the few CTTS systems that had been analysed synoptically over several consecutive rotational periods. Its analysis suggested a highly dynamical star-disk interaction mediated by the stellar magnetic field, as predicted by magneto-hydrodynamical simulations of young accreting systems.

NGC 2264 is a well studied young stellar cluster and we were therefore able to obtain a good estimate of cluster membership among the observed stars. We could also separate accreting from non-accreting systems, using indicators such as $H\alpha$ equivalent width, $H\alpha$ width at 10% intensity and $U - V$ excess. A total of 97 CTTS that belong to the cluster were observed by Corot. Among those, 26 were classified as possible AA Tau-like systems, which we defined as systems that present periodical light curve variability with an almost constant maximum interrupted by minima that can vary both in depth and width from one rotational cycle to the other. This type of light curve is believed to be due mostly to obscuration by material from the inner disk region. The data analysis shows that the AA Tau-type of light curve appears to be quite common among CTTS, which opens a good perspective to study the interaction between the stellar magnetic field and the inner disk region.

Corot and Spitzer IRAC data of 56 CTTSs were compared and it was shown that both datasets give coherent information about the evolution of the inner circumstellar disk. We used the α_{IRAC} index, the inclination of the spectral energy distribution between 3.6 μm and 8 μm, to classify the inner disk region. CTTS that present $\alpha_{IRAC} < -2.56$, naked photosphere systems, also presented spot-like light curves with no hint of obscuration by circumstellar material. As the α_{IRAC} index increases, corresponding to systems with anemic and thick disks, the percentage of spot-like light curves decreases rapidly and most of the Corot light curves present clear signs of obscuration by circumstellar material.

Highlights of Astronomy, Volume 15
XXVIIth IAU General Assembly, August 2009
Ian F. Corbett, ed.

© International Astronomical Union 2010
doi:10.1017/S1743921310011269

Evolution of protoplanetary disks

Cornelis P. Dullemond[1]

[1] Max-Planck-Institut für Astronomie
Königstuhl 17, 69117, Heidelberg, Germany
email: dullemon@mpia.de

In this review I gave an overview of the structure and evolution of protoplanetary disks, and how the evolution of dust affects this. This is an important topic because these determine the conditions under which planets are formed, or were formed in our solar system 4.5 billion years ago.

Nearly all protoplanetary disks display either dust features in emission or no dust features at all. This is strong evidence that the shape of these disks is flaring, i.e. the surface of the disk has a "bowl shape" such that the light of the star can illuminate the disk and create a warm surface layer on top of it (e.g. Chiang & Goldreich (1997)). This irradiation process is the main source of heating in most parts of the disk, except perhaps at relatively early stages and in the regions relatively close to the star. In those regions the midplane temperature can be dominated by heating via the dissipation of accretion energy. In the course of a few million years, however, the accretion rate in the disk drops and this source of midplane heating decays. As a result of this the snow line, which due to the accretional heating may initially be very far out (50 AU or so) will creep inward until it reaches about 1 AU (Davis 2005). Only when the gas and dust dissipate, after 3 to 10 million years, the snow line will move again further out because the disk becomes optically thin, and the dust can no longer keep the midplane cold.

The viscous evolution of a protoplanetary disk involves accretion as well as spreading of the disk. This has been described originally by Lynden-Bell & Pringle (1974), but it is still believed to be true. However, the mass of the disk decays only very slowly in the later stages of the evolution of the disk. It may take many tens of Myr to render the disk optically thin, while observations show that disks are destroyed in about 3 to 10 Myrs (e.g. Haisch *et al.* 2001; Carpenter *et al.* 2005). A possible solution to this problem comes from photoevaporation. The process of photoevaporation of disks by the Extreme UV (EUV) photons of the central star has been described in the context of massive stars by Hollenbach *et al.* (1994). For lower mass stars, Clarke *et al.* (2001) showed that EUV photoevaporation will drill a hole in the disk at about a few AU once the evolution of the disk has made the accretion rate to drop below about $\dot{M} \sim 10^{-9} M_{\odot}/\mathrm{yr}$. Once this hole is drilled, the disk is quickly destroyed through an inside-out evaporation process (Alexander *et al.* 2006). However, if the initial mass of the disk is large (say, 0.1 $M_{\odot}$), then it may take quite a while before this point is reached (over 10 Myr). Again, this appears to be problematic in the light of the observations. Gorti & Hollenbach (2009) have shown that another kind of UV photons, far-UV (FUV) photons, have a deeper penetration depth in the disk and overall contain more luminosity than the EUV photons. Also these photons tend to attack the outer regions of the disk (50-100 AU or so), which is the location where the mass reservoir of the disk is. This new photoevaporation process is therefore able to remove much more mass per second from the disk than EUV photoevaporation can. The viscous evolution in fact helps this process. As the disk wants to spread, it conveys material outward toward this 50-100 AU evaporation radius, where this material will be removed, leaving room for more material to be conveyed which also gets removed. This means that the combination of outward drift of the gas and the photoevaporation by FUV photons is very effective in destroying the disk (Gorti, Dullemond & Hollenbach in press). It is, however, very hard to model this process correctly. In the Gorti *et al.* model the gas temperature is calculated in great detail, but the evaporation process that follows is computed using simple analytic formulae. It is clear that the next step is to include the detailed heating/cooling physics into a 2-D axisymmetric hydrodynamics model where the 2-D dynamics of the disk is treated self-consistently, and thus the complex 2-D motion of the gas can be properly followed from the surface of the disk toward the sonic surface. Once such models are made with reasonably reliable detail, the results may be used to predict disk life times, which can be compared to observed life times and which are important ingredients in models of planet formation.

The models of photoevaporation of disks may also provide clues to the mysterious large inner holes observed in a subset of protoplanetary disks. Lately these objects are called "transition disks", as this is inspired by the idea that these disks form the phase between classical T Tauri stars and debris disks. Some authors also call them "cold disks", since the lack of hot dust in these inner holes renders the remaining emission "cold" (long wavelengths). The inference of the existence of these holes is usually done from the lack of cold dust in the spectral energy distribution (SED), but there are some examples where spatially resolved observations seem to confirm this (e.g. TW Hydra, Hughes *et al.* 2007; Ratzka *et al.* 2007). There is a subclass of these disks which have indeed such large inner holes, but still a minor amount of hot dust very close to the star. In these objects there is a clear lack of dust at temperatures between 100 K and 1000 K or thereabout, but there is hot dust (at about 1000-1500 K, i.e. residing around 0.1 AU around a T Tauri star) and there is cold dust (at temperatures below about 100 K, i.e. beyond about 20 AU or so). Also here these holes are mainly inferred from the SED, but also here some image evidence has been obtained (Brown *et al.* 2009).

The inner holes created by EUV photons might be an explanation for *some* of these sources. But Gorti *et al.* (in press) have shown that also FUV photoevaporation can drill a hole in the disk, and at much earlier times. So it is unclear which of the two mechanisms of photoevaporation would then be responsible for the observed transition disks.

References

Alexander, R. D., Clarke, C. C., & Pringle, J. E. 2006, *MNRAS* 369, 229

Brown J. M., Blake G. A., Qi, C., Dullemond, C. P., Wilner D. J., & Williams J. P. 2009, *ApJ* 704, 496

Carpenter, J. M., Wolf, S., Schreyer, K., Launhardt, R., & Henning, T. 2005, *AJ* 129, 1049

Chiang, E. I. & Goldreich, P. 1997, *ApJ* 490, 368 *ApJ* 490, 368

Clarke, C. J., Gendrin, A., & Sotomayor, M. 2001, *MNRAS* 328, 485

Davis, S. S. 2005, *ApJ* 620, 994

Gorti, U. & Hollenbach, D. 2009, *ApJ* 690, 1539

Haisch, K. E., Lada, E. A., & Lada, C. J. 2001, *ApJ* 553:L153.

Hollenbach, D., Johnstone, D., Lizano, S., & Shu, F. 1994, *ApJ* 428, 654

Hughes, A. M., Wilner, D. J., Calvet, N., D'Alessio, P., Claussen, M. J., & Hogerheijde, M. R. 2007, *ApJ* 664, 536.

Lynden-Bell, D. & Pringle, J. E. 1974, *MNRAS* 168, 603

Ratzka, T., Leinert, C., Henning, T., Bouwman, J., Dullemond, C. P., & Jaffe, W. 2007, *A&A* 471, 173

Highlights of Astronomy, Volume 15
XXVIIth IAU General Assembly, August 2009
Ian F. Corbett, ed.

© International Astronomical Union 2010
doi:10.1017/S1743921310011270

Proto-planetary disks with CARMA: sub-arsecond observations at millimeter wavelengths

Isella Andrea[1], John M. Carpenter[1], Laura Perez[1] and Anneila I. Sargent[1]

[1]Department of Astronomy, California Institute of Technology, MC 249-17, Pasadena, CA 91125. email:isella@astro.caltech.edu

Using the Combined Array for Research in Millimeter-wave Astronomy (CARMA) we observed several proto-planetary disks in the dust continuum emission at 1.3 and 2.8 mm (Isella *et al.* 2009a, 2009b). The observations have angular resolution between 0.15 and 0.7 arcsecond, corresponding to spatial scales spanning from about the orbit of Saturn up to about the orbital radius of Pluto. The observed disks are characterized by a variety of radial profiles for the dust density. We observe inner disk clearing as well as smooth density profiles, suggesting that disks may form, or evolve, in different ways. Despite that, we find that the characteristic disk radius is correlated with the stellar age increasing from 20 AU to 100 AU over about 5 Myr. Interpreting our results in terms of the temporal evolution of a viscous α-disk, we estimate that (i) at the beginning of the disk evolution about 60% of the circumstellar material was located inside radii of 25-40 AU, (ii) that disks formed with masses from 0.05 to 0.4 solar masses and (iii) that the viscous timescale at the disk initial radius is about 0.1-0.3 Myr. Viscous disk models tightly link the surface density $\Sigma(R)$ with the radial profile of the disk viscosity $\nu(R) \propto R^\gamma$. We find values of γ ranging from -0.8 to 0.8, suggesting that the viscosity dependence on the orbital radius can be very different in the observed disks. We demonstrate that the similarity solution for the surface density for $\gamma < 0$ can explain the properties of some "transitional" disks without requiring discontinuities in the disk surface density. In the case of LkCa 15, a smooth distribution of material from few stellar radii to about 240 AU can produce both the observed SED and the spatially resolved continuum emission at millimeter wavelengths. For two sources, RY Tau and DG Tau, we observed the dust emission with a resolution as high as 0.15 arcsecond, which corresponds to a spatial scale of 20 AU at the distance of the two stars. The achieved angular resolution is a factor 2 higher than any existing observation of circumstellar disks at the same wavelengths and enable us to investigate the disk structure with unprecedent details. In particular, we present a first attempt to derive the radial profile of the slope of the dust opacity β . We find mean values of β of 0.5 and 0.7 for DG Tau and RY Tau respectively and we exclude that β may vary by more than ± 0.4 between 20-70 AU. This implies that the circumstellar dust has a maximum grain size between 10 μm and few centimeters.

References

Isella, A., Carpenter, J. M., & Sargent, A. I. 2009a, *ApJ* 701, 260

Isella, A., Carpenter, J. M., & Sargent, A. I. 2009b, in preparation.

Highlights of Astronomy, Volume 15
XXVIIth IAU General Assembly, August 2009
Ian F. Corbett, ed.

© International Astronomical Union 2010
doi:10.1017/S1743921310011282

Gas Evolution in the Planet-Forming Region of Disks

Ilaria Pascucci[1,2]

[1] Space Telescope Science Institute
3700 San Martin Drive, Baltimore, MD 21218, USA
email: pascucci@stsci.edu

[2] Dept. of Physics & Astronomy, Johns Hopkins University
3400 N. Charles Street, Baltimore, MD 21218, USA
email: pascucci@pha.jhu.edu

Abstract. The timescale over which gas-rich disks disperse profoundly affects not only the formation of giant planets but also the habitability of terrestrial planets. In this contributed talk we presented new atomic and molecular diagnostics that can be used to trace the dispersal of gas at disk radii where planets form. We also showed the first observational evidence for photoevaporation driven by the central star and discussed the efficiency of this disk dispersal mechanism.

Keywords. accretion disks, line: identification, line: profiles, techniques: spectroscopic, planetary systems: formation, planetary systems: protoplanetary disks

1. New Gas Line Diagnostics and Disk Photoevaporation

In spite of its relevance to planet formation, little is known about the gas content of protoplanetary disks at the location where terrestrial and giant planets form. The Spitzer Space Telescope is providing the first view: several atomic and molecular lines, likely probing the planet-forming region, have been detected for the first time toward young sun-like stars. In this contribution, we briefly summarize how newly discovered mid-infrared gas lines can shed light on planet formation and disk dispersal mechanisms:

- [Ne II] at 12.81μm. This line was first discovered in 2007 in Spitzer spectra of several protoplanetary disks (Pascucci *et al.* 2007, Lahuis *et al.* 2007). Models of disks irradiated by stellar X-rays and/or EUV photons (Glassgold *et al.* 2007, Gorti & Hollenbach 2009) suggest that [Ne II] emission arises from the hot ($\sim$5,000 K) disk atmosphere out to about $\sim$15 AU. Follow-up ground-based high-resolution observations are on the way and confirm the disk origin in sources with no strong jets/outflows (Najita *et al.* 2009, Pascucci *et al.* 2009). In a few cases, the line profiles and peak emissions are consistent with disk gas being photoevaporated by the central star (Pascucci & Sterzik 2009). These results are in agreement with theoretical expectations that EUV-driven photoevaporation occurs at later times in the disk evolution and the mass loss rates are only modest.

- Water and OH transitions. Spitzer spectra of protoplanetary disks around young accreting stars show a forest of water and OH lines (Carr & Najita 2008, Salyk *et al.* 2008). Disks that have already evacuated the inner regions of small dust grains (also called transition disks) lack strong molecular lines in their infrared spectra (Najita *et al.* in prep.). This could result from low excitation due to low stellar accretion rates or physical depletion of gas in the inner disks, possibly due to an already formed giant planet

- Transitions from organic molecules. Simple organic molecules like HCN and C_2H_2 have been now detected toward many protoplanetary disks (Pascucci *et al.* 2009). The

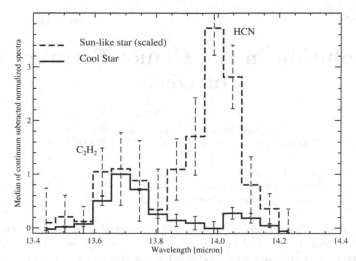

Figure 1. Median of continuum-subtracted and normalized spectra for the T Tauri star (Sun-like star, dashed line) and the very low-mass stars/brown dwarf (Cool star, solid line) samples presenting C_2H_2 and/or HCN emission bands. If brown dwarfs had the same flux ratio of HCN vs. C_2H_2 as the T Tauri stars do, HCN emission would have been easily detected toward them.

detection rate statistics and the line flux ratios of these two molecules show a striking difference between the sample of disks around T Tauri stars and very low-mass stars/brown dwarfs (see Fig. 1), demonstrating a significant underabundance of HCN relative to C_2H_2 in the disk surface of brown dwarfs. These results show how different stellar radiation fields can impact the synthesis of organic molecules and the following chemical disk network.

These recent discoveries with the Spitzer Space Telescope as well as the upcoming Herschel Space Observatory promise a bright future for gas disk studies. Using these different gas line diagnostics it will be possible to provide a definitive statement on the evolution of the gas at disk radii where planet formation takes place, understand whether gas and dust evolve in tandem, and explore whether the early accretion of water on terrestrial planets is a viable scenario.

References

Carr, J. & Najita, J. 2008, *Science* 319, 1504
Glassgold, A., Najita, J., & Igea, J. 2007, *ApJ* 656, 515
Gorti, U. & Hollenbach, D. 2009, *ApJ*, 690, 1539
Lahuis, F. *et al.* 2007, *ApJ* 665, 492
Najita, J. *et al.* 2009, *ApJ* 697, 957
Pascucci, I. *et al.* 2007, *ApJ* 663, 383
Pascucci, I. *et al.* 2009, *ApJ* 696, 143
Pascucci, I. & Sterzik, M. 2009, *ApJ* 702, 724
Salyk, C. *et al.* 2008, *ApJ* (Letters) 676, L49

Highlights of Astronomy, Volume 15
XXVIIth IAU General Assembly, August 2009
Ian F. Corbett, ed.

© International Astronomical Union 2010
doi:10.1017/S1743921310011294

Resolving structure in the HD100546 disk - signatures of planet building?

S. T. Maddison[1], C. M. Wright[2], D. Lommen[3], M. Burton[4], D. Wilner[5] and T. Bourke[5]

[1] Centre for Astrophysics & Supercomputing, Swinburne University, Australia,
email: smaddison@swin.edu.au

[2] UNSW@ADFA, Canberra, Australia
[3] Leiden Observatory, The Netherlands
[4] Department of Physics, UNSW, Australia
[5] Harvard-Smithsonian Centre for Astrophysics, USA

Several lines of evidence suggest that planet formation may be well underway within the circumstellar disk of the enigmatic Herbig Be star HD100546, including a cleared inner cavity, spiral structure, and similar dust mineralogy as seen in our own solar system. To learn more about the processes occurring in this disk we have conducted a multi-frequency observing program with the Australia Telescope Compact Array (ATCA).

We find the millimetre slope of the spectral energy distribution (SED) is $\alpha \sim 2.3$ (using fluxes integrated over the entire extent of the emission), suggesting a dust opacity index $\beta \sim 0.3$. From spatially resolved and temporally stable emission over $\gtrsim 3$ years, we find that the flux at 3, 7 and 16 mm is dominated by thermal emission from dust grains up to several tens of centimetre in size. At all millimetre wavelengths the peak emission is centered $\lesssim 0.5''$ west of the optical stellar position. There is however structure in the emission, especially at 3 mm where we detect a discrete clump about an arcsecond to the south-west coincident with features seen in HST ACS images, and a second clump about 2.5 arcsec to the north. Subarcsecond-resolution 3 mm images show a deficit of emission at the centre of the disk, in agreement with the inner cavity seen in infrared observations. We determine a dust disk FWHM of ~50–100 AU and mass of ~ 10 $M_{\rm Earth}$.

The longer wavelength 3.5 and 6.2 cm emission is relatively stable on the time scale of months and years, and we find the centimetre SED is consistent with emission arising from free-free processes. The 3.5 cm emission is elongated orthogonal to the millimetre disk emission, suggestive of a wind, with the data indicating a wind mass loss rate of $\sim 10^{-8}$ $M_\odot$/yr.

We also present $HCO^+(1{\rightarrow}0)$ line data which demonstrates the presence of dense molecular gas in the disk, supported by our ^{12}CO (4→3) and (7→6) data from NANTEN2. The line profile is double-peaked, with component velocities at 3.5 and 7.0 km/s, in agreement with APEX $^{12}CO(3{\rightarrow}2)$ data. Each component is coincident with the position of HD100546, but with a slight spatial offset approximately along the disk major axis. If interpreted as Keplerian rotation, the radius of the molecular gas emission is ~350 AU, with the SE side approaching and the NW side receding from Earth.

Our new results show that the pebble-sized grains in the disk of HD100546 are amongst the largest yet observed, and combined with the cleared inner dust cavity might suggest that planet formation is indeed well underway, or that grain sizes are so large that they are becoming invisible at millimetre wavelength. The detection of a large molecular gas disk and a wind, along with ultraviolet accretion signatures, would suggest that the system is still quite immature. Perhaps the gas is needed to allow grains to grow to such large sizes in the inner disk.

Highlights of Astronomy, Volume 15
XXVIIth IAU General Assembly, August 2009
Ian F. Corbett, ed.

High-energy radiation and particles in the environments of young stellar objects

Manuel Güdel

Institute of Astronomy, ETH Zurich, 8093 Zurich, Switzerland
email: guedel@astro.phys.ethz.ch

Abstract. High-energy radiation and particles profoundly affect circumstellar disk gas and solids. We discuss stellar high-energy sources and summarize their effects on circumstellar disks.

Keywords. Young stellar objects, X-rays, high-energy particles, ionization

1. Circumstellar disks subject to high-energy radiation and particles

Stellar X-rays may heat circumstellar disks to > 1000 K at AU distances, as suggested by, e.g. H_2 $2.12\mu m$ (Bary *et al.* 2003) and CO fundamental+overtone emission (Najita *et al.* 2003). X-rays may also ionize disks more efficiently than cosmic rays at similar distances, thus driving accretion through the magnetorotational instability (Glassgold *et al.* 1997). They may further dominate disk photoevaporation in the 10–40 AU range, resulting in mass loss rates of order 10^{-9} $M_\odot$ yr^{-1} (Ercolano *et al.* 2009).

What high-energy photons do reach the disk surface at all? Average neutral gas column densities around classical T Tauri stars (CTTS, $N_H \approx 10^{21} - 10^{22}$ cm^{-2}, Güdel *et al.* 2007a) typically exceed those of weak-lined T Tauri stars, suggesting excess gas located relatively close to the star. This gas easily absorbs EUV radiation, as do disk winds (Hollenbach & Gorti 2009), questioning the role of EUV radiation in CTTS disks. Absorption may be even more severe in some strong accretors where accretion streams seem to absorb all X-rays below $\approx 1 - 2$ keV (Güdel *et al.* 2007b).

Increasing X-ray absorption with increasing disk inclination provides, nevertheless, direct evidence for disk ionization (Kastner *et al.* 2005). Further support comes from fluorescent 6.4 keV Kα emission from "cold" iron at the disk surface during strong flares (Imanishi *et al.* 2001). Fluorescence may reach extremely high levels in protostars even outside any obvious flaring (Skinner *et al.* 2007).

The [Ne II] $12.81\mu m$ line, frequently detected in *Spitzer* spectra (Pascucci *et al.* 2007), may be another diagnostic for X-ray/EUV disk irradiation (Glassgold *et al.* 2007). A statistically significant correlation with the stellar X-ray luminosity is, however, dominated by scatter (Figure 1a); obviously, a number of further parameters (e.g., disk properties) are relevant. For example, the presence of jets leads to a large increase in [Ne II] luminosity (Fig. 1b) as also shown in spatially resolved observations (van Boekel *et al.* 2009).

Glauser *et al.* (2009) found dust crystallinity to anti-correlate with the central star's X-ray luminosity. Although X-rays carry insufficient momentum to induce lattice displacements, energetic (several keV) particles in the stellar wind are held responsible for amorphizing circumstellar dust.

2. What high-energy sources?

Apart from "traditional" magnetic coronae, additional high-energy sources have been identified. While accretion "suppresses" coronal X-rays by a factor of ≈ 2 (Telleschi *et al.*

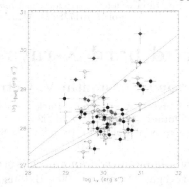 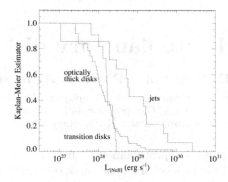

Figure 1. *Left (a):* [Ne II] luminosity vs. X-ray luminosity for CTTS; jet sources are shown by diamonds, disks without jets by circles (filled: detections; open: upper limits). *Right (b):* Kaplan-Meier estimator for the [Ne II] luminosity of CTTS disks without jets, CTTS jet sources, and transition disks. From Güdel *et al.* (2009).

2007), it adds a "soft excess" at cool (coronal) temperatures (Güdel & Telleschi 2007), perhaps due to an interaction between accretion streams and the corona. High densities inferred from X-ray line triplets (Kastner *et al.* 2002) have successfully been modeled in terms of shocks at the footpoints of accretion streams (Günther *et al.* 2007). CTTS jets also emit X-rays close to the star (Güdel *et al.* 2007b). Their "lamp-post" arrangement may provide ideal illumination of disk surfaces, avoiding absorption by accretion streams.

Large flares are of interest as they produce hard X-rays and may eject more energetic particles. Disks may thus be ionized more efficiently to deeper levels (Ilgner & Nelson 2006). However, as the energetics of X-ray emission appear to be dominated by the large population of *small* flares (Audard *et al.* 2000), hard X-rays should be continuously present, adding as yet unrecognized ionization power to the circumstellar environment.

References

Audard, M., Güdel, M., Drake, J. J., & Kashyap, V. L. 2000, *ApJ* 541, 396
Bary, J. S., Weintraub, D. A., & Kastner, J. H. 2003, *ApJ* 586, 1136
Ercolano, B., Clarke, C. J., & Drake, J. J. 2009, *ApJ* 699, 1639
Glassgold, A. E., Najita, J. R., & Igea, J. 1997, *ApJ* 480, 344
Glassgold, A. E., Najita, J. R., & Igea, J. 2007, *ApJ* 656, 515
Glauser, A. *et al.* 2009, *A&A*, in press
Güdel, M. & Telleschi, A. 2007, *A&A* 474, L25
Güdel, M. *et al.* 2007a, *A&A* 468, 353
Güdel, M., *et al.* 2007b, *A&A* 468, 515
Güdel *et al.* 2009, *A&A*, submitted
Günther, H. M., Schmitt, J. H. M. M., Robrade, J., & Liefke, C. 2007, A&A, 466, 1111
Hollenbach, D. & Gorti, U. 2009, *ApJ*, 703, 1203
Ilgner, M. & Nelson, R. P. 2006, *A&A* 455, 731
Imanishi, K., Koyama, K., & Tsuboi, Y. 2001, *ApJ* 557, 747
Kastner, J. H. *et al.* 2002, *ApJ* 567, 434
Kastner, J. H. *et al.* 2005, *ApJS* 160, 511
Najita, J., Carr, J. S., & Mathieu, R. D. 2003, *ApJ*, 589, 931
Pascucci, I. *et al.* 2007, *ApJ*, 663, 383
Skinner, S. L., Simmons, A. E., Audard, M., & Güdel, M. 2007, *ApJ* 658, 1144
Telleschi, A., Güdel, M., Briggs, K. R., Audard, M., & Palla, F. 2007, *A&A* 468, 425
van Boekel, R., Güdel, M., Henning, Th., Lahuis, F., & Pantin, E. 2009, *A&A*, 497, 137

Highlights of Astronomy, Volume 15
XXVIIth IAU General Assembly, August 2009
Ian F. Corbett, ed.
© International Astronomical Union 2010
doi:10.1017/S1743921310011312

Protoplanetary disks and hard X-rays

Eric D. Feigelson[1], Philip J. Armitage[2,3] and Konstantin V. Getman[1]

[1] Dept. of Astronomy and Astrophysics, Penn State University, University Park PA 16802
[2] JILA, Campus Box 440, University of Colorado, Boulder CO 80309
[3] Dept. of Astrophysical and Planetary Sciences, University of Colorado, Boulder CO 80309

The physics of protoplanetary disks and the early stages of planet formation is strongly affected by the level of ionization of the largely-neutral gas (Armitage 2009; Balbus 2009). Where the ionization fraction is above some limit around $\sim 10^{-12}$, the magnetorotational instability (MRI) will ensue and the gas will become turbulent. The presence or absence of disk turbulence at various locations and times has profound implications for viscosity, accretion, dust settling, protoplanet migration and other physical processes. The dominant source of ionization is very likely X-rays from the host star (Glassgold *et al.* 2000). X-ray emission is elevated in all pre-main sequence stars primarily due to the magnetic reconnection flares similar to, but much more powerful and frequent than, flares on the surface of the contemporary Sun (Feigelson *et al.* 2007).

Most theoretical studies of the ionization of X-rays into the circumstellar disk assume simplistic models for the X-ray spectrum with relatively soft spectra. But hard X-rays penetrate much deeper into the disk and will be critical for determining the extent of the Dead Zone where turbulence is absent. Ilgner & Nelson (2006) show that flares with plasma temperatures $kT = 7$ keV produce ~ 1000 times higher ionization at the disk midplane than flares with $kT = 3$ keV. Getman *et al.* (2008) study hundreds of strong X-ray flares in Orion Nebula stars and find that peak temperatures range from 1 to > 20 keV; the latter are nicknamed 'superhot' flares. Recalling that a significant flux of photons will be present at even higher energies, we are motivated to investigate the effect of hard X-ray irradiation on protoplanetary disks.

We have constructed a time-dependent protoplanetary disk model subject to X-ray ionization with layered Active and Dead Zones. X-ray spectrum, disk viscosity, recombination rate, and the critical Reynolds number for onset of the MRI are model parameters. For all X-ray inputs, the model reproduces observed accretion rates in pre-main sequence stars, declining from 10^{-7} $M_\odot$/yr at ages ~ 0.1 Myr to $\sim 10^{-10}$ at ~ 3 Myr. However, the presence of hard X-ray irradiation considerably changes disk structure, increasing surface density in the Dead Zone. This may promote the formation of planets in the Dead Zone.

References

Armitage, P. 2009, *Astrophysics of Planet Formation* (Cambridge)
Balbus, S. A. 2009 in: P. Garcia, (ed.), *Physical Processes in Circumstellar Disks Around Young Stars* (Chicago)
Feigelson, E., Townsley, L., Güdel, M., & Stassun, K. 2007 in: B. Reipurth *et al.* (eds.), *Protostars and Planets V* (Arizona), p. 313
Getman, K. V., Feigelson, E. D., Broos, P. S., Micela, G., & Garmire, G. P. 2008, *Astrophys. J.* 688, 418
Glassgold, A. E., Feigelson, E. D., & Montmerle, T. 2000 in: V. Mannings, *et al.* (eds.), *Protostars and Planets IV* (Arizona), p. 429
Ilgner, M. & Nelson, R. P. 2006, *Astron. Astrophys.* 455, 731

Highlights of Astronomy, Volume 15
XXVIIth IAU General Assembly, August 2009
Ian F. Corbett, ed.

© International Astronomical Union 2010
doi:10.1017/S1743921310011324

[Ne II] and X-ray emission from ρ Ophiuchi young stellar objects

E. Flaccomio, B. Stelzer, S. Sciortino, I. Pillitteri and G. Micela

INAF - Osservatorio Astronomico di Palermo Piazza del Parlamento, 1, 90134, Italy

Circumstellar disks are mostly made of gas. Constraining the spatial and thermal structure of the gas, and its time evolution, is crucial to understand the star- and planet-formation processes. Models predict that the gas is affected by UV and X-ray radiation from the central young stellar object (YSO), but many uncertainties remain, e.g. whether the EUV emission actually reaches the disk or is absorbed by disk winds. The infrared [Ne II] and [Ne III] fine structure lines at $12.81\mu m$ and $15.55\mu m$ have been theoretically predicted to trace the circumstellar disk gas subject to X-ray heating and ionization.

Flaccomio *et al.* (2009, A&A, arXiv:0906.4700) present observational results for a sample of 28 YSOs in the ρ Ophiuchi star formation region for which good quality infrared spectra and X-ray data have been obtained, the former with the Spitzer IRS and the latter with the Deep Rho Ophiuchi XMM-Newton Observation (DROXO). The [Ne II] and the [Ne III] lines are detected in 10 and 1 cases, respectively. In Figure 1 we plot the [Ne II] line luminosities vs. X-ray luminosity and accretion rate. No correlation with X-ray emission is observed. The luminosity of the [Ne II] line for one star, and that of both the [Ne II] and [Ne III] lines for another one, match the predictions of published models of X-ray irradiated disks; for the remaining 8 objects they are 1-2 dex higher than predicted on the basis of their L_X. Class I YSOs have significantly higher [Ne II] luminosities with respect to Class II objects. The [Ne II] line correlates with mass accretion rate. These results might point toward a role of accretion-generated UV emission in exciting the disk gas, or to [Ne II] emission in shocks within accretion-powered winds and jets.

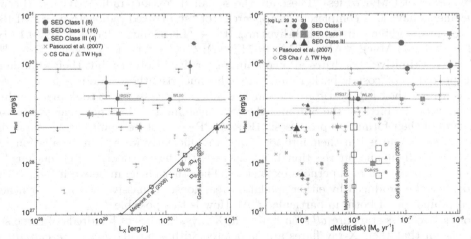

Figure 1. [Ne II] luminosity vs. L_X and $\dot{M}_{disk}$. Class I, II, and III objects are indicated with different symbols. In the right-hand panel symbol sizes are proportional to $\log L_X$. Also plotted here are 8 T-Tauri stars from Pascucci *et al.* (2007, ApJ, 663, 383) and Espaillat *et al.* (2007, ApJ, 664, L111) (CS Cha and TW Hya) and the model predictions of Meijerink *et al.* (2008, ApJ, 676, 518) and Gorti & Hollenbach (2008, ApJ, 683, 287, models 'A', 'B', and 'D').

Highlights of Astronomy, Volume 15
XXVIIth IAU General Assembly, August 2009
Ian F. Corbett, ed.
© International Astronomical Union 2010
doi:10.1017/S1743921310011336

Circumstellar disks in high-mass star environments: the early solar system

Thierry Montmerle[1], Matthieu Gounelle[2] and Georges Meynet[3]

[1]Laboratoire d'Astrophysique de Grenoble, France, [2]Muséum National d'Histoire Naturelle, Paris, France, [3]Observatoire de Genève, Switzerland
e-mail: montmerle@obs.ujf-grenoble.fr

Abstract. The early solar system represents the only case we have of a circumstellar disk that can be investigated "in situ" -albeit 4.6 Gyr after its formation. Meteorites studies give mounting evidence for an intense irradiation phase of the young circumsolar disk by energetic particles, and also for contamination by products of high-mass stellar and/or explosive nucleosynthesis. We thus discuss the conditions of the birth of the solar system in a high-mass star environment.

Keywords. early solar system, meteorites, nucleosynthesis, massive stars

---

The very early stages of the formation of the solar system took place while the young Sun was still surrounded by a circumstellar disk, some 4.6 Gyr ago. Although the growth from micron-sized particles (dust grains) to km-sized bodies ("planetesimals") remains a major unsolved problem in planet formation theory, collisions between small, undiffer-entiated bodies are still ongoing, and generate debris in the form of meteorites known today as "carbonaceous chondrites", that contain the earliest solids of the solar system.

The composition of these primitive meteorites is generally similar to the so-called "cosmic adundances", with a few important exceptions (see Lauretta & McSween 2006). They contain "calcium-aluminium-rich inclusions" ("CAIs"), which show an excess of elements that are the daughter products of the decay of radioactive isotopes with periods on the order of 1 Myr or less. These are the so-called "extinct radioactivities" (because the elements have long completely disappeared), or "short-lived radioactivities ("SLRs"). The parent-daughter pairs and half-lives are: ^{7}Be → ^{7}Li (52 days); ^{41}Ca → ^{41}K (0.1 Myr); ^{36}Cl → ^{36}S (0.3 Myr); ^{26}Al → ^{26}Mg (0.74 Myr); ^{10}B → ^{10}Be (1.5 Myr); ^{60}Fe → ^{60}Ni (2.7 Myr). Primitive chondrites also contain longer-lived isotopes, but their abundances are expected as a "cosmic background" of stellar nucleosynthesis in our galaxy.

In contrast, the SLRs testify to the nucleosynhetic processes that took place within a Myr or so of the birth of the Sun and the solar system. Two isotopes bear unequivocal signature of their formation processes: ^{7}Be and ^{10}Be cannot be coming from stars (since they are easily burned in their interiors) but are efficiently formed in "spallation reac-tions" (= in-flight nuclear reactions); in contrast, the heavy, neutron-rich nucleus ^{60}Fe can only be produced in supernova explosions. The other nuclei in between can, to vary-ing degrees, be produced by either spallation reactions, or massive star nucleosynthesis, or a combination of both, in particular ^{26}Al. How is this possible ?

Young low-mass stars are *all* seen flaring in X-rays. Studies of hundreds of these stars have shown that their X-ray flares are associated with solar-like *magnetic activity*, en-hanced by 3-4 orders of magnitude compared to the Sun ($L_X/L_{bol} \sim 10^{-4} - 10^{-3}$; e.g., Güdel 2004). Therefore, it is very reasonable to assume that, as is the case for solar flares, all young, low-mass stars accelerate particles in flares (notably protons, ^{3}He and α-particles), and are thus able to generate spallation reactions with the surrounding material -here dust grains (containing Si, C, O, Mg, etc.) in circumstellar disks. Along those lines, it has been shown that not only ^{7}Be and ^{10}Be, but also all the other SLRs (excluding ^{60}Fe) could be produced, within factors 2-3, by this single *generic* mechanism (e.g. Gounelle *et al.* 2006) -the young solar system thus being no exception.

However, low-mass stars can be formed either in loose clusters like Taurus-Auriga, or in "OB associations" of up to several thousand stars. If they are massive enough, the brightest stars turn supernova (hence contaminate their surroundings with the ^{60}Fe synthesized during the explosion) in only a few million years, i.e., while the associated low-mass stars may still possess dense circumstellar disks. Taking a disk e-folding lifetime of $\sim 1 - 2$ Myr (Cieza *et al.* 2007) puts a lower-limit of over $\sim 100 M_\odot$ to the mass of the exploding star. Therefore, it is logical to infer that, because ^{60}Fe was trapped in carbonaceous chondrites within ~ 1 Myr of the formation of the solar system, (i) a supernova from a very massive star exploded in its vicinity, (ii) the Sun was born in a rich OB association, such as the present-day Orion Nebula Cluster. However, one finds that such a "cosmic coincidence" is extremely unlikely (probability of a few percent at most, more likely $< 0.1\%$) (Williams & Gaidos 2007, Gounelle & Meibom 2008). The implication here would be that the Sun must have been born in a fairly exceptional environment. Note that the strong winds characteristic of massive stars before they explode can spread freshly synthesized ^{26}Al, after a time delay of $\gtrsim 1 - 2$ Myr to lift it from the stellar interior into the outer wind radiative acceleration zone (see Voss *et al.* 2009).

The above conclusion, however, rests on the assumption that the exploding massive star and the surrounding low-mass stars are more or less *coeval*. But it has long been known that star formation in OB associations takes place in successive bursts triggered by SN explosions. As a consequence, a given low-mass star may be born as a *second generation* and be exposed to the explosion of a first-generation massive star. A modern version of the "SN-triggered star-formation scenario" has been subject to extensive numerical simulations. In the so-called "turbulent convergent flow" model (see Hennebelle *et al.* 2007), molecular clouds form because they are compressed by stochastic SN explosions, hence can be contaminated by nucleosynthetic products like ^{26}Al and ^{60}Fe, and give birth to new stars. Little ^{26}Al will be left because of its short 0.73 Myr half-life, but ^{60}Fe, with its 2.7 Myr half-life, will not have decayed as significantly (Gounelle *et al.* 2009).

We conclude that our own circumstellar disk, that preceded the formation of the solar system, was subject to a series of events that could be of general relevance: (i) like all low-mass young stars, the flaring young Sun irradiated its inner circumsolar disk, and left traces of this irradiation in the form of radioactive Be isotopes and possibly heavier SLRs; (ii) the Sun was born in a massive-star environment: ^{60}Fe is likely due to explosions of massive stars of previous generation(s); ^{26}Al may be a combination of flare-induced spallation reactions and contamination by winds of massive stars of the same association, before these coeval stars explode as SN, too late to contaminate its disk.

References

Cieza, L. *et al.* 2007, ApJ, 667, 308

Gounelle, M. & Meibom, A. 2008, ApJ, 680, 781

Gounelle, M., Meibom, A., Hennebelle, P., & Inutsuka, S. 2009, ApJ, 694, L1

Gounelle, M., Shu, F., Shang, H., Glassgold, A. E., Rehm, K. E., & Lee, T. 2006, ApJ, 680, 781

Güdel, M. 2004, A&A Rev., 12, 71

Lauretta, D. L. & McSween, H. Y., Jr (Eds) 2006, *Meteorites and Early Solar System 2*, Arizona University Press, Tucson, 943 p.

Hennebelle, P., Mac Low, M. M., & Vazquez-Semadeni, E. 2007, arXiv:0711.2417v2

Voss, R. *et al.* 2009, A&A, 504, 531

Williams, J. & Gaidos, E. 2007, ApJ, 663, L33

Highlights of Astronomy, Volume 15
XXVIIth IAU General Assembly, August 2009
Ian F. Corbett, ed.

© International Astronomical Union 2010
doi:10.1017/S1743921310011348

Observational and numerical tests of jet models in young stars

S. Cabrit[1]

[1]LERMA, Observatoire de Paris et UMR 8112 du CNRS,
61 Av. de l'Observatoire, 75014 Paris. email: sylvie.cabrit@obspm.fr

Abstract. Jets are found in a wide range of accreting young stars, from brown dwarfs to massive protostars, but their launch region(s) and their role in angular momentum extraction are still debated. Many observational constraints exist on jet properties, including jet widths, kinematics along and across the jet, possible rotation signatures, ejection/accretion ratio, depletion and molecular counterparts. This contribution compares popular models, in particular disk winds, with these constraints and with MHD numerical simulations, highlighting a few open issues.

Keywords. accretion disks, MHD, stars: pre–main-sequence, ISM: jets and outflows

1. Stellar winds

He I profiles indicate accretion-powered inner "stellar" winds in $\geqslant 60\%$ of TTS (Kwan *et al.* 2007). However, launching from the stellar surface meets three caveats in explaining observed jets: (i) energetic difficulty to eject $>1\%$ of the accretion rate (Cranmer 2009, Ferreira *et al.* 2006), (ii) lack of dust, whereas iron and calcium depletion is measured in several jets (Podio *et al.* 2006, Dionatos *et al.* 2009), (iii) insufficient collimation (Bogovalov & Tsinganos 2001, Fendt 2009, Cabrit 2007); Other ejection sites thus seem required. More details may be found in the proceedings of IAU Symp. 243 and of the Conference *Protostellar jets in Context* (eds. T. Ray & K. Tsinganos).

2. Steady MHD winds from the disk surface

This mechanism is so far the only one where the difficult mass-loading problem from the accretion disk has been rigorously solved, and validated by numerical simulations (e.g. Casse & Ferreira 2000, Zanni *et al.* 2007, Salmeron *et al.* 2007). Synthetic predictions calculated for steady solutions compare well with high resolution observations such as: apparent jet widths (Garcia *et al.* 2001, Ray *et al.* 2007, Stute *et al.* 2009), maximum poloidal speeds and drop in velocity away from the jet axis (Cabrit 2007), and atomic jet rotation signatures (Pesenti *et al.* 2004). The latter require "warm" disk winds with a moderate magnetic lever arm $\lambda \leqslant 13$, launched out to a maximum radius of 0.1–3 AU, yielding an ejection to accretion ratio $\simeq 4\%$–13% in the observed range (Ferreira *et al.* 2006). However, T Tauri jet rotation signatures are still tentative due to possible contamination by shock asymmetries (e.g. Soker 2005). Disk winds could also naturally explain the presence of dust and molecules in/around jets (Panoglou *et al.* 2009)†. A caveat noted by Shu *et al.* (2008) is that a powerful disk wind induces transonic accretion; but since it occurs only out to a few AUs (limited by X-ray ionization of the disk surface), the disk lifetime is unaffected. The surface density in the inner "jet emitting" region is

---

† Note that the small launch radii < 0.1 AU inferred in molecular class 0 jets with SMA (Lee *et al.* 2008) are very uncertain since jet widths are unresolved, and rotation signatures are strongly suppressed in this case (Pesenti *et al.* 2004). ALMA will be crucial to settle this issue.

lowered, with possible impact on planet migration and disk photoevaporation, but the SED changes only moderately for typical accretion rates (Combet & Ferreira 2008). Infrared line profiles might offer a stronger test of disk wind extent in young stars.

3. The X-wind scenario

A scenario put forward by F. Shu and collaborators assumes that excess angular momentum from infalling matter is extracted by the funnel flow and transferred across corotation to power a steady centrifugal disk wind. A nice aspect is that for a terminal jet speed of 150 km s^{-1}, the required ejection/accretion ratio is 30%, close to observed. Adequate apparent jet collimation and excitation conditions can also be achieved (Shang *et al.* 2002). However, two key elements of the scenario are still unsolved: the angular momentum transfer across the X-region locked in corotation, since there is no shear to induce transport (and high magnetization could quench MRI turbulence); and how to load such a large mass-flux. MHD numerical simulations of star-disk interaction are so far unable to reproduce the X-wind conditions. The funnel flow spins *up* the star, rather than spin it down, and the conical wind from the inner disk edge is thus too slow (typically 50 km/s) compared to observed jets (Romanova *et al.* 2009). Stellar magnetic towers or sporadic reconnexion winds from the magnetosphere appear more promising to brake down the star (Ferreira *et al.* 2000, Zanni 2009, Romanova *et al.* 2009).

References

Bogovalov, S. & Tsinganos, K. 2001, MNRAS 325, 249

Cabrit, S. 2007, in IAU Symp. vol. 243, pp. 203–214

Casse, F. & Ferreira, J. 2000, A&A 361, 1178

Combet, C. & Ferreira, J. 2008, A&A 479, 481

Cranmer, S. R. 2009, ApJ 706, 824

Dionatos, O., Nisini, B., Garcia Lopez, R. *et al.* 2009, ApJ 692, 1

Fendt, C. 2009, ApJ 692, 346

Ferreira, J. & Pelletier, G., Appl, S. 2000, MNRAS 312, 387

Ferreira, J., Dougados, C., & Cabrit, S. 2006, A&A 453, 785

Garcia, P., Cabrit, S., Ferreira, J., & Binette, L. 2001, A&A 377, 609

Kwan, J., Edwards, S., & Fischer, W. 2007, ApJ 657, 897

Lee, C.-F. *et al.* 2008, ApJ 685, 1026

Panoglou, D., Cabrit, S., Pineau des Forêts, G., *et al.* 2009, submitted

Pesenti, N. *et al.* 2004, A&A 416, L9

Podio, L. *et al.* 2006, A&A 456, 189

Ray, T. P. *et al.* 2007, In *Protostars & Planets V*, B. Reipurth, D. Jewitt, and K. Keil (eds.), (University of Arizona Press, Tucson), pp. 231–244

Salmeron, R., Königl, A., & Wardle, M. 2007, MNRAS 375, 177

Romanova, M. M., Ustyugova, G. V., Koldoba, A. V., & Lovelace, R. V. E. 2009, MNRAS 399.1802

Shang, H., Glassgold, A. E., Shu, F. H., & Lizano, S. 2002, ApJ 564, 853

Shu, F. H., Lizano, S., Galli, D., Cai, M. J., & Mohanty, S. 2008, ApJ 682, L121

Soker, N. 2005, A&A 435, 125

Stute, M. *et al.* 2009, A&A in press

Zanni, C., Ferrari, A., Rosner, R., Bodo, G., & Massaglia, S. 2007, A&A 469, 811

Zanni, C. 2009, in *Protostellar jets in context*, (Spinger-Verlag, Berlin Heidelberg), eds. T. Ray and K. Tsinganos, in press

Highlights of Astronomy, Volume 15
XXVIIth IAU General Assembly, August 2009
Ian F. Corbett, ed.

© International Astronomical Union 2010
doi:10.1017/S174392131001135X

A movie of accretion/ejection of material in a high-mass YSO in Orion BN/KL at radii comparable to the Solar System

C. Goddi[1], L. Greenhill[1], E. Humphreys[1], L. Matthews[1] and C. Chandler[2]

[1]Harvard-Smithsonian Center for Astrophysics, 60 Garden Street, Cambridge, MA 02138

[2]National Radio Astronomy Observatory, P.O. Box O, Socorro, NM 87801

Around high-mass Young Stellar Objects (YSOs), outflows are expected to be launched and collimated by accretion disks inside radii of 100 AU. Strong observational constraints on disk-mediated accretion in this context have been scarce, largely owing to difficulties in probing the circumstellar gas at scales 10-100 AU around high-mass YSOs, which are on average distant (> 1 Kpc), form in clusters, and ignite quickly whilst still enshrouded in dusty envelopes. Radio Source I in Orion BN/KL is the nearest example of a high-mass YSO, and only one of three YSOs known to power SiO masers. Using VLA and VLBA observations of different SiO maser transitions, the KaLYPSO project (http://www.cfa.harvard.edu/kalypso/) aims to overcome past observational limitations by mapping the structure, 3-D velocity field, and dynamical evolution of the circumstellar gas within 1000 AU from Source I. Based on 19 epochs of VLBA observations of $v = 1, 2$ SiO masers over $\sim$2 years, we produced a movie of bulk gas flow tracing the compact disk and the base of the protostellar wind at radii < 100 AU from Source I. In addition, we have used the VLA to map 7mm SiO $v = 0$ emission and track proper motions over 10 years. We identify a narrowly collimated outflow with a mean motion of 18 km/s at radii 100-1000 AU, along a NE-SW axis perpendicular to that of the disk traced by the $v = 1, 2$ masers. The VLBA and VLA data exclude alternate models that place outflow from Source I along a NW-SE axis. The analysis of the complete (VLBA and VLA) dataset provides the most detailed evidence to date that high-mass star formation occurs via disk-mediated accretion.

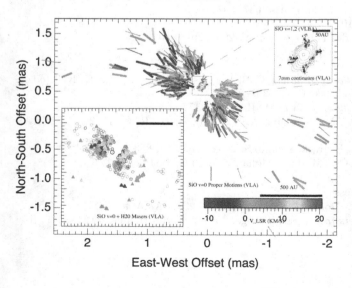

Figure 1. Disk/outflow in Orion Source I. *Main Frame)* Proper motions of SiO v=0 maser spots (*color arrows*) (4 VLA epochs over 10 years) identify a collimated, NE-SW oriented outflow (v_{mean} ~18 km/s, R ~100-1000 AU). *Top Right Inset)* VLA 7mm continuum emission (*red contours*) identifies an ionized disk around Source I (R ~50 AU). SiO v=1,2 maser emission (VLBA) (*black image*) traces a wide-angle, bipolar wind that emanates from the ionized disk. *Bottom Left Inset)* 1.3 cm water masers (VLA) overlies regions of SiO v=0 emission.

Highlights of Astronomy, Volume 15
XXVIIth IAU General Assembly, August 2009
Ian F. Corbett, ed.

© International Astronomical Union 2010
doi:10.1017/S1743921310011361

Warm and hot circumstellar gas in V1647 Ori during the 2008-2009 outburst

A. Carmona[1,*], M. Audard[1], M. van den Ancker[2], G. van der Plas[2,3], M. Goto[4] and D. Fedele[4]

[1]ISDC & Geneva Observatory, ch. d'Ecogia 16, 1290 Versoix, Switzerland
email: andres.carmona@unige.ch

[2]European Southern Observatory, Karl Schwarzschild Str 2, D-85748 Garching, Germany
[3]University of Amsterdam, Kruislaan 403, NL-1098 SJ Amsterdam, The Netherlands
[4]Max-Planck Institute for Astronomy, Königstuhl 17, D-69117 Heidelberg, Germany
*M.A and A.C acknowledge support from a Swiss National Science Foundation grant
(PP002–110504). Based on observations at the ESO-VLT (program ID: DDT 281C-5056).

Abstract. The pre-main sequence star V1647 Ori started a new outburst in August 2008. From October 2008 to February 2009 we monitored V1647 Ori, obtaining quasi-simultaneous VLT-CRIRES near-IR spectroscopy, VLT-VISIR mid-IR spectroscopy and VLT-FORS2 optical spectroscopy. We studied the evolution of H_2 and CO emission from hot and warm gas and $H\alpha$ and forbidden line-emission during the initial outburst phase of V1647 Ori. $H\alpha$ is observed in emission displaying P-Cygni profiles with blue-shifted absorption up to -700 km/s, suggesting the presence of a high velocity wind (Fig. 1a). [OI] emission at 6300 Å is observed displaying a blue-shifted emission shoulder, indicating the presence of material moving away from the star (Fig. 1b). We detect H_2 1-0 S(1) and CO (P4 to P14 and P30-P38) ro-vibrational lines centered at the velocity of the star at all epochs (Fig. 1c & d). This strongly suggests that the H_2 and CO emission originates from a disk and not from a warm outflow. The H_2 1–0 S(0) and 2-1 S(1) ro-vibrational lines at 2.22 and 2.24 μm and the pure-rotational H_2 0–0 S(1) and 0–0 S(2) lines at 17 and 12 μm were not detected in our spectra. Changes in the $H\alpha$ and [OI] profiles and the H_2 and CO emission observed do not correlate. We modeled the H_2 and CO line profiles assuming emission from a flat disk in keplerian rotation with line intensity decreasing with radius ($I \sim I_0 (R/R_{min})^{-\alpha}$). We found that the disk of V1647 Ori is observed nearly face-on and that the line emission is produced within a fraction of an AU of the star (Fig. 1d).

Keywords. stars: pre–main-sequence, circumstellar matter, individual (V1647 Ori)

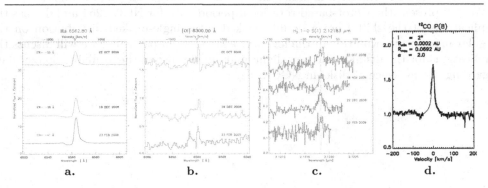

a.   b.   c.   d.

Figure 1. *Panels a, b & c:* $H\alpha$, [OI] and H_2 1-0 S(1) spectra observed. *Panel d:* Keplerian flat disk model of the CO P(8) line (Carmona *et al.* in preparation).

Highlights of Astronomy, Volume 15
XXVIIth IAU General Assembly, August 2009
Ian F. Corbett, ed.

© International Astronomical Union 2010
doi:10.1017/S1743921310011373

Stellar rotation at young ages: new results from Corot's monitoring NGC 2264

F. Favata[1], G. Micela[2], S. Alencar[3], S. Aigrain[4] and K. Zwintz[5]

[1]ESA

[2]Università di Palermo

[3]Departmento de Física, ICEx, UFMG, Belo Horizonte, MG 31270-901, Brazil

[4]University of Exeter

[5]Institute for Astronomy, University of Vienna, Austria

Stellar rotation at young ages: new results from Corot's Angular momentum is one of the driving forces in the early evolution of stars. Issues such as the coupling between the star and the accretion disk (the so-called disk regulation paradigm), are traced by the evolution of rotational momentum, but affect the star-forming process as a whole. One of the features observed in star-forming regions (e.g. ONC and NGC 2264) of age between 1 and few Myr, for masses above 0.25 solar masses, is a bimodality of the rotational period distribution, with a peak around 1 day and the other at around 4 to 7 days. This bimodality has been interpreted as the smoking gun of the disk-locking mechanism (with the fast rotators having lost their disk and the slow ones still being regulated by their disks).

We recently observed the NGC 2264 star-forming region (age 3 Myr) with the Corot space telescope, obtaining an uninterrupted 24 day long high accuracy photometric series. This allows the determination of rotational periods for $P < 12$ days with high accuracy and the removal of any biases due to the coverage imposed by ground-based observational campaigns. The Corot photometric campaign on NGC 2264 unambiguously shows that stars with $M > 0.25 M_{sun}$ have a single-peaked distribution (with a peak at 3-4 days) and no short period peak. The bimodal distribution that has been reported based on ground-based campaigns is not present in the Corot-based data, and the short-period peak in the distribution (centered around 1 day) appears to be spurious and due to aliasing effects between the actual rotational modulation and the observational window function.

We will present the Corot-based rotational period data for NGC 2264 and discuss the implications of the shift from a double-peaked to a single-peaked distribution on the angular momentum evolution in the few Myr age range. In particular we will discuss the implications for the disk-locking paradigm, making also use of the available Spitzer data regarding the presence of disks in NGC 2264.

Highlights of Astronomy, Volume 15
XXVIIth IAU General Assembly, August 2009
Ian F. Corbett, ed.

© International Astronomical Union 2010
doi:10.1017/S1743921310011385

Magnetospheric accretion & outflows in stars & brown dwarfs: theories and observational constraints

S. Mohanty[1]

[1]Imperial College, London

The manner in which young classical T Tauri stars (cTTs) and brown dwarfs accrete gas from their surrounding disks and simultaneously drive jets and outflows is central to star and planet formation and angular momentum evolution, but remains an ill-understood and hotly debated subject. One of the central concerns is the stellar field geometry: while analytic theories assume an idealized stellar dipole, T Tauri fields are observed to be complex multipolar beasts. I present an analytic generalization of the X-wind theory to include such fields. Independent of the precise field geometry, the generalized model makes a unique prediction about the relationship between various cTTs observables. I show that this prediction is supported by observations of accretion rate, hot spot size, stellar rotation and field strength from stellar to brown dwarf masses, including recent detailed spectropolarimetric measurements. I also discuss the unique insights offered by recent magnetic field measurements on accreting brown dwarfs: while they agree with the *accretion* theory above, they also pose a puzzle for *magnetic field generation* theory. Resolving this conundrum promises to illuminate our general picture of accretion and angular momentum transport in fully convective objects.

Editors Note: More details on this work have been published by Mohanty & Shu (2008).

References

Mohanty, S. & Shu, F. H. 2008, *ApJ* 687, 1323

Highlights of Astronomy, Volume 15
XXVIIth IAU General Assembly, August 2009
Ian F. Corbett, ed.
© International Astronomical Union 2010
doi:10.1017/S1743921310011397

Classical T Tauri-like Outflow Activity in the Brown Dwarf Mass Regime

E. T. Whelan[1], T. P. Ray[2], F. Bacciotti[3], L. Podio[2] and S. Randich[3]

[1]Laboratoire d'Astrophysique de l'Observatoire de Grenoble , [2]Dublin Institute for Advanced Studies, [3]INAF/Osservatorio Astrofisico di Arcetri

Since 2005 we have been analysing the spectra of brown dwarfs (BDs) using the technique of spectro-astrometry and to date have found 5 outflows driven by BDs. Our aim is to obtain an understanding of outflow activity in the BD mass regime and make a comprehensive comparison with low mass protostars, in particular the classical T Tauri stars (CTTSs). Table 1 summarises some results for the sources in our sample. Also see Whelan *et al.* (2009b) for a complete discussion and comparison with CTTSs. Some noteworthy results include the asymmetry in the ISO-217 bipolar outflow which is revealed in the relative brightness of the two lobes (red-shifted lobe is brighter) and the factor of two difference in radial velocity (the red-shifted lobe is faster). Asymmetries are common in jets from low mass protostars (0.1 Msun to 2 Msun) and the observation of a strong asymmetry at such a low mass supports the idea that BD outflow activity is scaled down from CTTSs. In addition, Whelan *et al.* (2009a) find a strong contribution to the Hα line emitted by LS-RCrA 1 and evidence of a dust hole in its disk. Using methods previously applied to CTTS Whelan *et al.* (2009b) estimate the mass outflow rate ($\dot{M}_{out}$) for LS-RCrA 1, ISO and ISO-Oph 102 $\dot{M}_{out}$ to be in the range 10^{-10} to 10^{-9} Msun yr^{-1} which is comparable to measured mass accretion rates.

Source	Mass (M_{JUP})	V_{rad} (kms^{-1})	Outflow PA (°)	Publication
ISO-217	80[1]	-20/30	202 ($\pm$) 8	Whelan *et al.* (2009b)
2MASS1207A	24[2]	-8/4		Whelan *et al.* (2007)
ISO-Oph 32	40[4]	-10-20	240 ($\pm$7)	Whelan *et al.* (2009b)
ISO-Oph 102	60[4]	-45	0	Whelan *et al.* (2005)
LS-RCrA 1	35-72[5]		15	Whelan *et al.* (2009a)

Table 1. BD candidates found to date to be driving outflows. In all cases the [OI]λ6300 line is the dominant line and V_{rad} is given here. For all sources where an outflow position angle (PA) is known (except ISO-Oph 102) this has been estimated from the spectro-astrometric analysis. 1-5 refer to the papers giving the mass estimates, 1=Muzerolle *et al.* (2005), 2=Mohanty *et al.* (2007), 3=Mohanty *et al.* (2004), 4=Natta *et al.* (2002) and 5=Barrado y Navascués *et al.* (2004).

References

Barrado y Navascués, D., Mohanty, S., & Jayawardhana, R. 2004, *ApJ*, 604, 284

Muzerolle, J. *et al.* 2005, *ApJ*, 625, 906

Mohanty, S., Jayawardhana, R., Huélamo, N., & Mamajek, E. 2007, *ApJ*, 657, 1064

Mohanty, S., Jayawardhana, R., & Basri, G. 2004, *ApJ*, 609, 885

Natta, A. *et al.* 2002, *A&A*, 393, 597

Whelan, E. T. *et al.* 2005, *NATURE*, 435, 652

Whelan, E. T. *et al.* 2007, *ApJL*, 659, L45

Whelan, E. T., Ray, T. P., & Bacciotti, F. 2009(a), *ApJL*, 691, L106

Whelan, E. T., Ray, T. P., Podio, L, Bacciotti, F., & Randich, S 2009(b), *ApJ*, submitted

Highlights of Astronomy, Volume 15
XXVIIth IAU General Assembly, August 2009
Ian F. Corbett, ed.

The early evolution of low mass stars and brown dwarfs

Isabelle Baraffe[1]

[1] École Normale Superiéure de Lyon

My talk will focus on the early evolution of low mass objects. I will discuss the main uncertainties on current evolutionary models and the effects of rotation, magnetic field and early accretion history on young object's structure. I will also present possible solutions to the well known spread in HRD observed in star formation regions for objects of a few Myr old.

Editors Note: More details on this work have been published in Baraffe, I., Chabrier, G., Gallardo, J., 2009, ApJ (Letters) 702, L27.

Highlights of Astronomy, Volume 15
XXVIIth IAU General Assembly, August 2009
Ian F. Corbett, ed.
© International Astronomical Union 2010
doi:10.1017/S1743921310011415

Brown Dwarf Model Atmospheres Based on Multi-Dimensional Radiation Hydrodynamics

France Allard and Bernd Freytag

Centre de Recherche Astrophysique de Lyon,

UMR 5574: CNRS, Université de Lyon, École Normale Supérieure de Lyon,
46 allée d'Italie, F-69364 Lyon Cedex 07, France
email: fallard@ens-lyon.fr

Abstract. The atmospheres of Brown Dwarfs (BDs) are the site of molecular opacities and cloud formation, and control their cooling rate, radius and brightness evolution. Brown dwarfs evolve from stellar-like properties (magnetic activity, spots, flares, mass loss) to planet-like properties (electron degeneracy of the interior, cloud formation, dynamical molecular transport) while retaining, due to their fully convective interior, larger rotational velocities ($\leqslant$ 30 km/s i.e. P < 4 hrs versus 11 hrs for Jupiter). Model atmospheres treating all this complexity are therefore essential to understand the evolution properties, and to interpret the observations of these objects. While the pure gas-phase based NextGen model atmospheres (Allard *et al.* 1997, Hauschildt *et al.* 1999) have allowed the understanding of the several populations of Very Low Mass Stars (VLMs), the AMES-Dusty models (Allard *et al.* 2001) based on equilibrium chemistry have reproduced some near-IR photometric properties of M and L-type brown dwarfs, and played a key role in the determination of the mass of brown dwarfs and Planetary Mass Objects (PMOs) in the eld and in young stellar clusters. In this paper, we present a new model atmosphere grid for VLMs, BDs, PMOs named BT-Settl, which includes a cloud model and dynamical molecular transport based on mixing information from 2D Radiation Hydrodynamic (RHD) simulations (Freytag *et al.* 2009). We also present the status of our 3D RHD simulations including rotation (Coriolis forces) of a cube on the surface of a brown dwarf. The BT-Settl model atmosphere grid will be available shortly via the Phoenix web simulator (http://phoenix.ens-lyon.fr/simulator/).

Keywords. stellar atmospheres, brown dwarfs, very low mass stars, radiation hydrodynamics

References

Allard, F., Hauschildt, P. H., Alexander, D. R., & Starrfield, S. 1997, ARA&A, 35, 137
Allard, F. *et al.* 2001, ApJ, 556, 357
Freytag, B., Allard, F., Ludwig, H.-G., Homeier, D., & Steffen, M. 2009, A&A, in press
Hauschildt, P. H., Allard, F., & Baron, E. 1999, ApJ, 512, 377

Highlights of Astronomy, Volume 15
XXVIIth IAU General Assembly, August 2009
Ian F. Corbett, ed.

© International Astronomical Union 2010
doi:10.1017/S1743921310011427

Physical Properties of Binary Brown Dwarfs

W. Brandner[1], M. Stumpf[1], R. Köhler[1], V. Joergens[1], F. Hormuth[1],
K. Geißler[1], B. Goldman[1], T. Henning[1], H. Bouy[2] and E. Martin[2]

[1] Max Planck Institute for Astronomy

[2] Instituto de Astrofísica de Canarias, Spain

Astrometric observations of binary brown dwarfs yield dynamical masses of the components independently of theoretical models. We give an update on our long-term high-resolution spectroscopic and photometric monitoring programme of spatially resolved binary brown dwarfs using ground-based adaptive optics and the Hubble Space Telescope. We present current orbital fits, including refined dynamical mass estimate of the Kelu-1 AB system. The results seem to support the previously reported trend that evolutionary and atmospheric models might underestimate the mass of very-low-mass stars and brown dwarfs.

Alternatively, additional, thus far unresoled system components (i.e. a higher degree of multiplicity) could explain the unexpectedly high dynamical mass estimates. In the case of Kelu-1 AB, we present potential spectroscopic evidence for a third system component. The confirmation of this third system component would resolve the reported discrepancy between evolutionary models and dynamical mass estimates. This might make Kelu-1 AB the first pure brown dwarf triple system discovered so far.

Highlights of Astronomy, Volume 15
XXVIIth IAU General Assembly, August 2009
Ian F. Corbett, ed.

© International Astronomical Union 2010
doi:10.1017/S1743921310011439

Testing Models with Brown Dwarf Binaries

Trent J. Dupuy and Michael C. Liu

Institute for Astronomy, University of Hawai'i
2680 Woodlawn Drive, Honolulu, HI, USA

Abstract. We have been using Keck laser guide star adaptive optics to monitor the orbits of ultracool binaries, providing dynamical masses at lower luminosities and temperatures than previously available and enabling strong tests of theoretical models. (1) We find that model color–magnitude diagrams cannot reliably be used to infer masses as they do not accurately reproduce the colors of ultracool dwarfs of known mass. (2) Effective temperatures inferred from evolutionary model radii can be inconsistent with temperatures derived from fitting observed spectra with atmospheric models by at most 100–300 K. (3) For the single pair of field brown dwarfs with a precise mass (3%) *and* age determination ($\approx$25%), the measured luminosities are $\sim$2–3$\times$ higher than predicted by model cooling rates (masses inferred from L_{bol} and age are 20–30% larger than measured). Finally, as the sample of binaries with measured orbits grows, novel tests of brown dwarf formation theories are made possible (e.g., testing theoretical eccentricity distributions).

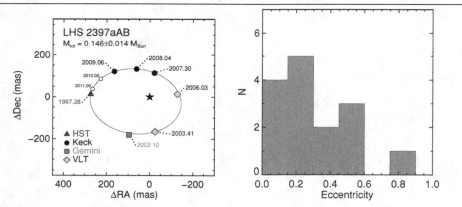

Figure 1. *Left:* Orbit of the M8+L7 binary LHS 2397aAB. *Right:* Eccentricity distribution for all nine published ultracool (>M6) binary orbits along with six unpublished orbits known to the author.

Left:Dupuy *et al.* (2009b) Dupuy, Liu, & Ireland
Right: Bouy *et al.* (2004) Bouy, Brandner, Martín, Delfosse, Allard, Baraffe, Forveille, & Demarco, Simon *et al.* (2006) Simon, Bender, & Prato, Seifahrt *et al.* (2008) Seifahrt, Röll, Neuhäuser, Reiners, Kerber, Käufl, Siebenmorgen, & Smette, Liu *et al.* (2008) Liu, Dupuy, & Ireland, Blake *et al.* (2008) Blake, Charbonneau, White, Torres, Marley, & Saumon, Cardoso *et al.* (2009) Cardoso, McCaughrean, King, Close, Scholz, Lenzen, Brandner, Lodieu, & Zinnecker, Dupuy *et al.* (2009a) Dupuy, Liu, & Ireland, Dupuy *et al.* (2009b) Dupuy, Liu, & Ireland, Dupuy *et al.* (2009c) Dupuy, Liu, & Bowler

References

Blake, C. H., *et al.* 2008, *Ap. Lett.*, 678, L125
Bouy, H., *et al.* 2004, *A&A*, 424, 213
Cardoso, C. V., *et al.* 2009, in AIP Conf. Ser. 1094, ed. E. Stempels, 509–512

Dupuy, T. J., Liu, M. C., & Ireland, M. J. 2009b, *ApJ*, 692, 729
—. 2009c, *ApJ*, 699, 168
Dupuy, T. J., Liu, M. C., & Bowler, B. P. 2009a, *ApJ*, in press (astro-ph/0909.4784)
Liu, M. C., Dupuy, T. J., & Ireland, M. J. 2008, *ApJ*, 689, 436
Seifahrt, A. *et al.* 2008, *A&A*, 484, 429
Simon, M., Bender, C., & Prato, L. 2006, *ApJ*, 644, 1183

Highlights of Astronomy, Volume 15
XXVIIth IAU General Assembly, August 2009
Ian F. Corbett, ed.

Observations of low mass companions to massive stars

H. Zinnecker[1]

[1] Astrophysikalisches Institut Potsdam

Massive stars are known to be multiple systems, often in tight, short-period OB stars binaries (SB1 and SB2, found by spectroscopic monitoring). However, little is known about low-mass companions to massive stars, such as A, F, and G stars with masses in the range of 1 to 3 solar masses. Yet systems of massive stars with wide low-mass companions (of the order of a few AU) must exist, for these are the progenitors of LMXB and HMXB (low-mass and high-mass X-ray binaries).

We discuss observational techniques to detect such high-mass/low-mass physical pairs, including long-baseline interferometry and astrometry (e.g. AMBER and PRIMA at the ESO VLTI on Paranal). The time has come to discover this new class of extreme contrast (1:100 to 1:10,000) binary systems, an effort similar to image exo-planets in the glare of solar-type main sequence stars.

Highlights of Astronomy, Volume 15
XXVIIth IAU General Assembly, August 2009
Ian F. Corbett, ed.

© International Astronomical Union 2010
doi:10.1017/S1743921310011452

Dynamical masses for the nearest brown dwarf binary: ε Indi Ba,b

C. V. Cardoso[1], M. J. McCaughrean[2,1], R. R. King[1], L. M. Close[3],
R.-D. Scholz[4], R. Lenzen[5], W. Brandner[5], N. Lodieu[6], H. Zinnecker[4],
R. Koehler[7] and Q. M. Konopacky[8]

[1]University of Exeter, [2]ESA, [3]Steward Observatory, [4]Astrophysikalisches Institut Potsdam,
[5]Max-Planck-Institut für Astronomie, [6]Instituto de Astrofísica de Canarias, [7]Landessternwarte
Zentrum für Astronomie Heidelberg, [8]UCLA,

Binary brown dwarfs are important because their dynamical masses can be determined in a model-independent way. If a main sequence star is also involved, the age and metallicity for the system can be determined, making it possible to break the sub-stellar mass-age degeneracy. The most suitable benchmark system for intermediate age T dwarfs is ε Indi Ba,b, two T dwarfs (spectral types T1 and T6; McCaughrean et al. (2004)) orbiting a K4.5V star, ε Indi A, at a projected separation of 1460AU. At a distance of 3.6224pc (HIPPARCOS distance to ε Indi A; van Leeuwen (2007)), these are the closest brown dwarfs to the Earth, and thus both components are bright and the system is well-resolved. The system has been monitored astrometrically with NACO and FORS2 on the VLT since June 2004 and August 2005, respectively, in order to determine the system and individual masses independent of evolutionary models. We have obtained a preliminary system mass of $121\pm1M_{\rm Jup}$. We have also analysed optical/near-IR spectra ($0.6-5.0\mu m$ at a resolution up to R~5000; King et al. (2009)) allowing us to determine bolometric luminosities, compare and calibrate evolutionary and atmospheric models of T dwarfs at an age of 4-8Gyr.

Acknowledgements: This work is funded by the EC FP6 Marie Curie RTN CONSTELLATION: MRTN-CT-2006-035890.

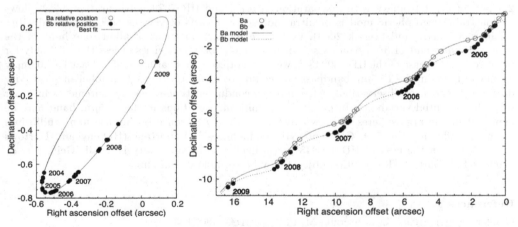

Figure 1. Left: Relative motion (NACO). Right: Absolute motion (FORS2).

References

McCaughrean, M. J., Close, L. M., Scholz, R.-D., Lenzen, R., et al. 2004, A&A 413, 1029
van Leeuwen, F. 2007, A&A 474, 653
King, R. R., McCaughrean, M. J., Homeier, D., Allard, F. et al. 2009, A&A in press

Highlights of Astronomy, Volume 15
XXVIIth IAU General Assembly, August 2009
Ian F. Corbett, ed.

Mid-Infrared Variability in Binary Brown Dwarfs

Michael F. Sterzik[1], Gael Chauvin[2], Kerstin Geißler[1,3] and Eric Pantin[4]

[1] European Southern Observatory, Casilla 19001, Santiago 19, Chile
email: msterzik@eso.org

[2] Laboratoire d' Astrophysique, Observatoire de Grenoble, 38041 Grenoble Cedex 9, France

[3] Max-Planck-Institut for Astronomy, Königstuhl 17, 69117 Heidelberg, Germany

[4] CEA/Saclay, DSM/DAPNIA/Service d' Astrophysique, 91191 Gif-sur-Yvette, France

Abstract. We have spatially resolved several nearby binary brown dwarfs and obtained mid-infrared photometry with VISIR at the VLT. In particular, we have monitored ε Indi B and HD 130948 in several narrow-band MIR filters. The 10.5μm band is a probe to constrain non-equilibrium chemistry in the atmosphere of cool brown dwarfs.

Keywords. stars: low-mass, brown dwarfs; binaries: close

1. Ground-based MIR measurements

Ground-based mid-IR imaging of binary brown dwarf systems with sub-arcsecond spatial resolution can complement high sensitivity, but low-spatial resolution space-based photometry as obtained e.g. with *Spitzer*. The spatially resolved photometry of the close (seperation 0.7") brown dwarf binary ε Indi Ba and Bb (Sterzik, Pantin, Hartung *et al.* 2005) and of three other brown dwarfs in binary systems, GJ 229 B (separation 7.8"), HD 130948 B (separation 2.6", B itself a L4 binary with a separation of 0.1") and HR 7329 B (separation 4.2") allows to constrain atmospheric models of ultra-cool brown dwarfs of various ages and metallicities (Geißler, Chauvin and Sterzik 2008). On-source integration times of about one hour in the 8.6μm, 10.5μm and 11.3μm bandpasses yield 3σ detection sensitivities of less then 1-2 mJy for point sources. In case of the HD 130948 B, we have noticed a flux variation of at least 1.7±0.6mJy within 48 hours in the 10.5μm bandpass and could not explain it through insufficient sensitivity during one epoch of observations. Therefore we conducted time-series measurements in order to probe potential variability. In particular, significant variations in the 10.5μm band may be expected in the atmospheres of brown dwarfs at the L/T transition in case non-equilibrium chemistry affecting the CO, CH_4 and NH_3 abundances is important (Hubeny and Burrows 2007). While in the case of HD 130948 B the likelihood of variability is small (Geißler *et al.* 2009), ε Indi Ba (a L/T transition object) may be variable in 10.5μm.

References

Geißler, K., Chauvin, G., & Sterzik, M. F. 2008, *A&A* 480, 193

Geißler, K., Sterzik, M. F., Chauvin, G., & Pantin, E. 2009, in: E. Stempels (ed.), *Cool Stars, Stellar Systems and the Sun*, 15th Cambridge Workshop (AIP), p. 521

Hubeny, I. & Burrows, A. 2007, *ApJ* 669, 1248

Sterzik, M. F., Pantin, E., Hartung, M., Huelamo, N., Kaufer, A., Käufl, H. U., Melo, C., Nürnberger, D., Siebenmorgen, R., & Smette, A. 2005, *A&A* 436, L39

Highlights of Astronomy, Volume 15
XXVIIth IAU General Assembly, August 2009
Ian F. Corbett, ed.

© International Astronomical Union 2010
doi:10.1017/S1743921310011476

Are pre-MS stars older than we thought?

Tim Naylor and N.J. Mayne

School of Physics, University of Exeter, EX4 4QL, UK

Abstract. We present a consistent age ordering for young clusters and groups determined using the contraction of stars through their pre-main-sequence phase. We compare these with ages derived from the evolution of the upper main-sequence stars, and find the upper MS ages are older by a factor 1.5 to 2. We show that increasing the binary fraction and number of equal-mass binaries amongst the O-stars compared to the rest of the MS cannot remove this discrepancy.

Keywords. methods: statistical, stars: formation, open clusters and associations: general

---

In Mayne *et al.* (2007) and Mayne & Naylor (2008) we developed an age-ordering for young stars and groups based on the luminosity of the pre-main-sequence. Table 1 shows the resulting ages, including those derived in subsequent papers. In Naylor (2009) we derived ages by fitting the change position in the colour-magnitude diagram of upper-MS stars as they evolve from the zero-age MS to the terminal-age MS. We found that the MS ages are a factor 1.5 to 2 longer than the ages derived from the PMS.

After my presentation Gaspard Duchêne pointed out that binarism amongst O-stars is much higher than in the rest of the MS. A higher binary fraction will shift the centroid of the combined single-star and binary-star sequences redwards, mimicking an older age and perhaps explaining the older MS ages. As the mass-ratio distribution is equally important, we tested this idea using the most extreme assumption we could reasonably make, the strong hypothesis of Lucy (2006), which we approximated as 25% of binaries evenly distributed over $0.95 < q < 1.0$, and 75% evenly distributed over $0.2 < q < 0.95$. Using this, and a binary fraction (restricted to $q > 0.2$) of 75% (e.g. Sana *et al.*, 2009) for all O-stars we find the ages of the clusters change by less than 5% compared with the results of Naylor (2009). Thus the discrepancy between the MS and PMS ages remains.

References

Jeffries, R. D., Naylor, T., Walter, F. M., Pozzo, M. P., & Devey, C. R. 2009, *MNRAS* 393, 538
Lucy, L. B. 2006, *A&A* 457, 629
Mayne, N. J. & Naylor, T. 2008, *MNRAS* 386, 261
Mayne, N. J., Naylor, T., Littlefair, S. P., Saunders, E. S., & Jeffries, R. D. 2007, *MNRAS* 375, 1220
Naylor, T. 2009, arXiv:0907.2307
Sana, H., Gosset, E., & Evans, C. J. 2009, arXiv:0909.0486

Table 1. Ages from PMS contraction

1Myr	2Myr	3Myr	4-5Myr	5-10Myr	10Myr	13Myr	40Myr
IC5146	ONC	λ Ori,	IC348	γ Vel[2]	NGC7160	h & χ Per	NGC2547
	NGC 6530	σ Ori,	Cep OB3b[1]				
		NGC2264	NGC2362				

[1] Littlefair *et al.* in prep.   [2] Jeffries *et al.* (2009)

Highlights of Astronomy, Volume 15
XXVIIth IAU General Assembly, August 2009
Ian F. Corbett, ed.

© International Astronomical Union 2010
doi:10.1017/S1743921310011488

Prospects for planet formation in multiple stellar systems

Gaspard Duchêne[1,2]

[1] Astronomy Department, UC Berkeley, 601 Campbell Hall, Berkeley CA 94720-3411, USA
email: gduchene@berkeley.edu

[2] Laboratoire d'Astrophysique de Grenoble, CNRS/UJF UMR 5571, 414 rue de la Piscine,
BP 53, 38041 Grenoble Cedex 9, France

Abstract. As I review here, planet formation in multiple stellar systems is far from exceptional. However, it appears that binaries with projected separation in the 5–100 AU range have different initial conditions and end result properties than wider systems, probably because they undergo different physical processes. In addition, very tight binaries, with projected separation less than a few AU, seem to fulfill all the known requirements to form planetary systems, suggesting that circumbinary planets are very likely to exist.

Keywords. binaries: general, planetary systems: formation

Planet formation around Sun-like stars appears to be a frequent phenomenon. The rapidly increasing number of known planets now enables detailed studies of their statistical properties, which should provide critical elements to disentangle the competing models of "core accretion" (Lissauer & Stevenson 2007) and "disk instability" (Durisen *et al.* 2007). In parallel, stellar multiplicity is an ubiquitous outcome of star formation (Duchêne *et al.* 2007), raising the question of the coexistence of these two phenomena. I present here a brief overview of several key trends connecting planet formation and stellar multiplicity that have been recently unearthed. An expanded version of this work will appear elsewhere (Duchêne, *in prep.*).

Studies of protoplanetary disks in pre-main sequence multiple systems have revealed that *the most massive component of a system is usually associated with the most massive and longest-lived disk* (see Monin *et al.* 2007). Considering disks located in multiple systems, *the properties of their innermost, planet-forming, regions are indistinguishable from those of single stars* (e.g., White & Ghez 2001, Pascucci *et al.* 2008). This suggests that the required initial conditions for planet formation are met equally well in the higher-mass component of a multiple system as in single stars, but that lower-mass components may have more adverse initial conditions to form planets. Beyond these overall properties, there is a clear trend for *disks in tighter binaries* ($a \lesssim 100$ *AU*) *to have a much shorter lifetime* (Cieza *et al.* 2009). In addition, it is a long-established fact that tighter binaries ($a \lesssim 300$ AU) have much lower (sub)millimeter fluxes than wider binaries and single stars alike (e.g., Andrews & Williams 2005). Both observations have been interpreted as the natural consequence of severe disk truncation in tight binaries, leading to a general belief that tight binaries are hard-pressed to form planetary systems at all. The situation is not necessarily as dark as it seems, however, as optical depth effects in dense, compact disks could explain their reduced thermal emission. It is also important to note that the few protoplanetary disks encompassing spectroscopic binaries currently known tend to have large (sub)millimeter fluxes (Jensen *et al.* 1996), similar to those around single stars, and could therefore offer fertile grounds for planets.

Moving on to the next evolutionary stage, let us focus on debris disks as proxies for newly formed planetary systems. *Overall, the debris disk phenomenon occurs at a very similar rate in binaries and single stars* (see Trilling *et al.* 2007). A proposed deficit of debris disks among intermediate separation (3–50 AU) binaries has not been confirmed (Plavchan *et al.* 2009), suggesting that *planetesimal disks form in binaries of all separations*. In particular, the shortest period binaries ($a \lesssim 3\,\mathrm{AU}$) frequently possess debris disks. In summary, the debris disks phenomenon appears to be indifferent to the presence and/or location of stellar companions.

Finally, let us review the connection between mature planetary systems and multiple stellar systems. Binary systems have long been excluded from planet searching surveys for practical reasons but more and more known examples are discovered in follow-up surveys of known exoplanet hosts. *Their multiplicity rate is only slightly lower than that of field stars despite selection biases* (Mugrauer & Neuhäuser 2009). Although the stellar companions tend to be at separations larger than 100 AU (Raghavan *et al.* 2006), there is a growing number of systems with projected separations as close as 20 AU, proving that *even a relatively tight companion does not preclude planet formation*. An early statistical analysis suggested a peculiar mass-period trend for planets in binary systems with respect to planets around single stars (Zucker & Mazeh 2002), but larger samples have now shown that this is not statistically significant. On the other hand, this larger sample has revealed that planets in tight binaries ($a \lesssim 100\,\mathrm{AU}$) all have $M_p > M_{Jup}$, whereas planets in wider binaries and single stars are roughly equally split above and below this threshold. In other words, *tight binaries only form high-mass planets*.

While there are caveats associated to the various studies listed here, not least the fact that each of them considers a different range of stellar masses, it is tempting to try and put together the various pieces of this puzzle. Overall, it seems clear that primary stars in "wide" binaries, say $a \gtrsim 100\,\mathrm{AU}$, evolve in the same manner as single stars, presumably through the "core accretion" scenario since disks survive for quite a long time (5–10 Myr). "Intermediate" binaries ($5 \lesssim a \lesssim 100\,\mathrm{AU}$ or so) have protoplanetary disks that are much denser and more compact, have a much shorter survival timescale, and yet form higher-mass planets on average. I propose that these various trends can be accounted for if planet formation in such binaries occurs via a violent process, such as the "disk instability" mechanism. Finally, while no such planet is known to date, the chances that planets exist around "tight" binaries ($a \lesssim 5\,\mathrm{AU}$) are quite high. While detecting such systems is a major observational challenge, it is certainly worth attempting to search for them with both direct and indirect techniques.

References

Andrews, S. M. & Williams, J. P. 2005, *ApJ*, 631, 1134
Cieza, L. A. *et al.* 2009, *ApJ* (Letters), 696, L84
Duchêne, G. *et al.* 2007, Protostars and Planets V, 379
Durisen, R. H. *et al.* 2007, in *Protostars and Planets V*, 607
Jensen, E. L. N., Mathieu, R. D., & Fuller, G. A. 1996, *ApJ*, 458, 312
Lissauer, J. J. & Stevenson, D. J. 2007, in *Protostars and Planets V*, 591
Monin, J.-L. *et al.* 2007, in *Protostars and Planets V*, 395
Mugrauer, M. & Neuhäuser, R. 2009, *A&A*, 494, 373
Pascucci, I. *et al.* 2008, *ApJ*, 673, 477
Plavchan, P. *et al.* 2009, *ApJ*, 698, 1068
Raghavan, D. *et al.* 2006, *ApJ*, 646, 523
Trilling, D. E. *et al.* 2007, *ApJ*, 658, 1289
White, R. J. & Ghez, A. M. 2001, *ApJ*, 556, 265
Zucker, S. & Mazeh, T. 2002, *ApJ* (Letters), 568, L113

Highlights of Astronomy, Volume 15
XXVIIth IAU General Assembly, August 2009
Ian F. Corbett, ed.

The Role of Multiplicity in Protoplanetary Disk Evolution

Adam L. Kraus[1] and Michael J. Ireland[2]

[1] Department of Astrophysics, California Institute of Technology, Pasadena, CA 91125, USA
email: alk@astro.caltech.edu

[2] School of Physics, University of Sydney, NSW 2006, Australia

Abstract. Interactions with close stellar or planetary companions can significantly influence the evolution and lifetime of protoplanetary disks. It has recently become possible to search for these companions, directly studying the role of multiplicity in protoplanetary disk evolution. We have described an ongoing survey to directly detect these stellar and planetary companions in nearby star-forming regions. Our program uses adaptive optics and sparse aperture mask interferometry to achieve typical contrast limits of $\Delta K = 5 - 6$ at the diffraction limit (5–8 M_{Jup} at 5–30 AU), while also detecting similar-flux binary companions at separations as low as 15 mas (2.5 AU). In most cases, our survey has found no evidence of companions (planetary or binary) among the well-known "transitional disk" systems; if the inner clearings are due to planet formation, as has been previously suggested, then this paucity places an upper limit on the mass of any resulting planet. Our survey also has uncovered many new binary systems, with the majority falling among the diskless (WTTS) population. This disparity suggests that disk evolution for close (5–30 AU) binary systems is very different from that for single stars. As we show in Figure 1, most circumbinary disks are cleared by ages of 1–2 Myr, while most circumstellar disks are not. These diskless binary systems have biased the disk frequency downward in previous studies. If we remove our new systems from those samples, we find that the disk fraction for single stars could be higher than was previously suggested.

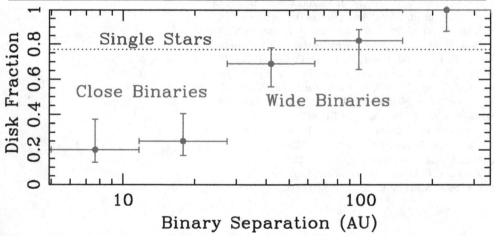

Figure 1. Disk fraction as a function of binary separation in Taurus-Auriga. Red points denote the disk fraction for several ranges of projected binary separations, while the blue dotted line shows the disk fraction for apparently single stars. Binary systems with separations of <30 AU have a significantly lower disk fraction than the population as a whole, whereas the disk fractions for wider binary systems and for single stars are indistinguishable.

Highlights of Astronomy, Volume 15
XXVIIth IAU General Assembly, August 2009
Ian F. Corbett, ed.

© International Astronomical Union 2010
doi:10.1017/S1743921310011506

Muti-technique observations and modelling of the gas and dust phases of protoplanetary disks

C. Pinte[1,2], F. Ménard[2], G. Duchêne[2,3] and J. C. Augereau[2]

[1] School of Physics, University of Exeter, Stocker Road, Exeter EX4 4QL, United Kingdom
[2] Laboratoire d'Astrophysique de Grenoble, CNRS/UJF UMR 5571, 414 rue de la Piscine, B.P. 53, F-38041 Grenoble Cedex 9, France
[3] Astronomy Dept, UC Berkeley, Berkeley CA 94720-3411, USA

A wide range of high-quality data is becoming available for protoplanetary disks. From these data sets many issues have already been addressed, such as constraining the large scale geometry of disks, finding evidence of dust grain evolution, as well as constraining the kinematics and physico-chemical conditions of the gas phase. Most of these results are based on models that emphasise fitting observations of either the dust component (SEDs or scattered light images or, more recently, interferometric visibilities), or the gas phase (resolved maps in molecular lines). In this contribution, we present a more global approach which aims at interpreting consistently the increasing amount of observational data in the framework of a single model, in order to to better characterize both the dust population and the gas disk properties, as well as their interactions. We present results of such modeling applied to a few disks (e.g. IM Lup, see Figure) with large observational data-sets available (scattered light images, polarisation maps, IR spectroscopy, X-ray spectrum, CO maps). These kinds of multi-wavelengths studies will become very powerful in the context of forthcoming instruments such as Herschel and ALMA.

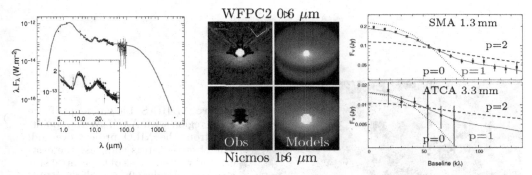

Figure 1. Multi-wavelength modelling of the observations of the disk surrounding IM Lupi. *Left panel:* SED. The mm spectral index indicates the presence of large mm-sized grains whereas the silicate bands indicate that micron-sized grains dominate in the disk surface, suggesting a stratified structure for the disk, with small grains at the surface and bigger grains close to the midplane. *Central panel:* the scattered light images at 0.6 and 1.6 μm give strong constraints on the disk geometry. *Right panel:* millimetre visibilities at 1.3 and 3.3 mm. In the right panel, several models with different surface density $\Sigma(r) \propto r^{-p}$ are presented, only the model with $p = 1$ reproduce the observations. The combination of these data allow strong constraints on the model parameters. From Pinte et al 2008, A&A, 489, 633.

Highlights of Astronomy, Volume 15
XXVIIth IAU General Assembly, August 2009
Ian F. Corbett, ed.

© International Astronomical Union 2010
doi:10.1017/S1743921310011518

First results from XILO: XMM-Newton Investigations in the Lambda Orionis star forming region

B. Stelzer[1], D. Barrado y Navascues[2], N. Huelamo[2], M. Morales-Calderon[2], A. Bayo[2]

[1] INAF - Osservatorio Astronomico di Palermo, 90134 Palermo, Italy
email: stelzer@astropa.unipa.it

[2] LAEFF - INTA, 28691 Villanueva de la Canada, Spain

The λ Orionis star formation region ($1 - 6\,$Myr, 400 pc) is a complex of star-forming clouds surrounded by a molecular ring with $\sim 5°$ radius which was probably formed by a supernova explosion (Dolan & Mathieu 2002). For a complete picture of star formation, believed to be determined by the supernova blast, the large-scale distribution of the pre-main sequence population in λ Ori needs to be examined. We have embarked on a multi-wavelength study (*XMM-Newton*/X-ray, CFHT/optical, *Spitzer*/IR) of selected areas within this intriguing star-forming complex that enables us to identify young stars and brown dwarfs. Our study comprises various areas within the cloud complex as shown in Fig.1. This data set is among the most extended X-ray surveys carried out with *XMM-Newton* in a coherent star-forming environment. The *XMM-Newton* observations combined with optical and IR data reveal the low-mass stellar population down to $\sim 0.4\,M_\odot$. For this mass-limited sample, our preliminary analysis confirms the anomalously low disk-fraction of the central star cluster Coll 69, the Eastern extension of its low-mass population pointing towards B 35, and the concentration of young stars in front of B 35. The analysis of the 'on-cloud field' of B 35 (white in the figure) will show if the cloud is currently forming stars. This will be crucial for determining the star-forming history in the whole λ Ori region.

Figure 1. IRAS $100\,\mu$m map of the λ Ori region with *XMM-Newton* pointings from the XILO project overlaid. The observed areas comprise the eastern and western sides of the central star cluster Collinder 69, the dark clouds B 30 & B 35, the 'bridge' connecting B 30 with Coll 69, the dark clouds LDN 1603 and LDN 1588 on the molecular ring, and a concentration of B stars near the north-east of the ring. The pointings from the original XILO project are shown in green, one observation yet to be executed is shown in red (field labelled 'B-stars'), and data added from the *XMM-Newton* archive are shown in white (field labelled 'B35E').

Highlights of Astronomy, Volume 15
XXVIIth IAU General Assembly, August 2009
Ian F. Corbett, ed.

© International Astronomical Union 2010
doi:10.1017/S174392131001152X

Stellar and brown dwarf properties from numerical simulations

Matthew R. Bate

School of Physics, University of Exeter
Stocker Road, Exeter EX4 4QL, United Kingdom
email: mbate@astro.ex.ac.uk

Abstract. We review the statistical properties of stars and brown dwarfs obtained from the first hydrodynamical simulation of star cluster formation to produce more than a thousand stars and brown dwarfs while simultaneously resolving the lowest mass brown dwarfs (those with masses set by the opacity limit for fragmentation), binaries with separations down to ~ 1 AU, and discs with radii greater than ~ 10 AU. In particular, we present the eccentricity distribution of the calculation's very-low-mass and brown dwarf binaries which has not been previously published.

Keywords. gravitation, hydrodynamics, stellar dynamics, methods: numerical, binaries: general, stars: formation, stars: low-mass, brown dwarfs, stars: luminosity function, mass function

---

Recently, Bate (2009a) published hydrodynamical simulations of the collapse and fragmentation of a 500 $M_\odot$ molecular cloud with a diameter of 0.8 pc, and a mean thermal Jeans mass of 1 $M_\odot$. Although initially spherical, the cloud was seeded with a supersonic divergence-free 'turbulent' velocity field with a power spectrum $P(k) \propto k^{-4}$. The turbulent energy was initially set equal in magnitude to the gravitational energy of the cloud and decayed during the calculations. A barotropic equation of state that mimics the heating of collapsing gas once it becomes optically thick to its own radiation was used, thus capturing the opacity limit for fragmentation. Once the central density of a protostar exceeded 10^{-10} g cm^{-3}, the object was replaced by a sink particle that accreted any gas approaching within a specified accretion radius. These objects began with masses of a few Jupiter masses, and subsequently accreted to become more massive brown dwarfs or stars. The same calculation was performed twice with two different accretion radii (5 AU and 0.5 AU). The former was run until 1.5 initial cloud free-fall times (285,000 yrs) by which time 1254 stars and brown dwarfs had formed. Due to shorter timesteps, the latter was only followed to 1.04 free-fall times, producing 258 objects. Comparison allowed the effect of different accretion radii on the statistical properties of the stars to be assessed. A brief summary of the statistical properties of the stars and brown dwarfs produced by the calculations is given below. In most cases, more detail can be found in Bate (2009a).

Multiplicity. The calculations produced stellar populations in which multiplicity is a strongly increasing function of primary mass with values in good agreement with observational surveys. In particular, the frequency of very-low-mass (VLM) binaries with primary masses of $0.03-0.10$ $M_\odot$ is found to be $19 \pm 5\%$ for the 0.5 AU accretion radius calculation, in good agreement with observations. We also note that multiplicity is predicted to continue decreasing with primary mass throughout the brown dwarf regime such that the predicted multiplicity of $0.01-0.03$ $M_\odot$ brown dwarfs is less than 7%.

Binary and multiple separations. The separations of binary and higher-order multiple systems are also found to depend on primary mass, with stellar systems having a median separation of 26 AU and VLM systems having a median separation of 10 AU. This tendency for lower mass binaries to have smaller separations is in agreement with observations, although observations indicate than VLM systems are even tighter than those produced in the calculations. However, we also note that the more accurate 0.5 AU accretion radius simulation produces tighter binaries than the 5 AU accretion radius simulation and also that the separations of VLM binaries decrease during the simulation.

Binary eccentricities. The eccentricity distribution of binaries is very sensitive to the accretion radius. The calculation using 5 AU accretion radii produced a severe excess of high eccentricity ($e > 0.7$) close (< 10 AU) systems. This was corrected when accretion radii of 0.5 AU were used,

M_1 $M_\odot$	M_2 $M_\odot$	Semi-major axis AU	Eccentricity
0.096	0.065	0.52	0.365
0.095	0.082	0.95	0.045
0.044	0.014	1.13	0.493
0.017	0.021	1.75	0.419
0.041	0.021	2.84	0.233
0.067	0.040	3.77	0.514
0.076	0.057	4.05	0.097
0.047	0.029	10.5	0.075
0.056	0.051	21.3	0.036
0.047	0.041	47.1	0.511
0.025	0.010	49.6	0.261
0.079	0.056	61.9	0.546
0.057	0.023	93.6	0.805
0.061	0.015	226	0.566
0.080	0.026	577	0.324
0.099	0.051	989	0.861

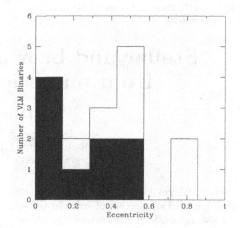

Figure 1. Masses and orbital properties of the 16 binary VLM objects produced by the star cluster formation calculation (Bate 2009a) using sink particle accretion radii of only 0.5 AU. The shaded histogram contains those binaries with semi-major axes less than 30 AU. These close systems are more likely to survive to join the field population and tend to have low eccentricities.

and the mean eccentricity is then in good agreement with observations. Bate (2009a) does not differentiate between stars and brown dwarfs when discussing eccentricities. However, at this meeting, preliminary results for the eccentricity distribution of VLM binaries were presented by Trent Dupuy (see the contribution by Dupuy, this volume). He finds VLM binaries tend to have low eccentricities (usually less than 0.5). In Fig. 1, we present the properties of the 16 VLM binaries produced by the 0.5 AU accretion radius calculation of Bate (2009a). In agreement with the new observational results, the simulated VLM binaries tend to have low eccentricities, particularly those with separations less than 30 AU which is typical of the field population.

Model deficiencies and conclusions. As discussed above, many of the stellar properties obtained from the simulations are in good agreement with observations. The implication is that the properties of binary and multiple systems are primarily determined by the combination of gravity and gas dynamics (i.e. dissipative gravitational dynamics). Other processes such as radiation transport and magnetic fields are not required to explain the origin of these properties. However, the hydrodynamical simulations do have two major deficiencies: they give a ratio of brown dwarfs to stars much higher than is observed (roughly 2:1) and there is a deficit of unequal-mass solar-type binaries. Recently, Bate (2009b) showed that including radiative feedback may correct the ratio of brown dwarfs to stars, but the origin of the unequal-mass solar-type binary deficit is still unknown.

References

Bate, M. R. 2009a, *MNRAS*, 392, 590
Bate, M. R. 2009b, *MNRAS*, 392, 1363

Highlights of Astronomy, Volume 15
XXVIIth IAU General Assembly, August 2009
Ian F. Corbett, ed.

© International Astronomical Union 2010
doi:10.1017/S1743921310011531

The formation of discs in clusters

Paul C. Clark[1]

[1]Institute for Theoretical Astrophysics, University of Heidelberg

We review the properties of the discs that form around 'sink particles' in smoothed particle hydrodynamics (SPH) simulations of cluster formation, similar to those of Bate *et al.* (2003) and Bonnell *et al.* (2004), and compare them to the observed properties of discs in nearby star-forming regions. Contrary to previous suggestions, discs can form and survive in such an environment, despite the chaotic effects of competitive accretion. We find the discs are typically massive, with ratios of disc mass to central object mass of around 0.1, or higher, being typical. Naturally, the evolution of these discs is dominated by gravitational torques, and the more massive examples exhibit strong $m = 2$ spiral modes. We also find that they can continuously grow over a period of 100,000 years, provided the central object is a single sink particle and the local density of sink particles is low. Discs that form around sink particles in the very centres of clusters tend to be shorter lived, but a single star can lose and gain a disc several times during the main accretion phase. However due to the nature of the turbulence in the cluster, the disc orientation can change dramatically over this time period, since disc-sink systems can accrete from counter-rotating envelopes. Since the competitive accretion process brings in material from large distances, the associated angular momentum can be higher than one would expect for an isolated star formation model. As such, we find that the discs are typically several hundred of AUs in extent, with the largest keplerian structures having radii of ~ 2000AU.

Highlights of Astronomy, Volume 15
XXVIIth IAU General Assembly, August 2009
Ian F. Corbett, ed.

© International Astronomical Union 2010
doi:10.1017/S1743921310011543

Special Session 7 - Sumary and Concluding remarks

Silvia H. P. Alencar[1]
and Jane Gregorio-Hetem[2]

[1] Departmento de Física, ICEx, UFMG, Belo Horizonte, MG 31270-901, Brazil
email: silvia@fisica.ufmg.br

[2] Universidade de São Paulo, IAG/USP, São Paulo, SP, 05508-090, Brazil

This was a very pleasant and interesting meeting on star formation. The debate run on freely and contributions were of a very high level, including the oral contributions of four exceptional graduate students.

We discussed star formation diagnostics over the full frequency range, going from radio to gamma rays. In the Special Session 7, high energy phenomena were often cited by researchers working on other frequencies. Most people became recently aware of the importance of high energy phenomena to star formation and disk evolution. The interesting issue is that it happened in recent years greatly due to Spitzer, an infrared telescope, after the detection of Neon lines in circumstellar disks of sun-like stars. It was said that, in what seemed to be Spitzer noise, raised a forest of water (so much sought) and Neon (not sought at all) lines in the spectra of the circumstellar disks of low mass stars.

From a number of talks we saw that brown dwarfs are just a scaled version of T Tauri stars. They have disks and accretion goes on with the same characteristic variability. They have jets, that are still quite hard to image, even in the VLT. They also form in binaries and, although eclipsing binaries seem to be hard to find, new observations of astrometric binaries are producing ways to test evolutionary models in the very low mass regime. Brown dwarf atmospheres remain a challenge that nevertheless evolves rapidly. We now need spectra of binaries with well determined masses to fully test the atmosphere models, which are beautiful, though quite complicated. It took a decade work to get to the actual state-of-the-art, but it looks to us like a good investment of time, given the results.

The first results of the CoRoT satellite on the star-forming region NGC 2264 were presented in the meeting, which includes the most detailed light curves up to now of young low mass stars. The rotation signatures are easy to measure and show substantial variations from ground-based results of the same region. The accretion signatures are quite impressive too, showing that, for a reasonable number of stars, we are able to probe the dynamic star-disk interaction.

Stellar ages in the pre-main sequence became though a lot more uncertain. It was suggested that they are not properly measured, which caused a lot of discussion. It was also shown that we may need to know the entire accretion history of an object to find out its age from an HR diagram in the Pre-Main-Sequence. This could explain the large dispersion observed in the HR diagram among young stars from the same star forming region, but it would also make precise age determination an almost impossible task in early stellar evolution.

The debate goes on among X-wind and disk-wind defenders, with much more detailed and complex models, including multipolar magnetic fields and MHD simulations on each side.

Finally, we saw that planet formation goes on in multiple stellar systems. Planets can form in circumstellar disks of wide binaries, as well as in circumbinary disks around tight companions. Tight binaries, however may have a significant influence on inner disk evolution.

From stars to planets and from gamma rays to radio wavelenghts, we discussed the formation and early evolution of star-disk systems. We would like to thank all the participants for making this a very special session.

Silvia Alencar and Jane Gregorio-Hetem

Part V. List of Poster papers

The Evolution of Disks, Protostars and the Young Cluster IRAS 20050+2730
  Nancy R Adams
Optical depth effects in the X-ray spectra of CTTSs
  Costanza Argiroffi
Energetic processes in young accreting stars with outbursts
  Marc Audard
Searching for Ionized Gas Tracers in Spitzer IRS Spectra of Young Stars in Taurus
  Carla Baldovin-Saavedra
Multi-Epoch Survey of 10 μm Silicate Variability in DG Tau and XZ Tau
  Jeffrey S. Bary
Modeling the X-ray emission from the nearest jets: HH 154 and DG Tau
  Rosaria Bonito
The Young Stellar Population in Orion OB1
  Cesar Briceño
The High Energy (UV/X-ray) Radiation Fields of the Young Stars GM Aurigae and
HD135344B and the Likely Effects on Their Transitional Disks
  Alexander Brown
An X-Ray Census of the Young Stars of Serpens
  Joanna M Brown
Rotation and Magnetic Fields in Fully Convective Stars: What simulations can tell us
  Matthew Browning
Day-night side cooling of the strongly irradiated planet
  Jan Budaj
Study of transitory disks between the protoplanetary and debris phases
  Carolina Chavero
Searching for DIBs in circumstellar environments of Herbig Ae/Be stars
  Cristiane Costa
The physical and chemical environment of a star-forming bright-rimmed cloud
  Alison Craigon
Tracing Outflows from Massive Young Stellar Objects through Masers and Mid-Infrared
Emission
  James M. De Buizer
The properties of pre-main sequence stars in the Magellanic Clouds
  Guido De Marchi
Resonant structures in Planetesimal Disk from system HD98800
  Rita C. Domingos
The disk and envelope structure of Class 0 protostars in Serpens
  Melissa Lanae Enoch
Dynamical study of mass accretion and outflow in the classical T Tauri star V354 Mon
  Nathalia Fonseca
Numerical results for the formation of the four giant planets of the Solar System
  Andrea Fortier
Identification of pre-stellar objects in the Rosette molecular cloud
  Diana R. G. Gama
Observing Gap Formation in the Dust Layer of a Protoplanetary Disk
  Jean-François Gonzalez

The 2008 Accretion Outburst of the Prototype EXOR EX Lupi
  Guy S. Stringfellow
Determination of the mass and temperature of the exoplanet candidate HD33636b using
VLTI+AMBER
  Ramarao Tata
SACY: Nurseries and Kindergartens in the solar neighborhood
  Carlos Alberto O. Torres
The effect of mass accretion on early stellar evolution
  Andrea Urban
The effects of the stellar wind on the magnetic field configuration of weak T Tauri stars
  Aline A. Vidotto
RV Crt, an intriguing PMS triple system
  Luiz Paulo R. Vaz
Confirmation of the nature of Herbig Ae/Be candidates
  Rodrigo Vieira
A Spectroscopic Study of Young Stellar Objects in the Serpens Cloud Core and NGC
1333
  Elaine M. Winston
Brown dwarf formation by fragmentation of protostellar discs
  A. Whitworth
Condensation in Brown Dwarf Atmospheres
  Soeren Witte
X-Ray and Infrared Emission from Young Stellar Objects near LkHalpha 101
  Scott J. Wolk
New tree-based gravity solver for octal tree AMR codes
  Richard Wunsch

Highlights of Astronomy, Volume 15
XXVIIth IAU General Assembly, August 2009
Ian F. Corbett, ed.

© International Astronomical Union 2010
doi:10.1017/S1743921310011555

Preface - SpS8 - The Galactic Plane

N. A. Walton, A. Damineli, co-chairs SOC, M. G. Hoare, J. E. Drew, editors

The first digital astronomical surveys emphasised exploration of the sky away from the crowded Galactic Plane. But now, increased computing power has made it possible to take on comprehensive surveying of the Galactic Plane even at high spatial resolution and down to faint magnitude limits. A number of ambitious wide-area surveys sampling high energies, optical wavelengths, the infrared, sub-millimetre and radio ranges are complete, in process, or about to begin. The goals of these surveys are as broad as Galactic science itself, but are mainly focused either on solving key problems in star formation and stellar evolution, or on mapping the complex substructures of the Galactic bulge and disk in order to see more clearly how the whole is constructed. This meeting brought together researchers directly involved in the many surveys, along with specialists in the observations and modelling of the ISM, stellar evolution, and the structure of the Galactic Disk and Bulge.

The programme of this Special Session, spread over 4 days, was grouped into ten topics. These were: 1. The new generation surveys; 2. Dust in the Milky Way; 3. Structure in the Galaxy; 4. Chemical picture of the Galaxy; 5. Disentangling the star formation process; 6. Mapping star formation in the Galactic disk; 7. Multi-object multi-wavelength galactic plane studies in the VO era; 8. End-states of stellar evolution; 9. Luminous evolved stars; 10. Perspective on the future.

The major threads running through the meeting were the preoccupations with identifying and locating structures within the Galactic Plane (disk and bulge), and what we are learning about its stellar populations and the star-forming process. The discussion of the structure did draw on the entire wavelength range from the high energies sampled by HESS up to the HI 21cm line work of GALFA. We heard about the progress being made in 3-D extinction mapping at optical and NIR wavelengths, and about the ongoing debate regarding the character of the disk's spiral arm structure and the bar in the bulge. The two stellar arm picture versus the four gaseous, star-forming arms was a recurrent theme throughout the meeting. Methanol masers were discussed both as phenomena linked to massive star formation and as targets for VLBI distance determinations. We also heard about the crucial matter of chemical abundance determinations from both stars and nebulae, based on optical data. The significant body of work stimulated by the Spitzer/GLIMPSE survey was very much in evidence during the meeting, as was the growing input from IPHAS. The future was anticipated in talks on LSST and Pan-STARRS.

It should go down on the record that Andrew Walsh wowed the audience with a magnificent sung (and versified) poster pitch. And it should also be mentioned that Anne Green won the poster competition and a bottle of cachaça with a high impact presentation on 'The southern sky at 843 MHz'. Specific thanks are due to SOC member, Gene Magnier for his enthusiastic help in the judging of the posters, alongside Nic and Janet. Last but not least, it is a great pleasure to acknowledge the financial support underpinning the Rio General Assembly as a whole, the efforts of the GA's LOC that made the venue work, and the active support of the members of the wider SOC in designing the programme for this interesting meeting.

Highlights of Astronomy, Volume 15
XXVIIth IAU General Assembly, August 2009
Ian F. Corbett, ed.

New and old roles for narrowband Hα

Janet E. Drew[1]

CAR/STRI, University of Hertfordshire, College Lane, Hatfield, AL10 9AB, UK

Abstract. Until recently, Hα has been seen as *the* tracer of ionized gas, picking out both star formation and the late stages of stellar evolution. This has been reaffirmed, spectacularly, by the recent WHAM and SHS surveys. But the advent of large-area digital detectors creates a new role for narrowband Hα as a direct, simultaneous, measure of intrinsic stellar colour and reddening when e.g. $r' - H\alpha$ is combined with a nearby broad band colour e.g. $r' - i'$. This new capability has been clearly demonstrated by the nearly-complete IPHAS survey.

The Hα line, observed in emission, has been recognised as a critical tracer of ionized gas for many decades now. Until recently, Hα survey data has been obtained using photographic media, deployed both in imaging surveys and in objective prism work delivering slitless low resolution spectra (the latter only being complete to $R \sim 12$). The transition to large-area digital.detectors, combined with increasingly capable database techniques, has wrought a major change. At the forefront of this have been three comprehensive surveys: the Wisconsin Hα Mapper (WHAM, Haffner *et al.* 2003), the Schmidt Hα Survey (SHS, Parker *et al.* 2005), and the INT/WFC Photometric Hα Survey (IPHAS, Drew *et al.* 2005, http://www.iphas.org). WHAM has migrated to the southern hemisphere, having already given us an extraordinarily dynamic image of the Warm Ionized Medium (WIM) across the northern sky. The SHS, on the other hand, is both an end and a beginning, as the last major photographic survey, and the first to take on the entire southern Galactic Plane at seeing-limited spatial resolution. Its major contribution to date has been a uniform planetary nebulae census (Parker *et al.* 2006).

A quite different role for narrowband Hα, now permitted by digital detectors, comes from sampling the stellar photospheric spectrum where either Hα presents as an absorption feature, or pseudo-emission between TiO bands is seen at K/M types. The net result is a monotonic sensitivity to stellar spectral type (or intrinsic colour), that is now being realised by the nearly-complete IPHAS survey of the northern Galactic Plane. As well as taking the sensitivity limit for the detection of emission line stars down to $m_r = 19.5$ (see Witham et al 2008), this ~1 arcsec resolution survey provides a colour-colour diagram ($r' - H\alpha$ versus $r' - i'$) that allows spectral types *and* reddenings to be assigned with high confidence to many millions of normal A–K stars This has many applications, including to 3-D extinction mapping (see e.g. Drew *et al.* 2008, Sale *et al.* 2009).

References

Drew J. E. *et al.* 2005, *MNRAS* 362, 753
Drew J. E. Greimel R., Irwin M. J., & Sale S. E. 2008, *MNRAS* 386, 1761
Haffner L. M. *et al.* 2003, *ApJS* 149, 405
Parker Q. A. *et al.* 2005, *MNRAS* 362, 6899
Parker Q. A. *et al.* 2006, *MNRAS* 373, 79
Sale S. E. *et al.* 2009, *MNRAS* 392, 497
Witham A. R. *et al.* 2008, *MNRAS* 369, 581

[1] *On behalf of the IPHAS Consortium*

Highlights of Astronomy, Volume 15
XXVIIth IAU General Assembly, August 2009
Ian F. Corbett, ed.

© International Astronomical Union 2010
doi:10.1017/S1743921310011579

Billions of stars: the near infrared view of the Plane with UKIDSS and VISTA

Philip W. Lucas[1,*] and David Samuel[1]

[1] Centre for Astrophysics, University of Hertfordshire, College Lane, Hatfield, AL10 9AB, UK

Abstract. The UKIDSS Galactic Plane Survey (GPS) is surveying the northern and equatorial plane in the J, H and K bands. Here we report initial results from searches for new clusters and star formation regions. 248 clusters have been detected by our Bayesian search, of which 127 are new. A visual inspection of the images is also proving successful. A cross match with *Spitzer*-GLIMPSE to find clusters of Young Stellar Objects is being attempted. No new globular clusters are detected except for two likely candidates already detected by Mercer *et al.* in GLIMPSE.

Keywords. surveys, stars: formation, (Galaxy:) open clusters and associations: general

The UKIDSS GPS is surveying the northern and equatorial plane at Galactic latitudes -5 deg$< b <$ 5 deg, in the J, H and K bands, see Lucas *et al.* (2008); Lawrence *et al.* (2007). Modal depth is K $\approx$ 18.0 in uncrowded fields and typical spatial resolution is 0.8 arcsec. Source confusion reduces these depths slightly in the midplane in the first Galactic quadrant, leading to a "needle in a haystack" problem when searching for clusters.

We are conducting a search for new clusters using 3 methods: (i) a Bayesian search for overdensities in the source catalogue; (ii) visual inspection of the jpeg "quick-look" images during quality control of the GPS; (iii) a cross match with *Spitzer*-GLIMPSE to detect Young Stellar Objects with red K-4.5 μm colours. The main aim of this search is to construct a large sample of pre-main sequence clusters and star formation regions (SFRs) in order to permit a statistical investigation of the star formation process. Recent successful searches for new clusters include those by (citeBica03, Mercer *et al.* (2005) and Froebrich, Scholz & Raftery (2007).

The Bayesian algorithm is similar to that of Mercer *et al.* The first scan includes only "extended" sources. This detects clusters and SFRs because nebulae and close stellar pairs and can appear as extended sources in the catalogue. 248 clusters have been detected so far in UKIDSS DR4, of which 127 are new. A large fraction are SFRs. The visual search is proving similarly successful, though there is overlap with the Bayesian list. We also detect many new SFRs with no associated cluster. Success in the GPS-GLIMPSE cross match must await efforts to improve the completeness of the catalogues in nebulous SFRs.

References

Bica, E., Dutra, C. M., Soares, J., & Barbuy, B. 2003, *A&A* 404, 223
Froebrich, D., Scholz, A., & Raftery, C. L. *MNRAS* 374, 399
Lawrence, A. *et al.* 2007, *MNRAS* 379, 1599
Lucas, P. W. *et al.* 2008, *MNRAS* 391, 136
Mercer, E. P., Clemens D.P. *et al.* 2005, *ApJ* 635, 560

*We thank the staff of UKIRT, the Cambridge Astronomical Survey Unit and Wide Field Astronomy Unit at Edinburgh University for making this project possible.

Highlights of Astronomy, Volume 15
XXVIIth IAU General Assembly, August 2009
Ian F. Corbett, ed.

© International Astronomical Union 2010
doi:10.1017/S1743921310011580

ATLASGAL, the APEX Telescope Large Area Survey of the Galaxy

F. Schuller[1], K. M. Menten[1], F. Wyrowski[1], H. Beuther[2], S. Bontemps[3], L. Bronfman[4], Y. Contreras[4], T. Henning[2], F. Motte[5], P. Schilke[6], M. Walmsley[7] and A. Zavagno[8]

[1]MPIfR, Bonn, Germany; email: schuller@mpifr-bonn.mpg.de

[2]MPIA, Heidelberg, Germany; [3]LAB, Bordeaux, France; [4]Univ. de Chile, Santiago, Chile

[5]CEA/DSM/IRFU/SAp, Saclay, France; [6]Universität zu Köln, Germany

[7]Osservatorio Astrofisico di Arcetri, Italy; [8]LAM, Marseille, France

Abstract. Submillimeter continuum emission traces high molecular column densities and, thus, dense cloud regions in which new stars are forming. Surveys of the Galactic plane in such emission have the potential of delivering an unbiased view of high-mass star formation throughout the Milky Way. Here we present the scope, current status and first results of ATLASGAL, an ongoing survey of the Galactic plane using the Large APEX Bolometer Camera (LABOCA) on the Atacama Pathfinder Experiment (APEX) telescope at the Chajnantor plateau in Chile. Aimed at mapping 360 square degrees at 870 μm, with a uniform sensitivity of 50 mJy/beam, this survey will provide the first unbiased sample of cold dusty clumps in the Galaxy at submillimeter wavelengths. These will be targets for molecular line follow-up observations and high resolution studies with ALMA and the EVLA.

Keywords. Surveys, submillimeter, stars: formation, ISM: structure, Galaxy: disk

---

The ATLASGAL project aims at mapping the cold dust content of the inner Galactic disk, in a uniform, systematic way. Observations are performed using the Large APEX Bolometer Camera (LABOCA), installed at the APEX 12 m telescope ((Güsten *et al.* 2006)) on a 5100 m high site in Chile. LABOCA is composed of 295 bolometers operating at 870 μm (Siringo *et al.* 2009). Our goals are to cover $\pm 60°$ in Galactic longitude, over $\pm 1.5°$ in latitude, to an r.m.s. of 50 mJy/beam. For typical cold dusty clumps, at 1–10 kpc distance, this corresponds to 5-σ detection limits of ~1 to a few 100 $M_\odot$.

Based on the first 95 deg^2 of data acquired in 2007, a preliminary catalogue of over 6000 compact sources brighter than 0.25 Jy/beam could be extracted; about 30% of them can be associated with infrared sources from the MSX catalogue (Price *et al.* 2001), but most sources have no bright infrared counterparts. They could correspond to the earliest stages of (massive) star formation.

The observations should be completed by the end of 2009; the raw data will become public one year after each observing period. Advanced data products (calibrated maps, source catalogues) will also be made available upon completion. A detailed description of the survey and early results are presented in Schuller *et al.* (2009).

References

Güsten, R., Nyman, L. Å., Schilke, P., Menten, K., Cesarsky, C., & Booth, R. 2006, *A&A* 454, L13

Price, S. D., Egan, M. P., Carey, S. J, Mizuno, D. R., & Kuchar, T. A. 2001, *AJ* 121, 2819

Schuller, F., Menten, K. M., Contreras, Y., *et al.* 2009, *A&A* in press

Siringo, G., Kreysa, E., Kovács, A., *et al.* 2009, *A&A* 497, 945

Highlights of Astronomy, Volume 15
XXVIIth IAU General Assembly, August 2009
Ian F. Corbett, ed.

CORNISH:
A 5 GHz VLA survey of the Galactic plane

Cormac R. Purcell[1], Melvin G. Hoare[2] and the CORNISH team

[1] Jodrell Bank Centre for Astrophysics, University of Manchester, Manchester M13 9PL UK.
email: `cormac.purcell@manchester.ac.uk`

[2] School of Physics & Astronomy, The University of Leeds Leeds, LS2 9JT, UK.

Abstract. The CORNISH (Co-Ordinated Radio 'N' Infrared Survey for High-mass star formation) project is the radio continuum part of a series of multi-wavelength surveys of the Galactic Plane that focus on the northern GLIMPSE-I region ($10° < l < 65°$, $|b| < 1°$) observed by the SPITZER satellite in the mid-infrared (Churchwell *et al.* 2009). CORNISH has delivered a complementary 5 GHz arcsecond resolution, radio-continuum survey to address key questions in high-mass star formation as well as many other areas of astrophysics.

Keywords. radio continuum: stars, stars: formation, surveys

The CORNISH survey was conducted in 2006 and 2007 with the Very Large Array in B and BnA configurations, yielding a uniform (< 10 percent variation), high-resolution ($\sim 1.5''$) map of the 110 square degree target area. With a root-mean-squared noise level of ~ 0.4 mJy/beam the survey is sufficiently sensitive to detect Ultra-Compact HII (UCHII) regions produced by B0 stars or earlier right across the Galaxy. In addition, the survey has furnished samples of a wide range of radio sources, including planetary nebulae, ionised winds from evolved massive stars, non-thermal emission from active stars, high energy sources, active galactic nuclei and radio galaxies.

Over 4000 radio sources have been detected above 7-σ, one quarter of which are unresolved and have a flat distribution in Galactic latitude. The distribution of sources with angular sizes greater than $2''$ peaks in the Galactic mid plane and exhibits infrared colours consistent with UCHII regions, planetary nebulae and evolved stars. We are in the process of identifying these sources via their GLIMPSE (Churchwell *et al.* 2009), MIPSGAL (Carey *et al.* 2009) and UKIDSS (Lawrence *et al.* 2007) colours, and via associations with other Galactic plane catalogues e.g., IPHAS (Drew *et al.* 2005) and the BU-FCRAO Galactic Ring Survey (Jackson *et al.* 2006). Pipeline reduced image data are now available to the community via the project website (www.ast.leeds.ac.uk/Cornish/) and a high-reliability catalogue is in preparation.

References

Churchwell, E. *et al.* 2009, PASP, 121, 213
Carey, S. J. *et al.* 2009, PASP, 121, 76
Lawrence, A., Warren, S. J., Almaini, O., *et al.* 2007, MNRAS, 379, 1599
Jackson, J. M., Rathborne, J. M., Shah, R. Y., *et al.* 2006, ApJ, 163, 145
Drew, J. E., Greimel, R., Irwin, M. J., *et al.* 2005, MNRAS, 362, 753

Highlights of Astronomy, Volume 15
XXVIIth IAU General Assembly, August 2009
Ian F. Corbett, ed.

Dust and the art of Galactic map making

**Douglas J. Marshall[1], Gilles Joncas[1], Anthony P. Jones[2],
Annie C. Robin[3], Céline Reylé[3] and Mathias Schultheis[3]**

[1]Département de physique, de génie physique et d'optique et Centre de recherche
en astrophysique du Québec, Université Laval, Québec, QC, G1V 0A6, Canada
email: douglas.marshall.1@ulaval.ca

[2]Institut d'Astrophysique Spatiale, bâtiment 121, Université Paris-XI, Orsay, 91405, France

[3]Observatoire de Besançon, Institut UTINAM, Université de Franche-Comté,
BP 1615, 25010 Besançon Cedex, France

Abstract. A Galactic model of stellar population synthesis is used along with a genetic algorithm to reconstruct the three dimensional dust distribution in the Milky Way. We have applied this technique towards over 1500 IRDC cloud candidates, for which we recovered distances and masses for 1259 of them. Aside from giving us the distance to the dust, the three dimensional extinction map also provides us with a temperature independent measure of its density. This new method is independent of any kinematical information, thus providing a new way to obtain information on the Galactic distribution of the ISM. It is a good complement to existing measures which are solely based on molecular gas kinematics as both methods are completely independent and both are affected by different systematics. It will be able to provide valuable distance information for use in the analysis and interpretation of far-infrared and sub-millimetre observations by Herschel and Planck. In the future it could be used with deeper stellar observations or observations at longer wavelengths in order to probe even higher density clouds and to even larger distances.

Keywords. ISM: clouds - dust, extinction - ISM: structure - Galaxy: structure

---

A three dimensional map of the dust distribution in the Milky Way has a number of uses. Apart from providing us with an alternate view on the structure of our home Galaxy, it can be put to many uses in order to gain a better understanding of the physical conditions in the ISM as well as helping us to understand the evolution of our Galaxy. In order to produce such a map, the Besançon model of Robin *et al.* (2003) is used along with a genetic algorithm as implemented by Charbonneau (1995). We have applied this new three dimensional extinction technique towards over 1500 IRDC cloud candidates from Simon *et al.* (2006a), for which we recovered distances and masses for 1259 of them, including over 1000 previously unmeasured clouds. The spatial distribution of the clouds is found to be concentrated in the molecular ring and along the spiral arms. The resulting three dimensional dust distribution can be used to help separate the dust emission signal that will be observed by Herschel and Planck into discrete components along the line of sight. In the future the 3d extinction method will be used with deeper stellar observations or observations at longer wavelengths in order to probe even higher density clouds and to even larger distances.

References

Charbonneau, P. 1995, ApJS, 101, 309
Robin, A. C., Reylé, C., Derrière, S., & Picaud, S. 2003, A&A, 409, 523
Simon, R., Jackson, J. M., Rathborne, J. M., & Chambers, E. T. 2006, ApJ, 639, 227

Highlights of Astronomy, Volume 15
XXVIIth IAU General Assembly, August 2009
Ian F. Corbett, ed.
© International Astronomical Union 2010
doi:10.1017/S1743921310011610

Exploring Extinction and Structure in the Milky Way Disk With 2MASS and *Spitzer*

G. Zasowski, S. R. Majewski, D. L. Nidever and R. Indebetouw

Department of Astronomy, University of Virginia
Box 400325, Charlottesville, VA, 22904 USA
email: **gailis@virginia.edu**

Abstract. We present new maps of the distribution of both dust and stars across the Galactic disk, based largely on an improved analysis of 2MASS and Spitzer-IRAC data. The infrared extinction law is rederived throughout the disk and we found strong longitudinal variations in both diffuse and dense environments that we incorporate in our analysis.

Keywords. dust, extinction — Galaxy: disk — Galaxy: structure — infrared: ISM

Many Galactic dust studies to date have found an IR extinction law that is nearly constant and universal in the diffuse interstellar medium (ISM; e.g., Indebetouw *et al.* 2005), but more recent work has demonstrated that in regions of dense ISM, such as dark cores, the MIR A_λ curve becomes shallower, likely due to dust grain growth (e.g., Weingartner & Draine 2001). Thus, we questioned whether the Galactic extinction law would change substantially through the various ISM environments in the disk, beyond the simple, frequently-used "dense"/"diffuse" paradigm.

For this study, we combined photometry from the MIR *Spitzer*/IRAC surveys (Benjamin *et al.* 2003), spanning ~150° of nearly contiguous disk longitude, with the NIR 2MASS Catalogue (Skrutskie *et al.* 2006), to obtain a consistent set of photometric data in 7 bands (1.2–8 μm). Using red clump stars, we have measured the relative extinction law along many lines of sight in the Galactic disk. We find strong, monotonic variations in A_λ/A_{K_s} as a function of galacto-centric angle, symmetric about the centre. This behaviour (a steepening extinction law at larger angles) persists even after the removal of known dense ISM, traced by ^{13}CO emission (GRS; Jackson *et al.* 2006), which suggests a secondary Galactic-scale dust property gradient (Zasowski *et al.* 2009).

We include these extinction law variations in a new technique for stellar reddening estimation, which uses long-baseline N/MIR colours to derive star-by-star extinction values more robustly than NIR-only techniques; in addition, this RJCE method (Rayleigh Jeans Color Excess method; Majewski, Nidever, & Zasowski, *in prep.*) preserves stellar type information to create reliably-cleaned mid-plane colour-magnitude diagrams. The preservation of stellar type and luminosity class information permits 3-D mapping of the stars and the intervening dust without reliance on a static Galactic model.

References

Benjamin, R. A. *et al.* 2003, *PASP*, 115, 953
Indebetouw, R. *et al.* 2005, *ApJ*, 619, 931
Jackson, J. M. *et al.* 2006, *ApJS*, 163, 145
Skrutskie, M. F. *et al.* 2006, *AJ*, 131, 1163
Weingartner, J. C. & Draine, B. T. 2001, *ApJ*, 548, 296
Zasowski, G. *et al.*, 2009, *ApJ, submitted*

Highlights of Astronomy, Volume 15
XXVIIth IAU General Assembly, August 2009
Ian F. Corbett, ed.

The Near-IR Extinction Law

Joseph J. Stead and Melvin G. Hoare

School of Physics and Astronomy, University of Leeds, Leeds, LS2 9JT, UK

Abstract. We show that the power-law slope of the near-IR extinction law is significantly steeper than previously thought. Simulated colour-colour diagrams including a stellar population synthesis, realistic extinction distribution along the line-of-sight and synthesis through the filter profiles are compared to data from the UKIDSS Galactic Plane Survey. The slope of extinction with wavelength is found to be 2.14±0.05 for total visual extinctions up to about 25 magnitudes and for a number of locations.

The typical value found in the literature for the slope of the near-IR extinction law, α, where $A_V \propto \lambda^{-\alpha}$, from photometric studies of stars is 1.7±0.1 (e.g. Draine (2003)). Here we use a new method to determine the slope of the extinction law using the new deep UKIDSS near-IR galactic plane survey data Lucas *et al.* (2008). J-H against H-K colour-colour diagrams for eight $60' \times 6'$ regions along the plane were constructed using only high-quality data. These were then compared to synthesized colour-colour data using the Besançon model of the galactic stellar population Robin *et al.* (2003) and a distribution of extinction along the line of sight based on those derived by Marshall *et al.* (2006).

Instead of adopting the usual straight reddening vector to redden the intrinsic colours of the stars, new curved reddening tracks appropriate for each stellar type were used. These reddening tracks were computed by taking model atmospheres, reddening them with a power-law of a given slope, and convolving with the UKIDSS filter profile. Tracks computed in this rigorous way are curved due to the changing effective wavelength with both spectral type and degree of reddening. These tracks are available at www.ast.leeds.ac.uk/RMS/ReddeningTracks.

To derive the slope of the extinction law synthetic colour-colour diagrams were constructed for a range of α values. A parameter representing the mean separation between the synthetic and observed data was calculated and minimized. Errors on the derived value of α were found from repeating the synthesis many times in a Monte Carlo calculation. The final value derived was $\alpha = 2.14\pm0.05$ and it was found not to vary significantly with galactic longitude between $27\,\mathrm{deg} < l < 100\,\mathrm{deg}$. Analysis using the same method with 2MASS data gave results which agreed within the errors. We believe our slope is larger than the previously derived values due to the choice of filter wavelength used to convert observed colour excess ratios to power-law slopes. Details of this work can be found in Stead & Hoare (2009). In future this will be applied to dereddening individual members of young clusters associated with massive young stellar objects from the RMS survey (see Lumsden *et al.*, this volume).

References

Draine, B. T. 2003, *ARAA*, 41, 241
Lucas, P. W. *et al.* 2008, *MNRAS*, 391, 136
Marshall, D. J., Robin, A. C., Reylé, C., Schultheis, M., & Picaud, S. 2006, *A&A*, 453, 635
Robin, A. C., Reylé, C., Derrière, S., & Picaud, S. 2003, *A&A*, 409, 523
Stead, J. J. & Hoare, M. G. 2009, *MNRAS*, 400, 731

Highlights of Astronomy, Volume 15
XXVIIth IAU General Assembly, August 2009
Ian F. Corbett, ed.

© International Astronomical Union 2010
doi:10.1017/S1743921310011634

High spatial resolution Galactic 3D extinction mapping with IPHAS

S. E. Sale[1], J. E. Drew[2] and the IPHAS Collaboration

[1] Astrophysics Group, Imperial College London, Blackett Laboratory, Prince Consort Road,
London SW7 2AZ, U.K.
email: **s.sale06@imperial.ac.uk**

[2] Centre for Astrophysics Research, STRI, University of Hertfordshire, College Lane Campus,
Hatfield, AL10 9AB, U.K.

Abstract. An algorithm, MEAD, is presented, which can map extinction in three dimensions, with fine distance and angular resolutions. MEAD is then employed when studying the structure of the outer Galaxy. We show that the Galaxy's radial density profile takes the form of a broken exponential, with density dropping off more steeply beyond a Galacto-centric radius of ~13 kpc.

Keywords. surveys, Galaxy: disk, Galaxy: structure, Galaxy: stellar content, ISM: dust, extinction, ISM: structure

---

MEAD, can determine intrinsic $(r' - i')$ colour, extinction, and distance for A0–K4 stars extracted from the IPHAS (**www.iphas.org**) $r'/i'/H\alpha$ photometric database. These data can be binned up to map extinction in three dimensions across the northern Galactic Plane with fine angular (~10 arcmin) and distance (~0.1 kpc) resolution to distances of up to 10 kpc.

Subsequently study the stellar density profile of the outer Galactic disc in the anti-centre direction. We select early A stars and compare observations with simulated photometry. By selecting A stars, we are appraising the properties of a population only ~100 Myrs old. We find the stellar density is well fit to an exponential with length scale of $(3020 \pm 120_{statistical} \pm 50_{R_\odot})$ pc, out to a galacto-centric radius of $R_T = (13.0 \pm 0.5_{statistical} \pm 0.2_{R_\odot})$ kpc. At larger radii the rate of decline appears to increase with the scale length dropping to (1200 ± 300) pc, see Fig.1. This result amounts to a refinement of the conclusions reached in previous studies that the stellar density profile is abruptly truncated.

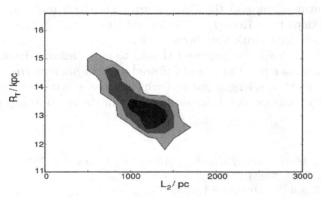

Figure 1. A contour plot showing the derived confidence limits on outer scale length and truncation radius. Contours are at 68%, 95% and 99%.

Highlights of Astronomy, Volume 15
XXVIIth IAU General Assembly, August 2009
Ian F. Corbett, ed.

The Disk of Our Galaxy

Delphine Russeil

LAM, Université de Provence,
F-13388, Marseille, France
email: delphine.russeil@oamp.fr

Abstract. We review our knowledge about the spiral structure of the disk of our Galaxy using tracers of star formation. These tracers reveal a 4-arm picture of the Galaxy.

Keywords. Galaxy: structure, disk

Multi-wavelength views (UV, CO, Hα) of spiral galaxies as M31, M51 and M83 show that the arms are well traced by young objects (OB stars, giant molecular clouds, HII regions). The mid-infrared Spitzer-24 μm image of M31 shows well traced arms, but in the near-infrared the arm contrast is low. For an observer inside the Galaxy, locating the arms can be confused by the numerous small filaments seen in Spitzer-8 μm images of M31 and M83. The HI emission, well known to extend far away the optical disk, is used to probe the external part of the disk. A new aspect comes from the recent 70, 100 and 160 μm Herschel images of M51 in the far-infrared (emission coming from active star-forming complexes) which show that spiral arms are well traced at these wavelengths.

To study the structure of our Galaxy we use the same tracers. Our approach is to use star-forming complexes (composed of the parental molecular cloud associated with the OB stars and their HII regions) as tracer of the spiral arms (Russeil (2003)). This method enables us to reduce the spatial and kinematic spread of objects belonging to the same complex. The multi-wavelength information (HII region recombination lines, parental cloud molecular lines and OB stars spectro-photometric distance) allows a better determination of the systemic velocity and distance. We then confirmed the 4-arm model and showed that the Carina-Sagittarius arm is a major optical arm. An updated view of the 4-arm structure is done by Hou *et al.* (2009) who also show a good connection with the HI arms seen in the outer part by Levine *et al.* (2006). From NIR star counts and FIR emission profiles, giving the arm tangencies, Benjamin (2008) and Drimmel (2000) find that the Scutum-Crux and the Perseus arms are the 2 major arms. But plotting the tangent directions from Hou *et al.* (2009) and Russeil (2003) one can see that these profiles can be compatible with the 4-arm models.

The arm locations should be improved thanks to new distance determination (GAIA stellar parallaxes, maser parallaxes and infrared spectro-photometric methods) and to recent/incoming multi-wavelength surveys which allow a systematic resolution of the distance ambiguity problem and a better definition of the star-forming complexes.

References

Benjamin, R. 2008, *Massive star Formation, ASP Conf. Series*, 387, 735
Drimmel, R. 2000, *A&A*, 358, L13
Hou, L., Han, J., & Shi, W. 2009, *A&A*, 499, 473
Levine, E., Blitz, L., & Heiles, C. 2006, *Science*, 312, 1773
Russeil, D. 2003, *A&A*, 397, 133

Highlights of Astronomy, Volume 15
XXVIIth IAU General Assembly, August 2009
Ian F. Corbett, ed.

Unveiling the Unseen: The Mid-IR Galactic Disk

Ed Churchwell

Dpeartment of Astronomy, University of Wisconsin, 475 N Charter St., Madison, Wisdonsin, USA

Abstract. The Spitzer mid-infrared (MIR) surveys, Galactic Legacy Infrared Mid-Plane Survey Extraordinaire (GLIMPSE) and MIPSGAL have revealed a new view of the disk of the Milky Way. Hallmarks of the Galactic disk at MIR wavelengths with spatial resolution $<2''$ are bubbles/HII regions, infrared dark clouds, young stellar objects (YSOs)/star formation regions, diffuse dust and extended polycyclic aromatic hydrocarbons (PAHs), and more than 100 million publically available archived stars with measured flux densities at 7 wavelengths and positions accurate to $0.1''$. At mid-IR wavelengths, the cool components in the Galaxy are preferentially bright and highlight physical processes that are not obvious at other wavelength regimes.

Three different types of phenomena seen in mid-IR surveys of the plane are discussed.

MIR Bubbles/HII Regions: Examples of both stellar wind-dominated and radiation dominated HII regions are found. They are easily distinguished in the mid-IR by whether the emission peaks at the location of the central star (radiation dominated) or whether there is a central wind-evacuated cavity around the central star(s) (wind-dominated). Both types of HII regions have warm dust within the ionized plasma traced by 24 μm emission. It is shown that dust should be blown out of the nebulae (large grains) or destroyed by sputtering (small grains) on short time scales relative to the age of wind-dominated HII regions. It is postulated that a continuous source of grains is required to understand the observed MIR emission. All HII regions are surrounded by bright PAH dominated 8.0 and 5.8 μm photo-dissociation shells.

YSOs/EGOs: The MIR signature of young stellar objects is a dusty accretion envelope that is particularly bright at 24 μm. The majority of the MIPSGAL 24 μm point sources and extended emission regions are Galactic YSOs, AGB stars, star forming clusters, and HII regions. Unlike the 8.0 μm emission, which primarily traces PAH emission and is extended throughout the inner Galactic plane, 24 μm emission is spotty and is confined to small regions around YSOs, young star clusters, and AGB stars. Extended green objects (EGOs), are so named because they have excess emission at 4.5 μm when designated as green in false color IRAC images. EGOs are believed to be very young protostars whose bipolar outflows have excited bright H_2 lines when crashing into the ambient interstellar medium. They are also strongly associated with methanol masers (both class I and II).

Infrared Dark Clouds (IRDCs): IRDCs represent the most opaque regions of dark molecular clouds. They are opaque at 8 μm ($A_V \gtrsim 100$ mag) and are seen in silhouette against the diffuse Galactic 8 μm background. These are the regions in molecular clouds where star formation is taking place, as demonstrated by the 24 μm point sources that are generally seen here. IRDCs provide information on the initial conditions for star formation.

Highlights of Astronomy, Volume 15
XXVIIth IAU General Assembly, August 2009
Ian F. Corbett, ed.

© International Astronomical Union 2010
doi:10.1017/S174392131001166X

I-GALFA: The Inner-Galaxy ALFA Low-Latitude H I Survey

Bon-Chul Koo[1], Steven J. Gibson[2,3], Ji-hyun Kang[1,2], Kevin A. Douglas[4,5], Geumsook Park[1], Joshua E. G. Peek[5], Eric J. Korpela[5], Carl E. Heiles[5], and Thomas M. Bania[6]

[1] Seoul National Univ., KOREA; [2] Arecibo Obs., USA; [3] Western Kentucky Univ., USA;
[4] Univ. of Exeter, UK; [5] Univ. of California - Berkeley, USA; [6] Boston Univ., USA

Abstract. The I-GALFA survey is mapping HI 21 cm emission in the inner parts of our Milky Way Galaxy using the Arecibo L-band Feed Array (ALFA). Examples of various H I features such as supershells and chimneys are shown.

The I-GALFA survey is mapping all the H I in the inner Galactic disk visible to the Arecibo 305m telescope within 10 degrees of the Galactic plane ($\ell = 32° < l < 77°$ at $b = 0°$). The survey, which will obtain $\sim 1.3 \times 10^6$ independent spectra, uses the 7-beam Arecibo L-Band Feed Array (ALFA) receiver and will be completed in September 2009. The survey data have a resolution of 3.'4, an RMS noise of ~ 0.25 K in 0.184 km s^{-1} channels covering LSR velocities of -750 to $+750$ km s^{-1}. Details of the observing and data reduction can be found in Peek & Heiles (2008). The data will be made publicly available when the calibrated and gridded cubes are completed. Further information on the I-GALFA project may be found at **www.naic.edu/~igalfa**.

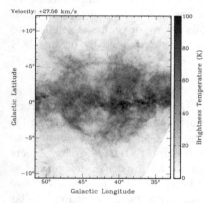
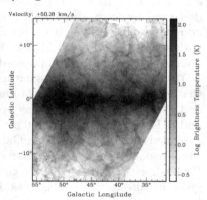

Figure 1: Partial I-GALFA H I line channel maps; more data are being added.
Left: Supershell in the Sagittarius spiral arm.
Right: Disk-halo clouds, chimneys, and worms.

Acknowledgements

We thank all members of the AO staff for the support of the I-GALFA survey. B.-C. K. is supported by the Korean Research Foundation under grant KRF-2008-313-0409. K. D. was supported by a Marie Curie fellowship. The Arecibo Observatory is part of the National Astronomy and Ionosphere Center, which is operated by Cornell University under a cooperative agreement with the U.S. National Science Foundation.

References

Peek, J. E. G. & Heiles, C. 2008, *astro-ph* arXiv:0810.1283v1

Highlights of Astronomy, Volume 15
XXVIIth IAU General Assembly, August 2009
Ian F. Corbett, ed.

© International Astronomical Union 2010
doi:10.1017/S1743921310011671

Abundance structure and chemical evolution of the Galactic disc

Thomas Bensby[1] and Sofia Feltzing[2]

[1] European Southern Observatory, Alonso de Cordova 3107, Vitacura, Santiago, Chile
email: **tbensby@eso.org**
[2] Lund Observatory, Box 43, SE-221 00 Lund, Sweden
email: **sofia@astro.lu.se**

Abstract. We have obtained high-resolution, high signal-to-noise spectra for 899 F and G dwarf stars in the Solar neighbourhood. The stars were selected on the basis of their kinematic properties to trace the thin and thick discs, the Hercules stream, and the metal-rich stellar halo. A significant number of stars with kinematic properties 'in between' the thin and thick discs were also observed to investigate in greater detail the dichotomy of the Galactic disc. All stars have been homogeneously analysed, using the exact same methods, atomic data, model atmospheres, etc., and also truly differentially to the Sun. Hence, the sample is likely to be free from internal errors, allowing us to, in a multi-dimensional space consisting of detailed elemental abundances, stellar ages, and the full three-dimensional space velocities, reveal very small differences between the stellar populations.

Keywords. stars: abundances, stars: kinematics, Galaxy: disk, Galaxy: evolution,

---

Compared to our previous studies of the Galactic thin and thick discs (Bensby *et al.* 2003, 2005) the current stellar sample is larger by a factor of ∼ 8. The figure shows the thin and thick disc abundance trends based on kinematical selection criteria only. The red full line in each plot is the running median from the thick disc stars, and the dashed blue line the running median from the thin disc stars. It is clear that there is separation between the two discs up to at least solar metallicities, signaling the dichotomy of the Galactic stellar disc, and that the two discs have had very different chemical histories. First results, based on this enlarged sample, regarding the origin of the Hercules stream and the metal-rich limit of the thick disc were published in Bensby *et al.* (2007a,b). The full data set will be published in the fourth quarter of 2009 where we will investigate in great detail the abundance structure and chemical evolution of the Galactic stellar disc.

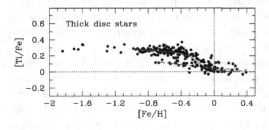 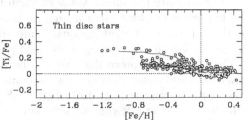

References

Bensby, T., Feltzing, S., & Lundström, I. 2003, *A&A*, 410, 527
Bensby, T., Feltzing, S., Lundström, I., & Ilyin, I. 2005, *A&A*, 415, 155
Bensby, T., Oey, M. S., Feltzing, S., & Gustafsson, B. 2007a, *ApJ*, 655, L89
Bensby, T., Zenn, A. R., Oey, M. S., & Feltzing, S. 2007b, *ApJ*, 663, L13

Highlights of Astronomy, Volume 15
XXVIIth IAU General Assembly, August 2009
Ian F. Corbett, ed.

Abundance gradients: tracing the chemical properties of the disk

Roberto D.D. Costa[1] and Walter J. Maciel[1]

[1]Depto. de Astronomia, IAG/USP, Rua do Matão 1226, 05508-090, São Paulo/SP, Brazil
email: roberto@astro.iag.usp.br, maciel@astro.iag.usp.br

Abstract. Abundance gradients are key parameters to constrain the chemical evolution of the galactic disk. In this review recent determinations for the radial gradient are described, including its slope as derived from different objects such as planetary nebulae, HII regions, cepheids, or B stars, and for different elements. Inner and outer limits for the radial gradient, as well as its time evolution, both related to the chemical evolution of the Galaxy, are also described. The possible existence of azimuthal and vertical gradients is also discussed.

Keywords. Galaxy: abundances, Galaxy: disk, Galaxy: evolution

1. Introduction

The existence of abundance gradients in the Milky Way has been known for a long time now. It can be traced using different objects such as HII regions, planetary nebulae, cepheids, B-stars or open clusters. Another important point that appears in both observational data and models is the temporal variation of the gradient, in the sense that younger objects display a flatter gradient. Results are less clear, however, when vertical or azimuthal abundance variations are investigated.

2. Summary of the results

Starting with early studies concerning the solar neighborhood, the radial gradient is now well established. Recent works such as Rudolph *et al.* (2006) or Costa & Maciel (2006) review this gradient, and despite some discrepancies concerning the true slope for each tracer, the results are qualitatively in agreement. They all show the gradient keeping the same slope from the inner edge of the disk up to nearly 10 kpc. Beyond this point results are less clear. Many of them indicate a flattening, which is consistent with the hypothesis of lower star formation rates at large galactocentric distances compared to those in more central regions. Time variation in the radial gradient appears when progenitors with different ages are used to trace it (for details, see Maciel *et al.* 2006).

There is a well known decrease in abundances when comparing thin disk, thick disk, inner halo and outer halo populations, and evidences of a *stricto sensu* vertical gradient now begin to appear (e.g., from the SDSS data). On the other hand, although some results indicate that different galactic longitudes may present distinct average abundances, currently it is not possible to define an azimuthal gradient throughout the galactic disk.

References

Costa, R. D. D. & Maciel, W. J. 2006, *Proc. IAU Symp. 234*, 243
Maciel, W. J., Lago, L. G., & Costa, R. D. D. 2006 *A&A* 453, 587
Rudolph, A. L., Fitch, M., Bell, G. R. *et al.* 2006, *ApJS* 162, 343

Highlights of Astronomy, Volume 15
XXVIIth IAU General Assembly, August 2009
Ian F. Corbett, ed.

© International Astronomical Union 2010
doi:10.1017/S1743921310011695

Physics and Structure of the Galactic disc(s)

Ralph A. Schönrich

Max Planck Institute for Astrophysics,
Garching, D-85741, Garching, Germany
email: rasch@mpa-garching.mpg.de

Abstract. The combination of classical models of chemical evolution with radial mixing and dynamics of the Galactic disc has opened alternative methods of understanding the detailed structure of the solar neighbourhood. I will show how radial mixing alters the views on chemical evolution as demonstrated by Schönrich & Binney (2009). I will explain how the model gives rise to a very natural division of the Galactic disc into a thick and a thin component, which can be examined in the light of detailed observational studies combining information on kinematics, chemistry and stellar ages.

Keywords. Galaxy: structure, Galaxy: evolution, Galaxy: abundances, Galaxy: kinematics and dynamics

---

Since the result of Sellwood & Binney (2002) it was known that from a theoretical point of view radial mixing must play a crucial role in the evolution of galactic discs. Schönrich & Binney (2009a) showed that by including radial mixing a chemical evolution model with only two free parameters gives a perfect fit to the full metallicity distribution provided by the Geneva-Copenhagen Survey (Nordström *et al.* 2004, Holmberg *et al.* 2007). In addition, no such fit is possible in a physically sound model without radial mixing. Without any further adaption the model provides - by the intrusion of stars from different galacto-centric radii into the solar neighbourhood - a natural explanation for the difference between theoretical predictions and observations of the time dependence of stellar velocity dispersions and especially for all the detailed links between chemistry and kinematics that were observed by Haywood (2008). It is predicted (cf. Schönrich & Binney 2009b) that no kinematic selection on a single disc component can be clean. The model further naturally gives rise to a thick disc created by mixing alone and not requiring any hiatus in star formation, matching all its properties to the observed precision. These are especially the abundance "gap" between high and low alpha element enrichment, the vertical density distribution as measured from the SDSS data by Jurić *et al.* (2008) and the detailed properties in age, metallicity and velocity space of the thin and thick disc under a kinematic selection as performed by Bensby *et al.* (2005). On these grounds there is no more convincing evidence for a hiatus of star formation in the past of the Galactic disc.

References

Bensby, T., Feltzing, S., Lundström, I., & Ilyin, I. 2005 *A&A*, 433, 185
Haywood, M. 2008, *MNRAS*, 388, 1175
Holmberg, J., Nordström, B., & Andersen, J. 2007 *A&A*, 475, 519
Jurić, M., Ivezić, Z., Brooks, A., *et al.* 2008, *ApJ*, 673, 864
Nordström, B., Mayor, M., Andersen, J., *et al.* 2004 *A&A*, 418, 989
Schönrich, R. & Binney, J 2009, *MNRAS*, 396, 203
Schönrich, R. & Binney, J 2009, *arXiv*, 0907.1899
Sellwood, J. A. & Binney, J. 2002, *MNRAS*, 336, 785

Highlights of Astronomy, Volume 15
XXVIIth IAU General Assembly, August 2009
Ian F. Corbett, ed.

Evolution of abundance gradients for galactic plane PNe

A.F. Kholtygin[1], Yu.V. Milanova
and V.V. Akimkin

Department of Astronomy, Saint-Petersburg State University, Saint-Petersburg, Russia
[1]email: afkholtygin@gmail.com

Abstract. The modern observations of planetary nebulae (PNe) are used to recalculate the element abundances for more than 150 PNe of the Milky Way and Magellanic Clouds. Basing on our data, we study the evolution of the abundance gradients for PNe in the thin disk and in the bulge.

A homogeneous set of the nebular parameters and element abundances for more than 300 galactic PNe (Milanova & Kholtygin 2009) was compiled using our calculations and literature data. These data were used to create a new catalogue of planetary nebulae (www.astro.spbu.ru/staff/afk/GalChemEvol) which contains the parameters of Galactic and Magellanic Cloud PNe. To study the evolution of the abundances in the Galactic disk we normalized the different distances found in literature catalogues using the well known distance to the Galactic center (Nikiforov 2004). The comparison of our data (Akimkin *et al.* 2009, in prep.) with those given by Stanghellini *et al.* 2008 is shown in Fig. 1.

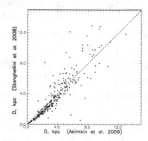

Figure 1. Our distances compared with those from Stanghellini *et al.* (2008)

We have found the regular increasing of the Ne and Cl abundance on the disk PNe with the age of the progenitor stars, but did not find clear evidence for the evolutionary changes of the radial abundances gradients for O and S. We compared the abundances for the different galactic subsystems and detected that abundance patterns are similar for bulge PNe and Peimbert's type II nebulae, for which progenitor stars belong to the population of the thin disk with the ages greater than 4-6 Gyrs. We suppose that the intense star formation both in the bulge and in the thin disk continued at least up to 4-6 Gyr ago.

References

Milanova, Yu. V. & Kholtygin A. F. 2009, Astronomy Letters, 35, 518
Nikiforov, I. 2004, ASP Conf. Series, 316
Stanghellini, L. Shaw, R. A., & Villaver, E. 2008, *ApJ* 689, 194

Highlights of Astronomy, Volume 15
XXVIIth IAU General Assembly, August 2009
Ian F. Corbett, ed.

Star formation simulations: caveats

Simon P. Goodwin

Department of Physics & Astronomy, University of Sheffield, Sheffield, S3 7RH, UK
email: s.goodwin@sheffield.ac.uk

Abstract. Star formation is such a huge problem, covering such a large range of physical scales and involving so many physical processes, that the results of simulations should always be taken with care.

Keywords. Stars:formation

Star formation covers such a huge range of physical scales and a wide variety of physical processes – all of which are important – that it is probably the single most difficult problem to address in astrophysics.

Star formation is a complex and chaotic process involving, to varying and uncertain degrees of importance, turbulence, magnetic fields (in particular non-ideal MHD), and feedback (mechanical and radiative, positive and negative). In addition, chemistry and radiative transfer play important roles in determining the thermodynamic properties of the gas (which in turn controls the collapse and fragmentation of the gas into multiple systems and clusters).

The range of physical scales required to understand star formation ranges from kpc or galactic scales (on which feedback-driven turbulence dominates?), to sub-au or Jupiter-mass scales (where gravity and magnetic fields are most important?). No simulation now, or in the foreseeable future can possibly cover the realistic formation of massive GMC complexes in galaxies and also the collapse and fragmentation on sub-au scales to stellar densities.

We have the further problem that what simulations tend to produce is not what observers actually observe. A particular problem is that observations do not directly observe the bulk of the gas in H_2 or He, but tracers such as dust or particular molecules (whose abundance varies with local conditions).

For good recent reviews of star formation theory I suggest McKee & Ostriker (2007), Klessen et al. (2009), as well as many chapters in 'Protostars & Planets V' (2007).

Current simulations can only probe a tiny fraction of this parameter space and so their results should always be taken with care. If simulations do not fit observations it does not mean that they are wrong, but if they do fit observations they could still be wrong.

References

Klessen, R. S., Krumholz, M. R., & Heitsch, F. 2009, arXiv:0906.4452
McKee, C. F. & Ostriker, E. C., 2007, ARAA, 45, 565

Highlights of Astronomy, Volume 15
XXVIIth IAU General Assembly, August 2009
Ian F. Corbett, ed.
© International Astronomical Union 2010
doi:10.1017/S1743921310011725

Deducing the Milky Way's Massive Cluster Population

M. M. Hanson[1], B. Popescu[1], S. S. Larsen[2] and V. D. Ivanov[3]

[1] Department of Physics, University of Cincinnati, Cincinnati, OH 45221, USA
[2] Astronomical Institute, University of Utrecht, Princetonplein 5, 3584 CC Utrecht, NL
[3] European Southern Observatory, Ave. Alonso de Cordova 3107, Santiago 19001, Chile

Abstract. Recent near-infrared surveys of the galactic plane have been used to identify new massive cluster candidates. Follow up study indicates about half are not true, gravitationally-bound clusters. These false positives are created by high density fields of unassociated stars, often due to a sight-line of reduced extinction. What is not so easy to estimate is the number of false negatives, clusters which exist but are not currently being detected by our surveys. In order to derive critical characteristics of the Milky Way's massive cluster population, such as cluster mass function and cluster lifetimes, one must be able to estimate the characteristics of these false negatives. Our group has taken on the daunting task of attempting such an estimate by first creating the stellar cluster imaging simulation program, MASSCLEAN. I will present our preliminary models and methods for deriving the biases of current searches.

Keywords. Galaxy: stellar content — open clusters and associations: general

What would we see if we could view our whole galaxy? With regards to the stellar clusters, van den Bergh & Lafontaine (1984) answered this by extrapolating the locally derived luminosity function of star clusters to our whole galaxy. This resulted in an estimate of 100 clusters with m $\sim 10^4$ M$_\odot$ and several clusters of m $\sim 10^5$ M$_\odot$. These early predictions have held up well with time. Using modern, infrared surveys, Ivanov *et al.* (2009) extended the observed local density of massive clusters to predict $\geqslant 81 \pm 21$ such clusters ($> 10^4$M$_\odot$) should exist in the Milky Way. Where are they all? Presently one cluster has a mass approaching 10^5 M$_\odot$ (Westerlund 1) and perhaps 10 additional clusters are known with mass $\sim 10^4$ M$_\odot$. Is this proof the cluster mass function in the Milky Way steepens or is truncated at the high mass end?

We believe even infrared surveys are plagued with complex biases which limit our ability to estimate what clusters might be missed. Estimating these false negatives is critical to deriving the mass function in the Milky Way. We have created a sophisticated stellar cluster simulation program, MASSCLEAN (Popescu & Hanson 2009) which when combined with current surveys, will be used to derive the biases of current search methods to differing cluster properties (age, mass, distance, extinction, core density, etc.) and as a function of location in the galaxy. MASSCLEAN will also be used to design more efficient stellar cluster searches.

This material was based upon work supported by the National Science Foundation under Grant No. 0607497 to the University of Cincinnati.

References

Ivanov, V. D., Messineo, M., Zhu, *et al.* 2009, *A&A*, submitted
Popescu, B. & Hanson, M. M. 2009, *AJ*, submitted
van den Bergh & Lafontaine 1984, *AJ*, 89, 1822

Highlights of Astronomy, Volume 15
XXVIIth IAU General Assembly, August 2009
Ian F. Corbett, ed.

© International Astronomical Union 2010
doi:10.1017/S1743921310011737

The Star Formation Activity of Molecular Clouds in the Galactic Plane

Joseph C. Mottram[1] and Christopher M. Brunt[1]

[1] School of Physics & Astronomy, University of Exeter,
Exeter, Devon, EX4 4QL, UK
email: joe@astro.ex.ac.uk, brunt@astro.ex.ac.uk

Abstract. It is well known that star formation takes place within molecular clouds. However, current observational surveys and investigations usually start by selecting a sample of sites where star formation is ongoing, thus biasing against those clouds and regions with little or no current formation activity. In an attempt to identify samples of clouds both with and without star formation, and to investigate their properties, we present an automated method for associating clouds identified in new 3D CO data with far-IR/sub-mm sources. Given the large number of surveys of the galactic plane currently planned, ongoing or being released, the methods used here may prove instructive in understanding how, where and under what conditions star formation takes place throughout our Galaxy. In addition, this will allow exploration of the properties of star forming regions on a range of spatial scales.

Keywords. ISM: Clouds, Stars: Formation, Catalogs, Surveys

We present an overview of new ^{12}CO (J=1-0) and ^{13}CO (J=1-0) mapping observations taken with SEQUOIA on the FCRAO between l of 55° and 102°, 141°and 195° with | b | $\leqslant$ 1°. These data have a spatial resolution of 45″ and a spectral resolution of ~ 0.15 km s^{-1}. Stray radiation results in a $\sim$ 30 % overestimate of the brightness temperatures, so we have therefore undertaken the first application of stay radiation correction to large scale galactic molecular line surveys. This results in these data being calibrated to within 10 %. We intend to release these data, along with the routines used to perform this correction, to the community in the near future. Those interested in obtaining the data sooner are encouraged to contact the authors.

We identify and catalogue clouds and clumps in the CO data using a threshold-based method, Brunt et al. (2003). We then assoicate sources to this catalogue with expectations that the association is a chance allignment below some threshold, as in Kerton & Brunt (2003). When associating clouds in a test region with BLAST (Chapin et al. (2008)) and RMS sources (Urquart et al. (2008)) the results produced agree well with the by-eye velocities assigned by those studies.

This method allows association of large samples of molecular clouds and sub-mm/far-IR data on reasonable timescales, which will be of particular value with SCUBA-2 and Herschel data. It will also allow us to study the star formation properties of clouds and sources on a statistical basis at a range of scales within the heirarchy of GMCs.

References

Brunt, C. M., Kerton, C. R., & Pomerleau, C. 2003, ApJS, 144, 47
Chapin, E. L., Ade, P. A. R., Bock, J. J., et al. 2008, ApJ, 681, 428
Kerton, C. R. & Brunt, C. M. 2003, A&A, 399, 1083
Urquhart, J. S., Hoare, M. G., Lumsden, S. L., et al. 2008, ASPC, 387, 381

Highlights of Astronomy, Volume 15
XXVIIth IAU General Assembly, August 2009
Ian F. Corbett, ed.

© International Astronomical Union 2010
doi:10.1017/S1743921310011749

A New Comprehensive Catalogue of Infrared Dark Clouds

G. A. Fuller[1] and N. Peretto[1]

[1] Jodrell Bank Centre for Astrophysics, School of Physics & Astronomy, Alan Turing Building,
University of Manchester, Oxford Road, Manchester, M13 9PL
email: G.Fuller@manchester.ac.uk

Abstract. To better characterise infrared dark clouds (IRDCs), and the star formation within them, a comprehensive catalogue of IRDCs has been constructed from the Spitzer GLIMPSE and MIPSGAL archival data. Mosaicing the individual survey blocks together, we have used a new extraction method to identify dark clouds up to 30′ in size, and produce a column density image of each cloud. In total the catalogue contains over 11,000 clouds, defined as connected regions with 8 micron optical depth > 0.35 (corresponding to column densities $> 10^{22}$ cm^{-2}). The extraction algorithm also identifies sub-structures (fragments) within each cloud. These Spitzer dark clouds (SDCs) range in mass from $10 M_\odot$ to $10^4 M_\odot$. About 80% of the SDCs were previously unidentified. Only $\sim 30\%$ of the SDCs are associated with $24 \mu m$ point-like sources, leaving the majority of these clouds with no apparent sign of star formation activity. This new catalogue provides an important new resource for future studies of the initial conditions of star formation in the Galaxy.

Keywords. Star formation, Molecular clouds

Infrared dark clouds seen in absorption against extended emission were first identified by Pérault et al. (1996) using ISO, and subsequently an extensive catalogue was produced from the MSX Galactic Plane survey (Simon et al. 2006). Using the Spitzer GLIMPSE and MIPSGAL data we have recently constructed a new catalogue of over 11,000 dark clouds identified from their $8 \mu m$ absorption using a new extraction algorithm which also identifies and characterises fragments within the clouds (Peretto & Fuller 2009). This catalogue, including a map of each cloud, is available online at www.irdarkclouds.org. The majority of these Spitzer dark clouds (SDCs) were previously unidentified, with only 20% of them appearing in the earlier MSX-based IRDC catalogue. Compared with the MSX clouds, the SDCs are more concentrated towards the Galactic Plane.

Inferring the mass of the clouds from their extinction and adopting a typical distance to a cloud of 4 kpc, the mass distribution of the clouds is found to be a power-law with an index indistinguishable from that for the mass distribution of CO clumps in molecular clouds (e.g. Kramer et al. 1998). However, the mass distribution for the fragments within the clouds has a log-normal form, indicative of a turbulent origin for these sub-structures (Hennebelle & Chabrier 2008).

References

Hennebelle, P. & Chabrier, G. 2008, *ApJ* 684, 395
Kramer, C., Stutzki, J., Rohrig, R., & Corneliussen, U. 1998, *A&A* 329, 249
Pérault, M., Omont, A., Simon, G., et al. 1996, *A&A* 315, 165
Peretto, N. & Fuller, G. A. 2009, *A&A* in press
Simon, R., Rathborne, J. M., Shah R. Y., Jackson, J. M., & Chambers, E. T. 2006, *ApJ* 653, 1325

Highlights of Astronomy, Volume 15
XXVIIth IAU General Assembly, August 2009
Ian F. Corbett, ed.

© International Astronomical Union 2010
doi:10.1017/S1743921310011750

The JCMT Legacy Survey: Mapping the Milky Way in the Submillimetre

Antonio Chrysostomou[1] & the members of the JLS teams

[1] Joint Astronomy Centre, 660 N. A'ohoku Place, Hilo, Hawaii, HI 96720, USA
email: a.chrysostomou@jach.hawaii.edu

Abstract. The JCMT Legacy Survey (JLS) is an ambitious programme of independent surveys to study our Galaxy and universe in the submillimetre ($\lambda = 450 - 850 \, \mu m$) from the summit of Mauna Kea, Hawaii. With its scientific breadth and unique spectral window, it is clear that the JLS will have a significant impact on star formation studies in the near future and beyond. Its complementarity with other surveys (e.g. Spitzer, Herschel) will make the JLS a very valuable resource for multi-wavelength studies for low and high-mass star formation across the Milky Way. The JLS is currently in its second year of operation.

The JLS comprises seven projects whose scientific objectives are to understand the formation and evolution of the planets, stars and galaxies in the Universe. Four projects will survey within the Milky Way, one will survey galaxies in the local universe, another will perform a deep cosmological survey, and there is an ambitious survey to map the whole submillimetre sky viewable from Mauna Kea. There are 3 projects specifically aimed at surveying the star formation content in our Galaxy: The Gould Belt Survey (GBS) will map the gas and dust content of all the major regions of low mass star formation within 500 pc of the Sun; the Spectral Legacy Survey (SLS) will obtain a chemical census from a range of known star formation environments; the JCMT Plane Survey (JPS) will obtain a deep (4 mJy at 850 μm) map of the spine of the Galactic plane.

Of these, the GBS and SLS have already begun (using HARP/ACSIS). Observations of the star formation content of the Serpens molecular cloud reveals in some detail how explosive star formation can be. Outflows can be seen breaking out in all directions from the central core (Hogerheijde *et al.* 2009) depositing significant energy and momentum into the natal cloud, a visual demonstration of the importance of feedback in star formation. In the Orion Bar, observations of key chemical species are revealing over and under abundances when compared to chemical models of photo-dissociation regions, forcing a refinement of these models to include processes such as depletion and evaporation from icy grain mantles (van der Wiel *et al.* 2009).

In 2010, we expect to start using SCUBA-2 for the JLS. The GBS will map all the star forming clouds within the Gould Belt as well as determining the magnetic field configuration in a few select clumps and cores. The JPS will initially map the GLIMPSE and FCRAO Outer Galaxy regions, before embarking on a complete, deep map of the Galactic plane (b= $\pm 0.5^{o}$). The JPS will detect a 1 $M_{\odot}$ object at 3 kpc and 40 $M_{\odot}$ objects out to 20 kpc, making the survey sensitive to almost all the massive star formation in the Galaxy. The SCUBA-2 "All-Sky" Survey (SASSy) will provide a shallower but broader survey of the Galactic Plane (b = $\pm 5^{o}$) down to 30 mJy at 850 μm before moving onto an ambitious phase of mapping the whole submillimetre sky viewable from Mauna Kea.

Highlights of Astronomy, Volume 15
XXVIIth IAU General Assembly, August 2009
Ian F. Corbett, ed.

Massive Star Formation Throughout the Galactic Disk

Stan Kurtz

Center for Radioastronomy and Astrophysics, National Autonomous University of Mexico,
Morelia, Michoacan 58089, Mexico
email: s.kurtz@crya.unam.mx

Abstract. High-mass star formation is manifestly a phenomenon of the Galactic Plane. The process begins with pre-stellar cores, evolves to proto-stellar objects, and culminates in massive main-sequence stars. Because massive young stellar objects are deeply embedded, the radio, sub-mm, and far/mid-infrared spectral windows are the most revealing. Galactic plane surveys at these wavelengths trace hot and cold molecular gas, interstellar masers, warm dust, and ionized gas that are present during star formation.

Keywords. surveys, stars: early type, stars: formation

Massive star formation is manifestly a phenomenon of the Galactic plane. All tracers of massive star formation are confined to the Galactic plane and are closely associated with the spiral arms.

The formation process of a high-mass star comprises four stages: 1) optically thin free-fall collapse within a molecular cloud; 2) a hydrostatic core accreting onto a protostellar object; 3) optically thick hot molecular cores; and 4) entry onto the main-sequence, accompanied by the formation of an ultracompact HII region. These four stages are characterised by: 1) cold, dusty molecular clouds; 2) warm, dusty molecular clumps; 3) hot, dusty molecular cores; and 4) compact, ionized gas.

Various Galactic Plane Surveys are well-suited to trace these distinct evolutionary states. Infrared dark clouds were discovered in *MSX* images and *SPITZER*, *AKARI* and *HERSCHEL* are well-suited to detect warm dust. The ATLASGAL and JCMT Legacy surveys cover substantial portions of the Galactic plane in the sub-millimetre regime, that probes cool dust emission. The Multibeam Methanol Survey traces maser emission sites, while CORNISH will be sensitive to all ultracompact and many hypercompact HII regions within the survey area.

Galactic Plane Surveys contribute to high-mass star formation studies in several ways. The most obvious is to address questions of a statistical nature. The more accurate estimates of the numbers of hot molecular cores, ultracompact HII regions, and other objects will allow improved lifetime estimates. Surveys can also contribute to more detailed studies. For example, hypercompact HII regions will not be resolved by any of the infrared, sub-mm, or centimetre-wave surveys, but just having the flux densities over this large a range will be useful to constrain models of their density distributions. However, the interpretation of survey data will not always be clear. In fact, we will need to learn some new astrophysics to properly interpret some survey data. An example is the Methanol Multibeam Survey which will locate star formation sites via methanol maser emission. Although these masers are thought to trace early stages of massive star formation, it is not well-known precisely what stage they trace, or for how long they are present. These and other questions must be addressed before we will be able to fully understand what the surveys are telling us.

Highlights of Astronomy, Volume 15
XXVIIth IAU General Assembly, August 2009
Ian F. Corbett, ed.

The Galactic star formation rate as seen by the *Spitzer Space Telescope*

Thomas P. Robitaille[1] and Barbara A. Whitney[2]

[1] Harvard-Smithsonian CfA, 60 Garden Street, Cambridge, MA, 02138, USA
[2] Space Science Institute, 4750 Walnut St. Suite 205, Boulder, CO 80301, USA

Abstract. We present preliminary results of a study to determine the star formation rate of the Galaxy using a census of young stellar objects (YSOs) in the *Spitzer*/GLIMPSE and MIPSGAL surveys, which cover nearly 300 square degrees of the Galactic mid-plane. We find a value of $1.7\,M_\odot/\text{yr}$, consistent with independent estimates.

Most existing estimates of the star formation rate of our Galaxy rely on tracers of the massive star population, such as the total ionising flux in HII regions (Smith *et al.* 1978; $5\,M_\odot/\text{yr}$), γ-rays from radioactive ^{26}Al, which traces nucleosynthesis in massive stars (Diehl *et al.* 2006; $4\,M_\odot/\text{yr}$), or free-free emission (Murray & Rahman 2009; $1.3\,M_\odot/\text{yr}$). The project described here aims to determine the Galactic star formation rate by directly counting YSOs detected in the *Spitzer*/GLIMPSE survey of the Galactic plane ((Benjamin *et al.* 2003)) and producing a model of Galaxy-wide star formation that reproduces the observed number of YSOs.

A synthetic population model was constructed by randomly sampling the position of young stars in the Galactic plane, weighted by a radially dependent star formation rate (Boissier & Prantzos 1999). The synthetic young stars are then randomly assigned masses from an IMF (Kroupa *et al.* 2001), ages from a constant star forming history, and corresponding colours and magnitudes using model SEDs (Robitaille *et al.* 2006). The magnitudes are scaled by the distance and extinction along the line of sight to the Sun, assuming a double exponential distribution for the dust (Misiriotis *et al.* 2006). Brightness and colour selection criteria matching those of a census of YSOs in the GLIMPSE survey (Robitaille *et al.* 2008) were applied, and the overall number of synthetic YSOs was adjusted until the number of 'selected' synthetic YSOs matched the observations. The resulting Galactic star formation rate is $1.7\,M_\odot/\text{yr}$, in agreement with existing estimates.

Support for this work was provided by NASA through the Spitzer Space Telescope Fellowship Program, through a contract issued by the Jet Propulsion Laboratory, California Institute of Technology under a contract with NASA.

References

Benjamin, R. A., *et al.* 2003, *PASP*, 115, 953
Boissier, S. & Prantzos, N. 1999, *MNRAS*, 307, 857
Diehl, R. *et al.* 2006, *Nature*, 439, 45
Kroupa, P. 2001, *MNRAS*, 322, 231
Misiriotis, A. *et al.* 2006, *A&A*, 459, 113
Murray, N. W. & Rahman, M. 2009, arXiv:0906.1026
Robitaille, T. P. *et al.* 2008, *AJ*, 136, 2413
Robitaille, T. P. *et al.* 2006, *ApJS*, 167, 256
Smith, L. F., Biermann, P., & Mezger, P. G. 1978, *A&A*, 66, 65

Highlights of Astronomy, Volume 15
XXVIIth IAU General Assembly, August 2009
Ian F. Corbett, ed.

© International Astronomical Union 2010
doi:10.1017/S1743921310011786

The Statistics and Galactic Properties of the Methanol Multibeam Survey

J. A. Green[1,*], J. L. Caswell[1], G. A. Fuller[2], A. Avison[2],
S. L. Breen[1,3], K. Brooks[1], M. G. Burton[4], A. Chrysostomou[5],
J. Cox[6], P. J. Diamond[2], S. P. Ellingsen[3], M. D. Gray[2], M. G. Hoare[7],
M. R. W. Masheder[8], N. M. McClure-Griffiths[1], M. Pestalozzi[5,11],
C. Phillips[1], L. Quinn[2], M. A. Thompson[5], M. A. Voronkov[1],
A. Walsh[9], D. Ward-Thompson[6], D. Wong-McSweeney[2], J. A. Yates[10]
and R. J. Cohen[2,*]

[1] Australia Telescope National Facility, CSIRO, PO Box 76, Epping, NSW 2121, Australia,
E-mail:james.green@csiro.au;
[2] Jodrell Bank Centre for Astrophysics, Alan Turing Building, University of Manchester,
Manchester, M13 9PL, UK; [3] School of Mathematics and Physics, University of Tasmania,
Private Bag 37, Hobart, TAS 7001, Australia; [4] School of Physics, University of New South
Wales, Sydney, NSW 2052, Australia; [5] Centre for Astrophysics Research, Science and
Technology Research Institute, University of Hertfordshire, College Lane, Hatfield, AL10 9AB,
UK; [6] Department of Physics and Astronomy, Cardiff University, 5 The Parade, Cardiff, CF24
3YB, UK; [7] School of Physics and Astronomy, University of Leeds, Leeds, LS2 9JT, UK;
[8] Astrophysics Group, Department of Physics, Bristol University, Tyndall Avenue, Bristol, BS8
1TL, UK; [9] School of Maths, Physics and IT, James Cook University, Townsville, QLD 4811,
Australia; [10] University College London, Department of Physics and Astronomy, Gower Street,
London, WC1E 6BT, UK; [11] Göteborgs Universitet Inkvissitutionen för Fysik, Göteborg,
Sweden

Abstract. The methanol multi-beam (MMB) survey has produced the largest and most complete catalogue of Galactic 6.7-GHz methanol masers to date. 6.7-GHz methanol masers are exclusively associated with high-mass star formation, and as such provide invaluable insight into the Galactic distribution and properties of high-mass star formation regions. I present the statistical properties of the MMB catalogue and, through the calculation of kinematic distances, investigate the resolution of distance ambiguities and explore the Galactic distribution.

Keywords. stars: formation, Masers, Surveys

6.7-GHz methanol masers provide an incredible tool to study both the properties of high-mass star formation regions and the structure of our Galaxy. The Methanol Multibeam (MMB) survey has recently completed its southern hemisphere observing with the Parkes Radio Telescope, covering over 60% of the Galactic plane and detecting in excess of 900 sources throughout the Galaxy. Factoring for the completeness of the survey gives a total population estimate of ∼1200 masers, in line with modelling of previous inhomogeneous surveys. The MMB sources have a narrow latitude distribution, peaking in longitude around ±20-30°, and have a flux density distribution peaking at around 1-2 Jy. Analysis of the distribution in longitude-velocity space shows 45 sources associated with the near and far 3-kpc arms. Kinematic distances to the MMB sources have been determined and the Galactic distribution analysed. Preliminary results suggest an overall galactocentric peak at 5-6 kpc with individual peaks at the positions tangential to the spiral arms.

* Deceased 2006 November 1.

Highlights of Astronomy, Volume 15
XXVIIth IAU General Assembly, August 2009
Ian F. Corbett, ed.

© International Astronomical Union 2010
doi:10.1017/S1743921310011798

The Red MSX Source Survey - Massive Star Formation in the Milky Way

Stuart Lumsden[1], Melvin Hoare[1], Ben Davis[1] and the RMS team[2]

[1] School of Physics and Astronomy, University of Leeds, Leeds, UK
email: s.l.lumsden@leeds.ac.uk
[2] http://www.ast.leeds.ac.uk/RMS

Abstract. We present the results of a Galaxy-wide survey for young massive stars still in the process of formation. Our data are consistent with a model in which the stars form through accretion disks with the overall Galactic star formation rate being $3\,M_\odot$ per year.

Keywords. Stars: formation, Stars: luminosity function, Galaxy: stellar content

We have used mid-infrared MSX and near-infrared 2MASS data to define a sample of massive young stellar objects (MYSOs) and compact HII regions (Lumsden *et al.* 2002). A programme of follow-up study has resulted in a final catalogue of about 500 of each type of object. We have successfully modelled the numbers we see in this population as a function of luminosity using an accretion based prescription for massive star formation (McKee & Tan 2003), and find good agreement between the predicted and observed numbers of MYSOs and compact HII regions given the known MSX sensitivity. The model

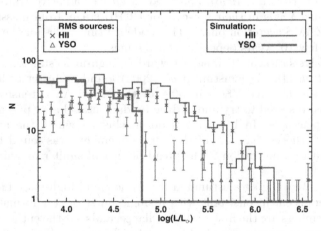

Figure 1. Luminosity distribution of the MYSOs (red points), compact HII regions (blue points) compared with the simulated population (solid lines)

References

Lumsden, S. L., Hoare, M. G., Oudmaijer, R. D., & Richards, D. 2002, *MNRAS* 336, 621
McKee, C. F. & Tan, J. C. 2003, *ApJ* 585, 850
Hosokawa, T. & Omukai, K. 2009, *ApJ*, 691, 823

Highlights of Astronomy, Volume 15
XXVIIth IAU General Assembly, August 2009
Ian F. Corbett, ed.

Star Formation Histories from Pan-Chromatic Infrared Continuum Surveys

Sergio Molinari[1]

[1] INAF - Istituto Fisica Spazio Interplanetario, Via Fosso del Cavaliere 100, 00133 Rome, Italy
email: molinari@ifsi-roma.inaf.it

Abstract. One of the currently most disputed issues in Star Formation is the timeline of the whole process. Is it a "slow" process of cloud assembly which, mediated by magnetic fields, evolve toward turbulence-supported clumps which are eventually super-critical to collapse, e.g. McKee & Tan (2003)? Or do clumps originate in already super-critical state in the post-shock regions of large-scale Galactic converging flows, e.g. Hartmann *et al.* (2001) with a rapid collapse in a crossing time or so (Elmegreen 2000)?

A pan-chromatic 1μm-1mm continuum view of cluster forming regions in their early stages offers access to the most massive members longward of 5-10μm, as well as the low-mass members which instead dominate the emission in the near-IR, offering an interesting potential in stimulating advances in theoretical modelling of clustered star formation, its history and rate.

Keywords. stars: formation

Molinari *et al.* (2008) used mid-IR to submm images to reconstruct the Spectral Energy Distribution (SED) and luminosity for the dominant YSOs in 42 high-mass star forming regions from Molinari *et al.* (1996), and used extensive radiative transfer modelling to disentangle massive YSOs in their hot-core or UCHII stage, from massive YSOs still in their active pre-ZAMS accretion phase. The modelled timescales for the formation are in the range 1 to a few 10^5 years depending on the mass.

In a recent analysis of near-IR images toward 26 high-mass star forming regions (also from Molinari *et al.* (1996)) Faustini *et al.* (2009) found a cluster of low mass objects associated with the massive YSOs in 80% of the cases. An extensive grid of Monte Carlo simulations was used to try and deduce fundamental cluster parameters, including indications on their ages. In less than a dozen clusters where the analysis delivered significant results, the median age of the cluster members was found between 1 and a few 10^6 years, with an indication for a spread of ages of similar magnitude within each cluster.

Taken at face value, the ages estimated for the low and high-mass members of young embedded clusters seem inconsistent with a single burst of star formation and suggest that low-mass members are the first to start collapse, and that the entire cluster formation process encompasses several dynamical timescales. Therefore this is difficult to reconcile with fast formation scenarios.

References

Elmegreen, B. G. 2000, ApJ 530, 277
Faustini, F., Molinari, S., Brand, J., & Testi, L. 2009, A&A 503, 801
Hartmann, L., Ballesteros-Paredes, J., & Bergin, E. A. 2001, ApJ 562, 852
McKee, C. F. & Tan, J. C. 2003, ApJ 585, 850
Molinari, S., Brand, J., Cesaroni, R., & Palla, F. 1996, A&A 308, 573
Molinari, S., Pezzuto, S., Cesaroni, R., Brand, J., Faustini, F., & Testi, L. 2008, A&A 481, 345

Highlights of Astronomy, Volume 15
XXVIIth IAU General Assembly, August 2009
Ian F. Corbett, ed.

© International Astronomical Union 2010
doi:10.1017/S1743921310011816

Virtual Observatory Access to the The IPHAS Data Releases

N. A. Walton[1], J. Drew[2], E. Gonzalez-Solares[1] and M. J. Irwin[1]

[1]Institute of Astronomy, University of Cambridge, Madingley Road, Cambridge, CB3 0HA, UK

[2]Centre for Astrophysics Research, Science and Technology Research Institute, University of Hertfordshire, College Lane, Hatfield, Hertfordshire, AL10 9AB, UK

Abstract. We highlight the IPHAS Data Releases and how access to the primary data products has been implemented through use of standard virtual observatory (VO) publishing interfaces as provided by the Astro- Grid system. The IPHAS Early Data release (EDR), is a photometric catalogue of more than 200 million unique objects, coupled with associated image data covering more than 1000 square degrees in three colours. These data represent the largest data sets to date published solely through Virtual Observatory interfaces.

The IPHAS project (www.iphas.org) is a large systematic optical/Hα survey being carried out with the 2.5-m Isaac Newton Telescope's Wide Field Imaging Camera of the entire northern galactic plane ($|b| < 5°$). The IPHAS Data Release (EDR) (Gonzalez-Solares *et al.* (2008)) (http://idr.iphas.org) is a photometric catalogue of more than 200 million unique objects, coupled with associated image data covering more than 1800 square degrees in three colours. Access to the primary data products has been implemented through use of standard virtual observatory (VO) publishing interfaces as provided by the AstroGrid system (www.astrogrid.org). The upcoming data releases, in particular the full 1st Data Release in Jan 2010, along with the data from VPHAS extension to the southern galactic plane will be published utilising the same interfaces.

Access, through the VO, is provided to the main IPHAS photometric catalogue, in addition to a number of common catalogues (such as 2MASS) which are of immediate relevance. The VO access allows for the user to simply perform a range of common science processes, for instance allowing improved combination of IPHAS and supplementary multi-wavelength data. For instance, cross-matching with 2MASS data in providing optical-IR colours of objects for extinction distance mapping, and cross matching with X-ray data leading to improved selections of CV samples and so forth. The publication of the IPHAS catalogues and image data represents the largest data sets to date published solely through Virtual Observatory interfaces.

Acknowledgements

This research has made use of data obtained using, or software provided by, the UK's AstroGrid Virtual Observatory Project, which was funded by the Science & Technology Facilities Council and through the EU's Framework 6 programme. The Isaac Newton Telescope, operated on the island of La Palma by the ING in the Spanish Observatorio del Roque de los Muchachos of the Instituto de Astrofisica de Canarias.

References

Gonalez-Solares, E. A., Walton, N. A., Greimel, R., *et al.* 2008, *MNRAS* 388, 89

Highlights of Astronomy, Volume 15
XXVIIth IAU General Assembly, August 2009
Ian F. Corbett, ed.

© International Astronomical Union 2010
doi:10.1017/S1743921310011828

A Close Look at the Galactic Ecosystem: the Canadian Galactic Plane Survey

Roland Kothes and Tom L. Landecker

NRC Canada, HIA, DRAO
P.O. Box 248, Penticton, British Columbia, V2A 6J9, Canada
email: roland.kothes@nrc-cnrc.gc.ca; tom.landecker@nrc-cnrc.gc.ca

Abstract. Energy and mass interchanges in the interstellar medium are driven by the life-cycle of stars. They appear in our own Galaxy over a broad range of scales, from point-like injection from stars to global input on the scale of spiral density waves. Understanding this Galactic Ecosystem requires observations of the different phases of the ISM over this large angular dynamic range. The Canadian Galactic Plane Survey (CGPS) is a project to combine radio, millimeter, and infrared surveys of the Galactic plane, providing arcminute-scale images of all major components of the interstellar medium over a large portion of the Galactic disk.

Keywords. atomic data, magnetic fields, polarization, surveys, ISM general, Galaxy: structure

---

The CGPS (Taylor *et al.*, 2003) combines multi-frequency surveys of the Galactic plane providing arcminute scale images of the ionized gas (radio continuum at 408 and 1420 MHz), magnetic fields (polarimetry at 1420 MHz), atomic gas (HI-line), molecular gas (CO(1-0) with the Five College Radio Astronomy Observatory), and dust (reprocessed IRAS surveys). The CGPS covers an area of 1260 square degrees within $52° \leqslant \ell \leqslant 192°$ and $-3.5° \leqslant b \leqslant +5.5°$. Final datasets are archived at the Canadian Astronomy Data Centre (webpage: http://www1.cadc-ccda.hia-iha.nrc-cnrc.gc.ca/cgps) and are freely available to the astronomical community.

Recent science highlights include the study of the large scale magnetic field in the Outer Galaxy (Brown *et al.*, 2003) and a new model for the density and velocity field of the Outer Galaxy traced by the distribution of HI, which can be exploited to determine more accurate distances to objects from systemic velocities (Foster & MacWilliams, 2006). The CGPS enabled a thorough study of HI self absorption features, which represent cold HI, recently compressed by the passage of a spiral shock, and on its way to forming molecules, which may be first step in forming stars (Gibson *et al.*, 2005). Normandeau, Taylor, & Dewdney (1996) discovered the "chimney", a conduit of radiation and material into the halo, driven by a cluster of massive stars at its base. Many new supernova remnants (SNRs) have been discovered through the CGPS (e.g. Kothes *et al.*, 2001a, Kothes *et al.*, 2006) and Kothes *et al.* (2001b) found an SNR, which is likely the result of triggered star formation, which closes the cycle and brings us back to the beginning.

References

Brown, J. C., Taylor, A. R., Wielebinski, R., & Müller, P. 2003 *ApJ*, 592, L29
Foster, T. & MacWilliams, J. 2006 *ApJ*, 644, 214
Gibson, S. J., Taylor, A. R., Higgs, L. A., Brunt, C. M., & Dewdney, P. E. 2005, *ApJ*, 626, 195
Kothes, R., Landecker, T.L., Foster, T., & Leahy, D.A. 2001, *A&A*, 376, 641
Kothes, R., Fedotov, K., Foster, T. J., & Uyanıker, B. 2006, *A&A*, 457, 1081
Kothes, R., Uyanıker, B., & Pineault, S. 2001, *ApJ*, 560, 236
Normandeau, M., Taylor, A. R., & Dewdney, P. E. 1996, *Nature*, 380, 687
Taylor, A. R., Gibosn, S. J., Peracaula, M., *et al.* 2003, *AJ*, 125, 3145

Highlights of Astronomy, Volume 15
XXVIIth IAU General Assembly, August 2009
Ian F. Corbett, ed.
© International Astronomical Union 2010
doi:10.1017/S174392131001183X

Massive Star Forming Regions in the Galactic Plane: A Comparative Study Using BGPS, Spitzer, & Optical/Near-IR Surveys

Guy S. Stringfellow[1] & the BGPS team

[1]CASA, University of Colorado, 389 UCB, Boulder, CO, USA, Guy.Stringfellow@colorado.edu

Abstract. The Bolocam Galactic Plane Survey (BGPS) is a 1.1 mm continuum survey that has detected more than 8300 clumps over a 170 square degree survey area in the Galactic plane. The full power of these data is realised only when considering the full complement of data spanning millimetre through x-ray wavelengths.

Keywords. Galaxy: structure, ISM: jets and outflows, ISM: structure, stars: formation, surveys

The search for highly obscured star forming regions has become possible through long-wavelength, large area, Galactic plane surveys using Spitzer and ground-based submillimeter (ATLASGAL; Schuller *et al.* 2009) and millimetre (Bolocam Galactic Plane Survey (BGPS); Aguirre *et al.* 2009 in press) surveys. BGPS is a 1.1 mm continuum survey that is contiguous over the range $-10.5° \leqslant l \leqslant 90.5°$ with $|b| \leqslant 0.5°$ and $75.5° \leqslant l \leqslant 87.5°$ with $|b| \leqslant 1.5°$. The BGPS survey has detected more than 8300 clumps over the entire 170 square degree survey area to a limiting non-uniform 5σ noise level in the range 30 to 60 mJy/beam (Rosolowsky *et al.* 2009). These clumps are believed to represent the earliest stages in the formation of massive stars.

For comparative analysis, Figure 1a shows a $\sim$32′ field-of-view of the BGPS survey centred on the DR21 and W75 complexes. The 1.1 mm emission traces active regions of cold, dense gas being heated by embedded massive stars, and matches well with the IRS and ERO objects.

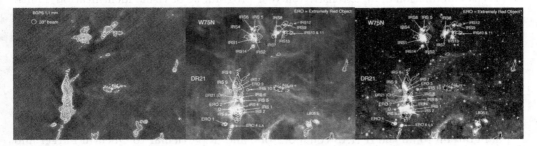

Figure 1. DR21/W75 massive star forming complexes: 1.1 mm dust continuum emission (*left*; contours: 0.2, 0.44, 0.96, 2.1, 4.6, 10 Jy/beam.), Spitzer 8 μm (*middle*) and 4.5 μm (*right*).

References

Aguirre, J. *et al.* 2009, *ApJ Suppl* in press
Rosolowsky, E. *et al.* 2009, *ApJ Suppl* in press
Schuller, F., Menten, K. M., Contreras, Y., *et al.* 2009, *A&A* in press

Highlights of Astronomy, Volume 15
XXVIIth IAU General Assembly, August 2009
Ian F. Corbett, ed.

© International Astronomical Union 2010
doi:10.1017/S1743921310011841

The population of Planetary Nebulae in the Milky Way

Romano L. M. Corradi

Instituto de Astrofísica de Canarias, E-38200 La Laguna, Tenerife, Spain

Departamento de Astrofísica, Universidad de La Laguna, E-38205 La Laguna, Tenerife, Spain
email: `rcorradi@iac.es`

Abstract. Extensive surveys of the Milky Way have provided a global view of the population of planetary nebulae in the Milky Way. Their chemical, dynamical, and morphological properties, the link with their red giant progenitors and white dwarf descendants, and the implications on our understanding of the chemical evolution of the Galaxy, are discussed.

PNe are considered to be tracers of the luminosity, mass, chemistry, and distance in almost any kind of stellar systems. However, recognizing these properties has raised new fundamental questions about the formation and evolution of PNe, among which: Which is the relevance of binarity in the formation of planetary nebulae? Do single stars form PNe at all? Which is the minimum stellar mass to form PNe? Why is the bright cutoff of the PN luminosity function universal? Which chemical elements are pristine in PNe, or how reliably can we adopt the (commonly used) PN oxygen abundances to assess the metallicity of the interstellar medium (ISM) at the time the PN progenitors were born?

Answering these questions relies on a better understanding of the global properties of PNe in our own galaxy, which is the only place where detailed studies of the physics of the nebulae and their central stars is possible. Two recent surface-brightness-limited Hα surveys of the Galactic Plane give us the possibility of making a big step forward in this direction, providing better PN counts, better distances, and better sampling of all the phases of PN evolution. The MASH project (Parker *et al.* 2006) is essentially complete, and is the culmination of an extensive ten year programme of visual identification, multi-wavelength comparison and spectroscopy of 1250 new PNe, which double the number of PNe previously known in the Galaxy. The search and confirmation of PNe in IPHAS (Miszalski *et al.* 2008) is in progress, and we estimate that another 500 to 1000 new PNe will be added in the Northern hemisphere, many of which will be located toward the anti-centre direction. This will substantially improve our knowledge of the properties of PNe and the ISM in the outer regions of the Galactic disc.

Acknowledgements

I am grateful to all the IPHAS collaborators, and in particular to Quentin Parker and David Frew from Macquarie University in Sydney for providing some of their results from MASH. I acknowledge funding from the Spanish AYA2007-66804, and from the Brasilian agency FAPERJ who partly supported my visit to Brasil.

References

Drew, J. E., Greimel, R., Irwin, M. J. *et al.* 2005, *MNRAS* 362, 753
Parker, Q. A., Acker, A., Frew, D. J. *et al.* 2006, *MNRAS* 373, 79
Miszalski, B., Parker, Q. A., Acker, A. *et al.* 2008, *MNRAS* 384, 525

Highlights of Astronomy, Volume 15
XXVIIth IAU General Assembly, August 2009
Ian F. Corbett, ed.

© International Astronomical Union 2010
doi:10.1017/S1743921310011853

The Galactic Pulsar Population

Duncan R. Lorimer

Department of Physics, 210 Hodges Hall, Morgantown, WV 26506, USA
email: duncan.lorimer@mail.wvu.edu

Abstract. Simulations of the Galactic pulsar population are reviewed. These include snapshot and time-dependent models of normal pulsars as well as binary and milli-second pulsars.

The radio pulsar population currently known totals over 1650 objects in our Galaxy, thanks largely to a number of large-scale surveys carried out with the Parkes radio telescope, as well as Green Bank, Arecibo and the Giant Metrewave telescope (see Lorimer 2008). Due to the complexities in describing the population and related selection effects analytically, the main approach to the problem is through a Monte Carlo simulation of the Galactic distribution of pulsars, propagation effects in the interstellar medium and the various large scale surveys that have been carried out. Two main types of simulations are typically used: (i) a snapshot model of the population at the current epoch; (ii) a time dependent model following the spin and kinematic evolution over time. The former approach requires fewer assumptions and is easy to optimize. The latter approach is ultimately more reliable, but has more model dependent results and covariances between various parameters. With these caveats in mind, Lorimer *et al.* (2006) have produced a snapshot model of the Galactic population of normal (i.e. non-millisecond) pulsars and find that their radial distribution closely follows the assumed distribution of free electrons and has a scale height around 330 pc. Faucher-Giguère & Kaspi (2006) have published a detailed time dependent model and find evidence for a log-normal luminosity function in which the luminosity depends upon spin parameters. They found that the birth velocity distribution is most likely distributed exponentially in each of the x, y and z components, with a mean three-dimensional speed of 380 km s^{-1}. Models of the millisecond and binary pulsar population are less well established, but it is now clear that they form a dynamically relaxed population with a smaller velocity dispersion and a larger scale height of 500 pc. A recent study by Story *et al.* (2007) reproduces many of the key features of the observable millisecond pulsar population. Kiel (2009) has made significant progress in the development of a self-consistent model of the millisecond and normal pulsar population which includes the effects of binary evolution. Models are available at `psrpop.phys.wvu.edu` and `psrpop.sourceforge.net`.

Acknowledgements

I thank the AAS for travel support. My research is supported by the National Science Foundation, the Research Corporation and West Virginia EPSCoR.

References

Faucher-Giguère, C. A. & Kaspi, V. M. 2006, ApJ, 643, 332
Kiel, P. 2009, PhD thesis, Swinburne University of Technology
Lorimer, D. R. *et al.* 2006, MNRAS, 372, 777
Lorimer, D. R. 2008, Living Rev. Relativity, 11
Story, S. R. *et al.* 2007, ApJ, 671, 713

Highlights of Astronomy, Volume 15
XXVIIth IAU General Assembly, August 2009
Ian F. Corbett, ed.

© International Astronomical Union 2010
doi:10.1017/S1743921310011865

The H.E.S.S. Galactic Plane Survey

Emma de Oña-Wilhelmi for the H.E.S.S. collaboration
email: emma@mpi-hd.mpg.de

Max-Planck-Institute für Kernphysik, P.O. Box 103980, D 69029 Heidelberg, Germany

Abstract. The H.E.S.S. Galactic Plane Survey (GPS) has revealed a large number of Galactic Sources, including Pulsar Wind Nebulae (PWN), Supernova Remnants (SNRs), giant molecular clouds, star formation regions and compact binary systems, as well as a number of unidentified objects, or dark sources, for which no obvious counterparts at other wavelengths have yet been found. We will review the latest results from the GPS observations and discuss the most interesting cases.

Keywords. gamma rays: observations

The H.E.S.S. telescope is located in Namibia at 23°16′17″ S 16°29′58″ E, at 1800 m above the sea level. H.E.S.S. is a system of four large (13 m diameter) imaging atmospheric Cherenkov telescopes, designed to detect gamma-rays in the very-high-energy (VHE; E > 100 GeV) domain. Each telescope has a mirror area of 107 m^2 and a total field of view of 5°, well suited to the study of extended sources. The system works in a coincidence mode. Its angular resolution reaches ∼ 5′per event and its sensitivity for a point-like source, is 2.0×10^{-13} ph cm^{-2}s^{-1} (1% of the Crab Nebula flux above 1 TeV) for a 5σ detection, within 25 hours observation time. The H.E.S.S. telescope has conducted an extension of the scan of the inner Galactic Plane Survey (GPS, Aharonian *et al.* (2005)), which was a major breakthrough in the Galactic field. This report can only give a brief overview; for details and further references, the reader is referred to the publications cited. The survey covers the as yet unexplored range in longitude between [-85°,60°] and [-2.5°,2.5°] in latitude. It has revealed more than two dozen new VHE sources, consisting of shell-type SNRs, PWN, X-ray binary systems, two putative young star clusters, as yet unidentified objects, the so-called dark sources, in which no obvious counterparts at other energy wavelengths are found (see e.g. Aharonian *et al.* (2006), Aharonian *et al.* (2008)). In the central 100 pc of the Milky Way diffuse emission is seen, and we are able to locate the Galactic Centre γ-ray source with a precision of 6″, consistent with the black hole Sgr A*, but excluding the nearby remnant Sgr A East. The sources line up along the Galactic plane, with a rms spread in latitude of about 0.3°, consistent with the scale height of the distribution of molecular gas and with the width of the distribution of supernova remnants and pulsars. Their Galactic origin is confirmed by the fact that nearly all sources are extended, with rms sizes up to 0.3°. The official H.E.S.S. Source Catalogue can be found at www.mpi-hd.mpg.de/hfm/HESS/pages/home/sources and includes all of the VHE γ-ray sources which were detected by H.E.S.S. and subsequently published in refereed journals.

References

Aharonian, F. A. *et al.* (H. E. S. S. Collaboration) 2005, Science, 307, 1938.
Aharonian F. A. *et al.* (H. E. S. S. Collaboration) 2006, ApJ, 636, 777.
Aharonian F. A. *et al.* (H. E. S. S. Collaboration), 2008, A&A, 477, 353.

Highlights of Astronomy, Volume 15
XXVIIth IAU General Assembly, August 2009
Ian F. Corbett, ed.

© International Astronomical Union 2010
doi:10.1017/S1743921310011877

Galactic plane structure in hard X-rays

A. Lutovinov[1], M.Revnivtsev[2,1] and R.Krivonos[2,1]

[1] Space Research Institute, Profsoyuznaya str. 84/32, Moscow, 117997, Russia
[2] Max-Plank Institute fur Astrophysics, Karl-Schwarzschild str. 1, Garching, Germany

Abstract. We study the structure of the Galaxy in the hard X-ray energy band ($\dot{\iota}$20 keV) using data from the INTEGRAL observatory. The increased sensitivity of the survey and the very deep observations performed during six years of the observatory operation allow us to detect about a hundred new sources. This significantly enlarges the sample of hard X-ray sources in the Galactic disk and bulge in a comparison with the previous studies.

We are continuing studies of the Galaxy structure in hard X-rays (>20 keV) using data from the INTEGRAL observatory. Observations and discoveries of a large number X-ray sources during more than 6 years of its operation in orbit have allowed us to improve significantly our knowledge about different populations of X-ray binaries, their properties and spatial distribution (Lutovinov *et al.* 2005, 2006, 2008; Krivonos *et al.* 2007).

At the moment (August, 2009) the total number of confidently detected sources with INTEGRAL amounts to 480, 80 of which are high-mass X-ray binaries (HMXB) and 92 are low-mass X-ray binaries (LMXB). We constructed the distribution of hard X-ray binaries in the Galaxy and found that they are concentrated towards the Galactic plane; however HMXBs and LMXBs have different vertical scale heights and spatial distribution along the plane, reflecting the age of stellar companions of these sources and their evolution. HMXBs are mostly located in spiral arms as a young galactic population, whilst LMXBs are concentrated to the Galactic Centre. It is worthy to note that the current INTEGRAL exposure and sensitivity reached in the Galactic plane allows us to detect hard X-ray sources with the luminosity of $L_X > 10^{35.5}$ erg s^{-1} practically throughout the entire Galaxy.

We also demonstrated that the spatial distribution of persistent LMXBs in the Galactic Centre/Galactic bulge region is consistent with a model of stellar mass distribution that includes the nuclear stellar disk component in the innermost degree of the Galaxy. The spatial distribution of transient LMXBs detected in the Galactic Centre region indicates an increased fraction of transient sources in the innermost degree of the Galaxy with respect to outer regions (Revnivtsev *et al.* 2008).

Acknowledgements

This work is supported by RFBR grant number 07-02-01051.

References

Lutovinov, A., Revnivtsev, M., Gilfanov, M., *et al.*, 2005, A&A 444, 821
Lutovinov, A., Revnivtsev, M., Gilfanov, M., & Sunyaev, R., 2006, IAUS 230, 340 (Eds. E.J.A. Meurs, G. Fabbiano; CUP)
Lutovinov, A., Revnivtsev, M., Gilfanov, M., & Sunyaev, R., 2008, Proceedings of the 6th IN-TEGRAL Workshop, 2007, ESA SP-622, p.241 (Eds. S.Grebenev, R.Sunyaev, C.Winkler)
Krivonos, R., Revnivtsev, M., Lutovinov, A., *et al.*, 2007, A&A 475, 775
Revnivtsev, M., Lutovinov, A., Churazov E., *et al.*, 2008, A&A 491, 209

Highlights of Astronomy, Volume 15
XXVIIth IAU General Assembly, August 2009
Ian F. Corbett, ed.

© International Astronomical Union 2010
doi:10.1017/S1743921310011889

Broad view on hard X-ray background emission of the Galaxy

Roman Krivonos[1,2], Mikhail Revnivtsev[2,3], Sergey Tsygankov[1,2], Eugene Churazov[1,2] and Rashid Sunyaev[1,2]

[1]Max Planck Institute for Astrophysics, Garching, Germany
email: krivonos@mpa-garching.mpg.de,
[2]Space Research Institute, Moscow, Russia
[3]Excellence Cluster Universe, Technische Universitat Muenchen, Garching, Germany

Abstract. The nature of the Galactic Ridge X-Ray Emission (GRXE) has been under scientific debate since its discovery more than 30 years ago. It is observed as extended emission along the Galactic disk. The question was: is GRXE truly diffuse or is it composed from a large number of unresolved point sources? Using near-infrared Galaxy maps measured with the DIRBE experiment and data from the INTEGRAL observatory, we show that the galactic background in the energy range 20-60 keV originates from the stellar population of the Galaxy, which is in contrast to the diffuse nature believed before (Krivonos *et al.*, 2007). Here we show preliminary results of studying the transition region from hard X-rays to gamma diffuse background of the Galaxy, revealing the broad band picture of Galactic Background emission.

Keywords. Galaxy: disk,Galaxy: bulge,Galaxy: stellar content,ISM: cosmic rays

---

In many high-energy and MeV- observations of the Galactic Plane the gamma-ray continuum emission was detected (see e.g. Kraushaar *et al.*, 1972) and is believed to originate from the interaction of cosmic rays with the interstellar medium. This suggests that at energies $> 100 - 200$ keV a change in the nature of the unresolved Galactic emission to cosmic-ray induced background should occur. This implies detection of the rising part of the Ridge spectrum in the transition region. Indications for this were found in INTEGRAL/SPI measurements (e.g. Bouchet *et al.*, 2008).

To investigate the Galactic background in the transition region between compact objects in X-rays, and truly diffuse (cosmic ray induced) in the MeV energy band, additional observations were scheduled of the Galactic disk ($l = -45°$). Here we report on the possible detection of the rising part of the Galactic background emission spectrum above 100 keV measured with the IBIS/ISGRI instrument onboard the INTEGRAL observatory. The spectrum's slope of 1.5 is compatible with that expected for the spectrum of cosmic-ray induced background emission and/or positron annihilation continuum. Subtracting a model of the high-energy rising part from the measured spectrum we obtained a GRXE spectrum at lower energies which is in good agreement with that measured in the Galactic center. This implies that the source population responsible for GRXE production is similar in the Galactic center and the plane in 17-100 keV energy band.

References

Krivonos R., Revnivtsev M., *et al.*, 2007, *A&A* 463, 957
Kraushaar *et al.*, 1972, ApJ, 177, 341
Bouchet *et al.*, 2008, The Astrophysical Journal, 679, 1315

Highlights of Astronomy, Volume 15
XXVIIth IAU General Assembly, August 2009
Ian F. Corbett, ed.

© International Astronomical Union 2010
doi:10.1017/S1743921310011890

Red clump giant stars as tracers of Galactic structure

M. López-Corredoira[1], A. Cabrera-Lavers[2], P. L. Hammersley[1], F. Garzón[1,3], T. J. Mahoney[1] and C. González-Fernández[1]

[1]Instituto de Astrofísica de Canarias, La Laguna (Tenerife, Spain)

[2]GRANTECAN, La Palma (Spain)

[3]Departamento de Astrofísica, Universidad de La Laguna (Tenerife, Spain)

Abstract. By isolating the red clump giant population in the color-magnitude diagrams and inverting their star counts, we can obtain directly the density distribution of the old stellar population along the line of sight. We have applied this method to several near infrared surveys and obtained information on the disc, bulge and long bar. The disc is well fitted by an exponential distribution in both the galactocentric distance and height, flared and warped in the outer parts, and with a deficit of stars in the inner in-plane regions. The long bar occupies these in-plane regions within $R < 3.9$ kpc, with approximate dimensions of 7.8 kpc×1.2 kpc× 0.2 kpc and a position angle of 40-45 deg. The bulge is a triaxial structure, possibly boxy, thicker and shorter than the long bar and with position angle of 10-30 deg.

Keywords. Galaxy: structure, infrared: stars

The red clump giants (approx. K0-2III stars) constitute a very prominent population in the near infrared color magnitude diagrams (CMDs). Their absolute magnitude is $M_K \approx -1.65$, and intrinsic color $J - K \approx 0.75$, so their property as standard candles can be used to derive the parameters of the Galactic structure. A method, explained in detail in López-Corredoira *et al.* (2002, Sect. 3), allows us to obtain the star density along a line of sight (l, b) from an analysis of near-infrared CMDs, such as m_K vs. $(J - K)$. First, the trace produced in the CMD by the red clump giants is identified. Second, we count the number of stars within a fixed width trace (we usually take 0.4 mag) as a function of the apparent magnitude. Third, since absolute magnitude and intrinsic color are known, we can obtain the extinction and density as a function of distance along the line of sight.

The method was applied to derive the parameters of the thin disc (López-Corredoira *et al.* 2002, 2004), thin+thick disc (Cabrera-Lavers *et al.* 2007a), long bar+thick bulge (Cabrera-Lavers *et al.* 2007b and references therein). The general obtained picture is that the disc is flared in the outer and inner parts, and without cut-off up to 16 kpc; and that the central 4 kpc of the Galaxy contains a double triaxial structure: long bar and thick bulge with position angles of 40-45 and 10-30 deg. respectively.

References

López-Corredoira, M., Cabrera-Lavers, A., Garzón, F., & Hammersley, P. L. 2002, *A&A* 394, 883

López-Corredoira, M., Cabrera-Lavers, A., Gerhard, O. E., & Garzón, F. 2004, *A&A* 421, 953

Cabrera-Lavers, A., Bilir, S., Ak, S., Yaz, E., & López-Corredoira, M. 2007a, *A&A* 464, 565

Cabrera-Lavers, A., Hammersley, P. L., González-Fernández, C., López-Corredoira, M., Garzón, F., & Mahoney, T. J. 2007b, *A&A* 465, 825

Highlights of Astronomy, Volume 15
XXVIIth IAU General Assembly, August 2009
Ian F. Corbett, ed.

© International Astronomical Union 2010
doi:10.1017/S1743921310011907

Long Period Variables as tracers of Galactic Structure

Martin A. T. Groenewegen

Royal Observatory of Belgium, Ringlaan 3, B-1180 Brussels, Belgium
email: marting@oma.be.

Abstract. Distances to a large sample of long period variables are derived using a PL-relation and a 3D model for the reddening. Their use as a tracer of galactic structure is discussed.

Almost all stars with initial masses $\lesssim 8$ $M_\odot$ will pass through the Asymptotic Giant Branch phase. One of the main characteristics of this phase is radial pulsation as Irregular (Irr), semi-regular (SR) or Mira variables. Many of the SR actually show well behaved light-curves, and so these stars and the Miras are called Long Period Variables (LPVs).

That Mira variables follow a period-luminosity (PL-) relation had been known for decades, but in the optical the scatter is large. The practical use increased with the extension to the K-band or bolometric magnitude, especially with sufficient phase coverage of the light-curve. Recently Whitelock *et al.* (2008) revisited the PL-relations in $M_{\rm bol}$ and $M_{\rm K}$ for Miras, including the Galactic PL-relation using the 2007 revised Hipparcos parallaxes and other observables as constraints. With the PL-relation as distance indicator one can make an attempt to study Galactic Structure using LPVs, see Figure 1.

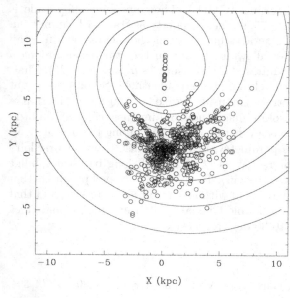

Figure 1. Location of 700 Miras, projected onto the Galactic Plane, with $|z| < 500$ pc. The Hou 4-spiral arm model is shown as reference. From the GCVS 4500 stars classified as Miras were selected, with periods in the range 50-400 d. Their K-band magnitudes were looked-up via SIMBAD (available for 1250). Using a PL-relation and a 3D model for the reddening (Groenewegen 2008) the distances were derived. I leave it up to the reader whether he/she believes that Miras trace Galactic Structure or not! The GCVS is clearly a very inhomogeneous source of information, so surveys like LSST and Skymapper will be essential in creating unbiased samples of LPVs.

References

Whitelock, P. A., Feast, M. W., & van Leeuwen, F. 2008, *mnras*, 386, 313
Groenewegen, M. A. T. 2008, *A&A*, 488, 25

Highlights of Astronomy, Volume 15
XXVIIth IAU General Assembly, August 2009
Ian F. Corbett, ed.

Galactic AGB stars from the IPHAS survey

N. J. Wright,[1] M. J. Barlow,[2] R. Greimel,[3] J. E. Drew[4] and M. Matsuura[2]

[1]Harvard-Smithsonian Center for Astrophysics, 60 Garden Street, Cambridge, MA 02138
[2]University College London, Gower Street, London WC1E 6BT, U.K.
[3]Institut für Physik, Karl-Franzen Universität Graz, Universitätsplatz 5, 8010 Graz, Austria
[4]University of Hertfordshire, College Lane, Hatfield, AL10 9AB, U.K.

Abstract. We present a photometric analysis of the properties of asymptotic giant branch stars identified in the INT Photometric H-alpha Survey (IPHAS) of the northern Galactic plane. Follow-up spectroscopy has revealed that the IPHAS (r - Ha) colour is a valuable diagnostic of the photospheric C/O ratio, and may be used to identify hundreds of carbon and S-type stars.

Keywords. stars: AGB and post-AGB - stars: carbon - infrared: stars - techniques: spectroscopic

---

Asymptotic giant branch (AGB) stars are one of the final evolutionary stages of all intermediate mass stars and one of the major sources of enriched material returned to the interstellar medium, including all s-process elements and the majority of carbon. The INT Photometric Hα Survey (IPHAS, Drew et al. 2005) is imaging the entire northern Galactic plane using Sloan r', i', and narrow-band Hα filters. One of the unique features of the IPHAS colour-colour diagram is that the main-sequence and giant branches are well separated at late spectral types, effectively allowing the AGB population across the entire northern Galactic plane to be identified and studied (Wright et al. 2008)

Follow-up spectroscopy on a number of optical and near-IR instruments has allowed this population to be studied in more detail. Wright et al. (2009) published a near-IR spectral library of AGB stars with a particular focus on very late-type sources. The spectral library includes spectra in all three near-IR bands as well as many variable objects and chemically evolved sources such as S-type and carbon stars. The library includes spectral classification sequences highlighting the various molecular features identified and discusses a number of rare features and the potential molecules responsible.

Finally we discuss a correlation between the IPHAS $(r'-H\alpha)$ colour and the C/O abundance index (Keenan & Boeshaar, 1980). Wright et al. (2009) found that the IPHAS $(r'-H\alpha)$ colour could be used to estimate C/O ratios for S-type stars and therefore determine their state of chemical evolution in the transition from O-rich (C/O < 1) to carbon rich (C/O > 1) via the intermediate S-type phase (C/O ~ 1). When combined with a near-IR colour the relationship has the potential to separate O-rich, S-type, and carbon stars across the Galactic plane based on their photometry alone, which will increase our understanding of AGB dredge-up mechanisms, galactic chemical evolution, and the structure and metallicity of the Galactic disk.

References

Drew, J. E., Greimel, R., Irwin, M. J. et al. 2005, *MNRAS*, 362, 753
Keenan, P. C. & Boeshaar, P. C. 1980 *ApJS*, 43, 379
Wright, N. J., Greimel, R., Barlow, M. J. et al. 2008, *MNRAS*, 390, 929
Wright, N. J., Barlow, M. J., Greimel, R. et al. 2009, *MNRAS*, in press

Highlights of Astronomy, Volume 15
XXVIIth IAU General Assembly, August 2009
Ian F. Corbett, ed.

New Galactic Wolf-Rayet Stars Discovered via 2MASS + *Spitzer*/GLIMPSE

Jon C. Mauerhan[1], Schuyler D. Van Dyk[1], Pat W. Morris[2]

[1] Spitzer Science Center, IPAC, California Institute of Technology, M/C 220-6, Pasadena, CA
91125, USA
email: mauerhan@ipac.caltech.edu

[2] NASA Herschel Science Center, IPAC, California Institute of Technology, M/C 100-22,
Pasadena, CA 91125, USA

Abstract. We have undertaken a campaign to expand the census of WRs throughout the Galaxy
using the 2MASS and Spitzer/GLIMPSE surveys. Free-free emission generated within the ion-
dense winds of WRs generates a significant infrared excess which forms the basis of an excellent
criterion for photometrically selecting WRs from the stellar field.

We report new results from our effort to identify obscured Wolf-Rayet stars (WRs) in
the Galaxy. Candidates were selected by their near-infrared (2MASS (Cutri *et al.* 2003)
and mid-infrared (*Spitzer*/GLIMPSE Churchwell *et al.* 2009) colour excesses, which are
consistent with free-free emission from ionised stellar winds and thermal excess from
hot dust. Mauerhan *et al.* (2009) have confirmed 12 new WRs in the Galactic disk,
including 9 WN type, and 3 WC; giving a total 27 discovered this way, including those
in Hadfield *et al.* (2007). We classify one of the new stars as a possible dust-producing
WC9d+OBI colliding-wind binary. A WC8 star in our sample appears to be a member of
the stellar cluster Danks 1, in contrast to the rest of the confirmed Wolf-Rayet stars that
generally do not appear to reside within dense stellar clusters. Therefore, the new WRs
are either (1) runaways from clusters, (2) members of cluster which are confused with
the background stellar field, or (3) formed in isolation. Follow-up observations to identify
early-type sibling stars in the vicinity of the new WRs will elucidate their origins.

Based on the WR infrared colour selection criterion from Hadfield *et al.* (2007) and
Mauerhan *et al.*(2009), the detection rate for WRs was ≈10%. Recently, we have expanded our selection criteria to include more highly reddened sources, while excluding
nearby contaminant stars more effectively (e. g., YSOs). This was made possible in part
by colour-magnitude restrictions, and the addition of a new colour selection sieve based
on [8]−[24] µm colours from the *Spitzer*/MIPSGAL survey (Carey *et al.* 2009). In addition, we have also cross-correlated our latest candidate WR sample with archival *Chandra*
and *XMM* point-source catalogues. With these new refinements, the most recent observations indicate that our WR detection rate has improved dramatically; 30 new WRs
have been discovered, 1/3 of which are serendipitous X-ray detections.

References

Carey, S. J. *et al.* 2009, PASP, 121, 76
Churchwell, E. *et al.* 2009, PASP, 121, 213
Cutri, R. M. *et al.* 2003, "Explanatory Supplement to the 2MASS All Sky Data Release"
Hadfield, L. J., van Dyk, S. D., Morris, P. W. *et al.* 2007, MNRAS, 376, 248
Mauerhan, J. C., van Dyk, S. D., & Morris, P. W. 2009, PASP, 121, 591

Highlights of Astronomy, Volume 15
XXVIIth IAU General Assembly, August 2009
Ian F. Corbett, ed.

© International Astronomical Union 2010
doi:10.1017/S1743921310011932

IPHAS A-type Stars with Mid-IR Excesses in Spitzer Surveys

Antonio S. Hales[1], Michael J. Barlow, Janet E. Drew, Yvonne C. Unruh, Robert Greimel, Michael J. Irwin and Eduardo González-Solares

[1] Joint ALMA Observatory, Av. El Golf 40, Piso 18, Santiago, Chile, email: ahales@alma.cl

Abstract. The Isaac Newton Photometric H-Alpha Survey (IPHAS) provides $(r'\text{-}H\alpha)\text{-}(r'-i')$ colors, which can be used to select AV0-5 Main Sequence star candidates (age∼20-200 Myr). By combining a sample of 23050 IPHAS-selected A-type stars with 2MASS, GLIMPSE and MIPSGAL photometry we searched for mid-infrared excesses attributable to dusty circumstellar disks. Positional cross-correlation yielded a sample of 2692 A-type stars, of which 0.6% were found to have 8-μm excesses above the expected photospheric values. The low fraction of main sequence stars with mid-IR excesses found in this work indicates that dust disks in the terrestrial planet zone of Main Sequence intermediate mass stars are rare. Dissipation mechanisms such as photo-evaporation, grain growth, collisional grinding or planet formation could possibly explain the depletion of dust detected in the inner regions of these disks.

Keywords. Stars: circumstellar matter; Planetary systems: proto-planetary disks

From the observed colors in the $(r'\text{-}H\alpha)\text{-}(r'-i')$ plane, we identify 23050 early A-type main sequence (MS) star candidates in the Isaac Newton Photometric H-Alpha Survey (IPHAS, Drew et al. 2005). Positional cross-correlation with the 2MASS and GLIMPSE catalogs led to a sample of 2692 candidate A-type stars with fully sampled 0.6 to 8 μm SEDs. Optical classification spectra obtained of 10 of the systems confirmed that all but one were main sequence A-type stars, the exception being an A5Ia super-giant. We searched for stars with 8-μm excesses by looking for unusually large $(K-8)$ colors. 17 stars (0.6%) were found to have 8-μm excesses above the expected photospheric value (Hales et al. 2009). Free-free emission from ionized gas can cause mid-IR excesses around B stars. However, these stars will also show prominent Hα emission (Uzpen et al. 2008), while our IPHAS-based color selection method ensures that stars in our sample have undetectable levels of Hα emission. Therefore the mid-IR excesses in our sample are most likely to originate in a disk of warm dust located close to the central stars (r < 1 AU). 10 out of the 17 excess stars had been covered by *Spitzer* MIPSGAL survey fields, of which 5 had detectable excesses at 24 μm. For sources with excesses detected in at least two mid-IR wavelength bands, blackbody fits to the excess SEDs yielded temperatures ranging from 270 to 650 K, and bolometric luminosity ratios comparable to those of warm debris-disk systems, similar to β Pictoris.

References

Drew, J. E. et al. 2005, *MNRAS*, 362, 753
Hales, A. S. et al., 2009, *ApJ*, 695, 75
Uzpen, B., Kobulnicky, H. A., Semler, D. R., Bensby, T., & Thom, C. 2008, *ApJ*, 685, 1157

Highlights of Astronomy, Volume 15
XXVIIth IAU General Assembly, August 2009
Ian F. Corbett, ed.

Gaia: A Stereoscopic Census of Our Galaxy

T. Prusti

European Space Agency, ESTEC, The Netherlands

Abstract. Gaia is a space astrometry mission, a broad survey project following the measurement and operational principles of Hipparcos. It will help solving one of the most difficult yet deeply fundamental challenges in modern astronomy: to create an extraordinarily precise three-dimensional map of about one billion stars throughout our Galaxy and beyond. In the process, it will map their three-dimensional motions, which encode the origin and subsequent evolution of the Galaxy. Through comprehensive photometric and spectroscopic classification, it will provide the detailed physical properties of each star observed: characterising their luminosity, temperature, gravity, and elemental composition. This massive stellar census will provide the basic observational data to tackle an enormous range of important problems related to the origin, structure, and evolutionary history of our Galaxy.

Within the Galactic plane the specific issue for Gaia is not surprisingly the high stellar density. While the astrometric instrument is expected to detect individually sources separated by no more than 0.1 arcsec, the photometric and spectroscopic instruments will suffer from crowding in the densest parts of the sky. This is due to the fact that the light is dispersed on the CCD. The maximum star densities are 1,000,000, 750,000, and 36,000 stars per square degree for astrometric, photometric and spectroscopic instruments respectively. In principle it is possible to operate Gaia in so called high density mode which will enable better completeness level even in regions with densities above a million stars per square degree. However, to achieve this a modification of the scanning law is required with a negative impact on overall sky homogeneity coverage. The precise use of the high density mode will be decided during the operational phase in order to balance the scientific benefits of an unbiased survey with the nominal scanning law and a better completeness in high density regions. As far as the large data quantities associated with the Galactic Plane are concerned, the planning of ground station time has been tuned to take this into account. When an operational day will contain a large fraction of Galactic Plane, Gaia will be served by two ground stations instead of a single one. This will ensure Gaia gathering a unique data set also for the Galactic Plane.

Currently Gaia is in the development phase with flight hardware being produced and the Data Processing Consortium preparing the software for handling the data. Gaia is scheduled for launch in spring 2012 with intermediate data releases anticipated a few years after the launch and final catalogue 2020.

Highlights of Astronomy, Volume 15
XXVIIth IAU General Assembly, August 2009
Ian F. Corbett, ed.

© International Astronomical Union 2010
doi:10.1017/S1743921310011956

Mapping the Milky Way with LSST

Željko Ivezić (for the LSST Collaboration)

Department of Astronomy, University of Washington, Seattle, WA 98155, USA
email: ivezic@astro.washington.edu

Abstract. The Large Synoptic Survey Telescope (LSST) is the most ambitious ground-based survey currently planned in the visible band. Mapping of the Milky Way is one of the four main science and design drivers. The main 20,000 sq.deg. large survey area will be imaged about 1000 times in six bands (*ugrizy*) during the anticipated 10 years of operations, with the first light expected in 2015. These data will result in databases including 10 billion galaxies and a similar number of stars, and will serve the majority of science programs. In the Milky Way context, these deep LSST data will enable unprecedented studies of distant stars in Galactic halo, as well as intrinsically faint nearby objects, such as white dwarfs and LTY dwarfs.

Keywords. Surveys, atlases, catalogs, astronomical data bases, astrometry, photometry

---

LSST will be a large, wide-field ground-based system designed to obtain multiple images covering the sky that is visible from Cerro Pachón in Northern Chile. The LSST design is driven by four main science themes: constraining dark energy and dark matter, taking an inventory of the Solar System, exploring the transient optical sky, and mapping the Milky Way. The current baseline design, with an 8.4m (6.7m effective) primary mirror, a 9.6 deg^2 field of view, and a 3,200 Megapixel camera, will allow about 10,000 square degrees of sky to be covered using pairs of 15-second exposures in two photometric bands every three nights on average. The system is designed to yield high image quality as well as superb astrometric and photometric accuracy. The survey area will include 30,000 deg^2 with $\delta < +34.5°$, and will be imaged multiple times in six bands, *ugrizy*, covering the wavelength range 320–1050 nm. About 90% of the observing time will be devoted to a deep-wide-fast survey mode which will observe a 20,000 deg^2 region about 1000 times in the six bands during the anticipated 10 years of operations. These data will result in databases including 10 billion galaxies and a similar number of stars, and will serve the majority of science programs. The remaining 10% of the observing time will be allocated to special programs such as Very Deep and Very Fast time domain surveys. More information about LSST can be obtained from www.lsst.org and Ivezić *et al.* (2008).

As a comparison, each 30-sec observation will be about 2 mag deeper than SDSS imaging, and the repeated observations will enable proper motion and trigonometric parallax measurements to $r = 24.5$, about 4-5 mag fainter limit than to be delivered by Gaia, and the coadded LSST map will reach $r = 27.5$. Due to Gaia's superb astrometric and photometric accuracy, and LSST's significantly deeper data, the two surveys are highly complementary: Gaia will map the Milky Way's disk with unprecedented detail, and LSST will extend this map all the way to the halo edge.

References

Ivezić, Ž., Tyson, J. A., Allsman, R. *et al.* 2008, arXiv:0805.2366

Highlights of Astronomy, Volume 15
XXVIIth IAU General Assembly, August 2009
Ian F. Corbett, ed.

The Pan-STARRS 3π Survey and the Brown Dwarf Factory

E. A. Magnier[1], M. Liu[1], B. Goldman[2], D. G. Monet[3],
K. C. Chambers[1] and N. Kaiser[1]

[1]IfA, U. Hawaii, 2680 Woodlawn Dr., Honolulu, HI 96822, USA email: eugene@ifa.hawaii.edu

[2]M.P.I.A., Königstul 17, D-69117 Heidelberg, Germany email: goldman@mpia-hd.mpg.de

[3]USNO, P.O. Box 1149, Flagstaff, AZ 86002, U.S.A email: dgm@nofs.navy.mil

Abstract. The Pan-STARRS 1 Telescope (PS1) is currently (2009 Aug) undergoing final commissioning efforts and starting to perform initial science observations for the PS1 survey mission. PS1 will greatly expand the known population of Brown Dwarfs, with discovery via photometry, proper-motion, and parallax.

PS1 uses a 1.8m primary mirror to image a 7 deg^2 field of view onto a 1.4 Gigapixel camera. PS1, located on the summit of Haleakala, Maui, was built by the Pan-STARRS Project and will be operated for the science survey mission by the PS1 Science Consortium (http://ps1sc.ifa.hawaii.edu). The Pan-STARRS Project, operating at the University of Hawaii, is working towards Pan-STARRS 4, a survey instrument to be deployed on Mauna Kea in Hawaii consisting of 4 identical 1.8m telescopes which observe the same portion of the sky at the same time. PS1 is a demonstration of the technology, and a major survey instrument on its own.

PS1 will perform a multi-level survey mission lasting at least 3 years, with wide ranging science goals. Roughly 56% of the time will be spent on the 3π Survey in which the full sky region observable from Hawaii will be repeatedly observed over the 3 years; most of the remainder of the time will go to repeated observations of various specific fields, and an additional portion will be used to search for potentially hazardous asteroids by observing along the Earth's orbit (the 'sweet-spot survey'). By the end of the mission, each patch in the 3π Survey region ($\delta > -30°$) will be observed 12 times in each of 5 filters, with expected detection limits of ($grizy = 23.2, 22.7, 22.6, 21.6, 20.1$).

Since 2009 April, PS1 has been operating with somewhat less than 50% of time for initial science observations. On-going engineering work is being performed to improve the image quality and operational efficiency. Even the modest science observations to date have resulted in a substantial survey data set. In a 30 day period in 2009 June & July, operating at this reduced rate, 3π observations have already covered over 25,000 deg^2 (single filter), equivalently to roughly 60% of the SDSS survey. The measured detection limit in y band depends on the image quality, but can reach the predicted 20.1 limit in appropriate conditions. Typical values in the survey data to date are in the range $y = 19.5 - 20.1$; the z-band detection limit is about 0.5 lower than expected.

We are learning how to search through the data set for brown dwarfs, starting with simple two-colour plots including 2MASS JHK photometry. At this stage, it is crucial to limit possible brown dwarf candidates to the most reliable objects by requiring two detections in each of the y and z filters. As multiple epochs become available, precision proper-motion and parallaxes will greatly improve the search efficiency.

Highlights of Astronomy, Volume 15
XXVIIth IAU General Assembly, August 2009
Ian F. Corbett, ed.

GASKAP: The Galactic ASKAP Survey

Snežana Stanimirović[1], John M. Dickey[2], Steven J. Gibson[3], José F. Gómez[4], Hiroshi Imai[5], Paul A. Jones[6,7] and Jacco Th. van Loon[8]

[1] Dept. of Astronomy, University of Wisconsin-Madison, 475 North Charter Street, Madison, WI 53706, USA; email: sstanimi@astro.wisc.edu

[2] School of Maths and Physics, University of Tasmania, Private Bag 37, Hobart, TAS 7001, Australia

[3] Dept. of Physics and Astronomy, Western Kentucky University, Bowling Green, KY 42101, USA

[4] Instituto de Astrofísica de Andalucía, CSIC, Apartado 3004, E-18080 Granada, Spain

[5] Dept. of Physics, Faculty of Science, Kagoshima University, 1-21-35 Korimoto, Kagoshima 890-0065, Japan

[6] School of Physics, University of New South Wales, Sydney, NSW 2052, Australia

[7] Dept. de Astronomía, Universidad de Chile, Casilla 36-D, Santiago, Chile

[8] Astrophysics Group, Lennard Jones Laboratories, Keele University, Staffordshire, ST5 5BG, UK

Abstract. The Galactic Australian SKA Pathfinder (GASKAP) survey is one of several key science projects with ASKAP, a new radio telescope being built in Australia as a technology demonstrator for the Square Kilometer Array (SKA). GASKAP aims to survey about 12,779 square degrees of the Galaxy and the Magellanic System, at high spectral resolution (0.2 km s^{-1}) and using several wavelengths: the λ21-cm HI line, the λ18-cm OH lines, and the comb of recombination lines around λ18-cm. The area covered by GASKAP includes all of the Galactic plane south of declination +40° with $|b| < 10°$, selected areas at higher latitudes covering important interstellar clouds in the disk and halo, the Large and Small Magellanic Clouds, and the Magellanic Bridge and Stream. Compared with previous surveys, GASKAP will achieve an order of magnitude or greater improvement in brightness sensitivity and resolution in various combinations of beam size and mapping speed matched to the astrophysical objectives.

Keywords. ISM: evolution, Galaxy: structure, Galaxy: halo, radio lines: ISM

Galaxy evolution begins at home! To make advances in the area of galaxy formation and evolution, GASKAP turns to our home neighbourhood (the Galaxy and the Magellanic Clouds), where essential physical processes can be studied in detail. The main scientific questions GASKAP will address are: What physical processes are responsible for converting ionised gas to atomic, then molecular clouds, and ultimately stars? How does the excitation temperature of the interstellar gas vary, and how are different interstellar medium phases mixed in different environments? How do feedback processes affect galaxy evolution? How do galaxies get their gas? What is the large-scale structure of present-day star formation? What can we learn by studying entire galactic systems in action? Being a key step toward the SKA, GASKAP will play an important role in training the scientific community for the future use of the SKA.

Highlights of Astronomy, Volume 15
XXVIIth IAU General Assembly, August 2009
Ian F Corbett, ed.

Marking the 400th Anniversary of Kepler's *Astronomia nova*

T. J. Mahoney[1],†

[1] Instituto de Astrofísica de Canarias, E38200 La Laguna, Tenerife, Spain
email: tjm[at]iac.es

Abstract. Special Session 9 of the XXVII General Assembly (11–14 August 2009, Rio de Janeiro) was devoted to the topic "Marking the 400th Anniversary of Kepler's *Astronomia nova*". During the two-and-a-half day meeting (spread over four days), there were nine invited and three contributed talks, a round-table discussion on the future of Kepler studies and an open session to propose the setting up of a Johannes Kepler Working Group under the aegis of the IAU.

Keywords. History and Philosophy of Astronomy: Johannes Kepler

---

1. Kepler's Cosmology

J. V. Field (Birkbeck College, University of London) discussed the geometrical basis of Kepler's cosmology. In Kepler's time, all Christian natural philosophers said God was a Geometer. Kepler went much further. He looked for the explanation of the observed structure of the Cosmos in Euclid's geometry. (In today's terms, this is to say that he saw the Universe as the expression of the nature of Euclidean space.) He found he also needed some geometry-derived arithmetic. However, his final cosmological model agrees closely with observations.

2. Why Should an Astronomer Today Care about Kepler?

Bruce Stephenson (Adler Planetarium) argued that astronomy is, and has been since the time of Isaac Newton, essentially a subfield of physics and questioned how this came to be.

In the pre-Newtonian world, astronomy was one of the seven liberal arts. The physical reality assumed to underlie astronomy consisted of rigid but transparent (hence invisible) spheres, whose rotations carried the visible planets in complex patterns around the centre of the Universe. In the mid-1500s Copernicus had proposed that the centre of the Universe was the Sun rather than the Earth.

Kepler, the first major astronomer to have come of age after Copernicus, believed that Tycho Brahe had disproved the existence of the rigid spheres once thought to carry the planets. Evidently, some force moved the huge massive bodies of the planets around the Sun, through empty space. From his earliest work, Kepler assumed the existence of such a force, and was able thereby to direct his astronomical work to better purpose. The planets all circled the Sun in the same direction, and all moved faster when close to the Sun, so he located the moving force in the Sun's body. A single planet changed its distance from the sun, so he hypothesized a quasi-magnetic force that alternately attracted and repelled planets. The details got messy, largely due to Kepler's assumption that moving

† Present address: Instituto de Astrofísica de Canarias, E38200 La Laguna, Tenerife, Spain.

a massive body required the application of force. Yet he persisted, eventually cobbling together physical theories that accounted for elliptical motion.

Historical study of such developments must begin by setting aside present knowledge of what we count as scientific truth, and evaluating past scientific work within the historical context in which it took place. Current science will someday be past science; it is not "true" in any eternal sense, but must be understood as a snapshot in a historical development

3. Trusting in Tycho: Kepler's Use of Tycho's Data in the *Astronomia nova*

Adam Mosley (Swansea University) described how Kepler's "new astronomy" depended upon his use of Tycho Brahe's observational data, and how he was led to abandon the hypothesis of circular planetary motion because of an eight-minute discrepancy between the longitudes of Mars predicted by his model and the longitudes derived from Tycho's observations. The talk addressed the questions of *why* Kepler placed so much trust in Tycho's data and to *what extent* his trust was warranted: it considered, therefore, not only Tycho's instruments and procedures, but also Kepler's knowledge of them and of the occasions on which Tycho's data had faced and met the challenges posed by other observers and critics. Mosley used sources other than the *Astronomia nova* (Kepler 1609), including Kepler's little-studied *Shieldbrearer of Tycho Brahe the Dane against the Anti-Tycho of Scipio Chiaramonti* (Kepler 1625), to suggest why, despite their intellectual differences and difficult working relationship, Kepler's justified respect for Tycho's observational legacy made the "new astronomy" possible.

4. The Mindset of Uniform Circular Motion

T. J. Mahoney (Instituto de Astrofísica de Canarias) argued that Kepler's immense achievement in establishing the ellipticity of planetary orbits could only be appreciated by understanding the ingrained nature of the paradigm of uniform circular motion. Few of Kepler's contemporaries accepted the idea of elliptical orbits; indeed, Galileo himself remained wedded to the circle throughout his life.

Mahoney mentioned Kepler's correspeondence with Frabricius concerning planetary orbits and Galileo's espousal of circular orbits as outlined in his *Diaglogue concerning Two Chief World Systems* (Galilei 1632).

5. Kepler's Laws: Some Myths Dispelled

A. E. L. Davis (Imperial College, London) said that Kepler is celebrated, above all else, for discovering the laws of planetary motion (Kepler 1609, 1619). These laws constitute the basis of our modern system of celestial dynamics. It would therefore, said Davis, come as a surprise to many people that Kepler himself had formulated the laws in a kinematic context, in which they were exactly true. He derived them simply by applying ancient Greek geometry in the traditional way. Davis went on to say that further notable features of Kepler's approach were his respect for observations and a conviction that his results should express physical reality. On the other hand, various erroneous opinions had been mis-attributed to him over the years, and these were rebutted by Davis.

6. The Kepler Map in Perspective

Ivan I. Shevchenko (Pulkovo Observatory) described the evolving role of the Kepler map in celestial mechanics. The Kepler map was derived by Petrosky (1986) and Chirikov & Vecheslavov (1986) in the framework of the restricted three-body and four-body problems in order to describe the long-term chaotic orbital behaviour of comets in perturbed, nearly parabolic motion and, in particular, that of Halley's Comet. It is a two-dimensional area-preserving map, describing a comets motion in terms of energy and time. Its second equation is based on Kepler's third law, hence the name of the map. Since the 1980s the Kepler map has become paradigmatic in a number of applications in celestial mechanics and atomic physics. It represents an important kind of general separatrix maps. Petrosky & Broucke (1988) used refined methods of mathematical physics to derive analytical expressions for its parameterization. These methods became available only in the 20th century, so it may seem that the map is inherently a very modern mathematical tool. Shevchenko showed that the Kepler map, including analytical formulae for its parameterization, can be derived by quite elementary methods and, although discovered so recently, it might well already have been derived in the first half of the 19th century. Without formulae for the parameterization it might have been derived even earlier. Shevchenko concluded by saying that one can state that the Kepler map is a direct consequence of Kepler's scientific work and legacy. Modern and perspective applications of the Kepler map and its modifications were discussed.

7. Kepler's *Astronomia nova* as a Rare Book

Jay M. Pasachoff (Williams College, Hopkins Observatory) spoke on Kepler's *Astronomia nova* (Kepler 1609) as of interest not only for its intellectual ideas but also for its physical embodiment as a book. Hundreds of first editions from 1609 are extant. Pasachoff discussed the physical format and variations of several copies of the first edition of *Astronomia nova* as well as of copies of the first editions of Kepler's *Mysterium cosmographicum* (1596), with its magnificent fold-out plate showing his conception of the Solar System as interspersed Platonic solids, and his *Harmonice mundi* (1619), which contained his third law.

8. Kepler's *Dream*

Jarosław Włodarczyk (Institute for the History of Science, Warsaw) discussed the first serious scientific study of the Moon, written by Johannes Kepler in his *Somnium (The Dream, or Posthumous Work on Lunar Astronomy)*, partly printed in Sagan in Silesia (1630) and completed in Frankfurt in the year 1634. *The Dream* combines a story which is fantasy, with a scientific treatment of lunar astronomy. The fantasy describes a journey to the Moon and is mixed with reflections whose importance has only become generally understood with the advent of the Space Age in the 20th century. The scientific part of the *Somnium*, said Włodarczyk, presents the astronomical phenomena that would be seen by an observer on the lunar surface. Kepler wrote *The Dream* with a clearly didactic intention: his perceptive description of celestial motions as seen from the Moon produced an ingenious argument on behalf of the Copernican theory. However, Kepler's analysis of many subtle effects arising from the motion of the Moon allowed him also to make discoveries that that have been neglected in the modern history of astronomy. Włodarczyk concluded by to urging all to re-read Kepler's *Somnium*.

9. The Great Synthesis: The Multifaceted New Astronomy of Johannes Kepler (1609)

Giora Hon (University of Haifa) noted that *Astronomia nova* (Kepler 1609) is one of the most revolutionary scientific texts ever written. . In this work Kepler developed an astronomical theory that departs fundamentally from the systems of Ptolemy and Copernicus, hence its distinctly approriate title. A comprehensive grasp of Kepler's astonishing achievements requires considering the conceptual, theological, metaphysical, epistemological, methodological and rhetorical elements that can be found in his astronomcal works. Moreover, one has to view Kepler not only as a mathematico–physical astronomer, but also as a designer of instruments and a practising observer.

One of the great innovations of the *Astronomia nova* is its explicit dependence on the science of optics. The declared goal of Kepler in his earlier publication, *Paralipomena to Witelo whereby The Optical Part of Astronomy is Treated* (Kepler 1604) was to solve difficulties and expose deceptive visual illusions which astronomers face when conducting astronomical observations with optical instruments.

What was the nature of the Keplerian revolution? At the centre of Kepler's revolutionary move is the transformation of theoretical astronomy, which was understood in terms of orbs (spherical shells to which the planets were attached) and models (called *hypotheses* at the time). Instead, Kepler introduced a single term: "orbit" (*orbita*); that is, the path of a planet in space resulting from the action of physical causes.

Kepler's *Astronomia nova*, concluded Hon, combines coherently and in a revolutionary way many layers of different kinds of knowledge that together offer a most powerful system of enquiry, whose scientific fruits we still enjoy today.

10. Galileo's Telescope and Kepler's Optics

Sven Dupré (Ghent University) said that the year 2009 jointly marks the 400th anniversary of the publication of Kepler's *Astronomia nova* as well as of Galileo's observations with the telescope published in *Sidereus nuncius* (Galilei 1610). His talk addresses the issue of the influence of Galileo's telescope on Kepler's work on optics. In *Paralipomena ad Vitellionem* (Kepler 1604) Kepler developed a new theory of vision; *Dioptice* (Kepler 1611) offered its reader an optical theory of the telescope. Is *Dioptrice* based solely on the concepts developed in *Paralipomena*, asked Dupré, or did Kepler learn something from his experiences with the telescope or Galileo's observations? The communication and relationship between Galileo and Kepler are often protrayed in terms of social failure and intellectual misunderstanding, from Kepler's mocking of Galileo's name to Galileo's rejection of Kepler's elliptical orbits. But this portrayal of the two men, said Dupré, misses their intense "conversations" in the period following Galileo's announcement of his telescopic observations.

11. "Third Man in the Middle": Kepler between Astronomy and Astrology

Sheila J. Rabin (St. Peter's College) explained how, combining the ideas of Copernicus and Tycho Brahe, Johannes Kepler concluded that the Universe was physical (Kepler 1610). That created a problem for his acceptance of astrology. Traditionally, belief in astrology had been grounded in the Aristotelian distinction between a physical sublunar world and the non-physical heavens. The celestial world could guide what happened on Earth because its lack of physicality made the celestial world superior. A physical

Universe meant that the celestial world lost its superiority and the consequent ability to guide us. Therefore, concluded Rabin, Kepler set about to reform astrology so that it would conform to his belief in the physical Universe.

12. Kepler and the Star of Bethlehem

Sidney Bludman (Universidad de Chile) questioned a number of astronomical explanations for the Star of Bethlehem. Dating Jesus' life is important for religion, for history and for understanding the goals and methods of ancient astronomy. he explained. The principal problem is to reduce naked eye observations of the Sun, Moon and planets from the concurrent Jewish lunar calendar to our Julian solar calendar. The Crucifixion took place on the 14th or 15th day after the new moon of Nisan, as determined by the first visibility of the lunar crescent in the evening sky. Allowing for atmospheric extinction by Rayleigh and Mie scattering and for stratospheric absorption, Bradley Schaeffer arrives at Julian dates AD 30 April 7 or AD 33 April 3, in agreement with established historical records.

The Star of Bethlehem is much more problematic, and may be entirely apocryphal. It is dated to about 1–10 BC, only by "historical" accounts, written about AD 80–90 by the evangelists Matthew and Luke, for Christian Jews gathered in Antioch. Nevertheless, astronomers since Kepler have tried to identify their brief gospel account, with some astronomical spectacle: an unusual planetary conjunction (7, 3, 2 BC), comet, eclipse, nova (March/April 5 BC), or supernova. All such astronomical interpretations fail to explain why the Magi saw the star in the east, but travelled west to Jerusalem; and why no one in Jerusalem sighted the spectacular event. Instead, Michael Molnar (1995, 1999) argues that the Greek texts needs to be read as an astrological horoscope, written to be understood non-scientifically, by Jews and their first century rulers. He argues for an astrological interpretation of a Sun-Moon-Jupiter-Mars-Venus alignment, in Ares (not Pisces), that would go unnoticed in Jerusalem, on 17 April 6 BC, but would give a regal horoscope, alarming King Herod. If the gospel account is anything more than a parable, Molnar's interpretation is supported by a contemporary Antioch coin (Molnar 1992), and is certainly more historical than any astronomical interpretation.

13. Round Table Discussion on the Future of Kepler Studies

One of the main purposes of the special session was to propose the setting up of a Johannes Kepler Working Group within the IAU. A round-table discussion was held among the invited historians to consider the following topics:

- Translation into English of Kepler's major works
- Translation into English of Kepler's correspondence
- Publication of an undergraduate-level textbook on Kepler's life and work
- Obtaining funding for future Kepler conferences
- Preparation of secondary-level educational material
- Representation of the WG at international level with other organizations
- Making enquiries concerning the future of the *Kepler-Kommission* offices and library

S. Rabin explained that Kepler's astrology gives insights into his views on nature, and that Kepler had difficulty in reconciling his astrology with his astronomy, so that further study of Kepler's astrology is crucial to understanding the decline of astrology.

B. Stephenson seconded Rabin's views and added that further study of Kepler's religious views, which were fundamental to his whole scientific outlook, was also required.

J. Włodarczyk pointed out that Kepler's theory of the moon was an important part of the history of selenography and needed to be pursued further. He also urged that the *Bayerische Akademie der Wissenschaften* (hosts of the now defunct *Kepler-Kommission*) be represented on the WG and stressed the importance of setting up a website for the WG.

G. Hon asked why we always speak of a Copernican revolution but never a Keplerian revolution. He recommended that Kepler studies be broaden to incorporate the philosophy of science.

A. E. L. Davis felt that making Kepler's name as well-known as Galileo's in the history of astronomy would result in a truer understanding of the subject. She also stressed the necessity to maintain access to the contents of the *Kepler-Kommission* library and the remaining unpublished manuscripts. Davis also advocated revision of the index of the *Johannes Keplers Gesammelte Werke* to make it more user-friendly.

S. Dupré said that a study of how *Dioptrice* was received by craftsmen and instrument builders was needed. An English translation of *Dioptrice*, including Kepler's marginal notes, was needed.

A. Mosley stressed the need for more translations into English of Kepler's works. He further indicated that investigation of the reception of Kepler's works and more studies of Kepler's contemporaries were vital, as was an up-to-date historiography of Kepler studies.

J. V. Field said that an English translation of *Stereometria* was needed and commented that the ongoing Copernicus project should help to shed light on Kepler.

14. Proposal for a Johannes Kepler Working Group

On the final day of the special session, the following persons drafted a formal proposal to Commission 41 for the setting up of a Johannes Kepler Working Group:

- T. J. Mahoney (Spain, IAU) Chair
- A. E. L. Davis (UK, IAU)
- S. Dupré (Belgium)
- J. V. Field (UK, IAU)
- E. Hoeg (Denmark, IAU, C41)
- G. Hon (Israel)
- A. Mosley (UK)
- J. M. Pasachoff (USA, IAU)
- J.-C. Pecker (France, IAU, C41)
- S. J. Rabin (USA)
- B. Stephenson (USA)
- J. Włodarczyk (Poland, IAU)
- G. Wolfschmidt (Germany, IAU, C41)

The aims of the proposed working group would be to promote Johannes Kepler awareness and studies through:

- Representations at international level with other organizations and institutions, specifically ICSU and its member organizations (e.g. the IUHPS)
- Promoting the digitization of Kepler's published works and unpublished manuscripts
- Promoting the maintenance of paper editions of the KGW (e.g. via print on demand)
- Establishing an IAU WG website dedicated to Johannes Kepler
- Establishing contacts with the *Bayerische Akademie der Wissenschaften* to secure continued access to the rich Kepler resources of the defunct *Kepler-Kommission*

- Producing and encouraging relevant publications (e.g. textbooks and monographs)
- Raising funding for projects relevant to Kepler studies

The proposal was handed presented to Clive Ruggles, Secretary of C41, immediately after the closure of the meeting.†

Acknowledgements

We gratefully acknowledge financial support from the IAU and the Royal Astronomical Society, without which this meeting would not have been possible.

References

Koch, W. 1983, *INP Preprint* 86–184

Galilei, Galileo, *Dialogo sopra due massimi sistemi del mondo* (Florence, 1632)

Galilei, Galileo, *Sidereus nuncius* (Venice, 1610)

Kepler, Johannes, *Ad Vitellionem paralipomena* (Prague, 1604)

Kepler, Johannes, *Astronomia nova* (Prague, 1609)

Kepler, Johannes, *Dioptrice* (Prague, 1611)

Kepler, Johannes, *Disertatio cum nuncio sidereo* (Prague, 1610)

Kepler, Johannes, *Harmonice mundi* (Linz, 1619)

Kepler, Johannes, *Mysterium cosmographicum* (Tübingen, 1596)

Kepler, Johannes, *Somnium* (Frankfurt, 1634)

Kepler, Johannes, *Tertius interveniens* (Prague, 1610)

Kepler, Johannes 1625, *Tychonis Brahei Dani hyperaspistes* (Frankfurt, 1625)

Molnar, Michael 1992, *S&T* 83, 37

Molnar, Michael 1995, *QJRAS* 36, 109

Molnar, Michael, *The Star of Bethlehem: The Legacy of the Magi* (New Jersey, 1999)

Petrosky, T. Y. 1986, *Phys. Lett.* A117, 328

Petrosky, T. Y. & Broucke, R. 1988, *Celest. Mech* 42, 53

† The proposal has since been accepted by Division XII on the unanimous recommendation of Commission 41.

Highlights of Astronomy, Volume 15
XXVIIth IAU General Assembly, August 2009
Ian F. Corbett, ed.

© International Astronomical Union 2010
doi:10.1017/S1743921310011993

SpS10-Next Generation Large Astronomical Facilties

Gerard F. Gilmore[1], Richard T. Schilizzi[2]

[1] Institute of Astronomy, University of Cambridge, UK,
[2] SKA Project

No manuscript was received for this Special Session.

Author Index - Part I Invited Discourses & Joint Discussions

Author Index - Part II Special Sessions

Printed in the United States
by Baker & Taylor Publisher Services